W9-BEH-697

ENHANCED

WebAssign

Increased Engagement.
Improved Outcomes.
Superior Service.

Exclusively from Cengage Learning, **Enhanced WebAssign®** combines the exceptional Mathematics content that you know and love with the most powerful and flexible online homework solution, **WebAssign**. **Enhanced WebAssign** engages students with immediate feedback, rich tutorial content, and interactive eBooks helping students to develop a deeper conceptual understanding of their subject matter. Online assignments can be built by selecting from thousands of text-specific problems or supplemented with problems from any Cengage Learning textbook in our collection.

With **Enhanced WebAssign**, you can

- Help students stay on task with the class by requiring regularly scheduled assignments using problems from the textbook.

- Provide students with access to a personal study plan to help them identify areas of weakness and offer remediation.

- Focus on teaching your course and not on grading assignments. Use **Enhanced WebAssign** item analysis to easily identify the problems that students are struggling with.

- Make the textbook a destination in the course by customizing your Cengage YouBook to include sharable notes and highlights, links to media resources and more.

- Design your course to meet the unique needs of traditional, lab-based or distance learning environments.

- Easily share or collaborate on assignments with other faculty or create a master course to help ensure a consistent student experience across multiple sections.

- Minimize the risk of cheating by offering algorithmic versions of problems to each student or fix assignment values to encourage group collaboration.

Learn more at www.cengage.com/ewa

TRANSFORM**I**NG **LEARNING** TRANSFORMING **LIVES**

CENGAGE
Learning

Ask the Authors!

We have taught math for many years. During that time, we have had students ask us a number of questions about mathematics and this course. Here you find some of the questions we have been asked most often, starting with the big one.

Dick Aufmann

Joanne Lockwood

Why do I have to take this course? You may have heard that *"Math is everywhere."* That is probably a slight exaggeration, but math does find its way into many disciplines. There are obvious places like engineering, science, and medicine. There are other disciplines such as business, social science, and political science where math may be less obvious but still essential. If you are going to be an artist, writer, or musician, the direct connection to math may be even less obvious. Even so, as art historians who have studied the Mona Lisa have shown, there is a connection to math. But, suppose you find these reasons not all that compelling. **There is still a reason to learn basic math skills: You will be a better consumer and be able to make better financial choices for you and your family.** For instance, is it better to buy a car or lease a car? Math can provide an answer.

I find math difficult. Why is that? It is true that some people, even very smart people, find math difficult. Some of this can be traced to previous math experiences. If your basic skills are lacking, it is more difficult to understand the math in a new math course. Some of the difficulty can be attributed to the ideas and concepts in math. They can be quite challenging to learn. Nonetheless, most of us can learn and understand the ideas in the math courses that are required for graduation. **If you want math to be less difficult, practice. When you have finished practicing, practice some more.** Ask an athlete, actor, singer, dancer, artist, doctor, skateboarder, or (name a profession) what it takes to become successful and the one common characteristic they all share is that they practiced—a lot.

Why is math important? As we mentioned earlier, math is found in many fields of study. There are, however, other reasons to take a math course. Primary among these reasons is to become a better problem solver. Math can help you learn critical thinking skills. It can help you develop a logical plan to solve a problem. Math can help you see relationships between ideas and to identify patterns. **When employers are asked what they look for in a new employee, being a problem solver is one of the highest ranked criteria.**

What do I need to do to pass this course? The most important thing you must do is to know and understand the requirements outlined by your instructor. These requirements are usually given to you in a syllabus. Once you know what is required, you can chart a course of action. Set time aside to study and do homework. If possible, choose your classes so that you have a free hour after your math class. Use this time to review your lecture notes, rework examples given by the instructor, and begin your homework. All of us eventually need help, so know where you can get assistance with this class. This means knowing your instructor's office hours, the hours of the math help center, and how to access available online resources. And finally, do not get behind. **Try to do some math EVERY day, even if it is for only 20 minutes.**

Introductory and Intermediate Algebra

AN APPLIED APPROACH

EDITION **6**

Annotated Instructor's Edition

Richard N. Aufmann

Palomar College

Joanne S. Lockwood

Nashua Community College

BROOKS/COLE
CENGAGE Learning

Australia • Brazil • Japan • Korea • Mexico • Singapore • Spain • United Kingdom • United States

BROOKS/COLE
CENGAGE Learning

*Introductory and Intermediate Algebra:
An Applied Approach*
Sixth Edition
Richard N. Aufmann, Joanne S. Lockwood

Senior Executive Editor: Charlie Van Wagner

Acquisitions Editor: Marc Bove

Developmental Editor: Danielle Derbenti

Assistant Editor: Lauren Crosby

Editorial Assistants: Jennifer Cordoba,
   Ryan Furtkamp

Media Editors: Heleny Wong, Guanglei Zhang

Brand Manager: Gordon Lee

Marketing Communications Manager:
   Jason LaChapelle

Content Project Manager: Cheryll Linthicum

Art Director: Vernon Boes

Manufacturing Planner: Becky Cross

Rights Acquisitions Specialist: Tom McDonough

Production and Composition Service:
   Graphic World Inc.

Photo Researcher: Bill Smith Group

Text Researcher: Pablo D'Stair

Copy Editor: Jean Bermingham

Illustrator: Graphic World Inc.

Text Designer: The Davis Group

Cover Designer: Irene Morris

Cover Image: Morris Design

© 2014, 2009 Brooks/Cole, Cengage Learning

ALL RIGHTS RESERVED. No part of this work covered by the copyright herein may be reproduced, transmitted, stored, or used in any form or by any means, graphic, electronic, or mechanical, including but not limited to photocopying, recording, scanning, digitizing, taping, Web distribution, information networks, or information storage and retrieval systems, except as permitted under Section 107 or 108 of the 1976 United States Copyright Act, without the prior written permission of the publisher.

For product information and technology assistance, contact us at
Cengage Learning Customer & Sales Support, 1-800-354-9706.

For permission to use material from this text or product,
submit all requests online at **www.cengage.com/permissions**.
Further permissions questions can be e-mailed to
permissionrequest@cengage.com.

Library of Congress Control Number: 2012955393

ISBN-13: 978-1-133-36541-9

ISBN-10: 1-133-36541-8

Brooks/Cole
20 Davis Drive
Belmont, CA 94002-3098
USA

Cengage Learning is a leading provider of customized learning solutions with office locations around the globe, including Singapore, the United Kingdom, Australia, Mexico, Brazil, and Japan. Locate your local office at **www.cengage.com/global**.

Cengage Learning products are represented in Canada by Nelson Education, Ltd.

To learn more about Brooks/Cole, visit **www.cengage.com/brookscole**.

Purchase any of our products at your local college store or at our preferred online store **www.CengageBrain.com**.

Printed in the United States of America
1 2 3 4 5 6 7 17 16 15 14 13

Brief Contents

Contents

CHAPTER

2

First-Degree Equations and Inequalities

75

CHAPTER

5

Systems of Linear Equations and Inequalities   297

CHAPTER

6

Polynomials   343

CHAPTER

7

Factoring   397

CHAPTER

8 Rational Expressions 449

CHAPTER

9 Exponents and Radicals 519

CHAPTER

10 Quadratic Equations   567

CHAPTER

11 Functions and Relations   615

CHAPTER

12 Exponential and Logarithmic Functions 669

CHAPTER

T Transitioning to Intermediate Algebra 721

Preface

mong the many questions we ask when we begin the process of revising a textbook, the most important is, "How can we improve the learning experience for the student?" We find answers to this question in a variety of ways, but most commonly by talking to students and instructors and by evaluating the written feedback we receive from instructors. Bearing this feedback in mind, our ultimate goal as we set out to create the sixth edition of *Introductory and Intermediate Algebra* was to provide students with more materials to help them better understand the underlying concepts presented in this course. As a result, we have made the following changes to the new edition.

New to this edition is the **Focus on Success** vignette that appears at the beginning of each chapter. **Focus on Success** offers practical tips for improving study habits and performance on tests and exams.

We now include an **Apply the Concept** box in many objectives in which a new concept is introduced. This feature gives an immediate real-world example of how that concept is applied. For instance, after linear functions are defined, there is an Apply the Concept example of using the linear function that converts Celsius temperatures to Fahrenheit temperatures.

The definition and key concept boxes have been enhanced in this edition; they now include examples to show how the general case translates to specific cases.

In each exercise set, the first group of exercises is now titled **Concept Check.** The **Concept Check** exercises focus on the concepts that lie behind the skills developed in the section. We consider an understanding of these concepts essential to a student's success in mastering the skills required to complete the exercises that follow.

Every chapter contains **Check Your Progress** exercises. This feature appears approximately mid-chapter and tests students' understanding of the concepts presented to that point in the chapter.

Critical Thinking exercises are included at the end of every exercise set. They may involve further exploration or analysis of the topic at hand. They may also integrate concepts introduced earlier in the text.

We trust that the new and enhanced features of the sixth edition will help students more successfully engage with the content. By narrowing the gap between the concrete and the abstract, between the real world and the theoretical, students should more plainly see that mastering the skills and topics presented is well within their reach and well worth the effort.

New to This Edition

- **Apply the Concept** boxes show how a just-introduced concept can be applied to real-world problems.

- **Concept Check** exercises appear at the beginning of each exercise set.

- Enhanced definition/key concept boxes now provide examples that illustrate how the general case applies to specific cases.

- The **Focus on Success** feature at the beginning of each chapter offers practical guidance to help students develop positive study habits.

- **Check Your Progress** exercises appear approximately mid-chapter and test students' understanding of the concepts presented thus far in the chapter.

- **In the News** articles within the exercise sets have been updated, as have application problems throughout the text.

- **Critical Thinking** exercises appear at the end of each exercise set.

- **Projects or Group Activities** are now included at the end of each exercise set.

- **Chapter A, AIM for Success,** now appears as the first chapter of the text. This chapter describes skills used by students who have been successful in this course. Topics include how to stay motivated, making a commitment to success, time management, and how to prepare for and take tests. A guide to the textbook is included to help students use its features effectively.

- More annotations have been added to the worked Examples, to more effectively explain the steps of the solutions.

- Many of the **Chapter Summaries** have been expanded to include more entries and more descriptive explanations.

Organizational Changes

We have made the following changes in order to improve the effectiveness of the textbook and enhance the student's learning experience.

- In Chapter 2, solution sets to inequalities involving "or" are now shown as the union of sets.

- In Chapter 5, Section 5.1 was revised; the concepts of independent, dependent, and inconsistent systems of equations are more clearly stated.

- Section 6.2 in Chapter 6 was expanded to place more emphasis on graphing polynomial functions. A table of the basic shapes of quadratic and cubic functions was added to give students an aid when graphing those functions.

- In Chapter 7, Section 7.5 on solving equations by factoring now includes a boxed list of the steps used in solving a quadratic equation by factoring.

- In Chapter 8, Section 8.2 was revised to provide a more thorough treatment of LCM.

- In Chapter 8, the objectives in Section 8.5 have been reorganized. Now in Objective 8.5A, students learn how to solve proportions and are introduced to applications of solving proportions. There are now more examples for the student to study, and the exercise set is longer. The student has more opportunity to master problem solving involving proportions. The topic of Objective 8.5B is solving problems involving similar triangles.

- The exposition in Section 8.7 has been expanded to provide more examples which apply the basic concepts of work problems. There is a gradual development of solving these problems. There is also more assistance in setting up the uniform motion problems in this section. There are several more work problems and uniform motion problems for the student to solve, and there is a wide variety of each type of word problem.

- In Chapter 9, Section 9.2 was separated into two sections: Section 9.2 Addition and Subtraction of Radical Expressions and Section 9.3 Multiplication and Division of Radical Expressions.

- Section 10.1 in Chapter 10 was consolidated into two objectives. Section 10.2 was expanded to include solving a quadratic equation by using the quadratic formula. The section on solving nonlinear inequalities was moved to the last section of the chapter. Consequently, applications of quadratic equations immediately follow the discussion on solving quadratic equations.

- Section 11.2 in Chapter 11 was rewritten to include graphing a function by using translations and graphing a function by using reflections.

- In Chapter 12, important concepts have been set off and highlighted. Now students can find these concepts easily and see basic examples of how the concepts are applied.

Take AIM and Succeed!

An Objective-Based Approach

Introductory and Intermediate Algebra: An Applied Approach is organized around a carefully constructed hierarchy of **objectives**. This "objective-based" approach provides an integrated learning path that enables you to find resources such as assessment tools (both within the text and online), videos, tutorials, and additional exercises for each objective in the text.

1 Each Chapter Opener outlines the learning **OBJECTIVES** that appear in each section of the chapter. The list of objectives serves as a resource to guide you in your study and review of the topics.

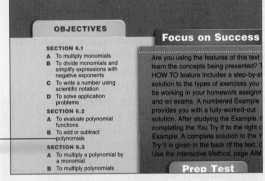

2 Taking the **PREP TEST** for each chapter will help you determine which topics you need to study more carefully and which topics you need only review. The **ANSWERS** to the **PREP TEST** provide references to the **OBJECTIVES** on which the exercises are based.

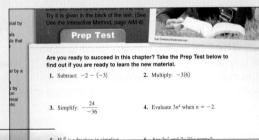

3 In every section, an **OBJECTIVE STATEMENT** introduces each new topic of discussion. Videos are available for each objective.

SECTION
6.1 Exponential Expressions

OBJECTIVE A *To multiply monomials*

A **monomial** is a number, a variable, or a product of a number and variables.

Point of Interest
Around A.D. 250, the monomial $3x^2$ shown at the right would have been written $\Delta^y 3$, or at least approximately

The examples at the right are monomials. The **degree of a monomial** is the sum of the expo-

x degree 1 $(x = x^1)$
$3x^2$ degree 2

4 Section exercises are keyed to **OBJECTIVE STATEMENTS.**

OBJECTIVE A *To multiply monomials*

For Exercises 7 to 35, simplify.

7. $(ab^3)(a^3b)$ 8. $(-2ab^4)(-3a^2b^4)$ 9. $(9xy^2)(-2x^2y^2)$ 10. $(x^2y)^2$

11. $(x^2y^4)^4$ 12. $(-2ab^2)^3$ 13. $(-3x^2y^3)^4$ 14. $(4a^2b^3)^3$

15. $(27a^2b^2)^2$ 16. $[(2ab)^3]^2$ 17. $[(2a^4b^3)^3]^2$ 18. $(xy)(x^2y)^4$

An Objective-Based Review

This "objective-based" approach continues through the end-of-chapter review and addresses a broad range of study styles by offering a **wide variety of review tools.**

 CHECK YOUR PROGRESS exercises appear approximately mid-chapter and test your understanding of the concepts presented up to that point in the chapter.

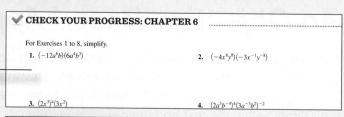

✓ CHECK YOUR PROGRESS: CHAPTER 6

For Exercises 1 to 8, simplify.
1. $(-12a^6b)(6a^4b^3)$
2. $(-4x^6y^8)(-3x^{-1}y^{-8})$
3. $(2x^3)^4(3x^2)$
4. $(2a^3b^{-4})^4(3a^{-3}b^2)^{-2}$

At the end of each chapter, you will find a **CHAPTER SUMMARY** containing **KEY WORDS** and **ESSENTIAL RULES AND PROCEDURES** presented in the chapter. Each entry includes an objective reference and a page reference that show where in the chapter the concept was introduced. An example demonstrating the concept is also included.

CHAPTER

6 Summary

Key Words	Examples
A monomial is a number, a variable, or a product of numbers and variables. [6.1A, p. 344]	5, y, and $8a^2b^2$ are monomials.

By completing the **CHAPTER REVIEW EXERCISES,** you can practice working on problems in an order that is different from the order in which they were presented in the chapter. The **ANSWER** to each Chapter Review exercise includes a reference to the objective on which the exercise is based. This reference will help you quickly identify where to go if you need further practice with a particular concept.

CHAPTER

6 Review Exercises

1. Add: $(12y^2 + 17y - 4) + (9y^2 - 13y + 3)$
2. Divide: $\dfrac{15x^2 + 2x - 2}{3x - 2}$
3. Simplify: $(2x^{-1}y^2z^5)^4(-3x^3yz^{-3})^2$
4. Expand: $(5y - 7)^2$

Each **CHAPTER TEST** is designed to simulate a typical test of the concepts covered in the chapter. Each **ANSWER** includes an objective reference as well as a reference to a numbered Example, You Try It, or HOW TO in the text that is similar to the given test question.

CHAPTER

6 TEST

1. Multiply: $2x(2x^2 - 3x)$
2. Use the Remainder Theorem to evaluate $P(x) = -x^3 + 4x - 8$ when $x = -2$.

CUMULATIVE REVIEW EXERCISES, which appear at the end of each chapter (beginning with Chapter 2), help you maintain previously learned skills. The **ANSWERS** include references to the section objectives on which the exercises are based.

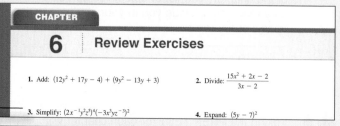

Cumulative Review Exercises

1. Simplify: $8 - 2[-3 - (-1)]^2 + 4$
2. Evaluate $\dfrac{2a - b}{b - c}$ when $a = 4$, $b = -2$, and $c = 6$.
3. Identify the property that justifies the statement $2x + (-2x) = 0$.
4. Simplify: $2x - 4[x - 2(3 - 2x) + 4]$

A **FINAL EXAM** is provided following the last chapter of the text. The Final Exam is designed to simulate a comprehensive exam covering all the concepts presented in the text. The **ANSWERS** to the Final Exam questions are provided in the appendix at the back of the text and include references to the section objectives on which the questions are based.

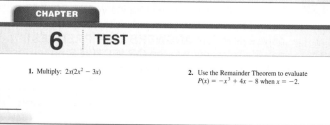

FINAL EXAM

1. Simplify: $12 - 8[3 - (-2)]^2 \div 5 - 3$
2. Evaluate $\dfrac{a^2 - b^2}{a - b}$ when $a = 3$ and $b = -4$.

Understanding the Concepts

Each of the following features is designed to give you a fuller understanding of the key concepts.

 CONCEPT CHECK exercises promote conceptual understanding. Completing these exercises will deepen your understanding of the concepts you are learning and provide the foundation you need to successfully complete the remaining exercises in the exercise set.

6.2 EXERCISES

✔ Concept Check

1. Identify each of the following as a monomial, a binomial, a trinomial, or none of these.
   a. $-3x^4 + 1$ b. $2x - 7$
   c. $3x^2y^5z$ d. $1 - 4x - x^2$
   e. $5z^4 - 2z^{-2} + 4$ f. 7

2. Write each polynomial in descending order.
   a. $3x - 7x^2 + 5$ b. $3x^4 - 7 - 2x + 4x^2$

3. What is the domain of a polynomial function?

4. Write the additive inverse of $4x^3 - 7x + 8$.

Definition/key concept boxes contain examples to illustrate how each definition or key concept is applied in practice.

Rule for Multiplying Exponential Expressions

If m and n are positive integers, then $x^m \cdot x^n = x^{m+n}$.

EXAMPLES
1. $x^4 \cdot x^7 = x^{4+7} = x^{11}$
2. $z \cdot z^6 = z^{1+6} = z^7$ • Recall that $z = z^1$.
3. $v^2 \cdot v^3 \cdot v^5 = v^{2+3+5} = v^{10}$
4. $(a^3b^5)(a^4b^7) = a^{3+4}b^{5+7} = a^7b^{12}$ • Add exponents on like bases.

TAKE NOTE boxes alert you to concepts that require special attention.

HOW TO 1   Factor: $4x^2 - 81y^2$

Write the binomial as the difference of two perfect squares.

$$4x^2 - 81y^2 = (2x)^2 - (9y)^2$$

The factors are the sum and difference of the square roots of the perfect squares.

$$= (2x + 9y)(2x - 9y)$$

📝 **Take Note**
Recall that, using FOIL,
$(x + 6)^2 = (x + 6)(x + 6)$
$= x^2 + 6x + 6x + 36$
$= x^2 + 12x + 36$

The square of a binomial is a **perfect-square trinomial.** Here are several examples.

Square of a Binomial	Perfect-Square Trinomial
$(x + 6)^2$	$x^2 + 12x + 36$
$(x - 7)^2$	$x^2 - 14x + 49$
$(a + b)^2$	$a^2 + 2ab + b^2$
$(a - b)^2$	$a^2 - 2ab + b^2$

Factoring a Perfect-Square Trinomial

$a^2 + 2ab + b^2 = (a + b)^2$ $a^2 - 2ab + b^2 = (a - b)^2$

EXAMPLES
1. $x^2 + 10x + 25 = (x + 5)^2$ 2. $y^2 - 12y + 36 = (y - 6)^2$

Unless otherwise noted, all content on this page is © Cengage Learning.

POINT OF INTEREST boxes, which relate to the topic under discussion, may be historical in nature or of general interest.

SECTION

6.1   Exponential Expressions

OBJECTIVE A   *To multiply monomials*

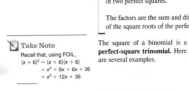 **Point of Interest**
Around A.D. 250, the monomial $3x^2$ shown at the right would have been written Δ^y3, or at least approximately like that. In A.D. 250, the symbol for 3 was not the one we use today.

A **monomial** is a number, a variable, or a product of a number and variables.

The examples at the right are monomials. The **degree of a monomial** is the sum of the exponents on the variables.

x	degree 1 $(x = x^1)$
$3x^2$	degree 2
$4x^2y$	degree 3
$6x^3y^4z^2$	degree 9

In this chapter, the variable n is considered a positive integer when used as an exponent.

x^n	degree n

Application of the Concepts

The section exercises offer many opportunities to put the concepts you are learning into practice.

 APPLY THE CONCEPT boxes illustrate how an arithmetic operation is applied to a real-world situation so that you understand how the operation is used in everyday life.

For instance, to evaluate $f(x) = x^2$ when $x = 4$, replace x by 4 and simplify.

$$f(x) = x^2$$
$$f(4) = 4^2 = 16$$

The *value* of the function is 16 when $x = 4$. An ordered pair of the function is $(4, 16)$.

> **APPLY THE CONCEPT**
>
> The height $s(t)$, in feet, of a ball above the ground t seconds after it is thrown upward at a velocity of 64 ft/s is given by $s(t) = -16t^2 + 64t + 4$. Find the height of the ball 1.5 s after it is released.
>
> To find the height, evaluate the function when $t = 1.5$.
>
> $s(t) = -16t^2 + 64t + 4$
> $s(1.5) = -16(1.5)^2 + 64(1.5) + 4$ • Evaluate the function when $t = 1.5$.
> $= -16(2.25) + 64(1.5) + 4$
> $= -36 + 96 + 4$
> $= 64$
>
> The ball is 64 ft above the ground 1.5 s after it is released.

THINK ABOUT IT exercises promote deeper conceptual understanding. Completing these exercises will expand your understanding of the concepts being addressed.

45. Does the graph of every straight line have a *y*-intercept? Explain.

46. Why is it not possible to graph an equation of the form $Ax + By = 0$ by using only the *x*- and *y*-intercepts?

CRITICAL THINKING exercises may involve further exploration or analysis of the topic at hand. They may also integrate concepts introduced earlier in the text.

Critical Thinking

58. Match each equation with its graph.

i. $y = -2x + 4$

ii. $y = 2x - 4$

iii. $y = 2$

iv. $2x + 4y = 0$

v. $y = \frac{1}{2}x + 4$

vi. $y = -\frac{1}{4}x - 2$

A. B. C. D. E. F.

Working through the application exercises that contain **REAL DATA** will prepare you to answer questions and solve problems that you encounter outside of class, using facts and information that you gather on your own.

IN THE NEWS exercises help you understand the importance of mathematics in our everyday world. These application exercises are based on information taken from popular media sources such as newspapers, magazines, and the Internet.

ated by $s = f(v) = 0.017v^2$. How far will a car skid after its brakes are applied if it is traveling at 60 mph?

47. **Automotive Technology** Read the news article at the right. Suppose you drive 12,000 mi per year and the price of gas is \$4.00 per gallon. Under these conditions, increasing your car's fuel efficiency by p percent can give you an annual cost savings, in dollars, of $S = f(p) = \frac{2400p}{1 + p}$. If you changed the tires on your car to low-rolling-resistance tires, what would be your minimum annual cost savings? Round to the nearest cent.

In the NEWS!

New Tires on a Roll to Lower Costs

Recent improvements in low-rolling-resistance tires have made these tires an increasingly popular choice for people who want to cut fuel costs. Low-rolling-resistance tires can raise a car's fuel efficiency by 5 to 7 percent.

Source: www.shopautoweek.com

48. **Airports** Airport administrators usually price airport parking at a rate that discour-

Hours Parked	Cost
$0 < t \le 1$	\$1.00

By completing the **WRITING** exercises, you will improve your communication skills while increasing your understanding of mathematical concepts.

Projects or Group Activities

58. There is an imaginary coordinate system on Earth that consists of *longitude* and *latitude*. Write a report on how location is determined on the surface of Earth. Include in your report the longitude and latitude coordinates of your school.

59. Describe the graph of all ordered pairs (x, y) that are 5 units from the origin.

60. Consider two distinct fixed points in a plane. Describe the graph of all points (x, y) that are equidistant from these fixed points.

Unless otherwise noted, all content on this page is © Cengage.

Focus on Study Skills

An emphasis on setting a foundation of good study habits is woven into the text.

UPDATED!

CHAPTER A, AIM FOR SUCCESS, outlines study skills that are used by students who have been successful in this course. By making Chapter A the first chapter of the text, the stage is set for a successful beginning to the course.

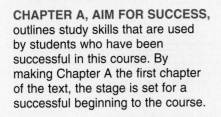

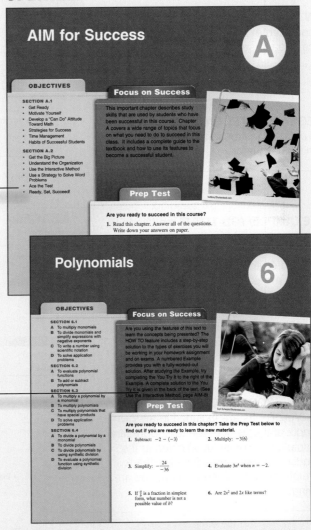

New ▶ FOCUS ON SUCCESS appears at the start of each Chapter Opener. These tips are designed to help you make the most of the text and your time as you progress through the course and prepare for tests and exams.

TIPS FOR SUCCESS boxes outline good study habits and function as reminders throughout the text.

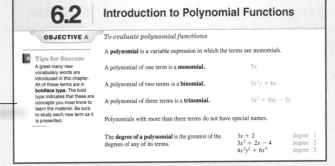

Focus on Skills and Problem Solving

The following features exemplify the emphasis on skills and the problem-solving process.

HOW TO examples provide solutions with detailed explanations for selected topics in each section.

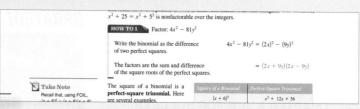

INTEGRATING TECHNOLOGY notes offer optional instruction in the use of a scientific calculator.

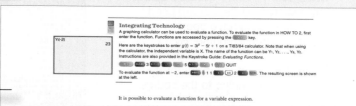

The **EXAMPLE/YOU TRY IT** matched pairs are designed to actively involve you in the learning process. The You Try Its are based on the Examples. These problems are paired so that you can easily refer to the steps in the Example as you work through the accompanying You Try It.

EXAMPLE 1

Factor: $25x^2 - 1$

Solution
$25x^2 - 1 = (5x)^2 - (1)^2$ • Difference of
$= (5x + 1)(5x - 1)$ two squares

YOU TRY IT 1

Factor: $x^2 - 36y^4$

Your solution

EXAMPLE 2

Factor: $4x^2 - 20x + 25$

Solution
$4x^2 - 20x + 25 = (2x - 5)^2$ • Perfect-square
trinomial

YOU TRY IT 2

Factor: $9x^2 + 12x + 4$

Your solution

EXAMPLE 3

Factor: $(x + y)^2 - 4$

Solution

YOU TRY IT 3

Factor: $(a + b)^2 - (a - b)^2$

Your solution

Complete, **WORKED-OUT SOLUTIONS** to the You Try Its are included in an appendix at the back of the text. Compare your solution to the solution given in the appendix to obtain immediate feedback and reinforcement of the concept you are studying.

SECTION 7.4

You Try It 1
$x^2 - 36y^4 = x^2 - (6y^2)^2$ • Difference of
$= (x + 6y^2)(x - 6y^2)$ two squares

You Try It 2
$9x^2 + 12x + 4 = (3x + 2)^2$ • Perfect-square
trinomial

You Try It 3
$(a + b)^2 - (a - b)^2$ • Difference of two squares
$= [(a + b) + (a - b)][(a + b) - (a - b)]$
$= (a + b + a - b)(a + b - a + b)$
$= (2a)(2b) = 4ab$

You Try It 4
$a^3b^3 - 27 = (ab)^3 - 3^3$ • Difference of two cubes
$= (ab - 3)(a^2b^2 + 3ab + 9)$

$2x - 3 = 0 \qquad 2x + 3 = 0$ • Principle of Zero Products
$2x = 3 \qquad 2x = -3$
$x = \dfrac{3}{2} \qquad x = -\dfrac{3}{2}$

The solutions are $\dfrac{3}{2}$ and $-\dfrac{3}{2}$.

You Try It 3
$(x + 2)(x - 7) = 52$
$x^2 - 5x - 14 = 52$
$x^2 - 5x - 66 = 0$
$(x + 6)(x - 11) = 0$
$x + 6 = 0 \qquad x - 11 = 0$ • Principle of Zero Products
$x = -6 \qquad x = 11$

The solutions are −6 and 11.

You Try It 4

The **PROBLEM-SOLVING APPROACH** used throughout the text emphasizes the importance of problem-solving strategies. Model strategies are presented as guides for you to follow as you attempt the You Try Its that accompany the numbered Examples.

EXAMPLE 9

How many miles does light travel in one day? The speed of light is 186,000 mi/s. Write the answer in scientific notation.

Strategy
To find the distance traveled:
• Write the speed of light in scientific notation.
• Write the number of seconds in one day in scientific notation.
• Use the equation $d = rt$, where r is the speed of light and t is the number of seconds in one

YOU TRY IT 9

The Roadrunner supercomputer from IBM can perform one arithmetic operation, called a FLOP (**FL**oating-point **OP**eration), in 9.74×10^{-16} s. How many arithmetic operations can be performed in 1 min? Write the answer in scientific notation.

Your strategy

UPDATED!

Projects or Group Activities

107. **Norman Window** A Norman window has the shape of a rectangle surmounted by a semicircle, as shown in the figure at the right. The exterior perimeter of the window in the figure is 50 ft. Find the height h and radius r that will maximize the area of the window.

Unless otherwise noted, all content on this page is © Cengage Learning.

PROJECTS OR GROUP ACTIVITIES appear at the end of each exercise set. Your instructor may assign these individually, or you may be asked to work through the activities in groups.

Additional Resources—
Get More from Your Textbook!

Instructor Resources

Annotated Instructor's Edition (AIE)
(ISBN 978-1-133-73434-3)

The Annotated Instructor's Edition features answers to all of the problems in the text, as well as an appendix denoting those problems that can be found in Enhanced WebAssign.

PowerLecture with Diploma®
(ISBN 978-1-285-42031-8)

This DVD provides the instructor with dynamic media tools for teaching. Create, deliver, and customize tests (both print and online) in minutes with Diploma's Computerized Testing featuring algorithmic equations. Easily build solution sets for homework or exams using Solution Builder's online solutions manual. Quickly and easily update your syllabus with the Syllabus Creator, which was created by the authors and contains the new edition's table of contents.

Complete Solutions Manual (ISBN 978-1-285-42033-2)
Authors: Ellena Reda, Dutchess Community College;
Pat Foard, South Plains College

The Complete Solutions Manual provides worked-out solutions to all of the problems in the text.

Instructor's Resource Binder with Appendix
(ISBN 978-1-285-42035-6)
Author: Maria H. Andersen, Muskegon Community College;
Appendices by Richard N. Aufmann, Palomar College, and
Joanne S. Lockwood, Nashua Community College

Each section of the main text is discussed in uniquely designed Teaching Guides that contain tips, examples, activities, worksheets, overheads, assessments, and solutions to all worksheets and activities.

Solution Builder

This online instructor database offers complete, worked-out solutions to all exercises in the text, allowing you to create customized, secure solutions printouts (in PDF format) matched exactly to the problems you assign in class. For more information, visit www.cengage.com/solutionbuilder.

Enhanced WebAssign®
Printed Access Card: 978-0-538-73810-1
Online Access Code: 978-1-285-18181-3

Exclusively from Cengage Learning, Enhanced WebAssign combines the exceptional mathematics content that you know and love with the most powerful online homework solution, WebAssign. Enhanced WebAssign engages students with immediate feedback, rich tutorial content, and interactive, fully customizable eBooks (YouBook), helping students to develop a deeper conceptual understanding of their subject matter. Online assignments can be built by selecting from thousands of text-specific problems or supplemented with problems from any Cengage Learning textbook.

Student Resources

Student Solutions Manual
(ISBN 978-1-285-41762-2)
Authors: Ellena Reda, Dutchess Community College;
Pat Foard, South Plains College

Go beyond answers and improve your grade! This manual provides worked-out, step-by-step solutions to the odd-numbered problems in the text. The Student Solutions Manual gives you the information you need to truly understand how the problems are solved.

Student Workbook (ISBN 978-1-285-42030-1)
Author: Maria H. Andersen, Muskegon Community College

Get a head start. The Student Workbook contains assessments, activities, and worksheets for classroom discussions, in-class activities, and group work.

AIM for Success Student Practice Sheets
(ISBN 978-1-285-42026-4)
Author: Christine S. Verity

AIM for Success Student Practice Sheets provide additional problems to help you learn the material.

Enhanced WebAssign
Printed Access Card: 978-0-538-73810-1
Online Access Code: 978-1-285-18181-3

Enhanced WebAssign (assigned by the instructor) provides you with instant feedback on homework assignments. This online homework system is easy to use and includes helpful links to textbook sections, video examples, and problem-specific tutorials.

Acknowledgments

The authors would like to thank the people who have reviewed the sixth edition and provided many valuable suggestions.

Becky Bradshaw, *Lake Superior College*
Harvey Cartine, *Warren County Community College*
Jim Dawson, *College of Southern Idaho*
Cindy Dickson, *College of Southern Idaho*
Estella G. Elliott, *College of Southern Idaho*
Stephen Ester, *Saint Petersburg College*
Cassie Firth, *Northern Oklahoma College*
Lori L. Grady, *University of Wisconsin–Whitewater*
Nicholas Grener, *California State University, East Bay*
Ryan Grossman, *Ivy Tech Community College–Indiana*
Autumn Hoover, *Angelo State University*
Pat Horacek, *Pensacola State College*
Kelly Jackson, *Camden County College*
Thomas Judge, *California State University, East Bay*
Katy Koe, *Lincoln College*
William Lind, *Bryant and Stratton College*
Renee Lustig, *LeCordon Bleu College of Culinary Arts*
David Maina, *Columbia College, Chicago*
Connie Meade, *College of Southern Idaho*
Eugenia M. Moreno, *Butte Community College*
Dan Quynh Nguyen, *California State University, East Bay*
Rod Oberdick, *Delaware Technical Community College*
Scott Phelps, *University of La Verne*
David Poock, *Davenport University*
Nolan Thomas Rice, *College of Southern Idaho*
Daria Santerre, *Norwalk Community College*
Patricia Shepherd, *Ivy Tech Community College*
Darlyn Thomas, *Hennepin Technical College*
Sherri Urcavich, *University of Wisconsin–Green Bay*
Dr. Pamela D. Walker, *Northwestern College*
Donna M. Weglarz, *Westwood College–DuPage*
Lisa Williams, *College of the Abermarle*
Solomon Lee Willis, *Cleveland Community College*
Jerry Jacob Woods, *Westwood College*
Chen Zhixiong, *New Jersey City University*

Special thanks go to Jean Bermingham for copyediting the manuscript and proofreading pages, to Ellena Reda and Pat Foard for preparing the solutions manuals, and to Lauri Semarne for her work in ensuring the accuracy of the text. We would also like to thank the many people at Cengage Learning who worked to guide the manuscript for the sixth edition from development through production.

Index of Applications

AIM for Success

Focus on Success

This important chapter describes study skills that are used by students who have been successful in this course. Chapter A covers a wide range of topics that focus on what you need to do to succeed in this class. It includes a complete guide to the textbook and how to use its features to become a successful student.

hxdbzxy/Shutterstock.com

Prep Test

Are you ready to succeed in this course?

1. Read this chapter. Answer all of the questions. Write down your answers on paper.

2. Write down your instructor's name.

3. Write down the classroom number.

4. Write down the days and times the class meets.

5. Bring your textbook, a notebook, and a pen or pencil to every class.

6. Be an active participant, not a passive observer.

A.1 | How to Succeed in This Course

Get Ready

We are committed to your success in learning mathematics and have developed many tools and resources to support you along the way.

DO YOU WANT TO EXCEL IN THIS COURSE?

Read on to learn about the skills you'll need and how best to use this book to get the results you want.

We have written this text in an *interactive* style. More about this later but, in short, this means that you are supposed to interact with the text. Do not just read the text! Work along with it. Ready? Let's begin!

WHY ARE YOU TAKING THIS COURSE?

Did you interact with the text, or did you just read the last question? Get some paper and a pencil or pen and answer the question. Really—you will have more success in math and other courses you take if you **actively participate.** Now, **interact.** Write down one reason you are taking this course.

Of course, we have no idea what you just wrote, but experience has shown us that many of you wrote something along the lines of "I have to take it to graduate" or "It is a prerequisite to another course I have to take" or "It is required for my major." Those reasons are perfectly fine. Every teacher has had to take courses that were not directly related to his or her major.

WHY DO YOU WANT TO SUCCEED IN THIS COURSE?

Think about why you want to succeed in this course. List the reasons here (not in your head . . . on the paper!):

One reason you may have listed is that math skills are important in order to be successful in your chosen career. That is certainly an important reason. Here are some other reasons.

- Math is a skill that applies across careers, which is certainly a benefit in our world of changing job requirements. A good foundation in math may enable you to more easily make a career change.
- Math can help you learn critical thinking skills, an attribute all employers want.
- Math can help you see relationships between ideas and identify patterns.

Intellistudies/Shutterstock.com

Motivate Yourself

You'll find many real-life problems in this book, relating to sports, money, cars, music, and more. We hope that these topics will help you understand how mathematics is used in everyday life. To learn all of the necessary skills and to understand how you can apply them to your life outside of this course, motivate yourself to learn.

One of the reasons we asked you why you are taking this course was to provide motivation for you to succeed. When there is a reason to do something, that task is easier to accomplish. We understand that you may not want to be taking this course but, to achieve your career goal, this is a necessary step. Let your career goal be your motivation for success.

MAKE THE COMMITMENT TO SUCCEED!

With practice, you will improve your math skills. Skeptical? Think about when you first learned to drive a car, ride a skateboard, dance, paint, surf, or any other talent that you now have. You may have felt self-conscious or concerned that you might fail. But with time and practice, you learned the skill.

List a situation in which you accomplished your goal by spending time practicing and perfecting your skills (such as learning to play the piano or to play basketball):

You do not get "good" at something by doing it once a week. **Practice** is the backbone of any successful endeavor—including math!

Develop a "Can Do" Attitude Toward Math

You can do math! When you first learned the skills you just listed above, you may not have done them well. With practice, you got better. With practice, you will get better at math. Stay focused, motivated, and committed to success.

We cannot emphasize enough how important it is to overcome the "I Can't Do Math" syndrome. If you listen to interviews of very successful athletes after a particularly bad performance, you will note that they focus on the positive aspects of what they did, not the negative. Sports psychologists encourage athletes always to be positive—to have a "can do" attitude. Develop this attitude toward math and you will succeed.

Change your conversation about mathematics. Do not say "I can't do math," "I hate math," or "Math is too hard." These comments just give you an excuse to fail. You don't want to fail, and we don't want you to fail. Write it down now: I can do math!

Strategies for Success

PREPARE TO SUCCEED

There are a number of things that may be worrisome to you as you begin a new semester. List some of those things now.

Take Note

Motivation alone won't lead to success. For example, suppose a person who cannot swim is rowed out to the middle of a lake and thrown overboard. That person has a lot of motivation to swim, but most likely will drown without some help. You'll need motivation *and* learning in order to succeed.

William Perugini/Shutterstock.com

Here are some of the concerns expressed by our students.

- **Tuition**
  Will I be able to afford school?
- **Job**
  I must work. Will my employer give me a schedule that will allow me to go to school?
- **Anxiety**
  Will I succeed?
- **Child care**
  What will I do with my kids while I'm in class or when I need to study?
- **Time**
  Will I be able to find the time to attend class and study?
- **Degree goals**
  How long will it take me to finish school and earn my degree?

These are all important and valid concerns. Whatever your concerns, acknowledge them. Choose an education path that allows you to accommodate your concerns. Make sure they don't prevent you from succeeding.

SELECT A COURSE

Many schools offer math assessment tests. These tests evaluate your present math skills. They don't evaluate how smart you are, so don't worry about your score on the test. If you are unsure about where you should start in the math curriculum, these tests can show you where to begin. You are better off starting at a level that is appropriate for you than starting with a more advanced class and then dropping it because you can't keep up. Dropping a class is a waste of time and money.

If you have difficulty with math, avoid short courses that compress the class into a few weeks. If you have struggled with math in the past, this environment does not give you the time to process math concepts. Similarly, avoid classes that meet once a week. The time delay between classes makes it difficult to make connections between concepts.

Some career goals require a number of math courses. If that is true of your major, try to take a math course every semester until you complete the requirements. Think about it this way. If you take, say, French I, and then wait two semesters before taking French II, you may forget a lot of material. Math is much the same. You must keep the concepts fresh in your mind.

Time Management

One of the most important requirements in completing any task is to acknowledge the amount of time it will take to finish the job successfully. Before a construction company starts to build a skyscraper, the company spends months looking at how much time each of the phases of construction will take. This is done so that resources can be allocated when appropriate. For instance, it would not make sense to schedule the electricians to run wiring until the walls are up.

MANAGE YOUR TIME!

We know how busy you are outside of school. Do you have a full-time or a part-time job? Do you have children? Do you visit your family often? Do you play school sports or participate in the school orchestra or theater company? It can be stressful to balance all of the important activities and responsibilities in your life. Creating a time management plan will help you schedule enough time to do everything you need to do. Let's get started.

wavebreakmedia ltd/Shutterstock.com

First, you need a calendar. You can use a daily planner, a calendar for a smartphone, or an online calendar, such as the ones offered by Google, MSN, or Yahoo. It is best to have a calendar on which you can fill in daily activities and be able to see a weekly or monthly view as well.

Start filling in your calendar now, even if it means stopping right here and finding a calendar. Some of the things you might include are:

- The hours each class meets
- Time for driving to and from work or school
- Leisure time, an important aspect of a healthy lifestyle
- Time for study. Plan at least one hour of study for each hour in class. This is a *minimum!*

- Time to eat
- Your work schedule
- Time for extracurricular activities such as sports, music lessons, or volunteer work
- Time for family and friends
- Time for sleep
- Time for exercise

Take Note

Be realistic about how much time you have. One gauge is that working 10 hours per week is approximately equivalent to taking one three-unit course. If your college considers 15 units a full load and you are working 10 hours per week, you should consider taking 12 units. The more you work, the fewer units you should take.

We really hope you did this. If not, please reconsider. One of the best pathways to success is understanding how much time it takes to succeed. When you finish your calendar, if it does not allow you enough time to stay physically and emotionally healthy, rethink some of your school or work activities. We don't want you to lose your job because you have to study math. On the other hand, we don't want you to fail in math because of your job.

If math is particularly difficult for you, consider taking fewer course units during the semesters you take math. This applies equally to any other subject that you may find difficult. There is no rule that you must finish college in four years. It is a myth—discard it now.

Now extend your calendar for the entire semester. Many of the entries will repeat, such as the time a class meets. In your extended calendar, include significant events that may disrupt your normal routine. These might include holidays, family outings, birthdays, anniversaries, or special events such as a concert or a football game. In addition to these events, be sure to include the dates of tests, the date of the final exam, and dates that projects or papers are due. These are all important semester events. Having them on your calendar will remind you that you need to make time for them.

CLASS TIME

To be successful, **attend class.** You should consider your commitment to attend class as serious as your commitment to your job or to keeping an appointment with a dear friend. It is difficult to overstate the importance of attending class. If you miss work, you don't get paid. If you miss class, you are not getting the full benefit of your tuition dollar. You are losing money.

If, by some unavoidable situation, you cannot attend class, find out as soon as possible what was covered in class. You might:

- Ask a friend for notes and the assignment.
- Contact your instructor and get the assignment. Missing class is no excuse for not being prepared for the next class.
- Determine whether there are online resources that you can use to help you with the topics and concepts that were discussed in the class you missed.

Going to class is important. Once you are there, **participate in class.** Stay involved and active. When your instructor asks a question, try to at least mentally answer the question. If you have a question, ask. Your instructor expects questions and wants you to understand the concept being discussed.

HOMEWORK TIME

In addition to attending class, you must **do homework.** Homework is the best way to reinforce the ideas presented in class. You should plan on at least one to two hours of

CandyBox Images/Shutterstock.com

homework and study for each hour you are in class. We've had many students tell us that one to two hours seems like a lot of time. That may be true, but if you want to attain your goals, you must be willing to devote the time to being successful in this math course.

You should schedule study time just as if it were class time. To do this, write down where and when you study best. For instance, do you study best at home, in the library, at the math center, under a tree, or somewhere else? Some psychologists who research successful study strategies suggest that just by varying where you study, you can increase the effectiveness of a study session. While you are considering where you prefer to study, also think about the time of day during which your study period will be most productive. Write down your thoughts.

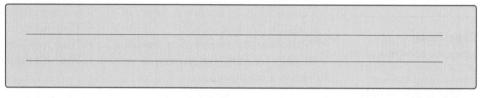

Look at what you have written, and be sure that you can consistently be in your favorite study environment at the time you have selected. Studying and homework are extremely important. Just as you should not miss class, **do not miss study time.**

Before we leave this important topic, we have a few suggestions. If at all possible, create a study hour right after class. The material will be fresh in your mind, and the immediate review, along with your homework, will help reinforce the concepts you are learning.

If you can't study right after class, make sure that you set aside some time *on the day of the class* to review notes and begin the homework. The longer you wait, the more difficult it will be to recall some of the important points covered during class. Study math in small chunks—one hour a day (perhaps not enough for most of us), every day, is better than seven hours in one sitting. If you are studying for an extended period of time, break up your study session by studying one subject for a while and then moving on to another subject. Try to alternate between similar or related courses. For instance, study math for a while, then science, and then back to math. Or study history for a while, then political science, and then back to history.

Meet some of the people in your class and try to **put together a study group.** The group could meet two or three times a week. During those meetings, you could quiz each other, prepare for a test, try to explain a concept to someone else in the group, or get help on a topic that is difficult for you.

After reading these suggestions, you may want to rethink where and when you study best. If so, do that now. Remember, however, that it is your individual style that is important. Choose what works for *you,* and stick to it.

Habits of Successful Students

Pattie Steib/Shutterstock.com

There are a number of habits that successful students use. Think about what these might be, and write them down.

What you have written is very important. The habits you have listed are probably the things you know you must do to succeed. Here is a list of some responses from successful students we have known.

- **Set priorities.** You will encounter many distractions during the semester. Do not allow them to prevent you from reaching your goal.

- **Take responsibility.** Your instructor, this textbook, tutors, math centers, and other resources are there to help you succeed. Ultimately, however, you must choose to learn. You must choose success.

- **Hang out with successful students.** Success breeds success. When you work and study with successful students, you are in an environment that will help you succeed. Seek out people who are committed to their goals.

- **Study regularly.** We have mentioned this before, but it is too important not to be repeated.

- **Self test.** Once every few days, select homework exercises from previous assignments and use them to test your understanding. Try to do these exercises without getting help from examples in the text. These self tests will help you gain confidence that you can do these types of problems on a test given in class.

- **Try different strategies.** If you read the text and are still having difficulty understanding a concept, consider going a step further. Contact the instructor or find a tutor. Many campuses have some free tutorial services. Go to the math or learning center. Consult another textbook. Be active and get the help you need.

- **Make flash cards.** This is one of the strategies that some math students do not think to try. Flash cards are a very important part of learning math. For instance, your instructor may use words or phrases such as *linear, quadratic, exponent, base, rational,* and many others. If you don't know the meanings of these words, you will not know what is being discussed.

- **Plod along.** Your education is not a race. The primary goal is to finish. Taking too many classes and then dropping some does not get you to the end any faster. Take only as many classes as you can successfully manage.

SECTION

A.2 How to Use This Text to Succeed in This Course

Helder Almeida/Shutterstock.com

Get the Big Picture

One of the major resources that you will have access to the entire semester is this text-book. We have written this text with you and your success in mind. The following is a guide to the features of this text that will help you succeed.

Actually, we want you to get the *really* big picture. Take a few minutes to read the table of contents. You may feel some anxiety about all the new concepts you will be learning. Try to think of this as an exciting opportunity to learn math. Now look through the entire book. Move quickly. Don't spend more than a few seconds on each page. Scan titles, look at pictures, and notice diagrams.

Getting this "big picture" view will help you see where this course is going. To reach your goal, it's important to get an idea of the steps you will need to take along the way.

As you look through the book, find topics that interest you. What's your preference? Racing? Sailing? TV? Amusement parks? Find the Index of Applications at the front of the book, and pull out three subjects that interest you. Write those topics here.

Understand the Organization

Look again at the Table of Contents. There are 12 chapters in this book. You'll see that every chapter is divided into sections, and each section contains a number of learning objectives. Each learning objective is labeled with a letter from A to G. Knowing how this book is organized will help you locate important topics and concepts as you're studying.

Before you start a new objective, take a few minutes to read the Objective Statement for that objective. Then, browse through the objective material. Especially note the words or phrases in bold type—these are important concepts that you'll need to know as you move along in the course. These words are good candidates for flash cards. If possible, include an example of the concept on the flash card, as shown at the left.

You will also see important concepts and rules set off in boxes. Here is one about multiplication. These rules are also good candidates for flash cards.

Flash Card

> **Rule for Multiplying Exponential Expressions**
>
> If m and n are positive integers, then $x^m \cdot x^n = x^{m+n}$.
>
> **Example:**
> $x^4 \cdot x^7 = x^{4+7} = x^{11}$

Rule for Multiplying Exponential Expressions

If m and n are positive integers, then $x^m \cdot x^n = x^{m+n}$.

EXAMPLES

1. $x^4 \cdot x^7 = x^{4+7} = x^{11}$
2. $z \cdot z^6 = z^{1+6} = z^7$ • Recall that $z = z^1$.
3. $v^2 \cdot v^3 \cdot v^5 = v^{2+3+5} = v^{10}$
4. $(a^3b^5)(a^4b^7) = a^{3+4}b^{5+7} = a^7b^{12}$ • Add exponents on like bases.

Leaf through Section 6.2 of Chapter 6. Write down words in bold and any concepts or rules that are displayed in boxes.

Use the Interactive Method

As we mentioned earlier, this textbook is based on an interactive approach. We want you to be actively involved in learning mathematics, and have given you many suggestions for getting "hands-on" with this book.

HOW TO Look on page 376. See the HOW TO 1? A HOW TO introduces a concept (in this case, dividing a polynomial by a monomial) and includes a step-by-step solution of the type of exercise you will find in the homework.

HOW TO 1   Divide and check: $\dfrac{16x^5 - 8x^3 + 4x}{2x}$

$$\dfrac{16x^5 - 8x^3 + 4x}{2x} = \dfrac{16x^5}{2x} - \dfrac{8x^3}{2x} + \dfrac{4x}{2x}$$

- Divide each term in the numerator by the denominator.

$$= 8x^4 - 4x^2 + 2$$

- Simplify each quotient.

Check:
$$2x(8x^4 - 4x^2 + 2) = 16x^5 - 8x^3 + 4x$$

- The quotient checks.

Grab paper and a pencil and work along as you're reading through a HOW TO. When you're done, get a clean sheet of paper. Write down the problem and try to complete the solution without looking at your notes or at the book. When you're done, check your answer. If you got it right, you're ready to move on.

Look through the text and find three instances of a HOW TO. Write the concept illustrated in each HOW TO here.

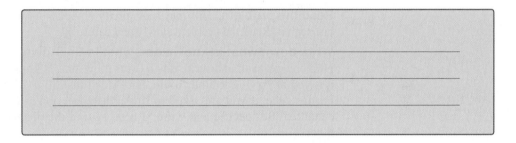

Example/You Try It Pair You'll need hands-on practice to succeed in mathematics. When we show you an example, work it out yourself, right beside the solution. Use the Example/You Try It pairs to get the practice you need.

Take a look at page 376. Example 1 and You Try It 1 are shown here.

EXAMPLE 1

Divide and check: $\dfrac{6x^3 - 3x^2 + 9x}{3x}$

Solution

$$\dfrac{6x^3 - 3x^2 + 9x}{3x}$$

$$= \dfrac{6x^3}{3x} - \dfrac{3x^2}{3x} + \dfrac{9x}{3x}$$

- Divide each term in the numerator by the denominator.

$$= 2x^2 - x + 3$$

- Simplify each quotient.

Check: $3x(2x^2 - x + 3) = 6x^3 - 3x^2 + 9x$

YOU TRY IT 1

Divide and check: $\dfrac{4x^3y + 8x^2y^2 - 4xy^3}{2xy}$

Your solution

Solution on p. S18

You'll see that each Example is fully worked out. Study the Example by carefully working through each step. Then, try to complete the You Try It. Use the solution to the Example as a model for solving the You Try It. If you get stuck, the solutions to the You Try Its are provided in the back of the book. There is a page number directly following the You Try It that shows you where you can find the completely-worked-out solution. Use the solution to get a hint for the step on which you are stuck. Then, try again!

When you've arrived at your solution, check your work against the solution in the back of the book. Turn to page S18 to see the solution for You Try It 1.

Remember that sometimes there is more than one way to solve a problem. But your answer should always match the answer we've given in the back of the book. If you have any questions about whether your method will always work, check with your instructor.

Use a Strategy to Solve Word Problems

Learning to solve word problems is one of the reasons you are studying math. This is where you combine all of the critical thinking skills you have learned to solve practical problems.

Try not to be intimidated by word problems. Basically, what you need is a strategy that will help you come up with the equation you will need to solve the problem. When you are looking at a word problem, try the following:

- **Read the problem.** This may seem pretty obvious, but we mean really **read** it. Don't just scan it. Read the problem slowly and carefully.

- **Write down what is known and what is unknown.** Now that you have read the problem, go back and write down everything that is known. Next, write down what it is you are trying to find. *Write* this—don't just think it! Be as specific as you can. For instance, if you are asked to find a distance, don't just write "I need to find the distance." Be specific and write "I need to find the distance between Earth and the moon."

- **Think of a method to find the unknown.** For instance, is there a formula that relates the known and unknown quantities? This is certainly the most difficult step. Eventually, you must write an equation to be solved.

- **Solve the equation.** Be careful as you solve the equation. There is no sense in getting to this point and then making a careless mistake. The unknown in most word problems will include a unit such as feet, dollars, or miles per hour. When you write your answer, include the unit. An answer such as 20 doesn't mean much. Is it 20 feet, 20 dollars, 20 miles per hour, or something else?

- **Check your solution.** Now that you have an answer, go back to the problem and ask yourself whether it makes sense. This is an important step. For instance, if, according to your answer, the cost of a car is $2.51, you know that something went wrong.

In this text, the solution of every word problem is broken down into two steps, **Strategy** and **Solution.** The Strategy consists of the first three steps discussed above. The Solution is the last two steps. Here is an Example from page 504 of the text. Because you have not yet studied the concepts involved in the problem, you may not be able to solve it. However, note the detail in the Strategy. When you do the You Try It following an Example, be sure to include your own Strategy.

When you have finished studying a section, **do the exercises your instructor has selected.** Math is not a spectator sport. You must practice every day. Do the homework and do not get behind.

EXAMPLE 1

The amount A of medication prescribed for a person is directly related to the person's weight W. For a 50-kilogram person, 2 ml of medication are prescribed. How many milliliters of medication are required for a person who weighs 75 kg?

Strategy

To find the required amount of medication:
- Write the basic direct variation equation, replace the variables by the given values, and solve for k.
- Write the direct variation equation, replacing k by its value. Substitute 75 for W and solve for A.

Solution

$A = kW$ • Direct variation equation

$2 = k \cdot 50$ • Replace A by 2 and W by 50.

$\dfrac{1}{25} = k$ • Solve for k.

$A = \dfrac{1}{25}W$ • Write the direct variation equation.

$A = \dfrac{1}{25} \cdot 75 = 3$ • Find A when $W = 75$.

The required amount of medication is 3 ml.

YOU TRY IT 1

The distance s a body falls from rest varies directly as the square of the time t of the fall. An object falls 64 ft in 2 s. How far will it fall in 5 s?

Your strategy

Your solution

Solution on p. S25

Ace the Test

There are a number of features in this text that will help you prepare for a test. These features will help you even more if you do just one simple thing: When you are doing your homework, go back to each previous homework assignment for the current chapter and rework two exercises. That's right—just *two* exercises. You will be surprised at how much better prepared you will be for a test by doing this.

Here are some additional aids to help you ace the test.

Chapter Summary Once you've completed a chapter, look at the Chapter Summary. The Chapter Summary is divided into two sections: **Key Words** and **Essential Rules and Procedures.** Flip to page 387 to see the Chapter Summary for Chapter 6. The summary shows all of the important topics covered in the chapter. Do you see the reference following each topic? This reference shows you the objective and page in the text where you can find more information on the concept.

Write down one Key Word and one Essential Rule or Procedure. Explain the meaning of the reference "6.1A, page 344."

Chapter Review Exercises Turn to page 390 to see the Chapter Review Exercises for Chapter 6. When you do the review exercises, you're giving yourself an important opportunity to test your understanding of the chapter. The answer to each review exercise is given at the back of the book, along with the objective the question relates to. When you're done with the Chapter Review Exercises, check your answers. If you had trouble with any of the questions, you can restudy the objectives and retry some of the exercises in those objectives for extra help.

Go to the Answer Section at the back of the text. Find the answers for the Chapter Review Exercises for Chapter 6. Write down the answer to Exercise 4. Explain the meaning of the reference "6.3C."

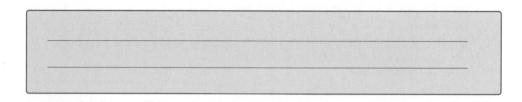

Chapter Test The Chapter Test for each chapter can be found after the Chapter Review Exercises and can be used to help you prepare for your exam. The answer to each question is given at the back of the book, along with both an objective reference and a reference to a HOW TO, Example, or You Try It that the question relates to. Think of these tests as "practice runs" for your in-class tests. Take the test in a quiet place, and try to work through it in the same amount of time that will be allowed for your actual exam.

The aids we have mentioned above will help you prepare for a test. You should begin your review *at least* two days before the test—three days is better. These aids will get you ready for the test.

Here are some suggestions to try while you are actually taking the test.

- **Try to relax.** We know that test situations make some students quite nervous or anxious. These feelings are normal. Try to stay calm and focused on what you know. If you have prepared as we have suggested, the answers will begin to come to you.
- **Scan the test.** Get a feeling for the big picture.
- **Read the directions carefully.** Make sure you answer each question fully.
- **Work the problems that are easiest for you first.** This will help you with your confidence and help reduce any nervous feelings you may have.

Ready, Set, Succeed!

hxdbzxy/Shutterstock.com

It takes hard work and commitment to succeed, but we know you can do it! Doing well in mathematics is just one step you'll take on your path to success. Good luck. We wish you success.

Real Numbers and Variable Expressions

1

OBJECTIVES

SECTION 1.1
- **A** To use inequality symbols with integers
- **B** To find the additive inverse and absolute value of a number
- **C** To add or subtract integers
- **D** To multiply or divide integers
- **E** To solve application problems

SECTION 1.2
- **A** To write a rational number as a decimal
- **B** To convert among percents, fractions, and decimals
- **C** To add or subtract rational numbers
- **D** To multiply or divide rational numbers
- **E** To evaluate exponential expressions
- **F** To simplify numerical radical expressions
- **G** To solve application problems

SECTION 1.3
- **A** To use the Order of Operations Agreement to simplify expressions

SECTION 1.4
- **A** To evaluate a variable expression
- **B** To simplify a variable expression using the Properties of Addition
- **C** To simplify a variable expression using the Properties of Multiplication
- **D** To simplify a variable expression using the Distributive Property
- **E** To translate a verbal expression into a variable expression

SECTION 1.5
- **A** To write a set using the roster method
- **B** To write and graph sets of real numbers

Focus on Success

Have you read Chapter A, AIM for Success? It describes study skills used by students who have been successful in their math courses. It gives you tips on how to stay motivated, how to manage your time, and how to prepare for exams. Chapter A also includes a complete guide to the textbook and how to use its features to be successful in this course. It starts on page AIM-1.

Katrina Brown/Shutterstock.com

Prep Test

Are you ready to succeed in this chapter? Take the Prep Test below to find out if you are ready to learn the new material.

1. What is 127.1649 rounded to the nearest hundredth?
127.16

2. Add: $3416 + 42{,}561 + 537$
46,514

3. Subtract: $5004 - 487$
4517

4. Multiply: 407×28
11,396

5. Divide: $11{,}684 \div 23$
508

6. What is the smallest number that both 8 and 12 divide evenly?
24

7. What is the greatest number that divides both 16 and 20 evenly?
4

8. Without using 1, write 21 as a product of two whole numbers.
$3 \cdot 7$

9. Represent the shaded portion of the figure as a fraction in simplest form.
$\frac{2}{5}$

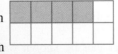

10. Which of the following, if any, is not possible?
(i) $6 + 0$ **(ii)** $6 - 0$
(iii) 6×0 **(iv)** $6 \div 0$
iv

INSTRUCTOR NOTE
These notes offer suggestions for presenting the material in the objective or provide information that may be helpful.

Unless otherwise noted, all content on this page is © Cengage Learning.

Valentyn Volkov/Shutterstock.com

SECTION

1.1   Introduction to Integers

OBJECTIVE A   *To use inequality symbols with integers*

The desire to group similar items seems to be a human characteristic. For instance, biologists place similar animals in groups called *species.* Nutritionists classify foods according to *food groups;* for example, pasta, crackers, and rice are among the foods in the bread group.

Mathematicians place objects with similar properties in groups called *sets.* A **set** is a collection of objects. The objects in a set are called the **elements of the set.**

The **roster method** of writing sets encloses a list of the elements in braces. Thus the set of sections within an orchestra is written {brass, percussion, string, woodwind}. When the elements of a set are listed, each element is listed only once. For instance, if the list of numbers 1, 2, 3, 2, 3 were placed in a set, the set would be {1, 2, 3}.

The symbol $\in$ means "is an element of." $2 \in B$ is read "2 is an element of set B."

Given $C = \{3, 5, 9\}$, then $3 \in C$, $5 \in C$, and $9 \in C$. $7 \notin C$ is read "7 is not an element of set C."

The numbers that we use to count objects, such as the students in a classroom or the horses on a ranch, are the *natural numbers.*

Natural numbers = $\{1, 2, 3, 4, 5, 6, 7, 8, 9, 10, \ldots\}$

The three dots mean that the list of natural numbers continues on and on and that there is no largest natural number.

Each natural number greater than 1 is a *prime* number or a *composite* number. A **prime number** is a natural number greater than 1 that is divisible (evenly) only by itself and 1. For example, 2, 3, 5, 7, 11, and 13 are the first six prime numbers. A natural number that is not a prime number is a **composite number.** The numbers 4, 6, 8, and 9 are the first four composite numbers.

The natural numbers do not have a symbol to denote the concept of none—for instance, the number of trees taller than 1000 feet. The *whole numbers* include zero and the natural numbers.

Whole numbers = $\{0, 1, 2, 3, 4, 5, 6, 7, 8, \ldots\}$

The whole numbers alone do not provide all the numbers that are useful in applications. For instance, a meteorologist needs numbers below zero and above zero.

Integers = $\{\ldots, -5, -4, -3, -2, -1, 0, 1, 2, 3, 4, 5, \ldots\}$

Each integer can be shown on a number line. The integers to the left of zero on the number line are called **negative integers.** The integers to the right of zero are called **positive integers,** or natural numbers. Zero is neither a positive nor a negative integer.

INSTRUCTOR NOTE

Margin notes entitled *Point of Interest, Take Note, Tips for Success,* and *Integrating Technology* are printed in the student text. The *Point of Interest* feature provides a historical note or mathematical fact of interest. The *Take Note* feature flags important information or provides assistance in understanding a concept. The *Tips for Success* feature offers suggestions on how to use this text and how to develop good study habits. The *Integrating Technology* feature describes some of the functions of a calculator.

Point of Interest

The Alexandrian astronomer Ptolemy began using *omicron, 0,* the first letter of the Greek word that means "nothing," as the symbol for zero in 150 A.D. It was not until the 13th century, however, that Fibonacci introduced 0 to the Western world as a placeholder so that we could distinguish, for example, 45 from 405.

INSTRUCTOR NOTE

Within the Microsoft PowerPoint® slides available with this text is a number line that extends from −10 to 10. It can be used to create a transparency on which to graph integers.

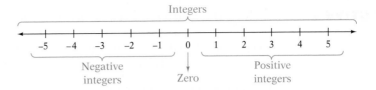
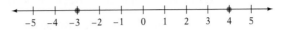

Point of Interest

The number zero was the cause of quite a controversy over the date on which the new millennium started. When our current calendar was created, numbering began with the year 1 because 0 had not yet been invented. Thus at the beginning of year 2, 1 year had elapsed; at the beginning of year 3, 2 years had elapsed; and so on. This means that at the beginning of year 2000, 1999 years had elapsed. It was not until the beginning of year 2001 that 2000 years had elapsed and a new millennium began.

The **graph** of an integer is shown by placing a heavy dot on the number line directly above the number. The graphs of -3 and 4 are shown on the number line below.

Consider the following sentences.

The quarterback threw the football and the receiver caught *it*.

A student purchased a computer and used *it* to write history papers.

In the first sentence, *it* is used to mean the football; in the second sentence, *it* means the computer. In language, the word *it* can stand for many different objects. Similarly, in mathematics, a letter of the alphabet can be used to stand for a number. Such a letter is called a **variable.** Variables are used in the following definition of inequality symbols.

Point of Interest

The symbols for "is less than" and "is greater than" were introduced by Thomas Harriot around 1630. Before that, ⊏ and ⊐ were used for $>$ and $<$, respectively.

Inequality Symbols

If a and b are two numbers and a is to the left of b on the number line, then a **is less than** b. This is written $a < b$.

If a and b are two numbers and a is to the right of b on the number line, then a **is greater than** b. This is written $a > b$.

EXAMPLES

1. $-2 < 1$ -2 is to the left of 1 on the number line. -2 is less than 1.

2. $3 > 0$ 3 is to the right of 0 on the number line. 3 is greater than 0.

The inequality symbols $\leq$ (is less than or equal to) and $\geq$ (is greater than or equal to) are also important. Note the examples below.

$$4 \leq 5 \text{ is a true statement because } 4 < 5.$$
$$5 \leq 5 \text{ is a true statement because } 5 = 5.$$

EXAMPLE 1

Let $y \in \{-7, 0, 6\}$. For which values of y is the inequality $y < 4$ a true statement?

Solution

Replace y by each of the elements of the set and determine whether the inequality is true.

$$y < 4$$
$$-7 < 4 \quad \text{True}$$
$$0 < 4 \quad \text{True}$$
$$6 < 4 \quad \text{False}$$

The inequality is true for -7 and 0.

YOU TRY IT 1

Let $z \in \{-10, -5, 6\}$. For which values of z is the inequality $z > -5$ a true statement?

Your solution

6

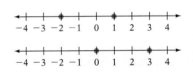

IN-CLASS EXAMPLES

1. Given $A = \{-9, -6, -3, 0, 3\}$, which elements of set A are:
   a. greater than -2? **0, 3**
   b. less than -5? **$-9, -6$**
   c. less than or equal to -6? **$-9, -6$**

Solution on p. S1

Unless otherwise noted, all content on this page is © Cengage Learning.

OBJECTIVE B | *To find the additive inverse and absolute value of a number*

The numbers 5 and -5 are the same distance from zero on the number line but on opposite sides of zero. The numbers 5 and -5 are called **additive inverses** or **opposites**.

The additive inverse (or opposite) of 5 is -5. The additive inverse of -5 is 5. The symbol for additive inverse is $-$.

$-(2)$ means the additive inverse of *positive* 2.

$-(-5)$ means the additive inverse of *negative* 5.

$-(2) = -2$

$-(-5) = 5$

The **absolute value** of a number is its distance from zero on the number line. The symbol for absolute value is $|\ \ |$.

Note from the figure above that the distance from 0 to 5 is 5. Therefore, $|5| = 5$. The figure also shows that the distance from 0 to -5 is 5. Therefore, $|-5| = 5$.

★ **Tips for Success**

One of the key instructional features of this text is the Example/You Try It pairs. Each Example is completely worked. You are to solve the You Try It problems. When you are ready, check your solution against the one given in the Solution section. The solution for You Try It 2 below is on page S1 (see the reference at the bottom right of the You Try It box). See *AIM for Success* in the Preface.

Absolute Value

The absolute value of a positive number is the number itself.
The absolute value of a negative number is the opposite of the number.
The absolute value of zero is zero.

EXAMPLES

1. $|6| = 6$ **2.** $|-8| = 8$ **3.** $|0| = 0$

HOW TO 1 Evaluate: $-|-12|$

$-|-12| = -12$ • The absolute value symbol does not affect the negative sign *in front of* the absolute value symbol.

EXAMPLE 2

Let $y \in \{-12, 0, 4\}$.
a. Determine $-y$, the additive inverse of y, for each element of the set.
b. Evaluate $|y|$ for each element of the set.

Solution

a. Replace y in $-y$ by each element of the set and determine the value of the expression.

$-y$
$-(-12) = 12$
$-(0) = 0$ • 0 is neither positive
$-(4) = -4$ nor negative.

b. Replace y in $|y|$ by each element of the set and determine the value of the expression.

$|y|$
$|-12| = 12$
$|0| = 0$
$|4| = 4$

YOU TRY IT 2

Let $d \in \{-11, 0, 8\}$.
a. Determine $-d$, the additive inverse of d, for each element of the set.
b. Evaluate $|d|$ for each element of the set.

Your solution

a. 11, 0, -8

b. 11, 0, 8

IN-CLASS EXAMPLES

2. Find the additive inverse of -64. **64**
3. Let $x \in \{-8, 0, 9\}$. For which values of x is $x > -2$ true? **0, 9**
4. a. Evaluate $|-14|$. **14**
   b. Evaluate $-|23|$. **-23**

Solution on p. S1

Unless otherwise noted, all content on this page is © Cengage Learning.

OBJECTIVE C *To add or subtract integers*

A number can be represented anywhere along the number line by an arrow. A positive number is represented by an arrow pointing to the right, and a negative number is represented by an arrow pointing to the left. The size of the number is represented by the length of the arrow.

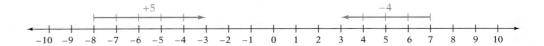

Addition is the process of finding the total of two numbers. The numbers being added are called **addends.** The total is called the **sum.** Addition of integers can be shown on the number line. To add integers, start at zero and draw, above the number line, an arrow representing the first number. At the tip of the first arrow, draw a second arrow representing the second number. The sum is below the tip of the second arrow.

$4 + 2 = 6$

$-4 + (-2) = -6$

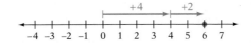

$-4 + 2 = -2$

$4 + (-2) = 2$

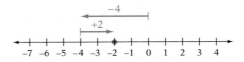

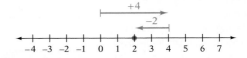

INSTRUCTOR NOTE

There are a number of models of addition of integers. A model that makes use of arrows on the number line is just one of them. Another suggestion is to use a checking account. If there is a balance of $25 in a checking account and a check is written for $30, the account will be overdrawn by $5 (−5).

Another model uses two colors of plastic chips, say blue for positive and red for negative, and the idea that a blue-red pair represents zero. To add −8 + 3, place 8 red and 3 blue chips in a circle. Make as many blue-red pairs as possible and remove them from the region. There are 5 red chips remaining, or −5.

The pattern for addition shown on the number lines above is summarized in the following rules for adding integers.

Addition of Integers

To add two numbers with the same sign, add the absolute values of the numbers. Then attach the sign of the addends.

To add two numbers with different signs, find the absolute value of each number. Subtract the smaller of the absolute values from the larger. Then attach the sign of the number with the larger absolute value.

EXAMPLES

1. Add: $-12 + (-26)$
$-12 + (-26) = -38$

• The signs are the same.
• Add the absolute values of the numbers $(12 + 26)$. Attach the sign of the addends.

2. Add: $-19 + 8$
$|-19| = 19; |8| = 8$
$19 - 8 = 11$
$-19 + 8 = -11$

• The signs are different.
• Find the absolute value of each number.
• Subtract the smaller absolute value from the larger.
• Attach the sign of the number with the larger absolute value.

Unless otherwise noted, all content on this page is © Cengage Learning.

Tips for Success
The HOW TO feature indicates an example with explanatory remarks. Using paper and pencil, you should work through the example. See *AIM for Success* in the Preface.

INSTRUCTOR NOTE
Sometimes students do not see the difference between a negative sign and a minus sign. Give examples such as $-3 - 6$, $-5 - (-3)$, and $8 - 12$, and ask students to distinguish the minus signs from the negative signs.

INSTRUCTOR NOTE
A subtraction model based on blue and red chips and similar to the addition model can be provided. Restrict the terms of the subtraction to, say, between -10 and 10, and start with 10 blue-red pairs in a circle. Because each blue-red pair represents zero, the circle just contains 10 zeros. To model $-3 - (-7)$, place 3 more red chips in the circle and remove (subtract) 7 red chips. Now pair as many blue and red chips as possible. There will be 4 blue chips without a red chip. Hence $-3 - (-7) = 4$.

HOW TO 2 Find the sum of -23, 47, -18, and -10.

Recall that a sum is the answer to an addition problem.

$$-23 + 47 + (-18) + (-10)$$
$$= 24 + (-18) + (-10)$$
$$= 6 + (-10) = -4$$

- To add more than two numbers, add the first two numbers. Then add the sum to the third number. Continue until all the numbers are added.

Look at the expressions below. Note that each expression equals the same number.

$$8 - 3 = 5 \quad \text{8 minus 3 is 5.}$$
$$8 + (-3) = 5 \quad \text{8 plus the opposite of 3 is 5.}$$

This example suggests the following.

Subtraction of Integers

To subtract one number from another, add the opposite of the second number to the first number.

EXAMPLES

1. Subtract: $-21 - (-40)$

$$-21 - (-40) = -21 + 40$$
- Rewrite the subtraction as addition of the opposite.
$$= 19$$
- Add.

2. Subtract: $15 - 51$

$$15 - 51 = 15 + (-51)$$
- Rewrite the subtraction as addition of the opposite.
$$= -36$$
- Add.

HOW TO 3 Subtract: $-12 - (-21) - 15$

$$-12 - (-21) - 15 = -12 + 21 + (-15)$$
$$= 9 + (-15) = -6$$

- Rewrite each subtraction as addition of the opposite. Then add.

HOW TO 4 Find the difference between -8 and 7.

A *difference* is the answer to a subtraction problem.

$$-8 - 7 = -8 + (-7)$$
$$= -15$$

- Rewrite the subtraction as addition of the opposite. Then add.

EXAMPLE 3

Add: $-52 + (-39)$

Solution

$-52 + (-39) = -91$

YOU TRY IT 3

Add: $100 + (-43)$

Your solution

57

EXAMPLE 4

Add: $37 + (-52) + (-14)$

Solution

$37 + (-52) + (-14) = -15 + (-14)$
$= -29$

YOU TRY IT 4

Add: $-51 + 42 + 17 + (-102)$

Your solution

-94

EXAMPLE 5

Find 11 more than -23.

Solution

$-23 + 11 = -12$

YOU TRY IT 5

Find -8 increased by 7.

Your solution

-1

EXAMPLE 6

Subtract: $-14 - 18 - (-21) - 4$

Solution

$-14 - 18 - (-21) - 4$
$= -14 + (-18) + 21 + (-4)$
$= -32 + 21 + (-4)$
$= -11 + (-4) = -15$

YOU TRY IT 6

Subtract: $-9 - (-12) - 17 - 4$

Your solution

-18

IN-CLASS EXAMPLES
Add.
5. $-43 + 75$ **32**
6. $39 + (-50) + (-13)$ **−24**
7. $-13 + 23 + (-7) + (-3)$ **0**
Subtract.
8. $13 - (-4)$ **17**
9. $-3 - 6 - (-12) - 9$ **−6**

EXAMPLE 7

Find 9 less than -4.

Solution

$-4 - 9 = -4 + (-9) = -13$

YOU TRY IT 7

Subtract -12 from -11.

Your solution

1

Solutions on p. S1

OBJECTIVE D *To multiply or divide integers*

 Point of Interest

The cross × was first used as a symbol for multiplication in 1631 in a book titled *The Key to Mathematics.* Also in that year, another book, *Practice of the Analytical Art,* advocated the use of a dot to indicate multiplication.

Several different symbols are used to indicate multiplication. The numbers being multiplied are called **factors;** for instance, 3 and 2 are factors in each of the examples at the right. The result is called the **product.** Note that when parentheses are used and there is no arithmetic symbol, the operation is multiplication.

$3 \times 2 = 6$
$3 \cdot 2 = 6$
$(3)(2) = 6$
$3(2) = 6$
$(3)2 = 6$

Multiplication is repeated addition of the same number. The product 3×5 is shown on the number line below.

5 is added 3 times.
$3 \times 5 = 5 + 5 + 5 = 15$

Now consider the product of a positive and a negative number.

-5 is added 3 times.
$$3(-5) = (-5) + (-5) + (-5) = -15$$

This example suggests that the product of a positive number and a negative number is negative. Here are a few more examples.

$$4(-7) = -28 \qquad -6 \cdot 7 = -42 \qquad (-8)7 = -56$$

Unless otherwise noted, all content on this page is © Cengage Learning.

To find the product of two negative numbers, look at the pattern at the right. As -5 multiplies a sequence of decreasing integers, the products increase by 5.

These numbers decrease by 1. → ← These numbers increase by 5.

$$-5 \times 3 = -15$$
$$-5 \times 2 = -10$$
$$-5 \times 1 = -5$$
$$-5 \times 0 = 0$$

The pattern can be continued by requiring that the product of two negative numbers be positive.

$$-5 \times (-1) = 5$$
$$-5 \times (-2) = 10$$
$$-5 \times (-3) = 15$$

INSTRUCTOR NOTE

A multiplication model based on blue and red chips and similar to that for addition and subtraction can be provided. Start with 10 blue-red pairs in a circle. Recall that because each blue-red pair represents zero, the circle just contains 10 zeros.

To model multiplication, the sign of the first factor determines whether we "put in" or "take out" chips, and the sign of the second factor determines which color chips we use.

To model $3(-2)$, put in 2 additional red chips three times. Now pair as many blue and red chips as possible. There will be 6 red chips without a blue chip: $3(-2) = -6$.

To model $-3(-2)$, start with 10 blue-red pairs. Take out 2 red chips three times. Now pair as many blue and red chips as possible, leaving 6 blue chips without a red chip. Hence, $-3(-2) = 6$.

Multiplication of Integers

To multiply two numbers with the same sign, multiply the absolute values of the numbers. The product is positive.

To multiply two numbers with different signs, multiply the absolute values of the numbers. The product is negative.

EXAMPLES

1. $-4(-8) = 32$ • The signs are the same. The product is positive.
2. $5(-12) = -60$ • The signs are different. The product is negative.

HOW TO 5 Find the product of -8 and -16.

A *product* is the answer to a multiplication problem.

$$-8(-16) = 128$$ • The signs are the same. The product is positive.

HOW TO 6 Multiply: $-2(5)(-7)(-4)$

$$-2(5)(-7)(-4) = -10(-7)(-4)$$
$$= 70(-4) = -280$$

• To multiply more than two numbers, multiply the first two. Then multiply the product by the third number. Continue until all the numbers are multiplied.

Consider the products shown at the right. Note that when there is an even number of negative factors, the product is positive. When there is an odd number of negative factors, the product is negative.

$$(-3)(-5) = 15$$
$$(-2)(-5)(-6) = -60$$
$$(-4)(-3)(-5)(-7) = 420$$
$$(-3)(-3)(-5)(-4)(-5) = -900$$
$$(-6)(-3)(-4)(-2)(-10)(-5) = 7200$$

Take Note

Think of the fraction bar as "divided by." Thus $\frac{8}{2}$ is 8 divided by 2. The number 2 is the **divisor.** The number 8 is the **dividend.** The result of the division, 4, is called the **quotient.**

This idea can be summarized by the following useful rule: The product of an even number of negative factors is positive; the product of an odd number of negative factors is negative.

For every division problem there is a related multiplication problem.

$$\frac{8}{2} = 4 \qquad \text{because} \qquad 4 \cdot 2 = 8.$$

Division   Related multiplication

This fact and the rules for multiplying integers can be used to illustrate the rules for dividing integers.

Note below that the quotient of two numbers with the same sign is positive.

$$\frac{12}{3} = 4 \text{ because } 4 \cdot 3 = 12. \qquad \frac{-12}{-3} = 4 \text{ because } 4(-3) = -12.$$

The next two examples illustrate that the quotient of two numbers with different signs is negative.

$$\frac{12}{-3} = -4 \text{ because } (-4)(-3) = 12. \qquad \frac{-12}{3} = -4 \text{ because } (-4)3 = -12.$$

Take Note

We can denote division using, for example,
$-18 \div (-2)$,
$-2\overline{)-18}$, or $\frac{-18}{-2}$.

INSTRUCTOR NOTE

The relationship between division and multiplication may help students better understand that division by zero is undefined. Introduce the students to a problem such as $12 \div 4$. When the students express confidence that $12 \div 4 = 3$, show that the answer can be "checked" by verifying that the product of the quotient and the divisor equals the dividend—that is, $3 \cdot 4 = 12$. Then introduce the problem $12 \div 0$ and ask what number, when multiplied by 0, produces the answer 12. Although students may guess several incorrect numbers, which can be rejected by actually carrying out the multiplication, they will conclude that there is no such number. In other words, division by zero is "undefined."

To relate division by zero to a real-life situation, show students that $6 \div 3 = 2$ means that if \$6 is divided among 3 people, each person receives \$2. Now, how can \$6 be divided among 0 people?

Division of Integers

To divide two numbers with the same sign, divide the absolute values of the numbers. The quotient is positive.

To divide two numbers with different signs, divide the absolute values of the numbers. The quotient is negative.

EXAMPLES

1. $-18 \div (-2) = 9$ • The signs are the same. The quotient is positive.

2. $-36 \div 9 = -4$ • The signs are different. The quotient is negative.

HOW TO 7   Find the quotient of -63 and -7.

A *quotient* is the answer to a division problem.

$$\frac{-63}{-7} = 9$$ • The signs are the same. The quotient is positive.

HOW TO 8   Simplify: $-\frac{-56}{7}$

$$-\frac{-56}{7} = -\left(\frac{-56}{7}\right) = -(-8) = 8$$

The properties of division are stated below. In these statements, the symbol $\neq$ is read "is not equal to."

Properties of Zero and One in Division

If $a \neq 0$, $\dfrac{0}{a} = 0$.   Zero divided by any number other than zero is zero.

If $a \neq 0$, $\dfrac{a}{a} = 1$.   Any number other than zero divided by itself is 1.

$\dfrac{a}{1} = a$ A number divided by 1 is the number.

$\dfrac{a}{0}$ is undefined.   Division by zero is not defined.

EXAMPLES

1. $\dfrac{0}{-5} = 0$ **2.** $\dfrac{-4}{-4} = 1$ **3.** $\dfrac{-7}{1} = -7$ **4.** $\dfrac{-12}{0}$ is undefined.

The fact that $\frac{-12}{3} = -4$, $\frac{12}{-3} = -4$, and $-\frac{12}{3} = -4$ suggests the following rule.

If a and b are integers, and $b \neq 0$, then $\frac{-a}{b} = \frac{a}{-b} = -\frac{a}{b}$.

EXAMPLE 8

Multiply: $(-3)4(-5)$

Solution

$(-3)4(-5) = (-12)(-5) = 60$

YOU TRY IT 8

Multiply: $8(-9)10$

Your solution

-720

EXAMPLE 9

Multiply: $12(-4)(-3)(-5)$

Solution

$12(-4)(-3)(-5) = (-48)(-3)(-5)$
$\qquad = 144(-5)$
$\qquad = -720$

YOU TRY IT 9

Multiply: $(-2)3(-8)7$

Your solution

336

IN-CLASS EXAMPLES
Multiply.
10. $4(-9)(-10)$ **360**
11. $-5(-3)(-1)(2)$ **−30**
Divide.
12. $(-120) \div (-20)$ **6**
13. $\frac{-24}{6}$ **−4**

EXAMPLE 10

Find the product of -13 and -9.

Solution

$-13(-9) = 117$

YOU TRY IT 10

What is -9 times 34?

Your solution

-306

14. Simplify: $-\frac{-36}{-3}$ **−12**

EXAMPLE 11

Divide: $(-120) \div (-8)$

Solution

$(-120) \div (-8) = 15$

YOU TRY IT 11

Divide: $(-135) \div (-9)$

Your solution

15

EXAMPLE 12

Divide: $\frac{95}{-5}$

Solution

$\frac{95}{-5} = -19$

YOU TRY IT 12

Divide: $\frac{-72}{4}$

Your solution

-18

EXAMPLE 13

Simplify: $-\frac{-81}{3}$

Solution

$-\frac{-81}{3} = -(-27) = 27$

YOU TRY IT 13

Simplify: $-\frac{36}{-12}$

Your solution

3

EXAMPLE 14

Find the quotient of 98 and -14.

Solution

$98 \div (-14) = -7$

YOU TRY IT 14

What is the ratio of -72 and -8?

Your solution

9

Solutions on p. S1

OBJECTIVE E   *To solve application problems*

An **export** is a good or service produced in one's own country and sold for consumption in another country. An **import** is a good or service consumed in one's own country that was bought from another country. A nation's **balance of trade** is the difference between the value of its exports and the value of its imports over a particular period of time.

A **favorable balance of trade** exists when the value of a nation's exports is greater than the value of its imports. In this case, the balance of trade is a positive number. An **unfavorable balance of trade** exists when the value of a nation's imports is greater than the value of its exports. In this case, the balance of trade is a negative number. An unfavorable balance of trade is referred to as a **trade deficit.** A trade deficit is considered unfavorable because more money is going out of the country to pay for imported goods than is coming into the country from sales of exported goods.

The U.S. government provides data on international trade. Statistics are reported monthly, quarterly, and annually. The following table lists the U.S. balance of trade for the years 2007 through 2011. Also shown is the total balance of trade for the five years.

Year	Balance of Trade (in billions of dollars)
2007	−697
2008	−698
2009	−379
2010	−495
2011	−560
Total	−2829

Source: www.census.gov

To determine the average annual trade deficit for the years 2007 through 2011, divide the sum of the balances of trade by the number of years (5).

$$-2829 \div 5 = -565.8$$

The average annual trade deficit for the years 2007 through 2011 was −$565.8 billion.

EXAMPLE 15

🌑 The average temperature on Mercury's sunlit side is 950°F. The average temperature on Mercury's dark side is −346°F. Find the difference between these two average temperatures.

Strategy

To find the difference, subtract the average temperature on the dark side (−346) from the average temperature on the sunlit side (950).

Solution

$$950 - (-346) = 950 + 346$$
$$= 1296$$

The difference between the average temperatures is 1296°F.

YOU TRY IT 15

The daily low temperatures (in degrees Celsius) during one week were recorded as −6°, −7°, 0°, −5°, −8°, −1°, and −1°. Find the average daily low temperature.

Your strategy

Your solution

−4°C

IN-CLASS EXAMPLES

15. The daily high temperatures (in degrees Celsius) for eight days in Nome, Alaska, were −8°, −5°, 2°, 7°, −6°, 1°, 3°, and −10°. Find the average daily high temperature.   **−2°C**

Solution on p. S1

Unless otherwise noted, all content on this page is © Cengage Learning.

1.1 EXERCISES

✔ Concept Check

SUGGESTED ASSIGNMENT
Exercises 1–8; Exercises 9–163, every other odd;
Exercises 165–179, odds
More challenging exercises: Exercises 180–183

1. Fill in the blank with *left* or *right*.

   a. On the number line, the number -7 is to the _____ left _____ of the number -5.

   b. On the number line, the number -1 is to the _____ right _____ of the number -8.

2. Fill in the blank with *positive* or *negative*.

   a. The additive inverse of a negative number is a _____ positive _____ number.

   b. The additive inverse of a positive number is a _____ negative _____ number.

3. The equation $|-10| = 10$ is read "the _absolute value_ of negative ten is ten."

4. ◣ Explain how to add two integers with the same sign.

5. ◣ Explain how to add two integers with different signs.

6. ◣ What is the difference between the terms *minus* and *negative*?

7. ◣ Explain how to subtract two integers.

8. Fill in the blank with *positive* or *negative*.

   a. The product or quotient of two numbers with the same sign is _____ positive _____.

   b. The product or quotient of two numbers with different signs is _____ negative _____.

OBJECTIVE A   *To use inequality symbols with integers*

For Exercises 9 to 18, place the correct symbol, $<$ or $>$, between the two numbers.

9. $8 > -6$ **10.** $-14 < 16$ **11.** $-12 < 1$ **12.** $35 > 28$ **13.** $42 > 19$

14. $-42 < 27$ **15.** $0 > -31$ **16.** $-17 < 0$ **17.** $53 > -46$ **18.** $-27 > -39$

19. Let $x \in \{-23, -18, -8, 0\}$. For which values of x is the inequality $x < -8$ a true statement?
$-23, -18$

20. Let $w \in \{-33, -24, -10, 0\}$. For which values of w is the inequality $w < -10$ a true statement?
$-33, -24$

21. Let $a \in \{-33, -15, 21, 37\}$. For which values of a is the inequality $a > -10$ a true statement?
$21, 37$

22. Let $v \in \{-27, -14, 14, 27\}$. For which values of v is the inequality $v > -15$ a true statement?
$-14, 14, 27$

◤ For Exercises 23 and 24, determine which of the following statements is true about n.
(i) n is positive.   **(ii)** n is negative.   **(iii)** n is zero.   **(iv)** n can be positive, negative, or zero.

23. The number n is to the right of the number 5 on the number line.   i

24. The number n is to the left of the number 5 on the number line.   iv

---

OBJECTIVE B   *To find the additive inverse and absolute value of a number*

For Exercises 25 to 30, find the additive inverse.

25. 4   **26.** 8   **27.** -9 **28.** -12 **29.** -28 **30.** -36
-4 -8 9   12   28   36

For Exercises 31 to 42, evaluate.

31. $-(-14)$
14

32. $-(-40)$
40

33. $-(77)$
-77

34. $-(39)$
-39

35. $-(0)$
0

36. $-(-13)$
13

37. $|-74|$
74

38. $|-96|$
96

39. $-|-82|$
-82

40. $-|-53|$
-53

41. $-|81|$
-81

42. $-|38|$
-38

For Exercises 43 to 50, place the correct symbol, $<$ or $>$, between the values of the two numbers.

43. $|-83| > |58|$

44. $|22| > |-19|$

45. $|43| < |-52|$

46. $|-71| < |-92|$

47. $|-68| > |-42|$

48. $|12| < |-31|$

49. $|-45| < |-61|$

50. $|-28| < |43|$

51. Let $p \in \{-19, 0, 28\}$. Evaluate $-p$ for each element of the set.
19, 0, -28

52. Let $q \in \{-34, 0, 31\}$. Evaluate $-q$ for each element of the set.
34, 0, -31

53. Let $x \in \{-45, 0, 17\}$. Evaluate $-|x|$ for each element of the set.
-45, 0, -17

54. Let $y \in \{-91, 0, 48\}$. Evaluate $-|y|$ for each element of the set.
-91, 0, -48

55. True or false? The absolute value of a negative number n is greater than n.
True

OBJECTIVE C   *To add or subtract integers*

56. Explain how to rewrite the subtraction $8 - (-6)$ as addition of the opposite.

For Exercises 57 to 107, add or subtract.

57. $-3 + (-8)$
-11

58. $-6 + (-9)$
-15

59. $-8 + 3$
-5

60. $-9 + 2$
-7

61. $-3 + (-80)$
-83

62. $-12 + (-1)$
-13

63. $-23 + (-23)$
-46

64. $-12 + (-12)$
-24

65. $16 + (-16)$
0

66. $-17 + 17$
0

67. $48 + (-53)$
-5

68. $19 + (-41)$
-22

69. $-17 + (-3) + 29$
9

70. $13 + 62 + (-38)$
37

71. $-3 + (-8) + 12$
1

72. $-27 + (-42) + (-18)$
-87

73. $16 - 8$
8

74. $12 - 3$
9

75. $7 - 14$
-7

76. $6 - 9$
-3

77. $-7 - 2$
-9

78. $-9 - 4$
-13

79. $7 - (-2)$
9

80. $3 - (-4)$
7

81. $-6 - (-3)$
-3

82. $-4 - (-2)$
-2

83. $6 - (-12)$
18

84. $-12 - 16$
-28

85. $13 + (-22) + 4 + (-5)$
-10

86. $-14 + (-3) + 7 + (-21)$
-31

87. $-16 + (-17) + (-18) + 10$
-41

88. $-25 + (-31) + 24 + 19$
-13

89. $26 + (-15) + (-11) + (-12)$
-12

90. $-32 + 40 + (-8) + (-19)$
-19

91. $-14 + (-15) + (-11) + 40$
0

92. $28 + (-19) + (-8) + (-1)$
0

93. $-4 - 3 - 2$
-9

94. $4 - 5 - 12$
-13

95. $12 - (-7) - 8$
11

96. $-12 - (-3) - (-15)$
6

97. $-19 - (-19) - 18$
-18

98. $-8 - (-8) - 14$
-14

99. $-17 - (-8) - (-9)$
0

100. $7 - 8 - (-1)$
0

101. $-30 - (-65) - 29 - 4$
2

102. $42 - (-82) - 65 - 7$
52

103. $-16 - 47 - 63 - 12$
-138

104. $42 - (-30) - 65 - (-11)$
18

105. $-47 - (-67) - 13 - 15$
-8

106. $-18 - 49 - (-84) - 27$
-10

107. $-19 - 17 - (-36) - 12$
-12

For Exercises 108 to 111, without finding the sum or difference, determine whether the sum or difference is positive or negative.

108. $812 + (-537)$
Positive

109. The sum of -57 and -31
Negative

110. $-25 - 52$
Negative

111. The difference between 8 and -5
Positive

OBJECTIVE D *To multiply or divide integers*

112. ✎ Name the operation in each expression. Justify your answer.
   a. $8(-7)$ **b.** $8 - 7$ **c.** $8 - (-7)$ **d.** $-xy$ **e.** $x(-y)$ **f.** $-x - y$

For Exercises 113 to 162, multiply or divide.

113. $(14)3$
42

114. $(17)6$
102

115. $-7 \cdot 4$
-28

116. $-8 \cdot 7$
-56

117. $(-12)(-5)$
60

118. $(-13)(-9)$
117

119. $-11(23)$
-253

120. $-8(21)$
-168

121. $(-17)14$
-238

122. $(-15)12$
-180

123. $6(-19)$
-114

124. $17(-13)$
-221

125. $12 \div (-6)$
-2

126. $18 \div (-3)$
-6

127. $(-72) \div (-9)$
8

128. $(-64) \div (-8)$
8

129. $-42 \div 6$
-7

130. $(-56) \div 8$
-7

131. $(-144) \div 12$
-12

132. $(-93) \div (-3)$
31

133. $48 \div (-8)$
-6

134. $57 \div (-3)$
-19

135. $\dfrac{-49}{7}$
-7

136. $\dfrac{-45}{5}$
-9

137. $\dfrac{-44}{-4}$
11

138. $\dfrac{-36}{-9}$
4

139. $\dfrac{98}{-7}$
-14

140. $\dfrac{85}{-5}$
-17

141. $-\dfrac{-120}{8}$
15

142. $-\dfrac{-72}{4}$
18

143. $-\dfrac{-80}{-5}$
-16

144. $-\dfrac{-114}{-6}$
-19

145. $0 \div (-9)$
0

146. $0 \div (-14)$
0

147. $\dfrac{-261}{9}$
-29

148. $\dfrac{-128}{4}$
-32

149. $9 \div 0$
Undefined

150. $(-21) \div 0$
Undefined

151. $\dfrac{132}{-12}$
-11

152. $\dfrac{250}{-25}$
-10

153. $\dfrac{0}{0}$
Undefined

154. $\dfrac{-58}{0}$
Undefined

155. $7(5)(-3)$
-105

156. $(-3)(-2)8$
48

157. $9(-7)(-4)$
252

158. $(-2)(6)(-4)$
48

159. $7(-2)(5)(-6)$
420

160. $(-3)7(-2)8$
336

161. $(-14)9(-11)0$
0

162. $(-13)(15)(-19)0$
0

163. ✎ You multiply four positive integers and three negative integers. Is the product positive or negative? Negative

OBJECTIVE E   *To solve application problems*

164. 🜨 **Temperature**   The news clipping at the right was written on February 11, 2008. The record low temperature for Minnesota is −51°C. Find the difference between the low temperature in International Falls on February 11, 2008, and the record low temperature for Minnesota.   11°C

165. 🜨 **Temperature**   The record high temperature in Illinois is 117°F. The record low temperature is −36°F. Find the difference between the record high and record low temperatures in Illinois.   153°F

166. Temperature   Find the temperature after a rise of 7°C from −8°C.   −1°C

167. Temperature   Find the temperature after a rise of 5°C from −19°C.   −14°C

🜨 **Geography**   The graph at the right shows Earth's three deepest ocean trenches and its three tallest mountains. Use this graph for Exercises 168 to 170.

168. What is the difference between the depth of the Philippine Trench and the depth of the Mariana Trench?   980 m

169. What is the difference between the height of Mt. Everest and the depth of the Mariana Trench?   20,370 m

170. 🐾 Could Mt. Everest fit in the Tonga Trench?   Yes

171. 🜨 **Golf Scores**   In golf, a player's score on a hole is 0 if the player completes the hole in *par*. **Par** is the number of strokes in which a golfer should complete a hole. In a golf match, scores are given both as a total number of strokes taken on all holes and as a value relative to par, such as −4 ("4 under par") or +2 ("2 over par").
   a. See the news clipping at the right. Convert each of Ken Duke's scores for the first three days into a score relative to par.   0, −4, −2
   b. In a golf tournament, a player's daily scores are added. Add Ken Duke's three daily scores to find his score, relative to par, for the first three days of the tournament.   −6
   c. Duke's score on the fourth day was 68. What was his final score, relative to par, for the four-day tournament?   −10

172. Meteorology   The daily low temperatures during one week were recorded as follows: 4°F, −5°F, 8°F, −1°F, −12°F, −14°F, −8°F. Find the average daily low temperature for the week.   −4°F

173. Meteorology   The daily high temperatures during one week were recorded as follows: −6°F, −11°F, 1°F, 5°F, −3°F, −9°F, −5°F. Find the average daily high temperature for the week.   −4°F

In the NEWS!

Minnesota Town Named "Icebox of the Nation"

In International Falls, Minnesota, the temperature fell to −40°C just days after the citizens received word that the town had won a federal trademark naming it the "Icebox of the Nation."
Source: news.yahoo.com

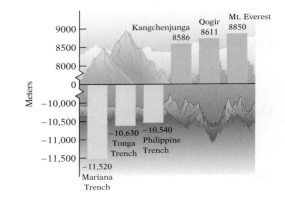

In the NEWS!

Duke Leads 2011 Nationwide Tour Championship

With scores of 72, 68, and 70 on his first three days, Ken Duke leads going into the last day of this four-day tournament. Par for the 18-hole golf course at Daniel Island in Charleston, South Carolina, is 72.
Source: www.pgatour.com

Unless otherwise noted, all content on this page is © Cengage Learning.

174. 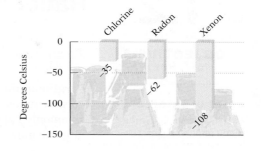 True or false? If five temperatures are all below 0°C, then the average of the five temperatures is also below 0°C. True

175. True or false? If the average of 10 temperatures is below 0°C, then all 10 temperatures are below 0°C. False

176. ● **Chemistry** The graph at the right shows the boiling points of three chemical elements. The boiling point of neon is seven times the highest boiling point shown in the graph.
   a. Without calculating the boiling point, determine whether the boiling point of neon is above 0°C or below 0°C. Below
   b. What is the boiling point of neon? −245°C

177. **Sports** The combined scores of the top 10 golfers in a tournament equaled −20 (20 under par). What was the average score of the 10 golfers? −2

178. **Games** During a card game of Hearts, Nick had a score of 11 points before his opponent "shot the moon," subtracting a score of 26 from Nick's total. What was Nick's score after his opponent shot the moon? −15 points

179. **Education** To discourage guessing on a multiple-choice exam, an instructor graded the test by giving 5 points for a correct answer, −2 points for an answer left blank, and −5 points for an incorrect answer. How many points did a student score who answered 20 questions correctly, answered 5 questions incorrectly, and left 2 questions blank? 71

Critical Thinking

180. The sum of two negative integers is −9. Find the integers. The integers can be −1 and −8, −2 and −7, −3 and −6, or −4 and −5.
181. Given the list of numbers −4, −3, −10, 9, and 15, find the largest difference that can be obtained by subtracting one number in the list from another. 25

182. If a and b are integers, is the expression $|a + b| = |a| + |b|$ always true, sometimes true, or never true? Sometimes true

183. Determine whether the statement is *true* or *false*.
   a. The product of a nonzero number and its opposite is negative. True
   b. The square of a negative number is a positive number. True

Projects or Group Activities

184. Make up three addition problems such that each problem involves one positive and one negative addend, and each problem has a sum of −4. Then describe a strategy for writing these problems.

185. Make up three subtraction problems such that each problem involves a negative number subtracted from a negative number, and each problem has a difference of −6. Then describe a strategy for writing these problems.

For Exercises 186 to 188, find the next three numbers in the pattern.

186. 5, −15, 45, −135, . . .
   405, −1215, 3645

187. −2, 4, −8, 16, . . .
   −32, 64, −128

188. −3, −12, −48, −192, . . .
   −768, −3072, −12,288

QUICK QUIZ
1. Place the correct symbol, < or >, between the two numbers.
   −6 −2 < **[1.1A]**
2. Given $B = \{-10, -5, 0, 5\}$, which elements of set B are less than −6?
   −10 [1.1A]
3. Find the additive inverse.
   a. 45 **−45**
   b. −27 **27 [1.1B]**
4. Evaluate.
   a. $|-16|$ **16**
   b. $-|8|$ **−8**
   c. $-|-30|$ **−30 [1.1B]**
Add.
5. −14 + (−26)
   −40 [1.1C]
6. 35 + (−10) + (−8)
   17 [1.1C]
Subtract.
7. −62 − (−53)
   −9 [1.1C]
8. 11 − 28 − (−9)
   −8 [1.1C]
Multiply.
9. −5(−6) **30 [1.1D]**
10. 15(−3)(10)
   −450 [1.1D]
Divide.
11. −48 ÷ 8 **−6 [1.1D]**
12. −27 ÷ (−9) **3 [1.1D]**
13. The daily low temperatures during one week were as follows: 4°, −6°, 8°, −2°, −9°, −11°, −5°. Find the average daily low temperature for the week. **−3° [1.1E]**

Unless otherwise noted, all content on this page is © Cengage Learning.

1.2   Rational and Irrational Numbers

OBJECTIVE A   *To write a rational number as a decimal*

Point of Interest

As early as A.D. 630, the Hindu mathematician Brahmagupta wrote a fraction as one number over another separated by a space. The Arab mathematician al Hassar (around A.D. 1050) was the first to show a fraction with a horizontal bar separating the numerator and denominator.

INSTRUCTOR NOTE

The term *rational* is related to the word *ratio*. A rational number is one that can be expressed as the ratio of two integers.

Point of Interest

Simon Stevin (1548–1620) was the first to name decimal numbers. He wrote the number 2.345 as 2 0 3 1 4 2 5 3. He called the whole number part the *commencement;* the tenths digit was *prime,* the hundredths digit was *second,* the thousandths digit was *third,* and so on.

IN-CLASS EXAMPLES

1. Write $\frac{3}{8}$ as a decimal.
   0.375

2. Write $\frac{7}{15}$ as a decimal. Place a bar over the repeating digits.   **$0.4\overline{6}$**

A **rational number** is the quotient of two integers. A rational number written in this way is commonly called a fraction. Some examples of rational numbers are shown at the right.

$$\frac{3}{4}, \quad \frac{-4}{9}, \quad \frac{15}{-4}, \quad \frac{8}{1}, \quad -\frac{5}{6}$$

Rational Numbers

A **rational number** is a number that can be written in the form $\frac{a}{b}$, where a and b are integers and $b \neq 0$.

EXAMPLES OF RATIONAL NUMBERS

1. $\frac{4}{5}$ 2. $\frac{-2}{13}$ 3. $\frac{9}{-5}$

Because an integer can be written as the quotient of the integer and 1, every integer is a rational number.

$$6 = \frac{6}{1} \qquad -8 = \frac{-8}{1}$$

A number written in **decimal notation** is also a rational number.

three-tenths $0.3 = \frac{3}{10}$ forty-three thousandths $0.043 = \frac{43}{1000}$

A rational number written as a fraction can be written in decimal notation by dividing the numerator of the fraction by the denominator. Think of the fraction bar as meaning "divided by."

HOW TO 1

Write $\frac{5}{8}$ as a decimal.

$$\begin{array}{r} 0.625 \\ 8{\overline{\smash{\big)}\,5.000}} \\ -4\,8 \\ \hline 20 \\ -16 \\ \hline 40 \\ -40 \\ \hline 0 \end{array}$$

• Divide the numerator, 5, by the denominator, 8.

When the remainder is zero, the decimal is called a **terminating decimal.** The decimal 0.625 is a terminating decimal.

$$\frac{5}{8} = 0.625$$

HOW TO 2

Write $\frac{4}{11}$ as a decimal.

$$\begin{array}{r} 0.3636 \\ 11{\overline{\smash{\big)}\,4.0000}} \\ -3\,3 \\ \hline 70 \\ -66 \\ \hline 40 \\ -33 \\ \hline 70 \\ -66 \\ \hline 4 \end{array}$$

• Divide the numerator, 4, by the denominator, 11.

No matter how long we continue to divide, the remainder is never zero. The decimal $0.\overline{36}$ is a **repeating decimal.** The bar over the 36 indicates that these digits repeat.

$$\frac{4}{11} = 0.\overline{36}$$

EXAMPLE 1

Write $\frac{8}{11}$ as a decimal. Place a bar over the repeating digits of the decimal.

Solution

$$\frac{8}{11} = 8 \div 11 = 0.7272\ldots = 0.\overline{72}$$

YOU TRY IT 1

Write $\frac{4}{9}$ as a decimal. Place a bar over the repeating digits of the decimal.

Your solution

$0.\overline{4}$

Solution on p. S1

OBJECTIVE B   *To convert among percents, fractions, and decimals*

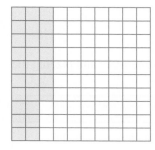

Percent means "parts of 100." Thus 27% means 27 parts of 100.

In applied problems involving percent, it may be necessary to rewrite a percent as a fraction or as a decimal, or to rewrite a fraction or a decimal as a percent.

To write a percent as a fraction, remove the percent sign and multiply by $\frac{1}{100}$.

$$27\% = 27\left(\frac{1}{100}\right) = \frac{27}{100}$$

To write a percent as a decimal, remove the percent sign and multiply by 0.01.

$$33\% \quad = \quad 33(0.01) \quad = \quad 0.33$$

Move the decimal point two places to the left. Then remove the percent sign.

To write a fraction as a percent, multiply by 100%. For example, $\frac{5}{8}$ is changed to a percent as follows:

$$\frac{5}{8} = \frac{5}{8}(100\%) = \frac{500}{8}\% = 62.5\%, \qquad \text{or} \qquad 62\frac{1}{2}\%$$

To write a decimal as a percent, multiply by 100%.

$$0.82 \quad = \quad 0.82(100\%) \quad = \quad 82\%$$

Move the decimal point two places to the right. Then write the percent sign.

 Take Note

The decimal equivalent of 100% is 1. Therefore, multiplying by 100% is the same as multiplying by 1 and does not change the value of the fraction.

$$\frac{5}{8} = \frac{5}{8}(1) = \frac{5}{8}(100\%)$$

EXAMPLE 2

Write 130% as a fraction and as a decimal.

Solution

$$130\% = 130\left(\frac{1}{100}\right) = \frac{130}{100} = \frac{13}{10}$$

$$130\% = 130(0.01) = 1.30$$

YOU TRY IT 2

Write 125% as a fraction and as a decimal.

Your solution

$\frac{5}{4}$, 1.25

Solution on p. S1

Unless otherwise noted, all content on this page is © Cengage Learning.

EXAMPLE 3

Write $\frac{5}{6}$ as a percent.

Solution

$$\frac{5}{6} = \frac{5}{6}(100\%) = \frac{500}{6}\% = 83\frac{1}{3}\%$$

EXAMPLE 4

Write 0.092 as a percent.

Solution

$$0.092 = 0.092(100\%) = 9.2\%$$

YOU TRY IT 3

Write $\frac{1}{3}$ as a percent.

Your solution

$33\frac{1}{3}\%$

YOU TRY IT 4

Write 0.043 as a percent.

Your solution

4.3%

IN-CLASS EXAMPLES

3. Write $8\frac{1}{3}\%$ as a fraction.

$\frac{1}{12}$

4. Write $\frac{7}{8}$ as a percent.

87.5% or $87\frac{1}{2}\%$

5. Write 1.5 as a percent.
150%

6. Write 40% as a decimal.
0.4

Solutions on p. S2

OBJECTIVE C   *To add or subtract rational numbers*

Fractions with the same denominator are added by adding the numerators and placing the sum over the common denominator.

IN-CLASS EXAMPLES
Simplify.

7. $-\frac{3}{4} + \left(-\frac{1}{6}\right)$ $-\frac{11}{12}$
8. $3.8 - 7.4$ **−3.6**
9. $6.2 - (-4.61)$ **10.81**

> **Addition of Fractions**
>
> To add two fractions with the same denominator, add the numerators and place the sum over the common denominator.
>
> $$\frac{a}{c} + \frac{b}{c} = \frac{a+b}{c}$$

To add fractions with different denominators, first rewrite the fractions as equivalent fractions with a common denominator. Then add the fractions.

The least common denominator is the **least common multiple** (LCM) of the denominators. This is the smallest number that is a multiple of each of the denominators.

HOW TO 3   Add: $-\frac{5}{6} + \frac{3}{10}$

The LCM of 6 and 10 is 30. Rewrite each fraction as an equivalent fraction with a denominator of 30. Then add the fractions.

$$-\frac{5}{6} + \frac{3}{10} = -\frac{5}{6}\cdot\frac{5}{5} + \frac{3}{10}\cdot\frac{3}{3} = -\frac{25}{30} + \frac{9}{30} = \frac{-25+9}{30} = \frac{-16}{30} = -\frac{8}{15}$$

To subtract fractions with the same denominator, subtract the numerators and place the difference over the common denominator.

HOW TO 4   Subtract: $-\frac{4}{9} - \left(-\frac{7}{12}\right)$

Rewrite subtraction as addition of the opposite. The LCM of 9 and 12 is 36. Rewrite each fraction as an equivalent fraction with a denominator of 36.

$$-\frac{4}{9} - \left(-\frac{7}{12}\right) = -\frac{4}{9} + \frac{7}{12} = -\frac{16}{36} + \frac{21}{36} = \frac{-16+21}{36} = \frac{5}{36}$$

Take Note

You can find the LCM by multiplying the denominators and then dividing by the *common factor* of the two denominators. In the case of 6 and 10, 6 · 10 = 60. Now divide by 2, the common factor of 6 and 10.

$$60 \div 2 = 30$$

Take Note

The least common multiple of the denominators is frequently called the **least common denominator** (LCD).

To add or subtract decimals, write the numbers so that the decimal points are in a vertical line. Then proceed as in the addition or subtraction of integers. Write the decimal point in the answer directly below the decimal points in the problem.

HOW TO 5 Add: $-114.039 + 84.76$

$|-114.039| = 114.039$
$|84.76| = 84.76$

• The signs are different. Find the absolute value of each number.

$$\begin{array}{r} 114.039 \\ -84.760 \\ \hline 29.279 \end{array}$$

• Subtract the smaller of these absolute values from the larger.

$-114.039 + 84.76 = -29.279$

• Attach the sign of the number with the larger absolute value. Because $|-114.039| > |84.76|$, use the sign of -114.039.

EXAMPLE 5

Simplify: $-\dfrac{3}{4} + \dfrac{1}{6} - \dfrac{5}{8}$

Solution The LCM of 4, 6, and 8 is 24.

$$-\frac{3}{4} + \frac{1}{6} - \frac{5}{8} = -\frac{18}{24} + \frac{4}{24} - \frac{15}{24}$$
$$= \frac{-18 + 4 - 15}{24}$$
$$= \frac{-29}{24} = -\frac{29}{24}$$

YOU TRY IT 5

Simplify: $-\dfrac{7}{8} - \dfrac{5}{6} + \dfrac{3}{4}$

Your solution
$-\dfrac{23}{24}$

EXAMPLE 6

Subtract: $42.987 - 98.61$

Solution $42.987 - 98.61$
$= 42.987 + (-98.61)$
$= -55.623$

YOU TRY IT 6

Subtract: $16.127 - 67.91$

Your solution
-51.783

Solutions on p. S2

OBJECTIVE D *To multiply or divide rational numbers*

The product of two fractions is the product of the numerators divided by the product of the denominators.

$$\frac{a}{b} \cdot \frac{c}{d} = \frac{ac}{bd}$$

IN-CLASS EXAMPLES

Simplify.

10. $-\dfrac{5}{6}\left(\dfrac{3}{10}\right)$ $-\dfrac{1}{4}$

11. $-6.8(-2.1)$ **14.28**

12. $9.44 \div (-8)$ **−1.18**

13. $-\dfrac{5}{6} \div \dfrac{5}{9}$ $-\dfrac{3}{2}$

HOW TO 6 Multiply: $\dfrac{3}{8} \cdot \dfrac{12}{17}$

$$\frac{3}{8} \cdot \frac{12}{17} = \frac{3 \cdot 12}{8 \cdot 17}$$

• Multiply the numerators. Multiply the denominators.

$$= \frac{3 \cdot \overset{1}{\cancel{2}} \cdot \overset{1}{\cancel{2}} \cdot 3}{2 \cdot \underset{1}{\cancel{2}} \cdot \underset{1}{\cancel{2}} \cdot 17}$$

• Write the prime factorization of each factor. Divide by the common factors.

$$= \frac{9}{34}$$

• Multiply the factors in the numerator and the factors in the denominator.

To divide fractions, invert the divisor. Then multiply the fractions.

Take Note

To invert the divisor means to write its reciprocal. The reciprocal of $\frac{18}{25}$ is $\frac{25}{18}$.

HOW TO 7 Divide: $\dfrac{3}{10} \div \left(-\dfrac{18}{25}\right)$

The signs are different. The quotient is negative.

$$\frac{3}{10} \div \left(-\frac{18}{25}\right) = -\left(\frac{3}{10} \div \frac{18}{25}\right) = -\left(\frac{3}{10} \cdot \frac{25}{18}\right)$$

$$= -\left(\frac{3 \cdot 25}{10 \cdot 18}\right) = -\left(\frac{\overset{1}{\cancel{3}} \cdot \overset{1}{\cancel{5}} \cdot 5}{2 \cdot \underset{1}{\cancel{5}} \cdot 2 \cdot \underset{1}{\cancel{3}} \cdot 3}\right) = -\frac{5}{12}$$

To multiply decimals, multiply as with integers. Write the decimal point in the product so that the number of decimal places in the product equals the sum of the numbers of decimal places in the factors.

HOW TO 8 Multiply: $-6.89(0.00035)$

$$
\begin{array}{r}
6.89 \\
\times\ 0.00035 \\
\hline
3445 \\
2067 \\
\hline
0.0024115
\end{array}
$$

2 decimal places
5 decimal places

7 decimal places

• Multiply the absolute values.

$-6.89(0.00035) = -0.0024115$ • The signs are different. The product is negative.

To divide decimals, move the decimal point in the divisor to the right to make the divisor a whole number. Move the decimal point in the dividend the same number of places to the right. Place the decimal point in the quotient directly above the decimal point in the dividend. Then divide as with whole numbers.

Take Note

The symbol ≈ is used to indicate that the quotient is an approximate value that has been rounded off.

HOW TO 9 Divide: $1.32 \div 0.27$. Round to the nearest tenth.

$$
\begin{array}{r}
4.88 \approx 4.9 \\
0.27\overline{)\ 1.32.00} \\
-1\ 08 \\
\hline
240 \\
-216 \\
\hline
240 \\
-216 \\
\hline
24
\end{array}
$$

• Move the decimal point 2 places to the right in the divisor and then in the dividend. Place the decimal point in the quotient directly above the decimal point in the dividend.

EXAMPLE 7

Divide: $-\dfrac{5}{8} \div \left(-\dfrac{5}{40}\right)$

Solution

The quotient is positive.

$$-\frac{5}{8} \div \left(-\frac{5}{40}\right) = \frac{5}{8} \div \frac{5}{40} = \frac{5}{8} \cdot \frac{40}{5} = \frac{5 \cdot 40}{8 \cdot 5}$$

$$= \frac{\overset{1}{\cancel{5}} \cdot \overset{1}{\cancel{2}} \cdot \overset{1}{\cancel{2}} \cdot \overset{1}{\cancel{2}} \cdot 5}{2 \cdot 2 \cdot 2 \cdot \underset{1}{\cancel{5}}} = \frac{5}{1} = 5$$

YOU TRY IT 7

Divide: $-\dfrac{3}{8} \div \left(-\dfrac{5}{12}\right)$

Your solution

$\dfrac{9}{10}$

Solution on p. S2

EXAMPLE 8

Multiply: $-4.29(8.2)$

Solution

The product is negative.

$$
\begin{array}{r}
4.29 \\
\times\ \ 8.2 \\
\hline
858 \\
3432\ \ \\
\hline
35.178
\end{array}
$$

$-4.29(8.2) = -35.178$

YOU TRY IT 8

Multiply: $-5.44(3.8)$

Your solution

-20.672

Solution on p. S2

OBJECTIVE E   *To evaluate exponential expressions*

 Point of Interest

René Descartes (1596–1650) was the first mathematician to use exponential notation extensively as it is used today. However, for some unknown reason, he always used *xx* for x^2.

Repeated multiplication of the same factor can be written using an exponent.

$$2 \cdot 2 \cdot 2 \cdot 2 \cdot 2 = 2^5 \longleftarrow \text{exponent}$$
$$\uparrow \rule{1.2cm}{0.4pt} \text{base}$$

$$a \cdot a \cdot a \cdot a = a^4 \longleftarrow \text{exponent}$$
$$\uparrow \rule{1.2cm}{0.4pt} \text{base}$$

The **exponent** indicates how many times the factor, called the **base,** occurs in the multiplication. The multiplication $2 \cdot 2 \cdot 2 \cdot 2 \cdot 2$ is in **factored form.** The exponential expression 2^5 is in **exponential form.**

2^1 is read "2 to the first power" or just "2." Usually the exponent 1 is not written.

2^2 is read "2 to the second power" or "2 squared."

2^3 is read "2 to the third power" or "2 cubed."

2^4 is read "2 to the fourth power."

a^4 is read "*a* to the fourth power."

IN-CLASS EXAMPLES
14. Evaluate $(-6)^2$.   **36**
15. Evaluate -6^2.   **−36**
16. Evaluate $-3^2 \cdot 5^2$.   **−225**

The first three natural-number powers can be interpreted geometrically as length, area, and volume, respectively.

$4^1 = 4$
Length: 4 ft

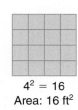

$4^2 = 16$
Area: 16 ft^2

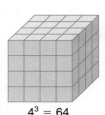

$4^3 = 64$
Volume: 64 ft^3

Unless otherwise noted, all content on this page is © Cengage Learning.

To evaluate an exponential expression, write each factor as many times as indicated by the exponent. Then multiply.

HOW TO 10 Evaluate $(-2)^4$.

$(-2)^4 = (-2)(-2)(-2)(-2)$ • Write (-2) as a factor 4 times.

$\qquad = 16$ • Multiply.

HOW TO 11 Evaluate -2^4.

$-2^4 = -(2 \cdot 2 \cdot 2 \cdot 2)$ • Write 2 as a factor 4 times.

$\qquad = -16$ • Multiply.

From these last two examples, note the difference between $(-2)^4$ and -2^4.

$$(-2)^4 = 16$$
$$-2^4 = -(2)^4 = -16$$

EXAMPLE 9

Evaluate -5^3.

Solution

$-5^3 = -(5 \cdot 5 \cdot 5) = -125$

YOU TRY IT 9

Evaluate -6^3.

Your solution

-216

EXAMPLE 10

Evaluate $(-4)^4$.

Solution

$(-4)^4 = (-4)(-4)(-4)(-4)$
$\qquad = 256$

YOU TRY IT 10

Evaluate $(-3)^4$.

Your solution

81

EXAMPLE 11

Evaluate $(-3)^2 \cdot 2^3$.

Solution $(-3)^2 \cdot 2^3 = (-3)(-3) \cdot (2)(2)(2)$
$\qquad\qquad\qquad = 9 \cdot 8 = 72$

YOU TRY IT 11

Evaluate $(3^3) \cdot (-2)^3$.

Your solution

-216

EXAMPLE 12

Evaluate $\left(-\dfrac{2}{3}\right)^3$.

Solution $\left(-\dfrac{2}{3}\right)^3 = \left(-\dfrac{2}{3}\right)\left(-\dfrac{2}{3}\right)\left(-\dfrac{2}{3}\right)$

$\qquad\qquad = -\dfrac{2 \cdot 2 \cdot 2}{3 \cdot 3 \cdot 3} = -\dfrac{8}{27}$

YOU TRY IT 12

Evaluate $\left(-\dfrac{2}{5}\right)^2$.

Your solution

$\dfrac{4}{25}$

EXAMPLE 13

Evaluate $-4(0.7)^2$.

Solution $-4(0.7)^2 = -4(0.7)(0.7)$
$\qquad\qquad\quad = -2.8(0.7) = -1.96$

YOU TRY IT 13

Evaluate $-3(0.3)^3$.

Your solution

-0.081

Solutions on p. S2

OBJECTIVE F   *To simplify numerical radical expressions*

A **square root** of a positive number x is a number whose square is x.

A square root of 16 is 4 because $4^2 = 16$.
A square root of 16 is -4 because $(-4)^2 = 16$.

Every positive number has two square roots, one positive and one negative. The symbol $\sqrt{}$, called a **radical sign,** is used to indicate the positive or **principal square root** of a number. For example, $\sqrt{16} = 4$ and $\sqrt{25} = 5$. The number under the radical sign is called the **radicand.**

When the negative square root of a number is to be found, a negative sign is placed in front of the radical. For example, $-\sqrt{16} = -4$ and $-\sqrt{25} = -5$.

The square of an integer is a **perfect square.** 49, 81, and 144 are examples of perfect squares.

$$7^2 = 49$$
$$9^2 = 81$$
$$12^2 = 144$$

The principal square root of an integer that is a perfect square is a positive integer.

$$\sqrt{49} = 7$$
$$\sqrt{81} = 9$$
$$\sqrt{144} = 12$$

> **Square Roots of Perfect Squares**
>
> $\sqrt{1} = 1$
> $\sqrt{4} = 2$
> $\sqrt{9} = 3$
> $\sqrt{16} = 4$
> $\sqrt{25} = 5$
> $\sqrt{36} = 6$
> $\sqrt{49} = 7$
> $\sqrt{64} = 8$
> $\sqrt{81} = 9$
> $\sqrt{100} = 10$
> $\sqrt{121} = 11$
> $\sqrt{144} = 12$

If a number is not a perfect square, its square root can only be approximated. For example, 2 and 7 are not perfect squares. The square roots of these numbers are called **irrational numbers.** Their decimal representations never terminate or repeat.

$$\sqrt{2} \approx 1.4142135\ldots \qquad \sqrt{7} \approx 2.6457513\ldots$$

IN-CLASS EXAMPLES

Simplify.

17. $\sqrt{675}$ $15\sqrt{3}$
18. $7\sqrt{48}$ $28\sqrt{3}$
19. $-3\sqrt{24}$ $-6\sqrt{6}$

Recall that rational numbers are fractions, such as $-\frac{6}{7}$ or $-\frac{10}{3}$, in which the numerator and denominator are integers. Rational numbers are also represented by repeating decimals, such as $0.25767676\ldots$, and by terminating decimals, such as 1.73. An irrational number is neither a repeating nor a terminating decimal. For instance, $2.45445444544445\ldots$ is an irrational number.

Real Numbers

The rational numbers and the irrational numbers taken together are called the **real numbers.**

 Take Note

Recall that a factor of a number divides the number evenly. For instance, 6 is a factor of 18. The perfect square 9 is also a factor of 18. It is a *perfect-square factor* of 18. The number 6 is not a perfect-square factor of 18 because 6 is not a perfect square.

Radical expressions that contain radicands that are not perfect squares are frequently written in simplest form. A radical expression is in *simplest form* when the radicand contains no factor greater than 1 that is a perfect square. For instance, $\sqrt{50}$ is not in simplest form because 25 is a perfect-square factor of 50. The radical expression $\sqrt{15}$ is in simplest form because there are no perfect-square factors of 15 that are greater than 1.

A knowledge of perfect squares and the Product Property of Square Roots are used to simplify radicands that are not perfect squares.

Unless otherwise noted, all content on this page is © Cengage Learning.

Take Note

From the example at the right, $\sqrt{72} = 6\sqrt{2}$. The two expressions are different representations of the same number. Using a calculator, we find that $\sqrt{72} \approx 8.485281$ and $6\sqrt{2} \approx 8.485281$.

The Product Property of Square Roots

If a and b are positive real numbers, then $\sqrt{ab} = \sqrt{a} \cdot \sqrt{b}$.

EXAMPLE

Simplify: $\sqrt{72}$

$\sqrt{72} = \sqrt{36 \cdot 2}$
- Write the radicand as the product of a perfect square and a factor that does not contain a perfect square.

$\quad\quad = \sqrt{36}\,\sqrt{2}$
- Use the Product Property of Square Roots to write the expression as a product.

$\quad\quad = 6\sqrt{2}$
- Simplify $\sqrt{36}$.

INSTRUCTOR NOTE

If students are having difficulty finding a perfect-square factor, have them write the prime factorization of the radicand. For example,

$\sqrt{288} = \sqrt{2^5 3^2} = \sqrt{2^4 3^2 \cdot 2}$
$\quad\quad = \sqrt{2^4 3^2}\,\sqrt{2}$
$\quad\quad = 2^2 \cdot 3\sqrt{2}$
$\quad\quad = 12\sqrt{2}$

Note that 72 must be written as the product of a perfect square and *a factor that does not contain a perfect square.* Therefore, it would not be correct to rewrite $\sqrt{72}$ as $\sqrt{9 \cdot 8}$ and simplify the expression as shown at the right. Although 9 is a perfect-square factor of 72, 8 also contains a perfect square factor $(8 = 4 \cdot 2)$. Therefore, $\sqrt{8}$ is not in simplest form. Remember to find the *largest* perfect-square factor of the radicand.

$\sqrt{72} = \sqrt{9 \cdot 8}$
$\quad\quad = \sqrt{9}\,\sqrt{8}$
$\quad\quad = 3\sqrt{8}$
Not in simplest form

HOW TO 12   Simplify: $\sqrt{-16}$

Because the square of any real number is positive, there is no real number whose square is -16. $\sqrt{-16}$ is not a real number.

EXAMPLE 14

Simplify: $3\sqrt{90}$

Solution

$3\sqrt{90} = 3\sqrt{9 \cdot 10}$
$\quad\quad = 3\sqrt{9}\,\sqrt{10}$
$\quad\quad = 3 \cdot 3\sqrt{10}$
$\quad\quad = 9\sqrt{10}$

YOU TRY IT 14

Simplify: $-5\sqrt{32}$

Your solution

$-20\sqrt{2}$

EXAMPLE 15

Simplify: $\sqrt{252}$

Solution

$\sqrt{252} = \sqrt{36 \cdot 7}$
$\quad\quad = \sqrt{36}\,\sqrt{7}$
$\quad\quad = 6\sqrt{7}$

YOU TRY IT 15

Simplify: $\sqrt{216}$

Your solution

$6\sqrt{6}$

OBJECTIVE G *To solve application problems*

One of the applications of percent is to express a portion of a total as a percent. For instance, a recent survey of 450 mall shoppers found that 270 preferred the mall closest to their home even though it did not have as much store variety as a mall farther from home. The percent of shoppers who preferred the mall closest to home can be found by converting a fraction to a percent.

$$\frac{\text{Portion preferring mall closest to home}}{\text{Total number surveyed}} = \frac{270}{450}$$

$$= 0.60 = 60\%$$

The Congressional Budget Office projected that the total surpluses for 2001 through 2011 would be $5.6 trillion. The number 5.6 trillion means

$$5.6 \times \underbrace{1,000,000,000,000}_{\text{1 trillion}} = 5,600,000,000,000$$

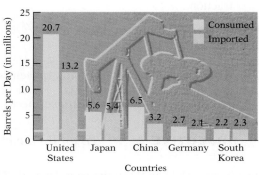 The graph at the right shows the numbers of barrels of oil per day, in millions, consumed by various countries and the numbers of barrels of oil per day, in millions, those countries import. Use this graph for Example 16 and You Try It 16.

Oil Consumption by Country
Source: IEA

EXAMPLE 16

Using the graph above, find the total number of barrels of oil consumed each day by the five countries shown.

Strategy

To find the total number of barrels:
• Read the numbers from the graph that correspond to oil consumption (20.7, 5.6, 6.5, 2.7, 2.2).
• Add the numbers.

Solution

$20.7 + 5.6 + 6.5 + 2.7 + 2.2 = 37.7$

A total of 37.7 million barrels of oil are consumed each day by the five countries shown.

YOU TRY IT 16

For the countries given in the graph above, find the difference between the total number of barrels of oil consumed each day and the total number imported each day.

Your strategy

Your solution

11.5 million barrels a day

IN-CLASS EXAMPLES

20. The lowest temperature ever recorded in Australia is −9.4°F. The highest temperature ever recorded is 128.0°F. (*Source:* National Climatic Data Center) Find the difference between these two extremes. **137.4°F**

Solution on p. S2

Unless otherwise noted, all content on this page is © Cengage Learning.

SUGGESTED ASSIGNMENT
Exercises 1–10; Exercises 11–177, every other odd;
Exercises 183–191, odds
More challenging exercises: Exercises 192–195

1.2 EXERCISES

✔ Concept Check

1. To write $\frac{3}{4}$ as a decimal, divide _____3_____ by _____4_____. The quotient is 0.75, which is a __terminating__ decimal.

2. To write a fraction as a percent, multiply the fraction by ___100%___.

3. To write a percent as a decimal, remove the percent sign and multiply by ____0.01____.

4. To add two fractions with the same denominator, add the __numerators__ and place the sum over the _common denominator_.

5. To subtract two fractions with different denominators, first rewrite the fractions as __equivalent__ fractions with a _common denominator_.

6. To multiply two fractions, first multiply the __numerators__. Then place the product over the product of the _denominators_.

7. To divide two fractions, multiply the first fraction by the __reciprocal__ of the second fraction.

8. In the expression $\sqrt{a}$, the symbol $\sqrt{\ }$ is called the __radical sign__, and a is called the __radicand__.

9. Write the sixth power of -5 in exponential form. $(-5)^6$

10. ◣ Explain why $2\sqrt{2}$ is in simplest form and $\sqrt{8}$ is not in simplest form.

OBJECTIVE A   *To write a rational number as a decimal*

For Exercises 11 to 22, write as a decimal. Place a bar over the repeating digits of a repeating decimal.

11. $\frac{1}{8}$ 0.125

12. $\frac{7}{8}$ 0.875

13. $\frac{2}{9}$ $0.\overline{2}$

14. $\frac{8}{9}$ $0.\overline{8}$

15. $\frac{1}{6}$ $0.1\overline{6}$

16. $\frac{5}{6}$ $0.8\overline{3}$

17. $\frac{9}{16}$ 0.5625

18. $\frac{15}{16}$ 0.9375

19. $\frac{7}{12}$ $0.58\overline{3}$

20. $\frac{11}{12}$ $0.91\overline{6}$

21. $\frac{21}{40}$ 0.525

22. $\frac{5}{11}$ $0.\overline{45}$

OBJECTIVE B   *To convert among percents, fractions, and decimals*

23. ◣ Explain why multiplying a number by 100% does not change its value.

For Exercises 24 to 33, write as a fraction and as a decimal.

24. 75% $\frac{3}{4}$, 0.75

25. 40% $\frac{2}{5}$, 0.40

26. 64% $\frac{16}{25}$, 0.64

27. 88% $\frac{22}{25}$, 0.88

28. 125% $\frac{5}{4}$, 1.25

29. 160% $\frac{8}{5}$, 1.6

30. 19% $\frac{19}{100}$, 0.19

31. 87% $\frac{87}{100}$, 0.87

32. 5% $\frac{1}{20}$, 0.05

33. 450% $\frac{9}{2}$, 4.50

For Exercises 34 to 43, write as a fraction.

34. $11\frac{1}{9}\%$ $\frac{1}{9}$ **35.** $4\frac{2}{7}\%$ $\frac{3}{70}$ **36.** $12\frac{1}{2}\%$ $\frac{1}{8}$ **37.** $37\frac{1}{2}\%$ $\frac{3}{8}$ **38.** $66\frac{2}{3}\%$ $\frac{2}{3}$

39. $\frac{1}{4}\%$ $\frac{1}{400}$ **40.** $\frac{1}{2}\%$ $\frac{1}{200}$ **41.** $6\frac{1}{4}\%$ $\frac{1}{16}$ **42.** $83\frac{1}{3}\%$ $\frac{5}{6}$ **43.** $5\frac{3}{4}\%$ $\frac{23}{400}$

For Exercises 44 to 53, write as a decimal.

44. 7.3%   0.073   **45.** 9.1%   0.091   **46.** 15.8%   0.158   **47.** 16.7%   0.167   **48.** 0.3%   0.003

49. 0.9%   0.009   **50.** 9.9%   0.099   **51.** 9.15%   0.0915   **52.** 121.2%   1.212   **53.** 18.23%   0.1823

For Exercises 54 to 73, write as a percent.

54. 0.15   15%   **55.** 0.37   37%   **56.** 0.05   5%   **57.** 0.02   2%   **58.** 0.175   17.5%

59. 0.125   12.5%   **60.** 1.15   115%   **61.** 1.36   136%   **62.** 0.008   0.8%   **63.** 0.004   0.4%

64. $\frac{27}{50}$ 54%   **65.** $\frac{83}{100}$ 83%   **66.** $\frac{1}{3}$ $33\frac{1}{3}\%$ **67.** $\frac{3}{8}$ $37\frac{1}{2}\%$ **68.** $\frac{5}{11}$ $45\frac{5}{11}\%$

69. $\frac{4}{9}$ $44\frac{4}{9}\%$ **70.** $\frac{7}{8}$ $87\frac{1}{2}\%$ **71.** $\frac{9}{20}$ 45%   **72.** $1\frac{2}{3}$ $166\frac{2}{3}\%$ **73.** $2\frac{1}{2}$ 250%

74. Does $\frac{4}{3}$ represent a percent greater than 100% or less than 100%?
Greater than 100%

75. Does 0.055 represent a percent greater than 1% or less than 1%?
Greater than 1%

OBJECTIVE C   *To add or subtract rational numbers*

For Exercises 76 to 97, add or subtract.

76. $-\frac{5}{6} - \frac{5}{9}$
$-\frac{25}{18}$

77. $-\frac{6}{13} + \frac{17}{26} = \frac{23}{39}$
$\frac{5}{26}$

78. $-\frac{7}{12} + \frac{5}{8}$
$\frac{1}{24}$

79. $\frac{5}{8} - \left(-\frac{3}{4}\right)$
$\frac{11}{8}$

80. $\frac{3}{5} - \frac{11}{12}$
$-\frac{19}{60}$

81. $\frac{11}{12} - \frac{5}{6}$
$\frac{1}{12}$

82. $-\frac{2}{3} - \left(-\frac{11}{18}\right)$
$-\frac{1}{18}$

83. $-\frac{5}{8} - \left(-\frac{11}{12}\right)$
$\frac{7}{24}$

84. $\dfrac{1}{3} + \dfrac{5}{6} - \dfrac{2}{9}$

$\dfrac{17}{18}$

85. $\dfrac{1}{2} - \dfrac{2}{3} + \dfrac{1}{6}$

0

86. $-\dfrac{5}{16} + \dfrac{3}{4} - \dfrac{7}{8}$

$-\dfrac{7}{16}$

87. $\dfrac{1}{2} - \dfrac{3}{8} - \left(-\dfrac{1}{4}\right)$

$\dfrac{3}{8}$

88. $-13.092 + 6.9$

-6.192

89. $2.54 - 3.6$

-1.06

90. $5.43 + 7.925$

13.355

91. $-16.92 - 6.925$

-23.845

92. $-3.87 + 8.546$

4.676

93. $6.9027 - 17.692$

-10.7893

94. $2.09 - 6.72 - 5.4$

-10.03

95. $-3.09 - 4.6 - (-27.3)$

19.61

96. $16.4 - (-3.09) - 7.93$

11.56

97. $2.66 - (-4.66) - 8.2$

-0.88

For Exercises 98 to 100, estimate each sum to the nearest integer. Do not find the exact sum.

98. $\dfrac{7}{8} + \dfrac{4}{5}$

2

99. $-0.125 + 1.25$

1

100. $-1.3 + 0.2$

-1

OBJECTIVE D *To multiply or divide rational numbers*

For Exercises 101 to 118, multiply or divide.

101. $\left(-\dfrac{3}{4}\right)\left(-\dfrac{8}{27}\right)$

$\dfrac{2}{9}$

102. $-\dfrac{1}{2}\left(\dfrac{8}{9}\right)$

$-\dfrac{4}{9}$

103. $\dfrac{5}{12}\left(-\dfrac{8}{15}\right)$

$-\dfrac{2}{9}$

104. $\dfrac{5}{8}\left(-\dfrac{7}{12}\right)\dfrac{16}{25}$

$-\dfrac{7}{30}$

105. $\left(\dfrac{5}{12}\right)\left(-\dfrac{8}{15}\right)\left(\dfrac{1}{3}\right)$

$-\dfrac{2}{27}$

106. $\dfrac{1}{2}\left(-\dfrac{3}{4}\right)\left(-\dfrac{5}{8}\right)$

$\dfrac{15}{64}$

107. $\dfrac{3}{8} \div \dfrac{1}{4}$

$\dfrac{3}{2}$

108. $\dfrac{5}{6} \div \left(-\dfrac{3}{4}\right)$

$-\dfrac{10}{9}$

109. $-\dfrac{5}{12} \div \dfrac{15}{32}$

$-\dfrac{8}{9}$

110. $\dfrac{1}{8} \div \left(-\dfrac{5}{12}\right)$

$-\dfrac{3}{10}$

111. $-\dfrac{4}{9} \div \left(-\dfrac{2}{3}\right)$

$\dfrac{2}{3}$

112. $-\dfrac{6}{11} \div \dfrac{4}{9}$

$-\dfrac{27}{22}$

113. $1.2(3.47)$

4.164

114. $(-0.8)6.2$

-4.96

115. $(-1.89)(-2.3)$

4.347

116. $(6.9)(-4.2)$
-28.98

117. $1.2(-0.5)(3.7)$
-2.22

118. $2.3(-0.6)(0.8)$
-1.104

For Exercises 119 to 124, divide. Round to the nearest hundredth.

119. $-1.27 \div (-1.7)$
0.75

120. $9.07 \div (-3.5)$
-2.59

121. $0.0976 \div 0.042$
2.32

122. $-6.904 \div 1.35$
-5.11

123. $-7.894 \div (-2.06)$
3.83

124. $-354.2086 \div 0.1719$
-2060.55

125. **a.** Without finding the product, determine whether $\frac{11}{13} \cdot \frac{50}{51}$ is greater than 1 or less than 1.   Less than 1
b. Without finding the quotient, determine whether $8.713 \div 7.2$ is greater than 1 or less than 1.   Greater than 1

OBJECTIVE E   *To evaluate exponential expressions*

For Exercises 126 to 144, evaluate.

126. 6^2
36

127. 7^4
2401

128. -7^2
-49

129. -4^3
-64

130. $(-3)^2$
9

131. $(-2)^3$
-8

132. $(-3)^4$
81

133. $(-5)^3$
-125

134. $\left(\frac{1}{2}\right)^2$
$\frac{1}{4}$

135. $\left(-\frac{3}{4}\right)^3$
$-\frac{27}{64}$

136. $(0.3)^2$
0.09

137. $(1.5)^3$
3.375

138. $\left(\frac{2}{3}\right)^2 \cdot 3^3$
12

139. $\left(-\frac{1}{2}\right)^3 \cdot 8$
-1

140. $(0.3)^3 \cdot 2^3$
0.216

141. $(-2) \cdot (-2)^2$
-8

142. $2^3 \cdot 3^3 \cdot (-4)$
-864

143. $(-3)^3 \cdot 5^2 \cdot 10$
-6750

144. $(-7) \cdot 4^2 \cdot 3^2$
-1008

For Exercises 145 to 148, without finding the product, state whether the given expression simplifies to a positive or a negative number.

145. $(-9)^7$
Negative

146. -8^6
Negative

147. $(-9^{10})(-5^4)$
Positive

148. $-(3^4)(-2^5)$
Positive

OBJECTIVE F   *To simplify numerical radical expressions*

For Exercises 149 to 172, simplify.

149. $\sqrt{16}$
4

150. $\sqrt{64}$
8

151. $\sqrt{49}$
7

152. $\sqrt{144}$
12

153. $\sqrt{32}$
$4\sqrt{2}$

154. $\sqrt{50}$
$5\sqrt{2}$

155. $\sqrt{8}$
$2\sqrt{2}$

156. $\sqrt{12}$
$2\sqrt{3}$

157. $6\sqrt{18}$
$18\sqrt{2}$

158. $-3\sqrt{48}$
$-12\sqrt{3}$

159. $5\sqrt{40}$
$10\sqrt{10}$

160. $2\sqrt{28}$
$4\sqrt{7}$

161. $\sqrt{15}$
$\sqrt{15}$

162. $\sqrt{21}$
$\sqrt{21}$

163. $\sqrt{29}$
$\sqrt{29}$

164. $\sqrt{13}$
$\sqrt{13}$

165. $-9\sqrt{72}$
$-54\sqrt{2}$

166. $11\sqrt{80}$
$44\sqrt{5}$

167. $\sqrt{45}$
$3\sqrt{5}$

168. $\sqrt{225}$
15

169. $\sqrt{0}$
0

170. $\sqrt{210}$
$\sqrt{210}$

171. $6\sqrt{128}$
$48\sqrt{2}$

172. $9\sqrt{288}$
$108\sqrt{2}$

For Exercises 173 to 178, find the decimal approximation to the nearest thousandth.

173. $\sqrt{240}$
15.492

174. $\sqrt{300}$
17.321

175. $\sqrt{288}$
16.971

176. $\sqrt{600}$
24.495

177. $\sqrt{256}$
16

178. $\sqrt{324}$
18

For Exercises 179 to 182, find consecutive integers m and n such that the given number is between m and n, or state that the given number is not a real number. Do not use a calculator.

179. $-\sqrt{115}$
−11 and −10

180. $-\sqrt{-90}$
Not a real number

181. $\sqrt{\sqrt{64}}$
2 and 3

182. $\sqrt{200}$
14 and 15

OBJECTIVE G   *To solve application problems*

183. ● **Meteorology**   On January 23, 1916, the temperature in Browing, Montana, was 6.67°C. On January 24, 1916, the temperature in Browing was −48.9°C. Find the difference between the temperatures in Browing on these two days.
55.57°C

184. ● **Meteorology**   On January 22, 1943, in Spearfish, South Dakota, the temperature fell from 12.22°C at 9 A.M. to −20°C at 9:27 A.M. How many degrees did the temperature fall during the 27-minute period?
32.22°C

185. ● **Chemistry**   The boiling point of oxygen is −182.962°C. Oxygen's melting point is −218.4°C. What is the difference between the boiling point and the melting point of oxygen?
35.438°C

186. ● **Chemistry**   The boiling point of nitrogen is −195.8°C, and the melting point is −209.86°C. Find the difference between the boiling point and the melting point of nitrogen.
14.06°C

QUICK QUIZ

1. Write $\dfrac{5}{16}$ as a decimal.
   0.3125   [1.2A]

2. Write $\dfrac{8}{11}$ as a decimal.
   $0.\overline{72}$ [1.2A]

3. Write 80% as a fraction and as a decimal.
   $\dfrac{4}{5}$, 0.8   [1.2B]

4. Write 0.9 and $\dfrac{7}{8}$ as percents.
   90%, 87.5%   [1.2B]

Simplify.
5. $5.63 - (-2.1)$
   7.73   [1.2C]

6. $-\dfrac{5}{6} - \left(-\dfrac{2}{9}\right)$
   $-\dfrac{11}{18}$ [1.2C]

(cont'd)

187. ● **Oil Production** Read the news clipping at the right.
    a. Find the difference between daily U.S. oil production in 2008 and in 1973.
       4.3 million barrels per day
    b. Calculate the predicted increase in U.S. oil production from 2008 to 2020.
       1.1 million barrels per day

188. **Interior Design** To reupholster a large sofa, an interior designer needs $12\frac{1}{2}$ yd of fabric that costs $5.43 per yard and $5\frac{3}{4}$ yd of fabric that costs $6.94 per yard. Find the total cost of the two fabrics.
    $107.78

189. **Food Science** A recipe calls for $\frac{3}{4}$ c of butter. If a chef wants to increase the recipe by one-half, how much butter should the chef use?
    $1\frac{1}{8}$ c

● **The Food Industry** The table at the right below shows the net weights of four different boxes of cereal. Use this table for Exercises 190 and 191.

190. Find the number of $\frac{3}{4}$-ounce servings in a box of Post Shredded Wheat.
    24 servings

191. Find the number of $1\frac{1}{2}$-ounce servings in a box of Kellogg's Honey Crunch Corn Flakes. 16 servings

Critical Thinking

192. Determine whether the statement is true or false.
    a. Every integer is a rational number.   True
    b. Every whole number is an integer.   True
    c. Every integer is a positive number.   False
    d. Every rational number is an integer.   False

193. **Number Problems**
    a. Find a rational number between 0.1 and 0.2.
    b. Find a rational number between 1 and 1.1.
    c. Find a rational number between 0 and 0.005.
    Answers will vary. For example, **a.** 0.15, **b.** 1.05, **c.** 0.001.

194. Find a rational number that is one-half the difference between $\frac{5}{11}$ and $\frac{4}{11}$. $\frac{1}{22}$

195. ◣ Given any two different rational numbers, is it always possible to find a rational number between the two given numbers? If so, explain how to find such a number. If not, give two rational numbers for which there is no rational number between them.

Projects or Group Activities

196. Use a calculator to determine the decimal representations of $\frac{17}{99}$, $\frac{45}{99}$, and $\frac{73}{99}$. Make a conjecture as to the decimal representation of $\frac{83}{99}$. Does your conjecture work for $\frac{33}{99}$? What about $\frac{1}{99}$?
    $\frac{17}{99} = 0.\overline{17}$; $\frac{45}{99} = 0.\overline{45}$; $\frac{73}{99} = 0.\overline{73}$; $\frac{83}{99} = 0.\overline{83}$; $\frac{33}{99} = 0.\overline{33} = 0.\overline{3}$, yes; $\frac{1}{99} = 0.\overline{01}$, yes

197. Find three natural numbers a, b, and c such that $\frac{1}{a} + \frac{1}{b} + \frac{1}{c}$ is a natural number.
    $a = 2$, $b = 3$, $c = 6$

Unless otherwise noted, all content on this page is © Cengage Learning.

In the NEWS!

U.S. Oil Production Expected to Grow

U.S. oil production was on the decline. Production went from 9.2 million barrels per day in 1973 to 4.9 million barrels per day in 2008. However, high oil prices in 2010 made it economical to extract crude out of oil-shale deposits. Now experts forecast production of 6.0 million barrels per day by 2020.
Source: Time, March 21, 2011

Cereal	Net Weight
Kellogg's Honey Crunch Corn Flakes	24 oz
Nabisco Instant Cream of Wheat	28 oz
Post Shredded Wheat	18 oz
Quaker Oats	41 oz

Simplify.
7. $-9.3(12.7)$
    -118.11 **[1.2D]**
8. $15.33 \div (-7)$
    -2.19 **[1.2D]**
9. $-\frac{2}{3} \div \left(-\frac{5}{6}\right)$
    $\frac{4}{5}$ **[1.2D]**
10. Evaluate $(-5)^2$.
    25 **[1.2E]**
11. Evaluate $-4^2 \cdot 3^3$.
    -432 **[1.2E]**

Simplify.
12. $\sqrt{18}$ $3\sqrt{2}$ **[1.2F]**
13. $-6\sqrt{75}$
    $-30\sqrt{3}$ **[1.2F]**
14. The lowest temperature ever recorded in North America is $-81.4°$F. The highest temperature ever recorded is $134.0°$F. (*Source:* National Climatic Data Center) Find the difference between these two extremes.   **215.4°F [1.2G]**

1.3 The Order of Operations Agreement

OBJECTIVE A *To use the Order of Operations Agreement to simplify expressions*

Let's evaluate $2 + 3 \cdot 5$.

There are two arithmetic operations, addition and multiplication, in this expression. The operations could be performed in different orders. We could multiply first and then add, or we could add first and then multiply. To prevent there being more than one answer when simplifying a numerical expression, an Order of Operations Agreement has been established.

The Order of Operations Agreement

Step 1 Perform operations inside grouping symbols. Grouping symbols include parentheses ( ), brackets [ ], braces { }, the absolute value symbol | |, and the fraction bar.
Step 2 Simplify exponential expressions.
Step 3 Do multiplication and division as they occur from left to right.
Step 4 Do addition and subtraction as they occur from left to right.

EXAMPLE

Evaluate $12 - 24(8 - 5) \div 2^2$.

$12 - 24(8 - 5) \div 2^2 = 12 - 24(3) \div 2^2$ • **Perform operations inside grouping symbols.**

$= 12 - 24(3) \div 4$ • **Simplify exponential expressions.**

$= 12 - 72 \div 4$ • **Do multiplication and division as they occur from left to right.**

$= 12 - 18$

$= -6$ • **Do addition and subtraction as they occur from left to right.**

 Integrating Technology

See the Keystroke Guide: *Basic Operations* for instruction on using a calculator to evaluate a numerical expression.

One or more of the steps listed above may not be needed to evaluate an expression. In that case, proceed to the next step in the Order of Operations Agreement.

HOW TO 1 Evaluate $\frac{4 + 8}{2 + 1} - (3 - 1) + 2$.

$\dfrac{4 + 8}{2 + 1} - (3 - 1) + 2 = \dfrac{12}{3} - 2 + 2$ • **Perform operations above and below the fraction bar and inside parentheses.**

$= 4 - 2 + 2$ • **Do multiplication and division as they occur from left to right.**

$= 2 + 2$ • **Do addition and subtraction as they occur from left to right.**

$= 4$

When an expression has grouping symbols inside grouping symbols, perform the operations inside the inner grouping symbols first.

HOW TO 2 Evaluate $6 \div [4 - (6 - 8)] + 2^2$.

$6 \div [4 - (6 - 8)] + 2^2 = 6 \div [4 - (-2)] + 2^2$ • Perform operations inside grouping symbols.

$= 6 \div 6 + 2^2$
$= 6 \div 6 + 4$ • Simplify exponential expressions.

$= 1 + 4$ • Do multiplication and division as they occur from left to right.

$= 5$ • Do addition and subtraction as they occur from left to right.

EXAMPLE 1

Evaluate $4 - 3[4 - 2(6 - 3)] \div 2$.

Solution

$4 - 3[4 - 2(6 - 3)] \div 2$
$= 4 - 3[4 - 2 \cdot 3] \div 2$ • Perform operations inside grouping symbols.
$= 4 - 3[4 - 6] \div 2$
$= 4 - 3[-2] \div 2$
$= 4 + 6 \div 2$ • Do multiplication and division from left to right.

$= 4 + 3$
$= 7$ • Do addition and subtraction from left to right.

YOU TRY IT 1

Evaluate $18 - 5[8 - 2(2 - 5)] \div 10$.

Your solution

11

IN-CLASS EXAMPLES
Evaluate.
1. $10 - 3[5 - 2(9 - 4)] \div 5$ **13**
2. $36 \div (9 - 7)^2 + (-4)^2 \cdot 3$ **57**
3. $-\dfrac{7 + 8}{2 - (-1)} - 2 \cdot 3^2$ **−23**
4. $7 - 2(4 - 7)^3 - (-2)^3$ **69**

EXAMPLE 2

Evaluate $(1.75 - 1.3)^2 \div 0.025 + 6.1$.

Solution

$(1.75 - 1.3)^2 \div 0.025 + 6.1$
$= (0.45)^2 \div 0.025 + 6.1$ • Perform operations inside grouping symbols.

$= 0.2025 \div 0.025 + 6.1$ • Simplify exponential expressions.

$= 8.1 + 6.1$ • Do the division.
$= 14.2$ • Do the addition.

YOU TRY IT 2

Evaluate $(6.97 - 4.72)^2 \cdot 4.5 \div 0.05$.

Your solution

455.625

Solutions on p. S2

1.3 EXERCISES

✔ Concept Check

SUGGESTED ASSIGNMENT
Exercises 1 and 2; Exercises 3–35, odds
More challenging exercises: Exercises 36–40

1. 🔲 Why do we need an Order of Operations Agreement?

2. 🔲 Describe each step in the Order of Operations Agreement.

OBJECTIVE A   *To use the Order of Operations Agreement to simplify expressions*

For Exercises 3 to 35, evaluate by using the Order of Operations Agreement.

3. $4 - 8 \div 2$
0

4. $2^2 \cdot 3 - 3$
9

5. $2(3 - 4) - (-3)^2$
-11

6. $16 - 32 \div 2^3$
12

7. $24 - 18 \div 3 + 2$
20

8. $8 - (-3)^2 - (-2)$
1

9. $8 - 2(3)^2$
-10

10. $16 - 16 \cdot 2 \div 4$
8

11. $12 + 16 \div 4 \cdot 2$
20

12. $16 - 2 \cdot 4^2$
-16

13. $27 - 18 \div (-3^2)$
29

14. $4 + 12 \div 3 \cdot 2$
12

15. $16 + 15 \div (-5) - 2$
11

16. $14 - 2^2 - (4 - 7)$
13

17. $14 - 2^2 - |4 - 7|$
7

18. $10 - |5 - 8| + 2^3$
15

19. $3 - 2[8 - (3 - 2)]$
-11

20. $-2^2 + 4[16 \div (3 - 5)]$
-36

21. $6 + \dfrac{16 - 4}{2^2 + 2} - 2$
6

22. $24 \div \dfrac{3^2}{8 - 5} - (-5)$
13

23. $18 \div |9 - 2^3| + (-3)$
15

24. $96 \div 2[12 + (6 - 2)] - 3^2$
759

25. $4[16 - (7 - 1)] \div 10$
4

26. $18 \div 2 - 4^2 - (-3)^2$
-16

27. $20 \div (10 - 2^3) + (-5)$
5

28. $16 - 3(8 - 3)^2 \div 5$
1

29. $4(-8) \div [2(7 - 3)^2]$
-1

30. $\dfrac{(-10) + (-2)}{6^2 - 30} \div |2 - 4|$
-1

31. $16 - 4 \cdot \dfrac{3^3 - 7}{2^3 + 2} - (-2)^2$
4

32. $(0.2)^2 \cdot (-0.5) + 1.72$
1.7

33. $0.3(1.7 - 4.8) + (1.2)^2$
0.51

34. $(1.8)^2 - 2.52 \div 1.8$
1.84

35. $(1.65 - 1.05)^2 \div 0.4 + 0.8$
1.7

36. Which expression is equivalent to $15 + 15 \div 3 - 4^2$?
(i) $30 \div 3 - 16$ **(ii)** $15 + 5 - 16$ **(iii)** $15 + 5 + 16$ **(iv)** $15 + 15 \div (-1)^2$
ii

Critical Thinking

37. Find two fractions between $\frac{2}{3}$ and $\frac{3}{4}$. (There is more than one answer to this question.)
Answers will vary. For example, $\dfrac{17}{24}$ and $\dfrac{33}{48}$.

38. A **magic square** is one in which the numbers in every row, column, and diagonal sum to the same number. Complete the magic square at the right.

$\frac{2}{3}$	$-\frac{1}{6}$	0
$-\frac{1}{2}$	$\frac{1}{6}$	$\frac{5}{6}$
$\frac{1}{3}$	$\frac{1}{2}$	$-\frac{1}{3}$

39. For each part below, find a rational number r that satisfies the condition.
a. $r^2 < r$ **b.** $r^2 = r$ **c.** $r^2 > r$
Answers will vary. For example, **a.** $\dfrac{1}{2}$, **b.** 1, **c.** 2.

40. The following was offered as the simplification of $6 + 2(4 - 9)$.
$$6 + 2(4 - 9) = 6 + 2(-5)$$
$$= 8(-5)$$
$$= -40$$

If this is a correct simplification, write yes for the answer. If it is incorrect, write no and explain the incorrect step.

Projects or Group Activities

41. In which column is the number 1 million, column A, B, or C?

A	B	C
1	8	27
64	125	216
.	.	.
.	.	.
.	.	.

Column A: $1{,}000{,}000 = 100^3$

QUICK QUIZ
Evaluate.
1. $3 \cdot 4^2$ **48 [1.3A]**
2. $5 - 2(7 - 10)$
   11 [1.3A]
3. $2^4 - 3(2 - 3^2)^2$
   -131 [1.3A]
4. $\dfrac{3^2 - 2}{2 + 5(2^2 - 3)}$
   1 [1.3A]
5. $\dfrac{6 - 3(8 - 6)}{3^2 - 4^2}$
   0 [1.3A]

✔ CHECK YOUR PROGRESS: CHAPTER 1 ....

1. Use the roster method to write the set of positive integers less than 9.   {1, 2, 3, 4, 5, 6, 7, 8}   [1.1A]

2. Let $A \in \{-7, 0, 2, 5\}$. For which elements of A is the inequality $A < 1$ a true statement?
$-7, 0$ [1.1A]

3. Find the additive inverse of -13.   13   [1.1B]

4. Evaluate $|-44|$ and $-|-18|$.   44; -18 [1.1B]

5. Place the correct symbol, $<$ or $>$, between the two expressions.
$|31|$ $>$ $|-13|$ [1.1B]

6. Add: $-47 + 23$ -24 [1.1C]

INSTRUCTOR NOTE
The notation [1.1A] following the answer to Exercise 1 indicates the objective the student should review if that question is answered incorrectly. The notation [1.1A] means Chapter 1, Section 1, Objective A. This notation is used following every answer in all of the Prep Tests (except Chapter 1), Check Your Progress exercises, Chapter Review Exercises, Chapter Tests, and Cumulative Reviews throughout the text.

7. Subtract: $-11 - (-27)$ 16   [1.1C]

8. Add: $-32 + 40 + (-9)$ -1 [1.1C]

9. Subtract: $42 - (-82) - 65 - 7$ 52   [1.1C]

10. Multiply: $16(-2)$ -32 [1.1D]

11. Multiply: $-9(7)(-5)$ 315   [1.1D]

12. Divide: $250 \div (-25)$ -10 [1.1D]

13. Divide: $-\dfrac{-80}{-5}$ -16 [1.1D]

14. Divide: $\dfrac{-58}{0}$ Undefined   [1.1D]

15. Write $\dfrac{11}{16}$ as a decimal.   0.6875   [1.2A]

16. Write $\dfrac{7}{11}$ as a decimal. Place a bar over any repeating digits. $0.\overline{63}$ [1.2A]

17. Write 45% as a fraction and as a decimal.
$\dfrac{9}{20}$; 0.45   [1.2B]

18. Write $14\frac{1}{2}\%$ as a fraction.
$\dfrac{29}{200}$ [1.2B]

19. Write $\frac{7}{8}$ as a percent.   87.5%   [1.2B]

20. Write 0.08 as a percent.   8%   [1.2B]

21. Add: $\frac{5}{6} + \frac{3}{18}$ 1   [1.2C]

22. Subtract: $\frac{3}{24} - \frac{1}{6}$ $-\frac{1}{24}$ [1.2C]

23. Simplify: $-18.39 + 4.9 - 23.7$ -37.19 [1.2C]

24. Multiply: $\frac{5}{8}\left(-\frac{9}{12}\right)\left(\frac{16}{25}\right)$ $-\frac{3}{2}$ [1.2D]

25. Divide: $-\frac{6}{11} \div \frac{9}{4}$ $-\frac{8}{33}$ [1.2D]

26. Multiply: $-1.6(0.2)$ -0.32 [1.2D]

27. Simplify: $3\sqrt{18}$ $6\sqrt{2}$ [1.2F]

28. Simplify: $\sqrt{27}$ $3\sqrt{3}$ [1.2F]

29. Evaluate: $-3^2 \cdot (-2)^4$ -144 [1.2E]

30. Evaluate: $5 - 4[3 - 2(7 - 1)] \div 9$ 9   [1.3A]

31. Evaluate: $-4 \cdot 2^3 - \dfrac{1 - 13}{2^2 \cdot 3}$ -31 [1.3A]

32. Evaluate: $(8 - 3^2)^6 + (2 \cdot 3 - 7)^9$ 0   [1.3A]

33. Temperature   Find the temperature after a rise of 8°C from −3°C.   5°C   [1.1E]

34. Temperature   The daily low temperatures (in degrees Celsius) during one week were recorded as −8°, −12°, 0°, −4°, 5°, −7°, and −9°. Find the average daily low temperature for the week.   −5°C   [1.1E]

35. Temperature   If the temperature rose 20.3°F during the day and reached a high of 15.7°F, at what temperature did the day begin?   −4.6°F   [1.2G]

© iStockphoto.com/Denis Jr. Tangney

1.4   Variable Expressions

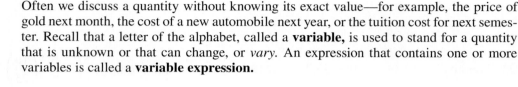

OBJECTIVE A   *To evaluate a variable expression*

 Tips for Success

Before you begin a new chapter, you should take some time to review previously learned skills. One way to do this is to complete the Prep Test. See page 75. This test focuses on the particular skills that will be required for the new chapter.

Often we discuss a quantity without knowing its exact value—for example, the price of gold next month, the cost of a new automobile next year, or the tuition cost for next semester. Recall that a letter of the alphabet, called a **variable,** is used to stand for a quantity that is unknown or that can change, or *vary.* An expression that contains one or more variables is called a **variable expression.**

A variable expression is shown at the right. The expression can be rewritten by writing subtraction as the addition of the opposite.

$$3x^2 - 5y + 2xy - x - 7$$
$$3x^2 + (-5y) + 2xy + (-x) + (-7)$$

 Point of Interest

Historical manuscripts indicate that mathematics is at least 4000 years old. Yet it was only 400 years ago that mathematicians started using variables to stand for numbers. The idea that a letter can stand for some number was a critical turning point in mathematics.

Today, *x* is used by most nations as the standard letter for a single unknown. In fact, *x*-rays were so named because the scientists who discovered them did not know what they were and thus labeled them the "unknown rays" or *x*-rays.

Note that the expression has five addends. The **terms** of a variable expression are the addends of the expression. The expression has five terms.

Five terms

$$\underbrace{3x^2 \quad - \quad 5y \quad + \quad 2xy \quad - \quad x}_{\text{Variable terms}} \quad \underbrace{- \quad 7}_{\substack{\text{Constant} \\ \text{term}}}$$

The terms $3x^2$, $-5y$, $2xy$, and $-x$ are **variable terms.**

The term -7 is a **constant term,** or simply a **constant.**

Each variable term is composed of a **numerical coefficient** and a **variable part** (the variable or variables and their exponents).

When the numerical coefficient is 1 or -1, the 1 is usually not written ($x = 1x$ and $-x = -1x$).

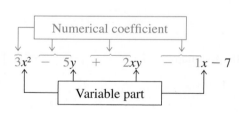

 Integrating Technology

See the Keystroke Guide: *Evaluating Variable Expressions* for instructions on using a graphing calculator to evaluate variable expressions.

Replacing each variable by its value and then simplifying the resulting numerical expression is called **evaluating a variable expression.**

HOW TO 1   Evaluate $ab - b^2$ when $a = 2$ and $b = -3$.

Replace each variable in the expression by its value. Then use the Order of Operations Agreement to simplify the resulting numerical expression.

$$ab - b^2$$
$$2(-3) - (-3)^2 = -6 - 9$$
$$= -15$$

When $a = 2$ and $b = -3$, the value of $ab - b^2$ is -15.

INSTRUCTOR NOTE

Emphasize that the result of evaluating a variable expression is one number.

EXAMPLE 1

Evaluate $\frac{a^2 - b^2}{a - b}$ when $a = 3$ and $b = -4$.

Solution

$$\frac{a^2 - b^2}{a - b}$$

$$\frac{3^2 - (-4)^2}{3 - (-4)} = \frac{9 - 16}{3 - (-4)} \qquad \bullet \ a = 3, b = -4$$

$$= \frac{-7}{7} = -1$$

YOU TRY IT 1

Evaluate $\frac{a^2 + b^2}{a + b}$ when $a = 5$ and $b = -3$.

Your solution

17

IN-CLASS EXAMPLES

1. Name the variable terms of the expression $3b^3 - 4b - 2$.
   $3b^3, -4b$
2. Evaluate $3a^2 - 4ab$ when $a = 5$ and $b = -4$. **155**
3. Evaluate $\frac{x^3 + y^3}{x + y}$ when $x = 2$ and $y = -3$. **19**

EXAMPLE 2

Evaluate $x^2 - 3(x - y) - z^2$ when $x = 2$, $y = -1$, and $z = 3$.

Solution

$$x^2 - 3(x - y) - z^2$$

$$2^2 - 3[2 - (-1)] - 3^2 \qquad \bullet \ x = 2, y = -1, z = 3$$

$$= 2^2 - 3(3) - 3^2$$

$$= 4 - 3(3) - 9$$

$$= 4 - 9 - 9$$

$$= -5 - 9$$

$$= -14$$

YOU TRY IT 2

Evaluate $x^3 - 2(x + y) + z^2$ when $x = 2$, $y = -4$, and $z = -3$.

Your solution

21

4. Evaluate $a^2 - 5(a - 2b) - c^2$ when $a = -3$, $b = 2$, and $c = -1$. **43**

Solutions on p. S2

OBJECTIVE B *To simplify a variable expression using the Properties of Addition*

 Take Note

Here is an example of the Distributive Property using just numbers.

$$2(5 + 9) = 2(5) + 2(9)$$
$$= 10 + 18 = 28$$

This is the same result we would obtain using the Order of Operations Agreement.

$$2(5 + 9) = 2(14) = 28$$

The usefulness of the Distributive Property will become more apparent as we explore variable expressions.

Like terms of a variable expression are terms with the same variable part. (Because $x^2 = x \cdot x$, x^2 and x are not like terms.)

Constant terms are like terms. 4 and 9 are like terms.

$$\underbrace{3x}_{} + \underbrace{4}_{} - \underbrace{7x}_{} + \underbrace{9}_{} - x^2$$

Like terms

Like terms

To simplify a variable expression, we use the Distributive Property to add the numerical coefficients of like variable terms. The variable part remains unchanged.

The Distributive Property

If a, b, and c are real numbers, then $a(b + c) = ab + ac$ or $(b + c)a = ba + ca$.

By the Distributive Property, the term outside the parentheses is multiplied by each term inside the parentheses.

EXAMPLES

1. $2(3 + 4) = 2 \cdot 3 + 2 \cdot 4$
$\qquad 2(7) = 6 + 8$
$\qquad 14 = 14$

2. $(4 + 5)2 = 4 \cdot 2 + 5 \cdot 2$
$\qquad (9)2 = 8 + 10$
$\qquad 18 = 18$

The Distributive Property in the form $(b + c)a = ba + ca$ is used to simplify a variable expression.

To simplify $2x + 3x$, use the Distributive Property to add the numerical coefficients of the like variable terms. This is called **combining like terms.**

$$2x + 3x = (2 + 3)x$$
$$= 5x$$

HOW TO 2   Simplify: $5y - 11y$

$$5y - 11y = (5 - 11)y$$
$$= -6y$$ • **Use the Distributive Property.**

HOW TO 3   Simplify: $5 + 7p$

The terms 5 and $7p$ are not like terms.

The expression $5 + 7p$ is in simplest form.

The following Properties of Addition are used to simplify variable expressions.

 Take Note

Simplifying an expression means combining like terms. The constant term 5 and the variable term $7p$ are not like terms and therefore cannot be combined.

INSTRUCTOR NOTE

Combining like terms can be related to many everyday experiences. For instance, 5 bricks plus 7 bricks is 12 bricks. But 5 bricks plus 7 nails is 5 bricks plus 7 nails. Students need to be constantly reminded that algebra is a reflection of our experiences, not some arbitrarily made up system of rules.

The Associative Property of Addition

If a, b, and c are real numbers, then $(a + b) + c = a + (b + c)$.

When three or more terms are added, the terms can be grouped (with parentheses, for example) in any order; the sum is the same.

EXAMPLES

1. $(5 + 7) + 15 = 5 + (7 + 15)$
$12 + 15 = 5 + 22$
$27 = 27$

2. $(3x + 5x) + 9x = 3x + (5x + 9x)$
$8x + 9x = 3x + 14x$
$17x = 17x$

The Commutative Property of Addition

If a and b are real numbers, then $a + b = b + a$.

When two terms are added, the terms can be added in either order; the sum is the same.

EXAMPLES

1. $15 + (-28) = (-28) + 15$
$-13 = -13$

2. $2x + (-4x) = -4x + 2x$
$-2x = -2x$

The Addition Property of Zero

If a is a real number, then $a + 0 = a$ and $0 + a = a$.

The sum of a term and zero is the term.

EXAMPLES

1. $-9 + 0 = -9$ and $0 + (-9) = -9$

2. $0 + 5x = 5x$ and $5x + 0 = 5x$

The Inverse Property of Addition

If a is a real number, then $a + (-a) = 0$ and $(-a) + a = 0$.

The sum of a term and its additive inverse (or opposite) is zero.

EXAMPLES

1. $8 + (-8) = 0$ and $-8 + 8 = 0$

2. $-7x + 7x = 0$ and $7x + (-7x) = 0$

HOW TO 4   Simplify: $8x + 4y - 8x + y$

$$8x + 4y - 8x + y$$
$$= (8x - 8x) + (4y + y)$$

$$= 0 + 5y = 5y$$

- Use the Commutative and Associative Properties of Addition to rearrange and group like terms.
- Combine like terms.

IN-CLASS EXAMPLES

Simplify.

5. $4a - 5b - 3a + 2b$
   $a - 3b$

6. $y^2 + 2 + 9y^2 - 14$
   $10y^2 - 12$

HOW TO 5   Simplify: $4x^2 + 5x - 6x^2 - 2x + 1$

$$4x^2 + 5x - 6x^2 - 2x + 1$$
$$= (4x^2 - 6x^2) + (5x - 2x) + 1$$

$$= -2x^2 + 3x + 1$$

- Use the Commutative and Associative Properties of Addition to rearrange and group like terms.
- Combine like terms.

EXAMPLE 3

Simplify: $3x + 4y - 10x + 7y$

Solution

$3x + 4y - 10x + 7y = -7x + 11y$

YOU TRY IT 3

Simplify: $3a - 2b - 5a + 6b$

Your solution

$-2a + 4b$

EXAMPLE 4

Simplify: $x^2 - 7 + 4x^2 - 16$

Solution

$x^2 - 7 + 4x^2 - 16 = 5x^2 - 23$

YOU TRY IT 4

Simplify: $-3y^2 + 7 + 8y^2 - 14$

Your solution

$5y^2 - 7$

Solutions on p. S2

OBJECTIVE C   *To simplify a variable expression using the Properties of Multiplication*

In simplifying variable expressions, the following Properties of Multiplication are used.

The Associative Property of Multiplication

If a, b, and c are real numbers, then $(ab)c = a(bc)$.

When three or more factors are multiplied, the factors can be grouped in any order; the product is the same.

EXAMPLES

1. $3(5 \cdot 6) = (3 \cdot 5)6$
   $3(30) = (15)6$
   $90 = 90$

2. $2(3x) = (2 \cdot 3)x$
   $= 6x$

📋 **Take Note**

The Associative Property of Multiplication allows us to multiply a coefficient by a number. Without this property, the expression $2(3x)$ could not be simplified.

Take Note
The Commutative Property of Multiplication allows us to rearrange factors. This property, along with the Associative Property of Multiplication, enables us to simplify some variable expressions.

The Commutative Property of Multiplication

If a and b are real numbers, then $ab = ba$.

Two factors can be multiplied in either order; the product is the same.

EXAMPLES

1. $5(-7) = -7(5)$
    $-35 = -35$

2. $(5x) \cdot 3 = 3 \cdot (5x)$ • Commutative Property
                                            of Multiplication
         $= (3 \cdot 5)x$ • Associative Property
                                of Multiplication
         $= 15x$

The Multiplication Property of One

If a is a real number, then $a \cdot 1 = a$ and $1 \cdot a = a$.

The product of a term and 1 is the term.

EXAMPLES

1. $9 \cdot 1 = 9$

2. $(8x) \cdot 1 = 8x$

The Inverse Property of Multiplication

If a is a real number and a is not equal to zero, then $a \cdot \dfrac{1}{a} = 1$ and $\dfrac{1}{a} \cdot a = 1$.

$\dfrac{1}{a}$ is called the **reciprocal** of a. $\dfrac{1}{a}$ is also called the **multiplicative inverse** of a.

The product of a number and its reciprocal is 1.

EXAMPLES

1. $7 \cdot \dfrac{1}{7} = 1$ and $\dfrac{1}{7} \cdot 7 = 1$

2. $x \cdot \dfrac{1}{x} = 1$ and $\dfrac{1}{x} \cdot x = 1,\quad x \neq 0$

 Take Note
In example (2), we must state that $x \neq 0$ because division by zero is undefined.

The multiplication properties are used to simplify variable expressions.

HOW TO 6   Simplify: $2(-x)$

$2(-x) = 2(-1 \cdot x)$ • Use the Associative Property of
$\qquad = [2(-1)]x$ Multiplication to group factors.
$\qquad = -2x$

INSTRUCTOR NOTE
Simplifying expressions such as these prepares the students for solving equations.

HOW TO 7   Simplify: $\dfrac{3}{2}\left(\dfrac{2x}{3}\right)$

$\dfrac{3}{2}\left(\dfrac{2x}{3}\right) = \dfrac{3}{2}\left(\dfrac{2}{3}x\right)$ • Note that $\dfrac{2x}{3} = \dfrac{2}{3}x$.

$\qquad = \left(\dfrac{3}{2} \cdot \dfrac{2}{3}\right)x$ • Use the Associative Property of
                                                               Multiplication to group factors.

$\qquad = 1 \cdot x$

$\qquad = x$

HOW TO 8 Simplify: $(16x)2$

$(16x)2 = 2(16x)$
$= (2 \cdot 16)x$
$= 32x$

• Use the Commutative and Associative Properties of Multiplication to rearrange and group factors.

EXAMPLE 5

Simplify: $-2(3x^2)$

Solution

$-2(3x^2) = (-2 \cdot 3)x^2 = -6x^2$

YOU TRY IT 5

Simplify: $-5(4y^2)$

Your solution

$-20y^2$

EXAMPLE 6

Simplify: $-5(-10x)$

Solution

$-5(-10x) = [(-5)(-10)]x = 50x$

YOU TRY IT 6

Simplify: $-7(-2a)$

Your solution

$14a$

EXAMPLE 7

Simplify: $-\dfrac{3}{4}\left(\dfrac{2}{3}x\right)$

Solution

$\left(-\dfrac{3}{4}\right)\left(\dfrac{2}{3}x\right) = \left(-\dfrac{3}{4} \cdot \dfrac{2}{3}\right)x = -\dfrac{1}{2}x$

YOU TRY IT 7

Simplify: $-\dfrac{3}{5}\left(-\dfrac{7}{9}a\right)$

Your solution

$\dfrac{7}{15}a$

IN-CLASS EXAMPLES
Simplify.
7. $-6(4y^2)$ $-24y^2$
8. $-3(-12b)$ $36b$
9. $(5a)(-6)$ $-30a$

Solutions on p. S3

OBJECTIVE D *To simplify a variable expression using the Distributive Property*

Recall that the Distributive Property states that if a, b, and c are real numbers, then

$$a(b + c) = ab + ac$$

The Distributive Property is used to remove parentheses from a variable expression.

HOW TO 9 Simplify: $3(2x + 7)$

$3(2x + 7) = 3(2x) + 3(7)$
$= 6x + 21$

• Use the Distributive Property. Multiply each term inside the parentheses by 3.

HOW TO 10 Simplify: $-5(4x + 6)$

$-5(4x + 6) = -5(4x) + (-5)(6)$
$= -20x - 30$

• Use the Distributive Property.

INSTRUCTOR NOTE

An expression such as $-(5x - 2)$ can be thought of as either the opposite of $5x - 2$ or $-1(5x - 2)$.

HOW TO 11 Simplify: $-(2x - 4)$

$$-(2x - 4) = -1(2x - 4)$$
$$= -1(2x) - (-1)(4)$$ • Use the Distributive Property.
$$= -2x + 4$$

From HOW TO 11, note that when a negative sign immediately precedes the parentheses, the sign of each term inside the parentheses is changed.

HOW TO 12 Simplify: $-\frac{1}{2}(8x - 12y)$

$$-\frac{1}{2}(8x - 12y) = -\frac{1}{2}(8x) - \left(-\frac{1}{2}\right)(12y)$$ • Use the Distributive Property.
$$= -4x + 6y$$

HOW TO 13 Simplify: $4(x - y) - 2(-3x + 6y)$

$$4(x - y) - 2(-3x + 6y)$$
$$= 4x - 4y + 6x - 12y$$ • Use the Distributive Property.
$$= 10x - 16y$$ • Combine like terms.

An extension of the Distributive Property is used when an expression contains more than two terms.

HOW TO 14 Simplify: $3(4x - 2y - z)$

$$3(4x - 2y - z) = 3(4x) - 3(2y) - 3(z)$$ • Use the Distributive Property.
$$= 12x - 6y - 3z$$

EXAMPLE 8

Simplify: $(2x - 6)2$

Solution
Use the Distributive Property.

$(2x - 6)2 = 4x - 12$

EXAMPLE 9

Simplify: $-3(-5a + 7b)$

Solution
Use the Distributive Property.

$-3(-5a + 7b) = 15a - 21b$

YOU TRY IT 8

Simplify: $(3a - 1)5$

Your solution
$15a - 5$

IN-CLASS EXAMPLES
Simplify.
10. $4(3 - 9x)$ **12 − 36x**
11. $(6 + 7y)8$ **48 + 56y**
12. $-9(-5a + 2b)$ **45a − 18b**

YOU TRY IT 9

Simplify: $-8(-2a + 7b)$

Your solution
$16a - 56b$

13. $4(y^2 - 3y + 7)$
$4y^2 - 12y + 28$
14. $-5(a^2 + 8a - 6)$
$-5a^2 - 40a + 30$

Solutions on p. S3

EXAMPLE 10

Simplify: $3(x^2 - x - 5)$

Solution

$3(x^2 - x - 5) = 3x^2 - 3x - 15$

YOU TRY IT 10

Simplify: $3(12x^2 - x + 8)$

Your solution

$36x^2 - 3x + 24$

EXAMPLE 11

Simplify: $2x - 3(2x - 7y)$

Solution

$2x - 3(2x - 7y)$
$= 2x - 6x + 21y$ • **Distributive Property**
$= -4x + 21y$ • **Combine like terms.**

YOU TRY IT 11

Simplify: $3y - 2(y - 7x)$

Your solution

$y + 14x$

EXAMPLE 12

Simplify: $7(x - 2y) - (-x - 2y)$

Solution

$7(x - 2y) - (-x - 2y)$
$= 7x - 14y + x + 2y$ • **Distributive Property**
$= 8x - 12y$ • **Combine like terms.**

YOU TRY IT 12

Simplify: $-2(x - 2y) - (-x + 3y)$

Your solution

$-x + y$

EXAMPLE 13

Simplify: $2x - 3[2x - 3(x + 7)]$

Solution

$2x - 3[2x - 3(x + 7)]$
$= 2x - 3[2x - 3x - 21]$
$= 2x - 3[-x - 21]$
$= 2x + 3x + 63$
$= 5x + 63$

YOU TRY IT 13

Simplify: $3y - 2[x - 4(2 - 3y)]$

Your solution

$-2x - 21y + 16$

IN-CLASS EXAMPLES

Simplify.
15. $5a - 4(3a + 2b)$ **−7a − 8b**
16. $-6(c - d) - (-c - 3d)$
    −5c + 9d
17. $9w - 4[3w - 5(w + 6)]$
    17w + 120

Solutions on p. S3

OBJECTIVE E   *To translate a verbal expression into a variable expression*

 Tips for Success

Before the class meeting in which your professor begins a new section, you should read each objective statement for that section. Next, browse through the objective material. The purpose of browsing through the material is so that your brain will be prepared to accept and organize the new information when it is presented to you. See *AIM for Success* in the Preface.

One of the major skills required in applied mathematics is the ability to translate a verbal expression into a variable expression. This requires recognizing the verbal phrases that translate into mathematical operations. A partial list of the verbal phrases used to indicate the different mathematical operations is given on the next page.

HOW TO 15   Translate "14 less than the cube of x" into a variable expression.

14 <u>less than</u> the <u>cube</u> of x

• Identify the words that indicate the mathematical operations.

$x^3 - 14$

• Use the identified operations to write the variable expression.

Point of Interest

The way in which expressions are symbolized has changed over time. Here are how some of the expressions shown at the right may have appeared in the early 16th century.

R p. 9 for $x + 9$. The symbol R was used for a variable raised to the first power. The symbol p. was used for plus.

R m. 3 for $x - 3$. The symbol R was used for the variable. The symbol m. was used for minus.

The square of a variable was designated by Q, and the cube was designated by C. The expression $x^2 + x^3$ was written Q p. C.

INSTRUCTOR NOTE

In all of these phrases, whatever term is mentioned first may be written first, and whatever term is mentioned second is written second—with the exception of the phrases *less than* and *subtracted from*. They are the only phrases in this list that require a reversal of the terms. For example, 5 *added to x* may be written as $5 + x$, but 5 *less than x* is always written as $x - 5$.

Words or Phrases for Addition

added to	6 added to y	$y + 6$
more than	8 more than x	$x + 8$
the sum of	the sum of x and z	$x + z$
increased by	t increased by 9	$t + 9$
the total of	the total of 5 and d	$5 + d$
plus	b plus 17	$b + 17$

Words or Phrases for Subtraction

minus	x minus 2	$x - 2$
less than	7 less than t	$t - 7$
less	7 less t	$7 - t$
subtracted from	5 subtracted from d	$d - 5$
decreased by	m decreased by 3	$m - 3$
the difference between	the difference between y and 4	$y - 4$

Words or Phrases for Multiplication

times	10 times t	$10t$
of	one-half of x	$\frac{1}{2}x$
the product of	the product of y and z	yz
multiplied by	b multiplied by 11	$11b$
twice	twice n	$2n$

Phrases for Division

divided by	x divided by 12	$\frac{x}{12}$
the quotient of	the quotient of y and z	$\frac{y}{z}$
the ratio of	the ratio of t to 9	$\frac{t}{9}$

Phrases for Power

the square of	the square of x	x^2
the cube of	the cube of a	a^3

In most applications that involve translating phrases into variable expressions, the variable to be used is not given. To translate these phrases, a variable must be assigned to an unknown quantity before the variable expression can be written.

HOW TO 16   Translate "the sum of two consecutive integers" into a variable expression. Then simplify.

the first integer: n

the next consecutive integer: $n + 1$

- Assign a variable to one of the unknown quantities.
- Use the assigned variable to write an expression for any other unknown quantity.

$n + (n + 1)$

$(n + n) + 1$

$2n + 1$

- Use the assigned variable to write the variable expression.
- Simplify the variable expression.

Many applications in mathematics require that you identify the unknown quantity, assign a variable to that quantity, and then attempt to express other unknown quantities in terms of the variable.

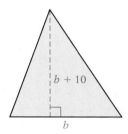

HOW TO 17   The height of a triangle is 10 ft longer than the base of the triangle. Express the height of the triangle in terms of the base of the triangle.

the base of the triangle: b

the height is 10 more than the base: $b + 10$

- Assign a variable to the base of the triangle.
- Express the height of the triangle in terms of b.

EXAMPLE 14

The length of a swimming pool is 4 ft less than two times the width. Express the length of the pool in terms of the width.

Solution

the width of the pool: w
the length is 4 ft less than two times the
    width: $2w - 4$

YOU TRY IT 14

The speed of a new jet plane is twice the speed of an older model. Express the speed of the new model in terms of the speed of the older model.

Your solution

Let s be the speed of the older model; $2s$

IN-CLASS EXAMPLES

18. The force of gravity on the moon is one-sixth the force of gravity on Earth. Represent the force of gravity on the moon in terms of the force of gravity on Earth.

    Force of gravity on Earth: F; $\frac{1}{6} F$

EXAMPLE 15

A banker divided $5000 between two accounts, one paying 10% annual interest and the second paying 8% annual interest. Express the amount invested in the 10% account in terms of the amount invested in the 8% account.

Solution

the amount invested at 8%: x
the amount invested at 10%: $5000 - x$

YOU TRY IT 15

A guitar string 6 ft long was cut into two pieces. Express the length of the shorter piece in terms of the length of the longer piece.

Your solution

Let y be the length of the longer piece; $6 - y$

Solutions on p. S3

Unless otherwise noted, all content on this page is © Cengage Learning.

1.4 EXERCISES

✔ **Concept Check**

SUGGESTED ASSIGNMENT
Exercises 1–14; Exercises 15–219, every other odd;
More challenging exercises: Exercises 221–224

For Exercises 1 to 3, name the terms of the variable expression. Then underline the constant term.

1. $2x^2 + 5x - 8$ $2x^2, 5x, \underline{-8}$ **2.** $-3n^2 - 4n + 7$ $-3n^2, -4n, \underline{7}$ **3.** $6 - a^4$ $-a^4, \underline{6}$

For Exercises 4 to 6, name the coefficients of the variable terms.

4. $x^2 - 9x + 2$ $1, -9$ **5.** $12a^2 - 8ab - b^2$ $12, -8, -1$ **6.** $n^3 - 4n^2 - n + 9$ $1, -4, -1$

7. The Inverse Property of Multiplication tells us that the product of a number and its
_____ is 1.   reciprocal (or multiplicative inverse)

8. The Inverse Property of Addition tells us that the sum of a number and its
_____ is 0.   opposite (or additive inverse)

9. ✎ What are *like terms*? Give an example of two like terms. Give an example of two terms that are not like terms.

10. ✎ Explain the meaning of the phrase "simplify a variable expression."

For each phrase in Exercises 11 and 12, identify the words that indicate mathematical operations.

11. twelve less than the quotient of x and negative two   less than, quotient

12. twenty subtracted from the product of eight and the cube of a number   subtracted from, product, cube

13. The sum of two numbers is 25. To express both numbers in terms of the same variable, let x represent one number. Then the other number is $\underline{25 - x}$.

14. The length of a rectangle is five times the width. To express the length and the width in terms of the same variable, let W represent the width. Then the length is $\underline{5W}$.

OBJECTIVE A   *To evaluate a variable expression*

For Exercises 15 to 23, evaluate the variable expression when $a = 2$, $b = 3$, and $c = -4$.

15. $6b \div (-a)$ -9 **16.** $bc \div (2a)$ -3 **17.** $b^2 - 4ac$ 41

18. $a^2 - b^2$ -5 **19.** $b^2 - c^2$ -7 **20.** $(a + b)^2$ 25

21. $a^2 + b^2$ 13 **22.** $2a - (c + a)^2$ 0 **23.** $\dfrac{5ab}{6} - 3cb$ 41

For Exercises 24 to 40, evaluate the variable expression when $a = -2$, $b = 4$, $c = -1$, and $d = 3$.

24. $\dfrac{d - b}{c}$ 1 **25.** $\dfrac{2d + b}{-a}$ 5 **26.** $\dfrac{b + 2d}{b}$ $\dfrac{5}{2}$ **27.** $\dfrac{b - d}{c - a}$ 1

28. $\dfrac{2c - d}{-ad}$ $-\dfrac{5}{6}$ **29.** $(b + d)^2 - 4a$ 57 **30.** $(d - a)^2 - 3c$ 28 **31.** $(d - a)^2 \div 5$ 5

32. $3(b - a) - bc$ 22

33. $\dfrac{b - 2a}{bc^2 - d}$ 8

34. $\dfrac{b^2 - a}{ad + 3c}$ -2

35. $\dfrac{1}{3}d^2 - \dfrac{3}{8}b^2$ -3

36. $\dfrac{5}{8}a^4 - c^2$ 9

37. $\dfrac{-4bc}{2a - b}$ -2

38. $-\dfrac{3}{4}b + \dfrac{1}{2}(ac + bd)$
  4

39. $-\dfrac{2}{3}d - \dfrac{1}{5}(bd - ac)$
  -4

40. $(b - a)^2 - (d - c)^2$
  20

For Exercises 41 to 44, without evaluating the expression, determine whether the expression is positive or negative when $a = -25$, $b = 67$, and $c = -82$.

41. $(c - a)(-b)$

Positive

42. $(a - c) + 3b$

Positive

43. $\dfrac{b + c}{abc}$

Negative

44. $\dfrac{ac}{-b^2}$

Negative

OBJECTIVE B   *To simplify a variable expression using the Properties of Addition*

For Exercises 45 to 76, simplify.

45. $6x + 8x$

  $14x$

46. $12x + 13x$

  $25x$

47. $9a - 4a$

  $5a$

48. $12a - 3a$

  $9a$

49. $7 - 3b$

  $7 - 3b$

50. $5 + 2a$

  $5 + 2a$

51. $-12a + 17a$

  $5a$

52. $-3a + 12a$

  $9a$

53. $-12xy + 17xy$

  $5xy$

54. $-15xy + 3xy$

  $-12xy$

55. $-3ab + 3ab$

  0

56. $-7ab + 7ab$

  0

57. $-\dfrac{1}{2}x - \dfrac{1}{3}x$

  $-\dfrac{5}{6}x$

58. $-\dfrac{2}{5}y + \dfrac{3}{10}y$

  $-\dfrac{1}{10}y$

59. $2.3x + 4.2x$

  $6.5x$

60. $6.1y - 9.2y$

  $-3.1y$

61. $x - 0.55x$

  $0.45x$

62. $0.65A - A$

  $-0.35A$

63. $5a - 3a + 5a$

  $7a$

64. $10a - 17a + 3a$

  $-4a$

65. $-5x^2 - 12x^2 + 3x^2$

  $-14x^2$

66. $-y^2 - 8y^2 + 7y^2$

  $-2y^2$

67. $\dfrac{3}{4}x - \dfrac{1}{3}x - \dfrac{7}{8}x$

  $-\dfrac{11}{24}x$

68. $-\dfrac{2}{5}a - \left(-\dfrac{3}{10}a\right) - \dfrac{11}{15}a$

  $-\dfrac{5}{6}a$

69. $7x - 3y + 10x$

  $17x - 3y$

70. $8y + 8x - 8y$

  $8x$

71. $3a + (-7b) - 5a + b$
  $-2a - 6b$

72. $-5b + 7a - 7b + 12a$
  $19a - 12b$

73. $3x + (-8y) - 10x + 4x$
  $-3x - 8y$

74. $3y + (-12x) - 7y + 2y$
  $-12x - 2y$

75. $x^2 - 7x + (-5x^2) + 5x$
  $-4x^2 - 2x$

76. $3x^2 + 5x - 10x^2 - 10x$
  $-7x^2 - 5x$

77. 🖋 Which of the following expressions are equivalent to $-10x - 10y - 10y - 10x$?
(i) 0   (ii) $-20y$ (iii) $-20x$ (iv) $-20x - 20y$ (v) $-20y - 20x$
iv, v

OBJECTIVE C   *To simplify a variable expression using the Properties of Multiplication*

For Exercises 78 to 117, simplify.

78. $4(3x)$
12x

79. $12(5x)$
60x

80. $-3(7a)$
$-21a$

81. $-2(5a)$
$-10a$

82. $-2(-3y)$
6y

83. $-5(-6y)$
30y

84. $(4x)2$
8x

85. $(6x)12$
72x

86. $(3a)(-2)$
$-6a$

87. $(7a)(-4)$
$-28a$

88. $(-3b)(-4)$
12b

89. $(-12b)(-9)$
108b

90. $-5(3x^2)$
$-15x^2$

91. $-8(7x^2)$
$-56x^2$

92. $\frac{1}{3}(3x^2)$
x^2

93. $\frac{1}{6}(6x^2)$
x^2

94. $\frac{1}{5}(5a)$
a

95. $\frac{1}{8}(8x)$
x

96. $-\frac{1}{2}(-2x)$
x

97. $-\frac{1}{4}(-4a)$
a

98. $-\frac{1}{7}(-7n)$
n

99. $-\frac{1}{9}(-9b)$
b

100. $(3x)\left(\frac{1}{3}\right)$
x

101. $(12x)\left(\frac{1}{12}\right)$
x

102. $(-6y)\left(-\frac{1}{6}\right)$
y

103. $(-10n)\left(-\frac{1}{10}\right)$
n

104. $\frac{1}{3}(9x)$
$3x$

105. $\frac{1}{7}(14x)$
$2x$

106. $-0.2(10x)$
$-2x$

107. $-0.25(8x)$
$-2x$

108. $-\frac{2}{3}(12a^2)$
$-8a^2$

109. $-\frac{5}{8}(24a^2)$
$-15a^2$

110. $-0.5(-16y)$
$8y$

111. $-0.75(-8y)$
$6y$

112. $(16y)\left(\frac{1}{4}\right)$
$4y$

113. $(33y)\left(\frac{1}{11}\right)$
$3y$

114. $(-6x)\left(\frac{1}{3}\right)$
$-2x$

115. $(-10x)\left(\frac{1}{5}\right)$
$-2x$

116. $(-8a)\left(-\frac{3}{4}\right)$
$6a$

117. $(21y)\left(-\frac{3}{7}\right)$
$-9y$

118. 🖋 After multiplying $\frac{2}{7}x^2$ by a proper fraction, is the coefficient of x^2 greater than 1 or less than 1?
Less than 1

OBJECTIVE D   *To simplify a variable expression using the Distributive Property*

For Exercises 119 to 157, simplify.

119. $2(4x - 3)$
$8x - 6$

120. $5(2x - 7)$
$10x - 35$

121. $-2(a + 7)$
$-2a - 14$

122. $-5(a + 16)$
$-5a - 80$

123. $-3(2y - 8)$
$-6y + 24$

124. $-5(3y - 7)$
$-15y + 35$

125. $-(x + 2)$
$-x - 2$

126. $-(x + 7)$
$-x - 7$

127. $(5 - 3b)7$
$35 - 21b$

128. $(10 - 7b)2$
$20 - 14b$

129. $\frac{1}{3}(6 - 15y)$
$2 - 5y$

130. $\frac{1}{2}(-8x + 4y)$
$-4x + 2y$

131. $3(5x^2 + 2x)$
$15x^2 + 6x$

132. $6(3x^2 + 2x)$
$18x^2 + 12x$

133. $-2(-y + 9)$
$2y - 18$

134. $-5(-2x + 7)$
$10x - 35$

135. $(-3x - 6)5$
$-15x - 30$

136. $(-2x + 7)7$
$-14x + 49$

137. $2(-3x^2 - 14)$
$-6x^2 - 28$

138. $5(-6x^2 - 3)$
$-30x^2 - 15$

139. $-3(2y^2 - 7)$
$-6y^2 + 21$

140. $-8(3y^2 - 12)$
$-24y^2 + 96$

141. $3(x^2 - y^2)$
$3x^2 - 3y^2$

142. $5(x^2 + y^2)$
$5x^2 + 5y^2$

143. $-\frac{2}{3}(6x - 18y)$
$-4x + 12y$

144. $-\frac{1}{2}(x - 4y)$
$-\frac{1}{2}x + 2y$

145. $-(6a^2 - 7b^2)$
$-6a^2 + 7b^2$

146. $3(x^2 + 2x - 6)$
$3x^2 + 6x - 18$

147. $4(x^2 - 3x + 5)$
$4x^2 - 12x + 20$

148. $-2(y^2 - 2y + 4)$
$-2y^2 + 4y - 8$

149. $\frac{3}{4}(2x - 6y + 8)$
$\frac{3}{2}x - \frac{9}{2}y + 6$

150. $-\frac{2}{3}(6x - 9y + 1)$
$-4x + 6y - \frac{2}{3}$

151. $4(-3a^2 - 5a + 7)$
$-12a^2 - 20a + 28$

152. $-5(-2x^2 - 3x + 7)$
$10x^2 + 15x - 35$

153. $-3(-4x^2 + 3x - 4)$
$12x^2 - 9x + 12$

154. $3(2x^2 + xy - 3y^2)$
$6x^2 + 3xy - 9y^2$

155. $5(2x^2 - 4xy - y^2)$
$10x^2 - 20xy - 5y^2$

156. $-(3a^2 + 5a - 4)$
$-3a^2 - 5a + 4$

157. $-(8b^2 - 6b + 9)$
$-8b^2 + 6b - 9$

158. After the expression $17x - 31$ is multiplied by a negative integer, is the constant term positive or negative?   Positive

159. Which of the following expressions is equivalent to $12 - 7(y - 9)$?
   (i) $5(y - 9)$ **(ii)** $12 - 7y - 63$ **(iii)** $12 - 7y + 63$ **(iv)** $12 - 7y - 9$
   iii

For Exercises 160 to 183, simplify.

160. $4x - 2(3x + 8)$
$-2x - 16$

161. $6a - (5a + 7)$
$a - 7$

162. $9 - 3(4y + 6)$
$-12y - 9$

163. $10 - (11x - 3)$
$-11x + 13$

164. $5n - (7 - 2n)$
$7n - 7$

165. $8 - (12 + 4y)$
$-4y - 4$

166. $3(x + 2) - 5(x - 7)$
$-2x + 41$

167. $2(x - 4) - 4(x + 2)$
$-2x - 16$

168. $12(y - 2) + 3(7 - 3y)$
$3y - 3$

169. $6(2y - 7) - (3 - 2y)$
$14y - 45$

170. $3(a - b) - (a + b)$
$2a - 4b$

171. $2(a + 2b) - (a - 3b)$
$a + 7b$

172. $4[x - 2(x - 3)]$
$-4x + 24$

173. $2[x + 2(x + 7)]$
$6x + 28$

174. $-2[3x + 2(4 - x)]$
$-2x - 16$

175. $-5[2x + 3(5 - x)]$
$5x - 75$

176. $-3[2x - (x + 7)]$
$-3x + 21$

177. $-2[3x - (5x - 2)]$
$4x - 4$

178. $2x - 3[x - (4 - x)]$
$-4x + 12$

179. $-7x + 3[x - (3 - 2x)]$
$2x - 9$

180. $-5x - 2[2x - 4(x + 7)] - 6$
$-x + 50$

181. $0.12(2x + 3) + x$
$1.24x + 0.36$

182. $0.05x + 0.02(4 - x)$
$0.03x + 0.08$

183. $0.03x + 0.04(1000 - x)$
$-0.01x + 40$

OBJECTIVE E   *To translate a verbal expression into a variable expression*

For Exercises 184 to 193, translate into a variable expression.

184. twelve minus a number
$12 - x$

185. a number divided by eighteen
$\dfrac{x}{18}$

186. two-thirds of a number
$\dfrac{2}{3}x$

187. twenty more than a number
$x + 20$

188. the quotient of twice a number and nine
$\dfrac{2x}{9}$

189. eight less than the product of eleven and a number
$11x - 8$

190. the quotient of fifteen and the sum of a number and twelve

$$\frac{15}{x + 12}$$

191. the difference between forty and the quotient of a number and twenty

$$40 - \frac{x}{20}$$

192. the quotient of five more than twice a number and the number

$$\frac{2x + 5}{x}$$

193. the sum of the square of a number and twice the number

$$x^2 + 2x$$

194. Which of the following phrases translate into the variable expression $32 - \frac{a}{7}$?
  (i) the difference between thirty-two and the quotient of a number and seven
  (ii) thirty-two decreased by the quotient of a number and seven
  (iii) thirty-two minus the ratio of a number to seven
  i, ii, iii

For Exercises 195 to 206, translate into a variable expression. Then simplify.

195. ten times the difference between a number and fifty
  $10(x - 50)$; $10x - 500$

196. nine less than the total of a number and two
  $(x + 2) - 9$; $x - 7$

197. the difference between a number and three more than the number
  $x - (x + 3)$; -3

198. four times the sum of a number and nineteen
  $4(x + 19)$; $4x + 76$

199. a number added to the difference between twice the number and four
  $(2x - 4) + x$; $3x - 4$

200. the product of five less than a number and seven
  $(x - 5)7$; $7x - 35$

201. a number added to the product of three and the number
  $3x + x$; $4x$

202. a number increased by the total of the number and nine
  $x + (x + 9)$; $2x + 9$

203. five more than the sum of a number and six
  $(x + 6) + 5$; $x + 11$

204. a number decreased by the difference between eight and the number
  $x - (8 - x)$; $2x - 8$

205. a number minus the sum of the number and ten
  $x - (x + 10)$; -10

206. two more than the total of a number and five
  $(x + 5) + 2$; $x + 7$

207. Museums   In a recent year, 3.8 million more people visited the Louvre in Paris than visited the Metropolitan Museum of Art in New York City. (*Sources:* The *Art Newspaper;* museums' accounts) Express the number of visitors to the Louvre in terms of the number of visitors to the Metropolitan Museum of Art. Let *M* be the number of visitors to the Metropolitan Museum of Art; $M + 3,800,000$

The Louvre

208. Astronomy   The diameter of Saturn's moon Rhea is 253 mi more than the diameter of Saturn's moon Dione. Express the diameter of Rhea in terms of the diameter of Dione. (*Source:* NASA)   Let *d* be the diameter of Dione; $d + 253$

209. ● **Noise Level**   The noise level of an ambulance siren is 10 decibels louder than that of a car horn. Express the noise level of an ambulance siren in terms of the noise level of a car horn. (*Source:* League for the Hard of Hearing)
Let *d* be the noise level, in decibels, of a car horn; $d + 10$

210. ● **Genetics**   The human genome contains 11,000 more genes than the round-worm genome. Express the number of genes in the human genome in terms of the number of genes in the roundworm genome. (*Source:* Celera, USA TODAY research)
Let *G* be the number of genes in the roundworm genome; $G + 11{,}000$

211. ● **Rock Band Tours**   See the news clipping at the right. Express Bruce Springsteen and the E Street Band's concert ticket sales in terms of U2's concert ticket sales.
Let *T* be U2's concert ticket sales; $T - 28{,}500{,}000$

In the NEWS!

U2 Concerts Top Annual Rankings in North America

The Irish rock band U2 performed the most popular concerts on the North American circuit this year. Bruce Springsteen and the E Street Band came in second, with $28.5 million less in ticket sales.
Source: new.music.yahoo.com

212. ● **Space Exploration**   A survey in *USA Today* reported that almost three-fourths of Americans think that money should be spent on exploration of Mars. Express the number of Americans who think that money should be spent on exploration of Mars in terms of the total number of Americans.
Let *N* be the total number of Americans; $\frac{3}{4}N$

213. ● **Biology**   According to the American Podiatric Medical Association, the bones in your foot account for one-fourth of all the bones in your body. Express the number of bones in your foot in terms of the number of bones in your body.
Let *N* be the number of bones in your body; $\frac{1}{4}N$

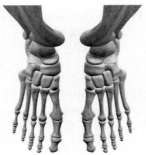

Aaliya Landholt/Shutterstock.com

214. **Football**   In football, the number of points awarded for a touchdown is three times the number of points awarded for a safety. Express the number of points awarded for a touchdown in terms of the number of points awarded for a safety.
Let *s* be the number of points awarded for a safety; $3s$

215. ● **Major League Sports**   See the news clipping at the right. Express the attendance at major league baseball games in terms of the attendance at major league basketball games.
Let *B* be the attendance at major league basketball games; $B + 50{,}000{,}000$

216. ● **Tax Refunds**   A recent survey conducted by Turbotax.com asked, "If you receive a tax refund, what will you do?" Forty-three percent of respondents said they would pay down their debt. (*Source: USA Today,* March 27, 2008) Express the number of people who would pay down their debt in terms of the number of people surveyed.   Let *N* be the number of people surveyed; $0.43N$

217. ● **Community Colleges**   According to the National Center for Education Statistics, 46% of U.S. undergraduate students attend two-year colleges. Express the number of U.S. undergraduate students who attend two-year colleges in terms of the number of U.S. undergraduate students.
Let *N* be the number of U.S. undergraduate students; $0.46N$

218. **Geometry**   The length of a rectangle is 5 m more than twice the width. Express the length of the rectangle in terms of the width.
Let *W* be the width of the rectangle; $2W + 5$

In the NEWS!

Over 70 Million Attend Major League Baseball Games

Among major league sports, attendance at major league baseball games topped attendance at other major league sporting events. Fifty million more people went to baseball games than went to basketball games. The attendance at football games and hockey games was even less than the attendance at basketball games.
Source: Time, December 28, 2010–January 4, 2010

219. Geometry In a triangle, the measure of the smallest angle is 10 degrees less than one-half the measure of the largest angle. Express the measure of the smallest angle in terms of the measure of the largest angle.

Let L be the measure of the largest angle; $\frac{1}{2}L - 10$

220. Wages An employee is paid $1172 per week plus $38 for each hour of overtime worked. Express the employee's weekly pay in terms of the number of hours of overtime worked. Let h be the number of hours of overtime worked; $1172 + 38h$

Critical Thinking

For Exercises 221 and 222, use the following situation: 83 more students enrolled in spring-term science classes than enrolled in fall-term science classes.

221. If s and $s + 83$ represent the quantities in this situation, what is s?
The number of students enrolled in fall-term science classes

222. If n and $n - 83$ represent the quantities in this situation, what is n?
The number of students enrolled in spring-term science classes

223. Metalwork A wire whose length is given as x inches is bent into a square. Express the length of a side of the square in terms of x.
$\frac{1}{4}x$

224. ⬤ Chemistry The chemical formula for glucose (sugar) is $C_6H_{12}O_6$. This formula means that there are 12 hydrogen atoms for every 6 carbon atoms and 6 oxygen atoms in each molecule of glucose (see the figure at the right). If x represents the number of atoms of oxygen in a pound of sugar, express the number of hydrogen atoms in the pound of sugar in terms of the number of oxygen atoms.
$2x$

225. ◥ Translate the expressions $5x + 8$ and $5(x + 8)$ into phrases.

Projects or Group Activities

226. Determine whether the statement is true or false. If the statement is false, give an example that illustrates that it is false.
a. Division is a commutative operation. False. For example, $8 \div 2 \neq 2 \div 8$
b. Division is an associative operation. False. For example, $(12 \div 4) \div 2 \neq 12 \div (4 \div 2)$
c. Subtraction is an associative operation. False. For example, $(9 - 2) - 3 \neq 9 - (2 - 3)$
d. Subtraction is a commutative operation. False. For example, $10 - 4 \neq 4 - 10$

227. Define an operation $\otimes$ as $a \otimes b = (a \cdot b) - (a + b)$.
For example, $7 \otimes 5 = (7 \cdot 5) - (7 + 5) = 35 - 12 = 23$.
a. Is $\otimes$ a commutative operation? Support your answer. Yes
b. Is $\otimes$ an associative operation? Support your answer. No

228. ◥ Give examples of two operations that occur in everyday experience that are not commutative (for example, putting on socks and then shoes).

229. Which of the following expressions are equivalent?
(i) $2x + 4(2x + 1)$
(ii) $x - (4 - 9x) + 8$
(iii) $7(x - 4) - 3(2x + 6)$
(iv) $3(2x + 8) + 4(x - 5)$
(v) $6 - 2[x + (3x - 4)] + 2(9x - 5)$
i, ii, iv, and v are equivalent; they are all equal to $10x + 4$.

Unless otherwise noted, all content on this page is © Cengage Learning.

QUICK QUIZ

1. Evaluate $2x - (y + z)^2$ when $x = 5$, $y = -3$, and $z = 6$. 1 [1.4A]

Simplify.
2. $9b - 5b$ $4b$ [1.4B]
3. $4y^2 + 3y - 6y^2 + 2y$
   $-2y^2 + 5y$ [1.4B]
4. $5(6a)$ $30a$ [1.4C]
5. $(-12c)\frac{2}{3}$
   $-8c$ [1.4C]
6. $8 - 5(3x + 2)$
   $-15x - 2$ [1.4D]
7. $3a - 5[a - (6 - a)]$
   $-7a + 30$ [1.4D]

Translate into a variable expression. Then simplify.
8. Five less than the total of a number and seven
   $(n + 7) - 5; n + 2$
   [1.4E]
9. A number decreased by the difference between six and the product of the number and four
   $n - (6 - 4n); 5n - 6$
   [1.4E]

1.5   Sets

OBJECTIVE A   *To write a set using the roster method*

Recall that a *set* is a collection of objects, which are called the *elements* of the set. The **roster method** of writing a set encloses a list of the elements in braces.

The set of the positive integers less than 5 is written $\{1, 2, 3, 4\}$.

HOW TO 1   Use the roster method to write the set of integers between 0 and 10.

$A = \{1, 2, 3, 4, 5, 6, 7, 8, 9\}$ • A set can be designated by a capital letter.
Note that 0 and 10 are not elements of the set.

HOW TO 2   Use the roster method to write the set of natural numbers.

$A = \{1, 2, 3, 4, \ldots\}$ • The three dots mean that the pattern of numbers continues without end.

INSTRUCTOR NOTE
Students want to write the empty set as $\{\varnothing\}$. It is very difficult to convince them that this is incorrect. Students, in general, have a difficult time with the idea that a set can be an element of another set.

The **empty set,** or **null set,** is the set that contains no elements. The symbol $\varnothing$ or $\{\ \}$ is used to represent the empty set.

The set of people who have run a 2-minute mile is the empty set.

Union and Intersection of Two Sets

The **union** of two sets, written $A \cup B$, is the set of all elements that belong to either set A *or* set B.

The **intersection** of two sets, written $A \cap B$, is the set that contains the elements that are common to both A and B.

EXAMPLES

Find $A \cup B$ and $A \cap B$, given $A = \{1, 2, 3, 4\}$ and $B = \{3, 4, 5, 6\}$.

$A \cup B = \{1, 2, 3, 4, 5, 6\}$ • The union of A and B contains all the elements of A and all the elements of B. Elements in both sets are listed only once.

$A \cap B = \{3, 4\}$ • The intersection of A and B contains the elements common to A and B.

EXAMPLE 1

Use the roster method to write the set of the odd positive integers less than 12.

Solution

$A = \{1, 3, 5, 7, 9, 11\}$

YOU TRY IT 1

Use the roster method to write the set of the odd negative integers greater than -10.

Your solution

$A = \{-9, -7, -5, -3, -1\}$

Solution on p. S3

EXAMPLE 2

Use the roster method to write the set of the even positive integers.

Solution

$A = \{2, 4, 6, \ldots\}$

EXAMPLE 3

Find $D \cup E$, given $D = \{6, 8, 10, 12\}$ and $E = \{-8, -6, 10, 12\}$.

Solution

$D \cup E = \{-8, -6, 6, 8, 10, 12\}$

EXAMPLE 4

Find $A \cap B$, given $A = \{5, 6, 9, 11\}$ and $B = \{5, 9, 13, 15\}$.

Solution

$A \cap B = \{5, 9\}$

EXAMPLE 5

Find $A \cap B$, given $A = \{1, 2, 3, 4\}$ and $B = \{8, 9, 10, 11\}$.

Solution

$A \cap B = \varnothing$

YOU TRY IT 2

Use the roster method to write the set of the odd positive integers.

Your solution

$A = \{1, 3, 5, \ldots\}$

YOU TRY IT 3

Find $A \cup B$, given $A = \{-2, -1, 0, 1, 2\}$ and $B = \{0, 1, 2, 3, 4\}$.

Your solution

$A \cup B = \{-2, -1, 0, 1, 2, 3, 4\}$

YOU TRY IT 4

Find $C \cap D$, given $C = \{10, 12, 14, 16\}$ and $D = \{10, 16, 20, 26\}$.

Your solution

$C \cap D = \{10, 16\}$

YOU TRY IT 5

Find $A \cap B$, given $A = \{-5, -4, -3, -2\}$ and $B = \{2, 3, 4, 5\}$.

Your solution

$A \cap B = \varnothing$

IN-CLASS EXAMPLES

1. Use the roster method to write the set of even integers between 20 and 30.
   $A = \{\mathbf{22, 24, 26, 28}\}$

2. Find $A \cup B$ and $A \cap B$, given $A = \{a, e, i, o, u\}$ and $B = \{a, b, c, d, e\}$.
   $\{\mathbf{a, b, c, d, e, i, o, u}\}; \{\mathbf{a, e}\}$

3. Find $C \cup D$ and $C \cap D$, given $C = \{10, 20, 30, 40\}$ and $D = \{5, 15, 25, 35\}$.
   $\{\mathbf{5, 10, 15, 20, 25, 30, 35, 40}\}; \varnothing$

Solutions on p. S3

OBJECTIVE B *To write and graph sets of real numbers*

 Point of Interest

The symbol $\in$ was first used in the book *Arithmeticae Principia,* published in 1889. It is the first letter of the Greek word $\varepsilon\sigma\tau\iota$, which means "is." The symbols for union and intersection were also introduced around the same time.

Another method of representing sets is called **set-builder notation.** This method of writing sets uses a rule to describe the elements of the set. Using set-builder notation, we represent the set of all positive integers less than 10 as

$\{x | x < 10, x \in \text{positive integers}\}$, which is read "the set of all positive integers x that are less than 10."

HOW TO 3 Use set-builder notation to write the set of integers less than or equal to 12.

$\{x | x \leq 12, x \in \text{integers}\}$ • This is read "the set of all integers x that are less than or equal to 12."

HOW TO 4 Use set-builder notation to write the set of real numbers greater than 4.

$\{x | x > 4, x \in \text{real numbers}\}$ • This is read "the set of all real numbers x that are greater than 4."

Take Note

Set-builder notation is mainly used to represent sets that have an infinite number of elements. The set $\{x|x > 4\}$ has an infinite number of elements and cannot be represented using the roster method.

For the remainder of this section, all variables will represent real numbers. Given this convention, $\{x|x > 4, x \in \text{real numbers}\}$ is written $\{x|x > 4\}$.

Some sets of real numbers that are written in set-builder notation can be written in **interval notation.** For instance, the interval notation $[-3, 2)$ represents the set of real numbers between -3 and 2. The bracket means that -3 is included in the set, and the parenthesis means that 2 is *not* included in the set. Using set-builder notation, the interval $[-3, 2)$ is written

$$\{x|-3 \leq x < 2\}$$ • This is read "the set of all real numbers x between -3 and 2, including -3 but excluding 2."

To indicate an interval that extends forever in the positive direction, we use the **infinity symbol, ∞;** to indicate an interval that extends forever in the negative direction, we use the **negative infinity symbol, $-\infty$.**

HOW TO 5   Write $\{x|x > 1\}$ in interval notation.

$\{x|x > 1\}$ is the set of real numbers greater than 1. This set extends forever in the positive direction. In interval notation, this set is written $(1, \infty)$.

HOW TO 6   Write $\{x|x \leq -2\}$ in interval notation.

$\{x|x \leq -2\}$ is the set of real numbers less than or equal to -2. This set extends forever in the negative direction. In interval notation, this set is written $(-\infty, -2]$.

When writing a set in interval notation, we always use a parenthesis to the right of ∞ and to the left of $-\infty$. Infinity is not a real number, so it cannot be represented as belonging to the set of real numbers by using a bracket.

HOW TO 7   Write $[1, 3]$ in set-builder notation.

This is the set of real numbers between 1 and 3, including 1 and 3. In set-builder notation, this set is written $\{x|1 \leq x \leq 3\}$.

We can graph sets of real numbers given in set-builder notation or in interval notation.

HOW TO 8   Graph: $(-\infty, -1)$

This is the set of real numbers less than -1, excluding -1. The parenthesis at -1 on the number line indicates that -1 is excluded from the set.

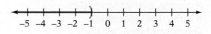

HOW TO 9   Graph: $\{x|x \geq 1\}$

This is the set of real numbers greater than or equal to 1. The bracket at 1 indicates that 1 is included in the set.

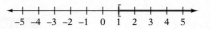

Unless otherwise noted, all content on this page is © Cengage Learning.

EXAMPLE 6

Write in interval notation.
a. $\{x|x \geq 2\}$ b. $\{x|0 \leq x \leq 1\}$

Solution

a. $\{x|x \geq 2\}$ is the set of real numbers greater than or equal to 2. This set extends forever in the positive direction. In interval notation, this set is written $[2, \infty)$.
b. $\{x|0 \leq x \leq 1\}$ is the set of real numbers between 0 and 1, including 0 and 1. In interval notation, this set is written $[0, 1]$.

YOU TRY IT 6

Write in interval notation.
a. $\{x|x \leq 3\}$ b. $\{x|-5 \leq x \leq -3\}$

Your solution

a. $(-\infty, 3]$ b. $[-5, -3]$

IN-CLASS EXAMPLES

4. Write $\{x|x \leq -5\}$ in interval notation. $(-\infty, -5]$
5. Write $(8, \infty)$ in set-builder notation. $\{x|x > 8\}$

EXAMPLE 7

Write in set-builder notation.
a. $(-\infty, 0]$ b. $(-3, 3)$

Solution

a. The interval $(-\infty, 0]$ is the set of real numbers less than or equal to 0. In set-builder notation, this set is written $\{x|x \leq 0\}$.
b. The interval $(-3, 3)$ is the set of real numbers between -3 and 3, excluding -3 and 3. In set-builder notation, this set is written $\{x|-3 < x < 3\}$.

YOU TRY IT 7

Write in set-builder notation.
a. $(-3, \infty)$ b. $[0, 4)$

Your solution

a. $\{x|x > -3\}$
b. $\{x|0 \leq x < 4\}$

6. Graph: $[-3, \infty)$

7. Graph: $\{x|0 \leq x \leq 5\}$

EXAMPLE 8

Graph.
a. $\{x|-2 < x < 1\}$ b. $\{x|x < 4\}$

Solution

a. The graph is the set of real numbers between -2 and 1, excluding -2 and 1. Use parentheses at -2 and 1.

b. The graph is the set of real numbers less than 4. Use a parenthesis at 4.

YOU TRY IT 8

Graph.
a. $\{x|-4 \leq x \leq 4\}$ b. $\{x|x > -3\}$

Your solution

a.

b.

EXAMPLE 9

Graph: $(-\infty, 5)$

Solution

The graph is the set of real numbers less than 5. Use a parenthesis at 5.

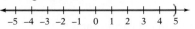

YOU TRY IT 9

Graph: $[2, 5]$

Your solution

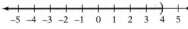

Solutions on p. S3

Unless otherwise noted, all content on this page is © Cengage Learning.

1.5 EXERCISES

✔ Concept Check

SUGGESTED ASSIGNMENT:
Exercises 1 and 2; Exercises 3–57, odds

1. The set $\{1, 2, 3\}$ is written using the ___roster___ method. The set $\{x|x < 4, x \in \text{positive integers}\}$ is written using ___set-builder___ notation. The set $[-3, 2]$ is written in ___interval___ notation.

2. 🖊 Explain how to find **a.** the union of two sets and **b.** the intersection of two sets.

OBJECTIVE A   *To write a set using the roster method*

For Exercises 3 to 6, use the roster method to write the set.

3. The integers between 15 and 22
$A = \{16, 17, 18, 19, 20, 21\}$

4. The integers between -10 and -4
$A = \{-9, -8, -7, -6, -5\}$

5. The odd integers between 8 and 18
$A = \{9, 11, 13, 15, 17\}$

6. The even integers between -11 and -1
$A = \{-10, -8, -6, -4, -2\}$

For Exercises 7 to 12, find $A \cup B$.

7. $A = \{3, 4, 5\}; B = \{4, 5, 6\}$
$A \cup B = \{3, 4, 5, 6\}$

8. $A = \{-3, -2, -1\}; B = \{-2, -1, 0\}$
$A \cup B = \{-3, -2, -1, 0\}$

9. $A = \{-10, -9, -8\}; B = \{8, 9, 10\}$
$A \cup B = \{-10, -9, -8, 8, 9, 10\}$

10. $A = \{m, n, p, q\}; B = \{m, n, o\}$
$A \cup B = \{m, n, o, p, q\}$

11. $A = \{1, 3, 7, 9\}; B = \{7, 9, 11, 13\}$
$A \cup B = \{1, 3, 7, 9, 11, 13\}$

12. $A = \{-3, -2, -1\}; B = \{-1, 1, 2\}$
$A \cup B = \{-3, -2, -1, 1, 2\}$

For Exercises 13 to 18, find $A \cap B$.

13. $A = \{3, 4, 5\}; B = \{4, 5, 6\}$
$A \cap B = \{4, 5\}$

14. $A = \{-4, -3, -2\}; B = \{-6, -5, -4\}$
$A \cap B = \{-4\}$

15. $A = \{-4, -3, -2\}; B = \{2, 3, 4\}$
$A \cap B = \varnothing$

16. $A = \{1, 2, 3, 4\}; B = \{1, 2, 3, 4\}$
$A \cap B = \{1, 2, 3, 4\}$

17. $A = \{a, b, c, d, e\}; B = \{c, d, e, f, g\}$
$A \cap B = \{c, d, e\}$

18. $A = \{m, n, o, p\}; B = \{k, l, m, n\}$
$A \cap B = \{m, n\}$

OBJECTIVE B   *To write and graph sets of real numbers*

For Exercises 19 to 24, use set-builder notation to write the set.

19. The negative integers greater than -5
$\{x|x > -5, x \in \text{negative integers}\}$

20. The positive integers less than 5
$\{x|x < 5, x \in \text{positive integers}\}$

21. The integers greater than 30
$\{x|x > 30, x \in \text{integers}\}$

22. The integers less than -70
$\{x|x < -70, x \in \text{integers}\}$

23. The real numbers greater than 8
$\{x|x > 8\}$

24. The real numbers less than 57
$\{x|x < 57\}$

For Exercises 25 to 33, write the set in interval notation.

25. $\{x|1 < x < 2\}$
$(1, 2)$

26. $\{x|-2 < x \leq 4\}$
$(-2, 4]$

27. $\{x|x > 3\}$
$(3, \infty)$

28. $\{x|x \leq 0\}$
$(-\infty, 0]$

29. $\{x|-4 \leq x < 5\}$
$[-4, 5)$

30. $\{x|-3 \leq x \leq 0\}$
$[-3, 0]$

31. $\{x|x \leq 2\}$
$(-\infty, 2]$

32. $\{x|x \geq -3\}$
$[-3, \infty)$

33. $\{x|-3 \leq x \leq 1\}$
$[-3, 1]$

For Exercises 34 to 42, write the interval in set-builder notation.

34. $[-4, 5]$
$\{x|-4 \leq x \leq 5\}$

35. $(-5, -3)$
$\{x|-5 < x < -3\}$

36. $(4, \infty)$
$\{x|x > 4\}$

37. $(-\infty, -2]$
$\{x|x \leq -2\}$

38. $(4, 9]$
$\{x|4 < x \leq 9\}$

39. $[-3, -2]$
$\{x|-3 \leq x \leq -2\}$

40. $[0, \infty)$
$\{x|x \geq 0\}$

41. $(-\infty, 6]$
$\{x|x \leq 6\}$

42. $(-\infty, \infty)$
$\{x|-\infty < x < \infty\}$

For Exercises 43 to 58, graph the set.

43. $[-5, 4]$

44. $(-3, 5]$

45. $\{x|x < 4\}$

46. $\{x|x \geq -3\}$

47. $\{x|x \leq -4\}$

48. $\{x|x > 0\}$

Unless otherwise noted, all content on this page is © Cengage Learning.

49. $(-\infty, 3]$

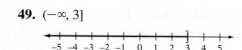

50. $(4, \infty)$

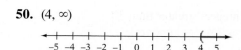

51. $[-1, 3)$

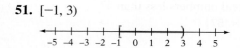

52. $(-3, 0]$

53. $\{x \mid -3 < x < 3\}$

54. $\{x \mid 0 \le x < 4\}$

55. $\{x \mid 2 \le x \le 4\}$

56. $\{x \mid -4 < x < 1\}$

57. $\{x \mid -\infty < x < \infty\}$

58. $(-\infty, \infty)$

59. How many elements are in the set given in interval notation as $(4, 4)$?
None

60. How many elements are in the set given by $\{x \mid 4 \le x \le 4\}$?
One

Critical Thinking

For Exercises 61 and 62, write an inequality that describes the situation.

61. To avoid shipping charges, one must spend a minimum m of $250.
$m \ge 250$

62. The temperature t never got above freezing ($32°F$).
$t \le 32$

63. True or false? If $A \cup B = A$, then $A \cap B = B$.
True

QUICK QUIZ

1. Use the roster method to write the set of integers between -6 and 0.
$\{-5, -4, -3, -2, -1\}$ **[1.5A]**

2. Find $C \cup D$ and $C \cap D$, given $C = \{-3, -2, -1, 0\}$ and $D = \{0, 1, 2, 3\}$.
$\{-3, -2, -1, 0, 1, 2, 3\};\ \{0\}$ **[1.5A]**

3. Write $\{x \mid x > -6\}$ in interval notation. $(-6, \infty)$ **[1.5B]**

4. Write $(-\infty, 7]$ in set-builder notation.
$\{x \mid x \le 7\}$ **[1.5B]**

5. Graph: $(-\infty, 4)$

  [1.5B]

6. Graph: $\{x \mid -4 < x < 2\}$

  [1.5B]

Projects or Group Activities

64. Make up sets A and B such that $A \cup B$ has three elements and $A \cap B$ has no elements. Write your sets using the roster method.
Answers will vary. For example, $A = \{1, 2\}$ and $B = \{3\}$.

65. Make up sets A and B such that $A \cup B$ has four elements and $A \cap B$ has four elements. Write your sets using the roster method.
Answers will vary. For example, $A = \{1, 2, 3, 4\}$ and $B = \{1, 2, 3, 4\}$.

66. Make up sets A and B such that $A \cup B$ has five elements and $A \cap B$ has two elements. Write your sets using the roster method.
Answers will vary. For example, $A = \{1, 2, 3\}$ and $B = \{1, 2, 4, 5\}$.

Unless otherwise noted, all content on this page is © Cengage Learning.

CHAPTER

1 | Summary

Key Words

Key Words	**Examples**

The set of **natural numbers** is $\{1, 2, 3, 4, 5, \ldots\}$.

The set of **whole numbers** is $\{0, 1, 2, 3, 4, 5, \ldots\}$.

The set of **integers** is $\{\ldots, -3, -2, -1, 0, 1, 2, 3, \ldots\}$. [1.1A, p. 2]

The number 1 is an element of the set of natural numbers, the set of whole numbers, and the set of integers.

A **prime number** is a natural number greater than 1 that is divisible (evenly) only by itself and 1. A natural number that is not a prime number is a **composite number**. [1.1A, p. 2]

2, 3, 5, 7, 11, and 13 are the first six prime numbers. 4, 6, 8, 9, 10, and 12 are the first six composite numbers.

A number a **is less than** a number b, written $a < b$, if a is to the left of b on a number line.

A number a **is greater than** a number b, written $a > b$, if a is to the right of b on a number line.

The symbol $\leq$ means **is less than or equal to**. The symbol $\geq$ means **is greater than or equal to**. [1.1A, p. 3]

$$-5 < -3 \qquad 9 > 0$$
$$3 \leq 3 \qquad 4 \leq 7$$
$$5 \geq 5 \qquad -6 \geq -9$$

Two numbers that are the same distance from zero on the number line but on opposite sides of zero are **opposite numbers** or **opposites**. The **additive inverse** of a number is the opposite of the number. [1.1B, p. 4]

7 and -7 are opposites.

$-\frac{3}{4}$ and $\frac{3}{4}$ are opposites.

The **absolute value** of a number is its distance from zero on the number line. [1.1B, p. 4]

$|5| = 5 \qquad |-2.3| = 2.3 \qquad |0| = 0$

A **rational number** (or fraction) is a number that can be written in the form $\frac{a}{b}$, where a and b are integers and $b \neq 0$. A rational number can be represented as a **terminating** or **repeating** **decimal**. [1.2A, p. 18]

$\frac{3}{8}$, $-\frac{9}{2}$, and 4 are rational numbers.

1.13 and $0.4\overline{73}$ are also rational numbers.

Percent means "parts of 100." [1.2B, p. 19]

72% means 72 of 100 equal parts.

An expression of the form a^n is in **exponential form**. The **base** is a and the **exponent** is n. [1.2E, p. 23]

5^4 is an exponential expression. The base is 5 and the exponent is 4.

A **square root** of a positive number x is a number whose square is x. The **principal square root** of a number is the positive square root.

The symbol $\sqrt{}$ is called a **radical sign** and is used to indicate the principal square root of a number. The **radicand** is the number under the radical sign. [1.2F, p. 25]

$\sqrt{25} = 5$
$-\sqrt{25} = -5$

The square of an integer is a **perfect square.** If a number is not a perfect square, its square root can only be approximated. [1.2F, p. 25]

$7^2 = 49$; 49 is a perfect square.

An **irrational number** is a number that has a decimal representation that never terminates or repeats. [1.2F, p. 25]

π, $\sqrt{2}$, and $1.34334333433334\ldots$ are irrational numbers.

The rational numbers and the irrational numbers taken together are the **real numbers.** [1.2F, p. 25]

$\frac{3}{8}$, $-\frac{9}{2}$, 4, 1.13, $0.4\overline{73}$, π, $\sqrt{2}$, and $1.34334333433334\ldots$ are real numbers.

A **variable** is a letter that is used for a quantity that is unknown or that can change. A **variable expression** is an expression that contains one or more variables. [1.4A, p. 40]

$4x + 2y - 6z$ is a variable expression. It contains the variables x, y, and z.

The **terms** of a variable expression are the addends of the expression. Each term is a **variable term** or a **constant term.** [1.4A, p. 40]

The expression $2a^2 - 3b^3 + 7$ has three terms: $2a^2$, $-3b^3$, and 7.

$2a^2$ and $-3b^3$ are variable terms.

7 is a constant term.

A variable term is composed of a **numerical coefficient** and a **variable part.** [1.4A, p. 40]

For the expression $-7x^3y^2$, -7 is the coefficient and x^3y^2 is the variable part.

In a variable expression, replacing each variable by its value and then simplifying the resulting numerical expression is called **evaluating the variable expression.** [1.4A, p. 40]

To evaluate $2ab - b^2$ when $a = 3$ and $b = -2$, replace a by 3 and b by -2 and then simplify the numerical expression.

$2(3)(-2) - (-2)^2 = -16$

Like terms of a variable expression are terms with the same variable part. Constant terms are like terms. [1.4B, p. 41]

For the expressions $3a^2 + 2b - 3$ and $2a^2 - 3a + 4$, $3a^2$ and $2a^2$ are like terms; -3 and 4 are like terms.

To simplify the sum of like variable terms, use the Distributive Property to add the numerical coefficients. This is called **combining like terms.** [1.4B, p. 42]

$5y + 3y = (5 + 3)y$
$= 8y$

The **multiplicative inverse** of a number is the **reciprocal** of the number. [1.4C, p. 44]

$\frac{3}{4}$ is the multiplicative inverse of $\frac{4}{3}$.

$-\frac{1}{4}$ is the multiplicative inverse of -4.

A **set** is a collection of objects, which are called the **elements** of the set.

The **roster method** of writing a set encloses a list of the elements in braces.

The **empty set** or **null set,** written $\varnothing$ or { }, is the set that contains no elements. [1.5A, p. 58]

Using the roster method, the set of integers from 1 to 8 is written

$A = \{1, 2, 3, 4, 5, 6, 7, 8\}$

The set of cars that can travel faster than 1000 mph is an empty set.

The **union** of two sets, written $A \cup B$, is the set that contains the elements of A and the elements of B. [1.5A, p. 58]

Let $A = \{2, 4, 6, 8\}$ and $B = \{0, 1, 2, 3, 4\}$.
Then $A \cup B = \{0, 1, 2, 3, 4, 6, 8\}$.

The **intersection** of two sets, written $A \cap B$, is the set that contains the elements that are common to both A and B. [1.5A, p. 58]

Let $A = \{2, 4, 6, 8\}$ and $B = \{0, 1, 2, 3, 4\}$.
Then $A \cap B = \{2, 4\}$.

Set-builder notation and **interval notation** are used to describe the elements of a set. [1.5B, pp. 59–60]

The set of real numbers greater than 2 is written in set-builder notation as $\{x | x > 2, x \in \text{real numbers}\}$.

In interval notation, the set of real numbers greater than 2 is written as $(2, \infty)$.

Essential Rules and Procedures	Examples
To add two numbers with the same sign, add the absolute values of the numbers. Then attach the sign of the addends. [1.1C, p. 5]	$7 + 15 = 22$ $-7 + (-15) = -22$
To add two numbers with different signs, find the absolute value of each number. Subtract the smaller of the two absolute values from the larger. Then attach the sign of the number with the larger absolute value. [1.1C, p. 5]	$7 + (-15) = -8$ $-7 + 15 = 8$
To subtract one number from another, add the opposite of the second number to the first number. [1.1C, p. 6]	$7 - 19 = 7 + (-19) = -12$ $-6 - (-13) = -6 + 13 = 7$
To multiply two numbers with the same sign, multiply the absolute values of the numbers. The product is positive. [1.1D, p. 8]	$7 \cdot 8 = 56$ $-7(-8) = 56$
To multiply two numbers with different signs, multiply the absolute values of the numbers. The product is negative. [1.1D, p. 8]	$-7 \cdot 8 = -56$ $7(-8) = -56$
To divide two numbers with the same sign, divide the absolute values of the numbers. The quotient is positive. [1.1D, p. 9]	$54 \div 9 = 6$ $(-54) \div (-9) = 6$
To divide two numbers with different signs, divide the absolute values of the numbers. The quotient is negative. [1.1D, p. 9]	$(-54) \div 9 = -6$ $54 \div (-9) = -6$
Properties of Zero and One in Division [1.1D, p. 9] If $a \neq 0$, $\frac{0}{a} = 0$. If $a \neq 0$, $\frac{a}{a} = 1$. $\frac{a}{1} = a$ $\frac{a}{0}$ is undefined.	$\frac{0}{-5} = 0$ $\frac{-12}{-12} = 1$ $\frac{7}{1} = 7$ $\frac{8}{0}$ is undefined.

To write a percent as a fraction, remove the percent sign and multiply by $\frac{1}{100}$. [1.2B, p. 19]

$$60\% = 60\left(\frac{1}{100}\right) = \frac{60}{100} = \frac{3}{5}$$

To write a percent as a decimal, remove the percent sign and multiply by 0.01. [1.2B, p. 19]

$$73\% = 73(0.01) = 0.73$$
$$1.3\% = 1.3(0.01) = 0.013$$

To write a decimal or a fraction as a percent, multiply by 100%. [1.2B, p. 19]

$$0.3 = 0.3(100\%) = 30\%$$

$$\frac{5}{8} = \frac{5}{8}(100\%) = \frac{500}{8}\% = 62.5\%$$

To add two fractions with the same denominator, add the numerators and place the sum over the common denominator. [1.2C, p. 20]

$$\frac{7}{10} + \frac{1}{10} = \frac{7+1}{10} = \frac{8}{10} = \frac{4}{5}$$

To subtract two fractions with the same denominator, subtract the numerators and place the difference over the common denominator. [1.2C, p. 20]

$$\frac{7}{10} - \frac{1}{10} = \frac{7-1}{10} = \frac{6}{10} = \frac{3}{5}$$

To multiply two fractions, place the product of the numerators over the product of the denominators. [1.2D, p. 21]

$$-\frac{2}{3} \cdot \frac{5}{6} = -\frac{2 \cdot 5}{3 \cdot 6} = -\frac{10}{18} = -\frac{5}{9}$$

To divide two fractions, multiply the dividend by the reciprocal of the divisor. [1.2D, p. 22]

$$-\frac{4}{5} \div \frac{2}{3} = -\frac{4}{5} \cdot \frac{3}{2} = -\frac{2 \cdot 2 \cdot 3}{5 \cdot 2} = -\frac{6}{5}$$

Product Property of Square Roots [1.2F, p. 26]
$$\sqrt{ab} = \sqrt{a} \cdot \sqrt{b}$$

$$\sqrt{50} = \sqrt{25 \cdot 2}$$
$$= \sqrt{25}\sqrt{2} = 5\sqrt{2}$$

Order of Operations Agreement [1.3A, p. 34]

Step 1 Perform operations inside grouping symbols. Grouping symbols include parentheses ( ), brackets [ ], braces { }, the absolute value symbol | |, and the fraction bar.

Step 2 Simplify exponential expressions.

Step 3 Do multiplication and division as they occur from left to right.

Step 4 Do addition and subtraction as they occur from left to right.

$$50 \div (-5)^2 + 2(7 - 16)$$
$$= 50 \div (-5)^2 + 2(-9)$$
$$= 50 \div 25 + 2(-9)$$
$$= 2 + (-18)$$
$$= -16$$

The Distributive Property [1.4B, p. 41]

If a, b, and c are real numbers, then

$a(b + c) = ab + ac$

or

$(b + c)a = ba + ca$

$5(4 + 7) = 5 \cdot 4 + 5 \cdot 7$
$= 20 + 35 = 55$

The Associative Property of Addition [1.4B, p. 42]

If a, b, and c are real numbers, then

$(a + b) + c = a + (b + c)$

$-4 + (2 + 7) = -4 + 9 = 5$
$(-4 + 2) + 7 = -2 + 7 = 5$

The Commutative Property of Addition [1.4B, p. 42]

If a and b are real numbers, then

$a + b = b + a$

$2 + 5 = 7$ and $5 + 2 = 7$

The Addition Property of Zero [1.4B, p. 42]

If a is a real number, then

$a + 0 = 0 + a = a$

$-8 + 0 = -8$ and $0 + (-8) = -8$

The Inverse Property of Addition [1.4B, p. 43]

If a is a real number, then

$a + (-a) = (-a) + a = 0$

$5 + (-5) = 0$ and $(-5) + 5 = 0$

The Associative Property of Multiplication [1.4C, p. 43]

If a, b, and c are real numbers, then

$(ab)c = a(bc)$

$-3 \cdot (5 \cdot 4) = -3(20) = -60$
$(-3 \cdot 5) \cdot 4 = -15 \cdot 4 = -60$

The Commutative Property of Multiplication [1.4C, p. 44]

If a and b are real numbers, then

$ab = ba$

$-3(7) = -21$ and $7(-3) = -21$

The Multiplication Property of One [1.4C, p. 44]

If a is a real number, then

$a \cdot 1 = 1 \cdot a = a$

$-3(1) = -3$ and $1(-3) = -3$

The Inverse Property of Multiplication [1.4C, p. 44]

If a is a real number and a is not equal to zero, then

$a \cdot \dfrac{1}{a} = \dfrac{1}{a} \cdot a = 1$

$-3 \cdot -\dfrac{1}{3} = 1$ and $-\dfrac{1}{3} \cdot -3 = 1$

1. Let $x \in \{-4, 0, 11\}$. For what values of x is the inequality $x < 1$ a true statement?

$-4, 0$ [1.1A]

2. Find the additive inverse of -4.

4   [1.1B]

3. Evaluate $-|-5|$.

-5 [1.1B]

4. Add: $-3 + (-12) + 6 + (-4)$

-13 [1.1C]

5. Subtract: $16 - (-3) - 18$

1   [1.1C]

6. Multiply: $(-6)(7)$

-42 [1.1D]

7. Divide: $-100 \div 5$

-20 [1.1D]

8. Write $\frac{7}{25}$ as a decimal.

0.28   [1.2A]

9. Write 6.2% as a decimal.

0.062   [1.2B]

10. Write $\frac{5}{8}$ as a percent.

62.5%   [1.2B]

11. Simplify: $\frac{1}{3} - \frac{1}{6} + \frac{5}{12}$

$\frac{7}{12}$ [1.2C]

12. Subtract: $5.17 - 6.238$

-1.068 [1.2C]

13. Divide: $-\frac{18}{35} \div \frac{17}{28}$

$-\frac{72}{85}$ [1.2D]

14. Multiply: $4.32(-1.07)$

-4.6224 [1.2D]

15. Evaluate $\left(-\frac{2}{3}\right)^4$.

$\frac{16}{81}$ [1.2E]

16. Simplify: $2\sqrt{36}$

12   [1.2F]

17. Simplify: $-3\sqrt{120}$

$-6\sqrt{30}$ [1.2F]

18. Evaluate $-3^2 + 4[18 + (12 - 20)]$.

31   [1.3A]

19. Evaluate $(b - a)^2 + c$ when $a = -2$, $b = 3$, and $c = 4$.
29   [1.4A]

20. Simplify: $6a - 4b + 2a$
$8a - 4b$ [1.4B]

21. Simplify: $-3(-12y)$
$36y$ [1.4C]

22. Simplify: $5(2x - 7)$
$10x - 35$ [1.4D]

23. Simplify: $-4(2x - 9) + 5(3x + 2)$
$7x + 46$ [1.4D]

24. Simplify: $5[2 - 3(6x - 1)]$
$-90x + 25$ [1.4D]

25. Use the roster method to write the set of odd positive integers less than 8.
$\{1, 3, 5, 7\}$ [1.5A]

26. Find $A \cap B$, given $A = \{1, 5, 9, 13\}$ and $B = \{1, 3, 5, 7, 9\}$.
$A \cap B = \{1, 5, 9\}$ [1.5A]

27. Graph $\{x \mid x > 3\}$.

-5 -4 -3 -2 -1 0 1 2 3 4 5   [1.5B]

28. Graph the interval $[1, 4]$.

-5 -4 -3 -2 -1 0 1 2 3 4 5   [1.5B]

29. Write the set $\{x \mid x > -4\}$ in interval notation.
$(-4, \infty)$ [1.5B]

30. Testing   To discourage random guessing on a multiple-choice exam, a professor assigns 6 points for a correct answer, -4 points for an incorrect answer, and -2 points for leaving a question blank. What is the score for a student who had 21 correct answers, had 5 incorrect answers, and left 4 questions blank?
98   [1.1E]

Should the penny be abolished?

31. 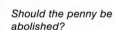 **Currency**   The graph at the right shows the responses of 2136 adults to the question "Would you favor or oppose abolishing the penny so that the nickel would be the lowest denomination of coin?" (*Source:* Harris Interactive) What percent of those surveyed opposed abolishing the penny? Round to the nearest tenth of a percent.
59.0%   [1.2G]

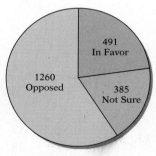

491
In Favor

1260
Opposed

385
Not Sure

Source: Harris Interactive

32. Translate "the difference between twice a number and one-half of the number" into a variable expression. Then simplify.
$2x - \dfrac{1}{2}x; \dfrac{3}{2}x$ [1.4E]

33. Baseball Cards   A baseball card collection contains five times as many National League player cards as American League player cards. Express the number of National League player cards in terms of the number of American League player cards.
Let A be the number of American League player cards; $5A$ [1.4E]

Unless otherwise noted, all content on this page is © Cengage Learning.

CHAPTER

1 | TEST

1. Place the correct symbol, $<$ or $>$, between the two numbers.

$-2 \; > \; -40$ [1.1A]

2. Find the opposite of -7.

7 [1.1B]

3. Evaluate $-|-4|$.

-4 [1.1B]

4. Subtract: $16 - 30$

-14 [1.1C]

5. Add: $-22 + 14 + (-8)$

-16 [1.1C]

6. Subtract: $16 - (-30) - 42$

4 [1.1C]

7. Divide: $-561 \div (-33)$

17 [1.1D]

8. Write $\frac{7}{9}$ as a decimal. Place a bar over the repeating digit of the decimal.

$0.\overline{7}$ [1.2A]

9. Write 45% as a fraction and as a decimal.

$\frac{9}{20}$, 0.45 [1.2B]

10. Add: $-\frac{2}{5} + \frac{7}{15}$

$\frac{1}{15}$ [1.2C]

11. Multiply: $6.02(-0.89)$

-5.3578 [1.2D]

12. Divide: $\frac{5}{12} \div \left(-\frac{5}{6}\right)$

$-\frac{1}{2}$ [1.2D]

13. Evaluate $\frac{3}{4} \cdot (4)^2$.

12 [1.2E]

14. Simplify: $-2\sqrt{45}$

$-6\sqrt{5}$ [1.2F]

15. Evaluate $16 \div 2[8 - 3(4 - 2)] + 1$.

17 [1.3A]

16. Evaluate $b^2 - 3ab$ when $a = 3$ and $b = -2$.

22 [1.4A]

17. Simplify: $3x - 5x + 7x$

$5x$ [1.4B]

18. Simplify: $\frac{1}{5}(10x)$

$2x$ [1.4C]

19. Simplify: $-3(2x^2 - 7y^2)$
$-6x^2 + 21y^2$ [1.4D]

20. Simplify: $2x - 3(x - 2)$
$-x + 6$ [1.4D]

21. Simplify: $2x + 3[4 - (3x - 7)]$

$-7x + 33$ [1.4D]

22. Use the roster method to write the set of integers between -3 and 4.
$\{-2, -1, 0, 1, 2, 3\}$ [1.5A]

23. Use set-builder notation to write the set of real numbers less than -3.
$\{x \mid x < -3, x \in \text{real numbers}\}$ [1.5B]

24. Find $A \cup B$ given $A = \{1, 3, 5, 7\}$ and $B = \{2, 4, 6, 8\}$.
$A \cup B = \{1, 2, 3, 4, 5, 6, 7, 8\}$ [1.5A]

25. Graph $\{x \mid x < 1\}$.

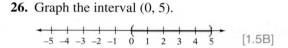

-5 -4 -3 -2 -1 0 1 2 3 4 5   [1.5B]

26. Graph the interval $(0, 5)$.

-5 -4 -3 -2 -1 0 1 2 3 4 5   [1.5B]

Year	Trade Balance
1980	−19.4
1981	−16.2
1982	−24.2
1983	−57.8
1984	−109.2
1985	−122.1
1986	−140.6
1987	−153.3
1988	−115.9
1989	−92.2
1990	−81.1
1991	−30.7
1992	−35.7
1993	−68.9
1994	−97.0
1995	−95.9
1996	−102.1
1997	−104.7
1998	−166.9
1999	−265.0
2000	−369.7

27. Translate "ten times the difference between a number and 3" into a variable expression. Then simplify.
$10(x - 3)$; $10x - 30$ [1.4E]

28. Baseball   The speed of a pitcher's fastball is twice the speed of the catcher's return throw. Express the speed of the fastball in terms of the speed of the return throw.
Let s be the speed of the catcher's return throw; $2s$ [1.4E]

29. ⬤ **Balance of Trade**   The table at the right shows the U.S. balance of trade, in billions of dollars, for the years 1980 to 2000 (*Source:* U.S. Dept. of Commerce).
a. In which years did the trade balance increase from the previous year?
'81, '88, '89, '90, '91, '95
b. Calculate the difference between the trade balance in 1990 and the trade balance in 2000. $288.6 billion
c. During which two consecutive years was the difference in the trade balance greatest?   1999 and 2000
d. How many times greater was the trade balance in 1990 than in 1980? Round to the nearest whole number.   4 times greater
e. Calculate the average trade balance per quarter for the year 2000.
−$92.425 billion [1.2G]

30. ⬤ **Temperature**   The lowest temperature recorded in North America is $-81.4°F$. The highest temperature recorded is $134.0°F$ (*Source:* National Climatic Data Center). Find the difference beween these two extremes.   215.4°F [1.2G]

Unless otherwise noted, all content on this page is © Cengage Learning.

First-Degree Equations and Inequalities

2

Focus on Success

Do you have trouble with word problems? Word problems show the variety of ways in which math can be used. The solution of every word problem can be broken down into two steps: Strategy and Solution. The Strategy consists of reading the problem, writing down what is known and unknown, and devising a plan to find the unknown. The Solution often consists of solving an equation and then checking the solution. (See Word Problems, page AIM-10).

Diego Cervo/Shutterstock.com

Prep Test

Are you ready to succeed in this chapter? Take the Prep Test below to find out if you are ready to learn the new material.

1. Write $\frac{9}{100}$ as a decimal.
0.09 [1.2B]

2. Write $\frac{3}{4}$ as a percent.
75% [1.2B]

3. Evaluate $3x^2 - 4x - 1$ when $x = -4$.
63 [1.4A]

4. Simplify: $R - 0.35R$
0.65R [1.4B]

5. Simplify: $\frac{1}{2}x + \frac{2}{3}x$
$\frac{7}{6}x$ [1.4B]

6. Simplify: $6x - 3(6 - x)$
9x − 18 [1.4D]

7. Simplify: $0.22(3x + 6) + x$
1.66x + 1.32 [1.4D]

8. Translate into a variable expression: "The difference between five and twice a number."
5 − 2n [1.4E]

9. Computers A new graphics card for computer games is five times faster than a graphics card made two years ago. Express the speed of the new card in terms of the speed of the old card.
Speed of old card: s
Speed of new card: 5s [1.4E]

10. Carpentry A board 5 ft long is cut into two pieces. If x represents the length of the longer piece, write an expression for the length of the shorter piece in terms of x.
5 − x [1.4E]

2.1 Introduction to Equations

OBJECTIVE A *To determine whether a given number is a solution of an equation*

Point of Interest

One of the most famous equations ever stated is $E = mc^2$. This equation, stated by Albert Einstein, shows that there is a relationship between mass m and energy E. As a side note, the chemical element einsteinium was named in honor of Einstein.

INSTRUCTOR NOTE

Remind students that determining whether a given number is a solution of an equation is a way to check an answer after the equation has been solved.

Take Note

The Order of Operations Agreement applies when evaluating $2(-2) + 5$ and $(-2)^2 - 3$.

An **equation** expresses the equality of two mathematical expressions. The expressions can be either numerical or variable expressions.

$$\left.\begin{array}{l} 9 + 3 = 12 \\ 3x - 2 = 10 \\ y^2 + 4 = 2y - 1 \\ z = 2 \end{array}\right\} \text{Equations}$$

The equation at the right is true if the variable is replaced by 5.

$$x + 8 = 13$$
$$5 + 8 = 13 \qquad \text{A true equation}$$

The equation is false if the variable is replaced by 7.

$$7 + 8 = 13 \qquad \text{A false equation}$$

A **solution of an equation** is a number that, when substituted for the variable, results in a true equation. 5 is a solution of the equation $x + 8 = 13$. 7 is not a solution of the equation $x + 8 = 13$.

HOW TO 1 Is -2 a solution of $2x + 5 = x^2 - 3$?

$$\begin{array}{c|c} 2x + 5 = x^2 - 3 \\ \hline 2(-2) + 5 & (-2)^2 - 3 \\ -4 + 5 & 4 - 3 \\ 1 = 1 \end{array}$$

Yes, -2 is a solution of the equation.

• Replace x by -2.
• Evaluate the numerical expressions.
• If the results are equal, -2 is a solution of the equation. If the results are not equal, -2 is not a solution of the equation.

EXAMPLE 1

Is -4 a solution of $4 + 5x = x^2 - 2x$?

Solution

$$\begin{array}{c|c} 4 + 5x = x^2 - 2x \\ \hline 4 + 5(-4) & (-4)^2 - 2(-4) \\ 4 + (-20) & 16 - (-8) \\ -16 \neq 24 \end{array}$$

• Replace x with -4.

($\neq$ means "is not equal to.")

No, -4 is not a solution.

YOU TRY IT 1

Is 5 a solution of $10x - x^2 = 3x - 10$?

Your solution

No

IN-CLASS EXAMPLES

1. Is -6 a solution of $4x + 3 = 2x - 9$? **Yes**

2. Is $-\dfrac{2}{3}$ a solution of $4 - 6x = 9x + 1$? **No**

Solution on p. S3

OBJECTIVE B *To solve an equation of the form $x + a = b$*

To **solve an equation** means to find a solution of the equation. The simplest equation to solve is an equation of the form *variable = constant,* because the constant is the solution.

The solution of the equation $x = 5$ is 5 because $5 = 5$ is a true equation.

Tips for Success

To learn mathematics, you must be an active participant. Listening and watching your professor do mathematics are not enough. Take notes in class, mentally think through every question your instructor asks, and try to answer it even if you are not called on to do so. Ask questions when you have them. See *AIM for Success* at the front of the book for other ways to be an active learner.

INSTRUCTOR NOTE

To help students see the value of the properties of equations, you might consider asking the students to find the solutions of equations such as the following by guessing. By the time they reach the last equation, they may have more appreciation for the properties.

$$x + 5 = 7$$
$$7 - x = 9$$
$$2x - 3 = 7$$
$$\frac{x}{2} - 1 = 3$$
$$5 - 3x = 27$$

 Take Note

An equation has some properties that are similar to those of a balance scale. For instance, if a balance scale is in balance and equal weights are added to each side of the scale, then the balance scale remains in balance. If an equation is true, then adding the same number to each side of the equation produces another true equation.

The solution of the equation at the right is 7 because $7 + 2 = 9$ is a true equation.

$$x + 2 = 9 \qquad 7 + 2 = 9$$

Note that if 4 is added to each side of the equation $x + 2 = 9$, the solution is still 7.

$$x + 2 = 9$$
$$x + 2 + 4 = 9 + 4$$
$$x + 6 = 13 \qquad 7 + 6 = 13$$

If -5 is added to each side of the equation $x + 2 = 9$, the solution is still 7.

$$x + 2 = 9$$
$$x + 2 + (-5) = 9 + (-5)$$
$$x - 3 = 4 \qquad 7 - 3 = 4$$

Equations that have the same solution are called **equivalent equations.** The equations $x + 2 = 9$, $x + 6 = 13$, and $x - 3 = 4$ are equivalent equations; each equation has 7 as its solution. These examples suggest that adding the same number to each side of an equation produces an equivalent equation. This is called the *Addition Property of Equations.*

Addition Property of Equations

The same number can be added to each side of an equation without changing its solution. In symbols, the equation $a = b$ has the same solution as the equation $a + c = b + c$.

EXAMPLE OF THIS PROPERTY

The equation $x - 3 = 7$ has the same solution as the equation $x - 3 + 3 = 7 + 3$.

In solving an equation, the goal is to rewrite the given equation in the form *variable = constant*. The Addition Property of Equations is used to remove a *term* from one side of the equation by adding the opposite of that term to each side of the equation.

> **HOW TO 2** Solve: $x - 4 = 2$
>
> $$x - 4 = 2$$ • The goal is to rewrite the equation in the form *variable = constant*.
>
> $$x - 4 + 4 = 2 + 4$$ • Add 4 to each side of the equation.
>
> $$x + 0 = 6$$ • Simplify.
>
> $$x = 6$$ • The equation is in the form *variable = constant*.
>
> *Check*: $\dfrac{x - 4 = 2}{6 - 4 \mid 2}$
>
> $$2 = 2 \quad \text{A true equation}$$
>
> The solution is 6.

Because subtraction is defined in terms of addition, the Addition Property of Equations also makes it possible to subtract the same number from each side of an equation without changing the solution of the equation.

Unless otherwise noted, all content on this page is © Cengage Learning.

INSTRUCTOR NOTE
The title of this objective contains the equation $x + a = b$. Ask students to give examples of this type of equation. Be sure they include $b = x + a$.

HOW TO 3   Solve: $y + \dfrac{3}{4} = \dfrac{1}{2}$

$y + \dfrac{3}{4} = \dfrac{1}{2}$ • The goal is to rewrite the equation in the form *variable = constant*.

$y + \dfrac{3}{4} - \dfrac{3}{4} = \dfrac{1}{2} - \dfrac{3}{4}$ • Subtract $\frac{3}{4}$ from each side of the equation.

$y + 0 = \dfrac{2}{4} - \dfrac{3}{4}$ • Simplify.

$y = -\dfrac{1}{4}$ • The equation is in the form *variable = constant*.

The solution is $-\frac{1}{4}$. You should check this solution.

EXAMPLE 2

Solve: $x + 15 = 23$

Solution

$x + 15 = 23$

$x + 15 - 15 = 23 - 15$ • Subtract 15 from each side.

$x + 0 = 8$ • Simplify each side.

$x = 8$ • Addition Property of Zero

The solution is 8.

YOU TRY IT 2

Solve: $26 = y - 14$

Your solution

40

IN-CLASS EXAMPLES
Solve.

3. $x - \dfrac{1}{4} = \dfrac{5}{6}$ $\dfrac{13}{12}$

4. $3 + x = 9$ 6

5. $5 = x + 5$ 0

Solution on p. S3

OBJECTIVE C   *To solve an equation of the form ax = b*

INSTRUCTOR NOTE
The requirement that each side be multiplied by a *nonzero* number is important. Later in the text, students will solve some equations by multiplying each side by a variable expression whose value may be zero. This can lead to extraneous solutions.

The solution of the equation at the right is 3 because $2 \cdot 3 = 6$ is a true equation.

Note that if each side of $2x = 6$ is multiplied by 5, the solution is still 3.

If each side of $2x = 6$ is multiplied by -4, the solution is still 3.

$2x = 6$ $2 \cdot 3 = 6$

$2x = 6$
$5(2x) = 5 \cdot 6$
$10x = 30$ $10 \cdot 3 = 30$

$2x = 6$
$(-4)(2x) = (-4)6$
$-8x = -24$ $-8 \cdot 3 = -24$

The equations $2x = 6$, $10x = 30$, and $-8x = -24$ are equivalent equations; each equation has 3 as its solution. These examples suggest that multiplying each side of an equation by the same nonzero number produces an equivalent equation.

Multiplication Property of Equations

Each side of an equation can be multiplied by the same nonzero number without changing the solution of the equation. In symbols, if $c \neq 0$, then the equation $a = b$ has the same solutions as the equation $ac = bc$.

EXAMPLE

The equation $3x = 21$ has the same solution as the equation $\frac{1}{3} \cdot 3x = \frac{1}{3} \cdot 21$.

INSTRUCTOR NOTE
Solutions to equations such as $ax = b$ have appeared in algebra texts for a long time. The problem below is an adaptation from Fibonacci's text *Liber Abaci*, which dates from 1202.

A merchant purchased 7 eggs for 1 denarius and sold them at a price of 5 eggs for 1 denarius. The merchant's profit was 18 denarii. How much did the merchant invest?

The resulting equation is

$$\frac{7}{5}x - x = 18$$

Solving this equation can serve as a classroom exercise. The solution is 45 denarii.

 Take Note

Remember to check the solution.

Check: $6x = 14$

$$6\left(\frac{7}{3}\right) \,\Big|\, 14$$

$$14 = 14$$

The Multiplication Property of Equations is used to remove a coefficient by multiplying each side of the equation by the reciprocal of the coefficient.

HOW TO 4   Solve: $\dfrac{3}{4}z = 9$

$\dfrac{3}{4}z = 9$ • The goal is to rewrite the equation in the form *variable* = *constant*.

$\dfrac{4}{3} \cdot \dfrac{3}{4}z = \dfrac{4}{3} \cdot 9$ • Multiply each side of the equation by $\dfrac{4}{3}$.

$1 \cdot z = 12$ • Simplify.

$z = 12$ • The equation is in the form *variable* = *constant*.

The solution is 12. You should check this solution.

Because division is defined in terms of multiplication, each side of an equation can be divided by the same nonzero number without changing the solution of the equation.

HOW TO 5   Solve: $6x = 14$

$6x = 14$ • The goal is to rewrite the equation in the form *variable* = *constant*.

$\dfrac{6x}{6} = \dfrac{14}{6}$ • Divide each side of the equation by 6.

$x = \dfrac{7}{3}$ • Simplify. The equation is in the form *variable* = *constant*.

The solution is $\dfrac{7}{3}$.

When using the Multiplication Property of Equations, multiply each side of the equation by the reciprocal of the coefficient when the coefficient is a fraction. Divide each side of the equation by the coefficient when the coefficient is an integer or a decimal.

EXAMPLE 3

Solve: $\dfrac{3x}{4} = -9$

Solution

$\dfrac{3x}{4} = -9$ • $\dfrac{3x}{4} = \dfrac{3}{4}x$

$\dfrac{4}{3} \cdot \dfrac{3}{4}x = \dfrac{4}{3}(-9)$ • Multiply each side by $\frac{4}{3}$.

$x = -12$

The solution is -12.

EXAMPLE 4

Solve: $5x - 9x = 12$

Solution

$5x - 9x = 12$

$-4x = 12$ • Combine like terms.

$\dfrac{-4x}{-4} = \dfrac{12}{-4}$ • Divide each side by -4.

$x = -3$

The solution is -3.

YOU TRY IT 3

Solve: $-\dfrac{2x}{5} = 6$

Your solution

-15

YOU TRY IT 4

Solve: $4x - 8x = 16$

Your solution

-4

IN-CLASS EXAMPLES
Solve.

6. $-\dfrac{5}{8}x = 25$ **−40**

7. $4y - 10y = -42$ **7**

8. $8 = \dfrac{3x}{4}$ $\dfrac{\mathbf{32}}{\mathbf{3}}$

9. $2z = 0$ **0**

Solutions on p. S3

OBJECTIVE D   *To solve application problems using the basic percent equation*

INSTRUCTOR NOTE
In application problems involving percent, substituting into the basic percent equation frequently leads to an equation of the form $ax = b$.

The simple interest equation and the percent mixture equation are introduced in this objective, and students are asked to use these equations to solve problems of the form $ax = b$. This will give them exposure to the concepts involved in solving interest problems and percent mixture problems prior to attempting the more difficult interest and mixture problems presented in Objectives 2.4B and 5.1C.

An equation that is used frequently in mathematics applications is the basic percent equation.

Basic Percent Equation

$$\text{Percent} \cdot \text{Base} = \text{Amount}$$
$$P \quad \cdot \quad B \quad = \quad A$$

In many application problems involving percent, the base follows the word *of.*

HOW TO 6   20% of what number is 30?

$P \cdot B = A$	• Use the basic percent equation.
$0.20B = 30$	• $P = 20\% = 0.20$, $A = 30$, and B is unknown.
$\dfrac{0.20B}{0.20} = \dfrac{30}{0.20}$	• Solve for B.
$B = 150$	

The number is 150.

In most cases, you should write the percent as a decimal before solving the basic percent equation, as in HOW TO 6. However, some percents are more easily written as fractions. For example,

$$33\frac{1}{3}\% = \frac{1}{3} \qquad 66\frac{2}{3}\% = \frac{2}{3} \qquad 16\frac{2}{3}\% = \frac{1}{6} \qquad 83\frac{1}{3}\% = \frac{5}{6}$$

 Take Note
We have written $P(80) = 70$ because that is the form of the basic percent equation. We could have written $80P = 70$. The important point is that each side of the equation is divided by 80, the coefficient of P.

HOW TO 7   70 is what percent of 80?

$P \cdot B = A$	• Use the basic percent equation.
$P(80) = 70$	• $B = 80$, $A = 70$, and P is unknown.
$\dfrac{P(80)}{80} = \dfrac{70}{80}$	• Solve for P.
$P = 0.875$	• The question asked for a percent.
$P = 87.5\%$	• Convert the decimal to a percent.

70 is 87.5% of 80.

©iStockphoto.com/photo25th

APPLY THE CONCEPT

 The world's production of cocoa for a recent year was 2928 metric tons. Of this, 1969 metric tons came from Africa. (*Source:* World Cocoa Foundation) What percent of the world's cocoa was produced in Africa? Round to the nearest tenth of a percent.

To find the percent, use the basic percent equation.

$$P \cdot B = A$$
$$P(2928) = 1969 \qquad \bullet \; B = 2928, A = 1969, \text{ and } P \text{ is unknown.}$$
$$P = \frac{1969}{2928} \approx 0.672$$

Approximately 67.2% of the world's cocoa was produced in Africa.

The simple interest that an investment earns is given by the **simple interest equation** $I = Prt$, where I is the simple interest, P is the principal, or amount invested, r is the simple interest rate, and t is the time.

APPLY THE CONCEPT

A $1500 investment has an annual simple interest rate of 7%. Find the simple interest earned on the investment after 18 months.

To find the interest, solve $I = Prt$ for I.

The time is given in months but the interest rate is an annual rate. Therefore, we must convert 18 months to years.   18 months $= \frac{18}{12}$ years $= 1.5$ years

$$I = Prt$$
$$I = 1500(0.07)(1.5) \qquad \bullet \; P = 1500, r = 0.07, t = 1.5$$
$$I = 157.5$$

The investment earned $157.50.

  Point of Interest

In the jewelry industry, the amount of gold in a piece of jewelry is measured in *karats*. Pure gold is 24 karats. A necklace that is 18 karats is $\frac{18}{24} = 0.75 = 75\%$ gold.

The amount of a substance in a solution can be given as a percent of the total solution. For instance, if a certain fruit juice drink is advertised as containing 27% cranberry juice, then 27% of the contents of the bottle must be cranberry juice.

The method for solving problems involving mixtures is based on the **percent mixture equation** $Q = Ar$, where Q is the quantity of a substance in the solution, A is the amount of the solution, and r is the percent concentration of the substance.

APPLY THE CONCEPT

The formula for a perfume requires that the concentration of jasmine be 1.2% of the total amount of perfume. How many ounces of jasmine are in a 2-ounce bottle of this perfume?

To find the number of ounces of jasmine, solve $Q = Ar$ for Q.

$$Q = Ar$$
$$Q = 2(0.012) \qquad \bullet \; A = \text{the amount of perfume} = 2 \text{ oz}$$
$$Q = 0.024 \qquad\qquad r = \text{the percent concentration} = 1.2\% = 0.012$$

There is 0.024 oz of jasmine in the perfume.

EXAMPLE 5

12 is $33\frac{1}{3}\%$ of what number?

Solution

$\quad P \cdot B = A$ • Use the basic percent equation.

$\quad \dfrac{1}{3}B = 12$ • $33\dfrac{1}{3}\% = \dfrac{1}{3}$

$\quad 3 \cdot \dfrac{1}{3}B = 3 \cdot 12$ • Multiply each side by 3.

$\quad\quad B = 36$

12 is $33\frac{1}{3}\%$ of 36.

YOU TRY IT 5

18 is $16\frac{2}{3}\%$ of what number?

Your solution

108

EXAMPLE 6

The data in the table below show the numbers of households (in millions) that downloaded music files for a three-month period in a recent year. (*Source:* NPD Group)

Month	April	May	June
Downloads	14.5	12.7	10.4

For the three-month period, what percent of the files were downloaded in May? Round to the nearest percent.

Strategy

To find the percent:
• Find the total number of files downloaded for the three-month period.
• Use the basic percent equation. B is the total number of files downloaded for the three-month period; $A = 12.7$, the number of files downloaded in May; P is unknown.

Solution

$14.5 + 12.7 + 10.4 = 37.6$

$\quad P \cdot B = A$ • Use the basic percent equation.

$P(37.6) = 12.7$ • $B = 37.6, A = 12.7$

$\quad P = \dfrac{12.7}{37.6} \approx 0.34$ • Divide each side by 37.6.

Approximately 34% of the files were downloaded in May.

YOU TRY IT 6

According to Wikipedia.org, 162.9 million people in the United States watched Super Bowl XLV. What percent of the U.S. population watched Super Bowl XLV? Use a figure of 310 million for the U.S. population. Round to the nearest tenth of a percent.

Your strategy

Your solution

52.5%

IN-CLASS EXAMPLES

10. 40 is $83\frac{1}{3}\%$ of what number?   **48**
11. A telephone bill of $27.25 consisted of charges for flat-rate service, charges for direct-dialed calls, and miscellaneous charges. Of this amount, $3.27 was for direct-dialed calls. What percent of the telephone bill was charges for direct-dialed calls?   **12%**
12. The annual simple interest rate on an account that earned $234 in interest in 18 months is 6.5%. Find the original principal.   **$2400**
13. Find the weight of a metal alloy that is 40% silver and contains 20 g of silver.   **50 g**

Solutions on p. S4

Unless otherwise noted, all content on this page is © Cengage Learning.

EXAMPLE 7

In April, Marshall Wardell was charged an interest fee of $8.72 on an unpaid credit card balance of $545. Find the annual interest rate on this credit card.

Strategy

The interest is $8.72. Therefore, $I = 8.72$. The unpaid balance is $545. This is the principal on which interest is calculated. Therefore, $P = 545$. The time is one month. Because the *annual* interest rate must be found and the time is given as one month, we write one month as $\frac{1}{12}$ year: $t = \frac{1}{12}$. To find the interest rate, solve $I = Prt$ for r.

Solution

$$I = Prt$$ • Use the simple interest equation.

$$8.72 = 545r\left(\frac{1}{12}\right)$$ • $I = 8.72, P = 545, t = \frac{1}{12}$

$$8.72 = \frac{545}{12}r$$

$$\frac{12}{545}(8.72) = \frac{12}{545}\left(\frac{545}{12}r\right)$$ • Multiply each side by the reciprocal of $\frac{545}{12}$.

$$0.192 = r$$

The annual interest rate is 19.2%.

YOU TRY IT 7

Clarissa Adams purchased a municipal bond for $1000. The bond earns an annual simple interest rate of 6.4%. How much must she deposit into an account that earns 8% annual simple interest so that the interest earned from each account after one year is the same?

Your strategy

Your solution

$800

EXAMPLE 8

To make a certain color of blue, 4 oz of cyan must be contained in 1 gal of paint. What is the percent concentration of cyan in the paint?

Strategy

The cyan is given in ounces and the amount of paint is given in gallons. We must convert ounces to gallons or gallons to ounces. For this problem, we will convert gallons to ounces: 1 gal = 128 oz. Solve $Q = Ar$ for r, with $Q = 4$ and $A = 128$.

Solution

$$Q = Ar$$ • Use the percent mixture equation.

$$4 = 128r$$ • $Q = 4, A = 128$

$$\frac{4}{128} = \frac{128r}{128}$$

$$0.03125 = r$$

The percent concentration of cyan is 3.125%.

YOU TRY IT 8

The concentration of sugar in a certain breakfast cereal is 25%. If there are 2 oz of sugar contained in a bowl of cereal, how many ounces of cereal are in the bowl?

Your strategy

Your solution

8 oz

Solutions on p. S4

OBJECTIVE E

To solve uniform motion problems

📝 **Take Note**

A car traveling in a *circle* at a constant speed of 45 mph is *not* in uniform motion because the direction of the car is always changing.

INSTRUCTOR NOTE

The equation $d = rt$ is introduced in this objective, and students are asked to use this equation to solve problems of the form $ax = b$. This will give them exposure to the concepts involved in solving uniform motion problems prior to attempting the more difficult motion problems presented in Objective 2.4C.

Any object that travels at a constant speed in a straight line is said to be in *uniform motion*. **Uniform motion** means that the speed and direction of an object do not change. For instance, a car traveling at a constant speed of 45 mph on a straight road is in uniform motion.

The solution of a uniform motion problem is based on the **uniform motion equation** $d = rt$, where d is the distance traveled, r is the rate of travel, and t is the time spent traveling. For instance, suppose a car travels at 50 mph for 3 h. Because the rate (50 mph) and time (3 h) are known, we can find the distance traveled by solving the equation $d = rt$ for d.

$$d = rt$$
$$d = 50(3) \qquad \bullet \ r = 50, t = 3$$
$$d = 150$$

The car travels a distance of 150 mi.

APPLY THE CONCEPT ·······································

A jogger runs 3 mi in 45 min. What is the rate of the jogger in miles per hour?

To find the rate of the jogger, solve the equation $d = rt$ for r.
The answer must be in miles per *hour* and the time is given in *minutes*.
Convert 45 min to hours: $45 \text{ min} = \frac{45}{60} \text{ h} = \frac{3}{4} \text{ h}$

$$d = rt$$
$$3 = r\left(\frac{3}{4}\right) \qquad \bullet \ d = 3, t = \frac{3}{4}$$
$$3 = \frac{3}{4}r$$
$$\left(\frac{4}{3}\right)3 = \left(\frac{4}{3}\right)\frac{3}{4}r \qquad \bullet \ \textbf{Multiply each side of the equation by the reciprocal of } \tfrac{3}{4}.$$
$$4 = r$$

The rate of the jogger is 4 mph.

···

If two objects are moving in opposite directions, then the rate at which the distance between them is increasing is the sum of the speeds of the two objects. For instance, in the diagram below, two cars start from the same point and travel in opposite directions. The distance between them is changing at the rate of 70 mph.

$$30 + 40 = 70 \text{ mph}$$

Unless otherwise noted, all content on this page is © Cengage Learning.

Similarly, if two objects are moving toward each other, the distance between them is decreasing at a rate that is equal to the sum of the speeds. The rate at which the two planes at the right are approaching one another is 800 mph.

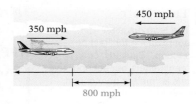

APPLY THE CONCEPT

Two cars start from the same point and move in opposite directions. The car moving west is traveling at 45 mph, and the car moving east is traveling at 60 mph. In how many hours will the cars be 210 mi apart?

To find the time, solve the equation $d = rt$ for t.

d = distance = 210 mi

The cars are moving in opposite directions, so the rate at which the distance between them is changing is the sum of the rates of the cars.

45 mph + 60 mph = 105 mph. Therefore, $r = 105$.

$$d = rt$$
$$210 = 105t \qquad \bullet \ d = 210, r = 105$$
$$\frac{210}{105} = \frac{105t}{105} \qquad \bullet \ \textbf{Divide each side of the equation by 105.}$$
$$2 = t$$

In 2 h, the cars will be 210 mi apart.

If a motorboat is on a river that is flowing at a rate of 4 mph, then the boat will float down the river at a speed of 4 mph when the motor is not on. Now suppose the motor is turned on and the power adjusted so that the boat would travel 10 mph without the aid of the current. Then, if the boat is moving with the current, its effective speed is the speed of the boat using power plus the speed of the current: 10 mph + 4 mph = 14 mph. (See the figure below.)

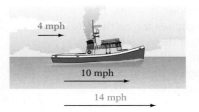

However, if the boat is moving against the current, the current slows the boat down. The effective speed of the boat is the speed of the boat using power minus the speed of the current: 10 mph − 4 mph = 6 mph. (See the figure below.)

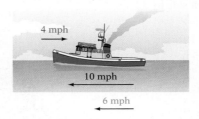

Unless otherwise noted, all content on this page is © Cengage Learning.

Take Note

The term ft/s is an abbreviation for "feet per second." Similarly, cm/s is "centimeters per second" and m/s is "meters per second."

There are other situations in which the preceding concepts may be applied.

APPLY THE CONCEPT

An airline passenger is walking between two airline terminals and decides to get on a moving sidewalk that is 150 ft long. If the passenger walks at a rate of 7 ft/s and the moving sidewalk moves at a rate of 9 ft/s, how long, in seconds, will it take for the passenger to walk from one end of the moving sidewalk to the other? Round to the nearest thousandth.

To find the time, solve the equation $d = rt$ for t.

d = distance = 150 ft

The passenger is traveling at 7 ft/s and the moving sidewalk is traveling at 9 ft/s.

The rate of the passenger is the sum of the two rates: 7 ft/s + 9 ft/s = 16 ft/s.

Therefore, $r = 16$.

$$d = rt$$
$$150 = 16t \quad \bullet \; d = 150, r = 16$$
$$\frac{150}{16} = \frac{16t}{16} \quad \bullet \; \textbf{Divide each side of the equation by 16.}$$
$$9.375 = t$$

It will take 9.375 s for the passenger to travel the length of the moving sidewalk.

EXAMPLE 9

Two cyclists start at the same time at opposite ends of an 80-mile course. One cyclist is traveling at 18 mph, and the second cyclist is traveling at 14 mph. How long after they begin cycling will they meet?

Strategy

The distance is 80 mi. Therefore, $d = 80$. The cyclists are moving toward each other, so the rate at which the distance between them is changing is the sum of the rates of the cyclists. The rate is 18 mph + 14 mph = 32 mph. Therefore, $r = 32$. To find the time, solve the equation $d = rt$ for t.

Solution

$$d = rt$$
$$80 = 32t \quad \bullet \; d = 80, r = 32$$
$$\frac{80}{32} = \frac{32t}{32} \quad \bullet \; \textbf{Divide each side by 32.}$$
$$2.5 = t$$

The cyclists will meet in 2.5 h.

YOU TRY IT 9

A plane that can normally travel at 250 mph in calm air is flying into a headwind of 25 mph. How far can the plane fly in 3 h?

Your strategy

Your solution

675 mi

IN-CLASS EXAMPLES

14. Ted leaves his house at 8:00 A.M. and arrives at work at 8:30 A.M. If the trip to work is 15 mi, determine Ted's average rate of speed.
30 mph

15. Joan leaves her house and travels at an average speed of 45 mph toward her cabin in the mountains. If the distance from her house to the cabin is 180 mi, how many hours will it take for Joan to arrive at her cabin if she stops one hour for lunch?   **5 h**

Solution on p. S4

2.1 EXERCISES

✔ Concept Check

SUGGESTED ASSIGNMENT
Exercises 1–10; Exercises 11–153, odds; Exercise 154; Exercises 155–167, odds

1. Label each of the following as either an expression or an equation.

   a. $3x + 7 = 9$ **b.** $3x + 7$ **c.** $4 - 6(y + 5)$ **d.** $a + b = 8$ **e.** $a + b - 8$

      Equation   Expression   Expression   Equation   Expression

2. ◣ What is the solution of the equation $x = 8$? Use your answer to explain why the goal in solving equations is to get the variable alone on one side of the equation.

3. Which of the following are equations of the form $x + a = b$? If an equation is of the form $x + a = b$, what would you do to solve the equation?

   (i) $d + 7.8 = -9.2$ **(ii)** $0.3 = t + 1.4$ **(iii)** $-9 = 3y$ **(iv)** $-8 + c = -5.6$

      i, ii, and iv are equations of the form $x + a = b$; you would subtract a from both sides.

4. Which of the following are equations of the form $ax = b$? If an equation is of the form $ax = b$, what would you do to solve the equation?

   (i) $3y = -12$ **(ii)** $2.4 = 0.6d$ **(iii)** $-5 = z - 10$ **(iv)** $-8c = -56$

      i, ii, and iv are equations of the form $ax = b$; you would divide both sides by a.

Identify the amount and the base.

5. 30 is 75% of 40.   Amount: 30, base: 40

6. 40% of 20 is 8.   Amount: 8, base: 20

Complete Exercises 7 and 8 by filling in the blanks with the correct number from the problem situation or with the word *unknown*.

[handwritten: P = percentage B = base A = amount]

7. Problem Situation: It rained on 24 of the 30 days of June. What percent of the days in June were rainy days?

   Using the formula $PB = A$, $P = \underline{\text{unknown}}$, $B = \underline{30}$, and $A = \underline{24}$.   unknown; 30; 24

8. Problem Situation: You bought a used car and made a down payment of 25% of the purchase price of $16,000. How much was the down payment?

   Using the formula $PB = A$, $P = \underline{25\%}$, $B = \underline{16,000}$, and $A = \underline{\text{unknown}}$.   25%; 16,000; unknown

9. Keith and Jennifer started at the same time and rode toward each other on a straight road. When they met, Keith had traveled 15 mi and Jennifer had traveled 10 mi. Who had the greater average speed?   Keith

10. Suppose you have a powerboat with the throttle set to move the boat at 8 mph in calm water. The rate of the current of a river is 4 mph.

   a. What is the speed of the boat when traveling on this river with the current?   12 mph

   b. What is the speed of the boat when traveling on this river against the current?   4 mph

[handwritten marginal notes: 8mph + 4 mph, 12mph, 8-4 = 4mph]

OBJECTIVE A   *To determine whether a given number is a solution of an equation*

11. Is 4 a solution of
$2x = 8$?
Yes

12. Is 3 a solution of
$y + 4 = 7$?
Yes

13. Is -1 a solution of
$2b - 1 = 3$?
No

14. Is -2 a solution of
$3a - 4 = 10$?
No

15. Is 1 a solution of
$4 - 2m = 3$?
No

16. Is 2 a solution of
$7 - 3n = 2$?
No

17. Is 5 a solution of
$2x + 5 = 3x$?
Yes

18. Is 4 a solution of
$3y - 4 = 2y$?
Yes

19. Is -2 a solution of
$3a + 2 = 2 - a$?
No

20. Is 3 a solution of
$z^2 + 1 = 4 + 3z$?
No

21. Is 2 a solution of
$2x^2 - 1 = 4x - 1$?
Yes

22. Is -1 a solution of
$y^2 - 1 = 4y + 3$?
No

23. Is $\frac{1}{2}$ a solution of
$4y + 1 = 3$?
Yes

24. Is $\frac{2}{5}$ a solution of
$5m + 1 = 10m - 3$?
No

25. Is $\frac{3}{4}$ a solution of
$8x - 1 = 12x + 3$?
No

26. 🐾 If A is a fixed number such that $A < 0$, is a solution of the equation $5x = A$
positive or negative?
Negative

OBJECTIVE B   *To solve an equation of the form $x + a = b$*

27. 🐾 ◤ Without solving the equation $x - \frac{11}{16} = \frac{19}{24}$,
determine whether x is less than or greater than
$\frac{19}{24}$. Explain your answer.

28. 🐾 ◤ Without solving the equation $x + \frac{13}{15} = -\frac{21}{43}$,
determine whether x is less than or greater than
$-\frac{21}{43}$. Explain your answer.

For Exercises 29 to 64, solve and check.

29. $x + 5 = 7$
2

30. $y + 3 = 9$
6

31. $b - 4 = 11$
15

32. $z - 6 = 10$
16

33. $2 + a = 8$
6

34. $5 + x = 12$
7

35. $n - 5 = -2$
3

36. $x - 6 = -5$
1

37. $b + 7 = 7$
0

38. $y - 5 = -5$
0

39. $z + 9 = 2$
-7

40. $n + 11 = 1$
-10

41. $10 + m = 3$
-7

42. $8 + x = 5$
-3

43. $9 + x = -3$
-12

44. $10 + y = -4$
-14

45. $2 = x + 7$
-5

46. $-8 = n + 1$
-9

47. $4 = m - 11$
15

48. $-6 = y - 5$
-1

49. $12 = 3 + w$
9

50. $-9 = 5 + x$
-14

51. $4 = -10 + b$
14

52. $-7 = -2 + x$
-5

53. $m + \frac{2}{3} = -\frac{1}{3}$
-1

54. $c + \frac{3}{4} = -\frac{1}{4}$
-1

55. $x - \frac{1}{2} = \frac{1}{2}$
1

56. $x - \frac{2}{5} = \frac{3}{5}$
1

57. $\frac{5}{8} + y = \frac{1}{8}$
$-\frac{1}{2}$

58. $\frac{4}{9} + a = -\frac{2}{9}$
$-\frac{2}{3}$

59. $-\frac{5}{6} = x - \frac{1}{4}$
$-\frac{7}{12}$

60. $-\frac{1}{4} = c - \frac{2}{3}$
$\frac{5}{12}$

61. $d + 1.3619 = 2.0148$ 0.6529

62. $w + 2.932 = 4.801$ 1.869

63. $6.149 = -3.108 + z$ 9.257

64. $5.237 = -2.014 + x$ 7.251

OBJECTIVE C   *To solve an equation of the form ax = b*

For Exercises 65 to 98, solve and check.

65. $5x = -15$
−3

66. $4y = -28$
−7

67. $3b = 0$
0

68. $2a = 0$
0

69. $-3x = 6$
−2

70. $-5m = 20$
−4

71. $-\dfrac{1}{6}n = -30$
180

72. $20 = \dfrac{1}{4}c$
80

73. $0 = -5x$
0

74. $0 = -8a$
0

75. $\dfrac{x}{3} = 2$
6

76. $\dfrac{x}{4} = 3$
12

77. $-\dfrac{y}{2} = 5$
−10

78. $-\dfrac{b}{3} = 6$
−18

79. $\dfrac{3}{4}y = 9$
12

80. $\dfrac{2}{5}x = 6$
15

81. $-\dfrac{2}{3}d = 8$
−12

82. $-\dfrac{3}{5}m = 12$
−20

83. $\dfrac{2n}{3} = 0$
0

84. $\dfrac{5x}{6} = 0$
0

85. $\dfrac{-3z}{8} = 9$
−24

86. $\dfrac{3x}{4} = 2$
$\dfrac{8}{3}$

87. $\dfrac{2}{9} = \dfrac{2}{3}y$
$\dfrac{1}{3}$

88. $-\dfrac{6}{7} = -\dfrac{3}{4}b$
$\dfrac{8}{7}$

89. $\dfrac{x}{1.46} = 3.25$
4.745

90. $\dfrac{z}{2.95} = -7.88$
−23.246

91. $3.47a = 7.1482$
2.06

92. $2.31m = 2.4255$
1.05

93. $2m + 5m = 49$
7

94. $5x + 2x = 14$
2

95. $3n + 2n = 20$
4

96. $7d - 4d = 9$
3

97. $10y - 3y = 21$
3

98. $2x - 5x = 9$
−3

For Exercises 99 to 102, suppose y is a positive integer. Determine whether x is positive or negative.

99. $15x = y$
Positive

100. $-6x = y$
Negative

101. $-\dfrac{1}{4}x = y$
Negative

102. $\dfrac{2}{9}x = -y$
Negative

OBJECTIVE D   *To solve application problems using the basic percent equation*

103. What is 35% of 80?
28

104. What percent of 8 is 0.5?
6.25%

105. Find 1.2% of 60.
0.72

106. 8 is what percent of 5?
160%

107. 125% of what is 80?
64

108. What percent of 20 is 30?
150%

109. 12 is what percent of 50?
24%

110. What percent of 125 is 50?
40%

111. Find 18% of 40.
7.2

112. What is 25% of 60?
15

113. 12% of what is 48?
400

114. 45% of what is 9?
20

115. What is $33\frac{1}{3}$% of 27?
9

116. Find $16\frac{2}{3}$% of 30.
5

117. What percent of 12 is 3?
25%

118. 10 is what percent of 15?
$66\frac{2}{3}$%

119. 12 is what percent of 6?
200%

120. 20 is what percent of 16?
125%

121. $5\frac{1}{4}$% of what is 21?
400

122. $37\frac{1}{2}$% of what is 15?
40

123. Find 15.4% of 50.
7.7

124. What is 18.5% of 46?
8.51

125. 1 is 0.5% of what?
200

126. 3 is 1.5% of what?
200

127. $\frac{3}{4}$% of what is 3?
400

128. $\frac{1}{2}$% of what is 3?
600

129. What is 250% of 12?
30

130. Without solving an equation, determine whether 40% of 80 is less than, equal to, or greater than 80% of 40.   Equal to

131. Without solving an equation, determine whether $\frac{1}{4}$% of 80 is less than, equal to, or greater than 25% of 80.   Less than

132. **Government**   To override a presidential veto, at least $66\frac{2}{3}$% of the Senate must vote to override the veto. There are 100 senators in the Senate. What is the minimum number of votes needed to override a veto?   67 votes

133. **Boston Marathon**   See the news clipping at the right. What percent of the participants who started the course finished the race? Round to the nearest tenth of a percent.   97.9%

134. **Natural Resources**   On average, a person uses 13.2 gal of water per day for showering. This is 17.8% of the total amount of water used per person per day in the average single-family home. Find the total amount of water used per person per day in the average single-family home. Round to the nearest whole number. (*Source:* American Water Works Association)   74 gal

135. **Safety**   Recently, the National Safety Council collected data on the leading causes of accidental death. The findings revealed that for people age 20, 30 died from a fall, 47 from fire, 200 from drowning, and 1950 from motor vehicle accidents. What percent of the accidental deaths were not attributed to motor vehicle accidents? Round to the nearest percent.   12%

In the NEWS!

Thousands Complete Boston Marathon

This year, there were 26,735 entrants in the Boston Marathon, the world's oldest annual marathon. Of those registered, 23,126 people started the race, and 22,629 people finished the 26.2-mile course.
Source: www.bostonmarathon.org

136. 🌑 **School Enrollment**   The circle graph at the right represents the U.S. population over 3 years old that is enrolled in school. To answer the question "How many people are enrolled in college or graduate school?," what additional piece of information is necessary?   Number of people 3 years old and older in the U.S. who are enrolled in school

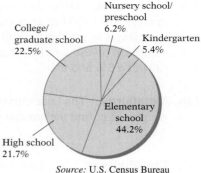

College/ graduate school 22.5%

Nursery school/ preschool 6.2%

Kindergarten 5.4%

Elementary school 44.2%

High school 21.7%

Source: U.S. Census Bureau

137. 🌑 **Travel**   According to the annual Summer Vacation Survey conducted by Myvesta, a nonprofit consumer education organization, the average summer vacation costs $2252. If $1850 of this amount is charged on a credit card, what percent of the vacation cost is charged? Round to the nearest tenth of a percent.   82.1%

INSTRUCTOR NOTE
See Exercise 137. Of those taking a vacation, 77.8% plan on using a credit card to pay for all or part of their trip, and 29.3% plan on taking three or more months to pay for their vacation charges. If the average vacation cost of $2252 is placed on a credit card and only the minimum payment is made, it will take almost 33 years to pay off the balance and will cost an additional $5687 in interest payments. That calculation assumes an 18% interest rate and a minimum monthly payment of 2%.

138. 🌑 **Energy**   The Energy Information Administration reports that if every U.S. household switched 4 h of lighting per day from incandescent bulbs to compact fluorescent bulbs, we would save 31.7 billion kilowatt-hours of electricity a year, or 33% of the total electricity used for home lighting. What is the total electricity used for home lighting in this country? Round to the nearest tenth of a billion.   96.1 billion kilowatt-hours

139. Investment   If Kachina Caron invested $1200 in a simple interest account and earned $72 in 8 months, what is the annual interest rate?
9%

140. Investment   How much money must Andrea invest for 2 years in an account that earns an annual simple interest rate of 8% if she wants to earn $300 from the investment?
$1875

141. Investment   Sal Boxer decided to divide a gift of $3000 into two different accounts. He placed $1000 in an account that earns an annual simple interest rate of 7.5%. The remaining money was placed in an account that earns an annual simple interest rate of 8.25%. How much interest will Sal earn from the two accounts after one year?
$240

142. Investment   If Americo invests $2500 at an 8% annual simple interest rate and Octavia invests $3000 at a 7% annual simple interest rate, which of the two will earn the greater amount of interest after one year?
Octavia

143. Investment   Makana invested $900 in a simple interest account that had an interest rate that was 1% more than that of her friend Marlys. If Marlys earned $51 after one year from an investment of $850, how much did Makana earn in one year?
$63

144. Investment   A $2000 investment at an annual simple interest rate of 6% earned as much interest after one year as another investment in an account that earns 8% simple interest. How much was invested at 8%?
$1500

Unless otherwise noted, all content on this page is © Cengage Learning.

145. **Investment** An investor placed $1000 in an account that earns 9% annual simple interest and $1000 in an account that earns 6% annual simple interest. If each investment is left in the account for the same period of time, is the interest rate on the combined investment less than 6%, between 6% and 9%, or greater than 9%?
Between 6% and 9%

146. **Metallurgy** The concentration of platinum in a necklace is 15%. If the necklace weighs 12 g, find the amount of platinum in the necklace.
1.8 g

147. **Dye Mixtures** A 250-milliliter solution of a fabric dye contains 5 ml of hydrogen peroxide. What is the percent concentration of the hydrogen peroxide?
2%

148. **Fabric Mixtures** A carpet is made with a blend of wool and other fibers. If the concentration of wool in the carpet is 75% and the carpet weighs 175 lb, how much wool is in the carpet?
131.25 lb

149. **Juice Mixtures** Apple Dan's 32-ounce apple-flavored fruit drink contains 8 oz of apple juice. A 40-ounce generic brand of an apple-flavored fruit drink contains 9 oz of apple juice. Which of the two brands has the greater concentration of apple juice?
Apple Dan

150. **Food Mixtures** Bakers use simple syrup in many of their recipes. Simple syrup is made by combining 500 g of sugar with 500 g of water and mixing it well until the sugar dissolves. What is the percent concentration of sugar in simple syrup?
50%

151. **Pharmacology** A pharmacist has 50 g of a topical cream that contains 75% glycerine. How many grams of the cream are not glycerine?
12.5 g

152. **Chemistry** A chemist has 100 ml of a solution that is 9% acetic acid. If the chemist adds 50 ml of pure water to this solution, what is the percent concentration of the resulting mixture?
6%

153. **Chemistry** A 500-gram salt-and-water solution contains 50 g of salt. This mixture is left in the open air, and 100 g of water evaporates from the solution. What is the percent concentration of salt in the remaining solution?
12.5%

OBJECTIVE E *To solve uniform motion problems*

154. Joe and John live 2 mi apart. They leave their houses at the same time and walk toward each other until they meet. Joe walks faster than John does.
   a. Is the distance walked by Joe less than, equal to, or greater than the distance walked by John?
   b. Is the time spent walking by Joe less than, equal to, or greater than the time spent walking by John?
   c. What is the total distance traveled by both Joe and John?
      a. Greater than   **b.** Equal to   **c.** 2 mi

155.  Morgan and Emma ride their bikes from Morgan's house to the store. Morgan begins biking 5 min before Emma begins. Emma bikes faster than Morgan and catches up with her just as they reach the store.
   a. Is the distance biked by Emma less than, equal to, or greater than the distance biked by Morgan?
   b. Is the time spent biking by Emma less than, equal to, or greater than the time spent biking by Morgan?
   a. Equal to   **b.** Less than

156. 🌐 **Trains** See the news clipping at the right. Find the time it will take the high-speed train to travel between the two cities. Round to the nearest tenth of an hour.   3.1 h

In the NEWS!

World's Fastest Train

China has unveiled the world's fastest rail link—a train that connects the cities of Guangzhou and Wuhan and can travel at speeds of up to 394.2 km/h. The distance between the two cities is 1069 km, and the train will travel that distance at an average speed of 350 km/h (217 mph). The head of the transport bureau at the Chinese railway ministry boasted, "It's the fastest train in operation in the world."
Source: news.yahoo.com

Bartłomiej Magierowski/Shutterstock.com

157. It takes a hospital dietician 40 min to drive from home to the hospital, a distance of 20 mi. What is the dietician's average rate of speed?
30 mph

158. As part of a training program for the Boston Marathon, a runner wants to build endurance by running at a rate of 9 mph for 20 min. How far will the runner travel in that time period?
3 mi

John Kropewnicki/Shutterstock.com

159. Marcella leaves home at 9:00 A.M. and drives to school, arriving at 9:45 A.M. If the distance between home and school is 27 mi, what is Marcella's average rate of speed?
36 mph

160. The Ride for Health Bicycle Club has chosen a 36-mile course for this Saturday's ride. If the riders plan on averaging 12 mph while they are riding, and they have a 1-hour lunch break planned, how long will it take them to complete the trip?
4 h

Arthur Eugene Preston/Shutterstock.com

161. Palmer's average running speed is 3 km/h faster than his walking speed. If Palmer can run around a 30-kilometer course in 2 h, how many hours would it take for Palmer to walk the same course?
2.5 h

162. A shopping mall has a moving sidewalk that takes shoppers from the shopping area to the parking garage, a distance of 250 ft. If your normal walking rate is 5 ft/s and the moving sidewalk is traveling at 3 ft/s, how many seconds would it take for you to walk from one end of the moving sidewalk to the other end?
31.25 s

163. Two joggers start at the same time from opposite ends of an 8-mile jogging trail and begin running toward each other. One jogger is running at a rate of 5 mph, and the other jogger is running at a rate of 7 mph. How long, in minutes, after they start will the two joggers meet?
40 min

164. 🔵 **sQuba**   See the news clipping at the right. Two sQubas are on opposite sides of a lake 1.6 mi wide. They start toward each other at the same time, one traveling on the surface of the water and the other traveling underwater. In how many minutes will the sQuba traveling on the surface of the water be directly above the sQuba traveling underwater? Assume they are traveling at top speed.
20 min

165. Two cyclists start from the same point at the same time and move in opposite directions. One cyclist is traveling at 8 mph, and the other cyclist is traveling at 9 mph. After 30 min, how far apart are the two cyclists?
8.5 mi

166. Petra and Celine can paddle their canoe at a rate of 10 mph in calm water. How long will it take them to travel 4 mi against the 2-mile-per-hour current of the river?
0.5 h

167. At 8:00 A.M., a train leaves a station and travels at a rate of 45 mph. At 9:00 A.M., a second train leaves the same station on the same track and travels in the direction of the first train at a speed of 60 mph. At 10:00 A.M., how far apart are the two trains?
30 mi

Critical Thinking

Solve.

168. $\dfrac{3y - 8y}{7} = 15$ -21

169. $\dfrac{2m + m}{5} = -9$ -15

170. $\dfrac{1}{\frac{1}{x}} + 8 = -19$ -27

171. $\dfrac{1}{\frac{1}{x}} = 5$ 5

172. $\dfrac{5}{7} - \dfrac{3}{7} = 6$ 21

173. $\dfrac{4}{\frac{3}{b}} = 8$ 6

174. Consumerism   Your bill for dinner, including a 7.25% sales tax, was $92.74. You want to leave a 15% tip on the cost of the dinner before the sales tax. Find the amount of the tip to the nearest dollar. $13

175. Business   A retailer decides to increase the original price of each item in the store by 10%. After the price increase, the retailer notices a significant drop in sales and so decides to reduce the current price of each item in the store by 10%. Are the prices back to the original prices? If not, are the prices lower or higher than the original prices?   No; lower

176. If a quantity increases by 100%, how many times its original value is the new value?   2 times

In the NEWS!

Underwater Driving —Not So Fast!
Swiss company Rinspeed, Inc., presented its new car, the sQuba, at the Geneva Auto Show. The sQuba can travel on land, on water, and underwater. With a new sQuba, you can expect top speeds of 77 mph when driving on land, 3 mph when driving on the surface of the water, and 1.8 mph when driving underwater!
Source: Seattle Times

QUICK QUIZ

1. Is $\frac{2}{3}$ a solution of $6x + 5 = 9$?   **Yes**   [2.1A]

Solve.

2. $a + 5 = -8$ **-13**   [2.1B]

3. $c + \dfrac{5}{6} = \dfrac{1}{3}$ **$-\dfrac{1}{2}$**   [2.1B]

4. $3x = -21$ **-7**   [2.1C]

5. $-12 = \dfrac{2}{3}x$ **-18**   [2.1C]

6. $8x - 3x = 30$ **6**   [2.1C]

7. 15% of what number is 30?   **200**   [2.1D]

8. A survey of chocolate truffle gourmets found that 77 people liked milk chocolate and 63 people liked dark chocolate. What percent of those surveyed liked milk chocolate?   **55%**   [2.1D]

9. One month, a credit card company charged $4.77 in interest on an unpaid credit card balance of $390. Find the annual interest rate on this credit card. Round to the nearest tenth of a percent.   **14.7%**   [2.1D]

10. Sweet Apple Drink contains 20% apple juice. Find the amount of Sweet Apple Drink in a bottle that contains 12.8 oz of apple juice.   **64 oz**   [2.1D]

11. Chu Min runs 1 mph faster than Sasha. If Chu Min can run 6 mi in 45 min, determine Sasha's running speed.   **7 mph**   [2.1E]

Unless otherwise noted, all content on this page is © Cengage Learning.

177. 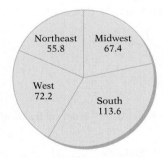 Employee A had an annual salary of $52,000, Employee B had an annual salary of $58,000, and Employee C had an annual salary of $56,000 before each employee was given a 5% raise. Which of the three employees now has the highest annual salary? Explain how you arrived at your answer.

178. Each of three employees earned an annual salary of $65,000 before Employee A was given a 3% raise, Employee B was given a 6% raise, and Employee C was given a 4.5% raise. Which of the three employees now has the highest annual salary? Explain how you arrived at your answer.

Projects or Group Activities

179. Make up an equation of the form $x + a = b$ that has 2 as a solution.   One possible answer is $x + 7 = 9$.

180. Make up an equation of the form $ax = b$ that has -2 as a solution.   One possible answer is $5x = -10$.

181. Two numbers form a "two-pair" if the sum of their reciprocals equals 2. For example, $\frac{8}{15}$ and 8 are a two-pair because $\frac{15}{8} + \frac{1}{8} = 2$. If two numbers a and b form a two-pair, and $a = \frac{7}{3}$, what is the value of b? $\frac{7}{11}$

182. Use the numbers 5, 10, and 15 to fill in the boxes in the equation $x + \square = \square - \square$.
  a. What is the largest solution possible?   0
  b. What is the smallest solution possible? -20

183. **U.S. Population**   The circle graph below shows the population of the United States, in millions, by region. (*Source:* U.S. Census Bureau)

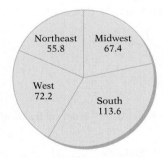

**U.S. Population by Region
(in millions of residents)**

  a. To the nearest tenth of a percent, what percent of the U.S. population lives in each region?   Northeast: 18.1%; Midwest: 21.8%; South: 36.8%; West: 23.4%
  b. Which region has the largest population? In which region does the largest percent of the population live?   South; South

According to the Census Bureau, California has the largest population of all the states, with 38 million. Wyoming, with 0.1683% of the U.S. population, has the least number of residents.
  c. What percent of the U.S. population lives in California? Round to the nearest tenth of a percent.   12.3%
  d. How many residents live in Wyoming? Round to the nearest ten thousand.   520,000 residents
  e. What percent of the U.S. population lives in the state you live in?   Answers will vary.

Unless otherwise noted, all content on this page is © Cengage Learning.

2.2   General Equations

OBJECTIVE A   *To solve an equation of the form $ax + b = c$*

In solving an equation of the form $ax + b = c$, the goal is to rewrite the equation in the form *variable = constant*. This requires the application of both the Addition and Multiplication Properties of Equations.

> **HOW TO 1**   Solve: $\dfrac{3}{4}x - 2 = -11$

The goal is to write the equation in the form *variable = constant*.

$$\frac{3}{4}x - 2 = -11$$

$$\frac{3}{4}x - 2 + 2 = -11 + 2 \qquad \bullet \text{ Add 2 to each side of the equation.}$$

$$\frac{3}{4}x = -9 \qquad \bullet \text{ Simplify.}$$

$$\frac{4}{3} \cdot \frac{3}{4}x = \frac{4}{3}(-9) \qquad \bullet \text{ Multiply each side of the equation by } \frac{4}{3}.$$

$$x = -12 \qquad \bullet \text{ The equation is in the form } variable = constant.$$

The solution is -12.

Take Note

Check:

$$\frac{3}{4}x - 2 = -11$$

$$\begin{array}{c|c} \frac{3}{4}(-12) - 2 & -11 \\ \hline -9 - 2 & -11 \\ -11 = -11 \end{array}$$

A true equation

Here is an example of solving an equation that contains more than one fraction.

> **HOW TO 2**   Solve: $\dfrac{2}{3}x + \dfrac{1}{2} = \dfrac{3}{4}$

$$\frac{2}{3}x + \frac{1}{2} = \frac{3}{4}$$

$$\frac{2}{3}x + \frac{1}{2} - \frac{1}{2} = \frac{3}{4} - \frac{1}{2} \qquad \bullet \text{ Subtract } \frac{1}{2} \text{ from each side of the equation.}$$

$$\frac{2}{3}x = \frac{1}{4} \qquad \bullet \text{ Simplify.}$$

$$\frac{3}{2}\left(\frac{2}{3}x\right) = \frac{3}{2}\left(\frac{1}{4}\right) \qquad \bullet \text{ Multiply each side of the equation by } \frac{3}{2}.$$

$$x = \frac{3}{8}$$

The solution is $\frac{3}{8}$.

INSTRUCTOR NOTE

On the next page, we solve the equation at the right by first *clearing denominators*. Introducing this method now will prepare students to solve equations that contain fractions with variables in the denominator.

It may be easier to solve an equation containing two or more fractions by multiplying each side of the equation by the least common multiple (LCM) of the denominators. For the equation above, the LCM of 3, 2, and 4 is 12. The LCM has the property that 3, 2, and 4 divide evenly into it. Therefore, if both sides of the equation are multiplied by 12, the denominators will divide evenly into 12. The result is an equation that does not contain any fractions. Multiplying each side of an equation that contains fractions by the LCM of the denominators is called **clearing denominators.** It is an alternative method, as we show in the next example, of solving an equation that contains fractions.

Take Note

This is the same example solved on the preceding page, but this time we are using the method of clearing denominators.

Observe that after we multiply both sides of the equation by the LCM of the denominators and then simplify, the equation no longer contains fractions.

Clearing denominators is a method of solving equations. The process applies only to equations, never to expressions.

HOW TO 3 Solve: $\dfrac{2}{3}x + \dfrac{1}{2} = \dfrac{3}{4}$

$$\frac{2}{3}x + \frac{1}{2} = \frac{3}{4}$$

$$12\left(\frac{2}{3}x + \frac{1}{2}\right) = 12\left(\frac{3}{4}\right)$$

• Multiply each side of the equation by 12, the LCM of 3, 2, and 4.

$$12\left(\frac{2}{3}x\right) + 12\left(\frac{1}{2}\right) = 12\left(\frac{3}{4}\right)$$

• Use the Distributive Property.

$$8x + 6 = 9$$

• Simplify.

$$8x + 6 - 6 = 9 - 6$$

• Subtract 6 from each side of the equation.

$$8x = 3$$

$$\frac{8x}{8} = \frac{3}{8}$$

• Divide each side of the equation by 8.

$$x = \frac{3}{8}$$

The solution is $\frac{3}{8}$.

INSTRUCTOR NOTE

One of the most common mistakes students make when solving an equation by clearing denominators is incorrectly applying the Distributive Property when multiplying both sides of the equation by the same number. We have shown this process in two steps to assist students.

Note that both methods give exactly the same solution. You may use either method to solve an equation containing fractions.

EXAMPLE 1

Solve: $3x - 7 = -5$

Solution

$$3x - 7 = -5$$

$$3x - 7 + 7 = -5 + 7$$

• Add 7 to each side.

$$3x = 2$$

$$\frac{3x}{3} = \frac{2}{3}$$

• Divide each side by 3.

$$x = \frac{2}{3}$$

The solution is $\frac{2}{3}$.

YOU TRY IT 1

Solve: $5x + 7 = 10$

Your solution

$\dfrac{3}{5}$

IN-CLASS EXAMPLES

Solve.

1. $8a + 3 = 10$ $\dfrac{7}{8}$

2. $7 = 12 + 5h$ -1

3. $\dfrac{2}{5}x - \dfrac{1}{4} = \dfrac{3}{2}$ $\dfrac{35}{8}$

4. $\dfrac{3}{8}x + \dfrac{2}{3} = \dfrac{5}{12}$ $-\dfrac{2}{3}$

5. $\dfrac{1}{3} - \dfrac{3}{5}x = \dfrac{1}{2}$ $-\dfrac{5}{18}$

EXAMPLE 2

Solve: $5 = 9 - 2x$

Solution

$$5 = 9 - 2x$$

$$5 - 9 = 9 - 9 - 2x$$

• Subtract 9 from each side.

$$-4 = -2x$$

$$\frac{-4}{-2} = \frac{-2x}{-2}$$

• Divide each side by −2.

$$2 = x$$

The solution is 2.

YOU TRY IT 2

Solve: $2 = 11 + 3x$

Your solution

-3

Solutions on p. S4

EXAMPLE 3

Solve: $\dfrac{2}{3} - \dfrac{x}{2} = \dfrac{3}{4}$

Solution

$$\dfrac{2}{3} - \dfrac{x}{2} = \dfrac{3}{4}$$

$$\dfrac{2}{3} - \dfrac{2}{3} - \dfrac{x}{2} = \dfrac{3}{4} - \dfrac{2}{3} \qquad \bullet \text{ Subtract } \tfrac{2}{3} \text{ from each side.}$$

$$-\dfrac{x}{2} = \dfrac{1}{12}$$

$$-2\left(-\dfrac{x}{2}\right) = -2\left(\dfrac{1}{12}\right) \qquad \bullet \text{ Multiply each side by } -2.$$

$$x = -\dfrac{1}{6}$$

The solution is $-\dfrac{1}{6}$.

YOU TRY IT 3

Solve: $\dfrac{5}{8} - \dfrac{2x}{3} = \dfrac{5}{4}$

Your solution

$-\dfrac{15}{16}$

EXAMPLE 4

Solve $\dfrac{4}{5}x - \dfrac{1}{2} = \dfrac{3}{4}$ by first clearing denominators.

Solution

The LCM of 5, 2, and 4 is 20.

$$\dfrac{4}{5}x - \dfrac{1}{2} = \dfrac{3}{4}$$

$$20\left(\dfrac{4}{5}x - \dfrac{1}{2}\right) = 20\left(\dfrac{3}{4}\right) \qquad \bullet \begin{array}{l}\textbf{Multiply each}\\ \textbf{side by 20.}\end{array}$$

$$20\left(\dfrac{4}{5}x\right) - 20\left(\dfrac{1}{2}\right) = 20\left(\dfrac{3}{4}\right) \qquad \bullet \begin{array}{l}\textbf{Use the Distributive}\\ \textbf{Property.}\end{array}$$

$$16x - 10 = 15$$

$$16x - 10 + 10 = 15 + 10 \qquad \bullet \begin{array}{l}\textbf{Add 10 to}\\ \textbf{each side.}\end{array}$$

$$16x = 25$$

$$\dfrac{16x}{16} = \dfrac{25}{16} \qquad \bullet \begin{array}{l}\textbf{Divide each}\\ \textbf{side by 16.}\end{array}$$

$$x = \dfrac{25}{16}$$

The solution is $\dfrac{25}{16}$.

YOU TRY IT 4

Solve $\dfrac{2}{3}x + 3 = \dfrac{7}{2}$ by first clearing denominators.

Your solution

$\dfrac{3}{4}$

INSTRUCTOR NOTE

One way to end this objective is to review the objective title, which is "to solve an equation of the form $ax + b = c$". If you ask students to name the variables of the equation, they may answer a, b, c, and x—and in a sense, that is true. However, as written symbolic math evolved, it became customary to think of letters at the beginning of the alphabet as constants and those at the end of the alphabet as variables. This kind of implicit understanding is often lost on students.

For this objective, the goal was to solve for x given a, b, and c. In the next objective, we solve $ax + b = cx + d$, again with the implicit understanding that a, b, c, and d are constants and coefficients.

Later in the text, we will introduce the equation $y = mx + b$, which also makes implicit assumptions about variables and constants. As students proceed through math courses, they will continually be exposed to the same kinds of understandings.

Solutions on pp. S4–S5

EXAMPLE 5

Solve: $2x + 4 - 5x = 10$

Solution

$$2x + 4 - 5x = 10$$

$$-3x + 4 = 10 \qquad \bullet \text{ Combine like terms.}$$

$$-3x + 4 - 4 = 10 - 4 \qquad \bullet \text{ Subtract 4 from each side.}$$

$$-3x = 6$$

$$\frac{-3x}{-3} = \frac{6}{-3} \qquad \bullet \text{ Divide each side by } -3.$$

$$x = -2$$

The solution is -2.

YOU TRY IT 5

Solve: $x - 5 + 4x = 25$

Your solution

6

Solution on p. S5

 OBJECTIVE B

To solve an equation of the form $ax + b = cx + d$

Tips for Success

Have you considered joining a study group? Getting together regularly with other students in the class to go over material and quiz each other can be very beneficial. See *AIM for Success* at the front of the book.

In solving an equation of the form $ax + b = cx + d$, the goal is to rewrite the equation in the form *variable = constant*. Begin by rewriting the equation so that there is only one variable term in the equation. Then rewrite the equation so that there is only one constant term.

HOW TO 4 Solve: $2x + 3 = 5x - 9$

$$2x + 3 = 5x - 9$$

$$2x - 5x + 3 = 5x - 5x - 9 \qquad \bullet \text{ Subtract } 5x \text{ from each side of the equation.}$$

$$-3x + 3 = -9 \qquad \bullet \text{ Simplify. There is only one variable term.}$$

$$-3x + 3 - 3 = -9 - 3 \qquad \bullet \text{ Subtract 3 from each side of the equation.}$$

$$-3x = -12 \qquad \bullet \text{ Simplify. There is only one constant term.}$$

$$\frac{-3x}{-3} = \frac{-12}{-3} \qquad \bullet \text{ Divide each side of the equation by } -3.$$

$$x = 4 \qquad \bullet \text{ The equation is in the form } \textit{variable = constant}.$$

The solution is 4. You should verify this by checking this solution.

EXAMPLE 6

Solve: $4x - 5 = 8x - 7$

Solution

$$4x - 5 = 8x - 7$$

$$4x - 8x - 5 = 8x - 8x - 7 \quad \bullet \text{ Subtract } 8x \text{ from each side.}$$

$$-4x - 5 = -7$$

$$-4x - 5 + 5 = -7 + 5 \quad \bullet \text{ Add 5 to each side.}$$

$$-4x = -2$$

$$\frac{-4x}{-4} = \frac{-2}{-4} \quad \bullet \text{ Divide each side by } -4.$$

$$x = \frac{1}{2}$$

The solution is $\frac{1}{2}$.

YOU TRY IT 6

Solve: $5x + 4 = 6 + 10x$

Your solution

$-\dfrac{2}{5}$

IN-CLASS EXAMPLES

Solve.

6. $5x - 4 = 3x - 10$ **−3**

7. $8x + 3 - 4x = 5 + x$ $\dfrac{2}{3}$

8. $3x - 7 = 5x - 7$ **0**

EXAMPLE 7

Solve: $3x + 4 - 5x = 2 - 4x$

Solution

$$3x + 4 - 5x = 2 - 4x$$

$$-2x + 4 = 2 - 4x \quad \bullet \text{ Combine like terms.}$$

$$-2x + 4 + 4x = 2 - 4x + 4x \quad \bullet \text{ Add } 4x \text{ to each side.}$$

$$2x + 4 = 2$$

$$2x + 4 - 4 = 2 - 4 \quad \bullet \text{ Subtract 4 from each side.}$$

$$2x = -2$$

$$\frac{2x}{2} = \frac{-2}{2} \quad \bullet \text{ Divide each side by 2.}$$

$$x = -1$$

The solution is -1.

YOU TRY IT 7

Solve: $5x - 10 - 3x = 6 - 4x$

Your solution

$\dfrac{8}{3}$

Solutions on p. S5

Unless otherwise noted, all content on this page is © Cengage Learning.

OBJECTIVE C | *To solve an equation containing parentheses*

INSTRUCTOR NOTE
Remind students that the goal is still *variable = constant*.

When an equation contains parentheses, one of the steps in solving the equation is to use the Distributive Property. The Distributive Property is used to remove parentheses from a variable expression.

HOW TO 5 Solve: $4 + 5(2x - 3) = 3(4x - 1)$

$$4 + 5(2x - 3) = 3(4x - 1)$$

$$4 + 10x - 15 = 12x - 3$$ • Use the Distributive Property. Then simplify.

$$10x - 11 = 12x - 3$$

$$10x - 12x - 11 = 12x - 12x - 3$$ • Subtract **12x** from each side of the equation.

$$-2x - 11 = -3$$ • Simplify.

$$-2x - 11 + 11 = -3 + 11$$ • Add **11** to each side of the equation.

$$-2x = 8$$ • Simplify.

$$\frac{-2x}{-2} = \frac{8}{-2}$$ • Divide each side of the equation by **−2**.

$$x = -4$$ • The equation is in the form *variable = constant*.

The solution is −4. You should verify this by checking this solution.

In the next example, we solve an equation containing parentheses and decimals.

HOW TO 6 Solve: $16 + 0.55x = 0.75(x + 20)$

$$16 + 0.55x = 0.75(x + 20)$$

$$16 + 0.55x = 0.75x + 15$$ • Use the Distributive Property.

$$16 + 0.55x - 0.75x = 0.75x - 0.75x + 15$$ • Subtract **0.75x** from each side of the equation.

$$16 - 0.20x = 15$$ • Simplify.

$$16 - 16 - 0.20x = 15 - 16$$ • Subtract **16** from each side of the equation.

$$-0.20x = -1$$ • Simplify.

$$\frac{-0.20x}{-0.20} = \frac{-1}{-0.20}$$ • Divide each side of the equation by **−0.20**.

$$x = 5$$ • The equation is in the form *variable = constant*.

The solution is 5.

EXAMPLE 8

Solve: $3x - 4(2 - x) = 3(x - 2) - 4$

Solution

$3x - 4(2 - x) = 3(x - 2) - 4$

$3x - 8 + 4x = 3x - 6 - 4$ • **Distributive Property**

$7x - 8 = 3x - 10$

$7x - 3x - 8 = 3x - 3x - 10$ • **Subtract 3x.**

$4x - 8 = -10$

$4x - 8 + 8 = -10 + 8$ • **Add 8.**

$4x = -2$

$\dfrac{4x}{4} = \dfrac{-2}{4}$ • **Divide by 4.**

$x = -\dfrac{1}{2}$

The solution is $-\frac{1}{2}$.

YOU TRY IT 8

Solve: $5x - 4(3 - 2x) = 2(3x - 2) + 6$

Your solution

2

IN-CLASS EXAMPLES

9. Solve: $9x - 3(2x + 5) = 4(5x + 2) - 6$ **−1**

10. Solve: $5[6 - 2(5x + 1)] = 8x - 9$ $\dfrac{1}{2}$

11. If $5x = 3x + 10$, evaluate $4x^2 - 10$.   **90**

EXAMPLE 9

Solve: $3[2 - 4(2x - 1)] = 4x - 10$

Solution

$3[2 - 4(2x - 1)] = 4x - 10$

$3[2 - 8x + 4] = 4x - 10$ • **Distributive Property**

$3[6 - 8x] = 4x - 10$

$18 - 24x = 4x - 10$ • **Distributive Property**

$18 - 24x - 4x = 4x - 4x - 10$ • **Subtract 4x.**

$18 - 28x = -10$

$18 - 18 - 28x = -10 - 18$ • **Subtract 18.**

$-28x = -28$

$\dfrac{-28x}{-28} = \dfrac{-28}{-28}$ • **Divide by −28.**

$x = 1$

The solution is 1.

YOU TRY IT 9

Solve: $-2[3x - 5(2x - 3)] = 3x - 8$

Your solution

2

OBJECTIVE D   *To solve application problems using formulas*

Take Note

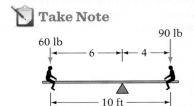

This system balances because

$F_1x = F_2(d - x)$
$60(6) = 90(10 - 6)$
$60(6) = 90(4)$
$360 = 360$

A lever system is shown at the right. It consists of a lever, or bar; a fulcrum; and two forces, F_1 and F_2. The distance d represents the length of the lever, x represents the distance from F_1 to the fulcrum, and $d - x$ represents the distance from F_2 to the fulcrum.

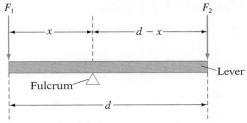

A principle of physics states that when the lever system balances, $F_1x = F_2(d - x)$.

EXAMPLE 10

A lever is 15 ft long. A force of 50 lb is applied to one end of the lever, and a force of 100 lb is applied to the other end. Where is the fulcrum located when the system balances?

Strategy

Make a drawing.

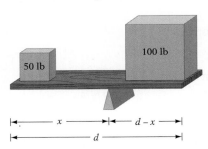

Given: $F_1 = 50$
$F_2 = 100$
$d = 15$
Unknown: x

Solution

$$F_1x = F_2(d - x)$$
$$50x = 100(15 - x) \quad \bullet \ F_1 = 50, F_2 = 100, d = 15$$
$$50x = 1500 - 100x$$
$$50x + 100x = 1500 - 100x + 100x \quad \bullet \ \text{Add } 100x.$$
$$150x = 1500$$
$$\frac{150x}{150} = \frac{1500}{150} \quad \bullet \ \text{Divide by 150.}$$
$$x = 10$$

The fulcrum is 10 ft from the 50-pound force.

YOU TRY IT 10

A lever is 25 ft long. A force of 45 lb is applied to one end of the lever, and a force of 80 lb is applied to the other end. Where is the location of the fulcrum when the system balances?

Your strategy

Your solution

16 ft from the 45-pound force

IN-CLASS EXAMPLES

12. A lever is 12 ft long. A force of 3 lb is applied to one end of the lever, and a force of 6 lb is applied to the other end. Where is the location of the fulcrum when the system balances?   **8 ft from the 3-pound force**

Solution on p. S5

Unless otherwise noted, all content on this page is © Cengage Learning.

2.2 EXERCISES

✓ Concept Check

SUGGESTED ASSIGNMENT
Exercises 1–8; Exercises 9–79,
odds; Exercises 89–157, odds

1. Match each equation with the first step in solving that equation.

a. $3x - 7 = 5$ **i.** Add 7 to each side.
b. $4x + 7 = -5$ **ii.** Add 5 to each side.
c. $7x - 5 = 2$ **iii.** Subtract 7 from each side.
d. $-7x + 5 = -2$ **iv.** Subtract 5 from each side.

a and i, b and iii, c and ii, d and iv

2. True or false? An equation of the form $ax + b = c$ cannot be solved if a is a negative number.   False

3. The first step in solving the equation $5 + 8x = 29$ is to subtract ___5___ from each side of the equation. The second step is to divide each side of the equation by ___8___.

4. To clear denominators from the equation $\frac{x}{9} + 2 = \frac{1}{6}$, multiply each side of the equation by ___18___, the least common multiple of the denominators 9 and 6.

5. True or false? The same variable term can be added to both sides of an equation without changing the solution of the equation.   True

6. True or false? The same variable term can be subtracted from both sides of an equation without changing the solution of the equation.   True

7. ✍ Describe the step that will enable you to rewrite the equation $2x - 3 = 7x + 12$ so that it has one variable term with a positive coefficient.   Subtract 2x from each side.

8. If you rewrite the equation $8 - y = y + 6$ so that it has one variable term on the left side of the equation, what will be the coefficient of the variable?   −2

OBJECTIVE A   *To solve an equation of the form $ax + b = c$*

For Exercises 9 to 80, solve and check.

9. $3x + 1 = 10$
3

10. $4y + 3 = 11$
2

11. $2a - 5 = 7$
6

12. $5m - 6 = 9$
3

13. $5 = 4x + 9$
−1

14. $2 = 5b + 12$
−2

15. $2x - 5 = -11$
−3

16. $3n - 7 = -19$
−4

17. $4 - 3w = -2$
2

18. $5 - 6x = -13$
3

19. $8 - 3t = 2$
2

20. $12 - 5x = 7$
1

21. $4a - 20 = 0$
5

22. $3y - 9 = 0$
3

23. $6 + 2b = 0$
−3

24. $10 + 5m = 0$
−2

25. $-2x + 5 = -7$
6

26. $-5d + 3 = -12$
3

27. $-1.2x + 3 = -0.6$
3

28. $-1.3 = -1.1y + 0.9$
2

29. $2 = 7 - 5a$
1

30. $3 = 11 - 4n$
2

31. $-35 = -6b + 1$
6

32. $-8x + 3 = -29$
4

33. $-3m - 21 = 0$
-7

34. $-5x - 30 = 0$
-6

35. $-4y + 15 = 15$
0

36. $-3x + 19 = 19$
0

37. $9 - 4x = 6$
$\dfrac{3}{4}$

38. $3t - 2 = 0$
$\dfrac{2}{3}$

39. $9x - 4 = 0$
$\dfrac{4}{9}$

40. $7 - 8z = 0$
$\dfrac{7}{8}$

41. $1 - 3x = 0$
$\dfrac{1}{3}$

42. $9d + 10 = 7$
$-\dfrac{1}{3}$

43. $12w + 11 = 5$
$-\dfrac{1}{2}$

44. $6y - 5 = -7$
$-\dfrac{1}{3}$

45. $8b - 3 = -9$
$-\dfrac{3}{4}$

46. $5 - 6m = 2$
$\dfrac{1}{2}$

47. $7 - 9a = 4$
$\dfrac{1}{3}$

48. $9 = -12c + 5$
$-\dfrac{1}{3}$

49. $10 = -18x + 7$
$-\dfrac{1}{6}$

50. $5y + \dfrac{3}{7} = \dfrac{3}{7}$
0

51. $9x + \dfrac{4}{5} = \dfrac{4}{5}$
0

52. $0.8 = 7d + 0.1$
0.1

53. $0.9 = 10x - 0.6$
0.15

54. $-6y + 5 = 13$
$-\dfrac{4}{3}$

55. $-4x + 3 = 9$
$-\dfrac{3}{2}$

56. $\dfrac{1}{2}a - 3 = 1$
8

57. $\dfrac{1}{3}m - 1 = 5$
18

58. $\dfrac{2}{5}y + 4 = 6$
5

59. $\dfrac{3}{4}n + 7 = 13$
8

60. $-\dfrac{2}{3}x + 1 = 7$
-9

61. $-\dfrac{3}{8}b + 4 = 10$
-16

62. $\dfrac{x}{4} - 6 = 1$
28

63. $\dfrac{y}{5} - 2 = 3$
25

64. $\dfrac{2x}{3} - 1 = 5$
9

65. $\dfrac{2}{3}x - \dfrac{5}{6} = -\dfrac{1}{3}$
$\dfrac{3}{4}$

66. $\dfrac{5}{4}x + \dfrac{2}{3} = \dfrac{1}{4}$
$-\dfrac{1}{3}$

67. $\dfrac{1}{2} - \dfrac{2}{3}x = \dfrac{1}{4}$
$\dfrac{3}{8}$

68. $\dfrac{3}{4} - \dfrac{3}{5}x = \dfrac{19}{20}$
$-\dfrac{1}{3}$

69. $\dfrac{3}{2} = \dfrac{5}{6} + \dfrac{3x}{8}$
$\dfrac{16}{9}$

70. $-\dfrac{1}{4} = \dfrac{5}{12} + \dfrac{5x}{6}$
$-\dfrac{4}{5}$

71. $\dfrac{11}{27} = \dfrac{4}{9} - \dfrac{2x}{3}$
$\dfrac{1}{18}$

72. $\dfrac{37}{24} = \dfrac{7}{8} - \dfrac{5x}{6}$
$-\dfrac{4}{5}$

73. $7 = \dfrac{2x}{5} + 4$

$\dfrac{15}{2}$

74. $5 - \dfrac{4c}{7} = 8$

$-\dfrac{21}{4}$

75. $7 - \dfrac{5}{9}y = 9$

$-\dfrac{18}{5}$

76. $6a + 3 + 2a = 11$

1

77. $5y + 9 + 2y = 23$

2

78. $7x - 4 - 2x = 6$

2

79. $11z - 3 - 7z = 9$

3

80. $2x - 6x + 1 = 9$

-2

For Exercises 81 to 84, without solving the equation, determine whether the solution is positive or negative.

81. $15x + 73 = -347$
Negative

82. $17 = 25 - 40a$
Positive

83. $290 + 51n = 187$
Negative

84. $-72 = -86y + 49$
Positive

85. Solve $3x + 4y = 13$ when $y = -2$.
$x = 7$

86. Solve $2x - 3y = 8$ when $y = 0$.
$x = 4$

87. If $4 - 5x = -1$, evaluate $x^2 - 3x + 1$.
-1

OBJECTIVE B *To solve an equation of the form $ax + b = cx + d$*

For Exercises 88 to 114, solve and check.

88. $8x + 5 = 4x + 13$
2

89. $6y + 2 = y + 17$
3

90. $5x - 4 = 2x + 5$
3

91. $13b - 1 = 4b - 19$
-2

92. $15x - 2 = 4x - 13$
-1

93. $7a - 5 = 2a - 20$
-3

94. $3x + 1 = 11 - 2x$
2

95. $n - 2 = 6 - 3n$
2

96. $2x - 3 = -11 - 2x$
-2

97. $4y - 2 = -16 - 3y$
-2

98. $0.2b + 3 = 0.5b + 12$
-30

99. $m + 0.4 = 3m + 0.8$
-0.2

100. $4y - 8 = y - 8$
0

101. $5a + 7 = 2a + 7$
0

102. $6 - 5x = 8 - 3x$
-1

103. $10 - 4n = 16 - n$
-2

104. $5 + 7x = 11 + 9x$
-3

105. $3 - 2y = 15 + 4y$
-2

106. $2x - 4 = 6x$
-1

107. $2b - 10 = 7b$
-2

108. $8m = 3m + 20$
4

109. $9y = 5y + 16$
4

110. $8b + 5 = 5b + 7$
$\dfrac{2}{3}$

111. $6y - 1 = 2y + 2$
$\dfrac{3}{4}$

112. $7x - 8 = x - 3$
$\dfrac{5}{6}$

113. $2y - 7 = -1 - 2y$
$\dfrac{3}{2}$

114. $2m - 1 = -6m + 5$
$\dfrac{3}{4}$

115. If $5x = 3x - 8$, evaluate $4x + 2$.
-14

116. If $7x + 3 = 5x - 7$, evaluate $3x - 2$.
-17

117. If $2 - 6a = 5 - 3a$, evaluate $4a^2 - 2a + 1$.
7

118. If $1 - 5c = 4 - 4c$, evaluate $3c^2 - 4c + 2$.
41

OBJECTIVE C *To solve an equation containing parentheses*

119. Without solving any of the equations, determine which of the following equations has the same solution as the equation $5 - 2(x - 1) = 8$.
(i) $3(x - 1) = 8$ **(ii)** $5 - 2x + 2 = 8$ **(iii)** $5 - 2x + 1 = 8$ ii

For Exercises 120 to 140, solve and check.

120. $5x + 2(x + 1) = 23$
3

121. $6y + 2(2y + 3) = 16$
1

122. $9n - 3(2n - 1) = 15$
4

123. $12x - 2(4x - 6) = 28$
4

124. $7a - (3a - 4) = 12$
2

125. $9m - 4(2m - 3) = 11$
-1

126. $5(3 - 2y) + 4y = 3$
2

127. $4(1 - 3x) + 7x = 9$
-1

128. $5y - 3 = 7 + 4(y - 2)$
2

129. $0.22(x + 6) = 0.2x + 1.8$
24

130. $0.05(4 - x) + 0.1x = 0.32$
2.4

131. $0.3x + 0.3(x + 10) = 300$
495

132. $2a - 5 = 4(3a + 1) - 2$
$-\dfrac{7}{10}$

133. $5 - (9 - 6x) = 2x - 2$
$\dfrac{1}{2}$

134. $7 - (5 - 8x) = 4x + 3$
$\dfrac{1}{4}$

135. $3[2 - 4(y - 1)] = 3(2y + 8)$
$-\dfrac{1}{3}$

136. $5[2 - (2x - 4)] = 2(5 - 3x)$
5

137. $3a + 2[2 + 3(a - 1)] = 2(3a + 4)$
$\dfrac{10}{3}$

138. $5 + 3[1 + 2(2x - 3)] = 6(x + 5)$
$\dfrac{20}{3}$

139. $-2[4 - (3b + 2)] = 5 - 2(3b + 6)$
$-\dfrac{1}{4}$

140. $-4[x - 2(2x - 3)] + 1 = 2x - 3$
2

141. If $4 - 3a = 7 - 2(2a + 5)$, evaluate $a^2 + 7a$.
0

142. If $9 - 5x = 12 - (6x + 7)$, evaluate $x^2 - 3x - 2$.
26

OBJECTIVE D   *To solve application problems using formulas*

Taxi Fares   The fare F to be charged a customer by a taxi company is calculated using the formula $F = 2.50 + 2.30(m - 1)$, where m is the number of miles traveled. Use this formula for Exercises 143 and 144.

143. A customer is charged $14.00. How many miles was the customer driven?   6 mi

144. A passenger is charged $20.90. Find the number of miles the passenger was driven.   9 mi

145. 🦵 **Physics**   Two people sit on a seesaw that is 8 ft long. The seesaw balances when the fulcrum is 3 ft from one of the people.
  a. How far is the fulcrum from the other person?
  b. Which person is heavier, the person who is 3 ft from the fulcrum or the other person?
  c. If the two people switch places, will the seesaw still balance?
       a. 5 ft   **b.** The person who is 3 ft from the fulcrum   **c.** No

Physics   For Exercises 146 to 151, use the lever system equation $F_1x = F_2(d - x)$.

146. A lever 10 ft long is used to move a 100-pound rock. The fulcrum is placed 2 ft from the rock. What force must be applied to the other end of the lever to move the rock?
25 lb

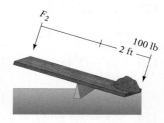

147. An adult and a child are on a seesaw 14 ft long. The adult weighs 175 lb and the child weighs 70 lb. How many feet from the child must the fulcrum be placed so that the seesaw balances?
10 ft

148. Two people are sitting 15 ft apart on a seesaw. One person weighs 180 lb. The second person weighs 120 lb. How far from the 180-pound person should the fulcrum be placed so that the seesaw balances?
6 ft

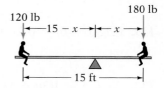

149. Two children are sitting on a seesaw that is 12 ft long. One child weighs 60 lb. The other child weighs 90 lb. How far from the 90-pound child should the fulcrum be placed so that the seesaw balances?
4.8 ft

150. In preparation for a stunt, two acrobats are standing on a plank 18 ft long. One acrobat weighs 128 lb and the second acrobat weighs 160 lb. How far from the 128-pound acrobat must the fulcrum be placed so that the acrobats are balanced on the plank?
10 ft

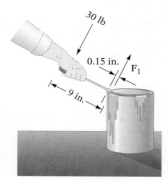

151. A screwdriver 9 in. long is used as a lever to open a can of paint. The tip of the screwdriver is placed under the lip of the can with the fulcrum 0.15 in. from the lip. A force of 30 lb is applied to the other end of the screwdriver. Find the force on the lip of the can.
1770 lb

Break-even Point   To determine the break-even point, or the number of units that must be sold so that no profit or loss occurs, an economist uses the formula $Px = Cx + F$, where P is the selling price per unit, x is the number of units that must be sold to break even, C is the cost to make each unit, and F is the fixed cost. Use this equation for Exercises 152 to 155.

152. A business analyst has determined that the selling price per unit for a laser printer is $1600. The cost to make one laser printer is $950, and the fixed cost is $211,250. Find the break-even point.
325 laser printers

Unless otherwise noted, all content on this page is © Cengage Learning.

153. A business analyst has determined that the selling price per unit for a gas barbecue is $325. The cost to make one gas barbecue is $175, and the fixed cost is $39,000. Find the break-even point.
260 barbecues

154. A manufacturer of headphones determines that the cost per unit for a pair of headphones is $38 and that the fixed cost is $24,400. The selling price for the headphones is $99. Find the break-even point.
400 headphones

155. A manufacturing engineer determines that the cost per unit for a soprano recorder is $12 and that the fixed cost is $19,240. The selling price for the recorder is $49. Find the break-even point.
520 recorders

Physiology   The oxygen consumption C, in millimeters per minute, of a small mammal at rest is related to the animal's weight m, in kilograms, by the equation $m = \frac{1}{6}(C - 5)$. Use this equation for Exercises 156 and 157.

156. What is the oxygen consumption of a mammal that weighs 10.4 kg?
67.4 ml/min

157. What is the oxygen consumption of a mammal that weighs 8.3 kg?
54.8 ml/min

Critical Thinking

For Exercises 158 to 161, solve. If the equation has no solution, write "No solution."

158. $3(2x - 1) - (6x - 4) = -9$ No solution

159. $\frac{1}{5}(25 - 10b) + 4 = \frac{1}{3}(9b - 15) - 6$ 4

160. $3[4(w + 2) - (w + 1)] = 5(2 + w)$ $-\frac{11}{4}$

161. $\frac{2(5x - 6) - 3(x - 4)}{7} = x + 2$ No solution

162. One-half of a certain number equals two-thirds of the same number. Find the number.
0

163. ◆ Does the sentence "Solve $3x - 4(x - 1)$" make sense? Why or why not?

164. ◆ The equation $x = x + 1$ has no solution, whereas the solution of the equation $2x + 3 = 3$ is zero. Is there a difference between no solution and a solution of zero? Explain your answer.

165. I am thinking of a number. When I subtract 4 from the number and then take 300% of the result, my new result is equal to the original number. What is the original number?   6

Projects or Group Activities

166. If $s = 5x - 3$ and $t = x + 4$, find the value of x for which $s = 3t - 1$.   7

167. The population of the town of Hampton increased by 10,000 people during the 1990s. In the first decade of the new millennium, the population of Hampton decreased by 10%, at which time the town had 6000 more people than at the beginning of the 1990s. Find Hampton's population at the beginning of the 1990s.
30,000 people

QUICK QUIZ

Solve.
1. $7b + 5 = 61$ 8   [2.2A]
2. $12 - 4c = 15$
   $-\frac{3}{4}$ [2.2A]
3. $\frac{4}{5}x + 3 = 7$ 5 [2.2A]
4. $-3 = 6m + 4 + m$
   -1 [2.2A]
5. $7x + 4 = 3x - 20$
   -6 [2.2B]
6. $4x + 5 = 23 - 2x$
   3   [2.2B]
7. $4x + 2(x + 3) = 3(x - 1)$
   -3 [2.2C]
8. Two people are sitting on a seesaw that is 9 ft long. One person weighs 120 lb. The second person weighs 150 lb. How far from the 120-pound person should the fulcrum be placed so that the seesaw balances?
   5 ft   [2.2D]

© iStockphoto.com/Rich Legg

2.3 Translating Sentences into Equations

OBJECTIVE A *To solve integer problems*

An equation states that two mathematical expressions are equal. Therefore, to **translate** a sentence into an equation, we must recognize the words or phrases that mean "equals." Some of these phrases are listed below.

$$\left.\begin{array}{l} \text{equals} \\ \text{is} \\ \text{is equal to} \\ \text{amounts to} \\ \text{represents} \end{array}\right\} \text{ translate to } =$$

Once the sentence is translated into an equation, the equation can be solved by rewriting it in the form *variable = constant*.

 Take Note

You can check the solution to a translation problem.

Check:

5 less than 18 is 13

$$\begin{array}{c|c} 18 - 5 & 13 \\ \hline & 13 = 13 \end{array}$$

HOW TO 1 Translate "five less than a number is thirteen" into an equation and solve.

Five less than a number	is	thirteen

- Find two verbal expressions for the same value.

The unknown number: n

- Assign a variable to the unknown number.

$$n - 5 \quad = \quad 13$$

- Write a mathematical expression for each verbal expression. Write the equals sign.

$$n - 5 + 5 = 13 + 5$$

$$n = 18$$

- Solve the equation.

The number is 18.

Recall that the integers are the numbers {..., −4, −3, −2, −1, 0, 1, 2, 3, 4, ...}. An **even integer** is an integer that is divisible by 2. Examples of even integers are −8, 0, and 22. An **odd integer** is an integer that is not divisible by 2. Examples of odd integers are −17, 1, and 39.

Consecutive integers are integers that follow one another in order. Examples of consecutive integers are shown at the right. (Assume that the variable *n* represents an integer.)	11, 12, 13 −8, −7, −6 $n, n + 1, n + 2$

Examples of **consecutive even integers** are shown at the right. (Assume that the variable *n* represents an even integer.)	24, 26, 28 −10, −8, −6 $n, n + 2, n + 4$

 Take Note

Both consecutive even and consecutive odd integers are represented using *n*, *n* + 2, *n* + 4, ...

Examples of **consecutive odd integers** are shown at the right. (Assume that the variable *n* represents an odd integer.)	19, 21, 23 −1, 1, 3 $n, n + 2, n + 4$

INSTRUCTOR NOTE
There is an opportunity, in solving problems that involve consecutive even or consecutive odd integers, to emphasize the importance of validating the solution. Give students the problem "The sum of two consecutive odd integers is eleven. Find the integers." In this case $x = 4.5$, which is not an odd integer. Therefore, the problem has no solution.

HOW TO 2   The sum of three consecutive odd integers is forty-five. Find the integers.

Strategy

- First odd integer: n
  Second odd integer: $n + 2$
  Third odd integer: $n + 4$
- The sum of the three odd integers is 45.

- Represent three consecutive odd integers.

Solution

$$n + (n + 2) + (n + 4) = 45$$
$$3n + 6 = 45$$
$$3n = 39$$
$$n = 13$$
$$n + 2 = 13 + 2 = 15$$
$$n + 4 = 13 + 4 = 17$$

- Write an equation.
- Solve the equation.

- The first odd integer is 13.
- Find the second odd integer.
- Find the third odd integer.

The three consecutive odd integers are 13, 15, and 17.

EXAMPLE 1

The sum of two numbers is sixteen. The difference between four times the smaller number and two is two more than twice the larger number. Find the two numbers.

Strategy

The difference between four times the smaller number and two	is	two more than twice the larger number

The smaller number: n
The larger number: $16 - n$

Solution

$$4n - 2 = 2(16 - n) + 2$$
$$4n - 2 = 32 - 2n + 2$$
$$4n - 2 = 34 - 2n$$
$$4n + 2n - 2 = 34 - 2n + 2n$$
$$6n - 2 = 34$$
$$6n - 2 + 2 = 34 + 2$$
$$6n = 36$$
$$\frac{6n}{6} = \frac{36}{6}$$
$$n = 6$$
$$16 - n = 16 - 6 = 10$$

- Distributive Property
- Combine like terms.
- Add $2n$ to each side.

- Add 2 to each side.

- Divide each side by 6.

- The smaller number is 6.
- Find the larger number.

The smaller number is 6.
The larger number is 10.

YOU TRY IT 1

The sum of two numbers is twelve. The total of three times the smaller number and six amounts to seven less than the product of four and the larger number. Find the two numbers.

Your strategy

Your solution

5, 7

IN-CLASS EXAMPLES

Translate into an equation and solve.
1. The sum of two numbers is twenty-five. The total of four times the smaller number and two is six less than the product of two and the larger number. Find the two numbers.
   $4x + 2 = 2(25 - x) - 6$; 7, 18
2. Three times the largest of three consecutive integers is ten more than the sum of the other two. Find the three integers.
   $3(n + 2) = n + (n + 1) + 10$; 5, 6, 7

Solution on pp. S5–S6

EXAMPLE 2

Find three consecutive even integers such that three times the second equals four more than the sum of the first and third.

Strategy

• First even integer: n
  Second even integer: $n + 2$
  Third even integer: $n + 4$
• Three times the second equals four more than the sum of the first and third.

Solution

$$3(n + 2) = n + (n + 4) + 4$$
$$3n + 6 = 2n + 8$$
$$3n - 2n + 6 = 2n - 2n + 8$$
$$n + 6 = 8$$
$$n = 2 \qquad \text{• The first even integer is 2.}$$
$$n + 2 = 2 + 2 = 4 \qquad \text{• Find the second even integer.}$$
$$n + 4 = 2 + 4 = 6 \qquad \text{• Find the third even integer.}$$

The three integers are 2, 4, and 6.

YOU TRY IT 2

Find three consecutive integers whose sum is negative six.

Your strategy

Your solution

$-3, -2, -1$

Solution on p. S6

OBJECTIVE B *To translate a sentence into an equation and solve*

EXAMPLE 3

A wallpaper hanger charges a fee of $25 plus $12 for each roll of wallpaper used in a room. If the total charge for hanging wallpaper is $97, how many rolls of wallpaper were used?

Strategy

To find the number of rolls of wallpaper used, write and solve an equation using n to represent the number of rolls of wallpaper used.

$25 plus $12 for each roll of wallpaper	is	$97

Solution

$$25 + 12n = 97$$
$$12n = 72 \qquad \text{• Subtract 25 from each side.}$$
$$\frac{12n}{12} = \frac{72}{12} \qquad \text{• Divide each side by 12.}$$
$$n = 6$$

6 rolls of wallpaper were used.

YOU TRY IT 3

The fee charged by a ticketing agency for a concert is $3.50 plus $17.50 for each ticket purchased. If your total charge for tickets is $161, how many tickets did you purchase?

Your strategy

Your solution

9 tickets

IN-CLASS EXAMPLES

3. An electric company charges $.07 for each of the first 249 kWh (kilowatt-hours) and $.14 for each kilowatt-hour over 249 kWh. Find the number of kilowatt-hours used by a family whose electric bill was $46.55. **457 kWh**

4. A piano wire 24 in. long is cut into two pieces. The length of the longer piece is 4 in. more than three times the length of the shorter piece. Find the length of each piece. **5 in., 19 in.**

Solution on p. S6

EXAMPLE 4

A board 20 ft long is cut into two pieces. Five times the length of the shorter piece is 2 ft more than twice the length of the longer piece. Find the length of each piece.

Strategy

Let x represent the length of the shorter piece. Then $20 - x$ represents the length of the longer piece.

Make a drawing.

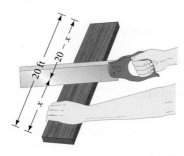

To find the lengths, write and solve an equation using x to represent the length of the shorter piece and $20 - x$ to represent the length of the longer piece.

Five times the length of the shorter piece	is	2 ft more than twice the length of the longer piece

Solution

$$5x = 2(20 - x) + 2$$
$$5x = 40 - 2x + 2 \quad \bullet \text{ Distributive Property}$$
$$5x = 42 - 2x \quad \bullet \text{ Combine like terms.}$$
$$5x + 2x = 42 - 2x + 2x \quad \bullet \text{ Add } 2x \text{ to each side.}$$
$$7x = 42$$
$$\frac{7x}{7} = \frac{42}{7} \quad \bullet \text{ Divide each side by 7.}$$
$$x = 6 \quad \bullet \text{ The shorter piece is 6 ft long.}$$
$$20 - x = 20 - 6 = 14 \quad \bullet \text{ Find the length of the longer piece.}$$

The length of the shorter piece is 6 ft.
The length of the longer piece is 14 ft.

YOU TRY IT 4

A wire 22 in. long is cut into two pieces. The length of the longer piece is 4 in. more than twice the length of the shorter piece. Find the length of each piece.

Your strategy

$22 - n$

$4n + 2$

INSTRUCTOR NOTE

Problems that begin with sentences such as "The sum of two numbers is twelve" and "A board 12 ft long is cut into two pieces" can have their two constituent parts represented identically, using n and $12 - n$.

Your solution

6 in., 16 in.

INSTRUCTOR NOTE

Some students have difficulty making the transition from the sentence "The sum of two numbers is sixteen" to the conclusion that if the first number is represented as n, the second number may be represented as $16 - n$. The confusion is due in part to the use of the word *sum* and the fact that the representation of the two numbers does not involve addition. It can be pointed out, however, that the sum of n and $(16 - n)$ is, in fact, 16.

Solution on p. S6

Unless otherwise noted, all content on this page is © Cengage Learning.

2.3 EXERCISES

✔ Concept Check

SUGGESTED ASSIGNMENT
Exercises 1–8; Exercises 9–47, odds

For Exercises 1 to 3, determine whether the statement is true or false.

1. When translating a sentence into an equation, we can use any variable to represent an unknown number.   True

2. An even integer is a multiple of 2.   True

3. Given the consecutive odd integers −5 and −3, the next consecutive odd integer is −1.   True

4. The sum of two numbers is 12.
   a. If x represents the larger number, represent the smaller number in terms of x.
   $12 - x$
   b. If x represents the smaller number, represent the larger number in terms of x.
   $12 - x$

5. When we translate a sentence into an equation, the word *is* translates into the __equals__ sign.

6. Integers that follow one another in order are called _____ integers.   consecutive

7. Two consecutive integers differ by __1__. Two consecutive even integers differ by __2__. Two consecutive odd integers differ by __2__.

8. The number of calories in a cup of low-fat milk is two-thirds the number of calories in a cup of whole milk. In this situation, let n represent the number of calories in a cup of __whole__ milk, and let $\frac{2}{3}n$ represent the number of calories in a cup of __low-fat__ milk.

OBJECTIVE A   *To solve integer problems*

For Exercises 9 to 24, translate into an equation and solve.

9. The difference between a number and fifteen is seven. Find the number.
   $x - 15 = 7; 22$

10. The sum of five and a number is three. Find the number.
    $5 + x = 3; -2$

11. The difference between nine and a number is seven. Find the number.
    $9 - x = 7; 2$

12. Three-fifths of a number is negative thirty. Find the number.
    $\frac{3}{5}x = -30; -50$

13. The difference between five and twice a number is one. Find the number.
    $5 - 2x = 1; 2$

14. Four more than three times a number is thirteen. Find the number.
    $3x + 4 = 13; 3$

15. The sum of twice a number and five is fifteen. Find the number.
    $2x + 5 = 15; 5$

16. The difference between nine times a number and six is twelve. Find the number.
    $9x - 6 = 12; 2$

17. Six less than four times a number is twenty-two. Find the number.
    $4x - 6 = 22; 7$

18. Four times the sum of twice a number and three is twelve. Find the number.
    $4(2x + 3) = 12; 0$

19. Three times the difference between four times a number and seven is fifteen. Find the number.
    $3(4x - 7) = 15; 3$

20. Twice the difference between a number and twenty-five is three times the number. Find the number.
    $2(x - 25) = 3x; -50$

21. The sum of two numbers is twenty. Three times the smaller is equal to two times the larger. Find the two numbers.
$3x = 2(20 - x)$; 8, 12

22. The sum of two numbers is fifteen. One less than three times the smaller is equal to the larger. Find the two numbers.
$3x - 1 = 15 - x$; 4, 11

23. The sum of two numbers is fourteen. The difference between two times the smaller and the larger is one. Find the two numbers.
$2x - (14 - x) = 1$; 5, 9

24. The sum of two numbers is eighteen. The total of three times the smaller and twice the larger is forty-four. Find the two numbers.
$3x + 2(18 - x) = 44$; 8, 10

25. The sum of three consecutive odd integers is fifty-one. Find the integers.
15, 17, 19

26. Find three consecutive even integers whose sum is negative eighteen.
$-8, -6, -4$

27. Find three consecutive odd integers such that three times the middle integer is one more than the sum of the first and third.
$-1, 1, 3$

28. Twice the smallest of three consecutive odd integers is seven more than the largest. Find the integers.
11, 13, 15

29. Find two consecutive even integers such that three times the first equals twice the second.
4, 6

30. Find two consecutive even integers such that four times the first is three times the second.
6, 8

31. The sum of two numbers is seven. Twice one number is four less than the other number. Which of the following equations does *not* represent this situation?
(i) $2(7 - x) = x - 4$ **(ii)** $2x = (7 - x) - 4$ **(iii)** $2n - 4 = 7 - n$
iii

OBJECTIVE B   *To translate a sentence into an equation and solve*

32. Depreciation   As a result of depreciation, the value of a car is now $19,200. This is three-fifths of its original value. Find the original value of the car. $32,000

33. Structures   The length of the Royal Gorge Bridge in Colorado is 320 m. This is one-fourth the length of the Golden Gate Bridge. Find the length of the Golden Gate Bridge.   1280 m

34. Nutrition   One slice of cheese pizza contains 290 calories. A medium-size orange has one-fifth that number of calories. How many calories are in a medium-size orange?   58 calories

35. History   John D. Rockefeller died in 1937. At the time of his death, Rockefeller had accumulated $1400 million, which was equal to one-sixty-fifth of the gross national product of the United States at that time. What was the U.S. gross national product in 1937? (*Source: The Wealthy 100: A Ranking of the Richest Americans, Past and Present*) $91 billion

36. Agriculture   A soil supplement that weighs 18 lb contains iron, potassium, and mulch. The supplement contains fifteen times as much mulch as iron and twice as much potassium as iron. Find the amount of mulch in the soil supplement.   15 lb

37. Geometry   An isosceles triangle has two sides of equal length. The length of the third side is 1 ft less than twice the length of one of the equal sides. Find the length of each side when the perimeter is 23 ft. (Recall that the perimeter of a figure is the distance around the figure.)   6 ft, 6 ft, 11 ft

The Golden Gate Bridge

John D. Rockefeller

38. **Geometry**   An isosceles triangle has two sides of equal length. The length of one of the equal sides is 2 m more than three times the length of the third side. If the perimeter is 46 m, find the length of each side. (Recall that the perimeter of a figure is the distance around the figure.)   20 m, 20 m, 6 m

 Point of Interest
The low-frequency pulses or whistles made by blue whales have been measured at up to 188 decibels, making them the loudest sounds produced by a living organism.

39. ● **Safety**   Loudness, or the intensity of sound, is measured in decibels. The sound level of a television is about 70 decibels, which is considered a safe hearing level. A food blender runs at 20 decibels higher than a TV, and a jet engine's decibel reading is 40 less than twice that of a blender. At this level, exposure can cause hearing loss. Find the intensity of the sound of a jet engine.   140 decibels

40. ● **Robots**   Kiva Systems, Inc., builds robots that companies can use to streamline order fulfillment operations in their warehouses. Salary and other benefits for one human warehouse worker can cost a company about $64,000 a year, an amount that is 103 times the company's yearly maintenance and operation costs for one robot. Find the yearly costs for a robot. Round to the nearest hundred. (*Source: The Boston Globe*) $600

Greenland

Iceland

41. ● **Geography**   Greenland, the largest island in the world, is 21 times larger than Iceland. The combined area of Greenland and Iceland is 880,000 mi². Find the area of Greenland.   840,000 mi²

42. **Consumerism**   The cost to replace a water pump in a sports car was $820. This included $375 for the water pump and $89 per hour for labor. How many hours of labor were required to replace the water pump?   5 h

43. **Utilities**   The cost of electricity in a certain city is $.09 for each of the first 300 kWh (kilowatt-hours) and $.15 for each kilowatt-hour over 300 kWh. Find the number of kilowatt-hours used by a family that receives a $59.25 electric bill.   515 kWh

44. **Labor Unions**   A union charges monthly dues of $4.00 plus $.25 for each hour worked during the month. A union member's dues for March were $46.00. How many hours did the union member work during the month of March?   168 h

45. **Business**   The cellular phone service for a business executive is $80 per month plus $.40 per minute of phone use over 900 min. For a month in which the executive's cellular phone bill was $100.40, how many minutes did the executive use the phone?   951 min

46. ● **Recycling**   Use the information in the article at the right to find how many tons of plastic drink bottles were stocked for sale in U.S. stores.   2.7 million tons

In the NEWS!

Americans' Unquenchable Thirst

Despite efforts to increase recycling, the 2.16 million tons of plastic drink bottles that ended up in landfills this year represent four-fifths of the plastic drink bottles stocked for sale in U.S. stores.

And Americans can't seem to get enough of bottled water. During a recent year, stores stocked 7.5 billion gallons of bottled water, an amount that is approximately the same as the volume of water that goes over Niagara Falls every 3 h.
Source: scienceline.org

Unless otherwise noted, all content on this page is © Cengage Learning.

Text Messaging   For Exercises 47 and 48, use the expression $2.99 + 0.15n$, which represents the total monthly text-messaging bill for n text messages over 300 in one month.

47. How much does the customer pay per text message over 300 messages? $.15

48. What is the fixed charge per month for the text-messaging service? $2.99

Critical Thinking

49. Metalwork   A wire 12 ft long is cut into two pieces. Each piece is bent into the shape of a square. The perimeter of the larger square is twice the perimeter of the smaller square. Find the perimeter of the larger square.   8 ft

50. The amount of liquid in a container triples every minute. The container becomes completely filled at 3:40 p.m. What fractional part of the container is filled at 3:39 P.M? $\frac{1}{3}$

51. Travel   A cyclist traveling at a constant speed completes three-fifths of a trip in $\frac{1}{2}$ h. In how many additional hours will the cyclist complete the entire trip? $\frac{1}{3}$ h

52. Business   During one day at an office, one-half of the amount of money in the petty cash drawer was used in the morning, and one-third of the remaining money was used in the afternoon, leaving $5 in the petty cash drawer at the end of the day. How much money was in the petty cash drawer at the start of the day? $15

53. Find four consecutive even integers whose sum is -36. $-12, -10, -8, -6$

54. Find four consecutive odd integers whose sum is -48. $-15, -13, -11, -9$

55. Find three consecutive odd integers such that the sum of the first and third integers is twice the second integer.   Any three consecutive odd integers

56. Find four consecutive integers such that the sum of the first and fourth integers equals the sum of the second and third integers.   Any four consecutive integers

Projects or Group Activities

Complete each statement with the word *even* or *odd*.

57. If k is an odd integer, then $k + 1$ is an __even__ integer.

58. If k is an odd integer, then $k - 2$ is an __odd__ integer.

59. If n is an integer, then $2n$ is an __even__ integer.

60. If m and n are even integers, then $m - n$ is an __even__ integer.

61. If m and n are even integers, then mn is an __even__ integer.

62. If m and n are odd integers, then $m + n$ is an __even__ integer.

63. If m and n are odd integers, then $m - n$ is an __even__ integer.

64. If m and n are odd integers, then mn is an __odd__ integer.

65. If m is an even integer and n is an odd integer, then $m - n$ is an __odd__ integer.

66. If m is an even integer and n is an odd integer, then $m + n$ is an __odd__ integer.

QUICK QUIZ
Translate into an equation and solve.
1. The product of five and a number is negative fifteen. Find the number.   **$5x = -15$; -3 [2.3A]**
2. Find three consecutive even integers such that four times the second is eight less than the third.   **$4(n + 2) = (n + 4) - 8$; $-4, -2, 0$ [2.3A]**
3. An investment of $8000 is divided into two accounts, one at a bank and one at a credit union. The value of the investment at the credit union is $1000 less than twice the value of the investment at the bank. Find the amount in each account.   **$3000 at the bank; $5000 at the credit union   [2.3B]**

Unless otherwise noted, all content on this page is © Cengage Learning.

✔ CHECK YOUR PROGRESS: CHAPTER 2

1. Is 3 a solution of $2a(a - 1) = 3a + 3$? Yes [2.1A]

2. Solve: $x + 7 = -4$ −11 [2.1B]

3. Solve: $-3y = -27$ 9 [2.1C]

4. What is 45% of 160? 72 [2.1D]

5. Solve: $6 - 4a = -10$ 4 [2.2A]

6. Is $-\frac{1}{4}$ a solution of $8t + 1 = -1$? Yes [2.1A]

7. Solve: $\frac{1}{6} + b = -\frac{1}{3}$ $-\frac{1}{2}$ [2.1B]

8. Solve: $5x - 4(3 - x) = 2(x - 1) - 3$ 1 [2.2C]

9. 18% of what number is 27? 150 [2.1D]

10. Solve: $6y + 5 - 8y = 3 - 4y$ −1 [2.2B]

11. Is 4 a solution of $x(x + 1) = x^2 + 5$? No [2.1A]

12. Solve: $84 = -16 + t$ 100 [2.1B]

13. Solve: $\frac{3}{4}c = \frac{3}{5}$ $\frac{4}{5}$ [2.1C]

14. Solve: $9 = \frac{1}{2}d - 5$ 28 [2.2A]

15. What percent of 170 is 42.5? 25% [2.1D]

16. Solve: $-\frac{8}{9} = -\frac{2}{3}y$ $\frac{4}{3}$ [2.1C]

17. Solve: $3n + 2(n - 4) = 7$ 3 [2.2C]

18. Solve: $3x - 8 = 5x + 6$ −7 [2.2B]

19. Solve: $2[3 - 5(x - 1)] = 7x - 1$ 1 [2.2C]

20. Solve: $18 = 2t$ 9 [2.1C]

21. Translate into an equation and solve: The quotient of fifteen and a number is negative three. Find the number. $\frac{15}{x} = -3$; −5 [2.3A]

22. Find four consecutive odd integers whose sum is 24. 3, 5, 7, 9 [2.3A]

23. ● **Health** According to *Health* magazine, the average American has increased his or her daily consumption of calories from 18 years ago by 11.6%. If the average daily consumption was 1970 calories 18 years ago, what is the average daily consumption today? Round to the nearest whole number. 2199 calories [2.1D]

24. **Recreation** K&B Tours offers a river trip that takes passengers from the K&B dock to a small island that is 24 mi away. The passengers spend 1 h at the island and then return to the K&B dock. If the speed of the boat is 10 mph in calm water and the rate of the current is 2 mph, how long does the trip last? 6 h [2.1E]

50 lb ← 3.5 ft → 60 lb

8 ft

25. **Physics** Two children are sitting 8 ft apart on a seesaw. One child weighs 60 lb and the second child weighs 50 lb. The fulcrum is 3.5 ft from the child weighing 60 lb. Is the seesaw balanced? Use the lever system equation $F_1 x = F_2(d - x)$. No [2.2D]

Unless otherwise noted, all content on this page is © Cengage Learning.

2.4 Mixture and Uniform Motion Problems

OBJECTIVE A *To solve value mixture problems*

A value mixture problem involves combining two ingredients that have different prices into a single blend. For example, a coffee merchant may blend two types of coffee into a single blend, or a candy manufacturer may combine two types of candy to sell as a variety pack.

The solution of a value mixture problem is based on the **value mixture equation** $AC = V$, where A is the amount of an ingredient, C is the cost per unit of the ingredient, and V is the value of the ingredient.

Take Note

The equation $AC = V$ is used to find the value of an ingredient. For example, the value of 4 lb of cashews costing $6 per pound is

$AC = V$
$4 \cdot \$6 = V$
$\$24 = V$

HOW TO 1 A coffee merchant wants to make 6 lb of a blend of coffee costing $5 per pound. The blend is made using a $7-per-pound grade and a $4-per-pound grade of coffee. How many pounds of each of these grades should be used?

Strategy for Solving a Value Mixture Problem

1. For each ingredient in the mixture, write a numerical or variable expression for the amount of the ingredient used, the unit cost of the ingredient, and the value of the amount used. For the blend, write a numerical or variable expression for the amount, the unit cost of the blend, and the value of the amount. The results can be recorded in a table.

The sum of the amounts is 6 lb.

Amount of $7 coffee: x
Amount of $4 coffee: $6 - x$

	Amount, A	·	Unit Cost, C	=	Value, V
$7 grade	x	·	7	=	$7x$
$4 grade	$6 - x$	·	4	=	$4(6 - x)$
$5 blend	6	·	5	=	$5(6)$

Take Note

Use the information given in the problem to fill in the amount and unit cost columns of the table. Fill in the value column by multiplying the two expressions you wrote in each row. Use the expressions in the last column to write the equation.

2. Determine how the values of the ingredients are related. Use the fact that the sum of the values of all the ingredients is equal to the value of the blend.

The sum of the values of the $7 grade and the $4 grade is equal to the value of the $5 blend.

$7x + 4(6 - x) = 5(6)$
$7x + 24 - 4x = 30$
$3x + 24 = 30$
$3x = 6$
$x = 2$

$6 - x = 6 - 2 = 4$ • Find the amount of the $4 grade coffee.

The merchant must use 2 lb of the $7 coffee and 4 lb of the $4 coffee.

Unless otherwise noted, all content on this page is © Cengage Learning.

EXAMPLE 1

How many ounces of a metal alloy that costs $4 an ounce must be mixed with 10 oz of an alloy that costs $6 an ounce to make a mixture that costs $4.32 an ounce?

Strategy

x oz
$4/oz

10 oz
$6/oz

• Ounces of $4 alloy: x

	Amount	Cost	Value
$4 alloy	x	4	$4x$
$6 alloy	10	6	$6(10)$
$4.32 mixture	$10 + x$	4.32	$4.32(10 + x)$

• The sum of the values before mixing equals the value after mixing.

Solution

$$4x + 6(10) = 4.32(10 + x)$$
• The sum of the values before mixing equals the value after mixing.

$$4x + 60 = 43.2 + 4.32x$$
$$-0.32x + 60 = 43.2$$
• Subtract $4.32x$ from each side.

$$-0.32x = -16.8$$
• Subtract 60 from each side.

$$x = 52.5$$
• Divide each side by -0.32.

52.5 oz of the $4 alloy must be used.

YOU TRY IT 1

A gardener has 20 lb of a lawn fertilizer that costs $.90 per pound. How many pounds of a fertilizer that costs $.75 per pound should be mixed with this 20 lb of lawn fertilizer to produce a mixture that costs $.85 per pound?

Your strategy

Your solution

10 lb

IN-CLASS EXAMPLES

1. A meatloaf mixture is made by combining ground turkey costing $1.49 per pound with ground beef costing $2.13 per pound. How many pounds of each should be used to make 8 lb of a meatloaf mixture that costs $1.89 per pound?
   Ground turkey: 3 lb; ground beef: 5 lb

Solution on p. S6

Unless otherwise noted, all content on this page is © Cengage Learning.

<table>
<tr><td>**OBJECTIVE B**</td><td>*To solve percent mixture problems*</td></tr>
</table>

INSTRUCTOR NOTE
These problems are similar to the value mixture problems. The sum of the quantities being mixed equals the desired quantity. Again, the last column of the table will be used to form the equation.

Recall from Section 2.1 that a percent mixture problem can be solved using the equation $Ar = Q$, where A is the amount of a solution, r is the percent concentration of a substance in the solution, and Q is the quantity of the substance in the solution.

For example, a 500-milliliter bottle is filled with a 4% solution of hydrogen peroxide.

$$Ar = Q$$
$$500\,(0.04) = Q$$
$$20 = Q$$

The bottle contains 20 ml of hydrogen peroxide.

HOW TO 2 How many gallons of a 20% salt solution must be mixed with 6 gal of a 30% salt solution to make a 22% salt solution?

> **Strategy for Solving a Percent Mixture Problem**
>
> 1. For each solution, write a numerical or variable expression for the amount of solution, the percent concentration, and the quantity of the substance in the solution. The results can be recorded in a table.

The unknown quantity of 20% solution: x

Take Note
Use the information given in the problem to fill in the amount and percent columns of the table. Fill in the quantity column by multiplying the two expressions you wrote in each row. Use the expressions in the last column to write the equation.

	Amount of Solution, A	·	Percent Concentration, r	=	Quantity of Substance, Q
20% solution	x	·	0.20	=	$0.20x$
30% solution	6	·	0.30	=	$0.30(6)$
22% solution	$x + 6$	·	0.22	=	$0.22(x + 6)$

> 2. Determine how the quantities of the substances in the solutions are related. Use the fact that the sum of the quantities of the substances being mixed is equal to the quantity of the substance after mixing.

The sum of the quantities of the substances in the 20% solution and the 30% solution is equal to the quantity of the substance in the 22% solution.

$$0.20x + 0.30(6) = 0.22(x + 6)$$
$$0.20x + 1.80 = 0.22x + 1.32$$
$$-0.02x + 1.80 = 1.32$$
$$-0.02x = -0.48$$
$$x = 24$$

24 gal of the 20% solution are required.

Unless otherwise noted, all content on this page is © Cengage Learning.

EXAMPLE 2

A chemist wishes to make 2 L of an 8% acid solution by mixing a 10% acid solution and a 5% acid solution. How many liters of each solution should the chemist use?

Strategy

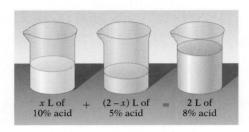

x L of 10% acid + $(2-x)$ L of 5% acid = 2 L of 8% acid

- Liters of 10% solution: x
  Liters of 5% solution: $2 - x$

	Amount	Percent	Quantity
10% solution	x	0.10	$0.10x$
5% solution	$2 - x$	0.05	$0.05(2 - x)$
8% solution	2	0.08	$0.08(2)$

- The sum of the quantities before mixing is equal to the quantity after mixing.

Solution

$0.10x + 0.05(2 - x) = 0.08(2)$ • The sum of the quantities before mixing equals the quantity after mixing.

$0.10x + 0.10 - 0.05x = 0.16$

$0.05x + 0.10 = 0.16$ • Combine like terms on the left side.

$0.05x = 0.06$ • Subtract 0.10 from each side.

$x = 1.2$ • Divide each side by 0.05.
Liters of 10% solution: 1.2

$2 - x = 2 - 1.2 = 0.8$ • Find the number of liters of 5% solution.

The chemist needs 1.2 L of the 10% solution and 0.8 L of the 5% solution.

YOU TRY IT 2

A pharmacist dilutes 5 L of a 12% solution with a 6% solution. How many liters of the 6% solution are added to make an 8% solution?

Your strategy

Your solution

10 L

IN-CLASS EXAMPLES

2. How many milliliters of a 20% acid solution must be mixed with 30 ml of an 80% acid solution to make a 32% acid solution?   **120 ml**

3. How many liters of pure water must be added to a 50-liter solution that contains 40% fruit juice to make a solution that is 30% fruit juice?

$16\frac{2}{3}$ **L**

Solution on pp. S6–S7

Unless otherwise noted, all content on this page is © Cengage Learning.

OBJECTIVE C   *To solve uniform motion problems*

INSTRUCTOR NOTE
Although this section
appears to cover three
different formulas ($AC = V$
for value mixture problems,
$Ar = Q$ for percent mixture
problems, and $rt = d$ for
uniform motion problems),
students may appreciate
knowing that all three types
of problems can be solved
using the generalized
formula $AR = T$, or Amount
times Rate equals Total.
Observant students may
notice that in all three
cases, the rate involves
the word *per*—for example,
per quart, *percent,* or *miles
per hour.* In all three cases,
amounts may be added and
totals may be added, but
rates may not.

Recall from Section 2.1 that an object traveling at a constant speed in a straight line is in *uniform motion*. The solution of a uniform motion problem is based on the equation $rt = d$, where r is the rate of travel, t is the time spent traveling, and d is the distance traveled.

HOW TO 3   A car leaves a town traveling at 40 mph. Two hours later, a second car leaves the same town, on the same road, traveling at 60 mph. In how many hours will the second car pass the first car?

> **Strategy for Solving a Uniform Motion Problem**
>
> 1. For each object, write a numerical or variable expression for the rate, time, and distance. The results can be recorded in a table.

The first car traveled 2 h longer than the second car.

Unknown time for the second car: t
Time for the first car: $t + 2$

	Rate, r	·	Time, t	=	Distance, d
First car	40	·	$t + 2$	=	$40(t + 2)$
Second car	60	·	t	=	$60t$

Take Note
Use the information given
in the problem to fill in the
rate and time columns of
the table. Fill in the distance
column by multiplying the
two expressions you wrote in
each row.

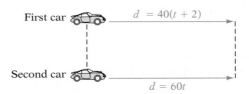

> 2. Determine how the distances traveled by the two objects are related. For example, the total distance traveled by both objects may be known, or it may be known that the two objects traveled the same distance.

INSTRUCTOR NOTE
One of the complications
of distance-rate problems
is that the variable may
not directly represent the
unknown. After doing a
problem similar to the one
at the right, show students a
problem similar to Example
4 on the next page. The
variable is time, but the
unknown is distance.

The two cars travel the same distance.

$$40(t + 2) = 60t$$
$$40t + 80 = 60t$$
$$80 = 20t$$
$$4 = t$$

The second car will pass the first car in 4 h.

EXAMPLE 3

Two cars, one traveling 10 mph faster than the other, start at the same time from the same point and travel in opposite directions. In 3 h, they are 300 mi apart. Find the rate of each car.

Strategy • Rate of first car: r
Rate of second car: $r + 10$

	Rate	Time	Distance
1st car	r	3	$3r$
2nd car	$r + 10$	3	$3(r + 10)$

• The total distance traveled by the two cars is 300 mi.

Solution

$3r + 3(r + 10) = 300$
$3r + 3r + 30 = 300$ • **Distributive Property**
$6r + 30 = 300$ • **Combine like terms.**
$6r = 270$ • **Subtract 30 from each side.**
$r = 45$ • **Divide each side by 6.**
$r + 10 = 45 + 10 = 55$ • **Find the rate of the second car.**

The first car is traveling at 45 mph.
The second car is traveling at 55 mph.

YOU TRY IT 3

Two trains, one traveling at twice the speed of the other, start at the same time on parallel tracks from stations that are 288 mi apart and travel toward each other. In 3 h, the trains pass each other. Find the rate of each train.

Your strategy

Your solution

32 mph; 64 mph

EXAMPLE 4

How far can the members of a bicycling club ride out into the country at a speed of 12 mph and return over the same road at 8 mph if they travel a total of 10 h?

Strategy • Time spent riding out: t
Time spent riding back: $10 - t$

	Rate	Time	Distance
Out	12	t	$12t$
Back	8	$10 - t$	$8(10 - t)$

• The distance out equals the distance back.

Solution

$12t = 8(10 - t)$
$12t = 80 - 8t$ • **Distributive Property**
$20t = 80$ • **Add 8t to each side.**
$t = 4$ (The time is 4 h.)   • **Divide each side by 20.**

The distance out $= 12t = 12(4) = 48$

The club can ride 48 mi into the country.

YOU TRY IT 4

A pilot flew out to a parcel of land and back in 5 h. The rate out was 150 mph, and the rate returning was 100 mph. How far away was the parcel of land?

Your strategy

Your solution

300 mi

IN-CLASS EXAMPLES

4. Two planes leave an airport at the same time and fly in opposite directions. One of the planes is flying at 450 mph, and the other plane is flying at 550 mph. In how many hours will they be 2000 mi apart? **2 h**

5. A cyclist starts on a course at 6 A.M. riding at 12 mph. An hour later, a second cyclist starts on the same course traveling at 18 mph. At what time will the second cyclist overtake the first cyclist? **9 A.M.**

Solutions on p. S7

Unless otherwise noted, all content on this page is © Cengage Learning.

2.4 EXERCISES

✔ **Concept Check**

SUGGESTED ASSIGNMENT
Exercises 1–12; Exercises 13–31, odds; Exercises 35–75, odds

1. The total value of a 7-pound bag of cat food that costs $1.50 per pound is __$10.50__.

2. If 8 L of a solvent costs $75 per liter, then the value of the 8 L of solvent is __$600__.

3. The cost per pound of a 5-pound bag of sugar that has a total value of $3.80 is __$.76__

4. A 250-milliliter bottle contains a solution that is 90% isopropyl alcohol. The amount of isopropyl alcohol in the solution is 250 (__0.90__) = __225__ ml.

5. A 500-gram can of legumes contains 20% navy beans. The can contains __100__ g of navy beans.

6. A 10-pound box of chocolate-covered cherries is 15% chocolate. There are __1.5__ lb of chocolate in the box. There are __8.5__ lb of cherries in the box.

For Exercises 7 to 9, determine whether the statement is true or false.

7. In the value mixture equation $V = AC$, the variable A represents the quantity of an ingredient.   True

8. Suppose we are mixing two salt solutions. Then the variable Q in the percent mixture equation $Q = Ar$ represents the amount of salt in a solution.   True

9. If we combine a 9% acid solution with a solution that is 4% acid, the resulting solution will be less than 4% acid.   False

10. ◣ Explain the meaning of each variable in the equation $V = AC$. Give an example of how this equation is used.

11. ◣ Explain the meaning of each variable in the equation $Q = Ar$. Give an example of how this equation is used.

12. ◣ Explain what each variable in the formula $d = rt$ represents.

OBJECTIVE A   *To solve value mixture problems*

13. At a veterinary clinic, a special high-protein dog food that costs $6.75 per pound is mixed with a vitamin supplement that costs $3.25 per pound. How many pounds of each should be used to make 5 lb of a mixture that costs $4.65 per pound? 2 lb of dog food; 3 lb of vitamin supplement

14. A goldsmith combined an alloy that costs $4.30 per ounce with an alloy that costs $1.80 per ounce. How many ounces of each were used to make a mixture of 200 oz costing $2.50 per ounce? $4.30 alloy: 56 oz; $1.80 alloy: 144 oz

15. How many pounds of chamomile tea that costs $18.20 per pound must be mixed with 12 lb of orange tea that costs $12.25 per pound to make a mixture that costs $14.63 per pound?   8 lb

© iStockphoto.com/agdalena Kucova

Andrea Skjold/Shutterstock.com

16. A wild birdseed mix is made by combining 100 lb of millet seed costing $.60 per pound with sunflower seeds costing $1.10 per pound. How many pounds of sunflower seeds are needed to make a mixture that costs $.70 per pound?   25 lb

17. Find the cost per pound of a coffee mixture made from 8 lb of coffee that costs $9.20 per pound and 12 lb of coffee that costs $5.50 per pound. $6.98

18. Find the cost per ounce of a mixture of 200 oz of a cologne that costs $7.50 per ounce and 500 oz of a cologne that costs $4.00 per ounce. $5.00

19. An herbalist has 30 oz of herbs costing $2 per ounce. How many ounces of herbs costing $1 per ounce should be mixed with these 30 oz of herbs to produce a mixture costing $1.60 per ounce?   20 oz

20. A snack food is made by mixing 5 lb of popcorn that costs $.80 per pound with caramel that costs $2.40 per pound. How much caramel is needed to make a mixture that costs $1.40 per pound?   3 lb

21. A grocery store offers a cheese sampler that includes a pepper cheddar cheese that costs $16 per kilogram and Pennsylvania Jack that costs $12 per kilogram. How many kilograms of each were used to make a 5-kilogram mixture that costs $13.20 per kilogram?   1.5 kg of pepper cheese; 3.5 kg of Pennsylvania Jack

22. A lumber company combined oak wood chips that cost $3.10 per pound with pine wood chips that cost $2.50 per pound. How many pounds of each were used to make an 80-pound mixture costing $2.65 per pound?   20 lb of oak chips; 60 lb of pine chips

23. The manager of a farmer's market has 500 lb of grain that costs $1.20 per pound. How many pounds of meal costing $.80 per pound should be mixed with the 500 lb of grain to produce a mixture that costs $1.05 per pound?   300 lb

24. A caterer made an ice cream punch by combining fruit juice that costs $4.50 per gallon with ice cream that cost $8.50 per gallon. How many gallons of each were used to make 100 gal of punch costing $5.50 per gallon? Fruit juice: 75 gal; ice cream: 25 gal

25. The manager of a specialty food store combined almonds that cost $6.50 per pound with walnuts that cost $5.50 per pound. How many pounds of each were used to make a 100-pound mixture that costs $5.87 per pound?   Almonds: 37 lb; walnuts: 63 lb

26. Find the cost per pound of a "house blend" of coffee that is made from 12 lb of Central American coffee that costs $8 per pound and 30 lb of South American coffee that costs $4.50 per pound. $5.50

27. Find the cost per pound of sugar-coated breakfast cereal made from 40 lb of sugar that costs $2.00 per pound and 120 lb of corn flakes that cost $1.20 per pound. $1.40

28. How many liters of a blue dye that costs $1.60 per liter must be mixed with 18 L of anil that costs $2.50 per liter to make a mixture that costs $1.90 per liter?   36 L

Unless otherwise noted, all content on this page is © Cengage Learning.

29. ● Tree Conservation   A town's parks department buys trees from the tree conservation program described in the news clipping at the right. The department spends $406 on 14 bundles of trees. How many bundles of seedlings and how many bundles of container-grown plants did the parks department buy?
8 bundles of seedlings, 6 bundles of container-grown plants

30. Find the cost per ounce of a gold alloy made from 25 oz of pure gold that costs $1282 per ounce and 40 oz of an alloy that costs $900 per ounce. $1046.92

31. Find the cost per ounce of a sunscreen made from 100 oz of a lotion that costs $2.50 per ounce and 50 oz of a lotion that costs $4.00 per ounce. $3

32. 🖾 A grocer mixes peanuts that cost $3 per pound with almonds that cost $7 per pound. Which of the following could be true about the cost C per pound of the mixture? There may be more than one correct answer.
  (i) $C = \$10$ **(ii)** $C > \$7$ **(iii)** $C < \$7$
  (iv) $C < \$3$ **(v)** $C > \$3$ **(vi)** $C = \$4$ iii, v, and vi

33. 🖾 A snack mix is made from 3 lb of sunflower seeds that cost S dollars per pound and 4 lb of raisins that cost R dollars per pound. Which expression gives the cost C per pound of the mixture?
  (i) $7(S + R)$ **(ii)** $3S + 4R$ **(iii)** $S + R$ **(iv)** $\dfrac{3S + 4R}{7}$ iv

In the NEWS!

Conservation Tree Planting Program Underway

The Kansas Forest Service is again offering its Conservation Tree Planting Program. Trees are sold in bundles of 25, in two sizes—seedlings cost $17 a bundle and larger container-grown plants cost $45 a bundle.
Source: Kansas Canopy

OBJECTIVE B   *To solve percent mixture problems*

34. Forty ounces of a 30% gold alloy are mixed with 60 oz of a 20% gold alloy. Find the percent concentration of the resulting gold alloy.   24%

35. One hundred ounces of juice that is 50% tomato juice are added to 200 oz of a vegetable juice that is 25% tomato juice. What is the percent concentration of tomato juice in the resulting mixture? $33\frac{1}{3}\%$

36. How many gallons of a 15% acid solution must be mixed with 5 gal of a 20% acid solution to make a 16% acid solution?   20 gal

37. How many pounds of a chicken feed that is 50% corn must be mixed with 400 lb of a feed that is 80% corn to make a chicken feed that is 75% corn?   80 lb

38. A rug is made by weaving 20 lb of yarn that is 50% wool with a yarn that is 25% wool. How many pounds of the yarn that is 25% wool must be used if the finished rug is to be 35% wool?   30 lb

39. Five gallons of a dark green latex paint that is 20% yellow paint are combined with a lighter green latex paint that is 40% yellow paint. How many gallons of the lighter green paint must be used to create a green paint that is 25% yellow paint? $1\frac{2}{3}$ gal

40. How many gallons of a plant food that is 9% nitrogen must be combined with another plant food that is 25% nitrogen to make 10 gal of a plant food that is 15% nitrogen?
6.25 gal

41. A chemist wants to make 50 ml of a 16% acid solution by mixing a 13% acid solution and an 18% acid solution. How many milliliters of each solution should the chemist use?
13% solution: 20 ml; 18% solution: 30 ml

42. Five grams of sugar are added to a 45-gram serving of a breakfast cereal that is 10% sugar. What is the percent concentration of sugar in the resulting mixture?   19%

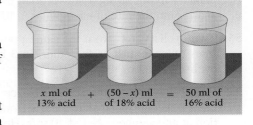

x ml of + $(50 - x)$ ml = 50 ml of
13% acid   of 18% acid   16% acid

Unless otherwise noted, all content on this page is © Cengage Learning.

43. Thirty ounces of pure silver are added to 50 oz of a silver alloy that is 20% silver. What is the percent concentration of the resulting alloy?   50%

44. To make the potpourri mixture sold at a florist shop, 70 oz of a potpourri that is 80% lavender are combined with a potpourri that is 60% lavender. The resulting potpourri is 74% lavender. How much of the potpourri that is 60% lavender is used?   30 oz

45. The manager of a garden shop mixes grass seed that is 40% rye grass with 40 lb of grass seed that is 60% rye grass to make a mixture that is 56% rye grass. How much of the 40% rye grass is used?   10 lb

46. A hair dye is made by blending a 7% hydrogen peroxide solution and a 4% hydrogen peroxide solution. How many milliliters of each are used to make a 300-milliliter solution that is 5% hydrogen peroxide?
7% solution: 100 ml; 4% solution: 200 ml

47. A clothing manufacturer has some pure silk thread and some thread that is 85% silk. How many kilograms of each must be woven together to make 75 kg of cloth that is 96% silk?
Pure silk thread: 55 kg; 85% silk thread: 20 kg

48. At a cosmetics company, 40 L of pure aloe cream are mixed with 50 L of a moisturizer that is 64% aloe. What is the percent concentration of aloe in the resulting mixture?   80%

49. ● **Ethanol Fuel**   See the news clipping at the right. *Gasohol* is a type of fuel made by mixing ethanol with gasoline. E10 is a fuel mixture of 10% ethanol and 90% gasoline. E20 contains 20% ethanol and 80% gasoline. How many gallons of ethanol must be added to 100 gal of E10 to make E20?   12.5 gal

50. A hair stylist combines 12 oz of shampoo that is 20% conditioner with an 8-ounce bottle of pure shampoo. What is the percent concentration of conditioner in the 20-ounce mixture?   12%

51. How many ounces of pure chocolate must be added to 150 oz of chocolate topping that is 50% chocolate to make a topping that is 75% chocolate?   150 oz

52. A recipe for a rice dish calls for 8 oz of pure wild rice and 12 oz of a rice mixture that is 20% wild rice. What is the percent concentration of wild rice in the 20-ounce mixture?   52%

53. ✎ True or false? A 10% salt solution can be combined with some amount of a 20% salt solution to create a 30% salt solution.   False

54. ✎ True or false? When *n* ounces of 100% acid are mixed with 2*n* ounces of pure water, the resulting mixture is a 50% acid solution.   False

> ## In the NEWS!
>
> ### Gasohol Reduces Harmful Emissions
>
> A new study indicates that using E20 fuel reduces carbon dioxide and hydrocarbon emissions, as compared with E10 blends or traditional gasoline.
> *Source:* www.sciencedaily.com

OBJECTIVE C   *To solve uniform motion problems*

55. Two small planes start from the same point and fly in opposite directions. The first plane is flying 25 mph slower than the second plane. In 2 h, the planes are 470 mi apart. Find the rate of each plane.   105 mph, 130 mph

|← ———— 470 mi ———— →|

56. Two cyclists start from the same point and ride in opposite directions. One cyclist rides twice as fast as the other. In 3 h, they are 81 mi apart. Find the rate of each cyclist.   9 mph, 18 mph

57. One speed skater starts across a frozen lake at an average speed of 8 m/s. Ten seconds later, a second speed skater starts from the same point and skates in the same direction at an average speed of 10 m/s. How many seconds after the second skater starts will the second skater overtake the first skater?   40 s

Unless otherwise noted, all content on this page is © Cengage Learning.

58. A long-distance runner starts on a course running at an average speed of 6 mph. Half an hour later, a second runner begins the same course at an average speed of 7 mph. How long after the second runner starts will the second runner overtake the first runner?   3 h

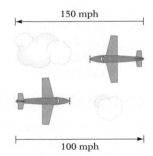

59. Michael Chan leaves a dock in his motorboat and travels at an average speed of 9 mph toward the Isle of Shoals, a small island off the coast of Massachusetts. Two hours later, a tour boat leaves the same dock and travels at an average speed of 18 mph toward the same island. How many hours after the tour boat leaves will Michael's boat be alongside the tour boat?   2 h

60. A jogger starts from one end of a 15-mile nature trail at 8:00 A.M. One hour later, a cyclist starts from the other end of the trail and rides toward the jogger. If the rate of the jogger is 6 mph and the rate of the cyclist is 9 mph, at what time will the two meet?   9:36 A.M.

61. An executive drove from home at an average speed of 30 mph to an airport where a helicopter was waiting. The executive boarded the helicopter and flew to the corporate offices at an average speed of 60 mph. The entire distance was 150 mi. The entire trip took 3 h. Find the distance from the airport to the corporate offices.
120 mi

62. A 555-mile, 5-hour plane trip was flown at two speeds. For the first part of the trip, the average speed was 105 mph. For the remainder of the trip, the average speed was 115 mph. How long did the plane fly at each speed?
2 h at 105 mph; 3 h at 115 mph

63. After a sailboat had been on the water for 3 h, a change in the wind direction reduced the average speed of the boat by 5 mph. The entire distance sailed was 57 mi. The total time spent sailing was 6 h. How far did the sailboat travel in the first 3 h?   36 mi

64. A stunt driver was needed at the production site of a Hollywood movie. The average speed of the stunt driver's flight to the site was 150 mph, and the average speed of the return trip was 100 mph. Find the distance of the round trip if the total flying time was 5 h.   600 mi

65. A passenger train leaves a train depot 2 h after a freight train leaves the same depot. The freight train is traveling 20 mph slower than the passenger train. Find the rate of each train if the passenger train overtakes the freight train in 3 h.
Passenger train: 50 mph; freight train: 30 mph

66. A car and a bus set out at 3 P.M. from the same point headed in the same direction. The average speed of the car is twice the average speed of the bus. In 2 h the car is 68 mi ahead of the bus. Find the rate of the car.   68 mph

67. A ship traveling east at 25 mph is 10 mi from a harbor when another ship leaves the harbor traveling east at 35 mph. How long does it take the second ship to catch up to the first ship?   1 h

68. At 10 A.M. a plane leaves Boston, Massachusetts, for Seattle, Washington, a distance of 3000 mi. One hour later a plane leaves Seattle for Boston. Both planes are traveling at a speed of 500 mph. How many hours after the plane leaves Seattle will the planes pass each other?   2.5 h

Unless otherwise noted, all content on this page is © Cengage Learning.

69.  **Bridges** See the news clipping at the right. Two cars, the first traveling 10 km/h faster than the second, start at the same time from opposite ends of the Hangzhou Bay Bridge and travel toward each other. The cars pass each other in 12 min. Find the rate of the faster car.   95 km/h

In the NEWS!

Longest Ocean-Crossing Bridge Opens to Public

The Hangzhou Bay Bridge is the longest ocean-crossing bridge in the world. It spans the Hangzhou Bay on the East China Sea and crosses the Qiantang River at the Yangtze River Delta. The S-shaped bridge connects Jiaxing to the north and Ningbo to the south. The bridge is 36 km long and has a speed limit of 100 km/h.

Source: www.roadtraffic-technology. com

70. At noon a train leaves Washington, D.C., headed for Charleston, South Carolina, a distance of 500 mi. The train travels at a speed of 60 mph. At 1 P.M. a second train leaves Charleston headed for Washington, D.C., traveling at 50 mph. How long after the train leaves Charleston will the two trains pass each other?   4 h

71. A race car driver starts along a 50-mile race course traveling at an average speed of 90 mph. Fifteen minutes later, a second driver starts along the same course at an average speed of 120 mph. Will the second car overtake the first car before the drivers reach the end of the course?   No

72. A bus traveled on a straight road for 2 h at an average speed that was 20 mph faster than its average speed on a winding road. The time spent on the winding road was 3 h. Find the average speed on the winding road if the total trip was 210 mi.   34 mph

73. A bus traveling at a rate of 60 mph overtakes a car traveling at a rate of 45 mph. If the car had a 1-hour head start, how far from the starting point does the bus overtake the car?   180 mi

74. A car traveling at 48 mph overtakes a cyclist who, riding at 12 mph, had a 3-hour head start. How far from the starting point does the car overtake the cyclist?   48 mi

75. A plane left Kennedy Airport on Tuesday morning for a 605 mile, 5-hour trip. For the first part of the trip, the average speed was 115 mph. For the remainder of the trip, the average speed was 125 mph. How long did the plane fly at each speed?   2 h at 115 mph; 3 h at 125 mph

Critical Thinking

76. Find the cost per ounce of a mixture of 30 oz of an alloy that costs $4.50 per ounce, 40 oz of an alloy that costs $3.50 per ounce, and 30 oz of an alloy that costs $3.00 per ounce. $3.65

77. A grocer combined walnuts that cost $5.60 per pound and cashews that cost $7.50 per pound with 20 lb of peanuts that cost $4.00 per pound. Find the amount of walnuts and the amount of cashews used to make a 50-pound mixture costing $5.72 per pound.   Walnuts: 10 lb; cashews: 20 lb

78. How many ounces of water evaporated from 50 oz of a 12% salt solution to produce a 15% salt solution?   10 oz

79. A chemist mixed pure acid with water to make 10 L of a 30% acid solution. How much pure acid and how much water did the chemist use?
Pure acid: 3 L; water: 7 L

80. How many grams of pure water must be added to 50 g of pure acid to make a solution that is 40% acid?   75 g

QUICK QUIZ

1. A trail mix is made by combining raisins that cost $4.20 per pound with granola that costs $2.20 per pound. How many pounds of each should be used to make 40 lb of trail mix that costs $2.75 per pound?   **Raisins: 11 lb; granola: 29 lb   [2.4A]**

2. How many pounds of wild birdseed that is 35% sunflower seeds must be combined with 10 lb of a birdseed mix that is 20% sunflower seeds to make a wild birdseed mixture that is 30% sunflower seeds?   **20 lb   [2.4B]**

81. Tickets to a performance by a community theater company cost $5.50 for adults and $2.75 for children. A total of 120 tickets were sold for $563.75. How many adults and how many children attended the performance?   85 adults, 35 children

82. A car and a cyclist start at at 10 A.M. from the same point, headed in the same direction. The average speed of the car is 5 mph more than three times the average speed of the cyclist. In 1.5 h, the car is 46.5 mi ahead of the cyclist. Find the rate of the cyclist.   13 mph

83. At 10 A.M., two campers left their campsite by canoe and paddled downstream at an average speed of 12 mph. They then turned around and paddled back upstream at an average rate of 4 mph. The total trip took 1 h. At what time did the campers turn around downstream?   10:15 A.M.

84. A truck leaves a depot at 11 A.M. and travels at a speed of 45 mph. At noon, a van leaves the same depot and travels the same route at a speed of 65 mph. At what time does the van overtake the truck?   2:15 P.M.

Projects or Group Activities

85. A radiator contains 15 gal of a 20% antifreeze solution. How many gallons must be drained from the radiator and replaced by pure antifreeze so that the radiator will contain 15 gal of a 40% antifreeze solution?   3.75 gal

86. When 5 oz of water are added to an acid solution, the new mixture is $33\frac{1}{3}\%$ acid. When 5 oz of pure acid are added to this new mixture, the resulting mixture is 50% acid. What was the percent concentration of acid in the original mixture?   50%

87. A bicyclist rides for 2 h at a speed of 10 mph and then returns at a speed of 20 mph. Find the cyclist's average speed for the trip. $13\frac{1}{3}$ mph

88. A car travels a 1-mile track at an average speed of 30 mph. At what average speed must the car travel the next mile so that the average speed for the 2 mi is 60 mph?   It is impossible to average 60 mph.

89. A mountain climber ascended a mountain at 0.5 mph and descended twice as fast. The trip took 12 h. How many miles was the round trip?   8 mi

90. ✏️ Explain why we look for patterns and relationships in mathematics. Include a discussion of the relationship between value mixture problems and percent mixture problems, and how understanding one of these can make it easier to understand the other. Also discuss why understanding how to solve the value mixture problems in this section can be helpful in solving Exercise 81.

Unless otherwise noted, all content on this page is © Cengage Learning.

3. A ship leaves a dock at 10 A.M. and travels south at 30 mph. One hour later, a second ship leaves the same dock and travels south at 50 mph. At what time does the second ship overtake the first ship?   **12:30 P.M.**   **[2.4C]**

Yu Lan/Shutterstock.com

2.5   First-Degree Inequalities

OBJECTIVE A   *To solve an inequality in one variable*

The **solution set of an inequality** is a set of numbers each element of which, when substituted for the variable, results in a true inequality.

The inequality at the right is true if the variable is replaced by (for instance) 3, -1.98, or $\frac{2}{3}$.

$$x - 1 < 4$$
$$3 - 1 < 4$$
$$-1.98 - 1 < 4$$
$$\frac{2}{3} - 1 < 4$$

> **Integrating Technology**
>
> See the Keystroke Guide: *Test* for instructions on using a graphing calculator to graph the solution set of an inequality.

There are many values of the variable x that will make the inequality $x - 1 < 4$ true. The solution set of the inequality is any number less than 5. The solution set can be written in set-builder notation as $\{x \mid x < 5\}$.

The graph of the solution set of $x - 1 < 4$ is shown at the right.

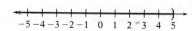

When solving an inequality, we use the **Addition and Multiplication Properties of Inequalities** to rewrite the inequality in the form *variable < constant* or in the form *variable > constant*.

Addition Property of Inequalities

The same term can be added to each side of an inequality without changing the solution set of the inequality. Symbolically, this is written

If $a < b$, then $a + c < b + c$.

If $a > b$, then $a + c > b + c$.

This property is also true for an inequality that contains $\leq$ or $\geq$.

The Addition Property of Inequalities is used to remove a term from one side of an inequality by adding the additive inverse of that term to each side of the inequality. Because subtraction is defined in terms of addition, the same number can be subtracted from each side of an inequality without changing the solution set of the inequality.

> **Take Note**
>
> The solution set of an inequality can be written in set-builder notation or in interval notation.

HOW TO 1   Solve and graph the solution set: $x + 2 \geq 4$

$$x + 2 \geq 4$$
$$x + 2 - 2 \geq 4 - 2 \qquad \bullet \text{ Subtract 2 from each side of the inequality.}$$
$$x \geq 2 \qquad\qquad\quad \bullet \text{ Simplify.}$$

The solution set is $\{x \mid x \geq 2\}$ or, in interval notation, $[2, \infty)$.

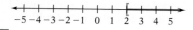

Unless otherwise noted, all content on this page is © Cengage Learning.

HOW TO 2 Solve: $3x - 4 < 2x - 1$
Write the solution set in set-builder notation.

$$3x - 4 < 2x - 1$$
$$3x - 4 - 2x < 2x - 1 - 2x \qquad \bullet \text{ Subtract } 2x \text{ from each side of the inequality.}$$
$$x - 4 < -1$$
$$x - 4 + 4 < -1 + 4 \qquad \bullet \text{ Add 4 to each side of the inequality.}$$
$$x < 3$$

The solution set is $\{x \mid x < 3\}$.

Care must be taken when multiplying each side of an inequality by a nonzero constant. The rule for multiplying each side by a *positive* number is different from the rule for multiplying each side by a *negative* number.

Take Note

$c > 0$ means c is a positive number. Note that the inequality symbol does not change.

$c < 0$ means c is a negative number. Note that the inequality symbol is reversed.

Multiplication Property of Inequalities

Rule 1 Each side of an inequality can be multiplied by the same *positive* constant without changing the solution set of the inequality. Symbolically, this is written

If $a < b$ and $c > 0$, then $ac < bc$.
If $a > b$ and $c > 0$, then $ac > bc$.

Rule 2 If each side of an inequality is multiplied by the same *negative* constant and the inequality symbol is reversed, then the solution set of the inequality is not changed. Symbolically, this is written

If $a < b$ and $c < 0$, then $ac > bc$.
If $a > b$ and $c < 0$, then $ac < bc$.

This property is also true for an inequality that contains $\leq$ or $\geq$.

Here are examples of this property.

Rule 1: Multiply by a *positive* number. **Rule 2:** Multiply by a *negative* number.

$2 < 5$	$3 > 2$	$2 < 5$	$3 > 2$
$2(4) < 5(4)$	$3(4) > 2(4)$	$2(-4) > 5(-4)$	$3(-4) < 2(-4)$
$8 < 20$	$12 > 8$	$-8 > -20$	$-12 < -8$

The Multiplication Property of Inequalities is used to remove a coefficient from one side of an inequality by multiplying each side of the inequality by the reciprocal of the coefficient.

Take Note

Each side of the inequality is divided by a negative number; the inequality symbol must be reversed.

HOW TO 3 Solve: $-3x > 9$
Write the solution set in interval notation.

$$-3x > 9$$
$$\frac{-3x}{-3} < \frac{9}{-3} \qquad \bullet \text{ Divide each side of the inequality by the coefficient } -3. \text{ Because } -3$$
$$\qquad\qquad\qquad \text{is a negative number, the inequality symbol must be reversed.}$$
$$x < -3$$

The solution set is $(-\infty, -3)$.

INSTRUCTOR NOTE
Students are often confused by the need to change the inequality symbol when multiplying or dividing by a negative number. Emphasize that it is what you multiply or divide *by* that is important. For instance:

$$2x < -8$$
$$\frac{2x}{2} < \frac{-8}{2}$$
$$x < -4$$

Here, we divide *by* 2. It does not matter that we are dividing *into* a negative number.

Here is another example:

$$-2x < 8$$
$$\frac{-2x}{-2} > \frac{8}{-2}$$
$$x > -4$$

Now we have divided *by* a negative number; the inequality symbol must be reversed.

Another source of confusion is problems of the following type: $-4x > 0$ and $3x < 0$. You might solve a few inequalities of this type in class, emphasizing that the Multiplication Property of Inequalities still applies.

HOW TO 4 Solve: $3x + 2 < -4$
Write the solution set in set-builder notation.

$$3x + 2 < -4$$
$$ 3x < -6 \qquad \text{• Subtract 2 from each side of the inequality.}$$
$$ \frac{3x}{3} < \frac{-6}{3} \qquad \text{• Divide each side of the inequality by the coefficient 3.}$$
$$ x < -2 \qquad \text{• Because 3 is a positive number, the inequality symbol remains the same.}$$

The solution set is $\{x \mid x < -2\}$.

HOW TO 5 Solve: $2x - 9 > 4x + 5$
Write the solution set in set-builder notation.

$$2x - 9 > 4x + 5$$
$$-2x - 9 > 5 \qquad \text{• Subtract } 4x \text{ from each side of the inequality.}$$
$$-2x > 14 \qquad \text{• Add 9 to each side of the inequality.}$$
$$\frac{-2x}{-2} < \frac{14}{-2} \qquad \text{• Divide each side of the inequality by the coefficient } -2.$$
$$\phantom{\frac{-2x}{-2}} x < -7 \qquad \text{Because } -2 \text{ is a negative number, reverse the inequality symbol.}$$

The solution set is $\{x \mid x < -7\}$.

HOW TO 6 Solve: $5(x - 2) \geq 9x - 3(2x - 4)$
Write the solution set in interval notation.

$$5(x - 2) \geq 9x - 3(2x - 4)$$
$$5x - 10 \geq 9x - 6x + 12 \qquad \text{• Use the Distributive Property to remove parentheses.}$$
$$5x - 10 \geq 3x + 12$$
$$2x - 10 \geq 12 \qquad \text{• Subtract } 3x \text{ from each side of the inequality.}$$
$$2x \geq 22 \qquad \text{• Add 10 to each side of the inequality.}$$
$$\frac{2x}{2} \geq \frac{22}{2} \qquad \text{• Divide each side of the inequality by the coefficient 2.}$$
$$x \geq 11$$

The solution set is $[11, \infty)$.

EXAMPLE 1

Solve and graph the solution set: $\dfrac{1}{6} - \dfrac{3}{4}x > \dfrac{11}{12}$

Write the solution set in set-builder notation.

Solution

$$\frac{1}{6} - \frac{3}{4}x > \frac{11}{12}$$
$$12\left(\frac{1}{6}\right) - 12\left(\frac{3}{4}x\right) > 12\left(\frac{11}{12}\right) \qquad \begin{array}{l} \text{• Clear fractions by} \\ \text{multiplying each side} \\ \text{of the inequality by 12.} \end{array}$$
$$2 - 9x > 11$$
$$-9x > 9 \qquad \text{• Subtract 2 from each side.}$$
$$\frac{-9x}{-9} < \frac{9}{-9} \qquad \text{• Divide each side by } -9.$$
$$x < -1$$

The solution set is $\{x \mid x < -1\}$.

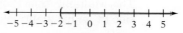

YOU TRY IT 1

Solve and graph the solution set: $2x - 1 < 6x + 7$
Write the solution set in set-builder notation.

Your solution

$\{x \mid x > -2\}$

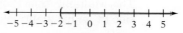

IN-CLASS EXAMPLES
Solve. Write the solution set in set-builder notation and in interval notation.
1. $x - 1 < 3$ $\{x \mid x < 4\}, (-\infty, 4)$
2. $-2x \leq 10$ $\{x \mid x \geq -5\}, [-5, \infty)$
3. $6 - 4x \geq 30$ $\{x \mid x \leq -6\}, (-\infty, -6]$
4. $5x - 3 > 2x + 6$ $\{x \mid x > 3\}, (3, \infty)$

Solution on p. S7

Unless otherwise noted, all content on this page is © Cengage Learning.

EXAMPLE 2

Solve: $3x - 5 \leq 3 - 2(3x + 1)$
Write the solution set in interval notation.

Solution

$3x - 5 \leq 3 - 2(3x + 1)$
$3x - 5 \leq 3 - 6x - 2$
$3x - 5 \leq 1 - 6x$
$9x - 5 \leq 1$
$\quad 9x \leq 6$
$\quad \dfrac{9x}{9} \leq \dfrac{6}{9}$
$\quad\quad x \leq \dfrac{2}{3}$

$\left(-\infty, \dfrac{2}{3} \right]$

YOU TRY IT 2

Solve: $6 - 3(2x + 1) \leq 8 - 4x$
Write the solution set in interval notation.

Your solution

$\left[-\dfrac{5}{2}, \infty \right)$

Solution on p. S7

OBJECTIVE B   *To solve a compound inequality*

INSTRUCTOR NOTE

The compound inequalities given in this section prepare students to solve the absolute value inequalities they will study in the next section.

A **compound inequality** is formed by joining two inequalities with a connective word such as *and* or *or*. The inequalities at the right are compound inequalities.

$2x < 4 \text{ and } 3x - 2 > -8$

$2x + 3 > 5 \text{ or } x + 2 < 5$

The solution set of a compound inequality with the connective word *and* is the set of all elements that are common to the solution sets of both inequalities. Therefore, it is the intersection of the solution sets of the two inequalities.

HOW TO 7   Solve: $2x < 6 \text{ and } 3x + 2 > -4$

$\begin{array}{lll} 2x < 6 & \text{and} & 3x + 2 > -4 \\ x < 3 & & 3x > -6 \quad \bullet \text{ Solve each inequality.} \\ \{x | x < 3\} & & x > -2 \\ & & \{x | x > -2\} \end{array}$

The solution set of a compound inequality with the word *and* is the intersection of the solution sets of the two inequalities.

$\{x | x < 3\} \cap \{x | x > -2\} = \{x | -2 < x < 3\}$ or, in interval notation, $(-2, 3)$.

HOW TO 8   Solve: $-3 < 2x + 1 < 5$

This inequality is equivalent to the compound inequality $-3 < 2x + 1$ and $2x + 1 < 5$.

$\begin{array}{lll} -3 < 2x + 1 & \text{and} & 2x + 1 < 5 \\ -4 < 2x & & 2x < 4 \quad \bullet \text{ Solve each inequality.} \\ -2 < x & & x < 2 \\ \{x | x > -2\} & & \{x | x < 2\} \end{array}$

The solution set of a compound inequality with the word *and* is the intersection of the solution sets of the two inequalities.

$\{x | x > -2\} \cap \{x | x < 2\} = \{x | -2 < x < 2\}$ or, in interval notation, $(-2, 2)$.

INSTRUCTOR NOTE
Have students solve the
compound inequality
$2x - 1 > 5$ and $3x - 2 < 1$.
The solution will require that
they find the intersection of
$\{x \mid x > 3\}$ and $\{x \mid x < 1\}$,
which is the empty set.
Suggest that another way
to think about the solution
is to ask, "What number is
greater than 3 *and* less than
1?" Because there is no such
number, the solution set is
the empty set.

Next, ask students whether
the solution set changes
if the word *and* in the
compound inequality is
replaced by *or*. The answer
is yes. The solution set
is now the union of the
two sets $\{x \mid x > 3\}$ and
$\{x \mid x < 1\}$. Another way to
think about this solution is
to say, "We are looking for a
number that is greater than 3
or less than 1."

There is an alternative method of solving the inequality in HOW TO 8.

HOW TO 9　Solve: $-3 < 2x + 1 < 5$

$$-3 < 2x + 1 < 5$$
$$-3 - 1 < 2x + 1 - 1 < 5 - 1$$
$$-4 < 2x < 4$$
$$\frac{-4}{2} < \frac{2x}{2} < \frac{4}{2}$$
$$-2 < x < 2$$

- Subtract 1 from each of the three parts of the inequality.

- Divide each of the three parts of the inequality by the coefficient 2.

The solution set is $\{x \mid -2 < x < 2\}$ or, in interval notation, $(-2, 2)$.

The solution set of a compound inequality with the connective word *or* is the union of the solution sets of the two inequalities.

HOW TO 10　Solve: $2x + 3 > 7$ or $4x - 1 < 3$

$$\begin{array}{ccc} 2x + 3 > 7 & \text{or} & 4x - 1 < 3 \\ 2x > 4 & & 4x < 4 \\ x > 2 & & x < 1 \\ \{x \mid x > 2\} & & \{x \mid x < 1\} \end{array}$$

- Solve each inequality.

The solution set of a compound inequality with the word *or* is the union of the solution sets of the two inequalities.

$$\{x \mid x > 2\} \cup \{x \mid x < 1\} \text{ or, in interval notation, } (-\infty, 1) \cup (2, \infty).$$

EXAMPLE 3

Solve: $1 < 3x - 5 < 4$
Write the solution set in interval notation.

Solution

$$1 < 3x - 5 < 4$$
$$1 + 5 < 3x - 5 + 5 < 4 + 5$$
$$6 < 3x < 9$$
$$\frac{6}{3} < \frac{3x}{3} < \frac{9}{3}$$
$$2 < x < 3$$
$$(2, 3)$$

- Add 5 to each of the three parts.

- Divide each of the three parts by 3.

YOU TRY IT 3

Solve: $-2 \le 5x + 3 \le 13$
Write the solution set in interval notation.

Your solution

$[-1, 2]$

IN-CLASS EXAMPLES
Solve. Write the solution set in
set-builder notation.
5.　$x + 1 \ge 6$ or $2x < 6$
　　$\{x \mid x \ge 5\} \cup \{x \mid x < 3\}$
6.　$4x - 3 > 5$ or $4x - 3 \le -15$
　　$\{x \mid x > 2\} \cup \{x \mid x \le -3\}$
Solve. Write the solution set in
interval notation.
7.　$2x < 8$ and $x + 3 > 1$　$(-2, 4)$
8.　$-1 < 2x + 5 < 9$　$(-3, 2)$

EXAMPLE 4

Solve: $11 - 2x > -3$ and $7 - 3x \le 4$
Write the solution set in set-builder notation.

Solution

$$\begin{array}{ccc} 11 - 2x > -3 & \text{and} & 7 - 3x \le 4 \\ -2x > -14 & & -3x \le -3 \\ x < 7 & & x \ge 1 \\ \{x \mid x < 7\} & & \{x \mid x \ge 1\} \end{array}$$
$$\{x \mid x < 7\} \cap \{x \mid x \ge 1\} = \{x \mid 1 \le x < 7\}$$

YOU TRY IT 4

Solve: $2 - 3x > 11$ or $5 + 2x > 7$
Write the solution set in set-builder notation.

Your solution

$\{x \mid x < -3\} \cup \{x \mid x > 1\}$

Solutions on p. S7

OBJECTIVE C   *To solve application problems*

EXAMPLE 5

A U.S. cellular phone company offers a golfer traveling to Ireland two plans. The first plan costs $5.99 per month with roaming rates of $.99 per minute. The second package has no monthly fee and roaming rates of $1.39 per minute. What minimum number of minutes must the golfer use in one month to make the first plan more economical than the second?

Strategy

To find the number of minutes, write and solve an inequality using x to represent the number of minutes of roaming time used by the golfer. Then the cost of the first plan is $5.99 + 0.99x$, and the cost of the second plan is $1.39x$.

Solution

Cost of plan 1 $<$ Cost of plan 2
$$5.99 + 0.99x < 1.39x$$
$$5.99 < 0.40x$$
$$14.975 < x$$

The golfer must use at least 15 min of roaming time.

EXAMPLE 6

Angeline is training for a triathlon. She wants to increase her training distance by 1 km each day and run a total of at least 50 km over the next five days. What is the minimum number of kilometers she must run on the first day to achieve her goal?

Strategy

To find the minimum number of kilometers to be run on the first day, write and solve an inequality using x to represent the number of kilometers run on the first day. Then the numbers of kilometers run on the next four days are $x + 1$, $x + 2$, $x + 3$, and $x + 4$.

Solution

Angeline must run at least 50 km over the five days.
$$x + (x + 1) + (x + 2) + (x + 3) + (x + 4) \geq 50$$
$$5x + 10 \geq 50$$
$$5x \geq 40$$
$$x \geq 8$$

Angeline must run at least 8 km on the first day.

YOU TRY IT 5

The base of a triangle is 12 in., and the height is $(x + 2)$ in. Express as an integer the maximum height of the triangle when the area is less than 50 in^2.

Your strategy

Your solution

8 in.

IN-CLASS EXAMPLES

9. Four times the difference between a number and six is less than or equal to six times the sum of the number and four. Find the smallest number that will satisfy the inequality.   **−24**

10. Company A rents a car for $36 per day and 10¢ for every mile driven. Company B rents a car for $68.50 per day with unlimited mileage. How many miles per day can you drive a Company A car if it is to cost you less than a Company B car?   **Less than 325 mi**

YOU TRY IT 6

An average score of 80 to 89 in a history course receives a B. Luisa Montez has grades of 72, 94, 83, and 70 on four exams. Find the range of scores on the fifth exam that will give Luisa a B for the course.

Your strategy

Your solution

$81 \leq N \leq 100$

Solutions on pp. S7–S8

2.5 EXERCISES

SUGGESTED ASSIGNMENT
Exercises 1–6; Exercises 7–51, every other odd; Exercises 55–109, odds;
More challenging exercises: Exercises 111–115

✔ Concept Check

1. State the Addition Property of Inequalities, and give numerical examples of its use.

2. State the Multiplication Property of Inequalities, and give numerical examples of its use.

3. Which numbers are solutions of the inequality $x + 7 \le -3$?
   (i) -17 **(ii)** 8   **(iii)** -10 **(iv)** 0
   i, iii

4. Which numbers are solutions of the inequality $2x - 1 > 5$?
   (i) 6   **(ii)** -4 **(iii)** 3   **(iv)** 5
   i, iv

5. Fill in the blank with $<$, $\le$, $>$, or $\ge$.
   If $-x > 0$, then x ____$<$____ 0.

6. **a.** Which set operation is used when a compound inequality is combined with *or*?
   b. Which set operation is used when a compound inequality is combined with *and*?

OBJECTIVE A   *To solve an inequality in one variable*

For Exercises 7 to 33, solve. Write the solution set in set-builder notation. For Exercises 7 to 12, graph the solution set.

7. $x - 3 < 2$ $\{x \mid x < 5\}$

8. $x + 4 \ge 2$ $\{x \mid x \ge -2\}$

9. $4x \le 8$ $\{x \mid x \le 2\}$

10. $6x > 12$ $\{x \mid x > 2\}$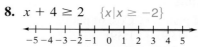

11. $-2x > 8$ $\{x \mid x < -4\}$

12. $-3x \le -9$ $\{x \mid x \ge 3\}$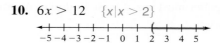

13. $3x - 1 > 2x + 2$
    $\{x \mid x > 3\}$

14. $5x + 2 \ge 4x - 1$
    $\{x \mid x \ge -3\}$

15. $2x - 1 > 7$
    $\{x \mid x > 4\}$

16. $3x + 2 < 8$
    $\{x \mid x < 2\}$

17. $5x - 2 \le 8$
    $\{x \mid x \le 2\}$

18. $4x + 3 \le -1$
    $\{x \mid x \le -1\}$

19. $6x + 3 > 4x - 1$
    $\{x \mid x > -2\}$

20. $7x + 4 < 2x - 6$
    $\{x \mid x < -2\}$

21. $8x + 1 \ge 2x + 13$
    $\{x \mid x \ge 2\}$

22. $5x - 4 < 2x + 5$
    $\{x \mid x < 3\}$

23. $4 - 3x < 10$
    $\{x \mid x > -2\}$

24. $2 - 5x > 7$
    $\{x \mid x < -1\}$

Unless otherwise noted, all content on this page is © Cengage Learning.

25. $7 - 2x \geq 1$
$\{x \mid x \leq 3\}$

26. $3 - 5x \leq 18$
$\{x \mid x \geq -3\}$

27. $-3 - 4x > -11$
$\{x \mid x < 2\}$

28. $-2 - x < 7$
$\{x \mid x > -9\}$

29. $4x - 2 < x - 11$
$\{x \mid x < -3\}$

30. $6x + 5 \leq x - 10$
$\{x \mid x \leq -3\}$

31. $x + 7 \geq 4x - 8$

$\{x \mid x \leq 5\}$

32. $3x + 1 \leq 7x - 15$

$\{x \mid x \geq 4\}$

33. $3x + 2 \leq 7x + 4$

$\left\{ x \,\middle|\, x \geq -\dfrac{1}{2} \right\}$

For Exercises 34 to 37, state whether the solution set of an inequality of the given form contains only negative numbers, only positive numbers, or both positive and negative numbers.

34. $x + n > a$, where both n and a are positive, and $n < a$ Only positive numbers

35. $nx > a$, where both n and a are negative   Both positive and negative numbers

36. $nx > a$, where n is negative and a is positive   Only negative numbers

37. $x - n > -a$, where both n and a are positive, and $n < a$ Both positive and negative numbers

For Exercises 38 to 53, solve. Write the solution set in interval notation.

38. $3x - 5 \geq -2x + 5$

$[2, \infty)$

39. $7x + 3 < 4x + 1$

$\left(-\infty, -\dfrac{2}{3} \right)$

40. $5x - 7 \leq x - 9$

$\left(-\infty, -\dfrac{1}{2} \right]$

41. $\dfrac{2}{3}x - \dfrac{3}{2} < \dfrac{7}{6} - \dfrac{1}{3}x$

$\left(-\infty, \dfrac{8}{3} \right)$

42. $\dfrac{7}{12}x - \dfrac{3}{2} < \dfrac{2}{3}x + \dfrac{5}{6}$

$(-28, \infty)$

43. $\dfrac{1}{2}x - \dfrac{3}{4} < \dfrac{7}{4}x - 2$

$(1, \infty)$

44. $6 - 2(x - 4) \leq 2x + 10$

$[1, \infty)$

45. $4(2x - 1) > 3x - 2(3x - 5)$

$\left(\dfrac{14}{11}, \infty \right)$

46. $2(1 - 3x) - 4 > 10 + 3(1 - x)$

$(-\infty, -5)$

47. $2 - 5(x + 1) \geq 3(x - 1) - 8$

$(-\infty, 1]$

48. $2 - 2(7 - 2x) < 3(3 - x)$

$(-\infty, 3)$

49. $3 + 2(x + 5) \geq x + 5(x + 1) + 1$

$\left(-\infty, \dfrac{7}{4} \right]$

50. $10 - 13(2 - x) < 5(3x - 2)$

$(-3, \infty)$

51. $3 - 4(x + 2) \leq 6 + 4(2x + 1)$

$\left[-\dfrac{5}{4}, \infty \right)$

52. $3x - 2(3x - 5) \leq 2 - 5(x - 4)$

$(-\infty, 6]$

53. $12 - 2(3x - 2) \geq 5x - 2(5 - x)$

$(-\infty, 2]$

OBJECTIVE B   *To solve a compound inequality*

For Exercises 54 to 67, solve. Write the solution set in interval notation.

54. $3x < 6$ and $x + 2 > 1$
$(-1, 2)$

55. $x - 3 \le 1$ and $2x \ge -4$
$[-2, 4]$

56. $x + 2 \ge 5$ or $3x \le 3$
$(-\infty, 1] \cup [3, \infty)$

57. $2x < 6$ or $x - 4 > 1$
$(-\infty, 3) \cup (5, \infty)$

58. $-2x > -8$ and $-3x < 6$
$(-2, 4)$

59. $\frac{1}{2}x > -2$ and $5x < 10$
$(-4, 2)$

60. $\frac{1}{3}x < -1$ or $2x > 0$
$(-\infty, -3) \cup (0, \infty)$

61. $\frac{2}{3}x > 4$ or $2x < -8$
$(-\infty, -4) \cup (6, \infty)$

62. $x + 4 \ge 5$ and $2x \ge 6$
$[3, \infty)$

63. $3x < -9$ and $x - 2 < 2$
$(-\infty, -3)$

64. $-5x > 10$ and $x + 1 > 6$
$\varnothing$

65. $2x - 3 > 1$ and $3x - 1 < 2$
$\varnothing$

66. $7x < 14$ and $1 - x < 4$
$(-3, 2)$

67. $4x + 1 < 5$ and $4x + 7 > -1$
$(-2, 1)$

For Exercises 68 to 71, state whether the inequality describes the empty set, all real numbers, two intervals of real numbers, or one interval of real numbers.

68. $x > -3$ and $x > 2$
One interval of real numbers

69. $x > -3$ or $x < 2$
All real numbers

70. $x < -3$ and $x > 2$
Empty set

71. $x < -3$ or $x > 2$
Two intervals of real numbers

For Exercises 72 to 91, solve. Write the solution set in set-builder notation.

72. $3x + 7 < 10$ or $2x - 1 > 5$
$\{x|x < 1\} \cup \{x|x > 3\}$

73. $6x - 2 < -14$ or $5x + 1 > 11$
$\{x|x < -2\} \cup \{x|x > 2\}$

74. $-5 < 3x + 4 < 16$
$\{x|-3 < x < 4\}$

75. $5 < 4x - 3 < 21$
$\{x|2 < x < 6\}$

76. $0 < 2x - 6 < 4$
$\{x|3 < x < 5\}$

77. $-2 < 3x + 7 < 1$
$\{x|-3 < x < -2\}$

78. $4x - 1 > 11$ or $4x - 1 \leq -11$

$\{x|x > 3\} \cup \left\{x \middle| x \leq -\dfrac{5}{2}\right\}$

79. $3x - 5 > 10$ or $3x - 5 < -10$

$\{x|x > 5\} \cup \left\{x \middle| x < -\dfrac{5}{3}\right\}$

80. $9x - 2 < 7$ and $3x - 5 > 10$

$\varnothing$

81. $8x + 2 \leq -14$ and $4x - 2 > 10$

$\varnothing$

82. $3x - 11 < 4$ or $4x + 9 \geq 1$

The set of real numbers

83. $5x + 12 \geq 2$ or $7x - 1 \leq 13$

The set of real numbers

84. $-6 \leq 5x + 14 \leq 24$

$\{x|-4 \leq x \leq 2\}$

85. $3 \leq 7x - 14 \leq 31$

$\left\{x \middle| \dfrac{17}{7} \leq x \leq \dfrac{45}{7}\right\}$

86. $3 - 2x > 7$ and $5x + 2 > -18$

$\{x|-4 < x < -2\}$

87. $1 - 3x < 16$ and $1 - 3x > -16$

$\left\{x \middle| -5 < x < \dfrac{17}{3}\right\}$

88. $5 - 4x > 21$ or $7x - 2 > 19$

$\{x|x < -4\} \cup \{x|x > 3\}$

89. $6x + 5 < -1$ or $1 - 2x < 7$

The set of real numbers

90. $3 - 7x \leq 31$ and $5 - 4x > 1$

$\{x|-4 \leq x < 1\}$

91. $9 - x \geq 7$ and $9 - 2x < 3$

$\varnothing$

OBJECTIVE C   *To solve application problems*

Exercises 92 to 95 make statements about temperatures t on a particular day. Match each statement with one of the following inequalities. Some inequalities may be used more than once.

$t > 21$	$t < 21$	$t \geq 21$	$t \leq 21$	$21 \leq t \leq 42$
$t < 42$	$t > 42$	$t \leq 42$	$t \geq 42$	$21 < t < 42$

92. The low temperature was 21°F.

$t \geq 21$

93. The temperature did not go above 42°F.

$t \leq 42$

94. The temperature ranged from 21°F to 42°F.

$21 \leq t \leq 42$

95. The high temperature was 42°F.

$t \leq 42$

96. Geometry The length of a rectangle is 2 ft more than four times the width. Express as an integer the maximum width of the rectangle when the perimeter is less than 34 ft.   2 ft

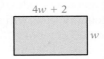

$4w + 2$

w

97. Geometry The length of a rectangle is 5 cm less than twice the width. Express as an integer the maximum width of the rectangle when the perimeter is less than 60 cm.   11 cm

Unless otherwise noted, all content on this page is © Cengage Learning.

98. ◐ **Aquariums** The following is a rule-of-thumb for making sure fish kept in an aquarium are not too crowded: The surface area of the water should be at least 12 times the total combined length of the fish kept in the aquarium. (*Source:* www.takomapet.com) Your 10-gallon aquarium has a water surface area of 288 in^2 and houses the following fish: one 2-inch odessa barb, three 1-inch gold tetra, three 1.75-inch cobra guppies, and five 1-inch neon tetra.
   a. Find the total combined length of all the fish in your aquarium.   15.25 in.
   b. Write and solve an inequality to find the greatest number n of 2-inch black hatchetfish that you can safely add to your aquarium without overcrowding the fish. $288 \geq 12(15.25 + 2n)$; 4 hatchetfish

99. **Advertising** To run an advertisement on a certain website, the website owner charges a setup fee of $250 and $12 per day to display the advertisement. If a marketing group has a budget of $1500 for an advertisement, what is the maximum number of days the advertisement can run on the site?
   104 days

100. **Consumerism** The entry fee to a state fair is $25 and includes five tickets for carnival rides at the fair. Additional tickets for carnival rides cost $1.50 each. If Alisha wants to spend a maximum of $45 for the entry fee and rides, how many additional carnival ride tickets can she purchase?
   13 tickets

101. **Consumerism** A homeowner has a budget of $100 to paint a room that has 320 ft^2 of wall space. Drop cloths, masking tape, and paint brushes cost $24. If 1 gal of paint will cover 100 ft^2 of wall space, what is the maximum cost per gallon of paint that the homeowner can pay?
   $19

102. **Temperature** The temperature range for a week was between 14°F and 77°F. Find the temperature range in degrees Celsius. Use the equation $F = \frac{9}{5}C + 32$.
   $-10° < C < 25°$

103. **Temperature** The temperature range for a week in a mountain town was between 0°C and 30°C. Find the temperature range in degrees Fahrenheit. Use the equation $C = \frac{5(F - 32)}{9}$.
   $32° < F < 86°$

104. **Compensation** You are a sales account executive earning $1200 per month plus 6% commission on the amount of sales. Your goal is to earn a minimum of $6000 per month. What amount of sales will enable you to earn $6000 or more per month?
   $80,000 or more

105. **Compensation** George Stoia earns $1000 per month plus 5% commission on the amount of sales. George's goal is to earn a minimum of $3200 per month. What amount of sales will enable George to earn $3200 or more per month?
   $44,000 or more

106. **Education** Some Chinese language students and their professors are planning a trip to China. Their goal is to practice their language skills and learn more about Chinese culture. In China, the group will be transported by small buses that can hold a maximum of 12 people. If 70 students and 10 professors are going on the trip, what is the minimum number of buses needed?
   7 buses

107. Business An organic juice company would like to increase its production of juice. The company's plan for the next five months is to increase the number of gallons of juice it produces each month by 400 gal. For the five-month period, the company wants to produce at least 8500 gal of juice. What is the minimum number of gallons of juice the company can produce in the first month if it is to achieve its goal?
900 gal

108. Education An average score of 90 or above in a history class receives an A grade. You have scores of 95, 89, and 81 on three exams. Find the range of scores on the fourth exam that will give you an A grade for the course.
$95 \leq N \leq 100$

109. Education An average of 70 to 79 in a mathematics class receives a C grade. A student has scores of 56, 91, 83, and 62 on four tests. Find the range of scores on the fifth test that will give the student a C for the course.
$58 \leq n \leq 100$

110. Hybrid Vehicles See the news clipping at the right. If a typical city bus has a fuel tank that holds 112 gal of diesel fuel, find the range of miles the buses in a city's fleet can travel on a full tank of fuel.
Between 392 mi and 560 mi

> **In the NEWS!**
>
> **Cities Introducing Hybrid Buses**
>
> More and more cities around the country are introducing diesel-electric hybrid buses to their public transportation bus fleets, achieving significant improvement in miles traveled on a single tank of fuel. While a city's conventional diesel buses may average as few as 3.5 mpg, the hybrids can average up to 5 mpg.
> *Sources:* www.boston.com, www.courier-journal.com, www.naplesnews.com, sanantonio.bizjournals.com

Critical Thinking

111. Let $-2 \leq x \leq 3$ and $a \leq 2x + 1 \leq b$.
    a. Find the largest possible value of a. -3
    b. Find the smallest possible value of b.   7

For Exercises 112 to 115, determine whether the statement is true or false.

112. If $a < b$ and $b > c$, then $c < a$.   False

113. If $a < b$ and $c > b$, then $a < b < c$.   True

114. If $a > b$, $c > d$, and $c < b$, then the smallest number is d.   True

115. If $a \leq b$ and $b \leq a$, then $a = b$.   True

QUICK QUIZ

Solve. Write the solution set in set-builder notation and in interval notation.
1. $2 - 7x \leq 16$
   $\{x \mid x \geq -2\}$, $[-2, \infty)$ **[2.5A]**
2. $6x + 1 \geq 4x - 5$
   $\{x \mid x \geq -3\}$, $[-3, \infty)$ **[2.5A]**
3. $x - 5 \leq 2$ and $3x > -12$
   $\{x \mid -4 < x \leq 7\}$, $(-4, 7]$ **[2.5B]**
4. $-2 < 3x + 1 < 10$
   $\{x \mid -1 < x < 3\}$, $(-1, 3)$ **[2.5B]**
5. $3x - 7 > 5$ or $3x - 7 < -10$
   $\{x \mid x < -1\} \cup \{x \mid x > 4\}$,
   $(-\infty, -1) \cup (4, \infty)$ **[2.5B]**
6. The length of a rectangle is 3 m less than four times the width. Express as an integer the maximum width of the rectangle when the perimeter is less than 74 m.
   7 m **[2.5C]**

Projects or Group Activities

116. Determine whether the following statements are always true, sometimes true, or never true.

    a. If $a > b$, then $-a < -b$.   Always true

    b. If $a < b$ and $a \neq 0$, $b \neq 0$, then $\frac{1}{a} < \frac{1}{b}$.   Sometimes true

    c. When dividing both sides of an inequality by an integer, we must reverse the inequality symbol.   Sometimes true

    d. If $a < 1$, then $a^2 < a$.   Sometimes true

    e. If $a < b < 0$ and $c < d < 0$, then $ac > bd$.   Always true

2.6 Absolute Value Equations and Inequalities

OBJECTIVE A   *To solve an absolute value equation*

Tips for Success

Before the class meeting in which your professor begins a new section, you should read each objective statement for that section. Next, browse through the objective material. The purpose of browsing through the material is to set the stage for your brain to accept and organize new information when it is presented to you. See *AIM for Success* in the Preface.

Recall that the *absolute value* of a number is its distance from zero on the number line. Distance is always a positive number or zero. Therefore, the absolute value of a number is always a positive number or zero.

The distance from 0 to 3 or from 0 to -3 is 3 units.

$$|3| = 3 \qquad |-3| = 3$$

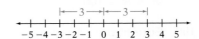

Absolute value can be used to represent the distance between any two points on the number line. The **distance between two points** on the number line is the absolute value of the difference between the coordinates of the two points.

The distance between point a and point b is given by $|b - a|$.

The distance between 4 and -3 on the number line is 7 units. Note that the order in which the coordinates are subtracted does not affect the distance.

$$\begin{aligned} \text{Distance} &= |-3 - 4| \\ &= |-7| \\ &= 7 \end{aligned} \qquad \begin{aligned} \text{Distance} &= |4 - (-3)| \\ &= |7| \\ &= 7 \end{aligned}$$

For any two numbers a and b, $|b - a| = |a - b|$.

An equation containing a variable within an absolute value symbol is called an **absolute value equation.** Here are three examples.

$$|x| = 3 \qquad\qquad |x + 2| = 8 \qquad\qquad |3x - 4| = 5x - 9$$

Take Note

You should always check your answers. Here is the check for examples (1), (2), and (3) at the right.

$$\begin{array}{c|c} |x| = 3 & |x| = 3 \\ \hline |-3| \mid 3 & |3| \mid 3 \\ 3 = 3 & 3 = 3 \end{array}$$

$$\begin{array}{c|c} |-x| = 8 & |-x| = 8 \\ \hline |-(-8)| \mid 8 & |-8| \mid 8 \\ |8| \mid 8 & 8 = 8 \\ 8 = 8 & \end{array}$$

$$\begin{array}{c|c} |x + 3| = 4 & |x + 3| = 4 \\ \hline |1 + 3| \mid 4 & |-7 + 3| \mid 4 \\ |4| \mid 4 & |-4| \mid 4 \\ 4 = 4 & 4 = 4 \end{array}$$

Solutions of an Absolute Value Equation

If $a > 0$ and $|x| = a$, then $x = -a$ or $x = a$. If $|x| = 0$, then $x = 0$. If $a < 0$, then $|x| = a$ has no solution.

EXAMPLES

1. If $|x| = 3$, then $x = -3$ or $x = 3$.
2. If $|-x| = 8$, then $x = -8$ or $x = 8$.
3. If $|x + 3| = 4$, then $x + 3 = 4$ or $x + 3 = -4$. The solution of $x + 3 = 4$ is 1. The solution of $x + 3 = -4$ is -7.
4. If $|z| = 0$, then $z = 0$.
5. If $|y| = -2$, then the equation has no solution. The absolute value of any number is greater than or equal to zero.

Unless otherwise noted, all content on this page is © Cengage Learning.

HOW TO 1   Solve: $|x + 2| = 8$

$$|x + 2| = 8$$
$$x + 2 = 8 \qquad x + 2 = -8$$ • Remove the absolute value sign and rewrite as two equations.
$$x = 6 \qquad\qquad x = -10$$ • Solve each equation.

Check:

$|x + 2| = 8$		$|x + 2| = 8$	
$|6 + 2|$	8	$|-10 + 2|$	8
$|8|$	8	$|-8|$	8
8 = 8		8 = 8	

The solutions are 6 and -10.

HOW TO 2   Solve: $|5 - 3x| - 8 = -4$

$$|5 - 3x| - 8 = -4$$
$$|5 - 3x| = 4$$ • Solve for the absolute value.
$$5 - 3x = 4 \qquad\qquad 5 - 3x = -4$$ • Remove the absolute value sign and rewrite as two equations.

$$-3x = -1 \qquad\qquad -3x = -9$$ • Solve each equation.
$$x = \frac{1}{3} \qquad\qquad x = 3$$

Check:

$|5 - 3x| - 8 = -4$		$|5 - 3x| - 8 = -4$	
$\left|5 - 3\left(\frac{1}{3}\right)\right| - 8$	-4	$|5 - 3(3)| - 8$	-4
$|5 - 1| - 8$	-4	$|5 - 9| - 8$	-4
$4 - 8$	-4	$4 - 8$	-4
$-4 = -4$		$-4 = -4$	

The solutions are $\frac{1}{3}$ and 3.

IN-CLASS EXAMPLES
Solve.
1. $|x + 3| = 1$ **$-2, -4$**
2. $|2 - 3x| = 2$ **$0, \frac{4}{3}$**
3. $|7 - y| - 2 = 2$ **3, 11**

EXAMPLE 1

Solve: $|2 - x| = 12$

Solution

$$|2 - x| = 12$$
$$2 - x = 12 \qquad 2 - x = -12$$
$$-x = 10 \qquad\quad -x = -14$$ • Subtract 2.
$$x = -10 \qquad\quad x = 14$$ • Multiply by -1.

The solutions are -10 and 14.

YOU TRY IT 1

Solve: $|2x - 3| = 5$

Your solution
4 and -1

EXAMPLE 2

Solve: $|2x| = -4$

Solution

$$|2x| = -4$$
There is no solution to this equation because the absolute value of a number must be nonnegative.

YOU TRY IT 2

Solve: $|x - 3| = -2$

Your solution
No solution

Solutions on p. S8

EXAMPLE 3

Solve: $3 - |2x - 4| = -5$

Solution

$$
\begin{aligned}
3 - |2x - 4| &= -5 \\
-|2x - 4| &= -8 \qquad \text{• Subtract 3.} \\
|2x - 4| &= 8 \qquad \text{• Multiply by } -1.
\end{aligned}
$$

$$
\begin{array}{ll}
2x - 4 = 8 & \quad 2x - 4 = -8 \\
\quad 2x = 12 & \qquad 2x = -4 \\
\qquad x = 6 & \qquad\ x = -2
\end{array}
$$

The solutions are 6 and -2.

YOU TRY IT 3

Solve: $5 - |3x + 5| = 3$

Your solution

-1 and $-\dfrac{7}{3}$

Solution on p. S8

OBJECTIVE B *To solve an absolute value inequality*

Recall that absolute value represents the distance between two points. For example, the solutions of the absolute value equation $|x - 1| = 3$ are the numbers whose distance from 1 is 3. Therefore, the solutions are -2 and 4. An **absolute value inequality** is an inequality that contains a variable within an absolute value symbol.

The solutions of the absolute value inequality $|x - 1| < 3$ are the numbers whose distance from 1 is less than 3. Therefore, the solutions are the numbers greater than -2 and less than 4. The solution set is $\{x | -2 < x < 4\}$.

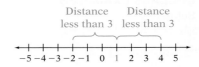

Distance less than 3   Distance less than 3

To solve an absolute value inequality of the form $|ax + b| < c$, solve the equivalent compound inequality $-c < ax + b < c$.

HOW TO 3 Solve: $|3x - 1| < 5$

$$
\begin{aligned}
|3x - 1| &< 5 \\
-5 &< 3x - 1 < 5 \qquad \text{• Solve the equivalent compound inequality.} \\
-5 + 1 &< 3x - 1 + 1 < 5 + 1 \\
-4 &< 3x < 6 \\
\frac{-4}{3} &< \frac{3x}{3} < \frac{6}{3} \\
-\frac{4}{3} &< x < 2
\end{aligned}
$$

The solution set is $\left\{x \,\middle|\, -\frac{4}{3} < x < 2\right\}$.

Take Note

In this objective, we will write all solution sets in set-builder notation.

The solutions of the absolute value inequality $|x + 1| > 2$ are the numbers whose distance from -1 is greater than 2. Therefore, the solutions are the numbers that are less than -3 or greater than 1. The solution set of $|x + 1| > 2$ is $\{x | x < -3\} \cup \{x | x > 1\}$.

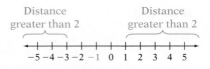

Distance greater than 2   Distance greater than 2

Unless otherwise noted, all content on this page is © Cengage Learning.

Take Note

Carefully observe the difference between the method used to solve $|ax + b| > c$ shown here and that used to solve $|ax + b| < c$ shown on the preceding page.

To solve an absolute value inequality of the form $|ax + b| > c$, solve the equivalent compound inequality $ax + b < -c$ or $ax + b > c$.

HOW TO 4   Solve: $|3 - 2x| > 1$

$$3 - 2x < -1 \quad \text{or} \quad 3 - 2x > 1$$
$$-2x < -4 \qquad\qquad -2x > -2 \qquad \bullet \text{ Solve each inequality.}$$
$$x > 2 \qquad\qquad\qquad x < 1$$
$$\{x | x > 2\} \qquad\qquad \{x | x < 1\}$$

The solution set of a compound inequality with the word *or* is the union of the solution sets of the two inequalities.

$$\{x | x > 2\} \cup \{x | x < 1\}$$

The rules for solving absolute value inequalities are summarized below.

IN-CLASS EXAMPLES

Solve.

4. $|x + 2| > 3$
   $\{x | x > 1\} \cup \{x | x < -5\}$
5. $|x - 3| \le 5$
   $\{x | -2 \le x \le 8\}$
6. $|10 - 2x| \le 0$
   $\{x | x = 5\}$
7. $|9 - 3x| \ge 0$
   $\{x | -\infty < x < \infty\}$

Solutions of Absolute Value Inequalities

To solve an absolute value inequality of the form $|ax + b| < c$, $c > 0$, solve the equivalent compound inequality $-c < ax + b < c$.

To solve an absolute value inequality of the form $|ax + b| > c$, solve the equivalent compound inequality $ax + b < -c$ or $ax + b > c$.

EXAMPLE 4

Solve: $|4x - 3| < 5$

Solution

Solve the equivalent compound inequality.

$$-5 < 4x - 3 < 5$$
$$-5 + 3 < 4x - 3 + 3 < 5 + 3$$
$$-2 < 4x < 8$$
$$\frac{-2}{4} < \frac{4x}{4} < \frac{8}{4}$$
$$-\frac{1}{2} < x < 2$$
$$\left\{ x \,\middle|\, -\frac{1}{2} < x < 2 \right\}$$

YOU TRY IT 4

Solve: $|3x + 2| < 8$

Your solution

$$\left\{ x \,\middle|\, -\frac{10}{3} < x < 2 \right\}$$

EXAMPLE 5

Solve: $|x - 3| < 0$

Solution

The absolute value of a number is greater than or equal to zero, since it measures the number's distance from zero on the number line. Therefore, the solution set of $|x - 3| < 0$ is the empty set.

YOU TRY IT 5

Solve: $|3x - 7| < 0$

Your solution

$\varnothing$

Solutions on p. S8

EXAMPLE 6

Solve: $|x + 4| > -2$

Solution

The absolute value of a number is greater than or equal to zero. Therefore, the solution set of $|x + 4| > -2$ is the set of real numbers.

YOU TRY IT 6

Solve: $|2x + 7| \geq -1$

Your solution

The set of real numbers

EXAMPLE 7

Solve: $|2x - 1| > 7$

Solution

Solve the equivalent compound inequality.

$$
\begin{array}{ll}
2x - 1 < -7 & \text{or} \quad 2x - 1 > 7 \\
\quad 2x < -6 & \qquad 2x > 8 \\
\quad\ \ x < -3 & \qquad\ \ x > 4 \\
\{x|x < -3\} & \{x|x > 4\}
\end{array}
$$

$\{x|x < -3\} \cup \{x|x > 4\}$

YOU TRY IT 7

Solve: $|5x + 3| > 8$

Your solution

$\left\{ x \middle| x < -\dfrac{11}{5} \right\} \cup \{x|x > 1\}$

Solutions on p. S8

OBJECTIVE C *To solve application problems*

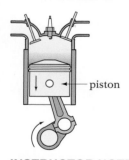

piston

INSTRUCTOR NOTE

Have students write absolute value inequalities from data you give them. For example:

a. An adult's normal body temperature is within 1°F of 98.6°F.
$|t - 98.6| \leq 1$

b. The net weight of a cereal box labeled "20 oz" must be within 0.45 oz of 20 oz.
$|w - 20| \leq 0.45$

The **tolerance** of a component, or part, is the amount by which it is acceptable for the component to vary from a given measurement. For example, the diameter of a piston may vary from the given measurement of 9 cm by 0.001 cm. This is written "9 cm ± 0.001 cm" and is read "9 centimeters plus or minus 0.001 centimeter." The maximum diameter, or **upper limit,** of the piston is 9 cm + 0.001 cm = 9.001 cm. The minimum diameter, or **lower limit,** is 9 cm − 0.001 cm = 8.999 cm.

The lower and upper limits of the diameter of the piston could also be found by solving the absolute value inequality $|d - 9| \leq 0.001$, where d is the diameter of the piston.

$$
\begin{aligned}
|d - 9| &\leq 0.001 \\
-0.001 \leq d - 9 &\leq 0.001 \\
-0.001 + 9 \leq d - 9 + 9 &\leq 0.001 + 9 \\
8.999 \leq d &\leq 9.001
\end{aligned}
$$

The lower and upper limits of the diameter of the piston are 8.999 cm and 9.001 cm.

Unless otherwise noted, all content on this page is © Cengage Learning.

EXAMPLE 8

The diameter of a piston for an automobile is $3\frac{5}{16}$ in., with a tolerance of $\frac{1}{64}$ in. Find the lower and upper limits of the diameter of the piston.

Strategy

To find the lower and upper limits of the diameter of the piston, let d represent the diameter of the piston, T the tolerance, and L the lower and upper limits of the diameter. Solve the absolute value inequality $|L - d| \le T$ for L.

Solution

$$|L - d| \le T$$

$$\left| L - 3\frac{5}{16} \right| \le \frac{1}{64}$$

$$-\frac{1}{64} \le L - 3\frac{5}{16} \le \frac{1}{64}$$

$$-\frac{1}{64} + 3\frac{5}{16} \le L - 3\frac{5}{16} + 3\frac{5}{16} \le \frac{1}{64} + 3\frac{5}{16}$$

$$3\frac{19}{64} \le L \le 3\frac{21}{64}$$

The lower and upper limits of the diameter of the piston are $3\frac{19}{64}$ in. and $3\frac{21}{64}$ in.

YOU TRY IT 8

A machinist must make a bushing that has a tolerance of 0.003 in. The diameter of the bushing is 2.55 in. Find the lower and upper limits of the diameter of the bushing.

IN-CLASS EXAMPLES

8. A machinist must make a bushing that has a tolerance of 0.004 in. The diameter of the bushing is 3.25 in. Find the lower and upper limits of the diameter of the bushing. **Lower limit: 3.246 in.; upper limit: 3.254 in.**

Your strategy

Your solution

Lower limit: 2.547 in.; upper limit: 2.553 in.

Solution on p. S8

Unless otherwise noted, all content on this page is © Cengage Learning.

2.6 EXERCISES

✔ Concept Check

SUGGESTED ASSIGNMENT
Exercises 1–16; Exercises 17–107, odds;
More challenging exercises: Exercises 109–115

1. Is 2 a solution of $|x - 8| = 6$?   Yes

2. Is -2 a solution of $|2x - 5| = 9$?   Yes

3. Is -1 a solution of $|3x - 4| = 7$?   Yes

4. Is 1 a solution of $|6x - 1| = -5$?   No

For Exercises 5 to 12, solve the absolute value equation.

5. $|x| = 7$
7 and -7

6. $|a| = 2$
2 and -2

7. $|-y| = 6$
6 and -6

8. $|-t| = 3$
3 and -3

9. $|x| = -4$
No solution

10. $|y| = -3$
No solution

11. $|-t| = -3$
No solution

12. $|-y| = -2$
No solution

13. Find the solution set of $|x| > 3$.
$\{x | x < -3\} \cup \{x | x > 3\}$

14. Find the solution set of $|x| \leq 5$.
$\{x | -5 \leq x \leq 5\}$

15. Write an absolute value inequality to represent all real numbers whose distance from 2 is less than 5.
$|x - 2| < 5$

16. Write an absolute value inequality to represent all real numbers whose distance from 4 is greater than 3.
$|x - 4| > 3$

OBJECTIVE A   *To solve an absolute value equation*

For Exercises 17 to 64, solve.

17. $|x + 2| = 3$
1 and -5

18. $|x + 5| = 2$
-3 and -7

19. $|y - 5| = 3$
8 and 2

20. $|y - 8| = 4$
12 and 4

21. $|a - 2| = 0$
2

22. $|a + 7| = 0$
-7

23. $|x - 2| = -4$
No solution

24. $|x + 8| = -2$
No solution

25. $|3 - 4x| = 9$
$-\dfrac{3}{2}$ and 3

26. $|2 - 5x| = 3$
$-\dfrac{1}{5}$ and 1

27. $|2x - 3| = 0$
$\dfrac{3}{2}$

28. $|5x + 5| = 0$
-1

29. $|3x - 2| = -4$
No solution

30. $|2x + 5| = -2$
No solution

31. $|x - 2| - 2 = 3$
7 and -3

32. $|x - 9| - 3 = 2$
14 and 4

33. $|3a + 2| - 4 = 4$
2 and $-\dfrac{10}{3}$

34. $|2a + 9| + 4 = 5$
-4 and -5

35. $|2 - y| + 3 = 4$
1 and 3

36. $|8 - y| - 3 = 1$
4 and 12

37. $|2x - 3| + 3 = 3$
$\dfrac{3}{2}$

38. $|4x - 7| - 5 = -5$
$\dfrac{7}{4}$

39. $|2x - 3| + 4 = -4$
No solution

40. $|3x - 2| + 1 = -1$
No solution

41. $|6x - 5| - 2 = 4$
$\dfrac{11}{6}$ and $-\dfrac{1}{6}$

42. $|4b + 3| - 2 = 7$
$\dfrac{3}{2}$ and -3

43. $|3t + 2| + 3 = 4$
$-\dfrac{1}{3}$ and -1

44. $|5x - 2| + 5 = 7$
$\dfrac{4}{5}$ and 0

45. $3 - |x - 4| = 5$
No solution

46. $2 - |x - 5| = 4$
No solution

47. $8 - |2x - 3| = 5$
3 and 0

48. $8 - |3x + 2| = 3$
1 and $-\dfrac{7}{3}$

49. $|2 - 3x| + 7 = 2$
No solution

50. $|1 - 5a| + 2 = 3$
0 and $\dfrac{2}{5}$

51. $|8 - 3x| - 3 = 2$
1 and $\dfrac{13}{3}$

52. $|6 - 5b| - 4 = 3$
$-\dfrac{1}{5}$ and $\dfrac{13}{5}$

53. $|2x - 8| + 12 = 2$
No solution

54. $|3x - 4| + 8 = 3$
No solution

55. $2 + |3x - 4| = 5$
$\dfrac{7}{3}$ and $\dfrac{1}{3}$

56. $5 + |2x + 1| = 8$
1 and -2

57. $5 - |2x + 1| = 5$
$-\dfrac{1}{2}$

58. $3 - |5x + 3| = 3$
$-\dfrac{3}{5}$

59. $6 - |2x + 4| = 3$
$-\dfrac{1}{2}$ and $-\dfrac{7}{2}$

60. $8 - |3x - 2| = 5$
$\dfrac{5}{3}$ and $-\dfrac{1}{3}$

61. $8 - |1 - 3x| = -1$
$-\dfrac{8}{3}$ and $\dfrac{10}{3}$

62. $3 - |3 - 5x| = -2$
$-\dfrac{2}{5}$ and $\dfrac{8}{5}$

63. $5 + |2 - x| = 3$
No solution

64. $6 + |3 - 2x| = 2$
No solution

For Exercises 65 to 68, assume that a and b are positive numbers such that $a < b$. State whether the given equation has no solution, two negative solutions, two positive solutions, or one positive and one negative solution.

65. $|x - b| = a$
Two positive solutions

66. $|x - b| = -a$
No solution

67. $|x + b| = a$
Two negative solutions

68. $|x + a| = b$
One positive and one negative solution

OBJECTIVE B   *To solve an absolute value inequality*

For Exercises 69 to 96, solve.

69. $|x + 1| > 2$
$\{x | x > 1\} \cup \{x | x < -3\}$

70. $|x - 2| > 1$
$\{x | x > 3\} \cup \{x | x < 1\}$

71. $|x - 5| \leq 1$
$\{x | 4 \leq x \leq 6\}$

72. $|x - 4| \leq 3$
$\{x | 1 \leq x \leq 7\}$

73. $|2 - x| \geq 3$
$\{x | x \geq 5\} \cup \{x | x \leq -1\}$

74. $|3 - x| \geq 2$
$\{x | x \leq 1\} \cup \{x | x \geq 5\}$

75. $|2x + 1| < 5$

$\{x | -3 < x < 2\}$

76. $|3x - 2| < 4$

$\left\{x \,\middle|\, -\dfrac{2}{3} < x < 2\right\}$

77. $|5x + 2| > 12$

$\{x | x > 2\} \cup \left\{x \,\middle|\, x < -\dfrac{14}{5}\right\}$

78. $|7x - 1| > 13$

$\{x | x > 2\} \cup \left\{x \,\middle|\, x < -\dfrac{12}{7}\right\}$

79. $|4x - 3| \le -2$

$\varnothing$

80. $|5x + 1| \le -4$

$\varnothing$

81. $|2x + 7| > -5$

The set of real numbers

82. $|3x - 1| > -4$

The set of real numbers

83. $|4 - 3x| \ge 5$

$\left\{x \,\middle|\, x \le -\dfrac{1}{3}\right\} \cup \{x | x \ge 3\}$

84. $|7 - 2x| > 9$

$\{x | x < -1\} \cup \{x | x > 8\}$

85. $|5 - 4x| \le 13$

$\left\{x \,\middle|\, -2 \le x \le \dfrac{9}{2}\right\}$

86. $|3 - 7x| < 17$

$\left\{x \,\middle|\, -2 < x < \dfrac{20}{7}\right\}$

87. $|6 - 3x| \le 0$

$\{x | x = 2\}$

88. $|10 - 5x| \ge 0$

The set of real numbers

89. $|2 - 9x| > 20$

$\{x | x < -2\} \cup \left\{x \,\middle|\, x > \dfrac{22}{9}\right\}$

90. $|5x - 1| < 16$

$\left\{x \,\middle|\, -3 < x < \dfrac{17}{5}\right\}$

91. $|2x - 3| + 2 < 8$

$\left\{x \,\middle|\, -\dfrac{3}{2} < x < \dfrac{9}{2}\right\}$

92. $|3x - 5| + 1 < 7$

$\left\{x \,\middle|\, -\dfrac{1}{3} < x < \dfrac{11}{3}\right\}$

93. $|2 - 5x| - 4 > -2$

$\{x | x < 0\} \cup \left\{x \,\middle|\, x > \dfrac{4}{5}\right\}$

94. $|4 - 2x| - 9 > -3$

$\{x | x < -1\} \cup \{x | x > 5\}$

95. $8 - |2x - 5| < 3$

$\{x | x > 5\} \cup \{x | x < 0\}$

96. $12 - |3x - 4| > 7$

$\left\{x \,\middle|\, -\dfrac{1}{3} < x < 3\right\}$

For Exercises 97 and 98, assume that a and b are positive numbers such that $a < b$. State whether the given inequality has no solution, all negative solutions, all positive solutions, or both positive and negative solutions.

97. $|x + b| < a$
All negative solutions

98. $|x + a| < b$
Both positive and negative solutions

OBJECTIVE C *To solve application problems*

99. A dosage of medicine may safely range from 2.8 ml to 3.2 ml. What is the desired dosage of the medicine? What is the tolerance?
3 ml; 0.2 ml

100. The tolerance, in inches, for the diameter of a piston is described by the absolute value inequality $|d - 5| \le 0.01$. What is the desired diameter of the piston? By how much can the actual diameter of the piston vary from the desired diameter?
5 in.; 0.01 in.

101. Mechanics The diameter of a bushing is 1.75 in. The bushing has a tolerance of 0.008 in. Find the lower and upper limits of the diameter of the bushing.
1.742 in.; 1.758 in.

← 1.75 in. →

102. Mechanics A machinist must make a bushing that has a tolerance of 0.004 in. The diameter of the bushing is 3.48 in. Find the lower and upper limits of the diameter of the bushing.
3.476 in.; 3.484 in.

Unless otherwise noted, all content on this page is © Cengage Learning.

103. Automobiles   The length of a piston rod for an automobile is $9\frac{5}{8}$ in. with a tolerance of $\frac{1}{32}$ in. Find the lower and upper limits of the length of the piston rod.

$9\frac{19}{32}$ in., $9\frac{21}{32}$ in.

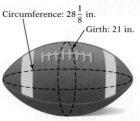

Circumference: $28\frac{1}{8}$ in.

Girth: 21 in.

|← Length: $11\frac{1}{32}$ in. →|

104. ● Football Manufacturing   An NCAA football must conform to the measurements shown in the diagram at the right, with tolerances of $\frac{1}{4}$ in. for the girth, $\frac{3}{8}$ in. for the circumference, and $\frac{5}{32}$ in. for the length. Find the lower and upper limits for **a.** the girth, **b.** the circumference, and **c.** the length of an NCAA football. (*Source:* www.ncaa.org)

a. $20\frac{3}{4}$ in., $21\frac{1}{4}$ in.   **b.** $27\frac{3}{4}$ in., $28\frac{1}{2}$ in.   **c.** $10\frac{7}{8}$ in., $11\frac{3}{16}$ in.

105. ● Political Polling   In a poll, the *margin of error* is a measure of the pollsters' confidence in their results. If the pollsters conduct the same poll many times, they expect that 95% of the time they will get results that fall within the margin of error of the reported results. Read the article at the right. For the poll described, the pollsters are 95% sure that the percent of American voters who felt the economy was the most important election issue lies between what lower and upper limits?

38% and 44%

106. ● Aquatic Environments   Different species of fish have different requirements for the temperature and pH of the water in which they live. The gold swordtail requires a temperature of 73°F plus or minus 9°F and a pH level of 7.65 plus or minus 0.65. Find the lower and upper limits of **a.** the temperature and **b.** the pH level of the water in which a gold swordtail lives. (*Source:* www.takomapet.com)

a. 64°F, 82°F   **b.** 7.0, 8.3

In the NEWS!

Economy Is Number-One Issue

A *Washington Post*/ABC News poll showed that 41% of American voters felt the economy was the most important election issue. The results of the poll had a margin of error of plus or minus 3 percentage points.

Source:
www.washingtonpost.com

Electronics   The tolerance of the resistors used in electronics is given as a percent. Use your calculator for Exercises 107 and 108.

107. 🖩 Find the lower and upper limits of a 29,000-ohm resistor with a 2% tolerance.

28,420 ohms; 29,580 ohms

108. 🖩 Find the lower and upper limits of a 15,000-ohm resistor with a 10% tolerance.

13,500 ohms; 16,500 ohms

Critical Thinking

109. For what values of the variable is the equation true? Write the solution set in set-builder notation.

a. $|x + 3| = x + 3$ $\{x | x \geq -3\}$ **b.** $|a - 4| = 4 - a$ $\{a | a \leq 4\}$

110. Replace the question mark with $\leq$, $\geq$, or $=$.

a. $|x + y| \,?\, |x| + |y|$ $\leq$ **b.** $|x - y| \,?\, |x| - |y|$ $\geq$
c. $||x| - |y|| \,?\, |x| - |y|$ $\geq$ **d.** $|xy| \,?\, |x||y|$ $=$

111. Let $|x| \leq 2$ and $|3x - 2| \leq a$. Find the smallest possible value of a.   4

Projects or Group Activities

For Exercises 112 to 115, solve the absolute value equation. Recall that if $|x| = a$, then $x = -a$ or $x = a$. For the equations below, a is an algebraic expression, so be careful with the negative sign. Remember to check your answers.

112. $|4x + 3| = 2x + 10$

$-\frac{13}{6}$ and $\frac{7}{2}$

113. $|3x - 4| = 2x + 10$

$-\frac{6}{5}$ and 14

114. $|x + 3| = 2x - 1$ 4

115. $|3x + 1| = 2x - 5$ No solution

Unless otherwise noted, all content on this page is © Cengage Learning.

QUICK QUIZ
Solve.
 1. $|3 - 2x| = 3$
    0, 3   **[2.6A]**
 2. $|2 - b| + 5 = 6$
    1, 3   **[2.6A]**
 3. $|x - 1| > 2$
    $\{x | x > 3\} \cup \{x | x < -1\}$
    [2.6B]
 4. $|x - 4| \leq 2$
    $\{x | 2 \leq x \leq 6\}$ **[2.6B]**
 5. A doctor has prescribed 3 cc of medicine for a patient. The tolerance is 0.05 cc. Find the lower and upper limits of the amount of medication to be given.   **Lower limit: 2.95 cc; upper limit: 3.05 cc**   **[2.6C]**

CHAPTER

2   Summary

Key Words	Examples
An **equation** expresses the equality of two mathematical expressions. [2.1A, p. 76]	$3 + 2(4x - 5) = x + 4$ is an equation.
A **solution of an equation** is a number that, when substituted for the variable, results in a true equation. [2.1A, p. 76]	-2 is a solution of $2 - 3x = 8$ because $2 - 3(-2) = 8$ is a true equation.
To **solve an equation** means to find a solution of the equation. The goal is to rewrite the equation in the form *variable* $=$ *constant*, because the constant is the solution. [2.1B, p. 76]	The equation $x = -3$ is in the form *variable* $=$ *constant*. The constant, -3, is the solution of the equation.
Consecutive integers follow one another in order. [2.3A, p. 110]	$5, 6, 7$ are consecutive integers. $-9, -8, -7$ are consecutive integers.
The **solution set of an inequality** is a set of numbers each element of which, when substituted in the inequality, results in a true inequality. [2.5A, p. 132]	Any number greater than 4 is a solution of the inequality $x > 4$.
A **compound inequality** is formed by joining two inequalities with a connective word such as **and** or **or**. [2.5B, p. 135]	$3x > 6$ and $2x + 5 < 7$ $2x + 1 < 3$ or $x + 2 > 4$
An **absolute value equation** is an equation that contains a variable within an absolute value symbol. [2.6A, p. 144]	$\lvert x - 2 \rvert = 3$
An **absolute value inequality** is an inequality that contains a variable within an absolute value symbol. [2.6B, p. 146]	$\lvert x - 4 \rvert < 5$ $\lvert 2x - 3 \rvert > 6$
The **tolerance** of a component or part is the amount by which it is acceptable for the component to vary from a given measurement. The maximum measurement is the **upper limit**. The minimum measurement is the **lower limit**. [2.6C, p. 148]	The diameter of a bushing is 1.5 in., with a tolerance of 0.005 in. The lower and upper limits of the diameter of the bushing are 1.5 in. $\pm$ 0.005 in.

Essential Rules and Procedures   # Examples

Addition Property of Equations [2.1B, p. 77]
The same number can be added to each side of an equation without changing the solution of the equation.

If $a = b$, then $a + c = b + c$.

Multiplication Property of Equations [2.1C, p. 78]
Each side of an equation can be multiplied by the same nonzero number without changing the solution of the equation.

If $a = b$ and $c \neq 0$, then $ac = bc$.

Basic Percent Equation [2.1D, p. 80]
Percent · Base = Amount
$$P \cdot B = A$$

30% of what number is 24?
$$PB = A$$
$$0.30B = 24$$
$$B = 80$$
30% of 80 is 24.

Simple Interest Equation [2.1D, p. 81]
Interest = Principal · Rate · Time
$$I = Prt$$

A credit card company charges an annual interest rate of 21% on the monthly unpaid balance on a card. Find the amount of interest charged on an unpaid balance of $232 for April.
$$I = Prt$$
$$I = 232(0.21)\left(\frac{1}{12}\right) = 4.06$$

Consecutive Integers [2.3A, p. 110]
$n, n + 1, n + 2, \ldots$

The sum of three consecutive integers is 33.
$$n + (n + 1) + (n + 2) = 33$$

Consecutive Even or Consecutive Odd Integers [2.3A, p. 110]
$n, n + 2, n + 4, \ldots$

The sum of three consecutive odd integers is 33.
$$n + (n + 2) + (n + 4) = 33$$

Value Mixture Equation [2.4A, p. 119]
Amount · Unit Cost = Value
$$AC = V$$

An herbalist has 30 oz of herbs costing $4 per ounce. How many ounces of herbs costing $2 per ounce should be mixed with the 30 oz to produce a mixture costing $3.20 per ounce?
$$30(4) + 2x = 3.20(30 + x)$$

Percent Mixture Equation [2.1D, p. 81; 2.4B, p. 121]
Quantity = Amount · Percent Concentration
$$Q = Ar$$

Forty ounces of a 30% gold alloy are mixed with 60 oz of a 20% gold alloy. Find the percent concentration of the resulting gold alloy.
$$0.30(40) + 0.20(60) = x(100)$$

Uniform Motion Equation [2.1E, p. 84; 2.4C, p. 123]

Distance = Rate · Time

$$d = rt$$

A boat traveled from a harbor to an island at an average speed of 20 mph. The average speed on the return trip was 15 mph. The total trip took 3.5 h. How long did it take for the boat to travel to the island?

$$20t = 15(3.5 - t)$$

Addition Property of Inequalities [2.5A, p. 132]

If $a > b$, then $a + c > b + c$.

If $a < b$, then $a + c < b + c$.

$$x + 3 > -2$$
$$x + 3 - 3 > -2 - 3$$
$$x > -5$$

Multiplication Property of Inequalities
[2.5A, p. 133]

Rule 1 If $a > b$ and $c > 0$, then $ac > bc$.
If $a < b$ and $c > 0$, then $ac < bc$.

$$3x > 12$$
$$\left(\frac{1}{3}\right)(3x) > \left(\frac{1}{3}\right)12$$
$$x > 4$$

Rule 2 If $a > b$ and $c < 0$, then $ac < bc$.
If $a < b$ and $c < 0$, then $ac > bc$.

$$-2x < 8$$
$$\frac{-2x}{-2} > \frac{8}{-2}$$
$$x > -4$$

Solutions of an Absolute Value Equation
[2.6A, p. 144]

If $a > 0$ and $|x| = a$, then $x = a$ or $x = -a$.

If $|x| = 0$, then $x = 0$.

If $a < 0$, then $|x| = a$ has no solution.

$$|x - 3| = 7$$

$x - 3 = 7$	$x - 3 = -7$
$x = 10$	$x = -4$

Solutions of Absolute Value Inequalities
[2.6B, p. 147]

To solve an absolute value inequality of the form $|ax + b| < c$, $c > 0$, solve the equivalent compound inequality $-c < ax + b < c$.

$$|x - 5| < 9$$
$$-9 < x - 5 < 9$$
$$-9 + 5 < x - 5 + 5 < 9 + 5$$
$$-4 < x < 14$$

To solve an absolute value inequality of the form $|ax + b| > c$, solve the equivalent compound inequality $ax + b < -c$ or $ax + b > c$.

$$|x - 5| > 9$$
$$x - 5 < -9 \quad \text{or} \quad x - 5 > 9$$
$$x < -4 \quad \text{or} \quad x > 14$$

$$\{x|x < -4\} \cup \{x|x > 14\}$$

2 | Review Exercises

1. Solve: $x + 3 = 24$
21   [2.1B]

2. Solve: $x + 5(3x - 20) = 10(x - 4)$
10   [2.2C]

3. Solve: $5x - 6 = 29$
7   [2.2A]

4. Is 3 a solution of $5x - 2 = 4x + 5$?
No   [2.1A]

5. Solve: $\dfrac{3}{5}a = 12$
20   [2.1C]

6. Solve: $3x - 7 > -2$
Write the solution set in interval notation.
$\left(\dfrac{5}{3}, \infty\right)$ [2.5A]

7. 30 is what percent of 12?
250%   [2.1D]

8. Solve: $5x + 3 = 10x - 17$
4   [2.2B]

9. Solve: $7 - [4 + 2(x - 3)] = 11(x + 2)$
−1   [2.2C]

10. Solve: $6 + |3x - 3| = 2$
No solution   [2.6A]

11. Solve: $|2x - 5| < 3$
$\{x \mid 1 < x < 4\}$ [2.6B]

12. Solve: $3x < 4$ and $x + 2 > -1$
Write the solution set in set-builder notation.
$\left\{x \middle| -3 < x < \dfrac{4}{3}\right\}$ [2.5B]

13. Solve: $3x - 2 > x - 4$ or $7x - 5 < 3x + 3$
Write the solution set in interval notation.
$(-\infty, \infty)$ [2.5B]

14. Solve: $|4x - 5| \geq 3$
$\left\{x \mid x \geq 2\right\} \cup \left\{x \mid x \leq \dfrac{1}{2}\right\}$ [2.6B]

15. Solve: $3y - 5 = 3 - 2y$
$\dfrac{8}{5}$ [2.2B]

16. Solve: $4x - 5 + x = 6x - 8$
3   [2.2B]

17. Solve: $3(x - 4) = -5(6 - x)$
9   [2.2C]

18. Solve: $\dfrac{3x - 2}{4} + 1 = \dfrac{2x - 3}{2}$
8   [2.2C]

19. Solve: $|5x + 8| = 0$
$-\dfrac{8}{5}$ [2.6A]

20. Solve: $|5x - 4| < -2$
$\varnothing$ [2.6B]

21. Physics   A lever is 12 ft long. At a distance of 2 ft from the fulcrum, a force of 120 lb is applied. How large a force must be applied to the other end so that the system will balance? Use the lever system equation $F_1 x = F_2(d - x)$.
24 lb   [2.2D]

22. Travel   A bus traveled on a level road for 2 h at an average speed that was 20 mph faster than its average speed on a winding road. The time spent on the winding road was 3 h. Find the average speed on the winding road if the total trip was 200 mi.
32 mph   [2.4C]

23. Mixtures   A health food store combined cranberry juice that cost $1.79 per quart with apple juice that cost $1.19 per quart. How many quarts of each were used to make 10 qt of cranapple juice costing $1.61 per quart?
Cranberry juice: 7 qt; apple juice: 3 qt   [2.4A]

24. Four times the second of three consecutive integers equals the sum of the first and third integers. Find the integers.
−1, 0, 1   [2.3A]

25. Translate "four less than the product of five and a number is sixteen" into an equation and solve.
$5n - 4 = 16$; 4   [2.3A]

26. 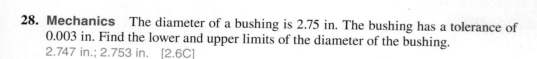 **Building Height**   The Empire State Building is 1472 ft tall. This is 654 ft less than twice the height of the Eiffel Tower. Find the height of the Eiffel Tower.
1063 ft   [2.3B]

27. Travel   A jet plane traveling at 600 mph overtakes a propeller-driven plane that had a 2-hour head start. The propeller-driven plane is traveling at 200 mph. How far from the starting point does the jet overtake the propeller-driven plane?
600 mi   [2.4C]

28. Mechanics   The diameter of a bushing is 2.75 in. The bushing has a tolerance of 0.003 in. Find the lower and upper limits of the diameter of the bushing.
2.747 in.; 2.753 in.   [2.6C]

29. Mixtures   A dairy owner mixed 5 gal of cream containing 30% butterfat with 8 gal of milk containing 4% butterfat. What is the percent of butterfat in the resulting mixture?
14%   [2.4B]

30. Uniform Motion   A ferry leaves a dock and travels to an island at an average speed of 16 mph. On the return trip, the ferry travels at an average speed of 12 mph. The total time for the trip is $2\frac{1}{3}$ h. How far is the island from the dock?
16 mi   [2.4C]

Songquan Deng/Shutterstock.com

Carolina K. Smith, M.D./ Shutterstock.com

CHAPTER

2   TEST

1. Solve: $3x - 2 = 5x + 8$
   −5   [2.2B]

2. Solve: $x - 3 = -8$
   −5   [2.1B]

3. Solve: $3x - 5 = -14$
   −3   [2.2A]

4. Solve: $4 - 2(3 - 2x) = 2(5 - x)$
   2   [2.2C]

5. Is −2 a solution of $x^2 - 3x = 2x - 6$?
   No   [2.1A]

6. Solve: $7 - 4x = -13$
   5   [2.2A]

7. What is 0.5% of 8?
   0.04   [2.1D]

8. Solve: $5x - 2(4x - 3) = 6x + 9$
   $-\dfrac{1}{3}$ [2.2C]

9. Solve: $5x + 3 - 7x = 2x - 5$
   2   [2.2B]

10. Solve: $\dfrac{3}{4}x = -9$
    −12   [2.1C]

11. Solve: $4 - 3(x + 2) < 2(2x + 3) - 1$
    Write the solution set in interval notation.
    $(-1, \infty)$ [2.5A]

12. Solve: $4x - 1 > 5$ or $2 - 3x < 8$
    Write the solution set in set-builder notation.
    $\{x | x > -2\}$ [2.5B]

13. Solve: $4 - 3x \geq 7$ and $2x + 3 \geq 7$
    Write the solution set in set-builder notation.
    $\varnothing$ [2.5B]

14. Solve: $|3 - 5x| = 12$
    3 and $-\dfrac{9}{5}$ [2.6A]

15. Solve: $2 - |2x - 5| = -7$
    7 and −2   [2.6A]

16. Solve: $|3x - 5| \leq 4$
    $\left\{ x \,\middle|\, \dfrac{1}{3} \leq x \leq 3 \right\}$ [2.6B]

17. **Mixtures**   A baker wants to make a 15-pound blend of flour that costs $.60 per pound. The blend is made using a rye flour that costs $.70 per pound and a wheat flour that costs $.40 per pound. How many pounds of each flour should be used?
    Rye: 10 lb; wheat: 5 lb   [2.4A]

18. Consecutive Integers   Find three consecutive even integers whose sum is 36.
10, 12, 14   [2.3A]

19. Chemistry   How many gallons of water must be mixed with 5 gal of a 20% salt solution to make a 16% salt solution?   1.25 gal   [2.4B]

20. Translate "the difference between three times a number and fifteen is twenty-seven" into an equation and solve.
$3x - 15 = 27$; 14   [2.3A]

21. Sports   A cross-country skier leaves a camp to explore a wilderness area. Two hours later a friend leaves the camp in a snowmobile, traveling 4 mph faster than the skier. The friend meets the skier 1 h later. Find the rate of the snowmobile.   6 mph   [2.4C]

Val Thoermer/Shutterstock.com

22. Business   A company makes 140 televisions per day. Three times the number of LCD rear-projection TVs made equals 20 less than the number of LCD flat-panel TVs made. Find the number of LCD flat-panel TVs made each day.
110 LCD flat-panel TVs   [2.3B]

23. The sum of two numbers is eighteen. The difference between four times the smaller number and seven is equal to the sum of two times the larger number and five. Find the two numbers.   8, 10   [2.3A]

24. Aviation   As part of flight training, a student pilot was required to fly to an airport and then return. The average speed to the airport was 90 mph, and the average speed returning was 120 mph. Find the distance between the two airports if the total flying time was 7 h.   360 mi   [2.4C]

25. Chemistry   A chemist mixes 100 g of water at 80°C with 50 g of water at 20°C. Find the final temperature of the water after mixing. Use the equation $m_1(T_1 - T) = m_2(T - T_2)$, where m_1 is the quantity of water at the hotter temperature, T_1 is the temperature of the hotter water, m_2 is the quantity of water at the cooler temperature, T_2 is the temperature of the cooler water, and T is the final temperature of the water after mixing.   60°C   [2.2D]

26. Consumerism   Gambelli Agency rents cars for $40 per day plus 25¢ for every mile driven. McDougal Rental rents cars for $58 per day with unlimited mileage. How many miles a day can you drive a Gambelli Agency car if it is to cost you less than a McDougal Rental car?
Less than 72 mi   [2.5C]

27. Mechanics   A machinist must make a bushing that has a tolerance of 0.002 in. The diameter of the bushing is 2.65 in. Find the lower and upper limits of the diameter of the bushing.
2.648 in.; 2.652 in.   [2.6C]

Cumulative Review Exercises

1. Subtract: $-6 - (-20) - 8$
6   [1.1C]

2. Multiply: $(-2)(-6)(-4)$
−48   [1.1D]

3. Subtract: $-\dfrac{5}{6} - \left(-\dfrac{7}{16}\right)$
$-\dfrac{19}{48}$ [1.2C]

4. Divide: $-\dfrac{7}{3} \div \dfrac{7}{6}$
−2   [1.2D]

5. Simplify: $-4^2 \cdot \left(-\dfrac{3}{2}\right)^3$
54   [1.2E]

6. Simplify: $25 - 3\dfrac{(5-2)^2}{2^3 + 1} - (-2)$
24   [1.3A]

7. Evaluate $3(a - c) - 2ab$ when $a = 2$, $b = 3$, and $c = -4$.
6   [1.4A]

8. Simplify: $3x - 8x + (-12x)$
−17x   [1.4B]

9. Simplify: $2a - (-3b) - 7a - 5b$
−5a − 2b   [1.4B]

10. Simplify: $(16x)\left(\dfrac{1}{8}\right)$
2x   [1.4C]

11. Simplify: $-4(-9y)$
36y   [1.4C]

12. Simplify: $-2(-x^2 - 3x + 2)$
2x² + 6x − 4   [1.4D]

13. Simplify: $-3[2x - 4(x - 3)] + 2$
6x − 34   [1.4D]

14. Find $A \cap B$ given $A = \{-4, -2, 0, 2\}$ and $B = \{-4, 0, 4, 8\}$.
$A \cap B = \{-4, 0\}$ [1.5A]

15. Graph: $\{x|x < 3\} \cap \{x|x > -2\}$

[1.5B]

16. Is -3 a solution of $x^2 + 6x + 9 = x + 3$?
Yes   [2.1A]

17. Find 32% of 60.
19.2   [2.1D]

18. Solve: $\dfrac{3}{5}x = -15$
−25   [2.1C]

19. Solve: $7x - 8 = -29$
−3   [2.2A]

20. Solve: $13 - 9x = -14$
3   [2.2A]

21. Solve: $8x - 3(4x - 5) = -2x - 11$

13   [2.2C]

22. 25% of what number is 30?

120   [2.1D]

23. Solve: $5x - 8 = 12x + 13$

−3   [2.2B]

24. Solve: $11 - 4x = 2x + 8$

$\dfrac{1}{2}$ [2.2B]

25. Solve: $3 - 2(2x - 1) \geq 3(2x - 2) + 1$

$\{x | x \leq 1\}$ [2.5A]

26. Solve: $3x + 2 \leq 5$ and $x + 5 \geq 1$

$\{x | -4 \leq x \leq 1\}$ [2.5B]

27. Solve: $|3 - 2x| = 5$

−1, 4   [2.6A]

28. Solve: $|3x - 1| > 5$

$\{x | x > 2\} \cup \left\{x | x < -\dfrac{4}{3}\right\}$ [2.6B]

29. Write 55% as a fraction.

$\dfrac{11}{20}$ [1.2B]

30. Write 1.03 as a percent.

103%   [1.2B]

31. Chemistry A chemist mixes 300 g of water at 75°C with 100 g of water at 15°C. Find the final temperature of the water after mixing. Use the equation $m_1(T_1 - T) = m_2(T - T_2)$, where m_1 is the quantity of water at the hotter temperature, T_1 is the temperature of the hotter water, m_2 is the quantity of water at the cooler temperature, T_2 is the temperature of the cooler water, and T is the final temperature of the water after mixing.   60°C   [2.2D]

32. Translate "The difference between twelve and the product of five and a number is negative eighteen" into an equation and solve.

$12 - 5x = -18$; 6   [2.3A]

33. Construction The area of a cement foundation of a house is 2000 ft². This is 200 ft² more than three times the area of the garage. Find the area of the garage.

600 ft²   [2.3B]

34. Mixtures How many pounds of an oat flour that costs $.80 per pound must be mixed with 40 lb of a wheat flour that costs $.50 per pound to make a blend that costs $.60 per pound?   20 lb   [2.4A]

35. Metallurgy How many grams of pure gold must be added to 100 g of a 20% gold alloy to make an alloy that is 36% gold?   25 g   [2.4B]

36. Sports A sprinter ran to the end of a track at an average rate of 8 m/s and then jogged back to the starting point at an average rate of 3 m/s. The sprinter took 55 s to run to the end of the track and jog back. Find the length of the track.

120 m   [2.4C]

Unless otherwise noted, all content on this page is © Cengage Learning.

Geometry

3

OBJECTIVES

SECTION 3.1
A To solve problems involving lines and angles
B To solve problems involving angles formed by intersecting lines
C To solve problems involving the angles of a triangle

SECTION 3.2
A To solve problems involving the perimeter of a geometric figure
B To solve problems involving the area of a geometric figure

SECTION 3.3
A To solve problems involving the volume of a solid
B To solve problems involving the surface area of a solid

Focus on Success

Did you make a time management plan when you started this course? If not, you can still benefit from doing so. Create a schedule that gives you enough time to do everything that you need to do. We want you to schedule enough time to study math each week so that you successfully complete this course. Once you have determined the hours during which you will study, consider your study time a commitment that you cannot break. (See Time Management, page AIM-4.)

wavebreakmedia ltd/Shutterstock.com

Prep Test

Are you ready to succeed in this chapter? Take the Prep Test below to find out if you are ready to learn the new material.

1. Simplify: $2(18) + 2(10)$ 56   [1.3A]

2. Solve: $x + 47 = 90$ 43   [2.1B]

3. Solve: $32 + 97 + x = 180$ 51   [2.1B]

4. Evaluate abc for $a = 2$, $b = 3.14$, and $c = 9$.   56.52   [1.4A]

5. Evaluate xyz^3 for $x = \frac{4}{3}$, $y = 3.14$, and $z = 3$.   113.04   [1.4A]

6. Evaluate $\frac{1}{2}a(b + c)$ when $a = 6$, $b = 25$, and $c = 15$.   120   [1.4A]

3.1   Introduction to Geometry

OBJECTIVE A   *To solve problems involving lines and angles*

Point of Interest

Geometry is one of the oldest branches of mathematics. Around 350 B.C., Euclid of Alexandria wrote *Elements,* which contained all of the known concepts of geometry. Euclid's contribution was to unify various concepts into a single deductive system that was based on a set of axioms.

The word *geometry* comes from the Greek words for *earth* and *measure*. In ancient Egypt, geometry was used by the Egyptians to measure land and to build structures such as the pyramids. Today geometry is used in many fields, such as physics, medicine, and geology. Geometry is also used in applied fields such as mechanical drawing and astronomy. Geometric forms are used in art and design.

Three basic concepts of geometry are point, line, and plane. A **point** is symbolized by drawing a dot. A **line** is determined by two distinct points and extends indefinitely in both directions, as the arrows on the line shown at the right indicate. This line contains points A and B and is represented by $\overleftrightarrow{AB}$. A line can also be represented by a single letter, such as ℓ.

A **ray** starts at a point and extends indefinitely in *one* direction. The point at which a ray starts is called the **endpoint** of the ray. The ray shown at the right is denoted by $\overrightarrow{AB}$. Point A is the endpoint of the ray.

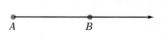

A **line segment** is part of a line and has two endpoints. The line segment shown at the right is denoted by $\overline{AB}$.

The distance between the endpoints of $\overline{AC}$ is denoted by AC. If B is a point on $\overline{AC}$, then AC (the distance from A to C) is the sum of AB (the distance from A to B) and BC (the distance from B to C).

$AC = AB + BC$

HOW TO 1   Given the figure above and the fact that $AB = 22$ cm and $AC = 31$ cm, find BC.

Write an equation for the distances between points on the line segment.	$AC = AB + BC$
Substitute the given distances for AB and AC into the equation.	$31 = 22 + BC$
Solve for BC.	$9 = BC$

$BC = 9$ cm

In this section we will be discussing figures that lie in a plane. A **plane** is a flat surface and can be pictured as a tabletop or blackboard that extends in all directions. Figures that lie in a plane are called **plane figures.**

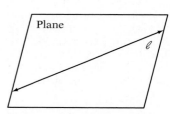

Unless otherwise noted, all content on this page is © Cengage Learning.

Lines in a plane can be intersecting or parallel. **Intersecting lines** cross at a point in the plane. **Parallel lines** never intersect. The distance between them is always the same.

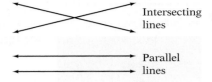

Intersecting lines

Parallel lines

The symbol ‖ means "is parallel to." In the figure at the right, $j \parallel k$ and $\overline{AB} \parallel \overline{CD}$. Note that j contains $\overline{AB}$ and k contains $\overline{CD}$. Parallel lines contain parallel line segments.

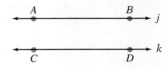

An **angle** is formed by two rays with the same endpoint. The **vertex** of the angle is the point at which the two rays meet. The rays are called the **sides** of the angle.

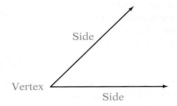

If A and C are points on rays r_1 and r_2, and B is the vertex, then the angle is called $\angle B$ or $\angle ABC$, where $\angle$ is the symbol for angle. Note that the angle is named by the vertex, or the vertex is the second point listed when the angle is named by giving three points. $\angle ABC$ could also be called $\angle CBA$.

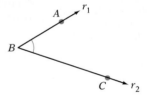

An angle can also be named by a variable written between the rays close to the vertex. In the figure at the right, $\angle x = \angle QRS$ and $\angle y = \angle SRT$. Note that in this figure, more than two rays meet at R. In this case, the vertex cannot be used to name an angle.

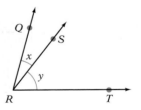

 Point of Interest

The Babylonians knew that Earth is in approximately the same position in the sky every 365 days. Historians suggest that one complete revolution of a circle is called 360° because 360 is the closest number to 365 that is divisible by many natural numbers.

An angle is measured in **degrees.** The symbol for degrees is a small raised circle, °. The angle formed by a ray rotating through a circle has a measure of 360° (360 degrees), probably because early Babylonians believed that Earth revolves around the sun in approximately 360 days.

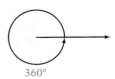

360°

Unless otherwise noted, all content on this page is © Cengage Learning.

Point of Interest

Sorin Colac/Shutterstock.com

The Leaning Tower of Pisa is the bell tower of the Cathedral in Pisa, Italy. Its construction began on August 9, 1173, and continued for about 200 years. The tower was designed to be vertical, but it started to lean during its construction. By 1350, it was 2.5° off from the vertical; by 1817, it was 5.1° off; and by 1990, it was 5.5° off. In 2001, work on the structure that returned its list to 5° was completed. (*Source: Time* magazine, June 25, 2001)

A **protractor** is used to measure an angle. Place the center of the protractor at the vertex of the angle with the edge of the protractor along a side of the angle. The angle shown in the figure below measures 58°.

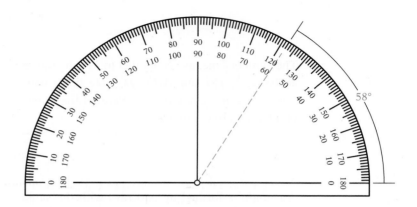

A 90° angle is called a **right angle.** The symbol ∟ represents a right angle.

Perpendicular lines are intersecting lines that form right angles.

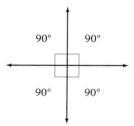

The symbol ⊥ means "is perpendicular to." In the figure at the right, $p \perp q$ and $\overline{AB} \perp \overline{CD}$. Note that line p contains $\overline{AB}$ and line q contains $\overline{CD}$. Perpendicular lines contain perpendicular line segments.

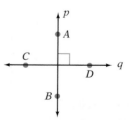

Complementary angles are two angles whose measures have the sum 90°.

$$\angle A + \angle B = 70° + 20° = 90°$$

$\angle A$ and $\angle B$ are complementary angles.

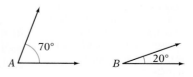

Unless otherwise noted, all content on this page is © Cengage Learning.

IN-CLASS EXAMPLES

1. Given $MN = 9$ cm,
   $OP = 10$ cm, and
   $MP = 25$ cm, find
   NO. **6 cm**

2. Find the complement of
   a 43° angle. **47°**
3. Find the supplement of
   a 57° angle. **123°**
4. Find the measure of
   $\angle a$. **59°**

A 180° angle is called a **straight angle.**

$\angle AOB$ is a straight angle.

Supplementary angles are two angles whose measures have the sum 180°.

$$\angle A + \angle B = 130° + 50° = 180°$$

$\angle A$ and $\angle B$ are supplementary angles.

An **acute angle** is an angle whose measure is between 0° and 90°. $\angle B$ above is an acute angle. An **obtuse angle** is an angle whose measure is between 90° and 180°. $\angle A$ above is an obtuse angle.

Two angles that share a common side are **adjacent angles.** In the figure at the right, $\angle DAC$ and $\angle CAB$ are adjacent angles. $\angle DAC = 45°$ and $\angle CAB = 55°$.

$$\angle DAB = \angle DAC + \angle CAB$$
$$= 45° + 55° = 100°$$

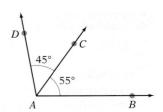

HOW TO 2   In the figure at the right, $\angle EDG = 80°$. $\angle FDG$ is three times the measure of $\angle EDF$. Find the measure of $\angle EDF$.

Let $x =$ the measure of $\angle EDF$. Then $3x =$ the measure of $\angle FDG$. Write an equation and solve for x, the measure of $\angle EDF$.

$\angle EDF = 20°$

$$\angle EDF + \angle FDG = \angle EDG$$
$$x + 3x = 80°$$
$$4x = 80°$$
$$x = 20°$$

EXAMPLE 1

Given $MN = 15$ mm, $NO = 18$ mm, and $MP = 48$ mm, find OP.

Solution

$MN + NO + OP = MP$ • $MN = 15$, $NO = 18$,
$15 + 18 + OP = 48$ $MP = 48$
$33 + OP = 48$ • Add 15 and 18.
$OP = 15$ • Subtract 33 from each side.

$OP = 15$ mm

YOU TRY IT 1

Given $QR = 24$ cm, $ST = 17$ cm, and $QT = 62$ cm, find RS.

Your solution

21 cm

Solution on p. S8

Unless otherwise noted, all content on this page is © Cengage Learning.

EXAMPLE 2

Given $XY = 9$ m and YZ is twice XY, find XZ.

$$\xleftarrow{\;\;\bullet\;\;\;\;\bullet\;\;\;\;\;\;\;\bullet\;\;\;}\rightarrow \ell$$
$$\quad X \quad\;\; Y \qquad\;\; Z$$

Solution

$XZ = XY + YZ$
$XZ = XY + 2(XY)$ • *YZ is twice XY.*
$XZ = 9 + 2(9)$ • *XY = 9*
$XZ = 9 + 18$
$XZ = 27$

$XZ = 27$ m

YOU TRY IT 2

Given $BC = 16$ ft and $AB = \frac{1}{4}(BC)$, find AC.

$$\xleftarrow{\;\;\bullet\;\;\bullet\;\;\;\;\;\;\;\;\bullet\;\;}\rightarrow \ell$$
$$\quad A \;\; B \qquad\;\; C$$

Your solution

20 ft

EXAMPLE 3

Find the complement of a 38° angle.

Strategy

Complementary angles are two angles whose sum is 90°. To find the complement, let x represent the complement of a 38° angle. Write an equation and solve for x.

Solution

$x + 38° = 90°$
$\quad\quad x = 52°$

The complement of a 38° angle is a 52° angle.

YOU TRY IT 3

Find the supplement of a 129° angle.

Your strategy

Your solution

51°

EXAMPLE 4

Find the measure of $\angle x$.

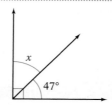

Strategy

To find the measure of $\angle x$, write an equation using the fact that the sum of the measure of $\angle x$ and 47° is 90°. Solve for $\angle x$.

Solution

$\angle x + 47° = 90°$
$\quad\quad \angle x = 43°$

The measure of $\angle x$ is 43°.

YOU TRY IT 4

Find the measure of $\angle a$.

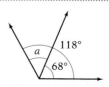

Your strategy

Your solution

50°

Solutions on pp. S8–S9

Unless otherwise noted, all content on this page is © Cengage Learning.

OBJECTIVE B *To solve problems involving angles formed by intersecting lines*

Point of Interest

Many cities in the New World, unlike those in Europe, were designed using rectangular street grids. Washington, D.C., was planned that way except that diagonal avenues were added, primarily for the purpose of enabling quick troop movement in the event that the city required defense. As an added precaution, monuments were constructed at major intersections so that attackers would not have a straight shot down a boulevard.

Four angles are formed by the intersection of two lines. If the two lines are perpendicular, each of the four angles is a right angle. If the two lines are not perpendicular, then two of the angles formed are acute angles and two of the angles are obtuse angles. The two acute angles are always opposite each other, and the two obtuse angles are always opposite each other.

In the figure at the right, $\angle w$ and $\angle y$ are acute angles. $\angle x$ and $\angle z$ are obtuse angles.

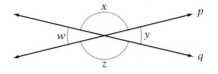

Two angles that are on opposite sides of the intersection of two lines are called **vertical angles.** Vertical angles have the same measure. $\angle w$ and $\angle y$ are vertical angles. $\angle x$ and $\angle z$ are vertical angles.

Vertical angles have the same measure.

$$\angle w = \angle y$$
$$\angle x = \angle z$$

Recall that two angles that share a common side are called **adjacent angles.** For the figure shown above, $\angle x$ and $\angle y$ are adjacent angles, as are $\angle y$ and $\angle z$, $\angle z$ and $\angle w$, and $\angle w$ and $\angle x$. Adjacent angles of intersecting lines are supplementary angles.

Adjacent angles of intersecting lines are supplementary angles.

$$\angle x + \angle y = 180°$$
$$\angle y + \angle z = 180°$$
$$\angle z + \angle w = 180°$$
$$\angle w + \angle x = 180°$$

HOW TO 3 Given that $\angle c = 65°$, find the measures of angles a, b, and d.

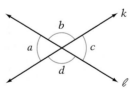

$\angle a = \angle c$ because $\angle a$ and $\angle c$ are vertical angles.

$\angle a = 65°$

$\angle b$ is supplementary to $\angle c$ because $\angle b$ and $\angle c$ are adjacent angles of intersecting lines.

$$\angle b + \angle c = 180°$$
$$\angle b + 65° = 180°$$
$$\angle b = 115°$$

$\angle d = \angle b$ because $\angle d$ and $\angle b$ are vertical angles.

$\angle d = 115°$

Unless otherwise noted, all content on this page is © Cengage Learning.

IN-CLASS EXAMPLES

5. Find x.   **14°**

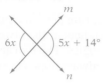

6. Given that $\ell_1 \parallel \ell_2$, find x.   **20°**

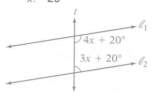

A line that intersects two other lines at different points is called a **transversal.**

If the lines cut by a transversal t are parallel lines and the transversal is perpendicular to the parallel lines, all eight angles formed are right angles.

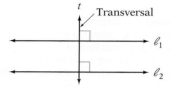

If the lines cut by a transversal t are parallel lines and the transversal is not perpendicular to the parallel lines, all four acute angles have the same measure and all four obtuse angles have the same measure. For the figure at the right,

$$\angle b = \angle d = \angle x = \angle z$$
$$\angle a = \angle c = \angle w = \angle y$$

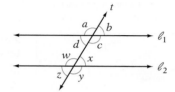

Alternate interior angles are two nonadjacent angles that are on opposite sides of the transversal and between the parallel lines. In the figure above, $\angle c$ and $\angle w$ are alternate interior angles; $\angle d$ and $\angle x$ are alternate interior angles. Alternate interior angles have the same measure.

Alternate interior angles have the same measure.

$$\angle c = \angle w$$
$$\angle d = \angle x$$

Alternate exterior angles are two nonadjacent angles that are on opposite sides of the transversal and outside the parallel lines. In the figure above, $\angle a$ and $\angle y$ are alternate exterior angles; $\angle b$ and $\angle z$ are alternate exterior angles. Alternate exterior angles have the same measure.

Alternate exterior angles have the same measure.

$$\angle a = \angle y$$
$$\angle b = \angle z$$

Corresponding angles are two angles that are on the same side of the transversal and are both acute angles or are both obtuse angles. For the figure above, the following pairs of angles are corresponding angles: $\angle a$ and $\angle w$, $\angle d$ and $\angle z$, $\angle b$ and $\angle x$, $\angle c$ and $\angle y$. Corresponding angles have the same measure.

Corresponding angles have the same measure.

$$\angle a = \angle w$$
$$\angle d = \angle z$$
$$\angle b = \angle x$$
$$\angle c = \angle y$$

Unless otherwise noted, all content on this page is © Cengage Learning.

HOW TO 4   Given that $\ell_1 \| \ell_2$ and $\angle c = 58°$, find the measures of $\angle f$, $\angle h$, and $\angle g$.

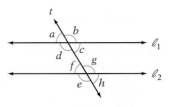

$\angle c$ and $\angle f$ are alternate interior angles.

$\angle f = \angle c = 58°$

$\angle c$ and $\angle h$ are corresponding angles.

$\angle h = \angle c = 58°$

$\angle g$ is supplementary to $\angle h$.

$\angle g + \angle h = 180°$
$\angle g + 58° = 180°$
$\angle g = 122°$

EXAMPLE 5

Find x.

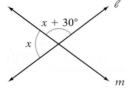

Strategy

The angles labeled are adjacent angles of intersecting lines and are therefore supplementary angles. To find x, write an equation and solve for x.

Solution

$x + (x + 30°) = 180°$ • **The sum of the measures**
$2x + 30° = 180°$ **of the two angles is 180°.**
$2x = 150°$
$x = 75°$

YOU TRY IT 5

Find x.

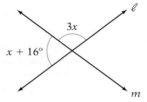

Your strategy

Your solution

41°

EXAMPLE 6

Given $\ell_1 \| \ell_2$, find x.

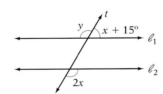

Strategy

$2x = y$ because alternate exterior angles have the same measure. $(x + 15°) + y = 180°$ because adjacent angles of intersecting lines are supplementary angles. Substitute $2x$ for y and solve for x.

Solution

$(x + 15°) + 2x = 180°$ • **The sum of the measures**
$3x + 15° = 180°$ **of the two angles is 180°.**
$3x = 165°$
$x = 55°$

YOU TRY IT 6

Given $\ell_1 \| \ell_2$, find x.

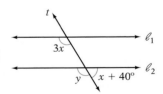

Your strategy

Your solution

35°

Solutions on p. S9

Unless otherwise noted, all content on this page is © Cengage Learning.

OBJECTIVE C *To solve problems involving the angles of a triangle*

IN-CLASS EXAMPLES

7. A triangle has a 110° angle and a 35° angle. Find the measure of the third angle of the triangle. **35°**

If the lines cut by a transversal are not parallel lines, the three lines will intersect at three points. In the figure at the right, the transversal t intersects lines p and q. The three lines intersect at points A, B, and C. These three points define three line segments, $\overline{AB}$, $\overline{BC}$, and $\overline{AC}$. The plane figure formed by these three line segments is called a **triangle.**

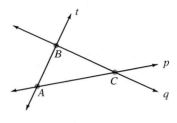

Each of the three points of intersection is the vertex of four angles. The angles within the region enclosed by the triangle are called **interior angles.** In the figure at the right, angles a, b, and c are interior angles. The sum of the measures of the interior angles of a triangle is 180°.

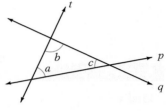

$$\angle a + \angle b + \angle c = 180°$$

The Sum of the Measures of the Interior Angles of a Triangle

The sum of the measures of the interior angles of a triangle is 180°.

EXAMPLE

The measures of two angles of a triangle are 40° and 60°. The measure of the third angle is:

$$40° + 60° + x = 180°$$
$$100° + x = 180°$$
$$x = 80°$$

An angle adjacent to an interior angle is an **exterior angle.** In the figure at the right, angles m and n are exterior angles for angle a. The sum of the measures of an interior and an exterior angle is 180°.

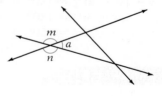

$$\angle a + \angle m = 180°$$
$$\angle a + \angle n = 180°$$

HOW TO 5 Given that $\angle c = 40°$ and $\angle d = 100°$, find the measure of $\angle e$.

$\angle d$ and $\angle b$ are supplementary angles.

$$\angle d + \angle b = 180°$$
$$100° + \angle b = 180°$$
$$\angle b = 80°$$

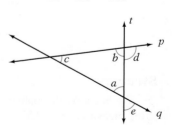

The sum of the interior angles is 180°.

$$\angle c + \angle b + \angle a = 180°$$
$$40° + 80° + \angle a = 180°$$
$$120° + \angle a = 180°$$
$$\angle a = 60°$$

$\angle a$ and $\angle e$ are vertical angles.

$$\angle e = \angle a = 60°$$

Unless otherwise noted, all content on this page is © Cengage Learning.

EXAMPLE 7

Given that $\angle y = 55°$, find the measures of angles a, b, and d.

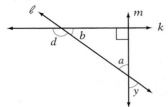

Strategy

• To find the measure of angle a, use the fact that $\angle a$ and $\angle y$ are vertical angles.
• To find the measure of angle b, use the fact that the sum of the measures of the interior angles of a triangle is 180°.
• To find the measure of angle d, use the fact that the sum of an interior and an exterior angle is 180°.

Solution

$\angle a = \angle y = 55°$ • $\angle a$ and $\angle y$ are vertical angles.

$\angle a + \angle b + 90° = 180°$ • The sum of the measures
$55° + \angle b + 90° = 180°$ of the interior angles of a
$\angle b + 145° = 180°$ triangle is 180°.
$\angle b = 35°$

$\angle d + \angle b = 180°$ • The sum of an interior and
$\angle d + 35° = 180°$ an exterior angle is 180°.
$\angle d = 145°$

YOU TRY IT 7

Given that $\angle a = 45°$ and $\angle x = 100°$, find the measures of angles b, c, and y.

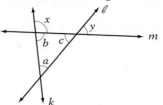

Your strategy

Your solution
$\angle b = 80°$, $\angle c = 55°$, $\angle y = 55°$

EXAMPLE 8

Two angles of a triangle measure 53° and 78°. Find the measure of the third angle.

Strategy

To find the measure of the third angle, use the fact that the sum of the measures of the interior angles of a triangle is 180°. Write an equation using x to represent the measure of the third angle. Solve the equation for x.

Solution

$x + 53° + 78° = 180°$
$x + 131° = 180°$
$x = 49°$

The measure of the third angle is 49°.

YOU TRY IT 8

One angle in a triangle is a right angle, and one angle measures 34°. Find the measure of the third angle.

Your strategy

Your solution
56°

Solutions on p. S9

Unless otherwise noted, all content on this page is © Cengage Learning.

3.1 EXERCISES

✔ **Concept Check**

SUGGESTED ASSIGNMENT
Exercises 1–10; Exercises 11–71, odds

For Exercises 1 to 4, fill in the blanks to complete an equation that can be used to find the value of x.

1.

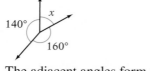

$AD = 12$ cm

AD is the sum of AB, BC, and CD, so
$\underline{\ 12\ } = \underline{\ 5\ } + \underline{\ x\ } + \underline{\ 4\ }$.

2. x is the supplement of a 113° angle.
The sum of an angle and its supplement is 180°,
so $x + \underline{\ 113°\ } = \underline{\ 180°\ }$.

3.

The adjacent angles form a circle, so
$x + \underline{\ 160°\ } + \underline{\ 140°\ } = \underline{\ 360°\ }$.

4.

$\angle LON$ is a right angle, so $x + \underline{\ 3x\ } = \underline{\ 90°\ }$.

For Exercises 5 to 8, use the diagram at the right, in which $\ell_1 \parallel \ell_2$. Fill in the blanks to complete an equation that models the given statement.

5. Vertical angles have the same measure. $\angle \underline{\ a\ } = \angle \underline{\ b\ }$

6. Alternate interior angles have the same measure. $\angle \underline{\ b\ } = \angle \underline{\ c\ }$

7. Adjacent angles of intersecting lines are supplementary angles. $\angle \underline{\ c\ } + \angle \underline{\ d\ } = \underline{\ 180°\ }$

8. Corresponding angles have the same measure. $\angle \underline{\ a\ } = \angle \underline{\ c\ }$

For Exercises 9 and 10, use the diagram at the right.

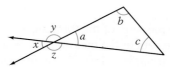

9. a. The angles that are interior angles are $\underline{\ \angle a, \angle b, \text{ and } \angle c\ }$.

b. The angles that are exterior angles are $\underline{\ \angle y \text{ and } \angle z\ }$.

c. The angle that is neither an interior nor an exterior angle of the triangle is $\underline{\ \angle x\ }$.

10. Complete the equations.

a. $\angle a + \underline{\ \angle b\ } + \underline{\ \angle c\ } = 180°$

b. $\angle a + \underline{\ \angle y\ } = 180°$ and $\angle a + \underline{\ \angle z\ } = 180°$

Unless otherwise noted, all content on this page is © Cengage Learning.

OBJECTIVE A   *To solve problems involving lines and angles*

For Exercises 11 to 16, use a protractor to measure the angle. State whether the angle is acute, obtuse, or right.

11. 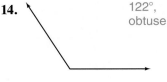   40°, acute

12. 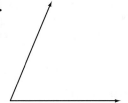   68°, acute

13. 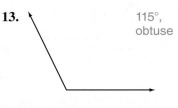   115°, obtuse

14. 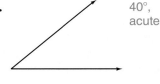   122°, obtuse

15. 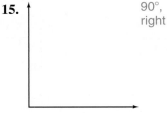   90°, right

16. 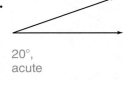   20°, acute

17. Find the complement of a 62° angle.   28°

18. Find the complement of a 31° angle.   59°

19. Find the supplement of a 162° angle.   18°

20. Find the supplement of a 72° angle.   108°

21. Given $AB = 12$ cm, $CD = 9$ cm, and $AD = 35$ cm, find the length of BC.
14 cm

22. Given $AB = 21$ mm, $BC = 14$ mm, and $AD = 54$ mm, find the length of CD.
19 mm

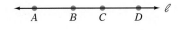

23. Given $QR = 7$ ft and RS is three times the length of QR, find the length of QS.
28 ft

24. Given $QR = 15$ in. and RS is twice the length of QR, find the length of QS.
45 in.

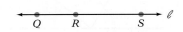

25. Given $EF = 20$ m and FG is $\frac{1}{2}$ the length of EF, find the length of EG.
30 m

26. Given $EF = 18$ cm and FG is $\frac{1}{3}$ the length of EF, find the length of EG.
24 cm

Unless otherwise noted, all content on this page is © Cengage Learning.

27. Given ∠LOM = 53° and ∠LON = 139°, find the measure of ∠MON.

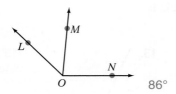

86°

28. Given ∠MON = 38° and ∠LON = 85°, find the measure of ∠LOM.

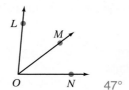

47°

Given that ∠LON is a right angle, find the measure of ∠x.

29.
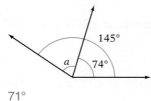
30°

30.
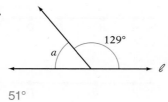
18°

31.

36°

32.
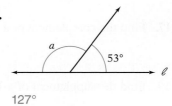
33°

Find the measure of ∠a.

33.
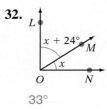
71°

34.
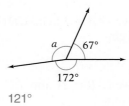
63°

35.
127°

36.
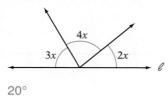
51°

37.
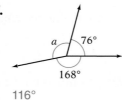
116°

38.
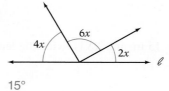
121°

For Exercises 39 to 44, find x.

39.
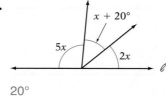
20°

40.
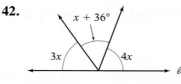
15°

41.
20°

42.
18°

Unless otherwise noted, all content on this page is © Cengage Learning.

43.

20°

44.

45°

45. Given that $\angle a = 51°$, find the measure of $\angle b$.

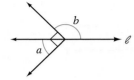

141°

46. Given that $\angle a = 38°$, find the measure of $\angle b$.

128°

47. 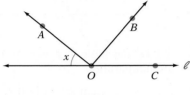 Given that $\overline{AO} \perp \overline{OB}$, express in terms of x the number of degrees in $\angle BOC$.

90° − x

48. 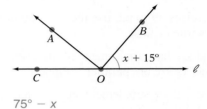 Given that $\overline{AO} \perp \overline{OB}$, express in terms of x the number of degrees in $\angle AOC$.

75° − x

OBJECTIVE B   *To solve problems involving angles formed by intersecting lines*

Find the measure of $\angle x$.

49.

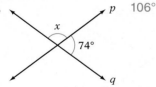   106°

50.

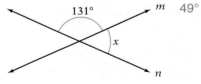   49°

Find x.

51.

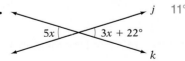   11°

52.

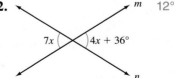   12°

Given that $\ell_1 \parallel \ell_2$, find the measures of angles a and b.

53.

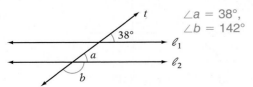 $\angle a = 38°$,
$\angle b = 142°$

54.

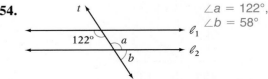 $\angle a = 122°$,
$\angle b = 58°$

Unless otherwise noted, all content on this page is © Cengage Learning.

55.

$\angle a = 47°$,
$\angle b = 133°$

56.

$\angle a = 44°$,
$\angle b = 136°$

Given that $\ell_1 \parallel \ell_2$, find x.

57.

20°

58.

47°

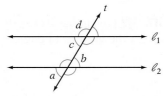

For Exercises 59 to 62, use the diagram at the right. Determine whether the statement is true or false.

59. $\angle b$ and $\angle c$ have the same measure even if ℓ_1 and ℓ_2 are not parallel.   False

60. $\angle a$ and $\angle b$ have the same measure even if ℓ_1 and ℓ_2 are not parallel.   True

61. $\angle a$ and $\angle d$ are supplementary if ℓ_1 and ℓ_2 are parallel.   True

62. $\angle c$ and $\angle d$ are supplementary only if ℓ_1 and ℓ_2 are parallel.   False

---

OBJECTIVE C *To solve problems involving the angles of a triangle*

63. Given that $\angle a = 95°$ and $\angle b = 70°$, find the measures of $\angle x$ and $\angle y$.

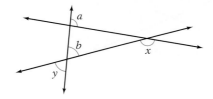

$\angle x = 155°$; $\angle y = 70°$

64. Given that $\angle a = 35°$ and $\angle b = 55°$, find the measures of $\angle x$ and $\angle y$.

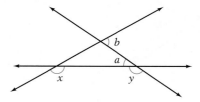

$\angle x = 160°$; $\angle y = 145°$

65. Given that $\angle y = 45°$, find the measures of $\angle a$ and $\angle b$.

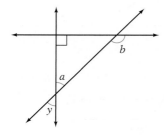

$\angle a = 45°$; $\angle b = 135°$

66. Given that $\angle y = 130°$, find the measures of $\angle a$ and $\angle b$.

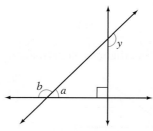

$\angle a = 40°$; $\angle b = 140°$

Unless otherwise noted, all content on this page is © Cengage Learning.

67. One angle in a triangle is a right angle, and one angle is equal to 30°. What is the measure of the third angle?   60°

68. A triangle has a 45° angle and a right angle. Find the measure of the third angle.   45°

69. Two angles of a triangle measure 42° and 103°. Find the measure of the third angle.   35°

70. A triangle has a 13° angle and a 65° angle. What is the measure of the third angle?   102°

71.  True or false? If one interior angle of a triangle is a right angle, then the other two interior angles of the triangle are complementary angles.   True

72.  True or false? If a triangle has an exterior angle that is a right angle, then the triangle also has an interior angle that is a right angle.   True

Critical Thinking

73. On a number line, the points A, B, C, and D have coordinates −2.5, 2, 5, and 3.5, respectively. Which of these points is halfway between two others?   D, or 3.5

74. Find the measure of the smaller angle between the hands of a clock when the time is 5 o'clock.   150°

75. **Forestry**   A forest ranger must determine the diameter of a redwood tree. Explain how the ranger could do this without cutting down the tree.

Projects or Group Activities

76. For the figure at the right, find the sum of the measures of angles x, y, and z.   360°

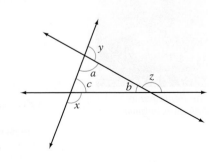

77. For the figure at the right, explain why $\angle a + \angle b = \angle x$. Write a rule that describes the relationship between an exterior angle of a triangle and the opposite interior angles. Use the rule to write an equation involving angles a, c, and z.

78. If $\overrightarrow{AB}$ and $\overrightarrow{CD}$ intersect at point O, and $\angle AOC = \angle BOC$, explain why $\overrightarrow{AB} \perp \overrightarrow{CD}$.

Unless otherwise noted, all content on this page is © Cengage Learning.

QUICK QUIZ

1. Given QR = 13 mm, RS = 11 mm, and QT = 32 mm, find ST.   **8 mm**   **[3.1A]**

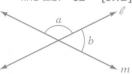

2. Find the complement of a 54° angle.   **36°**   **[3.1A]**

3. Given that $\angle a$ = 118°, find $\angle b$.   **62°**   **[3.1B]**

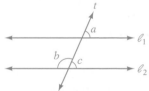

4. Given $\ell_1 \parallel \ell_2$ and $\angle c$ = 67°, find $\angle a$ and $\angle b$. $\angle a$ = **67°**, $\angle b$ = **113°**   **[3.1B]**

5. A triangle has a 21° angle and a 64° angle. Find the measure of the third angle of the triangle.   **95°**   **[3.1C]**

3.2 Plane Geometric Figures

OBJECTIVE A   *To solve problems involving the perimeter of a geometric figure*

A **polygon** is a closed figure determined by three or more line segments that lie in a plane. The line segments that form the polygon are called its **sides.** The figures below are examples of polygons.

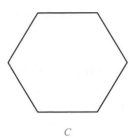
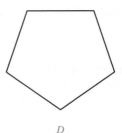
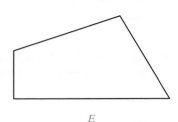

A   B   C   D   E

A **regular polygon** is one in which all sides have the same length and all angles have the same measure. The polygons in Figures *A*, *C*, and *D* above are regular polygons.

The name of a polygon is based on the number of its sides. The table below lists the names of polygons that have from 3 to 10 sides.

Point of Interest

Although a polygon is defined in terms of its sides, the word actually comes from the Latin word *polygonum*, meaning "many angles."

Number of Sides	Name of the Polygon
3	Triangle
4	Quadrilateral
5	Pentagon
6	Hexagon
7	Heptagon
8	Octagon
9	Nonagon
10	Decagon

Triangles and quadrilaterals are two of the most common types of polygons. Triangles are distinguished by the number of equal sides and also by the measures of their angles.

*The Pentagon in
Arlington, Virginia*

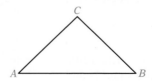
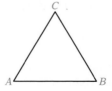
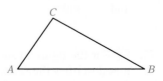

An **isosceles triangle** has two sides of equal length. The angles opposite the equal sides are of equal measure.

$AC = BC$
$\angle A = \angle B$

The three sides of an **equilateral triangle** are of equal length. The three angles are of equal measure.

$AB = BC = AC$
$\angle A = \angle B = \angle C$

A **scalene triangle** has no two sides of equal length. No two angles are of equal measure.

Unless otherwise noted, all content on this page is © Cengage Learning.

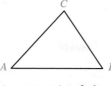

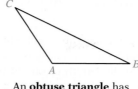

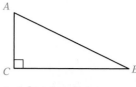

An **acute triangle** has three acute angles.

An **obtuse triangle** has an obtuse angle.

A **right triangle** has a right angle.

INSTRUCTOR NOTE

The diagram below shows the relationships among different types of quadrilaterals. A description of each quadrilateral is given within a drawing of that type of quadrilateral.

Quadrilaterals are also distinguished by their sides and angles, as shown below. Note that a rectangle, a square, and a rhombus are different forms of a parallelogram.

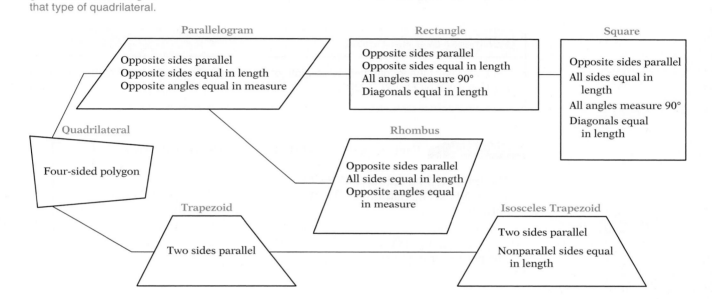

Parallelogram
Opposite sides parallel
Opposite sides equal in length
Opposite angles equal in measure

Rectangle
Opposite sides parallel
Opposite sides equal in length
All angles measure 90°
Diagonals equal in length

Square
Opposite sides parallel
All sides equal in length
All angles measure 90°
Diagonals equal in length

Quadrilateral
Four-sided polygon

Rhombus
Opposite sides parallel
All sides equal in length
Opposite angles equal in measure

Trapezoid
Two sides parallel

Isosceles Trapezoid
Two sides parallel
Nonparallel sides equal in length

The **perimeter** of a plane geometric figure is a measure of the distance around the figure. Perimeter is used, for example, in buying fencing for a lawn or determining how much baseboard is needed for a room.

The perimeter of a triangle is the sum of the lengths of the three sides.

Perimeter of a Triangle

Let a, b, and c be the lengths of the sides of a triangle. The perimeter P of the triangle is given by $P = a + b + c$.

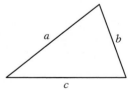

$P = a + b + c$

HOW TO 1   Find the perimeter of the triangle shown at the right.

$P = 5 + 7 + 10 = 22$

The perimeter is 22 ft.

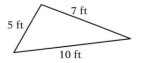

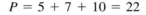

Unless otherwise noted, all content on this page is © Cengage Learning.

The perimeter of a quadrilateral is the sum of the lengths of its four sides.

A rectangle has four right angles and opposite sides of equal length. Usually the length L of a rectangle refers to the length of one of the longer sides of the rectangle, and the width W refers to the length of one of the shorter sides. The perimeter can then be represented $P = L + W + L + W$.

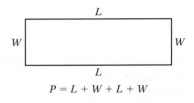

$$P = L + W + L + W$$

The formula for the perimeter of a rectangle is derived by combining like terms.

$$P = 2L + 2W$$

Perimeter of a Rectangle

Let L represent the length and W the width of a rectangle. The perimeter P of the rectangle is given by $P = 2L + 2W$.

HOW TO 2   Find the perimeter of the rectangle shown at the right.

The length is 5 m. Substitute 5 for L.
The width is 2 m. Substitute 2 for W.
Solve for P.

$$P = 2L + 2W$$
$$P = 2(5) + 2(2)$$
$$P = 10 + 4$$
$$P = 14$$

The perimeter is 14 m.

A square is a rectangle in which each side has the same length. If we let s represent the length of each side of a square, the perimeter of the square can be represented by $P = s + s + s + s$.

$$P = s + s + s + s$$

The formula for the perimeter of a square is derived by combining like terms.

$$P = 4s$$

Perimeter of a Square

Let s represent the length of a side of a square. The perimeter P of the square is given by $P = 4s$.

HOW TO 3   Find the perimeter of the square shown at the right.

$$P = 4s = 4(8) = 32$$

The perimeter is 32 in.

Unless otherwise noted, all content on this page is © Cengage Learning.

IN-CLASS EXAMPLES

1. Find the perimeter of a square for which each side is equal to 15 m.
   60 m
2. Find the perimeter of a rectangle with a length of 5 in. and a width of 1.4 in.   **12.8 in.**
3. Find the circumference of a circle with a radius of 11 cm. Round to the nearest hundredth.
   69.12 cm

A **circle** is a plane figure in which all points are the same distance from point O, called the **center** of the circle.

The **diameter** of a circle is a line segment across the circle through point O. AB is a diameter of the circle at the right. The variable d is used to designate the diameter of a circle.

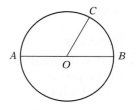

The **radius** of a circle is a line segment from the center of the circle to a point on the circle. OC is a radius of the circle at the right. The variable r is used to designate the radius of a circle.

The length of the diameter is twice the length of the radius. $d = 2r$ or $r = \dfrac{1}{2}d$

Point of Interest

Archimedes (c. 287–212 B.C.) is the person who calculated that $\pi \approx 3\frac{1}{7}$. He actually showed that $3\frac{10}{71} < \pi < 3\frac{1}{7}$. The approximation $3\frac{10}{71}$ is a more accurate approximation of π than $3\frac{1}{7}$, but it is more difficult to use.

The distance around a circle is called the **circumference.** The circumference C of a circle is equal to the product of π (pi) and the diameter. $C = \pi d$

Because $d = 2r$, the formula for the circumference can be written in terms of r. $C = 2\pi r$

The Circumference of a Circle

The circumference C of a circle with diameter d and radius r is given by $C = \pi d$ or $C = 2\pi r$.

The formula for circumference uses the number π, which is an irrational number. The value of π can be approximated by a fraction or by a decimal. $\pi \approx \dfrac{22}{7}$ or $\pi \approx 3.14$

The π key on a scientific calculator gives a closer approximation of π than 3.14. Use a scientific calculator to find approximate values in calculations involving π.

HOW TO 4   Find the circumference of a circle with a diameter of 6 in.

The diameter of the circle is given. Use the circumference formula that involves the diameter. $d = 6$

$$C = \pi d$$
$$C = \pi(6)$$

The exact circumference of the circle is 6π in. $C = 6\pi$

An approximate measure is found by using the π key on a calculator. $C \approx 18.85$

The approximate circumference is 18.85 in.

Integrating Technology

The π key on your calculator can be used to find decimal approximations to formulas that contain π. To perform the calculation at the right, enter 6 $\times$ π $=$.

Unless otherwise noted, all content on this page is © Cengage Learning.

EXAMPLE 1

A carpenter is designing a square patio with a perimeter of 44 ft. What is the length of each side?

Strategy

To find the length of each side, use the formula for the perimeter of a square. Substitute 44 for P and solve for s.

Solution

$P = 4s$ • The formula for the perimeter of a square
$44 = 4s$ • $P = 44$
$11 = s$

The length of each side of the patio is 11 ft.

YOU TRY IT 1

The infield of a softball field is a square with each side of length 60 ft. Find the perimeter of the infield.

Your strategy

Your solution

240 ft

EXAMPLE 2

The dimensions of a triangular sail are 18 ft, 11 ft, and 15 ft. What is the perimeter of the sail?

Strategy

To find the perimeter, use the formula for the perimeter of a triangle. Substitute 18 for a, 11 for b, and 15 for c. Solve for P.

Solution

$P = a + b + c$ • Use the formula for the
$P = 18 + 11 + 15$ perimeter of a triangle.
$P = 44$ $a = 18, b = 11, c = 15$

The perimeter of the sail is 44 ft.

YOU TRY IT 2

Find the length of decorative molding needed to edge the tops of the walls in a rectangular room that is 12 ft long and 8 ft wide.

Your strategy

Your solution

40 ft

EXAMPLE 3

Find the circumference of a circle with a radius of 15 cm. Round to the nearest hundredth.

Strategy

To find the circumference, use the circumference formula that involves the radius. An approximation is asked for; use the π key on a calculator.
$r = 15$

Solution

$C = 2\pi r = 2\pi(15) = 30\pi \approx 94.25$

The circumference is 94.25 cm.

YOU TRY IT 3

Find the circumference of a circle with a diameter of 9 in. Give the exact measure.

Your strategy

Your solution

9π in.

Solutions on p. S9

OBJECTIVE B

To solve problems involving the area of a geometric figure

Area is the amount of surface in a region. Area can be used to describe the size of, for example, a rug, a parking lot, a farm, or a national park. Area is measured in square units.

A square that measures 1 in. on each side has an area of 1 square inch, written 1 in^2.

A square that measures 1 cm on each side has an area of 1 square centimeter, written 1 cm^2.

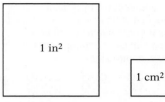

1 in^2

1 cm^2

Point of Interest

First Lady Michelle Obama's White House garden is a small, 1100 ft^2 plot of land, but it has yielded thousands of pounds of tomatoes, lettuce, eggplant, sweet potatoes, broccoli, cabbage, carrots, and other vegetables and herbs that have been used to feed people at the White House and at nearby homeless shelters.

Larger areas can be measured in square feet (ft^2), square meters (m^2), acres (43,560 ft^2), square miles (mi^2), or any other square unit.

The area of a geometric figure is the number of squares that are necessary to cover the figure. In the figures below, two rectangles have been drawn and covered with squares. In the figure on the left, 12 squares, each of area 1 cm^2, were used to cover the rectangle. The area of the rectangle is 12 cm^2. In the figure on the right, 6 squares, each of area 1 in^2, were used to cover the rectangle. The area of the rectangle is 6 in^2.

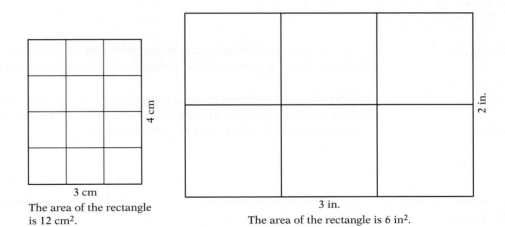

4 cm

3 cm

The area of the rectangle is 12 cm^2.

2 in.

3 in.

The area of the rectangle is 6 in^2.

Note from the above figures that the area of a rectangle can be found by multiplying the length of the rectangle by its width.

Area of a Rectangle

Let *L* represent the length and *W* the width of a rectangle. The area *A* of the rectangle is given by $A = LW$.

HOW TO 5 Find the area of the rectangle shown at the right.

$$A = LW = 11(7) = 77$$

The area is 77 m^2.

7 m

11 m

Unless otherwise noted, all content on this page is © Cengage Learning.

A square is a rectangle in which all sides are the same length. Therefore, both the length and the width of a square can be represented by s, and $A = LW = s \cdot s = s^2$.

Point of Interest

Figurate numbers are whole numbers that can be represented by regular geometric figures. For example, a square number is one that can be represented by a square array.

O   OO   OOO   OOOO
     OO   OOO   OOOO
           OOO   OOOO
                    OOOO

1   4   9   16

The square numbers are 1, 4, 9, 16, 25, ... They can be represented by 1^2, 2^2, 3^2, 4^2, 5^2, ...

INSTRUCTOR NOTE

Ask students to find the next three square numbers. For each number, have them form the square array, as well as represent the number in both standard form and exponential form.

Area of a Square
Let s represent the length of a side of a square. The area A of the square is given by $A = s^2$.

$A = s \cdot s = s^2$

HOW TO 6   Find the area of the square shown at the right.

$A = s^2 = 9^2 = 81$

The area is 81 mi^2.

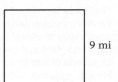

9 mi

Figure $ABCD$ is a parallelogram. BC is the **base** b of the parallelogram. AE, perpendicular to the base, is the **height** h of the parallelogram.

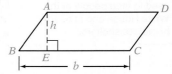

Any side of a parallelogram can be designated as the base. The corresponding height is found by drawing a line segment perpendicular to the base from the opposite side.

A rectangle can be formed from a parallelogram by cutting a right triangle from one end of the parallelogram and attaching it to the other end. The area of the resulting rectangle will equal the area of the original parallelogram.

Take Note

For a rectangle, $A = LW$.
For a parallelogram, $A = bh$.

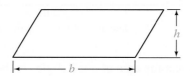

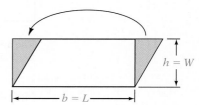

Area of a Parallelogram
Let b represent the length of the base and h the height of a parallelogram. The area A of the parallelogram is given by $A = bh$.

HOW TO 7   Find the area of the parallelogram shown at the right.

$A = bh = 12 \cdot 6 = 72$

The area is 72 m^2.

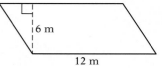

6 m

12 m

Unless otherwise noted, all content on this page is © Cengage Learning.

Luca Moi/Shutterstock.com

Point of Interest

A **glazier** is a person who cuts, fits, and installs glass, generally in doors and windows. Of particular challenge to a glazier are intricate stained glass window designs.

Figure *ABC* is a triangle. *AB* is the **base** *b* of the triangle. *CD*, perpendicular to the base, is the **height** *h* of the triangle.

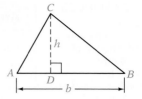

Any side of a triangle can be designated as the base. The corresponding height is found by drawing a line segment perpendicular to the base from the vertex opposite the base.

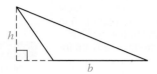

Consider the triangle with base *b* and height *h* shown at the right. By extending a line from *C* parallel to the base *AB* and equal in length to the base, a parallelogram is formed. The area of the parallelogram is *bh* and is twice the area of the triangle. Therefore, the area of the triangle is one-half the area of the parallelogram, or $\frac{1}{2}bh$.

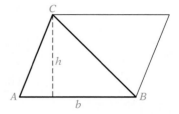

Area of a Triangle

Let *b* represent the length of the base and *h* the height of a triangle. The area *A* of the triangle is given by $A = \frac{1}{2}bh$.

HOW TO 8 Find the area of a triangle with a base of 18 cm and a height of 6 cm.

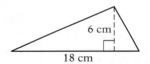

$$A = \frac{1}{2}bh = \frac{1}{2} \cdot 18 \cdot 6 = 54$$

The area is 54 cm².

Figure *ABCD* is a trapezoid. *AB* is one **base**, b_1, of the trapezoid, and *CD* is the other base, b_2. *AE*, perpendicular to the two bases, is the **height** *h*.

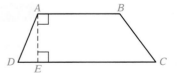

In the trapezoid at the right, the line segment *BD* divides the trapezoid into two triangles, *ABD* and *BCD*. In triangle *ABD*, b_1 is the base and *h* is the height. In triangle *BCD*, b_2 is the base and *h* is the height. The area of the trapezoid is the sum of the areas of the two triangles.

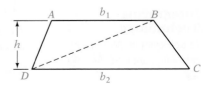

Area of trapezoid *ABCD* = area of triangle *ABD* + area of triangle *BCD*

$$= \frac{1}{2}b_1 h + \frac{1}{2}b_2 h = \frac{1}{2}h(b_1 + b_2)$$

Unless otherwise noted, all content on this page is © Cengage Learning.

IN-CLASS EXAMPLES
4. Find the area of a triangle with a base of 5 cm and a height of 1.6 cm.   **4 cm²**
5. Find the area of a square with a side of 8.5 ft.   **72.25 ft²**
6. Find the area of a circle with a diameter of 16 in. Round to the nearest hundredth.   **201.06 in²**

Area of a Trapezoid

Let b_1 and b_2 represent the lengths of the bases and h the height of a trapezoid. The area A of the trapezoid is given by $A = \frac{1}{2}h(b_1 + b_2)$.

HOW TO 9   Find the area of a trapezoid that has bases measuring 15 in. and 5 in. and a height of 8 in.

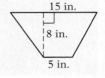

$$A = \frac{1}{2}h(b_1 + b_2)$$

$$= \frac{1}{2} \cdot 8(15 + 5) = 4(20) = 80$$

The area is 80 in².

The area of a circle is equal to the product of π and the square of the radius.

$A = \pi r^2$

 Tips for Success
You have now learned many different formulas for the perimeter and area of plane geometric figures. You will need to be able to recognize when to use each one. To test yourself, do the Chapter Review Exercises on page 213.

Area of a Circle

The area A of a circle with radius r is given by $A = \pi r^2$.

HOW TO 10   Find the area of a circle that has a radius of 6 cm.

Use the formula for the area of a circle. $r = 6$.

$$A = \pi r^2$$
$$A = \pi(6)^2$$
$$A = \pi(36)$$

The exact area of the circle is 36π cm².

$$A = 36\pi$$

An approximate measure is found by using the π key on a calculator.

$$A \approx 113.10$$

The approximate area of the circle is 113.10 cm².

 Integrating Technology
To approximate 36π on your calculator, enter 36 [×] [π] [=] .

For your reference, all of the formulas for the perimeter and area of the geometric figures presented in this section are listed in the Chapter Summary, which begins on page 210.

Unless otherwise noted, all content on this page is © Cengage Learning.

EXAMPLE 4

The parks and recreation department of a city plans to plant grass seed in a playground that has the shape of a trapezoid, as shown below. Each bag of grass seed will seed 1500 ft². How many bags of grass seed should the department purchase?

80 ft
64 ft
115 ft

Strategy

To find the number of bags to be purchased:
• Use the formula for the area of a trapezoid to find the area of the playground.
• Divide the area of the playground by the area one bag will seed (1500).

Solution

$A = \dfrac{1}{2}h(b_1 + b_2)$ • The formula for the area of a trapezoid

$A = \dfrac{1}{2} \cdot 64(80 + 115)$ • $h = 64, b_1 = 80, b_2 = 115$

$A = 6240$ • The area of the playground is 6240 ft².

$6240 \div 1500 = 4.16$

Because a portion of a fifth bag is needed, 5 bags of grass seed should be purchased.

YOU TRY IT 4

An interior designer decides to wallpaper two walls of a room. Each roll of wallpaper will cover 30 ft². Each wall measures 8 ft by 12 ft. How many rolls of wallpaper should be purchased?

Your strategy

Your solution

7 rolls

EXAMPLE 5

Find the area of a circle with a diameter of 5 ft. Give the exact measure.

Strategy

To find the area:
• Find the radius of the circle.
• Use the formula for the area of a circle. Leave the answer in terms of π.

Solution

$r = \dfrac{1}{2}d = \dfrac{1}{2}(5) = 2.5$ • Find the radius.

$A = \pi r^2 = \pi(2.5)^2 = \pi(6.25) = 6.25\pi$

The area of the circle is 6.25π ft².

YOU TRY IT 5

Find the area of a circle with a radius of 11 cm. Round to the nearest hundredth.

Your strategy

Your solution

380.13 cm²

Solutions on p. S10

Unless otherwise noted, all content on this page is © Cengage Learning.

3.2 EXERCISES

 Concept Check

SUGGESTED ASSIGNMENT
Exercises 1–12; Exercises 13–103, odds

Name each polygon.

1.

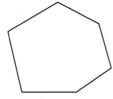

hexagon

2.

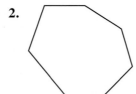

heptagon

3.

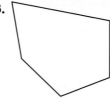

pentagon

4.

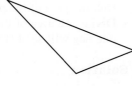

quadrilateral

Classify the triangle as isosceles, equilateral, or scalene.

5.

scalene

6.

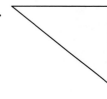

isosceles

7.

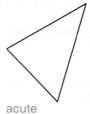

equilateral

8.

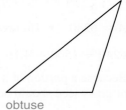

scalene

Classify the triangle as acute, obtuse, or right.

9.

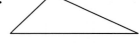

obtuse

10.

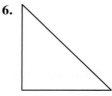

right

11.

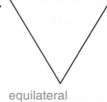

acute

12.

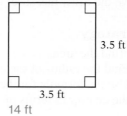

obtuse

OBJECTIVE A   *To solve problems involving the perimeter of a geometric figure*

Find the perimeter of the figure.

13.

12 in.   20 in.

24 in.

56 in.

14.

7 cm

11 cm

36 cm

15.

3.5 ft

3.5 ft

14 ft

16.

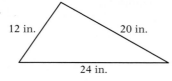

9 m

12 m   8 m

10 m

39 m

17.

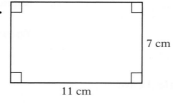

13 mi

10.5 mi

47 mi

18.

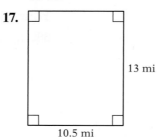

$2\frac{1}{2}$ in.

$2\frac{1}{2}$ in.

10 in.

Unless otherwise noted, all content on this page is © Cengage Learning.

For Exercises 19 to 24, find the circumference of the figure.
Give both the exact value and an approximation to the nearest hundredth.

19.   4 cm 8π cm;   25.13 cm

20.   12 m 24π m;   75.40 m

21.   5.5 mi 11π mi;   34.56 mi

22.   18 in. 18π in.;   56.55 in.

23.   17 ft 17π ft;   53.41 ft

24.   6.6 km 6.6π km;   20.73 km

25. The lengths of the three sides of a triangle are 3.8 cm, 5.2 cm, and 8.4 cm. Find the perimeter of the triangle.   17.4 cm

26. The lengths of the three sides of a triangle are 7.5 m, 6.1 m, and 4.9 m. Find the perimeter of the triangle.   18.5 m

27. The length of each of two sides of an isosceles triangle is $2\frac{1}{2}$ cm. The third side measures 3 cm. Find the perimeter of the triangle.   8 cm

28. The length of each side of an equilateral triangle is $4\frac{1}{2}$ in. Find the perimeter of the triangle.

$13\frac{1}{2}$ in.

29. A rectangle has a length of 8.5 m and a width of 3.5 m. Find the perimeter of the rectangle.   24 m

30. Find the perimeter of a rectangle that has a length of $5\frac{1}{2}$ ft and a width of 4 ft.   19 ft

31. Find the perimeter of a regular pentagon that measures 3.5 in. on each side.   17.5 in.

32. What is the perimeter of a regular hexagon that measures 8.5 cm on each side?
51 cm

33. The length of each side of a square is 12.2 cm. Find the perimeter of the square.
48.8 cm

34. Find the perimeter of a square that is 0.5 m on each side.   2 m

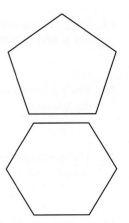

Unless otherwise noted, all content on this page is © Cengage Learning.

35. Find the circumference of a circle that has a diameter of 1.5 in. Give the exact value.
1.5π in.

36. The diameter of a circle is 4.2 ft. Find the circumference of the circle. Round to the nearest hundredth.   13.19 ft

37. The radius of a circle is 36 cm. Find the circumference of the circle. Round to the nearest hundredth.   226.19 cm

38. Find the circumference of a circle that has a radius of 2.5 m. Give the exact value.
5π m

39. How many feet of fencing should be purchased for a rectangular garden that is 18 ft long and 12 ft wide?   60 ft

40. How many meters of binding are required to bind the edge of a rectangular quilt that measures 3.5 m by 8.5 m?   24 m

41. Wall-to-wall carpeting is installed in a room that is 12 ft long and 10 ft wide. The edges of the carpet are nailed to the floor. Along how many feet must the carpet be nailed down?   44 ft

42. The length of a rectangular park is 55 yd. The width is 47 yd. How many yards of fencing are needed to surround the park?   204 yd

43. The perimeter of a rectangular playground is 440 ft. If the width is 100 ft, what is the length of the playground?   120 ft

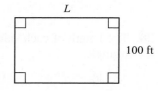

L

100 ft

44. A rectangular vegetable garden has a perimeter of 64 ft. The length of the garden is 20 ft. What is the width of the garden?   12 ft

45. Each of two sides of a triangular banner measures 18 in. If the perimeter of the banner is 46 in., what is the length of the third side of the banner?   10 in.

46. The perimeter of an equilateral triangle is 13.2 cm. What is the length of each side of the triangle?   4.4 cm

s

s   *s*

47. The perimeter of a square picture frame is 48 in. Find the length of each side of the frame.   12 in.

48. A square rug has a perimeter of 32 ft. Find the length of each edge of the rug.   8 ft

Unless otherwise noted, all content on this page is © Cengage Learning.

Solve. For Exercises 49 to 55, round to the nearest hundredth.

49. The circumference of a circle is 8 cm. Find the length of a diameter of the circle.
2.55 cm

50. The circumference of a circle is 15 in. Find the length of a radius of the circle.
2.39 in.

51. Find the length of molding needed to trim the edge of a circular table that is 4.2 ft in diameter.   13.19 ft

52. How much binding is needed to bind the edge of a circular rug that is 3 m in diameter?
9.42 m

53. A bicycle tire has a diameter of 24 in. How many feet does the bicycle travel when the wheel makes eight revolutions?   50.27 ft

54. A tricycle tire has a diameter of 12 in. How many feet does the tricycle travel when the wheel makes 12 revolutions?   37.70 ft

55. The distance from the surface of Earth to its center is 6356 km. What is the circumference of Earth?   39,935.93 km

56. Bias binding is to be sewed around the edge of a rectangular tablecloth measuring 72 in. by 45 in. If the bias binding comes in packages containing 15 ft of binding, how many packages of bias binding are needed for the tablecloth?   2 packages

57. Which has the greater perimeter, a square whose side measures 1 ft or a rectangle that has a length of 2 in. and a width of 1 in.?   A square whose side is 1 ft

58. The perimeter of an isosceles triangle is 54 ft. Let *s* be the length of one of the two equal sides. Is it possible for *s* to be 30 ft?   No

---

OBJECTIVE B   *To solve problems involving the area of a geometric figure*

Find the area of the figure.

59.

5 ft

12 ft

60 ft²

60.

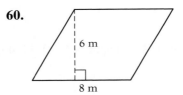

6 m

8 m

48 m²

61.

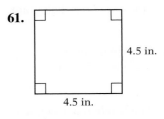

4.5 in.

4.5 in.

20.25 in²

Unless otherwise noted, all content on this page is © Cengage Learning.

62.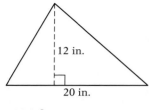

120 in²

63.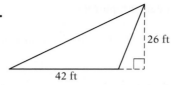

546 ft²

64.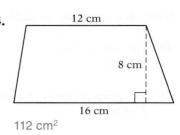

112 cm²

For Exercises 65 to 70, find the area of the figure.
Give both the exact value and an approximation to the nearest hundredth.

65.  16π cm²;
50.27 cm²

66.  144π m²;
452.39 m²

67. 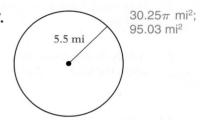 30.25π mi²;
95.03 mi²

68.  81π in²;
254.47 in²

69.  72.25π ft²;
226.98 ft²

70.  10.89π km²;
34.21 km²

71. The length of a side of a square is 12.5 cm. Find the area of the square. 156.25 cm²

72. Each side of a square measures $3\frac{1}{2}$ in. Find the area of the square. 12.25 in²

73. The length of a rectangle is 38 in., and the width is 15 in. Find the area of the rectangle. 570 in²

74. Find the area of a rectangle that has a length of 6.5 m and a width of 3.8 m. 24.7 m²

75. The length of the base of a parallelogram is 16 in., and the height is 12 in. Find the area of the parallelogram. 192 in²

76. The height of a parallelogram is 3.4 m, and the length of the base is 5.2 m. Find the area of the parallelogram. 17.68 m²

77. The length of the base of a triangle is 6 ft. The height is 4.5 ft. Find the area of the triangle. 13.5 ft²

78. The height of a triangle is 4.2 cm. The length of the base is 5 cm. Find the area of the triangle. 10.5 cm²

Unless otherwise noted, all content on this page is © Cengage Learning.

79. The length of one base of a trapezoid is 35 cm, and the length of the other base is 20 cm. If the height is 12 cm, what is the area of the trapezoid?   330 cm²

80. The height of a trapezoid is 5 in. The bases measure 16 in. and 18 in. Find the area of the trapezoid.   85 in²

81. The radius of a circle is 5 in. Find the area of the circle. Give the exact value. 25π in²

82. The diameter of a circle is 6.5 m. Find the area of the circle. Give the exact value. 10.5625π m²

83. ⬤ See the news clipping at the right. The nature reserve in Sankuru is about the size of Massachusetts. Consider Massachusetts a rectangle with a length of 150 mi and a width of 70 mi. Use these dimensions to approximate the area of the reserve in the Congo. 10,500 mi²

Ronald van der Beek/Shutterstock.com

In the NEWS!

Animal Sanctuary Established

The government of the Republic of Congo in Africa has set aside a vast expanse of land in the Sankuru Province to be used as a nature reserve. It will be a sanctuary for elephants; 11 species of primates, including the bonobos; and the okapi, a short-necked relative of the giraffe, which is on the endangered species list.
Source: www.time.com

84. The lens on the Hale telescope at Mount Palomar, California, has a diameter of 200 in. Find its area. Give the exact value.   10,000π in²

85. An irrigation system waters a circular field that has a 50-foot radius. Find the area watered by the irrigation system. Give the exact value.   2500π ft²

© Corbis Premium RF/Alamy

86. Find the area of a rectangular flower garden that measures 14 ft by 9 ft.   126 ft²

87. What is the area of a square patio that measures 8.5 m on each side?   72.25 m²

88. ⬤ See the news clipping at the right. What would be the cost of carpeting the entire living space if the cost of the carpet were $36 per square yard? $1,600,000

In the NEWS!

Billion-Dollar Home Built in Mumbai

The world's first billion-dollar home is a 27-story skyscraper in downtown Mumbai, India (formerly known as Bombay). It is 550 ft high with 400,000 ft² of living space.
Source: Forbes.com

89. Artificial turf is being used to cover a playing field. If the field is rectangular with a length of 100 yd and a width of 75 yd, how much artificial turf must be purchased to cover the field?   7500 yd²

90. A fabric wall hanging is to fill a space that measures 5 m by 3.5 m. Allowing for 0.1 m of the fabric to be folded back along each edge, how much fabric must be purchased for the wall hanging?   19.24 m²

Unless otherwise noted, all content on this page is © Cengage Learning.

91. The area of a rectangle is 300 in². If the length of the rectangle is 30 in., what is the width?   10 in.

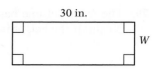

30 in.

W

92. The width of a rectangle is 12 ft. If the area is 312 ft², what is the length of the rectangle?   26 ft

93. The height of a triangle is 5 m. The area of the triangle is 50 m². Find the length of the base of the triangle.   20 m

94. The area of a parallelogram is 42 m². If the height of the parallelogram is 7 m, what is the length of the base?   6 m

95. You plan to stain the wooden deck attached to your house. The deck measures 10 ft by 8 ft. If a quart of stain will cover 50 ft², how many quarts of stain should you buy?   2 qt

96. You want to tile your kitchen floor. The floor measures 12 ft by 9 ft. How many tiles, each a square with sides of $1\frac{1}{2}$ ft, should you purchase for the job?   48 tiles

97. You are wallpapering two walls of a child's room, one measuring 9 ft by 8 ft and the other measuring 11 ft by 8 ft. The wallpaper costs $37 per roll, and each roll of wallpaper will cover 40 ft². What is the cost to wallpaper the two walls? $148

98. An urban renewal project involves reseeding a park that is in the shape of a square, 60 ft on each side. Each bag of grass seed costs $11.50 and will seed 1200 ft². How much money should be budgeted for buying grass seed for the park? $34.50

99. You want to rent a storage unit. You estimate that you will need 175 ft² of floor space. You see the ad shown at the right. You want to rent the smallest possible unit that will hold everything you want to store. Which of the six units pictured in the ad should you select?
10 × 20 unit

100. A circle has a radius of 8 in. Find the increase in area when the radius is increased by 2 in. Round to the nearest hundredth.   113.10 in²

101. A circle has a radius of 6 cm. Find the increase in area when the radius is doubled. Round to the nearest hundredth.   339.29 cm²

Unless otherwise noted, all content on this page is © Cengage Learning.

102. You want to install wall-to-wall carpeting in your living room, which measures 15 ft by 24 ft. If the cost of the carpet you would like to purchase is $31.90 per square yard, what is the cost of the carpeting for your living room? (*Hint:* $9 \text{ ft}^2 = 1 \text{ yd}^2$) $1276

103. You want to paint the walls of your bedroom. Two walls measure 15 ft by 9 ft, and the other two walls measure 12 ft by 9 ft. The paint you wish to purchase costs $29.98 per gallon, and each gallon will cover 400 ft^2 of wall. Find the total amount you will spend on paint. $59.96

104. A walkway 2 m wide surrounds a rectangular plot of grass. The plot is 30 m long and 20 m wide. What is the area of the walkway? 216 m^2

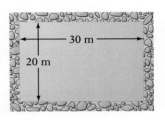

105. Pleated draperies for a window must be twice as wide as the width of the window. Draperies are being made for four windows, each 2 ft wide and 4 ft high. Since the drapes will fall slightly below the window sill and extra fabric will be needed for hemming the drapes, 1 ft must be added to the height of the window. How much material must be purchased to make the drapes? 80 ft^2

106. A circle has a radius of 5 cm.
  a. Can the exact area of the circle be A cm, where A is a decimal approximation?
  b. Can the exact area of the circle be $A \text{ cm}^2$, where A is a whole number?
  a. No; area is measured in square units.   **b.** No; A cannot be a whole number.

Critical Thinking

107. If both the length and the width of a rectangle are doubled, how many times larger is the area of the resulting rectangle?   4 times

108. Determine whether the statement is always true, sometimes true, or never true.
  a. If two triangles have the same perimeter, then the triangles have the same area.   Sometimes true
  b. If two rectangles have the same area, then the rectangles have the same perimeter.   Sometimes true

Projects or Group Activities

109. A rectangle has a perimeter of 20 units. What dimensions will result in a rectangle with the greatest possible area? Consider only whole-number dimensions.
  5 units by 5 units

110. Suppose a circle is cut into 16 equal pieces, which are then arranged as shown at the right. The figure formed resembles a parallelogram. What variable expression could describe the base of the parallelogram? What variable could describe its height? Explain how the formula for the area of a circle is derived from this approach.

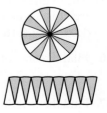

111. Prepare a report on the history of quilts in the United States. Find examples of quilt patterns that incorporate regular polygons.

Unless otherwise noted, all content on this page is © Cengage Learning.

QUICK QUIZ
1. Find the perimeter of a rectangle with a length of 3 m and a width of 0.75 m.   **7.5 m**   **[3.2A]**
2. Find the perimeter of a square for which each side measures 13.5 cm.   **54 cm**   **[3.2A]**
3. Find the length of a rubber gasket needed to fit around a circular porthole that has a 20-inch diameter. Round to the nearest hundredth.   **62.83 in.**   **[3.2A]**
4. Find the area of a triangle with a base of 10 ft and a height of 16 ft.   **80 ft²**   **[3.2B]**
5. Find the area of a rectangle with a length of 64 cm and a width of 22 cm.   **1408 cm²**   **[3.2B]**
6. Find the area of a circle with a diameter of 26 in. Round to the nearest hundredth.   **530.93 in²**   **[3.2B]**

✔ **CHECK YOUR PROGRESS: CHAPTER 3**

1. See the figure at the right. Given $BC = 15$ ft and $AB = \frac{1}{3}(BC)$, find AC.
   20 ft   [3.1A]

2. Find the supplement of a 12° angle.   168°   [3.1A]

3. See the figure at the right. Given that $\angle a = 42°$, find the measures of angles b, c, and d. $\angle b = 138°$, $\angle c = 42°$, $\angle d = 138°$ [3.1B]

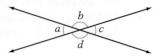

4. One angle in a right triangle measures 23°. Find the measure of the third angle.
   67°   [3.1C]

5. What is the area of a square quilt that measures 40 in. on each side?
   1600 in²   [3.2A]

6. See the figure at the right. Find x.   20°   [3.2B]

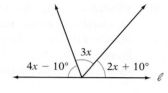

7. Find the circumference of a circle that has a diameter of 12 cm. Round to the nearest hundredth.   37.70 cm   [3.2A]

8. See the figure at the right. Given that $\ell_1 \parallel \ell_2$, find the measures of $\angle a$ and $\angle b$. $\angle a = 135°$, $\angle b = 45°$ [3.1B]

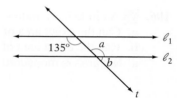

9. The height of a triangle is 8 m. The area of the triangle is 20 m². Find the length of the base of the triangle.   5 m   [3.2B]

10. Find the area of the parallelogram shown at the right.   98 m²   [3.2B]

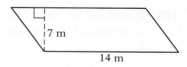

11. The perimeter of a square is 38 in. Find the length of a side of the square.
    9.5 in.   [3.2A]

12. See the figure at the right. Given that $\angle a = 72°$ and $\angle b = 48°$, find the measures of $\angle x$ and $\angle y$. $\angle x = 24°$, $\angle y = 156°$ [3.1C]

13. The width of a rectangle is 8 m. If the area is 128 m², what is the length of the rectangle?   16 m   [3.2B]

14. Find the perimeter of the triangle shown at the right.   36 in.   [3.2A]

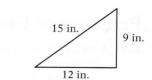

15. The diameter of a circle is 2.8 m. Find the area of the circle. Give both the exact value and an approximation to the nearest hundredth.
    1.96π m²; 6.16 m²   [3.2B]

16. Find the area of the trapezoid shown at the right.   72 cm²   [3.2B]

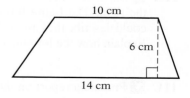

17. Find the length of decorative molding needed to edge the tops of the walls in a rectangular room that is 10 ft long and $8\frac{1}{2}$ ft wide.   37 ft   [3.2A]

Unless otherwise noted, all content on this page is © Cengage Learning.

SECTION

3.3 | Solids

OBJECTIVE A | *To solve problems involving the volume of a solid*

Geometric solids are figures in space. Five common geometric solids are the rectangular solid, the sphere, the cylinder, the cone, and the pyramid.

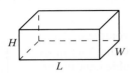

A **rectangular solid** is one in which all six sides, called **faces,** are rectangles. The variable L is used to represent the length of a rectangular solid, W its width, and H its height.

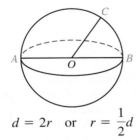

A **sphere** is a solid in which all points are the same distance from point O, called the **center** of the sphere. The **diameter** d of a sphere is a line across the sphere going through point O. The **radius** r is a line from the center to a point on the sphere. AB is a diameter and OC is a radius of the sphere shown at the left.

$$d = 2r \quad \text{or} \quad r = \frac{1}{2}d$$

The most common cylinder, called a **right circular cylinder,** is one in which the bases are circles and are perpendicular to the height of the cylinder. The variable r is used to represent the radius of a base of a cylinder, and h represents the height. In this text, only right circular cylinders are discussed.

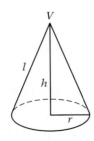

A **right circular cone** is obtained when one base of a right circular cylinder is shrunk to a point, called the **vertex** V. The variable r is used to represent the radius of the base of the cone, and h represents the height. The variable l is used to represent the **slant height,** which is the distance from a point on the circumference of the base to the vertex. In this text, only right circular cones are discussed.

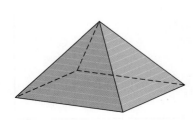

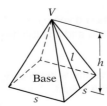

The base of a **regular pyramid** is a regular polygon, and the sides are isosceles triangles. The height h is the distance from the vertex V to the base and is perpendicular to the base. The variable l is used to represent the **slant height,** which is the height of one of the isosceles triangles on the face of the pyramid. The regular square pyramid at the left has a square base. This is the only type of pyramid discussed in this text.

Unless otherwise noted, all content on this page is © Cengage Learning.

A **cube** is a special type of rectangular solid. Each of the six faces of a cube is a square. The variable s is used to represent the length of one side of a cube.

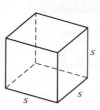

 Point of Interest

Originally, the human body was used as the standard of measure. A mouthful was used as a unit of measure in ancient Egypt; it was later referred to as a *half jigger*. In French, the word for *inch* is *pouce,* which means "thumb." A *span* was the distance from the tip of the outstretched thumb to the tip of the little finger. The *cubit* referred to the distance from the elbow to the end of the fingers. A *fathom* was the distance from the tip of the fingers on one hand to the tip of the fingers on the other hand when standing with arms fully extended from the sides. The *hand,* where 1 hand = 4 in., is still used today to measure horses.

Volume is a measure of the amount of space occupied by a geometric solid. Volume can be used to describe, for example, the amount of trash in a landfill, the amount of concrete poured for the foundation of a house, or the amount of water in a town's reservoir.

A cube that is 1 ft on each side has a volume of 1 cubic foot, which is written 1 ft^3. A cube that measures 1 cm on each side has a volume of 1 cubic centimeter, written 1 cm^3.

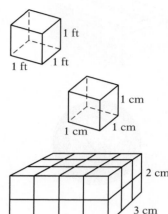

The volume of a solid is the number of cubes that are necessary to exactly fill the solid. The volume of the rectangular solid at the right is 24 cm^3 because it will hold exactly 24 cubes, each 1 cm on a side. Note that the volume can be found by multiplying the length times the width times the height.

The formulas for the volumes of the geometric solids described above are given below.

INSTRUCTOR NOTE

The difficulty students have distinguishing linear measure from square measure is compounded with volume measure. Ask students to give examples of things that would be measured in, for instance, feet, square feet, and cubic feet—for example, the length of a room, the area of the floor, and the volume of air in the room. Here are some more examples.
a. The distance across a lake, the area of the surface of the lake, and the volume of water in the lake.
b. The length of a driveway, the area of the driveway that needs to be resealed, and the volume of asphalt used to pave the driveway.

Volumes of Geometric Solids

The volume V of a **rectangular solid** with length L, width W, and height H is given by $V = LWH$.

The volume V of a **cube** with side s is given by $V = s^3$.

The volume V of a **sphere** with radius r is given by $V = \frac{4}{3}\pi r^3$.

The volume V of a **right circular cylinder** is given by $V = \pi r^2 h$, where r is the radius of the base and h is the height.

The volume V of a **right circular cone** is given by $V = \frac{1}{3}\pi r^2 h$, where r is the radius of the circular base and h is the height.

The volume V of a **regular square pyramid** is given by $V = \frac{1}{3}s^2 h$, where s is the length of a side of the base and h is the height.

Unless otherwise noted, all content on this page is © Cengage Learning.

HOW TO 1 Find the volume of a sphere with a diameter of 6 in.

First find the radius of the sphere.

$$r = \frac{1}{2}d = \frac{1}{2}(6) = 3$$

Use the formula for the volume of a sphere.

$$V = \frac{4}{3}\pi r^3$$

Substitute 3 for r.

$$V = \frac{4}{3}\pi (3)^3$$

$$V = \frac{4}{3}\pi (27)$$

The exact volume of the sphere is 36π in³.

$$V = 36\pi$$

An approximate measure can be found by using the π key on a calculator.

$$V \approx 113.10$$

The approximate volume is 113.10 in³.

EXAMPLE 1

The length of a rectangular solid is 5 m, the width is 3.2 m, and the height is 4 m. Find the volume of the solid.

Strategy

To find the volume, use the formula for the volume of a rectangular solid. $L = 5$, $W = 3.2$, $H = 4$

Solution

$V = LWH$

$V = 5(3.2)(4) = 64$

The volume of the rectangular solid is 64 m³.

YOU TRY IT 1

Find the volume of a cube that measures 2.5 m on a side.

Your strategy

Your solution

15.625 m³

IN-CLASS EXAMPLES

1. Find the volume of a rectangular solid with a length of 6 m, a width of 4 m, and a height of 4.5 m. **108 m³**
2. Find the volume of a sphere with a radius of 4 mm. Round to the nearest hundredth. **268.08 mm³**
3. Find the volume of a right circular cylinder with a radius of 15 cm and a height of 14 cm. Round to the nearest hundredth. **9896.02 cm³**

EXAMPLE 2

The radius of the base of a cone is 8 cm. The height is 12 cm. Find the volume of the cone. Round to the nearest hundredth.

Strategy

To find the volume, use the formula for the volume of a cone. An approximation is asked for; use the π key on a calculator. $r = 8$, $h = 12$

Solution

$$V = \frac{1}{3}\pi r^2 h$$

$$V = \frac{1}{3}\pi (8)^2(12) = \frac{1}{3}\pi (64)(12) = 256\pi \approx 804.25$$

The volume is approximately 804.25 cm³.

YOU TRY IT 2

The diameter of the base of a cylinder is 8 ft. The height of the cylinder is 22 ft. Find the exact volume of the cylinder.

Your strategy

Your solution

352π ft³

Solutions on p. S10

OBJECTIVE B   *To solve problems involving the surface area of a solid*

The **surface area** of a solid is the total area on the surface of the solid. Suppose you want to cover a geometric solid with wallpaper. The amount of wallpaper needed is equal to the surface area of the figure.

When a rectangular solid is cut open and flattened out, each face is a rectangle. The surface area *SA* of the rectangular solid is the sum of the areas of the six rectangles:

$$SA = LW + LH + WH + LW + WH + LH$$

which simplifies to

$$SA = 2LW + 2LH + 2WH$$

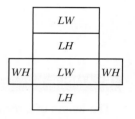

The surface area of a cube is the sum of the areas of the six faces of the cube. The area of each face is s^2. Therefore, the surface area *SA* of a cube is given by the formula $SA = 6s^2$.

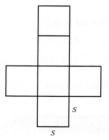

When a cylinder is cut open and flattened out, the top and bottom of the cylinder are circles. The side of the cylinder flattens out to a rectangle. The length of the rectangle is the circumference of the base, which is $2\pi r$; the width is *h*, the height of the cylinder. Therefore, the area of the rectangle is $2\pi rh$. The area of each circle is πr^2. The surface area *SA* of the cylinder is

$$SA = \pi r^2 + 2\pi rh + \pi r^2$$

which simplifies to

$$SA = 2\pi r^2 + 2\pi rh$$

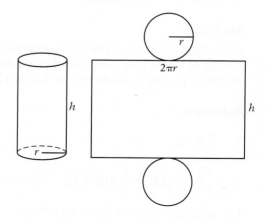

Unless otherwise noted, all content on this page is © Cengage Learning.

The surface area of a pyramid is the area of the base plus the area of the four isosceles triangles. A side of the square base is s; therefore, the area of the base is s^2. The slant height l is the height of each triangle, and s is the base of each triangle. The surface area SA of a pyramid is

$$SA = s^2 + 4\left(\frac{1}{2}sl\right)$$

which simplifies to

$$SA = s^2 + 2sl$$

Formulas for the surface areas of geometric solids are given below.

Surface Areas of Geometric Solids

The surface area SA of a **rectangular solid** with length L, width W, and height H is given by $SA = 2LW + 2LH + 2WH$.

The surface area SA of a **cube** with side s is given by $SA = 6s^2$.

The surface area SA of a **sphere** with radius r is given by $SA = 4\pi r^2$.

The surface area SA of a **right circular cylinder** is given by $SA = 2\pi r^2 + 2\pi rh$, where r is the radius of the base and h is the height.

The surface area SA of a **right circular cone** is given by $SA = \pi r^2 + \pi rl$, where r is the radius of the circular base and l is the slant height.

The surface area SA of a **regular pyramid** is given by $SA = s^2 + 2sl$, where s is the length of a side of the base and l is the slant height.

HOW TO 2   Find the surface area of a sphere with a diameter of 18 cm.

First find the radius of the sphere.

$$r = \frac{1}{2}d = \frac{1}{2}(18) = 9$$

Use the formula for the surface area of a sphere.
Substitute 9 for r.

$$SA = 4\pi r^2$$
$$SA = 4\pi (9)^2$$
$$SA = 4\pi (81)$$
$$SA = 324\pi$$

The exact surface area of the sphere is 324π cm^2.

An approximate measure can be found by using the π key on a calculator.

$$SA \approx 1017.88$$

The approximate surface area is 1017.88 cm^2.

Integrating Technology

To approximate 324π on your calculator, enter

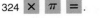 .

Unless otherwise noted, all content on this page is © Cengage Learning.

EXAMPLE 3

The diameter of the base of a cone is 5 m, and the slant height is 4 m. Find the surface area of the cone. Give the exact measure.

Strategy

To find the surface area of the cone:

- Find the radius of the base of the cone. $d = 5$
- Use the formula for the surface area of a cone. Leave the answer in terms of π.

Solution

$$r = \frac{1}{2}d = \frac{1}{2}(5) = 2.5 \qquad \bullet \text{ Find the radius.}$$

$$SA = \pi r^2 + \pi r l \qquad \bullet \text{ Formula for the surface}$$
$$\text{area of a cone}$$

$$SA = \pi (2.5)^2 + \pi (2.5)(4) \quad \bullet \ r = 2.5, l = 4$$
$$SA = \pi (6.25) + \pi (2.5)(4)$$
$$SA = 6.25\pi + 10\pi$$
$$SA = 16.25\pi$$

The surface area of the cone is 16.25π m^2.

YOU TRY IT 3

The diameter of the base of a cylinder is 6 ft, and the height is 8 ft. Find the surface area of the cylinder. Round to the nearest hundredth.

Your strategy

Your solution

207.35 ft^2

IN-CLASS EXAMPLES

4. Find the surface area of a rectangular solid with a length of 5 m, a width of 3 m, and a height of 2.5 m.   **70 m^2**
5. Find the surface area of a cube with a side of 5.5 mm.   **181.5 mm^2**
6. Find the surface area of a right circular cylinder with a radius of 10 cm and a height of 8 cm. Round to the nearest hundredth.   **1130.97 cm^2**

EXAMPLE 4

Find the surface area of a sphere with a diameter measuring 8 cm. Round to the nearest hundredth.

Strategy

To find the surface area:

- Find the radius of the sphere. $d = 8$
- Use the formula for the surface area of a sphere.
- An approximation is asked for; use the π key on a calculator.

Solution

$$r = \frac{1}{2}d = \frac{1}{2}(8) = 4 \qquad \bullet \text{ Find the radius.}$$

$$SA = 4\pi r^2 \qquad \bullet \text{ Formula for the surface}$$
$$\text{area of a sphere}$$

$$SA = 4\pi (4)^2 \qquad \bullet \ r = 4$$
$$SA = 4\pi (16) = 64\pi$$
$$SA \approx 201.06$$

The surface area of the sphere is 201.06 cm^2.

YOU TRY IT 4

Find the surface area of a cube with a side measuring 10 cm. Round to the nearest hundredth.

Your strategy

Your solution

600 cm^2

Solutions on p. S10

3.3 EXERCISES

✔ **Concept Check**

SUGGESTED ASSIGNMENT
Exercises 1–4; Exercises 5–29, odds; Exercises 30–36; Exercises 37–59, odds

1. Refer to Exercises 5 to 10 below. Fill in each blank with the name of the solid shown in the given exercise.

   a. Exercise 6 ___cone___   **b.** Exercise 8 ___cube___

   c. Exercise 9 ___sphere___   **d.** Exercise 10 ___cylinder___

2. Find the volume of the solid shown at the right.

   a. Use the formula for the volume of a ___pyramid___. $V = \dfrac{1}{3}s^2h$

   b. Replace s by ___6___ and h by ___7___. $V = \dfrac{1}{3}(\underline{}6\underline{})^2(\underline{}7\underline{})$

   c. Multiply. $V = \dfrac{1}{3}(\underline{}36\underline{})(7) = (\underline{}12\underline{})(7) = \underline{}84$

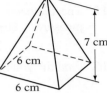
7 cm
6 cm
6 cm

   d. Fill in the blank with the correct unit: The volume of the pyramid is 84 ___cm³___.

3. To find the surface area of a pyramid with a slant height of 5 in. and a base with a side measuring 3 in., use the formula $SA = \underline{}s^2 + 2sl\underline{}$. Replace ___l___ by 5 and ___s___ by 3.

4. To find the surface area of a cylinder with a diameter of 12 cm and a height of 10 cm, use the formula $SA = \underline{}2\pi r^2 + 2\pi rh\underline{}$. Replace r by ___6___ and h by ___10___.

OBJECTIVE A   *To solve problems involving the volume of a solid*

For Exercises 5 to 10, find the volume of the figure. For calculations involving π, give both the exact value and an approximation to the nearest hundredth.

5.
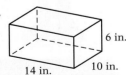
6 in.
14 in.   10 in.

840 in³

6.
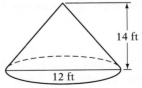
14 ft
12 ft

168π ft³; 527.79 ft³

7.
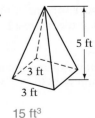
5 ft
3 ft
3 ft

15 ft³

8.
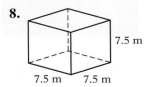
7.5 m
7.5 m   7.5 m

421.875 m³

9.

3 cm

4.5π cm³; 14.14 cm³

10.
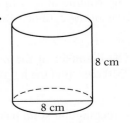
8 cm
8 cm

128π cm³; 402.12 cm³

Unless otherwise noted, all content on this page is © Cengage Learning.

11. A rectangular solid has a length of 6.8 m, a width of 2.5 m, and a height of 2 m. Find the volume of the solid.   34 m³

12. Find the volume of a rectangular solid that has a length of 4.5 ft, a width of 3 ft, and a height of 1.5 ft.   20.25 ft³

13. Find the volume of a cube whose side measures 2.5 in.   15.625 in³

14. The length of a side of a cube is 7 cm. Find the volume of the cube.   343 cm³

15. The diameter of a sphere is 6 ft. Find the volume of the sphere. Give the exact measure. 36π ft³

16. Find the volume of a sphere that has a radius of 1.2 m. Round to the nearest tenth.   7.2 m³

17. The diameter of the base of a cylinder is 24 cm. The height of the cylinder is 18 cm. Find the volume of the cylinder. Round to the nearest hundredth.   8143.01 cm³

18. The radius of the base of a cone is 5 in. The height of the cone is 9 in. Find the volume of the cone. Give the exact measure. 75π in³

19. The height of a cone is 15 cm. The diameter of the cone is 10 cm. Find the volume of the cone. Round to the nearest hundredth.   392.70 cm³

20. The length of a side of the base of a pyramid is 6 in., and the height is 10 in. Find the volume of the pyramid.   120 in³

21. The height of a pyramid is 8 m, and the length of a side of the base is 9 m. What is the volume of the pyramid?   216 m³

22. ● **The Statue of Liberty**   The index finger on the Statue of Liberty is 8 ft long. The circumference at the second joint is 3.5 ft. Use the formula for the volume of a cylinder to approximate the volume of the index finger on the Statue of Liberty. Round to the nearest hundredth.   7.80 ft³

23. The length of an aquarium is 18 in., and the width is 12 in. If the volume of the aquarium is 1836 in³, what is the height of the aquarium?   8.5 in.

24. The volume of a cylinder with a height of 10 in. is 502.4 in³. Find the radius of the base of the cylinder. Round to the nearest hundredth.   4.00 in.

25. The diameter of the base of a cylinder is 14 cm. If the volume of the cylinder is 2310 cm³, find the height of the cylinder. Round to the nearest hundredth.   15.01 cm

26. A rectangular solid has a square base and a height of 5 in. If the volume of the solid is 125 in³, find the length and the width.   Length: 5 in., width: 5 in.

Unless otherwise noted, all content on this page is © Cengage Learning.

27. Silos   A silo, which is in the shape of a cylinder, is 16 ft in diameter and has a height of 30 ft. The silo is three-fourths full. Find the volume of the portion of the silo that is not being used for storage. Round to the nearest hundredth.   1507.96 ft³

28. Storage Tanks   An oil storage tank, which is in the shape of a cylinder, is 4 m high and has a diameter of 6 m. The oil tank is two-thirds full. Find the number of cubic meters of oil in the tank. Round to the nearest hundredth.   75.40 m³

29. ● The Panama Canal   The Gatun Lock of the Panama Canal is 1000 ft long, 110 ft wide, and 60 ft deep. Find the volume of the lock in cubic feet.   6,600,000 ft³

Panama Canal

Construction   For Exercises 30 to 33, use the diagram at the right showing the concrete floor of a building. State whether the given expression can be used to calculate the volume of the concrete floor in cubic feet.

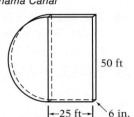

50 ft   ←25 ft→   6 in.

30. $(25)(50)(6) + (0.5)(3.14)(25^2)(6)$
No

31. $(25)(50)(0.5) + (0.5)(3.14)(25^2)(0.5)$
Yes

32. $0.5[(25)(50) + (0.5)(3.14)(25^2)]$
Yes

33. $(25)(50)(0.5) + (0.5)(3.14)(50^2)(0.5)$
No

34. Construction   Use the diagram for Exercises 30 to 33. At a cost of $10 per cubic foot, find the cost of having the floor poured. $11,156.25

35. ● Guacamole Consumption   See the news clipping at the right. What is the volume of the guacamole in cubic feet?   172,800 ft³

36. ● Guacamole Consumption   See the news clipping at the right. Assuming that each person eats 1 c of guacamole, how many people could be fed from the covered football field? $(1 \text{ ft}^3 = 59.84 \text{ pt})$
20,680,704 people

In the NEWS!

Super Bowl Win for Guacamole

Guacamole has become the dish of choice at Super Bowl parties. If all the guacamole eaten during the Super Bowl were piled onto a football field—that's a football field which, including end zones, is 360 ft long and 160 ft wide—it would cover the field to a depth of 3 ft!

Source: www.azcentral.com

OBJECTIVE B   *To solve problems involving the surface area of a solid*

Find the surface area of the figure.

37.
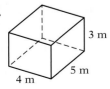
3 m   5 m   4 m

94 m²

38.
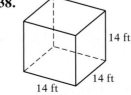
14 ft   14 ft   14 ft

1176 ft²

39.
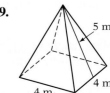
5 m   4 m   4 m

56 m²

Unless otherwise noted, all content on this page is © Cengage Learning.

For Exercises 40 to 42, find the surface area of the figure. Give both the exact value and an approximation to the nearest hundredth.

40.

2 cm

4π cm²; 12.57 cm²

41.

2 in.

6 in.

96π in²; 301.59 in²

42.

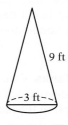

9 ft

3 ft

15.75π ft²; 49.48 ft²

43. The height of a rectangular solid is 5 ft, the length is 8 ft, and the width is 4 ft. Find the surface area of the solid.   184 ft²

44. The width of a rectangular solid is 32 cm, the length is 60 cm, and the height is 14 cm. What is the surface area of the solid?   6416 cm²

45. The side of a cube measures 3.4 m. Find the surface area of the cube.   69.36 m²

46. Find the surface area of a cube that has a side measuring 1.5 in.   13.5 in²

47. Find the surface area of a sphere with a diameter of 15 cm. Give the exact value. 225π cm²

48. The radius of a sphere is 2 in. Find the surface area of the sphere. Round to the nearest hundredth.   50.27 in²

49. The radius of the base of a cylinder is 4 in. The height of the cylinder is 12 in. Find the surface area of the cylinder. Round to the nearest hundredth.   402.12 in²

50. The diameter of the base of a cylinder is 1.8 m. The height of the cylinder is 0.7 m. Find the surface area of the cylinder. Give the exact value. 2.88π m²

51. The slant height of a cone is 2.5 ft. The radius of the base is 1.5 ft. Find the surface area of the cone. Give the exact value. 6π ft²

52. The diameter of the base of a cone is 21 in. The slant height is 16 in. What is the surface area of the cone? Round to the nearest hundredth.   874.15 in²

53. The length of a side of the base of a pyramid is 9 in., and the slant height is 12 in. Find the surface area of the pyramid.   297 in²

54. The slant height of a pyramid is 18 m, and the length of a side of the base is 16 m. What is the surface area of the pyramid?   832 m²

55. The surface area of a rectangular solid is 108 cm². The height of the solid is 4 cm, and the length is 6 cm. Find the width of the rectangular solid.   3 cm

Unless otherwise noted, all content on this page is © Cengage Learning.

56. The length of a rectangular solid is 12 ft. The width is 3 ft. If the surface area is 162 ft², find the height of the rectangular solid.   3 ft

57. Hot Air Balloons   A hot air balloon is in the shape of a sphere. Approximately how much fabric was used to construct the balloon if its diameter is 32 ft? Round to the nearest whole number.   3217 ft²

58. Paint   A can of paint will cover 300 ft². How many cans of paint should be purchased in order to paint a cylinder that has a height of 30 ft and a radius of 12 ft?
11 cans

59. Fish Tanks   How much glass is needed to make a fish tank that is 12 in. long, 8 in. wide, and 9 in. high? The fish tank is open at the top.   456 in²

QUICK QUIZ

1. Find the volume of a cube with a side of 8 ft.
512 ft³   [3.3A]

2. Find the volume of a right circular cylinder with a radius of 3 in. and a height of 5 in. Round to the nearest hundredth.
141.37 in³   [3.3A]

Critical Thinking

60. Half of a sphere is called a **hemisphere.** Derive formulas for the volume and surface area of a hemisphere.   See the *Solutions Manual.*

61. Determine whether the statement is always true, sometimes true, or never true.
a. The slant height of a regular pyramid is longer than the height.   Always true
b. The slant height of a cone is shorter than the height.   Never true
c. The four triangular faces of a regular pyramid are equilateral triangles.
Sometimes true

62. Refer to the cube and the pyramid shown at the right. Which equation correctly describes the relationship between the volume *C* of the cube and the volume *P* of the pyramid?

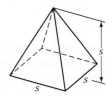

(i) $C = 2P$ **(ii)** $P = 3C$ **(iii)** $P = 2C$ **(iv)** $C = 3P$
iv

3. Find the volume of a sphere with a 5-foot diameter. Round to the nearest hundredth.   **65.45 ft³ [3.3A]**

63. A cone and a cylindrical can have the same radius and the same height. How many times should the cone be filled with water and the water poured into the cylindrical can to fill the cylindrical can (without its overflowing)?   Three times

4. Find the surface area of a rectangular solid with a length of 8 m, a width of 5 m, and a height of 3 m.   **158 m²   [3.3B]**

5. Find the surface area of a sphere with a diameter of 24 cm. Round to the nearest hundredth.
1809.56 cm²   [3.3B]

Projects or Group Activities

64. Prepare a report on the use of geometric forms in architecture. Include examples of both plane geometric figures and geometric solids.

65. Write a paper on the artist M. C. Escher. Explain how he used mathematics and geometry in his works.

Unless otherwise noted, all content on this page is © Cengage Learning.

3   Summary

Key Words	Examples

A **line** is determined by two distinct points and extends indefinitely in both directions. A **line segment** is part of a line and has two endpoints. **Parallel lines** never meet; the distance between them is always the same. **Perpendicular lines** are intersecting lines that form right angles. [3.1A, pp. 164–166]

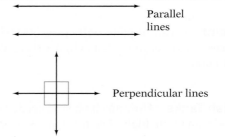

A **ray** starts at a point and extends indefinitely in one direction. The point at which a ray starts is the **endpoint** of the ray. An **angle** is formed by two rays with the same endpoint. The **vertex** of an angle is the point at which the two rays meet. An angle is measured in **degrees**. A 90° angle is a **right angle**. A 180° angle is a **straight angle**. An **acute angle** is an angle whose measure is between 0° and 90°. An **obtuse angle** is an angle whose measure is between 90° and 180°. **Complementary angles** are two angles whose measures have the sum 90°. **Supplementary angles** are two angles whose measures have the sum 180°. [3.1A, pp. 164–167]

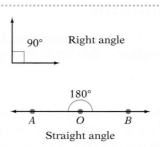

Two angles that are on opposite sides of the intersection of two lines are **vertical angles**; vertical angles have the same measure. Two angles that share a common side are **adjacent angles**; adjacent angles of intersecting lines are supplementary angles. [3.1A, p. 167; 3.1B, p. 169]

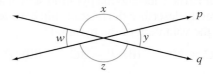

Angles *w* and *y* are vertical angles.

Angles *x* and *y* are adjacent angles.

A line that intersects two other lines at two different points is a **transversal**. If the lines cut by a transversal are parallel lines, equal angles are formed: **alternate interior angles, alternate exterior angles,** and **corresponding angles.** At the right, parallel lines ℓ_1 and ℓ_2 are cut by transversal *t*. All four acute angles have the same measure. All four obtuse angles have the same measure. [3.1B, p. 170]

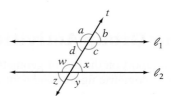

Unless otherwise noted, all content on this page is © Cengage Learning.

A **polygon** is a closed figure determined by three or more line segments. The line segments that form the polygon are its **sides**. A **regular polygon** is one in which all sides have the same length and all angles have the same measure. Polygons are classified by the number of sides. A **quadrilateral** is a four-sided polygon. A parallelogram, a rectangle, a square, a rhombus, and a trapezoid are all quadrilaterals. [3.2A, pp. 180–181]

Number of Sides	Name of the Polygon
3	Triangle
4	Quadrilateral
5	Pentagon
6	Hexagon
7	Heptagon
8	Octagon
9	Nonagon
10	Decagon

A **triangle** is a plane figure formed by three line segments. An **isosceles triangle** has two sides of equal length. The three sides of an **equilateral triangle** are of equal length. A **scalene triangle** has no two sides of equal length. An **acute triangle** has three acute angles. An **obtuse triangle** has one obtuse angle. A **right triangle** has a right angle. [3.1C, p. 172; 3.2A, pp. 180–181]

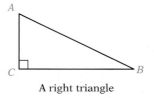

A right triangle

A **circle** is a plane figure in which all points are the same distance from the center of the circle. A **diameter** of a circle is a line segment across the circle through the center. A **radius** of a circle is a line segment from the center of the circle to a point on the circle. [3.2A, p. 183]

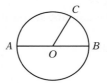

AB is a diameter of the circle.
OC is a radius.

Essential Rules and Procedures

Examples

Triangles [3.1C, p. 172]
The sum of the measures of the interior angles of a triangle is 180°. The sum of an interior and the corresponding exterior angle of a triangle is 180°.

In a right triangle, the measure of one acute angle is 12°. Find the measure of the other acute angle.

$$x + 12° + 90° = 180°$$
$$x + 102° = 180°$$
$$x = 78°$$

Unless otherwise noted, all content on this page is © Cengage Learning.

Formulas for Perimeter (the distance around a figure)
[3.2A, pp. 181–183]
Triangle: $P = a + b + c$
Rectangle: $P = 2L + 2W$
Square: $P = 4s$
Circumference of a circle: $C = \pi d$ or $C = 2\pi r$

The length of a rectangle is 8 m. The width is 5.5 m. Find the perimeter of the rectangle.

$P = 2L + 2W$

$P = 2(8) + 2(5.5)$

$P = 16 + 11$

$P = 27$

The perimeter is 27 m.

Formulas for Area (the amount of surface in a region) [3.2B, pp. 185–188]

Triangle: $A = \dfrac{1}{2}bh$

Rectangle: $A = LW$
Square: $A = s^2$
Circle: $A = \pi r^2$
Parallelogram: $A = bh$

Trapezoid: $A = \dfrac{1}{2}h(b_1 + b_2)$

The length of the base of a parallelogram is 12 cm, and the height is 4 cm. Find the area of the parallelogram.

$A = bh$

$A = 12(4)$

$A = 48$

The area is 48 cm².

Formulas for Volume (the amount of space inside a figure in space) [3.3A, p. 200]
Rectangular solid: $V = LWH$
Cube: $V = s^3$

Sphere: $V = \dfrac{4}{3}\pi r^3$

Right circular cylinder: $V = \pi r^2 h$

Right circular cone: $V = \dfrac{1}{3}\pi r^2 h$

Regular pyramid: $V = \dfrac{1}{3}s^2 h$

Find the volume of a cube that measures 3 in. on a side.

$V = s^3$

$V = 3^3$

$V = 27$

The volume is 27 in³.

Formulas for Surface Area (the total area on the surface of a solid) [3.3B, p. 203]
Rectangular solid: $SA = 2LW + 2LH + 2WH$
Cube: $SA = 6s^2$
Sphere: $SA = 4\pi r^2$
Right circular cylinder: $SA = 2\pi r^2 + 2\pi rh$
Right circular cone: $SA = \pi r^2 + \pi rl$
Regular pyramid: $SA = s^2 + 2sl$

Find the surface area of a sphere with a diameter of 10 cm. Give the exact value.

$r = \dfrac{1}{2}d = \dfrac{1}{2}(10) = 5$

$SA = 4\pi r^2$

$SA = 4\pi (5)^2$

$SA = 4\pi (25)$

$SA = 100\pi$

The surface area is 100π cm².

CHAPTER

3   Review Exercises

1. Given that $\angle a = 74°$ and $\angle b = 52°$, find the measures of angles x and y. $\angle x = 22°$, $\angle y = 158°$
                    [3.1C]

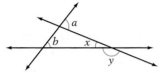

2. Find the perimeter of the rectangle in the figure below.   26 ft   [3.2A]

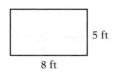

5 ft

8 ft

3. Find the volume of the figure.   168 in³   [3.3A]

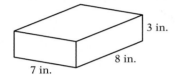

3 in.

8 in.

7 in.

4. Find the measure of $\angle x$.   68°   [3.1B]

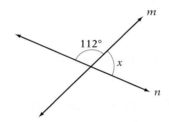

m

112°

x

n

5. Find the area of the circle shown below. Use 3.14 for π.   63.585 cm²   [3.2B]

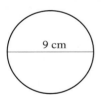

9 cm

6. Find the surface area of the figure. Round to the nearest hundredth.   125.66 m²
                    [3.3B]

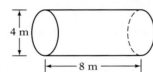

4 m

8 m

7. Given that $BC = 11$ cm and AB is three times the length of BC, find the length of AC.   44 cm   [3.1A]

A   B   C

8. Find x.   19°   [3.1A]

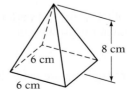

3x

4x   x + 28°

ℓ

9. Find the area of the figure.   32 in²   [3.2B]

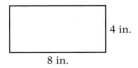

4 in.

8 in.

10. Find the volume of the figure.   96 cm³   [3.3A]

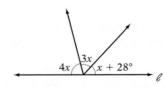

8 cm

6 cm

6 cm

Unless otherwise noted, all content on this page is © Cengage Learning.

11. Find the perimeter of the figure. 42 in. [3.2A]

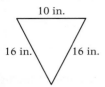

10 in.

16 in.   16 in.

12. Given that $\ell_1 \parallel \ell_2$, find the measures of angles *a* and *b*. $\angle a = 138°$, $\angle b = 42°$ [3.1B]

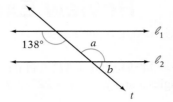

13. Find the surface area of the figure. 220 ft² [3.3B]

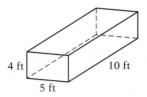

4 ft   10 ft

5 ft

14. Given $AB = 15$, $CD = 6$, and $AD = 24$, find *BC*. 3 [3.1A]

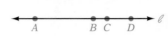

A   B C D ℓ

15. Find the volume of a cube whose side measures 3.5 in. 42.875 in³ [3.3A]

16. Find the supplement of a 32° angle. 148° [3.1A]

17. Find the volume of a rectangular solid with a length of 6.5 ft, a width of 2 ft, and a height of 3 ft. 39 ft³ [3.3A]

18. Two angles of a triangle measure 37° and 48°. Find the measure of the third angle. 95° [3.1C]

19. The height of a triangle is 7 cm. The area of the triangle is 28 cm². Find the length of the base of the triangle. 8 cm [3.2B]

20. Find the volume of a sphere that has a diameter of 12 mm. Give the exact value. 288π mm³ [3.3A]

21. Picture Frames The perimeter of a square picture frame is 86 cm. Find the length of each side of the frame. 21.5 cm [3.2A]

22. Paint A can of paint will cover 200 ft². How many cans of paint should be purchased in order to paint a cylinder that has a height of 15 ft and a radius of 6 ft? 4 cans [3.3B]

23. Fencing The length of a rectangular park is 56 yd. The width is 48 yd. How many yards of fencing are needed to surround the park? 208 yd [3.2A]

24. Patios What is the area of a square patio that measures 9.5 m on each side? 90.25 m² [3.2B]

25. Landscaping A walkway 2 m wide surrounds a rectangular plot of grass. The plot is 40 m long and 25 m wide. What is the area of the walkway? 276 m² [3.2B]

Unless otherwise noted, all content on this page is © Cengage Learning.

CHAPTER

3   TEST

1. Determine the volume of the rectangular solid shown below.

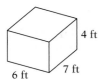

4 ft
6 ft
7 ft

168 ft³   [3.3A]

2. Given that ℓ_1 and ℓ_2 are parallel lines, determine the measure of angle a.

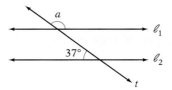

143°   [3.1B]

3. Determine the area of a rectangle with a length of 15 m and a width of 7.4 m.

111 m²   [3.2B]

4. Determine the area of a triangle whose base is 7 ft and whose height is 12 ft.

42 ft²   [3.2B]

5. Determine the exact volume of a right circular cone whose radius is 7 cm and whose height is 16 cm.

$\dfrac{784\pi}{3}$ cm³   [3.3A]

6. Determine the exact surface area of a pyramid whose square base is 3 m on each side and whose slant height is 11 m.

75 m²   [3.3B]

7. Determine the volume of the solid shown below. Round to the nearest hundredth.

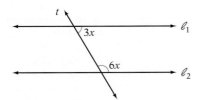

7 cm
30 cm

4618.14 cm³   [3.3A]

8. Determine the area of the trapezoid shown below.

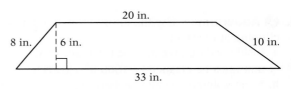

20 in.
8 in.   6 in.   10 in.
33 in.

159 in²   [3.2B]

9. Given that $\ell_1 \parallel \ell_2$, find x.   20°   [3.1B]

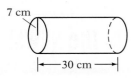

t
3x
6x
ℓ_1
ℓ_2

10. Determine the surface area of the figure shown below.

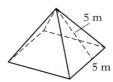

5 m
5 m

75 m²   [3.3B]

11. Find x.

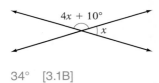

4x + 10°
x

34°   [3.1B]

12. Name the figure shown below.

Octagon   [3.2A]

Unless otherwise noted, all content on this page is © Cengage Learning.

13. Determine the exact surface area of the right circular cylinder shown below.

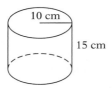

10 cm

15 cm

500π cm² [3.3B]

14. Determine the measure of angle a.

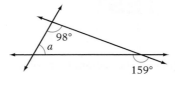

98°

a

159°

61° [3.1C]

15. Determine the perimeter of a square whose side is 5 m.
20 m [3.2A]

16. Determine the perimeter of a rectangle whose length is 8 cm and whose width is 5 cm.
26 cm [3.2A]

17. Find the supplement of a 41° angle.

139° [3.1A]

18. Two angles of a triangle measure 41° and 37°. Find the measure of the third angle.
102° [3.1C]

19. A right triangle has a 32° angle. Find the measures of the other two angles.
90° and 58° [3.1C]

20. Travel A bicycle tire has a diameter of 28 in. How many feet does the bicycle travel when the wheel makes 10 revolutions? Use 3.14 for π. Round to the nearest tenth of a foot.
73.3 ft [3.2A]

28 in.

21. 🔵 **Aquariums** Use the information in the news clipping at the right.
  a. Inside and out, what is the total area of window that must be cleaned?   1600 ft² [3.2B]
  b. What volume, in cubic inches, does the window in the exhibit fill?   1,440,000 in³ [3.3A]

Charles Crust/Danita Delimont.com "Danita Delimont Photography"/Newscom

In the NEWS!

Dive Shows at Aquarium

Visitors to the Window on Washington Waters exhibit at the Seattle Aquarium can choose from three daily dive shows. Divers in the 120,000-gallon tank talk to visitors through a 40-foot by 20-foot rectangular acrylic window that is 12.5 in. thick.

Source: Seattle Aquarium

22. Agriculture A silo in the shape of a cylinder is 9 ft in diameter and has a height of 18 ft. Find the volume of the silo. Use 3.14 for π.
1144.53 ft³ [3.3A]

23. Find the area of a right triangle with a base of 8 m and a height of 2.75 m.
11 m² [3.2B]

24. Forestry Find the cross-sectional area of a redwood tree that is 11 ft 6 in. in diameter. Use 3.14 for π. Round to the nearest hundredth.
103.82 ft² [3.2B]

iStockphoto.com/slobo

25. 🔵 **Alcatraz** Inmate cells at Alcatraz were 9 ft long and 5 ft wide. The height of a cell was 7 ft.
  a. Find the area of the floor of a cell at Alcatraz.   45 ft² [3.2B]
  b. Find the volume of a cell at Alcatraz.   315 ft³ [3.3A]

Unless otherwise noted, all content on this page is © Cengage Learning.

Cumulative Review Exercises

1. Let $x \in \{-3, 0, 1\}$. For what values of x is the inequality $x \leq 1$ a true statement?
$-3, 0, 1$ [1.1A]

2. Write 8.9% as a decimal.
0.089 [1.2B]

3. Write $\frac{7}{20}$ as a percent.
35% [1.2B]

4. Divide: $-\frac{4}{9} \div \frac{2}{3}$
$-\frac{2}{3}$ [1.2D]

5. Multiply: $5.7(-4.3)$
-24.51 [1.2D]

6. Simplify: $-\sqrt{125}$
$-5\sqrt{5}$ [1.2F]

7. Evaluate $5 - 3[10 + (5 - 6)^2]$.
-28 [1.3A]

8. Evaluate $a(b - c)^3$ when $a = -1$, $b = -2$, and $c = -4$.
-8 [1.4A]

9. Simplify: $5m + 3n - 8m$
$-3m + 3n$ [1.4B]

10. Simplify: $-7(-3y)$
$21y$ [1.4C]

11. Simplify: $4(3x + 2) - (5x - 1)$
$7x + 9$ [1.4D]

12. Use the roster method to write the set of negative integers greater than or equal to -2.
$\{-2, -1\}$ [1.5A]

13. Find $C \cup D$, given $C = \{0, 10, 20, 30\}$ and $D = \{-10, 0, 10\}$.
$C \cup D = \{-10, 0, 10, 20, 30\}$ [1.5A]

14. Graph: $x \leq 1$
[1.5B]

15. Solve: $4x + 2 = 6x - 8$
5 [2.2B]

16. Solve: $3(2x + 5) = 18$
$\frac{1}{2}$ [2.2C]

17. Solve: $4y - 3 \geq 6y + 5$
Write the solution set in interval notation.
$(-\infty, -4)$ [2.5A]

18. Solve: $8 - 4(3x + 5) \leq 6(x - 8)$
Write the solution set in set-builder notation.
$\{x | x \geq 2\}$ [2.5A]

Unless otherwise noted, all content on this page is © Cengage Learning.

19. Solve: $2x - 3 > 5$ or $x + 4 < 1$
$\{x|x < -3\} \cup \{x|x > 4\}$ [2.5B]

20. Solve: $-3 \le 2x - 7 \le 5$
$\{x|\ 2 \le x \le 6\}$ [2.5B]

21. Solve: $|3x - 1| = 2$
$1, -\dfrac{1}{3}$ [2.6A]

22. Solve: $|x - 8| \le 2$
$\{x|\ 6 \le x \le 10\}$ [2.6B]

23. Find the measure of $\angle x$.
131°   [3.1B]

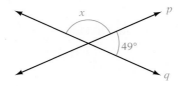

24. Translate "the difference between four times a number and ten is two" into an equation and solve.
$4x - 10 = 2; x = 3$ [2.3B]

25. Triangles   Two angles of a triangle measure 37° and 21°. Find the measure of the third angle of the triangle.
122°   [3.1C]

26. Investments   Michael deposits $5000 in an account that earns an annual simple interest rate of 4.5% and $2500 in an account that earns an annual simple interest rate of 3.5%. How much interest will Michael earn from the two accounts in one year?
$312.50   [2.1D]

27. Triangles   Two sides of an isosceles triangle measure 7.5 m. The perimeter of the triangle is 19.5 m. Find the measure of the third side of the triangle.
4.5 m   [3.2A]

28. Geometry   The volume of a box is 144 ft³. The length of the box is 12 ft, and the width is 4 ft. Find the height of the box.   3 ft   [3.3A]

29. Deep Sea Diving   The pressure P, in pounds per square inch, at a certain depth in the ocean can be approximated by the equation $P = 15 + \frac{1}{2}D$, where D is the depth in feet. Use this equation to find the depth when the pressure is 35 pounds per square inch.   40 ft   [2.2D]

30. Cell Phone   The monthly charge for your cell phone is $79 plus $.35 for every text message sent or received. In a month for which your cell phone bill was $86.70, how many text messages did you send or receive?
22 text messages   [2.3B]

Unless otherwise noted, all content on this page is © Cengage Learning.

Linear Functions and Inequalities in Two Variables

4

OBJECTIVES

SECTION 4.1
A To graph points in a rectangular coordinate system
B To determine ordered-pair solutions of an equation in two variables
C To graph a scatter diagram

SECTION 4.2
A To evaluate a function

SECTION 4.3
A To graph a linear function
B To graph an equation of the form $Ax + By = C$
C To find the x- and y-intercepts of a straight line
D To solve application problems

SECTION 4.4
A To find the slope of a line given two points
B To graph a line given a point and the slope

SECTION 4.5
A To find the equation of a line given a point and the slope
B To find the equation of a line given two points
C To solve application problems

SECTION 4.6
A To find parallel and perpendicular lines

SECTION 4.7
A To graph the solution set of an inequality in two variables

Focus on Success

Have you formed or are you part of a study group? Remember that a study group can be a great way to stay focused on succeeding in this course. You can support each other, get help and offer help on homework, and prepare for tests together. (See Homework Time, page AIM-5.)

© iStockphoto.com/Christopher Futcher

Prep Test

Are you ready to succeed in this chapter? Take the Prep Test below to find out if you are ready to learn the new material.

For Exercises 1 to 3, simplify.

1. $-4(x - 3)$
$-4x + 12$ [1.4D]

2. $\sqrt{(-6)^2 + (-8)^2}$
10 [1.2F]

3. $\dfrac{3 - (-5)}{2 - 6}$
-2 [1.3A]

4. Evaluate $-2x + 5$ for $x = -3$.
11 [1.4A]

5. Evaluate $\dfrac{2r}{r - 1}$ for $r = 5$.
2.5 [1.4A]

6. Evaluate $2p^3 - 3p + 4$ for $p = -1$.
5 [1.4A]

7. Evaluate $\dfrac{x_1 + x_2}{2}$ for $x_1 = 7$ and $x_2 = -5$.
1 [1.4A]

8. Given $3x - 4y = 12$, find the value of x when $y = 0$.
4 [2.2A]

4.1 The Rectangular Coordinate System

OBJECTIVE A *To graph points in a rectangular coordinate system*

INSTRUCTOR NOTE

Although Descartes is given credit for introducing analytic geometry, there were others working on the same concept, notably Pierre Fermat. Nowhere in Descartes's work is there a coordinate system as we draw it with two axes. Descartes did not use the word *coordinate* in his work. This word was introduced by Gottfried Leibnitz, who is also responsible for the use of the words *abscissa* and *ordinate*.

Before the 15th century, geometry and algebra were considered separate branches of mathematics. That all changed when René Descartes, a French mathematician who lived from 1596 to 1650, founded **analytic geometry.** In this geometry, a *coordinate system* is used to study relationships between variables.

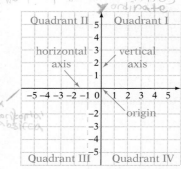

A **rectangular coordinate system** is formed by two number lines, one horizontal and one vertical, that intersect at the zero point of each line. The point of intersection is called the **origin.** The two lines are called **coordinate axes,** or simply **axes.** The axes determine a **plane,** which can be thought of as a large, flat sheet of paper. The two axes divide the plane into four regions called **quadrants,** which are numbered counterclockwise from I to IV.

Each point in the plane can be identified by a pair of numbers called an **ordered pair.** The first number of the pair measures a horizontal distance and is called the **abscissa.** The second number of the pair measures a vertical distance and is called the **ordinate.** The **coordinates of a point** are the numbers in the ordered pair associated with the point. The abscissa is also called the **first coordinate** of the ordered pair, and the ordinate is also called the **second coordinate** of the ordered pair.

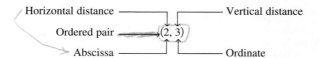

When drawing a rectangular coordinate system, we often label the horizontal axis *x* and the vertical axis *y*. In this case, the coordinate system is called an *xy***-coordinate system.** The coordinates of the points are given by ordered pairs (x, y), where the abscissa is called the *x***-coordinate** and the ordinate is called the *y***-coordinate.**

INSTRUCTOR NOTE

Within the Microsoft PowerPoint® slides available with this text is a blank coordinate grid. It can be used to create a transparency on which to plot points, equations, etc.

To **graph or plot a point in the plane,** place a dot at the location given by the ordered pair. The **graph of an ordered pair** (x, y) is the dot drawn at the coordinates of the point in the plane. The points whose coordinates are $(3, 4)$ and $(-2.5, -3)$ are graphed in the figures below.

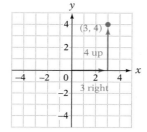

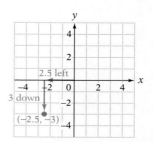

Unless otherwise noted, all content on this page is © Cengage Learning.

Take Note

This concept is very important. An **ordered pair** is a *pair* of coordinates, and the *order* in which the coordinates are listed is crucial.

The points whose coordinates are $(3, -1)$ and $(-1, 3)$ are graphed at the right. Note that the graphed points are in different locations. *The order of the coordinates in an ordered pair is important.*

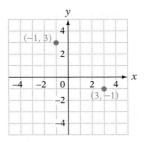

INSTRUCTOR NOTE

It may help students to think of an ordered pair as the address (location) of a point in the plane.

Each point in the plane is associated with an ordered pair, and each ordered pair is associated with a point in the plane. Although only the labels for integers are given on a coordinate grid, the graph of any ordered pair can be approximated. For example, the points whose coordinates are $(-2.3, 4.1)$ and $(\pi, 1)$ are shown on the graph at the right.

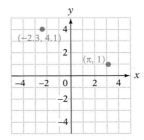

EXAMPLE 1

Graph the ordered pairs $(-2, -3)$, $(3, -2)$, $(0, -2)$, and $(3, 0)$.

Solution

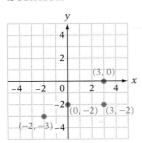

YOU TRY IT 1

Graph the ordered pairs $(-4, 1)$, $(3, -3)$, $(0, 4)$, and $(-3, 0)$.

Your solution

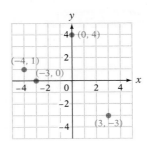

IN-CLASS EXAMPLES

1. Graph the ordered pairs $(5, 0)$, $(-4, 1)$, $(-1, 2)$, and $(-2, -4)$.

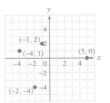

EXAMPLE 2

Give the coordinates of the points labeled A and B. Give the abscissa of point C and the ordinate of point D.

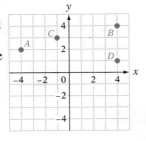

Solution

The coordinates of A are $(-4, 2)$.
The coordinates of B are $(4, 4)$.
The abscissa of C is -1.
The ordinate of D is 1.

YOU TRY IT 2

Give the coordinates of the points labeled A and B. Give the abscissa of point D and the ordinate of point C.

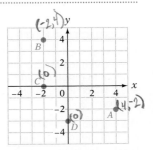

Your solution

$A(4, -2)$, $B(-2, 4)$
The abscissa of D is 0.
The ordinate of C is 0.

Solutions on p. S10

Unless otherwise noted, all content on this page is © Cengage Learning.

OBJECTIVE B   *To determine ordered-pair solutions of an equation in two variables*

An *xy*-coordinate system is used to study the relationship between two variables. Frequently this relationship is given by an equation. Examples of equations in two variables include

$$y = 2x - 3 \qquad 3x + 2y = 6 \qquad x^2 - y = 0$$

A **solution of an equation in two variables** is an ordered pair (x, y) whose coordinates make the equation a true statement.

HOW TO 1   Is $(-3, 7)$ a solution of the equation $y = -2x + 1$?

$$y = -2x + 1$$

7	$-2(-3) + 1$	• Replace x by -3; replace y by 7.
7	$6 + 1$	

$$7 = 7 \qquad \text{• The results are equal.}$$

Yes, $(-3, 7)$ is a solution of the equation $y = -2x + 1$.

Besides $(-3, 7)$, there are many other ordered-pair solutions of $y = -2x + 1$. For example, $(0, 1)$, $\left(-\frac{3}{2}, 4\right)$, and $(4, -7)$ are also solutions. In general, an equation in two variables has an infinite number of solutions. By choosing any value of x and substituting that value into the equation, we can calculate a corresponding value of y.

HOW TO 2   Find the ordered-pair solution of $y = \frac{2}{3}x - 3$ that corresponds to $x = 6$.

$$y = \frac{2}{3}x - 3$$

$$= \frac{2}{3}(6) - 3 \qquad \text{• Replace } x \text{ by 6.}$$

$$= 4 - 3 = 1 \qquad \text{• Simplify.}$$

The ordered-pair solution is $(6, 1)$.

The solutions of an equation in two variables can be graphed in an *xy*-coordinate system.

INSTRUCTOR NOTE

Problems such as HOW TO 3 are given to prepare the student to graph straight lines, which is the subject of Section 4.3.

We want to impress on the student the relationship between a solution of an equation and the pictorial representation of that solution, its graph.

HOW TO 3   Graph the ordered-pair solutions of $y = -2x + 1$ when $x = -2, -1, 0, 1,$ and 2.

Use the values of x to determine ordered-pair solutions of the equation. It is convenient to record these in a table.

x	$y = -2x + 1$	y	(x, y)
-2	$-2(-2) + 1$	5	$(-2, 5)$
-1	$-2(-1) + 1$	3	$(-1, 3)$
0	$-2(0) + 1$	1	$(0, 1)$
1	$-2(1) + 1$	-1	$(1, -1)$
2	$-2(2) + 1$	-3	$(2, -3)$

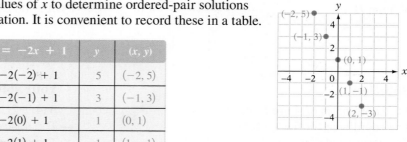

Unless otherwise noted, all content on this page is © Cengage Learning.

EXAMPLE 3

Is $(3, -2)$ a solution of $3x - 4y = 15$?

Solution

$$
\begin{array}{c|c}
3x - 4y = 15 & \\
\hline
3(3) - 4(-2) & 15 \\
9 + 8 & 15 \\
17 \neq 15 &
\end{array}
$$

• Replace x by 3 and y by -2.

No, $(3, -2)$ is not a solution of $3x - 4y = 15$.

YOU TRY IT 3

Is $(-2, 4)$ a solution of $x - 3y = -14$?

Your solution

Yes

EXAMPLE 4

Graph the ordered-pair solutions of $2x - 3y = 6$ when $x = -3, 0, 3$, and 6.

Solution

$$2x - 3y = 6$$ • Solve $2x - 3y = 6$ for y.
$$-3y = -2x + 6$$
$$y = \frac{2}{3}x - 2$$

Replace x in $y = \frac{2}{3}x - 2$ by $-3, 0, 3$, and 6.
For each value of x, determine the value of y.

x	$y = \dfrac{2}{3}x - 2$	y	(x, y)
-3	$\dfrac{2}{3}(-3) - 2$	-4	$(-3, -4)$
0	$\dfrac{2}{3}(0) - 2$	-2	$(0, -2)$
3	$\dfrac{2}{3}(3) - 2$	0	$(3, 0)$
6	$\dfrac{2}{3}(6) - 2$	2	$(6, 2)$

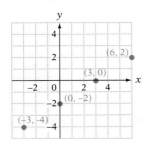

YOU TRY IT 4

Graph the ordered-pair solutions of $x + 2y = 4$ when $x = -4, -2, 0$, and 2.

Your solution

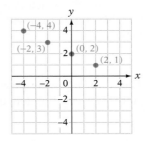

IN-CLASS EXAMPLES

2. Is $(0, -1)$ a solution of $2x - y = 1$? **Yes**
3. Is $(2, -2)$ a solution of $y = 3x + 8$? **No**
4. Graph the ordered-pair solutions of $y = -3x$ when $x = -1, 0$, and 1.

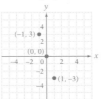

5. Graph the ordered-pair solutions of $3x - 4y = 4$ when $x = -4, 0$, and 4.

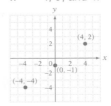

Solutions on pp. S10–S11

Unless otherwise noted, all content on this page is © Cengage Learning.

OBJECTIVE C   *To graph a scatter diagram*

IN-CLASS EXAMPLES

6. The table below shows the prices, in dollars, for a one-day admission ticket to an amusement park, for both an adult and a child, for a six-year period. Graph the scatter diagram for these data.

Adult Ticket, x	Child's Ticket, y
34	27
36	29
37	30
41	33
42	34
45	36

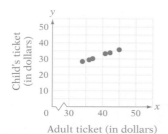

Child's ticket (in dollars) / Adult ticket (in dollars)

Discovering a relationship between two variables is an important task in the study of mathematics. These relationships occur in many forms and in a wide variety of applications. Here are some examples:

A botanist wants to know the relationship between the number of bushels of wheat yielded per acre and the amount of watering per acre.

An environmental scientist wants to know the relationship between the incidence of skin cancer and the amount of ozone in the atmosphere.

A business analyst wants to know the relationship between the price of a product and the number of products that are sold at that price.

A researcher may investigate the relationship between two variables by means of *regression analysis*, which is a branch of statistics. The study of the relationship between two variables may begin with a **scatter diagram,** which is a graph of the ordered pairs of the known data.

The following table gives data collected by a university registrar comparing the grade point averages of graduating high school seniors and their scores on a national test.

GPA, x	3.50	3.50	3.25	3.00	3.00	2.75	2.50	2.50	2.00	2.00	1.50
Test score, y	1500	1100	1200	1200	1000	1000	1000	900	800	900	700

The scatter diagram for these data is shown below.

INSTRUCTOR NOTE

This objective is an application of the concepts presented in Objective A; students are plotting points in the rectangular coordinate system. However, the grids are not restricted to the integers from −5 to 5.

Direct student attention to the note accompanying the graph in Example 6: "The jags indicate that a portion of the axis has been omitted."

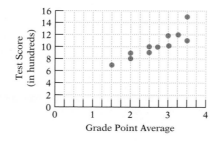

Test Score (in hundreds) / Grade Point Average

Each ordered pair represents the GPA and test score for a student. For example, the ordered pair $(2.75, 1000)$ indicates a student with a GPA of 2.75 who had a test score of 1000.

The dot on the scatter diagram at $(3, 12)$ represents the student with a GPA of 3.00 and a test score of 1200.

Unless otherwise noted, all content on this page is © Cengage Learning.

EXAMPLE 5

A nutritionist collected data on the number of grams of sugar and the number of grams of fiber in a 1-ounce serving of six brands of cereal. The data are recorded in the following table. Graph the scatter diagram for the data.

Sugar, x	6	8	6	5	7	5
Fiber, y	2	1	4	4	2	3

Solution

Graph the ordered pairs on the rectangular coordinate system. The horizontal axis represents the grams of sugar. The vertical axis represents the grams of fiber.

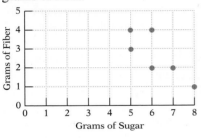

EXAMPLE 6

To test a heart medicine, a doctor measured the heart rates, in beats per minute, of five patients before and after they took the medication. The results are recorded in the scatter diagram. One patient's heart rate before taking the medication was 75 beats per minute. What was this patient's heart rate after taking the medication?

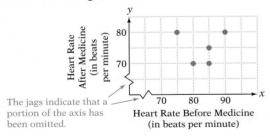

The jags indicate that a portion of the axis has been omitted.

Solution

Locate 75 beats per minute on the x-axis. Follow the vertical line from 75 to a point plotted in the diagram. Follow a horizontal line from that point to the y-axis. Read the number where the horizontal line intersects the y-axis.

The ordered pair is (75, 80), which indicates that the patient's heart rate before taking the medication was 75 and the heart rate after taking the medication was 80.

YOU TRY IT 5

A sports statistician collected data on the total number of yards gained and the total number of points scored by a college football team for six games. The data are recorded in the following table. Graph the scatter diagram for the data.

Yards, x	300	400	350	400	300	450
Points, y	18	24	14	21	21	30

Your solution

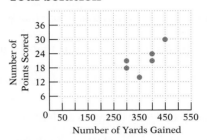

YOU TRY IT 6

A study by the FAA showed that narrow, over-the-wing emergency exit rows on airplanes slow passenger evacuation. The scatter diagram below shows the space between seats, in inches, and the evacuation time, in seconds, for a group of 35 passengers. What was the evacuation time when the space between seats was 20 in.?

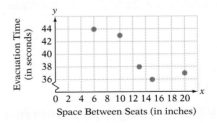

Your solution

37 s

Solutions on p. S11

Unless otherwise noted, all content on this page is © Cengage Learning.

4.1 EXERCISES

✔ Concept Check

SUGGESTED ASSIGNMENT
Exercises 1–12; Exercises 13–49, odds
More challenging exercises: Exercises 50–57

1. In which quadrant is the graph of $(-3, 4)$ located?   Quadrant II

2. In which quadrant is the graph of $(2, -5)$ located?   Quadrant IV

3. On which axis does the graph of $(0, -4)$ lie? y-axis

4. On which axis does the graph of $(-6, 0)$ lie? x-axis

5. Name any two points on a horizontal line that is 2 units above the x-axis.   Answers will vary. For example, $(-3, 2)$ and $(5, 2)$

6. Name any two points on a vertical line that is 3 units to the right of the y-axis.   Answers will vary. For example, $(3, -4)$ and $(3, 1)$

7. The point $(1, 3)$ lies on a vertical line. Name any other point on this line.   Answers will vary. Any point with x-coordinate 1 lies on the line.

Complete Exercises 8 and 9 by filling in each blank with the word *left, right, up,* or *down.*

8. To graph the point $(-1, 7)$, start at the origin and move 1 unit ____left____ and 7 units ____up____.

9. To graph the point $(5, -4)$, start at the origin and move 5 units ____right____ and 4 units ____down____.

10. Write as an ordered pair the coordinates of the point whose y-coordinate is 8 and whose x-coordinate is -7. $(-7, 8)$

11. Write as an ordered pair the coordinates of the point whose x-coordinate is 6 and whose y-coordinate is -5. $(6, -5)$

12. To decide whether the ordered pair $(1, 7)$ is a solution of the equation $y = 2x + 5$, substitute 1 for ____x____ and 7 for ____y____ to see whether the ordered pair $(1, 7)$ makes the equation $y = 2x + 5$ a true statement.

OBJECTIVE A   *To graph points in a rectangular coordinate system*

13. Graph $(-2, 1)$, $(3, -5)$, $(-2, 4)$, and $(0, 3)$.

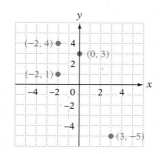

14. Graph $(5, -1)$, $(-3, -3)$, $(-1, 0)$, and $(1, -1)$.

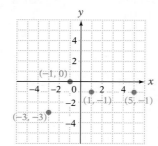

15. Graph $(0, 0)$, $(0, -5)$, $(-3, 0)$, and $(0, 2)$.

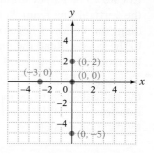

Unless otherwise noted, all content on this page is © Cengage Learning.

16. Graph $(-4, 5)$, $(-3, 1)$, $(3, -4)$, and $(5, 0)$.

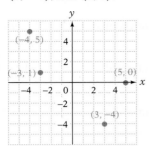

17. Graph $(-1, 4)$, $(-2, -3)$, $(0, 2)$, and $(4, 0)$.

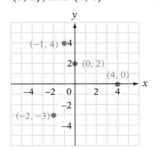

18. Graph $(5, 2)$, $(-4, -1)$, $(0, 0)$, and $(0, 3)$.

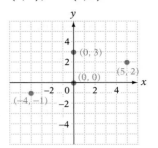

19. Find the coordinates of each of the points.

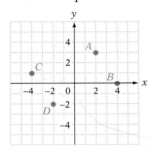

$A(2, 3)$, $B(4, 0)$, $C(-4, 1)$, $D(-2, -2)$

20. Find the coordinates of each of the points.

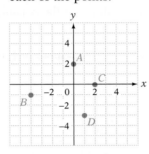

$A(0, 2)$, $B(-4, -1)$, $C(2, 0)$, $D(1, -3)$

21. Find the coordinates of each of the points.

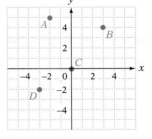

$A(-2, 5)$, $B(3, 4)$, $C(0, 0)$, $D(-3, -2)$

22. Find the coordinates of each of the points.

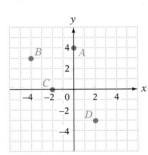

$A(0, 4)$, $B(-4, 3)$, $C(-2, 0)$, $D(2, -3)$

23. a. Name the abscissas of points A and C.
b. Name the ordinates of points B and D.

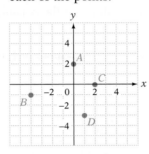

a. 2, -4 **b.** 1, -3

24. a. Name the abscissas of points A and C.
b. Name the ordinates of points B and D.

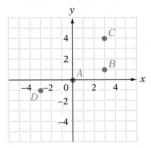

a. 0, 3   **b.** 1, -1

25. Let a and b be positive numbers such that $a < b$. In which quadrant is each point located?

a. (a, b) I   **b.** $(-a, b)$ II   **c.** $(b - a, -b)$ IV   **d.** $(a - b, -b - a)$ III

26. Let a and b be positive numbers. State whether the two given points lie on the x-axis, the y-axis, a horizontal line other than the x-axis, or a vertical line other than the y-axis.

a. $(-a, b)$ and $(-a, 0)$ **b.** $(a, 0)$ and $(-b, 0)$
    Vertical line other than y-axis x-axis

Unless otherwise noted, all content on this page is © Cengage Learning.

OBJECTIVE B *To determine ordered-pair solutions of an equation in two variables*

27. Is $(3, 4)$ a solution of $y = -x + 7$?   Yes

28. Is $(2, -3)$ a solution of $y = x + 5$?   No

29. Is $(-1, 2)$ a solution of $y = \frac{1}{2}x - 1$?   No

30. Is $(1, -3)$ a solution of $y = -2x - 1$?   Yes

31. Is $(4, 1)$ a solution of $2x - 5y = 4$?   No

32. Is $(-5, 3)$ a solution of $3x - 2y = 9$?   No

33. Suppose (x, y) is a solution of the equation $y = -3x + 6$, where $x > 2$. Is y positive or negative?
Negative

34. Suppose (x, y) is a solution of the equation $y = 4x - 8$, where $y > 0$. Is x less than or greater than 2?
Greater than

For Exercises 35 to 40, graph the ordered-pair solutions of the equation for the given values of x.

35. $y = 2x; x = -2, -1, 0, 2$

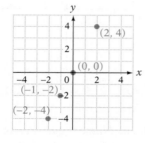

36. $y = -2x; x = -2, -1, 0, 2$

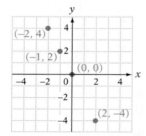

37. $y = \frac{2}{3}x + 1; x = -3, 0, 3$

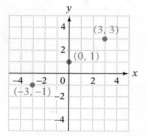

38. $y = -\frac{1}{3}x - 2; x = -3, 0, 3$

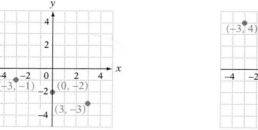

39. $2x + 3y = 6; x = -3, 0, 3$

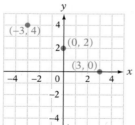

40. $x - 2y = 4; x = -2, 0, 2$

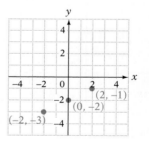

OBJECTIVE C *To graph a scatter diagram*

41. Employment The numbers of years of previous work experience and the monthly salaries of six people who completed a bachelor's degree in marketing are recorded in the following table. Graph the scatter diagram for these data.

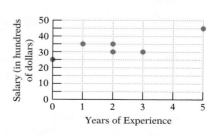

Years of experience, x	2	0	5	2	3	1
Salary (in hundreds), y	30	25	45	35	30	35

Unless otherwise noted, all content on this page is © Cengage Learning.

42. Physiology An exercise physiologist measured the time, in minutes, a person spent on a treadmill at a fast walk and the heart rate of that person. The results are recorded in the following table. Draw a scatter diagram for these data.

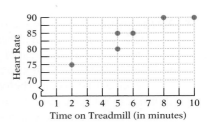

Time on treadmill, x	2	10	5	8	6	5
Heart rate, y	75	90	80	90	85	85

43. Criminology Sherlock Holmes solved a crime by recognizing a relationship between the length, in inches, of a person's stride and the height of that person in inches. The data for six people are recorded in the table below. Graph the scatter diagram for these data.

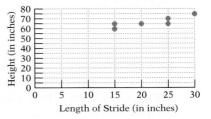

Length of stride, x	15	25	20	25	15	30
Height, y	60	70	65	65	65	75

44. ● World Cup Soccer Read the news clipping at the right. The table below shows the number of shots on goal and the number of goals scored for each of the 16 teams that played for a spot in the quarter-final round of the 2010 World Cup soccer tournament. Graph the scatter diagram for these data.

Shots on goal	8	4	6	6	8	6	7	5	9	3	6	2	7	2	8	3
Goals scored	2	1	1	2	4	1	3	1	2	1	3	0	0	0	1	0

In the NEWS!

World Cup Quarter-Final Matches Determined

Spain beat Portugal 1–0 to determine the last of the quarter-final match-ups in the 2010 World Cup. Spain's 8 shots on goal yielded just 1 goal, but that was enough to keep the Spaniards ahead of the Portuguese, who took only 3 shots on goal.
Source: www.nprstats.com

45. Track Meets The scatter diagram at the right shows the record times for races of different lengths at a junior high school track meet. What was the record time for the 800-meter race?   200 s

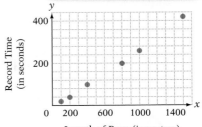

Length of Race (in meters)

For Exercises 46 and 47, refer to the scatter diagram in Exercise 45.

46. Suppose a new record is set for the 800-meter race. How will that change the scatter plot? Will the graph have the same number of points or an additional point?
The point at (800, 200) will be lowered to have a y-coordinate equal to the new record time. The graph will have the same number of points.

47. Suppose the school adds a 1200-meter race. How will that change the scatter plot? Will the graph have the same number of points or an additional point?
The point will be graphed with an x-coordinate of 1200 and a y-coordinate equal to the record time for the 1200-meter race. The graph will have an additional point.

QUICK QUIZ

1. Graph the ordered pairs $(5, -1)$, $(-4, -2)$, $(1, 0)$, and $(3, 4)$. **[4.1A]**

2. Is $(-3, 0)$ a solution of $y = 6x - 18$?
No **[4.1B]**

3. Graph the ordered-pair solutions of
$y = -\dfrac{2}{3}x - 2$ when
$x = -3, 0,$ and 3.
[4.1B]

Unless otherwise noted, all content on this page is © Cengage Learning.

48. ● **Incomes** To determine trends in income levels, economists use inflation-adjusted values. These numbers take into consideration how inflation affects purchasing power. The scatter diagram at the right shows the median income, rounded to the nearest thousand dollars, for selected years. The income is given in both actual dollars (before inflation) and 2008 inflation-adjusted dollars. (*Source*: U.S. Census Bureau) **a.** What was the adjusted median income when the annual median income was $44,000? **b.** What was the annual median income when the adjusted median income was $53,000?   **a.** $51,000   **b.** $42,000

Adjusted median income (in thousands of dollars) [y-axis: 49, 50, 51, 52, 53]

Annual median income (in thousands of dollars) [x-axis: 38 40 42 44 46 48 50 52 54]

49. ● **Fuel Efficiency** The American Council for an Energy-Efficient Economy releases rankings of environmentally friendly and unfriendly cars and trucks sold in the United States. The scatter diagram at the right shows the fuel usage, in miles per gallon, both in the city and on the highway, for twelve 2010-model vehicles ranked worst for the environment. **a.** What was the highway fuel usage, in miles per gallon, for the car that got 11 mpg in the city? **b.** What was city fuel usage, in miles per gallon, for the car that got 18 mpg on the highway?   **a.** 15 mpg   **b.** 13 mpg

Fuel usage, highway (in miles per gallon) [y-axis: 12, 13, 14, 15, 16, 17, 18, 19]

Fuel usage, city (in miles per gallon) [x-axis: 8 9 10 11 12 13 14 15]

Critical Thinking

For Exercises 50 to 52, find the distance from the given point to the horizontal axis.

50. $(-5, 1)$ 1 unit   **51.** $(3, -4)$ 4 units   **52.** $(-6, 0)$ 0 units

For Exercises 53 to 55, find the distance from the given point to the vertical axis.

53. $(-2, 4)$ 2 units   **54.** $(1, -3)$ 1 unit   **55.** $(5, 0)$ 5 units

56. Draw a line passing through every point whose abscissa equals its ordinate.

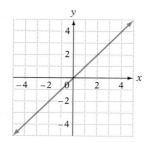

57. Draw a line passing through every point whose ordinate is the additive inverse of its abscissa.

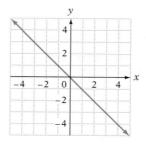

QUICK QUIZ
4. The following table shows the past and projected populations of baby boomers for the years 2000, 2031, and 2046. (*Source:* U.S. Census Bureau) Graph a scatter diagram for these data.   **[4.1C]**

Projects or Group Activities

58. ◤ There is an imaginary coordinate system on Earth that consists of *longitude* and *latitude*. Write a report on how location is determined on the surface of Earth. Include in your report the longitude and latitude coordinates of your school.

59. ◤ Describe the graph of all ordered pairs (x, y) that are 5 units from the origin.

60. ◤ Consider two distinct fixed points in a plane. Describe the graph of all points (x, y) that are equidistant from these fixed points.

Year	2000	2031	2046
Population in millions	79	51	19

Unless otherwise noted, all content on this page is © Cengage Learning.

4.2 | Introduction to Functions

OBJECTIVE A | *To evaluate a function*

Tips for Success

Have you considered joining a study group? Getting together regularly with other students in the class to go over material and quiz each other can be very beneficial. See *AIM for Success* in the Preface.

In mathematics and its applications, there are many times when it is necessary to investigate a relationship between two quantities. Here is a financial application: Consider a person who is planning to finance the purchase of a car. If the current interest rate for a 5-year loan is 5%, the equation that describes the relationship between the amount that is borrowed B and the monthly payment P is $P = 0.018871B$.

For each amount the purchaser may borrow (B), there is a certain monthly payment (P). The relationship between the amount borrowed and the payment can be recorded as a set of ordered pairs, where the first coordinate of each pair is the amount borrowed and the second coordinate is the monthly payment. Some of these ordered pairs are shown at the right.

$$0.018871B = P$$

(6000, 113.23)
(7000, 132.10)
(8000, 150.97)
(9000, 169.84)

A relationship between two quantities is not always given by an equation. The table at the right describes a grading scale that defines a relationship between a score on a test and a letter grade. For each score, the table assigns only one letter grade. The ordered pair $(84, B)$ indicates that a score of 84 receives a letter grade of B.

Score	Grade
90–100	A
80–89	B
70–79	C
60–69	D
0–59	F

The bar graph at the right shows the number of people who watched the Super Bowl for the years 2007 to 2012. The jagged line between 0 and 90 on the vertical axis indicates that a portion of the vertical axis has been omitted. The data in the graph can be written as a set of ordered pairs.

{(2007, 93.2), (2008, 97.4), (2009, 98.7), (2010, 106.5), (2011, 111.01), (2012, 111.35)}

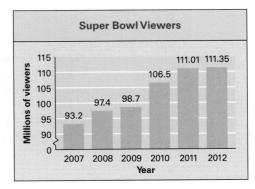

This set is a function. There are no two ordered pairs with the same first coordinate. The ordered pair (2010, 106.5) means that in 2010, the number of people who watched the Super Bowl was 106.5 million.

In each of the above examples, there is a rule (an equation, a table, or a graph) that determines a certain set of ordered pairs.

Unless otherwise noted, all content on this page is © Cengage Learning.

Definition of a Function

A **function** is a set of ordered pairs in which no two ordered pairs have the same first coordinate. The **domain** of a function is the set of first coordinates of the ordered pairs; the **range** of a function is the set of second coordinates of the ordered pairs.

EXAMPLES

1. {(1, 2), (2, 4), (3, 6), (4, 8)}
Domain = {1, 2, 3, 4}   Range = {2, 4, 6, 8}

2. {(−1, 0), (0, 0), (1, 0), (2, 0), (3, 0)}
Domain = {−1, 0, 1, 2, 3}   Range = {0}

Now consider the set of ordered pairs {(1, 2), (4, 5), (7, 8), (4, 6)}. This set of ordered pairs is *not* a function. There are two ordered pairs, (4, 5) and (4, 6), with the same first coordinate. This set of ordered pairs is called a *relation*. A **relation** is any set of ordered pairs. A function is a special type of relation. The concepts of domain and range apply to relations as well as to functions.

INSTRUCTOR NOTE
Domain and range are difficult concepts for most students. We are introducing the topic at this point. We will continually revisit the concepts of domain and range as we progress through the text.

HOW TO 1   Determine whether each set of ordered pairs is a function. State the domain and range.

A. {(2, 3), (4, 6), (6, 8), (10, 6)}
B. {(2, 2), (1, 1), (0, 0), (2, −2), (1, −1)}

A. No two ordered pairs have the same first element. The set of ordered pairs is a function. The domain is {2, 4, 6, 10}. The range is {3, 6, 8}.
B. The ordered pairs (2, 2) and (2, −2) have the same first coordinate. The set of ordered pairs is not a function. The domain is {0, 1, 2}. The range is {−2, −1, 0, 1, 2}.

For each element of the domain of a function there is a corresponding element in the range of the function. A possible diagram for the function in part A of HOW TO 1 is shown at the right. Each element of the domain is paired with exactly one element in the range. The diagram represents a function.

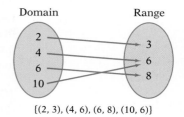

{(2, 3), (4, 6), (6, 8), (10, 6)}

A diagram for part B of HOW TO 1 is shown at the right. There are some elements in the domain that are paired with more than one element in the range. The diagram does not represent a function.

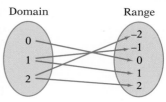

{(2, 2), (1, 1), (0, 0), (2, −2), (1, −1)}

Unless otherwise noted, all content on this page is © Cengage Learning.

Consider again the three examples of functions given on page 231. For the equation $0.018871B = P$, the domain is the possible amounts a consumer might borrow to purchase a car. Let's assume that the most a person would borrow is $50,000. Then the domain is $(B \mid 0 \le B \le 50,000)$. The range is all possible monthly payments. The largest monthly payment is $P = 0.018871(50,000) \approx 943.55$, so the range is $\{P \mid 0 \le P \le 943.55\}$.

For the grading-scale function, the domain is all possible test scores. The domain is $\{0, 1, 2, 3, \ldots, 97, 98, 99, 100\}$. The range is all possible grades. The range is $\{A, B, C, D, F\}$.

For a graph, the domain is represented on the horizontal axis and the range is represented on the vertical axis. For the graph of the Super Bowl data, the domain is the set of years. The domain is $\{2007, 2008, 2009, 2010, 2011, 2012\}$. The range is the number of people watching each year. The range is $\{93.2, 97.4, 98.7, 106.5, 111.01, 111.35\}$.

The **square function,** which pairs each real number with its square, can be defined by the equation

$$y = x^2$$

This equation states that for a given value of x in the domain, the corresponding value of y in the range is the square of x. For instance, if $x = 6$, then $y = 36$ and if $x = -7$, then $y = 49$. Because the value of y *depends* on the value of x, y is called the **dependent variable** and x is called the **independent variable.**

Take Note

The order in which the elements of a set are listed is not important. For instance,

$$\{a, b, c\} = \{b, a, c\}.$$

However, it is common practice to list domain and range elements in order from smallest to largest. Note that the elements of the domain of the grading-scale function were listed from smallest to largest.

Take Note

A pictorial representation of the square function is shown at the right. The function acts as a machine that changes a number from the domain into the square of the number.

INSTRUCTOR NOTE

To illustrate the importance of a function being single-valued, ask students whether a grading scale that allowed (92, A) and (92, C) would be a reasonable scale to use.

A function can be thought of as a rule that pairs one number with another number. For instance, the square function pairs a number with its square. The ordered pairs for the values shown at the right are $(-5, 25)$, $\left(\frac{3}{5}, \frac{9}{25}\right)$, $(0, 0)$, and $(3, 9)$. For this function, the second coordinate is the square of the first coordinate. If x represents the first coordinate, then the second coordinate is x^2 and the ordered pair is (x, x^2).

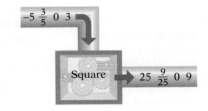

A function cannot have two ordered pairs with *different* second coordinates and the same first coordinate. However, a function may contain ordered pairs with the *same* second coordinate. For instance, the square function has the ordered pairs $(-3, 9)$ and $(3, 9)$; the second coordinates are the same but the first coordinates are different.

The **double function** pairs a number with twice that number. The ordered pairs for the values shown at the right are $(-5, -10)$, $\left(\frac{3}{5}, \frac{6}{5}\right)$, $(0, 0)$, and $(3, 6)$. For this function, the second coordinate is twice the first coordinate. If x represents the first coordinate, then the second coordinate is $2x$ and the ordered pair is $(x, 2x)$.

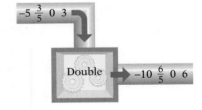

Not every equation in two variables defines a function. For instance, consider the equation

$$y^2 = x^2 + 9$$

Because

$$5^2 = 4^2 + 9 \qquad \text{and} \qquad (-5)^2 = 4^2 + 9$$

the ordered pairs $(4, 5)$ and $(4, -5)$ are both solutions of the equation. Consequently, there are two ordered pairs that have the same first coordinate (4) but *different* second coordinates $(5$ and $-5)$. Therefore, the equation does not define a function.

Unless otherwise noted, all content on this page is © Cengage Learning.

The phrase "y is a function of x," or the same phrase with different variables, is used to describe an equation in two variables that defines a function. To emphasize that the equation represents a function, **function notation** is used.

Just as the variable x is commonly used to represent a number, the letter f is commonly used to name a function. The square function is written in function notation as follows:

Take Note

The dependent variable y and the notation $f(x)$ can be used interchangeably.

This is the value of the function.
It is the number that is paired with x.

$$f(x) = x^2$$

The name of the function is f.   This is an algebraic expression that defines the relationship between the dependent and independent variables.

The symbol $f(x)$ is read "the value of f at x" or "f of x."

It is important to note that $f(x)$ does *not* mean f times x. The symbol $f(x)$ is the **value of the function** and represents the value of the dependent variable for a given value of the independent variable. We often write $y = f(x)$ to emphasize the relationship between the independent variable x and the dependent variable y. Remember that y and $f(x)$ are different symbols for the same number.

INSTRUCTOR NOTE

One way to help students evaluate functions is to write the equation of the function with open parentheses.

$g(t) = 3(t)^2 - 5(t) + 1$

$g(\ ) = 3(\ )^2 - 5(\ ) + 1$

To evaluate the function, fill each of the parentheses with the same number. Then simplify.

The letters used to represent a function are somewhat arbitrary. All of the following equations represent the same function.

$$\left.\begin{array}{l} f(x) = x^2 \\ s(t) = t^2 \\ P(v) = v^2 \end{array}\right\} \text{ Each equation represents the square function.}$$

The process of determining $f(x)$ for a given value of x is called **evaluating a function.** For instance, to evaluate $f(x) = x^2$ when $x = 4$, replace x by 4 and simplify.

$$f(x) = x^2$$
$$f(4) = 4^2 = 16$$

The *value* of the function is 16 when $x = 4$. An ordered pair of the function is $(4, 16)$.

APPLY THE CONCEPT ···

The height $s(t)$, in feet, of a ball above the ground t seconds after it is thrown upward at a velocity of 64 ft/s is given by $s(t) = -16t^2 + 64t + 4$. Find the height of the ball 1.5 s after it is released.

To find the height, evaluate the function when $t = 1.5$.

$$s(t) = -16t^2 + 64t + 4$$
$$s(1.5) = -16(1.5)^2 + 64(1.5) + 4 \qquad \text{• Evaluate the function when } t = 1.5.$$
$$= -16(2.25) + 64(1.5) + 4$$
$$= -36 + 96 + 4$$
$$= 64$$

The ball is 64 ft above the ground 1.5 s after it is released.

HOW TO 2 Evaluate $g(t) = 3t^2 - 5t + 1$ when $t = -2$.

$$g(t) = 3t^2 - 5t + 1$$
$$g(-2) = 3(-2)^2 - 5(-2) + 1 \qquad \bullet \text{ Replace } t \text{ by } -2 \text{ and then simplify.}$$
$$= 3(4) - 5(-2) + 1$$
$$= 12 + 10 + 1 = 23$$

When t is -2, the value of the function is 23.

Therefore, an ordered pair of the function is $(-2, 23)$.

Integrating Technology

A graphing calculator can be used to evaluate a function. To evaluate the function in HOW TO 2, first enter the function. Functions are accessed by pressing the key.

Here are the keystrokes to enter $g(t) = 3t^2 - 5t + 1$ on a TI83/84 calculator. Note that when using the calculator, the independent variable is X. The name of the function can be $Y_1, Y_2, \ldots, Y_9, Y_0$. Instructions are also provided in the Keystroke Guide: *Evaluating Functions*.

Y= CLEAR 3 X,T,θ,n x^2 — 5 X,T,θ,n + 1 2ND QUIT

To evaluate the function at -2, enter VARS ◊ 1 1 ((-) 2) ENTER. The resulting screen is shown at the left.

> Y₁(-2)
> 23

It is possible to evaluate a function for a variable expression.

HOW TO 3 Evaluate $P(z) = 3z - 7$ when $z = 3 + h$.

$$P(z) = 3z - 7$$
$$P(3 + h) = 3(3 + h) - 7 \qquad \bullet \text{ Replace } z \text{ by } 3 + h \text{ and then simplify.}$$
$$= 9 + 3h - 7$$
$$= 3h + 2$$

When z is $3 + h$, the value of the function is $3h + 2$.

Therefore, an ordered pair of the function is $(3 + h, 3h + 2)$.

IN-CLASS EXAMPLES

1. Find the domain and range of the function $\{(3, 10), (4, 13), (5, 16)\}$.
   D: {3, 4, 5};
   R: {10, 13, 16}
2. Evaluate $M(n) = 2n^2 - 3n - 5$ when $n = -1$. **0**
3. Evaluate $f(x) = 4x - 1$ for $x = 2a + 1$.
   8a + 3
4. What is the domain of $f(x) = \dfrac{3}{x - 6}$?
   {x | x ≠ 6}

When a function is represented by an equation, the domain of the function is all real numbers for which the value of the function is a real number. For instance:

- The domain of $f(x) = x^2$ is all real numbers, because the square of every real number is a real number. In set-builder notation, the domain is $\{x \mid -\infty < x < \infty\}$.

- The domain of $g(x) = \dfrac{1}{x - 2}$ is all real numbers except 2, because when $x = 2$, $g(2) = \dfrac{1}{2 - 2} = \dfrac{1}{0}$, which is not a real number. The domain is $\{x \mid x \neq 2\}$.

HOW TO 4 Find the domain of $f(x) = 2x^2 - 7x + 1$.

Because the value of $2x^2 - 7x + 1$ is a real number for any value of x, no values are excluded from the domain of $f(x) = 2x^2 - 7x + 1$. The domain of the function is all real numbers, or $\{x \mid -\infty < x < \infty\}$.

Unless otherwise noted, all content on this page is © Cengage Learning.

EXAMPLE 1

Find the domain and range of the function
$\{(5, 3), (9, 7), (13, 7), (17, 3)\}$.

Solution

The domain is the set of first coordinates. The range is the set of second coordinates.
The domain is $\{5, 9, 13, 17\}$.
The range is $\{3, 7\}$.

YOU TRY IT 1

Find the domain and range of the function
$\{(-1, 5), (3, 5), (4, 5), (6, 5)\}$.

Your solution

Domain: $\{-1, 3, 4, 6\}$
Range: $\{5\}$

EXAMPLE 2

Given $p(r) = 5r^3 - 6r - 2$, find $p(-3)$.

Solution

$$p(r) = 5r^3 - 6r - 2$$
$$\begin{aligned} p(-3) &= 5(-3)^3 - 6(-3) - 2 \\ &= 5(-27) + 18 - 2 \\ &= -135 + 18 - 2 \\ &= -119 \end{aligned}$$

YOU TRY IT 2

Evaluate $G(x) = \frac{3x}{x + 2}$ when $x = -4$.

Your solution

6

EXAMPLE 3

Evaluate $Q(r) = 2r + 5$ when $r = h + 3$.

Solution

$$\begin{aligned} Q(r) &= 2r + 5 \\ Q(h + 3) &= 2(h + 3) + 5 \\ &= 2h + 6 + 5 \\ &= 2h + 11 \end{aligned}$$

YOU TRY IT 3

Evaluate $f(x) = x^2 - 11$ when $x = 3h$.

Your solution

$9h^2 - 11$

EXAMPLE 4

What is the domain of $f(x) = \frac{2}{x - 5}$?

Solution

For $x = 5$, $f(x) = f(5) = \frac{2}{5 - 5} = \frac{2}{0}$, which is undefined. So, 5 is excluded from the domain of f. The domain is $\{x | x \neq 5\}$.

YOU TRY IT 4

What is the domain of $f(x) = 3x^2 - 5x + 2$?

Your solution

$\{x | -\infty < x < \infty\}$

Solutions on p. S11

4.2 EXERCISES

✔ **Concept Check**

SUGGESTED ASSIGNMENT
Exercises 1–6; Exercises 7–61, odds;
More challenging exercises: Exercises 63 and 64; 66–69

1. Given $f(4) = 5$, what is the value of the function f?   5

2. Given $y = f(x)$, what variable is the independent variable? What variable is the dependent variable? x, y

3. What are the domain and range of $\{(-3, 9), (-2, 4), (-1, 1), (1, 1), (2, 4), (3, 9)\}$?
   Domain: $\{-3, -2, -1, 1, 2, 3\}$; Range: $\{1, 4, 9\}$

4. Which of the following diagrams define a function?   i, ii, iv

(i)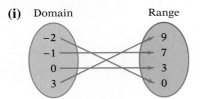

(ii)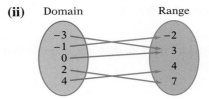

(iii)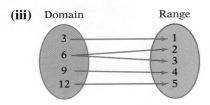

(iv)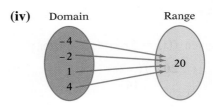

5. Does the ordered pair $(3, 9)$ belong to $f(x) = x^2$?   Yes

6. ◣ In your own words, explain what a function is.

OBJECTIVE A *To evaluate a function*

For Exercises 7 to 14, determine whether the set of ordered pairs is a function. State the domain and range.

7. $\{(0, 0), (2, 4), (3, 6), (4, 8), (5, 10)\}$
   Function; D = $\{0, 2, 3, 4, 5\}$, R = $\{0, 4, 6, 8, 10\}$

8. $\{(1, 3), (3, 5), (5, 7), (7, 9)\}$
   Function; D = $\{1, 3, 5, 7\}$, R = $\{3, 5, 7, 9\}$

9. $\{(-2, -1), (-4, -5), (0, -1), (3, 5)\}$
   Function; D = $\{-4, -2, 0, 3\}$, R = $\{-5, -1, 5\}$

10. $\{(-3, -1), (-1, -1), (0, 1), (2, 6)\}$
   Function; D = $\{-3, -1, 0, 2\}$, R = $\{-1, 1, 6\}$

11. $\{(-2, 3), (-1, 3), (0, -3), (1, 3), (2, 3)\}$
   Function; D = $\{-2, -1, 0, 1, 2\}$, R = $\{-3, 3\}$

12. $\{(0, 0), (1, 0), (2, 0), (3, 0), (4, 0)\}$
   Function; D = $\{0, 1, 2, 3, 4\}$, R = $\{0\}$

13. $\{(1, 1), (4, 2), (9, 3), (1, -1), (4, -2)\}$
   Not a function; D = $\{1, 4, 9\}$, R = $\{-2, -1, 1, 2, 3\}$

14. $\{(3, 1), (3, 2), (3, 3), (3, 4)\}$
   Not a function; D = $\{3\}$, R = $\{1, 2, 3, 4\}$

Unless otherwise noted, all content on this page is © Cengage Learning.

15. ● **Postal Rates**   Use the table in the article at the right.
    a. Does the table define a function?   Yes
    b. Given $w = 3.15$ oz, find p. $1.05
    c. Given $w = 2$ oz, find p. $.65
    d. How much does it cost to send a letter that weighs 0.25 oz? $.45

16. ● **Shipping**   The table at the right shows the cost to send a Zone 3 Priority Mail package using the U.S. Postal Service.
    a. Does this table define a function?   Yes
    b. Given $x = 3.54$ lb, find y. $8.15
    c. Given $x = 2$ lb, find y. $5.65
    d. How much does it cost to send a package that weighs 0.45 lb? $5.25

Weight in Pounds (x)	Cost (y)
$0 < x \le 1$	$5.25
$1 < x \le 2$	$5.65
$2 < x \le 3$	$6.95
$3 < x \le 4$	$8.15
$4 < x \le 5$	$9.35

In the NEWS!

First-Class Postage One Penny More

The U.S. Postal Service is raising postage rates for first-class mail by 1 cent. New rates for letters, shown below, go into effect on January 22.

Weight in Ounces, w	Postage in Dollars, p
$0 < w \le 1$	$.45
$1 < w \le 2$	$.65
$2 < w \le 3$	$.85
$3 < w \le 3.5$	$1.05

Sources:
www.washingtontimes.com,
www.usps.com

17. 🖐 True or false? If f is a function, then it is possible that $f(0) = -2$ and $f(3) = -2$.   True

18. 🖐 True or false? If f is a function, then it is possible that $f(4) = 3$ and $f(4) = 2$.   False

For Exercises 19 to 22, given $f(x) = 5x - 4$, evaluate.

19. $f(3)$
    11

20. $f(-2)$
    -14

21. $f(0)$
    -4

22. $f(-1)$
    -9

For Exercises 23 to 26, given $G(t) = 4 - 3t$, evaluate.

23. $G(0)$
    4

24. $G(-3)$
    13

25. $G(-2)$
    10

26. $G(4)$
    -8

For Exercises 27 to 30, given $q(r) = r^2 - 4$, evaluate.

27. $q(3)$
    5

28. $q(4)$
    12

29. $q(-2)$
    0

30. $q(-5)$
    21

For Exercises 31 to 34, given $F(x) = x^2 + 3x - 4$, evaluate.

31. $F(4)$
    24

32. $F(-4)$
    0

33. $F(-3)$
    -4

34. $F(-6)$
    14

For Exercises 35 to 38, given $H(p) = \dfrac{3p}{p + 2}$, evaluate.

35. $H(1)$
    1

36. $H(-3)$
    9

37. $H(t)$
    $\dfrac{3t}{t + 2}$

38. $H(v)$
    $\dfrac{3v}{v + 2}$

For Exercises 39 to 42, given $s(t) = t^3 - 3t + 4$, evaluate.

39. $s(-1)$
    6

40. $s(2)$
    6

41. $s(a)$
    $a^3 - 3a + 4$

42. $s(w)$
    $w^3 - 3w + 4$

43. Given $P(x) = 4x + 7$, write $P(-2 + h) - P(-2)$ in simplest form.
    $4h$

44. Given $G(t) = 9 - 2t$, write $G(-3 + h) - G(-3)$ in simplest form.
    $-2h$

Unless otherwise noted, all content on this page is © Cengage Learning.

45. Energy   The power a windmill can generate is a function of the velocity of the wind. The function can be approximated by $P = f(v) = 0.015v^3$, where P is the power in watts and v is the velocity of the wind in meters per second. How much power will be produced by a windmill when the velocity of the wind is 15 m/s?
50.625 watts

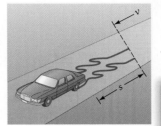

46. Automotive Technology   The distance s (in feet) a car will skid on a certain road surface after the brakes are applied is a function of the car's velocity v (in miles per hour). The function can be approximated by $s = f(v) = 0.017v^2$. How far will a car skid after its brakes are applied if it is traveling at 60 mph?
61.2 ft

47. ● **Automotive Technology**   Read the news article at the right. Suppose you drive 12,000 mi per year and the price of gas is $4.00 per gallon. Under these conditions, increasing your car's fuel efficiency by p percent can give you an annual cost savings, in dollars, of $S = f(p) = \frac{2400p}{1 + p}$. If you changed the tires on your car to low-rolling-resistance tires, what would be your minimum annual cost savings? Round to the nearest cent. $114.29

> ## In the NEWS!
>
> **New Tires on a Roll to Lower Costs**
>
> Recent improvements in low-rolling-resistance tires have made these tires an increasingly popular choice for people who want to cut fuel costs. Low-rolling-resistance tires can raise a car's fuel efficiency by 5 to 7 percent.
>
> *Source:* www.shopautoweek.com

48. Airports   Airport administrators usually price airport parking at a rate that discourages people from using the parking lot for long periods of time. The rate structure for an airport is given in the table at the right.
  a. Evaluate this function when $t = 2.5$ h. $6.50
  b. Evaluate this function when $t = 7$ h. $10.00

Hours Parked	Cost
$0 < t \le 1$	$1.00
$1 < t \le 2$	$3.00
$2 < t \le 4$	$6.50
$4 < t \le 7$	$10.00
$7 < t \le 12$	$14.00

49. Business   Game Engineering has just completed the programming and testing for a new computer game. The cost to manufacture and package the game depends on the number of units Game Engineering plans to sell. The table at the right shows the cost per game for various manufacturing quantities.
  a. Evaluate this function when $x = 7000$.
  b. Evaluate this function when $x = 20,000$.
  a. $4.75 per game   b. $4.00 per game

Number of Games Manufactured	Cost to Manufacture One Game
$0 < x \le 2500$	$6.00
$2500 < x \le 5000$	$5.50
$5000 < x \le 10,000$	$4.75
$10,000 < x \le 20,000$	$4.00
$20,000 < x \le 40,000$	$3.00

50. Real Estate   A real estate appraiser charges a fee that depends on the estimated value V of the property. The table at the right gives the fees charged for various estimated values of real estate.
  a. Evaluate this function when $V = \$5,000,000$. $3000
  b. Evaluate this function when $V = \$767,000$. $950

Value of Property	Appraisal Fee
$V < 100,000$	$350
$100,000 \le V < 500,000$	$525
$500,000 \le V < 1,000,000$	$950
$1,000,000 \le V < 5,000,000$	$2500
$5,000,000 \le V < 10,000,000$	$3000

Unless otherwise noted, all content on this page is © Cengage Learning.

For Exercises 51 to 62, what is the domain of the function? Write the answer in set-builder notation.

51. $f(x) = \dfrac{1}{x-1}$

$\{x \mid x \neq 1\}$

52. $g(x) = \dfrac{1}{x+4}$

$\{x \mid x \neq -4\}$

53. $f(x) = 3x + 2$

$\{x \mid -\infty < x < \infty\}$

54. $g(x) = 4 - 2x$

$\{x \mid -\infty < x < \infty\}$

55. $H(x) = \dfrac{1}{2}x^2$

$\{x \mid -\infty < x < \infty\}$

56. $f(x) = \dfrac{x-1}{x}$

$\{x \mid x \neq 0\}$

57. $g(x) = \dfrac{2x+5}{7}$

$\{x \mid -\infty < x < \infty\}$

58. $H(x) = x^2 - x + 1$

$\{x \mid -\infty < x < \infty\}$

59. $f(x) = 3x^2 + x + 4$

$\{x \mid -\infty < x < \infty\}$

60. $g(x) = \dfrac{3-5x}{5}$

$\{x \mid -\infty < x < \infty\}$

61. $H(x) = \dfrac{x-2}{x+2}$

$\{x \mid x \neq -2\}$

62. $h(x) = \dfrac{3-x}{6-x}$

$\{x \mid x \neq 6\}$

Critical Thinking

63. a. Find the set of ordered pairs (x, y) determined by the equation $y = x^3$, where $x \in \{-2, -1, 0, 1, 2\}$. $\{(-2, -8), (-1, -1), (0, 0), (1, 1), (2, 8)\}$

📝 **b.** Does the set of ordered pairs define a function? Why or why not?
Yes, the set defines a function because each member of the domain is assigned to exactly one member of the range.

64. a. Find the set of ordered pairs (x, y) determined by the equation $|y| = x$, where $x \in \{0, 1, 2, 3\}$. $\{(0, 0), (1, -1), (1, 1), (2, -2), (2, 2), (3, -3), (3, 3)\}$

📝 **b.** Does the set of ordered pairs define a function? Why or why not?
No, the set does not define a function because some values in the domain are assigned to two values in the range.

65. 📝 Explain the meanings of the words *relation* and *function*. Be sure to describe how the meanings of the two words differ.

QUICK QUIZ

1. Find the domain and range of the function $\{(2, 9), (3, 13), (4, 17)\}$.
   D: $\{2, 3, 4\}$;
   R: $\{9, 13, 17\}$ [4.2A]

2. Evaluate $f(z) = \dfrac{5z}{z+1}$ when $z = -2$.
   10   [4.2A]

3. Evaluate $h(x) = x^2 + x - 3$ when $x = -2a$.
   $4a^2 - 2a - 3$ [4.2A]

4. What is the domain of $f(x) = 3x^2 - 4x + 1$?
   $\{x \mid -\infty < x < \infty\}$
   [4.2A]

For Exercises 66 to 69, each graph defines a function.

66. Parachuting   The graph at the right shows the distance above the ground of a paratrooper after making a low-level training jump.

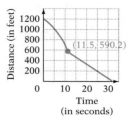

    a. The point with coordinates (11.5, 590.2) is on the graph. Write a sentence that explains the meaning of this ordered pair.

    b. Estimate the time from the beginning of the jump to the end of the jump.

       a. The paratrooper is 590.2 ft above the ground 11.5 s after beginning the jump.   **b.** 32 s

67. Parachuting   The graph at the right shows the descent speed of a paratrooper after making a low-level training jump.

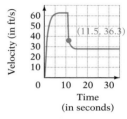

    a. The point with coordinates (11.5, 36.3) is on the graph. Write a sentence that explains the meaning of this ordered pair.

    b. Estimate the speed at which the paratrooper is falling 1 s after jumping from the plane.

       a. The speed of the paratrooper 11.5 s after beginning the jump is 36.3 ft/s.

       b. 30 ft/s

68. ⬤ **Clean Energy**   Use the graph in the article at the right.

    a. Estimate the amount of money that China invested in clean energy projects in 2005. $5 billion

    b. Estimate the amount of money that China invested in clean energy projects in 2010. $49 billion

69. Athletics   The graph at the right shows the decrease in the heart rate r (in beats per minute) of a runner t minutes after completing a race.

    a. Estimate the heart rate of a runner when $t = 5$ min.   110 beats/min

    b. Estimate the heart rate of a runner when $t = 20$ min.   75 beats/min

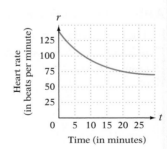

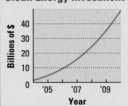

In the NEWS!

China Invests in Clean Energy

As China's demand for energy continues to rise, so does its involvement with clean energy projects.

Clean Energy Investment

Sources: The Boston Globe; Bloomberg New Energy Finance

Projects or Group Activities

70. ◣ Give a real-world example of a relation that is not a function. Is it possible to give an example of a function that is not a relation? If so, give one. If not, explain why it is not possible.

Unless otherwise noted, all content on this page is © Cengage Learning.

4.3   Linear Functions

OBJECTIVE A   *To graph a linear function*

Recall that the ordered pairs of a function can be written as $(x, f(x))$ or as (x, y). The **graph of a function** is a graph of the ordered pairs (x, y) that belong to the function. Certain functions have characteristic graphs. A function that can be written in the form $f(x) = mx + b$ (or $y = mx + b$) is called a **linear function** because its graph is a straight line.

Examples of linear functions are shown at the right. Note that the exponent on each variable is 1.

$$f(x) = 2x + 5 \qquad (m = 2, b = 5)$$
$$P(t) = 3t - 2 \qquad (m = 3, b = -2)$$
$$y = -2x \qquad (m = -2, b = 0)$$
$$y = -\frac{2}{3}x + 1 \qquad \left(m = -\frac{2}{3}, b = 1\right)$$
$$g(z) = z - 2 \qquad (m = 1, b = -2)$$

The equation $y = x^2 + 4x + 3$ is not a linear function because it includes a term with a variable squared. The equation $f(x) = \frac{3}{x - 2}$ is not a linear function because a variable occurs in the denominator. Another example of an equation that is not a linear function is $y = \sqrt{x} + 4$; this equation contains a variable within a radical and so is not a linear function.

There are many applications of linear functions.

APPLY THE CONCEPT ···

The linear function $F = T(C) = \frac{9}{5}C + 32$ is used to convert a Celsius temperature C to a Fahrenheit temperature F. Find the Fahrenheit temperature when the Celsius temperature is 20°C.

$$F = T(C) = \frac{9}{5}C + 32$$

$$= T(20) = \frac{9}{5}(20) + 32 \qquad \bullet \text{ Evaluate the function when } C = 20.$$

$$= 36 + 32 = 68$$

When the Celsius temperature is 20°C, the Fahrenheit temperature is 68°F.

···

Whether an equation is written as $f(x) = mx + b$ or as $y = mx + b$, the equation represents a linear function, and the graph of the equation is a straight line.

Because the graph of a linear function is a straight line, and a straight line is determined by two points, the graph of a linear function can be drawn by finding only two of the ordered pairs of the function. However, it is recommended that you find at least *three* ordered pairs to ensure accuracy.

Take Note

When the coefficient of *x* is a fraction, choose values of *x* that are multiples of the denominator of the fraction. This will result in coordinates that are integers.

INSTRUCTOR NOTE

Only two points are needed to graph a linear function. However, we recommend finding three points in order to ensure accuracy. If the three points plotted do not lie on a straight line, the student will know that an error in arithmetic or in plotting a point has been made.

HOW TO 1   Graph: $f(x) = -\dfrac{1}{2}x + 3$

x	$y = f(x)$
-4	5
0	3
2	2

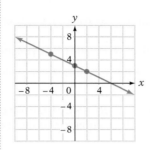

- Find at least three ordered pairs. Because the coefficient of *x* is a fraction with denominator 2, choosing values of *x* that are divisible by 2 simplifies the calculations. The ordered pairs can be displayed in a table.

- Graph the ordered pairs and draw a line through the points.

IN-CLASS EXAMPLES

1. Graph $y = 2x - 3$.
2. Graph $y = \dfrac{1}{2}x$.
3. Graph $y = -\dfrac{2}{3}x + 2$.

EXAMPLE 1

Graph: $f(x) = -\dfrac{3}{2}x - 3$

Solution

x	$y = f(x)$
0	-3
-2	0
-4	3

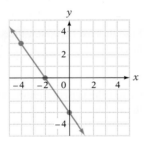

YOU TRY IT 1

Graph: $f(x) = \dfrac{3}{5}x - 4$

Your solution

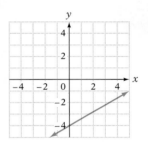

EXAMPLE 2

Graph: $y = \dfrac{2}{3}x$

Solution

x	y
0	0
3	2
-3	-2

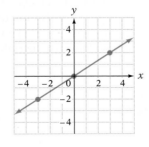

YOU TRY IT 2

Graph: $y = -\dfrac{3}{4}x$

Your solution

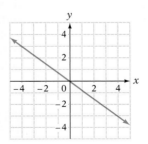

Solutions on p. S11

Unless otherwise noted, all content on this page is © Cengage Learning.

OBJECTIVE B   *To graph an equation of the form $Ax + By = C$*

INSTRUCTOR NOTE

There are many times when an equation in two variables must be solved for y before it can be graphed. For instance, most graphing calculators require that an equation be in this form. Rewriting linear equations in the form $y = mx + b$ is good practice for students.

Later we will graph the equation $Ax + By = C$ by using its intercepts.

The equation $Ax + By = C$, where A and B are coefficients and C is a constant, is also a *linear equation in two variables*. This equation can be written in the form $y = mx + b$.

HOW TO 2   Write $4x - 3y = 6$ in the form $y = mx + b$.

$$4x - 3y = 6$$
$$-3y = -4x + 6$$ • Subtract $4x$ from each side of the equation.
$$y = \frac{4}{3}x - 2$$ • Divide each side of the equation by -3. This is the form $y = mx + b$, with $m = \frac{4}{3}$ and $b = -2$.

To graph an equation of the form $Ax + By = C$, first solve the equation for y. Then follow the same procedure used to graph an equation of the form $y = mx + b$.

HOW TO 3   Graph: $3x + 2y = 6$

IN-CLASS EXAMPLES

  4. Graph $2x - y = 4$.
  5. Graph $5x - 2y = 10$.
  6. Graph $y = -3$.

$$3x + 2y = 6$$
$$2y = -3x + 6$$
$$y = -\frac{3}{2}x + 3$$ • Solve the equation for y.

x	y
0	3
2	0
4	−3

• Find at least three solutions. Choose multiples of 2 for x.

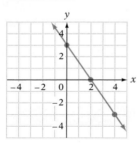

• Graph the ordered pairs and draw a line through the points.

If one of the coefficients A or B is zero, the graph of $Ax + By + C$ is a horizontal or vertical line. Consider the equation $y = -2$, where A, the coefficient of x, is 0. We can write this equation in *two* variables as $0 \cdot x + y = -2$. No matter what value of x is selected, $0 \cdot x = 0$. Therefore, y equals -2 for all values of x. The table below shows some ordered-pair solutions of $y = -2$. The graph is shown to the right of the table.

x	$0 \cdot x + y = -2$	y	(x, y)
−1	$0 \cdot (-1) + y = -2$	−2	$(-1, -2)$
0	$0 \cdot 0 + y = -2$	−2	$(0, -2)$
3	$0 \cdot 3 + y = -2$	−2	$(3, -2)$

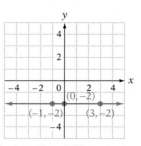

Unless otherwise noted, all content on this page is © Cengage Learning.

> ### Graph of $y = b$
>
> The **graph of $y = b$** is a horizontal line passing through the point with coordinates $(0, b)$.

HOW TO 4 Graph: $y + 4 = 0$

Solve for y.

$$y + 4 = 0$$
$$y = -4$$

The graph of $y = -4$ is a horizontal line passing through the point with coordinates $(0, -4)$.

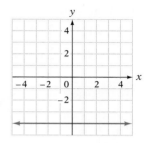

The equation $y = -2$ represents a function. Some of the ordered pairs of this function are $(-1, -2)$, $(0, -2)$, and $(3, -2)$. In function notation, we write $f(x) = -2$. This function is an example of a *constant function*. For every value of x, the value of the function is the constant -2. For instance, we have $f(17) = -2, f(\sqrt{2}) = -2$, and $f(\pi) = -2$.

> ### Constant Function
>
> A function given by $f(x) = b$, where b is a constant, is a **constant function.** The graph of a constant function is a horizontal line passing through the point with coordinates $(0, b)$.

Now consider the equation $x = 2$, where B, the coefficient of y, is zero. We write this equation in two variables by writing $x + 0 \cdot y = 2$. No matter what value of y is selected, $0 \cdot y = 0$. Therefore, x equals 2 for all values of y. The following table shows some ordered-pair solutions of $x = 2$. The graph is shown to the right of the table.

y	$x + 0 \cdot y = 2$	x	(x, y)
6	$x + 0 \cdot 6 = 2$	2	$(2, 6)$
1	$x + 0 \cdot 1 = 2$	2	$(2, 1)$
-4	$x + 0 \cdot (-4) = 2$	2	$(2, -4)$

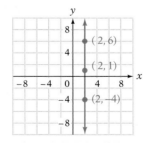

 Take Note

The equation $y = b$ represents a function. The equation $x = a$ does *not* represent a function. Remember, not all equations represent functions.

> ### Graph of $x = a$
>
> The **graph of $x = a$** is a vertical line passing through the point with coordinates $(a, 0)$.

Recall that a function is a set of ordered pairs in which no two ordered pairs have the same first coordinate and different second coordinates. Because $(2, 6)$, $(2, 1)$, and $(2, -4)$ are ordered pairs belonging to the equation $x = 2$, this equation does not represent a function, and its graph is not the graph of a function.

Unless otherwise noted, all content on this page is © Cengage Learning.

EXAMPLE 3

Graph: $2x + 3y = 9$

Solution

Solve the equation for y.

$2x + 3y = 9$
$\quad\quad 3y = -2x + 9$
$\quad\quad\ y = -\dfrac{2}{3}x + 3$

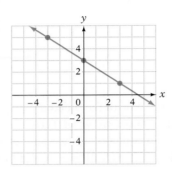

x	y
−3	5
0	3
3	1

YOU TRY IT 3

Graph: $-3x + 2y = 4$

Your solution

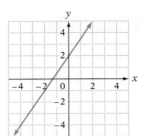

EXAMPLE 4

Graph: $x = -4$

Solution

The graph of an equation of the form $x = a$ is a vertical line passing through the point with coordinates $(a, 0)$.

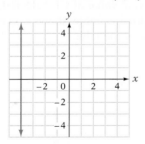

• The graph of $x = -4$ goes through the point with coordinates $(-4, 0)$.

YOU TRY IT 4

Graph: $y - 3 = 0$

Your solution

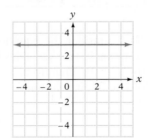

Solutions on p. S11

OBJECTIVE C   *To find the x- and y-intercepts of a straight line*

The graph of the equation $x - 2y = 4$ is shown at the right. The graph crosses the x-axis at the point with coordinates $(4, 0)$. This point is called the **x-intercept.** The graph crosses the y-axis at the point with coordinates $(0, -2)$. This point is called the **y-intercept.**

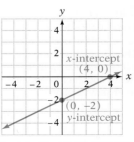

Unless otherwise noted, all content on this page is © Cengage Learning.

Take Note

The *x*-intercept occurs when $y = 0$.

The *y*-intercept occurs when $x = 0$.

HOW TO 5 Find the coordinates of the *x*- and *y*-intercepts of the graph of the equation $3x + 4y = -12$.

To find the *x*-intercept, let $y = 0$. (Any point on the *x*-axis has *y*-coordinate 0.)

$$3x + 4y = -12$$
$$3x + 4(0) = -12$$
$$3x = -12$$
$$x = -4$$

The *x*-intercept has coordinates $(-4, 0)$.

To find the *y*-intercept, let $x = 0$. (Any point on the *y*-axis has *x*-coordinate 0.)

$$3x + 4y = -12$$
$$3(0) + 4y = -12$$
$$4y = -12$$
$$y = -3$$

The *y*-intercept has coordinates $(0, -3)$.

HOW TO 6 Graph $3x - 2y = 6$ by using the *x*- and *y*-intercepts.

$$3x - 2y = 6$$
$$3x - 2(0) = 6$$
$$3x = 6$$
$$x = 2$$

• To find the *x*-intercept, let $y = 0$. Then solve for *x*.

The *x*-intercept has coordinates $(2, 0)$.

$$3x - 2y = 6$$
$$3(0) - 2y = 6$$
$$-2y = 6$$
$$y = -3$$

• To find the *y*-intercept, let $x = 0$. Then solve for *y*.

The *y*-intercept has coordinates $(0, -3)$.

Graph the *x*- and *y*-intercepts. Draw a line through the two points.

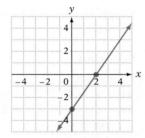

IN-CLASS EXAMPLES

Graph by using the *x*- and *y*-intercepts.

7. $x + 2y = 2$

8. $2x - 5y = -10$

9. Find the zero of $f(x) = 4x + 2.$ $\quad -\dfrac{1}{2}$

The graph of $f(x) = 2x - 4$ is shown at the right. Evaluating the function when $x = 2$, we have

$$f(x) = 2x - 4$$
$$f(2) = 2(2) - 4$$
$$f(2) = 0$$

2 is the value of *x* for which $f(x) = 0$. A value of *x* for which $f(x) = 0$ is called a *zero* of *f*.

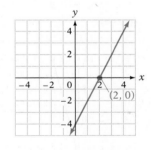

Note that the *x*-intercept of the graph has coordinates $(2, 0)$. The *x*-coordinate of the *x*-intercept is a zero of the function.

Zero of a Function

A value of *x* for which $f(x) = 0$ is called a **zero** of the function *f*.

EXAMPLES

1. Let $f(x) = 3x + 6$ and $x = -2$.
$f(-2) = 3(-2) + 6 = 0$
Because $f(-2) = 0$, -2 is a zero of *f*.

2. Let $f(x) = 2x - 6$ and $x = 0$.
$f(0) = 2(0) - 6 = -6 \neq 0$
Because $f(0) \neq 0$, 0 is not a zero of *f*.

To find a zero of a function, let $f(x) = 0$ and solve for *x*.

Unless otherwise noted, all content on this page is © Cengage Learning.

HOW TO 7 Find the zero of $f(x) = \frac{2}{3}x - 4$.

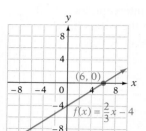

$$f(x) = \frac{2}{3}x - 4$$

$$0 = \frac{2}{3}x - 4 \qquad \bullet \text{ To find a zero of a function, let } f(x) = 0.$$

$$-\frac{2}{3}x = -4 \qquad \bullet \text{ Solve for } x.$$

$$x = 6$$

The zero is 6. The graph of f is shown at the left. Note that the x-coordinate of the x-intercept is the zero of f.

EXAMPLE 5

Graph $4x - y = 4$ by using the x- and y-intercepts.

Solution

x-intercept:

$4x - y = 4$

$4x - 0 = 4 \qquad \bullet \text{ Let } y = 0.$

$\quad 4x = 4$

$\quad\quad x = 1$

$(1, 0)$

y-intercept:

$\quad 4x - y = 4$

$4(0) - y = 4 \qquad \bullet \text{ Let } x = 0.$

$\quad\quad -y = 4$

$\quad\quad\quad y = -4$

$(0, -4)$

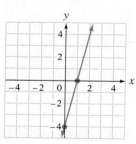

EXAMPLE 6

Find the zero of $h(t) = 4t + 6$.

Solution

$h(t) = 4t + 6$

$\quad 0 = 4t + 6 \qquad \bullet \text{ Let } h(t) = 0.$

$-4t = 6 \qquad\quad \bullet \text{ Solve for } t.$

$\quad t = -\frac{3}{2}$

The zero is $-\frac{3}{2}$.

YOU TRY IT 5

Graph $3x - y = 2$ by using the x- and y-intercepts.

Your solution

x-intercept: $\left(\frac{2}{3}, 0\right)$

y-intercept: $(0, -2)$

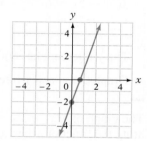

YOU TRY IT 6

Find the zero of $g(x) = 4 + \frac{2}{3}x$.

Your solution

-6

Solutions on pp. S11–S12

Unless otherwise noted, all content on this page is © Cengage Learning.

OBJECTIVE D   *To solve application problems*

There are a variety of applications of linear functions.

Take Note

In many applications, the domain of the variable is such that the equation makes sense. For this application, it would not be sensible to have values of *t* that are less than zero. This would indicate negative time! The number 10 is somewhat arbitrary, but after 10 min most people's heart rates would level off, and a linear function would no longer apply.

HOW TO 8   The heart rate *R* after *t* minutes for a person taking a brisk walk can be approximated by the equation $R = 2t + 72$. Graph this equation for $0 \leq t \leq 10$. The point with coordinates (5, 82) is on the graph. Write a sentence that describes the meaning of this ordered pair.

Find the values of *R* for $t = 0$ and $t = 10$. When $t = 0$, $R = 72$. When $t = 10$, $R = 84$.

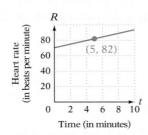

• **Graph (0, 72), (5, 82), and (10, 84). Draw the line segment that contains the three points.**

The ordered pair (5, 82) means that after 5 min, the person's heart rate is 82 beats per minute.

IN-CLASS EXAMPLES

10. A tennis instructor is paid $40 per hour for a private lesson. The equation that describes the total earnings *E* in dollars received by the instructor is $E = 40t$, where *t* is the total number of hours worked. Graph this equation for $0 \leq t \leq 30$. The point with coordinates (25, 1000) is on the graph. Write a sentence that describes the meaning of this ordered pair.   **The tennis instructor earns $1000 for working 25 h.**

EXAMPLE 7

An electronics technician charges $45 plus $1 per minute to repair defective wiring in a home or apartment. The equation that describes the total cost *C* to have defective wiring repaired is given by $C = t + 45$, where *t* is the number of minutes the technician works. Graph this equation for $0 \leq t \leq 60$. The point with coordinates (15, 60) is on the graph. Write a sentence that describes the meaning of this ordered pair.

Solution

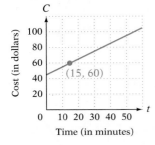

• **Graph $C = t + 45$.**
  When $t = 0$, $C = 45$.
  When $t = 50$, $C = 95$.

The ordered pair (15, 60) means that it costs $60 for the technician to work 15 min.

YOU TRY IT 7

The height *h* (in inches) of a person and the length *L* (in inches) of that person's stride while walking are related. The equation $h = \frac{3}{4}L + 50$ approximates this relationship. Graph this equation for $15 \leq L \leq 40$. The point with coordinates (32, 74) is on the graph. Write a sentence that describes the meaning of this ordered pair.

Your solution

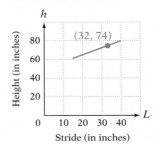

The ordered pair (32, 74) means that a person with a stride of 32 in. is 74 in. tall.

Solution on p. S12

Unless otherwise noted, all content on this page is © Cengage Learning.

4.3 EXERCISES

✔ Concept Check

SUGGESTED ASSIGNMENT
Exercises 1–4; Exercises 5–61, odds
More challenging exercises: Exercises 62–68

1. 🔲 When finding ordered pairs to graph a line, why do we recommend that you find at least three ordered pairs?

2. To graph points on the graph of $y = \frac{5}{3}x + 4$, it is helpful to choose values of x that are divisible by what number?   3

3. What is the x-coordinate of the y-intercept of a line? What is the y-coordinate of the x-intercept of a line?   0; 0

4. Is -3 a zero of $f(x) = 5x + 15$?   Yes

OBJECTIVE A   *To graph a linear function*

5. **Oceanography**   The linear function $P = f(d) = 0.097d + 1$ can be used to estimate the pressure P, in atmospheres (atm), on a vessel that is d meters below the surface of the ocean. Find the pressure on a vessel that is 500 m below the surface of the ocean. (*Note:* The pressure at sea level is 1 atm.)   49.5 atm

6. **Speed of Sound**   The speed of sound is affected by temperature and can be approximated by the linear function $S = f(T) = 0.6T + 340$, where S is the speed of sound in meters per second and T is the temperature in degrees Celsius. Find the speed of sound when the temperature is 25°C.   355 m/s

For Exercises 7 to 18, graph.

7. $y = x - 3$

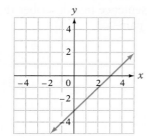

8. $y = -x + 2$

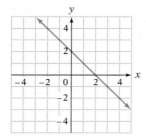

9. $y = -3x + 2$

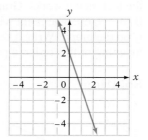

10. $y = -2x + 3$

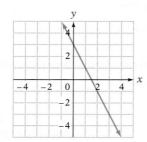

11. $f(x) = 3x - 4$

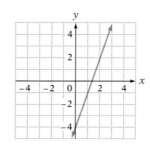

12. $f(x) = \frac{3}{2}x$

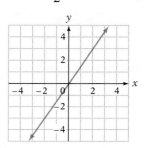

Unless otherwise noted, all content on this page is © Cengage Learning.

13. $f(x) = -\dfrac{2}{3}x$

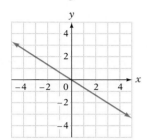

14. $f(x) = \dfrac{3}{4}x + 2$

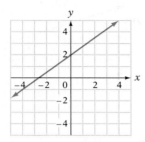

15. $y = \dfrac{2}{3}x - 4$

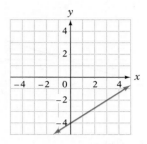

16. $y = -\dfrac{3}{2}x - 3$

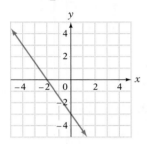

17. $f(x) = -\dfrac{1}{3}x + 2$

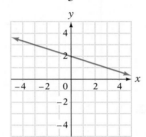

18. $f(x) = \dfrac{3}{5}x - 1$

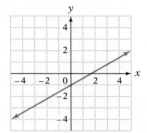

OBJECTIVE B   *To graph an equation of the form $Ax + By = C$*

For Exercises 19 to 33, graph.

19. $2x + y = -3$

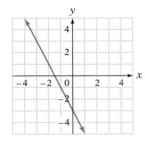

20. $2x - y = 3$

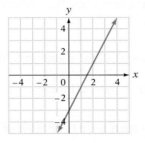

21. $x - 4y = 8$

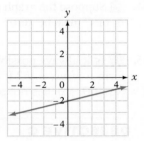

22. $2x + 5y = 10$

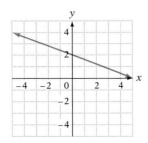

23. $4x + 3y = 12$

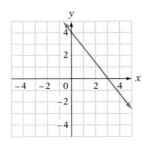

24. $2x - 5y = 10$

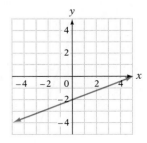

Unless otherwise noted, all content on this page is © Cengage Learning.

25. $x - 3y = 0$

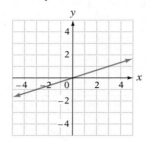

26. $x + 3y = 9$

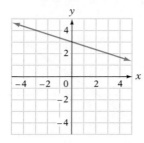

27. $y = -2$

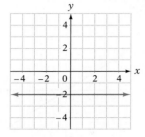

28. $f(x) = 3$

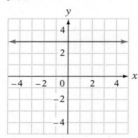

29. $x = -3$

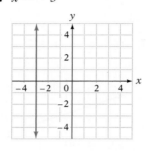

30. $x = 1$

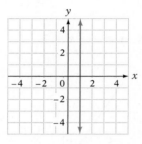

31. $3x - y = -2$

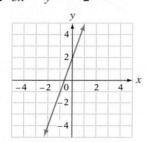

32. $2x - 3y = 12$

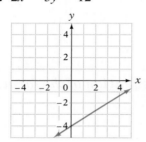

33. $3x - 2y = 8$

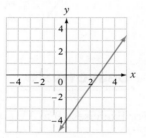

34. Suppose the graph of $Ax + By = C$ is a horizontal line. Which of the numbers A, B, or C must be zero? A

35. Is it always possible to solve $Ax + By = C$ for y and write the equation in the form $y = mx + b$? If not, explain.
No. If $B = 0$, it is not possible to solve $Ax + By = C$ for y.

OBJECTIVE C *To find the x- and y-intercepts of a straight line*

For Exercises 36 to 44, find the *x*- and *y*-intercepts, and graph.

36. $3x + y = 3$

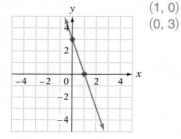

(1, 0)
(0, 3)

37. $x - 2y = -4$

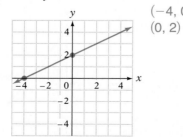

(−4, 0)
(0, 2)

38. $2x - y = 4$

(2, 0)
(0, −4)

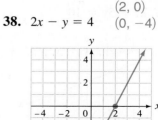

Unless otherwise noted, all content on this page is © Cengage Learning.

39. $2x - 3y = 9$

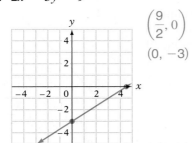

$\left(\dfrac{9}{2}, 0\right)$

$(0, -3)$

40. $4x - 3y = 8$

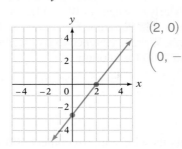

$(2, 0)$

$\left(0, -\dfrac{8}{3}\right)$

41. $2x + y = 3$ $\left(\dfrac{3}{2}, 0\right)$

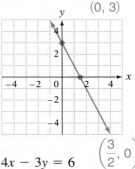

$(0, 3)$

42. $2x - 3y = 4$

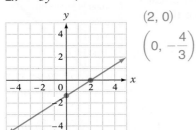

$(2, 0)$

$\left(0, -\dfrac{4}{3}\right)$

43. $3x + 2y = 4$

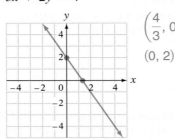

$\left(\dfrac{4}{3}, 0\right)$

$(0, 2)$

44. $4x - 3y = 6$ $\left(\dfrac{3}{2}, 0\right)$

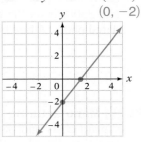

$(0, -2)$

45. Does the graph of every straight line have a *y*-intercept? Explain.
No. The graph of the equation $x = a$, $a \neq 0$, is a vertical line and has no *y*-intercept.

46. Why is it not possible to graph an equation of the form $Ax + By = 0$ by using only the *x*- and *y*-intercepts?   The *x*-intercept and *y*-intercept are both (0, 0), and at least two different points are needed to graph a line.

For Exercises 47 to 54, find the zero of the function.

47. $f(x) = 4x + 8$

-2

48. $h(x) = -3x - 9$

-3

49. $s(t) = \dfrac{3}{4}t - 9$

12

50. $f(t) = \dfrac{2}{5}t + 2$

-5

51. $f(x) = 4x$

0

52. $g(u) = -3u$

0

53. $g(x) = \dfrac{3}{2}x - 4$

$\dfrac{8}{3}$

54. $h(t) = 3 - \dfrac{4}{5}t$

$\dfrac{15}{4}$

OBJECTIVE D   *To solve application problems*

55. Biology   The heart of a ruby-throated hummingbird beats about 1200 times per minute while the bird is in flight. The equation $B = 1200t$ gives the total number of heartbeats B in t minutes of flight. How many times will this hummingbird's heart beat in 7 min of flight?   8400 times

56. Telecommunications   The monthly cost of a wireless telephone plan is \$39.99 for up to 450 min of calling time plus \$.45 per minute for each minute over 450 min. The equation that describes the cost of this plan is $C = 0.45x + 39.99$, where x is the number of minutes over 450. What is the cost of this plan if a person uses **a.** 398 min and **b.** 475 min of calling time?   **a.** \$39.99   **b.** \$51.24

Unless otherwise noted, all content on this page is © Cengage Learning.

57. ✎ **Compensation**   Marlys receives $11 per hour as a mathematics department tutor. The equation that describes Marlys's wages is $W = 11t$, where t is the number of hours she spends tutoring. Graph this equation for $0 \leq t \leq 20$. The point with coordinates (15, 165) is on the graph. Write a sentence that describes the meaning of this ordered pair.   Marlys receives $165 for tutoring 15 h.

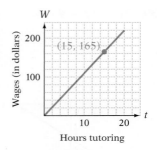

58. ✎ **Animal Science**   A bee beats its wings approximately 100 times per second. The equation that describes the total number of times a bee beats its wings is given by $B = 100t$. Graph this equation for $0 \leq t \leq 60$. The point with coordinates (35, 3500) is on the graph. Write a sentence that describes the meaning of this ordered pair.   A bee beats its wings 3500 times in 35 s.

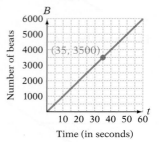

59. ✎ **Manufacturing**   The cost of manufacturing skis is $5000 for startup costs and $80 per pair of skis manufactured. The equation that describes the cost of manufacturing n pairs of skis is $C = 80n + 5000$. Graph this equation for $0 \leq n \leq 2000$. The point with coordinates (50, 9000) is on the graph. Write a sentence that describes the meaning of this ordered pair.
The cost of manufacturing 50 pairs of skis is $9000.

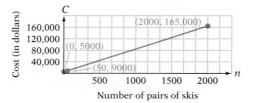

60. ◕ ✎ **Atomic Physics**   The Large Hadron Collider, or LHC, is a machine that is capable of accelerating a proton to a velocity that is 99.99% the speed of light. At this speed, a proton will travel approximately 0.98 ft in a billionth of a second (one nanosecond). (*Source:* news.yahoo.com) The equation $d = 0.98t$ gives the distance d, in feet, traveled by a proton in t nanoseconds. Graph this equation for $0 \leq t \leq 10$. The point with coordinates (4, 3.92) is on the graph. Write a sentence that explains the meaning of this ordered pair.
The proton travels 3.92 ft in 4 nanoseconds.

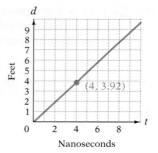

61. ◕ ✎ **Oceanography**   Read the article below about the small submarine *Alvin*.

In the NEWS!

Alvin, First Viewer of the *Titanic*, Gets an Upgrade

Alvin, the original Human Occupied Vehicle (HOV), is on vacation. Since 1964, scientists have used *Alvin* for deep-sea exploration. *Alvin* is able to descend at a rate of 30 m/min to a maximum depth of 4500 m below sea level. As part of a plan to create an HOV that can dive deeper and stay underwater longer, *Alvin* is now undergoing many structural and systems upgrades.
Source: Woods Hole Oceanographic Institute

The equation that describes *Alvin*'s depth D, in meters, is $D = -30t$, where t is the number of minutes *Alvin* has been descending. Graph this equation for $0 \leq t \leq 150$. The ordered pair (65, −1950) is on the graph. Write a sentence that describes the meaning of this ordered pair.   After 65 min, *Alvin* is 1950 m below sea level.

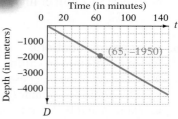

Unless otherwise noted, all content on this page is © Cengage Learning.

1. Graph $y = 3x - 6$.   [4.3A]   2. Graph $2x - 3y = 6$.
[4.3B]

Critical Thinking

62. ◣ Explain what the graph of an equation represents.

63. ◣ Explain how to graph the equation of a straight line by plotting points.

64. ◣ Explain how to graph the equation of a straight line by using its x- and y-intercepts.

65. ◣ Explain why you cannot graph the equation $4x + 3y = 0$ by using just its intercepts.

Projects or Group Activities

An equation of the form $\frac{x}{a} + \frac{y}{b} = 1$, where $a \neq 0$ and $b \neq 0$, is called the *intercept form of a straight line* because $(a, 0)$ and $(0, b)$ are the x- and y-intercepts of the graph of the equation. Graph the equations in Exercises 66 to 68.

Graph by using the x- and y-intercepts.
3. $3x + 2y = 6$ [4.3C]
4. A tree service charges $160 plus $175 for each tree removed. The equation that describes the cost C, in dollars, to remove n trees is $C = 175n + 160$. The ordered pair (20, 3660) is on the graph. Write a sentence that describes the meaning of this ordered pair.   **The tree service will charge $3660 to remove 20 trees.**   [4.3D]

66. $\dfrac{x}{3} + \dfrac{y}{5} = 1$

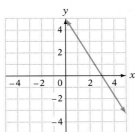

67. $\dfrac{x}{2} + \dfrac{y}{3} = 1$

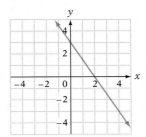

68. $\dfrac{x}{3} - \dfrac{y}{2} = 1$

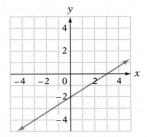

✔ CHECK YOUR PROGRESS: CHAPTER 4

1. Graph $(-4, 3)$, $(-2, -1)$, $(2, -1)$, and $(0, 3)$.   [4.1A]

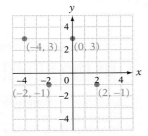

2. Graph $f(x) = -\frac{4}{5}x + 4$.   [4.3A]

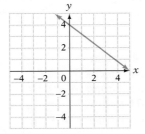

3. Graph $3x - 4y = 12$ by using the x- and y-intercepts.   [4.3C]

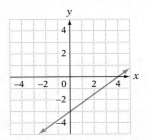

4. Find the ordered-pair solutions of $y = -3x + 2$ when $x = -1, 0, 1,$ and 2.
$(-1, 5)$, $(0, 2)$, $(1, -1)$, $(2, -4)$
[4.1B]

5. Describe the graph of $x = -5$.
A vertical line passing through $(-5, 0)$ [4.3B]

6. Find the domain and range of the relation $\{(-3, -2), (-2, -1), (-1, 0)\}$. Is the relation a function?
D: $\{-3, -2, -1\}$; R: $\{-2, -1, 0\}$; yes   [4.2A]

7. Evaluate $h(s) = s^2 + 3s$ when $h = -3$.   0   [4.2A]

8. Is $(-3, 0)$ a solution of $y = -\frac{1}{3}x - 1$?   Yes   [4.1B]

9. Does $y = x^2 - 4$, where $x \in \{-3, -1, 0, 1, 3\}$, define y as a function of x?
Yes   [4.2A]

Unless otherwise noted, all content on this page is © Cengage Learning.

10. Determine whether the set of ordered pairs is a function. State the domain and range.
$\{(0, 1), (1, 2), (2, 3), (-5, 3)\}$
Function; D: $\{-5, 0, 1, 2\}$; R: $\{1, 2, 3\}$ [4.2A]

11. What is the domain of $g(x) = \frac{x + 2}{2x}$?
$\{x | x \neq 0\}$ [4.2A]

12. Evaluate $s(t) = -3t^2 + 4t - 1$ when $t = -3$.
-40 [4.2A]

13. Evaluate $f(x) = 2x - 3$ when $x = 2 - a$.
$1 - 2a$ [4.2A]

For Exercises 14 to 17, graph the equation.

14. $f(x) = -\frac{3}{4}x + 2$

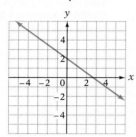

[4.3A]

15. $y = 2x - 3$

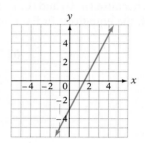

[4.3A]

16. $y + 5 = 0$

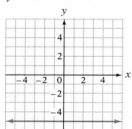

[4.3B]

17. $x = 4$

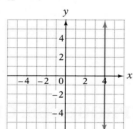

[4.3B]

18. Find the coordinates of the x- and y-intercepts of $4x - 5y = 20$. $(5, 0), (0, -4)$ [4.3C]

19. Find the zero of $g(t) = 3t + 6$.
-2 [4.3C]

20. ⬤ **Taxi Fares**   See the news clipping at the right. The equation $F = 2.80M + 2.20$ can be used to calculate the fare F, in dollars, for a ride of M miles. Graph this equation for values of M from 1 to 5. The point $(3, 10.6)$ is on the graph. Write a sentence that describes the meaning of this ordered pair.
A 3-mile taxi ride costs $10.60.   [4.3D]

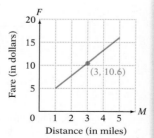

Fare (in dollars)

Distance (in miles)

In the NEWS!

Rate Hike for Boston Cab Rides

Taxi drivers soon will be raising their rates, perhaps in an effort to help pay for their required switch to hybrid vehicles by 2015. In the near future, a passenger will have to pay $5.00 for the first mile of a taxi ride and $2.80 for each additional mile.

Source: The Boston Globe

Unless otherwise noted, all content on this page is © Cengage Learning.

SECTION

4.4 Slope of a Straight Line

OBJECTIVE A   *To find the slope of a line given two points*

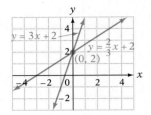

The graphs of $y = 3x + 2$ and $y = \frac{2}{3}x + 2$ are shown at the left. Each graph crosses the y-axis at the point $P(0, 2)$, but the graphs have different slants. The **slope** of a line is a measure of the slant of the line. The symbol for slope is m.

The slope of a line containing two points is the ratio of the change in the y values of the two points to the change in the x values. The line containing the points whose coordinates are $(-1, -3)$ and $(5, 2)$ is shown below.

The change in the y values is the difference between the y-coordinates of the two points.

$$\text{Change in } y = 2 - (-3) = 5$$

The change in the x values is the difference between the x-coordinates of the two points.

$$\text{Change in } x = 5 - (-1) = 6$$

The slope of the line between the two points is the ratio of the change in y to the change in x.

$$\text{Slope} = m = \frac{\text{change in } y}{\text{change in } x} = \frac{5}{6} \qquad m = \frac{2 - (-3)}{5 - (-1)} = \frac{5}{6}$$

In general, if $P_1(x_1, y_1)$ and $P_2(x_2, y_2)$ are two points on a line, then

$$\text{Change in } y = y_2 - y_1 \qquad \text{Change in } x = x_2 - x_1$$

Using these ideas, we can state a formula for slope.

Slope Formula

The slope of the line containing the two points $P_1(x_1, y_1)$ and $P_2(x_2, y_2)$ is given by

$$m = \frac{y_2 - y_1}{x_2 - x_1}, \, x_1 \neq x_2$$

Frequently, the Greek letter Δ (delta) is used to designate the change in a variable. Using this notation, we can write equations for the change in y and the change in x as follows:

$$\text{Change in } y = \Delta y = y_2 - y_1 \qquad \text{Change in } x = \Delta x = x_2 - x_1$$

Using this notation, the slope formula is written $m = \frac{\Delta y}{\Delta x}$.

Unless otherwise noted, all content on this page is © Cengage Learning.

Take Note

When finding the slope of the line between two points, it does not matter which point is called P_1 and which is called P_2. In HOW TO 1, we could have labeled the points $P_1(4, 5)$ and $P_2(-2, 0)$, reversing the names of P_1 and P_2. Then

$$m = \frac{y_2 - y_1}{x_2 - x_1}$$

$$m = \frac{0 - 5}{-2 - 4} = \frac{-5}{-6} = \frac{5}{6}$$

The result is the same.

HOW TO 1 Find the slope of the line passing through the points $P_1(-2, 0)$ and $P_2(4, 5)$.

From $P_1(-2, 0)$, we have $x_1 = -2$, $y_1 = 0$. From $P_2(4, 5)$, we have $x_2 = 4$, $y_2 = 5$. Now use the slope formula.

$$m = \frac{y_2 - y_1}{x_2 - x_1} = \frac{5 - 0}{4 - (-2)} = \frac{5}{6}$$

The slope of the line is $\frac{5}{6}$.

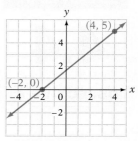
Positive slope

A line that slants upward to the right has a **positive slope.**

HOW TO 2 Find the slope of the line passing through the points $P_1(-3, 4)$ and $P_2(4, 2)$.

From $P_1(-3, 4)$, we have $x_1 = -3$, $y_1 = 4$. From $P_2(4, 2)$, we have $x_2 = 4$, $y_2 = 2$. Now use the slope formula.

$$m = \frac{y_2 - y_1}{x_2 - x_1} = \frac{2 - 4}{4 - (-3)} = \frac{-2}{7} = -\frac{2}{7}$$

The slope of the line is $-\frac{2}{7}$.

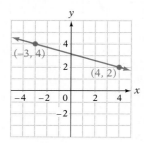
Negative slope

A line that slants downward to the right has a **negative slope.**

INSTRUCTOR NOTE

Students have a tendency to think that zero slope means *no slope*. Emphasize that a slope of zero refers to a line that is parallel to the *x*-axis; a line whose slope is undefined is parallel to the *y*-axis.

HOW TO 3 Find the slope of the line passing through the points $P_1(-2, 2)$ and $P_2(4, 2)$.

From $P_1(-2, 2)$, we have $x_1 = -2$, $y_1 = 2$. From $P_2(4, 2)$, we have $x_2 = 4$, $y_2 = 2$. Now use the slope formula.

$$m = \frac{y_2 - y_1}{x_2 - x_1} = \frac{2 - 2}{4 - (-2)} = \frac{0}{6} = 0$$

The slope of the line is 0.

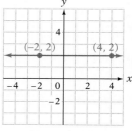
Zero slope

A horizontal line has **zero slope.**

★ **Tips for Success**

To learn mathematics, you must be an active participant. Listening to and watching your professor do mathematics is not enough. Take notes in class, mentally think through every question your instructor asks, and try to answer it even if you are not called on to do so. Ask questions when you have them. See *AIM for Success* in the Preface for other ways to be an active learner.

HOW TO 4 Find the slope of the line passing through the points $P_1(1, -2)$ and $P_2(1, 3)$.

From $P_1(1, -2)$, we have $x_1 = 1$, $y_1 = -2$. From $P_2(1, 3)$, we have $x_2 = 1$, $y_2 = 3$. Now use the slope formula.

$$m = \frac{y_2 - y_1}{x_2 - x_1} = \frac{3 - (-2)}{1 - 1} = \frac{5}{0}$$

$\frac{5}{0}$ is not a real number, because division by zero is undefined. The slope of the line is undefined.

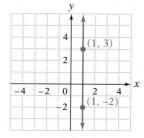
Undefined

A vertical line has **undefined slope.** A vertical line is also said to have **no slope.**

Unless otherwise noted, all content on this page is © Cengage Learning.

Point of Interest

One of the motivations for the discovery of calculus was the desire to solve a more complicated version of the distance-rate problem described at the right.

You may be familiar with twirling a ball on the end of a string. If you release the string, the ball flies off in a path as shown below.

The question that mathematicians tried to answer was essentially, "What is the slope of the line represented by the arrow?"

Answering questions similar to this led to the development of one aspect of calculus.

There are many applications of slope. Here are two examples.

The first record for the 1-mile run was recorded in 1865 in England. Richard Webster ran the mile in 4 min 36.5 s. His average speed was approximately 19 ft/s.

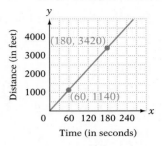

The graph at the right shows the distance Webster ran during that run. From the graph, note that after 60 s (1 min) he had traveled 1140 ft, and after 180 s (3 min) he had traveled 3420 ft.

We will choose $P_1(60, 1140)$ and $P_2(180, 3420)$. The slope of the line between these two points is

$$m = \frac{y_2 - y_1}{x_2 - x_1} = \frac{3420 - 1140}{180 - 60} = \frac{2280}{120} = 19$$

Note that the slope of the line is the same as Webster's average speed, 19 ft/s. Average speed is related to slope.

The resale value of a 2010 Chevrolet Corvette declines as the number of miles the car is driven increases, as shown in the graph at the right. From the graph, note that after the car is driven 25,000 mi, its value is \$34,400, and after the car is driven 50,000 mi, its value is \$33,000. (*Source:* Edmunds.com, December 2011)

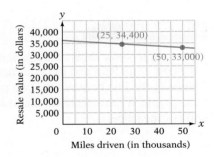

We will choose $P_1(25, 34{,}400)$ and $P_2(50, 33{,}000)$. The slope of the line between the two points is

IN-CLASS EXAMPLES

Find the slope of the line containing the points with the given coordinates.

1. $(4, -2)$, $(-1, 3)$ -1
2. $(-1, 2)$, $(5, 3)$ $\frac{1}{6}$
3. $(2, 3)$, $(4, -2)$ $-\frac{5}{2}$
4. $(1, 5)$, $(3, 5)$ 0
5. $(-2, -1)$, $(-2, 6)$ **Undefined**

$$m = \frac{y_2 - y_1}{x_2 - x_1} = \frac{33{,}000 - 34{,}400}{50 - 25} = \frac{-1400}{25} \approx -56$$

If we interpret negative slope as decreasing value, then the slope of the line represents the dollar decline in the value of the car for each 1000 mi driven. Thus the value of the car decreases by \$56 for each 1000 mi driven.

In general, any quantity that is expressed using the word *per* is represented mathematically as slope. In the first example, the slope represented the average speed, 19 ft/s. In the second example, the slope represented the rate at which the value of the car was decreasing, \$56 for each 1000 mi driven.

Unless otherwise noted, all content on this page is © Cengage Learning.

EXAMPLE 1

Find the slope of the line containing $P_1(2, -5)$ and $P_2(-4, 2)$.

Solution

$$m = \frac{y_2 - y_1}{x_2 - x_1} = \frac{2 - (-5)}{-4 - 2} = \frac{7}{-6}$$

The slope is $-\frac{7}{6}$.

YOU TRY IT 1

Find the slope of the line containing $P_1(4, -3)$ and $P_2(2, 7)$.

Your solution

-5

EXAMPLE 2

Find the slope of the line containing $P_1(-3, 4)$ and $P_2(5, 4)$.

Solution

$$m = \frac{y_2 - y_1}{x_2 - x_1} = \frac{4 - 4}{5 - (-3)} = \frac{0}{8} = 0$$

The slope of the line is zero.

YOU TRY IT 2

Find the slope of the line containing $P_1(6, -1)$ and $P_2(6, 7)$.

Your solution

The slope of the line is undefined.

EXAMPLE 3

The graph below shows the relationship between the cost of an item and the sales tax. Find the slope of the line between the two points shown on the graph. Write a sentence that states the meaning of the slope.

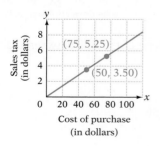

Solution

$$m = \frac{5.25 - 3.50}{75 - 50}$$

   • Choose $P_1(50, 3.50)$ and
     $P_2(75, 5.25)$.

$$= \frac{1.75}{25} = 0.07$$

A slope of 0.07 means the sales tax is $.07 per dollar.

YOU TRY IT 3

The graph below shows the decrease in the value of a recycling truck over 6 years. Find the slope of the line between the two points shown on the graph. Write a sentence that states the meaning of the slope.

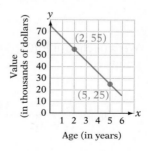

Your solution

$m = -10,000$
A slope of $-10,000$ means the value of the recycling truck is decreasing by $10,000 per year.

Solutions on p. S12

Unless otherwise noted, all content on this page is © Cengage Learning.

OBJECTIVE B | *To graph a line given a point and the slope*

The graph of the equation $y = -\frac{3}{4}x + 4$ is shown at the left. The points with coordinates $(-4, 7)$ and $(4, 1)$ are on the graph. The slope of the line is

$$m = \frac{7 - 1}{-4 - 4} = \frac{6}{-8} = -\frac{3}{4}$$

Note that the slope of the line is the coefficient of x in the equation of the line.

Recall that the y-intercept is found by replacing x by zero and solving for y.

$$y = -\frac{3}{4}x + 4 \qquad\qquad y = -\frac{3}{4}(0) + 4 = 4$$

The coordinates of the y-intercept are $(0, 4)$. Note that the y-coordinate of the y-intercept is the constant term of the equation of the line.

Slope-Intercept Form of a Straight Line

The equation $y = mx + b$ is called the **slope-intercept form** of a straight line. The slope of the line is m, the coefficient of x. The coordinates of the y-intercept are $(0, b)$.

When the equation of a straight line is in the form $y = mx + b$, its graph can be drawn by using the slope and y-intercept. First locate the y-intercept. Use the slope to find a second point on the line. Then draw a line through the two points.

HOW TO 5 Graph $y = \frac{5}{3}x - 4$ by using the slope and y-intercept.

The slope is the coefficient of x: $m = \frac{5}{3} = \frac{\text{change in } y}{\text{change in } x}$.
The coordinates of the y-intercept are $(0, -4)$.

Beginning at the y-intercept, which has coordinates $(0, -4)$, move up 5 units (change in y) and then right 3 units (change in x).

The point with coordinates $(3, 1)$ is a second point on the graph. Draw a line through the points with coordinates $(0, -4)$ and $(3, 1)$.

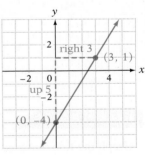

HOW TO 6 Graph $x + 2y = 4$ by using the slope and y-intercept.

Solve the equation for y.

$$x + 2y = 4$$
$$2y = -x + 4$$
$$y = -\frac{1}{2}x + 2 \quad \bullet \; m = -\frac{1}{2} = \frac{-1}{2},$$
$$y\text{-intercept} = (0, 2)$$

Beginning at the y-intercept, which has coordinates $(0, 2)$, move down 1 unit (change in y) and then right 2 units (change in x). The point with coordinates $(2, 1)$ is a second point on the graph. Draw a line through the points with coordinates $(0, 2)$ and $(2, 1)$.

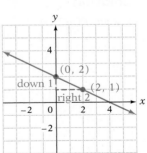

📋 **Take Note**

When graphing a line by using its slope and y-intercept, *always* start at the y-intercept.

INSTRUCTOR NOTE

For consistency, we always graph slope by moving up or down, depending on whether the slope is positive or negative, and then to the right. An interesting class activity is to graph the slope in other ways. For instance, a slope of $-\frac{1}{2}$ can be graphed by moving left 2 and up 1. It can also be graphed by moving right 1 and down $\frac{1}{2}$, right 4 and down 2, or in several other ways. This exercise will help reinforce the concept that slope measures the ratio of the change in y to the change in x.

Unless otherwise noted, all content on this page is © Cengage Learning.

The graph of a line can be drawn when any point on the line and the slope of the line are given.

Take Note

This HOW TO differs from the preceding two in that a point other than the *y*-intercept is used. In this case, start at the given point.

HOW TO 7   Graph the line that passes through $P(-4, -4)$ and has slope 2.

When the slope is an integer, write it as a fraction with denominator 1.

$$m = 2 = \frac{2}{1} = \frac{\text{change in } y}{\text{change in } x}$$

Beginning at the point $P(-4, -4)$, move up 2 units (change in *y*) and then right 1 unit (change in *x*). The point with coordinates $(-3, -2)$ is a second point on the graph. Draw a line through the points with coordinates $(-4, -4)$ and $(-3, -2)$.

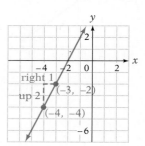

EXAMPLE 4

Graph $y = -\frac{3}{2}x + 4$ by using the slope and *y*-intercept.

Solution

$$m = -\frac{3}{2} = \frac{-3}{2}$$

y-intercept $= (0, 4)$

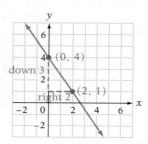

YOU TRY IT 4

Graph $2x + 3y = 6$ by using the slope and *y*-intercept.

Your solution

$$m = -\frac{2}{3}$$

y-intercept: $(0, 2)$

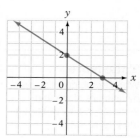

EXAMPLE 5

Graph the line that passes through $P(-2, 3)$ and has slope $-\frac{4}{3}$.

Solution

Locate $P(-2, 3)$.

$$m = -\frac{4}{3} = \frac{-4}{3}$$

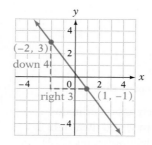

YOU TRY IT 5

Graph the line that passes through $P(-3, -2)$ and has slope 3.

Your solution

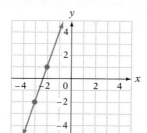

IN-CLASS EXAMPLES

Graph by using the slope and *y*-intercept.

6. $y = \frac{2}{3}x - 2$

7. $y = 2x - 3$

8. $y = -\frac{3}{4}x$

9. Graph the line that passes through $P(-3, -2)$ and has slope 3.

Solutions on p. S12

Unless otherwise noted, all content on this page is © Cengage Learning.

4.4 EXERCISES

✔ Concept Check

SUGGESTED ASSIGNMENT
Exercises 1–6; Exercises 7–57, odds;
More challenging exercises: Exercises 63–66

For Exercises 1 and 2, complete the sentence with *increases* or *decreases*.

1. If a line has positive slope, then as x increases, y ___increases___.

2. If a line has negative slope, then as x ___increases___, y decreases.

For Exercises 3 to 6, determine the slope and y-intercept of the graph of the equation.

3. $y = 3x + 4$

   3; (0, 4)

4. $y = -\dfrac{2}{3}x$

   $-\dfrac{2}{3}$; (0, 0)

5. $y = 1 - x$

   -1; (0, 1)

6. $y = \dfrac{4x}{3} - 2$

   $\dfrac{4}{3}$; (0, −2)

OBJECTIVE A   *To find the slope of a line given two points*

For Exercises 7 to 24, find the slope of the line containing the given points.

7. $P_1(1, 3), P_2(3, 1)$ -1

8. $P_1(2, 3), P_2(5, 1)$ $-\dfrac{2}{3}$

9. $P_1(-1, 4), P_2(2, 5)$ $\dfrac{1}{3}$

10. $P_1(3, -2), P_2(1, 4)$ -3

11. $P_1(-1, 3), P_2(-4, 5)$ $-\dfrac{2}{3}$

12. $P_1(-1, -2), P_2(-3, 2)$ -2

13. $P_1(0, 3), P_2(4, 0)$ $-\dfrac{3}{4}$

14. $P_1(-2, 0), P_2(0, 3)$ $\dfrac{3}{2}$

15. $P_1(2, 4), P_2(2, -2)$
   Undefined

16. $P_1(4, 1), P_2(4, -3)$
   Undefined

17. $P_1(2, 5), P_2(-3, -2)$ $\dfrac{7}{5}$

18. $P_1(4, 1), P_2(-1, -2)$ $\dfrac{3}{5}$

19. $P_1(2, 3), P_2(-1, 3)$ 0

20. $P_1(3, 4), P_2(0, 4)$ 0

21. $P_1(0, 4), P_2(-2, 5)$ $-\dfrac{1}{2}$

22. $P_1(-2, 3), P_2(-2, 5)$
   Undefined

23. $P_1(-3, -1), P_2(-3, 4)$
   Undefined

24. $P_1(-2, -5), P_2(-4, -1)$ -2

25. 🗒 Let l be a line passing through the points $P(a, b)$ and $Q(c, d)$. Which two of a, b, c, and d are equal if the slope of l is undefined? a and c

26. 🗒 Let l be a line passing through the points $P(a, b)$ and $Q(c, d)$. Which two of a, b, c, and d are equal if the slope of l is zero? b and d

27. 🖊 **Travel** The graph below shows the relationship between the distance traveled by a motorist and the time of travel. Find the slope of the line between the two points shown on the graph. Write a sentence that states the meaning of the slope.

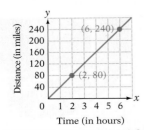

Time (in hours)

$m = 40$
The average speed of the motorist is 40 mph.

28. 🖊 **Media** The graph below shows the number of people subscribing to a sports magazine of increasing popularity. Find the slope of the line between the two points shown on the graph. Write a sentence that states the meaning of the slope.

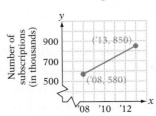

Year

$m = 54{,}000$
Each year, 54,000 subscribers are added.

Unless otherwise noted, all content on this page is © Cengage Learning.

29. 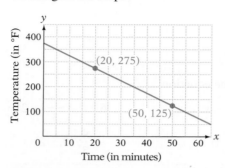 **Temperature**   The graph below shows the relationship between the temperature inside an oven and the time since the oven was turned off. Find the slope of the line. Write a sentence that states the meaning of the slope.

$m = -5$

The temperature of the oven is decreasing 5°/min.

30. **Home Maintenance**   The graph below shows the number of gallons of water remaining in a pool x minutes after a valve is opened to drain the pool. Find the slope of the line. Write a sentence that states the meaning of the slope.

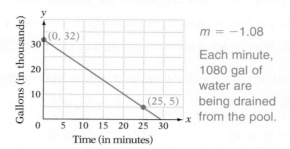

$m = -1.08$

Each minute, 1080 gal of water are being drained from the pool.

31. 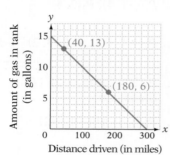 **Fuel Consumption**   The graph below shows how the amount of gas in the tank of a car decreases as the car is driven. Find the slope of the line. Write a sentence that states the meaning of the slope.

$m = -0.05$

For each mile the car is driven, approximately 0.05 gal of fuel is used.

32. **Meteorology**   The troposphere extends from Earth's surface to an elevation of about 11 km. The graph below shows the decrease in the temperature of the troposphere as altitude increases. Find the slope of the line. Write a sentence that states the meaning of the slope.

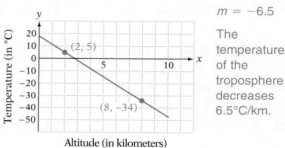

$m = -6.5$

The temperature of the troposphere decreases 6.5°C/km.

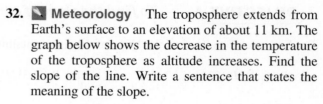

33. 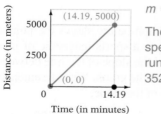 **Sports**   The graph below shows the relationship between distance and time for the world-record 5000-meter run set by Tirunesh Dibaba in 2008. Find the slope of the line between the two points shown on the graph. Round to the nearest tenth. Write a sentence that states the meaning of the slope.

$m \approx 352.4$

The average speed of the runner was 352.4 m/min.

34. 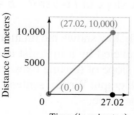 **Sports**   The graph below shows the relationship between distance and time for the Olympic-record 10,000-meter run set by Kenenisa Bekele in 2008. Find the slope of the line between the two points shown on the graph. Round to the nearest tenth. Write a sentence that states the meaning of the slope.

$m \approx 370.1$

The average speed of the runner was 370.1 m/min.

35. **Construction**   The American National Standards Institute (ANSI) states that the slope of a wheelchair ramp must not exceed $\frac{1}{12}$.

a. Does a ramp that is 6 in. high and 5 ft long meet the requirements of ANSI?   No
b. Does a ramp that is 12 in. high and 170 in. long meet the requirements of ANSI?   Yes

nicobatista/Shutterstock.com

36. ● **Solar Roof** Look at the butterfly roof design shown in the article below. Which side of the roof, the left or the right, has a slope of approximately 1? Right side

In the NEWS!

University of Maryland Wins Solar Decathlon

A team of students from the University of Maryland designed the winning solar house in this year's Solar Decathlon, sponsored by the U.S. Department of Energy. The winning home, named WaterShed, makes use of a butterfly roof, a design that combines two sections that slant toward each other at different angles and is ideal for the use of solar panels.

Source: www.news.cnet.com

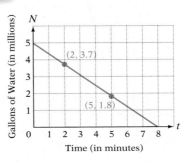

37. ◣ ● **Panama Canal** Ships in the Panama Canal are lowered through a series of locks. A ship is lowered as the water in a lock is discharged. The graph at the right shows the number of gallons of water N, in millions, remaining in a lock t minutes after the valves are opened to discharge the water. Find the slope of the line. Write a sentence that explains the meaning of the slope.

$m = -\dfrac{19}{30}$. Each minute, the water in the lock decreases by $0.6\overline{3}$ million gallons.

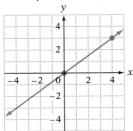

OBJECTIVE B *To graph a line given a point and the slope*

For Exercises 38 to 49, graph by using the slope and the y-intercept.

38. $y = \dfrac{2}{3}x - 3$

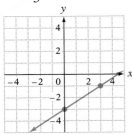

39. $y = \dfrac{1}{2}x + 2$

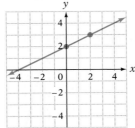

40. $y = \dfrac{3}{4}x$

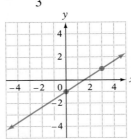

41. $y = -\dfrac{3}{2}x$

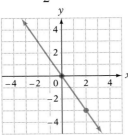

42. $y = \dfrac{2}{3}x - 1$

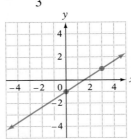

43. $y = -\dfrac{1}{2}x + 2$

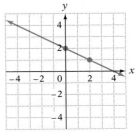

Unless otherwise noted, all content on this page is © Cengage Learning.

44. $y = -3x + 1$

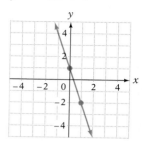

45. $y = 2x - 4$

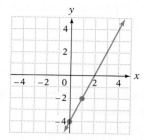

46. $4x + y = 2$

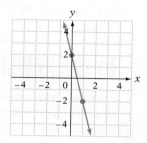

47. $4x - y = 1$

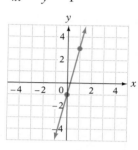

48. $3x + 2y = 8$

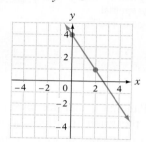

49. $x - 3y = 3$

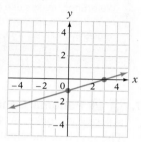

For Exercises 50 to 55, graph the line that passes through the given point and has the given slope.

50. $P(-2, -3)$; slope $\dfrac{5}{4}$

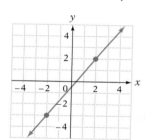

51. $P(-1, -3)$; slope $\dfrac{4}{3}$

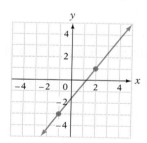

52. $P(2, -4)$; slope $-\dfrac{1}{2}$

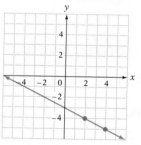

53. $P(-3, 0)$; slope -3

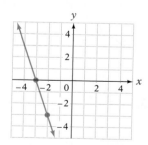

54. $P(1, 5)$; slope -4

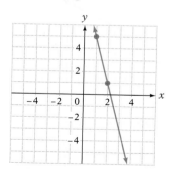

55. $P(-4, 1)$; slope $\dfrac{2}{3}$

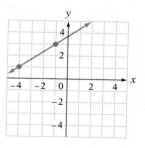

For Exercises 56 and 57, for the given conditions, state whether the graph of $Ax + By = C$ **a.** has its y-intercept above or below the x-axis, and **b.** has positive or negative slope.

56. A and C are positive numbers, and B is a negative number. **a.** Below **b.** Positive

57. A and B are positive numbers, and C is a negative number. **a.** Below **b.** Negative

Unless otherwise noted, all content on this page is © Cengage Learning.

Critical Thinking

58. Match each equation with its graph.

i. $y = -2x + 4$ D

ii. $y = 2x - 4$ C

iii. $y = 2$ B

iv. $2x + 4y = 0$ F

v. $y = \dfrac{1}{2}x + 4$ E

vi. $y = -\dfrac{1}{4}x - 2$ A

A. 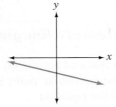   **B.** 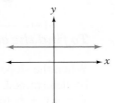   **C.**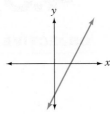

D. 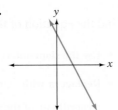   **E.** 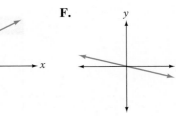   **F.**

For Exercises 59 to 62, complete the sentences using *increases* or *decreases* in the first blank and a positive number in the second blank.

59. If a line has a slope of 2, then the value of y __increases__ by __2__ as the value of x increases by 1.

60. If a line has a slope of -3, then the value of y __decreases__ by __3__ as the value of x increases by 1.

61. If a line has a slope of -2, then the value of y __increases__ by __2__ as the value of x decreases by 1.

62. If a line has a slope of 3, then the value of y __decreases__ by __3__ as the value of x decreases by 1.

QUICK QUIZ

Find the slope of the line containing the given points.

1. $P_1(1, -3)$, $P_2(5, -13)$

   $-\dfrac{5}{2}$ **[4.4A]**

2. $P_1(2, 3)$, $P_2(2, -1)$
   Undefined **[4.4A]**

3. Graph $y = \dfrac{1}{2}x - 1$ by using the slope and y-intercept. **[4.4B]**

Projects or Group Activities

63. Explain how you can use the slope of a line to determine whether three given points lie on the same line. Then use your procedure to determine whether each of the following sets of points lie on the same line.
  a. $(2, 5)$, $(-1, -1)$, $(3, 7)$ **b.** $(-1, 5)$, $(0, 3)$, $(-3, 4)$

For Exercises 64 to 66, determine the value of k such that the points whose coordinates are given lie on the same line.

64. $(3, 2)$, $(4, 6)$, $(5, k)$

$k = 10$

65. $(k, 1)$, $(0, -1)$, $(2, -2)$

$k = -4$

66. $(4, -1)$, $(3, -4)$, (k, k)

$k = \dfrac{13}{2}$

Unless otherwise noted, all content on this page is © Cengage Learning.

SECTION

4.5   Finding Equations of Lines

OBJECTIVE A   *To find the equation of a line given a point and the slope*

When the slope of a line and a point on the line are known, the equation of the line can be determined. If the particular point is the y-intercept, use the slope-intercept form, $y = mx + b$, to find the equation.

HOW TO 1   Find the equation of the line with y-intercept $P(0, 3)$ and slope $\frac{1}{2}$.

$$y = mx + b$$ • Use the slope-intercept form.

$$y = \frac{1}{2}x + 3$$ • Replace *m* with $\frac{1}{2}$, the given slope. Replace *b* with 3, the y-coordinate of the y-intercept.

The equation of the line is $y = \frac{1}{2}x + 3$.

One method of finding the equation of a line when the slope and *any* point on the line are known involves using the *point-slope formula*. This formula is derived from the formula for the slope of a line as follows.

Let $P_1(x_1, y_1)$ be the given point on the line, and let $P(x, y)$ be any other point on the line. See the graph at the left.

$$\frac{y - y_1}{x - x_1} = m$$ • Use the formula for the slope of a line.

$$\frac{y - y_1}{x - x_1}(x - x_1) = m(x - x_1)$$ • Multiply each side by $(x - x_1)$.

$$y - y_1 = m(x - x_1)$$ • Simplify.

Point-Slope Formula

Let *m* be the slope of a line, and let $P_1(x_1, y_1)$ be a point on the line. The equation of the line can be found from the **point-slope formula:**

$$y - y_1 = m(x - x_1)$$

HOW TO 2   Find the equation of the line that contains the point $P(4, -1)$ and has slope $-\frac{3}{4}$.

$$y - y_1 = m(x - x_1)$$ • Use the point-slope formula.

$$y - (-1) = \left(-\frac{3}{4}\right)(x - 4)$$ • $m = -\frac{3}{4}$, $(x_1, y_1) = (4, -1)$

$$y + 1 = -\frac{3}{4}x + 3$$ • Simplify.

$$y = -\frac{3}{4}x + 2$$ • Write the equation in the form $y = mx + b$.

INSTRUCTOR NOTE

Write the point-slope formula using parentheses as shown below. This will help some students make the correct substitutions, especially when a negative number is substituted.

$$y - y_1 = m(x - x_1)$$
$$\downarrow \quad \downarrow \qquad \downarrow$$
$$y - (\ ) = (\ )[x - (\ )]$$

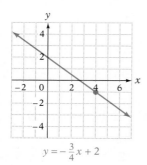

$$y = -\frac{3}{4}x + 2$$

Unless otherwise noted, all content on this page is © Cengage Learning.

IN-CLASS EXAMPLES

Find the equation of the line that contains the given point and has the given slope.

1. $P(0, 5)$, $m = -3$
   $y = -3x + 5$
2. $P(-2, 3)$, $m = -1$
   $y = -x + 1$
3. $P(4, -1)$, $m = \dfrac{3}{4}$
   $y = \dfrac{3}{4}x - 4$

HOW TO 3   Find the equation of the line that passes through the point $P(4, 3)$ and whose slope is undefined.

Because the slope is undefined, the point-slope formula cannot be used to find the equation. Instead, recall that when the slope of a line is undefined, the line is vertical. The equation of a vertical line is $x = a$, where a is the x-coordinate of the x-intercept. Because the line is vertical and passes through $P(4, 3)$, the x-intercept has coordinates $P(4, 0)$.
The equation of the line is $x = 4$.

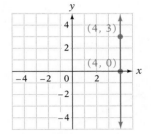

EXAMPLE 1

Find the equation of the line that contains the point $P(3, 0)$ and has slope -4.

Solution

$m = -4 \qquad (x_1, y_1) = (3, 0)$

$y - y_1 = m(x - x_1)$
$y - 0 = -4(x - 3)$
$\quad\; y = -4x + 12$

The equation of the line is $y = -4x + 12$.

YOU TRY IT 1

Find the equation of the line that contains the point $P(-3, -2)$ and has slope $-\frac{1}{3}$.

Your solution

$y = -\dfrac{1}{3}x - 3$

EXAMPLE 2

Find the equation of the line that contains the point $P(-2, 4)$ and has slope 2.

Solution

$m = 2 \qquad (x_1, y_1) = (-2, 4)$

$y - y_1 = m(x - x_1)$
$y - 4 = 2[x - (-2)]$
$y - 4 = 2(x + 2)$
$y - 4 = 2x + 4$
$\quad\; y = 2x + 8$

The equation of the line is $y = 2x + 8$.

YOU TRY IT 2

Find the equation of the line that contains the point $P(4, -3)$ and has slope -3.

Your solution

$y = -3x + 9$

Solutions on p. S12

Unless otherwise noted, all content on this page is © Cengage Learning.

OBJECTIVE B *To find the equation of a line given two points*

The point-slope formula and the formula for slope are used to find the equation of a line when two points are known.

HOW TO 4 Find the equation of the line containing $P_1(3, 2)$ and $P_2(-5, 6)$.

To use the point-slope formula, we must know the slope. Use the formula for slope to determine the slope of the line between the two given points.

$$m = \frac{y_2 - y_1}{x_2 - x_1} = \frac{6 - 2}{-5 - 3} = \frac{4}{-8} = -\frac{1}{2}$$

Now use the point-slope formula with $m = -\frac{1}{2}$ and $(x_1, y_1) = (3, 2)$.

$$y - y_1 = m(x - x_1)$$ • Use the point-slope formula.

$$y - 2 = \left(-\frac{1}{2}\right)(x - 3)$$ • $m = -\frac{1}{2}, (x_1, y_1) = (3, 2)$

$$y - 2 = -\frac{1}{2}x + \frac{3}{2}$$ • Simplify.

$$y = -\frac{1}{2}x + \frac{7}{2}$$ • Solve for y.

The equation of the line is $y = -\frac{1}{2}x + \frac{7}{2}$.

$y = -\frac{1}{2}x + \frac{7}{2}$

INSTRUCTOR NOTE
You might have students use (x_2, y_2) in the point-slope formula in HOW TO 4 and verify that the same equation results.

EXAMPLE 3

Find the equation of the line passing through the points $P_1(2, 3)$ and $P_2(4, 1)$.

Solution

$$m = \frac{y_2 - y_1}{x_2 - x_1} = \frac{1 - 3}{4 - 2} = \frac{-2}{2} = -1$$

$$y - y_1 = m(x - x_1)$$
$$y - 3 = -1(x - 2)$$
$$y - 3 = -x + 2$$
$$y = -x + 5$$

The equation of the line is $y = -x + 5$.

YOU TRY IT 3

Find the equation of the line passing through the points $P_1(2, 0)$ and $P_2(5, 3)$.

Your solution

$y = x - 2$

IN-CLASS EXAMPLES
Find the equation of the line that contains the given points.
4. $P_1(1, 3)$, $P_2(2, 4)$ $y = x + 2$
5. $P_1(0, -2)$, $P_2(-3, 2)$ $y = -\frac{4}{3}x - 2$
6. $P_1(3, -3)$, $P_2(2, -3)$ $y = -3$

EXAMPLE 4

Find the equation of the line containing $P_1(2, -3)$ and $P_2(2, 5)$.

Solution

$$m = \frac{y_2 - y_1}{x_2 - x_1} = \frac{5 - (-3)}{2 - 2} = \frac{8}{0}$$

The slope is undefined. The graph is a vertical line.

The equation of the line is $x = 2$.

YOU TRY IT 4

Find the equation of the line containing $P_1(2, 3)$ and $P_2(-5, 3)$.

Your solution

$y = 3$

Solutions on pp. S12–S13

Unless otherwise noted, all content on this page is © Cengage Learning.

OBJECTIVE C   *To solve application problems*

Linear functions can be used to model a variety of applications in science and business. For each application, data are collected and the independent and dependent variables are selected. Then a linear function that models the data is determined.

EXAMPLE 5

Suppose a manufacturer has determined that at a price of $150, consumers will purchase 1 million portable music players, and at a price of $125, consumers will purchase 1.25 million portable music players. Describe this situation with a linear function. Use the function to predict how many portable music players consumers will purchase if the price is $80.

Strategy

- Select the independent and dependent variables. Because you are trying to determine the number of portable music players, that quantity is the *dependent* variable, y. The price of a portable music player is the *independent* variable, x.
- From the given data, two ordered pairs are (150, 1) and (125, 1.25). (The ordinates are in millions of units.) Use these ordered pairs to determine the linear function.
- Evaluate the function when $x = 80$ to predict how many portable music players consumers will purchase if the price is $80.

Solution

Choose $P_1(150, 1)$ and $P_2(125, 1.25)$.

$$m = \frac{y_2 - y_1}{x_2 - x_1} = \frac{1.25 - 1}{125 - 150} = \frac{0.25}{-25} = -0.01$$

$$y - y_1 = m(x - x_1)$$

$$y - 1 = -0.01(x - 150)$$

$$y = -0.01x + 2.50$$

The linear function is $f(x) = -0.01x + 2.50$.

$$f(80) = -0.01(80) + 2.50 = 1.7$$

Consumers will purchase 1.7 million portable music players at a price of $80.

YOU TRY IT 5

Gabriel Daniel Fahrenheit invented the mercury thermometer in 1717. In terms of readings on this thermometer, water freezes at 32°F and boils at 212°F. In 1742, Anders Celsius invented the Celsius temperature scale. On this scale, water freezes at 0°C and boils at 100°C. Determine a linear function that can be used to predict the Celsius temperature when the Fahrenheit temperature is known.

Your strategy

Your solution

$$f(F) = \frac{5}{9}(F - 32)$$

IN-CLASS EXAMPLES

7. In 2000, there were 50,000 people 100 years old or older in the United States. Data from the Census Bureau show that the population of this group is expected to increase through the year 2020 at a rate of approximately 4250 people per year.
   a. Find a linear function that approximates the population, in millions, of people 100 years old or older in terms of the year.
      $f(x) = 4250x - 8,450,000$
   b. Use your function to approximate the population of this group of adults in 2015.   **113,750 people**

Solution on p. S13

4.5 EXERCISES

✔ **Concept Check**

SUGGESTED ASSIGNMENT
Exercises 1–4; Exercises 5–93, odds;
More challenging exercises: Exercises 95–103, odds

1. How many lines with a given slope can be drawn through a given point in the plane?   One

2. Given two points in the plane, how many lines can be drawn through the two points?   One

3. If you know the slope of a line, what other information would you need to know to find the equation of that line?   A point on the line

4. If you know a point on a line, what other information would you need to know to find the equation of that line?   Another point on the line or the slope of the line

OBJECTIVE A   *To find the equation of a line given a point and the slope*

5. ◩ Explain how to find the equation of a line given its slope and its y-intercept.

6. ◩ What is the point-slope formula and how is it used?

7. ◪ Through what point must the graph of the equation $y = mx$ pass?   (0, 0)

8. ◪ After you find an equation of a line given its slope and the coordinates of a point on the line, how can you determine whether you have the correct equation?   Check that the coefficient of x is the given slope. Check that the coordinates of the given point are a solution of your equation.

For Exercises 9 to 44, find the equation of the line that contains the given point and has the given slope.

9. $P(0, 5)$, $m = 2$

$y = 2x + 5$

10. $P(0, 3)$, $m = 1$

$y = x + 3$

11. $P(2, 3)$, $m = \dfrac{1}{2}$

$y = \dfrac{1}{2}x + 2$

12. $P(5, 1)$, $m = \dfrac{2}{3}$

$y = \dfrac{2}{3}x - \dfrac{7}{3}$

13. $P(-1, 4)$, $m = \dfrac{5}{4}$

$y = \dfrac{5}{4}x + \dfrac{21}{4}$

14. $P(-2, 1)$, $m = \dfrac{3}{2}$

$y = \dfrac{3}{2}x + 4$

15. $P(3, 0)$, $m = -\dfrac{5}{3}$

$y = -\dfrac{5}{3}x + 5$

16. $P(-2, 0)$, $m = \dfrac{3}{2}$

$y = \dfrac{3}{2}x + 3$

17. $P(2, 3)$, $m = -3$

$y = -3x + 9$

18. $P(1, 5)$, $m = -\dfrac{4}{5}$

$y = -\dfrac{4}{5}x + \dfrac{29}{5}$

19. $P(-1, 7)$, $m = -3$

$y = -3x + 4$

20. $P(-2, 4)$, $m = -4$

$y = -4x - 4$

21. $P(-1, -3)$, $m = \dfrac{2}{3}$

$y = \dfrac{2}{3}x - \dfrac{7}{3}$

22. $P(-2, -4)$, $m = \dfrac{1}{4}$

$y = \dfrac{1}{4}x - \dfrac{7}{2}$

23. $P(0, 0)$, $m = \dfrac{1}{2}$

$y = \dfrac{1}{2}x$

24. $P(0, 0)$, $m = \dfrac{3}{4}$

$y = \dfrac{3}{4}x$

25. $P(2, -3)$, $m = 3$

$y = 3x - 9$

26. $P(4, -5)$, $m = 2$

$y = 2x - 13$

27. $P(3, 5)$, $m = -\dfrac{2}{3}$

$y = -\dfrac{2}{3}x + 7$

28. $P(5, 1)$, $m = -\dfrac{4}{5}$

$y = -\dfrac{4}{5}x + 5$

29. $P(0, -3), m = -1$

$y = -x - 3$

30. $P(2, 0), m = \dfrac{5}{6}$

$y = \dfrac{5}{6}x - \dfrac{5}{3}$

31. $P(1, -4), m = \dfrac{7}{5}$

$y = \dfrac{7}{5}x - \dfrac{27}{5}$

32. $P(3, 5), m = -\dfrac{3}{7}$

$y = -\dfrac{3}{7}x + \dfrac{44}{7}$

33. $P(4, -1), m = -\dfrac{2}{5}$

$y = -\dfrac{2}{5}x + \dfrac{3}{5}$

34. $P(-3, 5), m = -\dfrac{1}{4}$

$y = -\dfrac{1}{4}x + \dfrac{17}{4}$

35. $P(3, -4)$, slope is undefined

$x = 3$

36. $P(-2, 5)$, slope is undefined

$x = -2$

37. $P(-2, -5), m = -\dfrac{5}{4}$

$y = -\dfrac{5}{4}x - \dfrac{15}{2}$

38. $P(-3, -2), m = -\dfrac{2}{3}$

$y = -\dfrac{2}{3}x - 4$

39. $P(-2, -3), m = 0$

$y = -3$

40. $P(-3, -2), m = 0$

$y = -2$

41. $P(4, -5), m = -2$

$y = -2x + 3$

42. $P(-3, 5), m = 3$

$y = 3x + 14$

43. $P(-5, -1)$, slope is undefined

$x = -5$

44. $P(0, 4)$, slope is undefined

$x = 0$

> **OBJECTIVE B**   *To find the equation of a line given two points*

45. After you find an equation of a line given the coordinates of two points on the line, how can you determine whether you have the correct equation?
Check that the coordinates of each given point are a solution of your equation.

46. If you are asked to find the equation of a line through two given points, does it matter which point is selected as (x_1, y_1) and which point is selected as (x_2, y_2)?
No. This is similar to finding the slope of a line between two points.

For Exercises 47 to 82, find the equation of the line that contains the given points.

47. $P_1(0, 2), P_2(3, 5)$

$y = x + 2$

48. $P_1(0, 4), P_2(1, 5)$

$y = x + 4$

49. $P_1(0, -3), P_2(-4, 5)$

$y = -2x - 3$

50. $P_1(0, -2), P_2(-3, 4)$

$y = -2x - 2$

51. $P_1(2, 3), P_2(5, 5)$

$y = \dfrac{2}{3}x + \dfrac{5}{3}$

52. $P_1(4, 1), P_2(6, 3)$

$y = x - 3$

53. $P_1(-1, 3), P_2(2, 4)$

$y = \dfrac{1}{3}x + \dfrac{10}{3}$

54. $P_1(-1, 1), P_2(4, 4)$

$y = \dfrac{3}{5}x + \dfrac{8}{5}$

55. $P_1(-1, -2), P_2(3, 4)$

$y = \dfrac{3}{2}x - \dfrac{1}{2}$

56. $P_1(-3, -1), P_2(2, 4)$

$y = x + 2$

57. $P_1(0, 3), P_2(2, 0)$

$y = -\dfrac{3}{2}x + 3$

58. $P_1(0, 4), P_2(2, 0)$

$y = -2x + 4$

59. $P_1(-3, -1), P_2(2, -1)$

$y = -1$

60. $P_1(-3, -5), P_2(4, -5)$

$y = -5$

61. $P_1(-2, -3), P_2(-1, -2)$

$y = x - 1$

62. $P_1(4, 1), P_2(3, -2)$

$y = 3x - 11$

63. $P_1(-2, 3), P_2(2, -1)$

$y = -x + 1$

64. $P_1(3, 1), P_2(-3, -2)$

$y = \dfrac{1}{2}x - \dfrac{1}{2}$

65. $P_1(2, 3), P_2(5, -5)$

$y = -\dfrac{8}{3}x + \dfrac{25}{3}$

66. $P_1(7, 2), P_2(4, 4)$

$y = -\dfrac{2}{3}x + \dfrac{20}{3}$

67. $P_1(2, 0), P_2(0, -1)$

$y = \dfrac{1}{2}x - 1$

68. $P_1(0, 4), P_2(-2, 0)$

$y = 2x + 4$

69. $P_1(3, -4), P_2(-2, -4)$

$y = -4$

70. $P_1(-3, 3), P_2(-2, 3)$

$y = 3$

71. $P_1(0, 0), P_2(4, 3)$

$y = \dfrac{3}{4}x$

72. $P_1(2, -5), P_2(0, 0)$

$y = -\dfrac{5}{2}x$

73. $P_1(2, -1), P_2(-1, 3)$

$y = -\dfrac{4}{3}x + \dfrac{5}{3}$

74. $P_1(3, -5), P_2(-2, 1)$

$y = -\dfrac{6}{5}x - \dfrac{7}{5}$

75. $P_1(-2, 5), P_2(-2, -5)$

$x = -2$

76. $P_1(3, 2), P_2(3, -4)$

$x = 3$

77. $P_1(2, 1), P_2(-2, -3)$

$y = x - 1$

78. $P_1(-3, -2), P_2(1, -4)$

$y = -\dfrac{1}{2}x - \dfrac{7}{2}$

79. $P_1(-4, -3), P_2(2, 5)$

$y = \dfrac{4}{3}x + \dfrac{7}{3}$

80. $P_1(4, 5), P_2(-4, 3)$

$y = \dfrac{1}{4}x + 4$

81. $P_1(0, 3), P_2(3, 0)$

$y = -x + 3$

82. $P_1(1, -3), P_2(-2, 4)$

$y = -\dfrac{7}{3}x - \dfrac{2}{3}$

OBJECTIVE C　　*To solve application problems*

83. Aviation　The pilot of a Boeing 777 jet takes off from Boston's Logan Airport, which is at sea level, and climbs to a cruising altitude of 32,000 ft at a constant rate of 1200 ft/min.
a. Write a linear function for the height of the plane in terms of the time after take-off.
b. Use your function to find the height of the plane 11 min after takeoff.

a. $f(x) = 1200x, 0 \le x \le 26\dfrac{2}{3}$　　**b.** 13,200 ft

Christopher Parypa/Shutterstock.com

84. Calories　A jogger running at 9 mph burns approximately 14 Calories per minute.
a. Write a linear function for the number of Calories burned by the jogger in terms of the number of minutes run.　　$f(x) = 14x$
b. Use your function to find the number of Calories the jogger has burned after jogging for 32 min.　　448 Calories

85. **● Ecology**   Use the information in the article at the right.
   a. Determine a linear function for the percent of hardwood trees at 2600 ft in terms of the year. $f(x) = 0.625x - 1170.5$
   b. Use your function to predict the percent of trees at 2600 ft that will be hardwoods in 2020.   92%

86. **Telecommunications**   A cellular phone company offers several different service options. One option, for people who plan on using the phone only in emergencies, costs the user $4.95 per month plus $.59 per minute for each minute the phone is used.
   a. Write a linear function for the monthly cost of the phone in terms of the number of minutes the phone is used. $f(x) = 0.59x + 4.95$
   b. Use your function to find the cost of using the cellular phone for 13 min in one month. $12.62

87. **Fuel Consumption**   The gas tank of a certain car contains 16 gal when the driver of the car begins a trip. For each mile driven by the driver, the amount of gas in the tank decreases by 0.032 gal.
   a. Write a linear function for the number of gallons of gas in the tank in terms of the number of miles driven. $f(x) = -0.032x + 16, 0 \le x \le 500$
   b. Use your function to find the number of gallons in the tank after driving 150 mi.   11.2 gal

88. **● Boiling Points**   At sea level, the boiling point of water is 100°C. At an altitude of 2 km, the boiling point of water is 93°C.
   a. Write a linear function for the boiling point of water in terms of the altitude above sea level. $f(x) = -3.5x + 100$
   b. Use your function to predict the boiling point of water on top of Mount Everest, which is approximately 8.85 km above sea level. Round to the nearest degree.   69°C

89. **Business**   A manufacturer of motorcycles has determined that 50,000 motorcycles per month can be sold at a price of $9000. At a price of $8750, the number of motorcycles sold per month increases to 55,000.
   a. Determine a linear function that predicts the number of motorcycles that will be sold each month at a given price. $f(x) = -20x + 230,000$
   b. Use this model to predict the number of motorcycles that will be sold at a price of $8500.   60,000 motorcycles

90. **Business**   A manufacturer of graphing calculators has determined that 10,000 calculators per week will be sold at a price of $95. At a price of $90, it is estimated that 12,000 calculators will be sold.
   a. Determine a linear function that predicts the number of calculators that will be sold each week at a given price. $f(x) = -400x + 48,000$
   b. Use this model to predict the number of calculators that will be sold each week at a price of $75.   18,000 calculators

91. **Calories**   There are approximately 126 Calories in a 2-ounce serving of lean hamburger and approximately 189 Calories in a 3-ounce serving.
   a. Determine a linear function for the number of Calories in lean hamburger in terms of the size of the serving. $f(x) = 63x$
   b. Use your function to estimate the number of Calories in a 5-ounce serving of lean hamburger.   315 Calories

92. **Compensation**   An account executive receives a base salary plus a commission. On $20,000 in monthly sales, the account executive receives $1800. On $50,000 in monthly sales, the account executive receives $3000.
   a. Determine a linear function that will yield the compensation of the sales executive for a given amount of monthly sales. $f(x) = 0.04x + 1000$
   b. Use this model to determine the account executive's compensation for $85,000 in monthly sales. $4400

In the NEWS!

Is Global Warming Moving Mountains?

In the mountains of Vermont, maples, beeches, and other hardwood trees that thrive in warm climates are gradually taking over areas that once supported more cold-loving trees, such as balsam and fir. Ecologists report that in 2004, 82% of the trees at an elevation of 2600 ft were hardwoods, as compared to only 57% in 1964.
Source: The Boston Globe

Mt. Everest

© iStockphoto.com/fotoVoyager

93. Refer to Exercise 90. Describe how you could use the linear function found in part (a) to find the price at which the manufacturer should sell the calculators in order to sell 15,000 calculators a week.   Substitute 15,000 for $f(x)$ and solve for x.

94. Refer to Exercise 92. Describe how you could use the linear function found in part (a) to find the monthly sales the executive would need to make to earn a commission of $6000 a month.   Substitute 6000 for $f(x)$ and solve for x.

95. Let f be a linear function. If $f(2) = 5$ and $f(0) = 3$, find $f(x)$.
$f(x) = x + 3$

96. Let f be a linear function. If $f(-3) = 4$ and $f(1) = -8$, find $f(x)$.
$f(x) = -3x - 5$

97. Let f be a linear function for which $f(1) = 3$ and $f(-1) = 5$. Determine $f(4)$.   0

98. Let f be a linear function for which $f(-3) = 2$ and $f(2) = 7$. Determine $f(0)$.   5

99. A line with slope $\frac{4}{3}$ passes through the point $P(3, 2)$.
   a. What is y when $x = -6$? -10
   b. What is x when $y = 6$?   6

100. A line with slope $-\frac{3}{4}$ passes through the point $P(8, -2)$.
   a. What is y when $x = -4$?   7
   b. What is x when $y = 1$?   4

Critical Thinking

101. A line contains the points $P_1(-3, 6)$ and $P_2(6, 0)$. Find the coordinates of three other points on this line.
Answers will vary. Possible answers are $(0, 4)$, $(3, 2)$, and $(9, -2)$.

102. A line contains the points $P_1(4, -1)$ and $P_2(2, 1)$. Find the coordinates of three other points on this line.
Answers will vary. Possible answers are $(0, 3)$, $(1, 2)$, and $(3, 0)$.

103. Find the equation of the line that passes through the midpoint of the line segment between $P_1(2, 5)$ and $P_2(-4, 1)$, and has slope -2.
$y = -2x + 1$

104. If $y = mx + b$, where m is a given constant, how does the graph of the equation change as the value of b changes?   Changing b moves the graph up or down.

Projects or Group Activities

105. Assume that the maximum speed your car will attain is a linear function of the steepness of the hill it is climbing or descending. If the steepness of the hill is 5° up (the road makes a 5° angle with the horizontal), your car's maximum speed is 77 km/h. If the steepness of the hill is 2° down, your car's maximum speed is 154 km/h. When your car's maximum speed is 99 km/h, how steep is the hill? State your answer in degrees, and note whether the steepness is up or down.   3° up

QUICK QUIZ
Find the equation of the line that contains the given point and has the given slope.

1. $P(0, 4)$, $m = -\frac{1}{2}$
   $y = -\frac{1}{2}x + 4$ [4.5A]

2. $P(2, 3)$, $m = -2$
   $y = -2x + 7$ [4.5A]

Find the equation of the line that contains the given points.

3. $(0, -1)$, $(-2, 3)$
   $y = -2x - 1$ [4.5B]

4. $(0, 3)$, $(2, 4)$
   $y = \frac{1}{2}x + 3$ [4.5B]

5. An Airbus 320 plane takes off from Denver International Airport, which is 5200 ft above sea level, and climbs to 30,000 ft at a constant rate of 1000 ft/min.
   a. Write a linear equation for the height of the plane in terms of the time after takeoff.
   $f(x) = 1000x + 5200$;
   $0 \le x \le 24.8$ [4.5C]
   b. Use your equation to find the height of the plane 8 min after takeoff.   **13,200 ft**
   [4.5C]

SECTION

4.6   Parallel and Perpendicular Lines

OBJECTIVE A   *To find parallel and perpendicular lines*

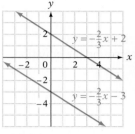

Figure 1

Two lines that have the same slope and different *y*-intercepts do not intersect and are called **parallel lines.**

Slopes of Parallel Lines

Two nonvertical lines with slopes of m_1 and m_2 are parallel if and only if $m_1 = m_2$. Vertical lines are parallel lines.

EXAMPLES

1. The graphs of $y = -\frac{2}{3}x + 2$ and $y = -\frac{2}{3}x - 3$ have the same slope, $-\frac{2}{3}$. The lines are parallel. See Figure 1 at the left.
2. The graphs of $x = 2$ and $x = 5$ are vertical lines. The lines are parallel. See Figure 2 at the left.

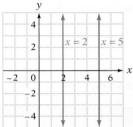

Figure 2

HOW TO 1   Is the line containing the points $P_1(-2, 1)$ and $P_2(-5, -1)$ parallel to the line that contains the points $Q_1(1, 0)$ and $Q_2(4, 2)$?

$$m_1 = \frac{-1 - 1}{-5 - (-2)} = \frac{-2}{-3} = \frac{2}{3}$$
• Find the slope of the line through $P_1(-2, 1)$ and $P_2(-5, -1)$.

$$m_2 = \frac{2 - 0}{4 - 1} = \frac{2}{3}$$
• Find the slope of the line through $Q_1(1, 0)$ and $Q_2(4, 2)$.

Because $m_1 = m_2$, the lines are parallel.

HOW TO 2   Find the equation of the line that contains the point $P(2, 3)$ and is parallel to the graph of $y = \frac{1}{2}x - 4$.

From the equation $y = \frac{1}{2}x - 4$, the slope of the given line is $\frac{1}{2}$. Because parallel lines have the same slope, the slope of the unknown line is also $\frac{1}{2}$.

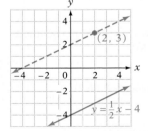

$$y - y_1 = m(x - x_1)$$
• Use the point-slope formula.

$$y - 3 = \frac{1}{2}(x - 2)$$
• $m = \frac{1}{2}$, $(x_1, y_1) = (2, 3)$

$$y - 3 = \frac{1}{2}x - 1$$
• Simplify.

$$y = \frac{1}{2}x + 2$$
• Write the equation in the form $y = mx + b$.

The equation of the line is $y = \frac{1}{2}x + 2$.

Unless otherwise noted, all content on this page is © Cengage Learning.

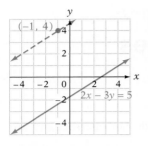

INSTRUCTOR NOTE

Encourage students to use the model given in the last section.

$$y - y_1 = m(x - x_1)$$
$$\downarrow \quad \downarrow \quad \downarrow$$
$$y - (\ ) = (\ )[x - (\ )]$$

If the lines are parallel, m is the same as the slope of the given line. If the lines are perpendicular, m is the negative reciprocal of the slope of the given line.

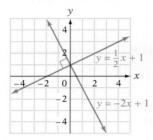

Figure 3

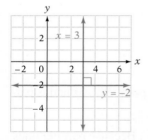

Figure 4

HOW TO 3 Find the equation of the line that contains the point $P(-1, 4)$ and is parallel to the graph of $2x - 3y = 5$.

Because the lines are parallel, the slope of the unknown line is the same as the slope of the given line. Solve $2x - 3y = 5$ for y and determine the slope of its graph.

$$2x - 3y = 5$$
$$-3y = -2x + 5$$
$$y = \frac{2}{3}x - \frac{5}{3}$$

The slope of the given line is $\frac{2}{3}$. Because the lines are parallel, this is the slope of the unknown line. Use the point-slope formula to determine the equation.

$$y - y_1 = m(x - x_1)$$ • Use the point-slope formula.

$$y - 4 = \frac{2}{3}[x - (-1)]$$ • $m = \frac{2}{3}$, $(x_1, y_1) = (-1, 4)$

$$y - 4 = \frac{2}{3}x + \frac{2}{3}$$ • Simplify.

$$y = \frac{2}{3}x + \frac{14}{3}$$ • Write the equation in the form $y = mx + b$.

The equation of the line is $y = \frac{2}{3}x + \frac{14}{3}$.

Two lines that intersect at right angles are called **perpendicular lines.**

Slopes of Perpendicular Lines

Two nonvertical lines with slopes m_1 and m_2 are perpendicular if and only if the product of the slopes is -1. This can be written $m_1 \cdot m_2 = -1$. A vertical line is perpendicular to a horizontal line.

EXAMPLES

1. The slope of the graph of $y = \frac{1}{2}x + 1$ is $m_1 = \frac{1}{2}$. The slope of the graph of $y = -2x + 1$ is $m_2 = -2$. The product of the two slopes is $\frac{1}{2} \cdot (-2) = -1$. The lines are perpendicular. See Figure 3 at the left.
2. The graph of $x = 3$ is a vertical line, and the graph of $y = -2$ is a horizontal line. The lines are perpendicular. See Figure 4 at the left.

Solving $m_1 \cdot m_2 = -1$ for m_1 gives $m_1 = -\dfrac{1}{m_2}$. This last equation states that the slopes of perpendicular lines are **negative reciprocals** of each other.

HOW TO 4 Is the line that contains the points $P_1(4, 2)$ and $P_2(-2, 5)$ perpendicular to the line that contains the points $Q_1(-4, 3)$ and $Q_2(-3, 5)$?

$$m_1 = \frac{5 - 2}{-2 - 4} = \frac{3}{-6} = -\frac{1}{2}$$ • Find the slope of the line through $P_1(4, 2)$ and $P_2(-2, 5)$.

$$m_2 = \frac{5 - 3}{-3 - (-4)} = \frac{2}{1} = 2$$ • Find the slope of the line through $Q_1(-4, 3)$ and $Q_2(-3, 5)$.

$$m_1 \cdot m_2 = -\frac{1}{2}(2) = -1$$ • Find the product of the two slopes.

Because $m_1 \cdot m_2 = -1$, the lines are perpendicular.

Unless otherwise noted, all content on this page is © Cengage Learning.

HOW TO 5   Are the graphs of $3x + 4y = 8$ and $8x + 6y = 5$ perpendicular?

To determine whether the lines are perpendicular, solve each equation for y and find the slope of each line. Then use the equation $m_1 \cdot m_2 = -1$.

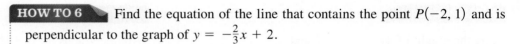

$$3x + 4y = 8 \qquad\qquad\qquad\qquad 8x + 6y = 5$$
$$4y = -3x + 8 \qquad\qquad\qquad 6y = -8x + 5$$
$$y = -\frac{3}{4}x + 2 \qquad\qquad\qquad y = -\frac{4}{3}x + \frac{5}{6}$$

$$m_1 = -\frac{3}{4} \qquad\qquad\qquad\qquad\quad m_2 = -\frac{4}{3}$$

$$m_1 \cdot m_2 = \left(-\frac{3}{4}\right)\left(-\frac{4}{3}\right) = 1$$

Because $m_1 \cdot m_2 = 1 \neq -1$, the lines are not perpendicular.

HOW TO 6   Find the equation of the line that contains the point $P(-2, 1)$ and is perpendicular to the graph of $y = -\frac{2}{3}x + 2$.

The slope of the given line is $-\frac{2}{3}$. The slope of the line perpendicular to the given line is the negative reciprocal of $-\frac{2}{3}$, which is $\frac{3}{2}$. Substitute this slope and the coordinates of the given point, $(-2, 1)$, into the point-slope formula.

$$y - y_1 = m(x - x_1)$$ • The point-slope formula

$$y - 1 = \frac{3}{2}[x - (-2)]$$ • $m = \frac{3}{2}$, $(x_1, y_1) = (-2, 1)$

$$y - 1 = \frac{3}{2}x + 3$$ • Simplify.

$$y = \frac{3}{2}x + 4$$ • Write the equation in the form $y = mx + b$.

The equation of the perpendicular line is $y = \frac{3}{2}x + 4$.

HOW TO 7   Find the equation of the line that contains the point $P(3, -4)$ and is perpendicular to the graph of $2x - y = -3$.

$$2x - y = -3$$ • Determine the slope of the given line by
$$-y = -2x - 3$$ solving the equation for y.
$$y = 2x + 3$$ • The slope is 2.

The slope of the line perpendicular to the given line is $-\frac{1}{2}$, the negative reciprocal of 2. Now use the point-slope formula to find the equation of the line.

$$y - y_1 = m(x - x_1)$$ • The point-slope formula

$$y - (-4) = -\frac{1}{2}(x - 3)$$ • $m = -\frac{1}{2}$, $(x_1, y_1) = (3, -4)$

$$y + 4 = -\frac{1}{2}x + \frac{3}{2}$$ • Simplify.

$$y = -\frac{1}{2}x - \frac{5}{2}$$ • Write the equation in the form $y = mx + b$.

The equation of the perpendicular line is $y = -\frac{1}{2}x - \frac{5}{2}$.

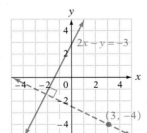

Unless otherwise noted, all content on this page is © Cengage Learning.

EXAMPLE 1

Is the line that contains the points $P_1(-4, 2)$ and $P_2(1, 6)$ parallel to the line that contains the points $Q_1(2, -4)$ and $Q_2(7, 0)$?

Solution

$$m_1 = \frac{6 - 2}{1 - (-4)} = \frac{4}{5}$$ • $(x_1, y_1) = (-4, 2)$, $(x_2, y_2) = (1, 6)$

$$m_2 = \frac{0 - (-4)}{7 - 2} = \frac{4}{5}$$ • $(x_1, y_1) = (2, -4)$, $(x_2, y_2) = (7, 0)$

$$m_1 = m_2 = \frac{4}{5}$$

The lines are parallel.

EXAMPLE 2

Are the graphs of $4x - y = -2$ and $x + 4y = -12$ perpendicular?

Solution

$$\begin{aligned} 4x - y &= -2 & x + 4y &= -12 \\ -y &= -4x - 2 & 4y &= -x - 12 \\ y &= 4x + 2 & y &= -\frac{1}{4}x - 3 \\ m_1 &= 4 & & \\ & & m_2 &= -\frac{1}{4} \end{aligned}$$

$$m_1 \cdot m_2 = 4\left(-\frac{1}{4}\right) = -1$$

The lines are perpendicular.

EXAMPLE 3

Find the equation of the line that contains the point $P(3, -1)$ and is parallel to the graph of $3x - 2y = 4$.

Solution

$$\begin{aligned} 3x - 2y &= 4 \\ -2y &= -3x + 4 \\ y &= \frac{3}{2}x - 2 \end{aligned}$$ • $m = \frac{3}{2}$

$$y - y_1 = m(x - x_1)$$

$$y - (-1) = \frac{3}{2}(x - 3)$$ • $(x_1, y_1) = (3, -1)$

$$y + 1 = \frac{3}{2}x - \frac{9}{2}$$

$$y = \frac{3}{2}x - \frac{11}{2}$$

The equation of the line is $y = \frac{3}{2}x - \frac{11}{2}$.

YOU TRY IT 1

Is the line that contains the points $P_1(-2, -3)$ and $P_2(7, 1)$ perpendicular to the line that contains the points $Q_1(4, 1)$ and $Q_2(6, -5)$?

Your solution

No

IN-CLASS EXAMPLES

1. Is the line that contains the points $P_1(-2, 3)$ and $P_2(5, -2)$ perpendicular to the line that contains the points $Q_1(2, 4)$ and $Q_2(-3, -3)$? **Yes**
2. Find the equation of the line containing the point $P(2, 1)$ and parallel to the graph of $3x + y = -1$. $y = -3x + 7$

YOU TRY IT 2

Are the graphs of $5x + 2y = 2$ and $5x + 2y = -6$ parallel?

Your solution

Yes

3. Find the equation of the line containing the point $P(1, -4)$ and perpendicular to the graph of $y = \frac{3}{2}x - 2$. $y = -\frac{2}{3}x - \frac{10}{3}$

YOU TRY IT 3

Find the equation of the line that contains the point $P(-2, 2)$ and is perpendicular to the graph of $x - 4y = 3$.

Your solution

$y = -4x - 6$

Solutions on p. S13

4.6 EXERCISES

✔ **Concept Check**

SUGGESTED ASSIGNMENT
Exercises 1–12; Exercises 13–33, odds;
More challenging exercises: Exercises 35–38

1. ◣ Explain how to determine whether the graphs of two lines are parallel.

2. ◣ Explain how to determine whether the graphs of two lines are perpendicular.

3. Complete the following sentence: Parallel lines have the same _____slope_____.

4. What is the negative reciprocal of $-\frac{2}{3}$? $\frac{3}{2}$

5. The slope of a line is -5. What is the slope of any line parallel to this line?

   -5

6. The slope of a line is 4. What is the slope of any line perpendicular to this line?

   $-\frac{1}{4}$

7. Give the slope of any line that is parallel to the graph of $y = -\frac{1}{3}x + 5$.

   $-\frac{1}{3}$

8. Give the slope of any line that is perpendicular to the graph of $y = \frac{3}{5}x + 2$.

   $-\frac{5}{3}$

9. Give the slope of any line that is perpendicular to the graph of $3x + 2y = 6$.

   $\frac{2}{3}$

10. Give the slope of any line that is parallel to the graph of $3x - 4y = 12$.

    $\frac{3}{4}$

11. Is the graph of $x = 4$ parallel to the graph of $x = -4$?

    Yes

12. Is the graph of $x = -2$ perpendicular to the graph of $y = 3$?

    Yes

OBJECTIVE A   *To find parallel and perpendicular lines*

13. Is the graph of $y = \frac{2}{3}x - 4$ parallel to the graph of $y = -\frac{3}{2}x - 4$?

    No

14. Is the graph of $y = -2x + \frac{2}{3}$ parallel to the graph of $y = -2x + 3$?

    Yes

15. Is the graph of $y = \frac{4}{3}x - 2$ perpendicular to the graph of $y = -\frac{3}{4}x + 2$?

    Yes

16. Is the graph of $y = \frac{1}{2}x + \frac{3}{2}$ perpendicular to the graph of $y = -\frac{1}{2}x + \frac{3}{2}$?

    No

17. Are the graphs of $2x + 3y = 2$ and $2x + 3y = -4$ parallel?

    Yes

18. Are the graphs of $2x - 4y = 3$ and $2x + 4y = -3$ parallel?

    No

19. Are the graphs of $x - 4y = 2$ and $4x + y = 8$ perpendicular?

    Yes

20. Are the graphs of $4x - 3y = 2$ and $4x + 3y = -7$ perpendicular?

    No

21. Is the line that contains the points $P_1(3, 2)$ and $P_2(1, 6)$ parallel to the line that contains the points $Q_1(-1, 3)$ and $Q_2(-1, -1)$?

    No

22. Is the line that contains the points $P_1(4, -3)$ and $P_2(2, 5)$ parallel to the line that contains the points $Q_1(-2, -3)$ and $Q_2(-4, 1)$?

    No

23. Is the line that contains the points $P_1(-3, 2)$ and $P_2(4, -1)$ perpendicular to the line that contains the points $Q_1(1, 3)$ and $Q_2(-2, -4)$?

    Yes

24. Is the line that contains the points $P_1(-1, 2)$ and $P_2(3, 4)$ perpendicular to the line that contains the points $Q_1(-1, 3)$ and $Q_2(-4, 1)$?

    No

25. Find the equation of the line that contains the point $P(3, -2)$ and is parallel to the graph of $y = 2x + 1$.
$y = 2x - 8$

26. Find the equation of the line that contains the point $P(-1, 3)$ and is parallel to the graph of $y = -x + 3$.
$y = -x + 2$

27. Find the equation of the line that contains the point $P(-2, -1)$ and is perpendicular to the graph of $y = -\frac{2}{3}x - 2$.
$y = \frac{3}{2}x + 2$

28. Find the equation of the line that contains the point $P(-4, 1)$ and is perpendicular to the graph of $y = 2x - 5$.
$y = -\frac{1}{2}x - 1$

29. Find the equation of the line containing the point $P(-2, -4)$ and parallel to the graph of $2x - 3y = 2$.
$y = \frac{2}{3}x - \frac{8}{3}$

30. Find the equation of the line containing the point $P(3, 2)$ and parallel to the graph of $3x + y = -3$.
$y = -3x + 11$

31. Find the equation of the line containing the point $P(4, 1)$ and perpendicular to the graph of $y = -3x + 4$.
$y = \frac{1}{3}x - \frac{1}{3}$

32. Find the equation of the line containing the point $P(2, -5)$ and perpendicular to the graph of $y = \frac{5}{2}x - 4$.
$y = -\frac{2}{5}x - \frac{21}{5}$

33. Find the equation of the line containing the point $P(-1, -3)$ and perpendicular to the graph of $3x - 5y = 2$.
$y = -\frac{5}{3}x - \frac{14}{3}$

34. Find the equation of the line containing the point $P(-1, 3)$ and perpendicular to the graph of $2x + 4y = -1$.
$y = 2x + 5$

Critical Thinking

A **perpendicular bisector** is a line that is perpendicular to a line segment and passes through the midpoint of the line segment. For Exercises 35 and 36, find the equation of the perpendicular bisector of the line segment with the given endpoints.

35. $P_1(3, 4), P_2(-1, 2)$
$y = -2x + 5$

36. $P_1(-3, 3), P_2(1, -7)$
$y = \frac{2}{5}x - \frac{8}{5}$

Projects or Group Activities

Physics For Exercises 37 and 38, suppose a ball is being twirled at the end of a string and the center of rotation is the origin of a coordinate system. If the string breaks, the initial path of the ball is on a line that is perpendicular to the radius of the circle.

37. Suppose the string breaks when the ball is at $P(6, 3)$. Find the equation of the line on which the initial path lies.
$y = -2x + 15$

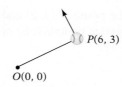
$P(6, 3)$
$O(0, 0)$

38. Suppose the string breaks when the ball is at $P(2, 8)$. Find the equation of the line on which the initial path lies.
$y = -\frac{1}{4}x + \frac{17}{2}$

QUICK QUIZ
1. Is the line that contains the points $P_1(2, 1)$ and $P_2(0, 5)$ parallel to the line that contains the points $Q_1(-2, 4)$ and $Q_2(-2, -2)$?
No [4.6A]
2. Find the equation of the line containing the point $P(-1, -3)$ and parallel to the graph of $x - 2y = 3$.
$y = \frac{1}{2}x - \frac{5}{2}$ **[4.6A]**
3. Find the equation of the line containing the point $P(3, 1)$ and perpendicular to the graph of $y = -2x + 5$.
$y = \frac{1}{2}x - \frac{1}{2}$ **[4.6A]**

Unless otherwise noted, all content on this page is © Cengage Learning.

SECTION

4.7 Inequalities in Two Variables

OBJECTIVE A *To graph the solution set of an inequality in two variables*

The graph of the linear equation $y = x - 1$ separates the plane into three sets: the set of points on the line, the set of points above the line, and the set of points below the line.

The point whose coordinates are (2, 1) is a solution of $y = x - 1$ and is a point on the line.

The point whose coordinates are (2, 4) is a solution of $y > x - 1$ and is a point above the line.

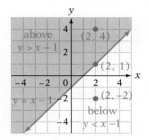

The point whose coordinates are $(2, -2)$ is a solution of $y < x - 1$ and is a point below the line.

The set of points on the line is the solution of the equation $y = x - 1$. The set of points above the line is the solution of the inequality $y > x - 1$. These points form a **half-plane.** The set of points below the line is the solution of the inequality $y < x - 1$. These points also form a half-plane.

An inequality of the form $y > mx + b$ or $Ax + By > C$ is a **linear inequality in two variables.** (The inequality symbol $>$ could be replaced by $\geq$, $<$, or $\leq$.) The solution set of a linear inequality in two variables is a half-plane.

HOW TO 1 illustrates the procedure for graphing the solution set of a linear inequality in two variables.

Take Note

When solving the inequality in HOW TO 1 for y, both sides of the inequality are divided by -4, so the inequality symbol must be reversed.

Take Note

As shown below, (0, 0) is a solution of the inequality in HOW TO 1.

$$y > \frac{3}{4}x - 3$$
$$0 > \frac{3}{4}(0) - 3$$
$$0 > 0 - 3$$
$$0 > -3$$

Because (0, 0) is a solution of the inequality, $P(0, 0)$ should be in the shaded region. The solution set as graphed is correct.

HOW TO 1 Graph the solution set of $3x - 4y < 12$.

$$3x - 4y < 12 \qquad \bullet \text{ Solve the inequality for } y.$$
$$-4y < -3x + 12$$
$$y > \frac{3}{4}x - 3$$

Change the inequality $y > \frac{3}{4}x - 3$ to the equality $y = \frac{3}{4}x - 3$, and graph the line.

If the inequality contains $\leq$ or $\geq$, the line belongs to the solution set and is shown by a *solid line*. If the inequality contains $<$ or $>$, the line is not part of the solution set and is shown by a *dashed line*.

If the inequality contains $>$ or $\geq$, shade the upper half-plane. If the inequality contains $<$ or $\leq$, shade the lower half-plane.

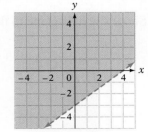

As a check, use the ordered pair (0, 0) to determine whether the correct region of the plane has been shaded. If (0, 0) is a solution of the inequality, then $P(0, 0)$ should be in the shaded region. If (0, 0) is not a solution of the inequality, then $P(0, 0)$ should not be in the shaded region.

Unless otherwise noted, all content on this page is © Cengage Learning.

Integrating Technology

See the Keystroke Guide: *Graphing Inequalities* for instructions on using a graphing calculator to graph the solution set of an inequality in two variables.

If the line passes through the point $P(0, 0)$, then another ordered pair, such as $(0, 1)$, must be used as a check.

From the graph of $y > \frac{3}{4}x - 3$, note that for a given value of x, more than one value of y can be paired with that value of x. For instance, $(4, 1)$, $(4, 3)$, $(5, 1)$, and $\left(5, \frac{9}{4}\right)$ are all ordered pairs that belong to the graph. Because there are ordered pairs with the same first coordinate and different second coordinates, the inequality does not represent a function. The inequality is a relation but not a function.

EXAMPLE 1

Graph the solution set of $x + 2y \le 4$.

Solution

$$x + 2y \le 4$$
$$2y \le -x + 4$$
$$y \le -\frac{1}{2}x + 2$$

Graph $y = -\frac{1}{2}x + 2$ as a solid line.

Shade the lower half-plane.

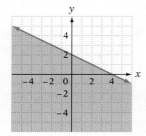

Check:

$$y \le -\frac{1}{2}x + 2$$
$$0 \le -\frac{1}{2}(0) + 2$$
$$0 \le 0 + 2$$
$$0 \le 2$$

The point $P(0, 0)$ should be in the shaded region.

YOU TRY IT 1

Graph the solution set of $x + 3y > 6$.

Your solution

IN-CLASS EXAMPLES
Graph the solution set.
1. $2x - 3y \ge 6$
2. $x + 2y > 6$
3. $y > 4$

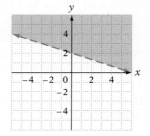

EXAMPLE 2

Graph the solution set of $x \ge -1$.

Solution

Graph $x = -1$ as a solid line.

Shade the half-plane to the right of the line.

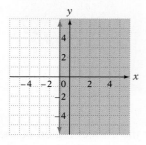

YOU TRY IT 2

Graph the solution set of $y < 2$.

Your solution

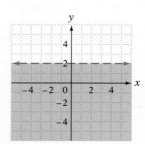

Solutions on p. S13

Unless otherwise noted, all content on this page is © Cengage Learning.

4.7 EXERCISES

Concept Check

SUGGESTED ASSIGNMENT
Exercises 1–6; Exercises 7–25, odds;
More challenging exercises: Exercises 27 and 28

1. What is a half-plane?

2. Explain a method you can use to check that the graph of a linear inequality in two variables has been shaded correctly.

3. Is (0, 0) a solution of $y > 2x - 7$?
   Yes

4. Is (0, 0) a solution of $y < 5x + 3$?
   Yes

5. Is (0, 0) a solution of $y \leq -\frac{2}{3}x - 8$?
   No

6. Is (0, 0) a solution of $y \geq -\frac{3}{4}x + 9$?
   No

OBJECTIVE A *To graph the solution set of an inequality in two variables*

For Exercises 7 to 24, graph the solution set.

7. $y \leq \frac{3}{2}x - 3$

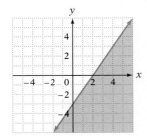

8. $y \geq \frac{4}{3}x - 4$

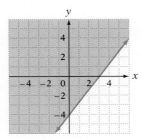

9. $y < -\frac{1}{3}x + 1$

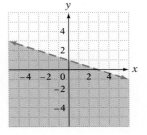

10. $y < \frac{3}{5}x - 3$

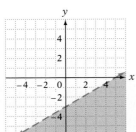

11. $4x - 5y > 10$

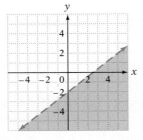

12. $4x + 3y < 9$

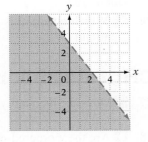

13. $x + 3y < 6$

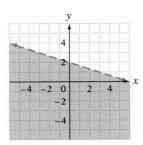

14. $2x - 5y \leq 10$

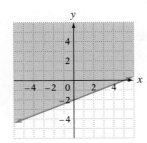

15. $2x + 3y \geq 6$

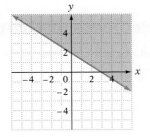

Unless otherwise noted, all content on this page is © Cengage Learning.

16. $3x + 2y < 4$

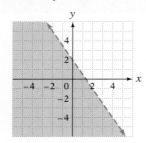

17. $-x + 2y > -8$

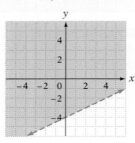

18. $-3x + 2y > 2$

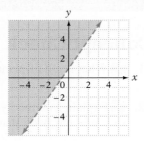

19. $y - 4 < 0$

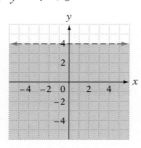

20. $x + 2 \geq 0$

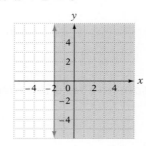

21. $6x + 5y < 15$

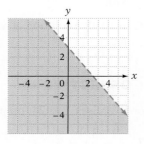

22. $3x - 5y < 10$

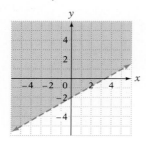

23. $-5x + 3y \geq -12$

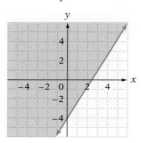

24. $3x + 4y \geq 12$

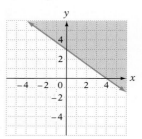

25. Which quadrant is represented by the two linear inequalities $x > 0$ and $y > 0$?
Quadrant I

26. Which quadrant is represented by the two linear inequalities $x < 0$ and $y < 0$?
Quadrant III

Critical Thinking

27. Does the inequality $y < 3x - 1$ represent a function? Explain your answer.

28. Are there ordered-pair solutions that satisfy both $y \leq x + 3$ and $y \geq -\frac{1}{2}x + 1$?
If so, give three such ordered-pair solutions. If not, explain why not.

Projects or Group Activities

29. Graph $|x| + |y| \leq 5$.

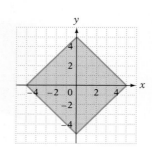

QUICK QUIZ
Graph the solution set.
1. $3x + 5y < 15$ [4.7A]
2. $2x - 5y \geq 10$ [4.7A]
3. $x \geq 2$ [4.7A]

Unless otherwise noted, all content on this page is © Cengage Learning.

CHAPTER

4 | Summary

Key Words

Examples

A **rectangular coordinate system** is formed by two number lines, one horizontal and one vertical, that intersect at the zero point of each line. The number lines that make up a rectangular coordinate system are called the **coordinate axes,** or simply **axes.** The **origin** is the point of intersection of the two coordinate axes. Generally, the horizontal axis is labeled the x-axis and the vertical axis is labeled the y-axis. The coordinate system divides the plane into four regions called **quadrants.** The **coordinates of a point** in the plane are given by an **ordered pair (x, y).** The first number in the ordered pair is called the **abscissa** or **x-coordinate.** The second number in the ordered pair is called the **ordinate** or **y-coordinate.** The **graph of an ordered pair (x, y)** is the dot drawn at the coordinates of the point in the plane.   [4.1A, p. 220]

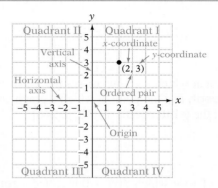

A **solution of an equation in two variables** is an ordered pair (x, y) that makes the equation a true statement.   [4.1B, p. 222]

The ordered pair $(-1, 1)$ is a solution of the equation $y = 2x + 3$ because when -1 is substituted for x and 1 is substituted for y, the result is a true equation.

A **scatter diagram** is a graph of a set of ordered pairs of data. [4.1C, p. 224]

A **function** is a set of ordered pairs in which no two ordered pairs have the same first coordinate and different second coordinates. The **domain** of a function is the set of the first coordinates of all the ordered pairs of the function. The **range** is the set of the second coordinates of all the ordered pairs of the function. [4.2A, p. 232]

$\{(2, 3), (3, 5), (5, 7), (6, 9)\}$
The domain is $\{2, 3, 5, 6\}$.
The range is $\{3, 5, 7, 9\}$.

A **relation** is any set of ordered pairs. [4.2A, p. 232]

$\{(2, 3), (2, 4), (3, 4), (5, 7)\}$

Function notation is used for those equations that represent functions. For the equation at the right, x is the **independent variable** and y is the **dependent variable.** The symbol $f(x)$ is the **value of the function** and represents the value of the dependent variable for a given value of the independent variable. [4.2A, pp. 233–234]

In function notation, $y = 3x + 7$ is written $f(x) = 3x + 7$.

The process of determining $f(x)$ for a given value of x is called **evaluating a function.** [4.2A, p. 234]

Evaluate $f(x) = 2x - 3$ when $x = 4$.

$f(x) = 2x - 3$
$f(4) = 2(4) - 3 = 5$

Unless otherwise noted, all content on this page is © Cengage Learning.

An equation of the form $y = mx + b$, where m and b are constants, is a **linear equation in two variables.** Using function notation, $f(x) = mx + b$ is called a **linear function.** A solution of a linear equation in two variables is an ordered pair (x, y) whose coordinates make the equation a true statement. The graph of a linear equation in two variables is a straight line. [4.3A, p. 242]

$y = 3x + 2$ is a linear equation in two variables; $m = 3$ and $b = 2$. Ordered-pair solutions of $y = 3x + 2$ are shown below, along with the graph of the equation.

x	y
1	5
0	2
−1	−1

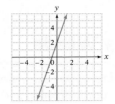

The point at which a graph crosses the x-axis is called the **x-intercept,** and the point at which a graph crosses the y-axis is called the **y-intercept.** [4.3C, p. 246]

The x-intercept of $x + y = 4$ has coordinates $(4, 0)$.
The y-intercept of $x + y = 4$ has coordinates $(0, 4)$.

A value of x for which $f(x) = 0$ is called a **zero** of the function. [4.3C, p. 247]

4 is a zero of $f(x) = 2x - 8$ because $f(4) = 2(4) - 8 = 0$.

The **slope** of a line is a measure of the slant, or tilt, of the line. The symbol for slope is m. A line that slants upward to the right has a **positive slope,** and a line that slants downward to the right has a **negative slope.** A horizontal line has **zero slope.** The slope of a vertical line is **undefined.** [4.4A, pp. 257–258]

The line $y = 2x - 3$ has a slope of 2 and slants upward to the right.
The line $y = -5x + 2$ has a slope of -5 and slants downward to the right.
The line $y = 4$ has a slope of 0.

An inequality of the form $y > mx + b$ or of the form $Ax + By > C$ is a **linear inequality in two variables.** (The symbol $>$ can be replaced by $\geq$, $<$, or $\leq$.) The solution set of an inequality in two variables is a **half-plane.** [4.7A, p. 283]

$4x - 3y < 12$ and $y \geq 2x + 6$ are linear inequalities in two variables.

Essential Rules and Procedures

Examples

Graph of $y = b$ [4.3B, p. 245]
The graph of $y = b$ is a horizontal line passing through the point with coordinates $(0, b)$.

The graph of $y = -5$ is a horizontal line passing through the point with coordinates $(0, -5)$.

Graph of a Constant Function [4.3B, p. 245]
A function given by $f(x) = b$, where b is a constant, is a **constant function.** The graph of the constant function is a horizontal line passing through the point with coordinates $(0, b)$.

The graph of $f(x) = -5$ is a horizontal line passing through the point with coordinates $(0, -5)$. Note that this is the same as the graph of $y = -5$.

Unless otherwise noted, all content on this page is © Cengage Learning.

Graph of $x = a$ [4.3B, p. 245]
The graph of $x = a$ is a vertical line passing through the point with coordinates $(a, 0)$.

The graph of $x = 4$ is a vertical line passing through the point with coordinates $(4, 0)$.

Finding Intercepts of Graphs of Linear Equations [4.3C, p. 247]
To find the x-intercept, let $y = 0$.
To find the y-intercept, let $x = 0$.
For any equation of the form $y = mx + b$, the y-intercept has coordinates $(0, b)$.

$$3x + 4y = 12$$

Let $y = 0$: $3x + 4(0) = 12$
$$3x = 12$$
$$x = 4$$
The x-intercept has coordinates $(4, 0)$.

Let $x = 0$: $3(0) + 4y = 12$
$$4y = 12$$
$$y = 3$$
The y-intercept has coordinates $(0, 3)$.

Slope Formula [4.4A, p. 257]
The slope of the line containing the two points $P_1(x_1, y_1)$ and $P_2(x_2, y_2)$ is given by $m = \frac{y_2 - y_1}{x_2 - x_1}$, $x_1 \neq x_2$.

$(x_1, y_1) = (-3, 2)$, $(x_2, y_2) = (1, 4)$

$$m = \frac{y_2 - y_1}{x_2 - x_1} = \frac{4 - 2}{1 - (-3)} = \frac{2}{4} = \frac{1}{2}$$

The slope of the line through $P_1(-3, 2)$ and $P_2(1, 4)$ is $\frac{1}{2}$.

Slope-Intercept Form of a Straight Line [4.4B, p. 261]
The equation $y = mx + b$ is called the *slope-intercept form* of a straight line. The slope of the line is m, the coefficient of x. The y-intercept has coordinates $(0, b)$.

For the equation $y = -3x + 2$, the slope is -3 and the coordinates of the y-intercept are $(0, 2)$.

Point-Slope Formula [4.5A, p. 268]
Let m be the slope of a line, and let $P_1(x_1, y_1)$ be a point on the line. The equation of the line can be found from the point-slope formula: $y - y_1 = m(x - x_1)$.

The equation of the line that passes through $P(4, 2)$ and has slope -3 is

$$y - y_1 = m(x - x_1)$$
$$y - 2 = -3(x - 4)$$
$$y - 2 = -3x + 12$$
$$y = -3x + 14$$

Slopes of Parallel Lines [4.6A, p. 277]
Two nonvertical lines with slopes of m_1 and m_2 are parallel if and only if $m_1 = m_2$. Vertical lines are parallel lines.

$y = 3x - 4$, $m_1 = 3$
$y = 3x + 2$, $m_2 = 3$
Because $m_1 = m_2$, the lines are parallel.

Slopes of Perpendicular Lines [4.6A, p. 278]
If m_1 and m_2 are the slopes of two lines, neither of which is vertical, then the lines are perpendicular if and only if $m_1 \cdot m_2 = -1$. A vertical line is perpendicular to a horizontal line.

$y = \frac{1}{2}x - 1$, $m_1 = \frac{1}{2}$

$y = -2x + 2$, $m_2 = -2$
Because $m_1 \cdot m_2 = -1$, the lines are perpendicular.

CHAPTER

4   Review Exercises

1. **a.** Graph the ordered pairs $(-2, 4)$ and $(3, -2)$.
   b. Name the abscissa of point A. -2
   c. Name the ordinate of point B. -4

   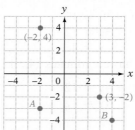

   [4.1A]

2. Graph the ordered-pair solutions of $y = -\frac{1}{2}x - 2$ when $x \in \{-4, -2, 0, 2\}$.

   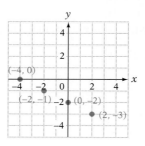

   [4.1B]

3. Determine the ordered-pair solution of $y = \frac{x}{x-2}$ that corresponds to $x = 4$.
   $(4, 2)$ [4.1B]

4. Given $P(x) = 3x + 4$, evaluate $P(-2)$ and $P(a)$.
   $-2; 3a + 4$ [4.2A]

5. Find the zero of $f(x) = -\frac{2}{3}x + 4$.   6   [4.3C]

6. What is the domain of $f(x) = \frac{x}{x+4}$?
   $\{x \mid x \neq -4\}$ [4.2A]

7. Determine the equation of the line that passes through the point $P(6, 1)$ and has slope $-\frac{5}{2}$.
   $y = -\frac{5}{2}x + 16$ [4.5A]

8. Find the domain and range of the function $\{(-1, 0), (0, 2), (1, 2), (5, 4)\}$.
   Domain = $\{-1, 0, 1, 5\}$; Range = $\{0, 2, 4\}$ [4.2A]

9. Graph $y = -\frac{2}{3}x - 2$ using the x- and y-intercepts.
   [4.3C]

   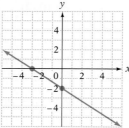

10. Graph $3x + 2y = -6$ using the x- and y-intercepts.
    [4.3C]

    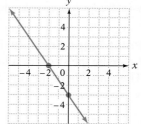

11. Graph: $y = -2x + 2$
    [4.3A]

    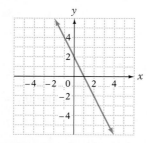

12. Graph: $4x - 3y = 12$
    [4.3B]

    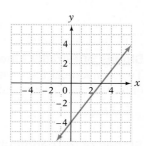

13. Find the slope of the line that contains the points $P_1(3, -2)$ and $P_2(-1, 2)$.
    -1 [4.4A]

14. Find the equation of the line that contains the point $P(-3, 4)$ and has slope $\frac{5}{2}$.
    $y = \frac{5}{2}x + \frac{23}{2}$ [4.5A]

15. Graph $5x + 3y = 15$.

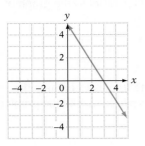

[4.3B]

16. Graph $3x - 2y = -6$.

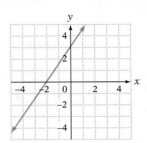

[4.3B]

17. Graph the line that passes through $P(-2, 3)$ and has slope $-\frac{1}{4}$.

[4.4B]

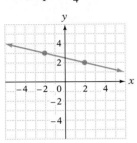

18. Graph the line that passes through the point $P(-1, 4)$ and has slope $-\frac{1}{3}$.

[4.4B]

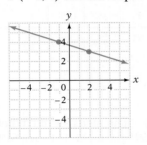

19. Find the equation of the line that contains the point $P(-2, 3)$ and is parallel to the graph of $y = -4x + 3$. $y = -4x - 5$ [4.6A]

20. Find the equation of the line that contains the point $P(-2, 3)$ and is perpendicular to the graph of $y = -\frac{2}{5}x - 3$.
$y = \frac{5}{2}x + 8$ [4.6A]

21. Graph $y = 1$.

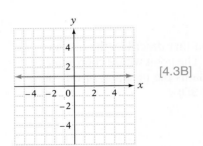

[4.3B]

22. Graph $x = -1$.

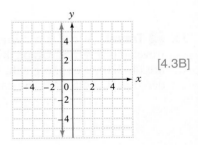

[4.3B]

23. Find the equation of the line that contains the point $P(-3, 3)$ and has slope $-\frac{2}{3}$.
$y = -\frac{2}{3}x + 1$ [4.5A]

24. Find the equation of the line that contains the points $P_1(-8, 2)$ and $P_2(4, 5)$.
$y = \frac{1}{4}x + 4$ [4.5B]

25. Graph the solution set of $y \geq 2x - 3$.

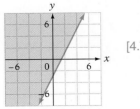

[4.7A]

26. Graph the solution set of $3x - 2y < 6$.

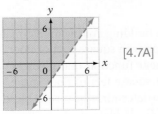

[4.7A]

Unless otherwise noted, all content on this page is © Cengage Learning.

27. Find the equation of the line that contains the points $P_1(-2, 4)$ and $P_2(4, -3)$.

$y = -\dfrac{7}{6}x + \dfrac{5}{3}$ [4.5B]

28. Find the equation of the line that contains the point $P(-2, -4)$ and is parallel to the graph of $4x - 2y = 7$.

$y = 2x$ [4.6A]

29. Find the equation of the line that contains the point $P(3, -2)$ and is parallel to the graph of $y = -3x + 4$.

$y = -3x + 7$ [4.6A]

30. Find the equation of the line that contains the point $P(2, 5)$ and is perpendicular to the graph of $y = -\dfrac{2}{3}x + 6$.

$y = \dfrac{3}{2}x + 2$ [4.6A]

31. Education   The math midterm scores and the final exam scores for six students are given in the following table. Graph the scatter diagram for these data.   [4.1C]

Midterm score, x	90	85	75	80	85	70
Final exam score, y	95	75	80	75	90	70

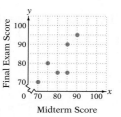

32. Hospitality Industry   The manager of a hotel determines that 200 rooms will be occupied if the rate is $95 per night. For each $10 increase in the rate, 10 fewer rooms will be occupied.

a. Determine a linear function that predicts the number of rooms that will be occupied at a given rate.

b. Use the model to predict occupancy when the rate is $120.

a. $f(x) = -x + 295, 0 \le x \le 295$ **b.** 175 rooms   [4.5C]

33. 🏳 **Travel**   A car is traveling at 55 mph. The equation that describes the distance traveled is $d = 55t$. Graph this equation for $0 \le t \le 6$. The point whose coordinates are (4, 220) is on the graph. Write a sentence that explains the meaning of this ordered pair.   After 4 h, the car has traveled 220 mi.   [4.3D]

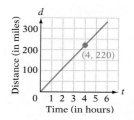

34. 🏳 **Manufacturing**   The graph at the right shows the relationship between the cost of manufacturing calculators and the number of calculators manufactured. Find the slope of the line between the two points shown on the graph. Write a sentence that states the meaning of the slope. $m = 20$; The manufacturing cost is $20 per calculator.   [4.4A]

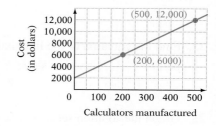

35. Construction   A building contractor estimates that the cost to build a new home is $25,000 plus $80 for each square foot of floor space.

a. Determine a linear function that gives the cost to build a house that contains a given number of square feet of floor space. $f(x) = 80x + 25,000$

b. Use the model to determine the cost to build a house that contains 2000 ft² of floor space. $185,000   [4.5C]

Unless otherwise noted, all content on this page is © Cengage Learning.

CHAPTER

4   TEST

1. Find the ordered-pair solution of $2x - 3y = 15$ corresponding to $x = 3$.

(3, −3)   [4.1B]

2. Graph the ordered-pair solutions of

$$y = -\frac{3}{2}x + 1 \text{ when } x = -2, 0, \text{ and } 4.$$

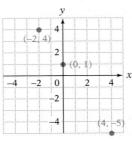

[4.1B]

3. Graph: $y = \dfrac{2}{3}x - 4$

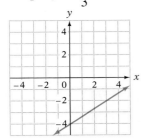

[4.3A]

4. Graph: $2x + 3y = -3$

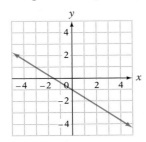

[4.3B]

5. Find the zero of $f(x) = 4x - 12$.

3   [4.3C]

6. Given $f(t) = t^2 + t$, find $f(2)$.

6   [4.2A]

7. Find the slope of the line that contains the points $P_1(-2, 3)$ and $P_2(4, 2)$.

$-\dfrac{1}{6}$ [4.4A]

8. Given $P(x) = 3x^2 - 2x + 1$, evaluate $P(2)$.

9   [4.2A]

9. Graph $2x - 3y = 6$ by using the x- and y-intercepts.

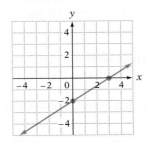

[4.3C]

10. Graph the line that passes through $P(-2, 3)$ and has slope $-\dfrac{3}{2}$.

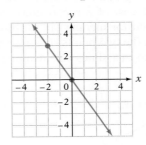

[4.4B]

11. Find the equation of the line that contains the point $P(-5, 2)$ and has slope $\dfrac{2}{5}$.

$y = \dfrac{2}{5}x + 4$ [4.5A]

12. What is the domain of $f(x) = \dfrac{2x + 1}{x}$?

$\{x \mid x \neq 0\}$ [4.2A]

Unless otherwise noted, all content on this page is © Cengage Learning.

13. Find the equation of the line that contains the points $P_1(3, -4)$ and $P_2(-2, 3)$.

$y = -\dfrac{7}{5}x + \dfrac{1}{5}$ [4.5B]

14. Find the x- and y-intercepts of $6x - 4y = 12$.

$(2, 0), (0, -3)$ [4.3C]

15. Find the domain and range of the function $\{(-4, 2), (-2, 2), (0, 0), (3, 5)\}$.

$D: \{-4, -2, 0, 3\}; \; R: \{0, 2, 5\}$ [4.2A]

16. Find the equation of the line that contains the point $P(1, 2)$ and is parallel to the graph of $y = -\dfrac{3}{2}x - 6$.

$y = -\dfrac{3}{2}x + \dfrac{7}{2}$ [4.6A]

17. Find the equation of the line that contains the point $P(-2, -3)$ and is perpendicular to the graph of $y = -\dfrac{1}{2}x - 3$.

$y = 2x + 1$ [4.6A]

18. Graph the solution set of $3x - 4y > 8$.

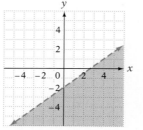

[4.7A]

19. 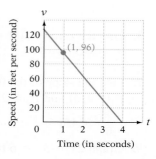 **Sports** The equation for the speed of a ball that is thrown straight up with an initial speed of 128 ft/s is $v = 128 - 32t$, where v is the speed of the ball after t seconds. Graph this equation for values of t from 0 to 4. The point whose coordinates are (1, 96) is on the graph. Write a sentence that describes the meaning of this ordered pair.

After 1 s, the ball is traveling 96 ft/s.   [4.3D]

20. **Depreciation** The graph below shows the relationship between the cost of a rental house and the depreciation allowed for income tax purposes. Find the slope between the two points shown on the graph. Write a sentence that states the meaning of the slope.

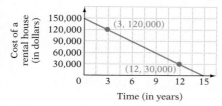

$m = -10,000$; The value of the house decreases by \$10,000 per year.   [4.4A]

21. **Summer Camp** The director of a baseball camp estimates that 100 students will enroll if the tuition is \$250. For each \$20 increase in tuition, six fewer students will enroll.

   a. Determine a linear function that predicts the number of students who will enroll at a given tuition.

   b. Use this model to predict enrollment when the tuition is \$300.

a. $f(x) = -\dfrac{3}{10}x + 175$ **b.** 85 students   [4.5C]

Unless otherwise noted, all content on this page is © Cengage Learning.

Cumulative Review Exercises

1. Identify the property that justifies the statement $(x + y) \cdot 2 = 2 \cdot (x + y)$.
Commutative Property of Multiplication   [1.4C]

2. Solve: $3 - \dfrac{x}{2} = \dfrac{3}{4}$
$\dfrac{9}{2}$ [2.2A]

3. Simplify: $3\sqrt{45}$
$9\sqrt{5}$ [1.2F]

4. Solve: $\dfrac{1 - 3x}{2} + \dfrac{7x - 2}{6} = \dfrac{4x + 2}{9}$
$-\dfrac{1}{14}$ [2.2C]

5. Solve: $x - 3 < -4$ or $2x + 2 > 3$
Write the solution set in set-builder notation.
$\{x \mid x < -1\} \cup \left\{x \mid x > \dfrac{1}{2}\right\}$ [2.5B]

6. Solve: $8 - |2x - 1| = 4$
$-\dfrac{3}{2}$ and $\dfrac{5}{2}$ [2.6A]

7. Solve: $|3x - 5| < 5$
$\left\{x \mid 0 < x < \dfrac{10}{3}\right\}$ [2.6B]

8. Simplify: $4 - 2(4 - 5)^3 + 2$
8   [1.3A]

9. Evaluate $(a - b)^2 \div (ab)$ when $a = 4$ and $b = -2$.
-4.5 [1.4A]

10. Graph: $\{x \mid x < -2\} \cup \{x \mid x > 0\}$
[1.5B]

11. Solve: $3x - 2[x - 3(2 - 3x)] = x - 7$
$\dfrac{19}{18}$ [2.2C]

12. Write $6\frac{2}{3}\%$ as a fraction.
$\dfrac{1}{15}$ [1.2B]

13. Solve: $3x - 1 < 4$ and $x - 2 > 2$
$\varnothing$ [2.5B]

14. Given $P(x) = x^2 + 5$, evaluate $P(-3)$.
14   [4.2A]

15. Find the ordered-pair solution of $y = -\dfrac{5}{4}x + 3$ that corresponds to $x = -8$.
$(-8, 13)$ [4.1B]

16. Find the slope of the line that contains the points $P_1(-1, 3)$ and $P_2(3, -4)$.
$-\dfrac{7}{4}$ [4.4A]

17. Find the equation of the line that contains the point $P(-1, 5)$ and has slope $\frac{3}{2}$.
$y = \dfrac{3}{2}x + \dfrac{13}{2}$ [4.5A]

18. Find the equation of the line that contains the points $P_1(4, -2)$ and $P_2(0, 3)$.
$y = -\dfrac{5}{4}x + 3$ [4.5B]

Unless otherwise noted, all content on this page is © Cengage Learning.

19. Find the equation of the line that contains the point $P(2, 4)$ and is parallel to the graph of $y = -\frac{3}{2}x + 2$.

$y = -\frac{3}{2}x + 7$ [4.6A]

20. Find the equation of the line that contains the point $P(4, 0)$ and is perpendicular to the graph of $3x - 2y = 5$.

$y = -\frac{2}{3}x + \frac{8}{3}$ [4.6A]

21. Find the zero of $f(x) = -2x + 6$.

3   [4.3C]

22. Graph $3x - 5y = 15$ by using the x- and y-intercepts.

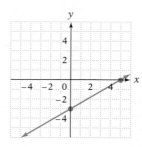

[4.3C]

23. Graph the line that passes through $P(-3, 1)$ and has slope $-\frac{3}{2}$.

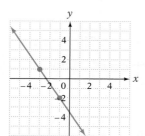

[4.4B]

24. Graph the solution set of $3x - 2y \geq 6$.

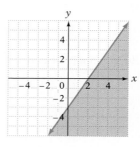

[4.7A]

25. **Uniform Motion** Two planes are 1800 mi apart and are traveling toward each other. One plane is traveling twice as fast as the other plane. The planes pass each other in 3 h. Find the speed of each plane.

First plane: 200 mph
Second plane: 400 mph   [2.4C]

26. **Mixtures** A grocer combines coffee costing $9 per pound with coffee costing $6 per pound. How many pounds of each should be used to make 60 lb of a blend costing $8 per pound?

40 lb of $9 coffee
20 lb of $6 coffee   [2.4A]

27. **Depreciation** The relationship between the value of a truck and the depreciation allowed for income tax purposes is shown in the graph at the right.

a. Write the equation for the line that represents the depreciated value of the truck.

b. 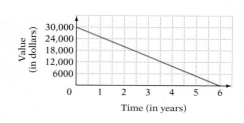 Write a sentence that states the meaning of the slope.

a. $y = -5000x + 30,000$
b. The value of the truck decreases by $5000 each year.   [4.5C]

Unless otherwise noted, all content on this page is © Cengage Learning.

Systems of Linear Equations and Inequalities

5

OBJECTIVES

SECTION 5.1

A To solve a system of linear equations by graphing

B To solve a system of linear equations by the substitution method

C To solve investment problems

SECTION 5.2

A To solve a system of two linear equations in two variables by the addition method

B To solve a system of three linear equations in three variables by the addition method

SECTION 5.3

A To solve rate-of-wind or rate-of-current problems

B To solve application problems

SECTION 5.4

A To graph the solution set of a system of linear inequalities

Focus on Success

Are you making attending class a priority? Remember that to be successful, you must attend class. You need to be in class to hear your instructor's explanations and instructions, as well as to ask questions when something is unclear. Most students who miss a class fall behind and then find it very difficult to catch up. (See Class Time, page AIM-5.)

Lisa F. Young/Shutterstock.com

Prep Test

Are you ready to succeed in this chapter? Take the Prep Test below to find out if you are ready to learn the new material.

1. Simplify: $10\left(\dfrac{3}{5}x + \dfrac{1}{2}y\right)$

$6x + 5y$ [1.4D]

2. Evaluate $3x + 2y - z$ for $x = -1$, $y = 4$, and $z = -2$.

7 [1.4A]

3. Given $3x - 2z = 4$, find the value of x when $z = -2$.

0 [2.2A]

4. Solve:
$3x + 4(-2x - 5) = -5$

-3 [2.2C]

5. Solve:
$0.45x + 0.06(-x + 4000) = 630$

1000 [2.2C]

6. Graph: $3x - 2y = 6$ [4.3B]

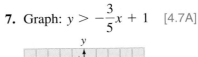

7. Graph: $y > -\dfrac{3}{5}x + 1$ [4.7A]

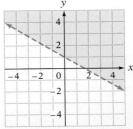

Unless otherwise noted, all content on this page is © Cengage Learning.

5.1   Solving Systems of Linear Equations by Graphing and by the Substitution Method

OBJECTIVE A   *To solve a system of linear equations by graphing*

A **system of equations** is two or more equations considered together. The system at the right is a system of two linear equations in two variables. The graphs of the equations are straight lines.

$$3x + 4y = 7$$
$$2x - 3y = 6$$

A **solution of a system of equations in two variables** is an ordered pair that is a solution of each equation of the system.

> **HOW TO 1**   Is $(3, -2)$ a solution of the system
> $$2x - 3y = 12$$
> $$5x + 2y = 11?$$

$$\begin{array}{r|l} 2x - 3y = 12 & \\ \hline 2(3) - 3(-2) & 12 \\ 6 - (-6) & 12 \\ 12 = 12 \end{array}$$
$$\begin{array}{r|l} 5x + 2y = 11 & \\ \hline 5(3) + 2(-2) & 11 \\ 15 + (-4) & 11 \\ 11 = 11 \end{array}$$

• **Replace x by 3 and y by -2.**

Yes, because $(3, -2)$ is a solution of each equation, it is a solution of the system of equations.

INSTRUCTOR NOTE
You might want to come back to HOW TO 1 after solving a system of equations by graphing. Graph $2x - 3y = 12$ and $5x + 2y = 11$ and show that the point of intersection is $(3, -2)$.

A solution of a system of linear equations can be found by graphing the equations of the system on the same set of coordinate axes. We will now look at three different systems of linear equations.

Consider the following system of equations:

$$x + 2y = 4$$
$$2x + y = -1$$

The graphs of the equations in this system are shown at the right. The lines intersect at a single point whose coordinates are $(-2, 3)$. Because this point lies on both lines, its coordinates give the ordered-pair solution of the system of equations. We can check this by substituting -2 for x and 3 for y in each equation of the system.

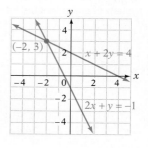

$$\begin{array}{r|l} x + 2y = 4 & \\ \hline -2 + 2(3) & 4 \\ -2 + 6 & 4 \\ 4 = 4 \end{array}$$
$$\begin{array}{r|l} 2x + y = -1 & \\ \hline 2(-2) + 3 & -1 \\ -4 + 3 & -1 \\ -1 = -1 \end{array}$$

• **Replace x by -2 and y by 3.**

The ordered pair $(-2, 3)$ is the solution of the system of equations.

When the graphs of the equations in a system of equations intersect at exactly one point, the system of equations is called an **independent system of equations.** The system of equations

$$x + 2y = 4$$
$$2x + y = -1$$

is an independent system of equations.

Unless otherwise noted, all content on this page is © Cengage Learning.

INSTRUCTOR NOTE
The idea of a system of equations is fairly difficult for students. You might motivate the discussion by first asking, "Find two numbers whose sum is 20." It will not take students long to realize that there are infinitely many solutions to this question. By guiding your students a little further, you can show that the solution set is the set of points whose coordinates satisfy $x + y = 20$.

Now ask, "Find two numbers whose difference is 4." This will provide the equation of another line.

Finally ask, "Find two numbers whose sum is 20 and whose difference is 4." You might have students guess and check until they find the solution (8 and 12). You can then show students that the solution represents the coordinates of the point of intersection of the two lines.

You can extend this number problem to illustrate inconsistent and dependent systems of equations as well. For example: "Find two numbers whose sum is 5 and whose sum is 8." There is no solution, and the graphs of the lines are parallel.

Here is another example. "Find two numbers whose sum is 10 such that the larger number is the difference between 10 and the smaller number." They are infinitely many solutions, and the graphs are the same line.

 Take Note

Keep in mind the differences among independent, dependent, and inconsistent systems of equations. You should be able to express your understanding of these terms by using graphs.

Now consider a second system of equations:

$$2x + 3y = 6$$
$$4x + 6y = -12$$

The graphs of the equations in this system are shown at the right. The graphs are parallel and therefore do not intersect. Because the lines do not intersect, the system of equations has no solution. A system of equations that has no solution is called an **inconsistent system of equations.**

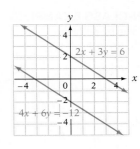

Finally, consider the following system of equations:

$$x - 2y = 4$$
$$2x - 4y = 8$$

The graphs of the equations in this system are shown at the right. In this case, the graph of one equation lies on top of the graph of the other. The graphs intersect at an infinite number of points, so there are an infinite number of solutions of the system of equations. Because each equation represents the same set of points, the solutions of the system of equations can be written using the ordered-pair solutions of either one of the equations. To do this, choose one of the equations in the system of equations. We have chosen the first equation.

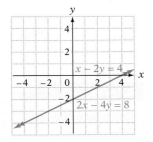

$$x - 2y = 4$$
$$-2y = -x + 4 \qquad \bullet \text{ Solve the equation for } y.$$

$$y = \frac{1}{2}x - 2$$

Write the ordered-pair solutions (x, y) as $\left(x, \frac{1}{2}x - 2\right)$. The ordered pairs $\left(x, \frac{1}{2}x - 2\right)$ are the solutions of the system of equations.

When the graphs of the equations in a system of equations intersect at infinitely many points, the system of equations is called a **dependent system of equations.** The system of equations

$$x - 2y = 4$$
$$2x - 4y = 8$$

is a dependent system of equations.

Summary of the Three Possibilities for a System of Linear Equations in Two Variables

1. The graphs intersect at one point.
   The solution of the system of equations is the ordered pair (x, y) whose coordinates name the point of intersection. The system of equations is independent.

2. The lines are parallel and never intersect.
   The system of equations has no solution. The system of equations is inconsistent.

3. The graphs are the same line, and they intersect at infinitely many points.
   There are infinitely many solutions of the system of equations. The system of equations is dependent.

Unless otherwise noted, all content on this page is © Cengage Learning.

IN-CLASS EXAMPLES
Solve by graphing.
1. $x + y = 5$
   $2x - y = 1$
   (2, 3)

2. $x - y = 2$
   $x + 3y = -2$
   (1, −1)

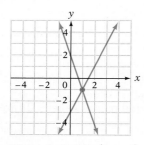

HOW TO 2 Solve by graphing: $2x - y = 3$
$4x - 2y = 6$

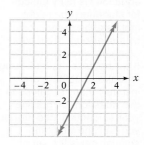

Graph each line.
The system of equations is dependent.
Solve one of the equations for y.

$$2x - y = 3$$
$$-y = -2x + 3$$
$$y = 2x - 3$$

The solutions are the ordered pairs $(x, 2x - 3)$.

EXAMPLE 1

Solve by graphing:
$2x - y = 3$
$3x + y = 2$

Solution

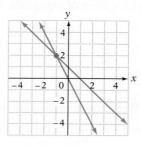

• **Find the coordinates of the point of intersection of the graphs of the equations.**

The solution is $(1, -1)$.

YOU TRY IT 1

Solve by graphing:
$x + y = 1$
$2x + y = 0$

Your solution

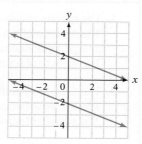

$(-1, 2)$

EXAMPLE 2

Solve by graphing:
$2x + 3y = 6$
$$y = -\frac{2}{3}x + 1$$

Solution

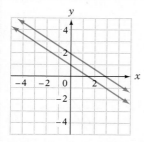

• **Graph the two equations.**

The lines are parallel and therefore do not intersect. The system of equations is inconsistent.
The system of equations has no solution.

YOU TRY IT 2

Solve by graphing:
$2x + 5y = 10$
$$y = -\frac{2}{5}x - 2$$

Your solution

The system of equations has no solution.

Solutions on pp. S13–S14

Unless otherwise noted, all content on this page is © Cengage Learning.

EXAMPLE 3

Solve by graphing:
$x - 2y = 6$
$$y = \frac{1}{2}x - 3$$

Solution

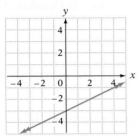

- **Graph the two equations.**

The system of equations is dependent. The solutions are the ordered pairs $\left(x, \frac{1}{2}x - 3\right)$.

YOU TRY IT 3

Solve by graphing:
$3x - 4y = 12$
$$y = \frac{3}{4}x - 3$$

Your solution

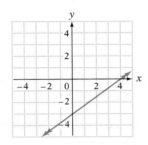

The solutions are the ordered pairs $(x, \frac{3}{4}x - 3)$.

Solution on p. S14

OBJECTIVE B

To solve a system of linear equations by the substitution method

INSTRUCTOR NOTE

When students evaluate a variable expression, they replace a variable with a constant. Here students are replacing a variable with a variable expression. Mentioning this connection will help some students grasp the substitution method of solving a system of equations. Also, have students imagine parentheses around the variable. This will help ensure that they apply the Distributive Property correctly.

When we solve a system of equations by graphing, we approximate the coordinates of a point of intersection. An algebraic method called the **substitution method** can be used to find an exact solution of a system of equations. To use the substitution method, we must write one of the equations of the system in terms of x or in terms of y.

HOW TO 3   Solve by the substitution method: (1) $3x + y = 5$
(2) $4x + 5y = 3$

(3)
$$3x + y = 5$$
$$y = -3x + 5$$

- **Solve Equation (1) for y.**
  This is Equation (3).

(2)
$$4x + 5y = 3$$
$$4x + 5(-3x + 5) = 3$$

- **This is Equation (2).**
- **Equation (3) states that $y = -3x + 5$. Substitute $-3x + 5$ for y in Equation (2).**

$$4x - 15x + 25 = 3$$
$$-11x + 25 = 3$$
$$-11x = -22$$
$$x = 2$$

- **Solve for x.**

(3)
$$y = -3x + 5$$
$$= -3(2) + 5$$
$$= -6 + 5$$
$$= -1$$

- **Substitute the value of x into Equation (3) to find the value of y.**

The solution is the ordered pair $(2, -1)$.

The graph of the system of equations is shown at the left. Note that the graphs intersect at the point whose coordinates are $(2, -1)$, the solution of the system of equations.

> **HOW TO 4**   Solve by the substitution method: (1) $6x + 2y = 8$
>   (2) $3x + y = 2$

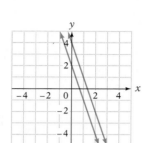

(3)
$$3x + y = 2$$
$$y = -3x + 2$$

- We will solve Equation (2) for y.
- This is Equation (3).

(1)
$$6x + 2y = 8$$
$$6x + 2(-3x + 2) = 8$$

- This is Equation (1).
- Equation (3) states that $y = -3x + 2$. Substitute $-3x + 2$ for y in Equation (1).

$$6x - 6x + 4 = 8$$
$$0x + 4 = 8$$
$$4 = 8$$

- Solve for x.

This is not a true equation. The system of equations is inconsistent. The system of equations has no solution.

The graph of the system of equations is shown at the left. Note that the lines are parallel.

EXAMPLE 4

Solve by substitution:
(1) $3x - 2y = 4$
(2) $-x + 4y = -3$

Solution

Solve Equation (2) for x.
$$-x + 4y = -3$$
$$-x = -4y - 3$$
$$x = 4y + 3 \qquad \bullet \text{ Equation (3)}$$

Substitute $4y + 3$ for x in Equation (1).
$$3x - 2y = 4 \qquad \bullet \text{ Equation (1)}$$
$$3(4y + 3) - 2y = 4 \qquad \bullet \; x = 4y + 3$$
$$12y + 9 - 2y = 4$$
$$10y + 9 = 4$$
$$10y = -5$$
$$y = -\frac{5}{10} = -\frac{1}{2}$$

Substitute the value of y into Equation (3).
$$x = 4y + 3 \qquad \bullet \text{ Equation (3)}$$
$$= 4\left(-\frac{1}{2}\right) + 3 \qquad \bullet \; y = -\frac{1}{2}$$
$$= -2 + 3 = 1$$

The solution is $\left(1, -\frac{1}{2}\right)$.

YOU TRY IT 4

Solve by substitution:
$$3x - y = 3$$
$$6x + 3y = -4$$

Your solution
$$\left(\frac{1}{3}, -2\right)$$

IN-CLASS EXAMPLES

Solve by substitution.

3. $y = x - 1$
     $2x + y = 5$ **(2, 1)**
4. $3x + y = 2$
     $2x + 3y = -8$ **(2, -4)**
5. $3x - y = 8$
     $y = 3x - 2$ **No solution**
6. $4x - 2y = 6$
     $y = 2x - 3$ **(x, 2x - 3)**
7. $x - 3y = -5$
     $2x - 5y = -9$ **(-2, 1)**

Unless otherwise noted, all content on this page is © Cengage Learning.

EXAMPLE 5

Solve by substitution and graph:
$$3x - 3y = 2$$
$$y = x + 2$$

Solution

$$3x - 3y = 2$$
$$3x - 3(x + 2) = 2 \qquad \bullet \; y = x + 2$$
$$3x - 3x - 6 = 2$$
$$-6 = 2$$

This is not a true equation. The system is
inconsistent. The system has no solution.

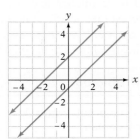

• **Graph the two equations.**

YOU TRY IT 5

Solve by substitution and graph:
$$y = 2x - 3$$
$$3x - 2y = 6$$

Your solution

$(0, -3)$

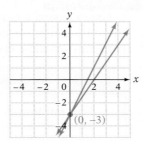

EXAMPLE 6

Solve by substitution and graph:
$$9x + 3y = 12$$
$$y = -3x + 4$$

Solution

$$9x + 3y = 12$$
$$9x + 3(-3x + 4) = 12 \qquad \bullet \; y = -3x + 4$$
$$9x - 9x + 12 = 12$$
$$12 = 12$$

This is a true equation. The system is
dependent. The solutions are the ordered
pairs $(x, -3x + 4)$.

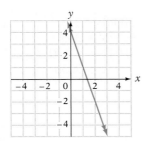

• **Graph the two equations.**

YOU TRY IT 6

Solve by substitution and graph:
$$6x - 3y = 6$$
$$2x - y = 2$$

Your solution

$(x, 2x - 2)$

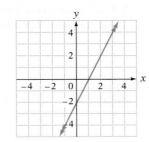

INSTRUCTOR NOTE

After presenting Example 6,
have students show that,
for example, if $x = 1$, then
$y = -3(1) + 4 = 1$. $(1, 1)$
is a solution of the system
of equations. If $x = -2$,
then $y = -3(-2) + 4 = 10$.
$(-2, 10)$ is a solution of the
system of equations.

Solutions on p. S14

Unless otherwise noted, all content on this page is © Cengage Learning.

OBJECTIVE C   *To solve investment problems*

The annual simple interest that an investment earns is given by the equation $Pr = I$, where P is the principal, or the amount invested, r is the simple interest rate, and I is the simple interest.

For instance, if you invest $500 at a simple interest rate of 5%, then the interest earned after one year is calculated as follows:

$$Pr = I$$
$$500(0.05) = I \qquad \text{• Replace } P \text{ by 500 and } r \text{ by 0.05 (5\%).}$$
$$25 = I \qquad \text{• Simplify.}$$

The amount of interest earned is $25.

Tips for Success
Note that solving a word problem includes stating a strategy and using the strategy to find a solution. If you have difficulty with a word problem, write down the known information. Be very specific. Write out a phrase or sentence that states what you are trying to find. See *AIM for Success* in the Preface.

HOW TO 5   You have a total of $5000 to invest in two simple interest accounts. On one account, a money market fund, the annual simple interest rate is 3.5%. On the second account, a bond fund, the annual simple interest rate is 7.5%. If you earn $245 per year from these two investments, how much do you have invested in each account?

Strategy for Solving Simple-Interest Investment Problems

1. For each amount invested, use the equation $Pr = I$. Write a numerical or variable expression for the principal, the interest rate, and the interest earned.

Amount invested at 3.5%: x
Amount invested at 7.5%: y

	Principal, P	·	Interest Rate, r	=	Interest Earned, I
Amount at 3.5%	x	·	0.035	=	$0.035x$
Amount at 7.5%	y	·	0.075	=	$0.075y$

2. Write a system of equations. One equation will express the relationship between the amounts invested. The second equation will express the relationship between the amounts of interest earned by the investments.

INSTRUCTOR NOTE
Students may not realize that investors do not always choose to put all of their money into the account with the greatest interest rate, because that account usually has the most risk. Placing money in different accounts allows the investor to diversify.

The total amount invested is $5000: $x + y = 5000$
The total annual interest earned is $245: $0.035x + 0.075y = 245$

Solve the system of equations.
(1) $\qquad x + y = 5000$
(2) $0.035x + 0.075y = 245$

Solve Equation (1) for y:
(3) $y = -x + 5000$
Substitute into Equation (2):
(2) $0.035x + 0.075(-x + 5000) = 245$
$$0.035x - 0.075x + 375 = 245$$
$$-0.04x = -130$$
$$x = 3250$$

Substitute the value of x into Equation (3) and solve for y.

$$y = -x + 5000$$
$$y = -3250 + 5000 = 1750$$

The amount invested at 3.5% is $3250.
The amount invested at 7.5% is $1750.

Unless otherwise noted, all content on this page is © Cengage Learning.

EXAMPLE 7

An investment of $4000 is made at an annual simple interest rate of 4.9%. How much additional money must be invested at an annual simple interest rate of 7.4% so that the total interest earned is 6.4% of the total investment?

Strategy

• Amount invested at 4.9%: $4000
• Amount invested at 7.4%: x
• Amount invested at 6.4%: y

	Principal	Rate	Interest
Amount at 4.9%	4000	0.049	0.049(4000)
Amount at 7.4%	x	0.074	0.074x
Amount at 6.4%	y	0.064	0.064y

• The amount invested at 6.4% (y) is $4000 more than the amount invested at 7.4% (x):
  $y = x + 4000$

• The sum of the interest earned at 4.9% and the interest earned at 7.4% equals the interest earned at 6.4%:
  $0.049(4000) + 0.074x = 0.064y$

Solution

(1) $\qquad\qquad y = x + 4000$
(2) $\quad 0.049(4000) + 0.074x = 0.064y$

Replace y in Equation (2) by $x + 4000$ from Equation (1). Then solve for x.

$$0.049(4000) + 0.074x = 0.064(x + 4000)$$
$$196 + 0.074x = 0.064x + 256$$
$$0.01x = 60$$
$$x = 6000$$

$6000 must be invested at an annual simple interest rate of 7.4%.

An investment club invested $13,600 in two simple interest accounts. On one account, the annual simple interest rate is 4.2%. On the other, the annual simple interest rate is 6%. How much should be invested in each account so that both accounts earn the same annual interest?

Your strategy

Your solution

$8000 at 4.2%; $5600 at 6%

IN-CLASS EXAMPLES

8. An investment of $12,000 is deposited in two simple interest accounts. On one account, the annual simple interest rate is 5.5%. On the other, the annual simple interest rate is 6.5%. How much should be invested in each account so that both accounts earn the same annual interest? **$6500 at 5.5%; $5500 at 6.5%**

9. A total of $10,000 is deposited in two accounts. On one account, the annual simple interest rate is 9.5%; on the other, the annual simple interest rate is 13%. How much is invested in each account if the total annual interest earned is $1090? **$6000 at 9.5%; $4000 at 13%**

Solution on pp. S14–S15

Unless otherwise noted, all content on this page is © Cengage Learning.

5.1 EXERCISES

SUGGESTED ASSIGNMENT
Exercises 1–8; Exercises 9–61, odds;
More challenging exercises: Exercises 63–65

✔ **Concept Check**

For Exercises 1 to 3, determine whether the ordered pair is a solution of the system of equations.

1. $(2, 1)$; $x + y = 3$
$2x - 3y = 1$
Yes

2. $(-3, -5)$; $x + y = -8$
$2x + 5y = -31$
Yes

3. $(1, -1)$; $3x - y = 4$
$7x + 2y = -5$
No

For Exercises 4 to 6, state whether the system of equations represented by the graph is independent, inconsistent, or dependent.

4.

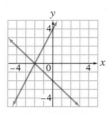

Independent

5.

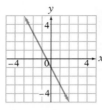

Dependent

6.

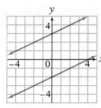

Inconsistent

7. What is the solution of the following system of equations? $\begin{aligned} x &= 4 \\ y &= -1 \end{aligned}$ $(4, -1)$

8. Fill in each blank with *equal* or *not equal*. Consider a system of two linear equations in two variables. For an independent system, the slopes of the lines are ___not equal___. For an inconsistent or dependent system, the slopes of the lines are ___equal___.

OBJECTIVE A   *To solve a system of linear equations by graphing*

For Exercises 9 to 23, solve by graphing.

9. $x + y = 2$
$x - y = 4$

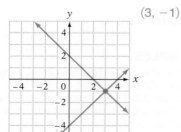
$(3, -1)$

10. $x + y = 1$
$3x - y = -5$

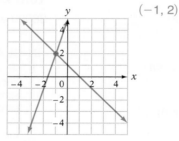
$(-1, 2)$

11. $x - y = -2$
$x + 2y = 10$

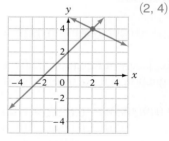
$(2, 4)$

12. $2x - y = 5$
$3x + y = 5$

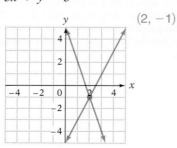
$(2, -1)$

13. $3x - 2y = 6$
$y = 3$

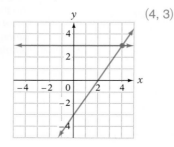
$(4, 3)$

14. $x = 4$
$3x - 2y = 4$

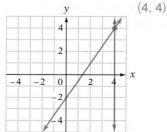
$(4, 4)$

Unless otherwise noted, all content on this page is © Cengage Learning.

15. $2x + 4y = 4$
   $-3x - 6y = -6$

$\left(x, -\dfrac{1}{2}x + 1\right)$

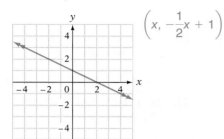

16. $2x - y = 6$
   $x - \dfrac{y}{2} = 3$

$(x, 2x - 6)$

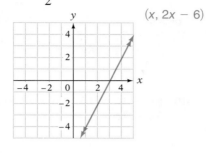

17. $x - y = 6$
   $x + y = 2$

$(4, -2)$

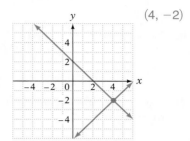

18. $2x + y = 2$
   $-x + y = 5$

$(-1, 4)$

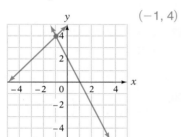

19. $y = x - 5$
   $2x + y = 4$

$(3, -2)$

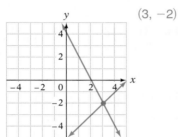

20. $2x + 3y = 6$
   $y = -\dfrac{2}{3}x + 1$

No solution

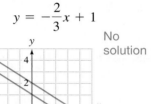

21. $y = \dfrac{1}{2}x - 2$
   $x - 2y = 8$

No solution

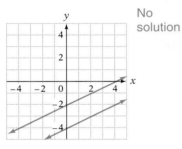

22. $3x - 2y = 6$
   $y = \dfrac{3}{2}x - 3$

$\left(x, \dfrac{3}{2}x - 3\right)$

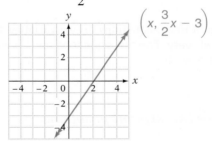

23. $2x - 5y = 10$
   $y = \dfrac{2}{5}x - 2$

$\left(x, \dfrac{2}{5}x - 2\right)$

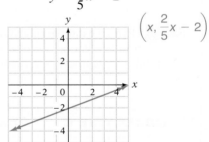

OBJECTIVE B   *To solve a system of linear equations by the substitution method*

For Exercises 24 to 50, solve by the substitution method.

24. $x = 3y + 1$ $(16, 5)$
   $x - 2y = 6$

25. $x = 2y - 3$ $(1, 2)$
   $3x + y = 5$

26. $5x - 2y = 9$ $(-1, -7)$
   $y = 3x - 4$

27. $4x - 3y = 2$ $\left(-\dfrac{5}{2}, -4\right)$
   $y = 2x + 1$

28. $x = 2y + 4$ $(-2, -3)$
   $4x + 3y = -17$

29. $3x - 2y = -11$ $(-1, 4)$
   $x = 2y - 9$

30. $5x + 4y = -1$ $(3, -4)$
   $y = 2 - 2x$

31. $3x + 2y = 4$ $(-2, 5)$
   $y = 1 - 2x$

32. $2x - 5y = -9$ $(3, 3)$
   $y = 9 - 2x$

Unless otherwise noted, all content on this page is © Cengage Learning.

33. $5x + 2y = 15$ (1, 5)
$x = 6 - y$

34. $7x - 3y = 3$ (0, −1)
$x = 2y + 2$

35. $3x - 4y = 6$ (2, 0)
$x = 3y + 2$

36. $2x + 2y = 7$ $\left(\dfrac{1}{2}, 3\right)$
$y = 4x + 1$

37. $3x + 7y = -5$ $\left(\dfrac{2}{3}, -1\right)$
$y = 6x - 5$

38. $3x + y = 5$ (1, 2)
$2x + 3y = 8$

39. $3x - y = 10$ No solution
$6x - 2y = 5$

40. $6x - 4y = 3$ No solution
$3x - 2y = 9$

41. $3x + 4y = 14$ (−2, 5)
$2x + y = 1$

42. $5x + 3y = 8$ (4, −4)
$3x + y = 8$

43. $3x + 5y = 0$ (0, 0)
$x - 4y = 0$

44. $2x - 7y = 0$ (0, 0)
$3x + y = 0$

45. $2x - 4y = 16$ $\left(x, \dfrac{1}{2}x - 4\right)$
$-x + 2y = -8$

46. $3x - 12y = -24$ $\left(x, \dfrac{1}{4}x + 2\right)$
$-x + 4y = 8$

47. $y = 3x + 2$ (1, 5)
$y = 2x + 3$

48. $y = 3x - 7$ (2, −1)
$y = 2x - 5$

49. $y = 3x + 1$ $\left(\dfrac{2}{3}, 3\right)$
$y = 6x - 1$

50. $y = 2x - 3$ $\left(\dfrac{1}{2}, -2\right)$
$y = 4x - 4$

51. The system of equations at the right is inconsistent.
What is the value of $\dfrac{a}{b}$? $\dfrac{2}{3}$

$4x - 6y = 7$
$ax - by = 9$

52. Give an example of a system of linear equations in two variables that has (0, 0) as its only solution.   Answers will vary. The two equations can be written in the form $y = ax$ and $y = bx$, where $a \neq b$.

OBJECTIVE C   *To solve investment problems*

For Exercises 53 and 54, use the system of equations shown at the right. The system models the investment of x dollars in one simple interest account and y dollars in a second simple interest account.

$x + y = \$6000$
$0.055x + 0.072y = \$391.20$

53. What are the interest rates on the two accounts?
5.5% and 7.2%

54. What is the total amount of money invested?
$6000

55. The Community Relief Charity Group is earning 3.5% simple interest on the $2800 it invested in a savings account. It also earns 4.2% simple interest on an insured bond fund. The annual interest earned from both accounts is $329. How much is invested in the insured bond fund? $5500

56. Two investments earn an annual income of $575. One investment earns an annual simple interest rate of 8.5%, and the other earns an annual simple interest rate of 6.4%. The total amount invested is $8000. How much is invested in each account?
$3000 at 8.5%; $5000 at 6.4%

57. An investment club invested $6000 at an annual simple interest rate of 4.0%. How much additional money must be invested at an annual simple interest rate of 6.5% so that the total annual interest earned will be 5% of the total investment? $4000

58. A company invested $30,000, putting part of it into a savings account that earned 3.2% annual simple interest and the remainder in a stock fund that earned 12.6% annual simple interest. If the investments earned $1665 annually, how much was invested in each account? $22,500 at 3.2%; $7500 at 12.6%

59. An account executive divided $42,000 between two simple interest accounts. On the tax-free account, the annual simple interest rate is 3.5%; on the money market fund, the annual simple interest rate is 4.5%. How much should be invested in each account so that both accounts earn the same annual interest? $23,625 at 3.5%; $18,375 at 4.5%

60. An investment club placed $33,000 into two simple interest accounts. On one account, the annual simple interest rate is 6.5%. On the other, the annual simple interest rate is 4.5%. How much should be invested in each account so that both accounts earn the same annual interest? $19,500 at 4.5%; $13,500 at 6.5%

61. The Cross Creek Investment Club decided to invest $16,000 in two bond funds. The first, a mutual bond fund, earns 4.5% annual simple interest. The second, a corporate bond fund, earns 8% annual simple interest. If the club earned $1070 from these two accounts, how much was invested in the mutual bond fund? $6000

62. Cabin Financial Service Group recommends that a client purchase for $10,000 a corporate bond that earns 5% annual simple interest. How much additional money must be placed in an investment that earns a simple interest rate of 3.5% so that the total annual interest earned from the two investments is 4% of the total investment? $20,000

Critical Thinking

63. For what values of k will the following system of equations be independent?

$2x + 3y = 6$
$2x + ky = 9$ Any real number except 3

64. If the following system of equations is inconsistent, how are the values of C and D related?

$3x - 4y = C$
$3x - 4y = D$ $C \neq D$

Projects or Group Activities

65. Write a system of equations in two variables that satisfies the given condition.
 a. The system of equations has $(-3, 5)$ as its only solution.
 b. The system of equations is a dependent system.
 c. The system of equations is an inconsistent system.   Answers will vary.

Unless otherwise noted, all content on this page is © Cengage Learning.

QUICK QUIZ

Solve by graphing.
1. $x + y = 3$
  $x - y = 1$

  $(2, 1)$ [5.1A]

Solve by substitution.
2. $y = -x + 3$
  $2x - y = 6$
  $(3, 0)$ [5.1B]
3. $x + 2y = 7$
  $2x - y = 4$
  $(3, 2)$ [5.1B]
4. $4x - y = 5$
  $y = 4x - 1$
  No solution   [5.1B]
5. An investment advisor deposits $38,000 into two simple interest accounts. On one account, the annual simple interest rate is 4.5%. On the other, the annual simple interest rate is 7.5%. How much should be invested in each account so that both accounts earn the same annual interest?   **$14,250 at 7.5%; $23,750 at 4.5%**   [5.1C]

5.2 Solving Systems of Linear Equations by the Addition Method

OBJECTIVE A *To solve a system of two linear equations in two variables by the addition method*

The **addition method** is an alternative method for solving a system of equations. This method is based on the Addition Property of Equations. We use the addition method when it is not convenient to solve one equation for one variable in terms of another.

 Point of Interest

There are records of Babylonian mathematicians solving systems of equations 3600 years ago. Here is a system of equations from that time (in our modern notation):

$$\frac{2}{3}x = \frac{1}{2}y + 500$$
$$x + y = 1800$$

We say *modern notation* for many reasons. Foremost is the fact that the use of variables did not become widespread until the 17th century. There are many other reasons: The equals sign had not been invented, 2 and 3 did not look like they do today, and zero had not even been considered as a possible number.

Note, for the system of equations at the right, the effect of adding Equation (2) to Equation (1). Because $-3y$ and $3y$ are additive inverses, adding the equations results in an equation with only one variable.

(1) $5x - 3y = 14$
(2) $2x + 3y = -7$
_____
$7x + 0y = 7$
$7x = 7$

The solution of the resulting equation is the first coordinate of the ordered-pair solution of the system.

$7x = 7$
$x = 1$

The second coordinate is found by substituting the value of x into Equation (1) or (2) and then solving for y. Equation (1) is used here.

(1) $5x - 3y = 14$
$5(1) - 3y = 14$
$5 - 3y = 14$
$-3y = 9$
$y = -3$

The solution is $(1, -3)$.

Sometimes each equation of a system of equations must be multiplied by a constant so that the coefficients of one of the variables are opposites.

HOW TO 1 Solve by the addition method:
(1) $3x + 4y = 2$
(2) $2x + 5y = -1$

To eliminate x, multiply Equation (1) by 2 and Equation (2) by -3. Note at the right how the constants are chosen.

$2(3x + 4y) = 2 \cdot 2$

$-3(2x + 5y) = -3(-1)$

• The negative is used so that the coefficients will be opposites.

$6x + 8y = 4$ • 2 times Equation (1).
$-6x - 15y = 3$ • -3 times Equation (2).
_____
$-7y = 7$ • Add the equations.
$y = -1$ • Solve for y.

 Tips for Success

Always check the proposed solution of a system of equations. For the system at the right:

$3x + 4y = 2$
_____
$3(2) + 4(-1)$ | 2
$6 - 4$ | 2
$2 = 2$

$2x + 5y = -1$
_____
$2(2) + 5(-1)$ | -1
$4 - 5$ | -1
$-1 = -1$

The solution checks.

Substitute the value of y into Equation (1) or Equation (2) and solve for x. Equation (1) will be used here.

(1) $3x + 4y = 2$
$3x + 4(-1) = 2$ • Substitute -1 for y.
$3x - 4 = 2$ • Solve for x.
$3x = 6$
$x = 2$

The solution is $(2, -1)$.

Take Note

The result of adding Equations (3) and (2) is $0 = 0$. It is not $x = 0$ and it is not $y = 0$. There is no variable in the equation $0 = 0$. This result does not indicate that the solution is $(0, 0)$; rather, it indicates a dependent system of equations.

INSTRUCTOR NOTE

Try approaching the solution set to HOW TO 2 by having students name an ordered-pair solution, then another, and another, and so on, until they suspect that there are infinitely many solutions. In order to name all the solutions, we must have variable coordinates in the ordered pair: $(x, 2x - 3)$.

HOW TO 2 Solve by the addition method:
(1) $2x - y = 3$
(2) $4x - 2y = 6$

To eliminate y, first multiply Equation (1) by -2.

(1) $-2(2x - y) = -2(3)$ • -2 times Equation (1).
(3) $-4x + 2y = -6$ • **This is Equation (3).**

Add Equation (3) to Equation (2).

(2) $4x - 2y = 6$
(3) $\underline{-4x + 2y = -6}$
$\, 0 = 0$

The equation $0 = 0$ indicates that the system of equations is dependent. This means that the graphs of the two lines are the same. Therefore, the solutions of the system of equations are the ordered-pair solutions of the equation of the line. Solve Equation (1) for y.

$2x - y = 3$
$-y = -2x + 3$
$y = 2x - 3$

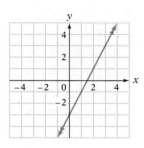

The ordered-pair solutions are $(x, 2x - 3)$, where $2x - 3$ is substituted for y.

INSTRUCTOR NOTE

The addition method is not clear to many students. Try showing them that this method is really a variation of the substitution method.

Using the very first system of equations shown on page 310, add $2x + 3y$ to each side of $5x - 3y = 14$.

$5x - 3y + (2x + 3y)$
$= 14 + (2x + 3y)$

Simplify the left side; substitute -7 for $2x + 3y$ on the right side. The result is $7x = 7$.

IN-CLASS EXAMPLES

Solve by the addition method.
1. $x - y = 6$
$3x + y = 10$ **(4, −2)**
2. $2x + y = 3$
$x + 2y = 0$ **(2, −1)**
3. $3x + 7y = 0$
$4x + 5y = 0$ **(0, 0)**
4. $2x - 3y = 9$
$4x - 6y = 4$
No solution

HOW TO 3 Solve by the addition method:
(1) $\dfrac{2}{3}x + \dfrac{1}{2}y = 4$

(2) $\dfrac{1}{4}x - \dfrac{3}{8}y = -\dfrac{3}{4}$

Clear fractions. Multiply each equation by the LCM of the denominators.

$6\left(\dfrac{2}{3}x + \dfrac{1}{2}y\right) = 6(4)$ • **The LCM of 3 and 2 is 6.**

$8\left(\dfrac{1}{4}x - \dfrac{3}{8}y\right) = 8\left(-\dfrac{3}{4}\right)$ • **The LCM of 4 and 8 is 8.**

$4x + 3y = 24$
$\underline{2x - 3y = -6}$ • **Eliminate y by adding the equations. Then solve for x.**
$6x = 18$
$x = 3$

$\dfrac{2}{3}x + \dfrac{1}{2}y = 4$ • **This is Equation (1).**

$\dfrac{2}{3}(3) + \dfrac{1}{2}y = 4$ • **Substitute $x = 3$ into Equation (1) and solve for y.**

$2 + \dfrac{1}{2}y = 4$

$\dfrac{1}{2}y = 2$

$y = 4$

The solution is $(3, 4)$.

Unless otherwise noted, all content on this page is © Cengage Learning.

EXAMPLE 1

Solve by the addition method:
(1) $3x - 2y = 2x + 5$
(2) $2x + 3y = -4$

Solution

Write Equation (1) in the form $Ax + By = C$.

$3x - 2y = 2x + 5$
$x - 2y = 5$

Solve the system:

$x - 2y = 5$
$2x + 3y = -4$

Eliminate x.

$-2(x - 2y) = -2(5)$
$2x + 3y = -4$

$-2x + 4y = -10$
$\underline{2x + 3y = -4}$

$7y = -14$ • Add the equations.
$y = -2$ • Solve for y.

Replace y in Equation (2).

$2x + 3y = -4$
$2x + 3(-2) = -4$
$2x - 6 = -4$
$2x = 2$
$x = 1$

The solution is $(1, -2)$.

EXAMPLE 2

Solve by the addition method:
(1) $4x - 8y = 36$
(2) $3x - 6y = 27$

Solution

Eliminate x.

$3(4x - 8y) = 3(36)$
$-4(3x - 6y) = -4(27)$

$12x - 24y = 108$
$\underline{-12x + 24y = -108}$
$ 0 = 0$ • Add the equations.

The system of equations is dependent.
Solve Equation (1) for y.

$4x - 8y = 36$
$-8y = -4x + 36$
$y = \dfrac{1}{2}x - \dfrac{9}{2}$

The solutions are the ordered pairs $\left(x, \dfrac{1}{2}x - \dfrac{9}{2}\right)$.

YOU TRY IT 1

Solve by the addition method:
$2x + 5y = 6$
$3x - 2y = 6x + 2$

Your solution

$(-2, 2)$

YOU TRY IT 2

Solve by the addition method:
$2x + y = 5$
$4x + 2y = 6$

Your solution

No solution

Solutions on p. S15

OBJECTIVE B

To solve a system of three linear equations in three variables by the addition method

An equation of the form $Ax + By + Cz = D$, where A, B, and C are the coefficients of the variables and D is a constant, is a **linear equation in three variables.** Examples of this type of equation are shown at the right.

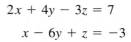

$$2x + 4y - 3z = 7$$
$$x - 6y + z = -3$$

Graphing an equation in three variables requires a third coordinate axis perpendicular to the xy-plane. The third axis is commonly called the z-axis. The result is a three-dimensional coordinate system called the ***xyz*-coordinate system.** To help visualize a three-dimensional coordinate system, think of a corner of a room: The floor is the xy-plane, one wall is the yz-plane, and the other wall is the xz-plane. A three-dimensional coordinate system is shown at the right.

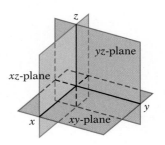

Each point in an xyz-coordinate system is the graph of an **ordered triple** (x, y, z). Graphing an ordered triple requires three moves, the first in the direction of the x-axis, the second in the direction of the y-axis, and the third in the direction of the z-axis. The graphs of the points with coordinates $(-4, 2, 3)$ and $(3, 4, -2)$ are shown at the right.

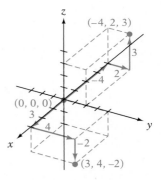

The graph of a linear equation in three variables is a plane. That is, if all the solutions of a linear equation in three variables were plotted in an xyz-coordinate system, the graph would look like a large piece of paper extending infinitely. The graph of $x + y + z = 3$ is shown at the right.

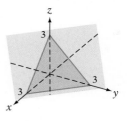

Point of Interest

In the early 1980s, Stephen Hoppe became interested in winning Monopoly strategies. Finding these strategies required solving a system that contained 123 equations in 123 variables!

Just as a solution of an equation in two variables is an ordered pair (x, y), a **solution of an equation in three variables** is an ordered triple (x, y, z). For example, the ordered triple $(2, 1, -3)$ is a solution of the equation $2x - y - 2z = 9$. The ordered triple $(1, 3, 2)$ is not a solution.

A **system of linear equations in three variables** is shown at the right. A **solution of a system of equations in three variables** is an ordered triple that is a solution of each equation of the system.

$$x - 2y + z = 6$$
$$3x + y - 2z = 2$$
$$2x - 3y + 5z = 1$$

Unless otherwise noted, all content on this page is © Cengage Learning.

INSTRUCTOR NOTE
Your book and a desk or
tabletop can be used to
model the various ways in
which planes intersect.

For a system of three equations in three variables to have a solution, the graphs of the equations must be three planes that intersect at a single point, must be three planes that intersect along a common line, or must be the same plane. These situations are shown in the figures that follow.

The three planes shown in Figure A intersect at a point. A system of equations represented by planes that intersect at a point is an independent system.

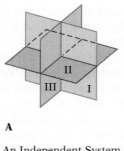

A

An Independent System
of Equations

The three planes shown in Figures B and C intersect along a common line. In Figure D, the three planes are all the same plane. The systems of equations represented by the planes in Figures B, C, and D are dependent systems.

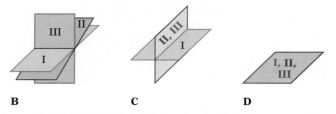

B   **C**   **D**

Dependent Systems of Equations

The systems of equations represented by the planes in the four figures below are inconsistent systems.

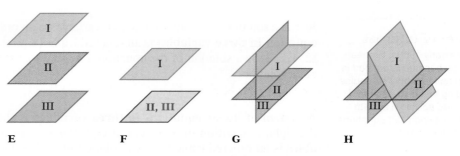

E   **F**   **G**   **H**

Inconsistent Systems of Equations

Unless otherwise noted, all content on this page is © Cengage Learning.

A system of linear equations in three variables can be solved by using the addition method. First, eliminate one variable from any two of the given equations. Then eliminate the same variable from any other two equations. The result will be a system of two equations in two variables. Solve this system by the addition method.

HOW TO 4   Solve:

$$\begin{aligned}(1) \quad & x + 4y - z = 10 \\ (2) \quad & 3x + 2y + z = 4 \\ (3) \quad & 2x - 3y + 2z = -7\end{aligned}$$

INSTRUCTOR NOTE
Show students that the goal of solving a system of three equations in three variables is to replace it with an equivalent system of two equations in two variables that can be solved using the techniques of the preceding objective.

Eliminate z from Equations (1) and (2) by adding the two equations. The result is Equation (4).

$$\begin{array}{l} x + 4y - z = 10 \\ \underline{3x + 2y + z = 4} \end{array}$$

(4) $\qquad 4x + 6y = 14$ • Add the equations.

Eliminate z from Equations (1) and (3). Multiply Equation (1) by 2 and add to Equation (3). The result is Equation (5).

$$\begin{array}{l} 2x + 8y - 2z = 20 \\ \underline{2x - 3y + 2z = -7} \end{array}$$

(5) $\qquad 4x + 5y = 13$

• 2 times Equation (1).
• This is Equation (3).
• Add the equations.

Using Equations (4) and (5), solve the system of two equations in two variables.

(4) $\qquad 4x + 6y = 14$
(5) $\qquad 4x + 5y = 13$

Eliminate x. Multiply Equation (5) by -1 and add to Equation (4).

$$\begin{array}{l} 4x + 6y = 14 \\ \underline{-4x - 5y = -13} \\ \qquad y = 1 \end{array}$$

• This is Equation (4).
• -1 times Equation (5).
• Add the equations.

Substitute the value of y into Equation (4) or Equation (5) and solve for x. Equation (4) is used here.

$$\begin{aligned} 4x + 6y &= 14 \\ 4x + 6(1) &= 14 \\ 4x + 6 &= 14 \\ 4x &= 8 \\ x &= 2 \end{aligned}$$

• This is Equation (4).
• $y = 1$

• Solve for x.

★ **Tips for Success**
Always check the proposed solution of a system of equations. For the system in HOW TO 4:

$$\begin{array}{l} x + 4y - z = 10 \\ \hline 2 + 4(1) - (-4) \ \big|\ 10 \\ \qquad\qquad\quad 10 = 10 \end{array}$$

$$\begin{array}{l} 3x + 2y + z = 4 \\ \hline 3(2) + 2(1) + (-4) \ \big|\ 4 \\ \qquad\qquad\qquad 4 = 4 \end{array}$$

$$\begin{array}{l} 2x - 3y + 2z = -7 \\ \hline 2(2) - 3(1) + 2(-4) \ \big|\ -7 \\ \qquad\qquad\qquad -7 = -7 \end{array}$$

The solution checks.

Substitute the value of y and the value of x into one of the equations in the original system, and solve for z. Equation (2) is used here.

$$\begin{aligned} 3x + 2y + z &= 4 \\ 3(2) + 2(1) + z &= 4 \\ 6 + 2 + z &= 4 \\ 8 + z &= 4 \\ z &= -4 \end{aligned}$$

• This is Equation (2).
• $x = 2, y = 1$

The solution is $(2, 1, -4)$.

IN-CLASS EXAMPLES

Solve by the addition method.

5. $x + 2y - z = 3$
    $2x - y + z = 3$
    $3x - 4y + 2z = -1$
    $(1, 3, 4)$

6. $2x - y + 4z = 7$
    $x + y + 2z = 2$
    $x + 2y - 2z = -1$
    $\left(2, -1, \dfrac{1}{2}\right)$

7. $x + 2y - z = 8$
    $2x + y + z = 1$
    $3x - y + 2z = -5$
    $(1, 2, -3)$

HOW TO 5   Solve: (1) $2x - 3y - z = 1$
                       (2) $x + 4y + 3z = 2$
                       (3) $4x - 6y - 2z = 5$

Eliminate x from Equations (1) and (2).

$$\begin{aligned} 2x - 3y - z &= 1 \\ -2x - 8y - 6z &= -4 \\ \hline -11y - 7z &= -3 \end{aligned}$$

- This is Equation (1).
- -2 times Equation (2).
- Add the equations.

Eliminate x from Equations (1) and (3).

$$\begin{aligned} -4x + 6y + 2z &= -2 \\ 4x - 6y - 2z &= 5 \\ \hline 0 &= 3 \end{aligned}$$

- -2 times Equation (1).
- This is Equation (3).
- Add the equations.

The equation $0 = 3$ is not a true equation. The system of equations is inconsistent and therefore has no solution.

HOW TO 6   Solve: (1) $3x - z = -1$
                       (2) $2y - 3z = 10$
                       (3) $x + 3y - z = 7$

Eliminate x from Equations (1) and (3). Multiply Equation (3) by -3 and add to Equation (1).

$$\begin{aligned} 3x \qquad - z &= -1 \\ -3x - 9y + 3z &= -21 \\ \hline -9y + 2z &= -22 \end{aligned}$$

(4)

- This is Equation (1).
- -3 times Equation (3).
- Add the equations.

Use Equations (2) and (4) to form a system of equations in two variables.

(2) $2y - 3z = 10$
(4) $-9y + 2z = -22$

Eliminate z. Multiply Equation (2) by 2 and Equation (4) by 3.

$$\begin{aligned} 4y - 6z &= 20 \\ -27y + 6z &= -66 \\ \hline -23y &= -46 \\ y &= 2 \end{aligned}$$

- 2 times Equation (2).
- 3 times Equation (4).
- Add the equations.
- Solve for y.

Substitute the value of y into Equation (2) or Equation (4) and solve for z. Equation (2) is used here.

(2)
$$\begin{aligned} 2y - 3z &= 10 \\ 2(2) - 3z &= 10 \\ 4 - 3z &= 10 \\ -3z &= 6 \\ z &= -2 \end{aligned}$$

- This is Equation (2).
- $y = 2$
- Solve for z.

Substitute the value of z into Equation (1) and solve for x.

(1)
$$\begin{aligned} 3x - z &= -1 \\ 3x - (-2) &= -1 \\ 3x + 2 &= -1 \\ 3x &= -3 \\ x &= -1 \end{aligned}$$

- This is Equation (1).
- $z = -2$
- Solve for x.

The solution is $(-1, 2, -2)$.

EXAMPLE 3

Solve: (1) $3x - y + 2z = 1$
         (2) $2x + 3y + 3z = 4$
         (3) $x + y - 4z = -9$

Solution

Eliminate y. Add Equations (1) and (3).

$$3x - y + 2z = 1$$
$$\underline{x + y - 4z = -9}$$
$$4x - 2z = -8$$

Simplify the resulting equation by multiplying each side of the equation by $\frac{1}{2}$.

$2x - z = -4$ • Equation (4)

Multiply Equation (1) by 3 and add to Equation (2).

$$9x - 3y + 6z = 3$$
$$\underline{2x + 3y + 3z = 4}$$
$$11x + 9z = 7$$ • Equation (5)

Solve the system of two equations formed by Equations (4) and (5).

(4) $2x - z = -4$
(5) $11x + 9z = 7$

Multiply Equation (4) by 9 and add to Equation (5). Solve for x.

$$18x - 9z = -36$$
$$\underline{11x + 9z = 7}$$
$$29x = -29$$
$$x = -1$$

Replace x by -1 in Equation (4). Solve for z.

$$2x - z = -4$$
$$2(-1) - z = -4$$
$$-2 - z = -4$$
$$-z = -2$$
$$z = 2$$

Replace x by -1 and z by 2 in Equation (3). Solve for y.

$$x + y - 4z = -9$$
$$-1 + y - 4(2) = -9$$
$$-9 + y = -9$$
$$y = 0$$

The solution is $(-1, 0, 2)$.

YOU TRY IT 3

Solve: $x - y + z = 6$
         $2x + 3y - z = 1$
         $x + 2y + 2z = 5$

Your solution

$(3, -1, 2)$

Solution on p. S15

5.2 EXERCISES

SUGGESTED ASSIGNMENT
Exercises 1 and 2; Exercises 3–69, odds;
More challenging exercises: Exercises 71, 73

✔ Concept Check

For Exercises 1 and 2, use the system of equations at the right.   (1) $5x - 7y = 9$
Fill in each blank with a number.   (2) $6x + 3y = 12$

1. To use the addition method to eliminate x, you could multiply Equation (1) by _____ and Equation (2) by _____ and then add the resulting equations.
   Answers will vary. Possible answers are 6 and -5.

2. To use the addition method to eliminate y, you could multiply Equation (1) by _____ and Equation (2) by _____ and then add the resulting equations.
   Answers will vary. Possible answers are 3 and 7.

OBJECTIVE A   *To solve a system of two linear equations in two variables by the addition method*

For Exercises 3 to 44, solve by the addition method.

3. $x - y = 5$
   $x + y = 7$

   (6, 1)

4. $x + y = 1$
   $2x - y = 5$

   (2, −1)

5. $3x + y = 4$
   $x + y = 2$

   (1, 1)

6. $x - 3y = 4$
   $x + 5y = -4$

   (1, −1)

7. $3x + y = 7$
   $x + 2y = 4$

   (2, 1)

8. $x - 2y = 7$
   $3x - 2y = 9$

   (1, −3)

9. $2x + 3y = -1$
   $x + 5y = 3$

   (−2, 1)

10. $x + 5y = 7$
    $2x + 7y = 8$

    (−3, 2)

11. $3x - y = 4$
    $6x - 2y = 8$

    $(x, 3x - 4)$

12. $x - 2y = -3$
    $-2x + 4y = 6$

    $\left(x, \dfrac{1}{2}x + \dfrac{3}{2}\right)$

13. $2x + 5y = 9$
    $4x - 7y = -16$

    $\left(-\dfrac{1}{2}, 2\right)$

14. $8x - 3y = 21$
    $4x + 5y = -9$

    $\left(\dfrac{3}{2}, -3\right)$

15. $4x - 6y = 5$
    $2x - 3y = 7$

    No solution

16. $3x + 6y = 7$
    $2x + 4y = 5$

    No solution

17. $3x - 5y = 7$
    $x - 2y = 3$

    (−1, −2)

18. $3x + 4y = 25$
    $2x + y = 10$

    (3, 4)

19. $x + 3y = 7$
    $-2x + 3y = 22$

    (−5, 4)

20. $2x - 3y = 14$
    $5x - 6y = 32$

    (4, −2)

21. $3x + 2y = 16$
$2x - 3y = -11$

$(2, 5)$

22. $2x - 5y = 13$
$5x + 3y = 17$

$(4, -1)$

23. $4x + 4y = 5$
$2x - 8y = -5$

$\left(\dfrac{1}{2}, \dfrac{3}{4}\right)$

24. $3x + 7y = 16$
$4x - 3y = 9$

$(3, 1)$

25. $5x + 4y = 0$
$3x + 7y = 0$

$(0, 0)$

26. $3x - 4y = 0$
$4x - 7y = 0$

$(0, 0)$

27. $5x + 2y = 1$
$2x + 3y = 7$

$(-1, 3)$

28. $3x + 5y = 16$
$5x - 7y = -4$

$(2, 2)$

29. $3x - 6y = 6$
$9x - 3y = 8$

$\left(\dfrac{2}{3}, -\dfrac{2}{3}\right)$

30. $\dfrac{2}{3}x - \dfrac{1}{2}y = 3$
$\dfrac{1}{3}x - \dfrac{1}{4}y = \dfrac{3}{2}$

$\left(x, \dfrac{4}{3}x - 6\right)$

31. $\dfrac{3}{4}x + \dfrac{1}{3}y = -\dfrac{1}{2}$
$\dfrac{1}{2}x - \dfrac{5}{6}y = -\dfrac{7}{2}$

$(-2, 3)$

32. $\dfrac{2}{5}x - \dfrac{1}{3}y = 1$
$\dfrac{3}{5}x + \dfrac{2}{3}y = 5$

$(5, 3)$

33. $\dfrac{5x}{6} + \dfrac{y}{3} = \dfrac{4}{3}$
$\dfrac{2x}{3} - \dfrac{y}{2} = \dfrac{11}{6}$

$(2, -1)$

34. $\dfrac{3x}{4} + \dfrac{2y}{5} = -\dfrac{3}{20}$
$\dfrac{3x}{2} - \dfrac{y}{4} = \dfrac{3}{4}$

$\left(\dfrac{1}{3}, -1\right)$

35. $\dfrac{2x}{5} - \dfrac{y}{2} = \dfrac{13}{2}$
$\dfrac{3x}{4} - \dfrac{y}{5} = \dfrac{17}{2}$

$(10, -5)$

36. $\dfrac{x}{2} + \dfrac{y}{3} = \dfrac{5}{12}$
$\dfrac{x}{2} - \dfrac{y}{3} = \dfrac{1}{12}$

$\left(\dfrac{1}{2}, \dfrac{1}{2}\right)$

37. $\dfrac{3x}{2} - \dfrac{y}{4} = -\dfrac{11}{12}$
$\dfrac{x}{3} - y = -\dfrac{5}{6}$

$\left(-\dfrac{1}{2}, \dfrac{2}{3}\right)$

38. $\dfrac{3x}{4} - \dfrac{2y}{3} = 0$
$\dfrac{5x}{4} - \dfrac{y}{3} = \dfrac{7}{12}$

$\left(\dfrac{2}{3}, \dfrac{3}{4}\right)$

39. $4x - 5y = 3y + 4$
$2x + 3y = 2x + 1$

$\left(\dfrac{5}{3}, \dfrac{1}{3}\right)$

40. $5x - 2y = 8x - 1$
$2x + 7y = 4y + 9$

$(-3, 5)$

41. $2x + 5y = 5x + 1$
$3x - 2y = 3y + 3$

No solution

42. $4x - 8y = 5$
$8x + 2y = 1$

$\left(\dfrac{1}{4}, -\dfrac{1}{2}\right)$

43. $5x + 2y = 2x + 1$
$2x - 3y = 3x + 2$

$(1, -1)$

44. $3x + 3y = y + 1$
$x + 3y = 9 - x$

$(-3, 5)$

OBJECTIVE B   *To solve a system of three linear equations in three variables by the addition method*

For Exercises 45 to 68, solve by the addition method.

45. $x + 2y - z = 1$
$2x - y + z = 6$
$x + 3y - z = 2$
$(2, 1, 3)$

46. $x + 3y + z = 6$
$3x + y - z = -2$
$2x + 2y - z = 1$
$(-1, 2, 1)$

47. $2x - y + 2z = 7$
$x + y + z = 2$
$3x - y + z = 6$
$(1, -1, 2)$

48. $x - 2y + z = 6$
$x + 3y + z = 16$
$3x - y - z = 12$
$(6, 2, 4)$

49. $3x + y = 5$
$3y - z = 2$
$x + z = 5$
$(1, 2, 4)$

50. $2y + z = 7$
$2x - z = 3$
$x - y = 3$
$(4, 1, 5)$

51. $x - y + z = 1$
$2x + 3y - z = 3$
$-x + 2y - 4z = 4$
$(2, -1, -2)$

52. $2x + y - 3z = 7$
$x - 2y + 3z = 1$
$3x + 4y - 3z = 13$
$(3, 1, 0)$

53. $2x + 3z = 5$
$3y + 2z = 3$
$3x + 4y = -10$
$(-2, -1, 3)$

54. $3x + 4z = 5$
$2y + 3z = 2$
$2x - 5y = 8$
$(-1, -2, 2)$

55. $2x + 4y - 2z = 3$
$x + 3y + 4z = 1$
$x + 2y - z = 4$
No solution

56. $x - 3y + 2z = 1$
$x - 2y + 3z = 5$
$2x - 6y + 4z = 3$
No solution

57. $2x + y - z = 5$
$x + 3y + z = 14$
$3x - y + 2z = 1$
$(1, 4, 1)$

58. $3x - y - 2z = 11$
$2x + y - 2z = 11$
$x + 3y - z = 8$
$(2, 1, -3)$

59. $3x + y - 2z = 2$
$x + 2y + 3z = 13$
$2x - 2y + 5z = 6$
$(1, 3, 2)$

60. $4x + 5y + z = 6$
$2x - y + 2z = 11$
$x + 2y + 2z = 6$
$(2, -1, 3)$

61. $2x - y + z = 6$
$3x + 2y + z = 4$
$x - 2y + 3z = 12$
$(1, -1, 3)$

62. $3x + 2y - 3z = 8$
$2x + 3y + 2z = 10$
$x + y - z = 2$
$(6, -2, 2)$

63. $3x - 2y + 3z = -4$
$2x + y - 3z = 2$
$3x + 4y + 5z = 8$
$(0, 2, 0)$

64. $3x - 3y + 4z = 6$
$4x - 5y + 2z = 10$
$x - 2y + 3z = 4$
$(0, -2, 0)$

65. $3x - y + 2z = 2$
$4x + 2y - 7z = 0$
$2x + 3y - 5z = 7$
$(1, 5, 2)$

66. $2x + 2y + 3z = 13$
$-3x + 4y - z = 5$
$5x - 3y + z = 2$
$(2, 3, 1)$

67. $2x - 3y + 7z = 0$
$x + 4y - 4z = -2$
$3x + 2y + 5z = 1$
$(-2, 1, 1)$

68. $5x + 3y - z = 5$
$3x - 2y + 4z = 13$
$4x + 3y + 5z = 22$
$(1, 1, 3)$

69. 🖎 For the following sentences, fill in the blank with one of the following phrases: **(i)** exactly one point, **(ii)** more than one point, or **(iii)** no points.
a. For an inconsistent system of linear equations in three variables, the planes representing the equations intersect at _____iii_____.
b. For a dependent system of linear equations in three variables, the planes representing the equations intersect at _____ii_____.
c. For an independent system of linear equations in three variables, the planes representing the equations intersect at _____i_____.

Critical Thinking

In Exercises 70 to 73, the systems are not systems of linear equations. However, each system can be solved by using a modification of the addition method. Solve each system of equations.

70. $\dfrac{1}{x} + \dfrac{2}{y} = 3$

$\dfrac{1}{x} - \dfrac{2}{y} = -1$

$(1, 1)$

71. $\dfrac{1}{x} - \dfrac{2}{y} = 3$

$\dfrac{2}{x} + \dfrac{3}{y} = -1$

$(1, -1)$

72. $\dfrac{3}{x} - \dfrac{5}{y} = -\dfrac{3}{2}$

$\dfrac{1}{x} - \dfrac{2}{y} = -\dfrac{2}{3}$

$(3, 2)$

73. $\dfrac{3}{x} + \dfrac{2}{y} = 1$

$\dfrac{2}{x} + \dfrac{4}{y} = -2$

$(1, -1)$

74. The point of intersection of the graphs of the equations $Ax + 3y = 6$ and $2x + By = -4$ is $(3, -2)$. Find A and B.
$A = 4, B = 5$

QUICK QUIZ
Solve by the addition method.

75. The point of intersection of the graphs of the equations $Ax + 3y + 2z = 8$, $2x + By - 3z = -12$, and $3x - 2y + Cz = 1$ is $(3, -2, 4)$. Find A, B, and C.
$A = 2, B = 3, C = -3$

Projects or Group Activities

76. The intersection of two distinct planes is a line. Let L be the line of intersection of the planes with equations $2x + y - z = 13$ and $x - 2y + z = 5$. If the point with coordinates $(x, 3, z)$ is on line L, find the value of $x - z$. 3

77. 🖎 Describe the graph of each of the following equations in an xyz-coordinate system.
a. $x = 3$ **b.** $y = 4$ **c.** $z = 2$ **d.** $y = x$

1. $x + y = 5$
$5x - y = 7$
$(2, 3)$ [5.2A]
2. $x - 4y = -7$
$3x - 2y = -11$
$(-3, 1)$ [5.2A]
3. $2x - y = 3$
$6x - 3y = 9$
$(x, 2x - 3)$ [5.2A]
4. $x - 2y - z = -5$
$5x - 4y + z = 5$
$2x + 3y + z = 9$
$(1, 1, 4)$ [5.2B]
5. $x - y + 2z = 3$
$3x + 2y + 2z = 2$
$2x + 3y - z = -4$
$(-2, 1, 3)$ [5.2B]

✔ CHECK YOUR PROGRESS: CHAPTER 5 ....

For Exercises 1 and 2, solve by graphing.

1. $2x - 3y = 5$

$x + 2y = -1$

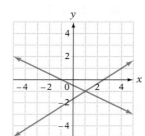

$(1, -1)$ [5.1A]

2. $2x + y = 6$

$x + \dfrac{y}{2} = 3$

 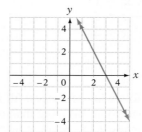

$(x, -2x + 6)$

[5.1A]

For Exercises 3 to 5, solve by the substitution method.

3. $4x - 2y = 16$
    $3x - y = 11$
    $(3, -2)$ [5.1B]

4. $9x + 12y = 11$
    $6x + 8y = 9$
    No solution   [5.1B]

5. $3x + 5y = 9$
    $2x - y = 7$
    $\left(\dfrac{44}{13}, -\dfrac{3}{13}\right)$ [5.1B]

For Exercises 6 to 10, solve by the addition method.

6. $3x + 5y = 14$
    $-2x + 3y = 16$

    $(-2, 4)$ [5.2A]

7. $2x + 5y = 10$
    $4x + 10y = 20$

    $\left(x, -\dfrac{2}{5}x + 2\right)$ [5.2A]

8. $x + 3y - 2z = -7$
    $2x + y + z = 6$
   $-3x - y + 3z = 4$
    $(2, -1, 3)$ [5.2B]

9. $4x + 5y - z = 22$
    $3x - 6y + 2z = -28$
    $x + 2y - 2z = 12$
    $(0, 4, -2)$ [5.2B]

10. $x - 3y + 2z = -2$
    $2x + y - z = 4$
    $x - y - 5z = 17$
    $(1, -1, -3)$ [5.2B]

11. Together, two investments earn an annual income of $1080. One investment earns an annual simple interest rate of 6%, and the second earns an annual simple interest rate of 4.5%. The total amount invested in the two accounts is $20,000. How much is invested in each account? $12,000 at 6%; $8000 at 4.5%   [5.1C]

Unless otherwise noted, all content on this page is © Cengage Learning.

5.3 | Application Problems

OBJECTIVE A   *To solve rate-of-wind or rate-of-current problems*

IN-CLASS EXAMPLES

1. Flying with the wind, a small plane flew 540 mi in 3 h. Flying against the wind, the plane could fly only 420 mi in the same amount of time. Find the rate of the plane in calm air and the rate of the wind.   **Plane: 160 mph; wind: 20 mph**

2. A cabin cruiser traveling with the current went 60 mi in 3 h. Traveling against the current, it took 5 h to go the same distance. Find the rate of the cabin cruiser in calm water and the rate of the current. **Cruiser: 16 mph; current: 4 mph**

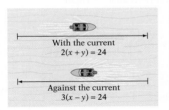

Solving motion problems that involve an object moving with or against a wind or current normally requires two variables.

HOW TO 1   A motorboat traveling with the current can go 24 mi in 2 h. Traveling against the current, it takes the boat 3 h to go the same distance. Find the rate of the motorboat in calm water and the rate of the current.

> **Strategy for Solving Rate-of-Wind or Rate-of-Current Problems**
>
> 1. Choose one variable to represent the rate of the object in calm conditions and a second variable to represent the rate of the wind or current. Using these variables, express the rate of the object with and against the wind or current. Use the equation $rt = d$ to write expressions for the distance traveled by the object. The results can be recorded in a table.

Rate of the boat in calm water: x
Rate of the current: y

	Rate		Time	=	Distance
With the current	$x + y$	$\cdot$	2	=	$2(x + y)$
Against the current	$x - y$	$\cdot$	3	=	$3(x - y)$

> 2. Determine how the expressions for distance are related.

The distance traveled with the current is 24 mi: $2(x + y) = 24$
The distance traveled against the current is 24 mi: $3(x - y) = 24$

Solve the system of equations.

$$2(x + y) = 24 \xrightarrow{\text{Multiply by } \frac{1}{2}.} \frac{1}{2} \cdot 2(x + y) = \frac{1}{2} \cdot 24 \longrightarrow x + y = 12$$

$$3(x - y) = 24 \xrightarrow{\text{Multiply by } \frac{1}{3}.} \frac{1}{3} \cdot 3(x - y) = \frac{1}{3} \cdot 24 \longrightarrow \underline{x - y = 8}$$

$$2x = 20 \quad \bullet \text{ Add the}$$
$$x = 10 \quad \text{equations.}$$

Replace x by 10 in the equation $x + y = 12$.
Solve for y.

$$x + y = 12$$
$$10 + y = 12$$
$$y = 2$$

The rate of the boat in calm water is 10 mph.
The rate of the current is 2 mph.

Unless otherwise noted, all content on this page is © Cengage Learning.

EXAMPLE 1

Flying with the wind, a plane flew 1000 mi in 5 h. Flying against the wind, the plane could fly only 500 mi in the same amount of time. Find the rate of the plane in calm air and the rate of the wind.

Strategy

• Rate of the plane in calm air: p
Rate of the wind: w

	Rate	Time	Distance
With wind	$p + w$	5	$5(p + w)$
Against wind	$p - w$	5	$5(p - w)$

• The distance traveled with the wind is 1000 mi:
$5(p + w) = 1000$
The distance traveled against the wind is 500 mi:
$5(p - w) = 500$

Solution

Solve the system of equations.

$$5(p + w) = 1000$$
$$5(p - w) = 500$$

$$\frac{1}{5} \cdot 5(p + w) = \frac{1}{5} \cdot 1000$$
$$\frac{1}{5} \cdot 5(p - w) = \frac{1}{5} \cdot 500$$

• Multiply each equation by $\frac{1}{5}$.

$$p + w = 200$$
$$\underline{p - w = 100}$$
$$2p = 300$$
$$p = 150$$

• Simplify.

• Add the equations.
• Solve for p.

$$p + w = 200$$
$$150 + w = 200$$
$$w = 50$$

• Substitute 150 for p.

The rate of the plane in calm air is 150 mph. The rate of the wind is 50 mph.

YOU TRY IT 1

A rowing team rowing with the current traveled 18 mi in 2 h. Rowing against the current, the team traveled 10 mi in 2 h. Find the rate of the rowing team in calm water and the rate of the current.

Your strategy

Your solution

Rate of rowing team in calm water: 7 mph
Rate of current: 2 mph

Solution on pp. S15–S16

Unless otherwise noted, all content on this page is © Cengage Learning.

OBJECTIVE B   *To solve application problems*

The application problems in this objective are varieties of problems solved earlier in the text. The strategy for each problem results in a system of equations.

IN-CLASS EXAMPLES

3. A sheet metal shop ordered 60 lb of tin and 30 lb of a zinc alloy for a total cost of $960. A second purchase, at the same prices, included 40 lb of tin and 70 lb of the zinc alloy. The total cost was $1540. Find the cost per pound of the tin and of the zinc alloy.   **Tin: $7/lb; zinc alloy: $18/lb**

HOW TO 2   A store owner purchased twenty 60-watt light bulbs and 30 fluorescent bulbs for a total cost of $80. A second purchase, at the same prices, included thirty 60-watt light bulbs and 10 fluorescent bulbs for a total cost of $50. Find the cost of a 60-watt bulb and that of a fluorescent bulb.

Strategy for Solving Application Problems
1. Choose a variable to represent each of the unknown quantities. Write numerical or variable expressions for all the remaining quantities. These results may be recorded in tables.

Cost of 60-watt bulb: b
Cost of fluorescent bulb: f

First Purchase

	Amount	·	Unit Cost	=	Value
60-watt	20	·	b	=	$20b$
Fluorescent	30	·	f	=	$30f$

Second Purchase

	Amount	·	Unit Cost	=	Value
60-watt	30	·	b	=	$30b$
Fluorescent	10	·	f	=	$10f$

2. Determine a system of equations. The strategies presented in the chapter on First-Degree Equations and Inequalities can be used to determine the relationships among the expressions in the tables. Each table will give one equation of the system of equations.

The total of the first purchase was $80: $20b + 30f = 80$
The total of the second purchase was $50: $30b + 10f = 50$

Solve the system of equations: (1) $20b + 30f = 80$
(2) $30b + 10f = 50$

$$\begin{aligned} 60b + 90f &= 240 \\ -60b - 20f &= -100 \\ \hline 70f &= 140 \\ f &= 2 \end{aligned}$$

• 3 times Equation (1).
• −2 times Equation (2).

$$\begin{aligned} 20b + 30f &= 80 \\ 20b + 30(2) &= 80 \\ 20b + 60 &= 80 \\ 20b &= 20 \\ b &= 1 \end{aligned}$$

• Replace f by 2 in Equation (1) and solve for b.

The cost of a 60-watt bulb was $1.00. The cost of a fluorescent bulb was $2.00.

Unless otherwise noted, all content on this page is © Cengage Learning.

EXAMPLE 2

A metallurgist has two alloys of stainless steel. Alloy I is 14% chromium and 6% nickel, and Alloy II is 18% chromium and 8% nickel. How many kilograms of each alloy should the metallurgist use to make a new stainless steel compound that contains 23 kg of chromium and 10 kg of nickel?

Strategy

- Kilograms of Alloy I: x
  Kilograms of Alloy II: y
  Use the percent mixture equation $Ar = Q$.

Kilograms of chromium:

	Amount	Percent	Quantity
Chromium in Alloy I	x	0.14	$0.14x$
Chromium in Alloy II	y	0.18	$0.18y$
Chromium in New Alloy			23

Kilograms of nickel:

	Amount	Percent	Quantity
Nickel in Alloy I	x	0.06	$0.06x$
Nickel in Alloy II	y	0.08	$0.08y$
Nickel in New Alloy			10

- The sum of the quantities of chromium in the two alloys equals the quantity of chromium in the new alloy: $0.14x + 0.18y = 23$.
- The sum of the quantities of nickel in the two alloys equals the quantity of nickel in the new alloy: $0.06x + 0.08y = 10$.

Solution

$$
\begin{aligned}
(1) \quad & 0.14x + 0.18y = 23 \\
(2) \quad & 0.06x + 0.08y = 10 \\
(3) \quad & 14x + 18y = 2300 \\
(4) \quad & 6x + 8y = 1000 \\
& 84x + 108y = 13{,}800 \\
& \underline{-84x - 112y = -14{,}000} \\
& \qquad\quad -4y = -200 \\
& \qquad\qquad\; y = 50 \\
(1) \quad & 0.14x + 0.18y = 23 \\
& 0.14x + 0.18(50) = 23 \\
& 0.14x + 9 = 23 \\
& 0.14x = 14 \\
& x = 100
\end{aligned}
$$

- Write a system of equations.
- Multiply each equation by 100 to remove decimals.
- Multiply Equation (3) by 6.
- Multiply Equation (4) by -14.
- Add the equations.
- Solve for y.

- Substitute 50 for y in Equation (1).

- Solve for x.

The metallurgist should use 100 kg of Alloy I and 50 kg of Alloy II.

YOU TRY IT 2

An investor has a total of $20,000 deposited in three different accounts, which earn annual interest rates of 9%, 7%, and 5%. The amount deposited in the 9% account is twice the amount in the 7% account. If the total annual interest earned by the three accounts is $1300, how much is invested in each account?

Your strategy

Your solution

9% account: $6000
7% account: $3000
5% account: $11,000

Solution on p. S16

Unless otherwise noted, all content on this page is © Cengage Learning.

5.3 EXERCISES

✔ Concept Check

SUGGESTED ASSIGNMENT
Exercises 1–4; Exercises 5–31, odds;
More challenging exercises: Exercises 32–34

1. The speed of a plane in calm air is 500 mph. If the plane is flying into a 50-mile-per-hour headwind, what is the speed of the plane relative to an observer on the ground?
450 mph

2. The speed of a boat in calm water is x miles per hour, and the speed of the current is y miles per hour. What is the rate of the boat going with the current?
$(x + y)$ miles per hour

3. A contractor bought 50 yd of nylon carpet for x dollars per yard and 100 yd of wool carpet for y dollars per yard. Express the total cost of the two purchases in terms of x and y. $50x + 100y$

4. A chemist has x grams of an alloy that is 20% silver and 10% gold, and y grams of an alloy that is 25% silver and 30% gold. Express the total number of grams of gold in the two alloys in terms of x and y. $0.10x + 0.30y$

OBJECTIVE A   *To solve rate-of-wind or rate-of-current problems*

5. 🖉 Traveling with the wind, a plane flies m miles in h hours. Traveling against the wind, the plane flies n miles in h hours. Is n less than, equal to, or greater than m?
Less than

6. 🖉 Traveling against the current, it takes a boat h hours to go m miles. Traveling with the current, the boat takes k hours to go m miles. Is h less than, equal to, or greater than k?   Greater than

7. A motorboat traveling with the current went 36 mi in 2 h. Traveling against the current, it took 3 h to go the same distance. Find the rate of the boat in calm water and the rate of the current.   Boat: 15 mph; current: 3 mph

8. A cabin cruiser traveling with the current went 45 mi in 3 h. Traveling against the current, it took 5 h to go the same distance. Find the rate of the cabin cruiser in calm water and the rate of the current.   Cabin cruiser: 12 mph; current: 3 mph

9. A jet plane flying with the wind went 2200 mi in 4 h. Flying against the wind, the plane could fly only 1820 mi in the same amount of time. Find the rate of the plane in calm air and the rate of the wind.   Plane: 502.5 mph; wind: 47.5 mph

10. Flying with the wind, a small plane flew 300 mi in 2 h. Flying against the wind, the plane could fly only 270 mi in the same amount of time. Find the rate of the plane in calm air and the rate of the wind.   Plane: 142.5 mph; wind: 7.5 mph

11. A rowing team rowing with the current traveled 20 km in 2 h. Rowing against the current, the team rowed 12 km in the same amount of time. Find the rate of the team in calm water and the rate of the current.   Team: 8 km/h; current: 2 km/h

John Kropewnicki/Shutterstock.com

12. A motorboat traveling with the current went 72 km in 3 h. Traveling against the current, the boat could go only 48 km in the same amount of time. Find the rate of the boat in calm water and the rate of the current.   Boat: 20 km/h; current: 4 km/h

13. A turboprop plane flying with the wind flew 800 mi in 4 h. Flying against the wind, the plane required 5 h to travel the same distance. Find the rate of the wind and the rate of the plane in calm air.   Plane: 180 mph; wind: 20 mph

14. Flying with the wind, a pilot flew 600 mi between two cities in 4 h. The return trip against the wind took 5 h. Find the rate of the plane in calm air and the rate of the wind.   Plane: 135 mph; wind: 15 mph

15. A plane flying with a tailwind flew 600 mi in 5 h. Flying against the wind, the plane required 6 h to fly the same distance. Find the rate of the plane in calm air and the rate of the wind.   Plane: 110 mph; wind: 10 mph

16. Flying with the wind, a plane flew 720 mi in 3 h. Against the wind, the plane required 4 h to fly the same distance. Find the rate of the plane in calm air and the rate of the wind.   Plane: 210 mph; wind: 30 mph

OBJECTIVE B   *To solve application problems*

17. 🗑 A coffee merchant's house blend contains 3 lb of dark roast coffee and 1 lb of light roast coffee. The merchant's breakfast blend contains 1 lb of dark roast coffee and 3 lb of light roast coffee. If the cost per pound of the house blend is greater than the cost per pound of the breakfast blend, is the cost per pound of the dark roast coffee less than, equal to, or greater than the cost per pound of the light roast coffee?
Greater than

18. 🗑 A chemist has two alloys of bronze, one that is 12% tin and 88% copper, and a second that is 10% tin and 90% copper. If these two alloys are melted and thoroughly mixed together, between what two values is the percent of tin in the mixture?
Between 10% and 12%

19. **Purchasing**   A carpenter purchased 60 ft of redwood and 80 ft of pine for a total cost of $286. A second purchase, at the same prices, included 100 ft of redwood and 60 ft of pine for a total cost of $396. Find the cost per foot of redwood and of pine.
Redwood: $3.30/ft; pine: $1.10/ft

20. **Business**   A merchant mixed 10 lb of cinnamon tea with 5 lb of spice tea. The 15-pound mixture cost $40. A second mixture included 12 lb of the cinnamon tea and 8 lb of the spice tea. The 20-pound mixture cost $54. Find the cost per pound of the cinnamon tea and of the spice tea.
Cinnamon: $2.50/lb; spice: $3/lb

21. **Purchasing**   A contractor buys 16 yd of nylon carpet and 20 yd of wool carpet for $1840. A second purchase, at the same prices, includes 18 yd of nylon carpet and 25 yd of wool carpet for $2200. Find the cost per yard of the wool carpet. $52/yd

Anna Hoychuk/Shutterstock.com

22. **Finances**   During one month, a homeowner used 500 units of electricity and 100 units of gas for a total cost of $352. The next month, 400 units of electricity and 150 units of gas were used for a total cost of $304. Find the cost per unit of gas.
$.32/unit

23. **Manufacturing**   A company manufactures both mountain bikes and trail bikes. The cost of materials for a mountain bike is $70, and the cost of materials for a trail bike is $50. The cost of labor to manufacture a mountain bike is $80, and the cost of labor to manufacture a trail bike is $40. During a week in which the company has budgeted $2500 for materials and $2600 for labor, how many mountain bikes does the company plan to manufacture?   25 mountain bikes

24. **Manufacturing**   A company manufactures both LCD and plasma televisions. The cost of materials for an LCD television is $125, and the cost of materials for a plasma TV is $150. The cost of labor to manufacture one LCD television is $80, and the cost of labor for one plasma television is $85. How many of each television can the manufacturer produce during a week in which $18,000 has been budgeted for materials and $10,750 has been budgeted for labor?   LCD: 60; plasma: 70

● **Fuel Economy**   For Exercises 25 and 26, use the information in the article at the right.

25. During one week, the owner of a hybrid car drove 394 mi and spent $34.74 on gasoline. How many miles did the owner drive in the city? On the highway?
City: 322 mi; highway: 72 mi

26. Gasoline for one week of driving cost the owner of a hybrid car $26.50. The owner would have spent $51.50 for gasoline to drive the same number of miles in a traditional car. How many miles did the owner drive in the city? On the highway?
City: 250 mi; highway: 50 mi

In the NEWS!

Hybrids Easier on the Pocketbook?

A hybrid car can make up for its high sticker price with savings at the pump. At current gas prices, here's a look at the cost per mile for one company's hybrid and traditional cars.

	Gasoline Cost per Mile	
Car Type	City ($/mi)	Highway ($/mi)
Hybrid	0.09	0.08
Traditional	0.18	0.13

Source: www.fueleconomy.gov

27. **Chemistry**   A chemist has two alloys, one of which is 10% gold and 15% lead, and the other of which is 30% gold and 40% lead. How many grams of each of the two alloys should be used to make an alloy that contains 60 g of gold and 88 g of lead?
First alloy: 480 g; second alloy: 40 g

28. **Health Science**   A pharmacist has two vitamin-supplement powders. The first powder is 20% vitamin B1 and 10% vitamin B2. The second is 15% vitamin B1 and 20% vitamin B2. How many milligrams of each powder should the pharmacist use to make a mixture that contains 130 mg of vitamin B1 and 80 mg of vitamin B2?
First powder: 560 mg; second powder: 120 mg

29. **Business**   On Monday, a computer manufacturing company sent out three shipments. The first order, which contained a bill for $114,000, was for 4 Model II, 6 Model VI, and 10 Model IX computers. The second shipment, which contained a bill for $72,000, was for 8 Model II, 3 Model VI, and 5 Model IX computers. The third shipment, which contained a bill for $81,000, was for 2 Model II, 9 Model VI, and 5 Model IX computers. What does the manufacturer charge for each Model VI computer? $4000

Unless otherwise noted, all content on this page is © Cengage Learning.

Image copyright Milevski Petar. Used under license from Shutterstock.com

30. **Purchasing**   A relief organization supplies blankets, cots, and lanterns to victims of fires, floods, and other natural disasters. One week, the organization purchased 15 blankets, 5 cots, and 10 lanterns for a total cost of $1250. The next week, at the same prices, the organization purchased 20 blankets, 10 cots, and 15 lanterns for a total cost of $2000. The third week, at the same prices, the organization purchased 10 blankets, 15 cots, and 5 lanterns for a total cost of $1625. Find the cost of one blanket, the cost of one cot, and the cost of one lantern.
$25 per blanket; $75 per cot; $50 per lantern

31. **Investments**   An investor has a total of $25,000 deposited in three different accounts, which earn annual interest rates of 8%, 6%, and 4%. The amount deposited in the 8% account is twice the amount in the 6% account. If the three accounts earn total annual interest of $1520, how much money is deposited in each account?
$10,400 at 8%; $5200 at 6%; $9400 at 4%

Critical Thinking

32. **Geometry**   Two angles are complementary. The measure of the larger angle is 9° more than eight times the measure of the smaller angle. Find the measures of the two angles. (Complementary angles are two angles whose sum is 90°.)
9° and 81°

33. **Geometry**   Two angles are supplementary. The measure of the larger angle is 40° more than three times the measure of the smaller angle. Find the measures of the two angles. (Supplementary angles are two angles whose sum is 180°.)
35° and 145°

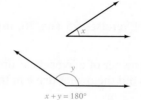

$x + y = 180°$

34. **Aviation**   A plane is flying the 3500 mi from New York City to London. The speed of the plane in calm air is 375 mph, and there is a 50-mile-per-hour tailwind. The *point of no return* is the point at which the flight time required to return to New York City is the same as the flight time to travel on to London. For this flight, how far from New York City is the point of no return? Round to the nearest whole number.
1517 mi

Projects or Group Activities

35. **Art**   A mobile is made by suspending three objects from a light rod that is 15 in. long, as shown below. The weight, in ounces, of each object is shown in the diagram. For the mobile to balance, the objects must be positioned so that $w_1d_1 = w_2d_2 + w_3d_3$. The artist wants d_3 to be three times d_2. Find the distances d_1, d_2, and d_3 such that the mobile will balance. $d_1 = 6$ in., $d_2 = 3$ in., $d_3 = 9$ in.

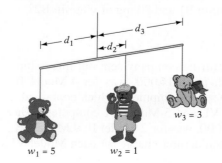

$w_1 = 5$ $w_2 = 1$ $w_3 = 3$

Unless otherwise noted, all content on this page is © Cengage Learning.

QUICK QUIZ

1. A jet plane flying with the wind went 2600 mi in 5 h. Flying against the wind, the plane could travel only 2200 mi in the same amount of time. Find the rate of the plane in calm air and the rate of the wind.   **Plane: 480 mph; wind: 40 mph**   **[5.3A]**

2. A motorboat traveling with the current went 80 mi in 4 h. Traveling against the current, it took 5 h to cover the same distance. Find the rate of the motorboat in calm water and the rate of the current.   **Motorboat: 18 mph; current: 2 mph**   **[5.3A]**

3. A restaurant manager buys 120 lb of hamburger and 60 lb of steak for a total cost of $720. A second purchase, at the same prices, includes 200 lb of hamburger and 80 lb of steak. The total cost is $1080. Find the cost of 1 lb of steak.   **$6/lb**   **[5.3B]**

SECTION

5.4 Solving Systems of Linear Inequalities

OBJECTIVE A *To graph the solution set of a system of linear inequalities*

Point of Interest

Large systems of linear inequalities containing over 100 inequalities have been used to solve application problems in such diverse areas as providing health care and hardening a nuclear missile silo.

Take Note

You can use a test point to check that the correct region has been designated as the solution set. We can see from the graph in HOW TO 1 that the point with coordinates (2, 4) is in the solution set. It is also, as shown below, a solution of each inequality of the system. This indicates that the solution set as graphed is correct.

$$2x - y \le 3$$
$$2(2) - (4) \le 3$$
$$0 \le 3 \quad \text{True}$$

$$3x + 2y > 8$$
$$3(2) + 2(4) > 8$$
$$14 > 8 \quad \text{True}$$

Integrating Technology

See the Keystroke Guide: *Graphing Inequalities* for instructions on using a graphing calculator to graph the solution set of a system of inequalities.

Two or more inequalities considered together are called a **system of inequalities.** The **solution set of a system of inequalities** is the intersection of the solution sets of the individual inequalities. To graph the solution set of a system of inequalities, first graph the solution set of each inequality. Recall that each of these solution sets is a half-plane. The solution set of the system of inequalities is the region of the plane represented by the intersection of the half-planes.

HOW TO 1 Graph the solution set: $2x - y \le 3$
$ 3x + 2y > 8$

Solve each inequality for *y*.

$$2x - y \le 3 \qquad\qquad 3x + 2y > 8$$
$$-y \le -2x + 3 \qquad\qquad 2y > -3x + 8$$
$$y \ge 2x - 3 \qquad\qquad y > -\frac{3}{2}x + 4$$

Graph $y = 2x - 3$ as a solid line. Because the inequality is $\ge$, shade above the line.

Graph $y = -\frac{3}{2}x + 4$ as a dashed line. Because the inequality is $>$, shade above the line.

The solution set of the system is the region of the plane that represents the intersection of the solution sets of the individual inequalities.

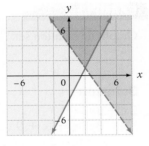

HOW TO 2 Graph the solution set: $-x + 2y \ge 4$
$ x - 2y \ge 6$

Solve each inequality for *y*.

$$-x + 2y \ge 4 \qquad\qquad x - 2y \ge 6$$
$$2y \ge x + 4 \qquad\qquad -2y \ge -x + 6$$
$$y \ge \frac{1}{2}x + 2 \qquad\qquad y \le \frac{1}{2}x - 3$$

Shade above the solid line graph of $y = \frac{1}{2}x + 2$.

Shade below the solid line graph of $y = \frac{1}{2}x - 3$.

Because the solution sets of the two inequalities do not intersect, the solution of the system is the empty set.

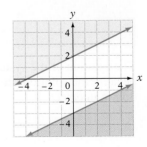

Unless otherwise noted, all content on this page is © Cengage Learning.

EXAMPLE 1

Graph the solution set: $y \geq x - 1$
$y < -2x$

Solution

Shade above the solid line graph of $y = x - 1$.
Shade below the dashed line graph of $y = -2x$.

The solution of the system is the intersection of
the solution sets of the individual inequalities.

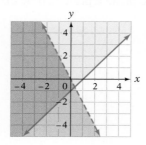

YOU TRY IT 1

Graph the solution set: $y \geq 2x - 3$
$y > -3x$

Your solution

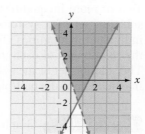

IN-CLASS EXAMPLES
Graph the solution set.
1. $3x - 2y < 12$
   $x + 3y < 6$

2. $x + 2y \leq 4$
   $x - y \leq 3$

EXAMPLE 2

Graph the solution set: $2x + 3y > 9$
$y < -\dfrac{2}{3}x + 1$

Solution

$2x + 3y > 9$
$\qquad 3y > -2x + 9$
$\qquad\quad y > -\dfrac{2}{3}x + 3$

Shade above the dashed line graph of
$y = -\dfrac{2}{3}x + 3$.

Shade below the dashed line graph of
$y = -\dfrac{2}{3}x + 1$.

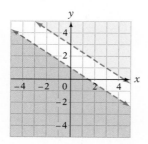

The solution of the system is the empty set
because the solution sets of the two inequalities
do not intersect.

YOU TRY IT 2

Graph the solution set: $3x + 4y > 12$
$y < \dfrac{3}{4}x - 1$

Your solution

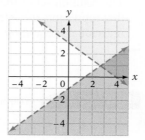

Solutions on p. S16

Unless otherwise noted, all content on this page is © Cengage Learning.

5.4 EXERCISES

✔ Concept Check

SUGGESTED ASSIGNMENT
Exercises 1 and 2; Exercises 3–17, odds;
More challenging exercises: Exercises 18–22

1. Which ordered pair is a solution of the system of inequalities shown at the right? $2x - y < 4$
   (i) $(5, 1)$ **(ii)** $(-3, -5)$ ii $x - 3y \geq 6$

2. Is the solution set of a system of inequalities the *union* or the *intersection* of the solution sets of the individual inequalities?   Intersection

---

OBJECTIVE A   *To graph the solution set of a system of linear inequalities*

For Exercises 3 to 17, graph the solution set.

3. $x - y \geq 3$
   $x + y \leq 5$

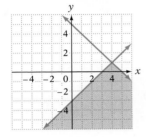

4. $2x - y < 4$
   $x + y < 5$

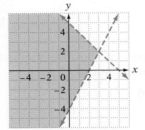

5. $3x - y < 3$
   $2x + y \geq 2$

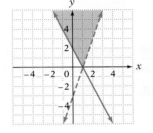

6. $x + 2y \leq 6$
   $x - y \leq 3$

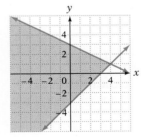

7. $2x + y \geq -2$
   $6x + 3y \leq 6$

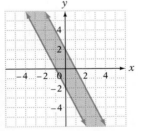

8. $x + y \geq 5$
   $3x + 3y \leq 6$ ∅

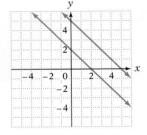

9. $3x - 2y < 6$
   $y \leq 3$

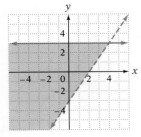

10. $x \leq 2$
   $3x + 2y > 4$

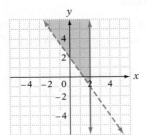

11. $x + 1 \geq 0$
   $y - 3 \leq 0$

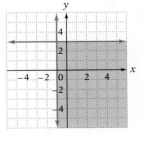

Unless otherwise noted, all content on this page is © Cengage Learning.

12. $x < 3$
$y < -2$

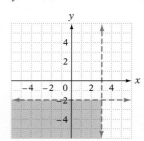

13. $2x + y \geq 4$
$3x - 2y < 6$

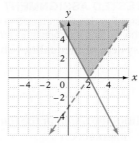

14. $5x - 2y \geq 10$
$3x + 2y \geq 6$

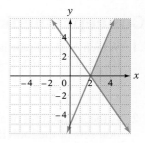

15. $x - 2y \leq 6$
$2x + 3y \leq 6$

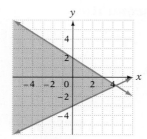

16. $x - 3y > 6$
$2x + y > 5$

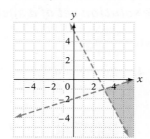

17. $x - 2y \leq 4$
$3x + 2y \leq 8$

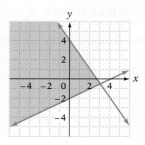

For Exercises 18 to 20, assume that a and b are positive numbers such that $a > b$. Describe the solution set of each system of inequalities.

18. $x + y > a$
$x + y > b$
Points above the
line $x + y = a$

19. $x + y < a$
$x + y > b$
Region between the
parallel lines $x + y = a$
and $x + y = b$

20. $x + y > a$
$x + y < b$
Empty set

Critical Thinking

For Exercises 21 and 22, graph the solution set.

21. $2x + 3y \leq 15$
$3x - y \leq 6$
$y \geq 0$

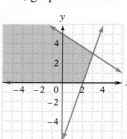

22. $2x - y \leq 4$
$3x + y < 1$
$y \leq 0$

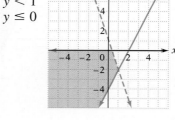

QUICK QUIZ
Graph the solution set.
1. $x + 2y \geq 4$
$4x + 3y < 12$

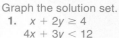

[5.4A]

2. $x - 3y \leq 6$
$2x + y \leq 4$

[5.4A]

Projects or Group Activities

23. A set of points in a plane is a **convex set** if each line segment connecting a pair of points in the set is contained completely within the set. Which of the following are convex sets?   ii, iii

(i) **(ii)**  **(iii)** **(iv)**

Unless otherwise noted, all content on this page is © Cengage Learning.

5 : Summary

Key Words	Examples
A **system of equations** is two or more equations considered together. A **solution of a system of equations in two variables** is an ordered pair that is a solution of each equation of the system. [5.1A, p. 298]	The solution of the system $$x + y = 2$$ $$x - y = 4$$ is the ordered pair $(3, -1)$. $(3, -1)$ is the only ordered pair that is a solution of both equations.
When the graphs of a system of equations intersect at exactly one point, the system is called an **independent system of equations.** [5.1A, p. 298]	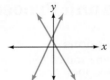
When the graphs of a system of equations do not intersect, the system has no solution and is called an **inconsistent system of equations.** [5.1A, p. 299]	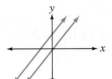
When the graphs of a system of equations coincide, the system is called a **dependent system of equations.** [5.1A, p. 299]	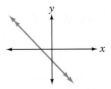
An equation of the form $Ax + By + Cz = D$, where A, B, and C are the coefficients of the variables and D is a constant, is a **linear equation in three variables**. A **solution of an equation in three variables** is an **ordered triple** (x, y, z). [5.2B, p. 313]	$3x + 2y - 5z = 12$ is a linear equation in three variables. One solution of this equation is the ordered triple $(0, 1, -2)$.

A **solution of a system of equations in three variables** is an ordered triple that is a solution of each equation of the system. [5.2B, p. 313]

The solution of the system

$$3x + y - 3z = 2$$
$$-x + 2y + 3z = 6$$
$$2x + 2y - 2z = 4$$

is the ordered triple $(1, 2, 1)$. $(1, 2, 1)$ is the only ordered triple that is a solution of all three equations.

Two or more inequalities considered together are called a **system of inequalities**. The **solution set of a system of inequalities** is the intersection of the solution sets of the individual inequalities. [5.4A, p. 331]

$x + y > 3$
$x - y > -2$

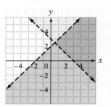

Essential Rules and Procedures

Examples

Solving Systems of Equations
A system of equations can be solved by:
1. Graphing [5.1A, pp. 298-299]

$y = \dfrac{1}{2}x + 2$

$y = \dfrac{5}{2}x - 2$

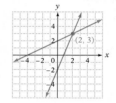

2. The substitution method [5.1B, p. 301]

(1) $2x - 3y = 4$
(2) $\qquad y = -x + 2$

Substitute $-x + 2$ for y in Equation (1).

$2x - 3(-x + 2) = 4$

3. The addition method [5.2A, p. 310]

$-2x + 3y = 7$
$\underline{2x - 5y = 2}$
$\qquad -2y = 9$ • Add the equations.

Annual Simple Interest Equation [5.1C, p. 304]
Principal · simple interest rate = simple interest
$Pr = I$

You have a total of $10,000 to invest in two simple interest accounts, one earning 4.5% annual simple interest and the other earning 5% annual simple interest. If you earn $485 per year in interest from these two investments, how much do you have invested in each account?

$x + y = 10,000$
$0.045x + 0.05y = 485$

Unless otherwise noted, all content on this page is © Cengage Learning.

CHAPTER

5   Review Exercises

1. Solve by substitution: $2x - 6y = 15$
$x = 4y + 8$

$\left(6, -\dfrac{1}{2}\right)$ [5.1B]

2. Solve by the addition method: $3x + 2y = 2$
$x + y = 3$

$(-4, 7)$ [5.2A]

3. Solve by graphing: $x + y = 3$
$3x - 2y = -6$

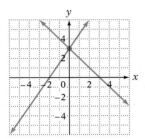

$(0, 3)$ [5.1A]

4. Solve by graphing: $2x - y = 4$
$y = 2x - 4$

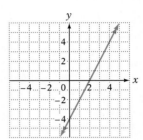

$(x, 2x - 4)$ [5.1A]

5. Solve by substitution: $3x + 12y = 18$
$x + 4y = 6$

$\left(x, -\dfrac{1}{4}x + \dfrac{3}{2}\right)$ [5.1B]

6. Solve by the addition method: $5x - 15y = 30$
$x - 3y = 6$

$\left(x, \dfrac{1}{3}x - 2\right)$ [5.2A]

7. Solve by the addition method:
$3x - 4y - 2z = 17$
$4x - 3y + 5z = 5$
$5x - 5y + 3z = 14$
$(3, -1, -2)$ [5.2B]

8. Solve by the addition method:
$3x + y = 13$
$2y + 3z = 5$
$x + 2z = 11$
$(5, -2, 3)$ [5.2B]

9. Is $(1, -2)$ a solution of the system of equations?
$6x + y = 4$
$2x - 5y = 12$
Yes   [5.1A]

10. Solve by substitution:
$2x - 4y = 11$
$y = 3x - 4$
$\left(\dfrac{1}{2}, -\dfrac{5}{2}\right)$ [5.1B]

Unless otherwise noted, all content on this page is © Cengage Learning.

11. Solve by substitution:
$$2x - y = 7$$
$$3x + 2y = 7$$
(3, −1)   [5.1B]

12. Solve by the addition method:
$$3x - 4y = 1$$
$$2x + 5y = 16$$
(3, 2)   [5.2A]

13. Solve:
$$x + y + z = 0$$
$$x + 2y + 3z = 5$$
$$2x + y + 2z = 3$$

(−1, −3, 4)   [5.2B]

14. Solve:
$$x + 3y + z = 6$$
$$2x + y - z = 12$$
$$x + 2y - z = 13$$
(2, 3, −5)   [5.2B]

15. Graph the solution set:
$$x + 3y \le 6$$
$$2x - y \ge 4$$ [5.4A]

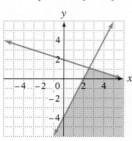

16. Graph the solution set:
$$2x + 4y \ge 8$$
$$x + y \le 3$$ [5.4A]

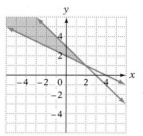

17. Boating   A cabin cruiser traveling with the current went 60 mi in 3 h. Traveling against the current, it took 5 h to go the same distance. Find the rate of the cabin cruiser in calm water and the rate of the current.
Cabin cruiser: 16 mph; current: 4 mph   [5.3A]

18. Aeronautics   A pilot flying with the wind flew 600 mi in 3 h. Flying against the wind, the pilot required 4 h to travel the same distance. Find the rate of the plane in calm air and the rate of the wind.   Plane: 175 mph; wind: 25 mph   [5.3A]

19. Ticket Sales   At a movie theater, admission tickets cost $5 for children and $8 for adults. The receipts for one Friday evening were $2500. The next day, three times as many children as the night before and only half the number of adults bought admission tickets, yet the receipts were still $2500. Find the number of children who attended on Friday evening.   100 children   [5.3B]

20. Investments   A trust administrator divides $20,000 between two accounts. One account earns an annual simple interest rate of 3%, and the second account earns an annual simple interest rate of 7%. The total annual income from the two accounts is $1200. How much is invested in each account?
$5000 at 3%; $15,000 at 7%   [5.1C]

Unless otherwise noted, all content on this page is © Cengage Learning.

CHAPTER

5 ┊ TEST

1. Solve by graphing: $2x - 3y = -6$
                              $2x - y = 2$

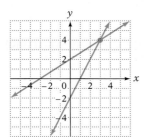

$(3, 4)$ [5.1A]

2. Solve by graphing: $x - 2y = -6$
                              $y = \dfrac{1}{2}x - 4$

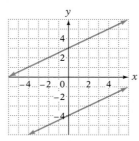

No solution   [5.1A]

3. Graph the solution set: $2x - y < 3$
                                  $4x + 3y < 11$

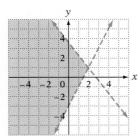

[5.4A]

4. Graph the solution set: $x + y > 2$
                                  $2x - y < -1$

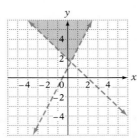

[5.4A]

5. Solve by substitution: $3x + 2y = 4$
                                 $x = 2y - 1$

$\left(\dfrac{3}{4}, \dfrac{7}{8}\right)$ [5.1B]

6. Solve by substitution: $5x + 2y = -23$
                                 $2x + y = -10$

$(-3, -4)$ [5.1B]

7. Solve by substitution: $y = 3x - 7$
                                 $y = -2x + 3$

$(2, -1)$ [5.1B]

8. Solve by the addition method:
     $3x + 4y = -2$
     $2x + 5y = 1$
     $(-2, 1)$ [5.2A]

9. Solve by the addition method:
     $4x - 6y = 5$
     $6x - 9y = 4$
     No solution   [5.2A]

10. Solve by the addition method:
     $3x - y = 2x + y - 1$
     $5x + 2y = y + 6$
     $(1, 1)$ [5.2A]

Unless otherwise noted, all content on this page is © Cengage Learning.

11. Solve by the addition method:
$$2x + 4y - z = 3$$
$$x + 2y + z = 5$$
$$4x + 8y - 2z = 7$$
No solution   [5.2B]

12. Solve by the addition method:
$$x - y - z = 5$$
$$2x + z = 2$$
$$3y - 2z = 1$$
$(2, -1, -2)$ [5.2B]

13. Solve by substitution:
$$x - y = 3$$
$$2x + y = -4$$
$\left(-\dfrac{1}{3}, -\dfrac{10}{3}\right)$ [5.1B]

14. Is $(2, -2)$ a solution of the system of equations?
$$5x + 2y = 6$$
$$3x + 5y = -4$$
Yes   [5.1A]

15. Solve:
$$x - y + z = 2$$
$$2x - y - z = 1$$
$$x + 2y - 3z = -4$$
$\left(\dfrac{1}{5}, -\dfrac{6}{5}, \dfrac{3}{5}\right)$ [5.2B]

16. Aeronautics   A plane flying with the wind went 350 mi in 2 h. The return trip, flying against the wind, took 2.8 h. Find the rate of the plane in calm air and the rate of the wind.   Plane: 150 mph; wind: 25 mph   [5.3A]

17. Purchasing   A clothing manufacturer purchased 60 yd of cotton and 90 yd of wool for a total cost of $1800. Another purchase, at the same prices, included 80 yd of cotton and 20 yd of wool for a total cost of $1000. Find the cost per yard of the cotton and of the wool.   Cotton: $9; wool: $14   [5.3B]

Thorsten Schmitt/Shutterstock.com

18. Investments   The annual interest earned on two investments is $549. One investment is in a 2.7% tax-free annual simple interest account, and the other investment is in a 5.1% annual simple interest CD. The total amount invested is $15,000. How much is invested in each account? $9000 at 2.7%; $6000 at 5.1%   [5.1C]

Cumulative Review Exercises

1. Solve: $\dfrac{3}{2}x - \dfrac{3}{8} + \dfrac{1}{4}x = \dfrac{7}{12}x - \dfrac{5}{6}$

$-\dfrac{11}{28}$ [2.2C]

2. Find the equation of the line that contains the points $P_1(2, -1)$ and $P_2(3, 4)$.
$y = 5x - 11$ [4.5B]

3. Simplify: $3[x - 2(5 - 2x) - 4x] + 6$
$3x - 24$ [1.4D]

4. Evaluate $a + bc \div 2$ when $a = 4$, $b = 8$, and $c = -2$.
-4 [1.4A]

5. Solve: $2x - 3 < 9$ or $5x - 1 < 4$
Write the solution set in interval notation.
$(-\infty, 6)$ [2.5B]

6. Solve: $|x - 2| - 4 < 2$
$\{x | -4 < x < 8\}$ [2.6B]

7. Solve: $|2x - 3| > 5$
$\{x | x < -1\} \cup \{x | x > 4\}$ [2.6B]

8. Given $f(x) = 3x^3 - 2x^2 + 1$, evaluate $f(-3)$.
-98 [4.2A]

9. What is the domain of $f(x) = 3x^2 - 2x$?
$\{x | -\infty < x < \infty\}$ [4.2A]

10. Given $F(x) = x^2 - 3$, find $F(2)$.
1 [4.2A]

11. Given $f(x) = 3x - 4$, write $f(2 + h) - f(2)$ in simplest form.
$3h$ [4.2A]

12. Graph the solution set of $\{x | x \le 2\} \cap \{x | x > -3\}$.

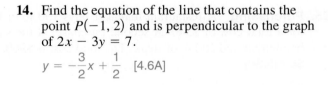

[1.5B]

13. Find the equation of the line that contains the point $P(-2, 3)$ and has slope $-\dfrac{2}{3}$.
$y = -\dfrac{2}{3}x + \dfrac{5}{3}$ [4.5A]

14. Find the equation of the line that contains the point $P(-1, 2)$ and is perpendicular to the graph of $2x - 3y = 7$.
$y = -\dfrac{3}{2}x + \dfrac{1}{2}$ [4.6A]

15. Graph $2x - 5y = 10$ by using the slope and y-intercept.   [4.4B]

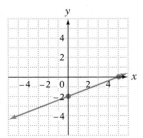

16. Graph the solution set of the inequality $3x - 4y \ge 8$.   [4.7A]

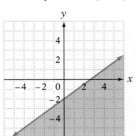

Unless otherwise noted, all content on this page is © Cengage Learning.

17. Solve by graphing.
$5x - 2y = 10$
$3x + 2y = 6$
(2, 0)   [5.1A]

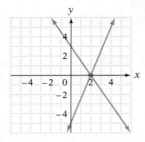

18. Graph the solution set.
$3x - 2y \geq 4$
$x + y < 3$
[5.4A]

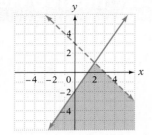

19. Solve by the addition method:
$3x + 2z = 1$
$2y - z = 1$
$x + 2y = 1$
(1, 0, −1)   [5.2B]

20. Solve:
$2x - y + z = 2$
$3x + y + 2z = 5$
$3x - y + 4z = 1$
(2, 1, −1)   [5.2B]

21. Solve by the addition method:
$4x - 3y = 17$
$3x - 2y = 12$
(2, −3)   [5.2B]

22. Solve by substitution:
$3x - 2y = 7$
$y = 2x - 1$
(−5, −11)   [5.1B]

23. Mixtures   How many milliliters of pure water must be added to 100 ml of a 4% salt solution to make a 2.5% salt solution?   60 ml   [2.4B]

24. Aeronautics   Flying with the wind, a small plane required 2 h to fly 150 mi. Against the wind, it took 3 h to fly the same distance. Find the rate of the wind.   12.5 mph   [5.3A]

25. Purchasing   A restaurant manager buys 100 lb of hamburger and 50 lb of steak for a total cost of $540. A second purchase, at the same prices, includes 150 lb of hamburger and 100 lb of steak. The total cost is $960. Find the cost of 1 lb of steak.
$6   [5.3B]

26. Electronics   Find the lower and upper limits of a 12,000-ohm resistor with a 15% tolerance.   Lower limit: 10,200 ohms; upper limit: 13,800 ohms   [2.6C]

27. Compensation   The graph shows the relationship between the monthly income and the sales of an account executive. Find the slope of the line between the two points shown on the graph. Write a sentence that states the meaning of the slope. $m = 40$
The commission rate of the executive is $40 for every $1000 worth of sales.   [4.4A]

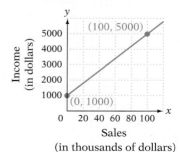

Unless otherwise noted, all content on this page is © Cengage Learning.

Polynomials

Supri Suharjoto/Shutterstock.com

6

Focus on Success

Are you using the features of this text to learn the concepts being presented? The HOW TO feature includes a step-by-step solution to the types of exercises you will be working in your homework assignment and on exams. A numbered Example provides you with a fully-worked-out solution. After studying the Example, try completing the You Try It to the right of the Example. A complete solution to the You Try It is given in the back of the text. (See Use the Interactive Method, page AIM-8)

Prep Test

Are you ready to succeed in this chapter? Take the Prep Test below to find out if you are ready to learn the new material.

1. Subtract: $-2 - (-3)$
1   [1.1C]

2. Multiply: $-3(6)$
-18 [1.1D]

3. Simplify: $-\dfrac{24}{-36}$
$\dfrac{2}{3}$ [1.1D]

4. Evaluate $3n^4$ when $n = -2$.
48   [1.4A]

5. If $\dfrac{a}{b}$ is a fraction in simplest form, what number is not a possible value of b?
0   [1.1D]

6. Are $2x^2$ and $2x$ like terms?
No   [1.4B]

7. Simplify:
$3x^2 - 4x + 1 + 2x^2 - 5x - 7$
$5x^2 - 9x - 6$ [1.4B]

8. Simplify: $-4y + 4y$
0   [1.4B]

9. Simplify: $-3(2x - 8)$
$-6x + 24$ [1.4D]

10. Simplify:
$3xy - 4y - 2(5xy - 7y)$
$-7xy + 10y$ [1.4D]

6.1 Exponential Expressions

OBJECTIVE A   *To multiply monomials*

 Point of Interest

Around A.D. 250, the monomial $3x^2$ shown at the right would have been written $\Delta^Y 3$, or at least approximately like that. In A.D. 250, the symbol for 3 was not the one we use today.

A **monomial** is a number, a variable, or a product of a number and variables.

The examples at the right are monomials. The **degree of a monomial** is the sum of the exponents on the variables.

x	degree 1 $(x = x^1)$
$3x^2$	degree 2
$4x^2y$	degree 3
$6x^3y^4z^2$	degree 9

In this chapter, the variable n is considered a positive integer when used as an exponent.

x^n degree n

The degree of a nonzero constant term is zero.

6   degree 0

The expression $5\sqrt{x}$ is not a monomial because $\sqrt{x}$ cannot be written as a product of variables. The expression $\frac{x}{y}$ is not a monomial because it is a quotient of variables.

The expression x^4 is an exponential expression. The exponent, 4, indicates the number of times the base, x, occurs as a factor.

INSTRUCTOR NOTE

You might prefer to have students develop the rule for multiplying exponential expressions themselves. Provide them with several expressions to simplify, such as

x^4x^3

y^2y^6

a^5a^7

Ask them to simplify the expressions by writing each expression in factored form and then writing the result with an exponent. After they have completed several examples, ask them to write a rule for x^mx^n. This same approach can be used to derive the Rule for Simplifying the Power of an Exponential Expression and the Rule for Dividing Exponential Expressions.

The product of exponential expressions with the *same base* can be simplified by writing each expression in factored form and writing the result with an exponent.

$$x^3 \cdot x^4 = \overbrace{(x \cdot x \cdot x)}^{3 \text{ factors}} \cdot \overbrace{(x \cdot x \cdot x \cdot x)}^{4 \text{ factors}}$$

$$\underbrace{}_{7 \text{ factors}}$$

$$= x^7$$

Note that adding the exponents results in the same product.

$$x^3 \cdot x^4 = x^{3+4} = x^7$$

Rule for Multiplying Exponential Expressions

If m and n are positive integers, then $x^m \cdot x^n = x^{m+n}$.

EXAMPLES

1. $x^4 \cdot x^7 = x^{4+7} = x^{11}$

2. $z \cdot z^6 = z^{1+6} = z^7$ • Recall that $z = z^1$.

3. $v^2 \cdot v^3 \cdot v^5 = v^{2+3+5} = v^{10}$

4. $(a^3b^5)(a^4b^7) = a^{3+4}b^{5+7} = a^7b^{12}$ • Add exponents on like bases.

HOW TO 1 Simplify: $(-4x^5y^3)(3xy^2)$

$(-4x^5y^3)(3xy^2) = (-4 \cdot 3)(x^5 \cdot x)(y^3 \cdot y^2)$
  • Use the Commutative and Associative Properties of Multiplication to rearrange and group factors.

$= -12(x^{5+1})(y^{3+2})$
  • Multiply variables with the same base by adding their exponents.

$= -12x^6y^5$
  • Simplify.

As shown below, the power of a monomial can be simplified by writing the power in factored form and then using the Rule for Multiplying Exponential Expressions. It can also be simplified by multiplying each exponent inside the parentheses by the exponent outside the parentheses.

$(a^2)^3 = a^2 \cdot a^2 \cdot a^2$
$= a^{2+2+2}$
$= a^6$

$(x^3y^4)^2 = (x^3y^4)(x^3y^4)$
$= x^{3+3}y^{4+4}$
$= x^6y^8$
  • Write in factored form. Then use the Rule for Multiplying Exponential Expressions.

$(a^2)^3 = a^{2\cdot3} = a^6$

$(x^3y^4)^2 = x^{3\cdot2}y^{4\cdot2} = x^6y^8$
  • Multiply each exponent inside the parentheses by the exponent outside the parentheses.

Rule for Simplifying the Power of an Exponential Expression

If m and n are integers, then $(x^m)^n = x^{mn}$.

EXAMPLES

1. $(x^4)^3 = x^{4\cdot3} = x^{12}$

2. $(z^5)^7 = z^{5\cdot7} = z^{35}$

Rule for Simplifying Powers of Products

If $m, n,$ and p are integers, then $(x^my^n)^p = x^{mp}y^{np}$.

EXAMPLES

1. $(x^2y^5)^3 = x^{2\cdot3}y^{5\cdot3} = x^6y^{15}$

2. $(4x^5)^3 = 4^{1\cdot3}x^{5\cdot3}$ • Recall that $4 = 4^1$.
$= 4^3x^{15} = 64x^{15}$

IN-CLASS EXAMPLES
Simplify.
  1. $(-2a^2b^3)(-4ab^2)$
    $8a^3b^5$
  2. $(3ab^2)^3$ $27a^3b^6$
  3. $(-2a^2b)^2(-3ab^2)^2$
    $36a^6b^6$

HOW TO 2 Simplify: $(2a^3b^4)^3$

$(2a^3b^4)^3 = 2^{1\cdot3}a^{3\cdot3}b^{4\cdot3}$
$= 2^3a^9b^{12}$
$= 8a^9b^{12}$
  • Use the Rule for Simplifying Powers of Products to multiply each exponent inside the parentheses by the exponent outside the parentheses.

EXAMPLE 1

Simplify: $(-4x^2y^5)^3$

Solution

$(-4x^2y^5)^3 = (-4)^3x^6y^{15}$ • $-4 = (-4)^1$
$= -64x^6y^{15}$

YOU TRY IT 1

Simplify: $(-3a^5b^7)^4$

Your solution

$81a^{20}b^{28}$

Solution on p. S16

EXAMPLE 2

Simplify: $(2xy^2)(-3xy^4)^3$

Solution

$$(2xy^2)(-3xy^4)^3 = (2xy^2)[(-3)^3 x^3 y^{12}]$$
$$= (2xy^2)(-27x^3 y^{12})$$
$$= -54x^4 y^{14}$$

YOU TRY IT 2

Simplify: $(-3a^2 b^4)(-2ab^3)^4$

Your solution

$-48a^6 b^{16}$

EXAMPLE 3

Simplify: $(-3x^3 y)^2 (-2x^3 y^5)^3$

Solution

$$(-3x^3 y)^2 (-2x^3 y^5)^3 = [(-3)^2 x^6 y^2][(-2)^3 x^9 y^{15}]$$
$$= [9x^6 y^2][-8x^9 y^{15}]$$
$$= -72x^{15} y^{17}$$

YOU TRY IT 3

Simplify: $(-4ab^4)^2 (2a^4 b^2)^4$

Your solution

$256a^{18} b^{16}$

Solutions on pp. S16–S17

OBJECTIVE B   *To divide monomials and simplify expressions with negative exponents*

The quotient of two exponential expressions with the same base can be simplified by writing each expression in factored form, dividing by the common factors, and then writing the result with an exponent.

$$\frac{x^5}{x^2} = \frac{\overset{1}{\cancel{x}} \cdot \overset{1}{\cancel{x}} \cdot x \cdot x \cdot x}{\underset{1}{\cancel{x}} \cdot \underset{1}{\cancel{x}}} = x^3$$

Note that subtracting the exponents gives the same result.

$$\frac{x^5}{x^2} = x^{5-2} = x^3$$

To divide two monomials with the same base, subtract the exponents of the like bases.

Rule for Dividing Exponential Expressions

If m and n are integers and $x \neq 0$, then $\dfrac{x^m}{x^n} = x^{m-n}$.

EXAMPLES

1. $\dfrac{x^9}{x^2} = x^{9-2} = x^7$

2. $\dfrac{a^4 b^5}{ab^3} = a^{4-1} b^{5-3} = a^3 b^2$ • Subtract exponents on like bases.

INSTRUCTOR NOTE

Have students copy this rule onto a piece of paper and then practice a few exercises, such as $\dfrac{a^9}{a^2}$ and $\dfrac{y^8}{y}$. It may also help to give $\dfrac{a^9}{b^5}$ as an exercise to emphasize that the bases must be the same in order to subtract the exponents.

HOW TO 3   Simplify: $\dfrac{6x^5 y^3}{8x^2 y}$

$$\frac{6x^5 y^3}{8x^2 y} = \frac{3x^{5-2} y^{3-1}}{4}$$ • Write $\dfrac{6}{8}$ in simplest form. Subtract the exponents on like bases.

$$= \frac{3x^3 y^2}{4}$$

Consider the expression $\dfrac{x^4}{x^4}$, $x \neq 0$. This expression can be simplified, as shown below, by subtracting exponents or dividing by common factors.

$$\dfrac{x^4}{x^4} = x^{4-4} = x^0 \qquad \dfrac{x^4}{x^4} = \dfrac{\overset{1}{\cancel{x}} \cdot \overset{1}{\cancel{x}} \cdot \overset{1}{\cancel{x}} \cdot \overset{1}{\cancel{x}}}{\underset{1}{\cancel{x}} \cdot \underset{1}{\cancel{x}} \cdot \underset{1}{\cancel{x}} \cdot \underset{1}{\cancel{x}}} = 1$$

The equations $\dfrac{x^4}{x^4} = x^0$ and $\dfrac{x^4}{x^4} = 1$ suggest the following definition of x^0.

Take Note

Pay special attention to example (3) at the right. The expression -15^0 is the *opposite* of 15^0. Note the difference between $-15^0 = -1$ and $(-15)^0 = 1$. Also note the difference between the expression in example (4) and the expression $(3z)^0 = 1$.

Definition of Zero as an Exponent

If $x \neq 0$, then $x^0 = 1$. The expression 0^0 is not defined.

EXAMPLES

Assume the value of the variable is not equal to zero.

1. $27^0 = 1$
2. $(-16z^3)^0 = 1$
3. $-15^0 = -(15)^0 = -1$
4. $3z^0 = 3 \cdot 1 = 3$

Point of Interest

In the 15th century, the expression $12^{2\overline{m}}$ was used to mean $12x^{-2}$. The use of $\overline{m}$ reflects an Italian influence, where m was used for "minus" and p was used for "plus." It was understood that $2\overline{m}$ referred to an unnamed variable. Isaac Newton, in the 17th century, advocated the use of a negative exponent, the symbol we use today.

Now consider the expression $\dfrac{x^4}{x^6}$, $x \neq 0$. This expression can be simplified, as shown below, by subtracting exponents or dividing by common factors.

$$\dfrac{x^4}{x^6} = x^{4-6} = x^{-2} \qquad \dfrac{x^4}{x^6} = \dfrac{\overset{1}{\cancel{x}} \cdot \overset{1}{\cancel{x}} \cdot \overset{1}{\cancel{x}} \cdot \overset{1}{\cancel{x}}}{\underset{1}{\cancel{x}} \cdot \underset{1}{\cancel{x}} \cdot \underset{1}{\cancel{x}} \cdot \underset{1}{\cancel{x}} \cdot x \cdot x} = \dfrac{1}{x^2}$$

The equations $\dfrac{x^4}{x^6} = x^{-2}$ and $\dfrac{x^4}{x^6} = \dfrac{1}{x^2}$ suggest that $x^{-2} = \dfrac{1}{x^2}$.

Definition of a Negative Exponent

If $x \neq 0$ and n is a positive integer, then $x^{-n} = \dfrac{1}{x^n}$ and $\dfrac{1}{x^{-n}} = x^n$.

EXAMPLES

Assume the value of the variable is not equal to zero.

1. $4^{-3} = \dfrac{1}{4^3} = \dfrac{1}{64}$

2. $2z^{-5} = \dfrac{2}{z^5}$

3. $(2z)^{-5} = \dfrac{1}{(2z)^5} = \dfrac{1}{32z^5}$

4. $\dfrac{7}{a^{-6}} = 7a^6$

Take Note

Note the difference between example (2) and example (3) in the box at the right.

An exponential expression is in simplest form when it is written with only positive exponents.

HOW TO 4   Simplify: $(3x^{-2}y^6)(-4x^5y^{-8})$

$$(3x^{-2}y^6)(-4x^5y^{-8}) = 3(-4)x^{-2+5}y^{6+(-8)}$$

- Multiply the coefficients. Add the exponents on like bases.

$$= -12x^3y^{-2}$$

$$= \frac{12x^3}{y^2}$$

- Write y^{-2} as $\frac{1}{y^2}$.

The expression $\left(\dfrac{x^3}{y^4}\right)^2$, $y \neq 0$, can be simplified by squaring $\dfrac{x^3}{y^4}$ or by multiplying each exponent in the quotient by the exponent outside the parentheses.

$$\left(\frac{x^3}{y^4}\right)^2 = \left(\frac{x^3}{y^4}\right)\left(\frac{x^3}{y^4}\right) = \frac{x^3 \cdot x^3}{y^4 \cdot y^4} = \frac{x^{3+3}}{y^{4+4}} = \frac{x^6}{y^8} \qquad \left(\frac{x^3}{y^4}\right)^2 = \frac{x^{3\cdot2}}{y^{4\cdot2}} = \frac{x^6}{y^8}$$

Rule for Simplifying Powers of Quotients

If m, n, and p are integers and $y \neq 0$, then $\left(\dfrac{x^m}{y^n}\right)^p = \dfrac{x^{mp}}{y^{np}}$.

EXAMPLES
Assume the value of the variable is not equal to zero.

1. $\left(\dfrac{a^2}{b^4}\right)^5 = \dfrac{a^{2\cdot5}}{b^{4\cdot5}} = \dfrac{a^{10}}{b^{20}}$

2. $\left(\dfrac{x^3}{y^2}\right)^{-4} = \dfrac{x^{3(-4)}}{y^{2(-4)}} = \dfrac{x^{-12}}{y^{-8}} = \dfrac{y^8}{x^{12}}$

3. $\left(\dfrac{2x^2y^4}{4x^4y^{-1}}\right)^3 = \left(\dfrac{x^{2-4}y^{4-(-1)}}{2}\right)^3$

   - Write $\dfrac{2}{4}$ in simplest form. Subtract exponents on like bases.

   $$= \left(\frac{x^{-2}y^5}{2}\right)^3$$

   - Simplify.

   $$= \frac{x^{-2(3)}y^{5(3)}}{2^3}$$

   - Use the Rule for Simplifying Powers of Quotients. Recall that $2 = 2^1$.

   $$= \left(\frac{x^{-6}y^{15}}{8}\right) = \frac{y^{15}}{8x^6}$$

Here is a summary of the rules of exponents presented in this section.

Rules of Exponents

If m, n, and p are integers, then

$$x^m \cdot x^n = x^{m+n} \qquad\qquad (x^m)^n = x^{mn} \qquad\qquad (x^my^n)^p = x^{mp}y^{np}$$

$$\frac{x^m}{x^n} = x^{m-n}, x \neq 0 \qquad\qquad \left(\frac{x^m}{y^n}\right)^p = \frac{x^{mp}}{y^{np}}, y \neq 0 \qquad\qquad x^{-n} = \frac{1}{x^n}, x \neq 0$$

$$x^0 = 1, x \neq 0$$

INSTRUCTOR NOTE
Examples such as HOW TO 5 are included to review the work on multiplying monomials and to demonstrate that negative exponents can be used with products of exponential expressions.

HOW TO 5 Simplify: $(3ab^{-4})(-2a^{-3}b^7)$

$$(3ab^{-4})(-2a^{-3}b^7) = [3 \cdot (-2)](a^{1+(-3)}b^{-4+7})$$
$$= -6a^{-2}b^3$$
$$= -\frac{6b^3}{a^2}$$

• When multiplying exponential expressions, add the exponents on like bases.

IN-CLASS EXAMPLES
Simplify.
4. $\dfrac{-24a^3b^9}{16a^6b^3}$ $-\dfrac{3b^6}{2a^3}$
5. $\dfrac{a^2}{3b^{-1}}$ $\dfrac{a^2b}{3}$
6. $(2x^{-3}y^3)^2$ $\dfrac{4y^6}{x^6}$

HOW TO 6 Simplify: $\dfrac{4a^{-2}b^5}{6a^5b^2}$

$$\frac{4a^{-2}b^5}{6a^5b^2} = \frac{2 \cdot 2a^{-2}b^5}{2 \cdot 3a^5b^2}$$
$$= \frac{2a^{-2-5}b^{5-2}}{3}$$
$$= \frac{2a^{-7}b^3}{3} = \frac{2b^3}{3a^7}$$

• Divide the coefficients by their common factor.
• Use the Rule for Dividing Exponential Expressions.
• Use the Definition of a Negative Exponent to rewrite the expression with a positive exponent.

HOW TO 7 Simplify: $\left[\dfrac{6m^2n^3}{8m^7n^2}\right]^{-3}$

INSTRUCTOR NOTE
There are a few ways to simplify the expression in HOW TO 7. Students could simplify it by starting as follows:
$$\left[\frac{6m^2n^3}{8m^7n^2}\right]^{-3} = \left[\frac{8m^7n^2}{6m^2n^3}\right]^3$$

$$\left[\frac{6m^2n^3}{8m^7n^2}\right]^{-3} = \left[\frac{3m^{2-7}n^{3-2}}{4}\right]^{-3}$$
$$= \left[\frac{3m^{-5}n}{4}\right]^{-3}$$
$$= \frac{3^{-3}m^{15}n^{-3}}{4^{-3}}$$
$$= \frac{4^3m^{15}}{3^3n^3} = \frac{64m^{15}}{27n^3}$$

• Simplify inside the brackets.
• Subtract the exponents.
• Use the Rule for Simplifying Powers of Quotients.
• Use the Definition of a Negative Exponent to rewrite the expression with positive exponents. Then simplify.

EXAMPLE 4

Simplify: $\dfrac{-28x^6z^{-3}}{42x^{-1}z^4}$

Solution
$$\frac{-28x^6z^{-3}}{42x^{-1}z^4} = -\frac{14 \cdot 2x^{6-(-1)}z^{-3-4}}{14 \cdot 3}$$
$$= -\frac{2x^7z^{-7}}{3} = -\frac{2x^7}{3z^7}$$

YOU TRY IT 4

Simplify: $\dfrac{20r^{-2}t^{-5}}{-16r^{-3}s^{-2}}$

Your solution
$$-\frac{5rs^2}{4t^5}$$

EXAMPLE 5

Simplify: $\dfrac{(3a^{-1}b^4)^{-3}}{(6^{-1}a^{-3}b^{-4})^3}$

Solution
$$\frac{(3a^{-1}b^4)^{-3}}{(6^{-1}a^{-3}b^{-4})^3} = \frac{3^{-3}a^3b^{-12}}{6^{-3}a^{-9}b^{-12}}$$
$$= \frac{6^3a^{12}b^0}{3^3} = \frac{216a^{12}}{27} = 8a^{12}$$

YOU TRY IT 5

Simplify: $\dfrac{(9u^{-6}v^4)^{-1}}{(6u^{-3}v^{-2})^{-2}}$

Your solution
$$\frac{4}{v^8}$$

Solutions on p. S17

To write a number using scientific notation

 Point of Interest

Astronomers measure the distance of some stars by using a unit called the *parsec*. One parsec is approximately 1.91×10^{13} mi.

Take Note

There are two steps involved in writing a number in scientific notation: (1) determine the number between 1 and 10, and (2) determine the exponent on 10.

INSTRUCTOR NOTE

For the example at the right, show students that $10^{-4} = \frac{1}{10^4} = \frac{1}{10,000} = 0.0001$. Stress that we are not changing the number, only writing the number in a different form.

Integer exponents are used to represent the very large and very small numbers encountered in the fields of science and engineering. For example, the mass of an electron is 0.00000000000000000000000000009 g. Numbers such as this are difficult to read and write, so a more convenient system for writing such numbers has been developed. It is called **scientific notation.**

To express a number in scientific notation, write the number as the product of a number between 1 and 10 and a power of 10. The form for scientific notation is $a \times 10^n$, where $1 \le a < 10$.

For numbers greater than 10, move the decimal point to the right of the first digit. The exponent n is positive and equal to the number of places the decimal point has been moved.

$$965,000 = 9.65 \times 10^5$$
$$3,600,000 = 3.6 \times 10^6$$
$$92,000,000,000 = 9.2 \times 10^{10}$$

For numbers less than 1, move the decimal point to the right of the first nonzero digit. The exponent n is negative. The absolute value of the exponent is equal to the number of places the decimal point has been moved.

$$0.0002 = 2 \times 10^{-4}$$
$$0.0000000974 = 9.74 \times 10^{-8}$$
$$0.000000000086 = 8.6 \times 10^{-11}$$

Converting a number written in scientific notation to decimal notation requires moving the decimal point.

When the exponent is positive, move the decimal point to the right the same number of places as the exponent.

$$1.32 \times 10^4 = 13,200$$
$$1.4 \times 10^8 = 140,000,000$$

When the exponent is negative, move the decimal point to the left the same number of places as the absolute value of the exponent.

$$1.32 \times 10^{-2} = 0.0132$$
$$1.4 \times 10^{-4} = 0.00014$$

Numerical calculations involving numbers that have more digits than a hand-held calculator is able to handle can be performed using scientific notation.

Integrating Technology

See the Keystroke Guide: *Scientific Notation* for instructions on entering a number that is in scientific notation into a graphing calculator.

HOW TO 8 Simplify: $\dfrac{220,000 \times 0.000000092}{0.0000011}$

$$\frac{220,000 \times 0.000000092}{0.0000011} = \frac{2.2 \times 10^5 \times 9.2 \times 10^{-8}}{1.1 \times 10^{-6}}$$

• Write the numbers in scientific notation.

$$= \frac{(2.2)(9.2) \times 10^{5 + (-8) - (-6)}}{1.1}$$

• Simplify.

$$= 18.4 \times 10^3 = 18,400$$

EXAMPLE 6

Write 0.000041 in scientific notation.

Solution $0.000041 = 4.1 \times 10^{-5}$

EXAMPLE 7

Write 3.3×10^7 in decimal notation.

Solution $3.3 \times 10^7 = 33,000,000$

YOU TRY IT 6

Write 942,000,000 in scientific notation.

Your solution 9.42×10^8

YOU TRY IT 7

Write 2.7×10^{-5} in decimal notation.

Your solution 0.000027

Solutions on p. S17

EXAMPLE 8

Simplify:
$$\frac{2,400,000,000 \times 0.0000063}{0.00009 \times 480}$$

Solution

$$\frac{2,400,000,000 \times 0.0000063}{0.00009 \times 480}$$

$$= \frac{2.4 \times 10^9 \times 6.3 \times 10^{-6}}{9 \times 10^{-5} \times 4.8 \times 10^2}$$

$$= \frac{(2.4)(6.3) \times 10^{9+(-6)-(-5)-2}}{(9)(4.8)}$$

$$= 0.35 \times 10^6 = 350,000$$

YOU TRY IT 8

Simplify:
$$\frac{5,600,000 \times 0.000000081}{900 \times 0.000000028}$$

Your solution

18,000

IN-CLASS EXAMPLES
Write in scientific notation.
7. 2,000,000,000 **2×10^9**
8. 0.000000016 **1.6×10^{-8}**
Write in decimal notation.
9. 7.4×10^8 **740,000,000**
10. 3.54×10^{-6} **0.00000354**
11. Simplify: $(0.0000076)(35,000,000)$ **266**

Solution on p. S17

OBJECTIVE D *To solve application problems*

EXAMPLE 9

● How many miles does light travel in one day? The speed of light is 186,000 mi/s. Write the answer in scientific notation.

Strategy

To find the distance traveled:
• Write the speed of light in scientific notation.
• Write the number of seconds in one day in scientific notation.
• Use the equation $d = rt$, where r is the speed of light and t is the number of seconds in one day.

Solution

$r = 186,000 = 1.86 \times 10^5$

$t = 24 \cdot 60 \cdot 60 = 86,400 = 8.64 \times 10^4$

$d = rt$
$d = (1.86 \times 10^5)(8.64 \times 10^4)$
$\quad = 1.86 \times 8.64 \times 10^9$
$\quad = 16.0704 \times 10^9$
$\quad = 1.60704 \times 10^{10}$

Light travels 1.60704×10^{10} mi in one day.

YOU TRY IT 9

● The Roadrunner supercomputer from IBM can perform one arithmetic operation, called a FLOP (**FL**oating-point **OP**eration), in 9.74×10^{-16} s. How many arithmetic operations can be performed in 1 min? Write the answer in scientific notation.

Your strategy

IN-CLASS EXAMPLES
Solve. Write the answer in scientific notation.
12. How many kilometers does light travel in 6 h? The speed of light is 300,000 km/s. **6.48×10^9 km**
13. A space vehicle travels 9.3×10^7 mi from Earth to Venus at an average velocity of 2.2×10^4 mph. How long does it take the space vehicle to reach Venus? **$\approx 4.2 \times 10^3$ h**

Your solution

6.18×10^{16} operations

Solution on p. S17

6.1 EXERGISES

✔ **Concept Check**

SUGGESTED ASSIGNMENT
Exercises 1–6; Exercises 7–117, odds;
More challenging exercises: Exercises 119–123, odds

1. Which of the following are monomials?   i, ii, iv

 (i) -8 **(ii)** $4x^2yz^5$ **(iii)** $3x + 1$ **(iv)** $\dfrac{x^2y^3}{4}$ **(v)** $\dfrac{4}{x}$

2. What is the Rule for Multiplying Exponential Expressions?   See page 344.

3. Does the Rule for Simplifying Powers of Products apply to $(x^2 + y^2)^3$? Why or why not?   No; the expression inside the parentheses is a sum, not a product.

4. If x is a positive number, is x^{-3} a positive or a negative number?   Positive

5. If $z \neq 0$, what are the values of $-z^0$ and z^0? $-1, 1$

6. Is the number 2.4055×10^{-5} less than or greater than 1?   Less than

---

OBJECTIVE A   *To multiply monomials*

For Exercises 7 to 35, simplify.

7. $(ab^3)(a^3b)$
 a^4b^4

8. $(-2ab^4)(-3a^2b^4)$
 $6a^3b^8$

9. $(9xy^2)(-2x^2y^2)$
 $-18x^3y^4$

10. $(x^2y)^2$
 x^4y^2

11. $(x^2y^4)^4$
 x^8y^{16}

12. $(-2ab^2)^3$
 $-8a^3b^6$

13. $(-3x^2y^3)^4$
 $81x^8y^{12}$

14. $(4a^2b^3)^3$
 $64a^6b^9$

15. $(27a^5b^3)^2$
 $729a^{10}b^6$

16. $[(2ab)^3]^2$
 $64a^6b^6$

17. $[(2a^4b^3)^3]^2$
 $64a^{24}b^{18}$

18. $(xy)(x^2y)^4$
 x^9y^5

19. $(x^2y^2)(xy^3)^3$
 x^5y^{11}

20. $(-4x^3y)^2(2x^2y)$
 $32x^8y^3$

21. $(-5ab)(3a^3b^2)^2$
 $-45a^7b^5$

22. $(-4r^2s^3)^3(2s^2)$
 $-128r^6s^{11}$

23. $(3x^5y)(-4x^3)^3$
 $-192x^{14}y$

24. $(4x^3z)(-3y^4z^5)$
 $-12x^3y^4z^6$

25. $(-6a^4b^2)(-7a^2c^5)$
 $42a^6b^2c^5$

26. $(4ab)^2(-2ab^2c^3)^3$
 $-128a^5b^8c^9$

27. $(-2ab^2)(-3a^4b^5)^3$
 $54a^{13}b^{17}$

28. $(2a^2b)^3(-3ab^4)^2$
 $72a^8b^{11}$

29. $(-3ab^3)^3(-2^2a^2b)^2$
 $-432a^7b^{11}$

30. $(-c^3)(-2a^2bc)(3a^2b)$
 $6a^4b^2c^4$

31. $(-2x^2y^3z)(3x^2yz^4)$
 $-6x^4y^4z^5$

32. $(x^2z^4)(2xyz^4)(-3x^3y^2)$
 $-6x^6y^3z^8$

33. $(2xy)(-3x^2yz)(x^2y^3z^3)$
 $-6x^5y^5z^4$

34. $(2a^2b)(-3ab^2c)(-b^3c^5)$
 $6a^3b^6c^6$

35. $(3b^5)(2ab^2)(-2ab^2c^2)$
 $-12a^2b^9c^2$

36. 📝 If $x^m \cdot x^n = x^9$ and $n = 3$, what is the value of m?   6

37. 📝 What is the value of n in the equation $2^{32} + 2^{32} = 2^n$?   33

OBJECTIVE B *To divide monomials and simplify expressions with negative exponents*

For Exercises 38 to 80, simplify.

38. 5^{-3}

$\dfrac{1}{125}$

39. 4^{-2}

$\dfrac{1}{16}$

40. $\dfrac{1}{3^{-6}}$

243

41. $\dfrac{1}{2^{-7}}$

128

42. x^{-8}

$\dfrac{1}{x^8}$

43. $3a^{-5}$

$\dfrac{3}{a^5}$

44. $\dfrac{2}{z^{-6}}$

$2z^6$

45. $\dfrac{1}{3a^{-7}}$

$\dfrac{a^7}{3}$

46. $\dfrac{2x^{-2}}{y^4}$ $\quad \dfrac{2}{x^2 y^4}$

47. $\dfrac{a^3}{4b^{-2}}$ $\quad \dfrac{a^3 b^2}{4}$

48. $x^{-3}y$ $\quad \dfrac{y}{x^3}$

49. xy^{-4} $\quad \dfrac{x}{y^4}$

50. $-5x^0$ $\quad -5$

51. $\dfrac{1}{2x^0}$ $\quad \dfrac{1}{2}$

52. $\dfrac{(2x)^0}{-2^3}$ $\quad -\dfrac{1}{8}$

53. $\dfrac{-3^{-2}}{(2y)^0}$ $\quad -\dfrac{1}{9}$

54. $(x^2 y^{-4})^2$

$\dfrac{x^4}{y^8}$

55. $(x^3 y^5)^{-2}$

$\dfrac{1}{x^6 y^{10}}$

56. $(2x^3 y^{-2})(-3x^{-4}y^3)$

$-\dfrac{6y}{x}$

57. $(-3a^{-4}b^{-5})(-5a^{-2}b^4)$

$\dfrac{15}{a^6 b}$

58. $(5m^{-3}n^2)^{-2}(10m^2n)$

$\dfrac{2m^8}{5n^3}$

59. $(4y^{-3}z^{-4})(-3y^3z^{-3})^{-2}$

$\dfrac{4z^2}{9y^9}$

60. $(-6mn^{-2})^3(-4m^{-3}n^{-1})^{-2}$

$-\dfrac{27m^9}{2n^4}$

61. $(4x^{-3}y^2)^{-3}(2xy^{-3})^4$

$\dfrac{x^{13}}{4y^{18}}$

62. $\dfrac{6a^4}{4a^6}$ $\quad \dfrac{3}{2a^2}$

63. $\dfrac{9x^5}{12x^8}$ $\quad \dfrac{3}{4x^3}$

64. $\dfrac{x^{17}y^5}{-x^7 y^{10}}$ $\quad -\dfrac{x^{10}}{y^5}$

65. $\dfrac{-6x^2 y}{12x^4 y}$ $\quad -\dfrac{1}{2x^2}$

66. $\dfrac{y^{-7}}{y^{-8}}$ $\quad y$

67. $\dfrac{y^{-2}}{y^6}$ $\quad \dfrac{1}{y^8}$

68. $\dfrac{a^{-1}b^{-3}}{a^4 b^{-5}}$ $\quad \dfrac{b^2}{a^5}$

69. $\dfrac{a^6 b^{-4}}{a^{-2}b^5}$ $\quad \dfrac{a^8}{b^9}$

70. $\dfrac{2x^2 y^4}{(3xy^2)^3}$ $\quad \dfrac{2}{27xy^2}$

71. $\dfrac{-3ab^2}{(9a^2 b^4)^3}$ $\quad -\dfrac{1}{243a^5 b^{10}}$

72. $\dfrac{(4x^2 y)^2}{(2xy^3)^3}$ $\quad \dfrac{2x}{y^7}$

73. $\dfrac{(3a^2 b)^3}{(-6ab^3)^2}$ $\quad \dfrac{3a^4}{4b^3}$

74. $\dfrac{(-4xy^3)^3}{(-2x^7 y)^4}$

$-\dfrac{4y^5}{x^{25}}$

75. $\dfrac{(-8x^2 y^2)^4}{(16x^3 y^7)^2}$

$\dfrac{16x^2}{y^6}$

76. $\dfrac{(-3x^3 y^{-2})^3}{(2xy^{-3})^{-2}}$

$-\dfrac{108x^{11}}{y^{12}}$

77. $\dfrac{(3a^4 b^{-2})^{-2}}{(2a^{-3}b)^3}$

$\dfrac{ab}{72}$

78. $\left(\dfrac{4^{-2}xy^{-3}}{x^{-3}y}\right)^3 \left(\dfrac{8^{-1}x^{-2}y}{x^4 y^{-1}}\right)^{-2}$

$\dfrac{x^{24}}{64y^{16}}$

79. $\left(\dfrac{9ab^{-2}}{8a^{-2}b}\right)^{-2}\left(\dfrac{3a^{-2}b}{2a^2 b^{-2}}\right)^3$

$\dfrac{8b^{15}}{3a^{18}}$

80. $\left(\dfrac{2ab^{-1}}{ab}\right)^{-1}\left(\dfrac{3a^{-2}b}{a^2 b^2}\right)^{-2}$

$\dfrac{a^8 b^4}{18}$

81. If $\dfrac{x^p}{x^q} = 1$, what is the value of $p - q$? $\quad 0$

82. If $m < n$, is the value of $\dfrac{2^m}{2^n}$ less than or greater than 1? $\quad$ Less than

OBJECTIVE C   *To write a number using scientific notation*

For Exercises 83 to 88, write in scientific notation.

83. 0.00000467
4.67×10^{-6}

84. 0.00000005
5×10^{-8}

85. 0.00000000017
1.7×10^{-10}

86. 4,300,000
4.3×10^{6}

87. 200,000,000,000
2×10^{11}

88. 9,800,000,000
9.8×10^{9}

For Exercises 89 to 94, write in decimal notation.

89. 1.23×10^{-7}
0.000000123

90. 6.2×10^{-12}
0.0000000000062

91. 8.2×10^{15}
8,200,000,000,000,000

92. 6.34×10^{5}
634,000

93. 3.9×10^{-2}
0.039

94. 4.35×10^{9}
4,350,000,000

For Exercises 95 to 106, simplify. Write the answer in decimal notation.

95. $(3 \times 10^{-12})(5 \times 10^{16})$
150,000

96. $(8.9 \times 10^{-5})(3.2 \times 10^{-6})$
0.0000000002848

97. $(0.0000065)(3,200,000,000,000)$
20,800,000

98. $(480,000)(0.0000000096)$
0.004608

99. $\dfrac{9 \times 10^{-3}}{6 \times 10^{5}}$
0.000000015

100. $\dfrac{2.7 \times 10^{4}}{3 \times 10^{-6}}$
9,000,000,000

101. $\dfrac{0.0089}{500,000,000}$
0.0000000000178

102. $\dfrac{0.000000346}{0.0000005}$
0.692

103. $\dfrac{(3.3 \times 10^{-11})(2.7 \times 10^{15})}{8.1 \times 10^{-3}}$
11,000,000

104. $\dfrac{(6.9 \times 10^{27})(8.2 \times 10^{-13})}{4.1 \times 10^{15}}$
1.38

105. $\dfrac{(0.00000004)(84,000)}{(0.0003)(1,400,000)}$
0.000008

106. $\dfrac{(720)(0.0000000039)}{(26,000,000,000)(0.018)}$
0.000000000000006

107. Is 5.27×10^{-6} less than zero or greater than zero?   Greater than

108. Place the correct symbol, $<$, $=$, or $>$, between the two numbers:
4.61×10^{5} __=__ 46.1×10^{4}.

OBJECTIVE D   *To solve application problems*

For Exercises 109 to 117, solve. Write the answer in scientific notation.

109. ● **Astronomy**   Our galaxy is estimated to be 5.6×10^{19} mi across. How many years would it take a spaceship traveling at 25,000 mph to cross the galaxy?
2.6×10^{11} years

110. ● **Astronomy**   How long does it take light to travel to Earth from the sun? The sun is 9.3×10^{7} mi from Earth, and light travels 1.86×10^{5} mi/s. 5×10^{2} s

111. ● **Physics**   The mass of an electron is 9.109×10^{-31} kg. The mass of a proton is 1.673×10^{-27} kg. How many times larger is the mass of a proton than the mass of an electron? 1.83664508×10^{3} times larger

The Milky Way

Viktar Malyshchyts/Shutterstock.com

112. ◑ **Astronomy** Use the information in the article at the right to determine the average number of miles traveled per day by *Curiosity* on its trip to Mars.
1.39×10^6 mi/day

113. ◑ **Astronomy** On the day the Mars rover *Curiosity* landed on Mars, NASA expected the distance from Mars to Earth to be 154 million miles and a radio signal to take 13.8 min to travel that distance. How fast does the radio signal travel?
1.12×10^7 mi/min

114. ◑ **The Federal Government** In 2012, the United States national debt was approximately 1.92×10^{13} dollars. How much would each American have to pay in order to pay off the debt? Use 3.13×10^8 as the number of U.S. citizens.
$\approx 6.13 \times 10^4$ dollars

115. ◑ **Geology** The mass of Earth is 5.9×10^{24} kg. The mass of the sun is 2×10^{30} kg. How many times larger is the mass of the sun than the mass of Earth?
3.3898305×10^5 times larger

116. ◑ **Forestry** Use the information in the article at the right. If every burned acre of Yellowstone Park had 12,000 lodgepole pine seedlings growing on it one year after the fire, how many new seedlings would be growing?
1.44×10^{11} seedlings

117. ◑ **Forestry** Use the information in the article at the right. Find the number of seeds released by the lodgepole pine trees for each surviving seedling.
$1.\overline{6} \times 10^2$ seeds

118. 🖎 One light-year is approximately 5.9×10^{12} mi and is defined as the distance light can travel in a vacuum in one year. One light-hour is approximately 6.7×10^8 mi.
True or false? $\dfrac{5.9 \times 10^{12}}{6.7 \times 10^8} \approx$ number of hours in one year
True

Critical Thinking

119. Evaluate 3^{x^2} for each value of *x*.
a. 2 81 **b.** 3 19,683 **c.** 0 1 **d.** -2 81

120. If *a* and *b* are positive real numbers and $a < b$, what is the relationship between a^{-1} and b^{-1}? $a^{-1} > b^{-1}$

For Exercises 121 to 124, simplify each expression. Assume that *m* and *n* are positive integers and that $x \neq 0$.

121. $x^{3n}x^{4n}$ x^{7n}

122. $(-2x^n y^m)(3x^n y^{2m})$ $-6x^{2n}y^{3m}$

123. $\dfrac{x^n y^{5m}}{x^{3n}y^m}$ $\dfrac{y^{4m}}{x^{2n}}$

124. $\left(\dfrac{3}{x^{3n}}\right)^{-2}$ $\dfrac{x^{6n}}{9}$

Projects or Group Activities

125. Simplify: **a.** $1 + [1 + (1 + 2^{-1})^{-1}]^{-1}$ $\dfrac{8}{5}$ **b.** $2 - [2 - (2 - 2^{-1})^{-1}]^{-1}$ $\dfrac{5}{4}$

126. Write $1 + \dfrac{1}{1 + \dfrac{1}{1 + \dfrac{1}{1 + x}}}$ on a single line by using negative exponents.
$1 + (1 + (1 + (1 + x)^{-1})^{-1})^{-1}$

In the NEWS!

Mission to Mars
Today NASA launched *Curiosity*, sending the new Mars rover on its 254-day, 354-million-mile journey to the red planet.
Source: Discovery News

In the NEWS!

Forest Fires Spread Seeds
The lodgepole pine is a tree that uses the intense heat of a fire to release its seeds from their cones. After a blaze that burned 12,000,000 acres of Yellowstone National Park, scientists counted 2 million lodgepole pine seeds on a single acre of the park. One year later, they returned to find 12,000 lodgepole pine seedlings growing.
Source: National Public Radio

QUICK QUIZ
Simplify.
1. $(8x^2y^2)(-3xy)$
   $-24x^3y^3$ [6.1A]
2. $(3ab^2)^2(-4a^2b)^2$
   $144a^6b^6$ [6.1A]
3. $\dfrac{42a^3b^{10}}{-35a^5b^9}$ $-\dfrac{6b}{5a^2}$ [6.1B]
4. $\dfrac{3x^{-1}}{x^6}$ $\dfrac{3}{x^7}$ [6.1B]
5. Write 0.00000029 in scientific notation.
   2.9×10^{-7} [6.1C]
   Write in decimal notation.
6. 5.34×10^6
   5,340,000 [6.1C]
7. 6×10^{-9}
   0.000000006 [6.1C]
8. The speed of light is 300,000,000 m/s. How many meters does light travel in 4 h? Write the answer in scientific notation. 4.32×10^{12} m [6.1D]

6.2 Introduction to Polynomial Functions

OBJECTIVE A *To evaluate polynomial functions*

Tips for Success

A great many new vocabulary words are introduced in this chapter. All of these terms are in **boldface type.** The bold type indicates that these are concepts you must know to learn the material. Be sure to study each new term as it is presented.

A **polynomial** is a variable expression in which the terms are monomials.

A polynomial of one term is a **monomial.**
$5x$

A polynomial of two terms is a **binomial.**
$5x^2y + 6x$

A polynomial of three terms is a **trinomial.**
$3x^2 + 9xy - 5y$

Polynomials with more than three terms do not have special names.

The **degree of a polynomial** is the greatest of the degrees of any of its terms.

$3x + 2$	degree 1
$3x^2 + 2x - 4$	degree 2
$4x^3y^2 + 6x^4$	degree 5

The terms of a polynomial in one variable are usually arranged so that the exponents on the variable decrease from left to right. This is called **descending order.**

$2x^2 - x + 8$

$3y^3 - 3y^2 + y - 12$

For a polynomial in more than one variable, *descending order* may refer to any one of the variables.

The polynomial at the right is shown first in descending order of the x variable and then in descending order of the y variable.

$2x^2 + 3xy + 5y^2$

$5y^2 + 3xy + 2x^2$

Polynomial functions have many applications in mathematics. In general, a **polynomial function** is an expression whose terms are monomials. The **linear function** given by $f(x) = mx + b$ is an example of a polynomial function. It is a polynomial function of degree 1. A second-degree polynomial function, called a **quadratic function,** is given by the equation $f(x) = ax^2 + bx + c$, $a \neq 0$. A third-degree polynomial function is called a **cubic function.**

HOW TO 1 Determine whether the function is a polynomial function.

A. $P(x) = 3x^{1/2} + 2x^2 - 3$
B. $T(x) = 3\sqrt{x} - 2x^2 - 3x + 2$
C. $R(x) = 14x^3 - \pi x^2 + 3x + 2$

A. This is not a polynomial function. A polynomial function does not have a variable raised to a fractional power.
B. This is not a polynomial function. A polynomial function does not have a variable expression within a radical.
C. This is a polynomial function. Each term is a monomial.

The **leading coefficient** of a polynomial function is the coefficient of the variable with the largest exponent. The **constant term** is the term without a variable.

> **HOW TO 2** Find the leading coefficient, the constant term, and the degree of the polynomial function $P(x) = 7x^4 - 3x^2 + 2x - 4$.
>
> The leading coefficient is 7, the constant term is -4, and the degree is 4.

To evaluate a polynomial function, replace the variable by its value and simplify.

Integrating Technology

See the Keystroke Guide: *Evaluating Functions* for instructions on using a graphing calculator to evaluate a function.

> **HOW TO 3** Given $P(x) = x^3 - 3x^2 + 4$, evaluate $P(-3)$.
>
> $$P(x) = x^3 - 3x^2 + 4$$
> $$P(-3) = (-3)^3 - 3(-3)^2 + 4 \qquad \bullet \text{ Substitute } -3 \text{ for } x \text{ and simplify.}$$
> $$= -27 - 27 + 4$$
> $$= -50$$

The graph of a linear function is a straight line and can be found by plotting just two points. The graph of a polynomial function of degree greater than 1 is a curve. Consequently, many points may have to be found before an accurate graph can be drawn.

Here is an example of graphing the cubic function $P(x) = x^3 - 2x^2 - 5x + 6$. Evaluating the function when $x = -2, -1, 0, 1, 2, 3,$ and 4 gives the graph in Figure 1 below. Evaluating for some noninteger values gives the graph in Figure 2. Finally, connecting the dots with a smooth curve gives the graph in Figure 3.

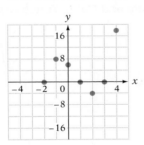

Figure 1

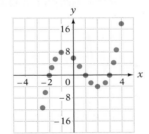

Figure 2

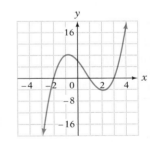

Figure 3

Integrating Technology

When graphing a polynomial function, you may have to plot a large number of points in order to achieve an accurate graph. Graphing utilities create graphs by evaluating the polynomial function at a large number of points and then connecting the points to form the graph. Here are some graphing calculator screens used to produce the graph of $P(x) = x^3 - 2x^2 - 5x + 6$ shown above. Note that the scales on the two axes are different. You may need to experiment with different window settings to produce an accurate graph of a polynomial function. For further assistance, refer to the Keystroke Appendix.

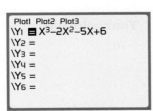

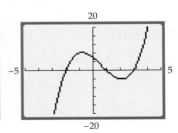

Unless otherwise noted, all content on this page is © Cengage Learning.

IN-CLASS EXAMPLES

1. Given
   $P(x) = 2x^2 - 4x + 5$,
   evaluate $P(3)$.   **11**
2. Given
   $f(x) = x^3 - 3x^2 + 6x - 7$,
   evaluate $f(-1)$.   **−17**
Graph.
3. $f(x) = -x^2 + 4x + 2$
4. $f(x) = x^3 - 4x$

HOW TO 4   Graph $P(x) = x^3 - 10$.

Evaluate the function for various values of x, plot the resulting ordered pairs, and then connect the dots to form the graph.

x	$y = f(x)$
−3	−37
−2	−18
−1	−11
0	−10
1	−9
2	−2
3	17
4	54

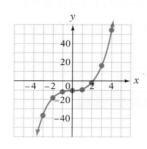

Note from the graph that the scale on the y-axis is different from the scale on the x-axis. When graphing polynomial functions, you may need to use different scales on the two axes.

The shape of the graph of a polynomial function depends on its degree. Knowing the basic shapes will help you create a graph more quickly. Below are the basic shapes of the graphs of quadratic and cubic polynomial functions.

Basic Graphs of Quadratic and Cubic Functions

Take Note

Recall that the graph of the linear function $f(x) = mx + b$ (not shown) is a straight line.

Quadratic Functions

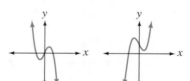

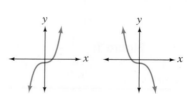

Cubic Functions

EXAMPLE 1

Given $P(x) = x^3 + 3x^2 - 2x + 8$, evaluate $P(-2)$.

Solution

$$P(x) = x^3 + 3x^2 - 2x + 8$$
$$P(-2) = (-2)^3 + 3(-2)^2 - 2(-2) + 8 \quad \bullet \text{ Replace}$$
$$= (-8) + 3(4) + 4 + 8 \qquad\qquad x \text{ by } -2.$$
$$= -8 + 12 + 4 + 8 \qquad\qquad \textbf{Simplify.}$$
$$= 16$$

YOU TRY IT 1

Given $R(x) = -2x^4 - 5x^3 + 2x - 8$, evaluate $R(2)$.

Your solution

−76

Solution on p. S17

Unless otherwise noted, all content on this page is © Cengage Learning.

EXAMPLE 2

Graph $f(x) = x^2 - 2$.

Solution

x	$y = f(x)$
-3	7
-2	2
-1	-1
0	-2
1	-1
2	2
3	7

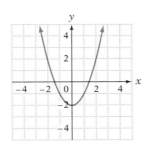

YOU TRY IT 2

Graph $f(x) = x^2 + 2x - 3$.

Your solution

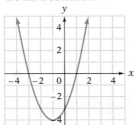

EXAMPLE 3

Graph $f(x) = x^3 - 1$.

Solution

x	$y = f(x)$
-2	-9
-1	-2
0	-1
1	0
2	7

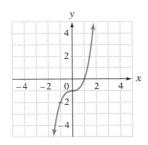

YOU TRY IT 3

Graph $f(x) = x^3 - 2x^2 - 11x + 12$.

Your solution

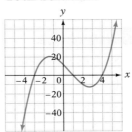

Solutions on p. S17

OBJECTIVE B   *To add or subtract polynomials*

Polynomials can be added by combining like terms. Either a vertical or a horizontal format can be used.

HOW TO 5   Add $(3x^2 + 2x - 7) + (7x^3 - 3 + 4x^2)$. Use a horizontal format.

Use the Commutative and Associative Properties of Addition to rearrange and group like terms.

$(3x^2 + 2x - 7) + (7x^3 - 3 + 4x^2)$
$= 7x^3 + (3x^2 + 4x^2) + 2x + (-7 - 3)$
$= 7x^3 + 7x^2 + 2x - 10$ • Combine like terms.

HOW TO 6   Add $(4x^2 + 5x - 3) + (7x^3 - 7x + 1) + (2x - 3x^2 + 4x^3 + 1)$.
Use a vertical format.

Arrange the terms of each polynomial in descending order, with like terms in the same column.

$$
\begin{array}{r}
4x^2 + 5x - 3 \\
7x^3 \qquad - 7x + 1 \\
4x^3 - 3x^2 + 2x + 1 \\
\hline
11x^3 + \ x^2 \qquad - 1
\end{array}
$$

• Add the terms in each column.

INSTRUCTOR NOTE

It may help to show students that adding polynomials is related to adding whole numbers. Adding the coefficients of like terms using a vertical format is similar to arranging whole numbers to be added in columns, with the units digits aligned.

Unless otherwise noted, all content on this page is © Cengage Learning.

Take Note

The additive inverse of a polynomial is that polynomial with the sign of every term changed.

Take Note

This is the same definition used for subtraction of integers: subtraction is addition of the opposite.

The **additive inverse of the polynomial** $x^2 + 5x - 4$ is $-(x^2 + 5x - 4)$.

To simplify the additive inverse of a polynomial, change the sign of every term inside the parentheses.

$$-(x^2 + 5x - 4) = -x^2 - 5x + 4$$

To subtract two polynomials, add the additive inverse of the second polynomial to the first.

HOW TO 7 Subtract $(3x^2 - 7xy + y^2) - (-4x^2 + 7xy - 3y^2)$.

Use a horizontal format.

Rewrite the subtraction as addition of the additive inverse.

$(3x^2 - 7xy + y^2) - (-4x^2 + 7xy - 3y^2)$
$= (3x^2 - 7xy + y^2) + (4x^2 - 7xy + 3y^2)$
$= 7x^2 - 14xy + 4y^2$ • Combine like terms.

IN-CLASS EXAMPLES

Simplify.

5. $(4x^2 + 3x - 5) + (x^2 - 7x + 10)$
   $5x^2 - 4x + 5$
6. $(5x^2 - x + 6) + (-2x^2 + 3x - 11)$
   $3x^2 + 2x - 5$
7. $(x^2 - 2x + 7) - (3x^2 - 4x + 7)$
   $-2x^2 + 2x$
8. $(4a^2 - 7a) - (-6a^2 + 5a - 7)$
   $10a^2 - 12a + 7$

HOW TO 8 Subtract $(6x^3 - 3x + 7) - (3x^2 - 5x + 12)$. Use a vertical format.

Rewrite the subtraction as addition of the additive inverse.

$(6x^3 - 3x + 7) - (3x^2 - 5x + 12) = (6x^3 - 3x + 7) + (-3x^2 + 5x - 12)$

Arrange the terms of each polynomial in descending order, with like terms in the same column.

$$
\begin{array}{r}
6x^3 \phantom{{}- 3x^2} - 3x + 7 \\
- 3x^2 + 5x - 12 \\
\hline
6x^3 - 3x^2 + 2x - 5
\end{array}
$$
• Combine the terms in each column.

Function notation can be used when adding or subtracting polynomials.

HOW TO 9 Given $P(x) = 3x^2 - 2x + 4$ and $R(x) = -5x^3 + 4x + 7$, find $P(x) + R(x)$.

$P(x) + R(x) = (3x^2 - 2x + 4) + (-5x^3 + 4x + 7)$
$= -5x^3 + 3x^2 + 2x + 11$

HOW TO 10 Given $P(x) = -5x^2 + 8x - 4$ and $R(x) = -3x^2 - 5x + 9$, find $P(x) - R(x)$.

$P(x) - R(x) = (-5x^2 + 8x - 4) - (-3x^2 - 5x + 9)$
$= (-5x^2 + 8x - 4) + (3x^2 + 5x - 9)$
$= -2x^2 + 13x - 13$

EXAMPLE 4

Add:
$(4x^2 - 3xy + 7y^2) + (-3x^2 + 7xy + y^2)$
Use a vertical format.

Solution

$$\begin{array}{r} 4x^2 - 3xy + 7y^2 \\ -3x^2 + 7xy + y^2 \\ \hline x^2 + 4xy + 8y^2 \end{array}$$

YOU TRY IT 4

Add:
$(-3x^2 - 4x + 9) + (-5x^2 - 7x + 1)$
Use a vertical format.

Your solution

$-8x^2 - 11x + 10$

EXAMPLE 5

Subtract:
$(3x^2 - 2x + 4) - (7x^2 + 3x - 12)$
Use a vertical format.

Solution

Add the additive inverse of $7x^2 + 3x - 12$ to $3x^2 - 2x + 4$.

$$\begin{array}{r} 3x^2 - 2x + 4 \\ -7x^2 - 3x + 12 \\ \hline -4x^2 - 5x + 16 \end{array}$$

YOU TRY IT 5

Subtract:
$(-5x^2 + 2x - 3) - (6x^2 + 3x - 7)$
Use a vertical format.

Your solution

$-11x^2 - x + 4$

EXAMPLE 6

Given $P(x) = -3x^2 + 2x - 6$ and
$R(x) = 4x^3 - 3x + 4$, find $P(x) + R(x)$.

Solution

$P(x) + R(x) = (-3x^2 + 2x - 6) + (4x^3 - 3x + 4)$
$= 4x^3 - 3x^2 - x - 2$

YOU TRY IT 6

Given $P(x) = 4x^3 - 3x^2 + 2$ and
$R(x) = -2x^2 + 2x - 3$, find $P(x) - R(x)$.

Your solution

$4x^3 - x^2 - 2x + 5$

Solutions on p. S17

6.2 EXERCISES

✔ **Concept Check**

SUGGESTED ASSIGNMENT
Exercises 1–10; Exercises 11–35, odds;
More challenging exercises: Exercises 37–42

1. Identify each of the following as a monomial, a binomial, a trinomial, or none of these.
   a. $-3x^4 + 1$ Binomial
   b. $2x - 7$ Binomial
   c. $3x^2y^5z$ Monomial
   d. $1 - 4x - x^2$ Trinomial
   e. $5z^4 - 2z^{-2} + 4$ None
   f. 7   Monomial

2. Write each polynomial in descending order.
   a. $3x - 7x^2 + 5$ $-7x^2 + 3x + 5$ **b.** $3x^4 - 7 - 2x + 4x^2$ $3x^4 + 4x^2 - 2x - 7$

3. What is the domain of a polynomial function?   All real numbers

4. Write the additive inverse of $4x^3 - 7x + 8$. $-4x^3 + 7x - 8$

For Exercises 5 to 10, indicate whether the expression defines a polynomial function. For those expressions that are polynomial functions: **a.** Identify the leading coefficient. **b.** Identify the constant term. **c.** State the degree.

5. $P(x) = -x^2 + 3x + 8$
   a. −1   b. 8   c. 2

6. $P(x) = 3x^4 - 3x - 7$
   a. 3   b. −7   c. 4

7. $f(x) = \sqrt{x} - x^2 + 2$
   Not a polynomial

8. $g(x) = -4x^5 - 3x^2 + x - \sqrt{7}$
   a. −4   b. $-\sqrt{7}$ c. 5

9. $P(x) = x^2 - 5x^4 - x^6$
   a. −1   b. 0   c. 6

10. $R(x) = \dfrac{1}{x} + 2$
    Not a polynomial

OBJECTIVE A   *To evaluate polynomial functions*

11. Given $P(x) = 3x^2 - 2x - 8$, evaluate $P(3)$.
    13

12. Given $P(x) = -3x^2 - 5x + 8$, evaluate $P(-5)$.
    −42

13. Given $R(x) = 2x^3 - 3x^2 + 4x - 2$, evaluate $R(2)$.
    10

14. Given $R(x) = -x^3 + 2x^2 - 3x + 4$, evaluate $R(-1)$.
    10

15. Given $f(x) = x^4 - 2x^2 - 10$, evaluate $f(-1)$.
    −11

16. Given $f(x) = x^5 - 2x^3 + 4x$, evaluate $f(2)$.
    24

For Exercises 17 to 22, graph.

17. $P(x) = x^2 - 3x - 3$

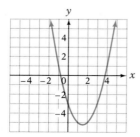

18. $P(x) = 2x^2 + 3x - 4$

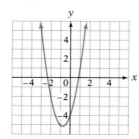

19. $P(x) = x^3 + 2$

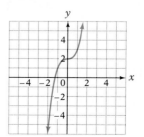

Unless otherwise noted, all content on this page is © Cengage Learning.

20. $f(x) = -x^3 - 10$

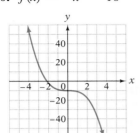

21. $f(x) = x^3 - 4x^2 - 4x + 16$

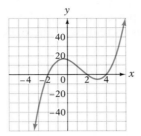

22. $f(x) = -x^3 - 3x^2 + 6x + 8$

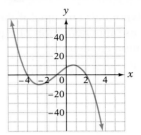

23. Suppose $f(x) = x^2$ and $g(x) = x^3$. For any number c between 0 and 1, is $f(c) - g(c) > 0$ or is $f(c) - g(c) < 0$? Suppose c is between -1 and 0. Is $f(c) - g(c) > 0$ or is $f(c) - g(c) < 0$? $f(c) - g(c) > 0; f(c) - g(c) < 0$

OBJECTIVE B   *To add or subtract polynomials*

24. If $P(x)$ and $Q(x)$ are polynomials of degree 3, is it possible for the sum of the two polynomials to have degree 2? If so, give an example. If not, explain why not.
Yes; one possible example is $P(x) = 5x^3 + 2x^2 - 6x + 1$ and $Q(x) = -5x^3 + 6x^2 + 8x - 3$.

For Exercises 25 to 28, simplify. Use a vertical format.

25. $(5x^2 + 2x - 7) + (x^2 - 8x + 12)$
$6x^2 - 6x + 5$

26. $(3x^2 - 2x + 7) + (-3x^2 + 2x - 12)$
-5

27. $(x^2 - 3x + 8) - (2x^2 - 3x + 7)$
$-x^2 + 1$

28. $(2x^2 + 3x - 7) - (5x^2 - 8x - 1)$
$-3x^2 + 11x - 6$

For Exercises 29 to 32, simplify. Use a horizontal format.

29. $(3y^2 - 7y) + (2y^2 - 8y + 2)$
$5y^2 - 15y + 2$

30. $(-2y^2 - 4y - 12) + (5y^2 - 5y)$
$3y^2 - 9y - 12$

31. $(2a^2 - 3a - 7) - (-5a^2 - 2a - 9)$
$7a^2 - a + 2$

32. $(3a^2 - 9a) - (-5a^2 + 7a - 6)$
$8a^2 - 16a + 6$

For Exercises 33 to 36, find the sum or difference of the polynomial functions.

33. $P(x) = x^2 - 3xy + y^2$ and
$R(x) = 2x^2 - 3y^2; P(x) + R(x)$
$3x^2 - 3xy - 2y^2$

34. $P(x) = 3x^4 - 2x + 1$ and
$R(x) = 3x^5 - 5x - 8; P(x) + R(x)$
$3x^5 + 3x^4 - 7x - 7$

35. $P(x) = 3x^2 + 2y^2$ and
$R(x) = -5x^2 + 2xy - 3y^2; P(x) - R(x)$
$8x^2 - 2xy + 5y^2$

36. $P(x) = 3x^4 - 3x^3 - x^2$ and
$R(x) = 3x^3 - 7x^2 + 2x; P(x) - R(x)$
$3x^4 - 6x^3 + 6x^2 - 2x$

Unless otherwise noted, all content on this page is © Cengage Learning.

Critical Thinking

Two polynomials are equal if the coefficients of like powers are equal. For Exercises 37 and 38, use this definition of equality of polynomials to find the value of k that makes the equation an identity.

37. $(2x^3 + 3x^2 + kx + 5) - (x^3 + x^2 - 5x - 2) = x^3 + 2x^2 + 3x + 7$ -2

QUICK QUIZ

1. Given $P(x) = 4x^2 - 3x + 6$, evaluate $P(2)$.   **16**   **[6.2A]**
2. Given $f(x) = 2x^3 + 4x^2 - 5x + 8$, evaluate $f(-2)$.   **18**   **[6.2A]**

Simplify.

3. $(6x^2 + 4x - 3) + (2x^2 - 5x + 7)$
   $8x^2 - x + 4$ **[6.2B]**
4. $(7x^2 - 5x + 6) + (-3x^2 + 4x - 10)$
   $4x^2 - x - 4$ **[6.2B]**
5. $(3x^2 + 2x - 6) - (4x^2 - 7x - 3)$
   $-x^2 + 9x - 3$ **[6.2B]**
6. $(2a^4 - 3a + 2) - (-4a - 7)$
   $2a^4 + a + 9$ **[6.2B]**

38. $(6x^3 + kx^2 - 2x - 1) - (4x^3 - 3x^2 + 1) = 2x^3 - x^2 - 2x - 2$ -4

39. If $P(-1) = -3$ and $P(x) = 4x^4 - 3x^2 + 6x + c$, find the value of c. 2

40. If $P(2) = 3$ and $P(x) = 2x^3 - 4x^2 - 2x + c$, find the value of c. 7

41. ◤ If $P(x)$ is a third-degree polynomial and $Q(x)$ is a fourth-degree polynomial, what can be said about the degree of $P(x) + Q(x)$? Give some examples of polynomials that support your answer.

42. Sports   The deflection D (in inches) of a beam that is uniformly loaded is given by the polynomial function $D(x) = 0.005x^4 - 0.1x^3 + 0.5x^2$, where x is the distance from one end of the beam. See the figure at the right. The maximum deflection occurs when x is the midpoint of the beam. Determine the maximum deflection for the beam in the diagram.   3.125 in.

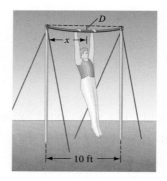

Projects or Group Activities

43. Graph $f(x) = x^2$, $g(x) = x^2 - 3$, and $h(x) = x^2 + 4$ on the same coordinate grid. From the graphs, make a conjecture about the shape and location of $k(x) = x^2 - 2$. Test your conjecture by graphing k.   The graph of k is the graph of f moved 2 units down.

44. Graph $f(x) = x^2$, $g(x) = (x - 3)^2$, and $h(x) = (x + 4)^2$ on the same coordinate grid. From the graphs, make a conjecture about the shape and location of $k(x) = (x - 2)^2$. Test your conjecture by graphing k.   The graph of k is the graph of f moved 2 units to the right.

Unless otherwise noted, all content on this page is © Cengage Learning.

✔ CHECK YOUR PROGRESS: CHAPTER 6

For Exercises 1 to 8, simplify.

1. $(-12a^6b)(6a^4b^3)$
$-72a^{10}b^4$ [6.1A]

2. $(-4x^6y^8)(-3x^{-1}y^{-8})$
$12x^5$ [6.1B]

3. $(2x^3)^4(3x^2)$

$48x^{14}$ [6.1A]

4. $(2a^3b^{-4})^4(3a^{-3}b^2)^{-2}$

$\dfrac{16a^{18}}{9b^{20}}$ [6.1B]

5. $\dfrac{x^4y}{xy^5}$

$\dfrac{x^3}{y^4}$ [6.1B]

6. $\dfrac{2x^{-3}}{4x^{-5}}$

$\dfrac{x^2}{2}$ [6.1B]

7. $\dfrac{3a^4b^2c^8}{6a^7b^{-2}c^8}$

$\dfrac{b^4}{2a^3}$ [6.1B]

8. $\dfrac{(3x^3y^{-2})^{-2}}{(2x^{-1}y^3)^3}$

$\dfrac{1}{72x^3y^5}$ [6.1B]

9. Write 0.000000683 in scientific notation.
6.83×10^{-7} [6.1C]

10. Write 2.607×10^{-3} in decimal notation.
0.002607 [6.1C]

11. Simplify: $\dfrac{35,000,000 \times 0.00642}{0.0007 \times 15}$
2,140,000 [6.1C]

12. Evaluate $P(x) = 4x^3 - 6x + 1$ when $x = -2$.

-19 [6.2A]

13. Add: $(3x^2 - 6x + 7) + (2x^2 + x - 9)$
$5x^2 - 5x - 2$ [6.2B]

14. Subtract: $(-5x^2 + 7x - 8) - (6x^2 + 7x - 7)$
$-11x^2 - 1$ [6.2B]

15. Graph $P(x) = x^3 - 4x$. [6.2A]

6.3 Multiplication of Polynomials

OBJECTIVE A   *To multiply a polynomial by a monomial*

To multiply a polynomial by a monomial, use the Distributive Property and the Rule for Multiplying Exponential Expressions.

IN-CLASS EXAMPLES
Simplify.
1. $-5y^2(3y - 4y^2)$
   $-15y^3 + 20y^4$
2. $3a + 2a(3 - a)$
   $9a - 2a^2$
3. $2a^2b(4a^2 - 3ab + 2b^2)$
   $8a^4b - 6a^3b^2 + 4a^2b^3$
4. $2x^2 - x[x - 3(2x - 1)]$
   $7x^2 - 3x$

HOW TO 1   Multiply: $-3x^2(2x^2 - 5x + 3)$

$-3x^2(2x^2 - 5x + 3)$

$= -3x^2(2x^2) - (-3x^2)(5x) + (-3x^2)(3)$ • Use the Distributive Property.

$= -6x^4 + 15x^3 - 9x^2$ • Use the Rule for Multiplying Exponential Expressions.

HOW TO 2   Simplify: $5x(3x - 6) + 3(4x - 2)$

$5x(3x - 6) + 3(4x - 2)$ • Use the Distributive Property.

$= 5x(3x) - 5x(6) + 3(4x) - 3(2)$

$= 15x^2 - 30x + 12x - 6$ • Simplify.

$= 15x^2 - 18x - 6$

HOW TO 3   Simplify: $2x^2 - 3x[2 - x(4x + 1) + 2]$

$2x^2 - 3x[2 - x(4x + 1) + 2]$

$= 2x^2 - 3x[2 - 4x^2 - x + 2]$ • Use the Distributive Property to remove the parentheses.

$= 2x^2 - 3x[-4x^2 - x + 4]$ • Simplify.

$= 2x^2 + 12x^3 + 3x^2 - 12x$ • Use the Distributive Property to remove the brackets.

$= 12x^3 + 5x^2 - 12x$ • Simplify.

EXAMPLE 1

Multiply: $(3a^2 - 2a + 4)(-3a)$

Solution

$(3a^2 - 2a + 4)(-3a)$

$= 3a^2(-3a) - 2a(-3a) + 4(-3a)$ • Use the Distributive Property.

$= -9a^3 + 6a^2 - 12a$

YOU TRY IT 1

Multiply: $(2b^2 - 7b - 8)(-5b)$

Your solution

$-10b^3 + 35b^2 + 40b$

Solution on p. S17

EXAMPLE 2

Simplify: $y - 3y[y - 2(3y - 6) + 2]$

Solution

$y - 3y[y - 2(3y - 6) + 2]$
$= y - 3y[y - 6y + 12 + 2]$
$= y - 3y[-5y + 14]$
$= y + 15y^2 - 42y$
$= 15y^2 - 41y$

YOU TRY IT 2

Simplify: $x^2 - 2x[x - x(4x - 5) + x^2]$

Your solution

$6x^3 - 11x^2$

Solution on p. S17

OBJECTIVE B *To multiply polynomials*

Multiplication of polynomials requires repeated application of the Distributive Property.

IN-CLASS EXAMPLES

Multiply.

5. $(3x - 2y)(2x + y)$
   $6x^2 - xy - 2y^2$
6. $(2x^2 - 3)(x^2 - 2)$
   $2x^4 - 7x^2 + 6$
7. $(x + 2)(x^2 - 3x - 6)$
   $x^3 - x^2 - 12x - 12$
8.
$(a + 3b)(2a^2 - 3ab - 4b^2)$
$2a^3 + 3a^2b - 13ab^2 - 12b^3$

HOW TO 4 Multiply: $(2x^2 - 5x + 1)(3x + 2)$

Use the Distributive Property to multiply the trinomial by each term of the binomial.

$(2x^2 - 5x + 1)(3x + 2) = (2x^2 - 5x + 1)3x + (2x^2 - 5x + 1)2$
$= (6x^3 - 15x^2 + 3x) + (4x^2 - 10x + 2)$
$= 6x^3 - 11x^2 - 7x + 2$

A convenient method of multiplying two polynomials is to use a vertical format similar to that used for multiplication of whole numbers.

HOW TO 5 Multiply $(2x^2 - 5x + 1)(3x + 2)$. Use a vertical format.

$$
\begin{array}{r}
2x^2 - 5x + 1 \\
3x + 2 \\
\hline
4x^2 - 10x + 2 \\
6x^3 - 15x^2 + 3x \\
\hline
6x^3 - 11x^2 - 7x + 2
\end{array}
$$

$= (2x^2 - 5x + 1)2$
$= (2x^2 - 5x + 1)3x$

It is frequently necessary to find the product of two binomials. The product can be found by using a method called **FOIL**, which is based on the Distributive Property. The letters of FOIL stand for **F**irst, **O**uter, **I**nner, and **L**ast.

 Take Note

FOIL is not really a different way of multiplying. It is based on the Distributive Property.

$(3x - 2)(2x + 5)$
$= 3x(2x + 5) - 2(2x + 5)$
$= 6x^2 + 15x - 4x - 10$
$= 6x^2 + 11x - 10$

FOIL is an efficient way of remembering how to do binomial multiplication.

Multiply: $(3x - 2)(2x + 5)$

Multiply the **F**irst terms.	$(3x - 2)(2x + 5)$	$3x \cdot 2x = 6x^2$
Multiply the **O**uter terms.	$(3x - 2)(2x + 5)$	$3x \cdot 5 = 15x$
Multiply the **I**nner terms.	$(3x - 2)(2x + 5)$	$-2 \cdot 2x = -4x$
Multiply the **L**ast terms.	$(3x - 2)(2x + 5)$	$-2 \cdot 5 = -10$

                                                    F   **O**   **I**   **L**

Add the products.	$(3x - 2)(2x + 5) =$	$6x^2 + 15x - 4x - 10$
Combine like terms.	$=$	$6x^2 + 11x - 10$

HOW TO 6   Multiply: $(6x - 5)(3x - 4)$

$$(6x - 5)(3x - 4) = 6x(3x) + 6x(-4) + (-5)(3x) + (-5)(-4)$$
$$= 18x^2 - 24x - 15x + 20$$
$$= 18x^2 - 39x + 20$$

 Take Note

The product of $x^2 - x - 12$ and $x + 2$ could have been found using a vertical format.

$$
\begin{array}{r}
x^2 - x - 12 \\
x + 2 \\
\hline
2x^2 - 2x - 24 \\
x^3 - x^2 - 12x \\
\hline
x^3 + x^2 - 14x - 24
\end{array}
$$

HOW TO 7   Multiply: $(x + 3)(x - 4)(x + 2)$

$(x + 3)(x - 4)(x + 2)$
$= (x^2 - x - 12)(x + 2)$ • Multiply $(x + 3)(x - 4)$.
$= x^2(x + 2) - x(x + 2) - 12(x + 2)$ • Use the Distributive Property.
$= x^3 + 2x^2 - x^2 - 2x - 12x - 24$
$= x^3 + x^2 - 14x - 24$ • Simplify.

EXAMPLE 3

Multiply $(4a^3 - 3a + 7)(a - 5)$.
Use a vertical format.

Solution

Keep like terms in the same column.

$$
\begin{array}{r}
4a^3 \quad\quad - 3a + 7 \\
a - 5 \\
\hline
-20a^3 \quad\quad + 15a - 35 \\
4a^4 \quad\quad - 3a^2 + 7a \\
\hline
4a^4 - 20a^3 - 3a^2 + 22a - 35
\end{array}
$$

• $-5(4a^3 - 3a + 7)$
• $a(4a^3 - 3a + 7)$

YOU TRY IT 3

Multiply $(-2b^2 + 5b - 4)(-3b + 2)$.
Use a vertical format.

Your solution

$6b^3 - 19b^2 + 22b - 8$

EXAMPLE 4

Multiply: $(5a - 3b)(2a + 7b)$

Solution

$(5a - 3b)(2a + 7b)$
$= 10a^2 + 35ab - 6ab - 21b^2$ • FOIL
$= 10a^2 + 29ab - 21b^2$

YOU TRY IT 4

Multiply: $(3x - 4)(2x - 3)$

Your solution

$6x^2 - 17x + 12$

EXAMPLE 5

Multiply: $(2x^2 - 3)(4x^2 + 1)$

Solution

$(2x^2 - 3)(4x^2 + 1)$
$= 8x^4 + 2x^2 - 12x^2 - 3$
$= 8x^4 - 10x^2 - 3$

YOU TRY IT 5

Multiply: $(3ab + 4)(5ab - 3)$

Your solution

$15a^2b^2 + 11ab - 12$

OBJECTIVE C   *To multiply polynomials that have special products*

Using FOIL, a pattern can be found for the **product of the sum and difference of two terms** [that is, a polynomial that can be expressed in the form $(a + b)(a - b)$] and for the **square of a binomial** [that is, a polynomial that can be expressed in the form $(a + b)^2$].

> Product of the Sum and Difference of Two Terms
> $$(a + b)(a - b) = a^2 - ab + ab - b^2$$
> $$= a^2 - b^2$$
>
> Square of the first term ————↑
> Square of the second term ————↑
>
> Square of a Binomial
> $$(a + b)^2 = (a + b)(a + b) = a^2 + ab + ab + b^2$$
> $$= a^2 + 2ab + b^2$$
>
> Square of the first term ————↑
> Twice the product of the two terms ————↑
> Square of the second term ————↑

IN-CLASS EXAMPLES
Simplify or expand.
9. $(2x - 1)(2x + 1)$
   $4x^2 - 1$
10. $(6a - b)(6a + b)$
   $36a^2 - b^2$
11. $(3a + 2)^2$
   $9a^2 + 12a + 4$
12. $(5x - 3y)^2$
   $25x^2 - 30xy + 9y^2$

 Take Note

The word *expand* is sometimes used to mean "multiply," especially when referring to a power of a binomial.

HOW TO 8   Multiply: $(4x + 3)(4x - 3)$

$(4x + 3)(4x - 3)$ is the sum and difference of the same two terms. The product is the difference of the squares of the terms.

$$(4x + 3)(4x - 3) = (4x)^2 - 3^2$$
$$= 16x^2 - 9$$

HOW TO 9   Expand: $(2x - 3y)^2$

$(2x - 3y)^2$ is the square of a binomial.

$$(2x - 3y)^2 = (2x)^2 + 2(2x)(-3y) + (-3y)^2$$
$$= 4x^2 - 12xy + 9y^2$$

EXAMPLE 6

Multiply: $(2a - 3)(2a + 3)$

Solution

$(2a - 3)(2a + 3)$ • The sum and difference
$= 4a^2 - 9$ of two terms

YOU TRY IT 6

Multiply: $(3x - 7)(3x + 7)$

Your solution

$9x^2 - 49$

EXAMPLE 7

Multiply: $(5x + y)(5x - y)$

Solution

$(5x + y)(5x - y)$ • The sum and difference
$= 25x^2 - y^2$ of two terms

YOU TRY IT 7

Multiply: $(2ab + 7)(2ab - 7)$

Your solution

$4a^2b^2 - 49$

Solutions on p. S18

EXAMPLE 8

Expand: $(2x + 7y)^2$

Solution

$(2x + 7y)^2 = 4x^2 + 28xy + 49y^2$ • The square of a binomial

YOU TRY IT 8

Expand: $(3x - 4y)^2$

Your solution

$9x^2 - 24xy + 16y^2$

EXAMPLE 9

Expand: $(3a^2 - b)^2$

Solution

$(3a^2 - b)^2 = 9a^4 - 6a^2b + b^2$ • The square of a binomial

YOU TRY IT 9

Expand: $(5xy + 4)^2$

Your solution

$25x^2y^2 + 40xy + 16$

Solutions on p. S18

OBJECTIVE D *To solve application problems*

EXAMPLE 10

The length of a rectangle is $(2x + 3)$ ft. The width is $(x - 5)$ ft. Find the area of the rectangle in terms of the variable x.

```
x - 5 [                    ]
           2x + 3
```

Strategy

To find the area, replace the variables L and W in the equation $A = L \cdot W$ by the given values, and solve for A.

Solution

$A = L \cdot W$
$A = (2x + 3)(x - 5)$
$ = 2x^2 - 10x + 3x - 15$ • FOIL
$ = 2x^2 - 7x - 15$

The area is $(2x^2 - 7x - 15)$ ft^2.

YOU TRY IT 10

The base of a triangle is $(2x + 6)$ ft. The height is $(x - 4)$ ft. Find the area of the triangle in terms of the variable x.

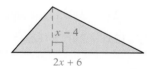

Your strategy

Your solution

$(x^2 - x - 12)$ ft^2

IN-CLASS EXAMPLES

13. The length of a rectangle is $(2x - 3)$ ft. The width is $(x + 3)$ ft. Find the area of the rectangle in terms of the variable x. **$(2x^2 + 3x - 9)$ ft^2**

14. The length of a side of a square is $(2x + 1)$ ft. Find the area of the square in terms of the variable x. **$(4x^2 + 4x + 1)$ ft^2**

15. The radius of a circle is $(4x + 3)$ in. Find the area of the circle in terms of the variable x. Use 3.14 for π. **$(50.24x^2 + 75.36x + 28.26)$ in^2**

Solution on p. S18

Unless otherwise noted, all content on this page is © Cengage Learning.

EXAMPLE 11

The corners are cut from a rectangular piece of cardboard measuring 8 in. by 12 in. The sides are folded up to make a box. Find the volume of the box in terms of the variable x, where x is the length of a side of the square cut from each corner of the rectangle.

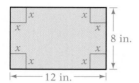

Strategy

Length of the box: $12 - 2x$
Width of the box: $8 - 2x$
Height of the box: x
To find the volume, replace the variables L, W, and H in the equation $V = L \cdot W \cdot H$, and solve for V.

Solution

$V = L \cdot W \cdot H$
$V = (12 - 2x)(8 - 2x)x$
$\quad = (96 - 24x - 16x + 4x^2)x \qquad \bullet \text{ FOIL}$
$\quad = (96 - 40x + 4x^2)x$
$\quad = 96x - 40x^2 + 4x^3$
$\quad = 4x^3 - 40x^2 + 96x$

The volume is $(4x^3 - 40x^2 + 96x)$ in^3.

YOU TRY IT 11

Find the volume of the solid shown in the diagram below. All dimensions are in feet.

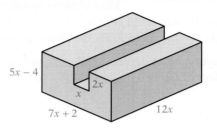

Your strategy

Your solution

$(396x^3 - 216x^2 - 96x)$ ft^3

EXAMPLE 12

The radius of a circle is $(3x - 2)$ cm. Find the area of the circle in terms of the variable x. Use 3.14 for π.

Strategy

To find the area, replace the variable r in the equation $A = \pi r^2$ by the given value, and solve for A.

Solution

$A = \pi r^2$
$A \approx 3.14(3x - 2)^2$
$\quad = 3.14(9x^2 - 12x + 4)$
$\quad = 28.26x^2 - 37.68x + 12.56$

The area is $(28.26x^2 - 37.68x + 12.56)$ cm^2.

YOU TRY IT 12

The radius of a circle is $(2x + 3)$ cm. Find the area of the circle in terms of the variable x. Use 3.14 for π.

Your strategy

Your solution

$(12.56x^2 + 37.68x + 28.26)$ cm^2

Solutions on p. S18

Unless otherwise noted, all content on this page is © Cengage Learning.

6.3 EXERCISES

SUGGESTED ASSIGNMENT
Exercises 1–4; Exercises 5–27, odds; 31–59, odds; 65–101, odds;
More challenging exercises: Exercises 103–108

✔ **Concept Check**

1. What is the first step when simplifying $5 + 2(2x + 1)$? Why?
   Multiply $2(2x + 1) = 4x + 2$. By the Order of Operations Agreement, do multiplication before addition.

2. If p is a polynomial of degree 2 and q is a polynomial of degree 3, what is the degree of the product of the two polynomials? 5

3. 📝 What is FOIL?

4. Determine whether the statement is true for all real numbers x and y.
   a. $(x - y)(x + y) = x^2 - y^2$ Yes   **b.** $(x + y)^2 = x^2 + y^2$ No

---

OBJECTIVE A   *To multiply a polynomial by a monomial*

For Exercises 5 to 28, multiply and simplify.

5. $2x(x - 3)$
   $2x^2 - 6x$

6. $2a(2a + 4)$
   $4a^2 + 8a$

7. $3x^2(2x^2 - x)$
   $6x^4 - 3x^3$

8. $-4y^2(4y - 6y^2)$
   $-16y^3 + 24y^4$

9. $3xy(2x - 3y)$
   $6x^2y - 9xy^2$

10. $-4ab(5a - 3b)$
    $-20a^2b + 12ab^2$

11. $-3xy^2(4x - 5y)$
    $-12x^2y^2 + 15xy^3$

12. $x - 2x(x - 2)$
    $-2x^2 + 5x$

13. $2b + 4b(2 - b)$
    $-4b^2 + 10b$

14. $-2y(3 - y) + 2y^2$
    $4y^2 - 6y$

15. $-2a^2(3a^2 - 2a + 3)$
    $-6a^4 + 4a^3 - 6a^2$

16. $4b(3b^3 - 12b^2 - 6)$
    $12b^4 - 48b^3 - 24b$

17. $(-3y^2 - 4y + 2)(y^2)$
    $-3y^4 - 4y^3 + 2y^2$

18. $(6b^4 - 5b^2 - 3)(-2b^3)$
    $-12b^7 + 10b^5 + 6b^3$

19. $-5x^2(4 - 3x + 3x^2 + 4x^3)$
    $-20x^2 + 15x^3 - 15x^4 - 20x^5$

20. $-2y^2(3 - 2y - 3y^2 + 2y^3)$
    $-6y^2 + 4y^3 + 6y^4 - 4y^5$

21. $-2x^2y(x^2 - 3xy + 2y^2)$
    $-2x^4y + 6x^3y^2 - 4x^2y^3$

22. $3ab^2(3a^2 - 2ab + 4b^2)$
    $9a^3b^2 - 6a^2b^3 + 12ab^4$

23. $5x^3 - 4x(2x^2 + 3x - 7)$
    $-3x^3 - 12x^2 + 28x$

24. $7a^3 - 2a(6a^2 - 5a - 3)$
    $-5a^3 + 10a^2 + 6a$

25. $2y^2 - y[3 - 2(y - 4) - y]$
    $5y^2 - 11y$

26. $3x^2 - x[x - 2(3x - 4)]$
    $8x^2 - 8x$

27. $2y - 3[y - 2y(y - 3) + 4y]$
    $6y^2 - 31y$

28. $4a^2 - 2a[3 - a(2 - a + a^2)]$
    $2a^4 - 2a^3 + 8a^2 - 6a$

29. Given $P(b) = 3b$ and $Q(b) = 3b^4 - 3b^2 + 8$, find $P(b) \cdot Q(b)$.
    $9b^5 - 9b^3 + 24b$

30. Given $P(x) = -2x^2$ and $Q(x) = 2x^2 - 3x - 7$, find $P(x) \cdot Q(x)$.
    $-4x^4 + 6x^3 + 14x^2$

OBJECTIVE B *To multiply polynomials*

For Exercises 31 to 60, multiply.

31. $(x - 2)(x + 7)$
$x^2 + 5x - 14$

32. $(y + 8)(y + 3)$
$y^2 + 11y + 24$

33. $(2y - 3)(4y + 7)$
$8y^2 + 2y - 21$

34. $(5x - 7)(3x - 8)$
$15x^2 - 61x + 56$

35. $(a + 3c)(4a - 5c)$
$4a^2 + 7ac - 15c^2$

36. $(2m - 3n)(5m + 4n)$
$10m^2 - 7mn - 12n^2$

37. $(5x - 7)(5x - 7)$
$25x^2 - 70x + 49$

38. $(5r + 2t)(5r - 2t)$
$25r^2 - 4t^2$

39. $2(2x - 3y)(2x + 5y)$
$8x^2 + 8xy - 30y^2$

40. $-3(7x - 3y)(2x - 9y)$
$-42x^2 + 207xy - 81y^2$

41. $(xy + 4)(xy - 3)$
$x^2y^2 + xy - 12$

42. $(xy - 5)(2xy + 7)$
$2x^2y^2 - 3xy - 35$

43. $(2x^2 - 5)(x^2 - 5)$
$2x^4 - 15x^2 + 25$

44. $(x^2 - 4)(x^2 - 6)$
$x^4 - 10x^2 + 24$

45. $(5x^2 - 5y)(2x^2 - y)$
$10x^4 - 15x^2y + 5y^2$

46. $(x^2 - 2y^2)(x^2 + 4y^2)$
$x^4 + 2x^2y^2 - 8y^4$

47. $(x + 5)(x^2 - 3x + 4)$
$x^3 + 2x^2 - 11x + 20$

48. $(a + 2)(a^2 - 3a + 7)$
$a^3 - a^2 + a + 14$

49. $(2a - 3b)(5a^2 - 6ab + 4b^2)$
$10a^3 - 27a^2b + 26ab^2 - 12b^3$

50. $(3a + b)(2a^2 - 5ab - 3b^2)$
$6a^3 - 13a^2b - 14ab^2 - 3b^3$

51. $(2x^3 + 3x^2 - 2x + 5)(2x - 3)$
$4x^4 - 13x^2 + 16x - 15$

52. $(3a^3 + 4a - 7)(4a - 2)$
$12a^4 - 6a^3 + 16a^2 - 36a + 14$

53. $(2x - 5)(2x^4 - 3x^3 - 2x + 9)$
$4x^5 - 16x^4 + 15x^3 - 4x^2 + 28x - 45$

54. $(2a - 5)(3a^4 - 3a^2 + 2a - 5)$
$6a^5 - 15a^4 - 6a^3 + 19a^2 - 20a + 25$

55. $(x^2 + 2x - 3)(x^2 - 5x + 7)$
$x^4 - 3x^3 - 6x^2 + 29x - 21$

56. $(x^2 - 3x + 1)(x^2 - 2x + 7)$
$x^4 - 5x^3 + 14x^2 - 23x + 7$

57. $(a - 2)(2a - 3)(a + 7)$
$2a^3 + 7a^2 - 43a + 42$

58. $(b - 3)(3b - 2)(b - 1)$
$3b^3 - 14b^2 + 17b - 6$

59. $(2x + 3)(x - 4)(3x + 5)$
$6x^3 - 5x^2 - 61x - 60$

60. $(3a - 5)(2a + 1)(a - 3)$
$6a^3 - 25a^2 + 16a + 15$

61. Given $P(y) = 2y^2 - 1$ and
$Q(y) = y^3 - 5y^2 - 3$, find $P(y) \cdot Q(y)$.
$2y^5 - 10y^4 - y^3 - y^2 + 3$

62. Given $P(b) = 2b^2 - 3$
and $Q(b) = 3b^2 - 3b + 6$, find $P(b) \cdot Q(b)$.
$6b^4 - 6b^3 + 3b^2 + 9b - 18$

63. 🖋 If $P(x)$ is a polynomial of degree m and $Q(x)$ is a polynomial of degree n, what is the degree of the product of the two polynomials?
mn

64. 🖋 Do all polynomials of degree 2 factor over the integers? If not, give an example of a polynomial of degree 2 that does not factor over the integers.
No; answers will vary. One possibility is $x^2 + 4x + 5$.

OBJECTIVE C   *To multiply polynomials that have special products*

For Exercises 65 to 88, simplify or expand.

65. $(3x - 2)(3x + 2)$
$9x^2 - 4$

66. $(4y + 1)(4y - 1)$
$16y^2 - 1$

67. $(6 - x)(6 + x)$
$36 - x^2$

68. $(10 + b)(10 - b)$
$100 - b^2$

69. $(2a - 3b)(2a + 3b)$
$4a^2 - 9b^2$

70. $(5x - 7y)(5x + 7y)$
$25x^2 - 49y^2$

71. $(3ab + 4)(3ab - 4)$
$9a^2b^2 - 16$

72. $(5xy - 8)(5xy + 8)$
$25x^2y^2 - 64$

73. $(x^2 + 1)(x^2 - 1)$
$x^4 - 1$

74. $(x^2 + y^2)(x^2 - y^2)$
$x^4 - y^4$

75. $(x - 5)^2$
$x^2 - 10x + 25$

76. $(y + 2)^2$
$y^2 + 4y + 4$

77. $(3a + 5b)^2$
$9a^2 + 30ab + 25b^2$

78. $(5x - 4y)^2$
$25x^2 - 40xy + 16y^2$

79. $(x^2 - 3)^2$
$x^4 - 6x^2 + 9$

80. $(x^2 + y^2)^2$
$x^4 + 2x^2y^2 + y^4$

81. $(2x^2 - 3y^2)^2$
$4x^4 - 12x^2y^2 + 9y^4$

82. $(2xy + 3)^2$
$4x^2y^2 + 12xy + 9$

83. $(3mn - 5)^2$
$9m^2n^2 - 30mn + 25$

84. $(2 - 7xy)^2$
$49x^2y^2 - 28xy + 4$

85. $y^2 - (x - y)^2$
$-x^2 + 2xy$

86. $a^2 + (a + b)^2$
$2a^2 + 2ab + b^2$

87. $(x - y)^2 - (x + y)^2$
$-4xy$

88. $(a + b)^2 + (a - b)^2$
$2a^2 + 2b^2$

89. True or false? $a^2 + b^2 = (a + b)(a + b)$
False

90. If $P(x)$ is a polynomial of degree 2 that factors as the difference of squares, is the coefficient of x in $P(x)$ **(i)** less than zero, **(ii)** equal to zero, **(iii)** greater than zero, or **(iv)** either less than or greater than zero?
ii

OBJECTIVE D   *To solve application problems*

91. If the measures of the width and length of the floor of a room are given in feet, what is the unit of measure of the area of the floor?
ft^2

92. If the measures of the width, length, and height of a box are given in meters, what is the unit of measure of the volume of the box?
m^3

93. Geometry The length of a rectangle is $(3x - 2)$ ft. The width is $(x + 4)$ ft. Find the area of the rectangle in terms of the variable x.
$(3x^2 + 10x - 8)$ ft^2

94. Geometry The base of a triangle is $(x - 4)$ ft. The height is $(3x + 2)$ ft. Find the area of the triangle in terms of the variable x.
$\left(\dfrac{3}{2}x^2 - 5x - 4\right)$ ft^2

95. Geometry   Find the area of the figure shown below. All dimensions given are in meters.
$(x^2 + 3x)$ m²

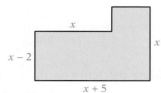

96. Geometry   Find the area of the figure shown below. All dimensions given are in feet.
$(x^2 + 12x + 16)$ ft²

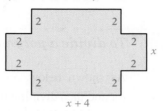

97. Geometry   The length of the side of a cube is $(x + 3)$ cm. Find the volume of the cube in terms of the variable x.
$(x^3 + 9x^2 + 27x + 27)$ cm³

98. Geometry   The length of a box is $(3x + 2)$ cm, the width is $(x - 4)$ cm, and the height is x cm. Find the volume of the box in terms of the variable x. $(3x^3 - 10x^2 - 8x)$ cm³

99. Geometry   Find the volume of the figure shown below. All dimensions given are in inches.
$(2x^3)$ in³

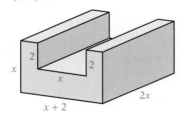

100. Geometry   Find the volume of the figure shown below. All dimensions given are in centimeters.
$(4x^3 + 32x^2 + 48x)$ cm³

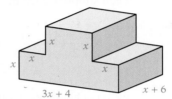

101. Geometry   The radius of a circle is $(5x + 4)$ in. Find the area of the circle in terms of the variable x. Use 3.14 for π.
$(78.5x^2 + 125.6x + 50.24)$ in²

102. Geometry   The radius of a circle is $(x - 2)$ in. Find the area of the circle in terms of the variable x. Use 3.14 for π.
$(3.14x^2 - 12.56x + 12.56)$ in²

Critical Thinking

103. For what value of k is the given equation an identity?
   a. $(3x - k)(2x + k) = 6x^2 + 5x - k^2$ $k = 5$
   b. $(4x + k)^2 = 16x^2 + 8x + k^2$ $k = 1$

104. Complete.
   a. If $m = n + 1$, then $\dfrac{a^m}{a^n} = $ _____ a _____ .

   b. If $m = n + 2$, then $\dfrac{a^m}{a^n} = $ _____ a^2 _____ .

105. What polynomial, when divided by $2x - 3$, has a quotient of $x + 7$?
$2x^2 + 11x - 21$

106. What polynomial, when divided by $x - 4$, has a quotient of $2x + 3$?
$2x^2 - 5x - 12$

Projects or Group Activities

107. Write two polynomials whose product is a polynomial of degree 3.
Answers will vary. For example, $(x + 3)(2x^2 - 1)$

108. Write two polynomials whose product is a polynomial of degree 4.
Answers will vary. For example, $(x^3 + 2x + 1)(x - 3)$

QUICK QUIZ
Multiply and simplify.
  1. $2x^2(3x^2 + x)$ $6x^4 + 2x^3$ [6.3A]
  2. $-3y(2 - y) + y^2$ $-6y + 4y^2$ [6.3A]
  3. $-3x^2y(x^2 - 2xy + 3y^2)$ $-3x^4y + 6x^3y^2 - 9x^2y^3$
    [6.3A]
  4. $(5x - 2y)(2x - 3y)$ $10x^2 - 19xy + 6y^2$ [6.3B]
  5. $(x - 1)(x^2 - 2x + 2)$ $x^3 - 3x^2 + 4x - 2$ [6.3B]

    Simplify or expand.
  6. $(3b - 8)(3b + 8)$
    $9b^2 - 64$ [6.3C]
  7. $(4x - 1)^2$
    $16x^2 - 8x + 1$ [6.3C]
  8. The length of a rectangle is $(2x + 4)$ ft. The width is $(x - 5)$ ft. Find the area of the rectangle in terms of the variable x.
    $(2x^2 - 6x - 20)$ ft²
    [6.3D]

Unless otherwise noted, all content on this page is © Cengage Learning.

SECTION

6.4 Division of Polynomials

OBJECTIVE A *To divide a polynomial by a monomial*

As shown below, $\frac{6 + 4}{2}$ can be simplified by first adding the terms in the numerator and then dividing the result by the denominator. It can also be simplified by first dividing each term in the numerator by the denominator and then adding the results.

$$\frac{6 + 4}{2} = \frac{10}{2} = 5 \qquad \frac{6 + 4}{2} = \frac{6}{2} + \frac{4}{2} = 3 + 2 = 5$$

It is the second method that is used to divide a polynomial by a monomial: Divide each term in the numerator by the denominator, and then write the sum of the quotients.

To divide $\frac{6x^2 + 4x}{2x}$, divide each term of the polynomial $6x^2 + 4x$ by the monomial $2x$. Then simplify each quotient.

Take Note

Recall that the fraction bar can be read "divided by."

$$\frac{6x^2 + 4x}{2x} = \frac{6x^2}{2x} + \frac{4x}{2x} \qquad \bullet \text{ Divide each term in the numerator by the denominator.}$$

$$= 3x + 2 \qquad \bullet \text{ Simplify each quotient.}$$

We can check this quotient by multiplying it by the divisor.

$$2x(3x + 2) = 6x^2 + 4x \qquad \bullet \text{ The product is the dividend. The quotient checks.}$$

HOW TO 1 Divide and check: $\dfrac{16x^5 - 8x^3 + 4x}{2x}$

IN-CLASS EXAMPLES

1. Divide and check:
   $\dfrac{3a^2 + 2a}{a}$ **3a + 2**

2. Divide and check:
   $\dfrac{x^6 - 3x^4 - x^2}{x^2}$
   $x^4 - 3x^2 - 1$

$$\frac{16x^5 - 8x^3 + 4x}{2x} = \frac{16x^5}{2x} - \frac{8x^3}{2x} + \frac{4x}{2x} \qquad \bullet \text{ Divide each term in the numerator by the denominator.}$$

$$= 8x^4 - 4x^2 + 2 \qquad \bullet \text{ Simplify each quotient.}$$

Check:
$$2x(8x^4 - 4x^2 + 2) = 16x^5 - 8x^3 + 4x \qquad \bullet \text{ The quotient checks.}$$

EXAMPLE 1

Divide and check: $\dfrac{6x^3 - 3x^2 + 9x}{3x}$

Solution

$$\frac{6x^3 - 3x^2 + 9x}{3x}$$

$$= \frac{6x^3}{3x} - \frac{3x^2}{3x} + \frac{9x}{3x} \qquad \bullet \text{ Divide each term in the numerator by the denominator.}$$

$$= 2x^2 - x + 3 \qquad \bullet \text{ Simplify each quotient.}$$

Check: $3x(2x^2 - x + 3) = 6x^3 - 3x^2 + 9x$

YOU TRY IT 1

Divide and check: $\dfrac{4x^3y + 8x^2y^2 - 4xy^3}{2xy}$

Your solution

$2x^2 + 4xy - 2y^2$

Solution on p. S18

OBJECTIVE B *To divide polynomials*

The division method illustrated in Objective A is appropriate only when the divisor is a monomial. To divide two polynomials in which the divisor is not a monomial, use a method similar to that used for division of whole numbers.

To check division of polynomials, use

$$\textbf{Dividend = (quotient × divisor) + remainder}$$

INSTRUCTOR NOTE
Begin this objective by showing an example of long division of whole numbers—for instance, 863 ÷ 57.

IN-CLASS EXAMPLES
Divide using long division.
3.
$(6x^2 + x - 5) \div (3x + 2)$
$2x - 1 - \dfrac{3}{3x + 2}$
4.
$(x^3 - 2x^2 + 3) \div (x - 3)$
$x^2 + x + 3 + \dfrac{12}{x - 3}$
5. $\dfrac{3x^3 - 7x^2 - 22x + 8}{3x - 1}$
$x^2 - 2x - 8$

HOW TO 2 Divide: $(x^2 + 5x - 7) \div (x + 3)$

Step 1

$$\begin{array}{r} x \phantom{{}+3x^2+5x-7} \\ x + 3 \overline{)x^2 + 5x - 7} \\ \underline{x^2 + 3x} \phantom{{}-7} \downarrow \\ 2x - 7 \end{array}$$

Think: $x\overline{)x^2} = \dfrac{x^2}{x} = x$

Multiply: $x(x + 3) = x^2 + 3x$

Subtract: $(x^2 + 5x) - (x^2 + 3x) = 2x$

Step 2

$$\begin{array}{r} x + 2 \\ x + 3 \overline{)x^2 + 5x - 7} \\ \underline{x^2 + 3x} \\ 2x - 7 \\ \underline{2x + 6} \\ -13 \end{array}$$

Think: $x\overline{)2x} = \dfrac{2x}{x} = 2$

Multiply: $2(x + 3) = 2x + 6$

Subtract: $(2x - 7) - (2x + 6) = -13$
The remainder is -13.

Check: $(x + 2)(x + 3) + (-13) = x^2 + 3x + 2x + 6 - 13 = x^2 + 5x - 7$

$$(x^2 + 5x - 7) \div (x + 3) = x + 2 - \dfrac{13}{x + 3}$$

INSTRUCTOR NOTE
After you have completed this objective, a possible extra-credit problem might be to divide

$2x - 4\overline{)3x^3 - 5x^2 - 6x + 8}$

$\dfrac{3}{2}x^2 + \dfrac{1}{2}x - 2$

HOW TO 3 Divide: $\dfrac{6 - 6x^2 + 4x^3}{2x + 3}$

Arrange the terms of each polynomial in descending order. Note that there is no term containing the first power of x in $4x^3 - 6x^2 + 6$. Insert a zero as $0x$ for the missing term so that like terms will be in the same columns.

$$\begin{array}{r} 2x^2 - 6x + 9 \\ 2x + 3 \overline{)4x^3 - 6x^2 + 0x + 6} \\ \underline{4x^3 + 6x^2} \\ -12x^2 + 0x \\ \underline{-12x^2 - 18x} \\ 18x + 6 \\ \underline{18x + 27} \\ -21 \end{array}$$

$$\dfrac{4x^3 - 6x^2 + 6}{2x + 3} = 2x^2 - 6x + 9 - \dfrac{21}{2x + 3}$$

EXAMPLE 2

Divide: $\dfrac{12x^2 - 11x + 10}{4x - 5}$

Solution

$$
\begin{array}{r}
3x + 1 \\
4x - 5 \overline{)12x^2 - 11x + 10} \\
\underline{12x^2 - 15x } \\
4x + 10 \\
\underline{4x - 5} \\
15
\end{array}
$$

$\dfrac{12x^2 - 11x + 10}{4x - 5} = 3x + 1 + \dfrac{15}{4x - 5}$

YOU TRY IT 2

Divide: $\dfrac{15x^2 + 17x - 20}{3x + 4}$

Your solution

$5x - 1 - \dfrac{16}{3x + 4}$

EXAMPLE 3

Divide: $\dfrac{x^3 + 1}{x + 1}$

Solution

$$
\begin{array}{r}
x^2 - x + 1 \\
x + 1 \overline{)x^3 + 0x^2 + 0x + 1} \\
\underline{x^3 + x^2 } \\
- x^2 + 0x \\
\underline{- x^2 - x} \\
x + 1 \\
\underline{x + 1} \\
0
\end{array}
$$

• **Insert zeros for the missing terms.**

$\dfrac{x^3 + 1}{x + 1} = x^2 - x + 1$

YOU TRY IT 3

Divide: $\dfrac{3x^3 + 8x^2 - 6x + 2}{3x - 1}$

Your solution

$x^2 + 3x - 1 + \dfrac{1}{3x - 1}$

EXAMPLE 4

Divide:
$(2x^4 - 7x^3 + 3x^2 + 4x - 5) \div (x^2 - 2x - 2)$

Solution

$$
\begin{array}{r}
2x^2 - 3x + 1 \\
x^2 - 2x - 2 \overline{)2x^4 - 7x^3 + 3x^2 + 4x - 5} \\
\underline{2x^4 - 4x^3 - 4x^2 } \\
- 3x^3 + 7x^2 + 4x \\
\underline{- 3x^3 + 6x^2 + 6x} \\
x^2 - 2x - 5 \\
\underline{x^2 - 2x - 2} \\
- 3
\end{array}
$$

$(2x^4 - 7x^3 + 3x^2 + 4x - 5) \div (x^2 - 2x - 2)$

$\quad = 2x^2 - 3x + 1 - \dfrac{3}{x^2 - 2x - 2}$

YOU TRY IT 4

Divide:
$(3x^4 - 11x^3 + 16x^2 - 16x + 8) \div (x^2 - 3x + 2)$

Your solution

$3x^2 - 2x + 4$

Solutions on pp. S18–S19

OBJECTIVE C   *To divide polynomials by using synthetic division*

 Tips for Success

An important element of success is practice. We cannot do anything well if we do not practice it repeatedly. Practice is crucial to success in mathematics. In this objective you are learning a new procedure, synthetic division. You will need to practice this procedure in order to be successful at it.

Synthetic division is a shorter method of dividing a polynomial by a binomial of the form $x - a$.

Divide $(3x^2 - 4x + 6) \div (x - 2)$ by using long division.

$$
\begin{array}{r}
3x + 2 \\
x - 2\overline{)3x^2 - 4x + 6} \\
\underline{3x^2 - 6x} \\
2x + 6 \\
\underline{2x - 4} \\
10
\end{array}
$$

$$(3x^2 - 4x + 6) \div (x - 2) = 3x + 2 + \frac{10}{x - 2}$$

The variables can be omitted because the position of a term indicates the power of the term.

$$
\begin{array}{r}
3 \quad\ 2 \\
-2\overline{)3 \ -4 \quad\ 6} \\
\underline{3 \ -6} \\
2 \quad\ 6 \\
\underline{2 \ -4} \\
10
\end{array}
$$

Each number shown in color is exactly the same as the number above it. Removing the colored numbers condenses the vertical spacing.

$$
\begin{array}{r}
3 \quad\ 2 \\
-2\overline{)3 \ -4 \quad\ 6} \\
\underline{-6 \ -4} \\
2 \quad 10
\end{array}
$$

The number in color in the top row is the same as the one in the bottom row. Writing the 3 from the top row in the bottom row allows the spacing to be condensed even further.

$$
\begin{array}{c|ccc}
-2 & 3 & -4 & 6 \\
& & -6 & -4 \\
\hline
& 3 & 2 & 10
\end{array}
$$

$\underbrace{\qquad\qquad}_{\text{Terms of the quotient}}$ $\underbrace{\qquad}_{\text{Remainder}}$

Because the degree of the dividend $(3x^2 - 4x + 6)$ is 2 and the degree of the divisor $(x - 2)$ is 1, the degree of the quotient is $2 - 1 = 1$. This means that, using the coefficients of the quotient given above, that quotient is $3x + 2$. The remainder is 10.

In general, the degree of the quotient of two polynomials is the difference between the degree of the dividend and the degree of the divisor.

By replacing the constant term in the divisor by its additive inverse, we may add rather than subtract terms. This is illustrated in HOW TO 4 on the next page.

IN-CLASS EXAMPLES

Divide by using synthetic division.

6. $(3x^2 + 11x + 6) \div (x + 3)$
   $3x + 2$

7. $(3x^2 - 6) \div (x - 2)$
   $3x + 6 + \dfrac{6}{x - 2}$

8. $(3x^3 + 2x^2 - 8x + 4) \div (x - 1)$
   $3x^2 + 5x - 3 + \dfrac{1}{x - 1}$

9. $\dfrac{2x^4 - 5x^3 - 8x^2 - 17x + 4}{x - 4}$
   $2x^3 + 3x^2 + 4x - 1$

HOW TO 4   Divide: $(3x^3 + 6x^2 - x - 2) \div (x + 3)$

The additive inverse of the binomial constant

Coefficients of the polynomial

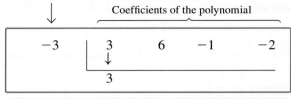

-3	3	6	-1	-2
	3			

• Bring down the 3.

-3	3	6	-1	-2
		-9		
	3	-3		

• Multiply $-3(3)$ and add the product to 6.

-3	3	6	-1	-2
		-9	9	
	3	-3	8	

• Multiply $-3(-3)$ and add the product to -1.

-3	3	6	-1	-2
		-9	9	-24
	3	-3	8	-26

• Multiply $-3(8)$ and add the product to -2.

Coefficients of the quotient   Remainder

The degree of the dividend is 3 and the degree of the divisor is 1. Therefore, the degree of the quotient is $3 - 1 = 2$.

$$(3x^3 + 6x^2 - x - 2) \div (x + 3) = 3x^2 - 3x + 8 - \frac{26}{x + 3}$$

INSTRUCTOR NOTE

Remind students that they can check an answer to a synthetic division problem in the same way that they check an answer to a long division problem.

HOW TO 5   Divide: $(2x^3 - x + 2) \div (x - 2)$

The additive inverse of the binomial constant

Coefficients of the polynomial

2	2	0	-1	2
	2			

• Insert a 0 for the missing x^2 term and bring down the 2.

2	2	0	-1	2
		4		
	2	4		

• Multiply $2(2)$ and add the product to 0.

2	2	0	-1	2
		4	8	
	2	4	7	

• Multiply $2(4)$ and add the product to -1.

2	2	0	-1	2
		4	8	14
	2	4	7	16

• Multiply $2(7)$ and add the product to 2.

Coefficients of the quotient   Remainder

$$(2x^3 - x + 2) \div (x - 2) = 2x^2 + 4x + 7 + \frac{16}{x - 2}$$

EXAMPLE 5

Divide: $(7 - 3x + 5x^2) \div (x - 1)$

Solution

Arrange the coefficients in decreasing powers of x.

```
1 |  5  -3   7
   |      5   2
      5   2   9
```

$(5x^2 - 3x + 7) \div (x - 1) = 5x + 2 + \dfrac{9}{x - 1}$

YOU TRY IT 5

Divide: $(6x^2 + 8x - 5) \div (x + 2)$

Your solution

$6x - 4 + \dfrac{3}{x + 2}$

EXAMPLE 6

Divide: $(2x^3 + 4x^2 - 3x + 12) \div (x + 4)$

Solution

```
-4 |  2   4   -3    12
   |     -8   16   -52
      2  -4   13   -40
```

$(2x^3 + 4x^2 - 3x + 12) \div (x + 4)$

$= 2x^2 - 4x + 13 - \dfrac{40}{x + 4}$

YOU TRY IT 6

Divide: $(5x^3 - 12x^2 - 8x + 16) \div (x - 2)$

Your solution

$5x^2 - 2x - 12 - \dfrac{8}{x - 2}$

EXAMPLE 7

Divide: $(3x^4 - 8x^2 + 2x + 1) \div (x + 2)$

Solution

Insert a zero for the missing x^3 term.

```
-2 |  3   0   -8    2    1
   |     -6   12   -8   12
      3  -6    4   -6   13
```

$(3x^4 - 8x^2 + 2x + 1) \div (x + 2)$

$= 3x^3 - 6x^2 + 4x - 6 + \dfrac{13}{x + 2}$

YOU TRY IT 7

Divide: $(2x^4 - 3x^3 - 8x^2 - 2) \div (x - 3)$

Your solution

$2x^3 + 3x^2 + x + 3 + \dfrac{7}{x - 3}$

Solutions on p. S19

OBJECTIVE D *To evaluate a polynomial function using synthetic division*

A polynomial can be evaluated by using synthetic division. Consider the polynomial $P(x) = 2x^4 - 3x^3 + 4x^2 - 5x + 1$. One way to evaluate the polynomial when $x = 2$ is to replace x by 2 and then simplify the resulting numerical expression.

$P(x) = 2x^4 - 3x^3 + 4x^2 - 5x + 1$

$P(2) = 2(2)^4 - 3(2)^3 + 4(2)^2 - 5(2) + 1$
$\quad\ \ = 2(16) - 3(8) + 4(4) - 5(2) + 1$
$\quad\ \ = 32 - 24 + 16 - 10 + 1$
$\quad\ \ = 15$

Now use synthetic division to divide $(2x^4 - 3x^3 + 4x^2 - 5x + 1)$ by $(x - 2)$.

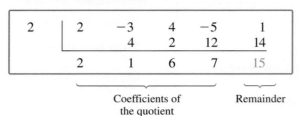

$$
\begin{array}{r|rrrrr}
2 & 2 & -3 & 4 & -5 & 1 \\
  &   & 4 & 2 & 12 & 14 \\
\hline
  & 2 & 1 & 6 & 7 & 15
\end{array}
$$

Coefficients of   Remainder
the quotient

Note that the remainder is 15, which is the same value as $P(2)$. This is not a coincidence. The following theorem states that this relationship is always true.

Remainder Theorem

If the polynomial $P(x)$ is divided by $x - a$, the remainder is $P(a)$.

HOW TO 6   Evaluate $P(x) = x^4 - 3x^2 + 4x - 5$ when $x = -2$ by using the Remainder Theorem.

The value at which the polynomial is evaluated

$$
\begin{array}{r|rrrrr}
-2 & 1 & 0 & -3 & 4 & -5 \\
   &   & -2 & 4 & -2 & -4 \\
\hline
   & 1 & -2 & 1 & 2 & -9
\end{array}
$$

- A **0** is inserted for the x^3 term.

⟵ The remainder

$P(-2) = -9$

IN-CLASS EXAMPLES

Use the Remainder Theorem to evaluate the polynomial.

10. $P(x) = 3x^2 - 4x + 5$; $P(3)$ **20**

11. $f(x) = x^3 - 6x^2 + 3x - 8$; $f(2)$ **−18**

EXAMPLE 8

Evaluate $P(x) = x^2 - 6x + 4$ when $x = 3$ by using the Remainder Theorem.

Solution

$$
\begin{array}{r|rrr}
3 & 1 & -6 & 4 \\
  &   & 3 & -9 \\
\hline
  & 1 & -3 & -5
\end{array}
$$

$P(3) = -5$

YOU TRY IT 8

Evaluate $P(x) = 2x^2 - 3x - 5$ when $x = 2$ by using the Remainder Theorem.

Your solution

$P(2) = -3$

EXAMPLE 9

Evaluate $P(x) = -x^4 + 3x^3 + 2x^2 - x - 5$ when $x = -2$ by using the Remainder Theorem.

Solution

$$
\begin{array}{r|rrrrr}
-2 & -1 & 3   & 2 & -1 & -5 \\
   &   & 2   & -10 & 16 & -30 \\
\hline
   & -1 & 5   & -8 & 15 & -35
\end{array}
$$

$P(-2) = -35$

YOU TRY IT 9

Evaluate $P(x) = 2x^3 - 5x^2 + 7$ when $x = -3$ by using the Remainder Theorem.

Your solution

$P(-3) = -92$

Solutions on p. S19

6.4 EXERCISES

✔ **Concept Check**

SUGGESTED ASSIGNMENT
Exercises 1–4; Exercises 5–39, odds; 43–61, odds; 65–83, odds;
More challenging exercises: Exercises 85–89

1. If $P(x)$ is a polynomial of degree 6 and $Q(x)$ is a polynomial of degree 2, what is the degree of the quotient of $P(x)$ and $Q(x)$?   4

2. Suppose r is the remainder when two polynomials are divided. Is the degree of r less than, equal to, or greater than the degree of the divisor?   Less than

3. ◣ What is the Remainder Theorem?

4. How can multiplication be used to check a division problem?
Use dividend = (quotient × divisor) + remainder.

OBJECTIVE A   *To divide a polynomial by a monomial*

For Exercises 5 to 16, divide and check.

5. $\dfrac{3x^2 - 6x}{3x}$

$x - 2$

6. $\dfrac{10y^2 - 6y}{2y}$

$5y - 3$

7. $\dfrac{5x^2 - 10x}{-5x}$

$-x + 2$

8. $\dfrac{3y^2 - 27y}{-3y}$

$-y + 9$

9. $\dfrac{5x^2y^2 + 10xy}{5xy}$

$xy + 2$

10. $\dfrac{8x^2y^2 - 24xy}{8xy}$

$xy - 3$

11. $\dfrac{x^3 + 3x^2 - 5x}{x}$

$x^2 + 3x - 5$

12. $\dfrac{a^3 - 5a^2 + 7a}{a}$

$a^2 - 5a + 7$

13. $\dfrac{9b^5 + 12b^4 + 6b^3}{3b^2}$

$3b^3 + 4b^2 + 2b$

14. $\dfrac{a^8 - 5a^5 - 3a^3}{a^2}$

$a^6 - 5a^3 - 3a$

15. $\dfrac{a^5b - 6a^3b + ab}{ab}$

$a^4 - 6a^2 + 1$

16. $\dfrac{5c^3d + 10c^2d^2 - 15cd^3}{5cd}$

$c^2 + 2cd - 3d^2$

17. ◔ If $\dfrac{P(x)}{3x} = 2x^2 + 7x - 5$, what is $P(x)$?

$6x^3 + 21x^2 - 15x$

18. ◔ If $\dfrac{6x^3 + 15x^2 - 24x}{ax} = 2x^2 + 5x - 8$,
what is a?

3

OBJECTIVE B   *To divide polynomials*

For Exercises 19 to 40, divide by using long division.

19. $(x^2 + 3x - 40) \div (x - 5)$

$x + 8$

20. $(x^2 - 14x + 24) \div (x - 2)$

$x - 12$

21. $(x^3 - 3x + 2) \div (x - 3)$

$x^2 + 3x + 6 + \dfrac{20}{x - 3}$

22. $(x^3 + 4x^2 - 8) \div (x + 4)$

$x^2 - \dfrac{8}{x + 4}$

23. $(6x^2 + 13x + 8) \div (2x + 1)$

$3x + 5 + \dfrac{3}{2x + 1}$

24. $(12x^2 + 13x - 14) \div (3x - 2)$

$4x + 7$

25. $(10x^2 + 9x - 5) \div (2x - 1)$

$5x + 7 + \dfrac{2}{2x - 1}$

26. $(18x^2 - 3x + 2) \div (3x + 2)$

$6x - 5 + \dfrac{12}{3x + 2}$

27. $(8x^3 - 9) \div (2x - 3)$

$4x^2 + 6x + 9 + \dfrac{18}{2x - 3}$

28. $(64x^3 + 4) \div (4x + 2)$

$16x^2 - 8x + 4 - \dfrac{4}{4x + 2}$

29. $(6x^4 - 13x^2 - 4) \div (2x^2 - 5)$

$3x^2 + 1 + \dfrac{1}{2x^2 - 5}$

30. $(12x^4 - 11x^2 + 10) \div (3x^2 + 1)$

$4x^2 - 5 + \dfrac{15}{3x^2 + 1}$

31. $\dfrac{-10 - 33x + 3x^3 - 8x^2}{3x + 1}$

$x^2 - 3x - 10$

32. $\dfrac{10 - 49x + 38x^2 - 8x^3}{1 - 4x}$

$2x^2 - 9x + 10$

33. $\dfrac{x^3 - 5x^2 + 7x - 4}{x - 3}$

$x^2 - 2x + 1 - \dfrac{1}{x - 3}$

34. $\dfrac{2x^3 - 3x^2 + 6x + 4}{2x + 1}$

$x^2 - 2x + 4$

35. $\dfrac{16x^2 - 13x^3 + 2x^4 + 20 - 9x}{x - 5}$

$2x^3 - 3x^2 + x - 4$

36. $\dfrac{x - x^2 + 5x^3 + 3x^4 - 2}{x + 2}$

$3x^3 - x^2 + x - 1$

37. $\dfrac{2x^3 + 4x^2 - x + 2}{x^2 + 2x - 1}$

$2x + \dfrac{x + 2}{x^2 + 2x - 1}$

38. $\dfrac{3x^3 - 2x^2 + 5x - 4}{x^2 - x + 3}$

$3x + 1 + \dfrac{-3x - 7}{x^2 - x + 3}$

39. $\dfrac{x^4 + 2x^3 - 3x^2 - 6x + 2}{x^2 - 2x - 1}$

$x^2 + 4x + 6 + \dfrac{10x + 8}{x^2 - 2x - 1}$

40. $\dfrac{x^4 - 3x^3 + 4x^2 - x + 1}{x^2 + x - 3}$

$x^2 - 4x + 11 + \dfrac{-24x + 34}{x^2 + x - 3}$

41. Given $Q(x) = 2x + 1$ and

$P(x) = 2x^3 + x^2 + 8x + 7$, find $\dfrac{P(x)}{Q(x)}$.

$x^2 + 4 + \dfrac{3}{2x + 1}$

42. Given $Q(x) = 3x - 2$ and

$P(x) = 3x^3 - 2x^2 + 3x - 5$, find $\dfrac{P(x)}{Q(x)}$.

$x^2 + 1 - \dfrac{3}{3x - 2}$

43. 🖐 If $\dfrac{6x^3 + 27x^2 + 18x - 30}{p(x)} = q(x)$,

what is $p(x) \cdot q(x)$?

$6x^3 + 27x^2 + 18x - 30$

44. 🖐 Suppose $\dfrac{4x^3 - 18x^2 + 6x + 18}{p(x)} = 2x - 3$.

Is it possible to find $p(x)$? If so, explain how.
If not, why not?

Yes. Divide $4x^3 - 18x^2 + 6x + 18$ by $2x - 3$.

OBJECTIVE C *To divide polynomials by using synthetic division*

45. 🖐 The display below shows the beginning of a synthetic division. What is the degree of the dividend?   3

```
4 |  2    -5    4    1
  |_____
```

46. 🖐 For the synthetic division shown below, what is the quotient and what is the remainder?

```
-2 |  3    7    4   -3
   |     -6   -2   -4
   |_____
      3    1    2   -7
```

Quotient: $3x^2 + x + 2$; remainder: $\dfrac{-7}{x + 2}$

For Exercises 47 to 62, divide by using synthetic division.

47. $(2x^2 - 6x - 8) \div (x + 1)$
$2x - 8$

48. $(3x^2 + 19x + 20) \div (x + 5)$
$3x + 4$

49. $(3x^2 - 14x + 16) \div (x - 2)$
$3x - 8$

50. $(4x^2 - 23x + 28) \div (x - 4)$
$4x - 7$

51. $(3x^2 - 4) \div (x - 1)$
$3x + 3 - \dfrac{1}{x - 1}$

52. $(4x^2 - 8) \div (x - 2)$
$4x + 8 + \dfrac{8}{x - 2}$

53. $(2x^3 - x^2 + 6x + 9) \div (x + 1)$
$2x^2 - 3x + 9$

54. $(3x^3 + 10x^2 + 6x - 4) \div (x + 2)$
$3x^2 + 4x - 2$

55. $(18 + x - 4x^3) \div (2 - x)$
$4x^2 + 8x + 15 + \dfrac{12}{x - 2}$

56. $(12 - 3x^2 + x^3) \div (x + 3)$
$x^2 - 6x + 18 - \dfrac{42}{x + 3}$

57. $(2x^3 + 5x^2 - 5x + 20) \div (x + 4)$
$2x^2 - 3x + 7 - \dfrac{8}{x + 4}$

58. $(5x^3 + 3x^2 - 17x + 6) \div (x + 2)$
$5x^2 - 7x - 3 + \dfrac{12}{x + 2}$

59. $\dfrac{5 + 5x - 8x^2 + 4x^3 - 3x^4}{2 - x}$
$3x^3 + 2x^2 + 12x + 19 + \dfrac{33}{x - 2}$

60. $\dfrac{3 - 13x - 5x^2 + 9x^3 - 2x^4}{3 - x}$
$2x^3 - 3x^2 - 4x + 1$

61. $\dfrac{3x^4 + 3x^3 - x^2 + 3x + 2}{x + 1}$
$3x^3 - x + 4 - \dfrac{2}{x + 1}$

62. $\dfrac{4x^4 + 12x^3 - x^2 - x + 2}{x + 3}$
$4x^3 - x + 2 - \dfrac{4}{x + 3}$

63. Given $Q(x) = x - 2$ and $P(x) = 3x^2 - 5x + 6$, find $\dfrac{P(x)}{Q(x)}$. $3x + 1 + \dfrac{8}{x - 2}$

64. Given $Q(x) = x + 5$ and $P(x) = 2x^2 + 7x - 12$, find $\dfrac{P(x)}{Q(x)}$. $2x - 3 + \dfrac{3}{x + 5}$

OBJECTIVE D   *To evaluate a polynomial function using synthetic division*

65. The result of a synthetic division is shown below. What is a first-degree polynomial factor of the dividend $p(x)$? $x - 3$

$$\begin{array}{r|rrrr} 3 & 1 & 1 & -7 & -15 \\ & & 3 & 12 & 15 \\ \hline & 1 & 4 & 5 & 0 \end{array}$$

66. The result of a synthetic division of $p(x)$ is shown below. What is $p(-3)$?   8

$$\begin{array}{r|rrrr} -3 & 1 & 2 & -7 & -4 \\ & & -3 & 3 & 12 \\ \hline & 1 & -1 & -4 & 8 \end{array}$$

For Exercises 67 to 84, use the Remainder Theorem to evaluate the polynomial function.

67. $P(x) = 2x^2 - 3x - 1; P(3)$
8

68. $Q(x) = 3x^2 - 5x - 1; Q(2)$
1

69. $R(x) = x^3 - 2x^2 + 3x - 1; R(4)$
43

70. $F(x) = x^3 + 4x^2 - 3x + 2; F(3)$
56

71. $P(z) = 2z^3 - 4z^2 + 3z - 1; P(-2)$
−39

72. $R(t) = 3t^3 + t^2 - 4t + 2; R(-3)$
−58

73. $Z(p) = 2p^3 - p^2 + 3; Z(-3)$
−60

74. $P(y) = 3y^3 + 2y^2 - 5; P(-2)$
−21

75. $Q(x) = x^4 + 3x^3 - 2x^2 + 4x - 9; Q(2)$
31

76. $Y(z) = z^4 - 2z^3 - 3z^2 - z + 7; Y(3)$
4

77. $F(x) = 2x^4 - x^3 + 2x - 5; F(-3)$
178

78. $Q(x) = x^4 - 2x^3 + 4x - 2; Q(-2)$
22

79. $P(x) = x^3 - 3; P(5)$
122

80. $S(t) = 4t^3 + 5; S(-4)$
−251

81. $R(t) = 4t^4 - 3t^2 + 5; R(-3)$
302

82. $P(z) = 2z^4 + z^2 - 3; P(-4)$
525

83. $Q(x) = x^5 - 4x^3 - 2x^2 + 5x - 2; Q(2)$
0

84. $R(x) = 2x^5 - x^3 + 4x - 1; R(-2)$
−65

Critical Thinking

For Exercises 85 to 88, for what value of k will the remainder be zero?

85. $(x^3 - 3x^2 - x + k) \div (x - 3)$ 3

86. $(x^3 - 2x^2 + x + k) \div (x - 2)$ −2

87. $(x^2 + kx - 6) \div (x - 3)$ −1

88. $(x^3 + kx + k - 1) \div (x - 1)$ 0

89. If $p(x)$ and $q(x)$ are polynomials of degree greater than zero, and the product of the two polynomials is of degree 6, what are the possible degrees of $p(x)$ and $q(x)$?
1, 5; 2, 4; 3, 3

90. Suppose $P(x) = x^4 + 2x^2 + 5$. Is there a real number a for which $P(a) = 0$? Explain.

Projects or Group Activities

The Factor Theorem is a result of the Remainder Theorem. The Factor Theorem states that a polynomial $P(x)$ has a factor $(x - c)$ if and only if $P(c) = 0$. In other words, a remainder of zero means that the divisor is a factor of the dividend. Use this fact for Exercises 91 and 92.

91. Determine whether $x + 5$ is a factor of $P(x) = x^4 + x^3 - 21x^2 - x + 20$.   Yes

92. Judging from your answer to Exercise 91, is −5 a zero of $P(x)$? Explain.   Yes

QUICK QUIZ

1. Divide and check:
$$\frac{20ab^2 - 10a^2b + 5ab}{5ab}$$
$4b - 2a + 1$ [6.4A]

Divide by using long division.
2. $(10x^2 + 14x - 15) \div (5x - 3)$
$2x + 4 - \frac{3}{5x - 3}$ [6.4B]

3. $(x^3 + 4x^2 - 6) \div (x + 2)$
$x^2 + 2x - 4 + \frac{2}{x + 2}$ [6.4B]

Divide by using synthetic division.
4. $(3x^2 + 6x - 24) \div (x + 4)$
$3x - 6$ [6.4C]

5. $(2x^3 + 3x^2 + 5x + 16) \div (x + 2)$
$2x^2 - x + 7 + \frac{2}{x + 2}$ [6.4C]

Use the Remainder Theorem to evaluate the polynomial function.
6. $R(x) = 4x^2 + 6x - 3$; $R(-2)$ 1   [6.4D]
7. $F(x) = x^3 - 5x^2 + 2x - 10$; $F(3)$ −22   [6.4D]

6 : Summary

Key Words	Examples
A **monomial** is a number, a variable, or a product of numbers and variables. [6.1A, p. 344]	5, y, and $8a^2b^2$ are monomials.
The **degree of a monomial** is the sum of the exponents on the variables. [6.1A, p. 344]	The degree of $8x^4y^5z$ is 10.
A **polynomial** is a variable expression in which the terms are monomials. [6.2A, p. 356]	$x^4 - 2xy - 32x + 8$ is a polynomial. The terms are x^4, $-2xy$, $-32x$, and 8.
A polynomial of one term is a **monomial,** a polynomial of two terms is a **binomial,** and a polynomial of three terms is a **trinomial.** [6.2A, p. 356]	$5x^4$ is a monomial. $6y^3 - 2y$ is a binomial. $2x^2 - 5x + 3$ is a trinomial.
The **degree of a polynomial** is the greatest of the degrees of any of its terms. [6.2A, p. 356]	The degree of the polynomial $x^3 + 3x^2y^2 - 4xy - 3$ is 4.
The terms of a polynomial in one variable are usually arranged so that the exponents on the variable decrease from left to right. This is called **descending order.** [6.2A, p. 356]	The polynomial $4x^3 + 5x^2 - x + 7$ is written in descending order.
A **polynomial function** is an expression whose terms are monomials. Polynomial functions include the **linear function** given by $f(x) = mx + b$; the **quadratic function** given by $f(x) = ax^2 + bx + c$, $a \neq 0$; and the **cubic function,** which is a third-degree polynomial function. The **leading coefficient** of a polynomial function is the coefficient of the variable with the largest exponent. The **constant term** is the term without a variable. [6.2A, pp. 356–357]	$f(x) = 5x - 4$ is a linear function. $f(x) = 3x^2 - 2x + 1$ is a quadratic function. 3 is the leading coefficient, and 1 is the constant term. $f(x) = x^3 - 1$ is a cubic function.

Essential Rules and Procedures	Examples
Rule for Multiplying Exponential Expressions [6.1A, p. 344] $x^m \cdot x^n = x^{m+n}$	$b^5 \cdot b^4 = b^{5+4} = b^9$

Rule for Simplifying the Power of an Exponential Expression [6.1A, p. 345]

$(x^m)^n = x^{mn}$

$(y^3)^7 = y^{3(7)} = y^{21}$

Rule for Simplifying Powers of Products [6.1A, p. 345]

$(x^m y^n)^p = x^{mp} y^{np}$

$(x^6 y^4 z^5)^2 = x^{6(2)} y^{4(2)} z^{5(2)} = x^{12} y^8 z^{10}$

Rule for Dividing Exponential Expressions [6.1B, p. 346]

For $x \neq 0$, $\dfrac{x^m}{x^n} = x^{m-n}$.

$\dfrac{y^8}{y^3} = y^{8-3} = y^5$

Definition of Zero as an Exponent [6.1B, p. 347]

For $x \neq 0$, $x^0 = 1$. The expression 0^0 is not defined.

$17^0 = 1$ $(5y)^0 = 1, y \neq 0$

Definition of a Negative Exponent [6.1B, p. 347]

For $x \neq 0$, $x^{-n} = \dfrac{1}{x^n}$ and $\dfrac{1}{x^{-n}} = x^n$.

$x^{-6} = \dfrac{1}{x^6}$ and $\dfrac{1}{x^{-6}} = x^6$

Rule for Simplifying Powers of Quotients [6.1B, p. 348]

For $y \neq 0$, $\left(\dfrac{x^m}{y^n}\right)^p = \dfrac{x^{mp}}{y^{np}}$.

$\left(\dfrac{x^2}{y^4}\right)^5 = \dfrac{x^{2 \cdot 5}}{y^{4 \cdot 5}} = \dfrac{x^{10}}{y^{20}}$

Scientific Notation [6.1C, p. 350]

To express a number in scientific notation, write it in the form $a \times 10^n$, where $1 \leq a < 10$ and n is an integer.

If the number is greater than 10, the exponent on 10 will be positive.

$367,000,000 = 3.67 \times 10^8$

If the number is less than 1, the exponent on 10 will be negative.

$0.0000059 = 5.9 \times 10^{-6}$

To change a number written in scientific notation to decimal notation, move the decimal point to the right if the exponent on 10 is positive and to the left if the exponent on 10 is negative. Move the decimal point the same number of places as the absolute value of the exponent.

$2.418 \times 10^7 = 24,180,000$
$9.06 \times 10^{-5} = 0.0000906$

To add polynomials, combine like terms, which means to add the coefficients of the like terms. [6.2B, p. 359]

$$(8x^2 + 2x - 9) + (-3x^2 + 5x - 7)$$
$$= (8x^2 - 3x^2) + (2x + 5x) + (-9 - 7)$$
$$= 5x^2 + 7x - 16$$

To write the additive inverse of a polynomial, change the sign of every term of the polynomial. [6.2B, p. 360]

The additive inverse of $-y^2 + 4y - 5$ is $y^2 - 4y + 5$.

To subtract two polynomials, add the additive inverse of the second polynomial to the first polynomial. [6.2B, p. 360]

$$(3y^2 - 8y + 6) - (-y^2 + 4y - 5)$$
$$= (3y^2 - 8y + 6) + (y^2 - 4y + 5)$$
$$= 4y^2 - 12y + 11$$

To multiply a polynomial by a monomial, use the Distributive Property and the Rule for Multiplying Exponential Expressions. [6.3A, p. 366]

$$-2x^3(4x^2 + 5x - 1)$$
$$= -8x^5 - 10x^4 + 2x^3$$

The FOIL Method [6.3B, p. 367]
The product of two binomials can be found by adding the products of the **First** terms, the **Outer** terms, the **Inner** terms, and the **Last** terms.

$$(4x + 3)(2x - 5)$$
$$= 8x^2 - 20x + 6x - 15$$
$$= 8x^2 - 14x - 15$$

To divide a polynomial by a monomial, divide each term of the polynomial by the monomial. [6.4A, p. 376]

$$\frac{12x^5 + 8x^3 - 6x}{4x^2} = 3x^3 + 2x - \frac{3}{2x}$$

Synthetic Division [6.4C, p. 379]
Synthetic division is a shorter method of dividing a polynomial by a binomial of the form $x - a$. This method uses only the coefficients of the variable terms.

$$(3x^3 - 9x - 5) \div (x - 2)$$

2	3	0	−9	−5
		6	12	6
	3	6	3	1

$$(3x^3 - 9x - 5) \div (x - 2)$$
$$= 3x^2 + 6x + 3 + \frac{1}{x - 2}$$

Remainder Theorem [6.4D, p. 382]
If the polynomial $P(x)$ is divided by $x - a$, the remainder is $P(a)$.

$$P(x) = x^3 - x^2 + x - 1$$

−2	1	−1	1	−1
		−2	6	−14
	1	−3	7	−15

$$P(-2) = -15$$

6 Review Exercises

1. Add: $(12y^2 + 17y - 4) + (9y^2 - 13y + 3)$
   $21y^2 + 4y - 1$ [6.2B]

2. Divide: $\dfrac{15x^2 + 2x - 2}{3x - 2}$
   $5x + 4 + \dfrac{6}{3x - 2}$ [6.4B]

3. Simplify: $(2x^{-1}y^2z^5)^4(-3x^3yz^{-3})^2$
   $144x^2y^{10}z^{14}$ [6.1B]

4. Expand: $(5y - 7)^2$
   $25y^2 - 70y + 49$ [6.3C]

5. Simplify: $\dfrac{a^{-1}b^3}{a^3b^{-3}}$
   $\dfrac{b^6}{a^4}$ [6.1B]

6. Use the Remainder Theorem to evaluate $P(x) = x^3 - 2x^2 + 3x - 5$ when $x = 2$.
   1 [6.4D]

7. Simplify: $(5x^2 - 8xy + 2y^2) - (x^2 - 3y^2)$
   $4x^2 - 8xy + 5y^2$ [6.2B]

8. Divide: $\dfrac{12b^7 + 36b^5 - 3b^3}{3b^3}$
   $4b^4 + 12b^2 - 1$ [6.4A]

9. Simplify: $\dfrac{3ab^4}{-6a^2b^4}$
   $-\dfrac{1}{2a}$ [6.1B]

10. Simplify: $(-2a^2b^4)(3ab^2)$
    $-6a^3b^6$ [6.1A]

11. Simplify: $\dfrac{8x^{12}}{12x^9}$
    $\dfrac{2x^3}{3}$ [6.1B]

12. Simplify: $\dfrac{4x^3 + 27x^2 + 10x + 2}{x + 6}$
    $4x^2 + 3x - 8 + \dfrac{50}{x + 6}$ [6.4B/6.4C]

13. Given $P(x) = 2x^3 - x + 7$, evaluate $P(-2)$.
    -7 [6.2A]

14. Subtract: $(13y^3 - 7y - 2) - (12y^2 - 2y - 1)$
    $13y^3 - 12y^2 - 5y - 1$ [6.2B]

15. Divide: $(b^3 - 2b^2 - 33b - 7) \div (b - 7)$
    $b^2 + 5b + 2 + \dfrac{7}{b - 7}$ [6.4B/6.4C]

16. Multiply: $4x^2y(3x^3y^2 + 2xy - 7y^3)$
    $12x^5y^3 + 8x^3y^2 - 28x^2y^4$ [6.3A]

17. Multiply: $(2a - b)(x - 2y)$
    $2ax - 4ay - bx + 2by$ [6.3B]

18. Multiply: $(2b - 3)(4b + 5)$
    $8b^2 - 2b - 15$ [6.3B]

19. Simplify: $5x^2 - 4x[x - 3(3x + 2) + x]$
    $33x^2 + 24x$ [6.3A]

20. Simplify: $(xy^5z^3)(x^3y^3z)$
    $x^4y^8z^4$ [6.1A]

21. Expand: $(4x - 3y)^2$
    $16x^2 - 24xy + 9y^2$ [6.3C]

22. Simplify: $\dfrac{x^4 - 4}{x - 4}$
    $x^3 + 4x^2 + 16x + 64 + \dfrac{252}{x - 4}$ [6.4B/6.4C]

23. Add: $(3x^2 - 2x - 6) + (-x^2 - 3x + 4)$
$2x^2 - 5x - 2$ [6.2B]

24. Multiply: $(5x^2yz^4)(2xy^3z^{-1})(7x^{-2}y^{-2}z^3)$
$70xy^2z^6$ [6.1B]

25. Simplify: $\dfrac{3x^4yz^{-1}}{-12xy^3z^2}$
$-\dfrac{x^3}{4y^2z^3}$ [6.1B]

26. Write 948,000,000 in scientific notation.
9.48×10^8 [6.1C]

27. Simplify: $\dfrac{3 \times 10^{-3}}{15 \times 10^2}$
2×10^{-6} [6.1C]

28. Use the Remainder Theorem to evaluate
$P(x) = -2x^3 + 2x^2 - 4$ when $x = -3$.
68   [6.4D]

29. Divide: $\dfrac{16x^5 - 8x^3 + 20x}{4x}$
$4x^4 - 2x^2 + 5$ [6.4A]

30. Divide: $\dfrac{12x^2 - 16x - 7}{6x + 1}$
$2x - 3 - \dfrac{4}{6x + 1}$ [6.4B]

31. Multiply: $a^{2n+3}(a^n - 5a + 2)$
$a^{3n+3} - 5a^{2n+4} + 2a^{2n+3}$ [6.3A]

32. Multiply: $(x + 6)(x^3 - 3x^2 - 5x + 1)$
$x^4 + 3x^3 - 23x^2 - 29x + 6$ [6.3B]

33. Multiply: $-2x(4x^2 + 7x - 9)$
$-8x^3 - 14x^2 + 18x$ [6.3A]

34. Multiply: $(3y^2 + 4y - 7)(2y + 3)$
$6y^3 + 17y^2 - 2y - 21$ [6.3B]

35. Simplify: $(-2u^3v^4)^4$
$16u^{12}v^{16}$ [6.1A]

36. Add: $(2x^3 + 7x^2 + x) + (2x^2 - 4x - 12)$
$2x^3 + 9x^2 - 3x - 12$ [6.2B]

37. Subtract: $(5x^2 - 2x - 1) - (3x^2 - 5x + 7)$
$2x^2 + 3x - 8$ [6.2B]

38. Multiply: $(a + 7)(a - 7)$
$a^2 - 49$ [6.3C]

39. Simplify: $(5a^7b^6)^2(4ab)$
$100a^{15}b^{13}$ [6.1A]

40. Write 1.46×10^7 in decimal notation.
14,600,000   [6.1C]

41. Simplify: $(-2x^3)^2(-3x^4)^3$
$-108x^{18}$ [6.1A]

42. Divide: $(6y^2 - 35y + 36) \div (3y - 4)$
$2y - 9$ [6.4B]

43. Evaluate: -4^{-2}
$-\dfrac{1}{16}$ [6.1B]

44. Multiply: $(5a - 7)(2a + 9)$
$10a^2 + 31a - 63$ [6.3B]

45. Divide: $\dfrac{7 - x - x^2}{x + 3}$
$-x + 2 + \dfrac{1}{x + 3}$ [6.4B/6.4C]

46. Write 0.000000127 in scientific notation.
1.27×10^{-7} [6.1C]

47. Divide: $\dfrac{16y^2 - 32y}{-4y}$
$-4y + 8$ [6.4A]

48. Simplify: $\dfrac{(2a^4b^{-3}c^2)^3}{(2a^3b^2c^{-1})^4}$
$\dfrac{c^{10}}{2b^{17}}$ [6.1B]

49. Multiply: $(x - 4)(3x + 2)(2x - 3)$
$6x^3 - 29x^2 + 14x + 24$ [6.3B]

50. Simplify: $(-3x^{-2}y^{-3})^{-2}$
$\dfrac{x^4y^6}{9}$ [6.1B]

51. Simplify: $(2a^{12}b^3)(-9b^2c^6)(3ac)$
$-54a^{13}b^5c^7$ [6.1A]

52. Multiply: $(5a + 2b)(5a - 2b)$
$25a^2 - 4b^2$ [6.3C]

53. Write 2.54×10^{-3} in decimal notation.
0.00254 [6.1C]

54. Multiply: $2ab^3(4a^2 - 2ab + 3b^2)$
$8a^3b^3 - 4a^2b^4 + 6ab^5$ [6.3A]

55. Graph $y = x^2 + 1$. [6.2A]

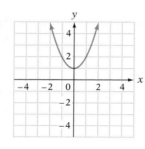

56. For the polynomial $P(x) = 3x^5 - 6x^2 + 7x + 8$:
 a. Identify the leading coefficient. 3
 b. Identify the constant term. 8
 c. State the degree. 5 [6.2A]

57.  **Physics** The mass of the moon is 3.7×10^{-8} times the mass of the sun. The mass of the sun is 2.19×10^{27} tons. Find the mass of the moon. Write the answer in scientific notation. 8.103×10^{19} tons [6.1D]

58. Geometry The side of a checkerboard is $(3x - 2)$ in. Express the area of the checkerboard in terms of the variable x.
$(9x^2 - 12x + 4)$ in² [6.3D]

The Moon

ravl/Shutterstock.com

59.  **Astronomy** The most distant object visible from Earth without the aid of a telescope is the Great Galaxy of Andromeda. It takes light from the Great Galaxy of Andromeda 2.2×10^6 years to travel to Earth. Light travels about 6.7×10^8 mph. How far from Earth is the Great Galaxy of Andromeda? Use a 365-day year. 1.291224×10^{19} mi [6.1D]

60. Geometry The length of a rectangle is $(5x + 3)$ cm. The width is $(2x - 7)$ cm. Find the area of the rectangle in terms of the variable x.
$(10x^2 - 29x - 21)$ cm² [6.3D]

Unless otherwise noted, all content on this page is © Cengage Learning.

CHAPTER

6 : TEST

1. Multiply: $2x(2x^2 - 3x)$
$4x^3 - 6x^2$ [6.3A]

2. Use the Remainder Theorem to evaluate $P(x) = -x^3 + 4x - 8$ when $x = -2$.
-8 [6.4D]

3. Simplify: $\dfrac{12x^2}{-3x^8}$
$-\dfrac{4}{x^6}$ [6.1B]

4. Simplify: $(-2xy^2)(3x^2y^4)$
$-6x^3y^6$ [6.1A]

5. Divide: $(x^2 + 1) \div (x + 1)$
$x - 1 + \dfrac{2}{x + 1}$ [6.4B/6.4C]

6. Multiply: $(x - 3)(x^2 - 4x + 5)$
$x^3 - 7x^2 + 17x - 15$ [6.3B]

7. Simplify: $(-2a^2b)^3$
$-8a^6b^3$ [6.1A]

8. Simplify: $\dfrac{(3x^{-2}y^3)^3}{3x^4y^{-1}}$
$\dfrac{9y^{10}}{x^{10}}$ [6.1B]

9. Multiply: $(a - 2b)(a + 5b)$
$a^2 + 3ab - 10b^2$ [6.3B]

10. Given $P(x) = 3x^2 - 8x + 1$, evaluate $P(2)$.
-3 [6.2A]

11. Divide using synthetic division:
$(x^2 + 6x - 7) \div (x - 1)$
$x + 7$ [6.4C]

12. Multiply: $-3y^2(-2y^2 + 3y - 6)$
$6y^4 - 9y^3 + 18y^2$ [6.3A]

13. Multiply: $(-2x^3 + x^2 - 7)(2x - 3)$
$-4x^4 + 8x^3 - 3x^2 - 14x + 21$ [6.3B]

14. Simplify: $(4y - 5)(4y + 5)$
$16y^2 - 25$ [6.3C]

15. Divide: $\dfrac{18x^5 + 9x^4 - 6x^3}{3x^2}$

$6x^3 + 3x^2 - 2x$ [6.4A]

16. Simplify: $\dfrac{2a^{-1}b}{2^{-2}a^{-2}b^{-3}}$

$8ab^4$ [6.1B]

17. Simplify: $\dfrac{(2a^{-4}b^2)^3}{4a^{-2}b^{-1}}$

$\dfrac{2b^7}{a^{10}}$ [6.1B]

18. Subtract: $(3a^2 - 2a - 7) - (5a^3 + 2a - 10)$

$-5a^3 + 3a^2 - 4a + 3$ [6.2B]

19. Simplify: $(2x - 5)^2$

$4x^2 - 20x + 25$ [6.3C]

20. Divide: $\dfrac{x^3 - 2x^2 - 5x + 7}{x + 3}$

$x^2 - 5x + 10 - \dfrac{23}{x + 3}$ [6.4B/6.4C]

21. Multiply: $(2x - 7y)(5x - 4y)$
$10x^2 - 43xy + 28y^2$ [6.3B]

22. Add: $(3x^3 - 2x^2 - 4) + (8x^2 - 8x + 7)$
$3x^3 + 6x^2 - 8x + 3$ [6.2B]

23. Write 0.00000000302 in scientific notation.
3.02×10^{-9} [6.1C]

24. Write the number of seconds in 10 weeks in scientific notation.
6.048×10^6 s   [6.1D]

25. Geometry   The radius of a circle is $(x - 5)$ m. Use the equation $A = \pi r^2$, where r is the radius, to find the area of the circle in terms of the variable x. Leave the answer in terms of π.
$(\pi x^2 - 10\pi x + 25\pi)$ m²   [6.3D]

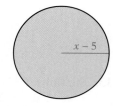

Unless otherwise noted, all content on this page is © Cengage Learning.

Cumulative Review Exercises

1. Simplify: $8 - 2[-3 - (-1)]^2 + 4$
  4   [1.3A]

2. Evaluate $\frac{2a - b}{b - c}$ when $a = 4$, $b = -2$, and $c = 6$.
  $-\frac{5}{4}$ [1.4A]

3. Identify the property that justifies the statement $2x + (-2x) = 0$.
  Inverse Property of Addition   [1.4B]

4. Simplify: $2x - 4[x - 2(3 - 2x) + 4]$
  $-18x + 8$ [1.4D]

5. Solve: $\frac{2}{3} - y = \frac{5}{6}$
  $-\frac{1}{6}$ [2.2A]

6. Solve: $8x - 3 - x = -6 + 3x - 8$
  $-\frac{11}{4}$ [2.2B]

7. Divide: $\frac{x^3 - 3}{x - 3}$
  $x^2 + 3x + 9 + \frac{24}{x - 3}$ [6.4B/6.4C]

8. Solve: $3 - |2 - 3x| = -2$
  -1 and $\frac{7}{3}$ [2.6A]

9. Given $P(x) = 3x^2 - 2x + 2$, evaluate $P(-2)$.
  18   [4.2A]

10. What is the domain of the function $f(x) = \frac{x + 1}{x + 2}$?
  $\{x | x \neq -2\}$ [4.2A]

11. Find the zero of the function given by $F(x) = 3x - 4$.
  $\frac{4}{3}$ [4.3C]

12. Find the slope of the line containing the points $P_1(-2, 3)$ and $P_2(4, 2)$.
  $-\frac{1}{6}$ [4.4A]

13. Find the equation of the line that contains the point $P(-1, 2)$ and has slope $-\frac{3}{2}$.
  $y = -\frac{3}{2}x + \frac{1}{2}$ [4.5A]

14. Find the equation of the line that contains the point $P(-2, 4)$ and is perpendicular to the line $3x + 2y = 4$.
  $y = \frac{2}{3}x + \frac{16}{3}$ [4.6A]

15. Solve by substitution:
  $2x - 3y = 4$
  $x + y = -3$
  $(-1, -2)$ [5.1B]

16. Solve by the addition method:
  $x - y + z = 0$
  $2x + y - 3z = -7$
  $-x + 2y + 2z = 5$
  $\left(-\frac{9}{7}, \frac{2}{7}, \frac{11}{7}\right)$ [5.2B]

17. Graph $3x - 4y = 12$ by using the x- and y-intercepts.

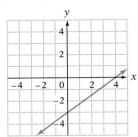

[4.3C]

18. Graph the solution set: $-3x + 2y < 6$

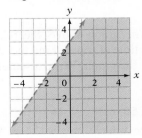

[4.7A]

Unless otherwise noted, all content on this page is © Cengage Learning.

19. Solve by graphing:

$$x - 2y = 3$$
$$-2x + y = -3$$

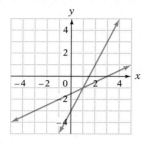

(1, −1) [5.1A]

20. Graph the solution set:

$$2x + y < 3$$
$$-2x + y \geq 1$$

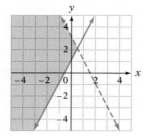

[5.4A]

21. Simplify: $(4a^{-2}b^3)(2a^3b^{-1})^{-2}$

$\dfrac{b^5}{a^8}$ [6.1B]

22. Simplify: $\dfrac{(5x^3y^{-3}z)^{-2}}{y^4z^{-2}}$

$\dfrac{y^2}{25x^6}$ [6.1B]

23. Simplify: $3 - (3 - 3^{-1})^{-1}$

$\dfrac{21}{8}$ [6.1B]

24. Multiply: $(2x + 3)(2x^2 - 3x + 1)$

$4x^3 - 7x + 3$ [6.3B]

25. Integers The sum of two integers is twenty-four. The difference between four times the smaller integer and nine is three less than twice the larger integer. Find the integers.

9, 15 [2.3A]

26. Investments An investor has a total of $12,000 invested in two simple interest accounts. On one account, the annual simple interest rate is 4%. On the second account, the annual simple interest rate is 4.5%. How much is invested in the 4% account if the total annual interest earned is $530?

$2000 [5.1C]

27. Uniform Motion Two bicyclists are 25 mi apart and are traveling toward each other. One cyclist is traveling at $\frac{2}{3}$ the rate of the other cyclist. The cyclists pass each other in 2 h. Find the rate of each cyclist.

Slower cyclist: 5 mph; faster cyclist: 7.5 mph [2.4C]

28. Mixtures How many ounces of pure silver that costs $360 per ounce must be mixed with 80 oz of an alloy that costs $120 per ounce to make a mixture that costs $200 per ounce?

40 oz [2.4A]

29. Uniform Motion The graph shows the relationship between the distance traveled and the time of travel. Find the slope of the line between the two points labeled on the graph. Write a sentence that states the meaning of the slope.

$m = 50$

The average speed is 50 mph. [4.4A]

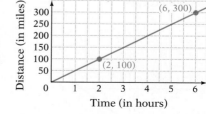

30. Astronomy A space vehicle travels 2.4×10^5 mi from Earth to the moon at an average velocity of 2×10^4 mph. How long does it take the vehicle to reach the moon?

12 h [6.1D]

Unless otherwise noted, all content on this page is © Cengage Learning.

Factoring

7

OBJECTIVES

SECTION 7.1
A To factor a monomial from a polynomial
B To factor by grouping

SECTION 7.2
A To factor a trinomial of the form $x^2 + bx + c$
B To factor completely

SECTION 7.3
A To factor a trinomial of the form $ax^2 + bx + c$ by using trial factors
B To factor a trinomial of the form $ax^2 + bx + c$ by grouping

SECTION 7.4
A To factor the difference of two perfect squares or a perfect-square trinomial
B To factor the sum or difference of two perfect cubes
C To factor a trinomial that is quadratic in form
D To factor completely

SECTION 7.5
A To solve equations by factoring
B To solve application problems

Focus on Success

Have you established a routine for doing your homework? If not, decide now where and when your study time is most productive. Perhaps it is at home, in the library, or in the math center, where you can get help as you need it. If possible, create a study hour right after class. The material will be fresh in your mind, and the immediate review, along with your homework, will reinforce the concepts you are learning. (See Homework Time, page AIM-5)

© iStockphoto.com/Chris Schmidt

Prep Test

Are you ready to succeed in this chapter? Take the Prep Test below to find out if you are ready to learn the new material.

1. Write 30 as a product of prime numbers.
$2 \cdot 3 \cdot 5$ [1.2D]

2. Simplify: $-3(4y - 5)$
$-12y + 15$ [1.4D]

3. Simplify: $-(a - b)$
$-a + b$ [1.4D]

4. Simplify: $2(a - b) - 5(a - b)$
$-3a + 3b$ [1.4D]

5. Solve: $4x = 0$
0 [2.1C]

6. Solve: $2x + 1 = 0$
$-\dfrac{1}{2}$ [2.2A]

7. Multiply: $(x + 4)(x - 6)$
$x^2 - 2x - 24$ [6.3B]

8. Multiply: $(2x - 5)(3x + 2)$
$6x^2 - 11x - 10$ [6.3B]

9. Simplify: $\dfrac{x^5}{x^2}$
x^3 [6.1B]

10. Simplify: $\dfrac{6x^4y^3}{2xy^2}$
$3x^3y$ [6.1B]

SECTION

7.1 Common Factors

OBJECTIVE A *To factor a monomial from a polynomial*

The **greatest common factor (GCF) of two or more numbers** is the largest common factor of the numbers. For example, the GCF of 9 and 12 is 3. The **greatest common factor (GCF) of two or more monomials** is the product of the GCF of the coefficients and the common variable factors.

$$6x^3y = 2 \cdot 3 \cdot x \cdot x \cdot x \cdot y$$
$$8x^2y^2 = 2 \cdot 2 \cdot 2 \cdot x \cdot x \cdot y \cdot y$$
$$\text{GCF} = 2 \cdot x \cdot x \cdot y = 2x^2y$$

Note that the exponent on each variable in the GCF is the same as the *smallest* exponent on that variable in either of the monomials.

The GCF of $6x^3y$ and $8x^2y^2$ is $2x^2y$.

HOW TO 1 Find the GCF of $12a^4b$ and $18a^2b^2c$.

The common variable factors are a^2 and b; c is not a common variable factor.

$$12a^4b = 2 \cdot 2 \cdot 3 \cdot a^4 \cdot b$$
$$18a^2b^2c = 2 \cdot 3 \cdot 3 \cdot a^2 \cdot b^2 \cdot c$$
$$\text{GCF} = 2 \cdot 3 \cdot a^2 \cdot b = 6a^2b$$

To **factor a polynomial** means to write the polynomial as a product of other polynomials. In the example at the right, $2x$ is the GCF of the terms $2x^2$ and $10x$.

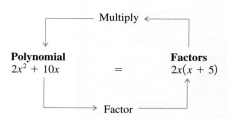

HOW TO 2 Factor: $5x^3 - 35x^2 + 10x$

Find the GCF of the terms of the polynomial.

$$5x^3 = 5 \cdot x^3$$
$$35x^2 = 5 \cdot 7 \cdot x^2$$
$$10x = 2 \cdot 5 \cdot x$$

The GCF is $5x$.

Rewrite the polynomial, expressing each term as a product with the GCF as one of the factors.

$$5x^3 - 35x^2 + 10x = 5x(x^2) + 5x(-7x) + 5x(2)$$
$$= 5x(x^2 - 7x + 2)$$

- Use the Distributive Property to write the polynomial as a product of factors.

 Take Note

At the right, the factors in parentheses are determined by dividing each term of the trinomial by the GCF, 5x.

$$\frac{5x^3}{5x} = x^2, \frac{-35x^2}{5x} = -7x, \text{ and}$$

$$\frac{10x}{5x} = 2$$

HOW TO 3   Factor: $21x^2y^3 - 6xy^5 + 15x^4y^2$

Find the GCF of the terms of the polynomial.

$$21x^2y^3 = 3 \cdot 7 \cdot x^2 \cdot y^3$$
$$6xy^5 = 2 \cdot 3 \cdot x \cdot y^5$$
$$15x^4y^2 = 3 \cdot 5 \cdot x^4 \cdot y^2$$

The GCF is $3xy^2$.

Rewrite the polynomial, expressing each term as a product with the GCF as one of the factors.

$$21x^2y^3 - 6xy^5 + 15x^4y^2$$
$$= 3xy^2(7xy) + 3xy^2(-2y^3) + 3xy^2(5x^3) \quad \bullet \text{ Use the Distributive Property to write}$$
$$= 3xy^2(7xy - 2y^3 + 5x^3) \qquad\qquad\qquad \text{the polynomial as a product of factors.}$$

EXAMPLE 1

Factor: $8x^2 + 2xy$

Solution
The GCF is $2x$.

$$8x^2 + 2xy = 2x(4x) + 2x(y) \quad \bullet \text{ Factor the GCF}$$
$$= 2x(4x + y) \qquad\qquad\quad \text{from each term.}$$

YOU TRY IT 1

Factor: $14a^2 - 21a^4b$

Your solution
$7a^2(2 - 3a^2b)$

EXAMPLE 2

Factor: $n^3 - 5n^2 + 2n$

Solution
The GCF is n.

$$n^3 - 5n^2 + 2n$$
$$= n(n^2) + n(-5n) + n(2) \quad \bullet \text{ Factor the GCF}$$
$$= n(n^2 - 5n + 2) \qquad\qquad\quad \text{from each term.}$$

YOU TRY IT 2

Factor: $27b^2 + 18b + 9$

Your solution
$9(3b^2 + 2b + 1)$

EXAMPLE 3

Factor: $16x^2y + 8x^4y^2 - 12x^4y^5$

Solution
The GCF is $4x^2y$.

$$16x^2y + 8x^4y^2 - 12x^4y^5$$
$$= 4x^2y(4) + 4x^2y(2x^2y) + 4x^2y(-3x^2y^4)$$
$$= 4x^2y(4 + 2x^2y - 3x^2y^4)$$

YOU TRY IT 3

Factor: $6x^4y^2 - 9x^3y^2 + 12x^2y^4$

Your solution
$3x^2y^2(2x^2 - 3x + 4y^2)$

IN-CLASS EXAMPLES
Factor.
1. $10y^2 - 15y^3z$ **$5y^2(2 - 3yz)$**
2. $12m^2 + 6m - 18$
   $6(2m^2 + m - 3)$
3. $20x^4y^3 - 30x^3y^4 + 40x^2y^5$
   $10x^2y^3(2x^2 - 3xy + 4y^2)$

Solutions on p. S19

OBJECTIVE B   *To factor by grouping*

A factor that has two terms is called a **binomial factor.** In the examples at the right, the binomials $a + b$ and $x - y$ are binomial factors.

$$2a(a + b)^2$$
$$3xy(x - y)$$

The Distributive Property is used to factor a common binomial factor from an expression.

The common binomial factor of the expression $6(x - 3) + y(x - 3)$ is $(x - 3)$. To factor the expression, use the Distributive Property to write the expression as a product of factors.

$$6\underline{(x - 3)} + y\underline{(x - 3)} = (x - 3)(6 + y)$$

Consider the following simplification of $-(a - b)$.

$$-(a - b) = -1(a - b) = -a + b = b - a$$

Thus
$$b - a = -(a - b)$$

This equation is sometimes used to factor a common binomial from an expression.

> **HOW TO 4**   Factor: $2x(x - y) + 5(y - x)$
>
> $2x(x - y) + 5(y - x) = 2x(x - y) - 5(x - y)$ • $5(y - x) = 5[(-1)(x - y)]$
> $\qquad\qquad\qquad\qquad\quad = (x - y)(2x - 5)$ $\qquad\qquad = -5(x - y)$

A polynomial can be **factored by grouping** if its terms can be grouped and factored in such a way that a common binomial factor is found.

INSTRUCTOR NOTE
Before you present factoring by grouping, have students practice inserting parentheses into expressions. Here are some suggestions.
$-a + 2b = -(a - 2b)$
$3x - 2y = -(-3x + 2y)$
$-4a - 3b = -(4a + 3b)$

> **HOW TO 5**   Factor: $ax + bx - ay - by$
>
> $ax + bx - ay - by = (ax + bx) - (ay + by)$ • Group the first two terms and the last two terms. Note that $-ay - by = -(ay + by)$.
> $\qquad\qquad\qquad\qquad = x(a + b) - y(a + b)$ • Factor each group.
> $\qquad\qquad\qquad\qquad = (a + b)(x - y)$ • Factor the GCF, $(a + b)$, from each group.
>
> *Check:* $(a + b)(x - y) = ax - ay + bx - by$
> $\qquad\qquad\qquad\quad = ax + bx - ay - by$

> **HOW TO 6**   Factor: $6x^2 - 9x - 4xy + 6y$
>
> $6x^2 - 9x - 4xy + 6y = (6x^2 - 9x) - (4xy - 6y)$ • Group the first two terms and the last two terms. Note that $-4xy + 6y = -(4xy - 6y)$.
> $\qquad\qquad\qquad\qquad\quad = 3x(2x - 3) - 2y(2x - 3)$ • Factor each group.
> $\qquad\qquad\qquad\qquad\quad = (2x - 3)(3x - 2y)$ • Factor the GCF, $(2x - 3)$, from each group.

EXAMPLE 4

Factor: $4x(3x - 2) - 7(3x - 2)$

Solution

$4x(3x - 2) - 7(3x - 2)$ • **$3x - 2$ is the common binomial factor.**

$= (3x - 2)(4x - 7)$

YOU TRY IT 4

Factor: $2y(5x - 2) - 3(2 - 5x)$

Your solution

$(5x - 2)(2y + 3)$

EXAMPLE 5

Factor: $9x^2 - 15x - 6xy + 10y$

Solution

$9x^2 - 15x - 6xy + 10y$

$= (9x^2 - 15x) - (6xy - 10y)$ • **$-6xy + 10y = -(6xy - 10y)$**

$= 3x(3x - 5) - 2y(3x - 5)$ • **$3x - 5$ is the common factor.**

$= (3x - 5)(3x - 2y)$

YOU TRY IT 5

Factor: $a^2 - 3a + 2ab - 6b$

Your solution

$(a - 3)(a + 2b)$

EXAMPLE 6

Factor: $3x^2y - 4x - 15xy + 20$

Solution

$3x^2y - 4x - 15xy + 20$

$= (3x^2y - 4x) - (15xy - 20)$ • **$-15xy + 20 = -(15xy - 20)$**

$= x(3xy - 4) - 5(3xy - 4)$ • **$3xy - 4$ is the common factor.**

$= (3xy - 4)(x - 5)$

YOU TRY IT 6

Factor: $2mn^2 - n + 8mn - 4$

Your solution

$(2mn - 1)(n + 4)$

EXAMPLE 7

Factor: $4ab - 6 + 3b - 2ab^2$

Solution

$4ab - 6 + 3b - 2ab^2$

$= (4ab - 6) + (3b - 2ab^2)$

$= 2(2ab - 3) + b(3 - 2ab)$

$= 2(2ab - 3) - b(2ab - 3)$ • **$3 - 2ab = -(2ab - 3)$**

$= (2ab - 3)(2 - b)$ • **$2ab - 3$ is the common factor.**

YOU TRY IT 7

Factor: $3xy - 9y - 12 + 4x$

Your solution

$(x - 3)(3y + 4)$

IN-CLASS EXAMPLES

Factor.

4. $6x(4x + 3) - 5(4x + 3)$
   $(4x + 3)(6x - 5)$

5. $8x^2 - 12x - 6xy + 9y$
   $(2x - 3)(4x - 3y)$

6. $7xy^2 - 3y + 14xy - 6$
   $(7xy - 3)(y + 2)$

7. $5xy - 9y - 18 + 10x$
   $(5x - 9)(y + 2)$

Solutions on p. S19

7.1 EXERCISES

✔ Concept Check

SUGGESTED ASSIGNMENT
Exercises 1–6; Exercises 7–39, odds; Exercises 43–65, odds

1. Name the greatest common factor of 4, 12, and 16. 4

2. Name the greatest common factor of x^3, x^5, and x^6. x^3

3. For the expression $x(2x - 1)$, name **a.** the monomial factor and **b.** the binomial factor. **a.** x **b.** $2x - 1$

4. Name the common binomial factor in the expression $5b(c - 6) + 8(c - 6)$. $c - 6$

5. Rewrite the expression $2x^3 - x^2 + 6x - 3$ by grouping the first two terms and the last two terms. $(2x^3 - x^2) + (6x - 3)$

6. ◥ Explain why the statement is true.
   a. The terms of the binomial $3x - 6$ have a common factor.
   b. The expression $3x^2 + 15$ is not in factored form.
   c. $5y - 7$ is a factor of $y(5y - 7)$.

OBJECTIVE A *To factor a monomial from a polynomial*

For Exercises 7 to 40, factor.

7. $5a + 5$
$5(a + 1)$

8. $7b - 7$
$7(b - 1)$

9. $16 - 8a^2$
$8(2 - a^2)$

10. $12 + 12y^2$
$12(1 + y^2)$

11. $8x + 12$
$4(2x + 3)$

12. $16a - 24$
$8(2a - 3)$

13. $7x^2 - 3x$
$x(7x - 3)$

14. $12y^2 - 5y$
$y(12y - 5)$

15. $3a^2 + 5a^5$
$a^2(3 + 5a^3)$

16. $6b^3 - 5b^2$
$b^2(6b - 5)$

17. $2x^4 - 4x$
$2x(x^3 - 2)$

18. $3y^4 - 9y$
$3y(y^3 - 3)$

19. $10x^4 - 12x^2$
$2x^2(5x^2 - 6)$

20. $12a^5 - 32a^2$
$4a^2(3a^3 - 8)$

21. $8a^8 - 4a^5$
$4a^5(2a^3 - 1)$

22. $16y^4 - 8y^7$
$8y^4(2 - y^3)$

23. $x^2y^2 - xy$
$xy(xy - 1)$

24. $a^2b^2 + ab$
$ab(ab + 1)$

25. $3x^2y^4 - 6xy$
$3xy(xy^3 - 2)$

26. $12a^2b^5 - 9ab$
$3ab(4ab^4 - 3)$

27. $3x^3 + 6x^2 + 9x$
$3x(x^2 + 2x + 3)$

28. $5y^3 - 20y^2 + 5y$
$5y(y^2 - 4y + 1)$

29. $2x^4 - 4x^3 + 6x^2$
$2x^2(x^2 - 2x + 3)$

30. $3y^4 - 9y^3 - 6y^2$
$3y^2(y^2 - 3y - 2)$

31. $2x^3 + 6x^2 - 14x$
$2x(x^2 + 3x - 7)$

32. $3y^3 - 9y^2 + 24y$
$3y(y^2 - 3y + 8)$

33. $2y^5 - 3y^4 + 7y^3$
$y^3(2y^2 - 3y + 7)$

34. $6a^5 - 3a^3 - 2a^2$
$a^2(6a^3 - 3a - 2)$

35. $x^3y - 3x^2y^2 + 7xy^3$
$xy(x^2 - 3xy + 7y^2)$

36. $2a^2b - 5a^2b^2 + 7ab^2$
$ab(2a - 5ab + 7b)$

37. $5y^3 + 10y^2 - 25y$
$5y(y^2 + 2y - 5)$

38. $4b^5 + 6b^3 - 12b$
$2b(2b^4 + 3b^2 - 6)$

39. $3a^2b^2 - 9ab^2 + 15b^2$
$3b^2(a^2 - 3a + 5)$

40. $8x^2y^2 - 4x^2y + x^2$
$x^2(8y^2 - 4y + 1)$

41. ◆ What is the GCF of the terms of the polynomial $x^a + x^b + x^c$ given that a, b, and c are all positive integers, and $a > b > c$? x^c

OBJECTIVE B   *To factor by grouping*

42. Use the three expressions at the right.
    a. Which expressions are equivalent to $x^2 - 5x + 6$? i, ii, iii
    b. Which expression can be factored by grouping? iii

(i) $x^2 - 15x + 10x + 6$
(ii) $x^2 - x - 4x + 6$
(iii) $x^2 - 2x - 3x + 6$

For Exercises 43 to 66, factor.

43. $x(b + 4) + 3(b + 4)$
$(b + 4)(x + 3)$

44. $y(a + z) + 7(a + z)$
$(a + z)(y + 7)$

45. $a(y - x) - b(y - x)$
$(y - x)(a - b)$

46. $3r(a - b) + s(a - b)$
$(a - b)(3r + s)$

47. $x(x - 2) + y(2 - x)$
$(x - 2)(x - y)$

48. $t(m - 7) + 7(7 - m)$
$(m - 7)(t - 7)$

49. $8c(2m - 3n) + (3n - 2m)$
$(2m - 3n)(8c - 1)$

50. $2y(4a + b) - (b + 4a)$
$(4a + b)(2y - 1)$

51. $x^2 + 2x + 2xy + 4y$
$(x + 2)(x + 2y)$

52. $x^2 - 3x + 4ax - 12a$
$(x - 3)(x + 4a)$

53. $p^2 - 2p - 3rp + 6r$
$(p - 2)(p - 3r)$

54. $t^2 + 4t - st - 4s$
$(t + 4)(t - s)$

55. $ab + 6b - 4a - 24$
$(a + 6)(b - 4)$

56. $xy - 5y - 2x + 10$
$(x - 5)(y - 2)$

57. $2z^2 - z + 2yz - y$
$(2z - 1)(z + y)$

58. $2y^2 - 10y + 7xy - 35x$
$(y - 5)(2y + 7x)$

59. $2x^2 - 5x - 6xy + 15y$
$(2x - 5)(x - 3y)$

60. $4a^2 + 5ab - 10b - 8a$
$(4a + 5b)(a - 2)$

61. $3y^2 - 6y - ay + 2a$
$(y - 2)(3y - a)$

62. $2ra + a^2 - 2r - a$
$(2r + a)(a - 1)$

63. $3xy - y^2 - y + 3x$
$(3x - y)(y + 1)$

64. $2ab - 3b^2 - 3b + 2a$
$(2a - 3b)(b + 1)$

65. $3st + t^2 - 2t - 6s$
$(3s + t)(t - 2)$

66. $4x^2 + 3xy - 12y - 16x$
$(4x + 3y)(x - 4)$

Critical Thinking

For Exercises 67 to 69, fill in the blank to make a true statement.

67. $a - 3 = $ _____ $(3 - a)$
-1

68. $2 - (x - y) = 2 + ($ _____ $)$
$y - x$

69. $4x + (3a - b) = 4x - ($ _____ $)$
$b - 3a$

Projects or Group Activities

70. Geometry Write an expression in factored form for the shaded portion in each of the following diagrams. Use the equation for the area of a rectangle $(A = LW)$ and the equation for the area of a circle $(A = \pi r^2)$.

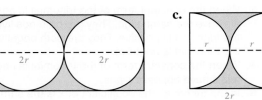

a.
$r^2(\pi - 2)$

b.
$2r^2(4 - \pi)$

c.
$r^2(4 - \pi)$

QUICK QUIZ
Factor.
1. $14x^2y^2 - 7xy$
   $7xy(2xy - 1)$ **[7.1A]**
2. $15r^2s + 20rs - 10rs^2$
   $5rs(3r + 4 - 2s)$ **[7.1A]**
3. $6a(b - 7) - 5(b - 7)$
   $(b - 7)(6a - 5)$ **[7.1B]**
4. $3xy - 9y + 2x - 6$
   $(x - 3)(3y + 2)$ **[7.1B]**
5. $2ay - a^2 - 3a + 6y$
   $(2y - a)(a + 3)$ **[7.1B]**

Unless otherwise noted, all content on this page is © Cengage Learning.

7.2 Factoring Polynomials of the Form $x^2 + bx + c$

OBJECTIVE A *To factor a trinomial of the form $x^2 + bx + c$*

Trinomials of the form $x^2 + bx + c$, where b and c are integers, are shown at the right.

$x^2 + 8x + 12; b = 8, c = 12$
$x^2 - 7x + 12; b = -7, c = 12$
$x^2 - 2x - 15; b = -2, c = -15$

To factor a trinomial of this form means to express the trinomial as the product of two binomials.

INSTRUCTOR NOTE
Students sometimes see factoring as unrelated to multiplication. Remind students that multiplying the binomial factors will give the polynomial.

Trinomials expressed as the product of binomials are shown at the right.

$x^2 + 8x + 12 = (x + 6)(x + 2)$
$x^2 - 7x + 12 = (x - 3)(x - 4)$
$x^2 - 2x - 15 = (x + 3)(x - 5)$

The method by which factors of a trinomial are found is based on FOIL. Consider the following binomial products, noting the relationship between the constant terms of the binomials and the terms of the trinomials.

The signs in the binomial factors are the same.

$(x + 6)(x + 2) = x^2 + 2x + 6x + (6)(2) = x^2 + 8x + 12$

Sum of 6 and 2
Product of 6 and 2

$(x - 3)(x - 4) = x^2 - 4x - 3x + (-3)(-4) = x^2 - 7x + 12$

Sum of -3 and -4
Product of -3 and -4

The signs in the binomial factors are opposites.

$(x + 3)(x - 5) = x^2 - 5x + 3x + (3)(-5) = x^2 - 2x - 15$

Sum of 3 and -5
Product of 3 and -5

$(x - 4)(x + 6) = x^2 + 6x - 4x + (-4)(6) = x^2 + 2x - 24$

Sum of -4 and 6
Product of -4 and 6

Factoring $x^2 + bx + c$: IMPORTANT RELATIONSHIPS

1. When the constant term of the trinomial is positive, the constant terms of the binomials have the same sign. They are both positive when the coefficient of the x term in the trinomial is positive. They are both negative when the coefficient of the x term in the trinomial is negative.
2. When the constant term of the trinomial is negative, the constant terms of the binomials have opposite signs.
3. In the trinomial, the coefficient of x is the sum of the constant terms of the binomials.
4. In the trinomial, the constant term is the product of the constant terms of the binomials.

HOW TO 1   Factor: $x^2 - 7x + 10$

Because the constant term is positive and the coefficient of x is negative, the binomial constants will be negative. Find two negative factors of 10 whose sum is -7.

Negative Factors of 10	Sum
$-1, -10$	-11
$-2, -5$	-7

• The results can be recorded in a table.

• These are the correct factors.

$x^2 - 7x + 10 = (x - 2)(x - 5)$ • Write the trinomial as a product of its factors.

You can check the proposed factorization by multiplying the two binomials.

Check: $(x - 2)(x - 5) = x^2 - 5x - 2x + 10 = x^2 - 7x + 10$

Take Note

Always check your proposed factorization to ensure accuracy.

HOW TO 2   Factor: $x^2 - 9x - 36$

The constant term is negative. The binomial constants will have opposite signs. Find two factors of -36 whose sum is -9.

Factors of -36	Sum
$+1, -36$	-35
$-1, +36$	35
$+2, -18$	-16
$-2, +18$	16
$+3, -12$	-9

• Once the correct factors are found, it is not necessary to try the remaining factors.

$x^2 - 9x - 36 = (x + 3)(x - 12)$ • Write the trinomial as a product of its factors.

INSTRUCTOR NOTE

The phrase *nonfactorable over the integers* may require additional examples. Explain that it does not mean that the polynomial does not factor; it means that it does not factor if *integers* are used. An analogy involving numbers may help. For instance, the only ways to write 7 as a product of integers are $1 \cdot 7$ or $(-1)(-7)$. However, $\frac{21}{5} \cdot \frac{5}{3} = 7$. The ability to factor depends on the types of numbers that can be used as factors.

For some trinomials, it is not possible to find integer factors of the constant term whose sum is the coefficient of the middle term. A polynomial that does not factor using only integers is **nonfactorable over the integers.**

HOW TO 3   Factor: $x^2 + 7x + 8$

Find two positive factors of 8 whose sum is 7.

Positive Factors of 8	Sum
$1, 8$	9
$2, 4$	6

• There are no positive integer factors of 8 whose sum is 7.

$x^2 + 7x + 8$ is nonfactorable over the integers.

Take Note

Just as 17 is a prime number, $x^2 + 7x + 8$ is a **prime polynomial.** Binomials of the form $x - a$ and $x + a$ are also prime polynomials.

EXAMPLE 1

Factor: $x^2 - 8x + 15$

Solution

Find two negative factors of 15 whose sum is -8.

Factors	Sum
$-1, -15$	-16
$-3, -5$	-8

$x^2 - 8x + 15 = (x - 3)(x - 5)$

YOU TRY IT 1

Factor: $x^2 + 9x + 20$

Your solution

$(x + 4)(x + 5)$

Solution on p. S19

Unless otherwise noted, all content on this page is © Cengage Learning.

EXAMPLE 2

Factor: $x^2 + 6x - 27$

Solution

Find two factors of
-27 whose sum is 6.

Factors	Sum
$+1, -27$	-26
$-1, +27$	26
$+3, -9$	-6
$-3, +9$	**6**

$x^2 + 6x - 27 = (x - 3)(x + 9)$

YOU TRY IT 2

Factor: $x^2 + 7x - 18$

Your solution
$(x + 9)(x - 2)$

IN-CLASS EXAMPLES
Factor.
1. $x^2 - 8x + 12$ $(x - 6)(x - 2)$
2. $x^2 + 8x + 12$ $(x + 6)(x + 2)$
3. $x^2 + 7x + 12$ $(x + 4)(x + 3)$
4. $x^2 - 4x - 12$ $(x - 6)(x + 2)$
5. $x^2 - 11x - 12$ $(x - 12)(x + 1)$

Solution on p. S20

OBJECTIVE B   *To factor completely*

 Take Note

The first step in *any* factoring problem is to determine whether the terms of the polynomial have a *common factor.* If they do, factor it out first.

INSTRUCTOR NOTE

When the terms of a polynomial contain a common factor, some students will attempt to find the common factor and the two binomial factors in one step; they rarely obtain the correct final answer. Encourage students to find only two factors at a time. Emphasize that the first step is to find the common factor and the resulting polynomial factor. The second step is to find the two binomial factors.

 Take Note

The terms $2y$ and $10y$ are placed in the binomials. This is necessary so that the middle term of the trinomial contains xy and the last term contains y^2.

A polynomial is **factored completely** when it is written as a product of factors that are nonfactorable over the integers.

HOW TO 4   Factor: $4y^3 - 4y^2 - 24y$

$4y^3 - 4y^2 - 24y = 4y(y^2) - 4y(y) - 4y(6)$ • The GCF is **4y.**

$= 4y(y^2 - y - 6)$ • Use the Distributive Property to factor out the GCF.

$= 4y(y + 2)(y - 3)$ • Factor $y^2 - y - 6$. The two factors of -6 whose sum is -1 are **2** and -3.

It is always possible to check a proposed factorization by multiplying the polynomial factors. Here is the check for HOW TO 4.

Check: $4y(y + 2)(y - 3) = 4y(y^2 - 3y + 2y - 6)$
$= 4y(y^2 - y - 6)$
$= 4y^3 - 4y^2 - 24y$ • This is the original polynomial.

HOW TO 5   Factor: $x^2 + 12xy + 20y^2$

There is no common factor.
Note that the variable part of the middle term is xy, and the variable part of the last term is y^2.

$x^2 + 12xy + 20y^2 = (x + 2y)(x + 10y)$ • The two factors of 20 whose sum is 12 are 2 and 10.

Note that the terms $2y$ and $10y$ are placed in the binomials. The following check shows why this is necessary.

Check: $(x + 2y)(x + 10y) = x^2 + 10xy + 2xy + 20y^2$
$= x^2 + 12xy + 20y^2$ • This is the original polynomial.

Unless otherwise noted, all content on this page is © Cengage Learning.

Take Note

When the coefficient of the highest power in a polynomial is negative, consider factoring out a negative GCF. Example 3 is another example of this technique.

HOW TO 6 Factor: $15 - 2x - x^2$

Because the coefficient of x^2 is -1, factor -1 from the trinomial, and then write the resulting trinomial in descending order.

$15 - 2x - x^2 = -(x^2 + 2x - 15)$ • $15 - 2x - x^2 = -1(-15 + 2x + x^2)$
$$= -(x^2 + 2x - 15)$$
$$= -(x + 5)(x - 3)$$ • Factor $x^2 + 2x - 15$. The two factors of -15 whose sum is 2 are 5 and -3.

Check: $-(x + 5)(x - 3) = -(x^2 + 2x - 15)$
$$= -x^2 - 2x + 15$$
$$= 15 - 2x - x^2$$ • This is the original polynomial.

EXAMPLE 3

Factor: $-3x^3 + 9x^2 + 12x$

Solution

The GCF is $-3x$. Factor out the GCF.

$-3x^3 + 9x^2 + 12x = -3x(x^2 - 3x - 4)$

Factor the trinomial $x^2 - 3x - 4$. Find two factors of -4 whose sum is -3.

Factors	Sum
$+1, -4$	-3

$-3x^3 + 9x^2 + 12x = -3x(x + 1)(x - 4)$

YOU TRY IT 3

Factor: $-2x^3 + 14x^2 - 12x$

Your solution

$-2x(x - 6)(x - 1)$

EXAMPLE 4

Factor: $4x^2 - 40xy + 84y^2$

Solution

The GCF is 4. Factor out the GCF.

$4x^2 - 40xy + 84y^2 = 4(x^2 - 10xy + 21y^2)$

Factor the trinomial $x^2 - 10xy + 21y^2$. Find two negative factors of 21 whose sum is -10.

Factors	Sum
$-1, -21$	-22
$-3, -7$	-10

$4x^2 - 40xy + 84y^2 = 4(x - 3y)(x - 7y)$

YOU TRY IT 4

Factor: $3x^2 - 9xy - 12y^2$

Your solution

$3(x + y)(x - 4y)$

IN-CLASS EXAMPLES

Factor.
6. $3x^2 - 6x - 72$ $3(x - 6)(x + 4)$
7. $2x^2 + 6x - 20$ $2(x + 5)(x - 2)$
8. $-4y^3 + 28y^2 - 48y$
   $-4y(y - 4)(y - 3)$
9. $5x^2 - 10xy - 15y^2$
   $5(x + y)(x - 3y)$

Solutions on p. S20

Unless otherwise noted, all content on this page is © Cengage Learning.

7.2 EXERCISES

SUGGESTED ASSIGNMENT
Exercises 1–6; Exercises 7–133, every other odd; More challenging exercises: Exercises 134, 136, 138

✔ **Concept Check**

1. The trinomial $x^2 - 8x + 7$ is of the form $x^2 + bx + c$. What is the value of b in the trinomial $x^2 - 8x + 7$? -8

2. Find two numbers whose sum is 9 and whose product is 14.   2 and 7

3. Find two numbers whose sum is 4 and whose product is -12. -2 and 6

4. When factoring a trinomial, if the constant term is positive, will the signs in the binomials be the same or different?   The same

5. When factoring a trinomial, if the constant term is negative, will the signs in the binomials be the same or different?   Different

6. What is the first step in factoring a trinomial?   Determine whether the terms of the trinomial have a common factor.

OBJECTIVE A   *To factor a trinomial of the form $x^2 + bx + c$*

For Exercises 7 to 75, factor.

7. $x^2 + 3x + 2$
$(x + 1)(x + 2)$

8. $x^2 + 5x + 6$
$(x + 2)(x + 3)$

9. $x^2 - x - 2$
$(x + 1)(x - 2)$

10. $x^2 + x - 6$
$(x + 3)(x - 2)$

11. $a^2 + a - 12$
$(a + 4)(a - 3)$

12. $a^2 - 2a - 35$
$(a + 5)(a - 7)$

13. $a^2 - 3a + 2$
$(a - 1)(a - 2)$

14. $a^2 - 5a + 4$
$(a - 1)(a - 4)$

15. $a^2 + a - 2$
$(a + 2)(a - 1)$

16. $a^2 - 2a - 3$
$(a + 1)(a - 3)$

17. $b^2 - 6b + 9$
$(b - 3)(b - 3)$

18. $b^2 + 8b + 16$
$(b + 4)(b + 4)$

19. $b^2 + 7b - 8$
$(b + 8)(b - 1)$

20. $y^2 - y - 6$
$(y + 2)(y - 3)$

21. $y^2 + 6y - 55$
$(y + 11)(y - 5)$

22. $z^2 - 4z - 45$
$(z + 5)(z - 9)$

23. $y^2 - 5y + 6$
$(y - 2)(y - 3)$

24. $y^2 - 8y + 15$
$(y - 3)(y - 5)$

25. $z^2 - 14z + 45$
$(z - 5)(z - 9)$

26. $z^2 - 14z + 49$
$(z - 7)(z - 7)$

27. $z^2 - 12z - 160$
$(z + 8)(z - 20)$

28. $p^2 + 2p - 35$
$(p + 7)(p - 5)$

29. $p^2 + 12p + 27$
$(p + 3)(p + 9)$

30. $p^2 - 6p + 8$
$(p - 2)(p - 4)$

31. $x^2 + 20x + 100$
$(x + 10)(x + 10)$

32. $x^2 + 9x - 70$
$(x + 14)(x - 5)$

33. $b^2 - b - 20$
$(b + 4)(b - 5)$

34. $b^2 + 3b - 40$
$(b + 8)(b - 5)$

35. $y^2 - 14y - 51$
$(y + 3)(y - 17)$

36. $y^2 - y - 72$
$(y + 8)(y - 9)$

37. $p^2 - 4p - 21$
$(p + 3)(p - 7)$

38. $p^2 + 16p + 39$
$(p + 3)(p + 13)$

39. $y^2 - 8y + 32$
Nonfactorable over
the integers

40. $y^2 - 9y + 81$
Nonfactorable over
the integers

41. $x^2 - 20x + 75$
$(x - 5)(x - 15)$

42. $x^2 - 12x + 11$
$(x - 11)(x - 1)$

43. $p^2 + 24p + 63$
$(p + 3)(p + 21)$

44. $x^2 - 15x + 56$
$(x - 7)(x - 8)$

45. $x^2 + 21x + 38$
$(x + 2)(x + 19)$

46. $x^2 + x - 56$
$(x + 8)(x - 7)$

47. $x^2 + 5x - 3$
Nonfactorable over
the integers

48. $a^2 - 21a - 7$
Nonfactorable over
the integers

49. $a^2 - 7a - 44$
$(a + 4)(a - 11)$

50. $a^2 - 15a + 36$
$(a - 3)(a - 12)$

51. $a^2 - 21a + 54$
$(a - 3)(a - 18)$

52. $z^2 - 9z - 136$
$(z + 8)(z - 17)$

53. $z^2 + 14z - 147$
$(z + 21)(z - 7)$

54. $c^2 - c - 90$
$(c + 9)(c - 10)$

55. $c^2 - 3c - 180$
$(c + 12)(c - 15)$

56. $z^2 + 15z + 44$
$(z + 4)(z + 11)$

57. $p^2 + 24p + 135$
$(p + 9)(p + 15)$

58. $c^2 + 19c + 34$
$(c + 2)(c + 17)$

59. $c^2 + 11c + 18$
$(c + 2)(c + 9)$

60. $x^2 - 4x - 96$
$(x + 8)(x - 12)$

61. $x^2 + 10x - 75$
$(x + 15)(x - 5)$

62. $x^2 - 22x + 112$
$(x - 8)(x - 14)$

63. $x^2 + 21x - 100$
$(x + 25)(x - 4)$

64. $b^2 + 8b - 105$
$(b + 15)(b - 7)$

65. $b^2 - 22b + 72$
$(b - 4)(b - 18)$

66. $a^2 - 9a - 36$
$(a + 3)(a - 12)$

67. $a^2 + 42a - 135$
$(a + 45)(a - 3)$

68. $b^2 - 23b + 102$
$(b - 6)(b - 17)$

69. $b^2 - 25b + 126$
$(b - 7)(b - 18)$

70. $a^2 + 27a + 72$
$(a + 3)(a + 24)$

71. $z^2 + 24z + 144$
$(z + 12)(z + 12)$

72. $x^2 + 25x + 156$
$(x + 12)(x + 13)$

73. $x^2 - 29x + 100$
$(x - 4)(x - 25)$

74. $x^2 - 10x - 96$
$(x + 6)(x - 16)$

75. $x^2 + 9x - 112$
$(x + 16)(x - 7)$

For Exercises 76 and 77, $x^2 + bx + c = (x + n)(x + m)$, where b and c are nonzero
and n and m are positive integers.

76. Is c positive or negative?
Positive

77. Is b positive or negative?
Positive

OBJECTIVE B *To factor completely*

For Exercises 78 to 131, factor completely.

78. $2x^2 + 6x + 4$
$2(x + 1)(x + 2)$

79. $3x^2 + 15x + 18$
$3(x + 2)(x + 3)$

80. $18 + 7x - x^2$
$-(x - 9)(x + 2)$

81. $12 - 4x - x^2$
$-(x + 6)(x - 2)$

82. $ab^2 + 2ab - 15a$
$a(b + 5)(b - 3)$

83. $ab^2 + 7ab - 8a$
$a(b + 8)(b - 1)$

84. $xy^2 - 5xy + 6x$
$x(y - 2)(y - 3)$

85. $xy^2 + 8xy + 15x$
$x(y + 3)(y + 5)$

86. $z^3 - 7z^2 + 12z$
$z(z - 3)(z - 4)$

87. $-2a^3 - 6a^2 - 4a$
$-2a(a + 2)(a + 1)$

88. $-3y^3 + 15y^2 - 18y$
$-3y(y - 3)(y - 2)$

89. $4y^3 + 12y^2 - 72y$
$4y(y + 6)(y - 3)$

90. $3x^2 + 3x - 36$
$3(x + 4)(x - 3)$

91. $2x^3 - 2x^2 + 4x$
$2x(x^2 - x + 2)$

92. $5z^2 - 15z - 140$
$5(z + 4)(z - 7)$

93. $6z^2 + 12z - 90$
$6(z + 5)(z - 3)$

94. $2a^3 + 8a^2 - 64a$
$2a(a + 8)(a - 4)$

95. $3a^3 - 9a^2 - 54a$
$3a(a + 3)(a - 6)$

96. $x^2 - 5xy + 6y^2$
$(x - 2y)(x - 3y)$

97. $x^2 + 4xy - 21y^2$
$(x + 7y)(x - 3y)$

98. $a^2 - 9ab + 20b^2$
$(a - 4b)(a - 5b)$

99. $a^2 - 15ab + 50b^2$
$(a - 5b)(a - 10b)$

100. $x^2 - 3xy - 28y^2$
$(x + 4y)(x - 7y)$

101. $s^2 + 2st - 48t^2$
$(s + 8t)(s - 6t)$

102. $y^2 - 15yz - 41z^2$
Nonfactorable over
the integers

103. $x^2 + 85xy + 36y^2$
Nonfactorable over
the integers

104. $z^4 - 12z^3 + 35z^2$
$z^2(z - 5)(z - 7)$

105. $z^4 + 2z^3 - 80z^2$
$z^2(z + 10)(z - 8)$

106. $b^4 - 22b^3 + 120b^2$
$b^2(b - 10)(b - 12)$

107. $b^4 - 3b^3 - 10b^2$
$b^2(b + 2)(b - 5)$

108. $2y^4 - 26y^3 - 96y^2$
$2y^2(y + 3)(y - 16)$

109. $3y^4 + 54y^3 + 135y^2$
$3y^2(y + 3)(y + 15)$

110. $-x^4 - 7x^3 + 8x^2$
$-x^2(x + 8)(x - 1)$

111. $-x^4 + 11x^3 + 12x^2$
$-x^2(x - 12)(x + 1)$

112. $4x^2y + 20xy - 56y$
$4y(x + 7)(x - 2)$

113. $3x^2y - 6xy - 45y$
$3y(x + 3)(x - 5)$

114. $c^3 + 18c^2 - 40c$
$c(c + 20)(c - 2)$

115. $-3x^3 + 36x^2 - 81x$
$-3x(x - 3)(x - 9)$

116. $-4x^3 - 4x^2 + 24x$
$-4x(x + 3)(x - 2)$

117. $x^2 - 8xy + 15y^2$
$(x - 3y)(x - 5y)$

118. $y^2 - 7xy - 8x^2$
$(y + x)(y - 8x)$

119. $a^2 - 13ab + 42b^2$
$(a - 6b)(a - 7b)$

120. $y^2 + 4yz - 21z^2$
$(y + 7z)(y - 3z)$

121. $y^2 + 8yz + 7z^2$
$(y + z)(y + 7z)$

122. $y^2 - 16yz + 15z^2$
$(y - z)(y - 15z)$

123. $3x^2y + 60xy - 63y$
$3y(x + 21)(x - 1)$

124. $4x^2y - 68xy - 72y$
$4y(x + 1)(x - 18)$

125. $3x^3 + 3x^2 - 36x$
$3x(x + 4)(x - 3)$

126. $4x^3 + 12x^2 - 160x$
$4x(x + 8)(x - 5)$

127. $2t^2 - 24ts + 70s^2$
$2(t - 5s)(t - 7s)$

128. $4a^2 - 40ab + 100b^2$
$4(a - 5b)(a - 5b)$

129. $3a^2 - 24ab - 99b^2$
$3(a + 3b)(a - 11b)$

130. $4x^3 + 8x^2y - 12xy^2$
$4x(x + 3y)(x - y)$

131. $5x^3 + 30x^2y + 40xy^2$
$5x(x + 2y)(x + 4y)$

132. State whether the trinomial has a factor of $x + 3$.
a. $3x^2 - 3x - 36$ **b.** $x^2y - xy - 12y$
  Yes   Yes

133. State whether the trinomial has a factor of $x + y$.
a. $2x^2 - 2xy - 4y^2$ **b.** $2x^2y - 4xy - 4y$
  Yes   No

Critical Thinking

134. If $a(x + 3) = x^2 + 2x - 3$, find a. $x - 1$

135. If $-2x^3 - 6x^2 - 4x = a(x + 1)(x + 2)$, find a. $-2x$

For Exercises 136 to 139, factor.

136. $20 + c^2 + 9c$ $(c + 4)(c + 5)$

137. $x^2y - 54y - 3xy$ $y(x + 6)(x - 9)$

138. $45a^2 + a^2b^2 - 14a^2b$ $a^2(b - 5)(b - 9)$

139. $12p^2 - 96p + 3p^3$ $3p(p + 8)(p - 4)$

QUICK QUIZ
Factor.
1. $x^2 + 5x - 24$ $(x + 8)(x - 3)$ **[7.2A]**
2. $x^2 - 13x + 36$ $(x - 9)(x - 4)$ **[7.2A]**
3. $3a^3 + 15a^2 + 18a$ $3a(a + 3)(a + 2)$ **[7.2B]**
4. $5x^3 - 15x^2y + 10xy^2$ $5x(x - y)(x - 2y)$ **[7.2B]**

Projects or Group Activities

For Exercises 140 to 143, find all integers k such that the trinomial can be factored over the integers.

140. $x^2 + kx + 35$
36, 12, −12, −36

141. $x^2 + kx + 18$
19, 11, 9, −9, −11, −19

142. $x^2 - kx + 21$
22, 10, −10, −22

143. $x^2 - kx + 14$
15, 9, −9, −15

For Exercises 144 to 149, determine the positive integer values of k for which the polynomial is factorable over the integers.

144. $y^2 + 4y + k$ 3, 4

145. $z^2 + 7z + k$ 6, 10, 12

146. $a^2 - 6a + k$ 5, 8, 9

147. $c^2 - 7c + k$ 6, 10, 12

148. $x^2 - 3x + k$ 2

149. $y^2 + 5y + k$ 4, 6

150. Exercises 144 to 149 included the requirement that $k > 0$. If k is allowed to be any integer, how many different values of k are possible for each polynomial? Explain your answer.

7.3 Factoring Polynomials of the Form $ax^2 + bx + c$

OBJECTIVE A   *To factor a trinomial of the form $ax^2 + bx + c$ by using trial factors*

INSTRUCTOR NOTE

The first objective of this section presents factoring by using trial factors. The second objective presents factoring by grouping. You may skip one of these objectives or do both.

Trinomials of the form $ax^2 + bx + c$, where a, b, and c are integers, are shown at the right.

$3x^2 - x + 4; a = 3, b = -1, c = 4$
$6x^2 + 2x - 3; a = 6, b = 2, c = -3$

These trinomials differ from those in the preceding section in that the coefficient of x^2 is not 1. There are various methods of factoring these trinomials. The method described in this objective is factoring by using trial factors.

To reduce the number of trial factors that must be considered, remember the following:

1. Use the signs of the constant term and the coefficient of x in the trinomial to determine the signs of the binomial factors. If the constant term is positive, the signs of the binomial factors will be the same as the sign of the coefficient of x in the trinomial. If the constant term is negative, the constant terms in the binomials will have opposite signs.

2. If the terms of the trinomial do not have a common factor, then the terms of each binomial factor will not have a common factor.

HOW TO 1   Factor: $2x^2 - 7x + 3$

The terms have no common factor. The constant term is positive. The coefficient of x is negative. The binomial constants will be negative.

Positive Factors of 2 (coefficient of x^2)	Negative Factors of 3 (constant term)
1, 2	−1, −3

Write trial factors. Use the **O**uter and **I**nner products of FOIL to determine the middle term, $-7x$, of the trinomial.

Trial Factors	Middle Term
$(x - 1)(2x - 3)$	$-3x - 2x = -5x$
$(x - 3)(2x - 1)$	$-x - 6x = -7x$

Write the factors of the trinomial.

$$2x^2 - 7x + 3 = (x - 3)(2x - 1)$$

HOW TO 2   Factor: $3x^2 + 14x + 15$

The terms have no common factor. The constant term is positive. The coefficient of x is positive. The binomial constants will be positive.

Positive Factors of 3 (coefficient of x^2)	Negative Factors of 15 (constant term)
1, 3	1, 15
	3, 5

Write trial factors. Use the **O**uter and **I**nner products of FOIL to determine the middle term, $14x$, of the trinomial.

Trial Factors	Middle Term
$(x + 1)(3x + 15)$	Common factor
$(x + 15)(3x + 1)$	$x + 45x = 46x$
$(x + 3)(3x + 5)$	$5x + 9x = 14x$
$(x + 5)(3x + 3)$	Common factor

Write the factors of the trinomial.

$$3x^2 + 14x + 15 = (x + 3)(3x + 5)$$

Unless otherwise noted, all content on this page is © Cengage Learning.

INSTRUCTOR NOTE
Remind students that they can eliminate the trial factors that have a common factor only when the terms of the polynomial do not have a common factor.

Take Note

For HOW TO 3, all the trial factors were listed. Once the correct factors have been found, however, the remaining trial factors can be omitted. For the examples and solutions in this text, all trial factors except those that have a common factor will be listed.

HOW TO 3   Factor: $6x^3 + 14x^2 - 12x$

Factor the GCF, $2x$, from the terms.

Factor the trinomial. The constant term is negative. The binomial constants will have opposite signs.

$$6x^3 + 14x^2 - 12x = 2x(3x^2 + 7x - 6)$$

Positive Factors of 3	Factors of −6
1, 3	1, −6
	−1, 6
	2, −3
	−2, 3

Write trial factors. Use the **O**uter and **I**nner products of FOIL to determine the middle term, $7x$, of the trinomial.

It is not necessary to test trial factors that have a common factor.

Trial Factors	Middle Term
$(x + 1)(3x - 6)$	Common factor
$(x - 6)(3x + 1)$	$x - 18x = -17x$
$(x - 1)(3x + 6)$	Common factor
$(x + 6)(3x - 1)$	$-x + 18x = 17x$
$(x + 2)(3x - 3)$	Common factor
$(x - 3)(3x + 2)$	$2x - 9x = -7x$
$(x - 2)(3x + 3)$	Common factor
$(x + 3)(3x - 2)$	$-2x + 9x = 7x$

Write the factors of the trinomial.

$$6x^3 + 14x^2 - 12x = 2x(x + 3)(3x - 2)$$

EXAMPLE 1

Factor: $3x^2 + x - 2$

Solution

Positive factors of 3: 1, 3

Factors of -2: 1, −2
                 −1, 2

Trial Factors	Middle Term
$(x + 1)(3x - 2)$	$-2x + 3x = \mathbf{x}$
$(x - 2)(3x + 1)$	$x - 6x = -5x$
$(x - 1)(3x + 2)$	$2x - 3x = -x$
$(x + 2)(3x - 1)$	$-x + 6x = 5x$

$3x^2 + x - 2 = (x + 1)(3x - 2)$

YOU TRY IT 1

Factor: $2x^2 - x - 3$

Your solution

$(x + 1)(2x - 3)$

EXAMPLE 2

Factor: $-12x^3 - 32x^2 + 12x$

Solution

$-12x^3 - 32x^2 + 12x = -4x(3x^2 + 8x - 3)$ • GCF

Factor $3x^2 + 8x - 3$.

Positive factors of 3: 1, 3

Factors of -3: 1, −3
                 −1, 3

Trial Factors	Middle Term
$(x - 3)(3x + 1)$	$x - 9x = -8x$
$(x + 3)(3x - 1)$	$-x + 9x = \mathbf{8x}$

$-12x^3 - 32x^2 + 12x = -4x(x + 3)(3x - 1)$

YOU TRY IT 2

Factor: $-45y^3 + 12y^2 + 12y$

Your solution

$-3y(3y - 2)(5y + 2)$

IN-CLASS EXAMPLES
Factor.
1. $5x^2 - 2x - 3$ $(5x + 3)(x - 1)$
2. $-12x^3 - 18x^2 + 30x$
   $-6x(2x + 5)(x - 1)$

Solutions on p. S20

OBJECTIVE B *To factor a trinomial of the form $ax^2 + bx + c$ by grouping*

In the preceding objective, trinomials of the form $ax^2 + bx + c$ were factored by using trial factors. In this objective, these trinomials will be factored by grouping.

To factor $ax^2 + bx + c$, first find two factors of $a \cdot c$ whose sum is b. Then use factoring by grouping to write the factorization of the trinomial.

HOW TO 4 Factor: $2x^2 + 13x + 15$

Find two positive factors of 30 $(a \cdot c = 2 \cdot 15 = 30)$ whose sum is 13.

Positive Factors of 30	Sum
1, 30	31
2, 15	17
3, 10	13

• Once the required sum has been found, the remaining factors need not be checked.

$$2x^2 + 13x + 15 = 2x^2 + 3x + 10x + 15$$

• Use the factors of 30 whose sum is 13 to write $13x$ as $3x + 10x$.

$$= (2x^2 + 3x) + (10x + 15)$$

• Factor by grouping.

$$= x(2x + 3) + 5(2x + 3)$$
$$= (2x + 3)(x + 5)$$

Check: $(2x + 3)(x + 5) = 2x^2 + 10x + 3x + 15$
$$= 2x^2 + 13x + 15$$

HOW TO 5 Factor: $6x^2 - 11x - 10$

Find two factors of -60 $[a \cdot c = 6(-10) = -60]$ whose sum is -11.

Factors of -60	Sum
1, -60	-59
-1, 60	59
2, -30	-28
-2, 30	28
3, -20	-17
-3, 20	17
4, -15	-11

$$6x^2 - 11x - 10 = 6x^2 + 4x - 15x - 10$$

• Use the factors of -60 whose sum is -11 to write $-11x$ as $4x - 15x$.

$$= (6x^2 + 4x) - (15x + 10)$$

• Factor by grouping. Recall that $-15x - 10 = -(15x + 10)$.

$$= 2x(3x + 2) - 5(3x + 2)$$
$$= (3x + 2)(2x - 5)$$

Check: $(3x + 2)(2x - 5) = 6x^2 - 15x + 4x - 10$
$$= 6x^2 - 11x - 10$$

Unless otherwise noted, all content on this page is © Cengage Learning.

HOW TO 6 ▶ Factor: $3x^2 - 2x - 4$

Find two factors of -12 $[a \cdot c = 3(-4) = -12]$ whose sum is -2.

Factors of -12	Sum
$1, -12$	-11
$-1, \ 12$	11
$2, -6$	-4
$-2, \ 6$	4
$3, -4$	-1
$-3, \ 4$	1

Because no integer factors of -12 have a sum of -2, $3x^2 - 2x - 4$ is nonfactorable over the integers.

EXAMPLE 3

Factor: $2x^2 + 19x - 10$

Solution

Factors of -20 [$2(-10)$]	Sum
$-1, 20$	**19**

$$\begin{aligned} 2x^2 + 19x - 10 &= 2x^2 - x + 20x - 10 \\ &= (2x^2 - x) + (20x - 10) \\ &= x(2x - 1) + 10(2x - 1) \\ &= (2x - 1)(x + 10) \end{aligned}$$

YOU TRY IT 3

Factor: $2a^2 + 13a - 7$

Your solution

$(2a - 1)(a + 7)$

EXAMPLE 4

Factor: $24x^2y - 76xy + 40y$

Solution

The GCF is $4y$.
$24x^2y - 76xy + 40y = 4y(6x^2 - 19x + 10)$
Factor $6x^2 - 19x + 10$.

Negative Factors of 60 [$6(10)$]	Sum
$-1, -60$	-61
$-2, -30$	-32
$-3, -20$	-23
$-4, -15$	-19

$$\begin{aligned} 6x^2 - 19x + 10 &= 6x^2 - 4x - 15x + 10 \\ &= (6x^2 - 4x) - (15x - 10) \\ &= 2x(3x - 2) - 5(3x - 2) \\ &= (3x - 2)(2x - 5) \end{aligned}$$

$$\begin{aligned} 24x^2y - 76xy + 40y &= 4y(6x^2 - 19x + 10) \\ &= 4y(3x - 2)(2x - 5) \end{aligned}$$

YOU TRY IT 4

Factor: $15x^3 + 40x^2 - 80x$

Your solution

$5x(3x - 4)(x + 4)$

IN-CLASS EXAMPLES

Factor.
3. $3x^2 + 7x + 4$ $(3x + 4)(x + 1)$
4. $72x^3 - 42x^2 - 72x$
   $6x(4x + 3)(3x - 4)$

Solutions on pp. S20–S21

Unless otherwise noted, all content on this page is © Cengage Learning.

7.3 EXERCISES

✔ **Concept Check**

If assigning factoring by using trial factors, the Suggested Assignment is Exercises 1–4; Exercises 9–73, odds; More challenging exercises: Exercises 139–147, odds

For Exercises 1 to 4, fill in the blank to make a true statement.

1. $6x^2 + 11x - 10 = (3x - 2)(\underline{\ 2x + 5\ })$

2. $40x^2 + 41x + 10 = (8x + 5)(\underline{\ 5x + 2\ })$

3. $20x^2 - 31x + 12 = (5x - 4)(\underline{\ 4x - 3\ })$

4. $12x^2 - 4x - 21 = (6x + 7)(\underline{\ 2x - 3\ })$

For Exercises 5 to 8, fill in the blanks.

If assigning factoring by grouping, the Suggested Assignment is Exercises 5–8; Exercises 77–133, odds; More challenging exercises: Exercises 139–147, odds

5. To factor $2x^2 - 5x + 2$ by grouping, find two numbers whose product is $\underline{\ 4\ }$ and whose sum is $\underline{\ -5\ }$.

6. To factor $3x^2 + 2x - 5$ by grouping, find two numbers whose product is $\underline{\ -15\ }$ and whose sum is $\underline{\ 2\ }$.

7. To factor $4x^2 - 8x + 3$ by grouping, $-8x$ must be written as $\underline{\ -2x - 6x\ }$.

8. To factor $6x^2 + 7x - 3$ by grouping, $7x$ must be written as $\underline{\ -2x + 9x\ }$.

OBJECTIVE A   *To factor a trinomial of the form $ax^2 + bx + c$ by using trial factors*

For Exercises 9 to 74, factor by using trial factors.

9. $2x^2 + 3x + 1$
$(x + 1)(2x + 1)$

10. $5x^2 + 6x + 1$
$(x + 1)(5x + 1)$

11. $2y^2 + 7y + 3$
$(y + 3)(2y + 1)$

12. $3y^2 + 7y + 2$
$(y + 2)(3y + 1)$

13. $2a^2 - 3a + 1$
$(a - 1)(2a - 1)$

14. $3a^2 - 4a + 1$
$(a - 1)(3a - 1)$

15. $2b^2 - 11b + 5$
$(b - 5)(2b - 1)$

16. $3b^2 - 13b + 4$
$(b - 4)(3b - 1)$

17. $2x^2 + x - 1$
$(x + 1)(2x - 1)$

18. $4x^2 - 3x - 1$
$(x - 1)(4x + 1)$

19. $2x^2 - 5x - 3$
$(x - 3)(2x + 1)$

20. $3x^2 + 5x - 2$
$(x + 2)(3x - 1)$

21. $2t^2 - t - 10$
$(t + 2)(2t - 5)$

22. $2t^2 + 5t - 12$
$(t + 4)(2t - 3)$

23. $3p^2 - 16p + 5$
$(p - 5)(3p - 1)$

24. $6p^2 + 5p + 1$
$(2p + 1)(3p + 1)$

25. $12y^2 - 7y + 1$
$(3y - 1)(4y - 1)$

26. $6y^2 - 5y + 1$
$(2y - 1)(3y - 1)$

27. $6z^2 - 7z + 3$
Nonfactorable
over the integers

28. $9z^2 + 3z + 2$
Nonfactorable
over the integers

29. $6t^2 - 11t + 4$
$(2t - 1)(3t - 4)$

30. $10t^2 + 11t + 3$
$(2t + 1)(5t + 3)$

31. $8x^2 + 33x + 4$
$(x + 4)(8x + 1)$

32. $7x^2 + 50x + 7$
$(x + 7)(7x + 1)$

33. $5x^2 - 62x - 7$
Nonfactorable
over the integers

34. $9x^2 - 13x - 4$
Nonfactorable
over the integers

35. $12y^2 + 19y + 5$
$(3y + 1)(4y + 5)$

36. $6b^2 - 19b + 15$
$(2b - 3)(3b - 5)$

37. $2z^2 - 27z - 14$
$(z - 14)(2z + 1)$

38. $4z^2 + 5z - 6$
$(z + 2)(4z - 3)$

39. $3p^2 + 22p - 16$
$(p + 8)(3p - 2)$

40. $7p^2 + 19p + 10$
$(p + 2)(7p + 5)$

41. $4x^2 + 6x + 2$
$2(x + 1)(2x + 1)$

42. $12x^2 + 33x - 9$
$3(x + 3)(4x - 1)$

43. $15y^2 - 50y + 35$
$5(y - 1)(3y - 7)$

44. $30y^2 + 10y - 20$
$10(y + 1)(3y - 2)$

45. $2x^3 - 11x^2 + 5x$
$x(x - 5)(2x - 1)$

46. $2x^3 - 3x^2 - 5x$
$x(x + 1)(2x - 5)$

47. $3a^2b - 16ab + 16b$
$b(a - 4)(3a - 4)$

48. $2a^2b - ab - 21b$
$b(a + 3)(2a - 7)$

49. $3z^2 + 95z + 10$
Nonfactorable over
the integers

50. $8z^2 - 36z + 1$
Nonfactorable over
the integers

51. $36x - 3x^2 - 3x^3$
$-3x(x + 4)(x - 3)$

52. $-2x^3 + 2x^2 + 4x$
$-2x(x - 2)(x + 1)$

53. $80y^2 - 36y + 4$
$4(4y - 1)(5y - 1)$

54. $24y^2 - 24y - 18$
$6(2y + 1)(2y - 3)$

55. $8z^3 + 14z^2 + 3z$
$z(2z + 3)(4z + 1)$

56. $6z^3 - 23z^2 + 20z$
$z(2z - 5)(3z - 4)$

57. $6x^2y - 11xy - 10y$
$y(2x - 5)(3x + 2)$

58. $8x^2y - 27xy + 9y$
$y(x - 3)(8x - 3)$

59. $10t^2 - 5t - 50$
$5(t + 2)(2t - 5)$

60. $16t^2 + 40t - 96$
$8(t + 4)(2t - 3)$

61. $3p^3 - 16p^2 + 5p$
$p(p - 5)(3p - 1)$

62. $6p^3 + 5p^2 + p$
$p(2p + 1)(3p + 1)$

63. $26z^2 + 98z - 24$
$2(z + 4)(13z - 3)$

64. $30z^2 - 87z + 30$
$3(2z - 5)(5z - 2)$

65. $10y^3 - 44y^2 + 16y$
$2y(y - 4)(5y - 2)$

66. $14y^3 + 94y^2 - 28y$
$2y(y + 7)(7y - 2)$

67. $4yz^3 + 5yz^2 - 6yz$
$yz(z + 2)(4z - 3)$

68. $12a^3 + 14a^2 - 48a$
$2a(2a - 3)(3a + 8)$

69. $42a^3 + 45a^2 - 27a$
$3a(2a + 3)(7a - 3)$

70. $36p^2 - 9p^3 - p^4$
$-p^2(p - 3)(p + 12)$

71. $9x^2y - 30xy^2 + 25y^3$
$y(3x - 5y)(3x - 5y)$

72. $8x^2y - 38xy^2 + 35y^3$
$y(2x - 7y)(4x - 5y)$

73. $9x^3y - 24x^2y^2 + 16xy^3$
$xy(3x - 4y)(3x - 4y)$

74. $9x^3y + 12x^2y + 4xy$
$xy(3x + 2)(3x + 2)$

For Exercises 75 and 76, let $(nx + p)$ and $(mx + q)$ be prime factors of the trinomial $ax^2 + bx + c$.

75. If n is even, must p be even or odd?   Odd

76. If p is even, must n be even or odd?   Odd

OBJECTIVE B   *To factor a trinomial of the form $ax^2 + bx + c$ by grouping*

For Exercises 77 to 133, factor by grouping.

77. $6x^2 - 17x + 12$
$(2x - 3)(3x - 4)$

78. $15x^2 - 19x + 6$
$(3x - 2)(5x - 3)$

79. $5b^2 + 33b - 14$
$(b + 7)(5b - 2)$

80. $8x^2 - 30x + 25$
$(2x - 5)(4x - 5)$

81. $6a^2 + 7a - 24$
$(2a - 3)(3a + 8)$

82. $14a^2 + 15a - 9$
$(2a + 3)(7a - 3)$

83. $4z^2 + 11z + 6$
$(z + 2)(4z + 3)$

84. $6z^2 - 25z + 14$
$(2z - 7)(3z - 2)$

85. $22p^2 + 51p - 10$
$(2p + 5)(11p - 2)$

86. $14p^2 - 41p + 15$
$(2p - 5)(7p - 3)$

87. $8y^2 + 17y + 9$
$(y + 1)(8y + 9)$

88. $12y^2 - 145y + 12$
$(y - 12)(12y - 1)$

89. $18t^2 - 9t - 5$
$(3t + 1)(6t - 5)$

90. $12t^2 + 28t - 5$
$(2t + 5)(6t - 1)$

91. $6b^2 + 71b - 12$
$(b + 12)(6b - 1)$

92. $8b^2 + 65b + 8$
$(b + 8)(8b + 1)$

93. $9x^2 + 12x + 4$
$(3x + 2)(3x + 2)$

94. $25x^2 - 30x + 9$
$(5x - 3)(5x - 3)$

95. $6b^2 - 13b + 6$
$(2b - 3)(3b - 2)$

96. $20b^2 + 37b + 15$
$(4b + 5)(5b + 3)$

97. $33b^2 + 34b - 35$
$(3b + 5)(11b - 7)$

98. $15b^2 - 43b + 22$
$(3b - 2)(5b - 11)$

99. $18y^2 - 39y + 20$
$(3y - 4)(6y - 5)$

100. $24y^2 + 41y + 12$
$(3y + 4)(8y + 3)$

101. $15a^2 + 26a - 21$
$(3a + 7)(5a - 3)$

102. $6a^2 + 23a + 21$
$(2a + 3)(3a + 7)$

103. $8y^2 - 26y + 15$
$(2y - 5)(4y - 3)$

104. $18y^2 - 27y + 4$
$(3y - 4)(6y - 1)$

105. $8z^2 + 2z - 15$
$(2z + 3)(4z - 5)$

106. $10z^2 + 3z - 4$
$(2z - 1)(5z + 4)$

107. $15x^2 - 82x + 24$
Nonfactorable
over the integers

108. $13z^2 + 49z - 8$
Nonfactorable
over the integers

109. $10z^2 - 29z + 10$
$(2z - 5)(5z - 2)$

110. $15z^2 - 44z + 32$
$(3z - 4)(5z - 8)$

111. $36z^2 + 72z + 35$
$(6z + 5)(6z + 7)$

112. $16z^2 + 8z - 35$
$(4z + 7)(4z - 5)$

113. $3x^2 + xy - 2y^2$
$(x + y)(3x - 2y)$

114. $6x^2 + 10xy + 4y^2$
$2(x + y)(3x + 2y)$

115. $3a^2 + 5ab - 2b^2$
$(a + 2b)(3a - b)$

116. $2a^2 - 9ab + 9b^2$
$(a - 3b)(2a - 3b)$

117. $4y^2 - 11yz + 6z^2$
$(y - 2z)(4y - 3z)$

118. $2y^2 + 7yz + 5z^2$
$(y + z)(2y + 5z)$

119. $28 + 3z - z^2$
$-(z - 7)(z + 4)$

120. $15 - 2z - z^2$
$-(z - 3)(z + 5)$

121. $8 - 7x - x^2$
$-(x - 1)(x + 8)$

122. $12 + 11x - x^2$
$-(x - 12)(x + 1)$

123. $9x^2 + 33x - 60$
$3(x + 5)(3x - 4)$

124. $16x^2 - 16x - 12$
$4(2x + 1)(2x - 3)$

125. $24x^2 - 52x + 24$
$4(2x - 3)(3x - 2)$

126. $60x^2 + 95x + 20$
$5(3x + 4)(4x + 1)$

127. $35a^4 + 9a^3 - 2a^2$
$a^2(5a + 2)(7a - 1)$

128. $15a^4 + 26a^3 + 7a^2$
$a^2(3a + 1)(5a + 7)$

129. $15b^2 - 115b + 70$
$5(b - 7)(3b - 2)$

130. $25b^2 + 35b - 30$
$5(b + 2)(5b - 3)$

131. $3x^2 - 26xy + 35y^2$
$(x - 7y)(3x - 5y)$

132. $4x^2 + 16xy + 15y^2$
$(2x + 3y)(2x + 5y)$

133. $216y^2 - 3y - 3$
$3(8y - 1)(9y + 1)$

For Exercises 134 to 137, information is given about the signs of b and c in the trinomial $ax^2 + bx + c$, where $a > 0$. If you want to factor $ax^2 + bx + c$ by grouping, you look for factors of ac whose sum is b. In each case, state whether the factors of ac should be two positive numbers, two negative numbers, or one positive and one negative number.

134. $b > 0$ and $c > 0$
Two positive

135. $b < 0$ and $c < 0$
One positive, one negative

136. $b < 0$ and $c > 0$
Two negative

137. $b > 0$ and $c < 0$
One positive, one negative

Critical Thinking

138. In your own words, explain how the signs of the last terms of the two binomial factors of a trinomial are determined.

For Exercises 139 to 147, factor.

139. $(x + 1)^2 - (x + 1) - 6$
$(x - 2)(x + 3)$

140. $(x - 2)^2 + 3(x - 2) + 2$
$x(x - 1)$

141. $(y + 3)^2 - 5(y + 3) + 6$
$y(y + 1)$

142. $2(y + 2)^2 - (y + 2) - 3$
$(2y + 1)(y + 3)$

143. $3(a + 2)^2 - (a + 2) - 4$
$(3a + 2)(a + 3)$

144. $4(y - 1)^2 - 7(y - 1) - 2$
$(4y - 3)(y - 3)$

145. $6y + 8y^3 - 26y^2$
$2y(y - 3)(4y - 1)$

146. $22p^2 - 3p^3 + 16p$
$p(2 + 3p)(8 - p)$

147. $a^3b - 24ab - 2a^2b$
$ab(a + 4)(a - 6)$

148. Given that $x + 2$ is a factor of $x^3 - 2x^2 - 5x + 6$, factor $x^3 - 2x^2 - 5x + 6$ completely. $(x + 2)(x - 3)(x - 1)$

Projects or Group Activities

For Exercises 149 to 154, find all integers k such that the trinomial can be factored over the integers.

149. $2x^2 + kx + 3$
$7, -7, 5, -5$

150. $2x^2 + kx - 3$
$5, -5, 1, -1$

151. $3x^2 + kx + 2$
$7, -7, 5, -5$

152. $3x^2 + kx - 2$
$5, -5, 1, -1$

153. $2x^2 + kx + 5$
$11, -11, 7, -7$

154. $2x^2 + kx - 5$
$9, -9, -3, 3$

QUICK QUIZ
Factor by using trial factors.
1. $2x^2 - 17x + 21$
   $(2x - 3)(x - 7)$ **[7.3A]**
2. $12x^3y + x^2y - 6xy$
   $xy(4x + 3)(3x - 2)$
   [7.3A]
Factor by grouping.
3. $10x^2 + x - 2$
   $(5x - 2)(2x + 1)$ **[7.3B]**
4. $12x^3y + 10x^2y - 8xy$
   $2xy(3x + 4)(2x - 1)$
   [7.3B]

155. **Geometry** The area of a rectangle is $(3x^2 + x - 2)$ ft². Find the dimensions of the rectangle in terms of the variable x. Given that $x > 0$, specify the dimension that is the length and the dimension that is the width. Can x be negative? Can $x = 0$? Explain your answers.

$A = 3x^2 + x - 2$

Unless otherwise noted, all content on this page is © Cengage Learning.

✔ CHECK YOUR PROGRESS: CHAPTER 7

1. Factor: $20b + 5$ $5(4b + 1)$ [7.1A]

2. Factor: $2x(7 + b) - y(b + 7)$
$(b + 7)(2x - y)$ [7.1B]

3. Factor: $x^2 + 20x + 100$ $(x + 10)(x + 10)$ [7.2A]

4. Factor: $x^2y - 2xy - 24y$
$y(x + 4)(x - 6)$ [7.2B]

5. Factor: $35 + 2x - x^2$ $-(x - 7)(x + 5)$ [7.2B]

6. Factor: $x^2 - 8x - 2$
Nonfactorable over the integers   [7.2A]

7. Factor: $21x^2 + 6xy - 49x - 14y$
$(7x + 2y)(3x - 7)$ [7.1B]

8. Factor: $6ab + 9a$ $3a(2b + 3)$ [7.1A]

9. Factor by using trial factors:
$5y^2 - 22y + 8$ $(y - 4)(5y - 2)$ [7.3A]

10. Factor by grouping:
$12x^2 + 31x + 9$ $(3x + 1)(4x + 9)$ [7.3B]

11. Factor: $9x - 5x^2$ $x(9 - 5x)$ [7.1A]

12. Factor: $2x^2 + x + 2xy + y$
$(2x + 1)(x + y)$ [7.1B]

13. Factor by grouping:
$8a^2 - 2ab - 3b^2$ $(2a + b)(4a - 3b)$ [7.3B]

14. Factor: $b^2 + 9b + 20$
$(b + 4)(b + 5)$ [7.2A]

15. Factor: $2a^3 + 24a^2 + 54a$
$2a(a + 9)(a + 3)$ [7.2B]

16. Factor by using trial factors:
$11a^2 - 54a - 5$ $(a - 5)(11a + 1)$ [7.3A]

17. Factor by grouping:
$360y^2 + 4y - 4$ $4(9y + 1)(10y - 1)$ [7.3B]

18. Factor: $14y^3 + 5y^2 + 11y$
$y(14y^2 + 5y + 11)$ [7.1A]

19. Factor: $x^2 - 7x + 10$ $(x - 2)(x - 5)$ [7.2A]

20. Factor: $x^2 + 8xy + 9y^2$
Nonfactorable over the integers   [7.2B]

21. Factor: $b^2 + 13b + 40$ $(b + 8)(b + 5)$ [7.2A]

22. Factor: $2x^2 - 5x - 6xy + 15y$
$(2x - 5)(x - 3y)$ [7.1B]

23. Factor: $x^2y - xy^3 + x^3y$
$xy(x - y^2 + x^2)$ [7.1A]

24. Factor by using trial factors:
$3b^2 + 16b + 16$ $(b + 4)(3b + 4)$ [7.3A]

25. Factor: $x^2 - 11x - 42$
$(x + 3)(x - 14)$ [7.2A]

7.4   Special Factoring

OBJECTIVE A   *To factor the difference of two perfect squares or a perfect-square trinomial*

The product of a term and itself is called a **perfect square.** The exponents on variables of perfect squares are always even numbers.

Term		Perfect Square
5	$5 \cdot 5 =$	25
x	$x \cdot x =$	x^2
$3y^4$	$3y^4 \cdot 3y^4 =$	$9y^8$

The **square root of a perfect square** is one of the two equal factors of the perfect square. $\sqrt{}$ is the symbol for square root. To find the exponent on the square root of a variable term, divide the exponent on the variable term by 2.

$$\sqrt{25} = 5$$
$$\sqrt{x^2} = x$$
$$\sqrt{9y^8} = 3y^4$$

The difference of two perfect squares factors as the sum and difference of the square roots of the perfect squares.

Factoring the Difference of Two Perfect Squares

$a^2 - b^2 = (a + b)(a - b)$

EXAMPLES

1. $x^2 - 25 = x^2 - 5^2 = (x + 5)(x - 5)$

2. $y^2 - 81 = y^2 - 9^2 = (y + 9)(y - 9)$

The sum of two squares, $a^2 + b^2$, does not factor over the integers. For instance, $x^2 + 25 = x^2 + 5^2$ is nonfactorable over the integers.

HOW TO 1   Factor: $4x^2 - 81y^2$

Write the binomial as the difference of two perfect squares.

$$4x^2 - 81y^2 = (2x)^2 - (9y)^2$$

The factors are the sum and difference of the square roots of the perfect squares.

$$= (2x + 9y)(2x - 9y)$$

 Take Note

Recall that, using FOIL,
$(x + 6)^2 = (x + 6)(x + 6)$
$ = x^2 + 6x + 6x + 36$
$ = x^2 + 12x + 36$

The square of a binomial is a **perfect-square trinomial.** Here are several examples.

Square of a Binomial	Perfect-Square Trinomial
$(x + 6)^2$	$x^2 + 12x + 36$
$(x - 7)^2$	$x^2 - 14x + 49$
$(a + b)^2$	$a^2 + 2ab + b^2$
$(a - b)^2$	$a^2 - 2ab + b^2$

Factoring a Perfect-Square Trinomial

$a^2 + 2ab + b^2 = (a + b)^2$ $\qquad$ $a^2 - 2ab + b^2 = (a - b)^2$

EXAMPLES

1. $x^2 + 10x + 25 = (x + 5)^2$ $\qquad$ **2.** $y^2 - 12y + 36 = (y - 6)^2$

Unless otherwise noted, all content on this page is © Cengage Learning.

In factoring a perfect-square trinomial, remember that the terms of the binomial are the square roots of the perfect squares of the trinomial. The sign of the binomial is the sign of the middle term of the trinomial.

HOW TO 2 Factor: $4x^2 + 12x + 9$

Because $4x^2$ is a perfect square $[4x^2 = (2x)^2]$ and 9 is a perfect square $(9 = 3^2)$, try factoring $4x^2 + 12x + 9$ as the square of a binomial.

$$4x^2 + 12x + 9 \overset{?}{=} (2x + 3)^2$$

Check: $(2x + 3)^2 = (2x + 3)(2x + 3) = 4x^2 + 6x + 6x + 9 = 4x^2 + 12x + 9$

The check verifies that $4x^2 + 12x + 9 = (2x + 3)^2$.

IN-CLASS EXAMPLES
Factor.
1. $25x^2 - 81$
   $(5x + 9)(5x - 9)$
2. $x^2 + 16$
   Nonfactorable
3. $36a^2 - 49b^2$
   $(6a + 7b)(6a - 7b)$
4. $y^2 - 8y + 16$
   $(y - 4)^2$
5. $25x^2 + 30x + 9$
   $(5x + 3)^2$

It is important to check a proposed factorization as we did above. HOW TO 3 illustrates the importance of this check.

HOW TO 3 Factor: $x^2 + 13x + 36$

Because x^2 is a perfect square and 36 is a perfect square, try factoring $x^2 + 13x + 36$ as the square of a binomial.

$$x^2 + 13x + 36 \overset{?}{=} (x + 6)^2$$

Check: $(x + 6)^2 = (x + 6)(x + 6) = x^2 + 6x + 6x + 36 = x^2 + 12x + 36$

INSTRUCTOR NOTE
Another example of a trinomial that appears to be a perfect-square trinomial but does not factor is $4x^2 - 17x + 25$.

In this case, the proposed factorization of $x^2 + 13x + 36$ does *not* check. Try another factorization. The numbers 4 and 9 are factors of 36 whose sum is 13.

$$x^2 + 13x + 36 = (x + 4)(x + 9)$$

EXAMPLE 1

Factor: $25x^2 - 1$

Solution

$25x^2 - 1 = (5x)^2 - (1)^2$ • Difference of
$= (5x + 1)(5x - 1)$ two squares

YOU TRY IT 1

Factor: $x^2 - 36y^4$

Your solution

$(x + 6y^2)(x - 6y^2)$

EXAMPLE 2

Factor: $4x^2 - 20x + 25$

Solution

$4x^2 - 20x + 25 = (2x - 5)^2$ • Perfect-square
trinomial

YOU TRY IT 2

Factor: $9x^2 + 12x + 4$

Your solution

$(3x + 2)^2$

EXAMPLE 3

Factor: $(x + y)^2 - 4$

Solution

$(x + y)^2 - 4$
$= (x + y)^2 - (2)^2$ • Difference of
$= (x + y + 2)(x + y - 2)$ two squares

YOU TRY IT 3

Factor: $(a + b)^2 - (a - b)^2$

Your solution

$4ab$

Solutions on p. S21

OBJECTIVE B | *To factor the sum or difference of two perfect cubes*

The product of the same three factors is called a **perfect cube.** The first seven perfect cube integers are:

$$1 = 1^3, 8 = 2^3, 27 = 3^3, 64 = 4^3, 125 = 5^3, 216 = 6^3, 343 = 7^3$$

A variable term is a perfect cube if the coefficient is a perfect cube and the exponent on each variable is divisible by 3. The table below shows some perfect-cube variable terms. Note that each exponent of the perfect cube is divisible by 3.

Term		Perfect Cube
x	$x \cdot x \cdot x = (x)^3 =$	x^3
$2y$	$2y \cdot 2y \cdot 2y = (2y)^3 =$	$8y^3$
$4x^2$	$4x^2 \cdot 4x^2 \cdot 4x^2 = (4x^2)^3 =$	$64x^6$
$3x^4y^3$	$3x^4y^3 \cdot 3x^4y^3 \cdot 3x^4y^3 = (3x^4y^3)^3 =$	$27x^{12}y^9$

The **cube root** of a perfect cube is one of the three equal factors of the perfect cube. $\sqrt[3]{}$ is the symbol for cube root. To find the exponents on the cube root of a perfect-cube variable expression, divide the exponents on the variables by 3.

$$\sqrt[3]{x^3} = x$$
$$\sqrt[3]{8y^3} = 2y$$
$$\sqrt[3]{64x^6} = 4x^2$$
$$\sqrt[3]{27x^{12}y^9} = 3x^4y^3$$

The following rules are used to factor the sum or difference of two perfect cubes.

Take Note

The first factoring formula is the result of finding the quotient

$$\frac{a^3 + b^3}{a + b} = a^2 - ab + b^2$$

Similarly,

$$\frac{a^3 - b^3}{a - b} = a^2 + ab + b^2$$

Factoring the Sum or Difference of Two Perfect Cubes

$$a^3 + b^3 = (a + b)(a^2 - ab + b^2)$$
$$a^3 - b^3 = (a - b)(a^2 + ab + b^2)$$

EXAMPLES

1. $x^3 + 8 = x^3 + 2^3 = (x + 2)(x^2 - 2x + 4)$
2. $z^3 - 64 = z^3 - 4^3 = (z - 4)(z^2 + 4z + 16)$

HOW TO 4 Factor: $64x^3 - 125$

$$64x^3 - 125 = (4x)^3 - 5^3$$

$$= (4x - 5)(16x^2 + 20x + 25)$$

Square of the first term of the binomial factor

Opposite of the product of the two terms

Square of the last term

• Write the binomial as the difference of two perfect cubes.
• The terms of the binomial factor are the cube roots of the perfect cubes. The sign of the binomial factor is the same as the sign of the given binomial. The trinomial factor is obtained from the binomial factor.

Take Note

Note the placement of the signs. The sign of the binomial factor is the same as the sign of the sum or difference of the perfect cubes. The first sign of the trinomial factor is the opposite of the sign of the binomial factor.

HOW TO 5   Factor: $m^3 + 64n^3$

$m^3 + 64n^3 = (m)^3 + (4n)^3$ • Write as the sum of two perfect cubes. $a = m$ and $b = 4n$

$\qquad = (m + 4n)(m^2 - 4mn + 16n^2)$ • Use $a^3 + b^3 = (a + b)(a^2 - ab + b^2)$.

HOW TO 6   Factor: $8x^3 - 27$

$8x^3 - 27 = (2x)^3 - 3^3$ • Write as the difference of two perfect cubes. $a = 2x$ and $b = 3$

$\qquad = (2x - 3)(4x^2 + 6x + 9)$ • Use $a^3 - b^3 = (a - b)(a^2 + ab + b^2)$.

EXAMPLE 4

Factor: $x^3y^3 - 1$

Solution

$x^3y^3 - 1 = (xy)^3 - 1^3$ • Difference of
$\qquad = (xy - 1)(x^2y^2 + xy + 1)$ two cubes

YOU TRY IT 4

Factor: $a^3b^3 - 27$

Your solution

$(ab - 3)(a^2b^2 + 3ab + 9)$

EXAMPLE 5

Factor: $64c^3 + d^3$

Solution

$64c^3 + d^3$
$= (4c)^3 + d^3$ • Sum of
$= (4c + d)(16c^2 - 4cd + d^2)$ two cubes

YOU TRY IT 5

Factor: $8x^3 + y^3z^3$

Your solution

$(2x + yz)(4x^2 - 2xyz + y^2z^2)$

EXAMPLE 6

Factor: $(x + y)^3 - x^3$

Solution

$(x + y)^3 - x^3$ • Difference of two cubes
$= [(x + y) - x][(x + y)^2 + x(x + y) + x^2]$
$= y(x^2 + 2xy + y^2 + x^2 + xy + x^2)$
$= y(3x^2 + 3xy + y^2)$

YOU TRY IT 6

Factor: $(x - y)^3 + (x + y)^3$

Your solution

$2x(x^2 + 3y^2)$

IN-CLASS EXAMPLES

Factor.
6. $x^3 + 64y^3$ $(x + 4y)(x^2 - 4xy + 16y^2)$
7. $8a^3 - b^3$ $(2a - b)(4a^2 + 2ab + b^2)$
8. $27x^3 - 1$ $(3x - 1)(9x^2 + 3x + 1)$

Solutions on p. S21

OBJECTIVE C   *To factor a trinomial that is quadratic in form*

Certain trinomials that are not quadratic can be expressed as quadratic trinomials by making suitable variable substitutions. A trinomial is **quadratic in form** if it can be written as $au^2 + bu + c$.

Take Note

An expression is quadratic in form if it can be written as $a(\ )^2 + b(\ ) + c$, where the same expression is placed in both sets of parentheses.

The expression $2x^6 - 7x^3 + 4$ is quadratic in form because
$$2x^6 - 7x^3 + 4$$
$$= 2(x^3)^2 - 7(x^3) + 4$$

The expression $5x^2y^2 + 3xy - 6$ is quadratic in form because
$$5x^2y^2 + 3xy - 6$$
$$= 5(xy)^2 + 3(xy) - 6$$

Trinomials That Are Quadratic in Form

A trinomial is **quadratic in form** if it can be written as $au^2 + bu + c$.

EXAMPLES

1. $2x^6 - 7x^3 + 4$
   Let $u = x^3$. Then $u^2 = (x^3)^2 = x^6$.
   $2x^6 - 7x^3 + 4 \Rightarrow 2u^2 - 7u + 4$
   $2x^6 - 7x^3 + 4$ is quadratic in form.
2. $5x^2y^2 + 3xy - 6$
   Let $u = xy$. Then $u^2 = (xy)^2 = x^2y^2$.
   $5x^2y^2 + 3xy - 6 \Rightarrow 5u^2 + 3u - 6$
   $5x^2y^2 + 3xy - 6$ is quadratic in form.

When we use this method to factor a trinomial that is quadratic in form, the variable part of the first term of each binomial factor will be u.

HOW TO 7   Factor: $x^4 + 5x^2 + 6$

$$x^4 + 5x^2 + 6 = u^2 + 5u + 6 \qquad \bullet \text{ Let } u = x^2.$$
$$= (u + 3)(u + 2) \qquad \bullet \text{ Factor.}$$
$$= (x^2 + 3)(x^2 + 2) \qquad \bullet \text{ Replace } u \text{ by } x^2.$$

Here is an example in which $u = \sqrt{x}$.

HOW TO 8   Factor: $x - 2\sqrt{x} - 15$

$$x - 2\sqrt{x} - 15 = u^2 - 2u - 15 \qquad \bullet \text{ Let } u = \sqrt{x}. \text{ Then } u^2 = x.$$
$$= (u - 5)(u + 3) \qquad \bullet \text{ Factor.}$$
$$= (\sqrt{x} - 5)(\sqrt{x} + 3) \qquad \bullet \text{ Replace } u \text{ by } \sqrt{x}.$$

IN-CLASS EXAMPLES

Factor.

9. $x^2y^2 - 9xy + 20$
   $(xy - 5)(xy - 4)$
10. $y^4 - 8y^2 - 48$
    $(y^2 - 12)(y^2 + 4)$
11. $a + 5\sqrt{a} - 36$
    $(\sqrt{a} + 9)(\sqrt{a} - 4)$

EXAMPLE 7

Factor: $6x^2y^2 - xy - 12$

Solution

$$6x^2y^2 - xy - 12$$
$$= 6u^2 - u - 12 \qquad \bullet \text{ Let } u = xy.$$
$$= (3u + 4)(2u - 3)$$
$$= (3xy + 4)(2xy - 3) \qquad \bullet \text{ Replace } u \text{ by } xy.$$

YOU TRY IT 7

Factor: $3x^4 + 4x^2 - 4$

Your solution

$(x^2 + 2)(3x^2 - 2)$

Solution on p. S21

OBJECTIVE D   *To factor completely*

Tips for Success
You now have completed all the lessons on factoring polynomials. You will need to be able to recognize all of the factoring patterns. To test yourself, try the Chapter 7 Review Exercises.

Take Note
Remember that you may have to factor more than once in order to write the polynomial as a product of *prime* factors.

General Factoring Strategy

When factoring a polynomial completely, ask the following questions about the polynomial.

1. Is there a common factor? If so, factor out the GCF.
2. If the polynomial is a binomial, is it the difference of two perfect squares, the sum of two perfect cubes, or the difference of two perfect cubes? If so, factor.
3. If the polynomial is a trinomial, is it a perfect-square trinomial or the product of two binomials? If so, factor.
4. Can the polynomial be factored by grouping? If so, factor.
5. Is each factor nonfactorable over the integers? If not, factor.

HOW TO 9   Factor: $64y^4 - 125y$

$$64y^4 - 125y = y(64y^3 - 125)$$
$$= y[(4y)^3 - 5^3]$$
$$= y(4y - 5)(16y^2 + 20y + 25)$$

- Factor out y, the GCF.
- Write the binomial as the difference of two perfect cubes. $a = 4y$ and $b = 5$
- Factor.

EXAMPLE 8

Factor: $6a^3 + 15a^2 - 36a$

Solution

$6a^3 + 15a^2 - 36a$
$= 3a(2a^2 + 5a - 12)$ • The GCF is $3a$.
$= 3a(2a - 3)(a + 4)$ • Factor the trinomial.

YOU TRY IT 8

Factor: $18x^3 - 6x^2 - 60x$

Your solution

$6x(3x + 5)(x - 2)$

EXAMPLE 9

Factor: $x^2y + 2x^2 - y - 2$

Solution

All four terms do not have a common factor. The polynomial is not a binomial or a trinomial. Try factoring by grouping.

$x^2y + 2x^2 - y - 2$
$= (x^2y + 2x^2) - (y + 2)$ • Factor by grouping.
$= x^2(y + 2) - (y + 2)$
$= (y + 2)(x^2 - 1)$ • $x^2 - 1$ is a difference
$= (y + 2)(x + 1)(x - 1)$ of squares.

YOU TRY IT 9

Factor: $4x - 4y - x^3 + x^2y$

Your solution

$(x - y)(2 + x)(2 - x)$

IN-CLASS EXAMPLES
Factor.
12. $x^4 - 7x^3 + 10x^2$ $x^2(x - 2)(x - 5)$
13. $18x^2 + 24x + 8$ $2(3x + 2)^2$
14. $x^4 - 8x$ $x(x - 2)(x^2 + 2x + 4)$
15. $x^4 - 81$ $(x^2 + 9)(x + 3)(x - 3)$
16. $x^3 + x^2 - 4x - 4$ $(x - 2)(x + 2)(x + 1)$

7.4 EXERCISES

✔ **Concept Check**

SUGGESTED ASSIGNMENT
Exercises 1–8; Exercises 9–115, odds;
More challenging exercises: Exercises 118–122

1. State whether each expression is a perfect square.
   a. $8x^4$ No
   b. $9x^{12}$ Yes
   c. $9x^9$ No
   d. $25x^{10}$ Yes

2. Name the square root of each expression.
   a. $16z^8$ $4z^4$
   b. $36d^{10}$ $6d^5$
   c. $81a^4b^6$ $9a^2b^3$
   d. $25m^2n^{12}$ $5mn^6$

3. State whether each expression is a difference of squares.
   a. $4x^2 - 9$ Yes
   b. $x^2 + 9$ No
   c. $x^2y^2 - 81$ Yes
   d. $16x^2 - 8$ No

4. State whether each expression is a perfect cube.
   a. $8x^6$ Yes
   b. $16x^3$ No
   c. $64x^6$ Yes
   d. $27x^9y^6$ Yes

5. Name the cube root of each expression.
   a. $8x^9$ $2x^3$
   b. $27y^{15}$ $3y^5$
   c. $64a^6b^{18}$ $4a^2b^6$
   d. $125c^{12}d^3$ $5c^4d$

6. State whether each expression is a sum or difference of perfect cubes.
   a. $8x^3 + 27$ Yes
   b. $(x + 8)^3$ No
   c. $(x - 27)^3$ No
   d. $x^3 - 64$ Yes

7. State whether each expression is quadratic in form.
   a. $2x^4 + x^2 - 6$ Yes
   b. $4x^4 + 9x + 25$ No
   c. $x^4 - 9$ Yes
   d. $3x - 5\sqrt{x} + 7$ Yes

8. Name the value of u such that each expression can be written in the form $au^2 + bu + c$.
   a. $x^4 - 6x^2 + 3$ **b.** $3x^4y^4 + 7x^2y^2 - 8$ **c.** $7x - 8\sqrt{x} + 1$ **d.** $x^6 + 6x^3 - 3$
   $u = x^2$ $u = x^2y^2$ $u = \sqrt{x}$ $u = x^3$

OBJECTIVE A   *To factor the difference of two perfect squares or a perfect-square trinomial*

For Exercises 9 to 40, factor.

9. $x^2 - 16$
   $(x + 4)(x - 4)$

10. $y^2 - 49$
    $(y + 7)(y - 7)$

11. $4x^2 - 1$
    $(2x + 1)(2x - 1)$

12. $81x^2 - 4$
    $(9x + 2)(9x - 2)$

13. $16x^2 - 121$
    $(4x + 11)(4x - 11)$

14. $49y^2 - 36$
    $(7y + 6)(7y - 6)$

15. $1 - 9a^2$
    $(1 + 3a)(1 - 3a)$

16. $16 - 81y^2$
    $(4 + 9y)(4 - 9y)$

17. $x^2y^2 - 100$
    $(xy + 10)(xy - 10)$

18. $a^2b^2 - 25$
    $(ab + 5)(ab - 5)$

19. $x^2 + 4$
    Nonfactorable

20. $a^2 + 16$
    Nonfactorable

21. $25 - a^2b^2$
    $(5 + ab)(5 - ab)$

22. $64 - x^2y^2$
    $(8 + xy)(8 - xy)$

23. $x^2 - 12x + 36$
    $(x - 6)^2$

24. $y^2 - 6y + 9$
    $(y - 3)^2$

25. $b^2 - 2b + 1$
    $(b - 1)^2$

26. $a^2 + 14a + 49$
    $(a + 7)^2$

27. $16x^2 - 40x + 25$
    $(4x - 5)^2$

28. $49x^2 + 28x + 4$
    $(7x + 2)^2$

29. $4a^2 + 4a - 1$
    Nonfactorable

30. $9x^2 + 12x - 4$
    Nonfactorable

31. $b^2 + 7b + 14$
    Nonfactorable

32. $y^2 - 5y + 25$
    Nonfactorable

33. $x^2 + 6xy + 9y^2$
$(x + 3y)^2$

34. $4x^2y^2 + 12xy + 9$
$(2xy + 3)^2$

35. $25a^2 - 40ab + 16b^2$
$(5a - 4b)^2$

36. $4a^2 - 36ab + 81b^2$
$(2a - 9b)^2$

37. $(x - 4)^2 - 9$
$(x - 7)(x - 1)$

38. $16 - (a - 3)^2$
$(7 - a)(1 + a)$

39. $(x - y)^2 - (a + b)^2$
$(x - y + a + b)(x - y - a - b)$

40. $(x - 2y)^2 - (x + y)^2$
$-3y(2x - y)$

OBJECTIVE B *To factor the sum or difference of two perfect cubes*

For Exercises 41 to 60, factor.

41. $x^3 - 27$
$(x - 3)(x^2 + 3x + 9)$

42. $y^3 + 125$
$(y + 5)(y^2 - 5y + 25)$

43. $8x^3 - 1$
$(2x - 1)(4x^2 + 2x + 1)$

44. $64a^3 + 27$
$(4a + 3)(16a^2 - 12a + 9)$

45. $x^3 - y^3$
$(x - y)(x^2 + xy + y^2)$

46. $x^3 - 8y^3$
$(x - 2y)(x^2 + 2xy + 4y^2)$

47. $m^3 + n^3$
$(m + n)(m^2 - mn + n^2)$

48. $27a^3 + b^3$
$(3a + b)(9a^2 - 3ab + b^2)$

49. $64x^3 + 1$
$(4x + 1)(16x^2 - 4x + 1)$

50. $1 - 125b^3$
$(1 - 5b)(1 + 5b + 25b^2)$

51. $27x^3 - 8y^3$
$(3x - 2y)(9x^2 + 6xy + 4y^2)$

52. $64x^3 + 27y^3$
$(4x + 3y)(16x^2 - 12xy + 9y^2)$

53. $x^3y^3 + 64$
$(xy + 4)(x^2y^2 - 4xy + 16)$

54. $8x^3y^3 + 27$
$(2xy + 3)(4x^2y^2 - 6xy + 9)$

55. $16x^3 - y^3$
Nonfactorable

56. $27x^3 - 8y^2$
Nonfactorable

57. $8x^3 - 9y^3$
Nonfactorable

58. $27a^3 - 16$
Nonfactorable

59. $125 - c^3$
$(5 - c)(25 + 5c + c^2)$

60. $27 - 64x^3$
$(3 - 4x)(9 + 12x + 16x^2)$

OBJECTIVE C *To factor a trinomial that is quadratic in form*

61. ✍ The expression $x - \sqrt{x} - 6$ is quadratic in form. Is the expression a polynomial? Explain.   No; polynomials cannot have square roots as variable terms.

62. ✍ The polynomial $x^4 - 2x^2 - 3$ is quadratic in form. Is the polynomial a quadratic polynomial? Explain.   No; it is a fourth-degree polynomial.

For Exercises 63 to 83, factor.

63. $x^2y^2 - 8xy + 15$
$(xy - 5)(xy - 3)$

64. $x^2y^2 - 8xy - 33$
$(xy + 3)(xy - 11)$

65. $x^2y^2 - 17xy + 60$
$(xy - 12)(xy - 5)$

66. $a^2b^2 + 10ab + 24$
$(ab + 6)(ab + 4)$

67. $x^4 - 9x^2 + 18$
$(x^2 - 6)(x^2 - 3)$

68. $y^4 - 6y^2 - 16$
$(y^2 + 2)(y^2 - 8)$

69. $b^4 - 13b^2 - 90$
$(b^2 + 5)(b^2 - 18)$

70. $a^4 + 14a^2 + 45$
$(a^2 + 9)(a^2 + 5)$

71. $x^4y^4 - 8x^2y^2 + 12$
$(x^2y^2 - 2)(x^2y^2 - 6)$

72. $a^4b^4 + 11a^2b^2 - 26$
$(a^2b^2 + 13)(a^2b^2 - 2)$

73. $x + 3\sqrt{x} + 2$
$(\sqrt{x} + 1)(\sqrt{x} + 2)$

74. $a - \sqrt{a} - 12$
$(\sqrt{a} + 3)(\sqrt{a} - 4)$

75. $3x^2y^2 - 14xy + 15$
$(3xy - 5)(xy - 3)$

76. $5x^2y^2 - 59xy + 44$
$(5xy - 4)(xy - 11)$

77. $6a^2b^2 - 23ab + 21$
$(2ab - 3)(3ab - 7)$

78. $10a^2b^2 + 3ab - 7$
$(ab + 1)(10ab - 7)$

79. $2x^4 - 13x^2 - 15$
$(x^2 + 1)(2x^2 - 15)$

80. $3x^4 + 20x^2 + 32$
$(3x^2 + 8)(x^2 + 4)$

81. $x^6 - x^3 - 6$
$(x^3 + 2)(x^3 - 3)$

82. $2z^6 + 7z^3 + 3$
$(2z^3 + 1)(z^3 + 3)$

83. $4x^2y^4 - 12xy^2 + 9$
$(2xy^2 - 3)^2$

OBJECTIVE D *To factor completely*

For Exercises 84 to 115, factor.

84. $5x^2 + 10x + 5$
$5(x + 1)^2$

85. $12x^2 - 36x + 27$
$3(2x - 3)^2$

86. $3x^4 - 81x$
$3x(x - 3)(x^2 + 3x + 9)$

87. $27a^4 - a$
$a(3a - 1)(9a^2 + 3a + 1)$

88. $7x^2 - 28$
$7(x + 2)(x - 2)$

89. $20x^2 - 5$
$5(2x + 1)(2x - 1)$

90. $y^4 - 10y^3 + 21y^2$
$y^2(y - 3)(y - 7)$

91. $y^5 + 6y^4 - 55y^3$
$y^3(y + 11)(y - 5)$

92. $x^4 - 16$
$(x^2 + 4)(x + 2)(x - 2)$

93. $16x^4 - 81$
$(4x^2 + 9)(2x + 3)(2x - 3)$

94. $8x^5 - 98x^3$
$2x^3(2x + 7)(2x - 7)$

95. $16a - 2a^4$
$2a(2 - a)(4 + 2a + a^2)$

96. $x^3y^3 - x^3$
$x^3(y - 1)(y^2 + y + 1)$

97. $a^3b^6 - b^3$
$b^3(ab - 1)(a^2b^2 + ab + 1)$

98. $x^6y^6 - x^3y^3$
$x^3y^3(xy - 1)(x^2y^2 + xy + 1)$

99. $8x^4 - 40x^3 + 50x^2$
$2x^2(2x - 5)^2$

100. $6x^5 + 74x^4 + 24x^3$
$2x^3(3x + 1)(x + 12)$

101. $x^4 - y^4$
$(x^2 + y^2)(x + y)(x - y)$

102. $16a^4 - b^4$
$(4a^2 + b^2)(2a + b)(2a - b)$

103. $x^6 + y^6$
$(x^2 + y^2)(x^4 - x^2y^2 + y^4)$

104. $x^4 - 5x^2 - 4$
Nonfactorable

105. $a^4 - 25a^2 - 144$
Nonfactorable

106. $3b^5 - 24b^2$
$3b^2(b - 2)(b^2 + 2b + 4)$

107. $16a^4 - 2a$
$2a(2a - 1)(4a^2 + 2a + 1)$

108. $x^4y^2 - 5x^3y^3 + 6x^2y^4$
$x^2y^2(x - 3y)(x - 2y)$

109. $a^4b^2 - 8a^3b^3 - 48a^2b^4$
$a^2b^2(a + 4b)(a - 12b)$

110. $x^3 - 2x^2 - x + 2$
$(x + 1)(x - 1)(x - 2)$

111. $x^3 - 2x^2 - 4x + 8$
$(x - 2)^2(x + 2)$

112. $4x^3 + 8x^2 - 9x - 18$
$(2x - 3)(2x + 3)(x + 2)$

113. $2x^3 + x^2 - 32x - 16$
$(x - 4)(x + 4)(2x + 1)$

114. $4x^2y^2 - 4x^2 - 9y^2 + 9$
$(2x + 3)(2x - 3)(y + 1)(y - 1)$

115. $4x^4 - x^2 - 4x^2y^2 + y^2$
$(2x + 1)(2x - 1)(x + y)(x - y)$

116. 🖊 What is the degree of the polynomial whose factored form is
$3x^3(x^2 + 4)(x - 2)(x + 3)$?   7

117. 🖊 What is the coefficient of x^6 when $4x^2(x^2 + 3)(x + 4)(2x - 5)$ is expanded and written as a polynomial?   8

Critical Thinking

For Exercises 118 and 119, find all integers k such that the trinomial is a perfect-square trinomial.

118. $4x^2 - kx + 25$ 20, −20

119. $9x^2 - kx + 1$ 6, −6

For Exercises 120 and 121, factor.

120. $(a - b)^3 - b^3$ $(a - 2b)(a^2 - ab + b^2)$

121. $a^3 + (a + b)^3$ $(2a + b)(a^2 + ab + b^2)$

122. Factor $x^4 + 64$. [*Suggestion:* Add and subtract $16x^2$ so that the expression becomes $(x^4 + 16x^2 + 64) - 16x^2$. Now factor the difference of two squares.]
$(x^2 - 4x + 8)(x^2 + 4x + 8)$

Projects or Group Activities

If you have not completed Exercises 91 and 92 from Section 6.4, you should complete them now.

123. Given that 3 is a zero of $P(x) = x^3 - x^2 - 3x - 9$, determine the factorization over the integers of $x^3 - x^2 - 3x - 9$. $(x - 3)(x^2 + 2x + 3)$

124. Given that −3 and 2 are zeros of $P(x) = x^4 + 2x^3 - 4x^2 - 5x - 6$, determine the factorization over the integers of $x^4 + 2x^3 - 4x^2 - 5x - 6$.
$(x + 3)(x - 2)(x^2 + x + 1)$

QUICK QUIZ
Factor.
1. $16x^2 - 49$
   (4x + 7)(4x − 7) [7.4A]
2. $a^2 + 25$
   Nonfactorable [7.4A]
3. $9x^2 - y^2$
   (3x + y)(3x − y) [7.4A]
4. $b^2 - 16b + 64$
   (b − 8)² [7.4A]
5. $125x^3 + 1$
   (5x + 1)(25x² − 5x + 1) [7.4B]
6. $x^2y^2 - 7xy + 10$
   (xy − 5)(xy − 2) [7.4C]
7. $x^4 - 14x^2 + 33$
   (x² − 11)(x² − 3) [7.4C]
8. $x^5 + 6x^4 - 27x^3$
   x³(x + 9)(x − 3) [7.4D]
9. $x^2y^3 - 8x^2$
   x²(y − 2)(y² + 2y + 4) [7.4D]
10. $5x^5 - 5x$
   5x(x² + 1)(x + 1)(x − 1) [7.4D]
11. $x^3 + 3x^2 - x - 3$
   (x + 1)(x − 1)(x + 3) [7.4D]

7.5 | Solving Equations

OBJECTIVE A | *To solve equations by factoring*

The Multiplication Property of Zero states that the product of a number and zero is zero. This property is stated below.

If a is a real number, then $a \cdot 0 = 0 \cdot a = 0$.

Now consider $a \cdot b = 0$. For this to be a true equation, either $a = 0$ or $b = 0$.

Principle of Zero Products

If the product of two factors is zero, then at least one of the factors must be zero.

If $a \cdot b = 0$, then $a = 0$ or $b = 0$.

The Principle of Zero Products is used to solve some equations.

HOW TO 1 Solve: $(x - 2)(x - 3) = 0$

$(x - 2)(x - 3) = 0$

If $(x - 2)(x - 3) = 0$, then $(x - 2) = 0$ or $(x - 3) = 0$.

$x - 2 = 0 \qquad x - 3 = 0$ • Let each factor equal zero (the Principle of Zero Products).

$x = 2 \qquad\qquad x = 3$ • Solve each equation for x.

Check:

$$
\begin{array}{c|c}
(x - 2)(x - 3) = 0 \\
\hline
(2 - 2)(2 - 3) & 0 \\
0(-1) & 0 \\
0 = 0 & \text{• A true equation}
\end{array}
\qquad
\begin{array}{c|c}
(x - 2)(x - 3) = 0 \\
\hline
(3 - 2)(3 - 3) & 0 \\
(1)(0) & 0 \\
0 = 0 & \text{• A true equation}
\end{array}
$$

The solutions are 2 and 3.

> **Take Note**
>
> $x - 2$ is equal to a number. $x - 3$ is equal to a number. In $(x - 2)(x - 3)$, two numbers are being multiplied. Since their product is zero, one of the numbers must be equal to zero. The number $x - 2$ is equal to 0 or the number $x - 3$ is equal to 0.

> **INSTRUCTOR NOTE**
>
> Quadratic equations are introduced here as an application of factoring. They can be omitted at this time. There is a complete discussion of the topic, including this material, later in the text.

An equation that can be written in the form $ax^2 + bx + c = 0$, $a \neq 0$, is a **quadratic equation.** A quadratic equation is in **standard form** when the polynomial is written in descending order and equal to zero. The quadratic equations at the right are in standard form.

$3x^2 + 2x + 1 = 0$
$a = 3, b = 2, c = 1$

$4x^2 - 3x + 2 = 0$
$a = 4, b = -3, c = 2$

A quadratic equation can be solved by using the Principle of Zero Products if the polynomial $ax^2 + bx + c$ is factorable.

HOW TO 2   Solve: $2x^2 + x = 6$

$$2x^2 + x = 6$$

$2x^2 + x - 6 = 0$ • Write the equation in standard form.

$(2x - 3)(x + 2) = 0$ • Factor.

$2x - 3 = 0 \qquad x + 2 = 0$ • Use the Principle of Zero Products.

$2x = 3 \qquad\qquad x = -2$ • Solve each equation for x.

$$x = \frac{3}{2}$$

Check: $\frac{3}{2}$ and -2 check as solutions.

The solutions are $\frac{3}{2}$ and -2.

HOW TO 2 illustrates the steps involved in solving a quadratic equation by factoring.

Steps in Solving a Quadratic Equation by Factoring

1. Write the equation in standard form.
2. Factor the polynominal.
3. Set each factor equal to zero.
4. Solve each equation for the variable.
5. Check the solutions.

EXAMPLE 1

Solve: $x(x - 3) = 0$

Solution

$x(x - 3) = 0$

$x = 0 \qquad x - 3 = 0$ • Use the Principle

$\qquad\qquad\quad x = 3$ of Zero Products.

The solutions are 0 and 3.

YOU TRY IT 1

Solve: $2x(x + 7) = 0$

Your solution

0, −7

EXAMPLE 2

Solve: $2x^2 - 50 = 0$

Solution

$$2x^2 - 50 = 0$$

$2(x^2 - 25) = 0$ • Factor out the GCF, 2.

$2(x + 5)(x - 5) = 0$ • Factor the difference
                       of two squares.

$x + 5 = 0 \qquad x - 5 = 0$ • Use the Principle

$x = -5 \qquad\quad x = 5$ of Zero Products.

The solutions are -5 and 5.

YOU TRY IT 2

Solve: $4x^2 - 9 = 0$

Your solution

$\dfrac{3}{2}, -\dfrac{3}{2}$

Solutions on p. S21

EXAMPLE 3

Solve: $(x - 3)(x - 10) = -10$

Solution

$(x - 3)(x - 10) = -10$

$x^2 - 13x + 30 = -10$ • Multiply $(x - 3)(x - 10)$.

$x^2 - 13x + 40 = 0$ • Add 10 to each side of the
equation. The equation is
now in standard form.

$(x - 8)(x - 5) = 0$

$x - 8 = 0 \qquad x - 5 = 0$

$\qquad x = 8 \qquad\qquad x = 5$

The solutions are 8 and 5.

YOU TRY IT 3

Solve: $(x + 2)(x - 7) = 52$

Your solution

$-6, 11$

IN-CLASS EXAMPLES

Solve.

1. $3x(x + 4) = 0$ **0, −4**

2. $16x^2 - 4 = 0$ $-\dfrac{1}{2}, \dfrac{1}{2}$

3. $(x + 3)(x - 5) = 9$ **6, −4**

Solution on p. S21

OBJECTIVE B   *To solve application problems*

EXAMPLE 4

The sum of the squares of two consecutive
positive even integers is equal to 100. Find the
two integers.

Strategy

First positive even integer: n
Second positive even integer: $n + 2$

The sum of the square of the first positive even
integer and the square of the second positive even
integer is 100.

Solution

$n^2 + (n + 2)^2 = 100$

$n^2 + n^2 + 4n + 4 = 100$

$2n^2 + 4n + 4 = 100$

$2n^2 + 4n - 96 = 0$ • Write the quadratic
equation in standard
form.

$2(n^2 + 2n - 48) = 0$

$2(n - 6)(n + 8) = 0$ • Factor.

$n - 6 = 0 \qquad n + 8 = 0$ • Principle of
Zero Products

$\quad n = 6 \qquad\qquad n = -8$

Because -8 is not a positive even integer, it is not
a solution.

$n = 6$

$n + 2 = 6 + 2 = 8$

The two integers are 6 and 8.

YOU TRY IT 4

The sum of the squares of two consecutive positive
integers is 61. Find the two integers.

Your strategy

Your solution

5, 6

IN-CLASS EXAMPLES

4. The sum of the squares of two
consecutive positive integers is
145. Find the two integers.
8, 9

5. The sum of the squares of
two consecutive negative odd
integers is 10. Find the two
integers.   **−3, −1**

6. A garden measures 12 ft by
16 ft. A uniform border around
the garden increases the total
area to 357 ft². What is the
width of the border?   **2.5 ft**

Solution on pp. S21–S22

EXAMPLE 5

A stone is thrown into a well with an initial speed of 4 ft/s. The well is 420 ft deep. How many seconds later will the stone hit the bottom of the well? Use the equation $d = vt + 16t^2$, where d is the distance in feet that the stone travels in t seconds when its initial speed is v feet per second.

Strategy

To find the time for the stone to drop to the bottom of the well, replace the variables d and v by their given values, and solve for t.

Solution

$$d = vt + 16t^2$$
$$420 = 4t + 16t^2$$
$$0 = -420 + 4t + 16t^2$$
$$0 = 16t^2 + 4t - 420 \quad \bullet \text{ Write the quadratic}$$
$$0 = 4(4t^2 + t - 105) \qquad \text{equation in standard form.}$$
$$0 = 4(4t + 21)(t - 5) \quad \bullet \text{ Factor.}$$

$$4t + 21 = 0 \qquad t - 5 = 0 \quad \bullet \text{ Principle of Zero}$$
$$4t = -21 \qquad t = 5 \qquad \text{Products}$$
$$t = -\frac{21}{4}$$

Because the time cannot be a negative number, $-\frac{21}{4}$ is not a solution.

The stone will hit the bottom of the well 5 s later.

YOU TRY IT 5

The length of a rectangle is 4 in. longer than twice the width. The area of the rectangle is 96 in². Find the length and width of the rectangle.

Your strategy

Your solution

Length: 16 in.; width: 6 in.

Solution on p. S22

7.5 EXERCISES

✔ Concept Check

SUGGESTED ASSIGNMENT
Exercises 1–4; Exercises 5–61, odds; Exercises 65–93, odds; More challenging exercises: Exercises 95–103, odds

1. Determine whether the equation is a quadratic equation.
   a. $2x^2 - 8 = 0$ Yes
   b. $2x - 8 = 0$ No
   c. $x^2 = 8x$ Yes

2. Write the equation in standard form.
   a. $x^2 + 4 = 4x$
      $x^2 - 4x + 4 = 0$
   b. $x + x^2 = 6$
      $x^2 + x - 6 = 0$

3. Can the equation be solved by using the Principle of Zero Products without first rewriting the equation?
   a. $4x(6x + 7) = 0$ Yes
   b. $0 = (4x - 5)(3x + 8)$ Yes
   c. $2x(x - 5) - 5 = 0$ No
   d. $(x - 7)(y + 3) = 0$ Yes
   e. $0 = (2x - 3)x + 3$ No
   f. $0 = (2x - 3)(x + 3)$ Yes

4. Fill in the blanks. If $(x + 5)(2x - 7) = 0$, then ___$x + 5$___ $= 0$ or
   ___$2x - 7$___ $= 0$.

OBJECTIVE A *To solve equations by factoring*

For Exercises 5 to 62, solve.

5. $(y + 3)(y + 2) = 0$
   $-3, -2$

6. $(y - 3)(y - 5) = 0$
   $3, 5$

7. $(z - 7)(z - 3) = 0$
   $7, 3$

8. $(z + 8)(z - 9) = 0$
   $-8, 9$

9. $x(x - 5) = 0$
   $0, 5$

10. $x(x + 2) = 0$
    $0, -2$

11. $a(a - 9) = 0$
    $0, 9$

12. $a(a + 12) = 0$
    $0, -12$

13. $y(2y + 3) = 0$
    $0, -\dfrac{3}{2}$

14. $t(4t - 7) = 0$
    $0, \dfrac{7}{4}$

15. $2a(3a - 2) = 0$
    $0, \dfrac{2}{3}$

16. $4b(2b + 5) = 0$
    $0, -\dfrac{5}{2}$

17. $(b + 2)(b - 5) = 0$
    $-2, 5$

18. $(b - 8)(b + 3) = 0$
    $8, -3$

19. $x^2 - 81 = 0$
    $9, -9$

20. $x^2 - 121 = 0$
    $11, -11$

21. $4x^2 - 49 = 0$
    $\dfrac{7}{2}, -\dfrac{7}{2}$

22. $16x^2 - 1 = 0$
    $\dfrac{1}{4}, -\dfrac{1}{4}$

23. $9x^2 - 1 = 0$
    $\dfrac{1}{3}, -\dfrac{1}{3}$

24. $16x^2 - 49 = 0$
    $\dfrac{7}{4}, -\dfrac{7}{4}$

25. $x^2 + 6x + 8 = 0$
    $-4, -2$

26. $x^2 - 8x + 15 = 0$
    $3, 5$

27. $z^2 + 5z - 14 = 0$
    $2, -7$

28. $z^2 + z - 72 = 0$
    $8, -9$

29. $2a^2 - 9a - 5 = 0$
$-\dfrac{1}{2}, 5$

30. $3a^2 + 14a + 8 = 0$
$-\dfrac{2}{3}, -4$

31. $6z^2 + 5z + 1 = 0$
$-\dfrac{1}{3}, -\dfrac{1}{2}$

32. $6y^2 - 19y + 15 = 0$
$\dfrac{5}{3}, \dfrac{3}{2}$

33. $x^2 - 3x = 0$
0, 3

34. $a^2 - 5a = 0$
0, 5

35. $x^2 - 7x = 0$
0, 7

36. $2a^2 - 8a = 0$
0, 4

37. $a^2 + 5a = -4$
$-1, -4$

38. $a^2 - 5a = 24$
$-3, 8$

39. $y^2 - 5y = -6$
2, 3

40. $y^2 - 7y = 8$
$-1, 8$

41. $2t^2 + 7t = 4$
$\dfrac{1}{2}, -4$

42. $3t^2 + t = 10$
$\dfrac{5}{3}, -2$

43. $3t^2 - 13t = -4$
$\dfrac{1}{3}, 4$

44. $5t^2 - 16t = -12$
$\dfrac{6}{5}, 2$

45. $x(x - 12) = -27$
3, 9

46. $x(x - 11) = 12$
12, -1

47. $y(y - 7) = 18$
9, -2

48. $y(y + 8) = -15$
$-3, -5$

49. $p(p + 3) = -2$
$-1, -2$

50. $p(p - 1) = 20$
5, -4

51. $y(y + 4) = 45$
5, -9

52. $y(y - 8) = -15$
3, 5

53. $x(x + 3) = 28$
4, -7

54. $p(p - 14) = 15$
15, -1

55. $(x + 8)(x - 3) = -30$
$-2, -3$

56. $(x + 4)(x - 1) = 14$
$-6, 3$

57. $(z - 5)(z + 4) = 52$
$-8, 9$

58. $(z - 8)(z + 4) = -35$
1, 3

59. $(z - 6)(z + 1) = -10$
1, 4

60. $(a + 3)(a + 4) = 72$
$-12, 5$

61. $(a - 4)(a + 7) = -18$
$-5, 2$

62. $(2x + 5)(x + 1) = -1$
$-\dfrac{3}{2}, -2$

For Exercises 63 and 64, the equation $ax^2 + bx + c = 0$, $a > 0$, is a quadratic equation that can be solved by factoring and then using the Principle of Zero Products.

63. If $ax^2 + bx + c = 0$ has one positive solution and one negative solution, is c greater than, less than, or equal to zero?   Less than

64. If zero is one solution of $ax^2 + bx + c = 0$, is c greater than, less than, or equal to zero?   Equal to

OBJECTIVE B *To solve application problems*

65. Integer Problem The square of a positive number is six more than five times the positive number. Find the number.   6

66. Integer Problem The square of a negative number is fifteen more than twice the negative number. Find the number.   −3

67. Integer Problem The sum of two numbers is six. The sum of the squares of the two numbers is twenty. Find the two numbers.   2, 4

68. Integer Problem The sum of two numbers is eight. The sum of the squares of the two numbers is thirty-four. Find the two numbers.   3, 5

For Exercises 69 and 70, use the following problem situation: The sum of the squares of two consecutive positive integers is 113. Find the two integers.

69. Which equation could be used to solve this problem?
(i) $x^2 + x^2 + 1 = 113$ **(ii)** $x^2 + (x + 1)^2 = 113$ **(iii)** $(x + x + 1)^2 = 113$
ii

70. Suppose the solutions of the correct equation in Exercise 69 are −8 and 7. Which solution should be eliminated, and why?
−8 should be eliminated because the problem specifies consecutive *positive* integers.

71. Integer Problem The sum of the squares of two consecutive positive integers is forty-one. Find the two integers.   4, 5

72. Integer Problem The sum of the squares of two consecutive positive even integers is one hundred. Find the two integers.   6, 8

73. Integer Problem The product of two consecutive positive integers is two hundred forty. Find the two integers.   15, 16

74. Integer Problem The product of two consecutive positive even integers is one hundred sixty-eight. Find the two integers.   12, 14

75. Geometry The length of the base of a triangle is three times the height. The area of the triangle is 54 ft². Find the base and height of the triangle.
Base: 18 ft; height: 6 ft

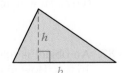

76. Geometry The height of a triangle is 4 m more than twice the length of the base. The area of the triangle is 35 m². Find the height of the triangle.   14 m

77. Geometry The length of a rectangle is 2 ft more than twice the width. The area is 144 ft². Find the length and width of the rectangle.   Length: 18 ft; width: 8 ft

78. Geometry The width of a rectangle is 5 ft less than the length. The area of the rectangle is 176 ft². Find the length and width of the rectangle.   Length: 16 ft; width: 11 ft

Unless otherwise noted, all content on this page is © Cengage Learning.

79. Geometry   The length of each side of a square is extended 4 m. The area of the resulting square is 64 m^2. Find the length of a side of the original square.   4 m

80. Geometry   The length of each side of a square is extended 2 cm. The area of the resulting square is 64 cm^2. Find the length of a side of the original square.   6 cm

81. Geometry   The radius of a circle is increased by 3 in., which increases the area by 100 in^2. Find the radius of the original circle. Round to the nearest hundredth.   3.81 in.

82. Geometry   The length of a rectangle is 5 cm, and the width is 3 cm. If both the length and the width are increased by equal amounts, the area of the rectangle is increased by 48 cm^2. Find the length and width of the larger rectangle.   7 cm by 9 cm

83. Geometry   The page of a book measures 6 in. by 9 in. A uniform border around the page leaves 28 in^2 for type. What are the dimensions of the type area?   4 in. by 7 in.

In the NEWS!

New Lane for Basketball Court

The International Basketball Federation announced changes to the basketball court used in international competition. The 3-second lane, currently a trapezoid, will be a rectangle 3 ft longer than it is wide, similar to the one used in NBA games.
Source: The New York Times

84. Geometry   A small garden measures 8 ft by 10 ft. A uniform border around the garden increases the total area to 143 ft^2. What is the width of the border?   1.5 ft

85. ◗ **Basketball**   See the news clipping at the right. If the area of the rectangular 3-second lane is 304 ft^2, find the width of the lane.   16 ft

Physics   For Exercises 86 and 87, use the formula $d = vt + 16t^2$, where d is the distance in feet, v is the initial velocity in feet per second, and t is the time in seconds.

86. An object is released from a plane at an altitude of 1600 ft. The initial velocity is 0 ft/s. How many seconds later will the object hit the ground?   10 s

87. An object is released from the top of a building 320 ft high. The initial velocity is 16 ft/s. How many seconds later will the object hit the ground?   4 s

Number Problems   For Exercises 88 and 89, use the formula $S = \dfrac{n^2 + n}{2}$, where S is the sum of the first n natural numbers.

88. How many consecutive natural numbers beginning with 1 will give a sum of 78?   12

89. How many consecutive natural numbers beginning with 1 will give a sum of 120?   15

Unless otherwise noted, all content on this page is © Cengage Learning.

SANDOR SZABO/EPA/Landov

Sports For Exercises 90 and 91, use the formula $N = \frac{t^2 - t}{2}$, where N is the number of basketball games that must be scheduled in a league with t teams if each team is to play every other team once.

90. A league has 28 games scheduled. How many teams are in the league if each team plays every other team once?   8 teams

91. A league has 45 games scheduled. How many teams are in the league if each team plays every other team once?   10 teams

Sports For Exercises 92 and 93, use the formula $h = vt - 16t^2$, where h is the height in feet that an object will attain (neglecting air resistance) in t seconds, and v is the initial velocity in feet per second.

92. A baseball player hits a "Baltimore chop," meaning the ball bounces off home plate after the batter hits it. The ball leaves home plate with an initial upward velocity of 32 ft/s. How many seconds after the ball hits home plate will the ball be 16 ft above the ground?   1 s

93. A golf ball is thrown onto a cement surface and rebounds straight up. The initial velocity of the rebound is 48 ft/s. How many seconds later will the golf ball return to the ground?   3 s

QUICK QUIZ
Solve.
1. $x^2 - 6x = 27$ **−3, 9** **[7.5A]**
2. $(x + 2)(x - 9) = -24$ **1, 6** **[7.5A]**

Critical Thinking

For Exercises 94 to 101, solve.

94. $2y(y + 4) = -5(y + 3)$ $-\dfrac{3}{2}, -5$

95. $2y(y + 4) = 3(y + 4)$ $\dfrac{3}{2}, -4$

96. $(a - 3)^2 = 36$ 9, −3

97. $(b + 5)^2 = 16$ −1, −9

98. $p^3 = 9p^2$ 0, 9

99. $p^3 = 7p^2$ 0, 7

100. $(2z - 3)(z + 5) = (z + 1)(z + 3)$ −6, 3

101. $(x + 3)(2x - 1) = (3 - x)(5 - 3x)$ 18, 1

102. Find $3n^2$ if $n(n + 5) = -4$.   48 or 3

103. Find $2n^3$ if $n(n + 3) = 4$.   2 or −128

104. ◤ Explain the error made in solving the equation at the right. Solve the equation correctly.

$$(x + 2)(x - 3) = 6$$
$$x + 2 = 6 \qquad x - 3 = 6$$
$$x = 4 \qquad x = 9$$

105. ◤ Explain the error made in solving the equation at the right. Solve the equation correctly.

$$x^2 = x$$
$$\frac{x^2}{x} = \frac{x}{x}$$
$$x = 1$$

3. The product of two consecutive even positive integers is 120. Find the integers.   **10, 12** **[7.5B]**

4. The length of a rectangle is four times the width. The area is 144 in². Find the length and width of the rectangle.   **Length: 24 in.; width: 6 in.** **[7.5B]**

Projects or Group Activities

106. **Geometry** The length of a rectangle is 7 cm, and the width is 4 cm. If both the length and the width are increased by equal amounts, the area of the rectangle is increased by 42 cm². Find the length and width of the larger rectangle.
Length: 10 cm; width: 7 cm

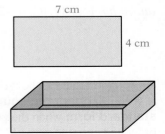

7 cm

4 cm

107. **Geometry** A rectangular piece of cardboard is 10 in. longer than it is wide. Squares 2 in. on a side are to be cut from each corner, and then the sides will be folded up to make an open box with a volume of 192 in³. Find the length and width of the piece of cardboard.   Length: 20 in.; width: 10 in.

108. Write an equation that has solutions 1, −2, and 3.
Answers will vary. For example, $(x - 1)(x + 2)(x - 3) = 0$

Unless otherwise noted, all content on this page is © Cengage Learning.

7 Summary

Key Words	Examples
To **factor a polynomial** means to write the polynomial as a product of other polynomials. [7.1A, p. 398]	To factor $x^2 + 3x + 2$ means to write it as the product $(x + 1)(x + 2)$.
A polynomial that does not factor using only integers is **nonfactorable over the integers.** [7.2A, p. 405]	The trinomial $x^2 + x + 4$ is nonfactorable over the integers. There are no integers whose product is 4 and whose sum is 1.
The **greatest common factor (GCF) of two or more monomials** is the product of the GCF of the coefficients and the common variable factors. [7.1A, p. 398]	The GCF of $8x^2y$ and $12xyz$ is $4xy$.
A factor that has two terms is called a **binomial factor.** [7.1B, p. 400]	$(x + 1)$ is a binomial factor of $3x(x + 1)$.
A polynomial is **factored completely** if it is written as a product of factors that are nonfactorable over the integers. [7.2B, p. 406]	The polynomial $3y^3 + 9y^2 - 12y$ is factored completely as $3y(y + 4)(y - 1)$.
The product of a term and itself is a **perfect square.** The **square root** of a perfect square is one of its two equal factors. [7.4A, p. 421]	$(5x)(5x) = 25x^2$; $25x^2$ is a perfect square. $\sqrt{25x^2} = 5x$
The product of the same three factors is a **perfect cube.** The **cube root** of a perfect cube is one of its three equal factors. [7.4B, p. 423]	$(2x)(2x)(2x) = 8x^3$; $8x^3$ is a perfect cube. $\sqrt[3]{8x^3} = 2x$
A trinomial is **quadratic in form** if it can be written as $au^2 + bu + c$. [7.4C, p. 425]	$6x^4 - 5x^2 - 4 = 6(x^2)^2 - 5(x^2) - 4$ $= 6u^2 - 5u - 4$
An equation that can be written in the form $ax^2 + bx + c = 0$, $a \neq 0$, is a **quadratic equation.** A quadratic equation is in **standard form** when the polynomial is written in descending order and equal to zero. [7.5A, p. 431]	The equation $2x^2 - 3x + 7 = 0$ is a quadratic equation in standard form.

Essential Rules and Procedures

Examples

Factoring by Grouping [7.1B, p. 400]
A polynomial can be factored by grouping if its terms can be grouped and factored in such a way that a common binomial factor is found.

$3a^2 - a - 15ab + 5b$
$= (3a^2 - a) - (15ab - 5b)$
$= a(3a - 1) - 5b(3a - 1)$
$= (3a - 1)(a - 5b)$

Factoring $x^2 + bx + c$: IMPORTANT RELATIONSHIPS [7.2A, p. 404]

1. When the constant term of the trinomial is positive, the constant terms of the binomials have the same sign. They are both positive when the coefficient of the x term in the trinomial is positive. They are both negative when the coefficient of the x term in the trinomial is negative.

$x^2 + 6x + 8 = (x + 4)(x + 2)$

$x^2 - 6x + 5 = (x - 5)(x - 1)$

2. When the constant term of the trinomial is negative, the constant terms of the binomials have opposite signs.

$x^2 - 4x - 21 = (x + 3)(x - 7)$

3. In the trinomial, the coefficient of x is the sum of the constant terms of the binomials.

In the three examples above, note that $6 = 4 + 2$, $-6 = -5 + (-1)$, and $-4 = 3 + (-7)$.

4. In the trinomial, the constant term is the product of the constant terms of the binomials.

In the three examples above, note that $8 = 4 \cdot 2$, $5 = -5(-1)$, and $-21 = 3(-7)$.

To Factor $ax^2 + bx + c$ by Grouping [7.3B, p. 414]
First find two factors of $a \cdot c$ whose sum is b. Then use factoring by grouping to write the factorization of the trinomial.

$3x^2 - 11x - 20$
$a \cdot c = 3(-20) = -60$
The product of 4 and -15 is -60.
The sum of 4 and -15 is -11.
$3x^2 + 4x - 15x - 20$
$= (3x^2 + 4x) - (15x + 20)$
$= x(3x + 4) - 5(3x + 4)$
$= (3x + 4)(x - 5)$

Factoring the Difference of Two Perfect Squares [7.4A, p. 421]
The difference of two perfect squares factors as the sum and difference of the square roots of the perfect squares.

$a^2 - b^2 = (a + b)(a - b)$

$x^2 - 64 = (x + 8)(x - 8)$
$4x^2 - 81 = (2x)^2 - 9^2$
$= (2x + 9)(2x - 9)$

Factoring a Perfect-Square Trinomial [7.4A, p. 421]
A perfect-square trinomial is the square of a binomial.

$a^2 + 2ab + b^2 = (a + b)^2$

$a^2 - 2ab + b^2 = (a - b)^2$

$x^2 + 14x + 49 = (x + 7)^2$

$x^2 - 10x + 25 = (x - 5)^2$

Factoring the Sum or Difference of Two Perfect Cubes
[7.4B, p. 423]

$a^3 + b^3 = (a + b)(a^2 - ab + b^2)$

$a^3 - b^3 = (a - b)(a^2 + ab + b^2)$

$x^3 + 64 = (x + 4)(x^2 - 4x + 16)$

$8b^3 - 1 = (2b - 1)(4b^2 + 2b + 1)$

To Factor Completely [7.4D, p. 426]
When factoring a polynomial completely, ask the following
questions about the polynomial.

1. Is there a common factor? If so, factor out the GCF.

2. If the polynomial is a binomial, is it the difference of two
   perfect squares, the sum of two perfect cubes, or the difference
   of two perfect cubes? If so, factor.

3. If the polynomial is a trinomial, is it a perfect-square trinomial
   or the product of two binomials? If so, factor.

4. Can the polynomial be factored by grouping? If so, factor.

5. Is each factor nonfactorable over the integers? If not, factor.

$$54x^3 - 6x = 6x(9x^2 - 1)$$
$$= 6x(3x + 1)(3x - 1)$$

Principle of Zero Products [7.5A, p. 431]
If the product of two factors is zero, then at least one of the factors
must be zero.

If $a \cdot b = 0$, then $a = 0$ or $b = 0$.

The Principle of Zero Products is used to solve a quadratic
equation by factoring.

$$x^2 + x = 12$$
$$x^2 + x - 12 = 0$$
$$(x - 3)(x + 4) = 0$$

$x - 3 = 0 \qquad x + 4 = 0$

$\qquad x = 3 \qquad\qquad x = -4$

7 | Review Exercises

1. Factor: $b^2 - 13b + 30$
$(b - 3)(b - 10)$ [7.2A]

2. Factor: $4x(x - 3) - 5(3 - x)$
$(x - 3)(4x + 5)$ [7.1B]

3. Factor $2x^2 - 5x + 6$ by using trial factors.
Nonfactorable over the integers [7.3A]

4. Factor: $21x^4y^4 + 23x^2y^2 + 6$
$(7x^2y^2 + 3)(3x^2y^2 + 2)$ [7.4C]

5. Factor: $14y^9 - 49y^6 + 7y^3$
$7y^3(2y^6 - 7y^3 + 1)$ [7.1A]

6. Factor: $y^2 + 5y - 36$
$(y - 4)(y + 9)$ [7.2A]

7. Factor $6x^2 - 29x + 28$ by using trial factors.
$(2x - 7)(3x - 4)$ [7.3A]

8. Factor: $12a^2b + 3ab^2$
$3ab(4a + b)$ [7.1A]

9. Factor: $a^6 - 100$
$(a^3 + 10)(a^3 - 10)$ [7.4A]

10. Factor: $n^4 - 2n^3 - 3n^2$
$n^2(n + 1)(n - 3)$ [7.2B]

11. Factor $12y^2 + 16y - 3$ by using trial factors.
$(6y - 1)(2y + 3)$ [7.3A]

12. Factor: $12b^3 - 58b^2 + 56b$
$2b(3b - 4)(2b - 7)$ [7.4D]

13. Factor: $9y^4 - 25z^2$
$(3y^2 + 5z)(3y^2 - 5z)$ [7.4A]

14. Factor: $c^2 + 8c + 12$
$(c + 6)(c + 2)$ [7.2A]

15. Factor $18a^2 - 3a - 10$ by grouping.
$(6a - 5)(3a + 2)$ [7.3B]

16. Solve: $4x^2 + 27x = 7$
$\dfrac{1}{4}, -7$ [7.5A]

17. Factor: $4x^3 - 20x^2 - 24x$
$4x(x - 6)(x + 1)$ [7.2B]

18. Factor: $64a^3 - 27b^3$
$(4a - 3b)(16a^2 + 12ab + 9b^2)$ [7.4B]

19. Factor $2a^2 - 19a - 60$ by grouping.
$(2a + 5)(a - 12)$ [7.3B]

20. Solve: $(x + 1)(x - 5) = 16$
$-3, 7$ [7.5A]

21. Factor: $21ax - 35bx - 10by + 6ay$
$(3a - 5b)(7x + 2y)$ [7.1B]

22. Factor: $36x^8 - 36x^4 + 5$
$(6x^4 - 1)(6x^4 - 5)$ [7.4C]

23. Factor: $10x^2 + 25x + 4xy + 10y$
$(2x + 5)(5x + 2y)$ [7.1B]

24. Factor: $5x^2 - 5x - 30$
$5(x + 2)(x - 3)$ [7.2B]

25. Factor: $3x^2 + 36x + 108$
$3(x + 6)^2$ [7.4D]

26. Factor $3x^2 - 17x + 10$ by grouping.
$(3x - 2)(x - 5)$ [7.3B]

27. Sports   The length of the field in field hockey is 20 yd less than twice the width of the field. The area of the field in field hockey is 6000 yd². Find the length and width of the field.   Length: 100 yd; width: 60 yd   [7.5B]

28. Image Projection   The size S of an image from a projector depends on the distance d of the screen from the projector and is given by $S = d^2$. Find the distance between the projector and the screen when the size of the picture is 400 ft².
20 ft   [7.5B]

© iStockphoto.com/Andrey Artykov

29. Photography   A rectangular photograph has dimensions 15 in. by 12 in. A picture frame around the photograph increases the total area to 270 in². What is the width of the frame?

1.5 in. or $1\frac{1}{2}$ in.   [7.5B]

30. Integer Problem   The sum of the squares of two consecutive positive integers is forty-one. Find the two integers.   4 and 5   [7.5B]

7 | TEST

1. Factor: $ab + 6a - 3b - 18$
$(b + 6)(a - 3)$ [7.1B]

2. Factor: $2y^4 - 14y^3 - 16y^2$
$2y^2(y + 1)(y - 8)$ [7.2B]

3. Factor $8x^2 + 20x - 48$ by grouping.
$4(x + 4)(2x - 3)$ [7.3B]

4. Factor $6x^2 + 19x + 8$ by using trial factors.
$(2x + 1)(3x + 8)$ [7.3A]

5. Factor: $a^2 - 19a + 48$
$(a - 3)(a - 16)$ [7.2A]

6. Factor: $6x^3 - 8x^2 + 10x$
$2x(3x^2 - 4x + 5)$ [7.1A]

7. Factor: $x^2 + 2x - 15$
$(x + 5)(x - 3)$ [7.2A]

8. Solve: $4x^2 - 1 = 0$
$\dfrac{1}{2}, -\dfrac{1}{2}$ [7.5A]

9. Factor: $5x^2 - 45x - 15$
$5(x^2 - 9x - 3)$ [7.1A]

10. Factor: $p^2 + 12p + 36$
$(p + 6)^2$ [7.4A]

11. Solve: $x(x - 8) = -15$
$3, 5$ [7.5A]

12. Factor: $3x^2 + 12xy + 12y^2$
$3(x + 2y)^2$ [7.4D]

13. Factor: $b^2 - 16$
$(b + 4)(b - 4)$ [7.4A]

14. Factor $6x^2y^2 + 9xy^2 + 3y^2$ by grouping.
$3y^2(2x + 1)(x + 1)$ [7.3B]

15. Factor: $27x^3 - 8$
$(3x - 2)(9x^2 + 6x + 4)$ [7.4B]

16. Factor: $6a^4 - 13a^2 - 5$
$(2a^2 - 5)(3a^2 + 1)$ [7.4C]

17. Factor: $x(p + 1) - (p + 1)$
$(p + 1)(x - 1)$ [7.1B]

18. Factor: $3a^2 - 75$
$3(a + 5)(a - 5)$ [7.4D]

19. Factor: $2x^2 + 4x - 5$
Nonfactorable over the integers   [7.3B]

20. Factor: $x^2 - 9x - 36$
$(x + 3)(x - 12)$ [7.2A]

21. Factor: $4a^2 - 12ab + 9b^2$
$(2a - 3b)^2$ [7.4A]

22. Factor: $4x^2 - 49y^2$
$(2x + 7y)(2x - 7y)$ [7.4A]

23. Solve: $(2a - 3)(a + 7) = 0$
$\dfrac{3}{2}, -7$ [7.5A]

24. Number Problem   The sum of two numbers is ten. The sum of the squares of the two numbers is fifty-eight. Find the two numbers.
3, 7   [7.5B]

25. Geometry   The length of a rectangle is 3 cm longer than twice the width. The area of the rectangle is 90 cm². Find the length and width of the rectangle.
Length: 15 cm; width: 6 cm   [7.5B]

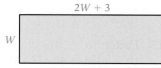

Unless otherwise noted, all content on this page is © Cengage Learning.

Cumulative Review Exercises

1. Subtract: $-2 - (-3) - 5 - (-11)$
7   [1.1C]

2. Simplify: $(3 - 7)^2 \div (-2) - 3 \cdot (-4)$
4   [1.3A]

3. Evaluate $-2a^2 \div (2b) - c$ when $a = -4$, $b = 2$, and $c = -1$.
-7 [1.4A]

4. Simplify: $-\dfrac{3}{4}(-20x^2)$
$15x^2$ [1.4C]

5. Simplify: $-2[4x - 2(3 - 2x) - 8x]$
12   [1.4D]

6. Solve: $-\dfrac{5}{7}x = -\dfrac{10}{21}$
$\dfrac{2}{3}$ [2.1C]

7. Solve: $3x - 2 = 12 - 5x$
$\dfrac{7}{4}$ [2.2B]

8. Solve: $-2 + 4[3x - 2(4 - x) - 3] = 4x + 2$
3   [2.2C]

9. 120% of what number is 54?
45   [2.1D]

10. Given $f(x) = -x^2 + 3x - 1$, find $f(2)$.
1   [4.2A]

11. Graph $y = \dfrac{1}{4}x + 3$.

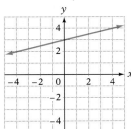

[4.3A]

12. Graph $5x + 3y = 15$.

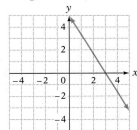

[4.3B]

13. Find the equation of the line that contains the point $(-3, 4)$ and has slope $\dfrac{2}{3}$.
$y = \dfrac{2}{3}x + 6$ [4.5A]

14. Solve by substitution: $8x - y = 2$
$y = 5x + 1$
$(1, 6)$ [5.1B]

15. Solve by the addition method:
$5x + 2y = -9$
$12x - 7y = 2$
$(-1, -2)$ [5.2A]

16. Simplify: $(-3a^3b^2)^2$
$9a^6b^4$ [6.1A]

Unless otherwise noted, all content on this page is © Cengage Learning.

17. Multiply: $(x + 2)(x^2 - 5x + 4)$
   $x^3 - 3x^2 - 6x + 8$ [6.3B]

18. Divide: $(8x^2 + 4x - 3) \div (2x - 3)$
   $4x + 8 + \dfrac{21}{2x - 3}$ [6.4B]

19. Simplify: $(x^{-4}y^3)^2$
   $\dfrac{y^6}{x^8}$ [6.1B]

20. Factor: $3a - 3b - ax + bx$
   $(a - b)(3 - x)$ [7.1B]

21. Factor: $15xy^2 - 20xy^4$
   $5xy^2(3 - 4y^2)$ [7.1A]

22. Factor: $x^2 - 5xy - 14y^2$
   $(x - 7y)(x + 2y)$ [7.2A]

23. Solve: $6x^2 + 60 = 39x$
   $\dfrac{5}{2}, 4$ [7.5A]

24. Factor: $18a^3 + 57a^2 + 30a$
   $3a(2a + 5)(3a + 2)$ [7.4D]

25. Factor: $36a^2 - 49b^2$
   $(6a - 7b)(6a + 7b)$ [7.4A]

26. Solve: $3x^2 + 19x - 14 = 0$
   $\dfrac{2}{3}, -7$ [7.5A]

27. Carpentry   A board 10 ft long is cut into two pieces. Four times the length of the shorter piece is 2 ft less than three times the length of the longer piece. Find the length of each piece.   4 ft, 6 ft   [2.3B]

28. Geometry   Given that lines ℓ_1 and ℓ_2 are parallel, find the measures of angles a and b.
$m\angle a = 72°; m\angle b = 108°$ [3.1B]

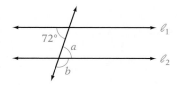

29. Travel   A family drove to a resort at an average speed of 42 mph and later returned over the same road at an average speed of 56 mph. Find the distance to the resort if the total driving time was 7 h.   168 mi   [2.4C]

30. Consecutive Integers   Find three consecutive even integers such that five times the middle integer is twelve more than twice the sum of the first and third integers.
10, 12, 14   [2.3A]

31. Geometry   The length of the base of a triangle is three times the height. The area of the triangle is 24 in². Find the length of the base of the triangle.   12 in.   [7.5B]

Unless otherwise noted, all content on this page is © Cengage Learning.

Rational Expressions

8

Focus on Success

Did you read Ask the Authors at the front of this text? If you did, then you know that the authors' advice is that you practice, practice, practice—and then practice some more. The more time you spend doing math outside of class, the more successful you will be in this course. (See Ask the Authors, page 0, and Make the Commitment to Succeed, page AIM-3.)

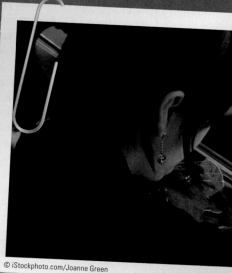

© iStockphoto.com/Joanne Green

Prep Test

Are you ready to succeed in this chapter? Take the Prep Test below to find out if you are ready to learn the new material.

1. Find the LCM of 10 and 25.
   50   [1.2C]

For Exercises 2 to 5, add, subtract, multiply, or divide.

2. $-\dfrac{3}{8} \cdot \dfrac{4}{9}$

   $-\dfrac{1}{6}$ [1.2D]

3. $-\dfrac{4}{5} \div \dfrac{8}{15}$

   $-\dfrac{3}{2}$ [1.2D]

4. $-\dfrac{5}{6} + \dfrac{7}{8}$

   $\dfrac{1}{24}$ [1.2C]

5. $-\dfrac{3}{8} - \left(-\dfrac{7}{12}\right)$

   $\dfrac{5}{24}$ [1.2C]

6. Evaluate $\dfrac{2x - 3}{x^2 - x + 1}$ for $x = 2$.

   $\dfrac{1}{3}$ [1.4A]

7. Solve: $4(2x + 1) = 3(x - 2)$

   -2 [2.2C]

8. Solve: $10\left(\dfrac{t}{2} + \dfrac{t}{5}\right) = 10(1)$

   $\dfrac{10}{7}$ [2.2C]

9. Two planes start from the same point and fly in opposite directions. The first plane is flying 20 mph slower than the second plane. In 2 h, the planes are 480 mi apart. Find the rate of each plane.   110 mph, 130 mph   [2.4C]

8.1 Multiplication and Division of Rational Expressions

OBJECTIVE A *To simplify a rational expression*

A fraction in which the numerator and denominator are polynomials is called a **rational expression.** Examples of rational expressions are shown at the right.

$$\frac{5}{z}, \qquad \frac{x^2 + 1}{2x - 1}, \qquad \frac{y^2 + y - 1}{4y^2 + 1}$$

Care must be exercised when evaluating a rational expression to ensure that the resulting denominator is not zero. Consider the rational expression at the right. The value of x cannot be 3 because the denominator would then be zero.

$$\frac{4x^2 - 9}{2x - 6}$$

$$\frac{4(3)^2 - 9}{2(3) - 6} = \frac{27}{0} \quad \text{Not a real number}$$

In the **simplest form of a rational expression,** the numerator and denominator have no common factors. The Multiplication Property of One is used to write a rational expression in simplest form.

INSTRUCTOR NOTE
Simplifying a rational expression is closely related to simplifying a rational number; the common factors are removed. Making this connection will help some students.

HOW TO 1 Simplify: $\dfrac{x^2 - 4}{x^2 - 2x - 8}$

$$\frac{x^2 - 4}{x^2 - 2x - 8} = \frac{(x - 2)(x + 2)}{(x - 4)(x + 2)}$$

• Factor the numerator and denominator.

$$= \frac{x - 2}{x - 4} \cdot \boxed{\frac{x + 2}{x + 2}} = \frac{x - 2}{x - 4} \cdot 1$$

$$= \frac{x - 2}{x - 4}, x \neq -2, 4$$

• The restrictions $x \neq -2$ and $x \neq 4$ are necessary to prevent division by zero.

This simplification is usually shown with slashes through the common factors:

$$\frac{x^2 - 4}{x^2 - 2x - 8} = \frac{(x - 2)\overset{1}{\cancel{(x + 2)}}}{(x - 4)\underset{1}{\cancel{(x + 2)}}}$$

• Factor the numerator and denominator.

$$= \frac{x - 2}{x - 4}, x \neq -2, 4$$

• Divide by the common factors. The restrictions $x \neq -2$ or 4 are necessary to prevent division by zero.

In summary, to simplify a rational expression, factor the numerator and denominator. Then divide the numerator and denominator by the common factors.

INSTRUCTOR NOTE
It is important to emphasize that the numerator and denominator must be written in factored form before simplifying. This will help students avoid errors such as
$$\frac{x^2 - x}{x^2} = \frac{\cancel{x^2} - x}{\cancel{x^2}} = 1 - x$$

HOW TO 2 Simplify: $\dfrac{10 + 3x - x^2}{x^2 - 4x - 5}$

$$\frac{10 + 3x - x^2}{x^2 - 4x - 5} = \frac{-(x^2 - 3x - 10)}{x^2 - 4x - 5}$$

• Because the coefficient of x^2 in the numerator is −1, factor −1 from the numerator.

$$= \frac{-\overset{1}{\cancel{(x - 5)}}(x + 2)}{\underset{1}{\cancel{(x - 5)}}(x + 1)}$$

• Factor the numerator and denominator. Divide by the common factors.

$$= -\frac{x + 2}{x + 1}, x \neq -1, 5$$

For the remaining examples, we will not list the restrictions on the variables that prevent division by zero and assume that the values of the variables are such that division by zero is not possible.

EXAMPLE 1

Simplify: $\dfrac{4x^3y^4}{6x^4y}$

Solution

$\dfrac{4x^3y^4}{6x^4y} = \dfrac{2y^3}{3x}$ • **Use the rules of exponents.**

YOU TRY IT 1

Simplify: $\dfrac{6x^5y}{12x^2y^3}$

Your solution

$\dfrac{x^3}{2y^2}$

EXAMPLE 2

Simplify: $\dfrac{x^2 + 2x - 15}{x^2 - 7x + 12}$

Solution

$\dfrac{x^2 + 2x - 15}{x^2 - 7x + 12} = \dfrac{(x + 5)\overset{1}{\cancel{(x - 3)}}}{\cancel{(x - 3)}(x - 4)} = \dfrac{x + 5}{x - 4}$

YOU TRY IT 2

Simplify: $\dfrac{x^2 + 4x - 12}{x^2 - 3x + 2}$

Your solution

$\dfrac{x + 6}{x - 1}$

EXAMPLE 3

Simplify: $\dfrac{9 - x^2}{x^2 + x - 12}$

Solution

$\dfrac{9 - x^2}{x^2 + x - 12} = \dfrac{\overset{-1}{\cancel{(3 - x)}}(3 + x)}{\cancel{(x - 3)}(x + 4)}$ • $\dfrac{3 - x}{x - 3} = \dfrac{-1(x - 3)}{x - 3}$

$= -\dfrac{x + 3}{x + 4}$ $= -1$

YOU TRY IT 3

Simplify: $\dfrac{x^2 + 2x - 24}{16 - x^2}$

Your solution

$-\dfrac{x + 6}{x + 4}$

IN-CLASS EXAMPLES

Simplify.

1. $\dfrac{18x^5y^2}{12xy^3}$ $\dfrac{3x^4}{2y}$

2. $\dfrac{x^2 - 1}{x^2 + 4x - 5}$ $\dfrac{x + 1}{x + 5}$

3. $\dfrac{a^2 - 2a}{4 - 2a}$ $-\dfrac{a}{2}$

Solutions on p. S22

OBJECTIVE B   *To multiply rational expressions*

The product of two fractions is a fraction whose numerator is the product of the numerators of the two fractions and whose denominator is the product of the denominators of the two fractions.

Multiplying Rational Expressions

To multiply two fractions, multiply the numerators and multiply the denominators.

$$\frac{a}{b} \cdot \frac{c}{d} = \frac{ac}{bd}$$

EXAMPLES

1. $\dfrac{2}{3} \cdot \dfrac{4}{5} = \dfrac{8}{15}$ 2. $\dfrac{3x}{y} \cdot \dfrac{2}{z} = \dfrac{6x}{yz}$ 3. $\dfrac{x + 2}{x} \cdot \dfrac{3}{x - 2} = \dfrac{3(x + 2)}{x(x - 2)}$

INSTRUCTOR NOTE
Remind students that
when they carry out the
multiplication step, writing
the product as a single
fraction, they should
leave the numerator and
denominator in factored
form. The simplified answer
also may be left in factored
form, as in Example 4.
(Note, however, that in
Objective 8.2B, students
will need to multiply out
the numerators when they
apply the skill of multiplying
rational expressions to the
skill of writing fractions in
terms of the LCM of their
denominators.)

HOW TO 3 Multiply: $\dfrac{x^2 + 3x}{x^2 - 3x - 4} \cdot \dfrac{x^2 - 5x + 4}{x^2 + 2x - 3}$

$$\dfrac{x^2 + 3x}{x^2 - 3x - 4} \cdot \dfrac{x^2 - 5x + 4}{x^2 + 2x - 3}$$

$$= \dfrac{x(x + 3)}{(x - 4)(x + 1)} \cdot \dfrac{(x - 4)(x - 1)}{(x + 3)(x - 1)}$$

• Factor the numerator and denominator of each fraction.

$$= \dfrac{x\cancel{(x + 3)}\cancel{(x - 4)}\cancel{(x - 1)}}{\cancel{(x - 4)}(x + 1)\cancel{(x + 3)}\cancel{(x - 1)}}$$

• Multiply. Then divide by the common factors.

$$= \dfrac{x}{x + 1}$$

• Write the answer in simplest form.

EXAMPLE 4

Multiply: $\dfrac{10x^2 - 15x}{12x - 8} \cdot \dfrac{3x - 2}{20x - 25}$

Solution

$$\dfrac{10x^2 - 15x}{12x - 8} \cdot \dfrac{3x - 2}{20x - 25}$$

$$= \dfrac{5x(2x - 3)}{4(3x - 2)} \cdot \dfrac{(3x - 2)}{5(4x - 5)}$$

• Factor.

$$= \dfrac{\cancel{5}x(2x - 3)\cancel{(3x - 2)}}{4\cancel{(3x - 2)}\cancel{5}(4x - 5)}$$

• Divide by the common factors.

$$= \dfrac{x(2x - 3)}{4(4x - 5)}$$

YOU TRY IT 4

Multiply: $\dfrac{12x^2 + 3x}{10x - 15} \cdot \dfrac{8x - 12}{9x + 18}$

Your solution

$$\dfrac{4x(4x + 1)}{15(x + 2)}$$

EXAMPLE 5

Multiply: $\dfrac{x^2 + x - 6}{x^2 + 7x + 12} \cdot \dfrac{x^2 + 3x - 4}{4 - x^2}$

Solution

$$\dfrac{x^2 + x - 6}{x^2 + 7x + 12} \cdot \dfrac{x^2 + 3x - 4}{4 - x^2}$$

$$= \dfrac{(x + 3)(x - 2)}{(x + 3)(x + 4)} \cdot \dfrac{(x + 4)(x - 1)}{(2 - x)(2 + x)}$$

• Factor.

$$= \dfrac{\cancel{(x + 3)}\overset{-1}{\cancel{(x - 2)}}\cancel{(x + 4)}(x - 1)}{\cancel{(x + 3)}\cancel{(x + 4)}\cancel{(2 - x)}(2 + x)}$$

• Divide by the common factors.

$$= -\dfrac{x - 1}{x + 2}$$

YOU TRY IT 5

Multiply: $\dfrac{x^2 + 2x - 15}{9 - x^2} \cdot \dfrac{x^2 - 3x - 18}{x^2 - 7x + 6}$

Your solution

$$-\dfrac{x + 5}{x - 1}$$

IN-CLASS EXAMPLES
Multiply.

4. $\dfrac{28a^5b^7}{5x^4} \cdot \dfrac{15x^2}{14ab}$ $\dfrac{6a^4b^6}{x^2}$

5. $\dfrac{6x^2 - 10x}{3 - 3x} \cdot \dfrac{x^2 - 1}{12x - 20}$ $-\dfrac{x(x + 1)}{6}$

OBJECTIVE C *To divide rational expressions*

The **reciprocal of a rational expression** is the rational expression with the numerator and denominator interchanged.

$$\text{Fraction} \begin{cases} \dfrac{a}{b} & \dfrac{b}{a} \\ x^2 = \dfrac{x^2}{1} & \dfrac{1}{x^2} \\ \dfrac{x+2}{x} & \dfrac{x}{x+2} \end{cases} \text{Reciprocal}$$

Dividing Rational Expressions

To divide two fractions, multiply the first faction by the reciprocal of the divisor.

$$\frac{a}{b} \div \frac{c}{d} = \frac{a}{b} \cdot \frac{d}{c} = \frac{ad}{bc}$$

EXAMPLES

1. $\dfrac{4}{x} \div \dfrac{y}{5} = \dfrac{4}{x} \cdot \dfrac{5}{y} = \dfrac{20}{xy}$

2. $\dfrac{x+4}{x} \div \dfrac{x-2}{4} = \dfrac{x+4}{x} \cdot \dfrac{4}{x-2} = \dfrac{4(x+4)}{x(x-2)}$

The basis for the division rule is shown at the right.

$$\frac{a}{b} \div \frac{c}{d} = \frac{\dfrac{a}{b}}{\dfrac{c}{d}} = \frac{\dfrac{a}{b} \cdot \dfrac{d}{c}}{\dfrac{c}{d} \cdot \dfrac{d}{c}} = \frac{\dfrac{a}{b} \cdot \dfrac{d}{c}}{1} = \frac{a}{b} \cdot \frac{d}{c}$$

EXAMPLE 6

Divide: $\dfrac{xy^2 - 3x^2y}{z^2} \div \dfrac{6x^2 - 2xy}{z^3}$

Solution

$\dfrac{xy^2 - 3x^2y}{z^2} \div \dfrac{6x^2 - 2xy}{z^3}$

$= \dfrac{xy^2 - 3x^2y}{z^2} \cdot \dfrac{z^3}{6x^2 - 2xy}$ • Multiply by the reciprocal.

$= \dfrac{xy\overset{-1}{\cancel{(y - 3x)}} \cdot z^3}{z^2 \cdot 2x\underset{1}{\cancel{(3x - y)}}} = -\dfrac{yz}{2}$

YOU TRY IT 6

Divide: $\dfrac{a^2}{4bc^2 - 2b^2c} \div \dfrac{a}{6bc - 3b^2}$

Your solution

$\dfrac{3a}{2c}$

EXAMPLE 7

Divide: $\dfrac{2x^2 + 5x + 2}{2x^2 + 3x - 2} \div \dfrac{3x^2 + 13x + 4}{2x^2 + 7x - 4}$

Solution

$\dfrac{2x^2 + 5x + 2}{2x^2 + 3x - 2} \div \dfrac{3x^2 + 13x + 4}{2x^2 + 7x - 4}$

$= \dfrac{2x^2 + 5x + 2}{2x^2 + 3x - 2} \cdot \dfrac{2x^2 + 7x - 4}{3x^2 + 13x + 4}$ • Multiply by the reciprocal.

$= \dfrac{(2x+1)\overset{1}{\cancel{(x+2)}} \cdot \overset{1}{\cancel{(2x-1)}}\overset{1}{\cancel{(x+4)}}}{\underset{1}{\cancel{(2x-1)}}\underset{1}{\cancel{(x+2)}} \cdot (3x+1)\underset{1}{\cancel{(x+4)}}} = \dfrac{2x+1}{3x+1}$

YOU TRY IT 7

Divide: $\dfrac{3x^2 + 26x + 16}{3x^2 - 7x - 6} \div \dfrac{2x^2 + 9x - 5}{x^2 + 2x - 15}$

Your solution

$\dfrac{x+8}{2x-1}$

IN-CLASS EXAMPLES

Divide.

6. $\dfrac{12a^5b}{7xy^4} \div \dfrac{9ab^3}{35xy^2}$ $\dfrac{20a^4}{3b^2y^2}$

7. $\dfrac{4x-8}{15 + 2x - x^2} \div \dfrac{x^2 - 4}{3x^2 - 15x}$

$-\dfrac{12x}{(x+3)(x+2)}$

Solutions on p. S22

8.1 EXERCISES

SUGGESTED ASSIGNMENT
Exercises 1–4; Exercises 5–55, odds;
Exercises 61–79, odds
More challenging exercises: Exercises 80–83

✔ Concept Check

1. 📝 What is a rational expression? Provide an example.

2. 📝 When is a rational expression in simplest form?

3. 📝 For the rational expression $\frac{x + 7}{x - 4}$, explain why the value of x cannot be 4.

4. 🔖 Why is the simplification at the right incorrect? $\dfrac{x + 3}{x} = \dfrac{\overset{1}{\cancel{x}} + 3}{\underset{1}{\cancel{x}}} = 4$

The numerator and denominator were divided by a term rather than by a factor.

OBJECTIVE A   *To simplify a rational expression*

For Exercises 5 to 32, simplify.

5. $\dfrac{9x^3}{12x^4}$

$\dfrac{3}{4x}$

6. $\dfrac{16x^2y}{24xy^3}$

$\dfrac{2x}{3y^2}$

7. $\dfrac{(x + 3)^2}{(x + 3)^3}$

$\dfrac{1}{x + 3}$

8. $\dfrac{(2x - 1)^5}{(2x - 1)^4}$

$2x - 1$

9. $\dfrac{3n - 4}{4 - 3n}$

-1

10. $\dfrac{5 - 2x}{2x - 5}$

-1

11. $\dfrac{6y(y + 2)}{9y^2(y + 2)}$

$\dfrac{2}{3y}$

12. $\dfrac{12x^2(3 - x)}{18x(3 - x)}$

$\dfrac{2x}{3}$

13. $\dfrac{6x(x - 5)}{8x^2(5 - x)}$

$-\dfrac{3}{4x}$

14. $\dfrac{14x^3(7 - 3x)}{21x(3x - 7)}$

$-\dfrac{2x^2}{3}$

15. $\dfrac{a^2 + 4a}{ab + 4b}$

$\dfrac{a}{b}$

16. $\dfrac{x^2 - 3x}{2x - 6}$

$\dfrac{x}{2}$

17. $\dfrac{4 - 6x}{3x^2 - 2x}$

$-\dfrac{2}{x}$

18. $\dfrac{5xy - 3y}{9 - 15x}$

$-\dfrac{y}{3}$

19. $\dfrac{y^2 - 3y + 2}{y^2 - 4y + 3}$

$\dfrac{y - 2}{y - 3}$

20. $\dfrac{x^2 + 5x + 6}{x^2 + 8x + 15}$

$\dfrac{x + 2}{x + 5}$

21. $\dfrac{x^2 + 3x - 10}{x^2 + 2x - 8}$

$\dfrac{x + 5}{x + 4}$

22. $\dfrac{a^2 + 7a - 8}{a^2 + 6a - 7}$

$\dfrac{a + 8}{a + 7}$

23. $\dfrac{x^2 + x - 12}{x^2 - 6x + 9}$

$\dfrac{x + 4}{x - 3}$

24. $\dfrac{x^2 + 8x + 16}{x^2 - 2x - 24}$

$\dfrac{x + 4}{x - 6}$

25. $\dfrac{x^2 - 3x - 10}{25 - x^2}$

$-\dfrac{x + 2}{x + 5}$

26. $\dfrac{4 - y^2}{y^2 - 3y - 10}$

$\dfrac{2 - y}{y - 5}$

27. $\dfrac{2x^3 + 2x^2 - 4x}{x^3 + 2x^2 - 3x}$

$\dfrac{2(x + 2)}{x + 3}$

28. $\dfrac{3x^3 - 12x}{6x^3 - 24x^2 + 24x}$

$\dfrac{x + 2}{2(x - 2)}$

29. $\dfrac{6x^2 - 7x + 2}{6x^2 + 5x - 6}$

$\dfrac{2x - 1}{2x + 3}$

30. $\dfrac{2n^2 - 9n + 4}{2n^2 - 5n - 12}$

$\dfrac{2n - 1}{2n + 3}$

31. $\dfrac{x^2 + 3x - 28}{24 - 2x - x^2}$

$-\dfrac{x + 7}{x + 6}$

32. $\dfrac{x^2 + 7x - 8}{1 + x - 2x^2}$

$\dfrac{x + 8}{2x + 1}$

OBJECTIVE B *To multiply rational expressions*

For Exercises 33 to 56, multiply.

33. $\dfrac{8x^2}{9y^3} \cdot \dfrac{3y^2}{4x^3}$

$\dfrac{2}{3xy}$

34. $\dfrac{14a^2b^3}{15x^5y^2} \cdot \dfrac{25x^3y}{16ab}$

$\dfrac{35ab^2}{24x^2y}$

35. $\dfrac{12x^3y^4}{7a^2b^3} \cdot \dfrac{14a^3b^4}{9x^2y^2}$

$\dfrac{8xy^2ab}{3}$

36. $\dfrac{18a^4b^2}{25x^2y^3} \cdot \dfrac{50x^5y^6}{27a^6b^2}$

$\dfrac{4x^3y^3}{3a^2}$

37. $\dfrac{3x - 6}{5x - 20} \cdot \dfrac{10x - 40}{27x - 54}$

$\dfrac{2}{9}$

38. $\dfrac{8x - 12}{14x + 7} \cdot \dfrac{42x + 21}{32x - 48}$

$\dfrac{3}{4}$

39. $\dfrac{3x^2 + 2x}{2xy - 3y} \cdot \dfrac{2xy^3 - 3y^3}{3x^3 + 2x^2}$

$\dfrac{y^2}{x}$

40. $\dfrac{4a^2x - 3a^2}{2by + 5b} \cdot \dfrac{2b^3y + 5b^3}{4ax - 3a}$

ab^2

41. $\dfrac{x^2 + 5x + 4}{x^3y^2} \cdot \dfrac{x^2y^3}{x^2 + 2x + 1}$

$\dfrac{y(x + 4)}{x(x + 1)}$

42. $\dfrac{x^2 + x - 2}{xy^2} \cdot \dfrac{x^3y}{x^2 + 5x + 6}$

$\dfrac{x^2(x - 1)}{y(x + 3)}$

43. $\dfrac{x^4y^2}{x^2 + 3x - 28} \cdot \dfrac{x^2 - 49}{xy^4}$

$\dfrac{x^3(x - 7)}{y^2(x - 4)}$

44. $\dfrac{x^5y^3}{x^2 + 13x + 30} \cdot \dfrac{x^2 + 2x - 3}{x^7y^2}$

$\dfrac{y(x - 1)}{x^2(x + 10)}$

45. $\dfrac{2x^2 - 5x}{2xy + y} \cdot \dfrac{2xy^2 + y^2}{5x^2 - 2x^3}$

$-\dfrac{y}{x}$

46. $\dfrac{3a^3 + 4a^2}{5ab - 3b} \cdot \dfrac{3b^3 - 5ab^3}{3a^2 + 4a}$

$-ab^2$

47. $\dfrac{x^2 - 2x - 24}{x^2 - 5x - 6} \cdot \dfrac{x^2 + 5x + 6}{x^2 + 6x + 8}$

$\dfrac{x + 3}{x + 1}$

48. $\dfrac{x^2 - 8x + 7}{x^2 + 3x - 4} \cdot \dfrac{x^2 + 3x - 10}{x^2 - 9x + 14}$

$\dfrac{x + 5}{x + 4}$

49. $\dfrac{x^2 + 2x - 35}{x^2 + 4x - 21} \cdot \dfrac{x^2 + 3x - 18}{x^2 + 9x + 18}$

$\dfrac{x - 5}{x + 3}$

50. $\dfrac{y^2 + y - 20}{y^2 + 2y - 15} \cdot \dfrac{y^2 + 4y - 21}{y^2 + 3y - 28}$

1

51. $\dfrac{x^2 - 3x - 4}{x^2 + 6x + 5} \cdot \dfrac{x^2 + 5x + 6}{8 + 2x - x^2}$

$-\dfrac{x + 3}{x + 5}$

52. $\dfrac{25 - n^2}{n^2 - 2n - 35} \cdot \dfrac{n^2 - 8n - 20}{n^2 - 3n - 10}$

$-\dfrac{n - 10}{n - 7}$

53. $\dfrac{16 + 6x - x^2}{x^2 - 10x - 24} \cdot \dfrac{x^2 - 6x - 27}{x^2 - 17x + 72}$

$-\dfrac{x + 3}{x - 12}$

54. $\dfrac{x^2 - 11x + 28}{x^2 - 13x + 42} \cdot \dfrac{x^2 + 7x + 10}{20 - x - x^2}$

$-\dfrac{x + 2}{x - 6}$

55. $\dfrac{2x^2 + 5x + 2}{2x^2 + 7x + 3} \cdot \dfrac{x^2 - 7x - 30}{x^2 - 6x - 40}$

$\dfrac{x + 2}{x + 4}$

56. $\dfrac{x^2 - 4x - 32}{x^2 - 8x - 48} \cdot \dfrac{3x^2 + 17x + 10}{3x^2 - 22x - 16}$

$\dfrac{x + 5}{x - 12}$

For Exercises 57 to 59, use the product $\dfrac{x^a}{y^b} \cdot \dfrac{y^c}{x^d}$, where a, b, c, and d are all positive integers.

57. If $a > d$ and $c > b$, what is the denominator of the simplified product?   1

58. If $a > d$ and $b > c$, which variable appears in the denominator of the simplified product? y

59. If $a < d$ and $b = c$, what is the numerator of the simplified product?   1

OBJECTIVE C   *To divide rational expressions*

For Exercises 60 to 79, divide.

60. $\dfrac{4x^2y^3}{15a^2b^3} \div \dfrac{6xy}{5a^3b^5}$

$\dfrac{2xy^2ab^2}{9}$

61. $\dfrac{9x^3y^4}{16a^4b^2} \div \dfrac{45x^4y^2}{14a^7b}$

$\dfrac{7a^3y^2}{40bx}$

62. $\dfrac{6x - 12}{8x + 32} \div \dfrac{18x - 36}{10x + 40}$

$\dfrac{5}{12}$

63. $\dfrac{28x + 14}{45x - 30} \div \dfrac{14x + 7}{30x - 20}$

$\dfrac{4}{3}$

64. $\dfrac{6x^3 + 7x^2}{12x - 3} \div \dfrac{6x^2 + 7x}{36x - 9}$

$3x$

65. $\dfrac{5a^2y + 3a^2}{2x^3 + 5x^2} \div \dfrac{10ay + 6a}{6x^3 + 15x^2}$

$\dfrac{3a}{2}$

66. $\dfrac{x^2 + 4x + 3}{x^2y} \div \dfrac{x^2 + 2x + 1}{xy^2}$

$\dfrac{y(x + 3)}{x(x + 1)}$

67. $\dfrac{x^3y^2}{x^2 - 3x - 10} \div \dfrac{xy^4}{x^2 - x - 20}$

$\dfrac{x^2(x + 4)}{y^2(x + 2)}$

68. $\dfrac{x^2 - 49}{x^4y^3} \div \dfrac{x^2 - 14x + 49}{x^4y^3}$

$\dfrac{x + 7}{x - 7}$

69. $\dfrac{x^2y^5}{x^2 - 11x + 30} \div \dfrac{xy^6}{x^2 - 7x + 10}$

$\dfrac{x(x - 2)}{y(x - 6)}$

70. $\dfrac{4ax - 8a}{c^2} \div \dfrac{2y - xy}{c^3}$

$-\dfrac{4ac}{y}$

71. $\dfrac{3x^2y - 9xy}{a^2b} \div \dfrac{3x^2 - x^3}{ab^2}$

$-\dfrac{3by}{ax}$

72. $\dfrac{x^2 - 5x + 6}{x^2 - 9x + 18} \div \dfrac{x^2 - 6x + 8}{x^2 - 9x + 20}$

$\dfrac{x - 5}{x - 6}$

73. $\dfrac{x^2 + 3x - 40}{x^2 + 2x - 35} \div \dfrac{x^2 + 2x - 48}{x^2 + 3x - 18}$

$\dfrac{(x + 6)(x - 3)}{(x + 7)(x - 6)}$

74. $\dfrac{x^2 + 2x - 15}{x^2 - 4x - 45} \div \dfrac{x^2 + x - 12}{x^2 - 5x - 36}$

1

75. $\dfrac{y^2 - y - 56}{y^2 + 8y + 7} \div \dfrac{y^2 - 13y + 40}{y^2 - 4y - 5}$

1

76. $\dfrac{8 + 2x - x^2}{x^2 + 7x + 10} \div \dfrac{x^2 - 11x + 28}{x^2 - x - 42}$

$-\dfrac{x + 6}{x + 5}$

77. $\dfrac{x^2 - x - 2}{x^2 - 7x + 10} \div \dfrac{x^2 - 3x - 4}{40 - 3x - x^2}$

$-\dfrac{x + 8}{x - 4}$

78. $\dfrac{2x^2 - 3x - 20}{2x^2 - 7x - 30} \div \dfrac{2x^2 - 5x - 12}{4x^2 + 12x + 9}$

$\dfrac{2x + 3}{x - 6}$

79. $\dfrac{6n^2 + 13n + 6}{4n^2 - 9} \div \dfrac{6n^2 + n - 2}{4n^2 - 1}$

$\dfrac{2n + 1}{2n - 3}$

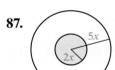 For Exercises 80 to 83, state whether the given division is equivalent to $\dfrac{x^2 - 3x - 4}{x^2 + 5x - 6}$.

80. $\dfrac{x - 4}{x + 6} \div \dfrac{x - 1}{x + 1}$

Yes

81. $\dfrac{x + 1}{x + 6} \div \dfrac{x - 1}{x - 4}$

Yes

82. $\dfrac{x + 1}{x - 1} \div \dfrac{x + 6}{x - 4}$

Yes

83. $\dfrac{x - 1}{x + 1} \div \dfrac{x - 4}{x + 6}$

No

Critical Thinking

For Exercises 84 to 86, name the values of x for which the rational expression is undefined. (*Hint:* Set the denominator equal to zero and solve for x.)

84. $\dfrac{x}{(x - 2)(x + 5)}$

2, −5

85. $\dfrac{x + 5}{x^2 - 4x - 5}$

5, −1

86. $\dfrac{3x - 8}{3x^2 - 10x - 8}$

$-\dfrac{2}{3}, 4$

Geometry   For Exercises 87 and 88, write in simplest form the ratio of the shaded area of the figure to the total area of the figure.

87.

$\dfrac{4}{25}$

88.

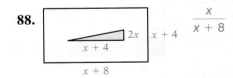

$\dfrac{x}{x + 8}$

89. Find two different pairs of rational expressions whose product is $\dfrac{2x^2 + 7x - 4}{3x^2 - 8x - 3}$.

$\dfrac{x + 4}{x - 3}$ and $\dfrac{2x - 1}{3x + 1}$ or $\dfrac{x + 4}{3x + 1}$ and $\dfrac{2x - 1}{x - 3}$

Projects or Group Activities

90. ◼ Given the expression $\dfrac{9}{x^2 + 1}$, choose some values of x and evaluate the expression for those values. Is it possible to choose a value of x for which the value of the expression is greater than 10? If so, give such a value. If not, explain why it is not possible.

91. ◼ Given the expression $\dfrac{1}{y - 3}$, choose some values of y and evaluate the expression for those values. Is it possible to choose a value of y for which the value of the expression is greater than 10,000,000? If so, give such a value. If not, explain why it is not possible.

QUICK QUIZ

Simplify.

1. $\dfrac{x^2 - x - 12}{x^2 + 9x + 18}$

$\dfrac{x - 4}{x + 6}$ [8.1A]

Multiply.

2. $\dfrac{x^2 - 5x + 4}{2x^2 + 5x - 3} \cdot \dfrac{x^2 - 9}{x^2 - 7x + 12}$

$\dfrac{x - 1}{2x - 1}$ [8.1B]

Divide.

3. $\dfrac{10x + 10}{3x - 6} \div \dfrac{2x + 2}{3xy - 6y}$

5y   [8.1C]

Unless otherwise noted, all content on this page is © Cengage Learning.

8.2 Addition and Subtraction of Rational Expressions

OBJECTIVE A *To rewrite rational expressions in terms of a common denominator*

Tips for Success

As you know, often in mathematics you learn one skill in order to perform another. This is true of this objective. You are learning to rewrite rational expressions in terms of a common denominator in order to add and subtract rational expressions in the next objective. To ensure success, be certain you understand this objective before studying the next.

INSTRUCTOR NOTE

The main purpose of this objective is to have students write rational expressions in terms of a common denominator. In the next objective, they will use this skill to add and subtract rational expressions.

Take Note

Our primary use of the LCM will be in adding or subtracting rational expressions with different denominators. When finding the LCM for this purpose, it is more useful *not* to multiply out the binomials of the LCM. See HOW TO 3.

IN-CLASS EXAMPLES

Write each fraction in terms of the LCM of the denominators.

1. $\dfrac{5}{12xy^2}, \dfrac{2}{3x^2y}$

$\dfrac{5x}{12x^2y^2}, \dfrac{8y}{12x^2y^2}$

2. $\dfrac{2x}{2x-1}, \dfrac{3x}{2x+1}$

$\dfrac{4x^2+2x}{(2x+1)(2x-1)},$

$\dfrac{6x^2-3x}{(x+1)(2x-1)}$

When adding or subtracting rational expressions, it is frequently necessary to express the rational expressions in terms of a common denominator. A common denominator is the least common multiple (LCM) of the denominators.

The **least common multiple (LCM) of two or more polynomials** is the polynomial of least degree that contains the factors of each polynomial. To find the LCM, begin by factoring each polynomial. The LCM must contain the factors of each polynomial.

HOW TO 1 Find the LCM of $6x^2y$ and $9xy^3$.

Factor each monomial and identify the common factors, shown below in red.

$6x^2y = 2 \cdot 3 \cdot x \cdot x \cdot y$

$9xy^3 = 3 \cdot 3 \cdot x \cdot y \cdot y \cdot y$

Create the LCM by first writing the common factors and then writing the remaining factors of each monomial (shown in blue).

$LCM = 3 \cdot x \cdot y \cdot 2 \cdot 3 \cdot x \cdot y \cdot y$

$\quad\quad = 18x^2y^3$

HOW TO 2 Find the LCM of $6x^2 + 30x$ and $4x^2 + 16x - 20$.

Factor each polynomial and identify the common factors, shown below in red.

$6x^2 + 30x = 6x(x + 5) = 2 \cdot 3 \cdot x(x + 5)$

$4x^2 + 16x - 20 = 4(x^2 + 4x - 5) = 2 \cdot 2(x - 1)(x + 5)$

Create the LCM by first writing the common factors and then writing the remaining factors of each polynomial (shown in blue).

$LCM = 2 \cdot (x + 5) \cdot 2 \cdot 3 \cdot x(x - 1)$

$\quad\quad = 12x(x + 5)(x - 1)$

HOW TO 3 Write the fractions $\dfrac{3x}{x-1}$ and $\dfrac{4}{2x+5}$ in terms of the LCM of the denominators.

First, find the LCM of $x - 1$ and $2x + 5$. Each polynomial is in simplest form, and there are no common factors. The LCM is the product of the two factors.

$LCM = (x - 1)(2x + 5)$

Write each fraction with the LCM as the denominator.

$\dfrac{3x}{x-1} \cdot \dfrac{2x+5}{2x+5} = \dfrac{3x(2x+5)}{(x-1)(2x+5)}$

$\quad\quad\quad\quad = \dfrac{6x^2+15x}{(x-1)(2x+5)}$

- **Multiply the numerator and denominator of $\dfrac{3x}{x-1}$ by $2x + 5$, because the product $(x-1)(2x+5)$ is the LCM.**

$\dfrac{4}{2x+5} \cdot \dfrac{x-1}{x-1} = \dfrac{4(x-1)}{(x-1)(2x+5)}$

$\quad\quad\quad\quad = \dfrac{4x-4}{(x-1)(2x+5)}$

- **Multiply the numerator and denominator of $\dfrac{4}{2x+5}$ by $x - 1$, because the product $(x-1)(2x+5)$ is the LCM.**

EXAMPLE 1

Write the fractions $\frac{x+2}{x^2-2x}$ and $\frac{5x-1}{3x-6}$ in terms of the LCM of the denominators.

Solution

$x^2 - 2x = x(x-2)$ • Find the LCM of the
$3x - 6 = 3(x-2)$ denominators.
LCM $= 3x(x-2)$

$\frac{x+2}{x^2-2x} = \frac{x+2}{x(x-2)} \cdot \frac{3}{3} = \frac{3(x+2)}{3x(x-2)} = \frac{3x+6}{3x(x-2)}$

$\frac{5x-1}{3x-6} = \frac{5x-1}{3(x-2)} \cdot \frac{x}{x} = \frac{x(5x-1)}{3x(x-2)} = \frac{5x^2-x}{3x(x-2)}$

YOU TRY IT 1

Write the fractions $\frac{2x}{2x-5}$ and $\frac{3}{x+4}$ in terms of the LCM of the denominators.

Your solution

$\frac{2x^2+8x}{(2x-5)(x+4)}$

$\frac{6x-15}{(2x-5)(x+4)}$

EXAMPLE 2

Write the fractions $\frac{2x-3}{3x-x^2}$ and $\frac{3x}{x^2-4x+3}$ in terms of the LCM of the denominators.

Solution

$3x - x^2 = x(3-x) = -x(x-3)$
$x^2 - 4x + 3 = (x-3)(x-1)$
The LCM is $x(x-3)(x-1)$.

$\frac{2x-3}{3x-x^2} = -\frac{2x-3}{x(x-3)} \cdot \frac{x-1}{x-1}$

$\qquad = -\frac{2x^2-5x+3}{x(x-3)(x-1)}$

$\frac{3x}{x^2-4x+3} = \frac{3x}{(x-3)(x-1)} \cdot \frac{x}{x}$

$\qquad = \frac{3x^2}{x(x-3)(x-1)}$

YOU TRY IT 2

Write the fractions $\frac{2x-7}{2x-x^2}$ and $\frac{3x-2}{3x^2-5x-2}$ in terms of the LCM of the denominators.

Your solution

$-\frac{6x^2-19x-7}{x(x-2)(3x+1)}$

$\frac{3x^2-2x}{x(x-2)(3x+1)}$

Solutions on p. S22

OBJECTIVE B *To add or subtract rational expressions*

Adding or Subtracting Rational Expressions

To add two rational expressions *with the same denominator*, add the numerators and place the sum over the common denominator. To subtract two rational expressions *with the same denominator*, subtract the numerators and place the difference over the common denominator.

EXAMPLE

$\frac{3x-5}{x^2-4} + \frac{x-3}{x^2-4} = \frac{(3x-5)+(x-3)}{x^2-4}$ • The denominators are the same. Add the numerators.

$\qquad = \frac{4x-8}{x^2-4} = \frac{4(x-2)}{(x+2)(x-2)}$ • Write the fraction in simplest form.

$\qquad = \frac{4}{x+2}$

Before two rational expressions with *different* denominators can be added or subtracted, each rational expression must be expressed in terms of a common denominator. A good common denominator to use is the LCM of the denominators, also called the **least common denominator (LCD).**

Take Note

Note the steps involved in adding or subtracting rational expressions with different denominators:

1. Find the LCM of the denominators.
2. Rewrite each fraction in terms of the common denominator.
3. Add or subtract the rational expressions.
4. Simplify the resulting sum or difference.

HOW TO 4 Simplify: $\dfrac{x}{x-3} - \dfrac{x+1}{x-2}$

The LCM is $(x-3)(x-2)$.

$$\dfrac{x}{x-3} - \dfrac{x+1}{x-2} = \dfrac{x}{x-3} \cdot \dfrac{x-2}{x-2} - \dfrac{x+1}{x-2} \cdot \dfrac{x-3}{x-3}$$

$$= \dfrac{x(x-2) - (x+1)(x-3)}{(x-3)(x-2)}$$

$$= \dfrac{(x^2 - 2x) - (x^2 - 2x - 3)}{(x-3)(x-2)}$$

$$= \dfrac{3}{(x-3)(x-2)}$$

• Find the LCM of the denominators.
• Express each fraction in terms of the LCM.
• Subtract the fractions.

• Simplify.

INSTRUCTOR NOTE

One reason that adding or subtracting rational expressions is difficult for students is the number of steps. Encourage students to develop an outline similar to the one above that they can use for the exercises in this objective.

HOW TO 5 Simplify: $\dfrac{3x}{2x-3} + \dfrac{3x+6}{2x^2 + x - 6}$

The LCM of $2x - 3$ and $2x^2 + x - 6$ is $(2x-3)(x+2)$.

$$\dfrac{3x}{2x-3} + \dfrac{3x+6}{2x^2 + x - 6}$$

$$= \dfrac{3x}{2x-3} \cdot \dfrac{x+2}{x+2} + \dfrac{3x+6}{(2x-3)(x+2)}$$

$$= \dfrac{3x(x+2) + (3x+6)}{(2x-3)(x+2)}$$

$$= \dfrac{(3x^2 + 6x) + (3x+6)}{(2x-3)(x+2)}$$

$$= \dfrac{3x^2 + 9x + 6}{(2x-3)(x+2)}$$

$$= \dfrac{3(x+2)(x+1)}{(2x-3)(x+2)} = \dfrac{3(x+1)}{2x-3}$$

• Find the LCM of the denominators.

• Express each fraction in terms of the LCM.

• Add the fractions.

• Simplify.

EXAMPLE 3

Simplify: $\dfrac{2}{x} - \dfrac{3}{x^2} + \dfrac{1}{xy}$

Solution

The LCM is $x^2 y$.

$$\dfrac{2}{x} - \dfrac{3}{x^2} + \dfrac{1}{xy} = \dfrac{2}{x} \cdot \dfrac{xy}{xy} - \dfrac{3}{x^2} \cdot \dfrac{y}{y} + \dfrac{1}{xy} \cdot \dfrac{x}{x}$$

$$= \dfrac{2xy}{x^2 y} - \dfrac{3y}{x^2 y} + \dfrac{x}{x^2 y}$$

$$= \dfrac{2xy - 3y + x}{x^2 y}$$

YOU TRY IT 3

Simplify: $\dfrac{2}{b} - \dfrac{1}{a} + \dfrac{4}{ab}$

Your solution

$\dfrac{2a - b + 4}{ab}$

Solution on p. S22

EXAMPLE 4

Simplify: $\dfrac{x}{2x - 4} - \dfrac{4 - x}{x^2 - 2x}$

Solution

$2x - 4 = \mathbf{2}(x - 2)$
$x^2 - 2x = x(x - 2)$

The LCM is $2x(x - 2)$.

$\dfrac{x}{2x - 4} - \dfrac{4 - x}{x^2 - 2x} = \dfrac{x}{\mathbf{2}(x - 2)} \cdot \dfrac{x}{x} - \dfrac{4 - x}{x(x - 2)} \cdot \dfrac{\mathbf{2}}{\mathbf{2}}$

$= \dfrac{x^2 - (4 - x)2}{\mathbf{2}x(x - 2)}$

$= \dfrac{x^2 - (8 - 2x)}{2x(x - 2)}$

$= \dfrac{x^2 + 2x - 8}{2x(x - 2)}$

$= \dfrac{(x + 4)(x - 2)}{2x(x - 2)}$

$= \dfrac{(x + 4)\cancel{(x - 2)}^{\,1}}{2x\cancel{(x - 2)}_{\,1}}$

$= \dfrac{x + 4}{2x}$

YOU TRY IT 4

Simplify: $\dfrac{a - 3}{a^2 - 5a} + \dfrac{a - 9}{a^2 - 25}$

Your solution

$\dfrac{2a + 3}{a(a + 5)}$

IN-CLASS EXAMPLES

Simplify.

3. $\dfrac{x}{x^2 - x - 2} - \dfrac{2}{x^2 - x - 2}$ $\dfrac{1}{x + 1}$

4. $\dfrac{4}{3x^2 y} - \dfrac{5}{6x} - \dfrac{7}{12xy}$ $\dfrac{16 - 10xy - 7x}{12x^2 y}$

5. $\dfrac{x}{x + 4} - \dfrac{4 - x}{x^2 - 16}$ $\dfrac{x + 1}{x + 4}$

6. $\dfrac{4x}{x - 2} - \dfrac{3x}{x - 3}$ $\dfrac{x(x - 6)}{(x - 2)(x - 3)}$

EXAMPLE 5

Simplify: $\dfrac{2x + 5}{x^2 + 5x - 6} - \dfrac{x - 3}{x^2 - 3x + 2}$

Solution

$x^2 + 5x - 6 = (x - 1)(x + 6)$
$x^2 - 3x + 2 = (x - 1)(x - 2)$

The LCM is $(x - 1)(x + 6)(x - 2)$.

$\dfrac{2x + 5}{x^2 + 5x - 6} - \dfrac{x - 3}{x^2 - 3x + 2}$

$= \dfrac{2x + 5}{(x - 1)(x + 6)} \cdot \dfrac{x - 2}{x - 2} - \dfrac{x - 3}{(x - 1)(x - 2)} \cdot \dfrac{x + 6}{x + 6}$

$= \dfrac{(2x + 5)(x - 2)}{(x - 1)(x + 6)(x - 2)} - \dfrac{(x - 3)(x + 6)}{(x - 1)(x + 6)(x - 2)}$

$= \dfrac{(2x + 5)(x - 2) - (x - 3)(x + 6)}{(x - 1)(x + 6)(x - 2)}$

$= \dfrac{(2x^2 + x - 10) - (x^2 + 3x - 18)}{(x - 1)(x + 6)(x - 2)}$

$= \dfrac{x^2 - 2x + 8}{(x - 1)(x + 6)(x - 2)}$

YOU TRY IT 5

Simplify: $\dfrac{2x}{x - 4} - \dfrac{x - 1}{x + 1} + \dfrac{2}{x^2 - 3x - 4}$

Your solution

$\dfrac{x^2 + 7x - 2}{(x - 4)(x + 1)}$

Solutions on p. S23

8.2 EXERCISES

SUGGESTED ASSIGNMENT
Exercises 1 and 2; Exercises 3–77, odds;
More challenging exercise: Exercise 78

✔ **Concept Check**

1. Find the LCM of each pair of polynomials.

a. $2x + 6, 7x + 21$
$14(x + 3)$

b. $x + 4, x - 6$
$(x + 4)(x - 6)$

c. $x^2 - 4, x^2 + 3x - 10$
$(x - 2)(x + 2)(x + 5)$

2. Add or subtract the rational expressions.

a. $\dfrac{2x}{3x + 12} + \dfrac{8}{3x + 12}$
$\dfrac{2}{3}$

b. $\dfrac{3x - 4}{x^2 + 4x - 12} - \dfrac{2x - 2}{x^2 + 4x - 12}$
$\dfrac{1}{x + 6}$

c. $\dfrac{x^2 + 4x - 9}{x^2 - 6x + 5} - \dfrac{2x - 6}{x^2 - 6x + 5}$
$\dfrac{x + 2}{x - 3}$

d. $\dfrac{x^2 + 4x - 5}{x^2 + 7x - 18} + \dfrac{1 - 4x}{x^2 + 7x - 18}$
$\dfrac{x + 2}{x + 9}$

OBJECTIVE A *To rewrite rational expressions in terms of a common denominator*

3. **a.** How many factors of a are in the LCM of $(a^2b)^3$ and a^4b^4?   Six
b. How many factors of b are in the LCM of $(a^2b)^3$ and a^4b^4?   Four

4. **a.** How many factors of $x - 4$ are in the LCM of $x^2 - x - 12$ and $x^2 - 8x + 16$?   Two
b. How many factors of $x - 4$ are in the LCM of $x^2 - x - 12$ and $x^2 - 16$?   One

For Exercises 5 to 27, write each fraction in terms of the LCM of the denominators.

5. $\dfrac{3}{4x^2y}, \dfrac{17}{12xy^4}$
$\dfrac{9y^3}{12x^2y^4}, \dfrac{17x}{12x^2y^4}$

6. $\dfrac{5}{16a^3b^3}, \dfrac{7}{30a^5b}$
$\dfrac{75a^2}{240a^5b^3}, \dfrac{56b^2}{240a^5b^3}$

7. $\dfrac{x - 2}{3x(x - 2)}, \dfrac{3}{6x^2}$
$\dfrac{2x^2 - 4x}{6x^2(x - 2)}, \dfrac{3x - 6}{6x^2(x - 2)}$

8. $\dfrac{5x - 1}{4x(2x + 1)}, \dfrac{2}{5x^3}$
$\dfrac{25x^3 - 5x^2}{20x^3(2x + 1)}, \dfrac{16x + 8}{20x^3(2x + 1)}$

9. $\dfrac{3x - 1}{2x(x - 5)}, -3x$
$\dfrac{3x - 1}{2x(x - 5)}, -\dfrac{6x^3 - 30x^2}{2x(x - 5)}$

10. $\dfrac{4x - 3}{3x(x - 2)}, 2x$
$\dfrac{4x - 3}{3x(x - 2)}, \dfrac{6x^3 - 12x^2}{3x(x - 2)}$

11. $\dfrac{3x}{2x - 3}, \dfrac{5x}{2x + 3}$
$\dfrac{6x^2 + 9x}{(2x + 3)(2x - 3)}, \dfrac{10x^2 - 15x}{(2x + 3)(2x - 3)}$

12. $\dfrac{2}{7y - 3}, \dfrac{-3}{7y + 3}$
$\dfrac{14y + 6}{(7y - 3)(7y + 3)}, \dfrac{-21y + 9}{(7y - 3)(7y + 3)}$

13. $\dfrac{2x}{x^2 - 9}, \dfrac{x + 1}{x - 3}$
$\dfrac{2x}{(x + 3)(x - 3)}, \dfrac{x^2 + 4x + 3}{(x + 3)(x - 3)}$

14. $\dfrac{3x}{16 - x^2}, \dfrac{2x}{16 - 4x}$
$\dfrac{12x}{4(4 + x)(4 - x)}, \dfrac{8x + 2x^2}{4(4 + x)(4 - x)}$

15. $\dfrac{3}{3x^2 - 12y^2}, \dfrac{5}{6x - 12y}$
$\dfrac{6}{6(x + 2y)(x - 2y)}, \dfrac{5x + 10y}{6(x + 2y)(x - 2y)}$

16. $\dfrac{2x}{x^2 - 36}, \dfrac{x - 1}{6x - 36}$
$\dfrac{12x}{6(x + 6)(x - 6)}, \dfrac{x^2 + 5x - 6}{6(x + 6)(x - 6)}$

17. $\dfrac{3x}{x^2 - 1}, \dfrac{5x}{x^2 - 2x + 1}$

$\dfrac{3x^2 - 3x}{(x + 1)(x - 1)^2}, \dfrac{5x^2 + 5x}{(x + 1)(x - 1)^2}$

18. $\dfrac{x^2 + 2}{x^3 - 1}, \dfrac{3}{x^2 + x + 1}$

$\dfrac{x^2 + 2}{(x - 1)(x^2 + x + 1)}, \dfrac{3x - 3}{(x - 1)(x^2 + x + 1)}$

19. $\dfrac{x - 3}{8 - x^3}, \dfrac{2}{4 + 2x + x^2}$

$-\dfrac{x - 3}{(x - 2)(x^2 + 2x + 4)},$

$\dfrac{2x - 4}{(x - 2)(x^2 + 2x + 4)}$

20. $\dfrac{2x}{x^2 + x - 6}, \dfrac{-4x}{x^2 + 5x + 6}$

$\dfrac{2x^2 + 4x}{(x + 3)(x - 2)(x + 2)}, -\dfrac{4x^2 - 8x}{(x + 3)(x - 2)(x + 2)}$

21. $\dfrac{2x}{x^2 + 2x - 3}, \dfrac{-x}{x^2 + 6x + 9}$

$\dfrac{2x^2 + 6x}{(x - 1)(x + 3)^2}, -\dfrac{x^2 - x}{(x - 1)(x + 3)^2}$

22. $\dfrac{3x}{2x^2 - x - 3}, \dfrac{-2x}{2x^2 - 11x + 12}$

$\dfrac{3x^2 - 12x}{(x + 1)(2x - 3)(x - 4)}, -\dfrac{2x^2 + 2x}{(x + 1)(2x - 3)(x - 4)}$

23. $\dfrac{-4x}{4x^2 - 16x + 15}, \dfrac{3x}{6x^2 - 19x + 10}$

$-\dfrac{12x^2 - 8x}{(2x - 3)(2x - 5)(3x - 2)}, \dfrac{6x^2 - 9x}{(2x - 3)(2x - 5)(3x - 2)}$

24. $\dfrac{3}{2x^2 + 5x - 12}, \dfrac{2x}{3 - 2x}, \dfrac{3x - 1}{x + 4}$

$\dfrac{3}{(2x - 3)(x + 4)}, -\dfrac{2x^2 + 8x}{(2x - 3)(x + 4)}, \dfrac{6x^2 - 11x + 3}{(2x - 3)(x + 4)}$

25. $\dfrac{5}{6x^2 - 17x + 12}, \dfrac{2x}{4 - 3x}, \dfrac{x + 1}{2x - 3}$

$\dfrac{5}{(3x - 4)(2x - 3)}, -\dfrac{4x^2 - 6x}{(3x - 4)(2x - 3)}, \dfrac{3x^2 - x - 4}{(3x - 4)(2x - 3)}$

26. $\dfrac{3x}{x - 4}, \dfrac{4}{x + 5}, \dfrac{x + 2}{20 - x - x^2}$

$\dfrac{3x^2 + 15x}{(x + 5)(x - 4)}, \dfrac{4x - 16}{(x + 5)(x - 4)}, -\dfrac{x + 2}{(x + 5)(x - 4)}$

27. $\dfrac{2x}{x - 3}, \dfrac{-2}{x + 5}, \dfrac{x - 1}{15 - 2x - x^2}$

$\dfrac{2x^2 + 10x}{(x + 5)(x - 3)}, -\dfrac{2x - 6}{(x + 5)(x - 3)}, -\dfrac{x - 1}{(x + 5)(x - 3)}$

OBJECTIVE B *To add or subtract rational expressions*

28. True or false? $\dfrac{1}{2x} + \dfrac{1}{3x} = \dfrac{1}{5x}$

False

29. True or false? $\dfrac{1}{x - 3} + \dfrac{1}{3 - x} = 0$

True

For Exercises 30 to 73, simplify.

30. $\dfrac{3}{2xy} - \dfrac{7}{2xy} - \dfrac{9}{2xy}$

$-\dfrac{13}{2xy}$

31. $-\dfrac{3}{4x^2} + \dfrac{8}{4x^2} - \dfrac{3}{4x^2}$

$\dfrac{1}{2x^2}$

32. $\dfrac{x}{x^2 - 3x + 2} - \dfrac{2}{x^2 - 3x + 2}$

$\dfrac{1}{x - 1}$

33. $\dfrac{3x}{3x^2 + x - 10} - \dfrac{5}{3x^2 + x - 10}$

$\dfrac{1}{x + 2}$

34. $\dfrac{3}{2x^2y} - \dfrac{8}{5x} - \dfrac{9}{10xy}$

$\dfrac{15 - 16xy - 9x}{10x^2y}$

35. $\dfrac{2}{5ab} - \dfrac{3}{10a^2b} + \dfrac{4}{15ab^2}$

$\dfrac{12ab - 9b + 8a}{30a^2b^2}$

36. $\dfrac{2}{3x} - \dfrac{3}{2xy} + \dfrac{4}{5xy} - \dfrac{5}{6x}$

$\dfrac{5y + 21}{30xy}$

37. $\dfrac{3}{4ab} - \dfrac{2}{5a} + \dfrac{3}{10b} - \dfrac{5}{8ab}$

$\dfrac{5 - 16b + 12a}{40ab}$

38. $\dfrac{2x - 1}{12x} - \dfrac{3x + 4}{9x}$

$-\dfrac{6x + 19}{36x}$

39. $\dfrac{3x-4}{6x} - \dfrac{2x-5}{4x}$

$\dfrac{7}{12x}$

40. $\dfrac{3x+2}{4x^2 y} - \dfrac{y-5}{6xy^2}$

$\dfrac{7xy + 6y + 10x}{12x^2 y^2}$

41. $\dfrac{2y-4}{5xy^2} + \dfrac{3-2x}{10x^2 y}$

$\dfrac{2xy - 8x + 3y}{10x^2 y^2}$

42. $\dfrac{2x}{x-3} - \dfrac{3x}{x-5}$

$-\dfrac{x(x+1)}{(x-3)(x-5)}$

43. $\dfrac{3a}{a-2} - \dfrac{5a}{a+1}$

$-\dfrac{a(2a-13)}{(a-2)(a+1)}$

44. $\dfrac{3}{2a-3} + \dfrac{2a}{3-2a}$

-1

45. $\dfrac{x}{2x-5} - \dfrac{2}{5x-2}$

$\dfrac{5x^2 - 6x + 10}{(2x-5)(5x-2)}$

46. $\dfrac{1}{x+h} - \dfrac{1}{h}$

$-\dfrac{x}{h(x+h)}$

47. $\dfrac{1}{a-b} + \dfrac{1}{b}$

$\dfrac{a}{b(a-b)}$

48. $\dfrac{2}{x} - 3 - \dfrac{10}{x-4}$

$-\dfrac{3x^2 - 4x + 8}{x(x-4)}$

49. $\dfrac{6a}{a-3} - 5 + \dfrac{3}{a}$

$\dfrac{a^2 + 18a - 9}{a(a-3)}$

50. $\dfrac{1}{2x-3} - \dfrac{5}{2x} + 1$

$\dfrac{4x^2 - 14x + 15}{2x(2x-3)}$

51. $\dfrac{5}{x} - \dfrac{5x}{5-6x} + 2$

$\dfrac{17x^2 + 20x - 25}{x(6x-5)}$

52. $\dfrac{3}{x^2-1} + \dfrac{2x}{x^2+2x+1}$

$\dfrac{2x^2 + x + 3}{(x-1)(x+1)^2}$

53. $\dfrac{1}{x^2-6x+9} - \dfrac{1}{x^2-9}$

$\dfrac{6}{(x+3)(x-3)^2}$

54. $\dfrac{x}{x+3} - \dfrac{3-x}{x^2-9}$

$\dfrac{x+1}{x+3}$

55. $\dfrac{1}{x+2} - \dfrac{3x}{x^2+4x+4}$

$-\dfrac{2(x-1)}{(x+2)^2}$

56. $\dfrac{2x-3}{x+5} - \dfrac{x^2-4x-19}{x^2+8x+15}$

$\dfrac{x+2}{x+3}$

57. $\dfrac{-3x^2+8x+2}{x^2+2x-8} - \dfrac{2x-5}{x+4}$

$-\dfrac{5x^2 - 17x + 8}{(x+4)(x-2)}$

58. $\dfrac{2x-2}{4x^2-9} - \dfrac{5}{3-2x}$

$\dfrac{12x+13}{(2x+3)(2x-3)}$

59. $\dfrac{x^2+4}{4x^2-36} - \dfrac{13}{x+3}$

$\dfrac{x^2 - 52x + 160}{4(x+3)(x-3)}$

60. $\dfrac{x-2}{x+1} - \dfrac{3-12x}{2x^2-x-3}$

$\dfrac{2x+3}{2x-3}$

61. $\dfrac{3x-4}{4x+1} + \dfrac{3x+6}{4x^2+9x+2}$

$\dfrac{3x-1}{4x+1}$

62. $\dfrac{x+1}{x^2+x-6} - \dfrac{x+2}{x^2+4x+3}$

$\dfrac{2x+5}{(x+3)(x-2)(x+1)}$

63. $\dfrac{x+1}{x^2+x-12} - \dfrac{x-3}{x^2+7x+12}$

$\dfrac{2(5x-3)}{(x+4)(x-3)(x+3)}$

64. $\dfrac{x^2 + 6x}{x^2 + 3x - 18} - \dfrac{2x - 1}{x + 6} + \dfrac{x - 2}{3 - x}$

$-\dfrac{2x^2 - 9x - 9}{(x + 6)(x - 3)}$

65. $\dfrac{2x^2 - 2x}{x^2 - 2x - 15} - \dfrac{2}{x + 3} + \dfrac{x}{5 - x}$

$\dfrac{x - 2}{x + 3}$

66. $\dfrac{7 - 4x}{2x^2 - 9x + 10} + \dfrac{x - 3}{x - 2} - \dfrac{x + 1}{2x - 5}$

$\dfrac{x - 12}{2x - 5}$

67. $\dfrac{x}{3x + 4} + \dfrac{3x + 2}{x - 5} - \dfrac{7x^2 + 24x + 28}{3x^2 - 11x - 20}$

1

68. $\dfrac{32x - 9}{2x^2 + 7x - 15} + \dfrac{x - 2}{3 - 2x} + \dfrac{3x + 2}{x + 5}$

$\dfrac{5x - 1}{2x - 3}$

69. $\dfrac{x + 1}{1 - 2x} - \dfrac{x + 3}{4x - 3} + \dfrac{10x^2 + 7x - 9}{8x^2 - 10x + 3}$

$\dfrac{x + 1}{2x - 1}$

70. $\dfrac{x^2}{x^3 - 8} - \dfrac{x + 2}{x^2 + 2x + 4}$

$\dfrac{4}{(x - 2)(x^2 + 2x + 4)}$

71. $\dfrac{2x}{4x^2 + 2x + 1} + \dfrac{4x + 1}{8x^3 - 1}$

$\dfrac{1}{2x - 1}$

72. $\dfrac{2x^2}{x^4 - 1} - \dfrac{1}{x^2 - 1} + \dfrac{1}{x^2 + 1}$

$\dfrac{2}{x^2 + 1}$

73. $\dfrac{x^2 - 12}{x^4 - 16} + \dfrac{1}{x^2 - 4} - \dfrac{1}{x^2 + 4}$

$\dfrac{1}{x^2 + 4}$

Critical Thinking

For Exercises 74 to 77, simplify.

74. $\dfrac{3}{x - 2} - \dfrac{x^2 + x}{2x^3 + 3x^2} \cdot \dfrac{2x^2 + x - 3}{x^2 + 3x + 2}$

$\dfrac{2x^2 + 9x - 2}{x(x - 2)(x + 2)}$

75. $\dfrac{x^2 - 4x + 4}{2x + 1} \cdot \dfrac{2x^2 + x}{x^3 - 4x} - \dfrac{3x - 2}{x + 1}$

$-\dfrac{2x^2 + 5x - 2}{(x + 2)(x + 1)}$

76. $\left(\dfrac{x - y}{x^2} - \dfrac{x - y}{y^2}\right) \div \dfrac{x^2 - y^2}{xy}$

$\dfrac{y - x}{xy}$

77. $\left(\dfrac{a - 2b}{b} + \dfrac{b}{a}\right) \div \left(\dfrac{b + a}{a} - \dfrac{2a}{b}\right)$

$\dfrac{b - a}{b + 2a}$

Projects or Group Activities

78. Let $f(x) = \dfrac{x}{x + 2}$, $g(x) = \dfrac{4}{x - 3}$, and $f(x) + g(x) = S(x)$. Then
$f(x) + g(x) = \dfrac{x}{x + 2} + \dfrac{4}{x - 3} = \dfrac{x^2 + x + 8}{x^2 - x - 6}$. Therefore, $S(x) = \dfrac{x^2 + x + 8}{x^2 - x - 6}$.
   a. Find $f(4)$, $g(4)$, and $S(4)$.
   b. Does $f(4) + g(4) = S(4)$?
   c. Does $f(x) + g(x) = S(x)$ for all values of x? Explain your answer.

   a. $\dfrac{2}{3}; 4; \dfrac{14}{3}$ **b.** Yes

   c. No. $f(x) + g(x) = S(x)$ for all values of x except $x = -2$ and $x = 3$.

QUICK QUIZ

Write each fraction in terms of the LCM of the denominators.

1. $\dfrac{6}{25ab}$, $\dfrac{9}{10a^2b^2}$

   $\dfrac{12ab}{50a^2b^2}$, $\dfrac{45}{50a^2b^2}$ **[8.2A]**

2. $\dfrac{5x}{x^2 - x - 6}$, $\dfrac{3x}{x^2 + 4x + 4}$

   $\dfrac{5x^2 + 10x}{(x - 3)(x + 2)^2}$, $\dfrac{3x^2 - 9x}{(x - 3)(x + 2)^2}$

   [8.2A]

Simplify.

3. $\dfrac{3}{4a^2b} - \dfrac{5}{3ab^2} - \dfrac{1}{12a^2b^2}$ $\dfrac{9b - 20a - 1}{12a^2b^2}$ **[8.2B]**

4. $\dfrac{7a}{a - 3} - \dfrac{5a}{a + 1}$ $\dfrac{2a(a + 11)}{(a - 3)(a + 1)}$ **[8.2B]**

5. $\dfrac{1}{x + 3} - \dfrac{2x}{x^2 + 6x + 9}$ $-\dfrac{x - 3}{(x + 3)^2}$ **[8.2B]**

8.3   Complex Fractions

OBJECTIVE A   *To simplify a complex fraction*

A **complex fraction** is a fraction in which the numerator or denominator contains one or more fractions. Examples of complex fractions are shown at the right.

$$\dfrac{5}{2 + \dfrac{1}{2}} \qquad \dfrac{5 + \dfrac{1}{y}}{5 - \dfrac{1}{y}} \qquad \dfrac{x + 4 + \dfrac{1}{x + 2}}{x - 2 + \dfrac{1}{x + 2}}$$

If the LCMs of the denominators in the numerator and denominator of a complex fraction are the same, the complex fraction can be simplified by multiplying the numerator and denominator by that expression.

 Take Note

Begin with the numerator of the complex fraction. The LCM of the denominators is x^2. Now consider the denominator of the complex fraction. The LCM of these denominators is also x^2.

HOW TO 1   Simplify: $\dfrac{1 + \dfrac{5}{x} - \dfrac{6}{x^2}}{1 + \dfrac{8}{x} + \dfrac{12}{x^2}}$

$$\dfrac{1 + \dfrac{5}{x} - \dfrac{6}{x^2}}{1 + \dfrac{8}{x} + \dfrac{12}{x^2}} = \dfrac{1 + \dfrac{5}{x} - \dfrac{6}{x^2}}{1 + \dfrac{8}{x} + \dfrac{12}{x^2}} \cdot \dfrac{x^2}{x^2}$$

• Multiply the numerator and denominator of the complex fraction by x^2.

$$= \dfrac{1 \cdot x^2 + \dfrac{5}{x} \cdot x^2 - \dfrac{6}{x^2} \cdot x^2}{1 \cdot x^2 + \dfrac{8}{x} \cdot x^2 + \dfrac{12}{x^2} \cdot x^2}$$

• Simplify.

$$= \dfrac{x^2 + 5x - 6}{x^2 + 8x + 12} = \dfrac{(x + 6)(x - 1)}{(x + 6)(x + 2)} = \dfrac{x - 1}{x + 2}$$

If the LCMs of the denominators in the numerator and denominator of a complex fraction are different, it may be easier to simplify the complex fraction by using a different approach.

 Take Note

Begin with the numerator of the complex fraction. The LCM of the denominators is $x - 3$. Now consider the denominator of the complex fraction. The LCM of the denominators is $x + 1$. These expressions are different.

Take Note

Either method of simplifying a complex fraction will always work. With experience, you will be able to decide which method works best for a particular complex fraction.

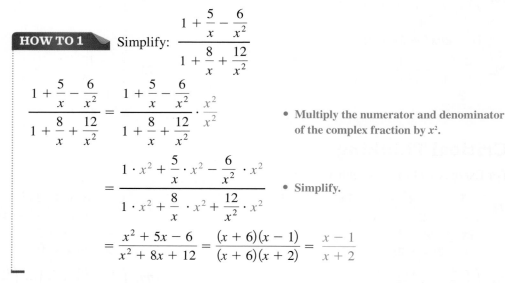

HOW TO 2   Simplify: $\dfrac{3 + \dfrac{10}{x - 3}}{3 - \dfrac{2}{x + 1}}$

$$\dfrac{3 + \dfrac{10}{x - 3}}{3 - \dfrac{2}{x + 1}} = \dfrac{\dfrac{3(x - 3)}{x - 3} + \dfrac{10}{x - 3}}{\dfrac{3(x + 1)}{x + 1} - \dfrac{2}{x + 1}}$$

• Simplify the numerator and denominator of the complex fraction by rewriting each as a single fraction.

$$= \dfrac{\dfrac{3x - 9}{x - 3} + \dfrac{10}{x - 3}}{\dfrac{3x + 3}{x + 1} - \dfrac{2}{x + 1}} = \dfrac{\dfrac{3x + 1}{x - 3}}{\dfrac{3x + 1}{x + 1}}$$

• Both the numerator and denominator of the complex fraction are now written as single fractions.

$$= \dfrac{3x + 1}{x - 3} \div \dfrac{3x + 1}{x + 1} = \dfrac{3x + 1}{x - 3} \cdot \dfrac{x + 1}{3x + 1} = \dfrac{x + 1}{x - 3}$$

• Divide the numerator by the denominator.

EXAMPLE 1

Simplify: $\dfrac{2x - 1 + \dfrac{7}{x + 4}}{3x - 8 + \dfrac{17}{x + 4}}$

Solution

The LCM of $x + 4$ and $x + 4$ is $x + 4$.

$$\dfrac{2x - 1 + \dfrac{7}{x + 4}}{3x - 8 + \dfrac{17}{x + 4}} = \dfrac{2x - 1 + \dfrac{7}{x + 4}}{3x - 8 + \dfrac{17}{x + 4}} \cdot \dfrac{x + 4}{x + 4}$$

$$= \dfrac{(2x - 1)(x + 4) + \dfrac{7}{x + 4}(x + 4)}{(3x - 8)(x + 4) + \dfrac{17}{x + 4}(x + 4)}$$

$$= \dfrac{2x^2 + 7x - 4 + 7}{3x^2 + 4x - 32 + 17}$$

$$= \dfrac{2x^2 + 7x + 3}{3x^2 + 4x - 15} = \dfrac{(2x + 1)(x + 3)}{(3x - 5)(x + 3)}$$

$$= \dfrac{(2x + 1)\cancel{(x + 3)}}{(3x - 5)\cancel{(x + 3)}} = \dfrac{2x + 1}{3x - 5}$$

YOU TRY IT 1

Simplify: $\dfrac{2x + 5 + \dfrac{14}{x - 3}}{4x + 16 + \dfrac{49}{x - 3}}$

Your solution

$\dfrac{x - 1}{2x + 1}$

IN-CLASS EXAMPLES

Simplify.

1. $\dfrac{2 + \dfrac{1}{x}}{4 - \dfrac{1}{x^2}}$ $\dfrac{x}{2x - 1}$

2. $\dfrac{1 + \dfrac{2}{x} - \dfrac{8}{x^2}}{1 + \dfrac{3}{x} - \dfrac{10}{x^2}}$ $\dfrac{x + 4}{x + 5}$

3. $\dfrac{x + 1 - \dfrac{4}{x - 2}}{x - 4 - \dfrac{24}{x - 2}}$ $\dfrac{x - 3}{x - 8}$

EXAMPLE 2

Simplify: $1 + \dfrac{a}{2 + \dfrac{1}{a}}$

Solution

The LCM of the denominators of a, 2, and $\frac{1}{a}$ is a.

$$1 + \dfrac{a}{2 + \dfrac{1}{a}} = 1 + \dfrac{a}{2 + \dfrac{1}{a}} \cdot \dfrac{a}{a}$$

$$= 1 + \dfrac{a \cdot a}{2 \cdot a + \dfrac{1}{a} \cdot a} = 1 + \dfrac{a^2}{2a + 1}$$

The LCM of the denominators of 1 and $\frac{a^2}{2a + 1}$
is $2a + 1$.

$$1 + \dfrac{a^2}{2a + 1} = 1 \cdot \dfrac{2a + 1}{2a + 1} + \dfrac{a^2}{2a + 1}$$

$$= \dfrac{2a + 1}{2a + 1} + \dfrac{a^2}{2a + 1} = \dfrac{2a + 1 + a^2}{2a + 1}$$

$$= \dfrac{a^2 + 2a + 1}{2a + 1} = \dfrac{(a + 1)^2}{2a + 1}$$

YOU TRY IT 2

Simplify: $2 - \dfrac{1}{2 - \dfrac{1}{x}}$

Your solution

$\dfrac{3x - 2}{2x - 1}$

8.3 EXERCISES

Concept Check

SUGGESTED ASSIGNMENT
Exercises 1–6; Exercises 7–43, odds;
More challenging exercises: Exercises 45–48

1. What is a complex fraction?

2. What is the LCD of the fractions in the complex fraction $\dfrac{1 - \dfrac{3}{x-3}}{\dfrac{2}{x} + 1}$? $x(x-3)$

For Exercises 3 to 6, simplify.

3. $\dfrac{2 - \dfrac{1}{3}}{4 + \dfrac{11}{3}}$ $\dfrac{5}{23}$

4. $\dfrac{3 + \dfrac{5}{2}}{8 - \dfrac{3}{2}}$ $\dfrac{11}{13}$

5. $\dfrac{3 - \dfrac{2}{3}}{5 + \dfrac{5}{6}}$ $\dfrac{2}{5}$

6. $\dfrac{5 - \dfrac{3}{4}}{2 + \dfrac{1}{2}}$ $\dfrac{17}{10}$

OBJECTIVE A *To simplify a complex fraction*

For Exercises 7 to 42, simplify.

7. $\dfrac{1 + \dfrac{1}{x}}{1 - \dfrac{1}{x^2}}$ $\dfrac{x}{x-1}$

8. $\dfrac{\dfrac{1}{y^2} - 1}{1 + \dfrac{1}{y}}$ $\dfrac{1-y}{y}$

9. $\dfrac{a-2}{\dfrac{4}{a} - a}$ $-\dfrac{a}{a+2}$

10. $\dfrac{\dfrac{25}{a} - a}{5 + a}$ $\dfrac{5-a}{a}$

11. $\dfrac{2 + \dfrac{1}{a}}{4 - \dfrac{1}{a^2}}$ $\dfrac{a}{2a-1}$

12. $\dfrac{9 - \dfrac{1}{x^2}}{3 + \dfrac{1}{x}}$ $\dfrac{3x-1}{x}$

13. $\dfrac{x - \dfrac{1}{x}}{x + \dfrac{1}{x}}$ $\dfrac{(x-1)(x+1)}{x^2+1}$

14. $\dfrac{a - \dfrac{1}{a}}{\dfrac{1}{a} + a}$ $\dfrac{(a-1)(a+1)}{a^2+1}$

15. $\dfrac{\dfrac{1}{a^2} - \dfrac{1}{a}}{\dfrac{1}{a^2} + \dfrac{1}{a}}$ $-\dfrac{a-1}{a+1}$

16. $\dfrac{\dfrac{1}{b} + \dfrac{1}{2}}{\dfrac{4}{b^2} - 1}$ $\dfrac{b}{2(2-b)}$

17. $\dfrac{2 - \dfrac{4}{x+2}}{5 - \dfrac{10}{x+2}}$ $\dfrac{2}{5}$

18. $\dfrac{4 + \dfrac{12}{2x-3}}{5 + \dfrac{15}{2x-3}}$ $\dfrac{4}{5}$

19. $\dfrac{\dfrac{3}{2a-3} + 2}{\dfrac{-6}{2a-3} - 4}$ $-\dfrac{1}{2}$

20. $\dfrac{\dfrac{-5}{b-5} - 3}{\dfrac{10}{b-5} + 6}$ $-\dfrac{1}{2}$

21. $\dfrac{1 - \dfrac{1}{x-4}}{1 - \dfrac{6}{x+1}}$ $\dfrac{x+1}{x-4}$

22. $\dfrac{1 + \dfrac{3}{x+2}}{1 + \dfrac{6}{x-1}}$ $\dfrac{x-1}{x+2}$

23. $\dfrac{1 - \dfrac{2}{x-3}}{1 + \dfrac{3}{2-x}}$ $\dfrac{x-2}{x-3}$

24. $\dfrac{1 + \dfrac{x}{x+1}}{1 + \dfrac{x-1}{x+2}}$ $\dfrac{x+2}{x+1}$

25. $\dfrac{x - 4 + \dfrac{9}{2x+3}}{x + 3 - \dfrac{5}{2x+3}}$ $\dfrac{x-3}{x+4}$

26. $\dfrac{2x - 3 - \dfrac{10}{4x-5}}{3x + 2 + \dfrac{11}{4x-5}}$ $\dfrac{2x-5}{3x-1}$

27. $\dfrac{x - 3 + \dfrac{10}{x + 4}}{x + 7 + \dfrac{16}{x - 3}}$
$\dfrac{x + 2}{x + 5}$

28. $\dfrac{x + 9 + \dfrac{30}{x - 2}}{x - 1 + \dfrac{8}{x + 5}}$
$\dfrac{x + 4}{x + 1}$

29. $\dfrac{1 - \dfrac{1}{x} - \dfrac{6}{x^2}}{1 - \dfrac{4}{x} + \dfrac{3}{x^2}}$
$\dfrac{x + 2}{x - 1}$

30. $\dfrac{1 - \dfrac{3}{x} - \dfrac{10}{x^2}}{1 + \dfrac{11}{x} + \dfrac{18}{x^2}}$
$\dfrac{x - 5}{x + 9}$

31. $\dfrac{1 + \dfrac{1}{x} - \dfrac{12}{x^2}}{\dfrac{9}{x^2} + \dfrac{3}{x} - 2}$
$-\dfrac{x + 4}{2x + 3}$

32. $\dfrac{\dfrac{15}{x^2} - \dfrac{2}{x} - 1}{\dfrac{4}{x^2} - \dfrac{5}{x} + 4}$
$-\dfrac{(x - 3)(x + 5)}{4x^2 - 5x + 4}$

33. $\dfrac{\dfrac{1}{y^2} - \dfrac{1}{xy} - \dfrac{2}{x^2}}{\dfrac{1}{y^2} - \dfrac{3}{xy} + \dfrac{2}{x^2}}$
$\dfrac{x + y}{x - y}$

34. $\dfrac{\dfrac{2}{b^2} - \dfrac{5}{ab} - \dfrac{3}{a^2}}{\dfrac{2}{b^2} + \dfrac{7}{ab} + \dfrac{3}{a^2}}$
$\dfrac{a - 3b}{a + 3b}$

35. $\dfrac{\dfrac{x}{x + 1} - \dfrac{1}{x}}{\dfrac{x}{x + 1} + \dfrac{1}{x}}$
$\dfrac{x^2 - x - 1}{x^2 + x + 1}$

36. $\dfrac{\dfrac{2a}{a - 1} - \dfrac{3}{a}}{\dfrac{1}{a - 1} + \dfrac{2}{a}}$
$\dfrac{2a^2 - 3a + 3}{3a - 2}$

37. $\dfrac{\dfrac{1}{a} - \dfrac{3}{a - 2}}{\dfrac{2}{a} + \dfrac{5}{a - 2}}$
$-\dfrac{2(a + 1)}{7a - 4}$

38. $\dfrac{\dfrac{2}{b} - \dfrac{5}{b + 3}}{\dfrac{3}{b} + \dfrac{3}{b + 3}}$
$\dfrac{b - 2}{2b + 3}$

39. $\dfrac{\dfrac{x - 1}{x + 1} - \dfrac{x + 1}{x - 1}}{\dfrac{x - 1}{x + 1} + \dfrac{x + 1}{x - 1}}$
$-\dfrac{2x}{x^2 + 1}$

40. $\dfrac{\dfrac{y}{y + 2} - \dfrac{y}{y - 2}}{\dfrac{y}{y + 2} + \dfrac{y}{y - 2}}$
$-\dfrac{2}{y}$

41. $a + \dfrac{a}{a + \dfrac{1}{a}}$
$\dfrac{a(a^2 + a + 1)}{a^2 + 1}$

42. $a - \dfrac{a}{1 - \dfrac{a}{1 - a}}$
$-\dfrac{a^2}{1 - 2a}$

43. 🐾 In simplest form, what is the reciprocal of the complex fraction $\dfrac{1}{1 - \dfrac{1}{a}}$? $\dfrac{a - 1}{a}$

44. 🐾 The denominator of a complex fraction is the reciprocal of its numerator. Which of the following is the simplified form of the complex fraction?
(i) 1 **(ii)** the square of the numerator of the complex fraction
(iii) the square of the denominator of the complex fraction   ii

QUICK QUIZ
Simplify.

1. $\dfrac{\dfrac{9}{y^2} - 1}{1 + \dfrac{3}{y}}$ $\dfrac{3 - y}{y}$ [8.3A]

Critical Thinking

2. $\dfrac{8 + \dfrac{6}{x} - \dfrac{5}{x^2}}{4 - \dfrac{3}{x} - \dfrac{10}{x^2}}$ $\dfrac{2x - 1}{x - 2}$ [8.3A]

3. $\dfrac{x + 2 - \dfrac{4}{x + 5}}{x - 1 - \dfrac{7}{x + 5}}$ $\dfrac{x + 1}{x - 2}$ [8.3A]

For Exercises 45 to 48, simplify.

45. $\dfrac{x^{-1}}{y^{-1}} + \dfrac{y}{x}$
$\dfrac{2y}{x}$

46. $\dfrac{x^{-1} + y^{-1}}{x^{-1} - y^{-1}}$
$\dfrac{y + x}{y - x}$

47. $\dfrac{\dfrac{1}{x + h} - \dfrac{1}{x}}{h}$
$-\dfrac{1}{x(x + h)}$

48. $\dfrac{\dfrac{1}{(x + h)^2} - \dfrac{1}{x^2}}{h}$
$-\dfrac{2x + h}{x^2(x + h)^2}$

Projects or Group Activities

49. Car Loans The interest rate on a car loan affects the monthly payment. The function that relates the monthly payment for a 5-year (60-month) loan to the monthly interest rate is shown at the right, where x is the monthly interest rate (as a decimal), C is the loan amount, and $P(x)$ is the monthly payment.

$P(x) = \dfrac{Cx}{\left[1 - \dfrac{1}{(x + 1)^{60}}\right]}$

a. Simplify the complex fraction. $P(x) = \dfrac{Cx(x + 1)^{60}}{(x + 1)^{60} - 1}$

b. Use a calculator to determine the monthly payment for a car loan of $20,000 at an annual interest rate of 8%. Round to the nearest cent. $405.53

8.4 Solving Equations Containing Fractions

OBJECTIVE A
To solve an equation containing fractions

INSTRUCTOR NOTE
If you did not cover Section 7.5 on solving quadratic equations by factoring, skip Example 1 and You Try It 1 on the next page. Also skip Exercises 34 to 40 in the Section 8.4 Exercises.

Recall that to solve an equation containing fractions, clear denominators by multiplying each side of the equation by the LCM of the denominators. Then solve for the variable.

HOW TO 1 Solve: $\dfrac{3x-1}{4}+\dfrac{2}{3}=\dfrac{7}{6}$

$$\dfrac{3x-1}{4}+\dfrac{2}{3}=\dfrac{7}{6}$$

$$12\left(\dfrac{3x-1}{4}+\dfrac{2}{3}\right)=12\cdot\dfrac{7}{6}$$ • The LCM is 12. To clear denominators, multiply each side of the equation by the LCM.

$$12\left(\dfrac{3x-1}{4}\right)+12\cdot\dfrac{2}{3}=12\cdot\dfrac{7}{6}$$ • Simplify by using the Distributive Property and the Properties of Fractions.

$$\dfrac{\overset{3}{\cancel{12}}}{1}\left(\dfrac{3x-1}{\cancel{4}}\right)+\dfrac{\overset{4}{\cancel{12}}}{1}\cdot\dfrac{2}{\cancel{3}}=\dfrac{\overset{2}{\cancel{12}}}{1}\cdot\dfrac{7}{\cancel{6}}$$

$$9x-3+8=14$$ • Solve for x.
$$9x+5=14$$
$$9x=9$$
$$x=1$$

1 checks as a solution. The solution is 1.

Occasionally, a value that appears to be a solution of an equation will make one of the denominators zero. In such a case, that value is not a solution of the equation.

HOW TO 2 Solve: $\dfrac{2x}{x-2}=1+\dfrac{4}{x-2}$

Take Note
HOW TO 2 at the right illustrates the importance of checking a solution of a rational equation when each side is multiplied by a variable expression. As shown in this example, a proposed solution may not check when it is substituted into the original equation.

$$\dfrac{2x}{x-2}=1+\dfrac{4}{x-2}$$

$$(x-2)\dfrac{2x}{x-2}=(x-2)\left(1+\dfrac{4}{x-2}\right)$$ • The LCM is $x-2$. Multiply each side of the equation by the LCM.

$$(x-2)\dfrac{2x}{x-2}=(x-2)\cdot1+(x-2)\dfrac{4}{x-2}$$ • Simplify by using the Distributive Property and the Properties of Fractions.

$$\dfrac{\cancel{(x-2)}}{1}\cdot\dfrac{2x}{\cancel{x-2}}=(x-2)\cdot1+\dfrac{\cancel{(x-2)}}{1}\cdot\dfrac{4}{\cancel{x-2}}$$

$$2x=x-2+4$$ • Solve for x.
$$2x=x+2$$
$$x=2$$

When x is replaced by 2, the denominators of $\dfrac{2x}{x-2}$ and $\dfrac{4}{x-2}$ are zero. Therefore, the equation has no solution.

EXAMPLE 1

Solve: $\dfrac{x}{x+4} = \dfrac{2}{x}$

Solution

The LCM is $x(x+4)$.

$$\dfrac{x}{x+4} = \dfrac{2}{x}$$

$$x(x+4)\left(\dfrac{x}{x+4}\right) = x(x+4)\left(\dfrac{2}{x}\right)$$ • Multiply by the LCM.

$$\dfrac{x\overset{1}{\cancel{(x+4)}}}{1} \cdot \dfrac{x}{\cancel{x+4}} = \dfrac{\overset{1}{\cancel{x}}(x+4)}{1} \cdot \dfrac{2}{\cancel{x}}$$ • Divide by the common factors.

$$x^2 = (x+4)2$$ • Simplify.
$$x^2 = 2x + 8$$

Solve the quadratic equation by factoring.

$$x^2 - 2x - 8 = 0$$ • Write in standard form.
$$(x-4)(x+2) = 0$$ • Factor.
$$x - 4 = 0 \qquad x + 2 = 0$$ • Principle of Zero
$$x = 4 \qquad\quad x = -2$$ Products

Both 4 and −2 check as solutions.
The solutions are 4 and −2.

YOU TRY IT 1

Solve: $\dfrac{x}{x+6} = \dfrac{3}{x}$

Your solution

−3, 6

EXAMPLE 2

Solve: $\dfrac{3x}{x-4} = 5 + \dfrac{12}{x-4}$

Solution

The LCM is $x - 4$.

$$\dfrac{3x}{x-4} = 5 + \dfrac{12}{x-4}$$

$$(x-4)\left(\dfrac{3x}{x-4}\right) = (x-4)\left(5 + \dfrac{12}{x-4}\right)$$ • Clear denominators.

$$\dfrac{\overset{1}{\cancel{(x-4)}}}{1} \cdot \dfrac{3x}{\cancel{x-4}} = (x-4)5 + \dfrac{\overset{1}{\cancel{(x-4)}}}{1} \cdot \dfrac{12}{\cancel{x-4}}$$

$$3x = (x-4)5 + 12$$ • Solve for x.
$$3x = 5x - 20 + 12$$
$$3x = 5x - 8$$
$$-2x = -8$$
$$x = 4$$

4 does not check as a solution.
The equation has no solution.

YOU TRY IT 2

Solve: $\dfrac{5x}{x+2} = 3 - \dfrac{10}{x+2}$

Your solution

No solution

IN-CLASS EXAMPLES

Solve.

1. $\dfrac{8}{4x-3} = -4$ \quad $\dfrac{1}{4}$

2. $\dfrac{3x}{2x+1} + \dfrac{1}{x+2} = \dfrac{4}{x+2}$ \quad 1, −1

Solutions on p. S23

8.4 EXERCISES

✔ Concept Check

SUGGESTED ASSIGNMENT
Exercises 1–4; Exercises 9–39, odds

1. The process of clearing denominators in an equation containing fractions is an application of which property of equations?   Multiplication Property of Equations

2. If the denominator of a fraction is $x + 3$, for what value of x is the fraction undefined?
   -3

3. Explain why you can clear denominators in part (a) below but not in part (b).

   a. $\dfrac{x}{2} + \dfrac{1}{3} = \dfrac{5}{2}$ **b.** $\dfrac{x}{2} + \dfrac{1}{3} + \dfrac{5}{2}$ We can clear denominators in an *equation*, as in part (a), but not in an *expression*, as in part (b).

4. ◤ After solving an equation containing fractions, why must we check the solution?

OBJECTIVE A   *To solve an equation containing fractions*

◤ When a proposed solution of a rational equation does not check in the original equation, it is because the proposed solution results in an expression that involves division by zero. For Exercises 5 to 7, state the values of x that would result in division by zero when substituted into the original equation.

5. $\dfrac{6x}{x + 1} - \dfrac{x}{x - 2} = 4$

   $-1, 2$

6. $\dfrac{1}{x + 5} = \dfrac{x}{x - 3} + \dfrac{2}{x^2 + 2x - 15}$

   $-5, 3$

7. $\dfrac{3}{x - 9} = \dfrac{1}{x^2 - 9x} + 2$

   $0, 9$

For Exercises 8 to 40, solve.

8. $\dfrac{2x}{3} - \dfrac{5}{2} = -\dfrac{1}{2}$

   3

9. $\dfrac{x}{3} - \dfrac{1}{4} = \dfrac{1}{12}$

   1

10. $\dfrac{x}{3} - \dfrac{1}{4} = \dfrac{x}{4} - \dfrac{1}{6}$

    1

11. $\dfrac{2y}{9} - \dfrac{1}{6} = \dfrac{y}{9} + \dfrac{1}{6}$

    3

12. $\dfrac{2x - 5}{8} + \dfrac{1}{4} = \dfrac{x}{8} + \dfrac{3}{4}$

    9

13. $\dfrac{3x + 4}{12} - \dfrac{1}{3} = \dfrac{5x + 2}{12} - \dfrac{1}{2}$

    2

14. $\dfrac{6}{2a + 1} = 2$

    1

15. $\dfrac{12}{3x - 2} = 3$

    2

16. $\dfrac{9}{2x - 5} = -2$

    $\dfrac{1}{4}$

17. $\dfrac{6}{4 - 3x} = 3$

    $\dfrac{2}{3}$

18. $2 + \dfrac{5}{x} = 7$

    1

19. $3 + \dfrac{8}{n} = 5$

    4

20. $1 - \dfrac{9}{x} = 4$

    -3

21. $3 - \dfrac{12}{x} = 7$

    -3

22. $\dfrac{2}{y} + 5 = 9$

    $\dfrac{1}{2}$

23. $\dfrac{6}{x} + 3 = 11$

$\dfrac{3}{4}$

24. $\dfrac{3}{x - 2} = \dfrac{4}{x}$

8

25. $\dfrac{5}{x + 3} = \dfrac{3}{x - 1}$

7

26. $\dfrac{2}{3x - 1} = \dfrac{3}{4x + 1}$

5

27. $\dfrac{5}{3x - 4} = \dfrac{-3}{1 - 2x}$

−7

28. $\dfrac{-3}{2x + 5} = \dfrac{2}{x - 1}$

−1

29. $\dfrac{4}{5y - 1} = \dfrac{2}{2y - 1}$

−1

30. $\dfrac{4x}{x - 4} + 5 = \dfrac{5x}{x - 4}$

5

31. $\dfrac{2x}{x + 2} - 5 = \dfrac{7x}{x + 2}$

−1

32. $2 + \dfrac{3}{a - 3} = \dfrac{a}{a - 3}$

No solution

33. $\dfrac{x}{x + 4} = 3 - \dfrac{4}{x + 4}$

No solution

34. $\dfrac{x}{x - 1} = \dfrac{8}{x + 2}$

2, 4

35. $\dfrac{x}{x + 12} = \dfrac{1}{x + 5}$

2, −6

36. $\dfrac{2x}{x + 4} = \dfrac{3}{x - 1}$

$-\dfrac{3}{2}$, 4

37. $\dfrac{5}{3n - 8} = \dfrac{n}{n + 2}$

$-\dfrac{2}{3}$, 5

38. $\dfrac{x}{x + 4} = \dfrac{11}{x^2 - 16} + 2$

−7, 3

39. $x - \dfrac{6}{x - 3} = \dfrac{2x}{x - 3}$

−1, 6

40. $\dfrac{8}{r} + \dfrac{3}{r - 1} = 3$

$\dfrac{2}{3}$, 4

Critical Thinking

For Exercises 41 to 44, solve.

41. $\dfrac{3}{5}y - \dfrac{1}{3}(1 - y) = \dfrac{2y - 5}{15}$

0

42. $\dfrac{3}{4}a = \dfrac{1}{2}(3 - a) + \dfrac{a - 2}{4}$

1

43. $\dfrac{x + 1}{x^2 + x - 2} = \dfrac{x + 2}{x^2 - 1} + \dfrac{3}{x + 2}$

$0, -\dfrac{2}{3}$

44. $\dfrac{y + 2}{y^2 - y - 2} + \dfrac{y + 1}{y^2 - 4} = \dfrac{1}{y + 1}$

−3

Projects or Group Activities

Intensity of Illumination   You are already aware that the standard unit of length in the metric system is the meter (m). You may not know that the standard unit of light intensity is the **candela (cd)**.

The rate at which light falls on a 1-square-unit area of surface is called the **intensity of illumination**. Intensity of illumination is measured in **lumens (lm)**. A lumen is defined in the following illustration.

Picture a source of light equal to 1 cd positioned at the center of a hollow sphere that has a radius of 1 m. The rate at which light falls on 1 m² of the inner surface of the sphere is equal to 1 lm. If a light source equal to 4 cd is positioned at the center of the sphere, then each square meter of the inner surface receives four times as much illumination, or 4 lm.

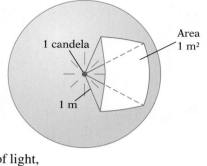

Light rays diverge as they leave a light source. The light that falls on an area of 1 m² at a distance of 1 m from the light source spreads out over an area of 4 m² when it is 2 m from the source. The same light spreads out over an area of 9 m² when it is 3 m from the light source, and over an area of 16 m² when it is 4 m from the light source. Therefore, as a surface moves farther away from the source of light, the intensity of illumination on the surface decreases from its value at 1 m to $\left(\frac{1}{2}\right)^2$, or $\frac{1}{4}$, that value at 2 m; to $\left(\frac{1}{3}\right)^2$, or $\frac{1}{9}$, that value at 3 m; and to $\left(\frac{1}{4}\right)^2$, or $\frac{1}{16}$, that value at 4 m.

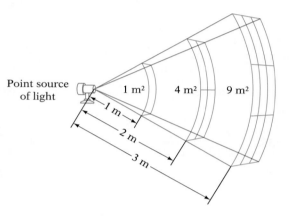

The formula for the intensity of illumination is

$$I = \frac{s}{r^2}$$

where I is the intensity of illumination in lumens, s is the intensity of the light source in candelas, and r is the distance in meters between the light source and the illuminated surface.

Example   A 30-candela lamp is positioned 0.5 m above a desk. Find the illumination on the desk.

$$I = \frac{s}{r^2}$$

$$I = \frac{30}{(0.5)^2} = 120$$

The illumination on the desk is 120 lm.

QUICK QUIZ

Solve.

1. $5 - \frac{8}{x} = 1$ 2   [8.4A]

2. $3 - \frac{4}{x+2} = \frac{2x}{x+2}$ **No solution**   [8.4A]

45. A 100-candela light is hanging 5 m above a floor. What is the intensity of the illumination on the floor beneath the light?   4 lm

46. A 25-candela source of light is positioned 2 m above a desk. Find the intensity of illumination on the desk.   6.25 lm

47. How strong a light source is needed to cast 20 lm of light on a surface 4 m from the source?   320 candela

48. How strong a light source is needed to cast 80 lm of light on a surface 5 m from the source?   2000 candela

49. How far from the desk surface must a 40-candela light source be positioned if the desired intensity of illumination is 10 lm?   2 m

50. Find the distance between a 36-candela light source and a surface if the intensity of illumination on the surface is 0.01 lm.   60 m

Unless otherwise noted, all content on this page is © Cengage Learning.

✔ CHECK YOUR PROGRESS: CHAPTER 8

1. Simplify: $\dfrac{x^2 - 2x - 8}{x^2 - 8x + 16}$

   $\dfrac{x + 2}{x - 4}$ [8.1A]

2. Simplify: $\dfrac{2x^2 - 11x - 40}{6x^2 - x - 40}$

   $\dfrac{x - 8}{3x - 8}$ [8.1A]

3. Multiply: $\dfrac{x^2 - 3x - 18}{x^2 - 5x - 24} \cdot \dfrac{x^2 - 2x - 15}{x^2 + 12x + 27}$

   $\dfrac{(x - 6)(x - 5)}{(x + 9)(x - 8)}$ [8.1B]

4. Multiply: $\dfrac{x^2 + x - 72}{x^2 + 14x + 45} \cdot \dfrac{2x^2 + 15x + 25}{3x^2 - 15x + 72}$

   $\dfrac{2x + 5}{3x + 9}$ [8.1B]

5. Divide: $\dfrac{6x^3 y^2}{18a^4 b} \div \dfrac{3xy}{9a^2 b^5}$

   $\dfrac{b^4 x^2 y}{a^2}$ [8.1C]

6. Divide: $\dfrac{3x^2 + 17x - 28}{x^2 + 2x - 15} \div \dfrac{12x^2 - 13x - 4}{x^2 - 6x + 9}$

   $\dfrac{(x + 7)(x - 3)}{(4x + 1)(x + 5)}$ [8.1C]

7. Find the LCM of $x^2 + 4x$ and $x^2 + 9x + 20$.

   $x(x + 4)(x + 5)$ [8.2A]

8. Find the LCM of $x^2 - 4$ and $x^2 + 2x - 8$.

   $(x - 2)(x + 2)(x + 4)$ [8.2A]

9. Add: $\dfrac{x + 6}{x - 1} + \dfrac{4x + 5}{x - 4}$

   $\dfrac{5x^2 + 3x - 29}{(x - 1)(x - 4)}$ [8.2B]

10. Add: $\dfrac{x + 9}{x - 3} + \dfrac{3x + 4}{x^2 - 12x + 27}$

    $\dfrac{x^2 + 3x - 77}{(x - 3)(x - 9)}$ [8.2B]

11. Subtract: $\dfrac{x + 9}{3x + 4} - \dfrac{x + 3}{x + 1}$

    $-\dfrac{2x^2 + 3x + 3}{(3x + 4)(x + 1)}$ [8.2B]

12. Subtract: $\dfrac{x - 8}{3x^2 + 20x - 63} - \dfrac{x + 2}{3x - 7}$

    $-\dfrac{x^2 + 10x + 26}{(x + 9)(3x - 7)}$ [8.2B]

13. Simplify: $\dfrac{1 + \dfrac{3}{x}}{1 - \dfrac{9}{x^2}}$

    $\dfrac{x}{x - 3}$ [8.3A]

14. Simplify: $\dfrac{\dfrac{7}{x - 3} - \dfrac{2}{3x}}{\dfrac{5}{3x} + \dfrac{1}{x - 3}}$

    $\dfrac{19x + 6}{8x - 15}$ [8.3A]

15. Solve: $\dfrac{5}{y + 3} - 2 = \dfrac{7}{y + 3}$

    -4 [8.4A]

16. Solve: $5 + \dfrac{8}{a - 2} = \dfrac{4a}{a - 2}$

    No solution [8.4A]

SECTION

8.5 Ratio and Proportion

OBJECTIVE A   *To solve a proportion*

 Point of Interest

The Women's Restroom Equity Bill was signed by New York City Mayor Michael Bloomberg and approved unanimously by the NYC Council. This bill requires that women's and men's bathroom stalls in bars, sports arenas, theaters, and highway service areas be in a ratio of 2 to 1. Nicknamed "potty parity," this legislation attempts to shorten the long lines at ladies rooms throughout the city.

Quantities such as 4 meters, 15 seconds, and 8 gallons are number quantities written with units. In these examples, the units are meters, seconds, and gallons.

A **ratio** is the quotient of two quantities that have the same unit.

The length of a living room is 16 ft and the width is 12 ft. The ratio of the length to the width is written

$$\frac{16 \text{ ft}}{12 \text{ ft}} = \frac{16}{12} = \frac{4}{3}$$ A ratio is in simplest form when the two numbers do not have a common factor. Note that the units are not written.

A **rate** is the quotient of two quantities that have different units.

There are 2 lb of salt in 8 gal of water. The salt-to-water rate is

$$\frac{2 \text{ lb}}{8 \text{ gal}} = \frac{1 \text{ lb}}{4 \text{ gal}}$$ A rate is in simplest form when the two numbers do not have a common factor. The units are written as part of the rate.

A **proportion** is an equation that states the equality of two ratios or rates. Examples of proportions are shown at the right.

$$\frac{30 \text{ mi}}{4 \text{ h}} = \frac{15 \text{ mi}}{2 \text{ h}} \qquad \frac{4}{6} = \frac{8}{12} \qquad \frac{3}{4} = \frac{x}{8}$$

HOW TO 1   Solve the proportion $\frac{4}{x} = \frac{2}{3}$.

$$\frac{4}{x} = \frac{2}{3}$$

$$3x\left(\frac{4}{x}\right) = 3x\left(\frac{2}{3}\right)$$ • The LCM of the denominators is $3x$. To clear denominators, multiply each side of the proportion by the LCM.

$$12 = 2x$$ • Solve the equation for x.

$$6 = x$$

The solution is 6.

APPLY THE CONCEPT

Nine ceramic tiles are required to tile a 4-square-foot area. At this rate, how many square feet can be tiled using 270 ceramic tiles?

To find the total area that 270 ceramic tiles will cover, write and solve a proportion using x to represent the number of square feet that 270 tiles will cover.

$$\frac{4}{9} = \frac{x}{270}$$ • The numerators represent square feet covered. The denominators represent numbers of tiles.

$$270\left(\frac{4}{9}\right) = 270\left(\frac{x}{270}\right)$$ • Multiply by the LCM of the denominators.

$$120 = x$$

A 120-square-foot area can be tiled using 270 tiles.

Athanasia Nomikou/Shutterstock.com

EXAMPLE 1

Solve: **A.** $\dfrac{8}{x+3} = \dfrac{4}{x}$ **B.** $\dfrac{6}{x+4} = \dfrac{12}{5x-13}$

Solution

A.
$$\frac{8}{x+3} = \frac{4}{x}$$

$$x(x+3)\frac{8}{x+3} = x(x+3)\frac{4}{x} \quad \bullet \text{ Clear denominators.}$$

$$8x = 4(x+3) \quad \bullet \text{ Solve for } x.$$

$$8x = 4x + 12$$

$$4x = 12$$

$$x = 3$$

The solution is 3.

B.
$$\frac{6}{x+4} = \frac{12}{5x-13}$$

$$(5x-13)(x+4)\frac{6}{x+4} = (5x-13)(x+4)\frac{12}{5x-13}$$

$$(5x-13)6 = (x+4)12$$

$$30x - 78 = 12x + 48$$

$$18x - 78 = 48$$

$$18x = 126$$

$$x = 7$$

The solution is 7.

YOU TRY IT 1

Solve. **A.** $\dfrac{2}{x+3} = \dfrac{6}{5x+5}$ **B.** $\dfrac{5}{2x-3} = \dfrac{10}{x+3}$

Your solution

A. 2 B. 3

IN-CLASS EXAMPLES

Solve.

1. $\dfrac{10}{x} = \dfrac{5}{7}$ **14**

2. $\dfrac{3}{8} = \dfrac{12}{x-2}$ **34**

3. $\dfrac{x-5}{4} = \dfrac{5x-1}{11}$ $-\dfrac{17}{3}$

4. Biologists catch, tag, and release 125 trout in a pond. Later, 200 trout are caught and checked for tags. Forty of these trout are found to have tags. Estimate the total trout population of the pond. **625 trout**

EXAMPLE 2

The monthly loan payment for a car is $28.35 for each $1000 borrowed. At this rate, find the monthly payment for a $6000 car loan.

Strategy

To find the monthly payment, write and solve a proportion using P to represent the monthly car payment.

Solution

$$\frac{28.35}{1000} = \frac{P}{6000} \quad \bullet \text{ Write a proportion.}$$

$$6000\left(\frac{28.35}{1000}\right) = 6000\left(\frac{P}{6000}\right) \quad \bullet \text{ Clear denominators.}$$

$$170.10 = P$$

The monthly payment is $170.10.

YOU TRY IT 2

Three ounces of medication are required for a 120-pound adult. At this rate, how many ounces of medication are required for a 180-pound adult?

Your strategy

Your solution

4.5 oz

Solutions on p. S24

magicinfoto/Shutterstock.com

OBJECTIVE B   *To solve problems involving similar triangles*

Similar objects have the same shape but not necessarily the same size. A tennis ball is similar to a basketball. A model ship is similar to an actual ship.

Similar objects have corresponding parts; for example, the rudder on the model ship corresponds to the rudder on the actual ship. The relationship between the sizes of each of the corresponding parts can be written as a ratio, and each ratio will be the same.

If the rudder on the model ship is $\frac{1}{100}$ the size of the rudder on the actual ship, then the model wheelhouse is $\frac{1}{100}$ the size of the actual wheelhouse, the width of the model is $\frac{1}{100}$ the width of the actual ship, and so on.

The two triangles ABC and DEF shown at the right are similar. Side $\overline{AB}$ corresponds to $\overline{DE}$, side $\overline{BC}$ corresponds to $\overline{EF}$, and side $\overline{AC}$ corresponds to $\overline{DF}$. The height $\overline{CH}$ corresponds to the height $\overline{FK}$. The ratios of corresponding parts of similar triangles are equal.

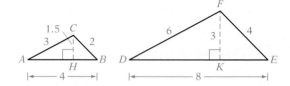

$$\frac{AB}{DE} = \frac{4}{8} = \frac{1}{2}, \qquad \frac{AC}{DF} = \frac{3}{6} = \frac{1}{2}, \qquad \frac{BC}{EF} = \frac{2}{4} = \frac{1}{2}, \qquad \text{and} \qquad \frac{CH}{FK} = \frac{1.5}{3} = \frac{1}{2}$$

Because the ratios of corresponding parts are equal, three proportions can be formed using the sides of the triangles.

$$\frac{AB}{DE} = \frac{AC}{DF}, \qquad \frac{AB}{DE} = \frac{BC}{EF}, \qquad \text{and} \qquad \frac{AC}{DF} = \frac{BC}{EF}$$

Three proportions can also be formed by using the sides and heights of the triangles.

$$\frac{AB}{DE} = \frac{CH}{FK}, \qquad \frac{AC}{DF} = \frac{CH}{FK}, \qquad \text{and} \qquad \frac{BC}{EF} = \frac{CH}{FK}$$

HOW TO 2   Triangles ABC and DEF at the right are similar. Find the area of triangle ABC.

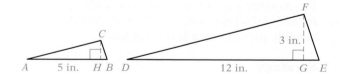

$\dfrac{AB}{DE} = \dfrac{CH}{FG}$ • Solve a proportion to find the height of triangle ABC.

$\dfrac{5}{12} = \dfrac{CH}{3}$ • $AB = 5$, $DE = 12$, and $FG = 3$.

$12 \cdot \dfrac{5}{12} = 12 \cdot \dfrac{CH}{3}$ • To clear denominators, multiply each side of the proportion by 12.

$5 = 4(CH)$ • Solve for CH.

$1.25 = CH$ • The height is 1.25 in. The base is 5 in.

$A = \dfrac{1}{2}bh = \dfrac{1}{2}(5)(1.25) = 3.125$ • Use the formula for the area of a triangle.

The area of triangle ABC is 3.125 in².

Unless otherwise noted, all content on this page is © Cengage Learning.

The measures of the corresponding angles of similar triangles are equal. Therefore, for the similar triangles in HOW TO 2,

$$m\angle A = m\angle D, \quad m\angle B = m\angle E, \quad \text{and} \quad m\angle C = m\angle F$$

It is also true that if the measures of the three angles of one triangle are equal, respectively, to the measures of the three angles of another triangle, then the two triangles are similar.

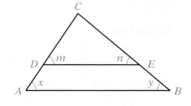

Take Note

Vertical angles of intersecting lines, corresponding angles of parallel lines, and angles of a triangle are discussed in Section 3.1.

A line $\overline{DE}$ is drawn parallel to the base $\overline{AB}$ in the triangle at the right. $m\angle x = m\angle m$ and $m\angle y = m\angle n$ because corresponding angles are equal. $m\angle C = m\angle C$; thus the measures of the three angles of triangle DEC are equal, respectively, to the measures of the three angles of triangle ABC. Triangle DEC is similar to triangle ABC.

The sum of the measures of the three angles of a triangle is 180°. If two angles of one triangle are equal in measure to two angles of another triangle, then the third angles must be equal in measure. Thus we can say that if two angles of one triangle are equal in measure to two angles of another triangle, then the two triangles are similar.

HOW TO 3   The line segments $\overline{AB}$ and $\overline{CD}$ intersect at point O in the figure at the right. Angles C and D are right angles. Find DO.

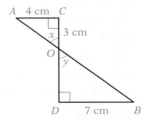

First we must determine whether triangle AOC is similar to triangle BOD.

$m\angle C = m\angle D$ because they are right angles.

$m\angle x = m\angle y$ because they are vertical angles.

Triangle AOC is similar to triangle BOD because two angles of one triangle are equal in measure to two angles of the other triangle.

$$\frac{AC}{DB} = \frac{CO}{DO}$$ • Use a proportion to find the length of the unknown side.

$$\frac{4}{7} = \frac{3}{DO}$$ • $AC = 4$, $CO = 3$, and $DB = 7$.

$$7(DO)\frac{4}{7} = 7(DO)\frac{3}{DO}$$ • To clear denominators, multiply each side of the proportion by $7(DO)$.

$$4(DO) = 7(3)$$ • Solve for DO.

$$4(DO) = 21$$

$$DO = 5.25$$

DO is 5.25 cm.

Unless otherwise noted, all content on this page is © Cengage Learning.

EXAMPLE 3

Triangles *ABC* and *DEF* are similar. Find *AC*.

 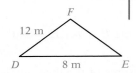

Strategy

To find the length of *AC*, write and solve a proportion.

Solution

$$\frac{AC}{DF} = \frac{AB}{DE}$$ • Write a proportion.

$$\frac{AC}{12} = \frac{5}{8}$$ • $DF = 12, AB = 5, DE = 8$

$$24\left(\frac{AC}{12}\right) = 24\left(\frac{5}{8}\right)$$ • Multiply each side by 24.

$$2(AC) = 15$$ • Solve for *AC*.
$$AC = 7.5$$

AC is 7.5 m.

YOU TRY IT 3

Triangles *ABC* and *DEF* are similar. Find *DE*.

 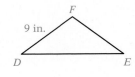

Your strategy

Your solution

16.2 in.

EXAMPLE 4

In the figure, $\overline{AB}$ is parallel to $\overline{DC}$.
$AB = 12$ m,
$DC = 4$ m,
and $AC = 18$ m. Find *CO*.

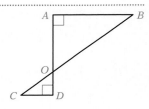

Strategy

Triangle *AOB* is similar to triangle *COD*. Solve a proportion to find the length of *CO*. Let *x* represent the length of *CO*, and let $18 - x$ represent the length of *AO*.

Solution

$$\frac{DC}{AB} = \frac{CO}{AO}$$ • Write a proportion.

$$\frac{4}{12} = \frac{x}{18 - x}$$ • Substitute.

$$12(18 - x) \cdot \frac{4}{12} = 12(18 - x) \cdot \frac{x}{18 - x}$$ • Clear denominators.

$$4(18 - x) = 12x$$ • Solve for *x*.
$$72 - 4x = 12x$$
$$72 = 16x$$
$$4.5 = x$$

CO is 4.5 m.

YOU TRY IT 4

In the figure, $\overline{AB}$ is parallel to $\overline{DC}$.
$AB = 10$ cm,
$CD = 4$ cm, and
$DO = 3$ cm. Find the area of triangle *AOB*.

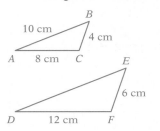

Your strategy

Your solution

37.5 cm²

IN-CLASS EXAMPLES

5. Triangles *ABC* and *DEF* are similar. Find the perimeter of triangle *DEF*. **33 cm**

Solutions on p. S24

Unless otherwise noted, all content on this page is © Cengage Learning.

8.5 EXERCISES

✔ **Concept Check**

SUGGESTED ASSIGNMENT
Exercises 1–6; Exercises 7–59, odds; Exercise 63; More challenging exercise: Exercise 66

1. Explain the difference between a ratio and a rate.   A ratio is the quotient of two quantities that have the same unit. A rate is the quotient of two quantities that have different units.

2. Explain the difference between a ratio and a proportion.   A ratio is a *quotient* (a fraction). A proportion is an *equation* that states the equality of two ratios (or rates).

3. Identify each of the following as a ratio or a rate. Then write it in simplest form.

a. $\dfrac{50 \text{ ft}}{4 \text{ s}}$ Rate, $\dfrac{25 \text{ ft}}{2 \text{ s}}$ **b.** $\dfrac{28 \text{ in.}}{21 \text{ in.}}$ Ratio, $\dfrac{4}{3}$ **c.** $\dfrac{20 \text{ mi}}{2 \text{ h}}$ Rate, $\dfrac{10 \text{ mi}}{1 \text{ h}}$ **d.** $\dfrac{3 \text{ gal}}{18 \text{ gal}}$ Ratio, $\dfrac{1}{6}$

For Exercises 4 and 5, use the pair of similar triangles shown at the right. Triangle *PQR* is similar to triangle *XYZ*.

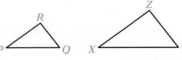

4. a. The corresponding part for side $\overline{RP}$ is __ZX__ .

b. The corresponding part for side $\overline{YX}$ is __QP__ .

c. The corresponding part for $\angle X$ is __$\angle P$__ .

5. a. Complete this proportion: $\dfrac{QR}{YZ} = \dfrac{PR}{XZ}$.

b. Complete this equality: $\angle Z = $ __$\angle R$__ .

6. In the diagram at the right, $\overline{BD}$ is parallel to $\overline{AE}$.

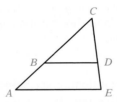

a. Triangle *CBD* is similar to triangle __CAE__ .

b. Complete this proportion: $\dfrac{BD}{AE} = \dfrac{CD}{CE}$.

OBJECTIVE A   *To solve a proportion*

For Exercises 7 to 21, solve.

7. $\dfrac{x}{12} = \dfrac{3}{4}$

9

8. $\dfrac{6}{x} = \dfrac{2}{3}$

9

9. $\dfrac{4}{9} = \dfrac{x}{27}$

12

10. $\dfrac{16}{9} = \dfrac{64}{x}$

36

11. $\dfrac{x+3}{12} = \dfrac{5}{6}$

7

12. $\dfrac{3}{5} = \dfrac{x-4}{10}$

10

13. $\dfrac{18}{x+4} = \dfrac{9}{5}$

6

14. $\dfrac{2}{11} = \dfrac{20}{x-3}$

113

15. $\dfrac{2}{x} = \dfrac{4}{x+1}$

1

16. $\dfrac{16}{x-2} = \dfrac{8}{x}$

−2

17. $\dfrac{x+3}{4} = \dfrac{x}{8}$

−6

18. $\dfrac{x-6}{3} = \dfrac{x}{5}$

15

Unless otherwise noted, all content on this page is © Cengage Learning.

19. $\dfrac{2}{x-1} = \dfrac{6}{2x+1}$

4

20. $\dfrac{9}{x+2} = \dfrac{3}{x-2}$

4

21. $\dfrac{2x}{7} = \dfrac{x-2}{14}$

$-\dfrac{2}{3}$

22. 🐾 True or false? (Assume that a, b, c, and d do not equal zero.)

a. If $\dfrac{a}{b} = \dfrac{c}{d}$, then $\dfrac{d}{b} = \dfrac{c}{a}$.   True

b. If $\dfrac{a}{b} = \dfrac{c}{d}$, then $\dfrac{b}{a} = \dfrac{d}{c}$.   True

23. Elections   An exit poll showed that 4 out of every 7 voters cast a ballot in favor of an amendment to a city charter. At this rate, how many people voted in favor of the amendment if 35,000 people voted?   20,000 people

24. Business   A quality control inspector found 3 defective transistors in a shipment of 500 transistors. At this rate, how many transistors would be defective in a shipment of 2000 transistors?   12 transistors

25. ● Health Insurance   See the news clipping at the right. How many Americans do not have health insurance? Use a figure of 300 million for the population of the United States.   45 million Americans

26. ● Poverty   See the news clipping at the right. How many American children live in poverty? Use a figure of 75 million for the number of children living in the United States.   12.5 million children

27. Construction   An air conditioning specialist recommends 2 air vents for every 300 ft² of floor space. At this rate, how many air vents are required for an office building of 21,000 ft²?   140 air vents

28. Television   In a city of 25,000 homes, a survey was taken to determine the number with Wi-Fi access. Of the 300 homes surveyed, 210 had Wi-Fi access. Estimate the number of homes in the city that have Wi-Fi access.   17,500 homes

● Fossils   For Exercises 29 and 30, use the information in the article at the right. Assume that all scorpions have approximately the same ratio of claw length to body length.

29. Estimate the length, in feet, of the longest known prehistoric sea scorpion's claw prior to the discovery of the new fossil. Round to the nearest hundredth.   1.23 ft

30. Today, scorpions range in length from about 0.5 in. to about 8 in. Estimate the length, in inches, of a claw of a 7-inch scorpion. Round to the nearest hundredth. (*Hint:* Convert 8.2 ft to inches.)   1.28 in.

31. Conservation   As part of a conservation effort for a lake, 40 fish were caught, tagged, and then released. Later, 80 fish were caught from the lake. Four of these 80 fish were found to have tags. Estimate the number of fish in the lake.   800 fish

In the NEWS!

Room for Improvement

According to a U.N. publication, the United States ranks 13th in the world in the area of human development. Government data show that in regard to health, 3 in 20 Americans do not have health insurance. With respect to standard of living, 1 in 6 American children lives in poverty.

Sources: www.undp.org, www.census.gov, Human Development Report 2009; Income, Poverty, and Health Insurance Coverage in the United States, 2008

In the NEWS!

390-Million-Year-Old Scorpion Fossil Found

Scientists have announced the unearthing of the largest fossil sea scorpion claw ever discovered. Based on the 18-inch claw length, scientists estimate that the scorpion would have measured 8.2 ft in length. Prior to this discovery, the longest known prehistoric sea scorpion was estimated to be 6.7 ft long.

Source: news.nationalgeographic.com

32. **Cooking**   A simple syrup is made by dissolving 2 c of sugar in $\frac{2}{3}$ c of boiling water. At this rate, how many cups of sugar are required for 2 c of boiling water?   6 c

33. **Energy**   The lighting for a billboard is provided by solar energy. If 3 energy panels generate 10 watts of power, how many panels are needed to provide 600 watts of power?   180 panels

34. **Business**   A company will accept a shipment of 10,000 computer disks if there are 2 or fewer defects in a sample of 100 randomly chosen disks. Assume that there are 300 defective disks in the shipment and that the rate of defective disks in the sample is the same as the rate in the shipment. Will the shipment be accepted?   No

35. **Business**   A company will accept a shipment of 20,000 precision bearings if there are 3 or fewer defects in a sample of 100 randomly chosen bearings. Assume that there are 400 defective bearings in the shipment and that the rate of defective bearings in the sample is the same as the rate in the shipment. Will the shipment be accepted?   Yes

36. **Art**   Leonardo da Vinci measured various distances on the human body in order to make accurate drawings. He determined that in general, the ratio of the kneeling height of a person to his or her standing height is $\frac{3}{4}$. Using this ratio, determine the standing height of a person who has a kneeling height of 48 in.   64 in.

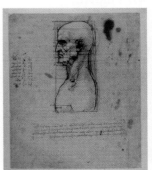

37. **Art**   In one of Leonardo da Vinci's notebooks, he wrote that ". . . from the top to the bottom of the chin is the sixth part of a face, and it is the fifty-fourth part of the man." Suppose the distance from the top to the bottom of a person's chin is 1.25 in. Using da Vinci's measurements, find the height of the person.   67.5 in.

Cameraphoto Arte, Venice/Art Resource, NY

38. **Cartography**   On a map, two cities are $2\frac{5}{8}$ in. apart. If $\frac{3}{8}$ in. on the map represents 25 mi, find the number of miles between the two cities.   175 mi

39. **Cartography**   On a map, two cities are $5\frac{5}{8}$ in. apart. If $\frac{3}{4}$ in. on the map represents 100 mi, find the number of miles between the two cities.   750 mi

40. The scale on a map shows that a distance of 3 cm on the map represents an actual distance of 10 mi. Would a distance of 8 cm on the map represent an actual distance that is greater than 30 mi or less than 30 mi?   Less than

41. **Rocketry**   The engine of a small rocket burns 170,000 lb of fuel in 1 min. At this rate, how many pounds of fuel does the engine burn in 45 s?   127,500 lb

42. **Construction**   To conserve energy and still allow for as much natural lighting as possible, an architect suggests that the ratio of the area of a window to the area of the total wall surface be 5 to 12. Using this ratio, determine the recommended area of a window to be installed in a wall that measures 8 ft by 12 ft.   40 ft²

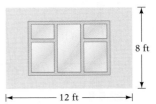

Unless otherwise noted, all content on this page is © Cengage Learning.

43. Paint Mixtures   A green paint is created by mixing 3 parts of yellow with every 5 parts of blue. How many gallons of yellow paint are needed to make 60 gal of this green paint?   22.5 gal

44. Food Industry   A soft drink is made by mixing 4 parts of carbonated water with every 3 parts of syrup. How many milliliters of carbonated water are in 280 ml of soft drink?   160 ml

45. Agriculture   A 50-acre field yields 1100 bushels of wheat annually. How many additional acres must be planted so that the annual yield will be 1320 bushels?
10 additional acres

46. Catering   A caterer estimates that 5 gal of coffee will serve 50 people. How much additional coffee is necessary to serve 70 people?   2 additional gallons

---

OBJECTIVE B   *To solve problems involving similar triangles*

Triangles *ABC* and *DEF* in Exercises 47 to 54 are similar. Round answers to the nearest tenth.

47. Find *AC*.

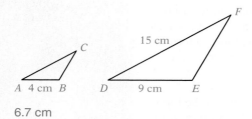

6.7 cm

48. Find *DE*.

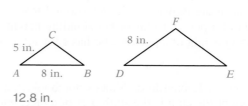

12.8 in.

49. Find the height of triangle *ABC*.

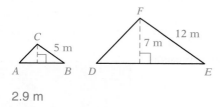

2.9 m

50. Find the height of triangle *DEF*.

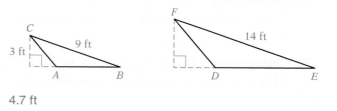

4.7 ft

51. Find the perimeter of triangle *DEF*.

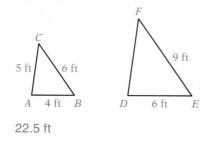

22.5 ft

52. Find the perimeter of triangle *ABC*.

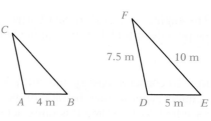

18 m

Unless otherwise noted, all content on this page is © Cengage Learning.

53. Find the area of triangle *ABC*.

 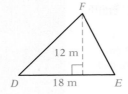

48 m²

54. Find the area of triangle *ABC*.

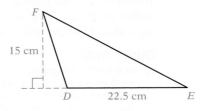

48 cm²

55. Given that $\overline{BD} \parallel \overline{AE}$, *BD* measures 5 cm, *AE* measures 8 cm, and *AC* measures 10 cm, find *BC*.

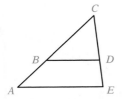

6.25 cm

56. Given that $\overline{AC} \parallel \overline{DE}$, *BD* measures 8 m, *AD* measures 12 m, and *BE* measures 6 m, find *BC*.

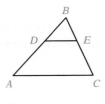

15 m

57. Given that $\overline{DE} \parallel \overline{AC}$, *DE* measures 6 in., *AC* measures 10 in., and *AB* measures 15 in., find *DA*.

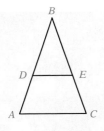

6 in.

58. Given that $\overline{MP}$ and $\overline{NQ}$ intersect at *O*, *NO* measures 25 ft, *MO* measures 20 ft, and *PO* measures 8 ft, find *QO*.

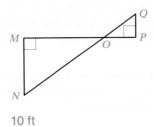

10 ft

59. Given that $\overline{MP}$ and $\overline{NQ}$ intersect at *O*, *NO* measures 24 cm, *MN* measures 10 cm, *MP* measures 39 cm, and *QO* measures 12 cm, find *OP*.

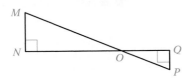

13 cm

60. Given that $\overline{MQ}$ and $\overline{NP}$ intersect at *O*, *NO* measures 12 m, *MN* measures 9 m, *PQ* measures 3 m, and *MQ* measures 20 m, find the perimeter of triangle *OPQ*.

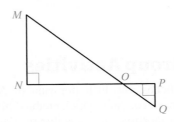

12 m

61. 🖊 True or false? The ratio of the perimeters of two similar triangles is the same as the ratio of their corresponding sides. True

62. 🖊 True or false? The ratio of the areas of two similar triangles is the same as the ratio of their corresponding sides. False

Unless otherwise noted, all content on this page is © Cengage Learning.

QUICK QUIZ

Solve.

1. $\dfrac{5}{y + 14} = \dfrac{7}{y}$

   −49 [8.5A]

2. $\dfrac{3}{2} = \dfrac{12}{x - 8}$ 16 [8.5A]

63. Similar triangles can be used as an indirect way of measuring inaccessible distances. The diagram at the right represents a river of width *DC*. The triangles *AOB* and *DOC* are similar. The distances *AB*, *BO*, and *OC* can be measured. Find the width of the river.   35 m

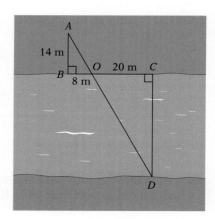

QUICK QUIZ

3. In a wildlife preserve, 10 elk were captured, tagged, and then released. Later, 15 elk were captured and 2 were found to have tags. Estimate the number of elk in the preserve.
75 elk   [8.5A]

4. Given that $\overline{DE} \parallel \overline{AC}$, *BD* = 2 cm, *DA* = 5 cm, and *BE* = 3 cm, find *BC*.   **10.5 cm   [8.5B]**

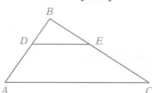

64. The sun's rays cast a shadow as shown in the diagram at the right. Find the height of the flagpole. Write the answer in terms of feet.   14.375 ft

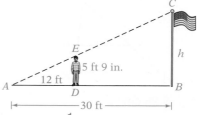

Critical Thinking

65. Number Problem   The sum of a number and its reciprocal is $\frac{26}{5}$. Find the number.   5 or $\frac{1}{5}$

66. Lotteries   Three people put their money together to buy lottery tickets. The first person put in $25, the second person put in $30, and the third person put in $35. One of the tickets was a winning ticket. If the winning ticket paid $4.5 million, what was the first person's share of the winnings? $1.25 million

67. Sports   A basketball player has made 5 out of every 6 foul shots attempted. If 42 foul shots were missed in the player's career, how many foul shots were made in the player's career?   210 foul shots

68. Fundraising   No one belongs to both the Math Club and the Photography Club, but the two clubs join to hold a car wash. Ten members of the Math Club and 6 members of the Photography Club participate. The profits from the car wash are $120. If each club's profits are proportional to the number of members participating, what share of the profits does the Math Club receive? $75

Dennis Ku/Shutterstock.com

Projects or Group Activities

69. History   Eratosthenes, the fifth librarian of Alexandria (230 B.C.), was familiar with certain astronomical data, which enabled him to calculate the circumference of Earth by using a proportion. He knew that on a midsummer day, the sun was directly overhead at Syene, as shown in the diagram. At the same time, at Alexandria, the sun was at a 7.5° angle from the zenith. The distance from Syene to Alexandria was 5000 stadia, or about 520 mi. Eratosthenes reasoned that the ratio of the 7.5° angle to one revolution was equal to the ratio of the arc length of 520 mi to the circumference of Earth. From this, he wrote and solved a proportion.
a. What did Eratosthenes calculate to be the circumference of Earth?   24,960 mi
b. Find the difference between his calculation and the accepted value of 24,874 mi.   86 mi

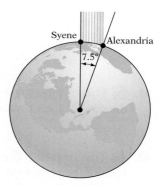

Unless otherwise noted, all content on this page is © Cengage Learning.

SECTION

8.6 Literal Equations

> **OBJECTIVE A** *To solve a literal equation for one of the variables*

INSTRUCTOR NOTE
Example 4 on the next page will be difficult for students. Before doing that example or one similar to it, remind students that when solving the equation $2x + 3x = 10$, they are using the Distributive Property to combine $2x$ and $3x$:

$2x + 3x = (2 + 3)x = 5x$

Now each side of the equation can be divided by 5.

For the equation $ax + bx = 10$, the procedure is exactly the same except that $a + b$ does not simplify further:

$ax + bx = (a + b)x$

Now each side of the equation can be divided by $(a + b)$.

A **literal equation** is an equation that contains more than one variable. Examples of literal equations are shown at the right.

$$2x + 3y = 6$$
$$4w - 2x + z = 0$$

Formulas are used to express relationships among physical quantities. A **formula** is a literal equation that states a rule about measurements. Examples of formulas are shown at the right.

$$\frac{1}{R_1} + \frac{1}{R_2} = \frac{1}{R} \quad \text{(Physics)}$$
$$s = a + (n - 1)d \quad \text{(Mathematics)}$$
$$A = P + Prt \quad \text{(Business)}$$

The Addition and Multiplication Properties can be used to solve a literal equation for one of the variables. The goal is to rewrite the equation so that the variable being solved for is alone on one side of the equation and all the other numbers and variables are on the other side.

HOW TO 1 Solve $A = P(1 + i)$ for i.

The goal is to rewrite the equation so that i is on one side of the equation and all other variables are on the other side.

$$A = P(1 + i)$$
$$A = P + Pi \quad \bullet \text{ Use the Distributive Property to remove parentheses.}$$
$$A - P = P - P + Pi \quad \bullet \text{ Subtract } P \text{ from each side of the equation.}$$
$$A - P = Pi$$
$$\frac{A - P}{P} = \frac{Pi}{P} \quad \bullet \text{ Divide each side of the equation by } P.$$
$$\frac{A - P}{P} = i$$

EXAMPLE 1

Solve $3x - 4y = 12$ for y.

Solution

$$3x - 4y = 12$$
$$3x - 3x - 4y = -3x + 12 \quad \bullet \text{ Subtract } 3x.$$
$$-4y = -3x + 12$$
$$\frac{-4y}{-4} = \frac{-3x + 12}{-4} \quad \bullet \text{ Divide by } -4.$$
$$y = \frac{3}{4}x - 3$$

YOU TRY IT 1

Solve $5x - 2y = 10$ for y.

Your solution

$$y = \frac{5}{2}x - 5$$

Solution on p. S24

EXAMPLE 2

Solve $I = \dfrac{E}{R + r}$ for R.

Solution

$$I = \frac{E}{R + r}$$

$$(R + r)I = (R + r)\frac{E}{R + r} \qquad \bullet \text{ Multiply by } (R + r).$$

$$RI + rI = E$$

$$RI + rI - rI = E - rI \qquad \bullet \text{ Subtract } rI.$$

$$RI = E - rI$$

$$\frac{RI}{I} = \frac{E - rI}{I} \qquad \bullet \text{ Divide by } I.$$

$$R = \frac{E - rI}{I}$$

YOU TRY IT 2

Solve $s = \dfrac{A + L}{2}$ for L.

Your solution

$L = 2s - A$

IN-CLASS EXAMPLES

1. Solve for b: $A = \dfrac{1}{2}bh$
   $b = \dfrac{2A}{h}$

2. Solve for x: $3x + 8y = 9$
   $x = -\dfrac{8}{3}y + 3$

3. Solve for y: $7x - y = 12$
   $y = 7x - 12$

EXAMPLE 3

Solve $L = a(1 + ct)$ for c.

Solution

$$L = a(1 + ct)$$

$$L = a + act \qquad \bullet \text{ Distributive Property}$$

$$L - a = a - a + act \qquad \bullet \text{ Subtract } a.$$

$$L - a = act$$

$$\frac{L - a}{at} = \frac{act}{at} \qquad \bullet \text{ Divide by } at.$$

$$\frac{L - a}{at} = c$$

YOU TRY IT 3

Solve $S = a + (n - 1)d$ for n.

Your solution

$n = \dfrac{S - a + d}{d}$

EXAMPLE 4

Solve $S = C - rC$ for C.

Solution

$$S = C - rC$$

$$S = (1 - r)C \qquad \bullet \text{ Factor.}$$

$$\frac{S}{1 - r} = \frac{(1 - r)C}{1 - r} \qquad \bullet \text{ Divide by } (1 - r).$$

$$\frac{S}{1 - r} = C$$

YOU TRY IT 4

Solve $S = rS + C$ for S.

Your solution

$S = \dfrac{C}{1 - r}$

Solutions on pp. S24–S25

8.6 EXERCISES

✔ Concept Check

SUGGESTED ASSIGNMENT
Exercises 1–4; Exercises 5–39, odds

For Exercises 1 and 2, determine whether the statement is true or false.

1. Literal equations are solved using the same properties of equations that are used to solve equations in one variable.   True

2. In solving a literal equation, the goal is to get the variable being solved for alone on one side of the equation and all numbers and other variables on the other side of the equation.   True

3. In solving $I = \dfrac{E}{R + r}$ for R, the goal is to get ___R___ alone on one side of the equation.

4. In solving $L = a(1 + ct)$ for c, the goal is to get ___c___ alone on one side of the equation.

> **OBJECTIVE A**   *To solve a literal equation for one of the variables*

For Exercises 5 to 20, solve the formula for the given variable.

5. $d = rt; t$ (Physics)

$$t = \frac{d}{r}$$

6. $E = IR; R$ (Physics)

$$R = \frac{E}{I}$$

7. $PV = nRT; T$ (Chemistry)

$$T = \frac{PV}{nR}$$

8. $A = bh; h$ (Geometry)

$$h = \frac{A}{b}$$

9. $P = 2l + 2w; l$ (Geometry)

$$l = \frac{P - 2w}{2}$$

10. $F = \dfrac{9}{5}C + 32; C$ (Temperature conversion)

$$C = \frac{5F - 160}{9}$$

11. $A = \dfrac{1}{2}h(b_1 + b_2); b_1$ (Geometry)

$$b_1 = \frac{2A - hb_2}{h}$$

12. $s = a(x - vt); t$ (Physics)

$$t = -\frac{s - ax}{av}$$

13. $V = \dfrac{1}{3}Ah; h$ (Geometry)

$$h = \frac{3V}{A}$$

14. $P = R - C; C$ (Business)

$$C = R - P$$

15. $R = \dfrac{C - S}{t}; S$ (Business)

$$S = C - Rt$$

16. $P = \dfrac{R - C}{n}; R$ (Business)

$$R = Pn + C$$

17. $A = P + Prt; P$ (Business)

$$P = \frac{A}{1 + rt}$$

18. $T = fm - gm; m$ (Engineering)

$$m = \frac{T}{f - g}$$

19. $A = Sw + w; w$ (Physics)

$$w = \frac{A}{S + 1}$$

20. $a = S - Sr; S$ (Mathematics)

$$S = \frac{a}{1 - r}$$

For Exercises 21 to 32, solve for y.

21. $3x + y = 10$

$y = -3x + 10$

22. $2x + y = 5$

$y = -2x + 5$

23. $4x - y = 3$

$y = 4x - 3$

24. $5x - y = 7$

$y = 5x - 7$

25. $3x + 2y = 6$

$y = -\dfrac{3}{2}x + 3$

26. $2x + 3y = 9$

$y = -\dfrac{2}{3}x + 3$

27. $2x - 5y = 10$

$y = \dfrac{2}{5}x - 2$

28. $5x - 2y = 4$

$y = \dfrac{5}{2}x - 2$

29. $2x + 7y = 14$

$y = -\dfrac{2}{7}x + 2$

30. $6x - 5y = 10$

$y = \dfrac{6}{5}x - 2$

31. $x + 3y = 6$

$y = -\dfrac{1}{3}x + 2$

32. $x + 2y = 8$

$y = -\dfrac{1}{2}x + 4$

For Exercises 33 to 40, solve for x.

33. $x + 3y = 6$

$x = -3y + 6$

34. $x + 6y = 10$

$x = -6y + 10$

35. $3x - y = 3$

$x = \dfrac{1}{3}y + 1$

36. $2x - y = 6$

$x = \dfrac{1}{2}y + 3$

37. $2x + 5y = 10$

$x = -\dfrac{5}{2}y + 5$

38. $4x + 3y = 12$

$x = -\dfrac{3}{4}y + 3$

39. $x - 2y + 1 = 0$

$x = 2y - 1$

40. $x - 4y - 3 = 0$

$x = 4y + 3$

41. Two students are working with the equation $A = P(1 + i)$. State whether the two students' answers are equivalent.

a. When asked to solve the equation for i, one student answered $i = \dfrac{A}{P} - 1$ and the other student answered $i = \dfrac{A - P}{P}$.   Yes

b. When asked to solve the equation for i, one student answered $i = -\dfrac{P - A}{P}$ and the other student answered $i = \dfrac{A - P}{P}$.   Yes

Critical Thinking

42. Solve for x: $cx - y = bx + 5$ $x = \dfrac{y + 5}{c - b}$

43. Solve the physics formula $\dfrac{1}{R_1} + \dfrac{1}{R_2} = \dfrac{1}{R}$ for R_2. $R_2 = \dfrac{RR_1}{R_1 - R}$

QUICK QUIZ

1. Use the equation $4x + y = 8$.
  a. Solve for y. $y = -4x + 8$
  b. Solve for x. $x = \dfrac{8 - y}{4}$ or
      $x = -\dfrac{1}{4}y + 2$ [8.6A]

2. Solve $P = C + Cr$ for r.
      $r = \dfrac{P - C}{C}$ [8.6A]

Projects or Group Activities

Business   Break-even analysis is a method used to determine the sales volume required for a company to "break even," or experience neither a profit nor a loss on the sale of its product. The break-even point represents the number of units that must be made and sold for income from sales to equal the cost of producing the product. The break-even point can be calculated using the formula $B = \dfrac{F}{S - V}$, where F is the fixed costs, S is the selling price per unit, and V is the variable costs per unit.

44. a. Solve the formula $B = \dfrac{F}{S - V}$ for S. $S = \dfrac{F + BV}{B}$

  b. Use your answer to part (a) to find the selling price per button pinhole video spycam required for a company to break even. The fixed costs are $15,000, the variable costs per spycam are $60, and the company plans to make and sell 200 spycams. $135

  c. Use your answer to part (a) to find the selling price per spy camera lighter required for a company to break even. The fixed costs are $18,000, the variable costs per lighter are $65, and the company plans to make and sell 600 lighters. $95

© Andrew Twort/Alamy

SECTION 8.7 | Application Problems

OBJECTIVE A | *To solve work problems*

If a painter can paint a room in 4 h, then in 1 h the painter can paint $\frac{1}{4}$ of the room. The painter's rate of work is $\frac{1}{4}$ of the room each hour. The **rate of work** is the part of a task that is completed in 1 unit of time.

A pipe can fill a tank in 30 min. This pipe can fill $\frac{1}{30}$ of the tank in 1 min. The rate of work is $\frac{1}{30}$ of the tank each minute. If a second pipe can fill the tank in x min, the rate of work for the second pipe is $\frac{1}{x}$ of the tank each minute.

In solving a work problem, the goal is to determine the time it takes to complete a task. The basic equation that is used to solve work problems is

Rate of work × time worked = part of task completed

Apply the Basic Concepts of Work Problems

EXAMPLE A   A faucet can fill a sink in 6 min. What fraction of the sink will the faucet fill in 5 min?

SOLUTION   The faucet can fill $\frac{1}{6}$ of the sink in 1 min. The rate of work is $\frac{1}{6}$ of the sink each minute.

Rate of work × time worked = part of task completed

$$\frac{1}{6} \times 5 = \frac{5}{6}$$

The faucet will fill $\frac{5}{6}$ of the sink in 5 min.

EXAMPLE B   Emily and Ian raked the yard in 40 min. It would have taken Ian 60 min to rake the yard by himself. What fraction of the yard did Ian rake?

SOLUTION   Ian can rake the yard in 60 min. His rate of work is $\frac{1}{60}$ of the yard each minute. The amount of time Ian raked was 40 min.

Rate of work × time worked = part of task completed

$$\frac{1}{60} \times 40 = \frac{40}{60} = \frac{2}{3}$$

Ian raked $\frac{2}{3}$ of the yard.

EXAMPLE C Sue and Ron wallpapered a room in 8 h. Sue wallpapered $\frac{3}{5}$ of the room. What fraction of the room did Ron wallpaper?

SOLUTION The sum of the part of the task completed by Ron and the part of the task completed by Sue is 1.

Let x = the part of the task completed by Ron.

Part of the task   Part of the task
completed by Sue + completed by Ron = 1

$$\frac{3}{5} + x = 1$$

$$x = \frac{2}{5} \quad \bullet \text{ Subtract } \frac{3}{5}.$$

Ron wallpapered $\frac{2}{5}$ of the room.

Try Concept Check Exercises 2 to 9 on Page 496.

Tips for Success

Note in the examples in this section that solving a word problem includes stating a strategy and using the strategy to find a solution. If you have difficulty with a word problem, write down the known information. Be very specific. Write out a phrase or sentence that states what you are trying to find. See *AIM for Success* at the front of the book.

Take Note

Use the information given in the problem to fill in the "Rate" and "Time" columns of the table. Fill in the "Part Completed" column by multiplying the two expressions you wrote in each row.

HOW TO 1 A painter can paint a wall in 20 min. The painter's apprentice can paint the same wall in 30 min. How long will it take to paint the wall if the painter and the apprentice work together?

Strategy for Solving a Work Problem

1. For each person or machine, write a numerical or variable expression for the rate of work, the time worked, and the part of the task completed. The results can be recorded in a table.

Unknown time to paint the wall working together: t

	Rate of Work	·	*Time Worked*	=	*Part of Task Completed*
Painter	$\frac{1}{20}$	·	t	=	$\frac{t}{20}$
Apprentice	$\frac{1}{30}$	·	t	=	$\frac{t}{30}$

2. Determine how the parts of the task completed are related. Use the fact that the sum of the parts of the task completed must equal 1, the complete task.

$$\frac{t}{20} + \frac{t}{30} = 1$$

• The sum of the part of the task completed by the painter and the part of the task completed by the apprentice is 1.

$$60\left(\frac{t}{20} + \frac{t}{30}\right) = 60 \cdot 1$$

• Multiply by the LCM of 20 and 30.

$$3t + 2t = 60$$

• Distributive Property

$$5t = 60$$

$$t = 12$$

Working together, the painter and the apprentice will paint the wall in 12 min.

Unless otherwise noted, all content on this page is © Cengage Learning.

EXAMPLE 1

A small water pipe takes three times longer to fill a tank than does a large water pipe. With both pipes open, it takes 4 h to fill the tank. Find the time it would take the small pipe, working alone, to fill the tank.

Strategy

• Time for large pipe to fill the tank: t
  Time for small pipe to fill the tank: $3t$

Fills tank
in $3t$ hours

Fills tank
in t hours

Fills $\frac{4}{3t}$ of the
tank in 4 hours

Fills $\frac{4}{t}$ of the
tank in 4 hours

	Rate	Time	Part
Small pipe	$\frac{1}{3t}$	4	$\frac{4}{3t}$
Large pipe	$\frac{1}{t}$	4	$\frac{4}{t}$

• The sum of the parts of the task completed by each pipe must equal 1.

Solution

$$\frac{4}{3t} + \frac{4}{t} = 1$$

$$3t\left(\frac{4}{3t} + \frac{4}{t}\right) = 3t \cdot 1 \quad \bullet \text{ Multiply by the LCM of } 3t \text{ and } t.$$

$$4 + 12 = 3t \quad \bullet \text{ Distributive Property}$$

$$16 = 3t$$

$$\frac{16}{3} = t \quad \bullet \text{ Time for large pipe to fill the tank}$$

$$3t = 3\left(\frac{16}{3}\right) = 16 \quad \bullet \text{ Time for small pipe to fill the tank}$$

The small pipe, working alone, takes 16 h to fill the tank.

YOU TRY IT 1

Two computer printers that work at the same rate are working together to print the payroll checks for a large corporation. After they work together for 2 h, one of the printers fails. The second printer requires 3 h more to complete the payroll checks. Find the time it would take one printer, working alone, to print the payroll.

Your strategy

Your solution

7 h

IN-CLASS EXAMPLES

1. It takes Erin Shaw 30 min to shovel the snow off her driveway by herself. Working by himself, Erin's son takes 45 min to clear the driveway of snow. If they work together, how long will it take Erin and her son to shovel all the snow off their driveway? **18 min**

Solution on p. S25

Unless otherwise noted, all content on this page is © Cengage Learning.

OBJECTIVE B

INSTRUCTOR NOTE
If you did not cover Section
7.5 on solving quadratic
equations by factoring,
skip Example 2 on the next
page. Also skip Exercises
44, 45, 50, and 63 in the
Section 8.7 Exercises.

To use rational expressions to solve uniform motion problems

A car that travels constantly in a straight line at 30 mph is in uniform motion. **Uniform motion** means that the speed and direction of an object do not change.

The basic equation used to solve uniform motion problems is

Distance = rate × time

An alternative form of this equation can be written by solving the equation for time.

$$\frac{\textbf{Distance}}{\textbf{Rate}} = \textbf{time}$$

This form of the equation is useful when the total time of travel for two objects or the time of travel between two points is known.

HOW TO 2   The speed of a boat in still water is 20 mph. The boat traveled 75 mi down a river in the same amount of time it took to travel 45 mi up the river. Find the rate of the river's current.

Strategy for Solving a Uniform Motion Problem

1. For each object, write a numerical or variable expression for the distance, rate, and time. The results can be recorded in a table.

The unknown rate of the river's current: r

Take Note
Use the information given
in the problem to fill in
the "Distance" and "Rate"
columns of the table. Fill in
the "Time" column by dividing
the two expressions you
wrote in each row.

	Distance	÷	Rate	=	Time
Down river	75	÷	$20 + r$	=	$\frac{75}{20 + r}$
Up river	45	÷	$20 - r$	=	$\frac{45}{20 - r}$

2. Determine how the times traveled by each object are related. For example, it may be known that the times are equal, or the total time may be known.

$$\frac{75}{20 + r} = \frac{45}{20 - r}$$
$$(20 + r)(20 - r)\frac{75}{20 + r} = (20 + r)(20 - r)\frac{45}{20 - r}$$
$$(20 - r)75 = (20 + r)45$$
$$1500 - 75r = 900 + 45r$$
$$-120r = -600$$
$$r = 5$$

- The time down the river is equal to the time up the river.
- Multiply by the LCM of the denominators.

- Distributive Property

The rate of the river's current is 5 mph.

Unless otherwise noted, all content on this page is © Cengage Learning.

EXAMPLE 2

A cyclist rode the first 20 mi of a trip at a constant rate. For the next 16 mi, the cyclist reduced the speed by 2 mph. The total time for the 36 mi was 4 h. Find the rate of the cyclist for each leg of the trip.

Strategy

*   Rate for the first 20 mi: r
    Rate for the next 16 mi: $r - 2$

	Distance	Rate	Time
First 20 mi	20	r	$\dfrac{20}{r}$
Next 16 mi	16	$r - 2$	$\dfrac{16}{r - 2}$

*   The total time for the trip was 4 h.

Solution

$$\frac{20}{r} + \frac{16}{r - 2} = 4$$

*   The total time was **4 h.**

$$r(r - 2)\left[\frac{20}{r} + \frac{16}{r - 2}\right] = r(r - 2) \cdot 4$$

*   **Multiply by the LCM of the denominators.**

$$(r - 2)20 + 16r = 4r^2 - 8r$$

*   **Distributive Property**

$$20r - 40 + 16r = 4r^2 - 8r$$

$$36r - 40 = 4r^2 - 8r$$

Solve the quadratic equation by factoring.

$$0 = 4r^2 - 44r + 40$$ • **Standard form**

$$0 = 4(r^2 - 11r + 10)$$

$$0 = 4(r - 10)(r - 1)$$ • **Factor.**

$r - 10 = 0 \qquad r - 1 = 0$ • **Principle of**
$\qquad r = 10 \qquad\quad r = 1$ **Zero Products**

The solution $r = 1$ mph is not possible, because the rate on the last 16 mi would then be -1 mph.

10 mph was the rate for the first 20 mi.
8 mph was the rate for the next 16 mi.

YOU TRY IT 2

The total time it took for a sailboat to sail across a lake 6 km wide and back was 2 h. The rate sailing back was three times the rate sailing across. Find the rate sailing out across the lake.

Your strategy

Your solution

4 km/h

IN-CLASS EXAMPLES

2. Lorenzo's bicycling rate is six times as fast as his walking rate. On his bicycle, he can complete a 9-mile route in $2\frac{1}{2}$ h less time than it takes him to walk the same route. Find Lorenzo's walking rate and his cycling rate.   **3 mph, 18 mph**

Solution on p. S25

Unless otherwise noted, all content on this page is © Cengage Learning.

8.7 EXERCISES

✔ **Concept Check**

SUGGESTED ASSIGNMENT
Exercises 1–14; Exercises 15–33, odds; Exercises 37–57, odds

1. ◨ Explain the meaning of the phrase "rate of work."

For Exercises 2 to 4, fill in the blank to make a true statement.

2. If it takes a janitorial crew 5 h to clean a company's offices, then in x hours the crew has completed _____ of the job. $\frac{x}{5}$

3. If it takes an automotive crew x minutes to service a car, then the rate of work is _____ of the job each minute. $\frac{1}{x}$

4. Two people completed a job. If one person completed $\frac{t}{30}$ of the job and the other person completed $\frac{t}{20}$ of the job, then $\frac{t}{30} + \frac{t}{20} = $ _____. 1

5. If Jen can paint a wall in 30 min and Amelia can paint the same wall in 45 min, who has the greater rate of work? Jen

6. It takes Pat 3 h to mow the lawn.
   a. What is Pat's rate of work? $\frac{1}{3}$ of the job per hour
   b. What fraction of the lawn can Pat mow in 2 h? $\frac{2}{3}$

7. It takes Chris x hours to lay a tile floor.
   a. What is Chris's rate of work? $\frac{1}{x}$ of the job per hour
   b. What fraction of the floor can Chris lay in 3 h? $\frac{3}{x}$

8. Dawn and Hugh painted a fence together in 8 h. It would have taken Hugh 12 h to paint the fence by himself.
   a. What fraction of the fence did Hugh paint? $\frac{2}{3}$
   b. What fraction of the fence did Dawn paint? $\frac{1}{3}$

9. Together, two printers printed a company's advertising brochures in h hours. The faster printer could have printed the brochures in 5 h. What fraction of the brochures did the faster printer print? $\frac{5}{h}$

10. If a plane flies 300 mph in calm air and the rate of the wind is r miles per hour, then the rate of the plane flying with the wind can be represented as ___$300 + r$___, and the rate of the plane flying against the wind can be represented as ___$300 - r$___.

11. Suppose you have a powerboat with the throttle set to move the boat at 8 mph in calm water, and the rate of the current is 4 mph. **a.** What is the speed of the boat when traveling with the current? **b.** What is the speed of the boat when traveling against the current? **a.** 12 mph **b.** 4 mph

12. The speed of a plane is 500 mph. There is a headwind of 50 mph. What is the speed of the plane relative to an observer on the ground?   450 mph

OBJECTIVE A   *To solve work problems*

13. One electrician can complete a wiring job in 10 h. It would take the electrician's assistant 12 h to complete the same wiring job. Let t represent the amount of time it would take the electrician and the assistant to complete the job if they worked together. Complete the following table.

	Rate of Work	·	Time Worked	=	Part of Task Completed
Electrician	$\frac{1}{10}$	·	t	=	$\frac{t}{10}$
Assistant	$\frac{1}{12}$	·	t	=	$\frac{t}{12}$

14. Refer to the situation presented in Exercise 13. When the wiring job is finished, the "part of task completed" is the whole task, so the sum of the parts completed by the electrician and by the assistant is ___1___. Use this fact and the expressions in the table in Exercise 13 to write an equation that can be solved to find the amount of time it would take for the electrician and the assistant to complete the job working together:

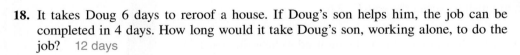

$$\frac{t}{10} + \frac{t}{12} = 1.$$

15. A park has two sprinklers that are used to fill a fountain. One sprinkler can fill the fountain in 3 h, whereas the second sprinkler can fill the fountain in 6 h. How long will it take to fill the fountain with both sprinklers operating?   2 h

16. One grocery clerk can stock a shelf in 20 min. A second clerk requires 30 min to stock the same shelf. How long would it take to stock the shelf if the two clerks worked together?   12 min

17. One person with a skiploader requires 12 h to transfer a large quantity of earth. With a larger skiploader, the same amount of earth can be transferred in 4 h. How long would it take to transfer the earth if both skiploaders were operated together?   3 h

18. It takes Doug 6 days to reroof a house. If Doug's son helps him, the job can be completed in 4 days. How long would it take Doug's son, working alone, to do the job?   12 days

19. One computer can solve a complex prime factorization problem in 75 h. A second computer can solve the same problem in 50 h. How long would it take both computers, working together, to solve the problem?   30 h

20. A new machine makes 10,000 aluminum cans three times faster than an older machine. With both machines operating, it takes 9 h to make 10,000 cans. How long would it take the new machine, working alone, to make 10,000 cans?   12 h

Unless otherwise noted, all content on this page is © Cengage Learning.

© RJH_CATALOG /Alamy

21. A small air conditioner can cool a room 5°F in 60 min. A larger air conditioner can cool the room 5°F in 40 min. How long would it take to cool the room 5°F with both air conditioners working?   24 min

22. One printing press can print the first edition of a book in 55 min. A second printing press requires 66 min to print the same number of copies. How long would it take to print the first edition of the book with both presses operating?   30 min

23. Two welders working together can complete a job in 6 h. One of the welders, working alone, can complete the task in 10 h. How long would it take the second welder, working alone, to complete the task?   15 h

24. Working together, Pat and Chris can reseal a driveway in 6 h. Working alone, Pat can reseal the driveway in 15 h. How long would it take Chris, working alone, to reseal the driveway?   10 h

25. Two oil pipelines can fill a small tank in 30 min. One of the pipelines, working alone, would require 45 min to fill the tank. How long would it take the second pipeline, working alone, to fill the tank?   90 min

26. A cement mason can construct a retaining wall in 8 h. A second mason requires 12 h to do the same job. After working alone for 4 h, the first mason quits. How long will it take the second mason to complete the wall?   6 h

27. With two reapers operating, a field can be harvested in 1 h. If only the newer reaper is used, the crop can be harvested in 1.5 h. How long would it take to harvest the field using only the older reaper?   3 h

28. A manufacturer of prefabricated homes has the company's employees work in teams. Team 1 can erect the Silvercrest model in 15 h. Team 2 can erect the same model in 10 h. How long would it take for Team 1 and Team 2, working together, to erect the Silvercrest model home?   6 h

29. One technician can wire a security alarm in 4 h, whereas it takes 6 h for a second technician to do the same job. After working alone for 2 h, the first technician quits. How long will it take the second technician to complete the wiring?   3 h

30. A wallpaper hanger requires 2 h to hang the wallpaper on one wall of a room. A second wallpaper hanger requires 4 h to hang the same amount of wallpaper. The first wallpaper hanger works alone for 1 h and then quits. How long will it take the second hanger, working alone, to finish papering the wall?   2 h

31. A large heating unit and a small heating unit are being used to heat the water in a pool. The large unit, working alone, requires 8 h to heat the pool. After both units have been operating for 2 h, the large unit is turned off. The small unit requires 9 more hours to heat the pool. How long would it take the small unit, working alone, to heat the pool? $14\frac{2}{3}$ h

32. Two machines fill cereal boxes at the same rate. After the two machines work together for 7 h, one machine breaks down. The second machine requires 14 more hours to finish filling the boxes. How long would it have taken one of the machines, working alone, to fill the boxes?   28 h

33. A mechanic requires 2 h to repair a transmission, whereas an apprentice requires 6 h to make the same repairs. The mechanic worked alone for 1 h and then stopped. How long will it take the apprentice, working alone, to complete the repairs?   3 h

34. A large drain and a small drain are opened to drain a pool. The large drain can empty the pool in 6 h. After both drains have been open for 1 h, the large drain becomes clogged and is closed. The small drain remains open and requires 9 more hours to empty the pool. How long would it have taken the small drain, working alone, to empty the pool?   12 h

35. It takes Sam h hours to rake the yard, and it takes Emma k hours to rake the yard, where $h > k$. Let t be the amount of time it takes Sam and Emma to rake the yard together. Is t less than k, between k and h, or greater than k?   Less than k

36. Zachary and Eli picked a row of peas together in m minutes. It would have taken Zachary n minutes to pick the row of peas by himself. What fraction of the row of peas did Zachary pick? What fraction of the row of peas did Eli pick? $\dfrac{m}{n}; \dfrac{n-m}{n}$

OBJECTIVE B   *To use rational expressions to solve uniform motion problems*

For Exercises 37 and 38, use the following problem situation: A plane can fly 380 mph in calm air. In the time it takes the plane to fly 1440 mi against a headwind, it could fly 1600 mi with the wind.

37. a. Let r represent the rate of the wind. Complete the following table.

	Distance	÷	Rate	=	Time
Against the wind	1440	÷	$380 - r$	=	$\dfrac{1440}{380 - r}$
With the wind	1600	÷	$380 + r$	=	$\dfrac{1600}{380 + r}$

b. Use the relationship between the expressions in the last column of the table to write an equation that can be solved to find the rate of the wind: _____ = _____.
$$\dfrac{1440}{380 - r} = \dfrac{1600}{380 + r}$$

38. Use the equation from part (b) of Exercise 37.

a. Explain the meanings of $380 - r$ and $380 + r$ in terms of the problem situation.

b. Explain the meanings of $\dfrac{1440}{380 - r}$ and $\dfrac{1600}{380 + r}$ in terms of the problem situation.

39. A camper drove 80 mi to a recreational area and then hiked 4 mi into the woods. The rate of the camper while driving was ten times the rate while hiking. The total time spent hiking and driving was 3 h. Find the rate at which the camper hiked.   4 mph

40. The president of a company traveled 1800 mi by jet and 300 mi on a prop plane. The rate of the jet was four times the rate of the prop plane. The entire trip took 5 h. Find the rate of the jet.   600 mph

Unless otherwise noted, all content on this page is © Cengage Learning.

Sudheer Sakthan/Shutterstock.com

41. To assess the damage done by a fire, a forest ranger traveled 1080 mi by jet and then an additional 180 mi by helicopter. The rate of the jet was four times the rate of the helicopter. The entire trip took 5 h. Find the rate of the jet. 360 mph

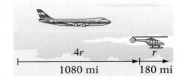

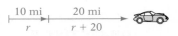
1080 mi   180 mi

42. An engineer traveled 165 mi by car and then an additional 660 mi by plane. The rate of the plane was four times the rate of the car. The total trip took 6 h. Find the rate of the car. 55 mph

43. After sailing 15 mi, a sailor changed direction and increased the boat's speed by 2 mph. An additional 19 mi was sailed at the increased speed. The total sailing time was 4 h. Find the rate of the boat for the first 15 mi. 7.5 mph

44. On a recent trip, a trucker traveled 330 mi at a constant rate. Because of road conditions, the trucker then reduced the speed by 25 mph. An additional 30 mi was traveled at the reduced rate. The entire trip took 7 h. Find the rate of the trucker for the first 330 mi. 55 mph

45. Commuting from work to home, a lab technician traveled 10 mi at a constant rate through congested traffic. Upon reaching the expressway, the technician increased the speed by 20 mph. An additional 20 mi was traveled at the increased speed. The total time for the trip was 1 h. At what rate did the technician travel through the congested traffic? 20 mph

10 mi   20 mi
r $r + 20$

46. As part of a conditioning program, a jogger ran 8 mi in the same amount of time it took a cyclist to ride 20 mi. The rate of the cyclist was 12 mph faster than the rate of the jogger. Find the rate of the jogger and the rate of the cyclist.
Jogger: 8 mph; cyclist: 20 mph

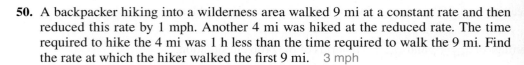
8 mi
r

20 mi
$r + 12$

47. In calm water, the rate of a small rental motorboat is 15 mph. The rate of the current on the river is 3 mph. How far down the river can a family travel and still return the boat in 3 h? 21.6 mi

48. The rate of a small aircraft in calm air is 125 mph. If the wind is currently blowing south at a rate of 15 mph, how far north can a pilot fly the plane and return it within 2 h? 123.2 mi

49. The speed of a boat in still water is 20 mph. The Jacksons traveled 75 mi down the Woodset River in this boat in the same amount of time it took them to return 45 mi up the river. Find the rate of the river's current. 5 mph

50. A backpacker hiking into a wilderness area walked 9 mi at a constant rate and then reduced this rate by 1 mph. Another 4 mi was hiked at the reduced rate. The time required to hike the 4 mi was 1 h less than the time required to walk the 9 mi. Find the rate at which the hiker walked the first 9 mi. 3 mph

51. An express train traveled 600 mi in the same amount of time it took a freight train to travel 360 mi. The rate of the express train was 20 mph faster than the rate of the freight train. Find the rate of each train.
Freight train: 30 mph; express train: 50 mph

52. A twin-engine plane flies 800 mi in the same amount of time it takes a single-engine plane to fly 600 mi. The rate of the twin-engine plane is 50 mph faster than the rate of the single-engine plane. Find the rate of the twin-engine plane. 200 mph

Unless otherwise noted, all content on this page is © Cengage Learning.

© iStockphoto.com/Craig Hansen

53. A small motor on a fishing boat can move the boat at a rate of 6 mph in calm water. Traveling with the current, the boat can travel 24 mi in the same amount of time it takes to travel 12 mi against the current. Find the rate of the current. 2 mph

54. A car is traveling at a rate that is 36 mph faster than the rate of a cyclist. The car travels 384 mi in the same amount of time it takes the cyclist to travel 96 mi. Find the rate of the car. 48 mph

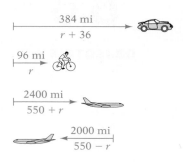

55. A commercial jet can fly 550 mph in calm air. Traveling with the jet stream, the plane can fly 2400 mi in the same amount of time it takes to fly 2000 mi against the jet stream. Find the rate of the jet stream. 50 mph

56. A cruise ship can sail 28 mph in calm water. Sailing with the Gulf Stream, the ship can sail 170 mi in the same amount of time it takes to sail 110 mi against the Gulf Stream. Find the rate of the Gulf Stream. 6 mph

57. Rowing with the current of a river, a rowing team can row 25 mi in the same amount of time it takes to row 15 mi against the current. The rate of the rowing team in calm water is 20 mph. Find the rate of the current. 5 mph

58. A plane can fly 180 mph in calm air. Flying with the wind, the plane can fly 600 mi in the same amount of time it takes to fly 480 mi against the wind. Find the rate of the wind. 20 mph

Critical Thinking

59. Work Problem   One pipe can fill a tank in 2 h, a second pipe can fill the tank in 4 h, and a third pipe can fill the tank in 5 h. How long would it take to fill the tank with all three pipes operating? $1\frac{1}{19}$ h

60. Work Problem   A mason can construct a retaining wall in 10 h. The mason's experienced apprentice can do the same job in 15 h. How long would it take the mason's novice apprentice to do the job if, working together, all three can complete the wall in 5 h? 30 h

61. Uniform Motion   An Outing Club traveled 18 mi by canoe and then hiked 3 mi. The rate by canoe was three times the rate on foot. The time spent walking was 1 h less than the time spent canoeing. Find the amount of time spent traveling by canoe. 2 h

62. Uniform Motion   A motorist drove 120 mi before running out of gas and walking 4 mi to a gas station. The motorist's driving rate was ten times the walking rate. The time spent walking was 2 h less than the time spent driving. How long did it take for the motorist to drive the 120 mi? 3 h

Projects or Group Activities

63. Uniform Motion   Because of bad weather, a bus driver reduced the usual speed along a 150-mile bus route by 10 mph. The bus arrived only 30 min later than its usual arrival time. How fast does the bus usually travel? 60 mph

64. Work Problem   A construction project must be completed in 15 days. Twenty-five workers did one-half of the job in 10 days. Working at the same rate, how many workers are needed to complete the job on schedule? 50 workers

QUICK QUIZ
1. One hose can fill a child's backyard pool in 24 min. A larger hose can fill the pool in 12 min. If the two hoses are used together, how long will it take to fill the pool? **8 min [8.7A]**
2. A plane flies 460 mph in calm air. Flying with the wind, the plane can travel 1560 mi in the same amount of time it takes to travel 1200 mi against the wind. Find the rate of the wind.
   60 mph [8.7B]

Unless otherwise noted, all content on this page is © Cengage Learning.

8.8   Variation

OBJECTIVE A   *To solve variation problems*

Direct variation describes a special function that can be expressed as the equation $y = kx^n$, where k is a constant and n is a positive number. The equation $y = kx^n$ is read "y varies directly as x to the nth" or "y is directly proportional to x to the nth." The constant k is called the **constant of variation** or the **constant of proportionality.**

A surveyor earns $43 per hour. The total wage w of the surveyor varies directly as the number of hours worked h. The direct variation equation is $w = 43h$. The constant of variation is 43, and the value of n is 1.

The distance s, in feet, that an object falls varies directly as the square of the time t, in seconds, that it falls. The direct variation equation is $s = kt^2$. The constant of variation is k, and the value of n is 2. If the object is dropped on Earth, $k = 16$; if the object is dropped on the moon, $k = 2.7$.

Many geometry formulas are expressed as direct variations. The circumference C of a circle is directly proportional to its diameter d. The direct variation equation is $C = \pi d$. The constant of proportionality is π, and the value of n is 1.

The area A of a circle varies directly as the square of its radius r. The direct variation equation is $A = \pi r^2$. The constant of proportionality is π, and the value of n is 2.

HOW TO 1   Given that V varies directly as r and that $V = 20$ when $r = 4$, find the constant of variation and the variation equation.

$V = kr$	• Write the basic direct variation equation.
$20 = k \cdot 4$	• Replace V and r by the given values. Then solve for k.
$5 = k$	• This is the constant of variation.
$V = 5r$	• Write the direct variation equation by substituting the value of k into the basic direct variation equation.

HOW TO 2   The tension T in a spring varies directly as the distance x it is stretched. If $T = 8$ lb when $x = 2$ in., find T when $x = 4$ in.

$T = kx$	• Write the basic direct variation equation.
$8 = k \cdot 2$	• Replace T and x by the given values.
$4 = k$	• Solve for the constant of variation.
$T = 4x$	• Write the direct variation equation.

To find T when $x = 4$ in., substitute 4 for x in the equation and solve for T.

$T = 4x$

$T = 4 \cdot 4 = 16$

The tension is 16 lb.

Inverse variation describes a function that can be written as the equation $y = \frac{k}{x^n}$, where k is a constant and n is a positive number. The equation $y = \frac{k}{x^n}$ is read "y varies inversely as x to the nth" or "y is inversely proportional to x to the nth."

The time t it takes a car to travel 100 mi varies inversely as the speed r of the car. The inverse variation equation is $t = \frac{100}{r}$. The variation constant is 100, and the value of n is 1.

The intensity I of a light source varies inversely as the square of the distance d from the source. The inverse variation equation is $I = \frac{k}{d^2}$. The constant of variation depends on the medium through which the light travels (air, water, glass), and the value of n is 2.

Point of Interest

The equation given in HOW TO 3 is important to concert sound engineers. Without additional speakers and reverberation, the sound intensity for someone about 20 rows back at this concert would be about 25 dB, the sound intensity of normal conversation.

HOW TO 3 The intensity I, in decibels (dB), of sound varies inversely as the square of the distance d from the source. If the intensity of an open-air concert is 110 dB in the front row, 10 ft from the band, find the variation equation.

$I = \frac{k}{d^2}$ • Write the basic inverse variation equation.

$110 = \frac{k}{10^2}$ • Replace I and d by the given values.

$11{,}000 = k$ • This is the constant of variation.

$I = \frac{11{,}000}{d^2}$ • This is the inverse variation equation.

IN-CLASS EXAMPLES

1. The distance d a spring will stretch varies directly as the force f applied to the spring. If a force of 12 lb is required to stretch a spring 6 in., what force is required to stretch the spring 10 in.? **20 lb**
2. At a constant temperature, the pressure P of a gas varies inversely as the volume V. If the pressure of a gas is 75 lb/in² when the volume is 200 ft³, find the pressure when the volume is 450 ft³. **$33.\overline{3}$ lb/in²**

HOW TO 4 The length L of a rectangle of fixed area is inversely proportional to the width w. If $L = 6$ ft when $w = 2$ ft, find L when $w = 3$ ft.

$L = \frac{k}{w}$ • Write the basic inverse variation equation.

$6 = \frac{k}{2}$ • Replace L and w by the given values.

$12 = k$ • Solve for the constant of variation.

$L = \frac{12}{w}$ • Write the inverse variation equation.

To find L when $w = 3$ ft, substitute 3 for w in the equation and solve for L.

$L = \frac{12}{w} = \frac{12}{3} = 4$

The length is 4 ft.

Joint variation is variation in which a variable varies directly as the product of two or more other variables. A joint variation can be expressed as the equation $z = kxy$, where k is a constant. The equation $z = kxy$ is read "z varies jointly as x and y."

For example, the area A of a triangle varies jointly as the base b and the height h. The joint variation equation is written $A = \frac{1}{2}bh$. The constant of variation is $\frac{1}{2}$.

A **combined variation** is a variation in which two or more types of variation occur at the same time. For example, in physics, the volume V of a gas varies directly as the temperature T and inversely as the pressure P. This combined variation is written $V = \frac{kT}{P}$.

© Ted Foxx/Alamy

HOW TO 5   A ball is being twirled on the end of a string. The tension T in the string is directly proportional to the square of the speed v of the ball and inversely proportional to the length r of the string. If the tension is 96 lb when the length of the string is 0.5 ft and the speed is 4 ft/s, find the tension when the length of the string is 1 ft and the speed is 5 ft/s.

$T = \dfrac{kv^2}{r}$ • Write the basic combined variation equation.

$96 = \dfrac{k \cdot 4^2}{0.5}$ • Replace T, v, and r by the given values.

$96 = \dfrac{k \cdot 16}{0.5}$

$96 = k \cdot 32$ • Solve for the constant of variation.
$3 = k$

$T = \dfrac{3v^2}{r}$ • Write the combined variation equation.

To find T when $r = 1$ ft and $v = 5$ ft/s, substitute 1 for r and 5 for v, and solve for T.

$T = \dfrac{3v^2}{r} = \dfrac{3 \cdot 5^2}{1} = 3 \cdot 25 = 75$

The tension is 75 lb.

EXAMPLE 1

The amount A of medication prescribed for a person is directly related to the person's weight W. For a 50-kilogram person, 2 ml of medication are prescribed. How many milliliters of medication are required for a person who weighs 75 kg?

Strategy

To find the required amount of medication:
- Write the basic direct variation equation, replace the variables by the given values, and solve for k.
- Write the direct variation equation, replacing k by its value. Substitute 75 for W and solve for A.

Solution

$A = kW$ • Direct variation equation
$2 = k \cdot 50$ • Replace A by 2 and W by 50.

$\dfrac{1}{25} = k$ • Solve for k.

$A = \dfrac{1}{25}W$ • Write the direct variation equation.

$A = \dfrac{1}{25} \cdot 75 = 3$ • Find A when $W = 75$.

The required amount of medication is 3 ml.

YOU TRY IT 1

The distance s a body falls from rest varies directly as the square of the time t of the fall. An object falls 64 ft in 2 s. How far will it fall in 5 s?

Your strategy

Your solution

400 ft

Solution on p. S25

EXAMPLE 2

A company that produces gaming laptop computers has determined that the number of laptops it can sell s is inversely proportional to the price P of the laptop. Two thousand laptops can be sold when the price is $2500. How many laptops can be sold when the price of a laptop is $2000?

Strategy

To find the number of laptops:
- Write the basic inverse variation equation, replace the variables by the given values, and solve for k.
- Write the inverse variation equation, replacing k by its value. Substitute 2000 for P and solve for s.

Solution

$$s = \frac{k}{P} \qquad \text{• Inverse variation equation}$$

$$2000 = \frac{k}{2500} \qquad \begin{array}{l}\text{• Replace } s \text{ by 2000 and} \\ P \text{ by 2500.}\end{array}$$

$$5{,}000{,}000 = k$$

$$s = \frac{5{,}000{,}000}{P} = \frac{5{,}000{,}000}{2000} = 2500 \qquad \begin{array}{l}\text{• } k = 5{,}000{,}000, \\ P = 2000\end{array}$$

At $2000 each, 2500 laptops can be sold.

YOU TRY IT 2

The resistance R to the flow of electric current in a wire of fixed length is inversely proportional to the square of the diameter d of the wire. If a wire of diameter 0.01 cm has a resistance of 0.5 ohm, what is the resistance in a wire that is 0.02 cm in diameter?

Your strategy

Your solution

0.125 ohm

EXAMPLE 3

The pressure P of a gas varies directly as the temperature T and inversely as the volume V. When $T = 50°$ and $V = 275$ in^3, $P = 20$ lb/in^2. Find the pressure of a gas when $T = 60°$ and $V = 250$ in^3.

Strategy

To find the pressure:
- Write the basic combined variation equation, replace the variables by the given values, and solve for k.
- Write the combined variation equation, replacing k by its value. Substitute 60 for T and 250 for V, and solve for P.

Solution

$$P = \frac{kT}{V} \qquad \begin{array}{l}\text{• Combined variation} \\ \text{equation}\end{array}$$

$$20 = \frac{k \cdot 50}{275} \qquad \begin{array}{l}\text{• Replace } P \text{ by 20, } T \text{ by} \\ \text{50, and } V \text{ by 275.}\end{array}$$

$$110 = k$$

$$P = \frac{110T}{V} = \frac{110 \cdot 60}{250} = 26.4 \qquad \begin{array}{l}\text{• } k = 110, T = 60, \\ V = 250\end{array}$$

The pressure is 26.4 lb/in^2.

YOU TRY IT 3

The strength s of a beam varies jointly as its width W and the square of its depth d and inversely as its length L. If the strength of a beam 2 in. wide, 12 in. deep, and 12 ft long is 1200 lb, find the strength of a beam 4 in. wide, 8 in. deep, and 16 ft long.

Your strategy

Your solution

800 lb

Solutions on pp. S25–S26

8.8 EXERCISES

SUGGESTED ASSIGNMENT
Exercises 1–4; Exercises 5–19, odds;
More challenging exercises: Exercises 21 and 22

✔ **Concept Check**

For Exercises 1 to 4, write the statement as an equation using k as the constant of variation.

1. y varies directly as x. $y = kx$

2. y varies inversely as x. $y = \dfrac{k}{x}$

3. z varies jointly as x and y. $z = kxy$

4. z varies directly as x and inversely as the square of y. $z = \dfrac{kx}{y^2}$

OBJECTIVE A *To solve variation problems*

5. Business The profit P realized by a company varies directly as the number of products it sells s. If a company makes a profit of $4000 on the sale of 250 products, what is the profit when the company sells 5000 products? $80,000

6. Compensation The income I of a computer analyst varies directly as the number of hours h worked. If the analyst earns $336 for working 8 h, how much will the analyst earn for working 36 h? $1512

7. Recreation The pressure p on a diver in the water varies directly as the depth d. If the pressure is 4.5 lb/in^2 when the depth is 10 ft, what is the pressure when the depth is 15 ft? 6.75 lb/in^2

8. Physics The distance d that a spring will stretch varies directly as the force f applied to the spring. If a force of 6 lb is required to stretch a spring 3 in., what force is required to stretch the spring 4 in.? 8 lb

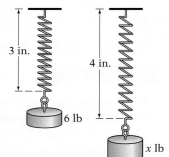

9. Physics The distance d an object falls is directly proportional to the square of the time t of the fall. If an object falls 144 ft in 3 s, how far will it fall in 10 s? 1600 ft

10. Physics The period p of a pendulum, or the time it takes the pendulum to make one complete swing, varies directly as the square root of the length L of the pendulum. If the period of a pendulum is 1.5 s when the length is 2 ft, find the period when the length is 5 ft. Round to the nearest hundredth. 2.37 s

11. Computer Science Parallel processing is the use of more than one computer to solve a problem. The time T it takes to solve a problem is inversely proportional to the number n of computers used. If it takes one computer 500 s to solve a problem, how long would it take five computers to solve the same problem? 100 s

12. Safety The stopping distance s of a car varies directly as the square of its speed v. If a car traveling at 30 mph requires 63 ft to stop, find the stopping distance for a car traveling at 55 mph. 211.75 ft

13. Sailing The load L, in pounds, on a sail varies directly as the square of the wind speed v, in miles per hour. If the load on a sail is 640 lb when the wind speed is 20 mph, what is the load on the sail when the wind speed is 15 mph? 360 lb

14. Whirlpools The speed v of the current in a whirlpool varies inversely as the distance d from the whirlpool's center. The Old Sow whirlpool, located off the coast of eastern Canada, is one of the most powerful whirlpools on Earth. At a distance of 10 ft from the center of the whirlpool, the speed of the current is about 2.5 ft/s. What is the speed of the current 2 ft from the center? 12.5 ft/s

Image Courtesy of Jim Lowe of Eastport, Maine

© iStockphoto.com/technotr

Unless otherwise noted, all content on this page is © Cengage Learning.

15. ● **Oil Spill**   Read the article at the right about the Deepwater Horizon oil spill of 2010. If the well leaked oil at the same rate throughout the duration of the spill, then the total amount of oil leaked would be directly proportional to the number of days the oil had been leaking. Using the data in the article, estimate how many barrels of oil leaked during the 86 days of the spill. Round to the nearest tenth of a million.
4.8 million barrels

16. **Architecture**   The heat loss H, in watts, through a single-pane window varies jointly as the area A and the difference T between the inside and outside temperatures. If the heat loss is 6 watts for a window with an area of 1.5 m² and a temperature difference of 2°C, what is the heat loss for a window with an area of 2 m² and a temperature difference of 5°C?   20 watts

17. **Electronics**   The current I in a wire varies directly as the voltage v and inversely as the resistance r. If the current is 10 amps when the voltage is 110 volts and the resistance is 11 ohms, find the current when the voltage is 180 volts and the resistance is 24 ohms.   7.5 amps

18. **Magnetism**   The repulsive force f between the north poles of two magnets is inversely proportional to the square of the distance d between them. If the repulsive force is 20 lb when the distance is 4 in., find the repulsive force when the distance is 2 in.   80 lb

19. **Light**   The intensity I of a light source is inversely proportional to the square of the distance d from the source. If the intensity is 12 foot-candles at a distance of 10 ft, what is the intensity when the distance is 5 ft?   48 foot-candles

20. **Mechanics**   The speed v of a gear varies inversely as the number of teeth t. If a gear that has 45 teeth makes 24 revolutions per minute (rpm), how many revolutions per minute will a gear that has 36 teeth make?   30 rpm

Critical Thinking

For Exercises 21 and 22, complete using the word *directly* or *inversely*.

21. If the area of a rectangle is held constant, the length of the rectangle varies ____inversely____ as the width.

22. If the length of a rectangle is held constant, the area of the rectangle varies ____directly____ as the width.

Projects or Group Activities

23. In the direct variation equation $y = kx$, what is the effect on y when x is doubled?
y is doubled.

24. In the inverse variation equation $y = \frac{k}{x}$, what is the effect on x when y is doubled?
x is halved.

25. If y varies inversely as the square of x, what is the effect on y when x is doubled?
y is divided by 4.

Unless otherwise noted, all content on this page is © Cengage Learning.

In the NEWS!

Oil Leak Finally Capped

On June 15, 56 days after the oil spill, government estimates of the total amount of oil leaked into the Gulf of Mexico stood at 3.1 million barrels. One month later, the well has been successfully capped, and finally, after 86 days, oil no longer leaks into the ocean waters.
Source: nytimes.com

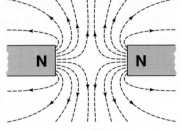

QUICK QUIZ

1. The pressure P on a diver in the water varies directly as the depth d. If the pressure is 6 lb/in² when the depth is 12 ft, what is the pressure when the depth is 16 ft?   **8 lb/in²** [8.8A]

2. The speed v of a gear varies inversely as the number of teeth t. If a gear that has 36 teeth makes 30 revolutions per minute, how many revolutions per minute will a gear that has 54 teeth make?
20 rpm [8.8A]

CHAPTER

8 | Summary

Key Words	Examples
A **rational expression** is a fraction in which the numerator and denominator are polynomials. A rational expression is in **simplest form** when the numerator and denominator have no common factors. [8.1A, p. 450]	$\frac{2x + 1}{x^2 + 4}$ is a rational expression in simplest form.
The **reciprocal of a rational expression** is the rational expression with the numerator and denominator interchanged. [8.1C, p. 453]	The reciprocal of $\frac{3x - y}{x + 4}$ is $\frac{x + 4}{3x - y}$.
The **least common multiple (LCM) of two or more polynomials** is the polynomial of least degree that contains all the factors of each polynomial. [8.2A, p. 458]	The LCM of $3x^2 - 6x$ and $x^2 - 4$ is $3x(x - 2)(x + 2)$, because it contains the factors of $3x^2 - 6x = 3x(x - 2)$ and the factors of $x^2 - 4 = (x - 2)(x + 2)$.
A **complex fraction** is a fraction in which the numerator or denominator contains one or more fractions. [8.3A, p. 466]	$\dfrac{x - \dfrac{2}{x + 1}}{1 - \dfrac{4}{x}}$ is a complex fraction.
A **ratio** is the quotient of two quantities that have the same unit. A **rate** is the quotient of two quantities that have different units. [8.5A, p. 476]	$\frac{9}{4}$ is a ratio. $\frac{60 \text{ m}}{12 \text{ s}}$ is a rate.
A **proportion** is an equation that states the equality of two ratios or rates. [8.5A, p. 476]	$\frac{3}{8} = \frac{12}{32}$ and $\frac{x \text{ ft}}{12 \text{ s}} = \frac{15 \text{ ft}}{160 \text{ s}}$ are proportions.
A **literal equation** is an equation that contains more than one variable. A **formula** is a literal equation that states a rule about measurements. [8.6A, p. 487]	$3x - 4y = 12$ is a literal equation. $A = LW$ is a literal equation that is also the formula for the area of a rectangle.
Direct variation is a function that can be expressed as the equation $y = kx^n$, where k is a constant called the **constant of variation**, and n is a positive number. [8.8A, p. 502]	$F = kv^2$ is a formula that gives drag force F created by an object with velocity v.
Inverse variation is a function that can be expressed as the equation $y = \frac{k}{x^n}$, where k is a constant and n is a positive number. [8.8A, p. 503]	$I = \frac{k}{d^2}$ gives the intensity of a light source at a distance d from the source.

Joint variation is a variation in which a variable varies directly as the product of two or more variables. A joint variation can be expressed as the equation $z = kxy$, where k is a constant. [8.8A, p. 503]

$C = kAT$ is a formula for the cost of insulation, where A is the area to be insulated and T is the thickness of the insulation.

A **combined variation** is a variation in which two or more types of variation occur at the same time. [8.8A, p. 503]

$V = \frac{kT}{P}$ is a formula that states that the volume of a gas is directly proportional to the temperature and inversely proportional to the pressure.

Essential Rules and Procedures

Examples

Simplifying Rational Expressions [8.1A, p. 450]
Factor the numerator and denominator. Divide the numerator and denominator by the common factors.

$$\frac{x^2 - 3x - 10}{x^2 - 25} = \frac{(x + 2)(x - 5)}{(x + 5)(x - 5)}$$
$$= \frac{x + 2}{x + 5}$$

Multiplying Rational Expressions [8.1B, p. 451]
Multiply the numerators. Multiply the denominators. Write the answer in simplest form.

$$\frac{a}{b} \cdot \frac{c}{d} = \frac{ac}{bd}$$

$$\frac{x^2 - 3x}{x^2 + x} \cdot \frac{x^2 + 5x + 4}{x^2 - 4x + 3}$$
$$= \frac{x(x - 3)}{x(x + 1)} \cdot \frac{(x + 1)(x + 4)}{(x - 3)(x - 1)}$$
$$= \frac{x(x - 3)(x + 1)(x + 4)}{x(x + 1)(x - 3)(x - 1)} = \frac{x + 4}{x - 1}$$

Dividing Rational Expressions [8.1C, p. 453]
Multiply the first fraction by the reciprocal of the divisor. Write the answer in simplest form.

$$\frac{a}{b} \div \frac{c}{d} = \frac{a}{b} \cdot \frac{d}{c} = \frac{ad}{bc}$$

$$\frac{4x + 16}{3x - 6} \div \frac{x^2 + 6x + 8}{x^2 - 4}$$
$$= \frac{4x + 16}{3x - 6} \cdot \frac{x^2 - 4}{x^2 + 6x + 8}$$
$$= \frac{4(x + 4)}{3(x - 2)} \cdot \frac{(x - 2)(x + 2)}{(x + 4)(x + 2)} = \frac{4}{3}$$

Adding and Subtracting Rational Expressions [8.2B, p. 459]

1. Find the LCM of the denominators.
2. Write each fraction as an equivalent fraction using the LCM as the denominator.
3. Add or subtract the numerators and place the result over the common denominator.
4. Write the answer in simplest form.

$$\frac{a}{b} + \frac{c}{b} = \frac{a + c}{b} \qquad \frac{a}{b} - \frac{c}{b} = \frac{a - c}{b}$$

$$\frac{x}{x + 1} - \frac{x + 3}{x - 2}$$
$$= \frac{x}{x + 1} \cdot \frac{x - 2}{x - 2} - \frac{x + 3}{x - 2} \cdot \frac{x + 1}{x + 1}$$
$$= \frac{x(x - 2)}{(x + 1)(x - 2)} - \frac{(x + 3)(x + 1)}{(x + 1)(x - 2)}$$
$$= \frac{x(x - 2) - (x + 3)(x + 1)}{(x + 1)(x - 2)}$$
$$= \frac{(x^2 - 2x) - (x^2 + 4x + 3)}{(x + 1)(x - 2)}$$
$$= \frac{-6x - 3}{(x + 1)(x - 2)}$$

Simplifying Complex Fractions [8.3A, p. 466]

Method 1: Multiply by 1 in the form $\dfrac{\text{LCM}}{\text{LCM}}$.

1. Determine the LCM of the denominators of the fractions in the numerator and denominator of the complex fraction.

2. Multiply the numerator and denominator of the complex fraction by the LCM.

3. Simplify.

Method 1:
$$\frac{\dfrac{1}{x} + \dfrac{1}{y}}{\dfrac{1}{x} - \dfrac{1}{y}} = \frac{\dfrac{1}{x} + \dfrac{1}{y}}{\dfrac{1}{x} - \dfrac{1}{y}} \cdot \frac{xy}{xy}$$

$$= \frac{\dfrac{1}{x} \cdot xy + \dfrac{1}{y} \cdot xy}{\dfrac{1}{x} \cdot xy - \dfrac{1}{y} \cdot xy}$$

$$= \frac{y + x}{y - x}$$

Method 2: Multiply the numerator by the reciprocal of the denominator.

1. Simplify the numerator to a single fraction and simplify the denominator to a single fraction.

2. Using the rule for dividing fractions, multiply the numerator by the reciprocal of the denominator.

3. Simplify.

Method 2:
$$\frac{\dfrac{1}{x} + \dfrac{1}{y}}{\dfrac{1}{x} - \dfrac{1}{y}} = \frac{\dfrac{y + x}{xy}}{\dfrac{y - x}{xy}}$$

$$= \frac{y + x}{xy} \cdot \frac{xy}{y - x}$$

$$= \frac{y + x}{y - x}$$

Solving Equations Containing Fractions [8.4A, p. 470]

Clear denominators by multiplying each side of the equation by the LCM of the denominators. Then solve for the variable.

$$\frac{1}{2a} = \frac{2}{a} - \frac{3}{8}$$

$$8a\left(\frac{1}{2a}\right) = 8a\left(\frac{2}{a}\right) - 8a\left(\frac{3}{8}\right)$$

$$4 = 16 - 3a$$

$$-12 = -3a$$

$$4 = a$$

Similar Triangles [8.5B, pp. 478–479]

Similar triangles have the same shape but not necessarily the same size. The ratios of corresponding sides of similar triangles are equal. The measures of the corresponding angles of similar triangles are equal.

Triangles *ABC* and *DFE* are similar triangles. The ratios of corresponding parts are equal to $\frac{2}{3}$.

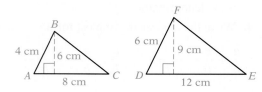

If two angles of one triangle are equal in measure to two angles of another triangle, then the two triangles are similar.

Triangles *AOB* and *COD* are similar because $m\angle AOB = m\angle COD$ and $m\angle B = m\angle D$.

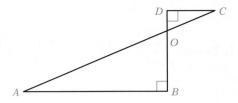

Solving Literal Equations [8.6A, p. 487]

Rewrite the equation so that the letter being solved for is alone on one side of the equation and all numbers and other variables are on the other side.

Solve $2x + ax = 5$ for x.

$$2x + ax = 5$$
$$x(2 + a) = 5$$
$$\frac{x(2 + a)}{2 + a} = \frac{5}{2 + a}$$
$$x = \frac{5}{2 + a}$$

Work Problems [8.7A, p. 491]

Rate of work $\times$ time worked = part of task completed

Pat can do a certain job in 3 h. Chris can do the same job in 5 h. How long would it take Pat and Chris, working together, to get the job done?

$$\frac{t}{3} + \frac{t}{5} = 1$$

Uniform Motion Problems with Rational Expressions
[8.7B, p. 494]

$$\frac{\text{Distance}}{\text{Rate}} = \text{time}$$

Train A's speed is 15 mph faster than train B's speed. Train A travels 150 mi in the same amount of time it takes train B to travel 120 mi. Find the rate of train B.

$$\frac{120}{r} = \frac{150}{r + 15}$$

Unless otherwise noted, all content on this page is © Cengage Learning.

8 | Review Exercises

1. Divide: $\dfrac{6a^2b^7}{25x^3y} \div \dfrac{12a^3b^4}{5x^2y^2}$ $\dfrac{b^3y}{10ax}$ [8.1C]

2. Add: $\dfrac{x+7}{15x} + \dfrac{x-2}{20x}$ $\dfrac{7x+22}{60x}$ [8.2B]

3. Multiply: $\dfrac{3x^3 + 9x^2}{6xy^2 - 18y^2} \cdot \dfrac{4xy^3 - 12y^3}{5x^2 + 15x}$ $\dfrac{2xy}{5}$ [8.1B]

4. Divide: $\dfrac{2x(x-y)}{x^2y(x+y)} \div \dfrac{3(x-y)}{x^2y^2}$ $\dfrac{2xy}{3(x+y)}$ [8.1C]

5. Simplify: $\dfrac{x - \dfrac{16}{5x-2}}{3x - 4 - \dfrac{88}{5x-2}}$ $\dfrac{x-2}{3x-10}$ [8.3A]

6. Simplify: $\dfrac{x^2 + x - 30}{15 + 2x - x^2}$ $-\dfrac{x+6}{x+3}$ [8.1A]

7. Simplify: $\dfrac{16x^5y^3}{24xy^{10}}$ $\dfrac{2x^4}{3y^7}$ [8.1A]

8. Solve: $\dfrac{20}{x+2} = \dfrac{5}{16}$ 62 [8.5A]

9. Divide: $\dfrac{10 - 23y + 12y^2}{6y^2 - y - 5} \div \dfrac{4y^2 - 13y + 10}{18y^2 + 3y - 10}$
$\dfrac{(3y-2)^2}{(y-1)(y-2)}$ [8.1C]

10. Solve $3ax - x = 5$ for x.

$x = \dfrac{5}{3a-1}$ [8.6A]

11. Solve: $\dfrac{2}{x} + \dfrac{3}{4} = 1$ 8 [8.4A]

12. Add: $\dfrac{x}{y} + \dfrac{3}{x}$ $\dfrac{x^2 + 3y}{xy}$ [8.2B]

13. Solve $5x + 4y = 20$ for y. $y = -\dfrac{5}{4}x + 5$ [8.6A]

14. Multiply: $\dfrac{8ab^2}{15x^3y} \cdot \dfrac{5xy^4}{16a^2b}$ $\dfrac{by^3}{6ax^2}$ [8.1B]

15. Simplify: $\dfrac{1 - \dfrac{1}{x}}{1 - \dfrac{8x-7}{x^2}}$ $\dfrac{x}{x-7}$ [8.3A]

16. Write each fraction in terms of the LCM of the denominators.

$\dfrac{x}{12x^2 + 16x - 3}$, $\dfrac{4x^2}{6x^2 + 7x - 3}$

$\dfrac{3x^2 - x}{(2x+3)(6x-1)(3x-1)}$, $\dfrac{24x^3 - 4x^2}{(2x+3)(6x-1)(3x-1)}$ [8.2A]

17. Solve $T = 2(ab + bc + ca)$ for a.

$a = \dfrac{T - 2bc}{2b + 2c}$ [8.6A]

18. Solve: $\dfrac{5}{7} + \dfrac{x}{2} = 2 - \dfrac{x}{7}$ 2 [8.4A]

19. Simplify: $\dfrac{2 + \dfrac{1}{x}}{3 - \dfrac{2}{x}}$ $\dfrac{2x + 1}{3x - 2}$ [8.3A]

20. Subtract: $\dfrac{2x}{x - 5} - \dfrac{x + 1}{x - 2}$ $\dfrac{x^2 + 5}{(x - 5)(x - 2)}$ [8.2B]

21. Solve $i = \dfrac{100m}{c}$ for c. $c = \dfrac{100m}{i}$ [8.6A]

22. Solve: $\dfrac{x + 8}{x + 4} = 1 + \dfrac{5}{x + 4}$ No solution [8.4A]

23. Divide: $\dfrac{20x^2 - 45x}{6x^3 + 4x^2} \div \dfrac{40x^3 - 90x^2}{12x^2 + 8x}$ $\dfrac{1}{x^2}$ [8.1C]

24. Add: $\dfrac{2y}{5y - 7} + \dfrac{3}{7 - 5y}$ $\dfrac{2y - 3}{5y - 7}$ [8.2B]

25. Subtract: $\dfrac{5x + 3}{2x^2 + 5x - 3} - \dfrac{3x + 4}{2x^2 + 5x - 3}$

$\dfrac{1}{x + 3}$ [8.2B]

26. Find the LCM of $10x^2 - 11x + 3$ and $20x^2 - 17x + 3$.

$(5x - 3)(2x - 1)(4x - 1)$ [8.2A]

27. Solve $4x + 9y = 18$ for y.

$y = -\dfrac{4}{9}x + 2$ [8.6A]

28. Multiply: $\dfrac{2x^2 - 5x - 3}{3x^2 - 7x - 6} \cdot \dfrac{3x^2 + 8x + 4}{x^2 + 4x + 4}$

$\dfrac{2x + 1}{x + 2}$ [8.1B]

29. Solve: $\dfrac{20}{2x + 3} = \dfrac{17x}{2x + 3} - 5$

5 [8.4A]

30. Add: $\dfrac{x - 1}{x + 2} + \dfrac{3x - 2}{5 - x} + \dfrac{5x^2 + 15x - 11}{x^2 - 3x - 10}$

$\dfrac{3x - 1}{x - 5}$ [8.2B]

31. Solve: $\dfrac{6}{x - 7} = \dfrac{8}{x - 6}$

10 [8.5A]

32. Solve: $\dfrac{3}{20} = \dfrac{x}{80}$

12 [8.5A]

33. Geometry   Given that $\overline{MP}$ and $\overline{NQ}$ intersect at O, NQ measures 25 cm, MO measures 6 cm, and PO measures 9 cm, find QO.   15 cm   [8.5B]

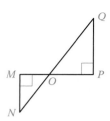

34. Geometry   Triangles ABC and DEF are similar triangles. Find the area of triangle DEF.
$\dfrac{256}{3}$ in²   [8.5B]

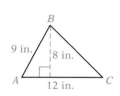

35. Work Problem   One hose can fill a pool in 15 h. A second hose can fill the pool in 10 h. How long would it take to fill the pool using both hoses?   6 h   [8.7A]

36. Uniform Motion   A car travels 315 mi in the same amount of time it takes a bus to travel 245 mi. The rate of the car is 10 mph faster than that of the bus. Find the rate of the car.   45 mph   [8.7B]

37. Uniform Motion   The rate of a jet is 400 mph in calm air. Traveling with the wind, the jet can fly 2100 mi in the same amount of time it takes to fly 1900 mi against the wind. Find the rate of the wind.   20 mph   [8.7B]

38. Baseball   A pitcher's earned run average (ERA) is the average number of runs allowed in 9 innings of pitching. If a pitcher allows 15 runs in 100 innings, find the pitcher's ERA.   1.35   [8.5A]

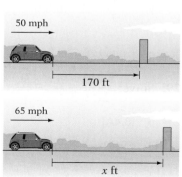

iStockphoto.com/jtroudt

39. Electronics   The current I in an electric circuit varies inversely as the resistance R. If the current in the circuit is 4 amps when the resistance is 50 ohms, find the current in the circuit when the resistance is 100 ohms.
2 amps   [8.8A]

40. Cartography   On a certain map, 2.5 in. represents 10 mi. How many miles would be represented by 12 in.?
48 mi   [8.5A]

41. Safety   The stopping distance s of a car varies directly as the square of the speed v of the car. For a car traveling at 50 mph, the stopping distance is 170 ft. Find the stopping distance for a car traveling at 65 mph.
287.3 ft   [8.8A]

42. Work   An electrician requires 65 min to install a ceiling fan. The electrician and an apprentice, working together, take 40 min to install the fan. How long would it take the apprentice, working alone, to install the ceiling fan?
104 min   [8.7A]

Unless otherwise noted, all content on this page is © Cengage Learning.

CHAPTER

8 | TEST

1. Subtract: $\dfrac{x}{x+3} - \dfrac{2x-5}{x^2+x-6}$

$\dfrac{x^2-4x+5}{(x-2)(x+3)}$ [8.2B]

2. Solve: $\dfrac{3}{x+4} = \dfrac{5}{x+6}$

-1 [8.5A]

3. Multiply: $\dfrac{x^2+2x-3}{x^2+6x+9} \cdot \dfrac{2x^2-11x+5}{2x^2+3x-5}$

$\dfrac{(x-5)(2x-1)}{(x+3)(2x+5)}$ [8.1B]

4. Simplify: $\dfrac{16x^5y}{24x^2y^4}$

$\dfrac{2x^3}{3y^3}$ [8.1A]

5. Solve $d = s + rt$ for t.

$t = \dfrac{d-s}{r}$ [8.6A]

6. Solve: $\dfrac{6}{x} - 2 = 1$

2 [8.4A]

7. Simplify: $\dfrac{x^2+4x-5}{1-x^2}$

$-\dfrac{x+5}{x+1}$ [8.1A]

8. Find the LCM of $6x - 3$ and $2x^2 + x - 1$.

$3(2x-1)(x+1)$ [8.2A]

9. Subtract: $\dfrac{2}{2x-1} - \dfrac{3}{3x+1}$

$\dfrac{5}{(2x-1)(3x+1)}$ [8.2B]

10. Divide: $\dfrac{x^2+3x+2}{x^2+5x+4} \div \dfrac{x^2-x-6}{x^2+2x-15}$

$\dfrac{x+5}{x+4}$ [8.1C]

11. Simplify: $\dfrac{1 + \dfrac{1}{x} - \dfrac{12}{x^2}}{1 + \dfrac{2}{x} - \dfrac{8}{x^2}}$

$\dfrac{x-3}{x-2}$ [8.3A]

12. Write each fraction in terms of the LCM of the denominators.

$\dfrac{3}{x^2-2x}, \dfrac{x}{x^2-4}$

$\dfrac{3x+6}{x(x-2)(x+2)}, \dfrac{x^2}{x(x-2)(x+2)}$ [8.2A]

13. **Subtract:** $\dfrac{2x}{x^2 + 3x - 10} - \dfrac{4}{x^2 + 3x - 10}$

$\dfrac{2}{x + 5}$ [8.2B]

14. **Solve** $3x - 8y = 16$ for y.

$y = \dfrac{3}{8}x - 2$ [8.6A]

15. **Solve:** $\dfrac{2x}{x + 1} - 3 = \dfrac{-2}{x + 1}$

No solution   [8.4A]

16. **Multiply:** $\dfrac{x^3 y^4}{x^2 - 4x + 4} \cdot \dfrac{x^2 - x - 2}{x^6 y^4}$

$\dfrac{x + 1}{x^3(x - 2)}$ [8.1B]

17. **Geometry**   Given that $\overline{AE} \parallel \overline{BD}$, AB measures 5 ft, ED measures 8 ft, and BC measures 3 ft, find CE.
12.8 ft   [8.5B]

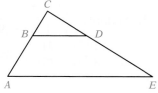

18. **Chemistry**   A saltwater solution is formed by mixing 4 lb of salt with 10 gal of water. At this rate, how many pounds of salt are required for 15 gal of water?
6 lb   [8.5A]

19. **Work Problem**   One pipe can fill a pool in 6 h, whereas a second pipe requires 12 h to fill the pool. How long would it take to fill the pool with both pipes turned on?
4 h   [8.7A]

20. **Uniform Motion**   A small plane can fly at 110 mph in calm air. Flying with the wind, the plane can fly 260 mi in the same amount of time it takes to fly 180 mi against the wind. Find the rate of the wind.
20 mph   [8.7B]

21. **Landscaping**   A landscape architect uses three sprinklers for each 200 ft² of lawn. At this rate, how many sprinklers are needed for a 3600-square-foot lawn?   54 sprinklers   [8.5A]

iStockphoto.com/Peter Eckhardt

22. **Sound Intensity**   The intensity I, in decibels, of a sound is inversely proportional to the square of the distance d, in meters, from the source. If the intensity is 50 decibels at a distance of 8 m from the source, what is the intensity at a distance of 5 m from the source?
128 decibels   [8.8A]

Unless otherwise noted, all content on this page is © Cengage Learning.

Cumulative Review Exercises

1. Evaluate: $\left(\dfrac{2}{3}\right)^2 \div \left(\dfrac{3}{2} - \dfrac{2}{3}\right) + \dfrac{1}{2}$

$\dfrac{31}{30}$ [1.3A]

2. Evaluate $-a^2 + (a - b)^2$ when $a = -2$ and $b = 3$.

21   [1.4A]

3. Simplify: $-2x - (-3y) + 7x - 5y$
5x − 2y   [1.4B]

4. Simplify: $2[3x - 7(x - 3) - 8]$
−8x + 26   [1.4D]

5. Solve: $4 - \dfrac{2}{3}x = 7$

$-\dfrac{9}{2}$ [2.2A]

6. Solve: $3[x - 2(x - 3)] = 2(3 - 2x)$

−12   [2.2C]

7. Find $16\dfrac{2}{3}\%$ of 60.

10   [2.1D]

8. Solve: $x - 3(1 - 2x) \geq 1 - 4(3 - 2x)$

$\{x | x \leq 8\}$ [2.5A]

9. Find the equation of the line that contains the point $(-2, -1)$ and is parallel to the line $3x - 2y = 6$.

$y = \dfrac{3}{2}x + 2$ [4.6A]

10. Solve: $2x - y + z = 2$
$3x + y - 2z = 9$
$x - y + z = 0$
(2, 1, −1)   [5.2B]

11. Simplify: $(a^2b^5)(ab^2)$
a^3b^7 [6.1A]

12. Multiply: $(a - 3b)(a + 4b)$
$a^2 + ab - 12b^2$ [6.3B]

13. Divide: $(x^3 - 8) \div (x - 2)$
$x^2 + 2x + 4$ [6.4B]

14. Factor: $12x^2 - x - 1$
$(4x + 1)(3x - 1)$ [7.3A/7.3B]

15. Given $P(x) = \dfrac{x - 1}{2x - 3}$, find $P(-2)$.
$\dfrac{3}{7}$ [4.2A]

16. Factor: $2a^3 + 7a^2 - 15a$
$a(2a - 3)(a + 5)$ [7.3A/7.3B]

17. Factor: $4b^2 - 100$

$4(b + 5)(b - 5)$ [7.4D]

18. Solve: $(x + 3)(2x - 5) = 0$

$-3, \dfrac{5}{2}$ [7.5A]

19. Simplify: $\dfrac{12x^4y^2}{18xy^7}$
$\dfrac{2x^3}{3y^5}$ [8.1A]

20. Simplify: $\dfrac{x^2 - 7x + 10}{25 - x^2}$
$-\dfrac{x - 2}{x + 5}$ [8.1A]

21. Divide: $\dfrac{x^2 - x - 56}{x^2 + 8x + 7} \div \dfrac{x^2 - 13x + 40}{x^2 - 4x - 5}$

1 [8.1C]

22. Subtract: $\dfrac{2}{2x - 1} - \dfrac{1}{x + 1}$

$\dfrac{3}{(2x - 1)(x + 1)}$ [8.2B]

23. Simplify: $\dfrac{1 - \dfrac{2}{x} - \dfrac{15}{x^2}}{1 - \dfrac{25}{x^2}}$

$\dfrac{x + 3}{x + 5}$ [8.3A]

24. Solve: $\dfrac{3x}{x - 3} - 2 = \dfrac{10}{x - 3}$

4 [8.4A]

25. Solve: $\dfrac{2}{x - 2} = \dfrac{12}{x + 3}$

3 [8.5A]

26. Solve $f = v + at$ for t.

$t = \dfrac{f - v}{a}$ [8.6A]

27. Integer Problem   Translate "the difference between five times a number and thirteen is the opposite of eight" into an equation and solve. $5x - 13 = -8$; $x = 1$ [2.3A]

28. Metallurgy   A silversmith mixes 60 g of an alloy that is 40% silver with 120 g of another silver alloy. The resulting alloy is 60% silver. Find the percent of silver in the 120-gram alloy.   70% [2.4B]

29. Insurance   A life insurance policy costs $16 for every $1000 of coverage. At this rate, how much money would a policy for $5000 cost? $80 [8.5A]

30. Electronics   The electrical resistance r of a cable varies directly as its length l and inversely as the square of its diameter d. If a cable 16,000 ft long and $\frac{1}{4}$ in. in diameter has a resistance of 3.2 ohms, what is the resistance of a cable that is 8000 ft long and $\frac{1}{2}$ in. in diameter?
0.4 ohm [8.6A]

31. Work Problem   One water pipe can fill a tank in 9 min, whereas a second pipe requires 18 min to fill the tank. How long would it take both pipes, working together, to fill the tank?   6 min [8.7A]

32. Uniform Motion   The rower of a boat can row at a rate of 5 mph in calm water. Traveling with the current, the boat travels 14 mi in the same amount of time it takes to travel 6 mi against the current. Find the rate of the current.   2 mph [8.7B]

© iStockphoto.com/Ashok Rodrigues

Exponents and Radicals

OBJECTIVES

SECTION 9.1
A To simplify expressions with rational exponents
B To write exponential expressions as radical expressions and to write radical expressions as exponential expressions
C To simplify radical expressions that are roots of perfect powers

SECTION 9.2
A To simplify radical expressions
B To add or subtract radical expressions

SECTION 9.3
A To multiply radical expressions
B To divide radical expressions

SECTION 9.4
A To solve a radical equation
B To solve application problems

SECTION 9.5
A To simplify a complex number
B To add or subtract complex numbers
C To multiply complex numbers
D To divide complex numbers

Focus on Success

There are a number of features in this text that will help you to be prepared for a test. Start with the Chapter Summary. The Chapter Summary describes the important topics covered in the chapter. Do the Chapter Review Exercises. If you have trouble with any of the questions, restudy the objectives the questions are taken from. Take the Chapter Test, working on it as if it were an actual exam. (See Ace the Test, page AIM-11.)

© iStockphoto.com/Mehmet Yunus Yeşil

Prep Test

Are you ready to succeed in this chapter? Take the Prep Test below to find out if you are ready to learn the new material.

1. Complete: $48 = ? \cdot 3$
16 [1.1D]

For Exercises 2 to 6, simplify.

2. 2^5
32 [1.2E]

3. $6\left(\dfrac{3}{2}\right)$
9 [1.2D]

4. $\dfrac{1}{2} - \dfrac{2}{3} + \dfrac{1}{4}$
$\dfrac{1}{12}$ [1.2C]

5. $(3 - 7x) - (4 - 2x)$
$-5x - 1$ [1.4D]

6. $\dfrac{3x^5y^6}{12x^4y}$
$\dfrac{xy^5}{4}$ [6.1B]

7. Expand: $(3x - 2)^2$
$9x^2 - 12x + 4$ [6.3C]

For Exercises 8 and 9, multiply.

8. $(2 + 4x)(5 - 3x)$
$-12x^2 + 14x + 10$ [6.3B]

9. $(6x - 1)(6x + 1)$
$36x^2 - 1$ [6.3C]

10. Solve: $x^2 - 14x - 5 = 10$
$-1, 15$ [7.5A]

SECTION

9.1 Rational Exponents and Radical Expressions

OBJECTIVE A   *To simplify expressions with rational exponents*

 Point of Interest

Nicolas Chuquet (c. 1475), a French physician, wrote an algebra text in which he used a notation for expressions with fractional exponents. He wrote $R^2 6$ to mean $6^{1/2}$ and $R^3 15$ to mean $15^{1/3}$. This was an improvement over earlier notations that used words for these expressions.

In this section, the definition of an exponent is extended beyond integers so that any rational number can be used as an exponent. The definition is expressed in such a way that the Rules of Exponents hold true for rational exponents.

Consider the expression $(a^{1/n})^n$ for $a > 0$ and n a positive integer. Now simplify, assuming the Rule for Simplifying the Power of an Exponential Expression is true.

$$(a^{1/n})^n = a^{\frac{1}{n} \cdot n} = a^1 = a$$

Because $(a^{1/n})^n = a$, the number $a^{1/n}$ is the number whose nth power is a.

If $a > 0$ and n is a positive number, then $a^{1/n}$ is called the nth root of a.

$25^{1/2} = 5$ because $(5)^2 = 25$.

$8^{1/3} = 2$ because $(2)^3 = 8$.

 **Integrating Technology**

A calculator can be used to evaluate expressions with rational exponents. For example, to evaluate the expression at the right, press

⊙ 27 ^ (1

÷ 3) ENTER . The display reads −3.

In the expression $a^{1/n}$, if a is a negative number and n is a positive even integer, then $a^{1/n}$ is not a real number.

$(-4)^{1/2}$ is not a real number, because there is no real number whose second power is -4.

When n is a positive odd integer, a can be a positive or a negative number.

$(-27)^{1/3} = -3$ because $(-3)^3 = -27$.

Using the definition of $a^{1/n}$ and the Rules of Exponents, it is possible to define any exponential expression that contains a rational exponent.

Rule for Rational Exponents

If m and n are positive integers and $a^{1/n}$ is a real number, then
$$a^{m/n} = (a^{1/n})^m$$

The expression $a^{m/n}$ can also be written $a^{m/n} = a^{m \cdot \frac{1}{n}} = (a^m)^{1/n}$. However, rewriting $a^{m/n}$ as $(a^m)^{1/n}$ is not as useful as rewriting it as $(a^{1/n})^m$. See the Take Note at the top of the next page.

INSTRUCTOR NOTE

The introduction to rational exponents excludes complex numbers. These numbers are discussed later in the chapter.

As shown above, expressions that contain rational exponents do not always represent real numbers when the base of the exponential expression is a negative number. For this reason, **all variables in this chapter represent positive numbers unless otherwise stated.**

Take Note

Although we can simplify an expression by rewriting $a^{m/n}$ in the form $(a^m)^{1/n}$, it is usually easier to simplify the form $(a^{1/n})^m$. For instance, simplifying $(27^{1/3})^2$ is easier than simplifying $(27^2)^{1/3}$.

Take Note

Note that $32^{-2/5} = \dfrac{1}{4}$, a positive number. A negative exponent does not affect the sign of a number.

IN-CLASS EXAMPLES

Simplify.

1. $16^{3/4}$ **8**
2. $8^{-2/3}$ $\dfrac{1}{4}$
3. $(-8)^{3/2}$
   Not a real number
4. $x^{1/4} \cdot x^{3/4}$ **x**
5. $\dfrac{a^{5/2}}{a^{1/2}}$ **a^2**
6. $(x^{-4}y^8)^{-1/2}$ $\dfrac{x^2}{y^4}$

Tips for Success

Remember that a HOW TO indicates a worked-out example. Using paper and pencil, work through the example. See *AIM for Success* in the Preface.

HOW TO 1   Simplify: $27^{2/3}$

$27^{2/3} = (3^3)^{2/3}$ • Rewrite 27 as 3^3.

$\qquad = 3^{3(2/3)} = 3^2$ • Multiply the exponents.

$\qquad = 9$

HOW TO 2   Simplify: $32^{-2/5}$

$32^{-2/5} = (2^5)^{-2/5}$ • Rewrite 32 as 2^5.

$\qquad = 2^{-2} = \dfrac{1}{2^2}$ • Multiply the exponents. Then use the Definition of Negative Exponents.

$\qquad = \dfrac{1}{4}$ • Simplify.

HOW TO 3   Simplify: $a^{1/2} \cdot a^{2/3} \cdot a^{-1/4}$

$a^{1/2} \cdot a^{2/3} \cdot a^{-1/4} = a^{1/2 + 2/3 - 1/4}$ • Use the Rule for Multiplying Exponential Expressions.

$\qquad = a^{6/12 + 8/12 - 3/12}$

$\qquad = a^{11/12}$ • Simplify.

HOW TO 4   Simplify: $(x^6y^4)^{3/2}$

$(x^6y^4)^{3/2} = x^{6(3/2)}y^{4(3/2)}$ • Use the Rule for Simplifying Powers of Products.

$\qquad = x^9y^6$ • Simplify.

HOW TO 5   Simplify: $3x^{-3/4}(2x^{11/4} - x^{-1/4})$

$3x^{-3/4}(2x^{11/4} - x^{-1/4}) = 3x^{-3/4}(2x^{11/4}) - 3x^{-3/4}(x^{-1/4})$ • Use the Distributive Property to remove parentheses.

$\qquad = 6x^{-3/4 + 11/4} - 3x^{-3/4 + (-1/4)}$ • Use the Rule for Multiplying Exponential Expressions.

$\qquad = 6x^{8/4} - 3x^{-4/4} = 6x^2 - 3x^{-1}$ • Simplify.

$\qquad = 6x^2 - \dfrac{3}{x}$

HOW TO 6   Simplify: $\left(\dfrac{8a^3b^{-4}}{64a^{-9}b^2}\right)^{2/3}$

$\left(\dfrac{8a^3b^{-4}}{64a^{-9}b^2}\right)^{2/3} = \left(\dfrac{2^3a^3b^{-4}}{2^6a^{-9}b^2}\right)^{2/3}$ • Rewrite 8 as 2^3 and 64 as 2^6.

$\qquad = (2^{-3}a^{12}b^{-6})^{2/3}$ • Use the Rule for Dividing Exponential Expressions.

$\qquad = 2^{-2}a^8b^{-4}$ • Use the Rule for Simplifying Powers of Products.

$\qquad = \dfrac{a^8}{2^2b^4} = \dfrac{a^8}{4b^4}$ • Use the Definition of Negative Exponents and simplify.

EXAMPLE 1

Simplify: $64^{-2/3}$

Solution $64^{-2/3} = (2^6)^{-2/3}$ • $64 = 2^6$

$\qquad\qquad\quad = 2^{-4} = \dfrac{1}{2^4} = \dfrac{1}{16}$

YOU TRY IT 1

Simplify: $16^{-3/4}$

Your solution $\dfrac{1}{8}$

EXAMPLE 2

Simplify: $(x^{1/2}y^{-3/2}z^{1/4})^{-3/2}$

Solution

$(x^{1/2}y^{-3/2}z^{1/4})^{-3/2}$

$= x^{-3/4}y^{9/4}z^{-3/8}$ • Use the Rule for Simplifying
$\qquad\qquad\qquad\qquad$ Powers of Products.

$= \dfrac{y^{9/4}}{x^{3/4}z^{3/8}}$

YOU TRY IT 2

Simplify: $(x^{3/4}y^{1/2}z^{-2/3})^{-4/3}$

Your solution

$\dfrac{z^{8/9}}{xy^{2/3}}$

EXAMPLE 3

Simplify: $\dfrac{x^{1/2}y^{-5/4}}{x^{-4/3}y^{1/3}}$

Solution

$\dfrac{x^{1/2}y^{-5/4}}{x^{-4/3}y^{1/3}}$

$= x^{3/6-(-8/6)}y^{-15/12-4/12}$ • Use the Rule for Dividing
$\qquad\qquad\qquad\qquad$ Exponential Expressions.

$= x^{11/6}y^{-19/12} = \dfrac{x^{11/6}}{y^{19/12}}$

YOU TRY IT 3

Simplify: $\left(\dfrac{16a^{-2}b^{4/3}}{9a^4b^{-2/3}}\right)^{-1/2}$

Your solution

$\dfrac{3a^3}{4b}$

Solutions on p. S26

OBJECTIVE B   *To write exponential expressions as radical expressions and to write radical expressions as exponential expressions*

Point of Interest

The radical sign was introduced in 1525 by Christoff Rudolff in a book called *Coss.* He modified the symbol to indicate square roots, cube roots, and fourth roots. The idea of using an index, as we do in our modern notation, did not occur until some years later.

Recall that $a^{1/n}$ is the nth root of a. The expression $\sqrt[n]{a}$ is another symbol for the nth root of a.

If a is a real number, then $a^{1/n} = \sqrt[n]{a}$.

In the expression $\sqrt[n]{a}$, the symbol $\sqrt{}$ is called a **radical,** n is the **index** of the radical, and a is the **radicand.** When $n = 2$, the radical expression represents a square root and the index 2 is usually not written.

An exponential expression with a rational exponent can be written as a radical expression.

Writing Exponential Expressions as Radical Expressions

If $a^{1/n}$ is a real number, then $a^{1/n} = \sqrt[n]{a}$ and $a^{m/n} = a^{m \cdot 1/n} = (a^m)^{1/n} = \sqrt[n]{a^m}$.

The expression $a^{m/n}$ can also be written $a^{m/n} = (a^{1/n})^m = (\sqrt[n]{a})^m$.

The exponential expression at the right has been written as a radical expression.

$$y^{2/3} = (y^2)^{1/3}$$
$$= \sqrt[3]{y^2}$$

The radical expressions at the right have been written as exponential expressions.

$$\sqrt[5]{x^6} = (x^6)^{1/5} = x^{6/5}$$
$$\sqrt{17} = 17^{1/2}$$

HOW TO 7 Write $(5x)^{2/5}$ as a radical expression.

$$(5x)^{2/5} = \sqrt[5]{(5x)^2}$$

- The denominator of the rational exponent is the index of the radical. The numerator is the power of the radicand.

$$= \sqrt[5]{25x^2}$$

- Simplify.

IN-CLASS EXAMPLES

7. Write the exponential expression $3^{1/2}$ as a radical expression. $\sqrt{3}$

8. Write the radical expression $\sqrt[3]{x^2}$ as an exponential expression. $x^{2/3}$

HOW TO 8 Write $\sqrt[3]{x^4}$ as an exponential expression with a rational exponent.

$$\sqrt[3]{x^4} = (x^4)^{1/3}$$

- The index of the radical is the denominator of the rational exponent. The power of the radicand is the numerator of the rational exponent.

$$= x^{4/3}$$

- Simplify.

HOW TO 9 Write $\sqrt[3]{a^3 + b^3}$ as an exponential expression with a rational exponent.

$$\sqrt[3]{a^3 + b^3} = (a^3 + b^3)^{1/3}$$

Note that $(a^3 + b^3)^{1/3} \neq a + b$.

EXAMPLE 4

Write $(3x)^{5/4}$ as a radical expression.

Solution $(3x)^{5/4} = \sqrt[4]{(3x)^5} = \sqrt[4]{243x^5}$

YOU TRY IT 4

Write $(2x^3)^{3/4}$ as a radical expression.

Your solution $\sqrt[4]{8x^9}$

EXAMPLE 5

Write $-2x^{2/3}$ as a radical expression.

Solution $-2x^{2/3} = -2(x^2)^{1/3} = -2\sqrt[3]{x^2}$

YOU TRY IT 5

Write $-5a^{5/6}$ as a radical expression.

Your solution $-5\sqrt[6]{a^5}$

EXAMPLE 6

Write $\sqrt[4]{3a}$ as an exponential expression.

Solution $\sqrt[4]{3a} = (3a)^{1/4}$

YOU TRY IT 6

Write $\sqrt[3]{3ab}$ as an exponential expression.

Your solution $(3ab)^{1/3}$

EXAMPLE 7

Write $\sqrt{a^2 - b^2}$ as an exponential expression.

Solution $\sqrt{a^2 - b^2} = (a^2 - b^2)^{1/2}$

YOU TRY IT 7

Write $\sqrt[4]{x^4 + y^4}$ as an exponential expression.

Your solution $(x^4 + y^4)^{1/4}$

Solutions on p. S26

OBJECTIVE C *To simplify radical expressions that are roots of perfect powers*

INSTRUCTOR NOTE

All of the expressions in this objective are perfect powers. Simplifying these expressions will prepare students for the next section.

Every positive number has two square roots, one a positive number and one a negative number. For example, because $(5)^2 = 25$ and $(-5)^2 = 25$, there are two square roots of 25: 5 and -5.

The symbol $\sqrt{}$ is used to indicate the positive square root, or **principal square root.** To indicate the negative square root of a number, a negative sign is placed in front of the radical.

$$\sqrt{25} = 5$$

$$-\sqrt{25} = -5$$

The square root of zero is zero.

$$\sqrt{0} = 0$$

The square root of a negative number is not a real number because the square of a real number must be positive.

$\sqrt{-25}$ is not a real
number.

Note that

$$\sqrt{(-5)^2} = \sqrt{25} = 5 \text{ and } \sqrt{5^2} = \sqrt{25} = 5$$

This pattern is true for all real numbers and is stated as the following result.

For any real number a, $\sqrt{a^2} = |a|$ and $-\sqrt{a^2} = -|a|$. If a is a positive real number, then $\sqrt{a^2} = a$ and $(\sqrt{a})^2 = a$.

Besides square roots, we can also determine cube roots, fourth roots, and so on.

Integrating Technology

See the Keystroke Guide: *Radical Expressions* for instructions on using a graphing calculator to evaluate a numerical radical expression.

$\sqrt[3]{8} = 2$, because $2^3 = 8$.   • **The cube root of a positive number is positive.**

$\sqrt[3]{-8} = -2$, because $(-2)^3 = -8$.   • **The cube root of a negative number is negative.**

$\sqrt[4]{625} = 5$, because $5^4 = 625$.

$\sqrt[5]{243} = 3$, because $3^5 = 243$.

The following properties hold true when finding the nth root of a real number.

If n is an even integer, then $\sqrt[n]{a^n} = |a|$ and $-\sqrt[n]{a^n} = -|a|$. If n is an odd integer, then $\sqrt[n]{a^n} = a$.

For example,

$$\sqrt[6]{y^6} = |y| \qquad -\sqrt[12]{x^{12}} = -|x| \qquad \sqrt[5]{b^5} = b$$

For the remainder of this chapter, we will assume that variable expressions inside a radical represent positive numbers. Therefore, it is not necessary to use the absolute value signs.

Take Note

Note that when the index is an even natural number, the nth root requires absolute value symbols.

$\sqrt[6]{y^6} = |y|$ but $\sqrt[5]{y^5} = y$

Because we stated that variables within radicals represent *positive* numbers, we will omit the absolute value symbols when writing an answer.

HOW TO 10   Simplify: $\sqrt[4]{x^4 y^8}$

$\sqrt[4]{x^4 y^8} = (x^4 y^8)^{1/4}$ • **The radicand is a perfect fourth power because the exponents on the variables are divisible by 4. Write the radical expression as an exponential expression.**

$= xy^2$ • **Use the Rule for Simplifying Powers of Products.**

HOW TO 11 Simplify: $\sqrt[3]{125c^9d^6}$

$\sqrt[3]{125c^9d^6} = (5^3c^9d^6)^{1/3}$
  • The radicand is a perfect cube because 125 is a perfect cube ($125 = 5^3$) and all the exponents on the variables are divisible by 3.

$= 5c^3d^2$
  • Use the Rule for Simplifying Powers of Products.

IN-CLASS EXAMPLES
Simplify.
9. $\sqrt[3]{x^6y^9}$ x^2y^3
10. $\sqrt[3]{-8x^9y^{21}}$ $-2x^3y^7$
11. $\sqrt{-25x^2y^4}$
   Not a real number
12. $-\sqrt[4]{a^{16}b^{24}}$ $-a^4b^6$

Note that a variable expression is a perfect power if the exponents on the factors are evenly divisible by the index of the radical.

The chart below shows roots of perfect powers. Knowledge of these roots is very helpful when simplifying radical expressions.

Square Roots		Cube Roots	Fourth Roots	Fifth Roots
$\sqrt{1} = 1$ $\sqrt{36} = 6$		$\sqrt[3]{1} = 1$	$\sqrt[4]{1} = 1$	$\sqrt[5]{1} = 1$
$\sqrt{4} = 2$ $\sqrt{49} = 7$		$\sqrt[3]{8} = 2$	$\sqrt[4]{16} = 2$	$\sqrt[5]{32} = 2$
$\sqrt{9} = 3$ $\sqrt{64} = 8$		$\sqrt[3]{27} = 3$	$\sqrt[4]{81} = 3$	$\sqrt[5]{243} = 3$
$\sqrt{16} = 4$ $\sqrt{81} = 9$		$\sqrt[3]{64} = 4$	$\sqrt[4]{256} = 4$	
$\sqrt{25} = 5$ $\sqrt{100} = 10$		$\sqrt[3]{125} = 5$	$\sqrt[4]{625} = 5$	

Take Note

From the chart, $\sqrt[5]{243} = 3$, which means that $3^5 = 243$. From this we know that $(-3)^5 = -243$, which means $\sqrt[5]{-243} = -3$.

HOW TO 12 Simplify: $\sqrt[5]{-243x^5y^{15}}$

$\sqrt[5]{-243x^5y^{15}} = -3xy^3$
  • From the chart, 243 is a perfect fifth power, and each exponent is divisible by 5. Therefore, the radicand is a perfect fifth power.

EXAMPLE 8

Simplify: $\sqrt[3]{-125a^6b^9}$

Solution

The radicand is a perfect cube.

$\sqrt[3]{-125a^6b^9} = -5a^2b^3$ • Divide each exponent by 3.

YOU TRY IT 8

Simplify: $\sqrt[3]{-8x^{12}y^3}$

Your solution

$-2x^4y$

EXAMPLE 9

Simplify: $-\sqrt[4]{16a^4b^8}$

Solution

The radicand is a perfect fourth power.

$-\sqrt[4]{16a^4b^8} = -2ab^2$ • Divide each exponent by 4.

YOU TRY IT 9

Simplify: $-\sqrt[4]{81x^{12}y^8}$

Your solution

$-3x^3y^2$

Solutions on p. S26

9.1 EXERCISES

✔ Concept Check

SUGGESTED ASSIGNMENT
Exercises 1–8; Exercises 9–139, odds;
More challenging exercises: Exercises 138 and 139

For Exercises 1 to 4, find the value of a.

1. $a^{1/3} = 5$ 125   **2.** $a^{1/4} = 3$ 81   **3.** $a^{1/5} = -2$ -32 **4.** $a^{1/3} = -2$ -8

For Exercises 5 to 8, simplify.

5. $\sqrt{81}$ 9   **6.** $\sqrt[3]{64}$ 4   **7.** $\sqrt[4]{81}$ 3   **8.** $\sqrt[5]{32}$ 2

OBJECTIVE A   *To simplify expressions with rational exponents*

9. 🗫 Which of the following are not real numbers?
(i) $(-7)^{1/2}$ (ii) $(-7)^{1/3}$ (iii) $(-7)^{1/4}$ (iv) $(-7)^{1/5}$ i, iii

10. 🗫 If $x^{1/n} = a$, what is a^n?   x

For Exercises 11 to 78, simplify.

11. $8^{1/3}$

2

12. $16^{1/2}$

4

13. $9^{3/2}$

27

14. $25^{3/2}$

125

15. $27^{-2/3}$

$\dfrac{1}{9}$

16. $64^{-1/3}$

$\dfrac{1}{4}$

17. $32^{2/5}$

4

18. $16^{3/4}$

8

19. $(-25)^{5/2}$

Not a real number

20. $(-36)^{1/4}$

Not a real number

21. $\left(\dfrac{25}{49}\right)^{-3/2}$

$\dfrac{343}{125}$

22. $\left(\dfrac{8}{27}\right)^{-2/3}$

$\dfrac{9}{4}$

23. $x^{1/2}x^{1/2}$

x

24. $a^{1/3}a^{5/3}$

a^2

25. $y^{-1/4}y^{3/4}$

$y^{1/2}$

26. $x^{2/5} \cdot x^{-4/5}$

$\dfrac{1}{x^{2/5}}$

27. $x^{-2/3} \cdot x^{3/4}$

$x^{1/12}$

28. $x \cdot x^{-1/2}$

$x^{1/2}$

29. $a^{1/3} \cdot a^{3/4} \cdot a^{-1/2}$

$a^{7/12}$

30. $y^{-1/6} \cdot y^{2/3} \cdot y^{1/2}$

y

31. $\dfrac{a^{1/2}}{a^{3/2}}$

$\dfrac{1}{a}$

32. $\dfrac{b^{1/3}}{b^{4/3}}$

$\dfrac{1}{b}$

33. $\dfrac{y^{-3/4}}{y^{1/4}}$

$\dfrac{1}{y}$

34. $\dfrac{x^{-3/5}}{x^{1/5}}$

$\dfrac{1}{x^{4/5}}$

35. $\dfrac{y^{2/3}}{y^{-5/6}}$

$y^{3/2}$

36. $\dfrac{b^{3/4}}{b^{-3/2}}$

$b^{9/4}$

37. $(x^2)^{-1/2}$

$\dfrac{1}{x}$

38. $(a^8)^{-3/4}$

$\dfrac{1}{a^6}$

39. $(x^{-2/3})^6$

$\dfrac{1}{x^4}$

40. $(y^{-5/6})^{12}$

$\dfrac{1}{y^{10}}$

41. $(a^{-1/2})^{-2}$

a

42. $(b^{-2/3})^{-6}$

b^4

43. $(x^{-3/8})^{-4/5}$

$x^{3/10}$

44. $(y^{-3/2})^{-2/9}$

$y^{1/3}$

45. $(a^{1/2} \cdot a)^2$

a^3

46. $(b^{2/3} \cdot b^{1/6})^6$

b^5

47. $(x^{-1/2} \cdot x^{3/4})^{-2}$

$\dfrac{1}{x^{1/2}}$

48. $(a^{1/2} \cdot a^{-2})^3$

$\dfrac{1}{a^{9/2}}$

49. $(y^{-1/2} \cdot y^{2/3})^{2/3}$

$y^{1/9}$

50. $(b^{-2/3} \cdot b^{1/4})^{-4/3}$

$b^{5/9}$

51. $(x^{-3}y^6)^{-1/3}$

$\dfrac{x}{y^2}$

52. $(a^2b^{-6})^{-1/2}$

$\dfrac{b^3}{a}$

53. $(x^{-2}y^{1/3})^{-3/4}$

$\dfrac{x^{3/2}}{y^{1/4}}$

54. $(a^{-2/3}b^{2/3})^{3/2}$

$\dfrac{b}{a}$

55. $\left(\dfrac{x^{1/2}}{y^2}\right)^4$

$\dfrac{x^2}{y^8}$

56. $\left(\dfrac{b^{-3/4}}{a^{-1/2}}\right)^8$

$\dfrac{a^4}{b^6}$

57. $\dfrac{x^{1/4}\cdot x^{-1/2}}{x^{2/3}}$

$\dfrac{1}{x^{11/12}}$

58. $\dfrac{b^{1/2}\cdot b^{-3/4}}{b^{1/4}}$

$\dfrac{1}{b^{1/2}}$

59. $\left(\dfrac{y^{2/3}\cdot y^{-5/6}}{y^{1/9}}\right)^9$

$\dfrac{1}{y^{5/2}}$

60. $\left(\dfrac{a^{1/3}\cdot a^{-2/3}}{a^{1/2}}\right)^4$

$\dfrac{1}{a^{10/3}}$

61. $\left(\dfrac{b^2\cdot b^{-3/4}}{b^{-1/2}}\right)^{-1/2}$

$\dfrac{1}{b^{7/8}}$

62. $\dfrac{(x^{-5/6}\cdot x^3)^{-2/3}}{x^{4/3}}$

$\dfrac{1}{x^{25/9}}$

63. $(a^{2/3}b^2)^6(a^3b^3)^{1/3}$

a^5b^{13}

64. $(x^3y^{-1/2})^{-2}(x^{-3}y^2)^{1/6}$

$\dfrac{y^{4/3}}{x^{13/2}}$

65. $(16x^{-2}y^4)^{-1/2}(xy^{1/2})$

$\dfrac{x^2}{4y^{3/2}}$

66. $(27s^3t^{-6})^{1/3}(s^{-1/3}t^{5/6})^6$

$\dfrac{3t^3}{s}$

67. $(x^{-2/3}y^{-3})^3(27x^{-3}y^6)^{-1/3}$

$\dfrac{1}{3xy^{11}}$

68. $(9x^{-2/3}y^{4/3})^{-1/2}(x^{2/3}y^{-8/3})^{1/2}$

$\dfrac{x^{2/3}}{3y^2}$

69. $\dfrac{(4a^{4/3}b^{-2})^{-1/2}}{(a^{1/6}b^{-3/2})^2}$

$\dfrac{b^4}{2a}$

70. $\dfrac{(4x^{-2}y^4)^{1/2}}{(8x^6y^{-3/2})^{-2/3}}$

$8x^3y$

71. $\left(\dfrac{x^{1/2}y^{-3/4}}{y^{2/3}}\right)^{-6}$

$\dfrac{y^{17/2}}{x^3}$

72. $\left(\dfrac{x^{1/2}y^{-5/4}}{y^{-3/4}}\right)^{-4}$

$\dfrac{y^2}{x^2}$

73. $\left(\dfrac{b^{-3}}{64a^{-1/2}}\right)^{-2/3}$

$\dfrac{16b^2}{a^{1/3}}$

74. $\left(\dfrac{49c^{5/3}}{a^{-1/4}b^{5/6}}\right)^{-3/2}$

$\dfrac{b^{5/4}}{343a^{3/8}c^{5/2}}$

75. $y^{3/2}(y^{1/2}-y^{-1/2})$

y^2-y

76. $y^{3/5}(y^{2/5}+y^{-3/5})$

$y+1$

77. $a^{-1/4}(a^{5/4}-a^{9/4})$

$a-a^2$

78. $x^{4/3}(x^{2/3}+x^{-1/3})$

x^2+x

OBJECTIVE B   *To write exponential expressions as radical expressions and to write radical expressions as exponential expressions*

79. 🐾 True or false? $8x^{1/3} = 2\sqrt[3]{x}$
False

80. 🐾 True or false? $(\sqrt[3]{x})^5 = (x^5)^{1/3}$
True

For Exercises 81 to 96, rewrite the exponential expression as a radical expression.

81. $3^{1/4}$
$\sqrt[4]{3}$

82. $5^{1/2}$
$\sqrt{5}$

83. $a^{3/2}$
$\sqrt{a^3}$

84. $b^{4/3}$
$\sqrt[3]{b^4}$

85. $(2t)^{5/2}$
$\sqrt{32t^5}$

86. $(3x)^{2/3}$
$\sqrt[3]{9x^2}$

87. $-2x^{2/3}$
$-2\sqrt[3]{x^2}$

88. $-3a^{2/5}$
$-3\sqrt[5]{a^2}$

89. $(a^2b)^{2/3}$
$\sqrt[3]{a^4b^2}$

90. $(x^2y^3)^{3/4}$
$\sqrt[4]{x^6y^9}$

91. $(a^2b^4)^{3/5}$
$\sqrt[5]{a^6b^{12}}$

92. $(a^3b^7)^{3/2}$
$\sqrt{a^9b^{21}}$

93. $(4x-3)^{3/4}$
$\sqrt[4]{(4x-3)^3}$

94. $(3x-2)^{1/3}$
$\sqrt[3]{3x-2}$

95. $x^{-2/3}$
$\dfrac{1}{\sqrt[3]{x^2}}$

96. $b^{-3/4}$
$\dfrac{1}{\sqrt[4]{b^3}}$

For Exercises 97 to 112, rewrite the radical expression as an exponential expression.

97. $\sqrt{14}$
$14^{1/2}$

98. $\sqrt{7}$
$7^{1/2}$

99. $\sqrt[3]{x}$
$x^{1/3}$

100. $\sqrt[4]{y}$
$y^{1/4}$

101. $\sqrt[3]{x^4}$
$x^{4/3}$

102. $\sqrt[4]{a^3}$
$a^{3/4}$

103. $\sqrt[5]{b^3}$
$b^{3/5}$

104. $\sqrt[4]{b^5}$
$b^{5/4}$

105. $\sqrt[3]{2x^2}$
$(2x^2)^{1/3}$

106. $\sqrt[5]{4y^7}$
$(4y^7)^{1/5}$

107. $-\sqrt{3x^5}$
$-(3x^5)^{1/2}$

108. $-\sqrt[4]{4x^5}$
$-(4x^5)^{1/4}$

109. $3x\sqrt[3]{y^2}$
$3xy^{2/3}$

110. $2y\sqrt{x^3}$
$2x^{3/2}y$

111. $\sqrt{a^2-2}$
$(a^2-2)^{1/2}$

112. $\sqrt{3-y^2}$
$(3-y^2)^{1/2}$

OBJECTIVE C   *To simplify radical expressions that are roots of perfect powers*

🐾 For Exercises 113 to 116, assume that x is a negative real number. State whether the expression simplifies to a positive number, a negative number, or a number that is not a real number.

113. $-\sqrt[3]{8x^{15}}$
Positive

114. $-\sqrt{9x^8}$
Negative

115. $\sqrt{-4x^{12}}$
Not a real number

116. $\sqrt[3]{-27x^9}$
Positive

For Exercises 117 to 140, simplify.

117. $\sqrt{x^{16}}$
x^8

118. $\sqrt{y^{14}}$
y^7

119. $-\sqrt{x^8}$
$-x^4$

120. $-\sqrt{a^6}$
$-a^3$

121. $\sqrt[3]{x^3y^9}$
xy^3

122. $\sqrt[3]{a^6b^{12}}$
a^2b^4

123. $-\sqrt[3]{x^{15}y^3}$
$-x^5y$

124. $-\sqrt[3]{a^9b^9}$
$-a^3b^3$

125. $\sqrt{16a^4b^{12}}$
$4a^2b^6$

126. $\sqrt{25x^8y^2}$
$5x^4y$

127. $\sqrt{-16x^4y^2}$
Not a real number

128. $\sqrt{-9a^4b^8}$
Not a real number

129. $\sqrt[3]{27x^9}$
$3x^3$

130. $\sqrt[3]{8a^{21}b^6}$
$2a^7b^2$

131. $\sqrt[3]{-64x^9y^{12}}$
$-4x^3y^4$

132. $\sqrt[3]{-27a^3b^{15}}$
$-3ab^5$

133. $-\sqrt[4]{x^8y^{12}}$
$-x^2y^3$

134. $-\sqrt[4]{a^{16}b^4}$
$-a^4b$

135. $\sqrt[5]{x^{20}y^{10}}$
x^4y^2

136. $\sqrt[5]{a^5b^{25}}$
ab^5

137. $\sqrt[4]{81x^4y^{20}}$
$3xy^5$

138. $\sqrt[4]{16a^8b^{20}}$
$2a^2b^5$

139. $\sqrt[5]{32a^5b^{10}}$
$2ab^2$

140. $\sqrt[5]{-32x^{15}y^{20}}$
$-2x^3y^4$

Critical Thinking

141. 🖊 If x is any real number, is $\sqrt{x^2} = x$ always true? Show why or why not.

142. Simplify.

a. $\sqrt[3]{\sqrt{x^6}}$ x

b. $\sqrt[4]{\sqrt{a^8}}$ a

c. $\sqrt{\sqrt{81y^8}}$ $3y^2$

d. $\sqrt{\sqrt[n]{a^{4n}}}$ a^2

e. $\sqrt[n]{\sqrt{b^{6n}}}$ b^3

f. $\sqrt{\sqrt[3]{x^{12}y^{24}}}$ x^2y^4

Projects or Group Activities

A **continued fraction** is a complex fraction, of the form shown at the right, that continues indefinitely. By stopping the process at some point, a continued fraction can be used to approximate the square root of a natural number.

$$a_0 + \cfrac{b_0}{a_1 + \cfrac{b_1}{a_2 + \cfrac{b_2}{a_3 + \ldots}}}$$

143. Show that $\sqrt{2} \approx 1 + \cfrac{1}{2 + \cfrac{1}{2 + \cfrac{1}{2 + \cfrac{1}{2}}}}$

144. Show that $\sqrt{3} \approx 1 + \cfrac{1}{1 + \cfrac{1}{2 + \cfrac{1}{1 + \cfrac{1}{2 + 1}}}}$

QUICK QUIZ

Simplify.

1. $16^{3/2}$ **64**   **[9.1A]**

2. $(a^4b^8)^{3/4}$ a^3b^6 **[9.1A]**

3. $\left(\dfrac{x^{1/3}}{y^{-2}}\right)^6$ x^2y^{12} **[9.1A]**

4. Rewrite $x^{3/4}$ as a radical expression.
$\sqrt[4]{x^3}$ **[9.1B]**

5. Rewrite $\sqrt[5]{a^4}$ as an exponential expression.
$a^{4/5}$ **[9.1B]**

Simplify.

6. $\sqrt[4]{a^8b^{12}}$ a^2b^3 **[9.1C]**

7. $\sqrt[3]{-x^{12}y^{24}}$
$-x^4y^8$ **[9.1C]**

8. $\sqrt{-81x^4y^6}$ **Not a real number**   **[9.1C]**

9.2

Addition and Subtraction of Radical Expressions

OBJECTIVE A | *To simplify radical expressions*

 Point of Interest

The Latin expression for irrational numbers was *numerus surdus,* which literally means "inaudible number." A prominent 16th-century mathematician wrote of irrational numbers, ". . . just as an infinite number is not a number, so an irrational number is not a true number, but lies hidden in some sort of cloud of infinity." In 1872, Richard Dedekind wrote a paper that established the first logical treatment of irrational numbers.

If a number is not a perfect power, its root can only be approximated; examples include $\sqrt{5}$ and $\sqrt[3]{3}$. These numbers are irrational numbers. Their decimal representations never terminate or repeat.

$$\sqrt{5} = 2.2360679\ldots \qquad \sqrt[3]{3} = 1.4422495\ldots$$

A radical expression is in simplest form when the radicand contains no factor that is a perfect power. The Product Property of Radicals is used to simplify radical expressions whose radicands are not perfect powers.

The Product Property of Radicals

If $\sqrt[n]{a}$ and $\sqrt[n]{b}$ are positive real numbers, then $\sqrt[n]{a} \cdot \sqrt[n]{b} = \sqrt[n]{ab}$.

INSTRUCTOR NOTE

Provide numerical examples of the Product Property of Radicals. For example:

$\sqrt{4 \cdot 9} = \sqrt{4} \cdot \sqrt{9}$ or
$\sqrt[3]{8 \cdot 64} = \sqrt[3]{8} \cdot \sqrt[3]{64}$

HOW TO 1 Simplify: $\sqrt{48}$

$\sqrt{48} = \sqrt{16 \cdot 3}$ • Write the radicand as the product of a perfect square and a factor that does not contain a perfect square.

$\quad = \sqrt{16}\,\sqrt{3}$ • Use the Product Property of Radicals to write the expression as a product.

$\quad = 4\sqrt{3}$ • Simplify $\sqrt{16}$.

Note that 48 must be written as the product of a perfect square and *a factor that does not contain a perfect square.* Therefore, it would not be correct to rewrite $\sqrt{48}$ as $\sqrt{4 \cdot 12}$ and simplify the expression as shown at the right. Although 4 is a perfect square factor of 48, 12 contains a perfect square $(12 = 4 \cdot 3)$ and thus $\sqrt{12}$ can be simplified further. Remember to find the *largest* perfect power that is a factor of the radicand.

$\sqrt{48} = \sqrt{4 \cdot 12}$

$\quad = \sqrt{4}\,\sqrt{12}$

$\quad = 2\sqrt{12}$

Not in simplest form

HOW TO 2 Simplify: $\sqrt{18x^2y^3}$

$\sqrt{18x^2y^3} = \sqrt{9x^2y^2 \cdot 2y}$ • Write the radicand as the product of a perfect square and factors that do not contain a perfect square.

$\quad = \sqrt{9x^2y^2}\,\sqrt{2y}$ • Use the Product Property of Radicals to write the expression as a product.

$\quad = 3xy\sqrt{2y}$ • Simplify.

IN-CLASS EXAMPLES
Simplify.

1. $\sqrt{x^3y^4z^6}$ $xy^2z^3\sqrt{x}$
2. $\sqrt{27a^5b^4}$ $3a^2b^2\sqrt{3a}$
3. $\sqrt[3]{-250x^5y^7}$
   $-5xy^2\sqrt[3]{2x^2y}$
4. $\sqrt[4]{32x^8y^{10}}$
   $2x^2y^2\sqrt[4]{2y^2}$

HOW TO 3   Simplify: $\sqrt[3]{x^7}$

$\sqrt[3]{x^7} = \sqrt[3]{x^6 \cdot x}$

• Write the radicand as the product of a perfect cube and a factor that does not contain a perfect cube.

$= \sqrt[3]{x^6}\,\sqrt[3]{x}$

• Use the Product Property of Radicals to write the expression as a product.

$= x^2\sqrt[3]{x}$

• Simplify.

HOW TO 4   Simplify: $\sqrt[4]{32x^7}$

$\sqrt[4]{32x^7} = \sqrt[4]{16x^4(2x^3)}$

• Write the radicand as the product of a perfect fourth power and factors that do not contain a perfect fourth power.

$= \sqrt[4]{16x^4}\,\sqrt[4]{2x^3}$

• Use the Product Property of Radicals to write the expression as a product.

$= 2x\sqrt[4]{2x^3}$

• Simplify.

EXAMPLE 1

Simplify: $\sqrt[4]{x^9}$

Solution

$\sqrt[4]{x^9} = \sqrt[4]{x^8 \cdot x}$

$= \sqrt[4]{x^8}\,\sqrt[4]{x}$

$= x^2\sqrt[4]{x}$

• x^8 is a perfect fourth power.

YOU TRY IT 1

Simplify: $\sqrt[5]{x^7}$

Your solution

$x\sqrt[5]{x^2}$

EXAMPLE 2

Simplify: $\sqrt[3]{-27a^5b^{12}}$

Solution

$\sqrt[3]{-27a^5b^{12}}$

$= \sqrt[3]{-27a^3b^{12}(a^2)}$

$= \sqrt[3]{-27a^3b^{12}}\,\sqrt[3]{a^2}$

$= -3ab^4\sqrt[3]{a^2}$

• $-27a^3b^{12}$ is a perfect cube.

YOU TRY IT 2

Simplify: $\sqrt[3]{-64x^8y^{18}}$

Your solution

$-4x^2y^6\sqrt[3]{x^2}$

Solutions on p. S26

OBJECTIVE B   *To add or subtract radical expressions*

INSTRUCTOR NOTE

Mention to students that adding and subtracting radical expressions is similar to combining like terms.

The Distributive Property is used to simplify the sum or difference of radical expressions that have the same radicand and the same index. For example,

$$3\sqrt{5} + 8\sqrt{5} = (3+8)\sqrt{5} = 11\sqrt{5}$$

$$2\sqrt[3]{3x} - 9\sqrt[3]{3x} = (2-9)\sqrt[3]{3x} = -7\sqrt[3]{3x}$$

Radical expressions that are in simplest form and have unlike radicands or different indices cannot be simplified by the Distributive Property. The expressions below cannot be simplified by the Distributive Property.

$$3\sqrt[4]{2} - 6\sqrt[4]{3} \qquad\qquad 2\sqrt[4]{4x} + 3\sqrt[3]{4x}$$

IN-CLASS EXAMPLES

Simplify.

5. $\sqrt{32} - \sqrt{50}$ $-\sqrt{2}$

6. $2b\sqrt{12a^3b} - a\sqrt{192ab^3}$
   $-4ab\sqrt{3ab}$

7. $6\sqrt{75} - 5\sqrt{48} + 2\sqrt{32}$
   $10\sqrt{3} + 8\sqrt{2}$

8. $x\sqrt[3]{16y^4} + 4\sqrt[3]{54x^3y^4} -$
   $xy\sqrt[3]{250y}$ $9xy\sqrt[3]{2y}$

HOW TO 5   Simplify: $3\sqrt{32x^2} - 2x\sqrt{2} + \sqrt{128x^2}$

$3\sqrt{32x^2} - 2x\sqrt{2} + \sqrt{128x^2}$

$= 3\sqrt{16x^2}\sqrt{2} - 2x\sqrt{2} + \sqrt{64x^2}\sqrt{2}$

$= 3 \cdot 4x\sqrt{2} - 2x\sqrt{2} + 8x\sqrt{2}$

$= 12x\sqrt{2} - 2x\sqrt{2} + 8x\sqrt{2}$

$= 18x\sqrt{2}$

- First simplify each term. Then combine like terms by using the Distributive Property.

EXAMPLE 3

Simplify: $5b\sqrt[4]{32a^7b^5} - 2a\sqrt[4]{162a^3b^9}$

Solution

$5b\sqrt[4]{32a^7b^5} - 2a\sqrt[4]{162a^3b^9}$

$= 5b\sqrt[4]{16a^4b^4 \cdot 2a^3b} - 2a\sqrt[4]{81b^8 \cdot 2a^3b}$

$= 5b \cdot 2ab\sqrt[4]{2a^3b} - 2a \cdot 3b^2\sqrt[4]{2a^3b}$

$= 10ab^2\sqrt[4]{2a^3b} - 6ab^2\sqrt[4]{2a^3b}$

$= 4ab^2\sqrt[4]{2a^3b}$

YOU TRY IT 3

Simplify: $3xy\sqrt[3]{81x^5y} - \sqrt[3]{192x^8y^4}$

Your solution

$5x^2y\sqrt[3]{3x^2y}$

9.2 EXERCISES

✔ **Concept Check**

SUGGESTED ASSIGNMENT
Exercises 1–12; Exercises 13–51, odds;
More challenging exercises: Exercises 53–56

For Exercises 1 to 8, determine whether the expression is a perfect square, perfect cube, or neither of these. There may be more than one answer for each expression.

1. 12
neither

2. 8
perfect cube

3. 49
perfect square

4. 64
perfect square,
perfect cube

5. x^3
perfect cube

6. $9x^4$
perfect square

7. $27x^9$
perfect cube

8. $64x^3y^6$
perfect cube

For Exercises 9 to 12, determine whether the radical expression is in simplest form.

9. $\sqrt{24}$
No

10. $\sqrt[3]{49}$
Yes

11. $\sqrt[3]{12x^4}$
No

12. $\sqrt{26xy^5}$
No

OBJECTIVE A *To simplify radical expressions*

For Exercises 13 to 28, simplify.

13. $\sqrt{x^4y^3z^5}$
$x^2yz^2\sqrt{yz}$

14. $\sqrt{x^3y^6z^9}$
$xy^3z^4\sqrt{xz}$

15. $\sqrt{8a^3b^8}$
$2ab^4\sqrt{2a}$

16. $\sqrt{24a^9b^6}$
$2a^4b^3\sqrt{6a}$

17. $\sqrt{45x^2y^3z^5}$
$3xyz^2\sqrt{5yz}$

18. $\sqrt{60xy^7z^{12}}$
$2y^3z^6\sqrt{15xy}$

19. $\sqrt[4]{48x^4y^5z^6}$
$2xyz\sqrt[4]{3yz^2}$

20. $\sqrt[4]{162x^9y^8z^2}$
$3x^2y^2\sqrt[4]{2xz^2}$

21. $\sqrt[3]{a^{16}b^8}$
$a^5b^2\sqrt[3]{ab^2}$

22. $\sqrt[3]{a^5b^8}$
$ab^2\sqrt[3]{a^2b^2}$

23. $\sqrt[3]{-125x^2y^4}$
$-5y\sqrt[3]{x^2y}$

24. $\sqrt[3]{-216x^5y^9}$
$-6xy^3\sqrt[3]{x^2}$

25. $\sqrt[3]{a^4b^5c^6}$
$abc^2\sqrt[3]{ab^2}$

26. $\sqrt[3]{a^8b^{11}c^{15}}$
$a^2b^3c^5\sqrt[3]{a^2b^2}$

27. $\sqrt[4]{16x^9y^5}$
$2x^2y\sqrt[4]{xy}$

28. $\sqrt[4]{64x^8y^{10}}$
$2x^2y^2\sqrt[4]{4y^2}$

OBJECTIVE B *To add or subtract radical expressions*

29. 🖊 True or false? $\sqrt{9+a} + \sqrt{a+9} = 2\sqrt{a+9}$
True

30. 🖊 True or false? $\sqrt{9+a} = 3 + \sqrt{a}$
False

For Exercises 31 to 52, simplify.

31. $2\sqrt{x} - 8\sqrt{x}$
$-6\sqrt{x}$

32. $3\sqrt{y} + 12\sqrt{y}$
$15\sqrt{y}$

33. $\sqrt{8x} - \sqrt{32x}$
$-2\sqrt{2x}$

34. $\sqrt{27a} - \sqrt{8a}$
$3\sqrt{3a} - 2\sqrt{2a}$

35. $\sqrt{18b} + \sqrt{75b}$
$3\sqrt{2b} + 5\sqrt{3b}$

36. $2\sqrt{2x^3} + 4x\sqrt{8x}$
$10x\sqrt{2x}$

37. $3\sqrt{8x^2y^3} - 2x\sqrt{32y^3}$
$-2xy\sqrt{2y}$

38. $2\sqrt{32x^2y^3} - xy\sqrt{98y}$
$xy\sqrt{2y}$

39. $2a\sqrt{27ab^5} + 3b\sqrt{3a^3b}$
$6ab^2\sqrt{3ab} + 3ab\sqrt{3ab}$

40. $\sqrt[3]{128} + \sqrt[3]{250}$
$9\sqrt[3]{2}$

41. $\sqrt[3]{16} - \sqrt[3]{54}$
$-\sqrt[3]{2}$

42. $2\sqrt[3]{3a^4} - 3a\sqrt[3]{81a}$
$-7a\sqrt[3]{3a}$

43. $2b\sqrt[3]{16b^2} + \sqrt[3]{128b^5}$
$8b\sqrt[3]{2b^2}$

44. $3\sqrt[3]{x^5y^7} - 8xy\sqrt[3]{x^2y^4}$
$-5xy^2\sqrt[3]{x^2y}$

45. $3\sqrt[4]{32a^5} - a\sqrt[4]{162a}$
$3a\sqrt[4]{2a}$

46. $2a\sqrt[4]{16ab^5} + 3b\sqrt[4]{256a^5b}$
$16ab\sqrt[4]{ab}$

47. $2\sqrt{50} - 3\sqrt{125} + \sqrt{98}$
$17\sqrt{2} - 15\sqrt{5}$

48. $3\sqrt{108} - 2\sqrt{18} - 3\sqrt{48}$
$6\sqrt{3} - 6\sqrt{2}$

49. $\sqrt{9b^3} - \sqrt{25b^3} + \sqrt{49b^3}$
$5b\sqrt{b}$

50. $\sqrt{4x^7y^5} + 9x^2\sqrt{x^3y^5} - 5xy\sqrt{x^5y^3}$
$6x^3y^2\sqrt{xy}$

51. $2x\sqrt{8xy^2} - 3y\sqrt{32x^3} + \sqrt{4x^3y^3}$
$-8xy\sqrt{2x} + 2xy\sqrt{xy}$

52. $5a\sqrt{3a^3b} + 2a^2\sqrt{27ab} - 4\sqrt{75a^5b}$
$-9a^2\sqrt{3ab}$

Critical Thinking

For Exercises 53 to 56, simplify.

53. $\sqrt[3]{54xy^3} - 5\sqrt[3]{2xy^3} + y\sqrt[3]{128x}$
$2y\sqrt[3]{2x}$

54. $2\sqrt[3]{24x^3y^4} + 4x\sqrt[3]{81y^4} - 3y\sqrt[3]{24x^3y}$
$10xy\sqrt[3]{3y}$

55. $2a\sqrt[4]{32b^5} - 3b\sqrt[4]{162a^4b} + \sqrt[4]{2a^4b^5}$
$-4ab\sqrt[4]{2b}$

56. $6y\sqrt[4]{48x^5} - 2x\sqrt[4]{243xy^4} - 4\sqrt[4]{3x^5y^4}$
$2xy\sqrt[4]{3x}$

Projects or Group Activities

A **radical function** is one for which a variable has a fractional exponent or a variable underneath a radical. Examples of radical functions are shown at the right.

$f(x) = 2\sqrt[3]{x} + 4$
$g(x) = \sqrt{2x} + 4$

The domain of a radical function is the set of real numbers for which the radical expression is a real number. For instance, -4 is not in the domain of function g shown above because $g(-4) = \sqrt{2(-4) + 4} = \sqrt{-4}$, which is not a real number. Because the square root of a negative number is not a real number, the domain of g is all values of x for which $2x + 4 \geq 0$. Solving this inequality gives $x \geq -2$. In interval notation, the domain of g is $[-2, \infty)$.

Now consider $f(x) = 2\sqrt[3]{x} + 4$. If x is a negative number, say $x = -8$, then $\sqrt[3]{-8} = -2$, a real number. If x is a positive number, say $x = 27$, then $\sqrt[3]{27} = 3$, a real number. In this case, the domain of f is all real numbers. This suggests the following rule. If the index of a radical expression is an even number, then the radicand must be greater than or equal to zero to ensure that the value of the radical expression is a real number. If the index of a radical expression is an odd number, then the radicand may be a positive or negative number.

For Exercises 57 to 60, find the domain of the radical function. Write the answer in interval notation. Use a graphing calculator to graph the function. For assistance with graphing, see the *Keystroke Guide: Graph.*

QUICK QUIZ
Simplify.

1. $\sqrt{x^5y^6z^7}$
   $x^2y^3z^3\sqrt{xz}$ [9.2A]

2. $\sqrt{48a^7b^3}$
   $4a^3b\sqrt{3ab}$ [9.2A]

3. $\sqrt[3]{-128x^7y^9}$
   $-4x^2y^3\sqrt[3]{2x}$ [9.2A]

4. $\sqrt[4]{162a^3b^5}$
   $3b\sqrt[4]{2a^3b}$ [9.2A]

5. $\sqrt{27} - \sqrt{12}$
   $\sqrt{3}$ [9.2B]

6. $5x\sqrt{x^3y^5} - 3y\sqrt{x^5y^3}$
   $2x^2y^2\sqrt{xy}$ [9.2B]

7. $6\sqrt{28} - 3\sqrt{112} + 2\sqrt{63}$
   $6\sqrt{7}$ [9.2B]

8. $3\sqrt[3]{24x^4y} + x\sqrt[3]{192xy} - \sqrt[3]{375x^4y}$
   $5x\sqrt[3]{3xy}$ [9.2B]

57. $f(x) = \sqrt{3 - x}$
$(-\infty, 3]$

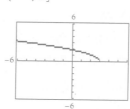

58. $f(x) = \sqrt[3]{x}$
$(-\infty, \infty)$

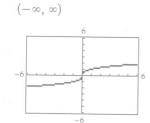

59. $f(x) = 2 + \sqrt[3]{1 - x}$
$(-\infty, \infty)$

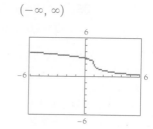

60. $f(x) = 2 - \sqrt{2x + 3}$
$\left[-\dfrac{3}{2}, \infty\right)$

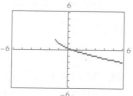

Unless otherwise noted, all content on this page is © Cengage Learning.

9.3 Multiplication and Division of Radical Expressions

OBJECTIVE A *To multiply radical expressions*

Recall the Product Property of Radicals: If $\sqrt[n]{a}$ and $\sqrt[n]{b}$ are real numbers, then $\sqrt[n]{a} \cdot \sqrt[n]{b} = \sqrt[n]{ab}$. This property is used to multiply radical expressions with the same index.

$$\sqrt{3x} \cdot \sqrt{5y} = \sqrt{15xy}$$

$$\sqrt[3]{4x^2z} \cdot \sqrt[3]{9y^2} = \sqrt[3]{36x^2y^2z}$$

$$\sqrt[4]{3xy} \cdot \sqrt[4]{9x^2y} = \sqrt[4]{27x^3y^2}$$

IN-CLASS EXAMPLES
Simplify.

1. $\sqrt{3xy^2}\,\sqrt{27xy^2}$ **$9xy^2$**

2. $\sqrt{5a}(\sqrt{20a} - \sqrt{a})$
   $10a - a\sqrt{5}$

3. $(\sqrt{x} - 4)^2$
   $x - 8\sqrt{x} + 16$

4. $(5\sqrt{x} + 2\sqrt{y})(5\sqrt{x} - 2\sqrt{y})$
   $25x - 4y$

HOW TO 1 Multiply: $\sqrt[3]{2a^5b}\,\sqrt[3]{16a^2b^2}$

$\sqrt[3]{2a^5b}\,\sqrt[3]{16a^2b^2} = \sqrt[3]{32a^7b^3}$
- Use the Product Property of Radicals to multiply the radicands.

$\qquad\qquad = \sqrt[3]{8a^6b^3 \cdot 4a}$
- Simplify.

$\qquad\qquad = 2a^2b\,\sqrt[3]{4a}$

HOW TO 2 Multiply: $\sqrt{2x}(\sqrt{8x} - \sqrt{3})$

$\sqrt{2x}(\sqrt{8x} - \sqrt{3}) = \sqrt{2x}(\sqrt{8x}) - \sqrt{2x}(\sqrt{3})$
- Use the Distributive Property.

$\qquad\qquad = \sqrt{16x^2} - \sqrt{6x}$
- Simplify.

$\qquad\qquad = 4x - \sqrt{6x}$

HOW TO 3 Multiply: $(2\sqrt{5} - 3)(3\sqrt{5} + 4)$

$(2\sqrt{5} - 3)(3\sqrt{5} + 4) = 6(\sqrt{5})^2 + 8\sqrt{5} - 9\sqrt{5} - 12$
- Use the FOIL method to multiply the numbers.

$\qquad\qquad = 30 + 8\sqrt{5} - 9\sqrt{5} - 12$

$\qquad\qquad = 18 - \sqrt{5}$
- Combine like terms.

HOW TO 4 Expand: $(5 - \sqrt{3x})^2$

$(5 - \sqrt{3x})^2 = (5 - \sqrt{3x})(5 - \sqrt{3x})$

$\qquad\qquad = 25 - 5\sqrt{3x} - 5\sqrt{3x} + (\sqrt{3x})^2$
- Use FOIL.

$\qquad\qquad = 25 - 10\sqrt{3x} + 3x$
- Combine like terms.

EXAMPLE 1

Multiply: $\sqrt{3}(\sqrt{15} - \sqrt{21})$

Solution

$\sqrt{3}(\sqrt{15} - \sqrt{21}) = \sqrt{45} - \sqrt{63}$
$\qquad\qquad\qquad = 3\sqrt{5} - 3\sqrt{7}$

YOU TRY IT 1

Multiply: $5\sqrt{2}(\sqrt{6} + \sqrt{24})$

Your solution

$30\sqrt{3}$

Solution on p. S26

EXAMPLE 2

Multiply: $(2 + 3\sqrt{5})(3 - \sqrt{5})$

Solution

$$
\begin{aligned}
(2 + 3\sqrt{5})(3 - \sqrt{5}) &= 6 - 2\sqrt{5} + 9\sqrt{5} - 3(\sqrt{5})^2 \\
&= 6 + 7\sqrt{5} - 3 \cdot 5 \\
&= 6 + 7\sqrt{5} - 15 \\
&= -9 + 7\sqrt{5}
\end{aligned}
$$

YOU TRY IT 2

Multiply: $(4 - 2\sqrt{7})(1 + 3\sqrt{7})$

Your solution

$-38 + 10\sqrt{7}$

EXAMPLE 3

Expand: $(3 - \sqrt{x + 1})^2$

Solution

$$
\begin{aligned}
(3 - \sqrt{x + 1})^2 &= (3 - \sqrt{x + 1})(3 - \sqrt{x + 1}) \\
&= 9 - 3\sqrt{x + 1} - 3\sqrt{x + 1} + (\sqrt{x + 1})^2 \\
&= 9 - 6\sqrt{x + 1} + (x + 1) \\
&= x - 6\sqrt{x + 1} + 10
\end{aligned}
$$

YOU TRY IT 3

Expand: $(4 - \sqrt{2x})^2$

Your solution

$2x - 8\sqrt{2x} + 16$

Solutions on p. S26

OBJECTIVE B *To divide radical expressions*

The Quotient Property of Radicals is used to divide radical expressions with the same index.

The Quotient Property of Radicals

If $\sqrt[n]{a}$ and $\sqrt[n]{b}$ are real numbers and $b \neq 0$, then $\dfrac{\sqrt[n]{a}}{\sqrt[n]{b}} = \sqrt[n]{\dfrac{a}{b}}$.

HOW TO 5 Divide: $\dfrac{\sqrt{5a^4b^7c^2}}{\sqrt{ab^3c}}$

$$
\begin{aligned}
\frac{\sqrt{5a^4b^7c^2}}{\sqrt{ab^3c}} &= \sqrt{\frac{5a^4b^7c^2}{ab^3c}} \\
&= \sqrt{5a^3b^4c} \\
&= \sqrt{a^2b^4 \cdot 5ac} = ab^2\sqrt{5ac}
\end{aligned}
$$

• Use the Quotient Property of Radicals.

• Simplify the radicand.

A radical expression is in simplest form when the denominator does not contain a radical. The radical expressions at the right are *not* in simplest form.

$$\frac{3}{\sqrt{7}} \qquad \frac{6}{\sqrt[3]{2x}} \qquad \frac{6}{3 - \sqrt{5}}$$

Not in simplest form

The procedure used to remove a radical expression from the denominator is called **rationalizing the denominator.** The idea is to multiply the numerator and denominator by an expression that will result in a denominator that is a perfect root of the index.

Take Note

$\dfrac{\sqrt{7}}{\sqrt{7}} = 1$. Therefore, we are multiplying by 1 and not changing the value of the expression.

HOW TO 6 Simplify: $\dfrac{3}{\sqrt{7}}$

$\dfrac{3}{\sqrt{7}} = \dfrac{3}{\sqrt{7}} \cdot \dfrac{\sqrt{7}}{\sqrt{7}}$

- Multiply the numerator and denominator by $\sqrt{7}$.

$= \dfrac{3\sqrt{7}}{\sqrt{49}} = \dfrac{3\sqrt{7}}{7}$

- $\sqrt{7} \cdot \sqrt{7} = \sqrt{49} = 7$. Because 49 is a perfect square, the denominator can now be written without a radical.

Take Note

Because the index of the radical is 3, we must multiply by a factor that will produce a perfect third power. Ask, "What must $4x$ be multiplied by to produce a perfect third power?"

$4x \cdot \, ? \, = 8x^3$
$4x \cdot 2x^2 = 8x^3$

We must multiply the numerator and denominator by $\sqrt[3]{2x^2}$.

HOW TO 7 Simplify: $\dfrac{6}{\sqrt[3]{4x}}$

$\dfrac{6}{\sqrt[3]{4x}} = \dfrac{6}{\sqrt[3]{4x}} \cdot \dfrac{\sqrt[3]{2x^2}}{\sqrt[3]{2x^2}}$

- Multiply the numerator and denominator by $\sqrt[3]{2x^2}$. See the Take Note at the left.

$= \dfrac{6\sqrt[3]{2x^2}}{\sqrt[3]{8x^3}} = \dfrac{6\sqrt[3]{2x^2}}{2x}$

- $\sqrt[3]{4x} \cdot \sqrt[3]{2x^2} = \sqrt[3]{8x^3} = 2x$. Because $8x^3$ is a perfect cube, the denominator can now be written without a radical.

$= \dfrac{3\sqrt[3]{2x^2}}{x}$

- Divide by the common factor.

To simplify a fraction that has a square-root radical expression with two terms in the denominator, multiply the numerator and denominator by the conjugate of the denominator.

Definition of Conjugate

The **conjugate** of $a + b$ is $a - b$, and the conjugate of $a - b$ is $a + b$.
The product of the conjugates is $(a + b)(a - b) = a^2 - b^2$.

EXAMPLES

1. The conjugate of $3 + \sqrt{7}$ is $3 - \sqrt{7}$. The product of the conjugates is
$$(3 + \sqrt{7})(3 - \sqrt{7}) = 3^2 - (\sqrt{7})^2 = 9 - 7 = 2$$

2. The conjugate of $-2 + 3\sqrt{2}$ is $-2 - 3\sqrt{2}$. The product of the conjugates is
$$(-2 + 3\sqrt{2})(-2 - 3\sqrt{2}) = (-2)^2 - (3\sqrt{2})^2 = 4 - (9 \cdot 2) = 4 - 18 = -14$$

3. The conjugate of $\sqrt{x} - \sqrt{y}$ is $\sqrt{x} + \sqrt{y}$. The product of the conjugates is
$$(\sqrt{x} - \sqrt{y})(\sqrt{x} + \sqrt{y}) = (\sqrt{x})^2 - (\sqrt{y})^2 = x - y$$

Take Note

The concept of conjugate is used in a number of different instances. Make sure you understand this idea.

The conjugate of $\sqrt{3} - 4$ is $\sqrt{3} + 4$.
The conjugate of $\sqrt{3} + 4$ is $\sqrt{3} - 4$.
The conjugate of $\sqrt{5a} + \sqrt{b}$ is $\sqrt{5a} - \sqrt{b}$.

INSTRUCTOR NOTE

Students need assurance that rationalizing a denominator is nothing more than multiplying an expression by 1. Have students evaluate $\dfrac{5}{\sqrt{2}}$ and $\dfrac{5\sqrt{2}}{2}$ with their calculators to see that the decimal representations displayed are the same.

HOW TO 8 Simplify: $\dfrac{6}{3 - \sqrt{5}}$

$\dfrac{6}{3 - \sqrt{5}} = \dfrac{6}{3 - \sqrt{5}} \cdot \dfrac{3 + \sqrt{5}}{3 + \sqrt{5}} = \dfrac{6(3 + \sqrt{5})}{3^2 - (\sqrt{5})^2}$

- Multiply the numerator and denominator by $3 + \sqrt{5}$, the conjugate of $3 - \sqrt{5}$.

$= \dfrac{6(3 + \sqrt{5})}{9 - 5} = \dfrac{6(3 + \sqrt{5})}{4}$

$= \dfrac{3(3 + \sqrt{5})}{2} = \dfrac{9 + 3\sqrt{5}}{2}$

EXAMPLE 4

Simplify: $\sqrt{\dfrac{3}{2x}}$

Solution

$\sqrt{\dfrac{3}{2x}} = \dfrac{\sqrt{3}}{\sqrt{2x}}$ • **Quotient Property of Radicals**

$= \dfrac{\sqrt{3}}{\sqrt{2x}} \cdot \dfrac{\sqrt{2x}}{\sqrt{2x}}$ • **Rationalize the denominator.**

$= \dfrac{\sqrt{6x}}{\sqrt{4x^2}} = \dfrac{\sqrt{6x}}{2x}$

YOU TRY IT 4

Simplify: $\sqrt{\dfrac{5}{6}}$

Your solution

$\dfrac{\sqrt{30}}{6}$

IN-CLASS EXAMPLES

Simplify.

5. $\dfrac{\sqrt{50x^3}}{\sqrt{2x}}$ $5x$ 8. $\dfrac{4}{\sqrt[3]{2x^2}}$ $\dfrac{2\sqrt[3]{4x}}{x}$

6. $\dfrac{8}{\sqrt{2a}}$ $\dfrac{4\sqrt{2a}}{a}$ 9. $\dfrac{\sqrt{32a^5b}}{\sqrt{12ab^3}}$ $\dfrac{2a^2\sqrt{6}}{3b}$

7. $\sqrt{\dfrac{x}{3}}$ $\dfrac{\sqrt{3x}}{3}$ 10. $\dfrac{2}{\sqrt{3}+2}$ $4-2\sqrt{3}$

EXAMPLE 5

Simplify: $\dfrac{6}{\sqrt[4]{9x}}$

Solution

Ask "What must $9x$ be multiplied by to produce a perfect fourth power?"

$9x \cdot ? = 81x^4$

$9x \cdot 9x^3 = 81x^4$

Multiply the numerator and denominator by $\sqrt[4]{9x^3}$.

$\dfrac{6}{\sqrt[4]{9x}} = \dfrac{6}{\sqrt[4]{9x}} \cdot \dfrac{\sqrt[4]{9x^3}}{\sqrt[4]{9x^3}} = \dfrac{6\sqrt[4]{9x^3}}{\sqrt[4]{81x^4}} = \dfrac{6\sqrt[4]{9x^3}}{3x} = \dfrac{2\sqrt[4]{9x^3}}{x}$

YOU TRY IT 5

Simplify: $\dfrac{3x}{\sqrt[3]{3x^2}}$

Your solution

$\sqrt[3]{9x}$

EXAMPLE 6

Simplify: $\dfrac{1-\sqrt{5}}{2+3\sqrt{5}}$

Solution

$\dfrac{1-\sqrt{5}}{2+3\sqrt{5}}$

$= \dfrac{1-\sqrt{5}}{2+3\sqrt{5}} \cdot \dfrac{2-3\sqrt{5}}{2-3\sqrt{5}}$ • **The conjugate of** $2+3\sqrt{5}$ **is** $2-3\sqrt{5}$.

$= \dfrac{2-3\sqrt{5}-2\sqrt{5}+3\sqrt{5}^2}{2^2-(3\sqrt{5})^2}$

$= \dfrac{2-5\sqrt{5}+3\cdot5}{4-9\cdot5}$

$= \dfrac{17-5\sqrt{5}}{-41} = \dfrac{-17+5\sqrt{5}}{41}$

YOU TRY IT 6

Simplify: $\dfrac{3}{2-\sqrt{x}}$

Your solution

$\dfrac{6+3\sqrt{x}}{4-x}$

Solutions on pp. S26–S27

9.3 EXERCISES

✔ **Concept Check**

SUGGESTED ASSIGNMENT
Exercises 1–8; Exercises 9–89, odds;
More challenging exercises: Exercises 91–93

1. What is the Product Property of Radicals?   If $\sqrt[n]{a}$ and $\sqrt[n]{b}$ are real numbers, then $\sqrt[n]{a} \cdot \sqrt[n]{b} = \sqrt[n]{ab}$.

2. What is the Quotient Property of Radicals?   If $\sqrt[n]{a}$ and $\sqrt[n]{b}$ are real numbers and $b \neq 0$, then $\dfrac{\sqrt[n]{a}}{\sqrt[n]{b}} = \sqrt[n]{\dfrac{a}{b}}$.

For Exercises 3 to 6, write the conjugate of the expression.

3. $3 + \sqrt{5}$
$3 - \sqrt{5}$

4. $-2 - \sqrt{7}$
$-2 + \sqrt{7}$

5. $4 - 3\sqrt{11}$
$4 + 3\sqrt{11}$

6. $-7\sqrt{3} + 6$
$-7\sqrt{3} - 6$

7. Which of the following can be simplified using the Product Property of Radicals?   i, iii
  (i) $\sqrt{5} \cdot \sqrt{23}$ **(ii)** $\sqrt[3]{7} \cdot \sqrt{7}$ **(iii)** $\sqrt[4]{3x^3y} \cdot \sqrt[4]{z}$ **(iv)** $\sqrt[4]{7xy^3} \cdot \sqrt[3]{7xy^3}$

8. Which of the following can be simplified using the Quotient Property of Radicals?   i, ii, iv
  (i) $\dfrac{\sqrt[3]{9}}{\sqrt[3]{11}}$ **(ii)** $\dfrac{\sqrt[4]{25}}{\sqrt[4]{5}}$ **(iii)** $\dfrac{\sqrt[3]{3xy^2}}{\sqrt{3xy^2}}$ **(iv)** $\dfrac{\sqrt{x^9}}{\sqrt{2x}}$

OBJECTIVE A   *To multiply radical expressions*

For Exercises 9 to 41, simplify.

9. $\sqrt{8}\,\sqrt{32}$
16

10. $\sqrt{14}\,\sqrt{35}$
$7\sqrt{10}$

11. $\sqrt[3]{4}\,\sqrt[3]{8}$
$2\sqrt[3]{4}$

12. $\sqrt[3]{6}\,\sqrt[3]{36}$
6

13. $\sqrt{x^2y^5}\,\sqrt{xy}$
$xy^3\sqrt{x}$

14. $\sqrt{a^3b}\,\sqrt{ab^4}$
$a^2b^2\sqrt{b}$

15. $\sqrt{2x^2y}\,\sqrt{32xy}$
$8xy\sqrt{x}$

16. $\sqrt{5x^3y}\,\sqrt{10x^3y^4}$
$5x^3y^2\sqrt{2y}$

17. $\sqrt[3]{x^2y}\,\sqrt[3]{16x^4y^2}$
$2x^2y\sqrt[3]{2}$

18. $\sqrt[3]{4a^2b^3}\,\sqrt[3]{8ab^5}$
$2ab^2\sqrt[3]{4b^2}$

19. $\sqrt[4]{12ab^3}\,\sqrt[4]{4a^5b^2}$
$2ab\sqrt[4]{3a^2b}$

20. $\sqrt[4]{36a^2b^4}\,\sqrt[4]{12a^5b^3}$
$2ab\sqrt[4]{27a^3b^3}$

21. $\sqrt{3}\,(\sqrt{27} - \sqrt{3})$
6

22. $\sqrt{10}\,(\sqrt{10} - \sqrt{5})$
$10 - 5\sqrt{2}$

23. $\sqrt{x}\,(\sqrt{x} - \sqrt{2})$
$x - \sqrt{2x}$

24. $\sqrt{y}\,(\sqrt{y} - \sqrt{5})$
$y - \sqrt{5y}$

25. $\sqrt{2x}\,(\sqrt{8x} - \sqrt{32})$
$4x - 8\sqrt{x}$

26. $\sqrt{3a}\,(\sqrt{27a^2} - \sqrt{a})$
$9a\sqrt{a} - a\sqrt{3}$

27. $(3 - 2\sqrt{5})(2 + \sqrt{5})$
$-4 - \sqrt{5}$

28. $(6 + 5\sqrt{2})(4 - 2\sqrt{2})$
$4 + 8\sqrt{2}$

29. $(-2 + \sqrt{7})(3 + 5\sqrt{7})$
$29 - 7\sqrt{7}$

30. $(5 - 2\sqrt{5})(7 - 3\sqrt{5})$
$65 - 29\sqrt{5}$

31. $(6 + 3\sqrt{2})(4 - 2\sqrt{2})$
12

32. $(10 - 3\sqrt{5})(3 + 2\sqrt{5})$
$11\sqrt{5}$

33. $(5 - 2\sqrt{7})(5 + 2\sqrt{7})$
  -3

34. $(4 + 2\sqrt{3})(4 - 2\sqrt{3})$
  4

35. $(3 - \sqrt{2x})(1 + 5\sqrt{2x})$
  $-10x + 14\sqrt{2x} + 3$

36. $(3 - 2\sqrt{x})^2$
  $4x - 12\sqrt{x} + 9$

37. $(2 + \sqrt{x})^2$
  $x + 4\sqrt{x} + 4$

38. $(\sqrt{2x} - 3)^2$
  $2x - 6\sqrt{2x} + 9$

39. $(\sqrt{3x} - 5)^2$
  $3x - 10\sqrt{3x} + 25$

40. $(5 - \sqrt{x + 2})^2$
  $x - 10\sqrt{x + 2} + 27$

41. $(4 - \sqrt{2x + 1})^2$
  $2x - 8\sqrt{2x + 1} + 17$

42. True or false? If $a > 0$, then
  $(\sqrt{a} - 1)(\sqrt{a} + 1) > a$.
  False

43. True or false? If $a > 0$, then
  $(\sqrt{a} + 1)^2 > a + 1$.
  True

OBJECTIVE B *To divide radical expressions*

44. When is a radical expression in simplest form?

45. Explain what it means to rationalize the denominator of a radical expression and how to do so.

For Exercises 46 to 49, by what expression should the numerator and denominator be multiplied in order to rationalize the denominator?

46. $\dfrac{1}{\sqrt{6}}$
  $\sqrt{6}$

47. $\dfrac{7}{\sqrt[3]{2x^5}}$
  $\sqrt[3]{4x}$

48. $\dfrac{8x}{\sqrt[4]{27x}}$
  $\sqrt[4]{3x^3}$

49. $\dfrac{4}{\sqrt{3} - x}$
  $\sqrt{3} + x$

For Exercises 50 to 90, simplify.

50. $\dfrac{\sqrt{32x^2}}{\sqrt{2x}}$
  $4\sqrt{x}$

51. $\dfrac{\sqrt{60y^4}}{\sqrt{12y}}$
  $y\sqrt{5y}$

52. $\dfrac{\sqrt{42a^3b^5}}{\sqrt{14a^2b}}$
  $b^2\sqrt{3a}$

53. $\dfrac{\sqrt{65ab^4}}{\sqrt{5ab}}$
  $b\sqrt{13b}$

54. $\dfrac{1}{\sqrt{5}}$
  $\dfrac{\sqrt{5}}{5}$

55. $\dfrac{1}{\sqrt{2}}$
  $\dfrac{\sqrt{2}}{2}$

56. $\dfrac{1}{\sqrt{2x}}$
  $\dfrac{\sqrt{2x}}{2x}$

57. $\dfrac{2}{\sqrt{3y}}$
  $\dfrac{2\sqrt{3y}}{3y}$

58. $\dfrac{5}{\sqrt{5x}}$
  $\dfrac{\sqrt{5x}}{x}$

59. $\dfrac{9}{\sqrt{3a}}$
  $\dfrac{3\sqrt{3a}}{a}$

60. $\sqrt{\dfrac{x}{5}}$
  $\dfrac{\sqrt{5x}}{5}$

61. $\sqrt{\dfrac{y}{2}}$
  $\dfrac{\sqrt{2y}}{2}$

62. $\dfrac{3}{\sqrt[3]{2}}$
  $\dfrac{3\sqrt[3]{4}}{2}$

63. $\dfrac{5}{\sqrt[3]{9}}$
  $\dfrac{5\sqrt[3]{3}}{3}$

64. $\dfrac{3}{\sqrt[3]{4x^2}}$
  $\dfrac{3\sqrt[3]{2x}}{2x}$

65. $\dfrac{5}{\sqrt[3]{3y}}$
  $\dfrac{5\sqrt[3]{9y^2}}{3y}$

66. $\dfrac{3}{\sqrt[4]{2x^3}}$

$\dfrac{3\sqrt[4]{8x}}{2x}$

67. $\dfrac{6x}{\sqrt[4]{9x}}$

$\dfrac{2\sqrt[4]{9x^3}}{3}$

68. $\dfrac{7}{\sqrt[5]{8x^2}}$

$\dfrac{7\sqrt[5]{4x^3}}{8x}$

69. $\dfrac{9x^2}{\sqrt[5]{27x}}$

$3x\sqrt[5]{9x^4}$

70. $\dfrac{\sqrt{40x^3y^2}}{\sqrt{80x^2y^3}}$

$\dfrac{\sqrt{2xy}}{2y}$

71. $\dfrac{\sqrt{15a^2b^5}}{\sqrt{30a^5b^3}}$

$\dfrac{b\sqrt{2a}}{2a^2}$

72. $\dfrac{\sqrt{24a^2b}}{\sqrt{18ab^4}}$

$\dfrac{2\sqrt{3ab}}{3b^2}$

73. $\dfrac{\sqrt{12x^3y}}{\sqrt{20x^4y}}$

$\dfrac{\sqrt{15x}}{5x}$

74. $\dfrac{5}{\sqrt{3}-2}$

$-10-5\sqrt{3}$

75. $\dfrac{-2}{1-\sqrt{2}}$

$2+2\sqrt{2}$

76. $\dfrac{-3}{2-\sqrt{3}}$

$-6-3\sqrt{3}$

77. $\dfrac{-4}{3-\sqrt{2}}$

$-\dfrac{12+4\sqrt{2}}{7}$

78. $\dfrac{2}{\sqrt{5}+2}$

$2\sqrt{5}-4$

79. $\dfrac{5}{2-\sqrt{7}}$

$-\dfrac{10+5\sqrt{7}}{3}$

80. $\dfrac{3}{\sqrt{y}-2}$

$\dfrac{3\sqrt{y}+6}{y-4}$

81. $\dfrac{-7}{\sqrt{x}-3}$

$-\dfrac{7\sqrt{x}+21}{x-9}$

82. $\dfrac{\sqrt{2}-\sqrt{3}}{\sqrt{2}+\sqrt{3}}$

$-5+2\sqrt{6}$

83. $\dfrac{\sqrt{3}+\sqrt{4}}{\sqrt{2}+\sqrt{3}}$

$-\sqrt{6}+3-2\sqrt{2}+2\sqrt{3}$

84. $\dfrac{2+3\sqrt{7}}{5-2\sqrt{7}}$

$-\dfrac{52+19\sqrt{7}}{3}$

85. $\dfrac{2+3\sqrt{5}}{1-\sqrt{5}}$

$-\dfrac{17+5\sqrt{5}}{4}$

86. $\dfrac{2\sqrt{3}-1}{3\sqrt{3}+2}$

$\dfrac{20-7\sqrt{3}}{23}$

87. $\dfrac{2\sqrt{a}-\sqrt{b}}{4\sqrt{a}+3\sqrt{b}}$

$\dfrac{8a-10\sqrt{ab}+3b}{16a-9b}$

88. $\dfrac{2\sqrt{x}-4}{\sqrt{x}+2}$

$\dfrac{2x-8\sqrt{x}+8}{x-4}$

89. $\dfrac{3\sqrt{y}-y}{\sqrt{y}+2y}$

$\dfrac{3-7\sqrt{y}+2y}{1-4y}$

90. $\dfrac{3\sqrt{x}-4\sqrt{y}}{3\sqrt{x}-2\sqrt{y}}$

$\dfrac{9x-6\sqrt{xy}-8y}{9x-4y}$

Critical Thinking

For Exercises 91 to 93, simplify.

91. $(\sqrt{8}-\sqrt{2})^3$ $2\sqrt{2}$

92. $(\sqrt{27}-\sqrt{3})^3$ $24\sqrt{3}$

93. $(\sqrt{2}-\sqrt{3})^2$ $5-2\sqrt{6}$

In some cases, it is necessary to rationalize the numerator of a radical expression rather than the denominator. In Exercises 94 and 95, rationalize the numerator.

94. $\dfrac{\sqrt{4+h}-2}{h}$ $\dfrac{1}{\sqrt{4+h}+2}$

95. $\dfrac{\sqrt{9+h}-3}{h}$ $\dfrac{1}{\sqrt{9+h}+3}$

Projects or Group Activities

The factorization formulas for the sum and difference of two perfect cubes can be used to simplify some radical expressions containing cube roots. Recall:

$$(a + b)(a^2 - ab + b^2) = a^3 + b^3$$
$$(a - b)(a^2 + ab + b^2) = a^3 - b^3$$

96. Using the sum of perfect cubes formula, show that
$$(3 + \sqrt[3]{2})(9 - 3\sqrt[3]{2} + \sqrt[3]{4}) = 29$$

97. Simplify $\dfrac{1}{3 + \sqrt[3]{2}}$ by multiplying the numerator and denominator by

$9 - 3\sqrt[3]{2} + \sqrt[3]{4}$. (*Hint:* See Exercise 96.) $\dfrac{9 - 3\sqrt[3]{2} + \sqrt[3]{4}}{29}$

98. Simplify: $\dfrac{4}{\sqrt[3]{5} - 1}$ $\sqrt[3]{25} + \sqrt[3]{5} + 1$

QUICK QUIZ

Simplify.

1. $\sqrt[4]{18a^3b^5}\ \sqrt[4]{6a^5b^3}$
   $a^2b^2\ \sqrt[4]{108}$ [9.3A]

2. $\sqrt{3x}(\sqrt{27x} - \sqrt{12})$
   $9x - 6\sqrt{x}$ [9.3A]

3. $(3\sqrt{2x} - \sqrt{y})(3\sqrt{2x} + \sqrt{y})$
   $18x - y$ [9.3A]

4. $\dfrac{\sqrt{48x^5}}{\sqrt{3x^2}}$ $4x\sqrt{x}$ [9.3B]

5. $\dfrac{7}{\sqrt{7x}}$ $\dfrac{\sqrt{7x}}{x}$ [9.3B]

6. $\dfrac{6}{\sqrt[3]{4y}}$ $\dfrac{3\sqrt[3]{2y^2}}{y}$ [9.3B]

7. $\dfrac{\sqrt{5} + \sqrt{3}}{\sqrt{5} - \sqrt{3}}$
   $4 + \sqrt{15}$ [9.3B]

✔ CHECK YOUR PROGRESS: CHAPTER 9

For Exercises 1 to 24, simplify.

1. $32^{4/5}$

   16   [9.1A]

2. $16^{-3/4}$

   $\dfrac{1}{8}$ [9.1A]

3. $\left(\dfrac{27}{8}\right)^{-1/3}$

   $\dfrac{2}{3}$ [9.1A]

4. $\left(\dfrac{64}{81}\right)^{-3/2}$

   $\dfrac{729}{512}$ [9.1A]

5. $x^{3/4} \cdot x^{-1/2}$

   $x^{1/4}$ [9.1A]

6. $(8x^6)^{2/3}$

   $4x^4$ [9.1A]

7. $\dfrac{z^{5/6}}{z^{3/4}}$

   $z^{1/12}$ [9.1A]

8. $\left(\dfrac{a^{-1/3}b^{3/2}}{c^{2/3}}\right)^{-6}$

   $\dfrac{a^2c^4}{b^9}$ [9.1A]

9. $\sqrt{9x^{12}}$

   $3x^6$ [9.1C]

10. $\sqrt[5]{-32a^5b^{15}}$

   $-2ab^3$ [9.1C]

11. $\sqrt{72a^3b^{10}}$

   $6ab^5\sqrt{2a}$ [9.2A]

12. $\sqrt[3]{16x^7y^3z^{11}}$

   $2x^2yz^3\sqrt[3]{2xy^3z^2}$ [9.2A]

13. $6\sqrt{8a^2b^3} - 4a\sqrt{32b^3}$
   $-4ab\sqrt{2b}$ [9.2B]

14. $3\sqrt{50} - 9\sqrt{72} + 6\sqrt{98}$
   $3\sqrt{2}$ [9.2B]

15. $x\sqrt{3x^3} + 2x^2\sqrt{27x} - \sqrt{75x^5}$
   $2x^2\sqrt{3x}$ [9.2B]

16. $(\sqrt[3]{4x^2y})(\sqrt[3]{6xy^4})$
   $2xy\sqrt[3]{3y^2}$ [9.3A]

17. $\sqrt{3x}(2\sqrt{6x^3} - \sqrt{12x})$
   $6x^2\sqrt{2} - 6x$ [9.3A]

18. $(2\sqrt{5} + 7)(3\sqrt{5} - 1)$
   $23 + 19\sqrt{5}$ [9.3A]

19. $(2\sqrt{x} - 3)^2$

   $4x - 12\sqrt{x} + 9$ [9.3A]

20. $\dfrac{6}{\sqrt{8}}$

   $\dfrac{3\sqrt{2}}{2}$ [9.3B]

21. $\sqrt[3]{\dfrac{3}{4}}$

   $\dfrac{\sqrt[3]{6}}{2}$ [9.3B]

22. $\dfrac{7}{2\sqrt{3} + 3}$

   $\dfrac{14\sqrt{3} - 21}{3}$ [9.3B]

23. $\dfrac{2\sqrt{x}}{\sqrt{x} - 2}$

   $\dfrac{2x + 4\sqrt{x}}{x - 4}$ [9.3B]

24. $\dfrac{3\sqrt{2} + 5}{2\sqrt{2} - 1}$

   $\dfrac{17 + 13\sqrt{2}}{7}$ [9.3B]

SECTION

9.4 Solving Equations Containing Radical Expressions

OBJECTIVE A *To solve a radical equation*

An equation that contains a variable expression in a radicand is called a **radical equation.**

$$\sqrt[3]{2x - 5} + x = 7$$
$$\sqrt{x + 1} - \sqrt{x} = 4$$

Radical equations

INSTRUCTOR NOTE

Some students may assume that the converse of this result is also true. That is, if $a^2 = b^2$, then $a = b$. Use $a = -4$ and $b = 4$ to show that $(-4)^2 = 4^2$ but $-4 \neq 4$. The fact that the converse is not true is what creates the potential for extraneous solutions.

The following property is used to solve a radical equation.

The Property of Raising Each Side of an Equation to a Power

If two numbers are equal, then the same powers of the numbers are equal.

If $a = b$, then $a^n = b^n$.

HOW TO 1 Solve: $\sqrt{x - 2} - 6 = 0$

$$\sqrt{x - 2} - 6 = 0$$
$$\sqrt{x - 2} = 6 \qquad \bullet \text{ Isolate the radical by adding 6 to each side of the equation.}$$
$$(\sqrt{x - 2})^2 = 6^2 \qquad \bullet \text{ Square each side of the equation.}$$
$$x - 2 = 36 \qquad \bullet \text{ Simplify and solve for } x.$$
$$x = 38$$

Check:
$$\begin{array}{c|c} \sqrt{x - 2} - 6 = 0 \\ \hline \sqrt{38 - 2} - 6 & 0 \\ \sqrt{36} - 6 & 0 \\ 6 - 6 & 0 \\ 0 = 0 \end{array}$$

38 checks as a solution. The solution is 38.

IN-CLASS EXAMPLES
Solve.
1. $\sqrt[3]{2x} = -2$ **−4**
2. $\sqrt{3x - 5} = 5$ **10**
3. $\sqrt[3]{x - 3} = 2$ **11**
4. $\sqrt[4]{3x + 1} = 4$ **85**
5. $\sqrt{2x - 1} - 1 = 2$ **5**

HOW TO 2 Solve: $\sqrt[3]{x + 2} = -3$

$$\sqrt[3]{x + 2} = -3$$
$$(\sqrt[3]{x + 2})^3 = (-3)^3 \qquad \bullet \text{ Cube each side of the equation.}$$
$$x + 2 = -27 \qquad \bullet \text{ Solve the resulting equation.}$$
$$x = -29$$

Check:
$$\begin{array}{c|c} \sqrt[3]{x + 2} = -3 \\ \hline \sqrt[3]{-29 + 2} & -3 \\ \sqrt[3]{-27} & -3 \\ -3 = -3 \end{array}$$

−29 checks as a solution. The solution is −29.

INSTRUCTOR NOTE

If you did not cover Section 7.5, skip HOW TO 3. This material is covered again in Chapter 10.

Raising each side of an equation to an *even* power may result in an equation that has extraneous solutions. Therefore, you *must* check proposed solutions of an equation if one of the steps in solving the equation is to raise each side to an even power.

 Take Note

Note that

$$(2 - \sqrt{x})^2$$
$$= (2 - \sqrt{x})(2 - \sqrt{x})$$
$$= 4 - 4\sqrt{x} + x$$

HOW TO 3   Solve: $\sqrt{2x - 1} + \sqrt{x} = 2$

$$\sqrt{2x - 1} + \sqrt{x} = 2$$

$$\sqrt{2x - 1} = 2 - \sqrt{x}$$ • Solve for one of the radical expressions.

$$(\sqrt{2x - 1})^2 = (2 - \sqrt{x})^2$$ • Square each side. Recall that $(a - b)^2 = a^2 - 2ab + b^2$.

$$2x - 1 = 4 - 4\sqrt{x} + x$$

$$x - 5 = -4\sqrt{x}$$

$$(x - 5)^2 = (-4\sqrt{x})^2$$ • Square each side again.

$$x^2 - 10x + 25 = 16x$$

$$x^2 - 26x + 25 = 0$$ • Solve the quadratic equation by factoring.

$$(x - 25)(x - 1) = 0$$

$$x = 25 \text{ or } x = 1$$

 Take Note

The proposed solutions of the equation were 1 and 25. However, 25 did not check as a solution. Therefore, it is an extraneous solution.

Check:

$\sqrt{2x - 1} + \sqrt{x} = 2$		$\sqrt{2x - 1} + \sqrt{x} = 2$	
$\sqrt{2(25) - 1} + \sqrt{25}$	2	$\sqrt{2(1) - 1} + \sqrt{1}$	2
$7 + 5$	2	$1 + 1$	2
$12 \neq 2$		$2 = 2$	

25 does not check as a solution. 1 checks as a solution. The solution is 1. See Take Note.

EXAMPLE 1

Solve: $\sqrt{x - 1} + \sqrt{x + 4} = 5$

Solution

$$\sqrt{x - 1} + \sqrt{x + 4} = 5$$

$$\sqrt{x + 4} = 5 - \sqrt{x - 1}$$ • Subtract $\sqrt{x - 1}$.

$$(\sqrt{x + 4})^2 = (5 - \sqrt{x - 1})^2$$ • Square each side.

$$x + 4 = 25 - 10\sqrt{x - 1} + x - 1$$

$$-20 = -10\sqrt{x - 1}$$

$$2 = \sqrt{x - 1}$$

$$2^2 = (\sqrt{x - 1})^2$$ • Square each side.

$$4 = x - 1$$

$$5 = x$$

The solution checks. The solution is 5.

EXAMPLE 2

Solve: $\sqrt[3]{3x - 1} = -4$

Solution

$$\sqrt[3]{3x - 1} = -4$$

$$(\sqrt[3]{3x - 1})^3 = (-4)^3$$ • Cube each side.

$$3x - 1 = -64$$

$$3x = -63$$

$$x = -21$$

The solution checks. The solution is -21.

YOU TRY IT 1

Solve: $\sqrt{x} - \sqrt{x + 5} = 1$

Your solution

The equation has no solution.

YOU TRY IT 2

Solve: $\sqrt[4]{x - 8} = 3$

Your solution

89

Solutions on p. S27

The Granger Collection, NYC—All rights reserved.

OBJECTIVE B *To solve application problems*

Pythagoras

A right triangle contains one 90° angle. The side opposite the 90° angle is called the hypotenuse. The other two sides are called legs.

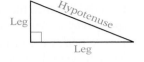

Pythagoras, a Greek mathematician, discovered that the square of the hypotenuse of a right triangle is equal to the sum of the squares of the two legs. Recall that this is called the Pythagorean Theorem.

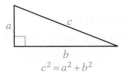

EXAMPLE 3

A ladder 20 ft long is leaning against a building. How high on the building will the ladder reach when the bottom of the ladder is 8 ft from the building? Round to the nearest tenth.

Strategy

To find the distance, use the Pythagorean Theorem. The hypotenuse is the length of the ladder. One leg is the distance from the bottom of the ladder to the base of the building. The distance along the building from the ground to the top of the ladder is the unknown leg.

Solution

$$c^2 = a^2 + b^2 \qquad \bullet \text{ Pythagorean Theorem}$$
$$20^2 = 8^2 + b^2 \qquad \bullet \text{ Replace } c \text{ by 20 and } a \text{ by 8.}$$
$$400 = 64 + b^2 \qquad \bullet \text{ Solve for } b.$$
$$336 = b^2$$

$$\sqrt{336} = \sqrt{b^2} \qquad \bullet \text{ Take the square root of each side.}$$

$$\sqrt{336} = b$$
$$18.3 \approx b$$

The distance is approximately 18.3 ft.

YOU TRY IT 3

Find the diagonal of a rectangle that is 6 cm long and 3 cm wide. Round to the nearest tenth.

IN-CLASS EXAMPLES

6. How far above the water would a submarine periscope have to be to locate a ship 5.6 mi away? The equation for the distance in miles that the lookout can see is $d = 1.4\sqrt{h}$, where h is the height in feet above the surface of the water. **16 ft**

Your strategy

Your solution

6.7 cm

Solution on p. S27

Unless otherwise noted, all content on this page is © Cengage Learning.

EXAMPLE 4

An object is dropped from a high building. Find the distance the object has fallen when its speed reaches 96 ft/s. Use the equation $v = \sqrt{64d}$, where v is the speed of the object in feet per second and d is the distance in feet.

Strategy

To find the distance the object has fallen, replace v in the equation with the given value and solve for d.

Solution

$$v = \sqrt{64d}$$
$$96 = \sqrt{64d} \quad \bullet \text{ Replace } v \text{ by 96.}$$
$$(96)^2 = (\sqrt{64d})^2 \quad \bullet \text{ Square each side.}$$
$$9216 = 64d$$
$$144 = d$$

The object has fallen 144 ft.

YOU TRY IT 4

How far above the water would a submarine periscope have to be to locate a ship 5.5 mi away? The equation for the distance in miles that the lookout can see is $d = \sqrt{1.5h}$, where h is the height in feet above the surface of the water. Round to the nearest hundredth.

Your strategy

Your solution

20.17 ft

EXAMPLE 5

Find the length of a pendulum that makes one swing in 1.5 s. The equation for the time of one swing is given by $T = 2\pi\sqrt{\dfrac{L}{32}}$, where T is the time in seconds and L is the length in feet. Use 3.14 for π. Round to the nearest hundredth.

Strategy

To find the length of the pendulum, replace T in the equation with the given value and solve for L.

Solution

$$T = 2\pi\sqrt{\frac{L}{32}}$$

$$1.5 = 2(3.14)\sqrt{\frac{L}{32}} \quad \bullet \text{ Replace } T \text{ by 1.5} \text{ and } \pi \text{ by 3.14.}$$

$$\frac{1.5}{2(3.14)} = \sqrt{\frac{L}{32}} \quad \bullet \text{ Divide each side by 2(3.14).}$$

$$\left[\frac{1.5}{2(3.14)}\right]^2 = \left(\sqrt{\frac{L}{32}}\right)^2 \quad \bullet \text{ Square each side.}$$

$$\left(\frac{1.5}{6.28}\right)^2 = \frac{L}{32} \quad \bullet \text{ Solve for } L. \text{ Multiply each side by 32.}$$

$$1.83 \approx L$$

The length of the pendulum is approximately 1.83 ft.

YOU TRY IT 5

Find the distance required for a car to reach a velocity of 88 ft/s when the acceleration is 22 ft/s². Use the equation $v = \sqrt{2as}$, where v is the velocity in feet per second, a is the acceleration, and s is the distance in feet.

Your strategy

Your solution

176 ft

Solutions on p. S27

9.4 EXERCISES

SUGGESTED ASSIGNMENT
Exercises 1–4; Exercises 5–35, odds;
More challenging exercises: Exercises 37 and 38

INSTRUCTOR NOTE
If you skipped Section 7.5, do not assign
Exercises 17, 18, and 23–25.

✔ **Concept Check**

For Exercises 1 and 2, determine whether the statement is sometimes true or always true.

1. If $a^2 = b^2$, then $a = b$.   Sometimes true

2. If $a^3 = b^3$, then $a = b$.   Always true

For Exercises 3 and 4, state whether 2 is a solution of the given equation.

3. $\sqrt{x + 2} - \sqrt{4x + 1} = 1$ No

4. $\sqrt{3x + 3} - \sqrt{x - 1} = 2$ Yes

OBJECTIVE A   *To solve a radical equation*

For Exercises 5 to 25, solve for x.

5. $\sqrt[3]{4x} = -2$
-2

6. $\sqrt[3]{6x} = -3$ $-\dfrac{9}{2}$

7. $\sqrt{3x - 2} = 5$
9

8. $\sqrt{3x + 9} - 12 = 0$
45

9. $\sqrt{4x - 3} + 9 = 4$
No solution

10. $\sqrt{2x - 5} + 4 = 1$
No solution

11. $\sqrt[3]{2x - 6} = 4$
35

12. $\sqrt[3]{x - 2} = 3$
29

13. $\sqrt[4]{3x + 2} = 5$
27

14. $\sqrt[4]{4x + 1} = 2$ $\dfrac{15}{4}$

15. $\sqrt[3]{2x - 3} + 5 = 2$
-12

16. $\sqrt[3]{x - 4} + 7 = 5$
-4

17. $4\sqrt{x - 2} + 2 = x + 3$
3 and 11

18. $2\sqrt{2x + 7} - 1 = x + 4$
-3 and 1

19. $\sqrt{x} + \sqrt{x - 5} = 5$
9

20. $\sqrt{x + 3} + \sqrt{x - 1} = 2$
1

21. $\sqrt{2x + 5} - \sqrt{2x} = 1$
2

22. $\sqrt{3x} - \sqrt{3x - 5} = 1$
3

23. $\sqrt{2x} - \sqrt{x - 1} = 1$
2

24. $\sqrt{2x - 5} + \sqrt{x + 1} = 3$
3

25. $\sqrt{2x + 2} + \sqrt{x} = 3$
1

26. ◼ Explain why the equation $\sqrt{x} = -4$ has no solution.

27. ◼ Without attempting to solve the equation, explain why $\sqrt{x} - \sqrt{x + 5} = 1$ has no solution. *Hint:* See Exercise 26.

OBJECTIVE B   *To solve application problems*

28. Physics   An object is dropped from a bridge. Find the distance the object has fallen when its speed reaches 100 ft/s. Use the equation $v = \sqrt{64d}$, where v is the speed in feet per second and d is the distance in feet.   156.25 ft

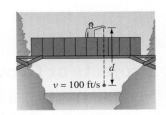

29. Physics   The time t, in seconds, it takes an object to fall a distance d, in feet, is given by the equation $t = \sqrt{\dfrac{2d}{g}}$, where g is the acceleration due to gravity. If an astronaut above the moon's surface drops an object, how far will it fall in 3 s? The acceleration on the moon's surface is 5.5 feet per second per second.   24.75 ft

30. Sailing   The total recommended area A, in square feet, of a sailboat's sails is given by $A = 16\sqrt[3]{d^2}$, where d is the displacement of the hull in cubic feet. If a sailboat has 400 ft² of sail, what is the displacement of the hull of the sailboat?   125 ft³

Unless otherwise noted, all content on this page is © Cengage Learning.

31. Water Tanks   A 6-foot-high conical water tank is filled to the top. When a valve at the bottom of the tank is opened, the height h, in feet, of the water in the tank is given by $h = (88.18 - 3.18t)^{2/5}$, where t is the time in seconds after the valve is opened. Find the height of the water 10 s after the valve is opened. How long will it take to empty the tank? Round answers to the nearest tenth.   5.0 ft; 27.7 s

32. Water Tanks   The velocity v, in feet per second, of the water pouring out of a small hole in the bottom of a cylindrical tank is given by $v = \sqrt{64h} + 10$, where h is the height, in feet, of the water in the tank. What is the height of the water in the tank when the velocity of the water leaving the tank is 14 ft/s? Round to the nearest tenth. 2.9 ft

33. Pendulums   Find the length of a pendulum that makes one swing in 3 s. The equation for the time of one swing of a pendulum is $T = 2\pi\sqrt{\dfrac{L}{32}}$, where T is the time in seconds and L is the length in feet. Round to the nearest hundredth.   7.30 ft

34. ◐ **Meteorology**   The sustained wind velocity v (in meters per second) in a hurricane is given by $v = 6.3\sqrt{1013 - p}$, where p is the air pressure in millibars (mb). Read the article at the right. What was the air pressure when Julia's winds were blowing at the velocity given in the article? What happens to wind speed in a hurricane as air pressure decreases?   892 mb; increases

35. Television   High definition television (HDTV) gives consumers a wider viewing area, more like a film screen in a theater. A regular television with a 27-inch diagonal measurement has a screen 16.2 in. tall. An HDTV screen with the same 16.2-inch height would have a diagonal measuring 33 in. How many inches wider is the HDTV screen? Round to the nearest hundredth.   7.15 in.

36. Construction   A carpenter inserts a 3-foot brace between two beams as shown. How far from the vertical beam will the brace reach? Round to the nearest tenth.
2.8 ft

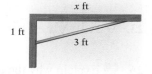

Critical Thinking

37. Moving Boxes   A moving box has a base that measures 2 ft by 3 ft, and the box is 4 ft tall. Find the length of the longest pole that could be placed in the box. Round to the nearest tenth.   5.4 ft

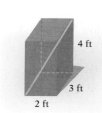

38. Geometry   Find the length of the side labeled x.
$x = \sqrt{6}$

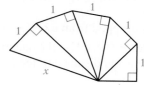

Projects or Group Activities

39. ◪ A 10-foot ladder is resting against a wall, with the bottom of the ladder 6 ft from the wall. The top of the ladder begins sliding down the wall at a constant rate of 2 ft/s. Is the bottom of the ladder sliding away from the wall at the same rate? Explain.
No.

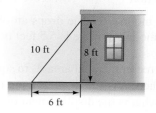

In the NEWS!

A Special Category of Hurricane

It hasn't happened since 1926—two Category 4 hurricanes existing at the same time over the Atlantic. Now, on the same day, Hurricanes Igor and Julia both have been rated Category 4, with Julia blowing the fiercer winds, at speeds of up to 69.3 m/s (155 mph).

Source: www.examiner.com/weather-in-baltimore

QUICK QUIZ

Solve.

1. $\sqrt{3y} = 9$ 27   **[9.4A]**
2. $\sqrt[3]{3x} = -6$
   −72   **[9.4A]**
3. $\sqrt[4]{2x - 1} = 3$
   41   **[9.4A]**
4. $\sqrt{3x - 6} - 4 = 2$
   14   **[9.4A]**
5. $\sqrt{x + 4} - \sqrt{x - 3} = 2$
   5   **[9.4A]**
6. An 18-foot ladder is leaning against a building. How high on the building will the ladder reach when the bottom of the ladder is 6 ft from the building? Round to the nearest tenth.
   17.0 ft   **[9.4B]**

Unless otherwise noted, all content on this page is © Cengage Learning.

SECTION

9.5 | Complex Numbers

OBJECTIVE A | *To simplify a complex number*

The radical expression $\sqrt{-4}$ is not a real number because there is no real number whose square is -4. However, the solution of an algebraic equation is sometimes the square root of a negative number.

For example, the equation $x^2 + 1 = 0$ does not have a real number solution because there is no real number whose square is -1.

$$x^2 + 1 = 0$$
$$x^2 = -1$$

Around the 17th century, a new number, called an **imaginary number,** was defined so that a negative number would have a square root. The letter i was chosen to represent the number whose square is -1.

$$i^2 = -1$$

An imaginary number is defined in terms of i.

 Point of Interest
The first written occurrence of an imaginary number was in a book published in 1545 by Hieronimo Cardan, where he wrote (in our modern notation) $5 + \sqrt{-15}$.
He went on to say that the number "is as refined as it is useless." It was not until the 20th century that applications of complex numbers were found.

Definition of $\sqrt{-a}$

If a is a positive real number, then $\sqrt{-a} = i\sqrt{a}$.

EXAMPLES

1. $\sqrt{-16} = i\sqrt{16} = 4i$ 2. $\sqrt{-12} = i\sqrt{12} = 2i\sqrt{3}$

3. $\sqrt{-23} = i\sqrt{23}$ 4. $\sqrt{-1} = i\sqrt{1} = i$

It is customary to write i in front of a radical to avoid confusing $\sqrt{a}\,i$ with $\sqrt{ai}$.

Definition of Complex Number

A **complex number** is a number of the form $a + bi$, where a and b are real numbers and $i = \sqrt{-1}$. The number a is the **real part** of the complex number, and b is the **imaginary part** of the complex number. A complex number written as $a + bi$ is in **standard form.**

EXAMPLES

1. $3 + 4i$ Real part is 3; imaginary part is 4.

2. $5 - 2i\sqrt{7}$ Real part is 5; imaginary part is $-2\sqrt{7}$.

3. 5 Real part is 5; imaginary part is 0, because $5 = 5 + 0i$.

4. $-4i$ Real part is 0; imaginary part is -4, because $-4i = 0 - 4i$.

5. $\dfrac{2 + 3i}{5} = \dfrac{2}{5} + \dfrac{3}{5}i$ Real part is $\dfrac{2}{5}$; imaginary part is $\dfrac{3}{5}$.

INSTRUCTOR NOTE

Complex numbers were not accepted by most mathematicians until late in the 19th century. Part of the difficulty was that there were no physical examples of these numbers.

Real Numbers
$a + 0i$

A *real number* is a complex number in which $b = 0$.

Complex numbers
$a + bi$

Imaginary Numbers
$0 + bi$

An *imaginary number* is a complex number in which $a = 0$.

HOW TO 1   Write $\dfrac{10 + \sqrt{-80}}{6}$ in standard form.

> **Tips for Success**
> Be sure you understand how to simplify expressions such as those in Example 1 and You Try It 1, as this skill is a prerequisite for solving quadratic equations in Chapter 10.

$$\frac{10 + \sqrt{-80}}{6} = \frac{10 + i\sqrt{-80}}{6}$$

$$= \frac{10 + 4i\sqrt{5}}{6}$$

$$= \frac{10}{6} + \frac{4\sqrt{5}}{6}i = \frac{5}{3} + \frac{2\sqrt{5}}{3}i$$

EXAMPLE 1

Simplify: $\sqrt{-80}$

Solution

$\sqrt{-80} = i\sqrt{80}$
$\quad = i\sqrt{16 \cdot 5} = 4i\sqrt{5}$

YOU TRY IT 1

Simplify: $\sqrt{-45}$

Your solution

$3i\sqrt{5}$

IN-CLASS EXAMPLES
Simplify.
1. $\sqrt{-49}$ **7i**
2. $\sqrt{-75}$ **5i√3**
3. $\sqrt{36} + \sqrt{-49}$ **6 + 7i**

EXAMPLE 2

Evaluate $-b + \sqrt{b^2 - 4ac}$ when $a = 2$, $b = -2$, and $c = 3$. Write the result as a complex number.

Solution

$-b + \sqrt{b^2 - 4ac}$
$-(-2) + \sqrt{(-2)^2 - 4(2)(3)}$
$\quad = 2 + \sqrt{4 - 24}$
$\quad = 2 + \sqrt{-20}$
$\quad = 2 + i\sqrt{20} = 2 + i\sqrt{4 \cdot 5}$
$\quad = 2 + 2i\sqrt{5}$

YOU TRY IT 2

Evaluate $-b + \sqrt{b^2 - 4ac}$ when $a = 1$, $b = 6$, and $c = 25$. Write the result as a complex number.

Your solution

$-6 + 8i$

Solutions on p. S27

OBJECTIVE B   *To add or subtract complex numbers*

Integrating Technology

See the Keystroke Guide: *Complex Numbers* for instructions on using a graphing calculator to perform operations on complex numbers.

> **Addition and Subtraction of Complex Numbers**
>
> To add two complex numbers, add the real parts and add the imaginary parts. To subtract two complex numbers, subtract the real parts and subtract the imaginary parts.
>
> $$(a + bi) + (c + di) = (a + c) + (b + d)i$$
> $$(a + bi) - (c + di) = (a - c) + (b - d)i$$

IN-CLASS EXAMPLES
Simplify.
4. $(3 + 5i) + (4 - 4i)$
 $7 + i$
5. $(-6 - 5i) - (5 - 9i)$
 $-11 + 4i$
6. $(7 - \sqrt{-9}) + (3 + \sqrt{-25})$
 $10 + 2i$
7. $(9 - \sqrt{-36}) - (3 + \sqrt{-81})$
 $6 - 15i$

HOW TO 2 Add: $(3 - 5i) + (2 + 3i)$

$(3 - 5i) + (2 + 3i)$
$= (3 + 2) + (-5 + 3)i$ • Add the real parts and add the imaginary parts of the complex number.
$= 5 - 2i$

HOW TO 3 Subtract: $(5 + 6i) - (7 - 3i)$

$(5 + 6i) - (7 - 3i)$
$= (5 - 7) + [6 - (-3)]i$ • Subtract the real parts and subtract the imaginary parts of the complex number.
$= -2 + 9i$

The additive inverse of the complex number $a + bi$ is $-a - bi$. The sum of these two numbers is zero.

$$(a + bi) + (-a - bi) = (a - a) + (b - b)i = 0 + 0i = 0$$

EXAMPLE 3

Simplify: $(3 + 2i) + (6 - 5i)$

Solution

$(3 + 2i) + (6 - 5i) = 9 - 3i$

YOU TRY IT 3

Simplify: $(-4 + 2i) - (6 - 8i)$

Your solution
$-10 + 10i$

EXAMPLE 4

Simplify: $(9 - \sqrt{-8}) - (5 + \sqrt{-32})$

Solution

$(9 - \sqrt{-8}) - (5 + \sqrt{-32})$
$= (9 - i\sqrt{8}) - (5 + i\sqrt{32})$
$= (9 - i\sqrt{4 \cdot 2}) - (5 + i\sqrt{16 \cdot 2})$
$= (9 - 2i\sqrt{2}) - (5 + 4i\sqrt{2})$
$= 4 - 6i\sqrt{2}$

YOU TRY IT 4

Simplify: $(16 - \sqrt{-45}) - (3 + \sqrt{-20})$

Your solution
$13 - 5i\sqrt{5}$

Solutions on p. S27

OBJECTIVE C *To multiply complex numbers*

When multiplying complex numbers, the term i^2 is often a part of the product. Recall that $i^2 = -1$.

HOW TO 4 Simplify: $2i \cdot 3i$

$2i \cdot 3i = 6i^2$ • Multiply the imaginary numbers.
$= 6(-1)$ • Replace i^2 by -1.
$= -6$ • Simplify.

Take Note

HOW TO 5 illustrates an important point. When working with an expression that contains a square root of a negative number, always rewrite the number as the product of a real number and *i* before continuing.

HOW TO 5   Simplify: $\sqrt{-6} \cdot \sqrt{-24}$

$$\sqrt{-6} \cdot \sqrt{-24} = i\sqrt{6} \cdot i\sqrt{24}$$ • Write each radical as the product of a real number and *i*.

$$= i^2\sqrt{144}$$ • Multiply the imaginary numbers.

$$= -\sqrt{144}$$ • Replace i^2 by -1.

$$= -12$$ • Simplify the radical expression.

Note from HOW TO 5 that it would have been incorrect to multiply the radicands of the two radical expressions. To illustrate,

$$\sqrt{-6} \cdot \sqrt{-24} = \sqrt{(-6)(-24)} = \sqrt{144} = 12,\ not\ -12$$

The Product Property of Radicals does *not* hold true when both radicands are negative and the index is an even number.

HOW TO 6   Simplify: $4i(3 - 2i)$

$$4i(3 - 2i) = 12i - 8i^2$$ • Use the Distributive Property to remove parentheses.

$$= 12i - 8(-1)$$ • Replace i^2 by -1.

$$= 8 + 12i$$ • Write the answer in the form $a + bi$.

IN-CLASS EXAMPLES
Simplify.
8. $(6i)(-7i)$ **42**
9. $\sqrt{-1}\ \sqrt{-16}$ **−4**
10. $2i(4 + i)$ **−2 + 8i**
11. $(7 - i)(3 + i)$ **22 + 4i**
12. $(2 + i)^2$ **3 + 4i**

To multiply two complex numbers, use FOIL.

HOW TO 7   Simplify: $(2 + 4i)(3 - 5i)$

$$(2 + 4i)(3 - 5i) = 6 - 10i + 12i - 20i^2$$ • Use the FOIL method to find the product.

$$= 6 + 2i - 20i^2$$

$$= 6 + 2i - 20(-1)$$ • Replace i^2 by -1.

$$= 26 + 2i$$ • Write the answer in the form $a + bi$.

EXAMPLE 5

Simplify: $(2i)(-5i)$

Solution

$$(2i)(-5i) = -10i^2$$
$$= (-10)(-1) = 10$$

YOU TRY IT 5

Simplify: $(-3i)(-10i)$

Your solution

−30

EXAMPLE 6

Simplify: $\sqrt{-10} \cdot \sqrt{-5}$

Solution

$$\sqrt{-10} \cdot \sqrt{-5} = i\sqrt{10} \cdot i\sqrt{5}$$
$$= i^2\sqrt{50}$$
$$= -\sqrt{25 \cdot 2} = -5\sqrt{2}$$

YOU TRY IT 6

Simplify: $-\sqrt{-8} \cdot \sqrt{-5}$

Your solution

$2\sqrt{10}$

Solutions on p. S27

EXAMPLE 7

Simplify: $3i(2 - 4i)$

Solution

$$3i(2 - 4i) = 6i - 12i^2$$
$$= 6i - 12(-1) \qquad \bullet \text{ Distributive Property}$$
$$= 12 + 6i$$

YOU TRY IT 7

Simplify: $-6i(3 + 4i)$

Your solution

$24 - 18i$

EXAMPLE 8

Simplify: $(3 - 4i)(2 + 5i)$

Solution

$$(3 - 4i)(2 + 5i) = 6 + 15i - 8i - 20i^2 \qquad \bullet \text{ FOIL}$$
$$= 6 + 7i - 20i^2$$
$$= 6 + 7i - 20(-1)$$
$$= 26 + 7i$$

YOU TRY IT 8

Simplify: $(4 - 3i)(2 - i)$

Your solution

$5 - 10i$

EXAMPLE 9

Expand: $(3 + 4i)^2$

Solution

$$(3 + 4i)^2 = (3 + 4i)(3 + 4i)$$
$$= 9 + 12i + 12i + 16i^2$$
$$= 9 + 24i + 16(-1)$$
$$= 9 + 24i - 16$$
$$= -7 + 24i$$

YOU TRY IT 9

Expand: $(5 - 3i)^2$

Your solution

$16 - 30i$

Solutions on pp. S27–S28

OBJECTIVE D *To divide complex numbers*

A rational expression containing one or more complex numbers is in simplest form when no imaginary number remains in the denominator.

HOW TO 8 Simplify: $\dfrac{2 - 3i}{2i}$

$$\frac{2 - 3i}{2i} = \frac{2 - 3i}{2i} \cdot \frac{i}{i} \qquad \bullet \text{ Multiply the numerator and denominator by } \frac{i}{i}.$$

$$= \frac{2i - 3i^2}{2i^2}$$

$$= \frac{2i - 3(-1)}{2(-1)} \qquad \bullet \text{ Replace } i^2 \text{ by } -1.$$

$$= \frac{3 + 2i}{-2} = -\frac{3}{2} - i \qquad \bullet \text{ Simplify. Write the answer in the form } a + bi.$$

IN-CLASS EXAMPLES
Simplify.

13. $\dfrac{3 - 2i}{6i}$ $\quad -\dfrac{1}{3} - \dfrac{1}{2}i$

14. $\dfrac{8}{5 + 4i}$ $\quad \dfrac{40}{41} - \dfrac{32}{41}i$

15. $\dfrac{1 - 2i}{2 + i}$ $\quad -i$

16. $\dfrac{\sqrt{-8}}{\sqrt{6} - \sqrt{-2}}$

$\quad -\dfrac{1}{2} + \dfrac{\sqrt{3}}{2}i$

To divide a complex number when the denominator is of the form $a + bi$, the *conjugate* of the denominator is used.

> ### Conjugate of a Complex Number
>
> The **conjugate** of $a + bi$ is $a - bi$, and the conjugate of $a - bi$ is $a + bi$. The product of the conjugates is $(a + bi)(a - bi) = a^2 + b^2$.
>
> **EXAMPLES**
>
> 1. The conjugate of $2 + 5i$ is $2 - 5i$. The product of the conjugates is
> $(2 + 5i)(2 - 5i) = 2^2 + 5^2 = 29$.
> 2. The conjugate of $3 - 4i$ is $3 + 4i$. The product of the conjugates is
> $(3 - 4i)(3 + 4i) = 3^2 + 4^2 = 25$.
> 3. The conjugate of $-5 + i$ is $-5 - i$. The product of the conjugates is
> $(-5 + i)(-5 - i) = (-5)^2 + 1^2 = 26$.

INSTRUCTOR NOTE
Students think that division of complex numbers is somehow different from division of other types of numbers. Show them that just as $\dfrac{12}{4} = 3$ because

$4 \cdot 3 = 12$, $\dfrac{3 + 2i}{1 + i} = \dfrac{5}{2} - \dfrac{1}{2}i$

because

$(1 + i)\left(\dfrac{5}{2} - \dfrac{1}{2}i\right) = 3 + 2i$.

HOW TO 9 Simplify: $\dfrac{3 + 2i}{1 + i}$

$\dfrac{3 + 2i}{1 + i} = \dfrac{3 + 2i}{1 + i} \cdot \dfrac{1 - i}{1 - i}$

• Multiply the numerator and denominator by the conjugate of $1 + i$, $1 - i$.

$\quad = \dfrac{3 - 3i + 2i - 2i^2}{1^2 + 1^2}$

$\quad = \dfrac{3 - i - 2(-1)}{1 + 1}$

• Replace i^2 by -1 and simplify.

$\quad = \dfrac{5 - i}{2} = \dfrac{5}{2} - \dfrac{1}{2}i$

• Write the answer in the form $a + bi$.

EXAMPLE 10

Simplify: $\dfrac{5 + 4i}{3i}$

Solution

$\dfrac{5 + 4i}{3i} = \dfrac{5 + 4i}{3i} \cdot \dfrac{i}{i} = \dfrac{5i + 4i^2}{3i^2}$

$\quad = \dfrac{5i + 4(-1)}{3(-1)} = \dfrac{-4 + 5i}{-3} = \dfrac{4}{3} - \dfrac{5}{3}i$

YOU TRY IT 10

Simplify: $\dfrac{2 - 3i}{4i}$

Your solution

$-\dfrac{3}{4} - \dfrac{1}{2}i$

EXAMPLE 11

Simplify: $\dfrac{5 - 3i}{4 + 2i}$

Solution

$\dfrac{5 - 3i}{4 + 2i} = \dfrac{5 - 3i}{4 + 2i} \cdot \dfrac{4 - 2i}{4 - 2i}$

$\quad = \dfrac{20 - 10i - 12i + 6i^2}{4^2 + 2^2}$

$\quad = \dfrac{20 - 22i + 6(-1)}{16 + 4}$

$\quad = \dfrac{14 - 22i}{20} = \dfrac{7 - 11i}{10} = \dfrac{7}{10} - \dfrac{11}{10}i$

YOU TRY IT 11

Simplify: $\dfrac{2 + 5i}{3 - 2i}$

Your solution

$-\dfrac{4}{13} + \dfrac{19}{13}i$

Solutions on p. S28

9.5 EXERCISES

✔ **Concept Check**

SUGGESTED ASSIGNMENT
Exercises 1–6; Exercises 7–75, odds;
More challenging exercises: Exercises 76–79

1. 🔲 What is an imaginary number? What is a complex number?

2. 🔲 Are all real numbers also complex numbers? Are all complex numbers also real numbers?

For Exercises 3 to 6, name the real part and the imaginary part of the complex number.

3. $3 + 7i$

3; 7

4. $2 - 3i$

2; −3

5. 7

7; 0

6. $4i$

0; 4

OBJECTIVE A *To simplify a complex number*

For Exercises 7 to 14, simplify.

7. $\sqrt{-25}$

$5i$

8. $\sqrt{-64}$

$8i$

9. $\sqrt{-98}$

$7i\sqrt{2}$

10. $\sqrt{-72}$

$6i\sqrt{2}$

11. $\dfrac{6 + \sqrt{-4}}{2}$

$3 + i$

12. $\dfrac{12 - \sqrt{-24}}{4}$

$3 - \dfrac{\sqrt{6}}{2}i$

13. $\dfrac{6 - 5\sqrt{-8}}{4}$

$\dfrac{3}{2} - \dfrac{5\sqrt{2}}{2}i$

14. $\dfrac{5 + 3\sqrt{-25}}{4}$

$\dfrac{5}{4} + \dfrac{15}{4}i$

For Exercises 15 to 26, evaluate $-b + \sqrt{b^2 - 4ac}$ for the given values of a, b, and c. Write the result as a complex number.

15. $a = 1, b = 4, c = 5$

$-4 + 2i$

16. $a = 1, b = -6, c = 13$

$6 + 4i$

17. $a = 2, b = -4, c = 10$

$4 + 8i$

18. $a = 4, b = -12, c = 45$

$12 + 24i$

19. $a = 3, b = -8, c = 6$

$8 + 2i\sqrt{2}$

20. $a = 3, b = -2, c = 9$

$2 + 2i\sqrt{26}$

21. $a = 4, b = 2, c = 7$

$-2 + 6i\sqrt{3}$

22. $a = 4, b = 5, c = 10$

$-5 + 3i\sqrt{15}$

23. $a = -2, b = 5, c = -6$

$-5 + i\sqrt{23}$

24. $a = -1, b = 4, c = -29$

$-4 + 10i$

25. $a = -3, b = 4, c = -6$

$-4 + 2i\sqrt{14}$

26. $a = -5, b = 1, c = -5$

$-1 + 3i\sqrt{11}$

OBJECTIVE B *To add or subtract complex numbers*

For Exercises 27 to 36, simplify.

27. $(2 + 4i) + (6 - 5i)$

$8 - i$

28. $(6 - 9i) + (4 + 2i)$

$10 - 7i$

29. $(-2 - 4i) - (6 - 8i)$

$-8 + 4i$

30. $(3 - 5i) + (8 - 2i)$
$11 - 7i$

31. $(8 - 2i) - (2 + 4i)$
$6 - 6i$

32. $(5 - 5i) - (11 - 6i)$
$-6 + i$

33. $5 + (6 - 4i)$
$11 - 4i$

34. $-7 + (3 + 5i)$
$-4 + 5i$

35. $3i - (6 + 5i)$
$-6 - 2i$

36. $(7 + 3i) - 8i$
$7 - 5i$

37. If the sum of two complex numbers is an imaginary number, what must be true of the complex numbers?
The real parts of the complex numbers are additive inverses.

38. If the sum of two complex numbers is a real number, what must be true of the complex numbers?
The imaginary parts of the complex numbers are additive inverses.

OBJECTIVE C   *To multiply complex numbers*

For Exercises 39 to 56, simplify.

39. $(7i)(-9i)$
63

40. $(-6i)(-4i)$
-24

41. $\sqrt{-2}\sqrt{-8}$
-4

42. $\sqrt{-5}\sqrt{-45}$
-15

43. $(5 + 2i)(5 - 2i)$
29

44. $(3 + 8i)(3 - 8i)$
73

45. $2i(6 + 2i)$
$-4 + 12i$

46. $-3i(4 - 5i)$
$-15 - 12i$

47. $-i(4 - 3i)$
$-3 - 4i$

48. $i(6 + 2i)$
$-2 + 6i$

49. $(5 - 2i)(3 + i)$
$17 - i$

50. $(2 - 4i)(2 - i)$
$-10i$

51. $(6 + 5i)(3 + 2i)$
$8 + 27i$

52. $(4 - 7i)(2 + 3i)$
$29 - 2i$

53. $(2 + 5i)^2$
$-21 + 20i$

54. $(3 - 4i)^2$
$-7 - 24i$

55. $\left(\dfrac{6}{5} + \dfrac{3}{5}i\right)\left(\dfrac{2}{3} - \dfrac{1}{3}i\right)$
1

56. $(2 - i)\left(\dfrac{2}{5} + \dfrac{1}{5}i\right)$
1

57. True or false? For all real numbers a and b, the product $(a + bi)(a - bi)$ is a positive real number.
True

58. Given that $\left(\dfrac{\sqrt{2}}{2} + \dfrac{\sqrt{2}}{2}i\right)^2 = i$ (try to verify this statement), what is a square root of i?
$\dfrac{\sqrt{2}}{2} + \dfrac{\sqrt{2}}{2}i$

OBJECTIVE D *To divide complex numbers*

For Exercises 59 to 73, simplify.

59. $\dfrac{3}{i}$

$-3i$

60. $\dfrac{4}{5i}$

$-\dfrac{4}{5}i$

61. $\dfrac{2-3i}{-4i}$

$\dfrac{3}{4}+\dfrac{1}{2}i$

62. $\dfrac{16+5i}{-3i}$

$-\dfrac{5}{3}+\dfrac{16}{3}i$

63. $\dfrac{4}{5+i}$

$\dfrac{10}{13}-\dfrac{2}{13}i$

64. $\dfrac{6}{5+2i}$

$\dfrac{30}{29}-\dfrac{12}{29}i$

65. $\dfrac{2}{2-i}$

$\dfrac{4}{5}+\dfrac{2}{5}i$

66. $\dfrac{5}{4-i}$

$\dfrac{20}{17}+\dfrac{5}{17}i$

67. $\dfrac{1-3i}{3+i}$

$-i$

68. $\dfrac{2+12i}{5+i}$

$\dfrac{11}{13}+\dfrac{29}{13}i$

69. $\dfrac{3i}{1+4i}$

$\dfrac{12}{17}+\dfrac{3}{17}i$

70. $\dfrac{-2i}{2-3i}$

$\dfrac{6}{13}-\dfrac{4}{13}i$

71. $\dfrac{2-3i}{3+i}$

$\dfrac{3}{10}-\dfrac{11}{10}i$

72. $\dfrac{3+5i}{1-i}$

$-1+4i$

73. $\dfrac{5+3i}{3-i}$

$\dfrac{6}{5}+\dfrac{7}{5}i$

74. True or false? The quotient of two imaginary numbers is an imaginary number.
False

75. True or false? The reciprocal of an imaginary number is an imaginary number.
True

Critical Thinking

For Exercises 76 to 79, determine whether the complex number is a solution of the equation.

76. $x^2-4x+13=0;\ 2+3i$
Yes

77. $x^2-10x+29=0;\ 5-3i$
No

78. $x^2-8x+19=0;\ 4-i\sqrt{3}$
Yes

79. $x^2-2x+4=0;\ 1-i\sqrt{3}$
Yes

Projects or Group Activities

80. Note the pattern when successive powers of i are simplified. Use the pattern to complete the statement.

$i^1=i$

$i^2=-1$

$i^3=i^2\cdot i=-i$

$i^4=i^2\cdot i^2=(-1)(-1)=1$

$i^5=i\cdot i^4=i(1)=i$

$i^6=i^2\cdot i^4=-1$

$i^7=i^3\cdot i^4=-i$

$i^8=i^4\cdot i^4=1$

When the exponent on i is a multiple of 4, the expression is equal to ____1____.

For Exercises 81 to 84, use the pattern in Exercise 80 to simplify the power of i.

81. i^{12}
1

82. i^{31}
$-i$

83. i^{57}
i

84. i^{82}
-1

QUICK QUIZ

1. Simplify: $\sqrt{-48}$
   $4i\sqrt{3}$ [9.5A]
2. Evaluate
   $-b+\sqrt{b^2-4ac}$ when
   $a=2, b=6,$ and $c=5$.
   $-6+2i$ [9.5A]

Simplify.
3. $(-3-3i)-(7-7i)$
   $-10+4i$ [9.5B]
4. $(-4+7i)-(6-8i)$
   $-10+15i$ [9.5B]
5. $(5-2i)(3+i)$
   $17-i$ [9.5C]
6. $(5-i)^2$
   $24-10i$ [9.5C]
7. $\dfrac{4-6i}{-2i}$ $3+2i$ [9.5D]
8. $\dfrac{5}{5+2i}$ $\dfrac{25}{29}-\dfrac{10}{29}i$
   [9.5D]
9. $\dfrac{1-5i}{5+i}$ $-i$ [9.5D]

CHAPTER

9 Summary

Key Words

Examples

$a^{1/n}$ is the **nth root of a.** [9.1A, p. 520]

$16^{1/2} = 4$ because $4^2 = 16$.

The expression $\sqrt[n]{a}$ is another symbol for the *n*th root of *a*. In the expression $\sqrt[n]{a}$, the symbol $\sqrt{}$ is called a **radical,** *n* is the **index** of the radical, and *a* is the **radicand.** [9.1B, p. 522]

$125^{1/3} = \sqrt[3]{125} = 5$
The index is 3, and the radicand is 125.

The symbol $\sqrt{}$ is used to indicate the positive square root, or **principal square root**, of a number. [9.1C, p. 524]

$\sqrt{16} = 4$
$-\sqrt{16} = -4$

The procedure used to remove a radical from the denominator of a radical expression is called **rationalizing the denominator.** [9.3B, p. 537]

$\dfrac{2}{1 - \sqrt{3}} = \dfrac{2}{1 - \sqrt{3}} \cdot \dfrac{1 + \sqrt{3}}{1 + \sqrt{3}}$

$= \dfrac{2(1 + \sqrt{3})}{(1 - \sqrt{3})(1 + \sqrt{3})}$

$= \dfrac{2 + 2\sqrt{3}}{1 - 3} = -1 - \sqrt{3}$

The expressions $a + b$ and $a - b$ are called **conjugates** of each other. The product of conjugates of the form $(a + b)(a - b)$ is $a^2 - b^2$. [9.3B, p. 537]

$(x + \sqrt{3})(x - \sqrt{3}) = x^2 - (\sqrt{3})^2$
$= x^2 - 3$

A **radical equation** is an equation that contains a variable expression in a radicand. [9.4A, p. 543]

$\sqrt{x - 2} - 3 = 6$ is a radical equation.

A **complex number** is a number of the form $a + bi$, where a and b are real numbers and $i = \sqrt{-1}$. For the complex number $a + bi$, a is the **real part** of the complex number and b is the **imaginary part** of the complex number. [9.5A, p. 549]

$3 + 2i$ is a complex number.
3 is the real part and 2 is the imaginary part of the complex number.

Essential Rules and Procedures

Examples

Rule for Rational Exponents [9.1A, p. 520]
If m and n are positive integers and $a^{1/n}$ is a real number, then $a^{m/n} = (a^{1/n})^m$.

$8^{2/3} = (8^{1/3})^2 = 2^2 = 4$

Definition of *n*th Root of *a* [9.1B, p. 522]

If *a* is a real number, then $a^{1/n} = \sqrt[n]{a}$.

$x^{1/3} = \sqrt[3]{x}$

Writing Exponential Expressions as Radical Expressions [9.1B, p. 522]

If $a^{1/n}$ is a real number, then $a^{m/n} = a^{m \cdot 1/n} = (a^m)^{1/n} = \sqrt[n]{a^m}$.

The expression $a^{m/n}$ can also be written $(\sqrt[n]{a})^m$.

$b^{3/4} = \sqrt[4]{b^3}$

$8^{2/3} = (\sqrt[3]{8})^2 = 2^2 = 4$

Product Property of Radicals [9.2A, p. 530]

If $\sqrt[n]{a}$ and $\sqrt[n]{b}$ are positive real numbers, then $\sqrt[n]{a} \cdot \sqrt[n]{b} = \sqrt[n]{ab}$.

$\sqrt{7} \cdot \sqrt{5} = \sqrt{7 \cdot 5} = \sqrt{35}$

Quotient Property of Radicals [9.3B, p. 536]

If $\sqrt[n]{a}$ and $\sqrt[n]{b}$ are positive real numbers and $b \neq 0$, then

$\dfrac{\sqrt[n]{a}}{\sqrt[n]{b}} = \sqrt[n]{\dfrac{a}{b}}$.

$\dfrac{\sqrt{42}}{\sqrt{6}} = \sqrt{\dfrac{42}{6}} = \sqrt{7}$

Property of Raising Each Side of an Equation to a Power [9.4A, p. 543]

If $a = b$, then $a^n = b^n$.

If $\sqrt{x} = 4$, then $(\sqrt{x})^2 = 4^2$. Thus, $x = 16$.

Pythagorean Theorem [9.4B, p. 545]

The square of the hypotenuse of a right triangle is equal to the sum of the squares of the two legs.

$c^2 = a^2 + b^2$

$5^2 = 3^2 + 4^2$

Definition of $\sqrt{-a}$ [9.5A, p. 549]

If *a* is a positive real number, then $\sqrt{-a} = i\sqrt{a}$.

$\sqrt{-8} = i\sqrt{8} = 2i\sqrt{2}$

Addition and Subtraction of Complex Numbers [9.5B, p. 550]

$(a + bi) + (c + di) = (a + c) + (b + d)i$
$(a + bi) - (c + di) = (a - c) + (b - d)i$

$(2 + 4i) + (3 + 6i)$
$= (2 + 3) + (4 + 6)i = 5 + 10i$
$(4 + 3i) - (7 + 4i)$
$= (4 - 7) + (3 - 4)i = -3 - i$

Multiplication of Complex Numbers [9.5C, p. 552]

To multiply two complex numbers, use FOIL.

$(2 - 3i)(5 + 4i)$
$= 10 + 8i - 15i - 12i^2$
$= 10 - 7i - 12(-1) = 22 - 7i$

Division of Complex Numbers [9.5D, p. 554]

To divide complex numbers, multiply the numerator and denominator by the conjugate of the denominator.

$\dfrac{9 - 7i}{3 + 2i} = \dfrac{9 - 7i}{3 + 2i} \cdot \dfrac{3 - 2i}{3 - 2i}$

$= \dfrac{13 - 39i}{13} = 1 - 3i$

Unless otherwise noted, all content on this page is © Cengage Learning.

CHAPTER

9 Review Exercises

1. Simplify: $(16x^{-4}y^{12})^{1/4}(100x^6y^{-2})^{1/2}$
$20x^2y^2$ [9.1A]

2. Solve: $\sqrt[4]{3x-5}=2$
7 [9.4A]

3. Multiply: $(6-5i)(4+3i)$
$39-2i$ [9.5C]

4. Rewrite $7y\sqrt[3]{x^2}$ as an exponential expression.
$7x^{2/3}y$ [9.1B]

5. Multiply: $(\sqrt{3}+8)(\sqrt{3}-2)$
$6\sqrt{3}-13$ [9.3A]

6. Solve: $\sqrt{4x+9}+10=11$
-2 [9.4A]

7. Divide: $\dfrac{x^{-3/2}}{x^{7/2}}$
$\dfrac{1}{x^5}$ [9.1A]

8. Simplify: $\dfrac{8}{\sqrt{3y}}$
$\dfrac{8\sqrt{3y}}{3y}$ [9.3B]

9. Simplify: $\sqrt[3]{-8a^6b^{12}}$
$-2a^2b^4$ [9.1C]

10. Simplify: $\sqrt{50a^4b^3}-ab\sqrt{18a^2b}$
$2a^2b\sqrt{2b}$ [9.2B]

11. Simplify: $\dfrac{14}{4-\sqrt{2}}$
$4+\sqrt{2}$ [9.3B]

12. Simplify: $\dfrac{5+2i}{3i}$
$\dfrac{2}{3}-\dfrac{5}{3}i$ [9.5D]

13. Simplify: $\sqrt{18a^3b^6}$
$3ab^3\sqrt{2a}$ [9.2A]

14. Subtract: $(17+8i)-(15-4i)$
$2+12i$ [9.5B]

15. Simplify: $3x\sqrt[3]{54x^8y^{10}}-2x^2y\sqrt[3]{16x^5y^7}$
$5x^3y^3\sqrt[3]{2x^2y}$ [9.2B]

16. Multiply: $\sqrt[3]{16x^4y}\,\sqrt[3]{4xy^5}$
$4xy^2\sqrt[3]{x^2}$ [9.3A]

17. Multiply: $i(3-7i)$
$7+3i$ [9.5C]

18. Simplify: $\dfrac{(4a^{-2/3}b^4)^{-1/2}}{(a^{-1/6}b^{3/2})^2}$
$\dfrac{a^{2/3}}{2b^5}$ [9.1A]

19. Simplify: $\sqrt[5]{-64a^8b^{12}}$
$-2ab^2\sqrt[5]{2a^3b^2}$ [9.2A]

20. Divide: $\dfrac{5 + 9i}{1 - i}$
$-2 + 7i$ [9.5D]

21. Multiply: $\sqrt{-12}\,\sqrt{-6}$
$-6\sqrt{2}$ [9.5C]

22. Solve: $\sqrt{x - 5} + \sqrt{x + 6} = 11$
30   [9.4A]

23. Simplify: $\sqrt[4]{81a^8b^{12}}$
$3a^2b^3$ [9.1C]

24. Simplify: $\dfrac{9}{\sqrt[3]{3x}}$
$\dfrac{3\sqrt[3]{9x^2}}{x}$ [9.3B]

25. Subtract: $(-8 + 3i) - (4 - 7i)$
$-12 + 10i$ [9.5B]

26. Expand: $(2 + \sqrt{2x - 1})^2$
$2x + 4\sqrt{2x - 1} + 3$ [9.3A]

27. Simplify: $4x\sqrt{12x^2y} + \sqrt{3x^4y} - x^2\sqrt{27y}$
$6x^2\sqrt{3y}$ [9.2B]

28. Simplify: $81^{-1/4}$
$\dfrac{1}{3}$ [9.1A]

29. Simplify: $(a^{16})^{-5/8}$
$\dfrac{1}{a^{10}}$ [9.1A]

30. Simplify: $-\sqrt{49x^6y^{16}}$
$-7x^3y^8$ [9.1C]

31. Rewrite $4a^{2/3}$ as a radical expression.
$4\sqrt[3]{a^2}$ [9.1B]

32. Simplify: $(9x^2y^4)^{-1/2}(x^6y^6)^{1/3}$
$\dfrac{x}{3}$ [9.1A]

33. Simplify: $\sqrt[4]{x^6y^8z^{10}}$
$xy^2z^2\sqrt[4]{x^2z^2}$ [9.2A]

34. Simplify: $\sqrt{54} + \sqrt{24}$
$5\sqrt{6}$ [9.2B]

35. Simplify: $\sqrt{48x^5y} - x\sqrt{80x^2y}$
$4x^2\sqrt{3xy} - 4x^2\sqrt{5y}$ [9.2B]

36. Simplify: $\sqrt{32}\,\sqrt{50}$
40   [9.3A]

37. Simplify: $\sqrt{3x}(3 + \sqrt{3x})$
$3x + 3\sqrt{3x}$ [9.3A]

38. Simplify: $\dfrac{\sqrt{125x^6}}{\sqrt{5x^3}}$
$5x\sqrt{x}$ [9.3B]

39. Simplify: $\dfrac{2 - 3\sqrt{7}}{6 - \sqrt{7}}$

$\dfrac{-9 - 16\sqrt{7}}{29}$ [9.3B]

40. Simplify: $\sqrt{-36}$

$6i$ [9.5A]

41. Evaluate $-b + \sqrt{b^2 - 4ac}$ when $a = 1$, $b = -8$, and $c = 25$.

$8 + 6i$ [9.5A]

42. Evaluate $-b + \sqrt{b^2 - 4ac}$ when $a = 1$, $b = 2$, and $c = 9$.

$-2 + 4i\sqrt{2}$ [9.5A]

43. Add: $(5 + 2i) + (4 - 3i)$

$9 - i$ [9.5B]

44. Simplify: $(3 + 2\sqrt{5})(3 - 2\sqrt{5})$

-11 [9.3A]

45. Simplify: $(3 - 9i) - 7$

$-4 - 9i$ [9.5B]

46. Expand: $(4 - i)^2$

$15 - 8i$ [9.5C]

47. Simplify: $\dfrac{-6}{i}$

$6i$ [9.5D]

48. Divide: $\dfrac{7}{2 - i}$

$\dfrac{14}{5} + \dfrac{7}{5}i$ [9.5D]

49. Solve: $\sqrt{2x - 7} + 2 = 5$

8 [9.4A]

50. Solve: $\sqrt[3]{9x} = -6$

-24 [9.4A]

51. Geometry   Find the width of a rectangle that has a diagonal of 13 in. and a length of 12 in.   5 in.   [9.4B]

52. Energy   The velocity of the wind determines the amount of power generated by a windmill. A typical equation for this relationship is $v = 4.05 \sqrt[3]{P}$, where v is the velocity in miles per hour and P is the power in watts. Find the amount of power generated by a 20-mile-per-hour wind. Round to the nearest whole number.

120 watts   [9.4B]

53. Automotive Technology   Find the distance required for a car to reach a velocity of 88 ft/s when the acceleration is 16 ft/s². Use the equation $v = \sqrt{2as}$, where v is the velocity in feet per second, a is the acceleration, and s is the distance in feet.

242 ft   [9.4B]

54. Home Maintenance   A 12-foot ladder is leaning against a house in preparation for washing the windows. How far from the house is the bottom of the ladder when the top of the ladder touches the house 10 ft above the ground? Round to the nearest hundredth.   6.63 ft   [9.4B]

Unless otherwise noted, all content on this page is © Cengage Learning.

CHAPTER

9 | TEST

1. Write $\dfrac{1}{2}\sqrt[4]{x^3}$ as an exponential expression.

   $\dfrac{1}{2}x^{3/4}$ [9.1B]

2. Simplify:

   $\sqrt[3]{54x^7y^3} - x\sqrt[3]{128x^4y^3} - x^2\sqrt[3]{2xy^3}$

   $-2x^2y\sqrt[3]{2x}$ [9.2B]

3. Write $3y^{2/5}$ as a radical expression.

   $3\sqrt[5]{y^2}$ [9.1B]

4. Multiply: $(2 + 5i)(4 - 2i)$

   $18 + 16i$ [9.5C]

5. Expand: $(3 - 2\sqrt{x})^2$

   $4x - 12\sqrt{x} + 9$ [9.3A]

6. Simplify: $\dfrac{r^{2/3}\, r^{-1}}{r^{-1/2}}$

   $r^{1/6}$ [9.1A]

7. Solve: $\sqrt{x + 12} - \sqrt{x} = 2$

   4 [9.4A]

8. Multiply: $\sqrt[4]{4a^5b^3}\ \sqrt[4]{8a^3b^7}$

   $2a^2b^2\sqrt[4]{2b^2}$ [9.3A]

9. Multiply: $\sqrt{3x}(\sqrt{x} - \sqrt{25x})$

   $-4x\sqrt{3}$ [9.3A]

10. Subtract: $(5 - 2i) - (8 - 4i)$

    $-3 + 2i$ [9.5B]

11. Simplify: $\sqrt{32x^4y^7}$

    $4x^2y^3\sqrt{2y}$ [9.2A]

12. Multiply: $(2\sqrt{3} + 4)(3\sqrt{3} - 1)$

    $14 + 10\sqrt{3}$ [9.3A]

13. Simplify: $\sqrt{-5}\ \sqrt{-20}$

    -10 [9.5C]

14. Simplify: $\dfrac{4 - 2\sqrt{5}}{2 - \sqrt{5}}$

    2 [9.3B]

15. Add: $\sqrt{18a^3} + a\sqrt{50a}$

$8a\sqrt{2a}$ [9.2B]

16. Multiply: $(\sqrt{a} - 3\sqrt{b})(2\sqrt{a} + 5\sqrt{b})$

$2a - \sqrt{ab} - 15b$ [9.3A]

17. Simplify: $\dfrac{(2x^{1/3}y^{-2/3})^6}{(x^{-4}y^8)^{1/4}}$

$\dfrac{64x^3}{y^6}$ [9.1A]

18. Simplify: $\dfrac{10x}{\sqrt[3]{5x^2}}$

$2\sqrt[3]{25x}$ [9.3B]

19. Simplify: $\dfrac{2 + 3i}{1 - 2i}$

$-\dfrac{4}{5} + \dfrac{7}{5}i$ [9.5D]

20. Solve: $\sqrt[3]{2x - 2} + 4 = 2$

-3 [9.4A]

21. Simplify: $\left(\dfrac{4a^4}{b^2}\right)^{-3/2}$

$\dfrac{b^3}{8a^6}$ [9.1A]

22. Simplify: $\sqrt[3]{27a^4b^3c^7}$

$3abc^2\sqrt[3]{ac}$ [9.2A]

23. Divide: $\dfrac{\sqrt{32x^5y}}{\sqrt{2xy^3}}$

$\dfrac{4x^2}{y}$ [9.3B]

24. Simplify: $\dfrac{5x}{\sqrt{5x}}$

$\sqrt{5x}$ [9.3B]

25. Physics An object is dropped from a high building. Find the distance the object has fallen when its speed reaches 192 ft/s. Use the equation $v = \sqrt{64d}$, where v is the speed of the object in feet per second and d is the distance in feet.

576 ft   [9.4B]

Cumulative Review Exercises

1. Simplify: $2^3 \cdot 3 - 4(3 - 4 \cdot 5)$
92 [1.3A]

2. Evaluate $4a^2b - a^3$ when $a = -2$ and $b = 3$.
56 [1.4A]

3. Simplify: $-3(4x - 1) - 2(1 - x)$
$-10x + 1$ [1.4D]

4. Solve: $5 - \dfrac{2}{3}x = 4$
$\dfrac{3}{2}$ [2.2A]

5. Solve: $2[4 - 2(3 - 2x)] = 4(1 - x)$
$\dfrac{2}{3}$ [2.2C]

6. Solve: $6x - 3(2x + 2) > 3 - 3(x + 2)$
Write the solution set in set-builder notation.
$\{x \mid x > 1\}$ [2.5A]

7. Solve: $2 + |4 - 3x| = 5$
$\dfrac{1}{3}, \dfrac{7}{3}$ [2.6A]

8. Solve: $|2x + 3| \leq 9$
$\{x \mid -6 \leq x \leq 3\}$ [2.6B]

9. Find the area of the triangle shown in the figure below.

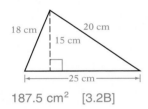

187.5 cm² [3.2B]

10. Find the volume of a rectangular solid with a length of 3.5 ft, a width of 2 ft, and a height of 2 ft.
14 ft³ [3.3A]

11. Graph $3x - 2y = -6$. State the slope and y-intercept.

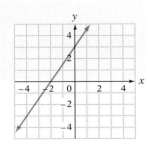

$m = \dfrac{3}{2}, b = 3$ [4.3B]

12. Graph the solution set of $3x + 2y \leq 4$.

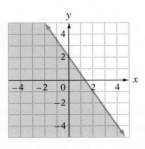

[4.7A]

13. Find the equation of the line that passes through the points $P_1(2, 3)$ and $P_2(-1, 2)$.
$y = \dfrac{1}{3}x + \dfrac{7}{3}$ [4.5B]

14. Solve by the addition method:
$$2x - y = 4$$
$$-2x + 3y = 0$$
$(3, 2)$ [5.2A]

15. Simplify: $(2^{-1}x^2y^{-6})(2^{-1}y^{-4})^{-2}$
$2x^2y^2$ [6.1B]

16. Factor: $81x^2 - y^2$
$(9x + y)(9x - y)$ [7.4A]

Unless otherwise noted, all content on this page is © Cengage Learning.

17. Factor: $x^5 + 2x^3 - 3x$

$x(x^2 + 3)(x + 1)(x - 1)$ [7.4D]

18. Solve $P = \dfrac{R - C}{n}$ for C.

$C = R - nP$ [8.6A]

19. Simplify: $\left(\dfrac{x^{-2/3}y^{1/2}}{y^{-1/3}}\right)^6$

$\dfrac{y^5}{x^4}$ [9.1A]

20. Subtract: $\sqrt{40x^3} - x\sqrt{90x}$

$-x\sqrt{10x}$ [9.2B]

21. Multiply: $(\sqrt{3} - 2)(\sqrt{3} - 5)$

$13 - 7\sqrt{3}$ [9.3A]

22. Simplify: $\dfrac{4}{\sqrt{6} - \sqrt{2}}$

$\sqrt{6} + \sqrt{2}$ [9.3B]

23. Simplify: $\dfrac{2i}{3 - i}$

$-\dfrac{1}{5} + \dfrac{3}{5}i$ [9.5D]

24. Solve: $\sqrt[3]{3x - 4} + 5 = 1$

-20 [9.4A]

25. The two triangles are similar triangles. Find the length of side DE.

27 m   [8.5C]

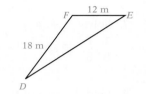

26. Investments   An investor has a total of $10,000 deposited in two simple interest accounts. On one account, the annual simple interest rate is 3.5%. On the second account, the annual simple interest rate is 4.5%. How much is invested in the 3.5% account if the total annual interest earned is $425?

$2500   [5.1C]

27. Uniform Motion   A sales executive traveled 25 mi by car and then an additional 625 mi by plane. The rate of the plane was five times greater than the rate of the car. The total time of the trip was 3 h. Find the rate of the plane.

250 mph   [8.7B]

28. ● **Astronomy**   How long does it take for light to travel from the moon to Earth when the moon is 232,500 mi from Earth? Light travels at a rate of 1.86×10^5 mi/s.

1.25 s   [6.1D]

29. Oceanography   How far above the water would a submarine periscope have to be to locate a ship 7 mi away? The equation for the distance in miles that the lookout can see is $d = \sqrt{1.5h}$, where h is the height in feet above the surface of the water. Round to the nearest tenth of a foot.

32.7 ft   [9.4B]

30. Investments   The graph shows the amount invested and the annual interest income from an investment. Find the slope of the line between the two points shown on the graph. Then write a sentence that states the meaning of the slope.

$m = 0.08$

The annual interest income is 8% of the investment.   [4.4A]

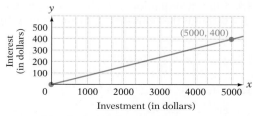

Unless otherwise noted, all content on this page is © Cengage Learning.

Quadratic Equations

OBJECTIVES

SECTION 10.1
A To solve a quadratic equation by factoring
B To solve a quadratic equation by taking square roots

SECTION 10.2
A To solve a quadratic equation by completing the square
B To solve a quadratic equation by using the quadratic formula

SECTION 10.3
A To solve an equation that is quadratic in form
B To solve a radical equation that is reducible to a quadratic equation
C To solve a rational equation that is reducible to a quadratic equation

SECTION 10.4
A To solve application problems

SECTION 10.5
A To solve a nonlinear inequality

Focus on Success

Do you get nervous before taking a math test? The more prepared you are, the less nervous you will be. We suggest you study the Chapter Summary. Then do the Chapter Review Exercises to test your understanding of the material in the chapter. If you have trouble with any of the questions, restudy the objectives the questions are taken from. Take the Chapter Test in a quiet place, working on it as if it were the actual exam. (See Ace the Test, page AIM-11.)

© shock/Shutterstock.com

Prep Test

Are you ready to succeed in this chapter? Take the Prep Test below to find out if you are ready to learn the new material.

1. Simplify: $\sqrt{18}$
   $3\sqrt{2}$ [1.2F]

2. Simplify: $\sqrt{-9}$
   $3i$ [9.5A]

3. Simplify: $\dfrac{3x - 2}{x - 1} - 1$
   $\dfrac{2x - 1}{x - 1}$ [8.2B]

4. Evaluate $b^2 - 4ac$ when $a = 2$, $b = -4$, and $c = 1$.
   8 [1.4A]

5. Is $4x^2 + 28x + 49$ a perfect square trinomial?
   Yes [7.4A]

6. Factor: $4x^2 - 4x + 1$
   $(2x - 1)^2$ [7.4A]

7. Factor: $9x^2 - 4$
   $(3x + 2)(3x - 2)$ [7.4A]

8. Graph $\{x \mid x < -1\} \cap \{x \mid x < 4\}$.

   $-5\ -4\ -3\ -2\ -1\ \ 0\ \ 1\ \ 2\ \ 3\ \ 4\ \ 5$
   [1.5B]

9. Solve: $x(x - 1) = x + 15$
   -3 and 5 [7.5A]

10. Solve: $\dfrac{4}{x - 3} = \dfrac{16}{x}$
    4 [8.5A]

Unless otherwise noted, all content on this page is © Cengage Learning.

10.1 Solving Quadratic Equations by Factoring or by Taking Square Roots

OBJECTIVE A *To solve a quadratic equation by factoring*

INSTRUCTOR NOTE

The material on solving a quadratic equation by factoring was covered earlier in the text. It is reviewed here for completeness.

Recall that a *quadratic equation* is an equation of the form $ax^2 + bx + c = 0$, where a and b are coefficients, c is a constant, and $a \neq 0$.

$$\text{Quadratic Equations} \begin{cases} 3x^2 - x + 2 = 0, & a = 3, & b = -1, & c = 2 \\ -x^2 + 4 = 0, & a = -1, & b = 0, & c = 4 \\ 6x^2 - 5x = 0, & a = 6, & b = -5, & c = 0 \end{cases}$$

A quadratic equation is in *standard form* when the polynomial is in descending order and equal to zero. Because the degree of the polynomial $ax^2 + bx + c$ is 2, a quadratic equation is also called a **second-degree equation.**

INSTRUCTOR NOTE

If you ask your students to write an equation such as $4x = 5x^2 - 3$ in standard form, some students may ask which side to make zero. Show them that it does not matter.

As we discussed earlier, quadratic equations sometimes can be solved by using the Principle of Zero Products. This method is reviewed here.

Principle of Zero Products

If $ab = 0$, then $a = 0$ or $b = 0$. This principle states that if the product of two factors is zero, then at least one of the factors must equal zero.

EXAMPLES

1. Suppose $5x = 0$. The factors are 5 and x. The product equals zero, so at least one of the factors must equal zero. Because $5 \neq 0$, we know that $x = 0$.
2. Suppose $-4(x - 2) = 0$. The factors are -4 and $x - 2$. The product equals zero, so at least one of the factors must equal zero. Because $-4 \neq 0$, we know that $x - 2 = 0$, which means $x = 2$.
3. Suppose $(x + 3)(x - 4) = 0$. The factors are $x + 3$ and $x - 4$. The product equals zero, so $x + 3 = 0$ or $x - 4 = 0$. If $x + 3 = 0$, then $x = -3$. If $x - 4 = 0$, then $x = 4$.

HOW TO 1 Solve by factoring: $3x^2 = 2 - 5x$

Take Note

Recall that the steps involved in solving a quadratic equation by factoring are:
1. Write the equation in standard form.
2. Factor.
3. Use the Principle of Zero Products to set each factor equal to zero.
4. Solve each equation.
5. Check the solutions.

$$3x^2 = 2 - 5x$$
$$3x^2 + 5x - 2 = 0$$ • Write the equation in standard form.
$$(3x - 1)(x + 2) = 0$$ • Factor.
$$3x - 1 = 0 \qquad x + 2 = 0$$ • Use the Principle of Zero Products to write two equations.
$$3x = 1 \qquad\qquad x = -2$$ • Solve each equation.
$$x = \frac{1}{3}$$

$\frac{1}{3}$ and -2 check as solutions. The solutions are $\frac{1}{3}$ and -2.

Take Note

When a quadratic equation has two solutions that are the same number, the solution is called a **double root** of the equation. The number 3 is a double root of $x^2 - 6x = -9$.

HOW TO 2   Solve by factoring: $x^2 - 6x = -9$

$$x^2 - 6x = -9$$
$$x^2 - 6x + 9 = 0$$
$$(x - 3)(x - 3) = 0$$

$x - 3 = 0 \qquad\qquad x - 3 = 0$
$\quad\ x = 3 \qquad\qquad\quad\ x = 3$

- Write the equation in standard form.
- Factor.
- Use the Principle of Zero Products.
- Solve each equation.

3 checks as a solution. The solution is 3.

EXAMPLE 1

Solve by factoring: $2x(x - 3) = x + 4$

Solution

$$2x(x - 3) = x + 4$$
$$2x^2 - 6x = x + 4$$
$$2x^2 - 7x - 4 = 0$$
$$(2x + 1)(x - 4) = 0$$

- Write in standard form.
- Solve by factoring.

$2x + 1 = 0 \qquad x - 4 = 0$
$\quad\ 2x = -1 \qquad\quad\ x = 4$
$\quad\ \ x = -\dfrac{1}{2}$

The solutions are $-\frac{1}{2}$ and 4.

YOU TRY IT 1

Solve by factoring: $2x^2 = 7x - 3$

Your solution

$\dfrac{1}{2}$ and 3

IN-CLASS EXAMPLES

Solve by factoring.
1. $x^2 - 2x = 15$ $-3, 5$
2. $x^2 + 12 = 8x$ $2, 6$
3. $4x^2 + 11x - 3 = 0$ $-3, \dfrac{1}{4}$
4. $x - 9 = x(x - 5)$ 3

Solution on p. S28

As shown below, the solutions of the equation $(x - r_1)(x - r_2) = 0$ are r_1 and r_2.

$$(x - r_1)(x - r_2) = 0$$

$x - r_1 = 0 \qquad\qquad\qquad x - r_2 = 0$
$\quad\ \ x = r_1 \qquad\qquad\qquad\quad\ x = r_2$

$Check:$

$$\begin{array}{c|c} (x - r_1)(x - r_2) = 0 & (x - r_1)(x - r_2) = 0 \\ \hline (r_1 - r_1)(r_1 - r_2) \;\big|\; 0 & (r_2 - r_1)(r_2 - r_2) \;\big|\; 0 \\ 0 \cdot (r_1 - r_2) \;\big|\; 0 & (r_2 - r_1) \cdot 0 \;\big|\; 0 \\ 0 = 0 & 0 = 0 \end{array}$$

Using the equation $(x - r_1)(x - r_2) = 0$ and the fact that r_1 and r_2 are solutions of this equation, it is possible to write a quadratic equation given its solutions.

HOW TO 3   Write a quadratic equation that has solutions 4 and -5.

$$(x - r_1)(x - r_2) = 0$$
$$(x - 4)[x - (-5)] = 0$$
$$(x - 4)(x + 5) = 0$$
$$x^2 + x - 20 = 0$$

- Replace r_1 by 4 and r_2 by -5.
- Simplify.
- Multiply.

IN-CLASS EXAMPLES
Write a quadratic equation with integer coefficients and the given numbers as solutions.

5. -1 and 5
$x^2 - 4x - 5 = 0$

6. -3 and 0 $\quad x^2 + 3x = 0$

7. $\frac{1}{3}$ and 2
$3x^2 - 7x + 2 = 0$

8. $-\frac{4}{5}$ and $\frac{2}{3}$
$15x^2 + 2x - 8 = 0$

HOW TO 4 Write a quadratic equation with integer coefficients and solutions $\frac{2}{3}$ and $\frac{1}{2}$.

$$(x - r_1)(x - r_2) = 0$$

$$\left(x - \frac{2}{3}\right)\left(x - \frac{1}{2}\right) = 0 \qquad \bullet \text{ Replace } r_1 \text{ by } \frac{2}{3} \text{ and } r_2 \text{ by } \frac{1}{2}.$$

$$x^2 - \frac{7}{6}x + \frac{1}{3} = 0 \qquad \bullet \text{ Multiply.}$$

$$6\left(x^2 - \frac{7}{6}x + \frac{1}{3}\right) = 6 \cdot 0 \qquad \bullet \text{ Multiply each side of the equation by the LCM of the denominators.}$$

$$6x^2 - 7x + 2 = 0$$

EXAMPLE 2

Write a quadratic equation with integer coefficients and solutions $\frac{1}{2}$ and -4.

Solution

$$(x - r_1)(x - r_2) = 0$$

$$\left(x - \frac{1}{2}\right)[x - (-4)] = 0 \qquad \bullet \; r_1 = \frac{1}{2}, r_2 = -4$$

$$\left(x - \frac{1}{2}\right)(x + 4) = 0$$

$$x^2 + \frac{7}{2}x - 2 = 0$$

$$2\left(x^2 + \frac{7}{2}x - 2\right) = 2 \cdot 0$$

$$2x^2 + 7x - 4 = 0$$

YOU TRY IT 2

Write a quadratic equation with integer coefficients and solutions 3 and $-\frac{1}{2}$.

Your solution

$2x^2 - 5x - 3 = 0$

Solution on p. S28

OBJECTIVE B *To solve a quadratic equation by taking square roots*

Recall that if x is a variable that can be positive or negative, then $\sqrt{x^2} = |x|$. This fact can be used to solve a quadratic equation by taking square roots.

HOW TO 5 Solve: $x^2 = 9$

$$x^2 = 9$$
$$\sqrt{x^2} = \sqrt{9}$$
$$|x| = 3$$

The solutions of $|x| = 3$ are -3 and 3. Therefore, the solutions of $x^2 = 9$ are -3 and 3.

A shortcut notation is used to represent the negative and positive of the same number. The notation ± 3 is read "plus or minus 3." Using this notation, we can write the solutions of the equation in HOW TO 5 as $x = \pm 3$.

HOW TO 6   Solve by taking square roots: $3x^2 = 54$

$$3x^2 = 54$$
$$x^2 = 18 \qquad \text{• Solve for } x^2.$$
$$\sqrt{x^2} = \sqrt{18} \qquad \text{• Take the square root of each side of the equation.}$$
$$x = \pm\sqrt{18} \qquad \text{• Simplify.}$$
$$x = \pm 3\sqrt{2}$$

The solutions are $3\sqrt{2}$ and $-3\sqrt{2}$.   • $3\sqrt{2}$ and $-3\sqrt{2}$ check as solutions.

Solving a quadratic equation by taking the square root of each side of the equation can lead to solutions that are complex numbers.

HOW TO 7   Solve by taking square roots: $2x^2 + 18 = 0$

$$2x^2 + 18 = 0$$
$$2x^2 = -18$$
$$x^2 = -9 \qquad \text{• Solve for } x^2.$$
$$\sqrt{x^2} = \sqrt{-9} \qquad \text{• Take the square root of each side of the equation.}$$
$$x = \pm\sqrt{-9} \qquad \text{• Simplify.}$$
$$x = \pm 3i$$

The solutions are $3i$ and $-3i$.

An equation containing the square of a binomial can be solved by taking square roots.

HOW TO 8   Solve by taking square roots: $(x + 2)^2 - 24 = 0$

$$(x + 2)^2 - 24 = 0$$
$$(x + 2)^2 = 24 \qquad \text{• Solve for } (x + 2)^2.$$
$$\sqrt{(x + 2)^2} = \sqrt{24} \qquad \text{• Take the square root of each}$$
$$x + 2 = \pm\sqrt{24} \qquad \text{side of the equation. Then}$$
$$x + 2 = \pm 2\sqrt{6} \qquad \text{simplify.}$$

$$x + 2 = 2\sqrt{6} \qquad x + 2 = -2\sqrt{6} \qquad \text{• Solve for } x.$$
$$x = -2 + 2\sqrt{6} \qquad x = -2 - 2\sqrt{6}$$

The solutions are $-2 + 2\sqrt{6}$ and $-2 - 2\sqrt{6}$.

Tips for Success

Always check the solution of an equation. Here is the check for HOW TO 7.

$2x^2 + 18 = 0$	
$2(3i)^2 + 18$	0
$2(-9) + 18$	0
$-18 + 18$	0
$0 = 0$	

$2x^2 + 18 = 0$	
$2(-3i)^2 + 18$	0
$2(-9) + 18$	0
$-18 + 18$	0
$0 = 0$	

EXAMPLE 3

Solve by taking square roots: $3(x - 2)^2 + 12 = 0$

Solution

$$3(x - 2)^2 + 12 = 0$$
$$3(x - 2)^2 = -12$$
$$(x - 2)^2 = -4 \qquad \text{• Solve for } (x - 2)^2.$$
$$\sqrt{(x - 2)^2} = \sqrt{-4} \qquad \text{• Take the square root}$$
$$x - 2 = \pm\sqrt{-4} \qquad \text{of each side of the}$$
$$x - 2 = \pm 2i \qquad \text{equation.}$$

$$x - 2 = 2i \qquad x - 2 = -2i \qquad \text{• Solve for } x.$$
$$x = 2 + 2i \qquad x = 2 - 2i$$

The solutions are $2 + 2i$ and $2 - 2i$.

YOU TRY IT 3

Solve by taking square roots: $2(x + 1)^2 - 24 = 0$

Your solution

$-1 + 2\sqrt{3}$ and $-1 - 2\sqrt{3}$

IN-CLASS EXAMPLES

Solve by taking square roots.

9. $4x^2 - 25 = 0 \qquad -\dfrac{5}{2}, \dfrac{5}{2}$

10. $y^2 + 4 = 0 \qquad -2i, 2i$

11. $3(y - 3)^2 = 12 \qquad 1, 5$

12. $(x + 3)^2 - 8 = 0 \qquad -3 + 2\sqrt{2}, -3 - 2\sqrt{2}$

13. $\left(t - \dfrac{1}{4}\right)^2 - 32 = 0 \qquad \dfrac{1 + 16\sqrt{2}}{4}, \dfrac{1 - 16\sqrt{2}}{4}$

Solution on p. S28

10.1 EXERGISES

✔ **Concept Check**

SUGGESTED ASSIGNMENT
Exercises 1–8; Exercises 9–39, every other odd;
Exercises 41–63, every other odd; Exercises 65–95, odds;
More challenging exercises: Exercises 97–107, odds

1. Which of the following are *not* quadratic equations?   i, iii

   (i) $4x + 5 = 0$ **(ii)** $x^2 = 3x - 5$ **(iii)** $x^2 + 4x - 5$ **(iv)** $2z^2 - \sqrt{5} = 2z$

2. Write each equation in standard form with the coefficient of x^2 positive.

   a. $3x^2 - 3 = 2x$ **b.** $4 - x^2 = 5x$ **c.** $5 + 4x = 3x^2$ **d.** $x^2 = 3$

   $3x^2 + 2x - 3 = 0$ $x^2 + 5x - 4 = 0$ $3x^2 - 4x - 5 = 0$ $x^2 - 3 = 0$

3. 🔖 State the Principle of Zero Products. How is it used to solve a quadratic equation?

4. 🔖 What is a double root of a quadratic equation?

For Exercises 5 to 8, solve the equation.

5. $5(x + 4) = 0$ 6. $(x - 3)(x + 2) = 0$ 7. $2x(x - 4) = 0$ 8. $(2x + 1)(3x - 4) = 0$

   -4 -2 and 3 0 and 4 $-\dfrac{1}{2}$ and $\dfrac{4}{3}$

OBJECTIVE A   *To solve a quadratic equation by factoring*

For Exercises 9 to 39, solve by factoring.

9. $x^2 - 4x = 0$

   0 and 4

10. $y^2 + 6y = 0$

    -6 and 0

11. $t^2 - 25 = 0$

    -5 and 5

12. $p^2 - 81 = 0$

    -9 and 9

13. $s^2 - s - 6 = 0$

    -2 and 3

14. $v^2 + 4v - 5 = 0$

    -5 and 1

15. $y^2 - 6y + 9 = 0$

    3

16. $x^2 + 10x + 25 = 0$

    -5

17. $9z^2 - 18z = 0$

    0 and 2

18. $4y^2 + 20y = 0$

    -5 and 0

19. $r^2 - 3r = 10$

    -2 and 5

20. $p^2 + 5p = 6$

    -6 and 1

21. $v^2 + 10 = 7v$

    2 and 5

22. $t^2 - 16 = 15t$

    -1 and 16

23. $2x^2 - 9x - 18 = 0$

    $-\dfrac{3}{2}$ and 6

24. $3y^2 - 4y - 4 = 0$

    $-\dfrac{2}{3}$ and 2

25. $4z^2 - 9z + 2 = 0$

    $\dfrac{1}{4}$ and 2

26. $2s^2 - 9s + 9 = 0$

    $\dfrac{3}{2}$ and 3

27. $3w^2 + 11w = 4$

    -4 and $\dfrac{1}{3}$

28. $2r^2 + r = 6$

    -2 and $\dfrac{3}{2}$

29. $6x^2 = 23x + 18$

    $-\dfrac{2}{3}$ and $\dfrac{9}{2}$

30. $6x^2 = 7x - 2$

$\dfrac{1}{2}$ and $\dfrac{2}{3}$

31. $4 - 15u - 4u^2 = 0$

-4 and $\dfrac{1}{4}$

32. $3 - 2y - 8y^2 = 0$

$-\dfrac{3}{4}$ and $\dfrac{1}{2}$

33. $x + 18 = x(x - 6)$

-2 and 9

34. $t + 24 = t(t + 6)$

-8 and 3

35. $4s(s + 3) = s - 6$

-2 and $-\dfrac{3}{4}$

36. $3v(v - 2) = 11v + 6$

$-\dfrac{1}{3}$ and 6

37. $u^2 - 2u + 4 = (2u - 3)(u + 2)$

-5 and 2

38. $(3v - 2)(2v + 1) = 3v^2 - 11v - 10$

-2 and $-\dfrac{4}{3}$

39. $(3x - 4)(x + 4) = x^2 - 3x - 28$

-4 and $-\dfrac{3}{2}$

For Exercises 40 to 63, write a quadratic equation with integer coefficients and the given numbers as solutions.

40. 3 and 1

$x^2 - 4x + 3 = 0$

41. 2 and 5

$x^2 - 7x + 10 = 0$

42. -1 and -3

$x^2 + 4x + 3 = 0$

43. -2 and -4

$x^2 + 6x + 8 = 0$

44. -2 and 5

$x^2 - 3x - 10 = 0$

45. 6 and -1

$x^2 - 5x - 6 = 0$

46. 5 and -5

$x^2 - 25 = 0$

47. 3 and -3

$x^2 - 9 = 0$

48. 2 and 2

$x^2 - 4x + 4 = 0$

49. 4 and 4

$x^2 - 8x + 16 = 0$

50. 0 and -2

$x^2 + 2x = 0$

51. 0 and 5

$x^2 - 5x = 0$

52. 2 and $\dfrac{2}{3}$

$3x^2 - 8x + 4 = 0$

53. 3 and $\dfrac{1}{2}$

$2x^2 - 7x + 3 = 0$

54. $-\dfrac{1}{2}$ and 5

$2x^2 - 9x - 5 = 0$

55. $-\dfrac{3}{4}$ and 2

$4x^2 - 5x - 6 = 0$

56. $-\dfrac{3}{2}$ and -1

$2x^2 + 5x + 3 = 0$

57. $-\dfrac{5}{3}$ and -2

$3x^2 + 11x + 10 = 0$

58. $\dfrac{3}{4}$ and $\dfrac{2}{3}$

$12x^2 - 17x + 6 = 0$

59. $\dfrac{1}{2}$ and $\dfrac{1}{3}$

$6x^2 - 5x + 1 = 0$

60. $\dfrac{3}{4}$ and $-\dfrac{3}{2}$

$8x^2 + 6x - 9 = 0$

61. $\dfrac{6}{5}$ and $-\dfrac{1}{2}$

$10x^2 - 7x - 6 = 0$

62. $-\dfrac{5}{6}$ and $-\dfrac{2}{3}$

$18x^2 + 27x + 10 = 0$

63. $-\dfrac{1}{4}$ and $-\dfrac{1}{2}$

$8x^2 + 6x + 1 = 0$

64. ✒ If u and v are solutions of $x^2 + bx + c = 0$, what are the values of b and c?

$b = -(u + v); c = uv$

65. ✒ If 0 is a solution of $ax^2 + bx + c = 0$, what is the value of c?

0

OBJECTIVE B   *To solve a quadratic equation by taking square roots*

For Exercises 66 to 92, solve by taking square roots.

66. $x^2 = 64$

-8 and 8

67. $y^2 = 49$

-7 and 7

68. $v^2 = -16$

$-4i$ and $4i$

69. $z^2 = -4$

$-2i$ and $2i$

70. $r^2 - 36 = 0$

-6 and 6

71. $s^2 - 4 = 0$

-2 and 2

72. $9x^2 - 16 = 0$

$-\dfrac{4}{3}$ and $\dfrac{4}{3}$

73. $4x^2 - 81 = 0$

$-\dfrac{9}{2}$ and $\dfrac{9}{2}$

74. $z^2 + 16 = 0$

$-4i$ and $4i$

75. $y^2 + 49 = 0$

$-7i$ and $7i$

76. $s^2 - 32 = 0$

$-4\sqrt{2}$ and $4\sqrt{2}$

77. $v^2 - 48 = 0$

$-4\sqrt{3}$ and $4\sqrt{3}$

78. $t^2 + 27 = 0$

$-3i\sqrt{3}$ and $3i\sqrt{3}$

79. $z^2 + 18 = 0$

$-3i\sqrt{2}$ and $3i\sqrt{2}$

80. $(x + 2)^2 = 25$

-7 and 3

81. $(x - 1)^2 = 36$

-5 and 7

82. $4(s - 2)^2 = 36$

-1 and 5

83. $5(z + 2)^2 = 125$

-7 and 3

84. $2(y - 3)^2 = 18$

0 and 6

85. $\left(v - \dfrac{1}{2}\right)^2 = \dfrac{1}{4}$

0 and 1

86. $\left(r + \dfrac{2}{3}\right)^2 = \dfrac{1}{9}$

-1 and $-\dfrac{1}{3}$

87. $(x + 5)^2 - 6 = 0$

$-5 - \sqrt{6}$ and $-5 + \sqrt{6}$

88. $(t - 1)^2 - 15 = 0$

$1 - \sqrt{15}$ and $1 + \sqrt{15}$

89. $(v - 3)^2 + 45 = 0$

$3 - 3i\sqrt{5}$ and $3 + 3i\sqrt{5}$

90. $(x + 5)^2 + 32 = 0$

$-5 - 4i\sqrt{2}$ and $-5 + 4i\sqrt{2}$

91. $\left(u + \dfrac{2}{3}\right)^2 - 18 = 0$

$-\dfrac{2 - 9\sqrt{2}}{3}$ and $-\dfrac{2 + 9\sqrt{2}}{3}$

92. $\left(z - \dfrac{1}{2}\right)^2 - 20 = 0$

$\dfrac{1 - 4\sqrt{5}}{2}$ and $\dfrac{1 + 4\sqrt{5}}{2}$

For Exercises 93 to 96, assume that a and b are positive real numbers. In each case, state how many real or complex number solutions the equation has.

93. $(x - a)^2 + b = 0$

Two complex solutions

94. $x^2 + a = 0$

Two complex solutions

95. $(x - a)^2 = 0$

Two equal real solutions

96. $(x - a)^2 - b = 0$

Two unequal real solutions

Critical Thinking

For Exercises 97 to 104, write a quadratic equation that has the given numbers as solutions.

97. $\sqrt{2}$ and $-\sqrt{2}$
$x^2 - 2 = 0$

98. $3\sqrt{2}$ and $-3\sqrt{2}$
$x^2 - 18 = 0$

99. i and $-i$
$x^2 + 1 = 0$

100. $2i\sqrt{3}$ and $-2i\sqrt{3}$
$x^2 + 12 = 0$

101. $3 - \sqrt{2}, 3 + \sqrt{2}$
$x^2 - 6x + 7 = 0$

102. $-4 - 2\sqrt{5}, -4 + 2\sqrt{5}$
$x^2 + 8x - 4 = 0$

103. $5 - i, 5 + i$
$x^2 - 10x + 26 = 0$

104. $-2 - 3i, -2 + 3i$
$x^2 + 4x + 13 = 0$

For Exercises 105 to 107, solve the equation by taking square roots.

105. $x^2 = \sqrt{7}$
$-\sqrt[4]{7}$ and $\sqrt[4]{7}$

106. $x^2 - \sqrt{5} = 0$
$-\sqrt[4]{5}$ and $\sqrt[4]{5}$

107. $x^2 - \sqrt[3]{2} = 0$
$-\sqrt[6]{2}$ and $\sqrt[6]{2}$

Projects or Group Activities

Checking Solutions of a Quadratic Equation

In Objective 10.1A, we created a quadratic equation from its roots. For instance, if 2 and -5 are roots of a quadratic equation, then the quadratic equation is

$$(x - 2)(x + 5) = 0, \text{ or } x^2 + 3x - 10 = 0$$

Notice that the coefficient of x, 3, is the *opposite* of the sum of the roots of the equation.

opposite of ——— sum of roots

$$-[2 + (-5)] = -[3] = 3$$

The constant term, -10, is the product of the roots.

product of roots

$$2(-5) = -10$$

This observation is true for all quadratic equations. If $x^2 + bx + c = 0$ is a quadratic equation with roots r_1 and r_2, then $b = -(r_1 + r_2)$ and $c = r_1 r_2$. For Exercises 108 to 113, use this fact to determine whether the given roots are solutions of the quadratic equation.

108. $x^2 - 12x - 28 = 0$
$r_1 = -2, r_2 = 14$
Yes

109. $x^2 + 6x + 5 = 0$
$r_1 = 1, r_2 = 5$
No

110. $x^2 + 6x + 7 = 0$
$r_1 = -3 - \sqrt{2}$,
$r_2 = -3 + \sqrt{2}$
Yes

111. $x^2 + 2x - 11 = 0$
$r_1 = -1 - 2\sqrt{3}$,
$r_2 = -1 + 2\sqrt{3}$
Yes

112. $x^2 - 8x + 17 = 0$
$r_1 = -4 + i$,
$r_2 = -4 - i$
No

113. $x^2 - 4x + 7 = 0$
$r_1 = 2 - i\sqrt{3}$,
$r_2 = 2 + i\sqrt{3}$
Yes

To check solutions by this method, the coefficient of x^2 must be 1. Suppose we want to check whether $-\frac{3}{5}$ and $\frac{2}{3}$ are solutions of the equation $15x^2 - x - 6 = 0$. First divide each side of the equation by 15, the coefficient of x^2. The resulting equation is $x^2 - \frac{1}{15}x - \frac{2}{5} = 0$. Note that the opposite of the sum of $-\frac{3}{5}$ and $\frac{2}{3}$ is $-\frac{1}{15}$. The product of $-\frac{3}{5}$ and $\frac{2}{3}$ is $-\frac{2}{5}$.

QUICK QUIZ

Solve by factoring.
1. $x^2 + 3x = 4$
$-4, 1$ **[10.1A]**
2. $x^2 - 12 = 11x$
$-1, 12$ **[10.1A]**
3. $3x^2 - 10x - 8 = 0$
$-\frac{2}{3}, 4$ **[10.1A]**
4. $x + 3 = x(x + 3)$
$-3, 1$ **[10.1A]**

Write a quadratic equation with integer coefficients and the given numbers as solutions.
5. -3 and 1
$x^2 + 2x - 3 = 0$
[10.1A]
6. -4 and 0
$x^2 + 4x = 0$ **[10.1A]**
7. $-\frac{2}{3}$ and 2
$3x^2 - 4x - 4 = 0$
[10.1A]

Solve by taking square roots.
8. $9x^2 - 25 = 0$
$-\frac{5}{3}, \frac{5}{3}$ **[10.1B]**
9. $s^2 + 50 = 0$ $5i\sqrt{2}$,
$-5i\sqrt{2}$ **[10.1B]**
10. $2(x + 2)^2 = 18$
$-5, 1$ **[10.1B]**
11. $(t - 2)^2 - 20 = 0$
$2 + 2\sqrt{5}$,
$2 - 2\sqrt{5}$ **[10.1B]**
12. $\left(y - \frac{3}{5}\right)^2 - 75 = 0$
$\frac{3 + 25\sqrt{3}}{5}$,
$\frac{3 - 25\sqrt{3}}{5}$ **[10.1B]**

SECTION

10.2

Solving Quadratic Equations by Completing the Square and by Using the Quadratic Formula

OBJECTIVE A   *To solve a quadratic equation by completing the square*

INSTRUCTOR NOTE

To say that students will find completing the square a daunting task is an understatement. It may be necessary to give quite a few examples before students can do this effectively. The Point of Interest below may help some students.

Recall that a perfect-square trinomial is the square of a binomial.

Perfect-Square Trinomial		Square of a Binomial
$x^2 + 8x + 16$	$=$	$(x + 4)^2$
$x^2 - 10x + 25$	$=$	$(x - 5)^2$
$x^2 - 5x + \dfrac{25}{4}$	$=$	$\left(x - \dfrac{5}{2}\right)^2$
$x^2 + 7x + \dfrac{49}{4}$	$=$	$\left(x + \dfrac{7}{2}\right)^2$

Point of Interest

Early attempts to solve quadratic equations were primarily geometric. The Persian mathematician al-Khowarizmi (c. A.D. 800) essentially completed a square of $x^2 + 12x$ as follows.

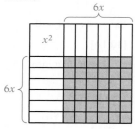

For each perfect-square trinomial, the square of $\frac{1}{2}$ of the coefficient of x equals the constant term.

$$\left(\frac{1}{2} \text{ coefficient of } x\right)^2 = \text{constant term}$$

$$x^2 + 8x + 16, \qquad \left(\frac{1}{2} \cdot 8\right)^2 = 16$$

$$x^2 - 10x + 25, \qquad \left[\frac{1}{2}(-10)\right]^2 = 25$$

$$x^2 - 5x + \frac{25}{4}, \qquad \left[\frac{1}{2}(-5)\right]^2 = \frac{25}{4}$$

$$x^2 + 7x + \frac{49}{4}, \qquad \left[\frac{1}{2}(7)\right]^2 = \frac{49}{4}$$

Adding to a binomial the constant term that makes it a perfect-square trinomial is called **completing the square.**

HOW TO 1   Complete the square on $x^2 + 12x$. Write the resulting perfect-square trinomial as the square of a binomial.

$$\left[\frac{1}{2}(12)\right]^2 = (6)^2 = 36$$ • **Find the constant term.**

$x^2 + 12x + 36$ • **Complete the square on $x^2 + 12x$ by adding the constant term.**

$x^2 + 12x + 36 = (x + 6)^2$ • **Write the resulting perfect-square trinomial as the square of a binomial.**

Unless otherwise noted, all content on this page is © Cengage Learning.

Tips for Success

This is a new skill and one that is difficult for many students. Be sure to do all you need to do in order to be successful at solving equations by completing the square: Read the introductory material, work through the HOW TO examples, study the paired Examples, do the You Try Its, and check your solutions against the ones given in the back of the book. See *AIM for Success* in the Preface.

HOW TO 2   Complete the square on $z^2 - 3z$. Write the resulting perfect-square trinomial as the square of a binomial.

$$\left[\frac{1}{2} \cdot (-3)\right]^2 = \left(-\frac{3}{2}\right)^2 = \frac{9}{4}$$ • Find the constant term.

$$z^2 - 3z + \frac{9}{4}$$ • Complete the square on $z^2 + 3z$ by adding the constant term.

$$z^2 - 3z + \frac{9}{4} = \left(z - \frac{3}{2}\right)^2$$ • Write the resulting perfect-square trinomial as the square of a binomial.

Any quadratic equation can be solved by completing the square.

Procedure for Solving a Quadratic Equation by Completing the Square

1. Write the equation in the form $ax^2 + bx = -c$.
2. Multiply both sides of the equation by $\dfrac{1}{a}$.
3. Complete the square on $x^2 + \dfrac{b}{a}x$. Add the number that completes the square to both sides of the equation.
4. Factor the perfect-square trinomial.
5. Take the square root of each side of the equation.
6. Solve the resulting equation for x.
7. Check the solutions.

HOW TO 3   Solve by completing the square: $x^2 - 6x - 15 = 0$

$$x^2 - 6x - 15 = 0$$
$$x^2 - 6x = 15$$ • Add 15 to each side of the equation.
$$x^2 - 6x + 9 = 15 + 9$$ • Complete the square. Add

$$\left[\frac{1}{2}(-6)\right]^2 = (-3)^2 = 9 \text{ to each side of the equation.}$$

$$(x - 3)^2 = 24$$ • Factor the perfect-square trinomial.

$$\sqrt{(x - 3)^2} = \sqrt{24}$$ • Take the square root of each side of the equation.

$$x - 3 = \pm\sqrt{24}$$ • Simplify.
$$x - 3 = \pm 2\sqrt{6}$$

$$x - 3 = 2\sqrt{6} \qquad x - 3 = -2\sqrt{6}$$ • Solve for x.
$$x = 3 + 2\sqrt{6} \qquad x = 3 - 2\sqrt{6}$$

Be sure to check the solutions. See the check at the left.

The solutions are $3 + 2\sqrt{6}$ and $3 - 2\sqrt{6}$.

Check:

$$\begin{array}{r|l} \hline x^2 - 6x - 15 = 0 \\ \hline (3 + 2\sqrt{6})^2 - 6(3 + 2\sqrt{6}) - 15 & 0 \\ 9 + 12\sqrt{6} + 24 - 18 - 12\sqrt{6} - 15 & 0 \\ & 0 = 0 \end{array}$$

$$\begin{array}{r|l} \hline x^2 - 6x - 15 = 0 \\ \hline (3 - 2\sqrt{6})^2 - 6(3 - 2\sqrt{6}) - 15 & 0 \\ 9 - 12\sqrt{6} + 24 - 18 + 12\sqrt{6} - 15 & 0 \\ & 0 = 0 \end{array}$$

In HOW TO 3 on the previous page, the solutions of the equation $x^2 - 6x - 15 = 0$ are $3 + 2\sqrt{6}$ and $3 - 2\sqrt{6}$. These are the exact solutions. However, in some situations it may be preferable to have decimal approximations of the solutions of a quadratic equation. Approximate solutions can be found by using a calculator and then rounding to the desired degree of accuracy.

$$3 + 2\sqrt{6} \approx 7.899 \text{ and } 3 - 2\sqrt{6} \approx -1.899$$

To the nearest thousandth, the approximate solutions of the equation $x^2 - 6x - 15 = 0$ are 7.899 and -1.899.

Integrating Technology

A graphing calculator can be used to check a proposed solution of a quadratic equation. For the equation in HOW TO 4, enter the expression $2x^2 - x - 2$ in Y_1. Then evaluate the function for $\dfrac{1 + \sqrt{17}}{4}$ and $\dfrac{1 - \sqrt{17}}{4}$. The value of each should be zero. Some typical graphing calculator screens are shown below.

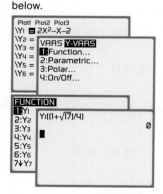

See the Keystroke Guide: *Evaluating Functions.*

HOW TO 4   Solve $2x^2 - x - 2 = 0$ by completing the square.

To complete the square on an expression, the coefficient of the squared term must be 1. After adding the constant term to each side of the equation, multiply each side of the equation by $\frac{1}{2}$.

$$2x^2 - x - 2 = 0$$
$$2x^2 - x = 2$$ • Add 2 to each side of the equation.
$$\frac{1}{2}(2x^2 - x) = \frac{1}{2} \cdot 2$$ • Multiply each side of the equation by $\frac{1}{2}$.
$$x^2 - \frac{1}{2}x = 1$$ • The coefficient of x^2 is now 1.
$$x^2 - \frac{1}{2}x + \frac{1}{16} = 1 + \frac{1}{16}$$ • Complete the square. Add $\left[\dfrac{1}{2}\left(-\dfrac{1}{2}\right)\right]^2 = \left(-\dfrac{1}{4}\right)^2 = \dfrac{1}{16}$ to each side of the equation.
$$\left(x - \frac{1}{4}\right)^2 = \frac{17}{16}$$ • Factor the perfect-square trinomial.
$$\sqrt{\left(x - \frac{1}{4}\right)^2} = \sqrt{\frac{17}{16}}$$ • Take the square root of each side of the equation.
$$x - \frac{1}{4} = \pm\frac{\sqrt{17}}{4}$$

$$x - \frac{1}{4} = \frac{\sqrt{17}}{4} \qquad x - \frac{1}{4} = -\frac{\sqrt{17}}{4}$$ • Solve for x.

$$x = \frac{1}{4} + \frac{\sqrt{17}}{4} \qquad x = \frac{1}{4} - \frac{\sqrt{17}}{4}$$

The solutions are $\dfrac{1 + \sqrt{17}}{4}$ and $\dfrac{1 - \sqrt{17}}{4}$.

Check the solutions. A check of one solution using a graphing calculator is shown at the left.

Unless otherwise noted, all content on this page is © Cengage Learning.

EXAMPLE 1

Solve by completing the square: $4x^2 - 8x + 1 = 0$

Solution

$4x^2 - 8x + 1 = 0$

$\quad 4x^2 - 8x = -1$ • Write in the form $ax^2 + bx = -c$.

$\dfrac{1}{4}(4x^2 - 8x) = \dfrac{1}{4}(-1)$ • Multiply each side by $\dfrac{1}{a}$.

$\quad x^2 - 2x = -\dfrac{1}{4}$

$x^2 - 2x + 1 = -\dfrac{1}{4} + 1$ • Complete the square.

$\quad (x - 1)^2 = \dfrac{3}{4}$ • Factor.

$\sqrt{(x - 1)^2} = \sqrt{\dfrac{3}{4}}$ • Take square roots.

$\quad x - 1 = \pm\dfrac{\sqrt{3}}{2}$

$x - 1 = \dfrac{\sqrt{3}}{2} \qquad x - 1 = -\dfrac{\sqrt{3}}{2}$ • Solve for x.

$x = 1 + \dfrac{\sqrt{3}}{2} \qquad x = 1 - \dfrac{\sqrt{3}}{2}$

$\quad = \dfrac{2 + \sqrt{3}}{2} \qquad\quad = \dfrac{2 - \sqrt{3}}{2}$

The solutions are $\dfrac{2 + \sqrt{3}}{2}$ and $\dfrac{2 - \sqrt{3}}{2}$.

YOU TRY IT 1

Solve by completing the square: $4x^2 - 4x - 1 = 0$

Your solution

$\dfrac{1 + \sqrt{2}}{2}$ and $\dfrac{1 - \sqrt{2}}{2}$

IN-CLASS EXAMPLES

Solve by completing the square.

1. $x^2 - 6x - 7 = 0$ $-1, 7$
2. $x^2 + 2x - 5 = 0$ $-1 + \sqrt{6}, -1 - \sqrt{6}$
3. $x^2 - 4x + 6 = 0$ $2 + i\sqrt{2}, 2 - i\sqrt{2}$
4. $4x^2 - 4x = 7$ $\dfrac{1 + 2\sqrt{2}}{2}, \dfrac{1 - 2\sqrt{2}}{2}$

EXAMPLE 2

Solve by completing the square: $x^2 + 4x + 5 = 0$

Solution

$x^2 + 4x + 5 = 0$

$\quad x^2 + 4x = -5$

$x^2 + 4x + 4 = -5 + 4$ • Complete the square.

$\quad (x + 2)^2 = -1$ • Factor.

$\sqrt{(x + 2)^2} = \sqrt{-1}$ • Take square roots.

$\quad x + 2 = \pm i$

$x + 2 = i \qquad x + 2 = -i$ • Solve for x.

$\quad x = -2 + i \qquad x = -2 - i$

The solutions are $-2 + i$ and $-2 - i$.

YOU TRY IT 2

Solve by completing the square: $x^2 + 4x + 8 = 0$

Your solution

$-2 + 2i$ and $-2 - 2i$

Solutions on p. S28

OBJECTIVE B *To solve a quadratic equation by using the quadratic formula*

A general formula known as the **quadratic formula** can be derived by applying the method of completing the square to the standard form of a quadratic equation. This formula can be used to solve any quadratic equation. The equation $ax^2 + bx + c = 0$ is solved by completing the square as follows.

$$ax^2 + bx + c = 0$$

Add the opposite of the constant term to each side of the equation.

$$ax^2 + bx + c + (-c) = 0 + (-c)$$

$$ax^2 + bx = -c$$

Multiply each side of the equation by the reciprocal of a, the coefficient of x^2.

$$\frac{1}{a}(ax^2 + bx) = \frac{1}{a}(-c)$$

$$x^2 + \frac{b}{a}x = -\frac{c}{a}$$

To complete the square, add $\left(\frac{1}{2} \cdot \frac{b}{a}\right)^2 = \frac{b^2}{4a^2}$ to each side of the equation.

$$x^2 + \frac{b}{a}x + \frac{b^2}{4a^2} = \frac{b^2}{4a^2} - \frac{c}{a}$$

Simplify the right side of the equation.

$$x^2 + \frac{b}{a}x + \frac{b^2}{4a^2} = \frac{b^2}{4a^2} - \left(\frac{c}{a} \cdot \frac{4a}{4a}\right)$$

$$x^2 + \frac{b}{a}x + \frac{b^2}{4a^2} = \frac{b^2}{4a^2} - \frac{4ac}{4a^2}$$

$$x^2 + \frac{b}{a}x + \frac{b^2}{4a^2} = \frac{b^2 - 4ac}{4a^2}$$

Factor the perfect-square trinomial on the left side of the equation.

$$\left(x + \frac{b}{2a}\right)^2 = \frac{b^2 - 4ac}{4a^2}$$

Take the square root of each side of the equation.

$$\sqrt{\left(x + \frac{b}{2a}\right)^2} = \sqrt{\frac{b^2 - 4ac}{4a^2}}$$

$$x + \frac{b}{2a} = \pm\frac{\sqrt{b^2 - 4ac}}{2a}$$

Solve for x.

$$x + \frac{b}{2a} = \frac{\sqrt{b^2 - 4ac}}{2a} \qquad x + \frac{b}{2a} = -\frac{\sqrt{b^2 - 4ac}}{2a}$$

$$x = -\frac{b}{2a} + \frac{\sqrt{b^2 - 4ac}}{2a} \qquad x = -\frac{b}{2a} - \frac{\sqrt{b^2 - 4ac}}{2a}$$

$$= \frac{-b + \sqrt{b^2 - 4ac}}{2a} \qquad\qquad = \frac{-b - \sqrt{b^2 - 4ac}}{2a}$$

⊚ Point of Interest

Although mathematicians have studied quadratic equations since around 500 B.C., it was not until the 18th century that the formula was written as it is today. Of further note, the word *quadratic* has the same Latin root as the word *square*.

The Quadratic Formula

The solutions of $ax^2 + bx + c = 0$, $a \neq 0$, are

$$\frac{-b + \sqrt{b^2 - 4ac}}{2a} \quad \text{and} \quad \frac{-b - \sqrt{b^2 - 4ac}}{2a}$$

The quadratic formula is frequently written as $x = \dfrac{-b \pm \sqrt{b^2 - 4ac}}{2a}$.

Take Note

The solutions of the quadratic equation in HOW TO 5 are rational numbers. When this happens, it means the equation could have been solved by factoring and using the Principle of Zero Products. This may be easier than applying the quadratic formula.

HOW TO 5   Solve by using the quadratic formula: $2x^2 + 5x + 3 = 0$

$$2x^2 + 5x + 3 = 0$$

$$x = \frac{-b \pm \sqrt{b^2 - 4ac}}{2a}$$

• The equation $2x^2 + 5x + 3 = 0$ is in standard form. $a = 2, b = 5, c = 3$

$$= \frac{-(5) \pm \sqrt{(5)^2 - 4(2)(3)}}{2(2)}$$

• Replace a, b, and c in the quadratic formula with these values.

$$= \frac{-5 \pm \sqrt{25 - 24}}{4}$$

$$= \frac{-5 \pm \sqrt{1}}{4}$$

$$= \frac{-5 \pm 1}{4}$$

$$x = \frac{-5 + 1}{4} = \frac{-4}{4} = -1 \qquad x = \frac{-5 - 1}{4} = \frac{-6}{4} = -\frac{3}{2}$$

The solutions are -1 and $-\frac{3}{2}$. Remember to check your solutions.

INSTRUCTOR NOTE

One of the difficulties students have when using the quadratic formula is in making correct substitutions. Have them first make sure that the equation is in standard form. Then have them circle the values of a, b, and c. They must remember that the sign that precedes a number is the sign of the number. For the equation in HOW TO 6, we have

$$\overset{a}{③}x^2 \overset{b}{\underset{}{\left(- 4\right)}}x \overset{c}{\underset{}{\left(- 6\right)}} = 0$$

Check:

$$3x^2 = 4x + 6$$

$3\left(\dfrac{2 + \sqrt{22}}{3}\right)^2$	$4\left(\dfrac{2 + \sqrt{22}}{3}\right) + 6$
$3\left(\dfrac{4 + 4\sqrt{22} + 22}{9}\right)$	$\dfrac{8}{3} + \dfrac{4\sqrt{22}}{3} + \dfrac{18}{3}$
$3\left(\dfrac{26 + 4\sqrt{22}}{9}\right)$	$\dfrac{26}{3} + \dfrac{4\sqrt{22}}{3}$
$\dfrac{26 + 4\sqrt{22}}{3}$	$= \dfrac{26 + 4\sqrt{22}}{3}$

HOW TO 6   Solve $3x^2 = 4x + 6$ by using the quadratic formula. Find the exact solutions, and approximate the solutions to the nearest thousandth.

$$3x^2 = 4x + 6$$

$$3x^2 - 4x - 6 = 0$$

• Write the equation in standard form. Subtract $4x$ and 6 from each side of the equation. $a = 3, b = -4, c = -6$

$$x = \frac{-b \pm \sqrt{b^2 - 4ac}}{2a}$$

$$= \frac{-(-4) \pm \sqrt{(-4)^2 - 4(3)(-6)}}{2(3)}$$

• Replace a, b, and c in the quadratic formula with these values.

$$= \frac{4 \pm \sqrt{16 - (-72)}}{6}$$

$$= \frac{4 \pm \sqrt{88}}{6}$$

$$= \frac{4 \pm 2\sqrt{22}}{6}$$

$$= \frac{2(2 \pm \sqrt{22})}{2 \cdot 3} = \frac{2 \pm \sqrt{22}}{3}$$

Check the solutions. A check of one of the solutions is shown at the left.

The exact solutions are $\dfrac{2 + \sqrt{22}}{3}$ and $\dfrac{2 - \sqrt{22}}{3}$.

$$\frac{2 + \sqrt{22}}{3} \approx 2.230 \qquad \frac{2 - \sqrt{22}}{3} \approx -0.897$$

To the nearest thousandth, the solutions are 2.230 and -0.897.

HOW TO 7 Solve by using the quadratic formula: $4x^2 = 8x - 13$

$$4x^2 = 8x - 13$$
$$4x^2 - 8x + 13 = 0$$ • Write the equation in standard form.

$$x = \frac{-b \pm \sqrt{b^2 - 4ac}}{2a}$$ • Use the quadratic formula.

$$= \frac{-(-8) \pm \sqrt{(-8)^2 - 4 \cdot 4 \cdot 13}}{2 \cdot 4}$$ • $a = 4, b = -8, c = 13$

$$= \frac{8 \pm \sqrt{64 - 208}}{8} = \frac{8 \pm \sqrt{-144}}{8}$$

$$= \frac{8 \pm 12i}{8} = \frac{2 \pm 3i}{2}$$

The solutions are $1 + \frac{3}{2}i$ and $1 - \frac{3}{2}i$.

IN-CLASS EXAMPLES

Solve by using the quadratic formula.

5. $x^2 + 3x - 10 = 0$
−5, 2

6. $x^2 - 4x - 30 = 0$
$2 + \sqrt{34}, 2 - \sqrt{34}$

7. $x^2 = 2x - 15$
$1 + i\sqrt{14}, 1 - i\sqrt{14}$

8. $4x^2 + 8x = 3$
$\frac{-2 + \sqrt{7}}{2}, \frac{-2 - \sqrt{7}}{2}$

Use the discriminant to determine whether the equation has two equal real number solutions, two unequal real number solutions, or two complex number solutions.

9. $2x^2 + x + 7 = 0$
Two complex

10. $3x^2 - x - 2 = 0$
Two unequal real

In the quadratic formula, the quantity $b^2 - 4ac$ is called the **discriminant.** When a, b, and c are real numbers, the discriminant determines whether a quadratic equation will have a double root, two real number solutions that are not equal, or two complex number solutions.

The Effect of the Discriminant on the Solutions of a Quadratic Equation

1. If $b^2 - 4ac = 0$, the equation has two equal real number solutions, a double root.
2. If $b^2 - 4ac > 0$, the equation has two unequal real number solutions.
3. If $b^2 - 4ac < 0$, the equation has two complex number solutions.

HOW TO 8 Use the discriminant to determine whether $x^2 - 4x - 5 = 0$ has two equal real number solutions, two unequal real number solutions, or two complex number solutions.

$$b^2 - 4ac$$
$$(-4)^2 - 4(1)(-5) = 16 + 20 = 36$$ • Evaluate the discriminant.
$$36 > 0$$ $a = 1, b = -4, c = -5$

Because $b^2 - 4ac > 0$, the equation has two unequal real number solutions.

EXAMPLE 3

Solve by using the quadratic formula:
$2x^2 - x + 5 = 0$

Solution
$$2x^2 - x + 5 = 0$$
$$x = \frac{-b \pm \sqrt{b^2 - 4ac}}{2a}$$
$$= \frac{-(-1) \pm \sqrt{(-1)^2 - 4(2)(5)}}{2 \cdot 2}$$ • $a = 2, b = -1,$
$$c = 5$$
$$= \frac{1 \pm \sqrt{1 - 40}}{4} = \frac{1 \pm \sqrt{-39}}{4} = \frac{1 \pm i\sqrt{39}}{4}$$

The solutions are $\frac{1}{4} + \frac{\sqrt{39}}{4}i$ and $\frac{1}{4} - \frac{\sqrt{39}}{4}i$.

YOU TRY IT 3

Solve by using the quadratic formula:
$x^2 = 2x - 10$

Your solution
$1 + 3i$ and $1 - 3i$

Solution on p. S29

EXAMPLE 4

Solve by using the quadratic formula:
$2x^2 = (x - 2)(x - 3)$

Solution

$2x^2 = (x - 2)(x - 3)$
$2x^2 = x^2 - 5x + 6$
$x^2 + 5x - 6 = 0$ • **Write in standard form.**
$a = 1, b = 5, c = -6$

$x = \dfrac{-b \pm \sqrt{b^2 - 4ac}}{2a}$

$= \dfrac{-5 \pm \sqrt{5^2 - 4(1)(-6)}}{2 \cdot 1}$

$= \dfrac{-5 \pm \sqrt{25 + 24}}{2}$

$= \dfrac{-5 \pm \sqrt{49}}{2}$

$= \dfrac{-5 \pm 7}{2}$

$x = \dfrac{-5 + 7}{2} \qquad x = \dfrac{-5 - 7}{2}$

$= \dfrac{2}{2} = 1 \qquad = \dfrac{-12}{2} = -6$

The solutions are 1 and -6.

YOU TRY IT 4

Solve by using the quadratic formula:
$4x^2 = 4x - 1$

Your solution

$\dfrac{1}{2}$

EXAMPLE 5

Use the discriminant to determine whether
$4x^2 - 2x + 5 = 0$ has two equal real number
solutions, two unequal real number solutions, or
two complex number solutions.

Solution

$a = 4, b = -2, c = 5$
$b^2 - 4ac = (-2)^2 - 4(4)(5)$
$\qquad\qquad = 4 - 80$
$\qquad\qquad = -76$
$-76 < 0$

Because the discriminant is less than zero, the
equation has two complex number solutions.

YOU TRY IT 5

Use the discriminant to determine whether
$3x^2 - x - 1 = 0$ has two equal real number solutions, two unequal real number solutions, or two
complex number solutions.

Your solution

Two unequal real number solutions

Solutions on p. S29

10.2 EXERCISES

✔ **Concept Check**

SUGGESTED ASSIGNMENT
Exercises 1–4; Exercises 5–89, odds;
More challenging exercises: Exercises 93–99, odds

1. State whether each expression is a perfect-square trinomial.

    a. $x^2 + 8x - 16$ No   **b.** $x^2 - 8x + 16$ Yes

    c. $y^2 - 7y + \dfrac{49}{4}$ Yes   **d.** $v^2 - 3v + \dfrac{9}{2}$ No

2. What term must be added to $x^2 - 10x$ to make the resulting expression a perfect square trinomial?   25

3. Can every quadratic equation be solved by using the quadratic formula?   Yes

4. If $ax^2 + bx + c = 0$, $a \neq 0$, what is $b^2 - 4ac$ called?   The discriminant

---

OBJECTIVE A   *To solve a quadratic equation by completing the square*

For Exercises 5 to 36, solve by completing the square.

5. $x^2 - 4x - 5 = 0$

    −1 and 5

6. $y^2 + 6y + 5 = 0$

    −5 and −1

7. $z^2 - 6z + 9 = 0$

    3

8. $u^2 + 10u + 25 = 0$

    −5

9. $r^2 + 4r - 7 = 0$

    $-2 - \sqrt{11}$ and $-2 + \sqrt{11}$

10. $s^2 + 6s - 1 = 0$

    $-3 - \sqrt{10}$ and $-3 + \sqrt{10}$

11. $x^2 - 6x + 7 = 0$

    $3 - \sqrt{2}$ and $3 + \sqrt{2}$

12. $y^2 + 8y + 13 = 0$

    $-4 - \sqrt{3}$ and $-4 + \sqrt{3}$

13. $p^2 = 3p - 1$

    $\dfrac{3 - \sqrt{5}}{2}$ and $\dfrac{3 + \sqrt{5}}{2}$

14. $r^2 - 5r = 2$

    $\dfrac{5 - \sqrt{33}}{2}$ and $\dfrac{5 + \sqrt{33}}{2}$

15. $y^2 - 6y = 4$

    $3 - \sqrt{13}$ and $3 + \sqrt{13}$

16. $w^2 = 2 - 4w$

    $-2 - \sqrt{6}$ and $-2 + \sqrt{6}$

17. $z^2 = z + 4$

    $\dfrac{1 - \sqrt{17}}{2}$ and $\dfrac{1 + \sqrt{17}}{2}$

18. $r^2 = 3r - 1$

    $\dfrac{3 - \sqrt{5}}{2}$ and $\dfrac{3 + \sqrt{5}}{2}$

19. $z^2 - 2z + 2 = 0$

    $1 - i$ and $1 + i$

20. $t^2 - 4t + 8 = 0$

    $2 - 2i$ and $2 + 2i$

21. $v^2 = 4v - 13$

    $2 - 3i$ and $2 + 3i$

22. $x^2 = 2x - 17$

    $1 - 4i$ and $1 + 4i$

23. $p^2 + 6p = -13$

$-3 - 2i$ and $-3 + 2i$

24. $x^2 + 4x = -20$

$-2 - 4i$ and $-2 + 4i$

25. $2s^2 = 4s + 5$

$\dfrac{2 - \sqrt{14}}{2}$ and $\dfrac{2 + \sqrt{14}}{2}$

26. $3u^2 = 6u + 1$

$\dfrac{3 - 2\sqrt{3}}{3}$ and $\dfrac{3 + 2\sqrt{3}}{3}$

27. $4x^2 - 4x + 5 = 0$

$\dfrac{1}{2} - i$ and $\dfrac{1}{2} + i$

28. $4t^2 - 4t + 17 = 0$

$\dfrac{1}{2} - 2i$ and $\dfrac{1}{2} + 2i$

29. $9x^2 - 6x + 2 = 0$

$\dfrac{1}{3} - \dfrac{1}{3}i$ and $\dfrac{1}{3} + \dfrac{1}{3}i$

30. $9y^2 - 12y + 13 = 0$

$\dfrac{2}{3} - i$ and $\dfrac{2}{3} + i$

31. $y - 2 = (y - 3)(y + 2)$

$1 - \sqrt{5}$ and $1 + \sqrt{5}$

32. $8s - 11 = (s - 4)(s - 2)$

$7 - \sqrt{30}$ and $7 + \sqrt{30}$

33. $6t - 2 = (2t - 3)(t - 1)$

$\dfrac{1}{2}$ and 5

34. $2z + 9 = (2z + 3)(z + 2)$

-3 and $\dfrac{1}{2}$

35. $(x - 4)(x + 1) = x - 3$

$2 - \sqrt{5}$ and $2 + \sqrt{5}$

36. $(y - 3)^2 = 2y + 10$

$4 - \sqrt{17}$ and $4 + \sqrt{17}$

For Exercises 37 to 40, solve by completing the square. Approximate the solutions to the nearest thousandth.

37. $z^2 + 2z = 4$
-3.236 and 1.236

38. $t^2 - 4t = 7$
-1.317 and 5.317

39. $2x^2 = 4x - 1$
0.293 and 1.707

40. $3y^2 = 5y - 1$
0.232 and 1.434

41. For what values of c does the equation $x^2 + 4x + c = 0$ have real number solutions?

$c \leq 4$

42. For what values of c does the equation $x^2 - 6x + c = 0$ have complex number solutions?

$c > 9$

OBJECTIVE B *To solve a quadratic equation by using the quadratic formula*

43. Write the quadratic formula. What does each variable in the formula represent?

44. Write the expression that appears under the radical symbol in the quadratic formula. What is this quantity called? What can it be used to determine?

For Exercises 45 to 77, solve by using the quadratic formula.

45. $x^2 - 3x - 10 = 0$

-2 and 5

46. $y^2 + 5y - 36 = 0$

-9 and 4

47. $x^2 - 8x + 9 = 0$

$4 - \sqrt{7}$ and $4 + \sqrt{7}$

48. $x^2 - 4x - 1 = 0$

$2 - \sqrt{5}$ and $2 + \sqrt{5}$

49. $v^2 = 6v + 19$

$3 - 2\sqrt{7}$ and $3 + 2\sqrt{7}$

50. $z^2 + 2z = 74$

$-1 - 5\sqrt{3}$ and $-1 + 5\sqrt{3}$

51. $x^2 = 14x - 4$

$7 - 3\sqrt{5}$ and $7 + 3\sqrt{5}$

52. $v^2 = 12v - 24$

$6 - 2\sqrt{3}$ and $6 + 2\sqrt{3}$

53. $2z^2 - 2z - 1 = 0$

$\dfrac{1 - \sqrt{3}}{2}$ and $\dfrac{1 + \sqrt{3}}{2}$

54. $9x^2 + 6x = 11$

$\dfrac{-1 - 2\sqrt{3}}{3}$ and $\dfrac{-1 + 2\sqrt{3}}{3}$

55. $4r^2 = 20r - 17$

$\dfrac{5 - 2\sqrt{2}}{2}$ and $\dfrac{5 + 2\sqrt{2}}{2}$

56. $9x^2 - 12x = 68$

$\dfrac{2 - 6\sqrt{2}}{3}$ and $\dfrac{2 + 6\sqrt{2}}{3}$

57. $z^2 + 2z + 2 = 0$

$-1 - i$ and $-1 + i$

58. $p^2 - 4p + 5 = 0$

$2 - i$ and $2 + i$

59. $y^2 - 2y + 5 = 0$

$1 - 2i$ and $1 + 2i$

60. $t^2 - 6t + 10 = 0$

$3 - i$ and $3 + i$

61. $4s - 13 = s^2$

$2 - 3i$ and $2 + 3i$

62. $-6x - 13 = x^2$

$-3 - 2i$ and $-3 + 2i$

63. $4x^2 - 4x + 33 = 0$

$\dfrac{1}{2} - 2i\sqrt{2}$ and $\dfrac{1}{2} + 2i\sqrt{2}$

64. $4x^2 + 89 = 12x$

$\dfrac{3}{2} - 2i\sqrt{5}$ and $\dfrac{3}{2} + 2i\sqrt{5}$

65. $6v + 71 = 9v^2$

$\dfrac{1 - 6\sqrt{2}}{3}$ and $\dfrac{1 + 6\sqrt{2}}{3}$

66. $47 - 4x = 4x^2$

$\dfrac{-1 - 4\sqrt{3}}{2}$ and $\dfrac{-1 + 4\sqrt{3}}{2}$

67. $2w^2 - w - 5 = w$

$\dfrac{1 - \sqrt{11}}{2}$ and $\dfrac{1 + \sqrt{11}}{2}$

68. $v^2 + 8v + 3 = -v^2$

$\dfrac{-4 - \sqrt{10}}{2}$ and $\dfrac{-4 + \sqrt{10}}{2}$

69. $3x^2 - x = x^2 - 5x + 6$

-3 and 1

70. $2x^2 - 15 = x^2 - 2x$

-5 and 3

71. $2x^2 + x = (x - 4)(x - 2)$

-8 and 1

72. $2x^2 - 3x - 1 = (x + 2)(x + 3)$

$4 - \sqrt{2}i$ and $4 + \sqrt{2}i$

73. $(2x + 1)(x + 2) = (x - 4)(x + 3)$

$-3 - i\sqrt{5}$ and $-3 + i\sqrt{5}$

74. $(3x - 1)(x - 1) = 6x^2 + x - 1$

-2 and $\dfrac{1}{3}$

75. $2x^2 - x = (x + 3)(x - 2)$

$1 - i\sqrt{5}$ and $1 + i\sqrt{5}$

76. $2x^2 + 3x + 1 = (x - 2)(x + 2)$

$-\dfrac{3}{2} - \dfrac{\sqrt{11}}{2}i$ and $-\dfrac{3}{2} + \dfrac{\sqrt{11}}{2}i$

77. $5t^2 - 5t + 7 = (t - 1)(t - 2)$

$\dfrac{1}{4} - \dfrac{\sqrt{19}}{4}i$ and $\dfrac{1}{4} + \dfrac{\sqrt{19}}{4}i$

For Exercises 78 to 83, solve by using the quadratic formula. Approximate the solutions to the nearest thousandth.

78. $x^2 - 6x - 6 = 0$
-0.873 and 6.873

79. $p^2 - 8p + 3 = 0$
0.394 and 7.606

80. $r^2 - 2r = 4$
-1.236 and 3.236

81. $w^2 + 4w = 1$
-4.236 and 0.236

82. $3t^2 = 7t + 1$
-0.135 and 2.468

83. $2y^2 = y + 5$
-1.351 and 1.851

For Exercises 84 to 89, use the discriminant to determine whether the quadratic equation has two equal real number solutions, two unequal real number solutions, or two complex number solutions.

84. $2z^2 - z + 5 = 0$
Two complex

85. $3y^2 + y + 1 = 0$
Two complex

86. $9x^2 - 12x + 4 = 0$
Two equal real

87. $4x^2 + 20x + 25 = 0$
Two equal real

88. $2v^2 - 3v - 1 = 0$
Two unequal real

89. $3w^2 + 3w - 2 = 0$
Two unequal real

90. 🐾 Suppose $a > 0$ and $c < 0$. Does the equation $ax^2 + bx + c = 0$ **(i)** always have real number solutions, **(ii)** never have real number solutions, or **(iii)** sometimes have real number solutions, depending on the value of b?
i

91. 🐾 If $a > 0$ and $c > 0$, what is the smallest value of b in $ax^2 + bx + c = 0$ that will guarantee that the equation will have real number solutions?
$\sqrt{4ac}$

Critical Thinking

For what values of p do the quadratic equations in Exercises 92 and 93 have two unequal real number solutions? Write the answer in set-builder notation.

92. $x^2 + 10x + p = 0$
$\{p \,|\, p < 25\}$

93. $x^2 - 6x + p = 0$
$\{p \,|\, p < 9\}$

For what values of p do the quadratic equations in Exercises 94 and 95 have two complex number solutions? Write the answer in interval notation.

94. $x^2 + 4x + p = 0$
$(4, \infty)$

95. $x^2 - 2x + p = 0$
$(1, \infty)$

For Exercises 96 and 97, i is the imaginary unit. Solve for x.

96. $x^2 + ix + 2 = 0$
$-2i$ and i

97. $x^2 - 2i + 15 = 0$
$-3i$ and $5i$

QUICK QUIZ

Solve by completing the square.

1. $x^2 + 4x + 3 = 0$
   $-3, -1$ **[10.2A]**

2. $2x^2 = 5 - x$
   $\dfrac{-1 + \sqrt{41}}{4}$,
   $\dfrac{-1 - \sqrt{41}}{4}$ **[10.2A]**

Solve by using the quadratic formula.

3. $x^2 + 3x - 18 = 0$
   $-6, 3$ **[10.2B]**

4. $x^2 = 4x - 6$
   $2 + i\sqrt{2}$,
   $2 - i\sqrt{2}$ **[10.2B]**

Projects or Group Activities

98. Sports   Assuming no air resistance, the height h, in feet, of a ball x feet from where it was hit by a batter can be approximated by $h = -0.0039x^2 + 1.1918x + 4$.
   a. Will the ball clear a fence that is 10 ft high and 300 ft away from the batter?
   Yes
   b. How far from the batter does the ball hit the ground? Round to the nearest foot.
   309 ft

99. Sports   After a baseball is hit, there are two equations that can be considered. One gives the height h (in feet) of the ball above the ground t seconds after it is hit. The second is the distance s (in feet) of the ball from home plate t seconds after it is hit. A model of this situation is given by $h = -16t^2 + 70t + 4$ and $s = 44.5t$. Using this model, determine whether the ball will clear a fence 325 ft from home plate.
No. The ball will have gone only 197.2 ft when it hits the ground.

© Aspen Photo/Shutterstock.com

10.3   Solving Equations That Are Reducible to Quadratic Equations

OBJECTIVE A   *To solve an equation that is quadratic in form*

Certain equations that are not quadratic can be expressed in quadratic form by making suitable substitutions.

QUADRATIC IN FORM

An equation is **quadratic in form** if it can be written in the form $au^2 + bu + c = 0$.

EXAMPLES

1. $x^4 - 4x^2 - 5 = 0$ is quadratic in form.
Let $u = x^2$. Then $u^2 = (x^2)^2 = x^4$. Replace x^4 with u^2 and x^2 with u.

$x^4 - 4x^2 - 5 = 0$
$u^2 - 4u - 5 = 0$ • $u = x^2, u^2 = x^4$

2. $4y - 3y^{1/2} + 6 = 0$ is quadratic in form.
Let $u = y^{1/2}$. Then $u^2 = (y^{1/2})^2 = y$. Replace y with u^2 and $y^{1/2}$ with u.

$4y - 3y^{1/2} + 6 = 0$
$4u^2 - 3u + 6 = 0$ • $u = y^{1/2}, u^2 = y$

The key to recognizing equations that are quadratic in form is as follows. When the equation is written in standard form, the exponent on one variable term is 2 times the exponent on the other variable term.

IN-CLASS EXAMPLES
Solve.
1. $x^4 - 10x^2 + 9 = 0$
   $-3, 3, -1, 1$
2. $x - 3x^{1/2} - 4 = 0$ 16
3. $x^4 - 15x^2 - 16 = 0$
   $-4, 4, -i, i$
4. $x^{2/3} - x^{1/3} - 2 = 0$
   $-1, 8$

HOW TO 1   Solve: $z + 7z^{1/2} - 18 = 0$

$z + 7z^{1/2} - 18 = 0$ • The exponent on z is 2 times the exponent on $z^{1/2}$. The equation is quadratic in form.

$u^2 + 7u - 18 = 0$ • Let $u = z^{1/2}$. Then $u^2 = z$.
$(u - 2)(u + 9) = 0$ • Solve by factoring.

$u - 2 = 0 \qquad u + 9 = 0$
$\quad u = 2 \qquad\quad u = -9$
$z^{1/2} = 2 \qquad z^{1/2} = -9$ • Replace u by $z^{1/2}$.
$(z^{1/2})^2 = 2^2 \qquad (z^{1/2})^2 = (-9)^2$
$\quad z = 4 \qquad\qquad z = 81$

Check:

$$
\begin{array}{r|l}
z + 7z^{1/2} - 18 = 0 & \\
\hline
4 + 7(4)^{1/2} - 18 & 0 \\
4 + 7 \cdot 2 - 18 & 0 \\
4 + 14 - 18 & 0 \\
& 0 = 0
\end{array}
\qquad
\begin{array}{r|l}
z + 7z^{1/2} - 18 = 0 & \\
\hline
81 + 7(81)^{1/2} - 18 & 0 \\
81 + 7 \cdot 9 - 18 & 0 \\
81 + 63 - 18 & 0 \\
& 126 \neq 0
\end{array}
$$

4 checks as a solution, but 81 does not check as a solution.

The solution is 4.

Take Note

When each side of an equation is squared, the resulting equation may have a solution that is not a solution of the original equation.

EXAMPLE 1

Solve: $x^4 + x^2 - 12 = 0$

Solution

$x^4 + x^2 - 12 = 0$ • The exponent on x^4 is 2 times the exponent on x^2. The equation is quadratic in form.

$u^2 + u - 12 = 0$ • Let $u = x^2$. Then $u^2 = x^4$.

$(u - 3)(u + 4) = 0$

$u - 3 = 0 \qquad u + 4 = 0$
$\quad u = 3 \qquad\qquad u = -4$

Replace u by x^2.

$x^2 = 3 \qquad\qquad x^2 = -4$ • Solve for x.
$\sqrt{x^2} = \sqrt{3} \qquad \sqrt{x^2} = \sqrt{-4}$
$\quad x = \pm\sqrt{3} \qquad\quad x = \pm 2i$

Check the solutions.

The solutions are $\sqrt{3}$, $-\sqrt{3}$, $2i$, and $-2i$.

YOU TRY IT 1

Solve: $x - 5x^{1/2} + 6 = 0$

Your solution

4 and 9

Solution on p. S29

OBJECTIVE B   *To solve a radical equation that is reducible to a quadratic equation*

Certain equations containing radicals can be expressed as quadratic equations.

 Take Note

In the third line of the solution in HOW TO 2, each side of the equation is squared. When this happens, you MUST check the resulting solutions. In this case, only one of the proposed solutions is actually a solution of the equation.

IN-CLASS EXAMPLES

Solve.

5. $\sqrt{x - 1} + x = 7$ **5**

6. $\sqrt{3w + 7} = w - 1$ **6**

7. $\sqrt{2x - 3} - 1 = \sqrt{x - 2}$

  2, 6

HOW TO 2   Solve: $\sqrt{x + 2} + 4 = x$

$\sqrt{x + 2} + 4 = x$

$\quad \sqrt{x + 2} = x - 4$ • Solve for the radical expression.

$(\sqrt{x + 2})^2 = (x - 4)^2$ • Square each side of the equation.

$\quad x + 2 = x^2 - 8x + 16$ • Simplify.

$\quad\quad 0 = x^2 - 9x + 14$ • Write the equation in standard form.

$\quad\quad 0 = (x - 7)(x - 2)$ • Factor.

$x - 7 = 0 \quad x - 2 = 0$ • Solve for x.
$\quad x = 7 \qquad\quad x = 2$

Check each solution.

Check:

$$\begin{array}{c|c} \sqrt{x + 2} + 4 = x \\ \hline \sqrt{7 + 2} + 4 & 7 \\ \sqrt{9} + 4 & 7 \\ 3 + 4 & 7 \\ 7 = 7 \end{array} \qquad \begin{array}{c|c} \sqrt{x + 2} + 4 = x \\ \hline \sqrt{2 + 2} + 4 & 2 \\ \sqrt{4} + 4 & 2 \\ 2 + 4 & 2 \\ 6 \neq 2 \end{array}$$

7 checks as a solution, but 2 does not check as a solution.

The solution is 7.

EXAMPLE 2

Solve: $\sqrt{7y - 3} + 3 = 2y$

Solution

$\sqrt{7y - 3} + 3 = 2y$

$\sqrt{7y - 3} = 2y - 3$ • Solve for the radical.

$(\sqrt{7y - 3})^2 = (2y - 3)^2$ • Square each side.

$7y - 3 = 4y^2 - 12y + 9$

$0 = 4y^2 - 19y + 12$ • Write in standard form.

$0 = (4y - 3)(y - 4)$ • Factor

$4y - 3 = 0 \qquad y - 4 = 0$ • Solve for y.

$4y = 3 \qquad\qquad y = 4$

$y = \dfrac{3}{4}$

4 checks as a solution.

$\frac{3}{4}$ does not check as a solution.

The solution is 4.

YOU TRY IT 2

Solve: $\sqrt{2x + 1} + x = 7$

Your solution

4

EXAMPLE 3

Solve: $\sqrt{2y + 1} - \sqrt{y} = 1$

Solution

$\sqrt{2y + 1} - \sqrt{y} = 1$

Solve for one of the radical expressions.

$\sqrt{2y + 1} = \sqrt{y} + 1$

$(\sqrt{2y + 1})^2 = (\sqrt{y} + 1)^2$ • Square each side.

$2y + 1 = y + 2\sqrt{y} + 1$

$y = 2\sqrt{y}$ • Solve for the radical.

$y^2 = (2\sqrt{y})^2$ • Square each side.

$y^2 = 4y$

$y^2 - 4y = 0$ • Write in standard form.

$y(y - 4) = 0$ • Factor.

$y = 0 \qquad y - 4 = 0$ • Solve for y.

$\qquad\qquad y = 4$

0 and 4 check as solutions.

The solutions are 0 and 4.

YOU TRY IT 3

Solve: $\sqrt{2x - 1} + \sqrt{x} = 2$

Your solution

1

Solutions on p. S29

OBJECTIVE C | *To solve a rational equation that is reducible to a quadratic equation*

After each side of a rational equation has been multiplied by the LCM of the denominators, the resulting equation may be a quadratic equation.

HOW TO 3   Solve: $\dfrac{1}{r} + \dfrac{1}{r+1} = \dfrac{3}{2}$

$$\frac{1}{r} + \frac{1}{r+1} = \frac{3}{2}$$

$$2r(r+1)\left(\frac{1}{r} + \frac{1}{r+1}\right) = 2r(r+1) \cdot \frac{3}{2}$$

- Multiply each side of the equation by the LCM of the denominators.

$$2(r+1) + 2r = r(r+1) \cdot 3$$
$$2r + 2 + 2r = 3r(r+1)$$
$$4r + 2 = 3r^2 + 3r$$
$$0 = 3r^2 - r - 2$$

- Write the equation in standard form.

$$0 = (3r+2)(r-1)$$

- Factor.

$$3r + 2 = 0 \qquad r - 1 = 0$$
$$3r = -2 \qquad\quad r = 1$$
$$r = -\frac{2}{3}$$

- Solve for r.

$-\frac{2}{3}$ and 1 check as solutions.

The solutions are $-\frac{2}{3}$ and 1.

Take Note

In the second line of the solution in HOW TO 3, each side of the equation is multiplied by a variable expression. When this happens, you MUST check the solutions. In this case, both proposed solutions are actual solutions. This is not always the case.

EXAMPLE 4

Solve: $\dfrac{9}{x-3} = 2x + 1$

Solution

$$\frac{9}{x-3} = 2x + 1$$

$$(x-3)\frac{9}{x-3} = (x-3)(2x+1)$$

- Multiply each side by $x - 3$.

$$9 = 2x^2 - 5x - 3$$
$$0 = 2x^2 - 5x - 12$$

- Write in standard form.

$$0 = (2x+3)(x-4)$$

- Factor.

$$2x + 3 = 0 \qquad x - 4 = 0$$
$$2x = -3 \qquad\quad x = 4$$
$$x = -\frac{3}{2}$$

- Solve for x.

The solutions are $-\frac{3}{2}$ and 4.

YOU TRY IT 4

Solve: $3y + \dfrac{25}{3y-2} = -8$

Your solution

-1

IN-CLASS EXAMPLES

Solve.

8. $x = \dfrac{9}{x-8}$ $-1, 9$

9. $\dfrac{x}{3x-4} = \dfrac{3}{x+2}$ 3, 4

10. $\dfrac{x}{x+2} = \dfrac{-4}{x-2}$ $-1 + i\sqrt{7},\ -1 - i\sqrt{7}$

11. $\dfrac{2x-3}{x-4} - x = 2$ $-1, 5$

12. $\dfrac{3x+4}{x+5} - 4x = 8$ $-\dfrac{9}{4}, -4$

13. $\dfrac{2}{x} - \dfrac{3}{2x+1} = 2$ $\dfrac{-1+\sqrt{33}}{8}, \dfrac{-1-\sqrt{33}}{8}$

Solution on p. S29

10.3 EXERCISES

SUGGESTED ASSIGNMENT
Exercises 1 and 2; Exercises 3–57, odds;
More challenging exercises: Exercises 59 and 60

✔ **Concept Check**

1. Which of the following equations are quadratic in form?   i, ii, iii, iv, v
   (i) $2x^4 - 6x^2 + 3 = 0$
   (ii) $x^6 - x^3 - 1 = 0$
   (iii) $2x + x^{1/2} - 5 = 0$
   (iv) $3x^{1/2} + 4x^{1/4} - 5 = 0$
   (v) $5x - 3\sqrt{x} - 5 = 0$
   (vi) $5x^4 + 3x - 6 = 0$

2. Is 12 a solution of $\sqrt{2x + 1} + x = 7$? No

OBJECTIVE A   *To solve an equation that is quadratic in form*

For Exercises 3 to 6, state whether the equation could be solved by writing it as a quadratic equation of the form $u^2 - 8u - 20 = 0$.

3. $x^{10} - 8x^5 - 20 = 0$
   Yes

4. $x^{16} - 8x^4 - 20 = 0$
   No

5. $x^{1/10} - 8x^{1/5} - 20 = 0$
   No

6. $x^{2/5} - 8x^{1/5} - 20 = 0$
   Yes

For Exercises 7 to 24, solve.

7. $x^4 - 13x^2 + 36 = 0$
   $-3, -2, 2, 3$

8. $y^4 - 5y^2 + 4 = 0$
   $-2, -1, 1, 2$

9. $z^4 - 6z^2 + 8 = 0$
   $-2, -\sqrt{2}, \sqrt{2}, 2$

10. $t^4 - 12t^2 + 27 = 0$
    $-3, -\sqrt{3}, \sqrt{3}, 3$

11. $p - 3p^{1/2} + 2 = 0$
    1 and 4

12. $v - 7v^{1/2} + 12 = 0$
    9 and 16

13. $x - x^{1/2} - 12 = 0$
    16

14. $w - 2w^{1/2} - 15 = 0$
    25

15. $z^4 + 3z^2 - 4 = 0$
    $-1, 1, -2i, 2i$

16. $y^4 + 5y^2 - 36 = 0$
    $-2, 2, -3i, 3i$

17. $x^4 + 12x^2 - 64 = 0$
    $-2, 2, -4i, 4i$

18. $x^4 - 81 = 0$
    $-3, 3, -3i, 3i$

19. $p + 2p^{1/2} - 24 = 0$
    16

20. $v + 3v^{1/2} - 4 = 0$
    1

21. $y^{2/3} - 9y^{1/3} + 8 = 0$
    1 and 512

22. $z^{2/3} - z^{1/3} - 6 = 0$
    -8 and 27

23. $9w^4 - 13w^2 + 4 = 0$
    $-1, -\dfrac{2}{3}, \dfrac{2}{3}, 1$

24. $4y^4 - 7y^2 - 36 = 0$
    $-2, 2, -\dfrac{3}{2}i, \dfrac{3}{2}i$

OBJECTIVE B   *To solve a radical equation that is reducible to a quadratic equation*

25. For which of Exercises 27 to 44 will the first step in solving the equation be to square each side of the equation?
    Exercises 30, 31, 32, 36, 37, 38, 40, 41, 44

26. For which of Exercises 27 to 44 will it be necessary to square each side of the equation twice?
    Exercises 39 to 44

For Exercises 27 to 44, solve.

27. $\sqrt{x + 1} + x = 5$

  3

28. $\sqrt{x - 4} + x = 6$

  5

29. $x = \sqrt{x + 6}$

  9

30. $\sqrt{2y - 1} = y - 2$

  5

31. $\sqrt{3w + 3} = w + 1$

  −1 and 2

32. $\sqrt{2s + 1} = s - 1$

  4

33. $\sqrt{4y + 1} - y = 1$

  0 and 2

34. $\sqrt{3s + 4} + 2s = 12$

  4

35. $\sqrt{10x + 5} - 2x = 1$

  $-\dfrac{1}{2}$ and 2

36. $\sqrt{t + 8} = 2t + 1$

  1

37. $\sqrt{p + 11} = 1 - p$

  −2

38. $x - 7 = \sqrt{x - 5}$

  9

39. $\sqrt{x - 1} - \sqrt{x} = -1$

  1

40. $\sqrt{y + 1} = \sqrt{y + 5}$

  4

41. $\sqrt{2x - 1} = 1 - \sqrt{x - 1}$

  1

42. $\sqrt{x + 6} + \sqrt{x + 2} = 2$

  −2

43. $\sqrt{t + 3} + \sqrt{2t + 7} = 1$

  −3

44. $\sqrt{5 - 2x} = \sqrt{2 - x} + 1$

  −2 and 2

OBJECTIVE C *To solve a rational equation that is reducible to a quadratic equation*

45. To solve Exercise 49, the first step will be to multiply each side of the equation by what expression? $y + 2$

46. To solve Exercise 53, the first step will be to multiply each side of the equation by what expression? $(x + 2)(x - 2)$

For Exercises 47 to 58, solve.

47. $x = \dfrac{10}{x - 9}$

  −1 and 10

48. $z = \dfrac{5}{z - 4}$

  −1 and 5

49. $\dfrac{y - 1}{y + 2} + y = 1$

  −3 and 1

50. $\dfrac{2p - 1}{p - 2} + p = 8$

  3 and 5

51. $\dfrac{3r + 2}{r + 2} - 2r = 1$

  −1 and 0

52. $\dfrac{2v + 3}{v + 4} + 3v = 4$

  $-\dfrac{13}{3}$ and 1

53. $\dfrac{1}{x + 2} + \dfrac{x}{x - 2} = \dfrac{x + 6}{x^2 - 4}$

  −4

54. $\dfrac{1}{x + 4} - \dfrac{x}{2 - x} = \dfrac{x + 10}{x^2 + 2x - 8}$

  −6

55. $\dfrac{16}{z - 2} + \dfrac{16}{z + 2} = 6$

  $-\dfrac{2}{3}$ and 6

56. $\dfrac{2}{y + 1} + \dfrac{1}{y - 1} = 1$

  0 and 3

57. $\dfrac{t}{t - 2} + \dfrac{2}{t - 1} = 4$

  $\dfrac{4}{3}$ and 3

58. $\dfrac{4t + 1}{t + 4} + \dfrac{3t - 1}{t + 1} = 2$

  $-\dfrac{11}{5}$ and 1

Critical Thinking

59. Solve: $(\sqrt{x} + 3)^2 - 4\sqrt{x} - 17 = 0$ (*Hint:* Let $u = \sqrt{x} + 3$.) 4

60. Solve: $(\sqrt{x} - 2)^2 - 5\sqrt{x} + 14 = 0$ (*Hint:* Let $u = \sqrt{x} - 2$.) 9 and 36

Projects or Group Activities

61. 🏈 **Sports** According to the Compton's Interactive Encyclopedia, the minimum dimensions of a football used in the National Football Association games are 10.875 in. long and 20.75 in. in circumference at the center. A possible model for the cross section of a football is given by $y = \pm 3.3041\sqrt{1 - \dfrac{x^2}{29.7366}}$, where x is the distance from the center of the football and y is the radius of the football at x.

a. What is the domain of the equation? $\{x \mid -\sqrt{29.7366} \le x \le \sqrt{29.7366}\}$

b. Graph $y = 3.3041\sqrt{1 - \dfrac{x^2}{29.7366}}$ and $y = -3.3041\sqrt{1 - \dfrac{x^2}{29.7366}}$ on the same coordinate axes. Explain why the $\pm$ symbol occurs in the equation.

c. Determine the radius of the football when x is 3 in. Round to the nearest ten-thousandth.   2.7592 in.

b. The $\pm$ symbol occurs in the equation so that the graph pictures the entire shape of the football.

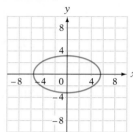

QUICK QUIZ
Solve.

1. $y^4 - 13y^2 + 36 = 0$
   $-2, 2, -3, 3$ [10.3A]
2. $x - 2x^{1/2} - 8 = 0$
   16 [10.3A]
3. $x^4 - 4x^2 - 5 = 0$
   $i, -i, \sqrt{5}, -\sqrt{5}$
   [10.3A]
4. $\sqrt{x + 2} + x = 10$
   7 [10.3B]
5. $\sqrt{x + 2} = \sqrt{x + 12}$
   4 [10.3B]
6. $\sqrt{3x + 4} - \sqrt{x} = 2$
   $0, 4$ [10.3B]
7. $x = \dfrac{6}{x - 5}$
   $-1, 6$ [10.3C]
8. $\dfrac{x}{x + 2} = \dfrac{3}{x - 2}$
   $6, -1$ [10.3C]
9. $\dfrac{x - 2}{x + 3} + x = 2$
   $-4, 2$ [10.3C]

✔ # CHECK YOUR PROGRESS: CHAPTER 10

For Exercises 1 and 2, solve the equation by factoring.

1. $3x^2 - 10x = 8$

$-\dfrac{2}{3}$ and 4 [10.1A]

2. $2x^2 + 3x = (x + 3)(x + 4)$

-2 and 6 [10.1A]

3. Find an equation with integer coefficients that has solutions -3 and $\frac{3}{5}$.

$5x^2 + 12x - 9 = 0$ [10.1A]

For Exercises 4 and 5, solve the quadratic equation by taking square roots.

4. $(x + 3)^2 = 20$
$-3 - 2\sqrt{5}$ and $-3 + 2\sqrt{5}$ [10.1B]

5. $(z - 4)^2 + 9 = 5$
$4 - 2i$ and $4 + 2i$ [10.1B]

For Exercises 6 and 7, solve the quadratic equation by completing the square.

6. $z^2 + 2x = 49$

$-1 - 5\sqrt{2}$ and $-1 + 5\sqrt{2}$ [10.2A]

7. $4x^2 + 12x + 21 = 0$

$-\dfrac{3}{2} - i\sqrt{3}$ and $-\dfrac{3}{2} + i\sqrt{3}$ [10.2A]

For Exercises 8 and 9, solve the quadratic equation by using the quadratic formula.

8. $4x^2 - 4x - 31 = 0$

$\dfrac{1 - 4\sqrt{2}}{2}$ and $\dfrac{1 + 4\sqrt{2}}{2}$ [10.2B]

9. $x^2 + 8x + 25 = 0$

$-4 - 3i$ and $-4 + 3i$ [10.2B]

For Exercises 10 to 12, solve the equation.

10. $x^4 + 8x^2 - 20 = 0$

$-\sqrt{2}, \sqrt{2}, -i\sqrt{10}, i\sqrt{10}$ [10.3A]

11. $\sqrt{2x + 1} - \sqrt{x + 1} = 2$

24 [10.3B]

12. $\dfrac{r}{r + 1} - \dfrac{2}{r} = \dfrac{3}{10}$

$-\dfrac{5}{7}$ and 4 [10.3C]

Unless otherwise noted, all content on this page is © Cengage Learning.

10.4   Applications of Quadratic Equations

OBJECTIVE A   *To solve application problems*

There are various applications of quadratic equations.

HOW TO 1   Assuming no air resistance, the height h, in feet, of a soccer ball x feet from where it was kicked at an angle of 45° to the ground can be given by $h = -\frac{32}{v^2}x^2 + x$, where v is the initial speed of the soccer ball in feet per second. If a soccer ball is kicked with an initial speed of 50 ft/s, how far from where the ball was kicked is the height of the ball 15 ft? Round to the nearest tenth.

Strategy

To find where the soccer ball will be 15 ft above the ground, use the equation $h = -\frac{32}{v^2}x^2 + x$. Substitute 15 for h and 50 for v, and then solve for x.

Solution

$$h = -\frac{32}{v^2}x^2 + x$$

$$15 = -\frac{32}{50^2}x^2 + x$$ • **Replace *h* by 15 and *v* by 50.**

$$0 = -0.0128x^2 + x - 15$$ • **Write the equation in standard form.**

$$x = \frac{-b \pm \sqrt{b^2 - 4ac}}{2a}$$ • **Solve by using the quadratic formula.**

$$x = \frac{-1 \pm \sqrt{1^2 - 4(-0.0128)(-15)}}{2(-0.0128)}$$ • ***a* = −0.0128, *b* = 1, *c* = −15**

$$= \frac{-1 \pm \sqrt{0.232}}{-0.0256} \approx \frac{-1 \pm 0.48166}{-0.0256}$$

$$x = \frac{-1 + 0.48166}{-0.0256} \quad \text{or} \quad x = \frac{-1 - 0.48166}{-0.0256}$$

$$= \frac{-0.51834}{-0.0256} \qquad\qquad = \frac{-1.48166}{-0.0256}$$

$$\approx 20.2 \qquad\qquad\qquad \approx 57.9$$

The ball is 15 ft above the ground when it is 20.2 ft and 57.9 ft from where it was kicked.

A drawing of the flight of the ball is shown below. Note that the ball is 15 ft high at two locations, 20.2 ft from the kicker and 57.9 ft from the kicker.

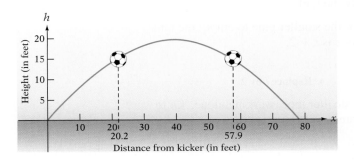

IN-CLASS EXAMPLES

1. An old pump requires 6 h longer to empty a pool than does a new pump. With both pumps working, the pool can be emptied in 4 h. Find the time required for the new pump, working alone, to empty the pool.   **6 h**

2. The rate of a jet in calm air is 225 mph. Flying with the wind, the jet can fly 1000 mi in 1 h less time than is required to make the return trip. Find the rate of the wind.   **25 mph**

3. The length of a rectangle is 2 ft less than three times the width. The area is 60 ft². Find the length of the rectangle. Round to the nearest hundredth.   **12.45 ft**

Unless otherwise noted, all content on this page is © Cengage Learning.

EXAMPLE 1

A small pipe takes 2 h longer to empty a tank than does a larger pipe. After the pipes work together for 1 h, the larger pipe becomes blocked. It takes the smaller pipe 1 more hour to empty the tank. How long would it take each pipe, working alone, to empty the tank?

Strategy

- This is a work problem.
- The unknown time for the larger pipe to empty the tank working alone: t
- The unknown time for the smaller pipe to empty the tank working alone: $t + 2$
- The larger pipe operates for 1 h. The smaller pipe operates for 2 h.

	Rate	·	Time	=	Part
Larger pipe	$\dfrac{1}{t}$	·	1	=	$\dfrac{1}{t}$
Smaller pipe	$\dfrac{1}{t + 2}$	·	2	=	$\dfrac{2}{t + 2}$

- The sum of the part of the task completed by the larger pipe and the part completed by the smaller pipe equals 1.

Solution

$$\frac{1}{t} + \frac{2}{t + 2} = 1$$

$$t(t + 2)\left(\frac{1}{t} + \frac{2}{t + 2}\right) = t(t + 2) \cdot 1$$

$$(t + 2) + 2t = t^2 + 2t$$

$$0 = t^2 - t - 2$$

$$0 = (t + 1)(t - 2)$$

$$t + 1 = 0 \qquad t - 2 = 0$$

$$t = -1 \qquad t = 2$$

Because time cannot be negative, $t = -1$ is not possible. It takes the larger pipe, working alone, 2 h to empty the tank.

The time for the smaller pipe to empty the tank working alone is $t + 2$.

$t + 2$
$2 + 2 = 4$ • **Replace t by 2.**

It takes the smaller pipe, working alone, 4 h to empty the tank.

It takes William 3 h longer than it does Olivia to detail a car. Working together, the friends can detail the car in 2 h. How long would it take William, working alone, to detail the car?

Your strategy

Your solution

6 h

Solution on p. S30

Unless otherwise noted, all content on this page is © Cengage Learning.

EXAMPLE 2

In 8 h, two campers rowed 15 mi down a river and then rowed back to their campsite. The rate of the river's current was 1 mph. Find the rate at which the campers row in still water.

Strategy

• Unknown rowing rate of the campers: r

	Distance	÷	Rate	=	Time
Down river	15	÷	$r + 1$	=	$\dfrac{15}{r+1}$
Up river	15	÷	$r - 1$	=	$\dfrac{15}{r-1}$

• The total time of the trip was 8 h.

Solution

$$\frac{15}{r+1} + \frac{15}{r-1} = 8$$

$$(r+1)(r-1)\left(\frac{15}{r+1} + \frac{15}{r-1}\right) = (r+1)(r-1) \cdot 8$$

$$15(r-1) + 15(r+1) = (r^2 - 1)8$$

$$30r = 8r^2 - 8$$

$$0 = 8r^2 - 30r - 8$$

$$0 = 2(4r+1)(r-4)$$

$$4r + 1 = 0 \qquad r - 4 = 0$$

$$r = -\frac{1}{4} \qquad r = 4$$

The rate cannot be negative, so the solution $-\frac{1}{4}$ is not possible. The rowing rate was 4 mph.

EXAMPLE 3

The length of a rectangle is 5 in. more than the width. The area is 36 in². Find the width.

Strategy

• Width of the rectangle: w
• Length of the rectangle: $w + 5$
• Use the formula $A = lw$.

Solution

$$A = lw$$

$$36 = (w+5)w$$

$$36 = w^2 + 5w$$

$$0 = w^2 + 5w - 36$$

$$0 = (w+9)(w-4)$$

$$w + 9 = 0 \qquad w - 4 = 0$$

$$w = -9 \qquad w = 4$$

The width cannot be negative. The width of the rectangle is 4 in.

YOU TRY IT 2

The rate of a jet in calm air is 250 mph. Flying with the wind, the jet can fly 1200 mi in 2 h less time than is required to make the return trip against the wind. Find the rate of the wind.

Your strategy

Your solution

50 mph

YOU TRY IT 3

The base of a triangle is 4 in. more than twice the height. The area is 35 in². Find the height of the triangle.

Your strategy

Your solution

5 in.

Solutions on p. S30

Unless otherwise noted, all content on this page is © Cengage Learning.

10.4 EXERCISES

SUGGESTED ASSIGNMENT
Exercises 1–4; Exercises 5–25, odds;
More challenging exercise: Exercise 27

✔ **Concept Check**

1. If the work on a project takes t hours to complete, what portion of the job is completed in 1 h? $\dfrac{1}{t}$

2. Suppose one person can do a job in 2 h and a second person can do the same job in 3 h. If they work together, will they complete the job in less than 2 h, in between 2 h and 3 h, or in more than 3h?   Less than 2 h

3. Let r be the rowing rate of a person in calm water. If the rate of a river's current is 2 mph, express the rate rowing down the river (with the current) and the rate rowing up the river (against the current) as variable expressions.   Down: $r + 2$; up: $r - 2$

4. Let r be the rate at which a plane can fly in calm air. Write a variable expression for the time it would take the plane to fly 500 mi into a headwind of 50 mph. Write a variable expression for the time it would take the plane to fly 500 mi with a tailwind of 50 mph.

    Headwind: $\dfrac{500}{r - 50}$; tailwind: $\dfrac{500}{r + 50}$

OBJECTIVE A *To solve application problems*

5. **Safety**   A car with good tire tread can stop in less distance than a car with poor tread. The formula for the stopping distance d, in feet, of a car with good tread on dry cement is approximated by $d = 0.04v^2 + 0.5v$, where v is the speed of the car. If the driver must be able to stop within 60 ft, what is the maximum safe speed, to the nearest mile per hour, of the car?   33 mph

6. **Rockets**   A model rocket is launched with an initial velocity of 200 ft/s. The height h, in feet, of the rocket t seconds after the launch is given by $h = -16t^2 + 200t$. How many seconds after the launch will the rocket be 300 ft above the ground? Round to the nearest hundredth of a second.   1.74 s and 10.76 s

7. **Physics**   The height of a projectile fired upward is given by $s = v_0 t - 16t^2$, where s is the height in feet, v_0 is the initial velocity, and t is the time in seconds. Find the time for a projectile with an initial velocity of 200 ft/s to return to Earth.   12.5 s

8. **Physics**   The depth d of a liquid in a bottle with a hole of area 0.5 cm² in its side can be approximated by $d = 0.0034t^2 - 0.52518t + 20$, where t is the time since a stopper was removed from the hole. When will the depth be 10 cm? Round to the nearest tenth of a second.   22.2 s

9. ⬤ **Sports**   The Water Cube was built in Beijing, China, to house the National Swimming Center for the 2008 Olympics. Although not actually a cube (its height is not equal to its length and width), the Water Cube is designed to look like a "cube" of water molecules. The volume of the 31-meter-high Water Cube is 971,199 m³. Find the length of a side of its square base. Recall that $V = LWH$.   177 m

10. 🔧 A large pipe and a small pipe, working together, can empty a pool in 4 h. Is it possible for the larger pipe, working alone, to empty the pool in less than 4 h?   No

© Nate A./Shutterstock.com

Marcel Lam/Corbis

Unless otherwise noted, all content on this page is © Cengage Learning.

11. **Tanks**   A small pipe can fill a tank in 6 min more time than it takes a larger pipe to fill the same tank. Working together, the pipes can fill the tank in 4 min. How long would it take each pipe, working alone, to fill the tank?
Smaller pipe: 12 min; larger pipe: 6 min

12. **Landscaping**   It takes a small sprinkler 16 min longer to soak a lawn than it takes a larger sprinkler. Working together, the sprinklers can soak the lawn in 6 min. How long would it take each sprinkler, working alone, to soak the lawn?
Larger sprinkler: 8 min; smaller sprinkler: 24 min

13. **Parallel Processing**   Parallel processing is the simultaneous use of more than one computer to run a program. Suppose one computer, working alone, takes 4 h longer than a second computer to run a program. After both computers work together for 1 h, the faster computer crashes. The slower computer continues working for another 2 h before completing the program. How long would it take the faster computer, working alone, to run the program?   2 h

14. **Payroll**   It takes one printer, working alone, 6 h longer to print a payroll than it takes a second printer. Working together, the printers can print the payroll in 4 h. How long would it take each printer, working alone, to print the payroll?
Slower printer: 12 h; faster printer: 6 h

15. **Flooring**   It takes an apprentice carpenter 2 h longer to install a section of floor than it does a more experienced carpenter. After the carpenters work together for 2 h, the experienced carpenter leaves for another job. It then takes the apprentice carpenter 2 h longer to complete the installation. How long would it take the apprentice, working alone, to install the floor? Round to the nearest tenth.   6.8 h

16. **Uniform Motion**   A ship made a trip of 100 mi in 8 h. The ship traveled the first 40 mi at a constant rate before increasing its speed by 5 mph. It traveled another 60 mi at the increased speed. Find the rate of the ship for the first 40 mi.   10 mph

17. ● **Uniform Motion**   The Concorde's speed in calm air was 1320 mph. Flying with the wind, the Concorde could fly from New York to London, a distance of approximately 4000 mi, in 0.5 h less time than was required to make the return trip. Find the rate of the wind to the nearest mile per hour.   108 mph

© Graham Bloomfield/Shutterstock.com

18. **Uniform Motion**   A car travels 120 mi. A second car, traveling 10 mph faster than the first car, makes the same trip in 1 h less time. Find the speed of each car.
First car: 30 mph; second car: 40 mph

19. ● **Air Force One**   The Air Force uses the designation VC-25 for the plane on which the president of the United States flies. When the president is on the plane, its call sign is Air Force One. The plane's speed in calm air is 630 mph. Flying with the jet stream, the plane can fly from Washington, D.C., to London, a distance of approximately 3660 mi, in 1.75 h less time than is required to make the return trip. Find the rate of the jet stream to the nearest mile per hour.   93 mph

© iStockphoto.com/David Birkbeck

20. **Uniform Motion**   For a portion of the Green River in Utah, the rate of the river's current is 4 mph. A tour guide can row 5 mi down this river and back in 3 h. Find the rowing rate of the guide in calm water.   6 mph

21. **Apartment rents** The manager of a 100-unit apartment complex is trying to decide what to charge for rent. Experience has shown that at a monthly rate of $1200, every unit will be occupied. For each $100 increase in the monthly rate, one additional unit will remain vacant. Find the number of units rented and the monthly rent if the total monthly rental revenue is $153,600.
(*Hint:* Revenue = monthly rent × number of units rented)
96 units rented at $1600/month; 16 units rented at $9600/month

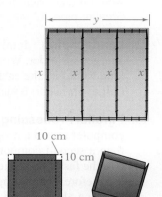

22. **Fencing** A rectangular enclosure for animals is fenced to produce three separate areas, as shown in the diagram at the right. If 800 ft of fencing is available and the total area to be enclosed is 18,750 ft², find the dimensions of the enclosure.
x = 75 ft and y = 250 ft, or x = 125 ft and y = 150 ft

10 cm

23. **Geometry** A square piece of cardboard is formed into a box by cutting 10-centimeter squares from each of the four corners and then folding up the sides, as shown in the figure. If the volume V of the box is to be 49,000 cm³, what size square piece of cardboard is needed? Recall that $V = LWH$. 90 cm by 90 cm

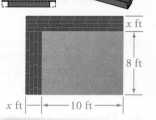

24. **Construction** A homeowner hires a mason to lay a brick border around a rectangular patio that measures 8 ft by 10 ft. If the total area of the patio and border is 168 ft², what is the width of the border? 4 ft

25. **Fencing** A dog trainer has 80 ft of fencing with which to create a rectangular work area for dogs. If the trainer wants to enclose an area of 300 ft², what will be the dimensions of the work area? 10 ft by 30 ft

26. ● **Sports** Read the article at the right.
   a. The screen in Michigan Stadium is a rectangle with length 9 ft less than twice its width. Find the length and width of the screen. Length: 85 ft; width: 47 ft
   b. The scoreboard structure is a rectangle with length 16 ft less than twice its width. Find the length and width of the scoreboard structure. Length: 108 ft; width: 62 ft

Critical Thinking

27. **Ice Cream** A perfectly spherical scoop of mint chocolate chip ice cream is placed in a cone, as shown in the figure. How far is the bottom of the scoop of ice cream from the bottom of the cone? Round to the nearest tenth. (*Hint:* A line segment from the center of the ice cream to the point at which the ice cream touches the cone is perpendicular to the edge of the cone.) 2.3 in.

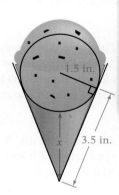

Projects or Group Activities

28. A 13-foot ladder rests against a building with the base of the ladder 5 ft from the base of the building as shown in the figure at the right.
   a. If the base of the ladder is moved 1 ft farther away from the wall, how many feet down does the top of the ladder move? Round to the nearest hundredth. 0.47 ft
   b. How much farther from its original position 5 ft from the wall should the base of the ladder be moved so that the top of the ladder moves down the same number of feet? 7 ft

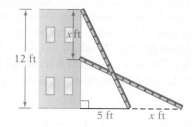

In the NEWS!

Wolverines on the Big Screen

Fans can now have a better view of Wolverine football, after the installation of 40% larger screens in Michigan Stadium's end zones. Each screen covers 3995 ft² of a new 6696-square-foot scoreboard structure.
Source: www.annarbor.com

QUICK QUIZ

1. A boat traveled 60 mi down a river and then returned. The total time for the round trip was 5 h, and the rate of the current was 5 mph. Find the rate of the boat in still water. **25 mph [10.4A]**

2. The height of a triangle is 3 in. less than the length of the base of the triangle. The area of the triangle is 80 in². Find the height of the triangle. Round to the nearest hundredth. **11.24 in. [10.4A]**

Unless otherwise noted, all content on this page is © Cengage Learning.

10.5   Quadratic Inequalities and Rational Inequalities

OBJECTIVE A   *To solve a nonlinear inequality*

A **quadratic inequality** is one that can be written in the form $ax^2 + bx + c < 0$ or $ax^2 + bx + c > 0$, where $a \neq 0$. The symbols $\leq$ and $\geq$ can also be used. The solution set of a quadratic inequality can be found by solving a compound inequality.

To solve $x^2 - 3x - 10 > 0$, first factor the trinomial.

$$x^2 - 3x - 10 > 0$$
$$(x + 2)(x - 5) > 0$$

There are two cases for which the product is positive: **(1)** both factors are positive, or **(2)** both factors are negative.

(1) $x + 2 > 0$ and $x - 5 > 0$
(2) $x + 2 < 0$ and $x - 5 < 0$

Solve each pair of compound inequalities.

(1) $x + 2 > 0$ and $x - 5 > 0$
         $x > -2$ $x > 5$
$\{x | x > -2\} \cap \{x | x > 5\} = \{x | x > 5\}$

(2) $x + 2 < 0$ and $x - 5 < 0$
         $x < -2$ $x < 5$
$\{x | x < -2\} \cap \{x | x < 5\} = \{x | x < -2\}$

Because cases (1) and (2) are connected by *or,* the solution set is the union of the solution sets of the individual inequalities.

$\{x | x > 5\} \cup \{x | x < -2\}$

IN-CLASS EXAMPLES
Solve.
 1. $x^2 - 5x + 6 \geq 0$
     $\{x | x \leq 2\} \cup \{x | x \geq 3\}$
 2. $(x - 1)(x + 1)(x - 2) < 0$
     $\{x | x < -1\} \cup \{x | 1 < x < 2\}$

Although the solution set of any quadratic inequality can be found by using the method outlined above, a graphical method is often easier to use.

HOW TO 1   Solve and graph the solution set of $x^2 - x - 6 < 0$.

Factor the trinomial.

$$x^2 - x - 6 < 0$$
$$(x - 3)(x + 2) < 0$$

On a number line, draw vertical lines indicating the numbers that make each factor equal to zero.

$x - 3 = 0$ $x + 2 = 0$
    $x = 3$ $x = -2$

For each factor, place plus signs above the regions where the factor is positive and minus signs where the factor is negative.

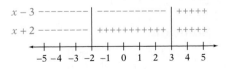

Because $x^2 - x - 6 < 0$, the solution set will be the regions where one factor is positive and the other factor is negative.

Write the solution set.

$\{x | -2 < x < 3\}$

The graph of the solution set of $x^2 - x - 6 < 0$ is shown at the right.

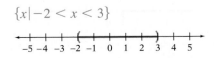

 Take Note

For each factor, choose a number in each region. For example: When $x = -4$, $x - 3$ is negative; when $x = 1$, $x - 3$ is negative; and when $x = 4$, $x - 3$ is positive. When $x = -4$, $x + 2$ is negative; when $x = 1$, $x + 2$ is positive; and when $x = 4$, $x + 2$ is positive.

Unless otherwise noted, all content on this page is © Cengage Learning.

HOW TO 2 Solve and graph the solution set of $(x - 2)(x + 1)(x - 4) > 0$.

On a number line, identify for each factor the regions where the factor is positive and those where the factor is negative.

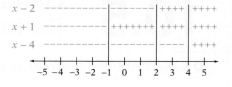

There are two regions where the product of the three factors is positive.

Write the solution set.

$$\{x|-1 < x < 2\} \cup \{x|x > 4\}$$

The graph of the solution set of $(x - 2)(x + 1)(x - 4) > 0$ is shown at the right.

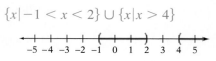

HOW TO 3 Solve: $\dfrac{2x - 5}{x - 4} \leq 1$

Rewrite the inequality so that zero appears on the right side of the inequality.

$$\frac{2x - 5}{x - 4} \leq 1$$

$$\frac{2x - 5}{x - 4} - 1 \leq 0$$

Simplify.

$$\frac{2x - 5}{x - 4} - \frac{x - 4}{x - 4} \leq 0$$

$$\frac{x - 1}{x - 4} \leq 0$$

On a number line, identify for each factor of the numerator and each factor of the denominator the regions where the factor is positive and those where the factor is negative.

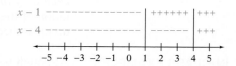

The region where the quotient of the two factors is negative is between 1 and 4.

Write the solution set.

$$\{x|1 \leq x < 4\}$$

Note that 1 is part of the solution set but 4 is not because the denominator of the rational expression is zero when $x = 4$.

EXAMPLE 1

Solve and graph the solution set of $2x^2 - x - 3 \geq 0$.

Solution

$$2x^2 - x - 3 \geq 0$$
$$(2x - 3)(x + 1) \geq 0$$

$$\{x|x \leq -1\} \cup \left\{x\middle|x \geq \frac{3}{2}\right\}$$

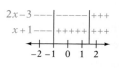

YOU TRY IT 1

Solve and graph the solution set of $2x^2 - x - 10 \leq 0$.

Your solution

$$\left\{x\middle|-2 \leq x \leq \frac{5}{2}\right\}$$

Solution on p. S30

Unless otherwise noted, all content on this page is © Cengage Learning.

10.5 EXERCISES

✔ Concept Check

SUGGESTED ASSIGNMENT
Exercises 1–6; Exercises 7–35, odds;
More challenging exercises: Exercises 37–41, odds

1. If $(x - 3)(x - 5) > 0$, what must be true of the values of $x - 3$ and $x - 5$?

2. For the inequality $\frac{x - 2}{x - 3} \leq 1$, which of the values 1, 2, and 3 is not a possible element of the solution set? Why?

For Exercises 3 to 6, for the given values of x, state whether the inequality is true or false.

3. $(x - 3)(x + 2)(x - 4) > 0$

   a. $x = 2$ **b.** $x = -2$ **c.** $x = -3$
     True   False   False

4. $(x + 5)(x - 6) \leq 0$

   a. $x < 6$ **b.** $-5 \leq x \leq 6$
     False   True

5. $\dfrac{(x + 1)(x - 5)}{x - 3} \geq 0$

   a. $x = 2$ **b.** $x = 3$ **c.** $x = -1$
     True   False   True

6. $\dfrac{x - 4}{x + 3} > 0$

   a. $x < -3$ **b.** $x > 4$
     True   True

OBJECTIVE A   *To solve a nonlinear inequality*

For Exercises 7 to 22, solve and graph the solution set.

7. $(x - 4)(x + 2) > 0$

$\{x | x < -2\} \cup \{x | x > 4\}$

8. $(x + 1)(x - 3) > 0$

$\{x | x < -1\} \cup \{x | x > 3\}$

9. $x^2 - 3x + 2 \geq 0$

$\{x | x \leq 1\} \cup \{x | x \geq 2\}$

10. $x^2 + 5x + 6 > 0$

$\{x | x < -3\} \cup \{x | x > -2\}$

11. $x^2 - x - 12 < 0$

$\{x | -3 < x < 4\}$

12. $x^2 + x - 20 < 0$

$\{x | -5 < x < 4\}$

Unless otherwise noted, all content on this page is © Cengage Learning.

13. $(x - 1)(x + 2)(x - 3) < 0$

$\{x|x < -2\} \cup \{x|1 < x < 3\}$

14. $(x + 4)(x - 2)(x + 1) > 0$

$\{x|-4 < x < -1\} \cup \{x|x > 2\}$

15. $(x + 4)(x - 2)(x - 1) \geq 0$

$\{x|-4 \leq x \leq 1\} \cup \{x|x \geq 2\}$

16. $(x - 1)(x + 5)(x - 2) \leq 0$

$\{x|x \leq -5\} \cup \{x|1 \leq x \leq 2\}$

17. $\dfrac{x - 4}{x + 2} > 0$

$\{x|x < -2\} \cup \{x|x > 4\}$

18. $\dfrac{x + 2}{x - 3} > 0$

$\{x|x < -2\} \cup \{x|x > 3\}$

19. $\dfrac{x - 3}{x + 1} \leq 0$

$\{x|-1 < x \leq 3\}$

20. $\dfrac{x - 1}{x} > 0$

$\{x|x < 0\} \cup \{x|x > 1\}$

21. $\dfrac{(x - 1)(x + 2)}{x - 3} \leq 0$

$\{x|x \leq -2\} \cup \{x|1 \leq x < 3\}$

22. $\dfrac{(x + 3)(x - 1)}{x - 2} \geq 0$

$\{x|-3 \leq x \leq 1\} \cup \{x|x > 2\}$

For Exercises 23 to 36, solve.

23. $x^2 - 16 > 0$

$\{x|x > 4\} \cup \{x|x < -4\}$

24. $x^2 - 4 \geq 0$

$\{x|x \leq -2\} \cup \{x|x \geq 2\}$

25. $x^2 - 9x \leq 36$

$\{x|-3 \leq x \leq 12\}$

26. $x^2 + 4x > 21$

$\{x|x < -7\} \cup \{x|x > 3\}$

27. $4x^2 - 8x + 3 < 0$

$\left\{x \middle| \dfrac{1}{2} < x < \dfrac{3}{2}\right\}$

28. $2x^2 + 11x + 12 \geq 0$

$\{x|x \leq -4\} \cup \left\{x \middle| x \geq -\dfrac{3}{2}\right\}$

Unless otherwise noted, all content on this page is © Cengage Learning.

29. $\dfrac{3}{x-1} < 2$

$\{x | x < 1\} \cup \left\{x | x > \dfrac{5}{2}\right\}$

30. $\dfrac{x}{(x-1)(x+2)} \geq 0$

$\{x | x > 1\} \cup \{x | -2 < x \leq 0\}$

31. $\dfrac{x-2}{(x+1)(x-1)} \leq 0$

$\{x | x < -1\} \cup \{x | 1 < x \leq 2\}$

32. $\dfrac{1}{x} < 2$

$\left\{x | x > \dfrac{1}{2}\right\} \cup \{x | x < 0\}$

33. $\dfrac{x}{2x-1} \geq 1$

$\left\{x | \dfrac{1}{2} < x \leq 1\right\}$

34. $\dfrac{x}{2x-3} \leq 1$

$\left\{x | x < \dfrac{3}{2}\right\} \cup \{x | x \geq 3\}$

35. $\dfrac{3}{x-5} > \dfrac{1}{x+1}$

$\{x | x > 5\} \cup \{x | -4 < x < -1\}$

36. $\dfrac{3}{x-2} > \dfrac{2}{x+2}$

$\{x | x > 2\} \cup \{x | -10 < x < -2\}$

QUICK QUIZ
Solve and graph the solution set.

1. $x^2 + 2x - 8 \leq 0$
  $\{x | -4 \leq x \leq 2\}$ **[10.5A]**

2. $(x+3)(x-1)(x+2) > 0$
  $\{x | x > 1\} \cup \{x | -3 < x < -2\}$
  [10.5A]

3. $\dfrac{x+1}{x-2} > 0$
  $\{x | x < -1\} \cup \{x | x > 2\}$
  [10.5A]

Critical Thinking

For Exercises 37 to 42, graph the solution set.

37. $(x-1)(x+3)(x-2)(x-4) \geq 0$

38. $(x+2)(x-3)(x+1)(x+4) > 0$

39. $(x^2 + 2x - 3)(x^2 + 3x + 2) \geq 0$

40. $(x^2 + 2x - 8)(x^2 - 2x - 3) < 0$

41. $\dfrac{x^2(3-x)(2x+1)}{(x+4)(x+2)} \geq 0$

42. $(x^2 + 1)(x^2 - 3x + 2) > 0$

Projects or Group Activities

43. You shoot an arrow into the air with an initial velocity of 70 m/s. The distance up, in meters, is given by $d = rt - 5t^2$, where t is the number of seconds since the arrow was shot and r is the initial velocity. Find the interval of time during which the arrow will be more than 200 m high.   Between 4 s and 10 s

Unless otherwise noted, all content on this page is © Cengage Learning.

10 Summary

Key Words	**Examples**
A **quadratic equation** is an equation of the form $ax^2 + bx + c = 0$, where $a \neq 0$. A quadratic equation is also called a **second-degree equation.** A quadratic equation is in **standard form** when the polynomial is in descending order and equal to zero. [10.1A, p. 568]	$3x^2 + 4x - 7 = 0$, $x^2 - 1 = 0$, and $4x^2 + 8x = 0$ are quadratic equations written in standard form.
For an equation of the form $ax^2 + bx + c = 0$, the quantity $b^2 - 4ac$ is called the **discriminant.** [10.2B, p. 582]	$2x^2 - 3x + 5 = 0$ $a = 2, b = -3, c = 5$ $b^2 - 4ac = (-3)^2 - 4(2)(5) = -31$
An equation is **quadratic in form** if it can be written as $au^2 + bu + c = 0$. [10.3A, p. 588]	$6x^4 - 5x^2 - 4 = 0$ $6u^2 - 5u - 4 = 0$ • $u = x^2, u^2 = x^4$ The equation is quadratic in form.
A **quadratic inequality** is one that can be written in the form $ax^2 + bx + c < 0$ or $ax^2 + bx + c > 0$, where $a \neq 0$. The symbols $\leq$ and $\geq$ can also be used. [10.5A, p. 601]	$3x^2 + 5x - 8 \leq 0$ is a quadratic inequality.

Essential Rules and Procedures	**Examples**
Principle of Zero Products [10.1A, p. 568] If a and b are real numbers and $ab = 0$, then $a = 0$ or $b = 0$.	$(x - 3)(x + 4) = 0$ $x - 3 = 0 \qquad x + 4 = 0$ $x = 3 \qquad\qquad x = -4$
To Solve a Quadratic Equation by Factoring [10.1A, p. 568] **1.** Write the equation in standard form. **2.** Factor the polynomial. **3.** Use the Principle of Zero Products to set each factor equal to zero. **4.** Solve each equation. **5.** Check the solutions.	$x^2 + 2x = 35$ $x^2 + 2x - 35 = 0$ $(x - 5)(x + 7) = 0$ $x - 5 = 0 \qquad x + 7 = 0$ $x = 5 \qquad\qquad x = -7$
To Write a Quadratic Equation Given Its Solutions [10.1A, p. 569] Use the equation $(x - r_1)(x - r_2) = 0$. Replace r_1 with one solution and r_2 with the other solution. Then multiply the two factors.	Write a quadratic equation that has solutions -3 and 6. $(x - r_1)(x - r_2) = 0$ $[x - (-3)](x - 6) = 0$ $(x + 3)(x - 6) = 0$ $x^2 - 3x - 18 = 0$

To Solve a Quadratic Equation by Taking Square Roots
[10.1B, pp. 570, 571]
1. Solve for x^2 or for $(x + a)^2$.
2. Take the square root of each side of the equation.
3. Simplify.
4. Check the solutions.

$$(x + 2)^2 - 9 = 0$$
$$(x + 2)^2 = 9$$
$$\sqrt{(x + 2)^2} = \sqrt{9}$$
$$x + 2 = \pm\sqrt{9}$$
$$x + 2 = \pm 3$$
$$x + 2 = 3 \qquad x + 2 = -3$$
$$x = 1 \qquad x = -5$$

To Complete the Square [10.2A, p. 576]
Add to a binomial of the form $x^2 + bx$ the square of $\frac{1}{2}$ of the coefficient of x, making it a perfect-square trinomial.

To complete the square on $x^2 - 8x$,
add $\left[\frac{1}{2}(-8)\right]^2 = 16$: $x^2 - 8x + 16$.

To Solve a Quadratic Equation by Completing the Square
[10.2A, p. 577]
1. Write the equation in the form $ax^2 + bx = -c$.
2. Multiply both sides of the equation by $\frac{1}{a}$.
3. Complete the square on $x^2 + \frac{b}{a}x$. Add the number that completes the square to both sides of the equation.
4. Factor the perfect-square trinomial.
5. Take the square root of each side of the equation.

6. Solve the resulting equation for x.
7. Check the solutions.

$$x^2 + 4x - 1 = 0$$
$$x^2 + 4x = 1$$
$$x^2 + 4x + 4 = 1 + 4$$
$$(x + 2)^2 = 5$$
$$\sqrt{(x + 2)^2} = \sqrt{5}$$
$$x + 2 = \pm\sqrt{5}$$
$$x + 2 = \sqrt{5} \qquad x + 2 = -\sqrt{5}$$
$$x = -2 + \sqrt{5} \qquad x = -2 - \sqrt{5}$$

Quadratic Formula [10.2B, p. 580]

The solutions of $ax^2 + bx + c = 0$, $a \neq 0$, are

$$x = \frac{-b \pm \sqrt{b^2 - 4ac}}{2a}.$$

$$2x^2 - 3x + 4 = 0$$
$$a = 2, b = -3, c = 4$$
$$x = \frac{-(-3) \pm \sqrt{(-3)^2 - 4(2)(4)}}{2(2)}$$
$$= \frac{3 \pm \sqrt{-23}}{4} = \frac{3 \pm i\sqrt{23}}{4}$$
$$= \frac{3}{4} \pm \frac{\sqrt{23}}{4}i$$

The Effect of the Discriminant on the Solutions of a Quadratic Equation [10.2B, p. 582]
1. If $b^2 - 4ac = 0$, the equation has two equal real number solutions, a double root.
2. If $b^2 - 4ac > 0$, the equation has two unequal real number solutions.
3. If $b^2 - 4ac < 0$, the equation has two complex number solutions.

$x^2 + 8x + 16 = 0$ has a double root because $b^2 - 4ac = 8^2 - 4(1)(16) = 0$.

$2x^2 + 3x - 5 = 0$ has two unequal real number solutions because $b^2 - 4ac = 3^2 - 4(2)(-5) = 49$.

$3x^2 + 2x + 4 = 0$ has two complex number solutions because $b^2 - 4ac = 2^2 - 4(3)(4) = -44$.

CHAPTER

10 Review Exercises

1. Solve by factoring: $2x^2 - 3x = 0$

0 and $\dfrac{3}{2}$ [10.1A]

2. Solve by factoring: $6x^2 + 9x = 6$

-2 and $\dfrac{1}{2}$ [10.1A]

3. Solve by taking square roots:
$x^2 = 48$

$-4\sqrt{3}$ and $4\sqrt{3}$ [10.1B]

4. Solve by taking square roots:
$$\left(x + \dfrac{1}{2}\right)^2 + 4 = 0$$

$-\dfrac{1}{2} - 2i$ and $-\dfrac{1}{2} + 2i$ [10.1B]

5. Solve by completing the square: $x^2 + 4x + 3 = 0$
-3 and -1 [10.2A]

6. Solve by completing the square:
$7x^2 - 14x + 3 = 0$

$\dfrac{7 - 2\sqrt{7}}{7}$ and $\dfrac{7 + 2\sqrt{7}}{7}$ [10.2A]

7. Solve by using the quadratic formula:
$12x^2 - 25x + 12 = 0$

$\dfrac{3}{4}$ and $\dfrac{4}{3}$ [10.2B]

8. Solve by using the quadratic formula:
$x^2 - x + 8 = 0$

$\dfrac{1}{2} - \dfrac{\sqrt{31}}{2}i$ and $\dfrac{1}{2} + \dfrac{\sqrt{31}}{2}i$ [10.2B]

9. Write a quadratic equation with integer coefficients that has solutions 0 and -3.
$x^2 + 3x = 0$ [10.1A]

10. Write a quadratic equation with integer coefficients that has solutions $\frac{3}{4}$ and $-\frac{2}{3}$.
$12x^2 - x - 6 = 0$ [10.1A]

11. Solve by completing the square: $x^2 - 2x + 8 = 0$
$1 - i\sqrt{7}$ and $1 + i\sqrt{7}$ [10.2A]

12. Solve by completing the square:
$(x - 2)(x + 3) = x - 10$
$-2i$ and $2i$ [10.2A]

13. Solve by using the quadratic formula:
$3x(x - 3) = 2x - 4$

$\dfrac{11 - \sqrt{73}}{6}$ and $\dfrac{11 + \sqrt{73}}{6}$ [10.2B]

14. Use the discriminant to determine whether $3x^2 - 5x + 3 = 0$ has two equal real number solutions, two unequal real number solutions, or two complex number solutions.
Two complex number solutions [10.2B]

15. Solve: $(x + 3)(2x - 5) < 0$
$\left\{x \mid -3 < x < \dfrac{5}{2}\right\}$ [10.5A]

16. Solve: $(x - 2)(x + 4)(2x + 3) \le 0$
$\{x \mid x \le -4\} \cup \left\{x \mid -\dfrac{3}{2} \le x \le 2\right\}$ [10.5A]

17. Solve: $x^{2/3} + x^{1/3} - 12 = 0$
−64 and 27   [10.3A]

18. Solve: $2(x - 1) + 3\sqrt{x - 1} - 2 = 0$
$\dfrac{5}{4}$ [10.3B]

19. Solve: $3x = \dfrac{9}{x - 2}$
−1 and 3   [10.3C]

20. Solve: $\dfrac{3x + 7}{x + 2} + x = 3$
−1   [10.3C]

21. Solve and graph the solution set:
$\dfrac{x - 2}{2x - 3} \geq 0$

$\left\{x\,\middle|\,x < \dfrac{3}{2}\right\} \cup \{x\,|\,x \geq 2\}$ [10.5A]

22. Solve and graph the solution set:
$\dfrac{(2x - 1)(x + 3)}{x - 4} \leq 0$

$\{x\,|\,x \leq -3\} \cup \left\{x\,\middle|\,\dfrac{1}{2} \leq x < 4\right\}$ [10.5A]

23. Solve: $x = \sqrt{x} + 2$
4   [10.3B]

24. Solve: $2x = \sqrt{5x + 24} + 3$
5   [10.3B]

25. Solve: $\dfrac{x - 2}{2x + 3} - \dfrac{x - 4}{x} = 2$
$\dfrac{-3 - \sqrt{249}}{10}$ and $\dfrac{-3 + \sqrt{249}}{10}$ [10.3C]

26. Solve: $1 - \dfrac{x + 4}{2 - x} = \dfrac{x - 3}{x + 2}$
$\dfrac{-11 - \sqrt{129}}{2}$ and $\dfrac{-11 + \sqrt{129}}{2}$ [10.3C]

27. Write a quadratic equation with integer coefficients that has solutions $\dfrac{1}{3}$ and −3.
$3x^2 + 8x - 3 = 0$ [10.1A]

28. Solve by factoring: $2x^2 + 9x = 5$
−5 and $\dfrac{1}{2}$ [10.1A]

29. Solve: $2(x + 1)^2 - 36 = 0$
$-1 - 3\sqrt{2}$ and $-1 + 3\sqrt{2}$ [10.1B]

30. Solve by using the quadratic formula:
$x^2 + 6x + 10 = 0$
$-3 - i$ and $-3 + i$ [10.2B]

31. Solve: $\dfrac{2}{x - 4} + 3 = \dfrac{x}{2x - 3}$
2 and 3   [10.3C]

32. Solve: $x^4 - 28x^2 + 75 = 0$
$-5, -\sqrt{3}, \sqrt{3}, 5$ [10.3A]

33. Solve: $\sqrt{2x - 1} + \sqrt{2x} = 3$
$\dfrac{25}{18}$ [10.3B]

34. Solve: $2x^{2/3} + 3x^{1/3} - 2 = 0$
−8 and $\dfrac{1}{8}$ [10.3A]

Unless otherwise noted, all content on this page is © Cengage Learning.

35. Solve: $\sqrt{3x - 2} + 4 = 3x$

2 [10.3B]

36. Solve by completing the square:
$x^2 - 10x + 7 = 0$

$5 - 3\sqrt{2}$ and $5 + 3\sqrt{2}$ [10.2A]

37. Solve: $\dfrac{2x}{x - 4} + \dfrac{6}{x + 1} = 11$

$-\dfrac{4}{9}$ and 5 [10.3C]

38. Solve by using the quadratic formula:
$9x^2 - 3x = 1$

$\dfrac{1 - \sqrt{5}}{6}$ and $\dfrac{1 + \sqrt{5}}{6}$ [10.2B]

39. Solve: $2x = 4 - 3\sqrt{x - 1}$

$\dfrac{5}{4}$ [10.3B]

40. Solve: $1 - \dfrac{x + 3}{3 - x} = \dfrac{x - 4}{x + 3}$

$\dfrac{-13 - \sqrt{217}}{2}$ and $\dfrac{-13 + \sqrt{217}}{2}$ [10.3C]

41. Use the discriminant to determine whether $2x^2 - 5x = 6$ has two equal real number solutions, two unequal real number solutions, or two complex number solutions.

Two unequal real number solutions [10.2B]

42. Solve: $x^2 - 3x \le 10$

$\{x \mid -2 \le x \le 5\}$ [10.5A]

43. Sports To prepare for an upcoming race, a sculling crew rowed 16 mi down a river and back in 6 h. If the rate of the river's current is 2 mph, find the sculling crew's rate of rowing in calm water. 6 mph [10.4A]

44. Geometry The length of a rectangle is 2 cm more than twice the width. The area of the rectangle is 60 cm^2. Find the length and width of the rectangle.

Length: 12 cm; width: 5 cm [10.4A]

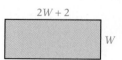

45. Computers An older computer requires 12 min longer to print the payroll than does a newer computer. Together the computers can print the payroll in 8 min. Find the time required for the new computer, working alone, to print the payroll.

12 min [10.4A]

46. Uniform Motion A car travels 200 mi. A second car, making the same trip, travels 10 mph faster than the first car and makes the trip in 1 h less time. Find the speed of each car. First car: 40 mph; second car: 50 mph [10.4A]

Unless otherwise noted, all content on this page is © Cengage Learning.

CHAPTER

10 | TEST

1. Solve by factoring: $3x^2 + 10x = 8$
   -4 and $\dfrac{2}{3}$ [10.1A]

2. Solve by factoring: $6x^2 - 5x - 6 = 0$
   $-\dfrac{2}{3}$ and $\dfrac{3}{2}$ [10.1A]

3. Write a quadratic equation with integer coefficients that has solutions 3 and -3.
   $x^2 - 9 = 0$ [10.1A]

4. Write a quadratic equation with integer coefficients that has solutions $\frac{1}{2}$ and -4.
   $2x^2 + 7x - 4 = 0$ [10.1A]

5. Solve by taking square roots: $3(x - 2)^2 - 24 = 0$
   $2 - 2\sqrt{2}$ and $2 + 2\sqrt{2}$ [10.1B]

6. Solve by completing the square: $x^2 - 6x - 2 = 0$
   $3 - \sqrt{11}$ and $3 + \sqrt{11}$ [10.2A]

7. Solve by completing the square:
   $3x^2 - 6x = 2$
   $\dfrac{3 - \sqrt{15}}{3}$ and $\dfrac{3 + \sqrt{15}}{3}$ [10.2A]

8. Solve by using the quadratic formula:
   $2x^2 - 2x = 1$
   $\dfrac{1 - \sqrt{3}}{2}$ and $\dfrac{1 + \sqrt{3}}{2}$ [10.2B]

9. Solve by using the quadratic formula:
   $x^2 + 4x + 12 = 0$
   $-2 - 2i\sqrt{2}$ and $-2 + 2i\sqrt{2}$ [10.2B]

10. Solve: $2x + 7x^{1/2} - 4 = 0$
    $\dfrac{1}{4}$ [10.3A]

11. Solve: $x^4 - 4x^2 + 3 = 0$
    $-\sqrt{3}, -1, 1, \sqrt{3}$ [10.3A]

12. Solve: $\sqrt{2x + 1} + 5 = 2x$
    4   [10.3B]

13. Solve: $\sqrt{x-2} = \sqrt{x} - 2$

No solution   [10.3B]

14. Solve: $\dfrac{2x}{x-3} + \dfrac{5}{x-1} = 1$

−9 and 2   [10.3C]

15. Solve and graph the solution set of $(x-2)(x+4)(x-4) < 0$.

$\{x \mid x < -4\} \cup \{x \mid 2 < x < 4\}$ [10.5A]

16. Solve and graph the solution set of $\dfrac{2x-3}{x+4} \le 0$.

$\left\{x \mid -4 < x \le \dfrac{3}{2}\right\}$ [10.5A]

17. Use the discriminant to determine whether $9x^2 + 24x = -16$ has two equal real number solutions, two unequal real number solutions, or two complex number solutions.   Two equal real number solutions   [10.2B]

18. Sports   A basketball player shoots at a basket that is 25 ft away. The height h, in feet, of the ball above the ground after t seconds is given by $h = -16t^2 + 32t + 6.5$. How many seconds after the ball is released does it hit the basket, which is 10 ft off the ground? Round to the nearest hundredth.   1.88 s   [10.4A]

19. Woodworking   It takes Clive 6 h longer to stain a bookcase than it does Cora. Working together, Clive and Cora can stain the bookcase in 4 h. Working alone, how long would it take Cora to stain the bookcase?   6 h   [10.4A]

20. Uniform Motion   The rate of a river's current is 2 mph. A canoeist paddled 6 mi down the river and back in 4 h. Find the paddling rate in calm water.   4 mph   [10.4A]

Unless otherwise noted, all content on this page is © Cengage Learning.

Cumulative Review Exercises

1. Evaluate $2a^2 - b^2 \div c^2$ when $a = 3$, $b = -4$, and $c = -2$.

14   [1.4A]

2. Find the volume of a cylinder with a height of 6 m and a radius of 3 m. Give the exact measure.

54π m³   [3.3A]

3. Find the slope of the line containing the points $P_1(3, -4)$ and $P_2(-1, 2)$.

$-\dfrac{3}{2}$ [4.4A]

4. Find the equation of the line that contains the point $P(1, 2)$ and is parallel to the graph of $x - y = 1$.

$y = x + 1$ [4.6A]

5. Factor: $-3x^3y + 6x^2y^2 - 9xy^3$

$-3xy(x^2 - 2xy + 3y^2)$ [7.1A]

6. Factor: $6x^2 - 7x - 20$

$(2x - 5)(3x + 4)$ [7.3A/7.3B]

7. Factor: $x^2 + xy - 2x - 2y$

$(x + y)(x - 2)$ [7.4D]

8. Divide: $(3x^3 - 13x^2 + 10) \div (3x - 4)$

$x^2 - 3x - 4 - \dfrac{6}{3x - 4}$ [6.4B]

9. Simplify: $\dfrac{x^2 + 2x + 1}{8x^2 + 8x} \cdot \dfrac{4x^3 - 4x^2}{x^2 - 1}$

$\dfrac{x}{2}$ [8.1B]

10. Triangles ABC and DEF are similar. Find the height of triangle DEF.

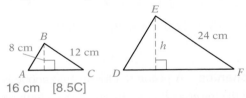

16 cm   [8.5C]

11. Solve $S = \frac{n}{2}(a + b)$ for b.

$b = \dfrac{2S - an}{n}$ [8.6A]

12. Simplify: $-2i(7 - 4i)$

$-8 - 14i$ [9.5C]

13. Simplify: $a^{-1/2}(a^{1/2} - a^{3/2})$

$1 - a$ [9.1A]

14. Simplify: $\dfrac{\sqrt[3]{8x^4y^5}}{\sqrt[3]{16xy^6}}$

$\dfrac{x\sqrt[3]{4y^2}}{2y}$ [9.3B]

15. Solve: $\dfrac{x}{x + 2} - \dfrac{4x}{x + 3} = 1$

$-\dfrac{3}{2}$ and -1 [10.3C]

16. Solve: $\dfrac{x}{2x + 3} - \dfrac{3}{4x^2 - 9} = \dfrac{x}{2x - 3}$

$-\dfrac{1}{2}$ [8.4A]

17. Solve: $x^4 - 6x^2 + 8 = 0$

$-2, -\sqrt{2}, \sqrt{2}, 2$ [10.3A]

18. Solve: $\sqrt{3x + 1} - 1 = x$

0 and 1   [10.3B]

Unless otherwise noted, all content on this page is © Cengage Learning.

19. Solve: $|3x - 2| < 8$

$\left\{ x \mid -2 < x < \dfrac{10}{3} \right\}$ [2.6B]

20. Find the x- and y-intercepts of the graph of $6x - 5y = 15$.

$\left(\dfrac{5}{2}, 0 \right)$ and $(0, -3)$ [4.3C]

21. Graph the solution set:

$x + y \le 3$
$2x - y < 4$

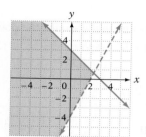

[5.4A]

22. Solve the system of equations.

$x + y + z = 2$
$-x + 2y - 3z = -9$
$x - 2y - 2z = -1$
$(1, -1, 2)$ [5.2B]

23. Given $f(x) = \dfrac{2x - 3}{x^2 - 1}$, find $f(-2)$.

$f(-2) = -\dfrac{7}{3}$ [4.2A]

24. Find the domain of the function

$f(x) = \dfrac{x - 2}{x^2 - 2x - 15}$.

$\{x \mid x \ne -3, 5\}$ [4.2A]

25. Solve and graph the solution set of $x^3 + x^2 - 6x < 0$.

$\{x \mid x < -3\} \cup \{x \mid 0 < x < 2\}$ [10.5A]

26. Solve and graph the solution set of

$\dfrac{(x - 1)(x - 5)}{x + 3} \ge 0$.

$\{x \mid -3 < x \le 1\} \cup \{x \mid x \ge 5\}$ [10.5A]

27. Mechanics   A piston rod for an automobile is $9\frac{3}{8}$ in. long, with a tolerance of $\frac{1}{64}$ in. Find the lower and upper limits of the length of the piston rod.

Lower limit: $9\dfrac{23}{64}$ in.; upper limit: $9\dfrac{25}{64}$ in.   [2.6C]

28. Geometry   The length of the base of a triangle is $(x + 8)$ ft. The height is $(2x - 4)$ ft. Find the area of the triangle in terms of the variable x.

$(x^2 + 6x - 16)$ ft²   [6.3D]

29. Use the discriminant to determine whether $2x^2 + 4x + 3 = 0$ has two equal real number solutions, two unequal real number solutions, or two complex number solutions.   Two complex number solutions   [10.2B]

30. Depreciation   The graph shows the relationship between the cost of a building and the depreciation allowed for income tax purposes. Find the slope of the line between the two points shown on the graph. Write a sentence that states the meaning of the slope.

$m = -\dfrac{25,000}{3}$; The building depreciates $\dfrac{\$25,000}{3}$, or about $8333, each year.   [4.4A]

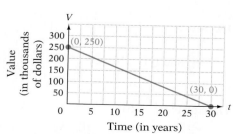

Unless otherwise noted, all content on this page is © Cengage Learning.

Functions and Relations

11

Focus on Success

What resources do you use when you need help in this course? You know to read and reread the text when you are having difficulty understanding a concept. Instructors are available to help you during their office hours. Most schools have a math center where students can get help. Some schools have a tutoring program. You might also ask a student who has been successful in this class for assistance. (See Habits of Successful Students, page AIM-6.)

© Monkey Business Images/Shutterstock.com

Prep Test

Are you ready to succeed in this chapter? Take the Prep Test below to find out if you are ready to learn the new material.

1. Evaluate $-\frac{b}{2a}$ for $b = -4$ and $a = 2$.

1 [1.4A]

2. Given $y = -x^2 + 2x + 1$, find the value of y when $x = -2$.

−7 [1.4A]

3. Given $f(x) = x^2 - 3x + 2$, find $f(-4)$.

30 [4.2A]

4. Evaluate $p(r) = r^2 - 5$ when $r = 2 + h$.

$h^2 + 4h - 1$ [4.2A]

5. Solve: $0 = 3x^2 - 7x - 6$

$-\frac{2}{3}$ and 3 [10.1A]

6. Solve by using the quadratic formula: $0 = x^2 - 4x + 1$

$2 - \sqrt{3}$ and $2 + \sqrt{3}$ [10.2B]

7. Solve $x = 2y + 4$ for y.

$y = \frac{1}{2}x - 2$ [8.6A]

8. Find the domain and range of the relation $\{(-2, 4), (3, 5), (4, 6), (6, 5)\}$. Is the relation a function?

D: $\{-2, 3, 4, 6\}$; R: $\{4, 5, 6\}$; Yes [4.2A]

9. What is the domain of $f(x) = \frac{3}{x - 8}$?

$\{x | x \neq 8\}$ [4.2A]

10. Graph: $x = -2$ [4.3B]

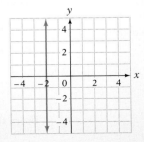

Unless otherwise noted, all content on this page is © Cengage Learning.

11.1 Properties of Quadratic Functions

OBJECTIVE A *To graph a quadratic function*

 Take Note

For a graph to represent a function, every value of *x* in the domain of the function must pair with *exactly one* value of *y*. If any value of *x* pairs with more than one value of *y*, the condition for a function is not met. The **vertical-line test** is used to determine whether a graph defines a function: A graph defines a function if any vertical line intersects the graph at no more than one point.

 Take Note

Sometimes the value of the independent variable is called the **input** because it is *put in* place of the independent variable. The result of evaluating the function is called the **output.**

Tables such as the one in HOW TO 1 are sometimes called **input/output tables.**

Recall that a linear function is one that can be expressed by the equation $f(x) = mx + b$. The graph of a linear function has certain characteristics. It is a straight line with slope *m* and *y*-intercept $P(0, b)$. A **quadratic function** is one that can be expressed by the equation $f(x) = ax^2 + bx + c$, $a \neq 0$. The graph of this function, called a **parabola,** also has certain characteristics. The graph of a quadratic function can be drawn by finding ordered pairs that belong to the function.

HOW TO 1 Graph $f(x) = x^2 - 2x - 3$.

By evaluating the function for various values of *x*, find enough ordered pairs to determine the shape of the graph.

x	$f(x) = x^2 - 2x - 3$	$f(x)$	(x, y)
-2	$f(-2) = (-2)^2 - 2(-2) - 3$	5	$(-2, 5)$
-1	$f(-1) = (-1)^2 - 2(-1) - 3$	0	$(-1, 0)$
0	$f(0) = (0)^2 - 2(0) - 3$	-3	$(0, -3)$
1	$f(1) = (1)^2 - 2(1) - 3$	-4	$(1, -4)$
2	$f(2) = (2)^2 - 2(2) - 3$	-3	$(2, -3)$
3	$f(3) = (3)^2 - 2(3) - 3$	0	$(3, 0)$
4	$f(4) = (4)^2 - 2(4) - 3$	5	$(4, 5)$

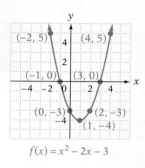

$f(x) = x^2 - 2x - 3$

 Take Note

In completing the square, 1 is both added and subtracted. Because $1 - 1 = 0$, the expression $x^2 - 2x - 3$ is not changed. Note that

$(x - 1)^2 - 4$
$= (x^2 - 2x + 1) - 4$
$= x^2 - 2x - 3$

which is the original expression.

Because the value of $f(x) = x^2 - 2x - 3$ is a real number for all values of *x*, the domain of *f* is all real numbers. From the graph, it appears that no value of *y* is less than -4. Thus the range is $\{y | y \geq -4\}$. The range can also be determined algebraically, as shown below, by completing the square.

$$f(x) = x^2 - 2x - 3$$
$$= (x^2 - 2x) - 3 \qquad \bullet \text{ Group the variable terms.}$$
$$= (x^2 - 2x + 1) - 1 - 3 \qquad \bullet \text{ Complete the square on } x^2 - 2x. \text{ Add and subtract}$$
$$\left[\frac{1}{2}(-2)\right]^2 = 1 \text{ to and from } x^2 - 2x.$$
$$= (x - 1)^2 - 4 \qquad \bullet \text{ Factor and combine like terms.}$$

Because the square of a real number is always nonnegative, we have

$$(x - 1)^2 \geq 0$$
$$(x - 1)^2 - 4 \geq -4 \qquad \bullet \text{ Subtract 4 from each side of the inequality.}$$
$$f(x) \geq -4 \qquad \bullet \; f(x) = (x - 1)^2 - 4$$
$$y \geq -4$$

From the last inequality, the range is $\{y | y \geq -4\}$.

Unless otherwise noted, all content on this page is © Cengage Learning.

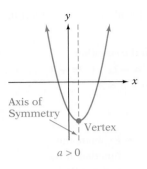

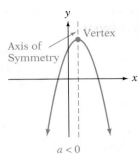

$a > 0$ $a < 0$

In general, the graph of $f(x) = ax^2 + bx + c, a \neq 0$, resembles a "cup" shape as shown at the left. When $a > 0$, the parabola opens up and the **vertex of the parabola** is the point with the least y-coordinate. When $a < 0$, the parabola opens down and the vertex is the point with the greatest y-coordinate. The **axis of symmetry of a parabola** is the vertical line that passes through the vertex of the parabola and is parallel to the y-axis. To understand the axis of symmetry, think of folding the graph along that vertical line. The two portions of the graph will match up.

The vertex and axis of symmetry of a parabola can be found by completing the square.

INSTRUCTOR NOTE

The argument used to determine the vertex is difficult for students to understand. One approach is to have students evaluate $x^2 + 3$ and $x^2 - 1$ for various values of x and discover that the least value of the expression occurs when $x = 0$. Then repeat this process for $(x + 2)^2 + 3$ and $(x + 2)^2 - 1$.

HOW TO 2 Find the coordinates of the vertex and the equation of the axis of symmetry for the graph of $F(x) = x^2 + 4x + 3$.

To find the coordinates of the vertex, complete the square.

$F(x) = x^2 + 4x + 3$

$\quad = (x^2 + 4x) + 3$ • Group the variable terms.

$\quad = (x^2 + 4x + 4) - 4 + 3$ • Complete the square on $x^2 + 4x$. Add and subtract $\left[\dfrac{1}{2}(4)\right]^2 = 4$ to and from $x^2 + 4x$.

$\quad = (x + 2)^2 - 1$ • Factor and combine like terms.

Because a, the coefficient of x^2, is positive $(a = 1)$, the parabola opens up and the vertex is the point with the least y-coordinate. Because $(x + 2)^2 \geq 0$ for all values of x, the least y-coordinate occurs when $(x + 2)^2 = 0$. The quantity $(x + 2)^2$ is equal to zero when $x = -2$. Therefore, the x-coordinate of the vertex is -2.

To find the y-coordinate of the vertex, evaluate the function at $x = -2$.

$F(x) = (x + 2)^2 - 1$

$F(-2) = (-2 + 2)^2 - 1 = -1$

The y-coordinate of the vertex is -1.

From the results above, the coordinates of the vertex are $(-2, -1)$.

The axis of symmetry is the vertical line that passes through the vertex. The equation of the vertical line that passes through the point with coordinates $(-2, -1)$ is $x = -2$. The equation of the axis of symmetry is $x = -2$. See the graph at the left.

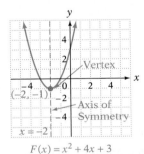

$F(x) = x^2 + 4x + 3$

IN-CLASS EXAMPLES

1. Find the vertex and axis of symmetry of the parabola given by the equation $y = x^2 - x - 2$. Then graph the equation.

Vertex: $\left(\dfrac{1}{2}, -\dfrac{9}{4}\right)$;

Axis of symmetry:

$x = \dfrac{1}{2}$

By following the process illustrated in HOW TO 2 and completing the square on $f(x) = ax^2 + bx + c$, we can find a formula for the coordinates of the vertex and the equation of the axis of symmetry of a parabola.

Vertex and Axis of Symmetry of a Parabola

Let $f(x) = ax^2 + bx + c$ be the equation of a parabola. The coordinates of the vertex are $\left(-\dfrac{b}{2a}, f\left(-\dfrac{b}{2a}\right)\right)$. The equation of the axis of symmetry is $x = -\dfrac{b}{2a}$.

Unless otherwise noted, all content on this page is © Cengage Learning.

HOW TO 3   Find the coordinates of the vertex of the parabola with equation $g(x) = -2x^2 + 3x + 1$. Then graph the equation.

x-coordinate of the vertex: $-\dfrac{b}{2a} = -\dfrac{3}{2(-2)} = \dfrac{3}{4}$

• From the equation
$g(x) = -2x^2 + 3x + 1$,
$a = -2, b = 3$.

y-coordinate of the vertex: $g(x) = -2x^2 + 3x + 1$

$$g\left(\frac{3}{4}\right) = -2\left(\frac{3}{4}\right)^2 + 3\left(\frac{3}{4}\right) + 1$$

$$g\left(\frac{3}{4}\right) = \frac{17}{8}$$

• Evaluate the function at the value of the x-coordinate of the vertex.

The coordinates of the vertex are $\left(\frac{3}{4}, \frac{17}{8}\right)$.

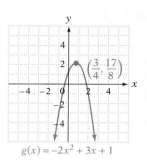

$g(x) = -2x^2 + 3x + 1$

Because a is negative, the graph opens down. Find a few ordered pairs that belong to the function and then sketch the graph, as shown at the left.

📋 **Take Note**

Once the coordinates of the vertex are found, the range of the quadratic function can be determined.

Once the y-coordinate of the vertex is known, the range of the function can be determined. Here, the graph of g opens down, so the y-coordinate of the vertex is the greatest value of y. Therefore, the range of g is $\left\{y \mid y \leq \frac{17}{8}\right\}$. The domain is $\{x \mid -\infty < x < \infty\}$.

EXAMPLE 1

Find the coordinates of the vertex and the equation of the axis of symmetry for the parabola with equation $y = -x^2 + 4x + 1$. Then graph the equation.

Solution

x-coordinate of vertex:

$-\dfrac{b}{2a} = -\dfrac{4}{2(-1)} = 2$

y-coordinate of vertex:

$y = -x^2 + 4x + 1$
$= -(2)^2 + 4(2) + 1 = 5$

Vertex: (2, 5)
Axis of symmetry: $x = 2$

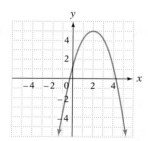

EXAMPLE 2

Find the domain and range of $f(x) = 0.5x^2 - 3$. Then graph the equation.

Solution

x-coordinate of vertex:

$-\dfrac{b}{2a} = -\dfrac{0}{2(0.5)} = 0$

y-coordinate of vertex:

$f(x) = 0.5x^2 - 3$
$f(0) = 0.5(0)^2 - 3 = -3$
Vertex: $(0, -3)$

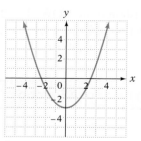

The domain is $\{x \mid -\infty < x < \infty\}$.
The range is $\{y \mid y \geq -3\}$.

YOU TRY IT 1

Find the coordinates of the vertex and the equation of the axis of symmetry for the parabola with equation $y = 4x^2 + 4x + 1$. Then graph the equation.

Your solution

Vertex: $\left(-\dfrac{1}{2}, 0\right)$

Axis of symmetry: $x = -\dfrac{1}{2}$

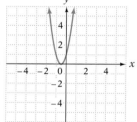

YOU TRY IT 2

Find the domain and range of
$f(x) = -x^2 - 2x - 1$. Then graph the equation.

Your solution

Domain: $\{x \mid -\infty < x < \infty\}$
Range: $\{y \mid y \leq 0\}$

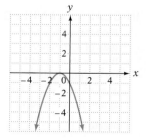

Solutions on p. S31

Unless otherwise noted, all content on this page is © Cengage Learning.

OBJECTIVE B   *To find the x-intercepts of a parabola*

INSTRUCTOR NOTE
Intercepts were discussed previously in the context of linear functions. Making a connection to that discussion may help students realize that this is not a new concept.

Recall that a point at which a graph crosses the *x*- or *y*-axis is called an *intercept* of the graph. The *x*-intercepts of the graph of an equation occur when $y = 0$; the *y*-intercepts occur when $x = 0$.

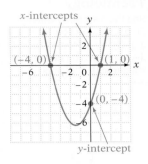

The graph of $y = x^2 + 3x - 4$ is shown at the right. The points whose coordinates are $(-4, 0)$ and $(1, 0)$ are the *x*-intercepts of the graph. The point whose coordinates are $(0, -4)$ is the *y*-intercept of the graph.

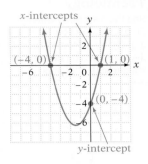

$y = 4x^2 - 4x + 1$

HOW TO 4   Find the coordinates of the *x*-intercepts of the graph of $y = 4x^2 - 4x + 1$.

To find the *x*-intercepts, let $y = 0$ and solve for *x*.

$y = 4x^2 - 4x + 1$
$0 = 4x^2 - 4x + 1$ • Let $y = 0$.
$0 = (2x - 1)(2x - 1)$ • Solve for *x* by factoring and using the Principle of Zero Products.

$2x - 1 = 0 \qquad 2x - 1 = 0$
$2x = 1 \qquad\quad\ 2x = 1$
$x = \dfrac{1}{2} \qquad\quad x = \dfrac{1}{2}$

The coordinates of the *x*-intercept are $\left(\dfrac{1}{2}, 0\right)$.

In HOW TO 4, the parabola has only one *x*-intercept. In this case, the parabola is said to be *tangent* to the *x*-axis at $x = \dfrac{1}{2}$.

HOW TO 5   Find the coordinates of the *x*-intercepts of the graph of $y = 2x^2 - x - 6$.

To find the *x*-intercepts, let $y = 0$ and solve for *x*.

$y = 2x^2 - x - 6$
$0 = 2x^2 - x - 6$ • Let $y = 0$.
$0 = (2x + 3)(x - 2)$ • Solve for *x* by factoring and using the Principle of Zero Products.

$2x + 3 = 0 \qquad x - 2 = 0$
$x = -\dfrac{3}{2} \qquad\quad x = 2$

The coordinates of the *x*-intercepts are $\left(-\dfrac{3}{2}, 0\right)$ and $(2, 0)$.

Take Note
A zero of a function is the *x*-coordinate of an *x*-intercept of the graph of the function. Because the *x*-intercepts of the graph of $f(x) = 2x^2 - x - 6$ are $\left(-\dfrac{3}{2}, 0\right)$ and $(2, 0)$, the zeros are $-\dfrac{3}{2}$ and 2.

If the equation $y = 2x^2 - x - 6$ in HOW TO 5 is written in function notation as $f(x) = 2x^2 - x - 6$, then to find the coordinates of the *x*-intercepts you would let $f(x) = 0$ and solve for *x*. Recall that a value of *x* for which $f(x) = 0$ is called a *zero* of the function. Thus $-\dfrac{3}{2}$ and 2 are zeros of $f(x) = 2x^2 - x - 6$.

Unless otherwise noted, all content on this page is © Cengage Learning.

Integrating Technology

See the Keystroke Guide: *Zero* for instructions on using a graphing calculator to find the zeros of a function or the *x*-intercepts of a graph.

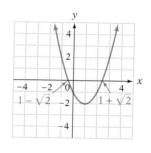

HOW TO 6   Find the zeros of $f(x) = x^2 - 2x - 1$.

To find the zeros, let $f(x) = 0$ and solve for x.

$$f(x) = x^2 - 2x - 1$$
$$0 = x^2 - 2x - 1$$

$$x = \frac{-b \pm \sqrt{b^2 - 4ac}}{2a}$$

• Because $x^2 - 2x - 1$ does not easily factor, use the quadratic formula to solve for x.

$$= \frac{-(-2) \pm \sqrt{(-2)^2 - 4(1)(-1)}}{2(1)}$$

• $a = 1, b = -2, c = -1$

$$= \frac{2 \pm \sqrt{4 + 4}}{2} = \frac{2 \pm \sqrt{8}}{2}$$

$$= \frac{2 \pm 2\sqrt{2}}{2} = 1 \pm \sqrt{2}$$

The zeros of the function are $1 - \sqrt{2}$ and $1 + \sqrt{2}$. The graph of $f(x) = x^2 - 2x - 1$ is shown at the left. Note that the zeros are the *x*-coordinates of the *x*-intercepts of the graph of *f*.

INSTRUCTOR NOTE

This is a difficult concept for students. Giving an additional simple example may help.

Find the *x*-intercept of $y = 2x - 6$.

Find the zero of $f(x) = 2x - 6$.

Solve $2x - 6 = 0$.

The preceding examples suggest that there is a relationship among the *x*-intercepts of the graph of a function, the zeros of the function, and the solutions of an equation of the function. In fact, those three concepts are different ways of discussing the same number. The choice depends on the focus of the discussion. If we are discussing graphing, then the intercepts are our focus; if we are discussing functions, then the zeros of the function are our focus; and if we are discussing equations, the solutions of the equation are our focus.

The *x*-axis is a real number line. Therefore, if a graph crosses the *x*-axis, it must have a real number zero. It is also true that if a graph does not cross the *x*-axis, it does not have a real number zero. The zeros in this case are complex numbers.

INSTRUCTOR NOTE

The discriminant was discussed in conjunction with the solution of a quadratic equation. It may be necessary to review that material at this time.

The graph of $f(x) = x^2 - 2x + 2$ is shown at the right. The graph has no *x*-intercepts. Thus *f* has no real zeros.

Using the quadratic formula, the complex number solutions of $x^2 - 2x + 2 = 0$ are $1 - i$ and $1 + i$. Thus the zeros of $f(x) = x^2 - 2x + 2$ are the complex numbers $1 - i$ and $1 + i$.

Recall that the *discriminant* of $ax^2 + bx + c$ is the expression $b^2 - 4ac$ and that this expression can be used to determine whether $ax^2 + bx + c = 0$ has zero, one, or two real number solutions. Because there is a connection between the solutions of $ax^2 + bx + c = 0$ and the *x*-intercepts of the graph of $y = ax^2 + bx + c$, the discriminant can be used to determine the number of *x*-intercepts of a parabola.

Tips for Success

The paragraph at the right begins with the word "Recall." This signals that the content refers to material presented earlier in the text. The ideas presented here will be more meaningful if you return to the discussion of the word *discriminant* on page 582 and review the concepts presented there.

The Effect of the Discriminant on the Number of *x*-Intercepts of the Graph of a Parabola

1. If $b^2 - 4ac = 0$, the parabola has one *x*-intercept.
2. If $b^2 - 4ac > 0$, the parabola has two *x*-intercepts.
3. If $b^2 - 4ac < 0$, the parabola has no *x*-intercepts.

Unless otherwise noted, all content on this page is © Cengage Learning.

HOW TO 7 Use the discriminant to determine the number of x-intercepts of the graph of the parabola with equation $y = 2x^2 - x + 2$.

$b^2 - 4ac$ • Evaluate the discriminant.
$(-1)^2 - 4(2)(2) = 1 - 16 = -15$ • $a = 2, b = -1, c = 2$
$-15 < 0$

The discriminant is less than zero. Therefore, the parabola has no x-intercepts.

EXAMPLE 3

Find the x-intercepts of the graph of $y = 2x^2 - 5x + 2$.

Solution

$y = 2x^2 - 5x + 2$
$0 = 2x^2 - 5x + 2$ • Let $y = 0$.
$0 = (2x - 1)(x - 2)$ • Solve for x by factoring.

$2x - 1 = 0 \qquad x - 2 = 0$
$\quad 2x = 1 \qquad\quad x = 2$
$\quad\ x = \dfrac{1}{2}$

The x-intercepts are $\left(\dfrac{1}{2}, 0\right)$ and $(2, 0)$.

YOU TRY IT 3

Find the x-intercepts of the graph of $y = x^2 + 3x + 4$.

Your solution

The graph has no x-intercepts.

EXAMPLE 4

Find the zeros of $f(x) = x^2 + 4x + 5$.

Solution

$f(x) = x^2 + 4x + 5$
$0 = x^2 + 4x + 5$ • Let $f(x) = 0$.

$x = \dfrac{-b \pm \sqrt{b^2 - 4ac}}{2a}$ • Use the quadratic formula.

$\quad = \dfrac{-4 \pm \sqrt{4^2 - 4(1)(5)}}{2(1)}$ • $a = 1, b = 4, c = 5$

$\quad = \dfrac{-4 \pm \sqrt{16 - 20}}{2} = \dfrac{-4 \pm \sqrt{-4}}{2}$

$\quad = \dfrac{-4 \pm 2i}{2} = -2 \pm i$

The zeros of the function are $-2 + i$ and $-2 - i$.

YOU TRY IT 4

Find the zeros of $g(x) = x^2 - x + 6$.

Your solution

$\dfrac{1}{2} + \dfrac{\sqrt{23}}{2}i$ and $\dfrac{1}{2} - \dfrac{\sqrt{23}}{2}i$

IN-CLASS EXAMPLES

2. Find the x-intercepts of the graph of $y = x^2 - 2x - 3$.
   $(3, 0), (-1, 0)$

3. Find the zeros of the quadratic function
   $f(x) = 2x^2 - 3x - 2$. $2, -\dfrac{1}{2}$

4. Use the discriminant to determine the number of x-intercepts of the graph of the equation.
   a. $y = 2x^2 - x - 1$ **Two x-intercepts**
   b. $y = x^2 + x + 2$ **No x-intercepts**

EXAMPLE 5

Use the discriminant to determine the number of x-intercepts of the graph of $y = x^2 - 6x + 9$.

Solution

$b^2 - 4ac$ • Evaluate the discriminant.
$(-6)^2 - 4(1)(9)$ • $a = 1, b = -6, c = 9$
$\quad = 36 - 36 = 0$

The discriminant is equal to zero.
The parabola has one x-intercept.

YOU TRY IT 5

Use the discriminant to determine the number of x-intercepts of the graph of $y = x^2 - x - 6$.

Your solution

Two x-intercepts

Solutions on p. S31

OBJECTIVE C   *To find the minimum or maximum of a quadratic function*

Tips for Success

After studying this objective, you should be able to describe in words the meaning of the minimum or maximum value of a quadratic function and calculate algebraically the minimum or maximum value of a quadratic function.

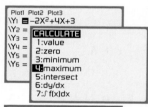

Integrating Technology

A graphing calculator can be used to approximate the maximum or minimum value of a function. Some typical screens for the function in HOW TO 8 are shown below. More information can be found in the Keystroke Guide: *Min and Max*.

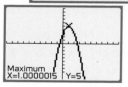

The graph of $f(x) = x^2 - 2x + 3$ is shown at the right. Because a is positive, the parabola opens up. The vertex of the parabola is the lowest point on the parabola. It is the point that has the minimum y-coordinate. This point represents the **minimum value of the function.**

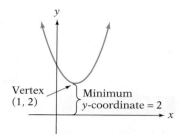

The graph of $f(x) = -x^2 + 2x + 1$ is shown at the right. Because a is negative, the parabola opens down. The vertex of the parabola is the highest point on the parabola. It is the point that has the maximum y-coordinate. This point represents the **maximum value of the function.**

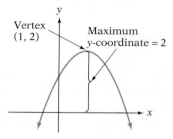

To find the minimum or maximum value of a quadratic function, first find the x-coordinate of the vertex. Then evaluate the function at that value.

HOW TO 8   Find the minimum or maximum value of $f(x) = -2x^2 + 4x + 3$.

$$x = -\frac{b}{2a} = -\frac{4}{2(-2)} = 1$$

- Find the x-coordinate of the vertex. $a = -2, b = 4$

$$f(x) = -2x^2 + 4x + 3$$
$$f(1) = -2(1)^2 + 4(1) + 3$$
$$= -2 + 4 + 3$$
$$= 5$$

- Evaluate the function at $x = 1$.

The maximum value of the function is 5.

- Because $a < 0$, the graph of f opens down. Therefore, the function has a maximum value.

EXAMPLE 6

Find the minimum or maximum value of $f(x) = 2x^2 - 3x + 1$.

- Find the x-coordinate of the vertex

Solution

$$x = -\frac{b}{2a} = -\frac{-3}{2(2)} = \frac{3}{4}$$

$$f(x) = 2x^2 - 3x + 1$$

- $x = \frac{3}{4}$

$$f\left(\frac{3}{4}\right) = 2\left(\frac{3}{4}\right)^2 - 3\left(\frac{3}{4}\right) + 1$$

$$= \frac{9}{8} - \frac{9}{4} + 1 = -\frac{1}{8}$$

Because a is positive ($a = 2$), the graph opens up. The function has a minimum value.

The minimum value of the function is $-\frac{1}{8}$.

YOU TRY IT 6

Find the minimum or maximum value of $f(x) = -3x^2 + 4x - 1$.

Your solution

Maximum value: $\frac{1}{3}$

IN-CLASS EXAMPLES

Find the minimum or maximum value of the quadratic function.

5. $f(x) = x^2 - 4x + 5$ **Minimum value: 1**

6. $f(x) = -2x^2 - 4x + 1$ **Maximum value: 3**

7. $f(x) = -3x^2 + 2x + 1$ **Maximum value:** $\frac{4}{3}$

Solution on p. S31

Unless otherwise noted, all content on this page is © Cengage Learning.

OBJECTIVE D   *To solve application problems*

CandyBox Images/Shutterstock.com

HOW TO 9   A carpenter is forming a rectangular floor for a storage shed. The perimeter of the rectangle is 44 ft. What dimensions of the rectangle will give the floor a maximum area? What is the maximum area?

We are given the perimeter of the rectangle, and we want to find the dimensions of the rectangle that will yield the maximum area for the floor. Use the equation for the perimeter of a rectangle.

$$P = 2L + 2W$$
$$44 = 2L + 2W \qquad \bullet \ P = 44$$
$$22 = L + W \qquad \bullet \ \text{Divide both sides of the equation by 2.}$$
$$22 - L = W \qquad \bullet \ \text{Solve the equation for } W.$$

Now use the equation for the area of a rectangle. Use substitution to express the area in terms of L.

$$A = LW$$
$$A = L(22 - L) \qquad \bullet \ \text{From the equation above, } W = 22 - L. \text{ Substitute } 22 - L \text{ for } W.$$
$$A = 22L - L^2 \qquad \bullet \ \text{The area of the rectangle is } 22L - L^2.$$

To find the length of the rectangle, find the L-coordinate of the vertex of the function $f(L) = -L^2 + 22L$.

$$L = -\frac{b}{2a} = -\frac{22}{2(-1)} = 11 \qquad \bullet \ \text{For the equation } f(L) = -L^2 + 22L,$$
$$a = -1 \text{ and } b = 22.$$

The length of the rectangle is 11 ft.

To find the width, replace L in $22 - L$ by the L-coordinate of the vertex and evaluate.

$$W = 22 - L$$
$$W = 22 - 11 = 11 \qquad \bullet \ \text{Replace } L \text{ by 11 and evaluate.}$$

The width of the rectangle is 11 ft.

The dimensions of the rectangle that will give the floor a maximum area are 11 ft by 11 ft.

To find the maximum area of the floor, evaluate $f(L) = -L^2 + 22L$ at the L-coordinate of the vertex.

$$f(L) = -L^2 + 22L$$
$$f(11) = -(11)^2 + 22(11) \qquad \bullet \ \text{Evaluate the function at } L = 11.$$
$$= -121 + 242 = 121$$

The maximum area of the floor is 121 ft².

Point of Interest

Calculus is a branch of mathematics that demonstrates, among other things, how to find the maximum or minimum values of functions other than quadratic functions. These are very important problems in applied mathematics. For instance, an automotive engineer wants to design a car whose shape will *minimize* the effects of air resistance. The same engineer tries to *maximize* the efficiency of the car's engine. Similarly, an economist may try to determine what business practices will *minimize* cost and *maximize* profit.

The graph of $f(L) = -L^2 + 22L$ is shown at the right. Note that the vertex of the parabola has coordinates $(11, 121)$. For any value of L less than 11, the area of the floor will be less than 121 ft². For any value of L greater than 11, the area of the floor will be less than 121 ft². The maximum value of the function is 121, and the maximum value occurs when $L = 11$.

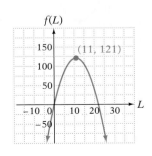

EXAMPLE 7

A mining company has determined that the cost c, in dollars per ton, of mining a mineral is given by $c(x) = 0.2x^2 - 2x + 12$, where x is the number of tons of the mineral mined. Find the number of tons of the mineral that should be mined to minimize the cost. What is the minimum cost per ton?

Strategy

- To find the number of tons of the mineral that should be mined to minimize the cost, find the x-coordinate of the vertex.
- To find the minimum cost, evaluate $c(x)$ at the x-coordinate of the vertex.

Solution

$c(x) = 0.2x^2 - 2x + 12$ • $a = 0.2, b = -2,$
                                    $c = 12$

$x = -\dfrac{b}{2a} = -\dfrac{(-2)}{2(0.2)} = 5$ • **Find the x-coordinate of the vertex.**

To minimize the cost, 5 tons of the mineral should be mined.

$c(x) = 0.2x^2 - 2x + 12$
$c(5) = 0.2(5)^2 - 2(5) + 12$ • $x = 5$
$\quad = 5 - 10 + 12$
$\quad = 7$

The minimum cost is $7 per ton.

Note: The graph of $c(x) = 0.2x^2 - 2x + 12$ is shown below. The vertex of the parabola has coordinates $(5, 7)$. For any value of x less than 5, the cost per ton is greater than $7. For any value of x greater than 5, the cost per ton is greater than $7. Seven is the minimum value of the function, and the minimum value occurs when $x = 5$.

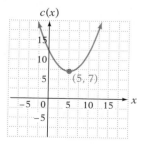

YOU TRY IT 7

The height s, in feet, of a ball thrown straight up is given by $s(t) = -16t^2 + 64t$, where t is the time in seconds. Find the time it takes the ball to reach its maximum height. What is the maximum height?

Your strategy

Your solution

2 s; 64 ft

IN-CLASS EXAMPLES

8. The height s, in feet, of a rocket after t seconds is given by the formula $s(t) = 192t - 16t^2$. Find the maximum height of the rocket. **576 ft**

9. Find two numbers whose sum is 40 and whose product is a maximum. **20 and 20**

10. A rectangle has a perimeter of 48 ft. What dimensions of the rectangle will produce a maximum area? What is the maximum area? **12 ft × 12 ft; 144 ft²**

Solution on p. S31

Unless otherwise noted, all content on this page is © Cengage Learning.

EXAMPLE 8

Find two numbers whose difference is 10 and whose product is a minimum. What is the minimum product of the two numbers?

YOU TRY IT 8

A rectangular fence is being constructed along a stream to enclose a picnic area. If 100 ft of fencing is available, what dimensions of the rectangle will produce the maximum area for picnicking?

Strategy

- Let x represent one number. Because the difference between the two numbers is 10,

$$x + 10$$

represents the other number.
[*Note:* $(x + 10) - (x) = 10$]
Then the product of the numbers is represented by

$$x(x + 10) = x^2 + 10x$$

- To find one of the two numbers, find the x-coordinate of the vertex of $f(x) = x^2 + 10x$.
- To find the other number, replace x in $x + 10$ by the x-coordinate of the vertex and evaluate.
- To find the minimum product, evaluate the function at the x-coordinate of the vertex.

Your strategy

Solution

$f(x) = x^2 + 10x$ • $a = 1, b = 10, c = 0$

$x = -\dfrac{b}{2a} = -\dfrac{10}{2(1)} = -5$ • One number is -5.

$x + 10$
$-5 + 10 = 5$ • The other number is 5.

The numbers are -5 and 5.

$\begin{aligned} f(x) &= x^2 + 10x \\ f(-5) &= (-5)^2 + 10(-5) \\ &= 25 - 50 \\ &= -25 \end{aligned}$ • The x-coordinate of the vertex is -5.

The minimum product of the two numbers is -25.

Your solution

25 ft by 50 ft

INSTRUCTOR NOTE

If you present Example 8 in class, ask students to name other pairs of numbers whose difference is 10. Find the product of each pair named to show that every product named is greater than the product of 5 and -5.

Solution on p. S32

Unless otherwise noted, all content on this page is © Cengage Learning.

11.1 EXERCISES

SUGGESTED ASSIGNMENT
Exercises 1–10; Exercises 11–103, odds;
More challenging exercises: Exercises 105 and 106

✔ **Concept Check**

1. Which of the following functions are quadratic functions?   i, iii

   (i) $f(x) = 3x^2 - x - 7$ **(ii)** $f(x) = \dfrac{1}{x^2}$ **(iii)** $f(x) = 1 - x^2$ **(iv)** $f(x) = 2x - 3$

2. ◥ What is the vertex of a parabola?   3. ◥ What is the axis of symmetry of a parabola?

4. The equation of the axis of symmetry of a parabola is $x = -2$, and the point $P(2, 5)$ is on the parabola. What are the coordinates of another point on the parabola? $(-6, 5)$

5. How many x-intercepts can the graph of $f(x) = ax^2 + bx + c$ have?   Zero, one, or two

6. How many y-intercepts does the graph of $f(x) = ax^2 + bx + c$ have?   One

7. The zeros of $f(x) = x^2 - 2x - 3$ are -1 and 3. What are the coordinates of the x-intercepts of the graph of f? $(-1, 0), (3, 0)$

8. The solutions of $x^2 + 3x - 4 = 0$ are -4 and 1. What are the coordinates of the x-intercepts of the graph of $f(x) = x^2 + 3x - 4$? $(-4, 0), (1, 0)$

9. ◥ What is the minimum or maximum value of a quadratic function?

10. State whether the function has a minimum or a maximum value.

   a. $g(x) = -2x^2 + 4x - 30$ **b.** $f(x) = x^2 + 15$ **c.** $f(x) = 4 - x^2$
       Maximum   Minimum   Maximum

OBJECTIVE A   *To graph a quadratic function*

For Exercises 11 to 25, find the coordinates of the vertex and the equation of the axis of symmetry for the parabola given by the equation. Then graph the equation.

11. $y = x^2 - 2x - 4$ **12.** $y = x^2 + 4x - 4$ **13.** $y = -x^2 + 2x - 3$

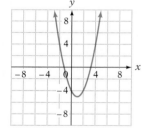   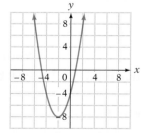   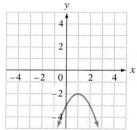

Vertex: $(1, -5)$ Vertex: $(-2, -8)$ Vertex: $(1, -2)$
Axis of symmetry: $x = 1$ Axis of symmetry: $x = -2$ Axis of symmetry: $x = 1$

14. $y = -x^2 + 4x - 5$ **15.** $f(x) = x^2 - x - 6$ **16.** $G(x) = x^2 - x - 2$

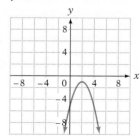   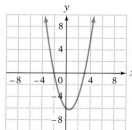   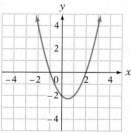

Vertex: $(2, -1)$ Vertex: $\left(\dfrac{1}{2}, -\dfrac{25}{4}\right)$ Vertex: $\left(\dfrac{1}{2}, -\dfrac{9}{4}\right)$

Axis of symmetry: $x = 2$ Axis of symmetry: $x = \dfrac{1}{2}$ Axis of symmetry: $x = \dfrac{1}{2}$

Unless otherwise noted, all content on this page is © Cengage Learning.

17. $F(x) = x^2 - 3x + 2$

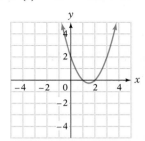

Vertex: $\left(\dfrac{3}{2}, -\dfrac{1}{4}\right)$

Axis of symmetry: $x = \dfrac{3}{2}$

18. $y = 2x^2 - 4x + 1$

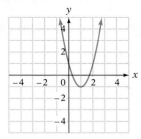

Vertex: $(1, -1)$

Axis of symmetry: $x = 1$

19. $y = -2x^2 + 6x$

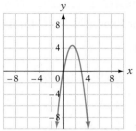

Vertex: $\left(\dfrac{3}{2}, \dfrac{9}{2}\right)$

Axis of symmetry: $x = \dfrac{3}{2}$

20. $y = \dfrac{1}{2}x^2 + 4$

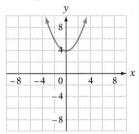

Vertex: $(0, 4)$

Axis of symmetry: $x = 0$

21. $y = -\dfrac{1}{4}x^2 - 1$

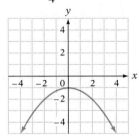

Vertex: $(0, -1)$

Axis of symmetry: $x = 0$

22. $h(x) = \dfrac{1}{2}x^2 - x + 1$

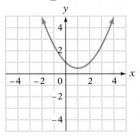

Vertex: $\left(1, \dfrac{1}{2}\right)$

Axis of symmetry: $x = 1$

23. $P(x) = -\dfrac{1}{2}x^2 + 2x - 3$

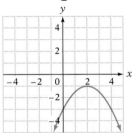

Vertex: $(2, -1)$

Axis of symmetry: $x = 2$

24. $y = \dfrac{1}{2}x^2 + 2x - 6$

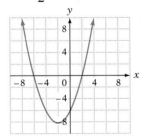

Vertex: $(-2, -8)$

Axis of symmetry: $x = -2$

25. $y = -\dfrac{1}{2}x^2 + x - 3$

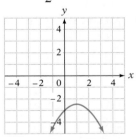

Vertex: $\left(1, -\dfrac{5}{2}\right)$

Axis of symmetry: $x = 1$

26. Is 3 in the range of the function given by $f(x) = (x + 2)^2 + 4$?   No

27. What are the domain and range of the function given by $f(x) = (x - 1)^2 + 2$?
Domain: $\{x \mid -\infty < x < \infty\}$; Range: $\{y \mid y \geq 2\}$

For Exercises 28 to 33, state the domain and range of the function.   For Exercises 28 to 33, the domain is $\{x \mid -\infty < x < \infty\}$.

28. $f(x) = 2x^2 - 4x - 5$
Range: $\{y \mid y \geq -7\}$

29. $f(x) = 2x^2 + 8x + 3$
Range: $\{y \mid y \geq -5\}$

30. $f(x) = -2x^2 - 3x + 2$
Range: $\left\{y \mid y \leq \dfrac{25}{8}\right\}$

31. $f(x) = -x^2 + 6x - 9$
Range: $\{y \mid y \leq 0\}$

32. $f(x) = -x^2 - 4x - 5$
Range: $\{y \mid y \leq -1\}$

33. $f(x) = x^2 + 4x - 3$
Range: $\{y \mid y \geq -7\}$

Unless otherwise noted, all content on this page is © Cengage Learning.

OBJECTIVE B *To find the x-intercepts of a parabola*

For Exercises 34 to 45, find the coordinates of the *x*-intercepts of the graph of the equation.

34. $y = x^2 - 4$

(2, 0), (−2, 0)

35. $y = x^2 - 9$

(3, 0), (−3, 0)

36. $y = 2x^2 - 4x$

(0, 0), (2, 0)

37. $y = 3x^2 + 6x$

(0, 0), (−2, 0)

38. $y = x^2 - x - 2$

(2, 0), (−1, 0)

39. $y = x^2 - 2x - 8$

(4, 0), (−2, 0)

40. $y = 2x^2 - x - 1$

$\left(-\dfrac{1}{2}, 0\right)$, (1, 0)

41. $y = 2x^2 - 5x - 3$

$\left(-\dfrac{1}{2}, 0\right)$, (3, 0)

42. $y = x^2 + 2x - 1$

(−1 + $\sqrt{2}$, 0), (−1 − $\sqrt{2}$, 0)

43. $y = x^2 + 4x - 3$

(−2 + $\sqrt{7}$, 0), (−2 − $\sqrt{7}$, 0)

44. $y = x^2 + 6x + 10$

No *x*-intercepts

45. $y = -x^2 - 4x - 5$

No *x*-intercepts

For Exercises 46 to 63, find the zeros of *f*.

46. $f(x) = x^2 + 3x + 2$

−2, −1

47. $f(x) = x^2 - 4x - 5$

−1, 5

48. $f(x) = 2x^2 - x - 3$

$-1, \dfrac{3}{2}$

49. $f(x) = 3x^2 - 2x - 8$

$-\dfrac{4}{3}, 2$

50. $f(x) = x^2 - 6x + 9$

3

51. $h(x) = 4x^2 - 4x + 1$

$\dfrac{1}{2}$

52. $f(x) = 2x^2 - 3x$

$0, \dfrac{3}{2}$

53. $f(x) = -3x^2 + 4x$

$0, \dfrac{4}{3}$

54. $f(x) = 2x^2 - 32$

−4, 4

55. $f(x) = -3x^2 + 12$

−2, 2

56. $f(x) = 2x^2 - 4$

−$\sqrt{2}$, $\sqrt{2}$

57. $f(x) = 2x^2 - 54$

−3$\sqrt{3}$, 3$\sqrt{3}$

58. $f(x) = x^2 - 4x + 1$

2 − $\sqrt{3}$, 2 + $\sqrt{3}$

59. $f(x) = x^2 - 2x - 17$

1 − 3$\sqrt{2}$, 1 + 3$\sqrt{2}$

60. $f(x) = x^2 - 6x + 4$

3 − $\sqrt{5}$, 3 + $\sqrt{5}$

61. $f(x) = x^2 + 4x + 5$

−2 − *i*, −2 + *i*

62. $f(x) = x^2 - 8x + 20$

4 − 2*i*, 4 + 2*i*

63. $f(x) = x^2 + 4x + 13$

−2 − 3*i*, −2 + 3*i*

For Exercises 64 to 75, use the discriminant to determine the number of x-intercepts of the graph of the equation.

64. $y = 2x^2 + 2x - 1$
Two

65. $y = -x^2 - x + 3$
Two

66. $y = x^2 - 8x + 16$
One

67. $y = x^2 - 10x + 25$
One

68. $y = -3x^2 - x - 2$
No x-intercepts

69. $y = -2x^2 + x - 1$
No x-intercepts

70. $y = -2x^2 + x + 1$
Two

71. $y = 4x^2 - x - 2$
Two

72. $y = 2x^2 + x + 1$
No x-intercepts

73. $y = 2x^2 + x + 4$
No x-intercepts

74. $y = -3x^2 + 2x - 8$
No x-intercepts

75. $y = 4x^2 + 2x - 5$
Two

76. Let $f(x) = (x - 3)^2 + a$. For what values of a will the graph of f have **a.** two x-intercepts, **b.** one x-intercept, and **c.** no x-intercepts?
a. $a < 0$ **b.** $a = 0$ **c.** $a > 0$

77. Let $f(x) = -(x + 1)^2 + a$. For what values of a will the graph of f have **a.** two x-intercepts, **b.** one x-intercept, and **c.** no x-intercepts?
a. $a > 0$ **b.** $a = 0$ **c.** $a < 0$

OBJECTIVE C   *To find the minimum or maximum of a quadratic function*

For Exercises 78 to 89, find the minimum or maximum value of the quadratic function.

78. $f(x) = x^2 - 2x + 3$
Minimum: 2

79. $f(x) = 2x^2 + 4x$
Minimum: -2

80. $f(x) = -2x^2 + 4x - 3$
Maximum: -1

81. $f(x) = -2x^2 + 4x - 5$
Maximum: -3

82. $f(x) = -2x^2 - 3x + 4$
Maximum: $\dfrac{41}{8}$

83. $f(x) = -2x^2 - 3x$
Maximum: $\dfrac{9}{8}$

84. $f(x) = 2x^2 + 3x - 8$
Minimum: $-\dfrac{73}{8}$

85. $f(x) = 3x^2 + 3x - 2$
Minimum: $-\dfrac{11}{4}$

86. $f(x) = -3x^2 + x - 6$
Maximum: $-\dfrac{71}{12}$

87. $f(x) = -x^2 - x + 2$
Maximum: $\dfrac{9}{4}$

88. $f(x) = x^2 - 5x + 3$
Minimum: $-\dfrac{13}{4}$

89. $f(x) = 3x^2 + 5x + 2$
Minimum: $-\dfrac{1}{12}$

OBJECTIVE D *To solve application problems*

90. Physics The height s, in feet, of a rock thrown upward at an initial speed of 64 ft/s from a cliff 50 ft above an ocean beach is given by the function $s(t) = -16t^2 + 64t + 50$, where t is the time in seconds. Find the maximum height above the beach that the rock will attain.   114 ft

91. Business A manufacturer of microwave ovens believes that the revenue R, in dollars, the company receives is related to the price P, in dollars, of an oven by the function $R(P) = 125P - 0.25P^2$. What price will give the maximum revenue? $250

92. Business A tour operator believes that the profit P, in dollars, from selling x tickets is given by $P(x) = 40x - 0.25x^2$. Using this model, what is the maximum profit the tour operator can expect? $1600

93. Mathematics Find two numbers whose sum is 20 and whose product is a maximum.   10 and 10

94. Mathematics Find two numbers whose difference is 14 and whose product is a minimum.   −7 and 7

95. ● Weightlessness Read the article at the right. Suppose the height h, in meters, of the airplane is modeled by the equation $h(t) = -1.42t^2 + 119t + 6000$, where t is the number of seconds elapsed since the plane entered its parabolic path. Show that the maximum height of the airplane is approximately the value given in the article. Round values to the nearest unit.   The vertex of the parabola is approximately (42, 8500), so the maximum height of the airplane is about 8500 m.

In the NEWS!

Zero-G Painting Sells Sky High

A painting completed while the artist floated weightless in zero gravity sold at auction for over $300,000. The artist flew in an airplane that produces a weightless environment by flying in a series of parabolic paths. The plane can reach a maximum height of about 8500 m above Earth.

Sources: BBC News, www.msnbc.msn.com, www.space-travellers.com

96. ● Weightlessness The airplane described in Exercise 95 is used to prepare Russian cosmonauts for their work in the weightless environment of space travel. NASA uses a similar technique to train American astronauts. Suppose the height h, in feet, of NASA's airplane is modeled by the equation $h(t) = -6.63t^2 + 431t + 25,000$, where t is the number of seconds elapsed since the plane entered its parabolic path. Find the maximum height of the airplane. Round to the nearest thousand feet.   32,000 ft

97. Fountains The height h, in feet, of a parabolic stream of water t seconds after it passes through a fountain nozzle can be approximated by $h(t) = -16t^2 + 30t$. Find the time at which the stream reaches its maximum height. What is the maximum height?   0.9375 s; 14.0625 ft

98. Motocross The height y, in feet, of a motocross jumper x feet from the jumping-off point can be approximated by $y(x) = -0.0032x^2 + x + 40$.

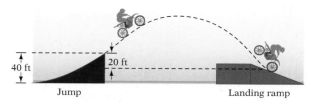

Jump   Landing ramp

40 ft   20 ft

What was the maximum height of the jumper above the ground?   118.125 ft

© Kristina Postnikova/Shutterstock.com

Unless otherwise noted, all content on this page is © Cengage Learning.

99. Apartment Rentals The manager of a 100-unit apartment complex is trying to decide what to charge for rent. Experience has shown that at a monthly rate of $1200, every unit will be occupied. For each $100 increase in the monthly rate, one additional unit will remain vacant. Find the number of units the manager should rent to maximize revenue. What is the maximum revenue?
(*Hint:* Revenue = monthly rent × number of units rented)   56 units; $313,600

100. Fencing A rectangular enclosure for animals is fenced to produce three separate feeding areas, as shown in the diagram at the right. If 800 ft of fencing is available, find the dimensions of the enclosure that will maximize the feeding area. What is the maximum area? $x = 100$ ft and $y = 200$ ft; 20,000 ft²

101. Ranching A rancher has 200 ft of fencing with which to build a rectangular corral alongside an existing fence. Determine the dimensions of the corral that will maximize the enclosed area.   Length: 100 ft; width: 50 ft

102. Construction A courtyard at the corner of two buildings is to be enclosed using 100 ft of redwood fencing. Find the dimensions of the courtyard that will maximize the area.   50 ft by 50 ft

103. Recreation A large lot in a park is going to be split into two softball fields, and each field will be enclosed with a fence. The parks and recreation department has 2100 ft of fencing to enclose the fields. What dimensions will enclose the greatest area?   350 ft by 525 ft

104. If the height h, in feet, of a ball t seconds after it has been tossed directly upward is given by $h(t) = -16(t - 2)^2 + 40$, what is the maximum height the ball will attain?   40 ft

Critical Thinking

105. The zeros of $f(x) = mx^2 + nx + 1$ are -2 and 3. What are the zeros of $g(x) = nx^2 + mx - 1$? $-2, 3$

106. What is the value of k if the vertex of the graph of $y = x^2 - 8x + k$ is a point on the x-axis?   16

Projects or Group Activities

107. Norman Window A Norman window has the shape of a rectangle surmounted by a semicircle, as shown in the figure at the right. The exterior perimeter of the window in the figure is 50 ft. Find the height h and radius r that will maximize the area of the window.

$h = r = \dfrac{50}{\pi + 4}$ ft or about 7.00 ft

Unless otherwise noted, all content on this page is © Cengage Learning.

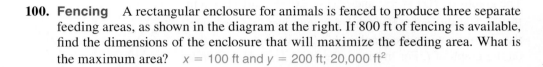

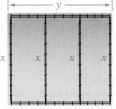

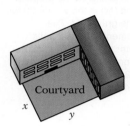

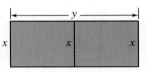

QUICK QUIZ

1. Find the vertex and axis of symmetry of the parabola given by the equation $y = x^2 + 2x - 3$. Then graph the equation.
   Vertex: $(-1, -4)$;
   Axis of symmetry:
   $x = -1$ [11.1A]

2. Find the coordinates of the x-intercepts of the graph of $y = x^2 - 3x - 18$.
   $(6, 0)$, $(-3, 0)$ [11.1B]

3. Find the zeros of the quadratic function $f(x) = 3x^2 - 4x + 1$.
   $1, \dfrac{1}{3}$ [11.1B]

4. Use the discriminant to determine the number of x-intercepts of the graph of $y = x^2 - 2x + 1$.
   One   [11.1B]

5. Find the minimum or maximum value of the quadratic function $f(x) = -2x^2 + 4x - 1$.
   Maximum: 1   [11.1C]

6. A manufacturer of DVD players believes that the revenue R, in dollars, the company receives is related to the price P of a DVD player by the function $R(P) = 180P - \dfrac{1}{3}P^2$. What price will give the maximum revenue?
   $270   [11.1D]

11.2 Translating and Reflecting Graphs

OBJECTIVE A — To graph by using translations

The graphs of $f(x) = \frac{1}{4}x^2$ and $h(x) = \frac{1}{4}x^2 + 2$ are shown in Figure 1 at the right. Note that for any given x-coordinate, the y-coordinate on the graph of h is 2 units higher than that on the graph of f. For instance, the point $(-2, 3)$ is 2 units higher than the point $(-2, 1)$. The graph of h is said to be a **vertical translation,** or **vertical shift,** of the graph of f 2 units up.

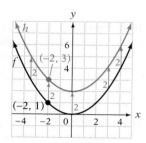

Figure 1

> **Take Note**
>
> Remember that $h(x)$ is the y-coordinate on the graph of h, and $f(x)$ is the y-coordinate on the graph of f. The statement "$h(x)$ is 2 units greater than $f(x)$" means the y-coordinate on the graph of h is 2 units greater than the y-coordinate on the graph of f.

Note that because $f(x) = \frac{1}{4}x^2$,

$$h(x) = \frac{1}{4}x^2 + 2$$
$$= f(x) + 2$$

Thus $h(x)$ is 2 units greater than $f(x)$. See the Take Note at the left.

Now consider the graphs of the functions $f(x) = \frac{1}{4}x^2$ and $g(x) = \frac{1}{4}x^2 - 3$, shown in Figure 2. Note that for a given x-coordinate, the y-coordinate on the graph of g is 3 units lower than that on the graph of f. The graph of g is a vertical translation of the graph of f 3 units down.

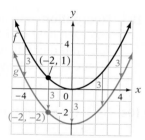

Figure 2

In Figure 1, the graph of $h(x) = \frac{1}{4}x^2 + 2$ is the graph of f shifted 2 units up, whereas in Figure 2, the graph of $g(x) = \frac{1}{4}x^2 - 3$ is the graph of f shifted 3 units down.

Vertical Translations

If f is a function and c is a positive constant, then

- The graph of $y = f(x) + c$ is the graph of $y = f(x)$ shifted c units up.
- The graph of $y = f(x) - c$ is the graph of $y = f(x)$ shifted c units down.

HOW TO 1 Given the graph of $y = f(x)$ shown below in black, graph $g(x) = f(x) + 3$.

The graph of $g(x) = f(x) + 3$ is the graph of $y = f(x)$ shifted 3 units up. The graph of g is shown in blue at the right.

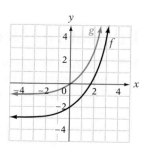

Unless otherwise noted, all content on this page is © Cengage Learning.

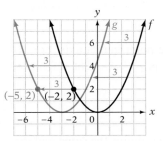

Figure 3

The graphs of $f(x) = \frac{1}{2}x^2$ and $g(x) = f(x + 3) = \frac{1}{2}(x + 3)^2$ are shown in Figure 3. Note that the graph of g is the graph of f shifted 3 units to the left. The graph of g is a **horizontal translation,** or **horizontal shift,** of the graph of f. In this situation, each x-coordinate is moved 3 units to the left, but the y-coordinate remains unchanged.

The graphs of $f(x) = \frac{1}{2}x^2$ and $g(x) = f(x - 2) = \frac{1}{2}(x - 2)^2$ are shown in Figure 4. Note that the graph of g is the graph of f shifted 2 units to the right.

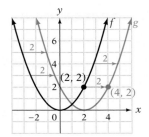

Figure 4

Integrating Technology

Given the equation of a function, a graphing calculator can be used to show a translation of the graph of the function. The equation of the function graphed in HOW TO 2 is $f(x) = |x|$. Enter this equation in Y_1. Then enter the equation for A in Y_2, as shown below. Press GRAPH.

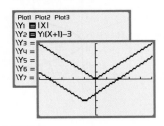

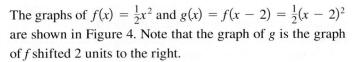

Horizontal Translations

If f is a function and c is a positive constant, then

- The graph of $y = f(x + c)$ is the graph of $y = f(x)$ shifted c units to the left.
- The graph of $y = f(x - c)$ is the graph of $y = f(x)$ shifted c units to the right.

It is possible for a graph to involve both a horizontal and a vertical translation.

HOW TO 2   Given the graph of $y = f(x)$ shown at the right in black, graph $A(x) = f(x + 1) - 3$.

The graph of A is a horizontal translation of f 1 unit to the left and a vertical translation of f 3 units down. The graph of A is shown in blue at the right.

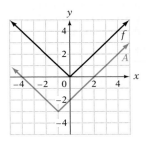

EXAMPLE 1

Given the graph of f shown at the right, graph $g(x) = f(x - 1) + 2$.

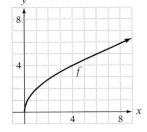

Solution

The graph of g is the graph of f shifted 1 unit to the right and 2 units up. The graph of g is shown in blue.

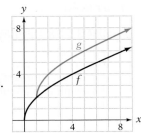

YOU TRY IT 1

Given the graph of h shown at the right, graph $K(x) = h(x + 3) - 2$.

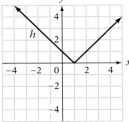

Your solution

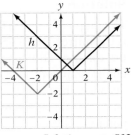

IN-CLASS EXAMPLES

1. Use the graph of f in HOW TO 2. Graph $h(x) = f(x) - 3$.
   Translate the graph of f 3 units down.
2. Use the graph of f in HOW TO 2. Graph $g(x) = f(x - 3) + 1$.
   Translate the graph of f 3 units right and 1 unit up.

Solution on p. S32

Unless otherwise noted, all content on this page is © Cengage Learning.

OBJECTIVE B

To graph by using reflections

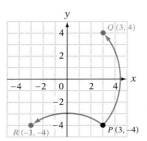

Figure 5

A reflection of a point about an axis is the mirror image of the point in that axis. For instance, in Figure 5, the reflection of $P(3, -4)$ about the x-axis is $Q(3, 4)$. The reflection of P about the y-axis is $R(-3, -4)$.

Note that when P is reflected about the x-axis, the x-coordinate remains the same and the y-coordinate changes sign. When P is reflected about the y-axis, the y-coordinate remains the same and the x-coordinate changes sign. The reflection of a graph about an axis is the mirror image of all the points on the graph.

Reflections of a Graph About an Axis

- The reflection of the graph of $y = f(x)$ about the x-axis is the graph of $y = -f(x)$.
- The reflection of the graph of $y = f(x)$ about the y-axis is the graph of $y = f(-x)$.

Take Note

The first definition in the box states that reflection about the x-axis changes the sign of the y-coordinate:

$y = -f(x)$.

The second definition states that reflection about the y-axis changes the sign of the x-coordinate:

$y = f(-x)$.

HOW TO 3 Given the graph of f shown in black, graph $y = -f(x)$ and $y = f(-x)$.

The graph of $y = -f(x)$ is the mirror image about the x-axis of the graph of f. The graph of $y = -f(x)$ is shown in blue. The graph of $y = f(-x)$ is the mirror image about the y-axis of the graph of f. The graph of $y = f(-x)$ is shown in red.

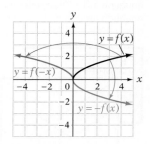

EXAMPLE 2

Given the graph of f shown at the right, graph $y = -f(x)$.

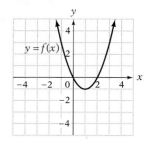

Solution

The graph of $y = -f(x)$ is the mirror image about the x-axis of all the points on the graph of f. For instance, the point $(1, -1)$ on $y = f(x)$ reflects to the point $(1, 1)$ on $y = -f(x)$.

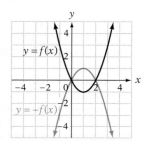

YOU TRY IT 2

Given the graph of f shown at the right, graph $y = f(-x)$.

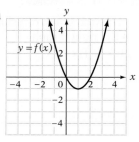

Your solution

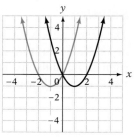

IN-CLASS EXAMPLES

3. Use the graph of h in You Try It 1. Graph $y = h(-x)$. **Reflect the graph of h about the y-axis.**

Solution on p. S32

Unless otherwise noted, all content on this page is © Cengage Learning.

11.2 EXERCISES

✔ Concept Check

SUGGESTED ASSIGNMENT
Exercises 1–6; Exercises 7–31, odds;
More challenging problems: Exercises 33–41, odds

For Exercises 1 to 4, complete the sentence.

1. If c is a positive constant, then the graph of $y = f(x + c)$ is the graph of $y = f(x)$ shifted c units ___left___.

2. If c is a positive constant, then the graph of $y = f(x - c)$ is the graph of $y = f(x)$ shifted c units ___right___.

3. If c is a positive constant, then the graph of $y = f(x) + c$ is the graph of $y = f(x)$ shifted c units ___up___.

4. If c is a positive constant, then the graph of $y = f(x) - c$ is the graph of $y = f(x)$ shifted c units ___down___.

5. What are the coordinates of $P(-3, -5)$ after it has been reflected about the x-axis? $(-3, 5)$

6. What are the coordinates of $P(2, 4)$ after it has been reflected about the y-axis? $(-2, 4)$

OBJECTIVE A   *To graph by using translations*

For Exercises 7 to 22, use translations to draw the graphs. You will use the graphs in Figures 1 to 5 below.

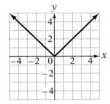
Figure 1

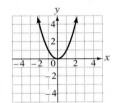
Figure 2

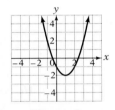
Figure 3

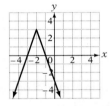
Figure 4

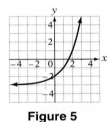
Figure 5

7. Given the graph of $y = f(x)$ in Figure 1 above, graph $g(x) = f(x) + 2$.

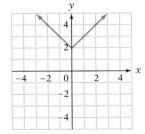

8. Given the graph of $y = f(x)$ in Figure 1 above, graph $g(x) = f(x) - 3$.

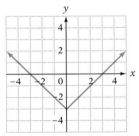

9. Given the graph of $y = f(x)$ in Figure 2 above, graph $g(x) = f(x) - 2$.

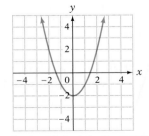

10. Given the graph of $y = f(x)$ in Figure 2 above, graph $g(x) = f(x) + 1$.

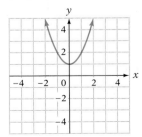

Unless otherwise noted, all content on this page is © Cengage Learning.
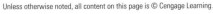

11. Given the graph of $y = f(x)$ in Figure 3 on page 635, graph $g(x) = f(x + 3)$.

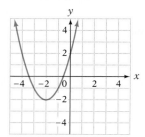

12. Given the graph of $y = f(x)$ in Figure 3 on page 635, graph $g(x) = f(x - 1)$.

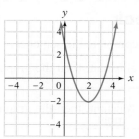

13. Given the graph of $y = f(x)$ in Figure 4 on page 635, graph $g(x) = f(x) + 1$.

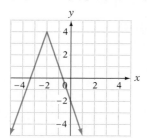

14. Given the graph of $y = f(x)$ in Figure 4 on page 635, graph $g(x) = f(x + 1)$.

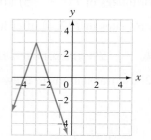

15. Given the graph of $y = f(x)$ in Figure 5 on page 635, graph $g(x) = f(x - 2)$.

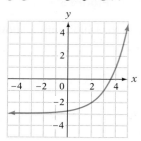

16. Given the graph of $y = f(x)$ in Figure 5 on page 635, graph $g(x) = f(x) - 2$.

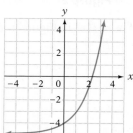

17. Given the graph of $y = f(x)$ in Figure 3 on page 635, graph $g(x) = f(x + 3) - 2$.

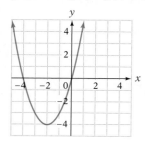

18. Given the graph of $y = f(x)$ in Figure 4 on page 635, graph $g(x) = f(x - 4) - 1$.

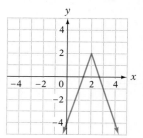

Unless otherwise noted, all content on this page is © Cengage Learning.

19. Given the graph of $y = f(x)$ in Figure 5 on page 635, graph $g(x) = f(x - 2) + 1$.

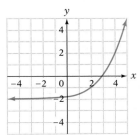

20. Given the graph of $y = f(x)$ in Figure 3 on page 635, graph $g(x) = f(x + 2) - 3$.

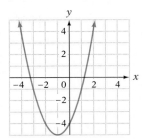

21. Given the graph of $y = f(x)$ in Figure 3 on page 635, graph $g(x) = f(x + 3) + 2$.

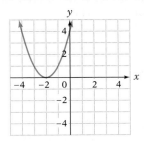

22. Given the graph of $y = f(x)$ in Figure 1 on page 635, graph $g(x) = f(x - 3) + 1$.

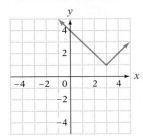

23. If $(0, 7)$ are the coordinates of the y-intercept of $y = f(x) - 2$, then what are the coordinates of the y-intercept of $y = f(x)$? $(0, 9)$

24. If $(5, 0)$ are the coordinates of an x-intercept of $y = f(x)$, then what are the coordinates of an x-intercept of $y = f(x - 3)$? $(8, 0)$

OBJECTIVE B   *To graph by using reflections*

For Exercises 25 to 32, use reflections to draw the graphs.

25. Given the graph of $y = f(x)$ in Figure 1 on page 635, graph $g(x) = -f(x)$.

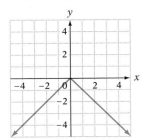

26. Given the graph of $y = f(x)$ in Figure 3 on page 635, graph $g(x) = -f(x)$.

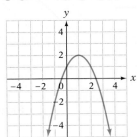

27. Given the graph of $y = f(x)$ in Figure 3 on page 635, graph $g(x) = f(-x)$.

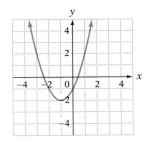

28. Given the graph of $y = f(x)$ in Figure 4 on page 635, graph $g(x) = f(-x)$.

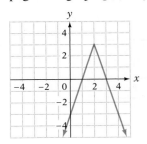

Unless otherwise noted, all content on this page is © Cengage Learning.

29. Given the graph of $y = f(x)$ in Figure 5 on page 635, graph $g(x) = -f(x)$.

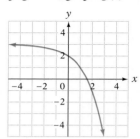

30. Given the graph of $y = f(x)$ in Figure 5 on page 635, graph $g(x) = f(-x)$.

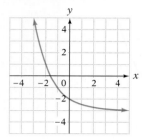

31. Given the graph of $y = f(x)$ in Figure 4 on page 635, graph $g(x) = -f(x)$.

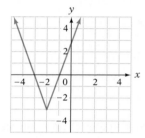

32. Given the graph of $y = f(x)$ in Figure 2 on page 635, graph $g(x) = -f(x)$.

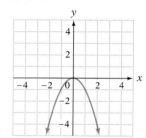

Critical Thinking

For Exercises 33 to 38, use the graph in Figure 6, called a *square wave,* and the graph in Figure 7, called a *sawtooth wave.*

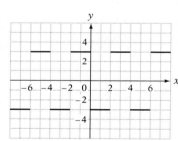 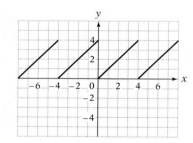

Figure 6   Figure 7

33. Given the graph of $y = f(x)$ in Figure 6 above, graph $g(x) = f(x) + 1$.

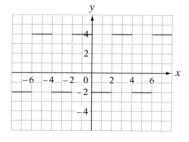

34. Given the graph of $y = f(x)$ in Figure 7 above, graph $g(x) = f(x) - 3$.

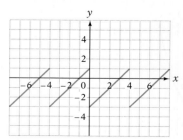

Unless otherwise noted, all content on this page is © Cengage Learning.

35. Given the graph of $y = f(x)$ in Figure 7 on page 638, graph $g(x) = f(x - 3)$.

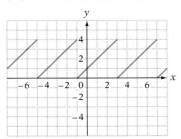

36. Given the graph of $y = f(x)$ in Figure 6 on page 638, graph $g(x) = f(x + 2)$.

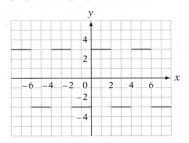

37. Given the graph of $y = f(x)$ in Figure 6 on page 638, graph $g(x) = -f(x)$.

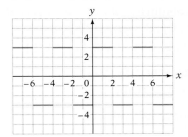

38. Given the graph of $y = f(x)$ in Figure 7 on page 638, graph $g(x) = f(-x)$.

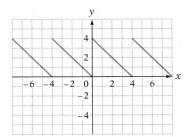

Projects or Group Activities

Translations and reflections of graphs can be combined. For Exercises 39 to 41, use the graph in Figure 8 at the right.

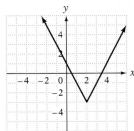

Figure 8

39. Given the graph of $y = f(x)$ in Figure 8, graph $g(x) = -f(x + 3)$.

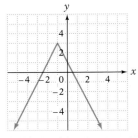

40. Given the graph of $y = f(x)$ in Figure 8, graph $g(x) = f(-x) + 3$.

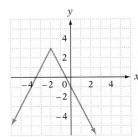

41. Given the graph of $y = f(x)$ in Figure 8, graph $g(x) = -f(-x)$.

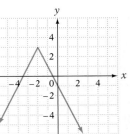

QUICK QUIZ

1. Given the graph of $y = f(x)$ in Figure 1 on page 635, graph $g(x) = f(x + 2)$. **Translate the graph of f 2 units left.   [11.2A]**

2. Given the graph of $y = f(x)$ in Figure 5 on page 635, graph $g(x) = f(x - 1) + 2$. **Translate the graph of f 1 unit right and 2 units up.   [11.2A]**

3. Given the graph of $y = f(x)$ in Figure 8 above, graph $y = -f(x)$. **Reflect the graph of f about the x-axis.   [11.2B]**

Unless otherwise noted, all content on this page is © Cengage Learning.

✔ CHECK YOUR PROGRESS: CHAPTER 11

For Exercises 1 and 2, find the coordinates of the vertex and the equation of the axis of symmetry for the parabola given by the equation. Then graph the equation.

1. $y = x^2 + 6x + 3$

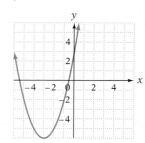

Vertex: $(-3, -6)$
Axis of symmetry:
$x = -3$ [11.1A]

2. $f(x) = -\dfrac{1}{2}x^2 + 2x + 3$

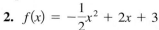
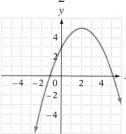

Vertex: $(2, 5)$
Axis of symmetry:
$x = 2$ [11.1A]

3. Find the coordinates of the x-intercepts of the graph of $y = 2x^2 + 5x - 3$.

$(-3, 0), \left(\dfrac{1}{2}, 0\right)$ [11.1B]

4. Find the coordinates of the x-intercepts of the graph of $f(x) = x^2 + 2x - 1$.

$(-1 - \sqrt{2}, 0), (-1 + \sqrt{2}, 0)$ [11.1B]

5. Find the zeros of $f(x) = x^2 + x - 30$.

$-6, 5$ [11.1B]

6. Find the zeros of $f(x) = x^2 - 4x + 40$.

$2 - 6i, 2 + 6i$ [11.1B]

7. Find the minimum value of $f(x) = x^2 - 4x - 32$.

-36 [11.1C]

8. Find the maximum value of $f(x) = x^2 - 4x + 8$.

4 [11.1C]

9. Use the graph of f shown below to graph $g(x) = f(x - 5)$.

[11.2A]

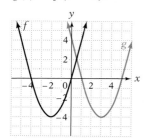

10. Use the graph of f shown below to graph $g(x) = f(x) + 5$.

[11.2A]

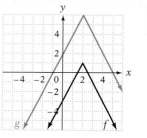

11. Use the graph of f shown below to graph $g(x) = -f(x)$.

[11.2B]

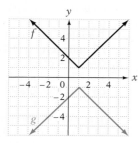

12. Use the graph of f shown below to graph $g(x) = f(-x)$.

[11.2B]

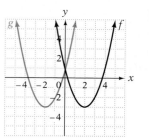

13. Physics The height s, in feet, of a rock thrown upward at an initial speed of 64 ft/s from a cliff 76 ft above an ocean beach is given by $s(t) = -16t^2 + 64t + 76$, where t is the time in seconds. Find the maximum height above the beach that the rock will attain.   140 ft   [11.1D]

Unless otherwise noted, all content on this page is © Cengage Learning.

11.3 | Algebra of Functions

OBJECTIVE A | *To perform operations on functions*

The operations of addition, subtraction, multiplication, and division of functions are defined as follows.

Operations on Functions

If f and g are functions and x is an element of the domain of each function, then

$$(f + g)(x) = f(x) + g(x) \qquad (f \cdot g)(x) = f(x) \cdot g(x)$$

$$(f - g)(x) = f(x) - g(x) \qquad \left(\frac{f}{g}\right)(x) = \frac{f(x)}{g(x)}, g(x) \neq 0$$

INSTRUCTOR NOTE

Definitions are usually difficult for students. After you have given some examples of operations on functions, you might point out the requirement in the definition that x must be an element of the domain of each function. For example, $(f + g)(1)$ is not a real number if $f(x) = x^2$ and $g(x) = \sqrt{x - 2}$.

HOW TO 1   Given $f(x) = x^2 + 1$ and $g(x) = 3x - 2$, find $(f + g)(3)$ and $(f \cdot g)(-1)$.

$$\begin{aligned}
(f + g)(3) &= f(3) + g(3) \\
&= [(3)^2 + 1] + [3(3) - 2] \\
&= 10 + 7 = 17
\end{aligned}$$

$$\begin{aligned}
(f \cdot g)(-1) &= f(-1) \cdot g(-1) \\
&= [(-1)^2 + 1] \cdot [3(-1) - 2] \\
&= 2 \cdot (-5) = -10
\end{aligned}$$

Using $f(x) = x^2 + 1$ and $g(x) = 3x - 2$ from **HOW TO 1**, let $S(x)$ be the sum of the two functions. Then

$$\begin{aligned}
S(x) = (f + g)(x) &= f(x) + g(x) \\
&= [x^2 + 1] + [3x - 2] \\
S(x) &= x^2 + 3x - 1
\end{aligned}$$

- **Definition of addition of functions**
- $f(x) = x^2 + 1, g(x) = 3x - 2$

Now evaluate $S(3)$.

$$\begin{aligned}
S(x) &= x^2 + 3x - 1 \\
S(3) &= (3)^2 + 3(3) - 1 \\
&= 9 + 9 - 1 \\
&= 17 = (f + g)(3)
\end{aligned}$$

Note that $S(3) = 17$ and $(f + g)(3) = 17$. This shows that adding $f(x) + g(x)$ and then evaluating is the same as evaluating $f(x)$ and $g(x)$ and then adding. The same is true for the other operations on functions. For instance, let $P(x)$ be the product of the functions $f(x) = x^2 + 1$ and $g(x) = 3x - 2$. Then

$$\begin{aligned}
P(x) = (f \cdot g)(x) &= f(x) \cdot g(x) \\
&= (x^2 + 1)(3x - 2) \\
&= 3x^3 - 2x^2 + 3x - 2
\end{aligned}$$

$$\begin{aligned}
P(-1) &= 3(-1)^3 - 2(-1)^2 + 3(-1) - 2 \\
&= -3 - 2 - 3 - 2 \\
&= -10
\end{aligned}$$

- **Note that $P(-1) = -10$ and $(f \cdot g)(-1) = -10$.**

IN-CLASS EXAMPLES
For $f(x) = x^2 - 4$ and
$g(x) = -3x + 1$, find:
1. $f(4) + g(4)$ **1**
2. $f(-1) - g(-1)$ **-7**
3. $(f \cdot g)(3)$ **-40**
4. $\left(\dfrac{g}{f}\right)(2)$ **Undefined**

HOW TO 2   Given $f(x) = 2x^2 - 5x + 3$ and $g(x) = x^2 - 1$, find $\left(\dfrac{f}{g}\right)(1)$.

$$\left(\frac{f}{g}\right)(1) = \frac{f(1)}{g(1)}$$

$$= \frac{2(1)^2 - 5(1) + 3}{(1)^2 - 1} = \frac{0}{0} \qquad \bullet \text{ Not a real number}$$

Because $\frac{0}{0}$ is undefined, the expression $\left(\dfrac{f}{g}\right)(1)$ cannot be evaluated.

EXAMPLE 1

Given $f(x) = x^2 - x + 1$ and $g(x) = x^3 - 4$, find $(f - g)(3)$.

Solution

$$
\begin{aligned}
(f - g)(3) &= f(3) - g(3) \\
&= (3^2 - 3 + 1) - (3^3 - 4) \\
&= 7 - 23 \\
&= -16
\end{aligned}
$$

$(f - g)(3) = -16$

YOU TRY IT 1

Given $f(x) = x^2 + 2x$ and $g(x) = 5x - 2$, find $(f + g)(-2)$.

Your solution

$(f + g)(-2) = -12$

EXAMPLE 2

Given $f(x) = x^2 + 2$ and $g(x) = 2x + 3$, find $(f \cdot g)(-2)$.

Solution

$$
\begin{aligned}
(f \cdot g)(-2) &= f(-2) \cdot g(-2) \\
&= [(-2)^2 + 2] \cdot [2(-2) + 3] \\
&= 6(-1) \\
&= -6
\end{aligned}
$$

$(f \cdot g)(-2) = -6$

YOU TRY IT 2

Given $f(x) = 4 - x^2$ and $g(x) = 3x - 4$, find $(f \cdot g)(3)$.

Your solution

$(f \cdot g)(3) = -25$

EXAMPLE 3

Given $f(x) = x^2 + 4x + 4$ and $g(x) = x^3 - 2$, find $\left(\dfrac{f}{g}\right)(3)$.

Solution

$$\left(\frac{f}{g}\right)(3) = \frac{f(3)}{g(3)}$$

$$= \frac{3^2 + 4(3) + 4}{3^3 - 2}$$

$$= \frac{25}{25}$$

$$= 1$$

$\left(\dfrac{f}{g}\right)(3) = 1$

YOU TRY IT 3

Given $f(x) = x^2 - 4$ and $g(x) = x^2 + 2x + 1$, find $\left(\dfrac{f}{g}\right)(4)$.

Your solution

$\left(\dfrac{f}{g}\right)(4) = \dfrac{12}{25}$

Solutions on p. S32

© Sky Light Pictures/Shutterstock.com

OBJECTIVE B   *To find the composition of two functions*

Composition of functions is another way functions can be combined. This method of combining functions uses the output of one function as the input for a second function.

Suppose a forest fire is started by lightning striking a tree, and the spread of the fire can be approximated by a circle whose radius r, in feet, is given by $r(t) = 24\sqrt{t}$, where t is the number of hours since the tree was struck by lightning. The area of the fire is the area of a circle and is given by the formula $A(r) = \pi r^2$. Because the area of the fire depends on the radius of the circle and the radius depends on the time since the tree was struck, there is a relationship between the area of the fire and time. This relationship can be found by evaluating the formula for the area of a circle using $r(t) = 24\sqrt{t}$.

$$A(r) = \pi r^2$$
$$A[r(t)] = \pi [r(t)]^2 \qquad \bullet \text{ Replace } r \text{ by } r(t).$$
$$= \pi [24\sqrt{t}]^2 \qquad \bullet \ r(t) = 24\sqrt{t}$$
$$= 576\pi t \qquad \bullet \text{ Simplify.}$$

The result is the function $A(t) = 576\pi t$, which gives the area of the fire in terms of the time since the lightning struck. For instance, when $t = 3$, we have

$$A(t) = 576\pi t$$
$$A(3) = 576\pi (3) \approx 5429$$

Three hours after the lightning strikes, the area of the fire is approximately 5429 ft^2.

The function above, formed by evaluating one function at another function, is referred to as the *composition* of A with r. The notation $A \circ r$ is used to denote this composition of functions. That is,

$$(A \circ r)(t) = 576\pi t$$

Definition of the Composition of Two Functions

Let f and g be two functions such that $g(x)$ is in the domain of f for all x in the domain of g. Then the **composition of the two functions,** denoted by $f \circ g$, is the function whose value at x is given by $(f \circ g)(x) = f[g(x)]$.

The function defined by $(f \circ g)(x)$ is also called the **composite** of f and g and represents a **composite function.** We read $(f \circ g)(x)$ or $f[g(x)]$ as "f of g of x."

The function machine at the right illustrates the composition of $g(x) = x^2$ and $f(x) = 2x$. Note how a composite function combines two functions. First, one function pairs an input with an output. That output is then used as the input for a second function, which in turn produces a final output.

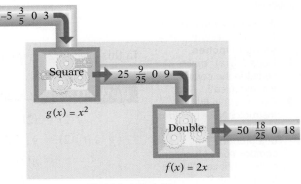

$$(f \circ g)(x) = f[g(x)]$$

Unless otherwise noted, all content on this page is © Cengage Learning.

Consider $f(x) = 3x - 1$ and $g(x) = x^2 + 1$. The expression $(f \circ g)(-2)$ or, equivalently, $f[g(-2)]$, means to evaluate the function f at $g(-2)$.

$$g(x) = x^2 + 1$$
$$g(-2) = (-2)^2 + 1 \qquad \bullet \text{ Evaluate } g \text{ at } -2.$$
$$g(-2) = 5$$
$$f(x) = 3x - 1$$
$$f(5) = 3(5) - 1 = 14 \qquad \bullet \text{ Evaluate } f \text{ at } g(-2) = 5.$$

If we apply our function machine analogy, the composition of functions looks something like the figure below.

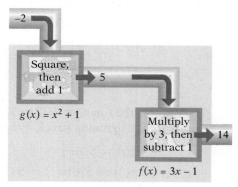

$$f[g(-2)] = 14$$

We can find a general expression for $f[g(x)]$ by evaluating f at $g(x)$. For instance, using $f(x) = 3x - 1$ and $g(x) = x^2 + 1$, we have

$$f(x) = 3x - 1$$
$$f[g(x)] = 3[g(x)] - 1 \qquad \bullet \text{ Replace } x \text{ by } g(x).$$
$$= 3[x^2 + 1] - 1 \qquad \bullet \text{ Replace } g(x) \text{ by } x^2 + 1.$$
$$= 3x^2 + 2 \qquad \bullet \text{ Simplify.}$$

The requirement in the definition of the composition of two functions that $g(x)$ be in the domain of f for all x in the domain of g is important. For instance, let

$$f(x) = \frac{1}{x - 1} \qquad \text{and} \qquad g(x) = 3x - 5$$

When $x = 2$,

$$g(2) = 3(2) - 5 = 1$$
$$f[g(2)] = f(1) = \frac{1}{1 - 1} = \frac{1}{0} \qquad \bullet \text{ This is not a real number.}$$

In this case, $g(2)$ is not in the domain of f. Thus the composition is not defined at 2.

> **HOW TO 3** Given $f(x) = x^3 - x + 1$ and $g(x) = 2x^2 - 10$, evaluate $(g \circ f)(2)$.
>
> $$f(2) = (2)^3 - (2) + 1 = 7$$
> $$(g \circ f)(2) = g[f(2)]$$
> $$= g(7)$$
> $$= 2(7)^2 - 10 = 88$$

Unless otherwise noted, all content on this page is © Cengage Learning.

INSTRUCTOR NOTE

You might present the functions $I(x) = 12x$, which converts x feet into I inches, and $F(y) = 3y$, which converts y yards into F feet. Ask students the following questions.

1. Explain the meaning of $I[F(y)]$. **It converts y yards to inches.**

2. Does $F[I(x)]$ make sense in the context of this exercise? **No**

3. Exercise 2 in this Instructor Note provides a real-world application of a property that composition of functions does not have. What is that property? **Commutativity**

IN-CLASS EXAMPLES
Given $f(x) = 3x - 2$ and
$g(x) = \dfrac{1}{2}x + 4$, evaluate
the composite function.
   5. $f[g(2)]$ **13**
   6. $g[f(0)]$ **3**

HOW TO 4   Given $f(x) = 3x - 2$ and $g(x) = x^2 - 2x$, find $(f \circ g)(x)$.

$$
\begin{aligned}
(f \circ g)(x) &= f[g(x)] \\
&= 3(x^2 - 2x) - 2 \\
&= 3x^2 - 6x - 2
\end{aligned}
$$

When we evaluate compositions of functions, the order in which the functions are applied is important. In the two diagrams below, the order in which the *square* function, $g(x) = x^2$, and the *double* function, $f(x) = 2x$, are applied is interchanged. Note that the final outputs are different. Therefore, $(f \circ g)(x) \neq (g \circ f)(x)$.

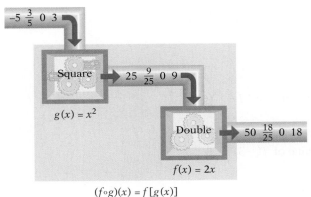

$(f \circ g)(x) = f[g(x)]$

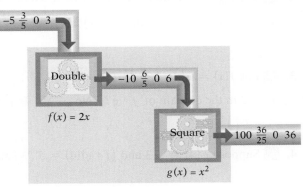

$(g \circ f)(x) = g[f(x)]$

EXAMPLE 4

Given $f(x) = x^2 - x$ and $g(x) = 3x - 2$, find $f[g(3)]$.

Solution

$$
\begin{aligned}
g(x) &= 3x - 2 \\
g(3) &= 3(3) - 2 = 9 - 2 = 7
\end{aligned}
$$
• Evaluate g at 3.

$$
\begin{aligned}
f(x) &= x^2 - x \\
f[g(3)] &= f(7) = 7^2 - 7 = 42
\end{aligned}
$$
• Evaluate f at $g(3) = 7$.

YOU TRY IT 4

Given $f(x) = 1 - 2x$ and $g(x) = x^2$, find $f[g(-1)]$.

Your solution

$f[g(-1)] = -1$

EXAMPLE 5

Given $s(t) = t^2 + 3t - 1$ and $v(t) = 2t + 1$, determine $s[v(t)]$.

Solution

$$
\begin{aligned}
s(t) &= t^2 + 3t - 1 \\
s[v(t)] &= (2t + 1)^2 + 3(2t + 1) - 1 \\
&= (4t^2 + 4t + 1) + 6t + 3 - 1 \\
&= 4t^2 + 10t + 3
\end{aligned}
$$
• Evaluate s at $v(t)$.

YOU TRY IT 5

Given $L(s) = s + 1$ and $M(s) = s^3 + 1$, determine $M[L(s)]$.

Your solution

$M[L(s)] = s^3 + 3s^2 + 3s + 2$

Solutions on p. S32

Unless otherwise noted, all content on this page is © Cengage Learning.

11.3 EXERCISES

✔ **Concept Check**

SUGGESTED ASSIGNMENT
Exercises 1–4; Exercises 5–59 odds;
More challenging exercises: Exercises 61–77, odds

1. 🔎 Let $f(x) = x^2 + 4$ and $g(x) = \sqrt{x} + 4$. For the given value of x, is it possible to find $(f + g)(x)$?

   a. 0   Yes   **b.** −4   Yes   **c.** 1   Yes   **d.** −5   No

2. 🔎 Let $s(1) = -5$ and $v(1) = 3$. Find each of the following.

   a. $(s + v)(1)$ −2   **b.** $(s - v)(1)$ −8   **c.** $(s \cdot v)(1)$ −15   **d.** $\left(\dfrac{s}{v}\right)(1)$ $-\dfrac{5}{3}$

3. 🔎 Let $f(x) = x^2 + 1$ and $g(x) = \dfrac{x}{x^2 - 4}$.

   a. Is 2 in the domain of $f \circ g$?   No   **b.** Is 2 in the domain of $g \circ f$?   Yes

4. 🔎 Suppose $g(4) = -3$ and $(f \circ g)(4) = 5$. What is the value of $f(-3)$?   5

OBJECTIVE A *To perform operations on functions*

For Exercises 5 to 12, let $f(x) = 2x^2 - 3$ and $g(x) = -2x + 4$. Find:

5. $f(2) - g(2)$

   5

6. $f(3) - g(3)$

   17

7. $f(0) + g(0)$

   1

8. $f(1) + g(1)$

   1

9. $(f \cdot g)(2)$

   0

10. $(f \cdot g)(-1)$

    −6

11. $\left(\dfrac{f}{g}\right)(4)$

    $-\dfrac{29}{4}$

12. $\left(\dfrac{g}{f}\right)(-3)$

    $\dfrac{2}{3}$

For Exercises 13 to 20, let $f(x) = 2x^2 + 3x - 1$ and $g(x) = 2x - 4$. Find:

13. $f(1) + g(1)$
    2

14. $f(-3) + g(-3)$
    −2

15. $f(4) - g(4)$
    39

16. $f(-2) - g(-2)$
    9

17. $(f \cdot g)(1)$
    −8

18. $(f \cdot g)(-2)$
    −8

19. $\left(\dfrac{f}{g}\right)(2)$
    Undefined

20. $(f \cdot g)\left(\dfrac{1}{2}\right)$
    −3

For Exercises 21 to 23, let $f(x) = x^2 + 3x - 5$ and $g(x) = x^3 - 2x + 3$. Find:

21. $f(2) - g(2)$

    −2

22. $(f \cdot g)(-3)$

    90

23. $\left(\dfrac{f}{g}\right)(-2)$

    7

OBJECTIVE B *To find the composition of two functions*

Given $f(x) = 2x - 3$ and $g(x) = 4x - 1$, evaluate the composite functions in Exercises 24 to 29.

24. $f[g(0)]$
-5

25. $g[f(0)]$
-13

26. $f[g(2)]$
11

27. $g[f(-2)]$
-29

28. $f[g(x)]$
$8x - 5$

29. $g[f(x)]$
$8x - 13$

Given $h(x) = 2x + 4$ and $f(x) = \frac{1}{2}x + 2$, evaluate the composite functions in Exercises 30 to 35.

30. $h[f(0)]$
8

31. $f[h(0)]$
4

32. $h[f(2)]$
10

33. $f[h(-1)]$
3

34. $h[f(x)]$
$x + 8$

35. $f[h(x)]$
$x + 4$

Given $g(x) = x^2 + 3$ and $h(x) = x - 2$, evaluate the composite functions in Exercises 36 to 41.

36. $g[h(0)]$
7

37. $h[g(0)]$
1

38. $g[h(4)]$
7

39. $h[g(-2)]$
5

40. $g[h(x)]$
$x^2 - 4x + 7$

41. $h[g(x)]$
$x^2 + 1$

Given $f(x) = x^2 + x + 1$ and $h(x) = 3x + 2$, evaluate the composite functions in Exercises 42 to 47.

42. $f[h(0)]$
7

43. $h[f(0)]$
5

44. $f[h(-1)]$
1

45. $h[f(-2)]$
11

46. $f[h(x)]$
$9x^2 + 15x + 7$

47. $h[f(x)]$
$3x^2 + 3x + 5$

Given $f(x) = x - 2$ and $g(x) = x^3$, evaluate the composite functions in Exercises 48 to 53.

48. $f[g(2)]$
6

49. $f[g(-1)]$
-3

50. $g[f(2)]$
0

51. $g[f(-1)]$
-27

52. $f[g(x)]$
$x^3 - 2$

53. $g[f(x)]$
$x^3 - 6x^2 + 12x - 8$

54. **Oil Spills**   Suppose the spread of an oil leak from a tanker can be approximated by a circle with the tanker at its center and radius r, in feet. The radius of the spill t hours after the beginning of the leak is given by $r(t) = 45t$.
   a. Find the area of the spill as a function of time. $A(t) = 2025\pi t^2$
   b. What is the area of the spill after 3 h? Round to the nearest whole number.
     57,256 ft^2

© Cheryl Casey/Shutterstock.com

55. **Manufacturing**   Suppose the manufacturing cost, in dollars, per digital camera is given by the function $M(x) = \frac{50x + 10{,}000}{x}$. A camera store will sell the cameras by marking up the manufacturing cost per camera, $M(x)$, by 60%.
   a. Express the selling price of a camera as a function of the number of cameras to be manufactured. That is, find $S \circ M$.
     $S[M(x)] = 80 + \dfrac{16{,}000}{x}$
   b. Find $(S \circ M)(5000)$. $\$83.20$
   c. ✎ Write a sentence that explains the meaning of the answer to part (b).
     When 5000 digital cameras are manufactured, the camera store sells each camera for $83.20.

56. **Manufacturing**   The number of electric scooters e that a factory can produce per day is a function of the number of hours h it operates and is given by $e(h) = 250h$, $0 \le h \le 10$. The daily cost c to manufacture e electric scooters is given by the function $c(e) = 0.05e^2 + 60e + 1000$.
   a. Find $(c \circ e)(h)$. $c[e(h)] = 3125h^2 + 15{,}000h + 1000$
   b. Evaluate $(c \circ e)(10)$. $\$463{,}500$
   c. ✎ Write a sentence that explains the meaning of the answer to part (b).
     If the plant operates for 10 h, the cost to produce the scooters is $463,500.

© mypokcik/Shutterstock.com

57. ◐ **Electric Cars**   Read the article at the right. The garage's income from the conversion of vehicles is given by $I(n) = 12{,}500n$, where n is the number of vehicles converted. The number of vehicles the garage has converted is given by $n(m) = 4m$, where m is the number of months.
   a. Find $(I \circ n)(m)$. $I[n(m)] = 50{,}000m$
   b. Evaluate $(I \circ n)(3)$. $\$150{,}000$
   c. ✎ Write a sentence that explains the meaning of the answer to part (b).
     The garage's income from 3 months' worth of conversions is $150,000.

58. **Electric Cars**   A company in Santa Cruz, California, sells kits for converting cars to electric power. The company's income, in dollars, from the sale of conversion kits is given by $I(n) = 10{,}000n$, where n is the number of conversion kits sold. The company sells an average of 9 kits per week.
   a. Find the income the company receives from conversion kits as a function of the number of weeks w. $I(w) = 90{,}000w$
   b. What is the company's income from the sale of conversion kits after one year? $\$4{,}680{,}000$

In the NEWS!

Going Electric

If you don't want to wait for car manufacturers to come out with a reasonably priced electric car, you can take your gasoline-powered vehicle to a garage right here in Walton, Kansas. The garage charges $12,500 per vehicle to convert cars and trucks to electric power and is able to convert four vehicles per month.

Source: Associated Press

59. **Automobile Rebates** A car dealership offers a $1500 rebate and a 10% discount off the price of a new car. Let p be the sticker price of a new car on the dealer's lot, r the price after the rebate, and d the discounted price. Then $r(p) = p - 1500$ and $d(p) = 0.90p$.

   a. Write a composite function for the dealer taking the rebate first and then the discount. $d[r(p)] = 0.90p - 1350$

   b. Write a composite function for the dealer taking the discount first and then the rebate. $r[d(p)] = 0.90p - 1500$

   c. Which composite function would you prefer the dealer use when you buy a new car? $r[d(p)]$ (The final price is less.)

Critical Thinking

Use the graphs of f and g shown at the right to determine the values of the composite functions in Exercises 60 to 65.

60. $f[g(-1)]$
    6

61. $g[f(1)]$
    0

62. $(g \circ f)(2)$
    -3

63. $(f \circ g)(3)$
    -2

64. $g[f(3)]$
    -4

65. $f[g(0)]$
    7

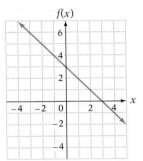

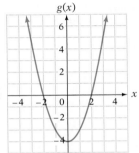

For Exercises 66 to 71, let $g(x) = x^2 - 1$. Find:

66. $g(2 + h)$
    $h^2 + 4h + 3$

67. $g(3 + h) - g(3)$
    $h^2 + 6h$

68. $g(-1 + h) - g(-1)$

    $h^2 - 2h$

69. $\dfrac{g(1 + h) - g(1)}{h}$

    $2 + h$

70. $\dfrac{g(-2 + h) - g(-2)}{h}$

    $-4 + h$

71. $\dfrac{g(a + h) - g(a)}{h}$

    $2a + h$

Projects or Group Activities

For Exercises 72 to 77, let $f(x) = 2x$, $g(x) = 3x - 1$, and $h(x) = x - 2$. Find:

72. $f(g[h(2)])$
    -2

73. $g(h[f(1)])$
    -1

74. $h(g[f(-1)])$
    -9

75. $f(h[g(0)])$
    -6

76. $f(g[h(x)])$
    $6x - 14$

77. $g(f[h(x)])$
    $6x - 13$

QUICK QUIZ

For $f(x) = 2x^2 + x - 4$ and $g(x) = -x + 3$, find:

1. $f(2) + g(2)$ **7**   **[11.3A]**
2. $f(3) - g(3)$
   17   **[11.3A]**
3. $(f \cdot g)(-1)$
   -12   **[11.3A]**
4. $\left(\dfrac{f}{g}\right)(-2)$ $\dfrac{2}{5}$ **[11.3A]**

Given $g(x) = 2x^2 - 1$ and $h(x) = 3x + 5$, evaluate the composite function.

5. $g[h(-1)]$ **7**   **[11.3B]**
6. $h[g(0)]$ **2**   **[11.3B]**

Unless otherwise noted, all content on this page is © Cengage Learning.

11.4 One-to-One and Inverse Functions

OBJECTIVE A *To determine whether a function is one-to-one*

INSTRUCTOR NOTE

It may help students to recall that a function is a set of ordered pairs in which no two ordered pairs have the same first coordinate. A 1–1 function is one in which no two ordered pairs have the same second coordinate either.

Recall that a function is a set of ordered pairs in which no two ordered pairs have the same first coordinate. This means that given any *x*, there is only one *y* that can be paired with that *x*. A **one-to-one function** satisfies the additional condition that given any *y*, there is only one *x* that can be paired with that *y*. One-to-one functions are commonly written as 1–1.

> **One-to-One Function**
>
> A function *f* is a 1–1 function if, for any *a* and *b* in the domain of *f*, $f(a) = f(b)$ implies $a = b$.

This definition states that if the *y*-coordinates of an ordered pair are equal, $f(a) = f(b)$, then the *x*-coordinates must be equal, $a = b$.

The function defined by $f(x) = 2x + 1$ is a 1–1 function. To show this, determine $f(a)$ and $f(b)$. Then form the equation $f(a) = f(b)$.

$$f(a) = 2a + 1 \qquad f(b) = 2b + 1$$

$$f(a) = f(b)$$
$$2a + 1 = 2b + 1$$
$$2a = 2b \qquad \bullet \text{ Subtract 1 from each side of the equation.}$$
$$a = b \qquad \bullet \text{ Divide each side of the equation by 2.}$$

Because $f(a) = f(b)$ implies $a = b$, the function is a 1–1 function.

Consider the function defined by $g(x) = x^2 - x$. Evaluate *g* at -2 and 3.

$$g(-2) = (-2)^2 - (-2) = 6 \qquad g(3) = 3^2 - 3 = 6$$

Note that $g(-2) = 6$ and $g(3) = 6$, but $-2 \neq 3$. Thus *g* is not a 1–1 function.

The graphs of $f(x) = 2x + 1$ and $g(x) = x^2 - x$ are shown at the right. Note that a horizontal line intersects the graph of *f* at no more than one point. However, a horizontal line intersects the graph of *g* at more than one point.

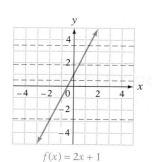

$f(x) = 2x + 1$

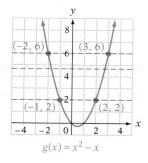

$g(x) = x^2 - x$

IN-CLASS EXAMPLE

1. Determine whether the graph represents the graph of a 1–1 function.

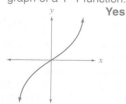

Yes

Note from the graph of *f*, that for each *y*-coordinate, there is only one *x*-coordinate. Thus *f* is a 1–1 function. From the graph of *g*, however, there are *two x*-coordinates for a given *y*-coordinate. For instance, $(-2, 6)$ and $(3, 6)$ are the coordinates of two points on the graph for which the *y*-coordinates are the same and the *x*-coordinates are different. Therefore, *g* is not a 1–1 function.

Unless otherwise noted, all content on this page is © Cengage Learning.

Horizontal-Line Test

The graph of a function represents the graph of a 1–1 function if any horizontal line intersects the graph at no more than one point.

EXAMPLE 1

Determine whether the graph shown below is the graph of a 1–1 function.

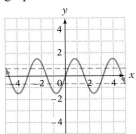

Solution

The dashed horizontal lines intersect the graph more than once. The graph is not the graph of a 1–1 function.

YOU TRY IT 1

Determine whether the graph shown below is the graph of a 1–1 function.

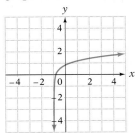

Your solution

Yes

Solution on p. S32

OBJECTIVE B   *To find the inverse of a function*

The **inverse** of a function f is the set of ordered pairs formed by reversing the coordinates of each ordered pair of f.

For instance, let $f(x) = 3x$ with domain $\{-2, -1, 0, 1, 2\}$. The ordered pairs of f are $\{(-2, -6), (-1, -3), (0, 0), (1, 3), (2, 6)\}$. Reversing the coordinates of the ordered pairs of f gives $\{(-6, -2), (-3, -1), (0, 0), (3, 1), (6, 2)\}$. This set of ordered pairs is a function and is called the *inverse function* of f. The inverse function of f is denoted by f^{-1}.

From the ordered pairs of f and f^{-1}, we have

Domain of f: $\{-2, -1, 0, 1, 2\}$ Range of f: $\{-6, -3, 0, 3, 6\}$

Domain of f^{-1}: $\{-6, -3, 0, 3, 6\}$ Range of f^{-1}: $\{-2, -1, 0, 1, 2\}$

The domain of f^{-1} is the range of f, and the range of f^{-1} is the domain of f.

 Take Note

It is important to note that $f^{-1}(x)$ is *not* the reciprocal of $f(x)$. $f^{-1}(x) = \dfrac{x}{3}$; the *reciprocal* of f is $\dfrac{1}{f(x)} = \dfrac{1}{3x}$.

Note that $f^{-1}(6) = 2$, whereas evaluating the reciprocal of f at $x = 6$ gives $\dfrac{1}{18}$.

Because the function f multiplies an element of the domain by 3, a reasonable guess would be that the *inverse* function would *divide* an element by 3. To test this, we can write $f^{-1}(x) = \dfrac{x}{3}$, where the symbol $f^{-1}(x)$ is read "f inverse of x." Using $\{-6, -3, 0, 3, 6\}$ as the domain of f^{-1}, the ordered pairs of f^{-1} are $\{(-6, -2), (-3, -1), (0, 0), (3, 1), (6, 2)\}$. This is the same result we obtained by interchanging the coordinates of the ordered pairs of f.

Unless otherwise noted, all content on this page is © Cengage Learning.

Now let $g(x) = x^2$ with domain $\{-2, -1, 0, 1, 2\}$. The ordered pairs of g are $\{(-2, 4), (-1, 1), (0, 0), (1, 1), (2, 4)\}$. Reversing the coordinates of the ordered pairs of g gives $\{(4, -2), (1, -1), (0, 0), (1, 1), (4, 2)\}$. This set of ordered pairs *is not* a function. There are ordered pairs with the same first coordinate and different second coordinates. This illustrates that not every function has an inverse function.

The graphs of $f(x) = 3x$ and $g(x) = x^2$, with the set of real numbers as the domain, are shown below.

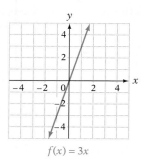

$f(x) = 3x$

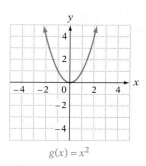

$g(x) = x^2$

By the horizontal-line test, f is a 1–1 function but g is not a 1–1 function.

Condition for an Inverse Function

A function f has an inverse function f^{-1} if and only if f is a 1–1 function.

HOW TO 1   Find the inverse of the function $\{(-2, -8), (-1, -1), (0, 0), (1, 1), (2, 8)\}$.

Reverse the coordinates of the ordered pairs of the function.

$$\{(-8, -2), (-1, -1), (0, 0), (1, 1), (8, 2)\}$$

Because no two ordered pairs have the same first coordinate, this is a function. The inverse function is $\{(-8, -2), (-1, -1), (0, 0), (1, 1), (8, 2)\}$.

We can use the fact that the inverse of a function is found by interchanging x and y to determine the inverse of a function that is given by an equation.

HOW TO 2   Find the inverse of $f(x) = 3x + 6$.

$$f(x) = 3x + 6$$
$$y = 3x + 6 \qquad \bullet \text{ Replace } f(x) \text{ by } y.$$
$$x = 3y + 6 \qquad \bullet \text{ Interchange } x \text{ and } y.$$
$$x - 6 = 3y \qquad \bullet \text{ Solve for } y.$$
$$\frac{1}{3}x - 2 = y$$
$$f^{-1}(x) = \frac{1}{3}x - 2 \qquad \bullet \text{ Replace } y \text{ by } f^{-1}(x).$$

The inverse of the function $f(x) = 3x + 6$ is $f^{-1}(x) = \frac{1}{3}x - 2$.

Unless otherwise noted, all content on this page is © Cengage Learning.

A special property relates the composition of a function and its inverse.

Composition of Inverse Functions Property

If f is a function and f^{-1} is its inverse function, then $f^{-1}[f(x)] = x$ and $f[f^{-1}(x)] = x$.

Take Note

Inverse functions can be likened to, for instance, multiplying a number by 5 and then dividing the result by 5; you will be back to the original number. Evaluating f at a number and then evaluating f^{-1} at the output of f returns you to the original number.

In words, $f^{-1}[f(x)] = x$ states that if $f(a) = b$, then $f^{-1}(b) = a$, the original number used to evaluate f. A similar statement can be made for $f[f^{-1}(x)] = x$.

For instance, consider $f(x) = 3x + 6$ and $f^{-1}(x) = \frac{1}{3}x - 2$ from HOW TO 2. Choose *any* value of x in the domain of f, say, $x = 4$. Then $f(4) = 3 \cdot 4 + 6 = 18$. Now evaluate f^{-1} at 18. $f^{-1}(18) = \frac{1}{3}(18) - 2 = 4$, the original value of x. The Composition of Inverse Functions Property can be used to determine whether two functions are inverses of each other.

HOW TO 3 Are $f(x) = 2x - 4$ and $g(x) = \frac{1}{2}x + 2$ inverses of each other?

To determine whether the functions are inverses, use the Composition of Inverse Functions Property.

$$f[g(x)] = 2\left(\frac{1}{2}x + 2\right) - 4 \qquad g[f(x)] = \frac{1}{2}(2x - 4) + 2$$
$$= x + 4 - 4 \qquad\qquad\qquad = x - 2 + 2$$
$$= x \qquad\qquad\qquad\qquad = x$$

Because $f[g(x)] = x$ and $g[f(x)] = x$, the functions are inverses of each other.

Inverse functions can be used to model many real-world situations.

HOW TO 4 The equation $F(x) = \frac{9}{5}x + 32$ converts a Celsius temperature x to a Fahrenheit temperature $F(x)$. Find the inverse of this function, and write a sentence that explains its meaning.

$$F(x) = \frac{9}{5}x + 32$$

$$y = \frac{9}{5}x + 32 \qquad \bullet \text{ Replace } F(x) \text{ by } y.$$

$$x = \frac{9}{5}y + 32 \qquad \bullet \text{ Interchange } x \text{ and } y.$$

$$x - 32 = \frac{9}{5}y \qquad \bullet \text{ Solve for } y.$$

$$\frac{5}{9}(x - 32) = y$$

$$F^{-1}(x) = \frac{5}{9}(x - 32) \qquad \bullet \text{ Replace } y \text{ by } F^{-1}(x).$$

The inverse function is $F^{-1}(x) = \frac{5}{9}(x - 32)$. This equation gives a formula for converting a Fahrenheit temperature to a Celsius temperature.

IN-CLASS EXAMPLES

Find $f^{-1}(x)$.

2. $f(x) = 2x - 6$

$f^{-1}(x) = \frac{1}{2}x + 3$

3. $f(x) = \frac{1}{3}x - 2$

$f^{-1}(x) = 3x + 6$

EXAMPLE 2

Find the inverse of $f(x) = 2x - 3$.

Solution

$$f(x) = 2x - 3$$
$$y = 2x - 3 \qquad \bullet \text{ Replace } f(x) \text{ by } y.$$
$$x = 2y - 3 \qquad \bullet \text{ Interchange } x \text{ and } y.$$
$$x + 3 = 2y \qquad \bullet \text{ Solve for } y.$$
$$\frac{x}{2} + \frac{3}{2} = y$$
$$f^{-1}(x) = \frac{x}{2} + \frac{3}{2} \qquad \bullet \text{ Replace } y \text{ by } f^{-1}(x).$$

The inverse of the function is given by
$f^{-1}(x) = \frac{x}{2} + \frac{3}{2}$.

YOU TRY IT 2

Find the inverse of $f(x) = \frac{1}{2}x + 4$.

Your solution

$f^{-1}(x) = 2x - 8$

EXAMPLE 3

Are $f(x) = 3x - 6$ and $g(x) = \frac{1}{3}x + 2$ inverses of each other?

Solution

$$f[g(x)] = 3\left(\frac{1}{3}x + 2\right) - 6 = x + 6 - 6 = x$$

$$g[f(x)] = \frac{1}{3}(3x - 6) + 2 = x - 2 + 2 = x$$

Yes, the functions are inverses of each other.

YOU TRY IT 3

Are $f(x) = 2x - 6$ and $g(x) = \frac{1}{2}x - 3$ inverses of each other?

Your solution

No

EXAMPLE 4

To convert feet to inches, we can use the formula $f(x) = 12x$, where x is the number of feet and $f(x)$ is the number of inches. Find $f^{-1}(x)$, and write a sentence that explains its meaning.

Solution

$$f(x) = 12x$$
$$y = 12x \qquad \bullet \text{ Replace } f(x) \text{ by } y.$$
$$x = 12y \qquad \bullet \text{ Interchange } x \text{ and } y.$$
$$\frac{x}{12} = y \qquad \bullet \text{ Solve for } y.$$
$$\frac{x}{12} = f^{-1}(x) \qquad \bullet \text{ Replace } y \text{ by } f^{-1}(x).$$

The inverse is $f^{-1}(x) = \frac{x}{12}$. This equation gives a formula for converting inches to feet.

YOU TRY IT 4

The speed of sound s, in feet per second, depends on the temperature of the air and can be approximated by $s(x) = \frac{3}{5}x + 1125$, where x is the air temperature in degrees Celsius. Find $s^{-1}(x)$, and write a sentence that explains its meaning.

Your solution

$s^{-1}(x) = \frac{5}{3}x - 1875$. This equation gives the temperature of the air given the speed of sound.

Solutions on p. S33

11.4 EXERCISES

SUGGESTED ASSIGNMENT
Exercises 1–8; Exercises 9–59, odds;
More challenging exercises: Exercises 61–77, odds

✔ Concept Check

1. What is a 1–1 function?

2. What is the horizontal-line test?

3. **a.** Suppose f is a function and $f(3) = 5$. Is it possible that there exists another value c for which $f(c) = 5$?   Yes
   b. Suppose f is a 1–1 function and $f(3) = 5$. Is it possible that there exists another value c for which $f(c) = 5$?   No

4. The domain of a 1–1 function f has n elements. How many elements are in the range of f? n

5. How are the ordered pairs of the inverse of a function related to the function?

6. Suppose the domain of f is $\{1, 2, 3, 4, 5\}$, and the range of f is $\{3, 7, 11\}$. Does f have an inverse function?   No

7. Suppose f is a 1–1 function and $f(3) = -2$. Which of the following is (are) possible?   **(i)** $f^{-1}(-2) = -2$ **(ii)** $f^{-1}(-2) = 3$ **(iii)** $f^{-1}(-2) = 6$ ii

8. Let f be a 1–1 function with $f(3) = 4$, $f(4) = 5$, and $f(6) = 7$. What is the value of **a.** $f^{-1}(5)$, **b.** $f^{-1}(4)$, and **c.** $f^{-1}(7)$?
   a. 4   b. 3   c. 6

OBJECTIVE A   *To determine whether a function is one-to-one*

For Exercises 9 to 20, determine whether the graph represents the graph of a 1–1 function.

9.
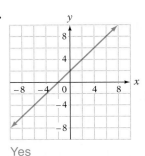
Yes

10.
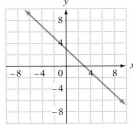
Yes

11.
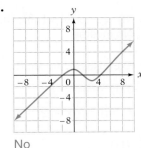
No

12.
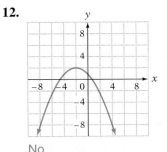
No

13.
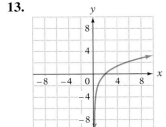
Yes

14.
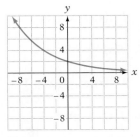
Yes

Unless otherwise noted, all content on this page is © Cengage Learning.

15.

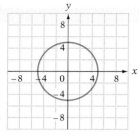

No

16.

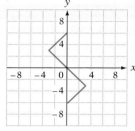

No

17.

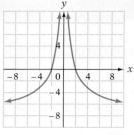

No

18.

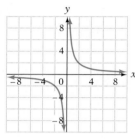

Yes

19.

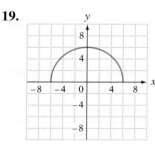

No

20.

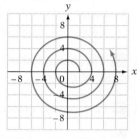

No

OBJECTIVE B *To find the inverse of a function*

For Exercises 21 to 28, find the inverse of the function. If the function does not have an inverse function, write "no inverse."

21. $\{(1, 0), (2, 3), (3, 8), (4, 15)\}$
$\{(0, 1), (3, 2), (8, 3), (15, 4)\}$

22. $\{(1, 0), (2, 1), (-1, 0), (-2, 0)\}$
No inverse

23. $\{(3, 5), (-3, -5), (2, 5), (-2, -5)\}$
No inverse

24. $\{(-5, -5), (-3, -1), (-1, 3), (1, 7)\}$
$\{(-5, -5), (-1, -3), (3, -1), (7, 1)\}$

25. $\{(0, -2), (-1, 5), (3, 3), (-4, 6)\}$
$\{(-2, 0), (5, -1), (3, 3), (6, -4)\}$

26. $\{(-2, -2), (0, 0), (2, 2), (4, 4)\}$
$\{(-2, -2), (0, 0), (2, 2), (4, 4)\}$

27. $\{(-2, -3), (-1, 3), (0, 3), (1, 3)\}$
No inverse

28. $\{(2, 0), (1, 0), (3, 0), (4, 0)\}$
No inverse

For Exercises 29 to 46, find $f^{-1}(x)$.

29. $f(x) = 4x - 8$
$f^{-1}(x) = \dfrac{1}{4}x + 2$

30. $f(x) = 3x + 6$
$f^{-1}(x) = \dfrac{1}{3}x - 2$

31. $f(x) = 2x + 4$
$f^{-1}(x) = \dfrac{1}{2}x - 2$

32. $f(x) = x - 5$

$f^{-1}(x) = x + 5$

33. $f(x) = \dfrac{1}{2}x - 1$
$f^{-1}(x) = 2x + 2$

34. $f(x) = \dfrac{1}{3}x + 2$
$f^{-1}(x) = 3x - 6$

35. $f(x) = -2x + 2$
$f^{-1}(x) = -\dfrac{1}{2}x + 1$

36. $f(x) = -3x - 9$
$f^{-1}(x) = -\dfrac{1}{3}x - 3$

37. $f(x) = \dfrac{2}{3}x + 4$
$f^{-1}(x) = \dfrac{3}{2}x - 6$

38. $f(x) = \dfrac{3}{4}x - 4$
$f^{-1}(x) = \dfrac{4}{3}x + \dfrac{16}{3}$

39. $f(x) = -\dfrac{1}{3}x + 1$
$f^{-1}(x) = -3x + 3$

40. $f(x) = -\dfrac{1}{2}x + 2$
$f^{-1}(x) = -2x + 4$

Unless otherwise noted, all content on this page is © Cengage Learning.

41. $f(x) = 2x - 5$

$f^{-1}(x) = \dfrac{1}{2}x + \dfrac{5}{2}$

42. $f(x) = 3x + 4$

$f^{-1}(x) = \dfrac{1}{3}x - \dfrac{4}{3}$

43. $f(x) = 5x - 2$

$f^{-1}(x) = \dfrac{1}{5}x + \dfrac{2}{5}$

44. $f(x) = 4x - 2$

$f^{-1}(x) = \dfrac{1}{4}x + \dfrac{1}{2}$

45. $f(x) = 6x - 3$

$f^{-1}(x) = \dfrac{1}{6}x + \dfrac{1}{2}$

46. $f(x) = -8x + 4$

$f^{-1}(x) = -\dfrac{1}{8}x + \dfrac{1}{2}$

For Exercises 47 to 49, let $f(x) = 3x - 5$. Find:

47. $f^{-1}(0)$

$\dfrac{5}{3}$

48. $f^{-1}(2)$

$\dfrac{7}{3}$

49. $f^{-1}(4)$

3

For Exercises 50 to 52, state whether the graph is the graph of a function. If it is the graph of a function, does the function have an inverse?

50.
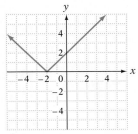
Yes; No

51.
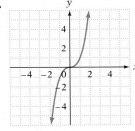
Yes; Yes

52.
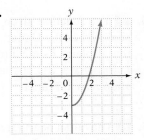
Yes; Yes

For Exercises 53 to 60, use the Composition of Inverse Functions Property to determine whether the functions are inverses.

53. $f(x) = 4x; g(x) = \dfrac{x}{4}$

Yes

54. $g(x) = x + 5; h(x) = x - 5$

Yes

55. $f(x) = 3x; h(x) = \dfrac{1}{3x}$

No

56. $h(x) = x + 2; g(x) = 2 - x$

No

57. $g(x) = 3x + 2; f(x) = \dfrac{1}{3}x - \dfrac{2}{3}$

Yes

58. $h(x) = 4x - 1; f(x) = \dfrac{1}{4}x + \dfrac{1}{4}$

Yes

59. $f(x) = \dfrac{1}{2}x - \dfrac{3}{2}; g(x) = 2x + 3$

Yes

60. $g(x) = -\dfrac{1}{2}x - \dfrac{1}{2}; h(x) = -2x + 1$

No

Unless otherwise noted, all content on this page is © Cengage Learning.

61. ◣ To convert ounces to pounds, we can use the formula $f(x) = \frac{x}{16}$, where x is the number of ounces and $f(x)$ is the number of pounds. Find $f^{-1}(x)$, and write a sentence that explains its meaning.

$f^{-1}(x) = 16x$; The inverse function converts pounds to ounces.

62. ◣ To convert pounds to kilograms, we can use the formula $f(x) = \frac{x}{2.2}$, where x is the number of pounds and $f(x)$ is the number of kilograms. Find $f^{-1}(x)$, and write a sentence that explains its meaning.

$f^{-1}(x) = 2.2x$; The inverse function converts kilograms to pounds.

63. ◣ The function given by $f(x) = x + 30$ converts a dress size x in the United States to a dress size $f(x)$ in France. Find the inverse of this function, and write a sentence that explains its meaning.

$f^{-1}(x) = x - 30$; The inverse function converts a dress size in France to a dress size in the United States.

64. ◣ The function given by $f(x) = x - 31$ converts a shoe size x in France to a shoe size $f(x)$ in the United States. Find the inverse of this function, and write a sentence that explains its meaning.

$f^{-1}(x) = x + 31$; The inverse function converts a shoe size in the United States to a shoe size in France.

65. ◣ The target heart rate $f(x)$ for a certain athlete is given by $f(x) = 90x + 65$, where x is the training intensity percent. Find the inverse of this function, and write a sentence that explains its meaning.

$f^{-1}(x) = \frac{1}{90}x - \frac{13}{18}$; The inverse function gives the intensity of training percent for a given target heart rate for the athlete.

66. ◣ A data messaging service charges $5 per month plus $.10 per message. The function given by $f(x) = 0.1x + 5$ gives the monthly cost $f(x)$ for sending x messages. Find the inverse of this function, and write a sentence that explains its meaning.

$f^{-1}(x) = 10x - 50$; The inverse function gives the number of messages sent given the monthly cost of the service.

67. ● **Currency Exchange** Read the article at the right. The function $f(x) = 120.381x$ represents the exchange rate in January 2007. For this function, the U.S. dollar is the base currency, which means x is in dollars and $f(x)$ is in yen. Find an equation for f^{-1}, which represents the exchange rate with Japanese yen as the base currency. *Note:* Exchange rates less than 1 are often given to six decimal places. $f^{-1}(x) = 0.008307x$

68. ● **Currency Exchange** Read the article at the right and refer to Exercise 67. Give the exchange rate between U.S. dollars and Japanese yen in January 2012 by writing two functions that are inverses of each other. Tell which currency is the base currency for each function. $f(x) = 76.9677x$, base dollar; $f^{-1}(x) = 0.012992x$, base yen

In the NEWS!

Tougher Exchange Rates for Exchange Students

The recession has made life harder for American students living in Tokyo, Japan. In January 2007, when one U.S. dollar was worth 120.381 yen, living in this very expensive city was affordable. Now, in January 2012, with an exchange rate of 76.9677 yen per dollar, American exchange students are feeling the pinch.

Source: www.temple-news.com, www.x-rates.com

Critical Thinking

Each of the tables in Exercises 69 and 70 defines a function. Is the inverse of the function a function? Explain your answer.

69. 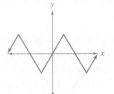 **Grading Scale**

Score	Grade
90–100	A
80–89	B
70–79	C
60–69	D
0–59	F

70. **First-Class Postage Rates**

Weight	Cost
$0 < w < 1$	$.45
$1 < w \leq 2$	$.65
$2 < w \leq 3$	$.85
$3 < w \leq 3.5$	$1.05

71. Is the inverse of a constant function a function? Explain your answer.

72. The graphs of all functions given by $f(x) = mx + b$, $m \neq 0$, are straight lines. Are all of these functions 1–1 functions? If so, explain why. If not, give an example of a linear function that is not 1–1.

Projects or Group Activities

Recall that interchanging the *x*- and *y*-coordinates of each ordered pair of a 1–1 function forms the inverse function. Given the graphs in Exercises 73 to 78, interchange the *x*- and *y*-coordinates of the given ordered pairs and then plot the new ordered pairs. Draw a smooth graph through the ordered pairs. The graph is the graph of the inverse function.

Unless otherwise noted, all content on this page is © Cengage Learning.

QUICK QUIZ

1. Determine whether the graph represents the graph of a 1–1 function.

    No   [11.4A]

Find $f^{-1}(x)$.

2. $f(x) = 4x + 8$

    $f^{-1}(x) = \dfrac{1}{4}x - 2$

    [11.4B]

3. $f(x) = \dfrac{1}{2}x + 1$

    $f^{-1}(x) = 2x - 2$

    [11.4B]

4. $f(x) = -6x + 3$

    $f^{-1}(x) = -\dfrac{1}{6}x + \dfrac{1}{2}$

    [11.4B]

73.

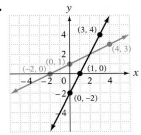

74.

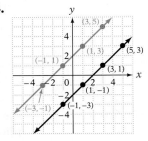

75.

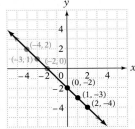

The inverse is the same graph.

76.

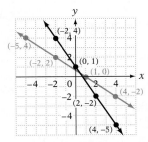

77.

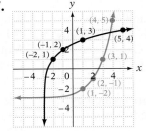

78.

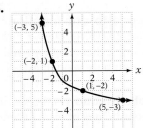

The inverse is the same graph.

11   Summary

Key Words

Examples

A **quadratic function** is one that can be expressed by the equation $f(x) = ax^2 + bx + c$, $a \neq 0$. [11.1A, p. 616]

$f(x) = 4x^2 + 3x - 5$ is a quadratic function.

The graph of a quadratic function is a **parabola**. When $a > 0$, the parabola opens up and the **vertex** of the parabola is the point with the least y-coordinate. When $a < 0$, the parabola opens down and the **vertex** of the parabola is the point with the largest y-coordinate. The **axis of symmetry** is the vertical line that passes through the vertex of the parabola and is parallel to the y-axis. [11.1A, pp. 616, 617]

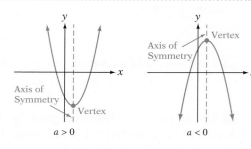

A point at which a graph crosses the x-axis is called an **x-intercept** of the graph. The x-intercepts occur when $y = 0$. [11.1B, p. 619]

$x^2 - 5x + 6 = 0$
$(x - 2)(x - 3) = 0$

$x - 2 = 0 \qquad x - 3 = 0$
$\qquad x = 2 \qquad\qquad x = 3$

The x-intercepts are $(2, 0)$ and $(3, 0)$.

A value of x for which $f(x) = 0$ is called a **zero** of the function. [11.1B, p. 619]

$f(x) = x^2 - 7x + 10$
$0 = x^2 - 7x + 10$
$0 = (x - 2)(x - 5)$

$x - 2 = 0 \qquad x - 5 = 0$
$\qquad x = 2 \qquad\qquad x = 5$

2 and 5 are zeros of the function.

The graph of a quadratic function has a **minimum value** if $a > 0$ and a **maximum value** if $a < 0$. [11.1C, p. 622]

The graph of $f(x) = 3x^2 - x + 4$ has a minimum value.
The graph of $f(x) = -x^2 - 6x + 5$ has a maximum value.

A function is a **one-to-one function** if, for any a and b in the domain of f, $f(a) = f(b)$ implies $a = b$. This means that given any y, there is only one x that can be paired with that y. One-to-one functions are commonly written as 1–1. [11.4A, p. 650]

A non-constant linear function is a 1–1 function.
A quadratic function is not a 1–1 function.

The **inverse of a function** is the set of ordered pairs formed by reversing the coordinates of each ordered pair of the function. [11.4B, p. 651]

The inverse of the function $\{(1, 2), (2, 4), (3, 6), (4, 8), (5, 10)\}$ is $\{(2, 1), (4, 2), (6, 3), (8, 4), (10, 5)\}$.

Unless otherwise noted, all content on this page is © Cengage Learning.

Essential Rules and Procedures

Examples

Vertex and Axis of Symmetry of a Parabola [11.1A, p. 617]

Let $f(x) = ax^2 + bx + c$ be the equation of a parabola. The coordinates of the vertex are $\left(-\frac{b}{2a}, f\left(-\frac{b}{2a}\right)\right)$. The equation of the axis of symmetry is $x = -\frac{b}{2a}$.

$f(x) = x^2 - 2x - 4$
$a = 1, b = -2$

$$-\frac{b}{2a} = -\frac{-2}{2(1)} = 1$$

$$f(1) = 1^2 - 2(1) - 4 = -5$$

The coordinates of the vertex are $(1, -5)$.

The equation of the axis of symmetry is $x = 1$.

Effect of the Discriminant on the Number of x-Intercepts of a Parabola [11.1B, p. 620]

1. If $b^2 - 4ac = 0$, the parabola has one x-intercept.
2. If $b^2 - 4ac > 0$, the parabola has two x-intercepts.
3. If $b^2 - 4ac < 0$, the parabola has no x-intercepts.

$x^2 + 8x + 16 = 0$ has one x-intercept because $b^2 - 4ac = 8^2 - 4(1)(16) = 0$.

$2x^2 + 3x - 5 = 0$ has two x-intercepts because
$b^2 - 4ac = 3^2 - 4(2)(-5) = 49$.

$3x^2 + 2x + 4 = 0$ has no x-intercepts because
$b^2 - 4ac = 2^2 - 4(3)(4) = -44$.

To Find the Minimum or Maximum Value of a Quadratic Function [11.1C, p. 622]

First find the x-coordinate of the vertex. Then evaluate the function at that value.

$f(x) = x^2 - 2x - 4$
$a = 1, b = -2$

$$-\frac{b}{2a} = -\frac{-2}{2(1)} = 1$$

$$f(1) = 1^2 - 2(1) - 4 = -5$$

$a > 0$. The minimum value of the function is -5.

Vertical Translations [11.2A, p. 632]

If f is a function and c is a positive constant, then

- The graph of $y = f(x) + c$ is the graph of $y = f(x)$ shifted c units up.
- The graph of $y = f(x) - c$ is the graph of $y = f(x)$ shifted c units down.

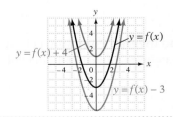

Horizontal Translations [11.2A, p. 633]

If f is a function and c is a positive constant, then

- The graph of $y = f(x + c)$ is the graph of $y = f(x)$ shifted c units to the left.
- The graph of $y = f(x - c)$ is the graph of $y = f(x)$ shifted c units to the right.

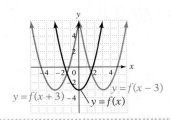

Unless otherwise noted, all content on this page is © Cengage Learning.

Reflections of a Graph About an Axis [11.2B, p. 634]

- The reflection of the graph of $y = f(x)$ about the x-axis is the graph of $y = -f(x)$.
- The reflection of the graph of $y = f(x)$ about the y-axis is the graph of $y = f(-x)$.

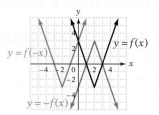

Operations on Functions [11.3A, p. 641]

If f and g are functions and x is an element of the domain of each function, then

$(f + g)(x) = f(x) + g(x)$
$(f - g)(x) = f(x) - g(x)$
$(f \cdot g)(x) = f(x) \cdot g(x)$
$\left(\dfrac{f}{g}\right)(x) = \dfrac{f(x)}{g(x)}, \; g(x) \neq 0$

Given $f(x) = x + 2$ and $g(x) = 2x$:
$(f + g)(4) = f(4) + g(4)$
$\qquad = (4 + 2) + 2(4) = 6 + 8$
$\qquad = 14$
$(f - g)(4) = f(4) - g(4)$
$\qquad = (4 + 2) - 2(4) = 6 - 8$
$\qquad = -2$
$(f \cdot g)(4) = f(4) \cdot g(4)$
$\qquad = (4 + 2) \cdot 2(4) = 6 \cdot 8$
$\qquad = 48$
$\left(\dfrac{f}{g}\right)(4) = \dfrac{f(4)}{g(4)} = \dfrac{4 + 2}{2(4)} = \dfrac{6}{8} = \dfrac{3}{4}$

Definition of the Composition of Two Functions [11.3B, p. 643]

The composition of two functions f and g, symbolized by $f \circ g$, is the function whose value at x is given by $(f \circ g)(x) = f[g(x)]$.

Given $f(x) = x - 4$ and $g(x) = 4x$:
$(f \circ g)(2) = f[g(2)]$
$\qquad = f(8)$ • $g(2) = 8$
$\qquad = 8 - 4 = 4$

Horizontal-Line Test [11.4A, p. 651]

The graph of a function represents the graph of a 1–1 function if any horizontal line intersects the graph at no more than one point.

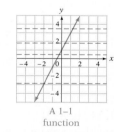

A 1–1
function

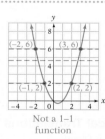

Not a 1–1
function

Condition for an Inverse Function [11.4B, p. 652]

A function f has an inverse function if and only if f is a 1–1 function.

The function $f(x) = x^2$ does not have an inverse function. When $y = 4$, $x = 2$ or -2; therefore, the function $f(x)$ is not a 1–1 function.

Composition of Inverse Functions Property [11.4B, p. 653]

$f^{-1}[f(x)] = x$ and $f[f^{-1}(x)] = x$

$f(x) = 2x - 3 \qquad f^{-1}(x) = \dfrac{1}{2}x + \dfrac{3}{2}$

$f^{-1}[f(x)] = \dfrac{1}{2}(2x - 3) + \dfrac{3}{2} = x$

$f[f^{-1}(x)] = 2\left(\dfrac{1}{2}x + \dfrac{3}{2}\right) - 3 = x$

Unless otherwise noted, all content on this page is © Cengage Learning.

CHAPTER

11 | Review Exercises

1. Use the graph of f shown below to graph $g(x) = f(x - 2)$.

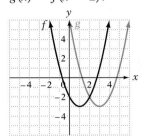

[11.2A]

2. Is the graph shown below the graph of a 1–1 function?

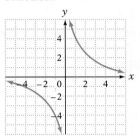

Yes   [11.4A]

3. Use the graph of f shown below to graph $g(x) = f(x - 3) + 2$.

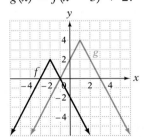

[11.2A]

4. Use the graph of f shown below to graph $g(x) = f(-x)$.

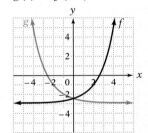

[11.2B]

5. Use the graph of f shown below to graph $g(x) = -f(x)$.

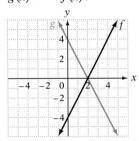

[11.2B]

6. Find the coordinates of the vertex and the equation of the axis of symmetry for the parabola with equation $y = x^2 - 2x + 3$. Then graph the equation.

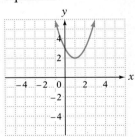

Vertex: (1, 2)
Axis of symmetry: $x = 1$ [11.1A]

7. Use the discriminant to determine the number of x-intercepts of the graph of $y = -3x^2 + 4x + 6$.
Two x-intercepts   [11.1B]

8. Use the discriminant to determine the number of x-intercepts of the graph of $f(x) = 2x^2 + x + 5$.
No x-intercepts   [11.1B]

9. Find the x-intercepts of the graph of $y = 3x^2 + 9x$.
(0, 0), (−3, 0)   [11.1B]

10. Find the x-intercepts of the graph of $f(x) = x^2 - 6x + 7$.
$(3 - \sqrt{2}, 0), (3 + \sqrt{2}, 0)$ [11.1B]

11. Find the zeros of $f(x) = 2x^2 - 7x - 15$.
$-\dfrac{3}{2}, 5$ [11.1B]

12. Find the zeros of $f(x) = x^2 - 2x + 10$.
$1 - 3i, 1 + 3i$ [11.1B]

Unless otherwise noted, all content on this page is © Cengage Learning.

13. Find the maximum value of
$f(x) = -2x^2 + 4x + 1$.
3 [11.1C]

14. Find the minimum value of $f(x) = x^2 - 7x + 8$.
$-\dfrac{17}{4}$ [11.1C]

15. Given $f(x) = x^2 + 4$ and $g(x) = 4x - 1$, find
$f[g(0)]$.
5 [11.3B]

16. Given $f(x) = 6x + 8$ and $g(x) = 4x + 2$, find
$g[f(-1)]$.
10 [11.3B]

17. Given $f(x) = 3x^2 - 4$ and $g(x) = 2x + 1$, find
$f[g(x)]$.
$12x^2 + 12x - 1$ [11.3B]

18. Given $f(x) = 2x^2 + x - 5$ and $g(x) = 3x - 1$,
find $g[f(x)]$.
$6x^2 + 3x - 16$ [11.3B]

For Exercises 19 to 22, use $f(x) = x^2 + 2x - 3$ and $g(x) = x^2 - 2$.

19. Evaluate: $(f + g)(2)$
7 [11.3A]

20. Evaluate: $(f - g)(-4)$
-9 [11.3A]

21. Evaluate: $(f \cdot g)(-4)$

70 [11.3A]

22. Evaluate: $\left(\dfrac{f}{g}\right)(3)$

$\dfrac{12}{7}$ [11.3A]

23. Find the inverse of $f(x) = -6x + 4$.

$f^{-1}(x) = -\dfrac{1}{6}x + \dfrac{2}{3}$ [11.4B]

24. Find the inverse of $f(x) = \frac{2}{3}x - 12$.

$f^{-1}(x) = \dfrac{3}{2}x + 18$ [11.4B]

25. Use the Composition of Inverse Functions Property to determine whether
$f(x) = -\frac{1}{4}x + \frac{5}{4}$ and $g(x) = -4x + 5$ are inverses of each other.
Yes [11.4B]

26. Use the Composition of Inverse Functions Property to determine whether $f(x) = \frac{1}{2}x$
and $g(x) = 2x + 1$ are inverses of each other.
No [11.4B]

27. ◣ **Deep-sea Diving** The pressure p, in pounds per square inch, on a scuba diver
x feet below the surface of the water can be approximated by $p(x) = 0.4x + 15$. Find
$p^{-1}(x)$, and write a sentence that explains its meaning.
$p^{-1}(x) = 2.5x - 37.5$; The inverse function gives the diver's depth below the surface
of the water for a given pressure on the diver. [11.4B]

28. **Business** The monthly profit P, in dollars, earned by a company from the sale of
x youth baseball gloves is given by $P(x) = -x^2 + 100x + 2500$. How many gloves
should be made each month to maximize profit? What is the maximum monthly
profit? 50 gloves; $5000 [11.1D]

29. **Geometry** The perimeter of a rectangle is 28 ft. What dimensions would give the
rectangle a maximum area? 7 ft by 7 ft [11.1D]

CHAPTER

11 | TEST

1. Find the coordinates of the vertex and the equation of the axis of symmetry for the parabola with equation $y = x^2 - 6x + 4$. Then graph the equation.

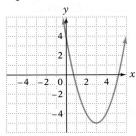

Vertex: $(3, -5)$
Axis of symmetry: $x = 3$ [11.1A]

2. Use the graph of f shown below to graph $g(x) = f(x + 3)$.

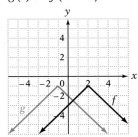

[11.2A]

3. Use the graph of f shown below to graph $g(x) = f(x + 5) - 3$.

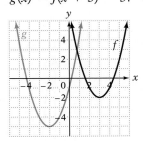

[11.2A]

4. Use the graph of f shown below to graph $g(x) = f(-x)$.

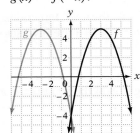

[11.2B]

5. Use the discriminant to determine the number of x-intercepts of the graph of $f(x) = 3x^2 - 4x + 6$.
No x-intercepts [11.1B]

6. Find the coordinates of the x-intercepts of the graph of $y = 3x^2 - 7x - 6$.
$\left(-\dfrac{2}{3}, 0\right)$, $(3, 0)$ [11.1B]

7. Find the maximum value of $f(x) = -x^2 + 8x - 7$.
9 [11.1C]

8. Find the domain and range of $f(x) = 2x^2 + 4x - 5$.
Domain: $\{x \mid -\infty < x < \infty\}$
Range: $\{y \mid y \geq -7\}$ [11.1A]

9. Given $f(x) = x^2 + 2x - 3$ and $g(x) = x^3 - 1$, find $(f - g)(2)$.
-2 [11.3A]

10. Given $f(x) = x^3 + 1$ and $g(x) = 2x - 3$, find $(f \cdot g)(-3)$.
234 [11.3A]

11. Given $f(x) = 4x - 5$ and $g(x) = x^2 + 3x + 4$, find $\left(\dfrac{f}{g}\right)(-2)$.

$-\dfrac{13}{2}$ [11.3A]

12. Given $f(x) = x^2 + 4$ and $g(x) = 2x^2 + 2x + 1$, find $(f - g)(-4)$.
-5 [11.3A]

Unless otherwise noted, all content on this page is © Cengage Learning.

13. Given $f(x) = 2x - 7$ and $g(x) = x^2 - 2x - 5$, find $f[g(2)]$.
−17 [11.3B]

14. Given $f(x) = x^2 + 1$ and $g(x) = x^2 + x + 1$, find $g[f(-2)]$.
31 [11.3B]

15. Given $f(x) = x^2 - 1$ and $g(x) = 3x + 2$, find $g[f(x)]$.
$3x^2 - 1$ [11.3B]

16. Given $f(x) = 2x^2 - 7$ and $g(x) = x - 1$, find $f[g(x)]$.
$2x^2 - 4x - 5$ [11.3B]

17. Does the following function have an inverse function?
$\{(1, 4), (2, 5), (3, 6), (4, 5), (5, 4)\}$
No [11.4B]

18. Find the inverse of the following function.
$\{(2, 6), (3, 5), (4, 4), (5, 3)\}$
$\{(6, 2), (5, 3), (4, 4), (3, 5)\}$ [11.4B]

19. Find the inverse of $f(x) = 4x - 2$.
$f^{-1}(x) = \dfrac{1}{4}x + \dfrac{1}{2}$ [11.4B]

20. Find the inverse of $f(x) = \dfrac{1}{4}x - 4$.
$f^{-1}(x) = 4x + 16$ [11.4B]

21. Use the Composition of Inverse Functions Property to determine whether $f(x) = \frac{1}{2}x + 2$ and $g(x) = 2x - 4$ are inverses of each other.
Yes [11.4B]

22. Use the Composition of Inverse Functions Property to determine whether $f(x) = \frac{2}{3}x - 2$ and $g(x) = \frac{3}{2}x + 3$ are inverses of each other.
Yes [11.4B]

23. Determine whether the graph at the right is the graph of a 1–1 function.
No [11.4A]

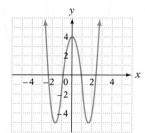

24. ✎ **Delivery Service** A company that delivers documents in a city via bicycle messenger charges $5 per message plus $1.25 per mile for the service. The total cost $C(x)$ to deliver a document to a location x miles away is given by $C(x) = 1.25x + 5$. Find $C^{-1}(x)$, and write a sentence that explains its meaning.
$C^{-1}(x) = 0.8x - 4$. The inverse function gives the number of miles to the location for a given cost. [11.4B]

25. **Business** The daily production cost $C(x)$, in dollars, to manufacture small speakers is given by $C(x) = x^2 - 50x + 675$, where x is the number of speakers produced per day. How many speakers should be made per day to minimize the daily production cost? What is the minimum daily production cost?
25 speakers; $50 [11.1D]

26. **Mathematics** Find two numbers whose sum is 28 and whose product is a maximum.
14 and 14 [11.1D]

27. **Geometry** The perimeter of a rectangle is 200 cm. What dimensions will give the rectangle a maximum area? What is the maximum area?
50 cm by 50 cm; 2500 cm² [11.1D]

Unless otherwise noted, all content on this page is © Cengage Learning.

Cumulative Review Exercises

1. Evaluate $-3a + \left|\dfrac{3b - ab}{3b - c}\right|$ when $a = 2$, $b = 2$, and $c = -2$.

$-\dfrac{23}{4}$ [1.4A]

2. Graph $\{x|x < -3\} \cap \{x|x > -4\}$.

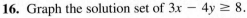 [1.5B]

3. Solve: $\dfrac{3x - 1}{6} - \dfrac{5 - x}{4} = \dfrac{5}{6}$

3   [2.2C]

4. Solve: $4x - 2 < -10$ or $3x - 1 > 8$
Write the solution set in set-builder notation.
$\{x|x < -2\} \cup \{x|x > 3\}$ [2.5B]

5. Solve: $|8 - 2x| \geq 0$
All real numbers   [2.6B]

6. Simplify: $\left(\dfrac{3a^3b}{2a}\right)^2\left(\dfrac{a^2}{-3b^2}\right)^3$

$-\dfrac{a^{10}}{12b^4}$ [6.1B]

7. Multiply: $(x - 4)(2x^2 + 4x - 1)$
$2x^3 - 4x^2 - 17x + 4$ [6.3B]

8. Solve by using the addition method:
$6x - 2y = -3$
$4x + y = 5$
$\left(\dfrac{1}{2}, 3\right)$ [5.2A]

9. Factor: $x^3y + x^2y^2 - 6xy^3$
$xy(x + 3y)(x - 2y)$ [7.2B]

10. Solve: $(b + 2)(b - 5) = 2b + 14$
-3 and 8   [7.5A]

11. Solve: $x^2 - 2x > 15$
$\{x|x < -3\} \cup \{x|x > 5\}$ [10.5A]

12. Subtract: $\dfrac{x^2 + 4x - 5}{2x^2 - 3x + 1} - \dfrac{x}{2x - 1}$

$\dfrac{5}{2x - 1}$ [8.2B]

13. Solve: $\dfrac{5}{x^2 + 7x + 12} = \dfrac{9}{x + 4} - \dfrac{2}{x + 3}$
-2 [8.4A]

14. Divide: $\dfrac{4 - 6i}{2i}$

$-3 - 2i$ [9.5D]

15. Graph $f(x) = \frac{1}{4}x^2$. Find the coordinates of the vertex and the equation of the axis of symmetry.

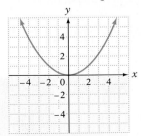

Vertex: $(0, 0)$
Axis of symmetry: $x = 0$ [11.1A]

16. Graph the solution set of $3x - 4y \geq 8$.

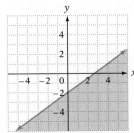

[4.7A]

Unless otherwise noted, all content on this page is © Cengage Learning.

17. Find the equation of the line containing the points $P_1(-3, 4)$ and $P_2(2, -6)$.
$y = -2x - 2$ [4.5B]

18. Find the equation of the line that contains the point $P(-3, 1)$ and is perpendicular to the graph of $2x - 3y = 6$.
$y = -\dfrac{3}{2}x - \dfrac{7}{2}$ [4.6A]

19. Solve: $3x^2 = 3x - 1$
$\dfrac{1}{2} - \dfrac{\sqrt{3}}{6}i$ and $\dfrac{1}{2} + \dfrac{\sqrt{3}}{6}i$ [10.2B]

20. Solve: $\sqrt{8x + 1} = 2x - 1$
3 [10.3B]

21. Find the minimum value of $f(x) = 2x^2 - 3$.
-3 [11.1C]

22. Find the zero of $f(x) = 3x - 4$.
$\dfrac{4}{3}$ [4.3C]

23. Is the following set of ordered pairs a function?
$\{(-3, 0), (-2, 0), (-1, 1), (0, 1)\}$

Yes [4.2A]

24. Solve: $\sqrt[3]{5x - 2} = 2$
2 [9.4A]

25. Given $g(x) = 3x - 5$ and $h(x) = \frac{1}{2}x + 4$, find $g[h(2)]$.
10 [11.3B]

26. Find the inverse of $f(x) = -3x + 9$.
$f^{-1}(x) = -\dfrac{1}{3}x + 3$ [11.4B]

27. Mixtures Find the cost per pound of a tea mixture made from 30 lb of tea costing $4.50 per pound and 45 lb of tea costing $3.60 per pound.
$3.96 [2.4A]

28. Mixtures How many pounds of an 80% copper alloy must be mixed with 50 lb of a 20% copper alloy to make an alloy that is 40% copper?
25 lb [2.4B]

29. Mixtures Six ounces of insecticide are mixed with 16 gal of water to make a spray for spraying an orange grove. How much additional insecticide is required if it is to be mixed with 28 gal of water?
4.5 oz [8.5B]

30. Tanks A large pipe can fill a tank in 8 min less time than it takes a smaller pipe to fill the same tank. Working together, the pipes can fill the tank in 3 min. How long would it take the larger pipe, working alone, to fill the tank?
4 min [10.4A]

31. Physics The distance d that a spring stretches varies directly as the force f used to stretch the spring. If a force of 50 lb can stretch a spring 30 in., how far can a force of 40 lb stretch the spring?
24 in. [8.8A]

32. Music The frequency of vibration f in a pipe in an open pipe organ varies inversely as the length L of the pipe. If the air in a pipe 2 m long vibrates 60 times per minute, find the frequency in a pipe that is 1.5 m long.
80 vibrations/min [8.8A]

© iStockphoto.com/Soubrette

Exponential and Logarithmic Functions

OBJECTIVES

SECTION 12.1
A To evaluate an exponential function
B To graph an exponential function

SECTION 12.2
A To find the logarithm of a number
B To use the Properties of Logarithms to simplify expressions containing logarithms
C To use the Change-of-Base Formula

SECTION 12.3
A To graph a logarithmic function

SECTION 12.4
A To solve an exponential equation
B To solve a logarithmic equation

SECTION 12.5
A To solve application problems

Focus on Success

The end of the semester is generally a very busy and stressful time. You may be dealing with the anxiety of taking final exams. You have covered a great deal of material in this course, and reviewing all of it may be daunting. You might begin by reviewing the Chapter Summary for each chapter that you were assigned during the term. Then take the Final Exam on page 717. The answer to each exercise is given at the back of the book. (See Ace the Test, page AIM-11.)

Prep Test

© iStockphoto.com/Pete Saloutos

Are you ready to succeed in this chapter? Take the Prep Test below to find out if you are ready to learn the new material.

1. Simplify: 3^{-2}

$\dfrac{1}{9}$ [6.1B]

2. Simplify: $\left(\dfrac{1}{2}\right)^{-4}$

16 [6.1B]

3. Complete: $\dfrac{1}{8} = 2^{?}$

-3 [6.1B]

4. Evaluate $f(x) = x^4 + x^3$ for $x = -1$ and $x = 3$.

0; 108 [4.2A/6.2A]

5. Solve: $3x + 7 = x - 5$

-6 [2.2B]

6. Solve: $16 = x^2 - 6x$

$-2, 8$ [10.1A]

7. Evaluate $A(1 + i)^n$ for $A = 5000$, $i = 0.04$, and $n = 6$. Round to the nearest hundredth.

6326.60 [1.4A]

8. Graph: $f(x) = x^2 - 1$

[11.1A]

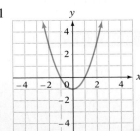

Unless otherwise noted, all content on this page is © Cengage Learning.

SECTION

12.1 Exponential Functions

OBJECTIVE A *To evaluate an exponential function*

The growth of a $500 savings account that earns 5% annual interest compounded daily is shown in the graph at the right. In approximately 14 years, the savings account contains approximately $1000, twice the initial amount. The growth of this savings account is an example of an exponential function.

The pressure of the atmosphere at a certain altitude is shown in the graph at the right. This is another example of an exponential function. From the graph, we read that the air pressure is approximately 6.5 lb/in² at an altitude of 20,000 ft.

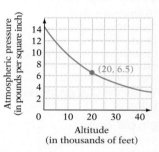

INSTRUCTOR NOTE

It is important for students to distinguish between $P(x) = x^2$ and $F(x) = 2^x$. The first is a polynomial function; the second is an exponential function. Polynomial functions have a variable base and a constant exponent, whereas exponential functions are characterized by a constant base and a variable exponent.

Definition of an Exponential Function

The **exponential function with base b** is defined by $f(x) = b^x$ where $b > 0$, $b \neq 1$, and x is any real number.

EXAMPLES

1. $f(x) = 3^x$ is an exponential function with base $b = 3$.

2. $g(x) = \left(\dfrac{2}{3}\right)^x$ is an exponential function with base $b = \dfrac{2}{3}$.

3. $h(x) = 2^{-x}$ is an exponential function with base $b = \dfrac{1}{2}$. To see this, rewrite $h(x)$ as follows.

$$h(x) = 2^{-x} = (2^{-1})^x = \left(\frac{1}{2}\right)^x$$

In the definition of an exponential function, b, the base, is required to be positive. If the base were a negative number, the value of the function would be a complex number for some values of x. For instance, the value of $f(x) = (-4)^x$ when $x = \frac{1}{2}$ is $f\left(\frac{1}{2}\right) = (-4)^{1/2} = \sqrt{-4} = 2i$. To avoid complex number values of a function, the base of the exponential function is always a positive number.

HOW TO 1 Evaluate $f(x) = 2^x$ at $x = 3$ and $x = -2$.

$f(3) = 2^3 = 8$ • Substitute 3 for x and simplify.

$f(-2) = 2^{-2} = \dfrac{1}{2^2} = \dfrac{1}{4}$ • Substitute −2 for x and simplify.

Unless otherwise noted, all content on this page is © Cengage Learning.

IN-CLASS EXAMPLES

1. Evaluate $f(x) = 3^{x+1}$ at $x = 2$ and $x = -3$.

   $27; \dfrac{1}{9}$

2. Given $f(x) = \left(\dfrac{1}{3}\right)^{2x}$, evaluate $f(0)$ and $f\left(\dfrac{1}{2}\right)$. $1; \dfrac{1}{3}$

APPLY THE CONCEPT

If \$250 is deposited into an account that earns 4% annual interest compounded quarterly, the value of the investment after n years is given by the exponential function $V(n) = 250(1.01)^{4n}$. The value of the investment after 8 years can be determined by evaluating the function at $n = 8$.

$$V(n) = 250(1.01)^{4n}$$
$$V(8) = 250(1.01)^{4(8)} = 250(1.01)^{32} \approx 343.74 \qquad \bullet\ n = 8$$

The value of the investment after 8 years is \$343.74.

To evaluate an exponential expression at an irrational number such as $\sqrt{2}$, we obtain an approximation to the value of the function by approximating the irrational number. For instance, the value of $f(x) = 4^x$ at $x = \sqrt{2}$ can be approximated by using an approximation of $\sqrt{2}$.

$$f(\sqrt{2}) = 4^{\sqrt{2}} \approx 4^{1.4142} \approx 7.1029$$

Because $f(x) = b^x$ ($b > 0$, $b \neq 1$) can be evaluated at both rational and irrational numbers, the domain of f is all real numbers. And because $b^x > 0$ for all values of x, the range of f is the positive real numbers.

A frequently used base in applications of exponential functions is an irrational number designated by e. **The number e is approximately 2.71828183.** It is an irrational number, so it has a nonterminating, nonrepeating decimal representation.

 Take Note

The natural exponential function is an extremely important function. It is used extensively in applied problems in virtually all disciplines, from archaeology to zoology. Leonhard Euler (1707–1783) was the first to use the letter e as the base of the natural exponential function.

Natural Exponential Function

The function defined by $f(x) = e^x$ is called the **natural exponential function.**

 Integrating Technology

A graphing calculator can be used to evaluate an exponential function at an irrational number, as shown at the left below. Use the e^x key to evaluate the natural exponential function. Evaluations of e^5 and $e^{-1.2}$ are shown at the right below.

INSTRUCTOR NOTE

If time permits, show students some of the remarkable relationships that exist among i, e, and π. For instance, $e^{\pi i} = -1$ and $i^{-i} = e^{\pi/2}$.

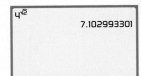

EXAMPLE 1

Evaluate $f(x) = \left(\dfrac{1}{2}\right)^x$ at $x = 2$ and $x = -3$.

Solution

$$f(2) = \left(\dfrac{1}{2}\right)^2 = \dfrac{1}{4} \qquad \bullet\ x = 2$$

$$f(-3) = \left(\dfrac{1}{2}\right)^{-3} = 2^3 = 8 \qquad \bullet\ x = -3$$

YOU TRY IT 1

Evaluate $f(x) = \left(\dfrac{2}{3}\right)^x$ at $x = 3$ and $x = -2$.

Your solution

$$f(3) = \dfrac{8}{27}$$

$$f(-2) = \dfrac{9}{4}$$

Solution on p. S33

Unless otherwise noted, all content on this page is © Cengage Learning.

EXAMPLE 2

Evaluate $f(x) = 2^{3x-1}$ at $x = 1$ and $x = -1$.

Solution

$$f(1) = 2^{3(1)-1} = 2^2 = 4 \qquad \bullet\; x = 1$$
$$f(-1) = 2^{3(-1)-1} = 2^{-4} = \frac{1}{2^4} = \frac{1}{16} \qquad \bullet\; x = -1$$

YOU TRY IT 2

Evaluate $f(x) = 2^{2x+1}$ at $x = 0$ and $x = -2$.

Your solution

$$f(0) = 2$$

$$f(-2) = \frac{1}{8}$$

EXAMPLE 3

Evaluate $f(x) = e^{2x}$ at $x = 1$ and $x = -1$.
Round to the nearest ten-thousandth.

Solution

$$f(1) = e^{2 \cdot 1} = e^2 \approx 7.3891 \qquad \bullet\; x = 1$$
$$f(-1) = e^{2(-1)} = e^{-2} \approx 0.1353 \qquad \bullet\; x = -1$$

YOU TRY IT 3

Evaluate $f(x) = e^{2x-1}$ at $x = 2$ and $x = -2$.
Round to the nearest ten-thousandth.

Your solution

$$f(2) \approx 20.0855$$
$$f(-2) \approx 0.0067$$

Solutions on p. S33

OBJECTIVE B *To graph an exponential function*

Some properties of an exponential function can be seen from its graph.

Integrating Technology

See the Keystroke Guide: *Graph* for instructions on using a graphing calculator to graph functions.

HOW TO 2 Graph $f(x) = 2^x$.

Think of this as the equation $y = 2^x$. Choose values of x and find the corresponding values of y. The results can be recorded in a table.

Graph the ordered pairs on a rectangular coordinate system. Connect the points with a smooth curve.

INSTRUCTOR NOTE

Graph $g(x) = x^2$ so that students can see the difference between the graph of $g(x) = x^2$ and the graph of $f(x) = 2^x$.

x	$f(x) = 2^x$	y	(x, y)
-2	$f(-2) = 2^{-2}$	$\frac{1}{4}$	$\left(-2, \frac{1}{4}\right)$
-1	$f(-1) = 2^{-1}$	$\frac{1}{2}$	$\left(-1, \frac{1}{2}\right)$
0	$f(0) = 2^0$	1	$(0, 1)$
1	$f(1) = 2^1$	2	$(1, 2)$
2	$f(2) = 2^2$	4	$(2, 4)$
3	$f(3) = 2^3$	8	$(3, 8)$

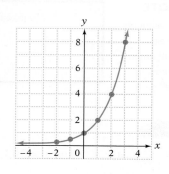

Note that any vertical line would intersect the graph in HOW TO 2 at only one point. Therefore, by the vertical-line test (page 616), the graph of $f(x) = 2^x$ is the graph of a function. Also note that any horizontal line would intersect the graph at only one point. Therefore, the graph of $f(x) = 2^x$ is the graph of a one-to-one function.

Unless otherwise noted, all content on this page is © Cengage Learning.

IN-CLASS EXAMPLES

3. Graph $f(x) = 3^{x-1}$.

4. Graph $f(x) = \left(\frac{1}{2}\right)^x + 2$.

HOW TO 3   Graph $f(x) = \left(\frac{2}{3}\right)^x$.

Think of this as the equation $y = \left(\frac{2}{3}\right)^x$. Choose values of x and find the corresponding values of y. Graph the ordered pairs on a rectangular coordinate system. You will need to approximate the y values as best you can. Connect the points with a smooth curve.

x	$f(x) = \left(\frac{2}{3}\right)^x$	y	(x, y)
-3	$f(-3) = \left(\frac{2}{3}\right)^{-3}$	$\frac{27}{8}$	$\left(-3, \frac{27}{8}\right)$
-2	$f(-2) = \left(\frac{2}{3}\right)^{-2}$	$\frac{9}{4}$	$\left(-2, \frac{9}{4}\right)$
-1	$f(-1) = \left(\frac{2}{3}\right)^{-1}$	$\frac{3}{2}$	$\left(-1, \frac{3}{2}\right)$
0	$f(0) = \left(\frac{2}{3}\right)^{0}$	1	$(0, 1)$
1	$f(1) = \left(\frac{2}{3}\right)^{1}$	$\frac{2}{3}$	$\left(1, \frac{2}{3}\right)$
2	$f(2) = \left(\frac{2}{3}\right)^{2}$	$\frac{4}{9}$	$\left(2, \frac{4}{9}\right)$

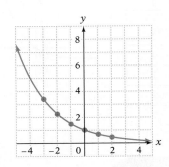

INSTRUCTOR NOTE

After you discuss the graph of the exponential function, you might give students the extra credit problem of graphing $f(x) = x^x$. A discussion of the domain of this function may help students gain a better understanding of domain.

HOW TO 4   Graph $f(x) = 2^{-x}$.

Think of this as the equation $y = 2^{-x}$. Choose values of x and find the corresponding values of y. Graph the ordered pairs on a rectangular coordinate system. Connect the points with a smooth curve.

x	$f(x) = 2^{-x}$	y	(x, y)
-3	$f(-3) = 2^{-(-3)} = 2^3$	8	$(-3, 8)$
-2	$f(-2) = 2^{-(-2)} = 2^2$	4	$(-2, 4)$
-1	$f(-1) = 2^{-(-1)} = 2^1$	2	$(-1, 2)$
0	$f(0) = 2^{-0} = 2^0$	1	$(0, 1)$
1	$f(1) = 2^{-1}$	$\frac{1}{2}$	$\left(1, \frac{1}{2}\right)$
2	$f(2) = 2^{-2}$	$\frac{1}{4}$	$\left(2, \frac{1}{4}\right)$

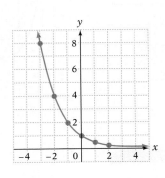

 Take Note

Applying the vertical-line and horizontal-line tests to the graphs in HOW TO 3 and HOW TO 4 reveals that

$f(x) = \left(\frac{2}{3}\right)^x$ and $f(x) = 2^{-x}$

are one-to-one functions.

EXAMPLE 4

Graph: $f(x) = 3^{1/2\,x - 1}$

Solution

x	$y = 3^{\frac{1}{2}x - 1}$
-2	$\frac{1}{9}$
0	$\frac{1}{3}$
2	1
4	3

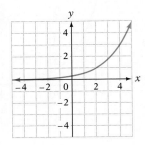

YOU TRY IT 4

Graph: $f(x) = 2^{-1/2\,x}$

Your solution

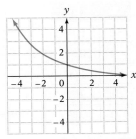

Solution on p. S33

Unless otherwise noted, all content on this page is © Cengage Learning.

EXAMPLE 5

Graph: $f(x) = 2^x - 1$

Solution

x	$y = 2^x - 1$
-2	$-\dfrac{3}{4}$
-1	$-\dfrac{1}{2}$
0	0
1	1
2	3
3	7

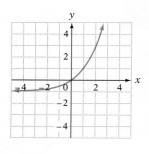

YOU TRY IT 5

Graph: $f(x) = 2^x + 1$

Your solution

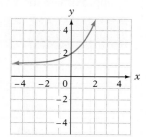

EXAMPLE 6

Graph: $f(x) = \left(\dfrac{1}{3}\right)^x - 2$

Solution

x	$y = \left(\dfrac{1}{3}\right)^x - 2$
-2	7
-1	1
0	-1
1	$-\dfrac{5}{3}$
2	$-\dfrac{17}{9}$

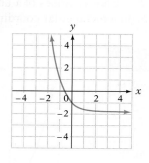

YOU TRY IT 6

Graph: $f(x) = 2^{-x} + 2$

Your solution

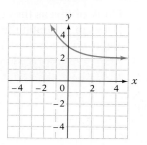

EXAMPLE 7

Graph: $f(x) = e^x$

Solution Use a calculator to approximate the values of y.

x	$y = e^x$
-2	0.14
-1	0.37
0	1.00
1	2.72
2	7.39

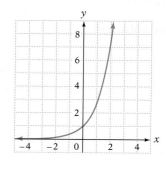

YOU TRY IT 7

Graph: $f(x) = e^{-x/2}$

Your solution

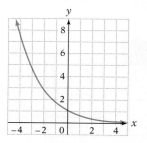

Solutions on p. S33

Unless otherwise noted, all content on this page is © Cengage Learning.

12.1 EXERCISES

✔ **Concept Check**

SUGGESTED ASSIGNMENT
Exercises 1–4; Exercises 5–35, odds;
More challenging exercises: Exercises 37–41 (A graphing calculator is required.)

1. ◤ What is an exponential function?

2. ◤ What is the natural exponential function?

3. Which of the following cannot be the base of an exponential function?

   (i) 7   **(ii)** $\dfrac{1}{4}$ **(iii)** -5 **(iv)** 0.01   **(v)** 1   **(vi)** $-\dfrac{2}{3}$

   iii, v, vi

4. State whether the function is an exponential function.

   a. $f(x) = x^3$ **b.** $f(x) = e^{-x}$ **c.** $f(x) = 1.5^{3x}$ **d.** $f(x) = \left(\dfrac{1}{2}\right)^{2x}$ **e.** $f(x) = 3x^{-2x}$

       No   Yes   Yes   Yes   No

OBJECTIVE A   *To evaluate an exponential function*

5. Given $f(x) = 3^x$, evaluate the following.
   a. $f(2)$ **b.** $f(0)$ **c.** $f(-2)$

   9   1 $\dfrac{1}{9}$

6. Given $H(x) = 2^x$, evaluate the following.
   a. $H(-3)$ **b.** $H(0)$ **c.** $H(2)$

   $\dfrac{1}{8}$ 1   4

7. Given $g(x) = 2^{x+1}$, evaluate the following.
   a. $g(3)$ **b.** $g(1)$ **c.** $g(-3)$

   16   4 $\dfrac{1}{4}$

8. Given $F(x) = 3^{x-2}$, evaluate the following.
   a. $F(-4)$ **b.** $F(-1)$ **c.** $F(0)$

   $\dfrac{1}{729}$ $\dfrac{1}{27}$ $\dfrac{1}{9}$

9. Given $P(x) = \left(\dfrac{1}{2}\right)^{2x}$, evaluate the following.
   a. $P(0)$ **b.** $P\left(\dfrac{3}{2}\right)$ **c.** $P(-2)$

   1          $\dfrac{1}{8}$ 16

10. Given $R(t) = \left(\dfrac{1}{3}\right)^{3t}$, evaluate the following.
    a. $R\left(-\dfrac{1}{3}\right)$ **b.** $R(1)$ **c.** $R(-2)$

    3         $\dfrac{1}{27}$ 729

11. Given $G(x) = e^{x/2}$, evaluate the following. Round to the nearest ten-thousandth.
    a. $G(4)$ **b.** $G(-2)$ **c.** $G\left(\dfrac{1}{2}\right)$

    7.3891   0.3679   1.2840

12. Given $f(x) = e^{2x}$, evaluate the following. Round to the nearest ten-thousandth.
    a. $f(-2)$ **b.** $f\left(-\dfrac{2}{3}\right)$ **c.** $f(2)$

    0.0183   0.2636   54.5982

13. Given $H(r) = e^{-r+3}$, evaluate the following. Round to the nearest ten-thousandth.
    a. $H(-1)$ **b.** $H(3)$ **c.** $H(5)$
        54.5982   1   0.1353

14. Given $P(t) = e^{-\frac{1}{2}t}$, evaluate the following. Round to the nearest ten-thousandth.
    a. $P(-3)$ **b.** $P(4)$ **c.** $P\left(\dfrac{1}{2}\right)$

        4.4817   0.1353   0.7788

15. Given $F(x) = 2^{x^2}$, evaluate the following.

    a. $F(2)$ **b.** $F(-2)$ **c.** $F\left(\dfrac{3}{4}\right)$

        16   16   1.4768

16. Given $Q(x) = 2^{-x^2}$, evaluate the following.

    a. $Q(3)$ **b.** $Q(-1)$ **c.** $Q(-2)$

     $\dfrac{1}{512}$ $\dfrac{1}{2}$ $\dfrac{1}{16}$

17. Given $f(x) = e^{-x^2/2}$, evaluate the following. Round to the nearest ten-thousandth.

    a. $f(-2)$ **b.** $f(2)$ **c.** $f(-3)$

     0.1353   0.1353   0.0111

18. Given $f(x) = e^{-2x} + 1$, evaluate the following. Round to the nearest ten-thousandth.

    a. $f(-1)$ **b.** $f(3)$ **c.** $f(-2)$

     8.3891   1.0025   55.5982

19. 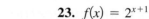 Suppose a and b are real numbers with $a < b$. If $f(x) = 2^{-x}$, then is $f(a) < f(b)$ or is $f(a) > f(b)$?

    $f(a) > f(b)$

20. Suppose that u and v are real numbers with $u < v$ and that $f(x) = b^x$ ($b > 0$, $b \neq 1$). If $f(u) < f(v)$, then is $0 < b < 1$ or is $b > 1$?

    $b > 1$

OBJECTIVE B   *To graph an exponential function*

For Exercises 21 to 32, graph the function.

21. $f(x) = 3^x$

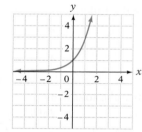

22. $f(x) = 3^{-x}$

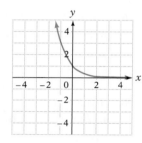

23. $f(x) = 2^{x+1}$

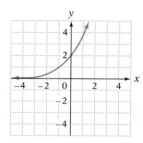

24. $f(x) = 2^{x-1}$

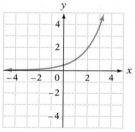

25. $f(x) = \left(\dfrac{1}{3}\right)^x$

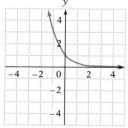

26. $f(x) = \left(\dfrac{2}{3}\right)^x$

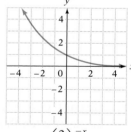

27. $f(x) = 2^{-x} + 1$

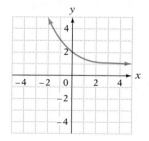

28. $f(x) = 2^x - 3$

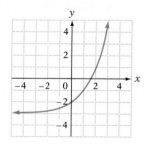

29. $f(x) = \left(\dfrac{1}{3}\right)^{-x}$

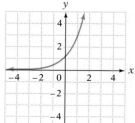

30. $f(x) = \left(\dfrac{3}{2}\right)^{-x}$

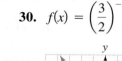

31. $f(x) = e^{x+1} - 1$

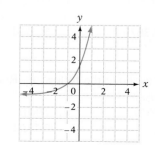

32. $f(x) = e^{-x} + 1$

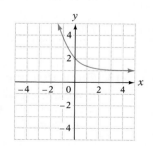

Unless otherwise noted, all content on this page is © Cengage Learning.

33. **Physics** If air resistance is ignored, the speed v, in feet per second, of an object t seconds after it has been dropped is given by $v(t) = 32t$. However, if air resistance is considered, then the speed depends on the mass (and on other things). For a certain mass, the speed of an object t seconds after it has been dropped is given by $v(t) = 32(1 - e^{-t})$. Find the speed of the object after 4 s. Round to the nearest hundredth. 31.41 ft/s

34. **Investments** The exponential function given by $F(n) = 500(1.00021918)^{365n}$ gives the value after n years of a \$500 investment in a certificate of deposit that earns 8% annual interest compounded daily. What is the value of the investment after 9 years? Round to the nearest cent. \$1027.14

35. Which of the following have the same graph?

(i) $f(x) = 3^x$ (ii) $f(x) = \left(\dfrac{1}{3}\right)^x$ (iii) $f(x) = \left(\dfrac{1}{3}\right)^{-x}$ (iv) $f(x) = 3^{-x}$

i and iii; ii and iv

36. Which of the following have the same graph?

(i) $f(x) = \left(\dfrac{1}{4}\right)^{-x}$ (ii) $f(x) = 4^{-x}$ (iii) $f(x) = 4^x$ (iv) $f(x) = \left(\dfrac{1}{4}\right)^{x}$

i and iii; ii and iv

Critical Thinking

For Exercises 37 to 39, use a graphing calculator to graph the function.

37. $P(x) = \left(\sqrt{3}\right)^x$

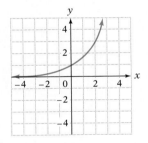

38. $Q(x) = \left(\sqrt{3}\right)^{-x}$

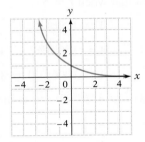

39. $f(x) = \pi^x$

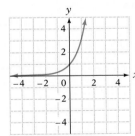

Projects or Group Activities

40. Evaluate $\left(1 + \dfrac{1}{n}\right)^n$ for $n = 100$, 1000, 10,000, and 100,000, and compare the results with the value of e, the base of the natural exponential function. On the basis of your evaluation, complete the following sentence: As n increases, $\left(1 + \dfrac{1}{n}\right)^n$ becomes closer to _____ e _____.

41. Evaluate $(1 + n)^{1/n}$ for $n = 0.01$, 0.001, 0.0001, and 0.00001. Compare the results with the value of e, the base of the natural exponential function. On the basis of your evaluation, complete the following sentence: As n decreases, $(1 + n)^{1/n}$ becomes closer to _____ e _____.

Unless otherwise noted, all content on this page is © Cengage Learning.

QUICK QUIZ

1. Evaluate $f(x) = 2^x$ at $x = 3$ and $x = -2$.
   $8; \dfrac{1}{4}$ **[12.1A]**

2. Given $f(x) = \left(\dfrac{1}{2}\right)^{x-1}$, evaluate $f(1)$ and $f(3)$.
   $1; \dfrac{1}{4}$ **[12.1A]**

3. Graph $f(x) = 3^{x+1}$.

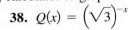

[12.1B]

4. Graph $f(x) = 2^{-x} + 2$.

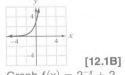

[12.1B]

SECTION

12.2 Introduction to Logarithms

OBJECTIVE A *To find the logarithm of a number*

INSTRUCTOR NOTE

To prepare students for the inverse function nature of the exponential and logarithmic functions, you can use other functions the students have already studied. For example:

(i) The halving function
$f(x) = \frac{1}{2}x$ and the doubling function
$g(x) = 2x$: $\frac{1}{2}(6) = 3$ and $2(3) = 6$;
$\frac{1}{2}(20) = 10$ and $2(10) = 20$

(ii) The square root function
$f(x) = \sqrt{x}$, $x \geq 0$, and the squaring function
$g(x) = x^2$:
$\sqrt{9} = 3$ and $3^2 = 9$;
$\sqrt{25} = 5$ and $5^2 = 25$

You might have students create tables of values for the inverse functions presented above.

Suppose a bacteria colony that originally contained 1000 bacteria doubles in size every hour. Then the table at the right shows the number of bacteria that would be in the colony after 1 h, 2 h, and 3 h.

Time (in hours)	Number of Bacteria
0	1000
1	2000
2	4000
3	8000

The exponential function $A = 1000(2^t)$, where A is the number of bacteria in the colony at time t, is a model of the growth of the colony. For instance, when $t = 3$ h, we have

$$A = 1000(2^t)$$
$$A = 1000(2^3) \qquad \bullet \text{ Replace } t \text{ by 3.}$$
$$A = 1000(8) = 8000$$

After 3 h, there are 8000 bacteria in the colony.

Now we ask, "How long will it take for there to be 32,000 bacteria in the colony?" To answer this question, we must solve the *exponential equation* $32{,}000 = 1000(2^t)$. By trial and error, we find that when $t = 5$,

$$A = 1000(2^t)$$
$$A = 1000(2^5) \qquad \bullet \text{ Replace } t \text{ by 5.}$$
$$A = 1000(32) = 32{,}000$$

After 5 h, there will be 32,000 bacteria in the colony.

Now suppose we want to know how long it will take before the colony reaches 50,000 bacteria. To answer that question, we must find t so that $50{,}000 = 1000(2^t)$. Using trial and error again, we find that

$$1000(2^5) = 32{,}000 \quad \text{and} \quad 1000(2^6) = 64{,}000$$

Because 50,000 is between 32,000 and 64,000, we conclude that t is between 5 h and 6 h. If we try $t = 5.5$ (halfway between 5 and 6), then

$$A = 1000(2^t)$$
$$A = 1000(2^{5.5}) \qquad \bullet \text{ Replace } t \text{ by 5.5.}$$
$$A \approx 1000(45.25) = 45{,}250$$

In 5.5 h, there are approximately 45,250 bacteria in the colony. Because this is less than 50,000, the value of t must be a little greater than 5.5.

Unless otherwise noted, all content on this page is © Cengage Learning.

We could continue to use trial and error to find the correct value of *t,* but it would be more efficient if we could just solve the exponential equation $50{,}000 = 1000(2^t)$ for *t.* If we follow the procedures for solving equations that were discussed earlier in the text, we have

$$50{,}000 = 1000(2^t)$$
$$50 = 2^t \qquad \text{• Divide each side of the equation by 1000.}$$

To proceed to the next step, it would be helpful to have a function that would find the power of 2 that produces 50.

Around the mid-sixteenth century, mathematicians created such a function, which we now call a *logarithmic function.* We write the solution of $50 = 2^t$ as $t = \log_2 50$. This is read "*t* equals the logarithm base 2 of 50" and it means "*t* equals the power of 2 that produces 50." When logarithms were first introduced, tables were used to find a numerical value of *t.* Today, a calculator is used. Using a calculator, we can approximate the value of *t* as 5.644. This means that $2^{5.644} \approx 50$.

The equivalence of the expressions $50 = 2^t$ and $t = \log_2 50$ are described in the following definition of **logarithm.**

Take Note

When we tried $t = 5.5$, we stated that the actual value of *t* must be greater than 5.5. Note that 5.644 is a little greater than 5.5.

Integrating Technology

Using a calculator, we can verify that $2^{5.644} \approx 50$. On a graphing calculator, press 2 ^ 5.644.

Take Note

Read $\log_b x$ as "the logarithm of *x*, base *b*" or "log base *b* of *x*."

Definition of Logarithm

For $x > 0$, $b > 0$, $b \ne 1$, $y = \log_b x$ is equivalent to $x = b^y$.

EXAMPLES

1. $4 = \log_2 16$ is equivalent to $16 = 2^4$.
2. $2 = \log_{2/3}\left(\dfrac{4}{9}\right)$ is equivalent to $\dfrac{4}{9} = \left(\dfrac{2}{3}\right)^2$.
3. $-1 = \log_{10}(0.1)$ is equivalent to $0.1 = 10^{-1}$.

Tips for Success

Be sure you can rewrite an exponential equation as a logarithmic equation and a logarithmic equation as an exponential equation. This relationship is very important.

HOW TO 1 Write $\log_3 81 = 4$ in exponential form.

$\log_3 81 = 4$ is equivalent to $3^4 = 81$.

INSTRUCTOR NOTE

Some students see the definition of logarithm as artificial. The following analogy may help.

Defining a logarithm as the inverse of the exponential function is similar to defining square root as the inverse of square. "If $49 = x^2$, what is *x*?" The answer is "the square root of 49."

Now consider the question "If $8 = 2^x$, what is *x*?" The answer is "the logarithm, base 2, of 8."

This analogy can be extended to motivate the use of tables or calculators.

"If $19 = x^2$, what is *x*?"
"If $21 = 10^x$, what is *x*?"

HOW TO 2 Write $10^{-2} = 0.01$ in logarithmic form.

$10^{-2} = 0.01$ is equivalent to $\log_{10}(0.01) = -2$.

It is important to note that the exponential function is a 1–1 function and thus has an inverse function. The inverse function of the exponential function is called a logarithm.

The 1–1 property of exponential functions can be used to evaluate some logarithms.

1–1 Property of Exponential Functions

Suppose $b > 0$, $b \ne 1$. If $b^u = b^v$, then $u = v$.

EXAMPLES

1. If $4^x = 4^3$, then $x = 3$.   2. If $3^{-2} = 3^x$, then $x = -2$.

IN-CLASS EXAMPLES

Write the exponential
equation in logarithmic
form.
  1. $2^3 = 8$ $\log_2 8 = 3$
  2. $b^x = d$ $\log_b d = x$
Write the logarithmic
equation in exponential
form.
  3. $\log_3 81 = 4$ $3^4 = 81$
  4. $\log_b c = y$ $b^y = c$

  5. Evaluate $\log_6 36$.   **2**
  6. Evaluate $\log_{10} 1$.   **0**
  7. Solve for x:
     $\log_2 x = 3$ **8**
  8. Solve for x: $\log_7 x = -2$
     $\dfrac{1}{49}$

HOW TO 3   Evaluate $\log_2 8$.

$\log_2 8 = x$ • Write an equation.

$\quad\ 8 = 2^x$ • Write the equation in its equivalent exponential form.

$\ 2^3 = 2^x$ • Write 8 as 2^3.

$\quad\ 3 = x$ • Use the 1–1 Property of Exponential Functions.

$\log_2 8 = 3$

HOW TO 4   Solve $\log_4 x = -2$ for x.

$\log_4 x = -2$

$\quad 4^{-2} = x$ • Write the equation in its equivalent exponential form.

$\quad \dfrac{1}{16} = x$ • Simplify the negative exponent.

The solution is $\dfrac{1}{16}$.

Integrating Technology

The logarithms of most numbers are irrational numbers. Therefore, the value displayed by a calculator is an approximation.

Logarithms to the base 10 are called **common logarithms.** Usually the base, 10, is omitted when writing the common logarithm of a number. Therefore, $\log_{10} x$ is written $\log x$. To find the common logarithm of most numbers, a calculator is necessary. A calculator was used to find the value of $\log 384$, shown below.

$$\log 384 \approx 2.5843312$$

When e (the base of the natural exponential function) is used as the base of a logarithm, the logarithm is referred to as the **natural logarithm** and is abbreviated $\ln x$. This is read "el en x." Use a calculator to approximate natural logarithms.

$$\ln 23 \approx 3.135494216$$

EXAMPLE 1

Evaluate: $\log_3 \dfrac{1}{9}$

Solution

$\log_3 \dfrac{1}{9} = x$ • Write an equation.

$\dfrac{1}{9} = 3^x$ • Write the equivalent exponential form.

$3^{-2} = 3^x$ • $\dfrac{1}{9} = 3^{-2}$

$-2 = x$ • The bases are the same. The exponents are equal.

$\log_3 \dfrac{1}{9} = -2$

YOU TRY IT 1

Evaluate: $\log_4 64$

Your solution
3

EXAMPLE 2

Solve for x: $\log_5 x = 2$

Solution $\log_5 x = 2$

$\qquad\quad 5^2 = x$ • Write the equivalent

$\qquad\quad 25 = x$ exponential form.

The solution is 25.

YOU TRY IT 2

Solve for x: $\log_2 x = -4$

Your solution
$\dfrac{1}{16}$

Solutions on p. S34

EXAMPLE 3

Solve $\log x = -1.5$ for x. Round to the nearest ten-thousandth.

Solution

$$\log x = -1.5$$
$$10^{-1.5} = x$$ • Write the equivalent exponential form.
$$0.0316 \approx x$$ • Use a calculator.

YOU TRY IT 3

Solve $\ln x = 3$ for x. Round to the nearest ten-thousandth.

Your solution

20.0855

Solution on p. S34

OBJECTIVE B *To use the Properties of Logarithms to simplify expressions containing logarithms*

 Take Note

The properties stated here can be proved using the definition of logarithm and the properties of exponents. For instance, to prove the Property of Log_b 1, let $\log_b 1 = x$. Then, by the definition of logarithm, $b^x = 1$. Because $b^0 = 1$, we have $x = 0$.

From the definition of a logarithm, a number of properties of logarithms can be stated.

Property of Log$_b$ 1

If $b > 0$, $b \neq 1$, then $\log_b 1 = 0$. That is, the logarithm base b of 1 is 0.

EXAMPLES

1. $\log_9 1 = 0$ 2. $\log_2 1 = 0$ 3. $\log_{3/4} 1 = 0$

Property of Log$_b$ b

If $b > 0$, $b \neq 1$, then $\log_b b = 1$. That is, the logarithm base b of b is 1.

EXAMPLES

1. $\log_7 7 = 1$ 2. $\log 10 = 1$ 3. $\log_{1/2}\left(\dfrac{1}{2}\right) = 1$

Point of Interest

Logarithms were developed independently by Jobst Burgi (1552–1632) and John Napier (1550–1617) as a means of simplifying the calculations of astronomers. The idea was to devise a method by which two numbers could be multiplied by performing additions. Napier is usually given credit for logarithms because he published his results first.

1–1 Property of Logarithms

Assume $b > 0$, $b \neq 1$, and x and y are positive numbers. If $\log_b x = \log_b y$, then $x = y$.

EXAMPLES

1. If $\log_b 7 = \log_b x$, then $x = 7$.   2. If $\log_b z = \log_b 9$, then $z = 9$.

The next property is a direct result of the fact that logarithmic functions and exponential functions are inverses of one another.

Inverse Property of Logarithms

If $b > 0$, $b \neq 1$, and $x > 0$, then $\log_b b^x = x$ and $b^{\log_b x} = x$.

EXAMPLES

1. $\log_6 6^{3x-7} = 3x - 7$ 2. $12^{\log_{12}(x^2+1)} = x^2 + 1$

John Napier

Because a logarithm is related to an exponential expression, logarithms have properties that are similar to those of exponential expressions. The proofs of the next three properties can be found in the Appendix.

The following property relates to the logarithm of the product of two numbers.

© INTERFOTO/Alamy

Take Note

Pay close attention to this property. Note, for instance, that this property states that $\log_3 (4p) = \log_3 4 + \log_3 p$. It also states that $\log_5 9 + \log_5 z = \log_5 (9z)$. It does *not* state any relationship regarding the expression $\log_b (x + y)$. **This expression cannot be simplified.**

Product Property of Logarithms

For any positive real numbers x, y, and b, $b \neq 1$, $\log_b (xy) = \log_b x + \log_b y$.

In words, this property states that the logarithm of the product of two numbers equals the sum of the logarithms of the two numbers.

HOW TO 5   Write $\log_b (6z)$ in expanded form.

$\log_b (6z) = \log_b 6 + \log_b z$ • Use the Product Property of Logarithms.

HOW TO 6   Write $\log_b 12 + \log_b r$ as a single logarithm.

$\log_b 12 + \log_b r = \log_b (12r)$ • Use the Product Property of Logarithms.

The Product Property of Logarithms can be extended to include the logarithm of the product of more than two factors. For instance,

$$\log_b (7rt) = \log_b 7 + \log_b r + \log_b t$$

The next property relates to the logarithm of the quotient of two numbers.

 Take Note

This property is used to rewrite expressions such as $\log_5 \left(\dfrac{m}{8}\right) = \log_5 m - \log_5 8$. It does *not* state any relationship regarding the expression $\dfrac{\log_b x}{\log_b y}$. **This expression cannot be simplified.**

Quotient Property of Logarithms

For any positive real numbers x, y, and b, $b \neq 1$, $\log_b \dfrac{x}{y} = \log_b x - \log_b y$.

In words, this property states that the logarithm of the quotient of two numbers equals the difference of the logarithms of the two numbers.

HOW TO 7   Write $\log_b \dfrac{p}{8}$ in expanded form.

$\log_b \dfrac{p}{8} = \log_b p - \log_b 8$ • Use the Quotient Property of Logarithms.

HOW TO 8   Write $\log_b y - \log_b v$ as a single logarithm.

$\log_b y - \log_b v = \log_b \dfrac{y}{v}$ • Use the Quotient Property of Logarithms.

Another property of logarithms is used to simplify the logarithm of a power of a number.

INSTRUCTOR NOTE

Examples such as these give students practice in applying the Properties of Logarithms. Facility with these operations will help students when they solve exponential and logarithmic equations later in the chapter.

Power Property of Logarithms

For any positive real numbers x and b, $b \neq 1$, and for any real number r, $\log_b x^r = r \log_b x$.

In words, this property states that the logarithm of the power of a number equals the power times the logarithm of the number.

HOW TO 9   Rewrite $\log_b x^3$ in terms of $\log_b x$.

$\log_b x^3 = 3 \log_b x$ • Use the Power Property of Logarithms.

IN-CLASS EXAMPLES
Write the logarithm in
expanded form.
9. $\log_2 (x^2 y^3)$
2 log$_2$ x + 3 log$_2$ y

10. $\log_5 \dfrac{x^3 y}{z^2}$
3 log$_5$ x + log$_5$ y − 2 log$_5$ z

11. $\log_4 \sqrt[3]{xy}$
$\dfrac{1}{3}$ **log$_4$ x +** $\dfrac{1}{3}$ **log$_4$ y**

Express as a single
logarithm with a coefficient
of 1.
12. $5(\log_3 x + \log_3 y)$
log$_3$ (x^{5}y^5)

13. $3(\log_5 x - \log_5 y)$
log$_5$ $\dfrac{x^3}{y^3}$

HOW TO 10 Rewrite $\frac{2}{3} \log_b x$ with a coefficient of 1.

$$\frac{2}{3} \log_b x = \log_b x^{2/3}$$ • Use the Power Property of Logarithms.

Here are some additional examples that combine some of the properties of logarithms.

HOW TO 11 Write $\log_b \dfrac{x^3}{y^2 z}$ in expanded form.

$$\log_b \frac{x^3}{y^2 z} = \log_b x^3 - \log_b y^2 z$$ • Use the Quotient Property of Logarithms.
$$= \log_b x^3 - (\log_b y^2 + \log_b z)$$ • Use the Product Property of Logarithms.
$$= 3 \log_b x - (2 \log_b y + \log_b z)$$ • Use the Power Property of Logarithms.
$$= 3 \log_b x - 2 \log_b y - \log_b z$$ • Write in simplest form.

HOW TO 12 Write $2 \log_b x + 4 \log_b y - \log_b z$ as a single logarithm with a coefficient of 1.

$$2 \log_b x + 4 \log_b y - \log_b z = \log_b x^2 + \log_b y^4 - \log_b z$$ • Use the Power Property.
$$= \log_b (x^2 y^4) - \log_b z$$ • Use the Product Property.
$$= \log_b \frac{x^2 y^4}{z}$$ • Use the Quotient Property.

EXAMPLE 4

Write $\log \sqrt{x^3 y}$ in expanded form.

Solution

$\log \sqrt{x^3 y}$
$= \log (x^3 y)^{1/2} = \dfrac{1}{2} \log (x^3 y)$ • Power Property
$= \dfrac{1}{2} (\log x^3 + \log y)$ • Product Property
$= \dfrac{1}{2} (3 \log x + \log y)$ • Power Property
$= \dfrac{3}{2} \log x + \dfrac{1}{2} \log y$ • Distributive Property

EXAMPLE 5

Write $\frac{1}{2}(\log_3 x - 3 \log_3 y + \log_3 z)$ as a single logarithm with a coefficient of 1.

Solution

$\dfrac{1}{2}(\log_3 x - 3 \log_3 y + \log_3 z)$
$= \dfrac{1}{2} (\log_3 x - \log_3 y^3 + \log_3 z)$
$= \dfrac{1}{2} \left(\log_3 \dfrac{x}{y^3} + \log_3 z \right)$
$= \dfrac{1}{2} \left(\log_3 \dfrac{xz}{y^3} \right) = \log_3 \left(\dfrac{xz}{y^3} \right)^{1/2} = \log_3 \sqrt{\dfrac{xz}{y^3}}$

YOU TRY IT 4

Write $\log_8 \sqrt[3]{xy^2}$ in expanded form.

Your solution

$\dfrac{1}{3} \log_8 x + \dfrac{2}{3} \log_8 y$

YOU TRY IT 5

Write $\frac{1}{3} (\log_4 x - 2 \log_4 y + \log_4 z)$ as a single logarithm with a coefficient of 1.

Your solution

$\log_4 \sqrt[3]{\dfrac{xz}{y^2}}$

Solutions on p. S34

To use the Change-of-Base Formula

Although only common logarithms and natural logarithms are programmed into calculators, the logarithms for other positive bases can be found.

Integrating Technology

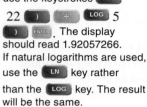

To evaluate $\dfrac{\log 22}{\log 5}$ using a graphing calculator, use the keystrokes **LOG** 22 **)** **÷** **LOG** 5 **)** **ENTER**. The display should read 1.92057266. If natural logarithms are used, use the **LN** key rather than the **LOG** key. The result will be the same.

HOW TO 13 Evaluate $\log_5 22$. Round to the nearest ten-thousandth.

$$\log_5 22 = x$$ • Write an equation.

$$5^x = 22$$ • Write the equation in its equivalent exponential form.

$$\log 5^x = \log 22$$ • Apply the common logarithm to each side of the equation.

$$x \log 5 = \log 22$$ • Use the Power Property of Logarithms.

$$x = \frac{\log 22}{\log 5}$$ • Exact answer

$$x \approx 1.9206$$ • Approximate answer

$$\log_5 22 \approx 1.9206$$

In the third step above, the natural logarithm, instead of the common logarithm, could have been applied to each side of the equation. The same result would have been obtained.

Using a procedure similar to the one used in HOW TO 13, we can derive a formula for changing bases.

Change-of-Base Formula

$$\log_a N = \frac{\log_b N}{\log_b a}$$

IN-CLASS EXAMPLES
Evaluate. Round to the nearest ten-thousandth.
14. $\log_2 6$ **2.5850**
15. $\log_5 20$ **1.8614**
16. $\log_3 7.5$ **1.8340**

HOW TO 14 Evaluate $\log_2 14$. Round to the nearest ten-thousandth.

$$\log_2 14 = \frac{\log 14}{\log 2} \approx 3.8074$$ • Use the Change-of-Base Formula with $N = 14$, $a = 2$, and $b = 10$.

In HOW TO 14, common logarithms were used. Here is the same example using natural logarithms. Note that the answers are the same.

$$\log_2 14 = \frac{\ln 14}{\ln 2} \approx 3.8074$$

EXAMPLE 6

Evaluate $\log_8 0.137$ by using natural logarithms. Round to the nearest ten-thousandth.

Solution $\log_8 0.137 = \dfrac{\ln 0.137}{\ln 8} \approx -0.9559$

YOU TRY IT 6

Evaluate $\log_3 0.834$ by using natural logarithms. Round to the nearest ten-thousandth.

Your solution -0.1652

EXAMPLE 7

Evaluate $\log_2 90.813$ by using common logarithms. Round to the nearest ten-thousandth.

Solution $\log_2 90.813 = \dfrac{\log 90.813}{\log 2} \approx 6.5048$

YOU TRY IT 7

Evaluate $\log_7 6.45$ by using common logarithms. Round to the nearest ten-thousandth.

Your solution 0.9579

Solutions on p. S34

12.2 EXERCISES

✔ **Concept Check**

SUGGESTED ASSIGNMENT
Exercises 1–16; Exercises 17–61, odds; Exercises 63–127, every other odd;
More challenging exercises: Exercises 129–134

1. ◤ What is a common logarithm?

2. ◤ What is a natural logarithm?

For Exercises 3 to 6, write the exponential equation in logarithmic form.

3. $5^2 = 25$

$\log_5 25 = 2$

4. $10^3 = 1000$

$\log 1000 = 3$

5. $4^{-2} = \dfrac{1}{16}$

$\log_4 \left(\dfrac{1}{16}\right) = -2$

6. $e^v = u$

$\ln_u = v$

For Exercises 7 to 10, write the logarithmic equation in exponential form.

7. $\log_3 81 = 4$

$3^4 = 81$

8. $\log 0.001 = -3$

$10^{-3} = 0.001$

9. $\ln p = q$

$e^q = p$

10. $\log_5 \left(\dfrac{1}{5}\right) = -1$

$5^{-1} = \dfrac{1}{5}$

For Exercises 11 to 16, determine whether the statement is true or false. Assume x and y are positive numbers.

11. $\log_5 (x + y) = \log_5 x + \log_5 y$
False

12. $\log_7 12 = \log_7 4 + \log_7 3$
True

13. $\ln e^x = x$
True

14. $\dfrac{\log 8}{\log 2} = \log 4$
False

15. $\log_3 10 - \log_3 19 = \log_3 \dfrac{10}{19}$
True

16. $\ln \dfrac{8}{2} = \ln 4$
True

OBJECTIVE A   *To find the logarithm of a number*

For Exercises 17 to 28, evaluate the expression.

17. $\log_3 81$
4

18. $\log_7 49$
2

19. $\log_2 128$
7

20. $\log_5 125$
3

21. $\log 100$
2

22. $\log 0.001$
−3

23. $\ln e^3$
3

24. $\ln e^2$
2

25. $\log_8 1$
0

26. $\log_3 243$
5

27. $\log_5 625$
4

28. $\log_2 64$
6

For Exercises 29 to 36, solve for x.

29. $\log_3 x = 2$
9

30. $\log_5 x = 1$
5

31. $\log_4 x = 3$
64

32. $\log_2 x = 6$
64

33. $\log_7 x = -1$
$\dfrac{1}{7}$

34. $\log_8 x = -2$
$\dfrac{1}{64}$

35. $\log_6 x = 0$
1

36. $\log_4 x = 0$
1

For Exercises 37 to 44, solve for x. Round to the nearest hundredth.

37. $\log x = 2.5$
316.23

38. $\log x = 3.2$
1584.89

39. $\log x = -1.75$
0.02

40. $\log x = -2.1$
0.01

41. $\ln x = 2$
7.39

42. $\ln x = 1.4$
4.06

43. $\ln x = -\dfrac{1}{2}$
0.61

44. $\ln x = -1.7$
0.18

45. ◤ If $\log_3 x > 0$, then $x >$ _____.
1

46. ◤ Why is there no real number value for $\log_5 (-2)$?
Suppose $\log_5 (-2) = x$. Then $5^x = -2$. However, $5^x > 0$ for all real numbers x.

OBJECTIVE B *To use the Properties of Logarithms to simplify expressions containing logarithms*

47. ◣ What is the Product Property of Logarithms?

48. ◣ What is the Quotient Property of Logarithms?

49. ◉ True or false? $\log_5(-2) + \log_5(-3) = \log_5[(-2)(-3)]$
$= \log_5 6$

False

50. ◉ For what values of x is the equation $\log_b x^2 = 2\log_b x$ true?

$x > 0$

For Exercises 51 to 62, evaluate the expression.

51. $\log_{12} 1$
0

52. $\ln 1$
0

53. $\ln e$
1

54. $\log_{10} 10$
1

55. $\log_3 3^x$
x

56. $8^{\log_8 p}$
p

57. $e^{\ln v}$
v

58. $\ln e^{3x}$
$3x$

59. $2^{\log_2(x^2+1)}$
$x^2 + 1$

60. $\log_4 4^{3x+1}$
$3x + 1$

61. $\log_5 5^{x^2-x-1}$
$x^2 - x - 1$

62. $8^{\log_8(3x-7)}$
$3x - 7$

For Exercises 63 to 86, write the logarithm in expanded form.

63. $\log_8(xz)$

$\log_8 x + \log_8 z$

64. $\log_7(rt)$

$\log_7 r + \log_7 t$

65. $\log_3 x^5$

$5\log_3 x$

66. $\log_2 y^7$

$7\log_2 y$

67. $\log_b \dfrac{r}{s}$

$\log_b r - \log_b s$

68. $\log_c \dfrac{z}{4}$

$\log_c z - \log_c 4$

69. $\log_3(x^2 y^6)$

$2\log_3 x + 6\log_3 y$

70. $\log_4(t^4 u^2)$

$4\log_4 t + 2\log_4 u$

71. $\log_7 \dfrac{u^3}{v^4}$

$3\log_7 u - 4\log_7 v$

72. $\log_{10} \dfrac{s^5}{t^2}$

$5\log_{10} s - 2\log_{10} t$

73. $\log_2(rs)^2$

$2\log_2 r + 2\log_2 s$

74. $\log_3(x^2 y)^3$

$6\log_3 x + 3\log_3 y$

75. $\ln(x^2 yz)$

$2\ln x + \ln y + \ln z$

76. $\ln(xy^2 z^3)$

$\ln x + 2\ln y + 3\ln z$

77. $\log_5 \dfrac{xy^2}{z^4}$

$\log_5 x + 2\log_5 y - 4\log_5 z$

78. $\log_b \dfrac{r^2 s}{t^3}$

$2\log_b r + \log_b s - 3\log_b t$

79. $\log_8 \dfrac{x^2}{yz^2}$

$2\log_8 x - \log_8 y - 2\log_8 z$

80. $\log_9 \dfrac{x}{y^2 z^3}$

$\log_9 x - 2\log_9 y - 3\log_9 z$

81. $\log_4 \sqrt{x^3 y}$

$\dfrac{3}{2}\log_4 x + \dfrac{1}{2}\log_4 y$

82. $\log_3 \sqrt{x^5 y^3}$

$\dfrac{5}{2}\log_3 x + \dfrac{3}{2}\log_3 y$

83. $\log_7 \sqrt{\dfrac{x^3}{y}}$

$\dfrac{3}{2}\log_7 x - \dfrac{1}{2}\log_7 y$

84. $\log_b \sqrt[3]{\dfrac{r^2}{t}}$

$\dfrac{2}{3}\log_b r - \dfrac{1}{3}\log_b t$

85. $\log_3 \dfrac{t}{\sqrt{x}}$

$\log_3 t - \dfrac{1}{2}\log_3 x$

86. $\log_4 \dfrac{x}{\sqrt{y^2 z}}$

$\log_4 x - \log_4 y - \dfrac{1}{2}\log_4 z$

For Exercises 87 to 110, write the expression as a single logarithm with a coefficient of 1.

87. $\log_3 x^3 + \log_3 y^2$

$\log_3 (x^3 y^2)$

88. $\log_7 x + \log_7 z^2$

$\log_7 (xz^2)$

89. $\ln x^4 - \ln y^2$

$\ln \dfrac{x^4}{y^2}$

90. $\ln x^2 - \ln y$

$\ln \dfrac{x^2}{y}$

91. $3 \log_7 x$

$\log_7 x^3$

92. $4 \log_8 y$

$\log_8 y^4$

93. $3 \ln x + 4 \ln y$

$\ln (x^3 y^4)$

94. $2 \ln x - 5 \ln y$

$\ln \dfrac{x^2}{y^5}$

95. $2(\log_4 x + \log_4 y)$

$\log_4 (x^2 y^2)$

96. $3(\log_5 r + \log_5 t)$

$\log_5 (r^3 t^3)$

97. $2 \log_3 x - \log_3 y + 2 \log_3 z$

$\log_3 \dfrac{x^2 z^2}{y}$

98. $4 \log_5 r - 3 \log_5 s + \log_5 t$

$\log_5 \dfrac{r^4 t}{s^3}$

99. $\ln x - (2 \ln y + \ln z)$

$\ln \dfrac{x}{y^2 z}$

100. $2 \log_b x - 3(\log_b y + \log_b z)$

$\log_b \dfrac{x^2}{y^3 z^3}$

101. $\dfrac{1}{2}(\log_6 x - \log_6 y)$

$\log_6 \sqrt{\dfrac{x}{y}}$

102. $\dfrac{1}{3}(\log_8 x - \log_8 y)$

$\log_8 \sqrt[3]{\dfrac{x}{y}}$

103. $2(\log_4 s - 2 \log_4 t + \log_4 r)$

$\log_4 \dfrac{s^2 r^2}{t^4}$

104. $3(\log_9 x + 2 \log_9 y - 2 \log_9 z)$

$\log_9 \dfrac{x^3 y^6}{z^6}$

105. $\ln x - 2(\ln y + \ln z)$

$\ln \dfrac{x}{y^2 z^2}$

106. $\ln t - 3(\ln u + \ln v)$

$\ln \dfrac{t}{u^3 v^3}$

107. $\dfrac{1}{2}(3 \log_4 x - 2 \log_4 y + \log_4 z)$

$\log_4 \sqrt{\dfrac{x^3 z}{y^2}}$

108. $\dfrac{1}{3}(4 \log_5 t - 5 \log_5 u - 7 \log_5 v)$

$\log_5 \sqrt[3]{\dfrac{t^4}{u^5 v^7}}$

109. $\dfrac{1}{2}\log_2 x - \dfrac{2}{3}\log_2 y + \dfrac{1}{2}\log_2 z$

$\log_2 \dfrac{\sqrt{xz}}{\sqrt[3]{y^2}}$

110. $\dfrac{2}{3}\log_3 x + \dfrac{1}{3}\log_3 y - \dfrac{1}{2}\log_3 z$

$\log_3 \dfrac{\sqrt[3]{x^2 y}}{\sqrt{z}}$

OBJECTIVE C   *To use the Change-of-Base Formula*

For exercises 111 to 126, evaluate the expression. Round to the nearest ten-thousandth.

111. $\log_8 6$

0.8617

112. $\log_4 8$

1.5000

113. $\log_5 30$

2.1133

114. $\log_6 28$

1.8597

115. $\log_3 0.5$

-0.6309

116. $\log_5 0.6$

-0.3174

117. $\log_7 1.7$

0.2727

118. $\log_6 3.2$

0.6492

119. $\log_5 15$
1.6826

120. $\log_3 25$
2.9299

121. $\log_{12} 120$
1.9266

122. $\log_9 90$
2.0480

123. $\log_4 2.55$
0.6752

124. $\log_8 6.42$
0.8942

125. $\log_5 67$
2.6125

126. $\log_8 35$
1.7098

127. Use the Change-of-Base Formula to write $\log_5 x$ in terms of $\log x$.
$$\log_5 x = \frac{\log x}{\log 5}$$

128. Use the Change-of-Base Formula to write $\log_5 x$ in terms of $\ln x$.
$$\log_5 x = \frac{\ln x}{\ln 5}$$

Critical Thinking

For Exercises 129 to 132, solve for x.

129. $\log_3(\log_3 x) = 2$
19,683

130. $\log_2(\log_2 16) = x$
2

131. $\log_2(\log_2 256) = x$
3

132. $\log_2(\log_4 x) = 3$
65,536

133. Suppose $x = 4$. What is wrong with the following application of the Product Property of Logarithms? $\log x + \log(x - 5) = \log[x(x - 5)]$
Because $x = 4$, $x - 5 = -1$. The logarithm of a negative number is undefined.

134. It is possible to change an exponential expression to an equivalent exponential expression with a different base. Write 2^{3x} with e as the base. $e^{(3 \ln 2)x}$

Projects or Group Activities

135. ● **Biology** To discuss the variety of species that live in a certain environment, a biologist needs a precise definition of *diversity*. Let $p_1, p_2, \ldots, p_n$ be the proportions of n species that live in an environment. The biological diversity D of this system is

$$D = -(p_1 \log_2 p_1 + p_2 \log_2 p_2 + \cdots + p_n \log_2 p_n)$$

The larger the value of D, the greater the diversity of the system. Suppose an ecosystem has exactly five different varieties of grass: rye (R), Bermuda (B), blue (L), fescue (F), and St. Augustine (A).

Table 1

R	B	L	F	A
$\frac{1}{5}$	$\frac{1}{5}$	$\frac{1}{5}$	$\frac{1}{5}$	$\frac{1}{5}$

Table 2

R	B	L	F	A
$\frac{1}{8}$	$\frac{3}{8}$	$\frac{1}{16}$	$\frac{1}{8}$	$\frac{5}{16}$

Table 3

R	B	L	F	A
0	$\frac{1}{4}$	0	0	$\frac{3}{4}$

Table 4

R	B	L	F	A
0	0	0	0	1

a. Calculate the diversity of this ecosystem if the proportions are as shown in Table 1. 2.3219281

b. Because Bermuda and St. Augustine are virulent grasses, after a time the proportions are as shown in Table 2. Does this system have more or less diversity than the one given in Table 1? Less

c. After an even longer period, the Bermuda and St. Augustine grasses completely overrun the environment, and the proportions are as in Table 3. Calculate the diversity of this system. (*Note:* For purposes of the diversity definition, $0 \log_2 0 = 0$.) Does it have more or less diversity than the system given in Table 2? Less

d. ◣ Finally, the St. Augustine overruns the Bermuda, and the proportions are as in Table 4. Calculate the diversity of this system. Write a sentence that explains your answer. 0; Because this system has only one species, there is no diversity in the system.

QUICK QUIZ

1. Write $4^2 = 16$ in logarithmic form.
$\log_4 16 = 2$ **[12.2A]**

2. Write $\log_{10} 0.001 = -3$ in exponential form.
$10^{-3} = 0.001$ **[12.2A]**

3. Evaluate $\log_9 81$.
2 **[12.2A]**

4. Solve for x: $\log_5 x = -2$
$\frac{1}{25}$ **[12.2A]**

5. Write $\log_8 (x^3yz^2)$ in expanded form.
$3 \log_8 x + \log_8 y + 2 \log_8 z$ **[12.2B]**

6. Express $3 \log_5 x - 4 \log_5 y$ as a single logarithm with a coefficient of 1.
$\log_5 \frac{x^3}{y^4}$ **[12.2B]**

Evaluate. Round to the nearest ten-thousandth.

7. $\log_4 14$ **1.9037** **[12.2C]**

8. $\log_2 8.8$ **3.1375** **[12.2C]**

Unless otherwise noted, all content on this page is © Cengage Learning.

SECTION

12.3 Graphs of Logarithmic Functions

OBJECTIVE A *To graph a logarithmic function*

 Point of Interest

Although logarithms were originally developed to assist with computations, logarithmic functions have a much broader use today. These functions are used in geology, acoustics, chemistry, and economics, for example.

The graph of a logarithmic function can be drawn by using the relationship between the exponential and logarithmic functions.

 Integrating Technology

To graph a logarithmic function (of base other than base 10 or base e) using a graphing calculator, use the Change-of-Base Formula. For instance, to graph $f(x) = \log_2 x + 1$, write $\log_2 x$ as $\dfrac{\log x}{\log 2}$. Now graph $Y1 = \dfrac{\log x}{\log 2} + 1$. The graph is shown below.

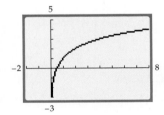

HOW TO 1 Graph: $f(x) = \log_2 x$

Think of $f(x) = \log_2 x$ as the equation $y = \log_2 x$.

Write the equivalent exponential equation.

$$f(x) = \log_2 x$$
$$y = \log_2 x$$
$$x = 2^y$$

Because the equation is solved for x in terms of y, it is easier to choose values of y and find the corresponding values of x. The results can be recorded in a table.

Graph the ordered pairs on a rectangular coordinate system.

Connect the points with a smooth curve.

$x = 2^y$	y
$\frac{1}{4}$	-2
$\frac{1}{2}$	-1
1	0
2	1
4	2

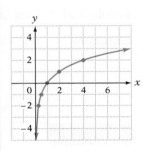

Applying the vertical-line and horizontal-line tests reveals that $f(x) = \log_2 x$ is a one-to-one function.

IN-CLASS EXAMPLES

1. Graph:
   $f(x) = \log_2 (2x + 1)$

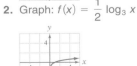

2. Graph: $f(x) = \dfrac{1}{2} \log_3 x$

HOW TO 2 Graph: $f(x) = \log_2 x + 1$

Think of $f(x) = \log_2 x + 1$ as the equation $y = \log_2 x + 1$.

Solve for $\log_2 x$.

Write the equivalent exponential equation.

Choose values of y and find the corresponding values of x.

Graph the ordered pairs on a rectangular coordinate system.

Connect the points with a smooth curve.

$$f(x) = \log_2 x + 1$$
$$y = \log_2 x + 1$$
$$y - 1 = \log_2 x$$
$$2^{y-1} = x$$

$x = 2^{y-1}$	y
$\frac{1}{4}$	-1
$\frac{1}{2}$	0
1	1
2	2
4	3

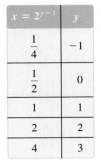
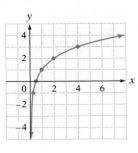

Unless otherwise noted, all content on this page is © Cengage Learning.

EXAMPLE 1

Graph: $f(x) = \log_3 x$

Solution
$$f(x) = \log_3 x$$
$$y = \log_3 x \quad \bullet \; f(x) = y$$
$$3^y = x \quad \bullet \; \textbf{Exponential form}$$

$x = 3^y$	y
$\dfrac{1}{9}$	-2
$\dfrac{1}{3}$	-1
1	0
3	1

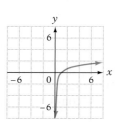

YOU TRY IT 1

Graph: $f(x) = \log_2 (x - 1)$

Your solution

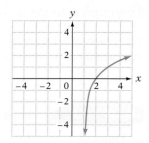

EXAMPLE 2

Graph: $f(x) = 2 \log_3 x$

Solution
$$f(x) = 2 \log_3 x$$
$$y = 2 \log_3 x \quad \bullet \; f(x) = y$$
$$\frac{y}{2} = \log_3 x \quad \bullet \; \textbf{Divide each side by 2.}$$
$$3^{y/2} = x \quad \bullet \; \textbf{Exponential form}$$

$x = 3^{y/2}$	y
$\dfrac{1}{9}$	-4
$\dfrac{1}{3}$	-2
1	0
3	2

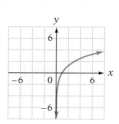

YOU TRY IT 2

Graph: $f(x) = \log_3 (2x)$

Your solution

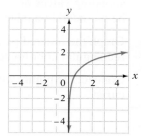

EXAMPLE 3

Graph: $f(x) = -\log_2 (x - 2)$

Solution
$$f(x) = -\log_2 (x - 2)$$
$$y = -\log_2 (x - 2) \quad \bullet \; f(x) = y$$
$$-y = \log_2 (x - 2) \quad \bullet \; \textbf{Multiply each side by } -1.$$
$$2^{-y} = x - 2 \quad \bullet \; \textbf{Exponential form}$$
$$2^{-y} + 2 = x$$

$x = 2^{-y} + 2$	y
6	-2
4	-1
3	0
$\dfrac{5}{2}$	1

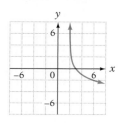

YOU TRY IT 3

Graph: $f(x) = -\log_3 (x + 1)$

Your solution

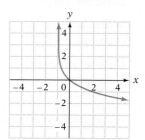

Solutions on p. S34

Unless otherwise noted, all content on this page is © Cengage Learning.

12.3 EXERCISES

✔ Concept Check

1. ◤ What is the relationship between the graphs of $x = 3^y$ and $y = \log_3 x$?

2. ◤ What is the relationship between the graphs of $y = 3^x$ and $y = \log_3 x$?

SUGGESTED ASSIGNMENT
Exercises 1–6; Exercises 7–19, odds;
More challenging exercises: Exercises 21–25
(A graphing calculator is required.)

For Exercises 3 to 6, write the equation in an equivalent form that does not include a logarithm.

3. $y = 3 \log_2 x$
   $x = 2^{y/3}$

4. $y = 4 \log_5 (x - 3)$
   $x = 5^{y/4} + 3$

5. $y = 3 \ln x + 2$
   $x = e^{(y-2)/3}$

6. $y = 2 \log x - 5$
   $x = 10^{(y+5)/2}$

OBJECTIVE A *To graph a logarithmic function*

For Exercises 7 to 18, graph the function.

7. $f(x) = \log_4 x$

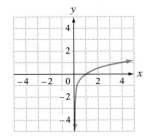

8. $f(x) = \log_2 (x + 1)$

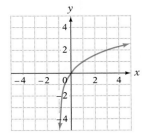

9. $f(x) = \log_3 (2x - 1)$

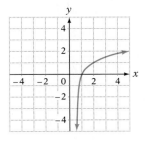

10. $f(x) = \log_2 \left(\dfrac{1}{2} x \right)$

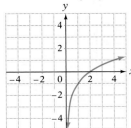

11. $f(x) = 3 \log_2 x$

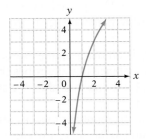

12. $f(x) = \dfrac{1}{2} \log_2 x$

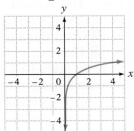

13. $f(x) = -\log_2 x$

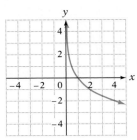

14. $f(x) = -\log_3 x$

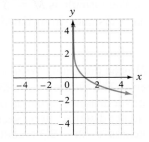

15. $f(x) = \log_2 (x - 1)$

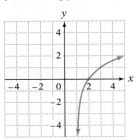

Unless otherwise noted, all content on this page is © Cengage Learning.

16. $f(x) = \log_3 (2 - x)$

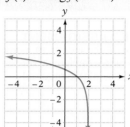

17. $f(x) = -\log_2 (x - 1)$

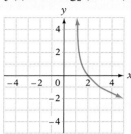

18. $f(x) = -\log_2 (1 - x)$

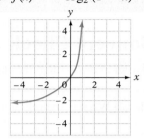

19. If $a \neq b$, at what point do the graphs of $y = \log_b x$ and $y = \log_a x$ intersect?
(1, 0)

20. If $a \neq b$, what is the difference between the y-coordinates of the graphs of $y = \log_b b$ and $y = \log_a a$?   0

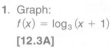

QUICK QUIZ
1. Graph:
   $f(x) = \log_3 (x + 1)$
   [12.3A]
2. Graph:
   $f(x) = -\log_2 (2x)$
   [12.3A]

Critical Thinking

For Exercises 21 to 23, use a graphing calculator to graph the function.

21. $f(x) = x - \log_2 (1 - x)$

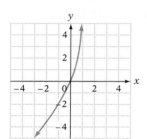

22. $f(x) = -\dfrac{1}{2} \log_2 x - 1$

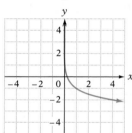

23. $f(x) = \dfrac{x}{2} - 2 \log_2 (x + 1)$

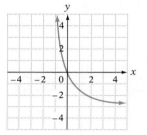

Projects or Group Activities

24. Memory A person's ability to recite a list of memorized words decreases over time. The equation $P = 100 - 30 \ln(t + 1)$ approximates the percent P of previously memorized words a person will remember after t weeks.
   a. Graph this equation.
   b. The point with coordinates (6, 42) is on the graph. (Coordinates are rounded to the nearest whole number.) Write a sentence that describes the meaning of this ordered pair.   The ordered pair (6, 42) means that after 6 weeks, a person will remember 42% of the words on the list.
   c. Use your graph from part (a) to estimate the number of weeks after which a person will remember only 25% of the list. Round to the nearest whole number.   11 weeks

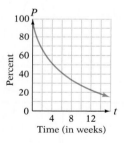

25. **Astronomy** Astronomers use the *distance modulus* of a star as a method of determining the star's distance from Earth. The formula is $M = 5 \log s - 5$, where M is the distance modulus and s is the star's distance from Earth in parsecs. (One parsec $\approx 1.9 \times 10^{13}$ mi)
   a. Graph the equation.
   b. The point with coordinates (25.1, 2) is on the graph. Write a sentence that describes the meaning of this ordered pair.   The ordered pair (25.1, 2) means that a star that is 25.1 parsecs from Earth has a distance modulus of 2.
   c. Use your graph from part (a) to estimate the distance, in parsecs, of a star with a distance modulus of –2. Round to the nearest tenth.   6.3 parsecs

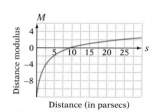

Unless otherwise noted, all content on this page is © Cengage Learning.

✓ CHECK YOUR PROGRESS: CHAPTER 12 ....

For Exercises 1 to 3, evaluate the function for the given value of x.

1. $f(x) = 3^x$; $x = 4$

81   [12.1A]

2. $f(x) = 2^{x-5}$; $x = 2$

$\dfrac{1}{8}$ [12.1A]

3. $f(x) = 4^{2x+3}$; $x = -2$

$\dfrac{1}{4}$ [12.1A]

For Exercises 4 to 6, graph the function.

4. $f(x) = 2^{x-2}$

[12.1B]

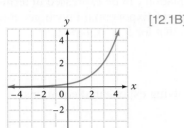

5. $f(x) = \log_2(x + 2)$

[12.3A]

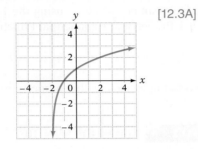

6. $f(x) = \left(\dfrac{2}{3}\right)^x + 1$

[12.1B]

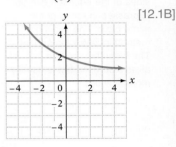

For Exercises 7 to 10, evaluate the expression.

7. $\log_3 81$

4   [12.2A]

8. $\log_4\left(\dfrac{1}{64}\right)$

−3   [12.2A]

9. $\log_5\left(\dfrac{1}{5}\right)$

−1   [12.2A]

10. $\log_7 7^{33}$

33   [12.2B]

For Exercises 11 to 14, solve for x.

11. $\log_5 x = 4$

625   [12.2A]

12. $\log_3 x = -3$

$\dfrac{1}{27}$ [12.2A]

13. $\log_7 x = 1$

0   [12.2A]

14. $\log x = -4$

0.0001   [12.2A]

For Exercises 15 to 17, expand the logarithmic expression.

15. $\log_7 (x^2 y^5)$

$2 \log_7 x + 5 \log_7 y$ [12.2B]

16. $\log_8 \dfrac{x}{y^3}$

$\log_8 x - 3 \log_8 y$ [12.2B]

17. $\log_3 \dfrac{x^2}{\sqrt{yz}}$

$2 \log_3 x - \dfrac{1}{2} \log_3 y - \dfrac{1}{2} \log_3 z$ [12.2B]

For Exercises 18 to 20, write the expression as a single logarithm with a coefficient of 1.

18. $3 \log_3 x - 4 \log_3 y$

$\log_3 \dfrac{x^3}{y^4}$ [12.2B]

19. $\ln x - (2 \ln y - 5 \ln z)$

$\ln \dfrac{xz^5}{y^2}$ [12.2B]

20. $\dfrac{1}{2}(\log x + \log y)$

$\log \sqrt{xy}$ [12.2B]

For Exercises 21 to 23, evaluate the expression. Round to the nearest ten-thousandth.

21. $\log_3 12$

2.2619   [12.2C]

22. $\log_5 0.1$

−1.4307   [12.2C]

23. $\log_7 5$

0.8271   [12.2C]

Unless otherwise noted, all content on this page is © Cengage Learning.

12.4 Solving Exponential and Logarithmic Equations

OBJECTIVE A *To solve an exponential equation*

INSTRUCTOR NOTE

Two methods for solving an exponential equation are presented here. The first method is used when both sides of the equation can be expressed in terms of the same base. The second method involves logarithms. The logarithm method can always be used but is usually more difficult.

An **exponential equation** is one in which a variable occurs in an exponent. The equations at the right are exponential equations.

$$6^{2x+1} = 6^{3x-2}$$
$$4^x = 3$$
$$2^{x+1} = 7$$

An exponential equation in which each side of the equation can be expressed in terms of the same base can be solved by using the 1–1 Property of Exponential Functions. Recall that the 1–1 Property of Exponential Functions states that for $b > 0$, $b \neq 1$,

$$\text{if } b^x = b^y, \text{ then } x = y.$$

In the two examples below, this property is used in solving exponential equations.

HOW TO 1 Solve: $10^{3x+5} = 10^{x-3}$

$$10^{3x+5} = 10^{x-3}$$
$$3x + 5 = x - 3 \qquad \bullet \text{ Use the 1–1 Property of Exponential Functions to equate the exponents.}$$
$$2x + 5 = -3 \qquad \bullet \text{ Solve the resulting equation.}$$
$$2x = -8$$
$$x = -4$$

IN-CLASS EXAMPLES

Solve for *x*. Round to the nearest ten-thousandth.

1. $6^{5x+3} = 6^{x-9}$ -3
2. $8^x = 4$ $\dfrac{2}{3}$
3. $5^x = 7$ **1.2091**
4. $3^{2x+1} = 15$ **0.7325**
5. $3^{2x-1} = 27^x$ -1

Check:

$$\begin{array}{c|c} 10^{3x+5} & = 10^{x-3} \\ \hline 10^{3(-4)+5} & 10^{-4-3} \\ 10^{-12+5} & 10^{-7} \\ 10^{-7} & = 10^{-7} \end{array}$$

The solution is -4.

 Take Note

The 1–1 Property of Exponential Functions requires that the bases be equal. For HOW TO 2 at the right, we can write $9 = 3^2$ and $27 = 3^3$. After simplifying, the bases of the exponential expressions are equal.

HOW TO 2 Solve: $9^{x+1} = 27^{x-1}$

$$9^{x+1} = 27^{x-1}$$
$$(3^2)^{x+1} = (3^3)^{x-1} \qquad \bullet \; 3^2 = 9; 3^3 = 27$$
$$3^{2x+2} = 3^{3x-3}$$
$$2x + 2 = 3x - 3 \qquad \bullet \text{ Use the 1–1 Property of Exponential Functions to equate the exponents.}$$
$$2 = x - 3 \qquad \bullet \text{ Solve for } x.$$
$$5 = x$$

Check:

$$\begin{array}{c|c} 9^{x+1} & = 27^{x-1} \\ \hline 9^{5+1} & 27^{5-1} \\ 9^6 & 27^4 \\ 531{,}441 & = 531{,}441 \end{array}$$

The solution is 5.

When both sides of an exponential equation cannot easily be expressed in terms of the same base, logarithms are used to solve the exponential equation.

HOW TO 3 Solve: $4^x = 7$. Round to the nearest ten-thousandth.

$$4^x = 7$$
$$\log 4^x = \log 7$$
$$x \log 4 = \log 7$$
$$x = \frac{\log 7}{\log 4} \approx 1.4037$$

- Take the common logarithm of each side of the equation.
- Rewrite the equation using the Properties of Logarithms.
- Solve for x.

The solution is 1.4037.

- Note that $\dfrac{\log 7}{\log 4} \neq \log 7 - \log 4$.

Integrating Technology

To evaluate $\dfrac{\log 7}{\log 4}$ on a scientific calculator, use the keystrokes

7 **LOG** **÷** 4 **LOG** **ENTER**

The display should read 1.4036775.

HOW TO 4 Solve: $3^{x+1} = 5$. Round to the nearest ten-thousandth.

$$3^{x+1} = 5$$
$$\log 3^{x+1} = \log 5$$
$$(x + 1)\log 3 = \log 5$$
$$x + 1 = \frac{\log 5}{\log 3}$$
$$x + 1 \approx 1.4650$$
$$x \approx 0.4650$$

- Take the common logarithm of each side of the equation.
- Rewrite the equation using the Properties of Logarithms.

- Solve for x.

The solution is 0.4650.

Tips for Success

Always check the solution of an equation, even when the solution is an approximation. For the equation in HOW TO 4:

$$3^{x+1} = 5$$
$$\begin{array}{c|c} 3^{0.4650+1} & 5 \\ 3^{1.4650} & 5 \\ 5.000145 \approx 5 \end{array}$$

EXAMPLE 1

Solve for n: $(1.1)^n = 2$. Round to the nearest ten-thousandth.

Solution

$$(1.1)^n = 2$$
$$\log (1.1)^n = \log 2$$
$$n \log 1.1 = \log 2$$
$$n = \frac{\log 2}{\log 1.1}$$
$$n \approx 7.2725$$

- Take the log of each side.
- Power Property

- Divide both sides by log 1.1.

The solution is 7.2725.

YOU TRY IT 1

Solve for n: $(1.06)^n = 1.5$. Round to the nearest ten-thousandth.

Your solution

6.9585

EXAMPLE 2

Solve for x: $3^{2x} = 4$. Round to the nearest ten-thousandth.

Solution

$$3^{2x} = 4$$
$$\log 3^{2x} = \log 4$$
$$2x \log 3 = \log 4$$
$$2x = \frac{\log 4}{\log 3}$$
$$2x \approx 1.2619$$
$$x \approx 0.6310$$

- Take the log of each side.
- Power Property

- Divide both sides by log 3.

- Divide both sides by 2.

The solution is 0.6310.

YOU TRY IT 2

Solve for x: $4^{3x} = 25$. Round to the nearest ten-thousandth.

Your solution

0.7740

Solutions on p. S35

To solve a logarithmic equation

The 1–1 Property of Logarithms from Section 12.2 can be used to solve some logarithmic equations.

Take Note

It is important to check a proposed solution of a logarithmic equation. Consider the equation $\log (2x - 3) = \log (x - 2)$.

$\log (2x - 3) = \log (x - 2)$
$2x - 3 = x - 2$
$x = 1$

Substituting 1 into the original equation gives $\log (-1) = \log (-1)$. However, logarithms of negative numbers are not defined. Therefore, the equation $\log (2x - 3) = \log (x - 2)$ has no solution.

HOW TO 5 Solve: $\log_5 (3x - 1) = \log_5 (7 - x)$

$\log_5 (3x - 1) = \log_5 (7 - x)$
$3x - 1 = 7 - x$ • Use the 1–1 Property of Logarithms.
$4x = 8$ • Solve for x.
$x = 2$

Because logarithms are defined only for positive numbers, we must check the solution.

Check: $\dfrac{\log_5 (3x - 1) = \log_5 (7 - x)}{\log_5 [3(2) - 1] \mid \log_5 (7 - 2)}$ • Replace x by 2.
$\log_5 (5) = \log_5 (5)$

The solution checks. The solution is 2.

Solving some logarithmic equations may require using several of the properties of logarithms.

HOW TO 6 Solve: $\log_3 6 - \log_3 (2x + 3) = \log_3 (x + 1)$

$\log_3 6 - \log_3 (2x + 3) = \log_3 (x + 1)$

$\log_3 \dfrac{6}{2x + 3} = \log_3 (x + 1)$ • Use the Quotient Property of Logarithms.

$\dfrac{6}{2x + 3} = x + 1$ • Use the 1–1 Property of Logarithms.

$6 = (2x + 3)(x + 1)$
$6 = 2x^2 + 5x + 3$
$0 = 2x^2 + 5x - 3$ • Write in standard form.
$0 = (2x - 1)(x + 3)$ • Factor and use the Principle of Zero Products.

$2x - 1 = 0 \qquad x + 3 = 0$
$x = \dfrac{1}{2} \qquad\qquad x = -3$

-3 does not check as a solution. The solution is $\frac{1}{2}$.

IN-CLASS EXAMPLES

Solve for x.
6. $\log_2 (x + 1) = 3$ 7
7. $\log_2 (x^2 - 2x) = 3$
   $-2, 4$
8. $\log_3 \left(\dfrac{2x}{x + 1}\right) = 1$ -3
9. $\log_2 24 - \log_2 (x^2 - 1)$
   $= \log_2 3$
   $-3, 3$

HOW TO 7 Solve: $\log_9 x + \log_9 (x - 8) = 1$

$\log_9 x + \log_9 (x - 8) = 1$

$\log_9 [x(x - 8)] = 1$ • Use the Product Property of Logarithms.
$9^1 = x(x - 8)$ • Write the equation in exponential form.
$9 = x^2 - 8x$
$0 = x^2 - 8x - 9$ • Write in standard form.
$0 = (x - 9)(x + 1)$ • Factor and use the Principle of Zero Products.

$x - 9 = 0 \qquad x + 1 = 0$
$x = 9 \qquad\quad x = -1$

-1 does not check as a solution. The solution is 9.

EXAMPLE 3

Solve for x: $\log_3 (2x - 1) = 2$

Solution

$\log_3 (2x - 1) = 2$
$3^2 = 2x - 1$ • **Write in exponential form.**
$9 = 2x - 1$
$10 = 2x$
$5 = x$

The solution is 5.

YOU TRY IT 3

Solve for x: $\log_4 (x^2 - 3x) = 1$

Your solution

-1 and 4

EXAMPLE 4

Solve for x:
$\log_2 (3x + 8) = \log_2 (2x + 2) + \log_2 (x - 2)$

Solution

$\log_2 (3x + 8) = \log_2 (2x + 2) + \log_2 (x - 2)$
$\log_2 (3x + 8) = \log_2 [(2x + 2)(x - 2)]$
$\log_2 (3x + 8) = \log_2 (2x^2 - 2x - 4)$
$3x + 8 = 2x^2 - 2x - 4$ • **1–1 Property of Logarithms**

$0 = 2x^2 - 5x - 12$
$0 = (2x + 3)(x - 4)$ • **Solve by factoring.**

$2x + 3 = 0 \qquad x - 4 = 0$
$x = -\dfrac{3}{2} \qquad x = 4$

$-\dfrac{3}{2}$ does not check as a solution. The solution is 4.

YOU TRY IT 4

Solve for x: $\log_3 x + \log_3 (x + 3) = \log_3 4$

Your solution

1

EXAMPLE 5

Solve for x: $\log_3 (5x + 4) - \log_3 (2x - 1) = 2$

Solution

$\log_3 (5x + 4) - \log_3 (2x - 1) = 2$

$\log_3 \dfrac{5x + 4}{2x - 1} = 2$ • **Quotient Property**

$3^2 = \dfrac{5x + 4}{2x - 1}$ • **Definition of logarithm**

$9 = \dfrac{5x + 4}{2x - 1}$

$(2x - 1)9 = (2x - 1)\left(\dfrac{5x + 4}{2x - 1}\right)$

$18x - 9 = 5x + 4$
$13x = 13$
$x = 1$

1 checks as a solution. The solution is 1.

YOU TRY IT 5

Solve for x: $\log_3 x + \log_3 (x + 6) = 3$

Your solution

3

Solutions on p. S35

12.4 EXERCISES

✔ Concept Check

SUGGESTED ASSIGNMENT
Exercises 1–4; Exercises 5–53, odds
More challenging exercises: Exercises 57, 59;
   Exercise 62 (A graphing calculator is required.)

1. ◣ What does the 1–1 Property of Exponential Functions state?

2. ◣ What does the 1–1 Property of Logarithms state?

3. ◔ Let $2^x = \frac{1}{3}$. Without solving the equation, is $x < 0$ or is $x > 0$? $x < 0$

4. ◔ Without solving, determine which of the following equations have no solution.
   (i) $5^{-x} = 6$ **(ii)** $5^x = -6$ **(iii)** $5^{-x} = -6$ **(iv)** $5^x = 6$ ii and iii

---

OBJECTIVE A   *To solve an exponential equation*

For Exercises 5 to 36, solve for x. If necessary, round to the nearest ten-thousandth.

5. $5^{4x-1} = 5^{x-2}$
   $-\dfrac{1}{3}$

6. $7^{4x-3} = 7^{2x+1}$
   2

7. $8^{x-4} = 8^{5x+8}$
   -3

8. $10^{4x-5} = 10^{x+4}$
   3

9. $9^x = 3^{x+1}$
   1

10. $2^{x-1} = 4^x$
    -1

11. $8^{x+2} = 16^x$
    6

12. $9^{3x} = 81^{x-4}$
    -8

13. $16^{2-x} = 32^{2x}$
    $\dfrac{4}{7}$

14. $27^{2x-3} = 81^{4-x}$
    $\dfrac{5}{2}$

15. $25^{3-x} = 125^{2x-1}$
    $\dfrac{9}{8}$

16. $8^{4x-7} = 64^{x-3}$
    $\dfrac{1}{2}$

17. $5^x = 6$
    1.1133

18. $7^x = 10$
    1.1833

19. $8^{x/4} = 0.4$
    -1.7626

20. $5^{x/2} = 0.5$
    -0.8614

21. $2^{3x} = 5$
    0.7740

22. $3^{6x} = 0.5$
    -0.1052

23. $2^{-x} = 7$
    -2.8074

24. $3^{-x} = 14$
    -2.4022

25. $2^{x-1} = 6$
    3.5850

26. $4^{x+1} = 9$
    0.5850

27. $3^{2x-1} = 4$
    1.1309

28. $4^{-x+2} = 12$
    0.2075

29. $\left(\dfrac{1}{2}\right)^{x+1} = 3$
    -2.5850

30. $\left(\dfrac{3}{5}\right)^{-2x} = 2$
    0.6785

31. $3 \cdot 2^x = 7$
    1.2224

32. $5 \cdot 3^{2-x} = 4$
    2.2031

33. $7 = 10\left(\dfrac{1}{2}\right)^{x/8}$
    4.1166

34. $8 = 15\left(\dfrac{1}{2}\right)^{x/22}$
    19.9516

35. $15 = 12e^{0.05x}$
    4.4629

36. $7 = 42e^{-3x}$
    0.5973

OBJECTIVE B   *To solve a logarithmic equation*

For Exercises 37 to 54, solve for x. If necessary, round to the nearest ten-thousandth.

37. $\log x = \log(1 - x)$
    $\dfrac{1}{2}$

38. $\ln(3x - 2) = \ln(x + 1)$
    $\dfrac{3}{2}$

39. $\ln(3x + 2) = \ln(5x + 4)$
    No solution

40. $\log_3(x - 2) = \log_3(2x)$
    No solution

41. $\log_2(8x) - \log_2(x^2 - 1) = \log_2 3$
    3

42. $\log_5(3x) - \log_5(x^2 - 1) = \log_5 2$
    2

43. $\log_9 x + \log_9 (2x - 3) = \log_9 2$ **44.** $\log_6 x + \log_6 (3x - 5) = \log_6 2$ **45.** $\log_2 (2x - 3) = 3$
2   2 $\dfrac{11}{2}$

46. $\log_4 (3x + 1) = 2$ **47.** $\ln (3x + 2) = 4$ **48.** $\ln (2x + 3) = -1$
5   17.5327   −1.3161

49. $\log_2 (x + 1) + \log_2 (x + 3) = 3$ **50.** $\log_4 (3x - 4) + \log_4 (x + 6) = 2$ **51.** $\log_5 (2x) - \log_5 (x - 1) = 1$
1   2 $\dfrac{5}{3}$

52. $\log_3 (3x) - \log_3 (2x - 1) = 2$ **53.** $\log_8 (6x) = \log_8 2 + \log_8 (x - 4)$ **54.** $\log_7 (5x) = \log_7 3 + \log_7 (2x + 1)$
$\dfrac{3}{5}$ No solution   No solution

55. 🖊 If u and v are positive numbers and $u < v$, then $\log u < \log v$. Use this fact to explain why $\log (x - 2) - \log x = 3$ has no solution. Do not solve the equation.
$x - 2 < x$, and therefore $\log (x - 2) < \log x$. This means that
$\log (x - 2) - \log x < 0$ and cannot equal the positive number 3.

56. 🖊 Sometimes it is easy to represent one number as a power of another. For instance, $81 = 3^4$. Although it is not as easy, show how to represent 81 as a power of 7.
$81 = 7^{\frac{\log 81}{\log 7}}$

QUICK QUIZ
Solve for x. For Exercises 2 and 3, round to the nearest ten-thousandth.
1. $3^{x-1} = 9^x$
   −1 **[12.4A]**
2. $7^x = 20$ **1.5395**
   [12.4A]
3. $2^{x+1} = 10$ **2.3219**
   [12.4A]
4. $\log_4 (x - 2) = 1$
   6 **[12.4B]**
5. $\log_5 x + \log_5 (2x - 1)$
   $= \log_5 1$ 1 **[12.4B]**

Critical Thinking

For Exercises 57 to 60, solve for x. Round to the nearest ten-thousandth.

57. $3^{x+1} = 2^{x-2}$ **58.** $2^{2x} = 5^{x+1}$

−6.1285   −7.2126

59. $7^{2x-1} = 3^{2x+3}$ **60.** $4^{3x+2} = 3^{2x-5}$

3.0932   −4.2136

Projects or Group Activities

61. Physics   A model for the distance s, in feet, that an object experiencing air resistance will fall in t seconds is given by $s = 312.5 \ln \dfrac{e^{0.32t} + e^{-0.32t}}{2}$.

   a. 🖩 Graph this equation. *Suggestion:* Use Xmin $= 0$, Xmax $= 4.5$, Ymin $= 0$, Ymax $= 140$, and Yscl $= 20$.
   b. Determine, to the nearest hundredth of a second, the time it takes for the object to fall 100 ft.
   2.64 s

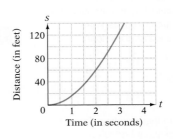

62. Physics   A model for the distance s, in feet, that an object experiencing air resistance will fall in t seconds is given by $s = 78 \ln \dfrac{e^{0.8t} + e^{-0.8t}}{2}$.

   a. 🖩 Graph this equation. *Suggestion:* Use Xmin $= 0$, Xmax $= 4.5$, Ymin $= 0$, Ymax $= 140$, and Yscl $= 20$.
   b. Determine, to the nearest hundredth of a second, the time it takes for the object to fall 125 ft.
   2.86 s

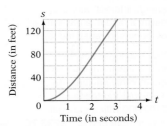

SECTION

12.5 Applications of Exponential and Logarithmic Functions

OBJECTIVE A *To solve application problems*

Point of Interest

C. Northcote Parkinson

Parkinson's Law, named after C. Northcote Parkinson, is sometimes stated as "A job will expand to fill the time allotted for the job." However, Parkinson actually said that in any new government administration, administrative employees will be added at the rate of about 5% to 6% per year. This is an example of exponential growth and means that a staff of 500 will grow to approximately 630 by the end of a 4-year term.

INSTRUCTOR NOTE

Another exponential function from finance that will be of interest to students is the function that is used to calculate the amount of an amortized loan payment, such as a car loan:

$$P = B\left[\frac{\dfrac{i}{12}}{1 - \left(1 + \dfrac{i}{12}\right)^{-n}}\right]$$

In this equation, B is the amount borrowed, i is the annual interest rate as a decimal, and n is the number of months to repay the loan.

You might ask students to use this formula to find the monthly payment on a 30-year mortgage of $100,000 with an annual interest rate of 8%. **$733.76**

A biologist places one single-celled bacterium in a culture. Each hour, every member of that particular species of bacteria divides into two bacteria. After 1 h, there will be two bacteria. After 2 h, each of those two bacteria will divide and there will be four bacteria. After 3 h, each of the four bacteria will divide and there will be eight bacteria.

The table at the right shows the number of bacteria in the culture after various intervals of time t, in hours. Values in this table could also be found by using the exponential equation $N = 2^t$.

Time, t	Number of Bacteria, N
0	1
1	2
2	4
3	8
4	16

The equation $N = 2^t$ is an example of an **exponential growth equation.** In general, any equation that can be written in the form $A = A_0 b^{kt}$, where A is the size at time t, A_0 is the initial size, $b > 1$, and k is a positive real number, is an exponential growth equation. These equations are important not only in population growth studies but also in physics, chemistry, psychology, and economics.

Recall that interest is the amount of money that one pays (or receives) for borrowing (or investing) money. **Compound interest** is interest that is computed not only on the original principal but also on the interest already earned. The compound interest formula is an exponential equation.

The **compound interest formula** is $A = P(1 + i)^n$, where P is the original value of an investment, i is the interest rate per compounding period, n is the total number of compounding periods, and A is the value of the investment after n periods.

HOW TO 1 An investment broker deposits $1000 into an account that earns 8% annual interest compounded quarterly. What is the value of the investment after 3 years?

$i = \dfrac{8\%}{4} = \dfrac{0.08}{4} = 0.02$
 • Find i, the interest rate per quarter. The quarterly rate is the annual rate divided by 4, the number of quarters in 1 year.

$n = 4 \cdot 3 = 12$
 • Find n, the number of compounding periods. The investment is compounded quarterly, 4 times a year, for 3 years.

$A = P(1 + i)^n$
$A = 1000(1 + 0.02)^{12}$
$A \approx 1268$
 • Use the compound interest formula.
 • Replace P, i, and n by their values.
 • Solve for A.

The value of the investment after 3 years is approximately $1268.

Unless otherwise noted, all content on this page is © Cengage Learning.

Exponential decay offers another example of an exponential equation. A common illustration of exponential decay is the decay of a radioactive element. One form of the equation that is used to model radioactive decay is $A = A_0(0.5)^{t/k}$, where A is the amount of the substance remaining after a time period t, A_0 is the initial amount of the radioactive material, and k is the *half-life* of the material. The **half-life** of a radioactive substance is the time it takes for one-half of the material to disintegrate.

The table at the right indicates the amount of an initial 10-microgram sample of tritium that remains after various intervals of time t in years. Note from the table that after each 12-year period of time, the amount of tritium is reduced by one-half. The half-life of tritium is 12 years. The equation that models the decay of the 10-microgram sample is $A = 10(0.5)^{t/12}$, where t is in years and A is in micrograms.

Time, t	Amount, A
0	10
12	5
24	2.5
36	1.25

HOW TO 2   Yttrium-90 is a radioactive isotope that is used to treat some cancers. A sample that originally contained 5 mg of yttrium-90 was measured again after 20 h and found to have 4 mg of yttrium-90. What is the half-life of yttrium-90? Round to the nearest whole number.

$$A = A_0(0.5)^{t/k}$$ • Use the half-life equation.

$$4 = 5(0.5)^{20/k}$$ • Replace A by **4**, A_0 by **5**, and t by **20**.

$$0.8 = (0.5)^{20/k}$$ • Divide each side by 5 to isolate the exponential expression.

$$\log(0.8) = \log(0.5)^{20/k}$$ • Take the common logarithm of each side.

$$\log(0.8) = \frac{20}{k}\log(0.5)$$ • Use the Power Property of Logarithms.

$$k\log(0.8) = 20\log(0.5)$$ • Solve for k.

$$k = \frac{20\log(0.5)}{\log(0.8)}$$ • Exact answer

$$k \approx 62.13$$ • Approximate answer

The half-life of yttrium-90 is approximately 62 h.

Unless otherwise noted, all content on this page is © Cengage Learning.

A method by which an archaeologist can measure the age of a bone is called *carbon dating*. Carbon dating is based on a radioactive isotope of carbon called carbon-14, which has a half-life of approximately 5570 years. The exponential decay equation is given by $A = A_0(0.5)^{t/5570}$, where A_0 is the original amount of carbon-14 present in the bone, t is the age of the bone in years, and A is the amount of carbon-14 present after t years.

HOW TO 3   A bone that originally contained 100 mg of carbon-14 now has 70 mg of carbon-14. What is the approximate age of the bone in years?

$$A = A_0\,(0.5)^{t/5570}$$ • Use the exponential decay equation.

$$70 = 100(0.5)^{t/5570}$$ • Replace A by **70** and A_0 by **100**, and solve for t.

$$0.7 = (0.5)^{t/5570}$$ • Divide each side by 100.

$$\log 0.7 = \log (0.5)^{t/5570}$$ • Take the common logarithm of each side of the equation.

$$\log 0.7 = \frac{t}{5570}\log 0.5$$ • Power Property

$$\frac{5570 \log 0.7}{\log 0.5} = t$$ • Multiply by 5570, and divide by log 0.5.

$$2866 \approx t$$

The bone is approximately 2866 years old.

Logarithmic functions are used to scale very large or very small numbers into numbers that are easier to comprehend. For instance, the *Richter scale magnitude* of an earthquake uses a logarithmic function to convert the intensity of shock waves I into a number M, which for most earthquakes is in the range of 0 to 10. The intensity I of an earthquake is often given in terms of the constant I_0, where I_0 is the intensity of the smallest earthquake, called a *zero-level earthquake*, that can be measured on a seismograph near the earthquake's epicenter. An earthquake with an intensity I has a Richter scale magnitude of $M = \log \dfrac{I}{I_0}$, where I_0 is the measure of the intensity of a zero-level earthquake.

Point of Interest

Charles F. Richter

The Richter scale was created by seismologist Charles F. Richter in 1935. Note that a tenfold increase in the intensity level of an earthquake increases the Richter scale magnitude of the earthquake by only 1.

HOW TO 4   The Richter scale magnitude of an earthquake that occurred on May, 2008, in Sichuan, China, was 7.9. Find the intensity of the earthquake in terms of I_0. Round to the nearest thousand.

$$M = \log \frac{I}{I_0}$$

$$7.9 = \log \frac{I}{I_0}$$ • Replace M by 7.9.

$$\frac{I}{I_0} = 10^{7.9}$$ • Write in exponential form.

$$I = 10^{7.9}I_0$$ • Solve for I.

$$I \approx 79{,}432{,}823I_0$$

The Sichuan earthquake had an intensity that was approximately 79,433,000 times the intensity of a zero-level earthquake.

IN-CLASS EXAMPLES

1. $5000 is invested at 6% annual interest compounded semiannually. In approximately how many years will the investment be worth twice the original amount? Use the compound interest formula $A = P(1 + i)^n$, where P is the original value of an investment, i is the interest rate per compounding period, n is the total number of compounding periods, and A is the value of the investment after n periods. **12 years**

2. Find the pH of a hydrogen chloride solution for which the hydrogen ion concentration is 5.2×10^{-5}. Use the equation pH $= -\log (H^+)$, where H^+ is the hydrogen ion concentration of a solution. Round to the nearest tenth. **4.3**

HOW TO 5 The San Francisco earthquake of 1906 measured 7.8 on the Richter scale. It is the earthquake of greatest magnitude recorded in the 48 contiguous states in modern history. The 2011 Fukushima earthquake off the coast of Honshu, Japan, measured 9.0 on the Richter scale. How many times stronger was the Fukushima earthquake than the San Francisco earthquake? Round to the nearest whole number. (*Source:* U.S. Geological Survey)

Let I_1 represent the intensity of the San Francisco earthquake, and let I_2 represent the intensity of the Fukushima earthquake. Use the equation $M = \log\left(\dfrac{I}{I_0}\right) = \log I - \log I_0$ to express the magnitude of each earthquake in terms of its intensity.

$9.0 = \log I_2 - \log I_0$ • The magnitude of the Fukushima earthquake was 9.0.

$\dfrac{7.8 = \log I_1 - \log I_0}{1.2 = \log I_2 - \log I_1}$ • The magnitude of the San Francisco earthquake was 7.8.
• Subtract the equations.

$1.2 = \log \dfrac{I_2}{I_1}$ • Use the Quotient Property of Logarithms.

$\dfrac{I_2}{I_1} = 10^{1.2}$ • Write the equivalent exponential equation.

$I_2 = 10^{1.2}\, I_1$ • Solve for I_2.

$I_2 \approx 15.85 I_1$

The Fukushima earthquake was approximately 16 times stronger than the San Francisco earthquake.

Point of Interest

Søren Sørensen

The pH scale was created by Danish biochemist Søren Sørensen in 1909 to measure the acidity of water used in the brewing of beer. pH is an abbreviation for *pondus hydrogenii*, which translates as "potential hydrogen."

The Oesper Collections, University of Cincinnati

A chemist measures the acidity or alkalinity of a solution by measuring the concentration of hydrogen ions, H^+, in the solution using the formula pH $= -\log (H^+)$. A neutral solution such as distilled water has a pH of 7, acids have a pH less than 7, and alkaline solutions (also called basic solutions) have a pH greater than 7.

HOW TO 6 Find the pH of orange juice that has a hydrogen ion concentration, H^+, of 2.9×10^{-4}. Round to the nearest tenth.

pH $= -\log (H^+)$

$= -\log (2.9 \times 10^{-4})$ • $H^+ = 2.9 \times 10^{-4}$

≈ 3.5376

The pH of the orange juice is approximately 3.5.

EXAMPLE 1

An investment of $3000 is placed into an account that earns 12% annual interest compounded monthly. In approximately how many years will the investment be worth twice the original amount?

Strategy

To find the time, solve the compound interest formula for n. Use $A = 6000$, $P = 3000$, and $i = \frac{12\%}{12} = \frac{0.12}{12} = 0.01$.

Solution

$$A = P(1 + i)^n$$
$$6000 = 3000(1 + 0.01)^n$$
$$6000 = 3000(1.01)^n$$
$$2 = (1.01)^n \qquad \bullet \text{ Divide by 3000.}$$
$$\log 2 = \log(1.01)^n \qquad \bullet \text{ Take the log of each side.}$$
$$\log 2 = n \log 1.01 \qquad \bullet \text{ Power Property}$$
$$\frac{\log 2}{\log 1.01} = n \qquad \bullet \text{ Divide each side by } \log 1.01.$$
$$70 \approx n$$

70 months $\div$ 12 $\approx$ 5.8 years

In approximately 6 years, the investment will be worth $6000.

YOU TRY IT 1

Find the hydrogen ion concentration, H^+, of vinegar that has a pH of 2.9.

Your strategy

Your solution
0.00126

EXAMPLE 2

Rhenium-186 is a radioactive isotope with a half-life of approximately 3.78 days. Rhenium-186 is sometimes used for pain management. If a patient receives a dose of 5 micrograms of rhenium-186, how long (in days) will it take for the patient's rhenium-186 level to reach 2 micrograms?

Strategy

- Use the equation $A = A_0 (0.5)^{t/k}$.
- Replace A with 2, A_0 with 5, and k with 3.78. Then solve for t.

Solution

$$A = A_0 (0.5)^{t/k}$$
$$2 = 5(0.5)^{t/3.78}$$
$$\frac{2}{5} = (0.5)^{t/3.78}$$
$$\log(0.4) = \frac{t}{3.78} \log(0.5)$$
$$\frac{3.78 \log(0.4)}{\log(0.5)} = t$$
$$4.997 \approx t$$

In approximately 5 days, there will be 2 micrograms of rhenium-186 remaining in the patient's body.

YOU TRY IT 2

The percent of light p, as a decimal, that passes through a substance of thickness d, in meters, is given by $\log p = -kd$. The value of k for a type of opaque glass is 6. How thick a piece of this glass is necessary so that 20% of the light passes through the glass?

Your strategy

Your solution
0.116 m

12.5 EXERCISES

✔ **Concept Check**

SUGGESTED ASSIGNMENT
Exercises 1 and 2; Exercises 3–31, odds

1. ◳ What is compound interest?

2. In each case, state whether $A = A_0 b^{kt}$ represents exponential *growth* or exponential *decay*.
    a. $0 < b < 1$ Decay   **b.** $b > 1$ Growth

OBJECTIVE A *To solve application problems*

Compound Interest For Exercises 3 to 6, use the compound interest formula $A = P(1 + i)^n$, where P is the original value of an investment, i is the interest rate per compounding period, n is the total number of compounding periods, and A is the value of the investment after n periods.

3. An investment broker deposits $1000 into an account that earns 8% annual interest compounded quarterly. What is the value of the investment after 2 years? Round to the nearest dollar. $1172

4. A financial advisor recommends that a client deposit $2500 into a fund that earns 7.5% annual interest compounded monthly. What will be the value of the investment after 3 years? Round to the nearest cent. $3128.62

5. To save for college tuition, the parents of a preschooler invest $5000 in a bond fund that earns 6% annual interest compounded monthly. In approximately how many years will the investment be worth $15,000?   18 years

6. A hospital administrator deposits $10,000 into an account that earns 6% annual interest compounded monthly. In approximately how many years will the investment be worth $15,000?   7 years

Radioactivity For Exercises 7 to 10, use the exponential decay equation $A = A_0 (0.5)^{t/k}$, where A is the amount of a radioactive material present after time t, k is the half-life of the radioactive substance, and A_0 is the original amount of the radioactive substance. Round to the nearest tenth.

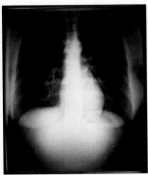

Fuse/Jupiter Images

7. An isotope of technetium is used to prepare images of internal body organs. The isotope has a half-life of about 6 h. A patient is injected with 30 mg of the isotope.
    a. What is the technetium level in the patient after 3 h?   21.2 mg
    b. How long (in hours) will it take for the technetium level to reach 20 mg?   3.5 h

8. Iodine-131 is an isotope that is used to study the functioning of the thyroid gland. This isotope has a half-life of approximately 8 days. A patient is given an injection that contains 8 micrograms of iodine-131.
    a. What is the iodine level in the patient after 5 days?   5.2 micrograms
    b. How long (in days) will it take for the iodine level to reach 5 micrograms?   5.4 days

9. A sample of promethium-147 (used in some paints) weighs 25 mg. One year later, the sample weighs 18.95 mg. What is the half-life of promethium-147, in years?
    2.5 years

10. Francium-223 is a very rare radioactive isotope discovered in 1939 by Marguerite Percy. A 3-microgram sample of francium-223 decays to 2.54 micrograms in 5 min. What is the half-life of francium-223, in minutes?   20.8 min

● **Seismology** For Exercises 11 to 14, use the Richter scale equation $M = \log \frac{I}{I_0}$, where M is the magnitude of an earthquake, I is the intensity of the shock waves, and I_0 is the measure of the intensity of a zero-level earthquake.

11. An earthquake in Japan on March 2, 1933, measured 8.9 on the Richter scale. Find the intensity of the earthquake in terms of I_0. Round to the nearest whole number.
794,328,235I_0

12. Read the article at the right about the April 13, 2010, earthquake in Qinghai province, China. Find the intensity of this earthquake, in terms of I_0, for each of the magnitudes reported in the article. Round to the nearest thousand.
Magnitude 7.1: 12,589,000I_0; magnitude 6.9: 7,943,000I_0

13. In 2008, a 6.9-magnitude earthquake occurred off the coast of Honshu, Japan. In the same year, a 6.4-magnitude earthquake occurred near Quetta, Pakistan. How many times stronger was the Honshu earthquake than the Quetta earthquake? Round to the nearest tenth.   3.2 times stronger

14. In 2008, a 6.4-magnitude earthquake occurred off the coast of Oregon. In the same year, a 7.2-magnitude earthquake occurred near Xinjiang, China. How many times stronger was the Xinjiang earthquake than the one off the coast of Oregon? Round to the nearest tenth.   6.3 times stronger

In the NEWS!

Earthquake Strikes Remote Area of China

Rescue workers continue to look for survivors of a severe earthquake that shook a remote area of western China. The China Earthquake Commission reported the quake at magnitude 7.1 on the Richter scale, while the U.S. Geological Survey recorded the quake at magnitude 6.9.

Source: www.msnbc.msn.com

● **Seismology** A *seismogram* is used to find the magnitude of an earthquake. Magnitude depends on a shock wave's amplitude A and the difference in time t between the arrival of a *primary wave* (p-wave) and a *secondary wave* (s-wave). The amplitude A of a wave is half the difference between its high and low points. For the seismogram at the right, A is 23 mm and t is 24 s. The equation is $M = \log A + 3 \log 8t - 2.92$. Use this equation for Exercises 15 to 17. Round to the nearest tenth.

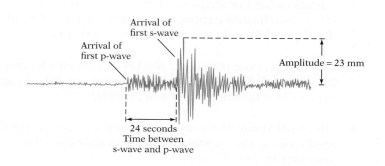

Arrival of first s-wave
Arrival of first p-wave
Amplitude = 23 mm
24 seconds
Time between s-wave and p-wave

15. Determine the magnitude of the earthquake for the seismogram shown above.   5.3

16. Find the magnitude of an earthquake that has a seismogram with an amplitude of 30 mm and for which t is 21 s.   5.2

17. Find the magnitude of an earthquake that has a seismogram with an amplitude of 28 mm and for which t is 28 s.   5.6

Chemistry For Exercises 18 to 21, use the equation $pH = -\log (H^+)$, where H^+ is the hydrogen ion concentration of a solution.

18. Find the pH of milk, which has a hydrogen ion concentration of 3.97×10^{-7}. Round to the nearest tenth.   6.4

19. Find the pH of a baking soda solution for which the hydrogen ion concentration is 3.98×10^{-9}. Round to the nearest tenth.   8.4

20. The pH of pure water is 7. What is the hydrogen ion concentration of pure water?
1×10^{-7}

21. Peanuts grow best in soils that have a pH between 5.3 and 6.6. What is the range of hydrogen ion concentrations for these soils? 2.5×10^{-7} to 5.0×10^{-6}

Unless otherwise noted, all content on this page is © Cengage Learning.

Sound   For Exercises 22 to 25, use the equation $D = 10(\log I + 16)$, where D is the number of decibels of a sound and I is the intensity of the sound in watts per square centimeter. Round decibels to the nearest whole number.

22. Find the number of decibels of normal conversation. The intensity of the sound of normal conversation is approximately 3.2×10^{-10} watts/cm².   65 decibels

23. ◖ The loudest sound made by any animal is made by the blue whale and can be heard from more than 500 mi away. The intensity of the sound is 630 watts/cm². Find the number of decibels of sound emitted by the blue whale.   188 decibels

24. Although pain thresholds for sound vary in humans, a decibel level of 125 decibels will produce pain (and even hearing loss) for most people. What is the intensity, in watts per square centimeter, of 125 decibels? 3.16×10^{-4} watts/cm²

25. The purr of a cat is approximately 25 decibels. What is the intensity, in watts per square centimeter, of 25 decibels? 3.16×10^{-14} watts/cm²

Light   For Exercises 26 and 27, use the equation $\log P = -kd$, which gives the percent P, as a decimal, of light passing through a substance of thickness d, in meters.

26. The value of k for a swimming pool is approximately 0.05. At what depth, in meters, will the percent of light be 75% of the light at the surface of the pool?   2.5 m

27. The constant k for a piece of blue stained glass is 20. What percent of light will pass through a piece of this glass that is 0.005 m thick?   79.4%

28. Earth Science   The atmospheric pressure changes as one rises above Earth's surface. At an altitude of h kilometers, where $0 < h < 80$, the pressure P in newtons per square centimeter is approximately modeled by the equation $P(h) = 10.13e^{-0.116h}$.
  a. What is the approximate pressure at 40 km above Earth's surface?
  b. What is the approximate pressure on Earth's surface?
  c. Does atmospheric pressure increase or decrease as one rises above Earth's surface?
    a. 0.098 newton/cm²   **b.** 10.13 newtons/cm²   **c.** Decreases

29. Chemistry   The intensity I of an x-ray after it passes through a material x centimeters thick is given by $I = I_0 e^{-kx}$, where I_0 is the initial intensity of the x-ray and k is a number that depends on the material. The constant k for copper is 3.2. Find the thickness of copper such that the intensity of an x-ray after passing through the copper is 25% of the original intensity. Round to the nearest tenth.   0.4 cm

30. Failure Rate   The probability P, as a percent, that a certain computer keyboard in a public library will last more than t years can be approximated by $P = 100e^{-1.2t}$.
  a. What is the probability that the keyboard will last more than 3 years? Round to the nearest tenth.   2.7%
  b. After how many years will the probability of keyboard failure be 90%? Round to the nearest tenth.   1.9 years

31. ◪ Assuming an exponential model is appropriate for each of the following, would the situation be modeled using exponential growth or exponential decay?
  a. The atmospheric pressure x meters above the surface of Earth   Decay
  b. The temperature of a roast t minutes after it is put in a hot oven   Growth
  c. The temperature of a hot cup of tea that is put in a refrigerator   Decay
  d. The spread of a contagious disease   Growth

© Ina Raschke/Shutterstock.com

Critical Thinking

32. Doubling Time Some investors want to know the "doubling time" of an investment—that is, the amount of time it will take for the investment to grow to twice its initial value. Suppose the value of an investment in an account earning an annual interest rate of 7% compounded daily grows according to the equation $A = A_0\left(1 + \frac{0.07}{365}\right)^{365t}$, where A_0 is the initial value of the investment and t is the time in years. Find the doubling time for the investment. Round to the nearest year.
10 years

33. Continuous Compounding In some scenarios, economists will use continuous compounding of an investment. The equation for continuous compounding is $A = A_0 e^{rt}$, where A is the present value of an initial investment of A_0 at an annual interest rate r (as a decimal), and t is the time in years.

a. Find the value of an investment of $5000 after 3 years if interest is compounded continuously at an annual rate of 6%. $5986.09

b. If an investor wants to grow a continuously compounded investment of $1000 to $1250 in 2 years, what interest rate must the investor receive? Round to the nearest tenth. 11.2%

34. ● **Uranium Dating** A tract of pinkish bedrock on the shore of Canada's Hudson Bay may contain the oldest known rock on Earth. Some scientists estimate the age of the rock at 4.28 billion years. The carbon-14 dating method does not work on rocks, so methods based on different radioactive elements are used. One such method uses uranium-235, which has a half-life of approximately 713 million years. Use the uranium-235 dating equation $A = A_0\left(\frac{1}{2}\right)^{t/713,000,000}$ to estimate the percent of uranium-235 remaining in the Hudson Bay rock. (*Hint:* Begin by finding $\frac{A}{A_0}$ when $t = 4.28$ billion years.) Round to the nearest tenth of a percent. 1.6%

© All Canada Photos/Alamy

Projects or Group Activities

35. ● **Steroid Use** When air resistance is considered, the height f, in feet, of a baseball x feet from home plate after being hit at a certain angle can be approximated by $f(x) = \left(\frac{0.5774v + 155.3}{v}\right)x + 565.3 \ln\left(\frac{v - 0.2747x}{v}\right) + 3.5$, where v is the speed of the ball in feet per second when it leaves the bat.

a. If $v = 160$ ft/s, show that a baseball will hit near the bottom of a 15-foot-high fence 375 ft from home plate. When $v = 160$, $f(375) = 0.43$ ft.

b. Read the article at the right. If the speed in part (a) is increased by 4%, show that the baseball will clear the fence by approximately 9 ft.

c. ▣ Use the value of v from part (b). Determine the greatest distance from home plate the 15-foot fence could be placed such that a ball hit with speed v would clear the fence. Round to the nearest foot. 385 ft
b. When $v = 166.4$, $f(375) = 24.4$ ft.

36. ▣ **Ball Flight** Air resistance plays a large role in the flight of a ball. If air resistance is ignored, then the height h, in feet, of a baseball x feet from home plate after being hit at the same angle as in Exercise 35 can be approximated by $h(x) = \frac{-21.33}{v^2}x^2 + 0.5774x + 3.5$. If the speed of a ball as it leaves the bat is 160 ft/s, how much farther will the ball travel before hitting the ground if air resistance is ignored than it would if air resistance were considered? Round to the nearest foot. 324 ft

QUICK QUIZ

1. The percent of light that will pass through a material is given by the equation $\log P = -kd$, where P is the percent of light passing through the material, k is a constant that depends on the material, and d is the thickness of the material in centimeters. The constant k for a piece of opaque glass that is 0.5 cm thick is 0.4. Find the percent of light that will pass through the glass. Round to the nearest percent. **63%** **[12.5A]**

In the NEWS!

Steroids Increase Number of Home Runs

Steroids can help batters hit 50 percent more home runs by boosting their muscle mass by just 10 percent, according to Roger Tobin of Tufts University. Calculations show that by acquiring 10 percent more muscle mass, a batter can swing about 5 percent faster, increasing the ball's speed as it leaves the bat by 4 percent.

Source: www.reuters.com

CHAPTER

12 : Summary

Key Words

Examples

A function of the form $f(x) = b^x$, $b > 0$, $b \neq 1$, is an **exponential function.** The number b is the *base* of the exponential function. [12.1A, p. 670]

$f(x) = 3^x$ is an exponential function. 3 is the base of the function.

The function defined by $f(x) = e^x$ is called the **natural exponential function.** [12.1A, p. 671]

$f(x) = 2e^{x-1}$ is a natural exponential function. e is an irrational number approximately equal to 2.71828183.

Because the exponential function is a 1–1 function, it has an inverse function, called a **logarithm.** The definition of logarithm is as follows: For $x > 0$, $b > 0$, $b \neq 1$, $y = \log_b x$ is equivalent to $x = b^y$. [12.2A, p. 679]

$\log_2 8 = 3$ is equivalent to $8 = 2^3$.

Logarithms with base 10 are called **common logarithms.** We usually omit the base, 10, when writing the common logarithm of a number. [12.2A, p. 680]

$\log_{10} 100 = 2$ is usually written $\log 100 = 2$.

When e (the base of the natural exponential function) is used as a base of a logarithm, the logarithm is referred to as a **natural logarithm** and is abbreviated ln x. [12.2A, p. 680]

$\log_e 100 \approx 4.61$ is usually written $\ln 100 \approx 4.61$.

An **exponential equation** is one in which a variable occurs in an exponent. [12.4A, p. 694]

$2^x = 12$ is an exponential equation.

An **exponential growth equation** is an equation that can be written in the form $A = A_0 b^{kt}$, where A is the size at time t, A_0 is the initial size, $b > 1$, and k is a positive real number. In an **exponential decay equation,** the base is between 0 and 1. [12.5A, pp. 700–702]

$P = 1000(1.03)^n$ is an exponential growth equation.
$A = 10(0.5)^x$ is an exponential decay equation.

Essential Rules and Procedures

Examples

1–1 Property of Exponential Functions [12.2A, p. 679]
For $b > 0$, $b \neq 1$, if $b^u = b^v$, then $u = v$.

If $b^x = b^5$, then $x = 5$.

Property of $\log_b$ 1 [12.2B, p. 681]
If $b > 0$, $b \neq 1$, then $\log_b 1 = 0$.

$\log_6 1 = 0$

Property of $\log_b$ b [12.2B, p. 681]
If $b > 0$, $b \neq 1$, then $\log_b b = 1$.

$\log_4 4 = 1$

1–1 Property of Logarithms [12.2B, p. 681]
For any positive real numbers x, y, and b, $b \neq 1$, if $\log_b x = \log_b y$, then $x = y$.

If $\log_5 (x - 2) = \log_5 3$, then $x - 2 = 3$.

Inverse Property of Logarithms [12.2B, p. 681]
For any positive real numbers x and b, $b \neq 1$, $b^{\log_b x} = x$ and $\log_b b^x = x$.

$\log_3 3^4 = 4$

Product Property of Logarithms
[12.2B, p. 682]
For any positive real numbers x, y, and b,
$b \neq 1$, $\log_b (xy) = \log_b x + \log_b y$.

$\log_b (3x) = \log_b 3 + \log_b x$

Quotient Property of Logarithms [12.2B, p. 682]
For any positive real numbers x, y, and b, $b \neq 1$,
$\log_b \dfrac{x}{y} = \log_b x - \log_b y$.

$\log_b \dfrac{x}{20} = \log_b x - \log_b 20$

Power Property of Logarithms [12.2B, p. 682]
For any positive real numbers x and b, $b \neq 1$, and for any real number r, $\log_b x^r = r \log_b x$.

$\log_b x^5 = 5 \log_b x$

Change-of-Base Formula [12.2C, p. 684]

$\log_a N = \dfrac{\log_b N}{\log_b a}$

$\log_3 12 = \dfrac{\log 12}{\log 3} \qquad \log_6 16 = \dfrac{\ln 16}{\ln 6}$

CHAPTER

12 | Review Exercises

1. Evaluate $f(x) = e^{x-2}$ at $x = 2$.
1 [12.1A]

2. Write $\log_5 25 = 2$ in exponential form.
$5^2 = 25$ [12.2A]

3. Graph: $f(x) = 3^{-x} + 2$

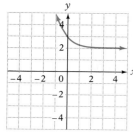

[12.1B]

4. Graph: $f(x) = \log_3 (x - 1)$

[12.3A]

5. Write $\log_3 \sqrt[5]{x^2 y^4}$ in expanded form.
$\dfrac{2}{5} \log_3 x + \dfrac{4}{5} \log_3 y$ [12.2B]

6. Write $2 \log_3 x - 5 \log_3 y$ as a single logarithm with a coefficient of 1.
$\log_3 \dfrac{x^2}{y^5}$ [12.2B]

Unless otherwise noted, all content on this page is © Cengage Learning.

7. Solve: $27^{2x+4} = 81^{x-3}$
-12 [12.4A]

8. Solve: $\log_5 \dfrac{7x+2}{3x} = 1$
$\dfrac{1}{4}$ [12.4B]

9. Find $\log_6 22$. Round to the nearest ten-thousandth.
1.7251   [12.2C]

10. Solve: $\log_2 x = 5$
32   [12.2A]

11. Solve: $\log_3 (x + 2) = 4$
79   [12.4B]

12. Solve: $\log_{10} x = 3$
1000   [12.2A]

13. Write $\dfrac{1}{3}(\log_7 x + 4 \log_7 y)$ as a single logarithm with a coefficient of 1.
$\log_7 \sqrt[3]{xy^4}$ [12.2B]

14. Write $\log_8 \sqrt{\dfrac{x^5}{y^3}}$ in expanded form.
$\dfrac{5}{2} \log_8 x - \dfrac{3}{2} \log_8 y$ [12.2B]

15. Write $2^5 = 32$ in logarithmic form.
$\log_2 32 = 5$ [12.2A]

16. Find $\log_3 1.6$. Round to the nearest ten-thousandth.
0.4278   [12.2C]

17. Solve $3^{x+2} = 5$. Round to the nearest thousandth.
-0.535 [12.4A]

18. Evaluate $f(x) = \left(\dfrac{2}{3}\right)^{x+2}$ at $x = -3$.
$\dfrac{3}{2}$ [12.1A]

19. Solve: $\log_2(x + 3) - \log_2(x - 1) = 3$
$\dfrac{11}{7}$ [12.4B]

20. Solve: $\log_3 (2x + 3) + \log_3 (x - 2) = 2$
3   [12.4B]

21. Graph: $f(x) = \left(\dfrac{2}{3}\right)^{x+1}$

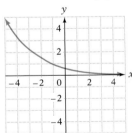

[12.1B]

22. Graph: $f(x) = \log_2 (2x - 1)$

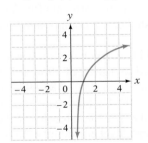

[12.3A]

23. Evaluate: $\log_6 36$
2   [12.2A]

24. Write $\dfrac{1}{3}(\log_2 x - \log_2 y)$ as a single logarithm with a coefficient of 1.
$\log_2 \sqrt[3]{\dfrac{x}{y}}$ [12.2B]

25. Solve for x: $9^{2x} = 3^{x+3}$
1   [12.4A]

26. Solve for x: $5 \cdot 3^{x/2} = 12$. Round to the nearest ten-thousandth.
1.5938   [12.4A]

Unless otherwise noted, all content on this page is © Cengage Learning.

27. Solve for x: $\log_5 x = -1$

  $\dfrac{1}{5}$ [12.2A]

28. Write $3^4 = 81$ in logarithmic form.

  $\log_3 81 = 4$ [12.2A]

29. Solve for x: $\log x + \log (x - 2) = \log 15$

  5 [12.4B]

30. Write $\log_5 \sqrt[3]{x^2 y}$ in expanded form.

  $\dfrac{2}{3} \log_5 x + \dfrac{1}{3} \log_5 y$ [12.2B]

31. Solve for x: $6e^{-2x} = 17$. Round to the nearest ten-thousandth.

  -0.5207 [12.4A]

32. Evaluate $f(x) = 7^{x+2}$ at $x = -3$.

  $\dfrac{1}{7}$ [12.1A]

33. Evaluate: $\log_2 16$

  4 [12.2A]

34. Solve for x: $\log_6 x = \log_6 2 + \log_6 (2x - 3)$

  2 [12.4B]

35. Evaluate $\log_2 5$. Round to the nearest ten-thousandth.

  2.3219 [12.2C]

36. Solve for x: $4^x = 8^{x-1}$

  3 [12.4A]

37. Solve for x: $\log_5 x = 4$

  625 [12.2A]

38. Write $3 \log_b x - 7 \log_b y$ as a single logarithm with a coefficient of 1.

  $\log_b \dfrac{x^3}{y^7}$ [12.2B]

39. Evaluate $f(x) = 5^{-x-1}$ at $x = -2$.

  5 [12.1A]

40. Solve $5^{x-2} = 7$ for x. Round to the nearest ten-thousandth.

  3.2091 [12.4A]

41. Investments Use the compound interest formula $P = A(1 + i)^n$, where A is the original value of an investment, i is the interest rate per compounding period, n is the number of compounding periods, and P is the value of the investment after n periods, to find the value of an investment after 2 years. The amount of the investment is $4000, and it is invested at 8% compounded monthly. Round to the nearest dollar.

  $4692 [12.5A]

42. ⬤ Seismology The earthquake of greatest magnitude ever recorded in the United States occurred in March of 1964 at Prince William Sound, Alaska. The intensity of the earthquake was $I = 1,584,893,192 I_0$. Find the Richter scale magnitude of the earthquake. Use the Richter scale equation $M = \log \dfrac{I}{I_0}$, where M is the magnitude of an earthquake, I is the intensity of the shock waves, and I_0 is the measure of the intensity of a zero-level earthquake. Round to the nearest tenth. 9.2 [12.5A]

43. Radioactivity Use the exponential decay equation $A = A_0 (0.5)^{t/k}$, where A is the amount of a radioactive material present after time t, k is the half-life of the radioactive material, and A_0 is the original amount of radioactive material, to find the half-life of a material that decays from 25 mg to 15 mg in 20 days. Round to the nearest whole number. 27 days [12.5A]

44. Sound The number of decibels D of a sound can be given by the equation $D = 10(\log I + 16)$, where I is the intensity of the sound measured in watts per square centimeter. Find the number of decibels of sound emitted from a busy street corner for which the intensity of the sound is 5×10^{-6} watts/cm^2. Round to the nearest decibel. 107 decibels [12.5A]

© iStockphoto.com/Cristian Baitg

CHAPTER

12 | TEST

1. Evaluate $f(x) = \left(\frac{2}{3}\right)^x$ at $x = 0$.

$f(0) = 1$ [12.1A]

2. Evaluate $f(x) = 3^{x+1}$ at $x = -2$.

$f(-2) = \frac{1}{3}$ [12.1A]

3. Graph: $f(x) = 2^x - 3$

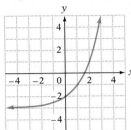

[12.1B]

4. Graph: $f(x) = 2^x + 2$

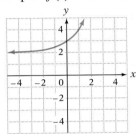

[12.1B]

5. Evaluate: $\log_4 16$

2   [12.2A]

6. Solve for x: $\log_3 x = -2$

$\frac{1}{9}$ [12.2A]

7. Graph: $f(x) = \log_2(2x)$

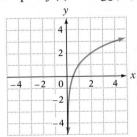

[12.3A]

8. Graph: $f(x) = \log_3(x + 1)$

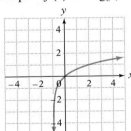

[12.3A]

9. Write $\log_6 \sqrt{xy^3}$ in expanded form.

$\frac{1}{2}\log_6 x + \frac{3}{2}\log_6 y$ [12.2B]

10. Write $\frac{1}{2}(\log_3 x - \log_3 y)$ as a single logarithm with a coefficient of 1.

$\log_3 \sqrt{\frac{x}{y}}$ [12.2B]

11. Write $\ln \frac{x}{\sqrt{z}}$ in expanded form.

$\ln x - \frac{1}{2}\ln z$ [12.2B]

12. Write $3\ln x - \ln y - \frac{1}{2}\ln z$ as a single logarithm with a coefficient of 1.

$\ln \frac{x^3}{y\sqrt{z}}$ [12.2B]

Unless otherwise noted, all content on this page is © Cengage Learning.

13. Solve for x: $3^{7x+1} = 3^{4x-5}$
  −2 [12.4A]

14. Solve for x: $8^x = 2^{x-6}$
  −3 [12.4A]

15. Solve for x: $3^x = 17$. Round to the nearest ten-thousandth.
  2.5789 [12.4A]

16. Solve for x: $\log x + \log(x-4) = \log 12$
  6 [12.4B]

17. Solve for x: $\log_6 x + \log_6(x-1) = 1$
  3 [12.4B]

18. Find $\log_5 9$. Round to the nearest ten-thousandth.
  1.3652 [12.2C]

19. Find $\log_3 19$. Round to the nearest ten-thousandth.
  2.6801 [12.2C]

20. Solve for x: $5^{2x-5} = 9$. Round to the nearest ten-thousandth.
  3.1826 [12.4A]

21. Solve for x: $2e^{x/4} = 9$. Round to the nearest ten-thousandth.
  6.0163 [12.4A]

22. Solve for x: $\log_5(30x) - \log_5(x+1) = 2$
  5 [12.4B]

23. Carbon Dating A shard from a vase originally contained 250 mg of carbon-14 and now contains 170 mg of carbon-14. Use the equation $A = A_0 (0.5)^{t/5570}$, where A_0 is the original amount of carbon-14 in the shard and A is the amount of carbon-14 in the shard t years later, to find the approximate age of the shard. Round to the nearest whole number.   3099 years old [12.5A]

24. Sound Use the decibel equation $D = 10(\log I + 16)$, where D is the decibel level and I is the intensity of a sound in watts per square centimeter, to find the intensity of a 75-decibel dial tone. 3.16×10^{-9} watts/cm² [12.5A]

25. Radioactivity Use the exponential decay equation $A = A_0 (0.5)^{t/k}$, where A is the amount of a radioactive material present after time t, k is the half-life of the material, and A_0 is the original amount of radioactive material, to find the half-life of a material that decays from 10 mg to 9 mg in 5 h. Round to the nearest whole number.   33 h [12.5A]

Cumulative Review Exercises

1. Solve: $4 - 2[x - 3(2 - 3x) - 4x] = 2x$

$\dfrac{8}{7}$ [2.2C]

2. Find the equation of the line that contains the point $P(2, -2)$ and is parallel to the graph of $2x - y = 5$.

$y = 2x - 6$ [4.6A]

3. Factor: $4x^4 + 7x^2 + 3$

$(4x^2 + 3)(x^2 + 1)$ [7.3A/7.3B]

4. Simplify: $\dfrac{1 - \dfrac{5}{x} + \dfrac{6}{x^2}}{1 + \dfrac{1}{x} - \dfrac{6}{x^2}}$

$\dfrac{x - 3}{x + 3}$ [8.3A]

5. Simplify: $\dfrac{\sqrt{xy}}{\sqrt{x} - \sqrt{y}}$

$\dfrac{x\sqrt{y} + y\sqrt{x}}{x - y}$ [9.3B]

6. Solve by completing the square: $x^2 - 4x - 6 = 0$

$2 - \sqrt{10}$ and $2 + \sqrt{10}$ [10.2A]

7. Write a quadratic equation with integer coefficients and solutions $\dfrac{1}{3}$ and -3.

$3x^2 + 8x - 3 = 0$ [10.1A]

8. Graph the solution set: $\begin{aligned} 2x - y &< 3 \\ x + y &< 1 \end{aligned}$

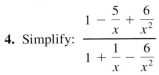

[5.4A]

9. Solve by the addition method:
$$\begin{aligned} 3x - y + z &= 3 \\ x + y + 4z &= 7 \\ 3x - 2y + 3z &= 8 \end{aligned}$$
$(0, -1, 2)$ [5.2B]

10. Subtract: $\dfrac{x - 4}{2 - x} - \dfrac{1 - 6x}{2x^2 - 7x + 6}$

$-\dfrac{2x^2 - 17x + 13}{(x - 2)(2x - 3)}$ [8.2B]

11. Solve: $x^2 + 4x - 5 \le 0$
Write the solution set in set-builder notation.

$\{x \mid -5 \le x \le 1\}$ [10.5A]

12. Solve: $|2x - 5| \le 3$

$\{x \mid 1 \le x \le 4\}$ [2.6B]

13. Graph: $f(x) = \left(\dfrac{1}{2}\right)^x + 1$

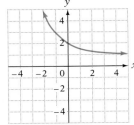

[12.1B]

14. Graph: $f(x) = \log_2 x - 1$

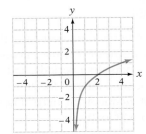

[12.3A]

Unless otherwise noted, all content on this page is © Cengage Learning.

15. Evaluate $f(x) = 2^{-x-1}$ at $x = -3$.
    4   [12.1A]

16. Solve for x: $\log_5 x = 3$
    125   [12.2A]

17. Write $3 \log_b x - 5 \log_b y$ as a single logarithm with a coefficient of 1.

$\log_b \dfrac{x^3}{y^5}$ [12.2B]

18. Find $\log_3 7$. Round to the nearest ten-thousandth.
    1.7712   [12.2C]

19. Solve for x: $4^{5x-2} = 4^{3x+2}$
    2   [12.4A]

20. Solve for x: $\log x + \log(2x + 3) = \log 2$

$\dfrac{1}{2}$ [12.4B]

21. Banking   A bank offers two types of business checking accounts. One account has a charge of $5 per month plus 2 cents per check. The second account has a charge of $2 per month plus 8 cents per check. How many checks can a customer who has the second type of account write if it is to cost the customer less than the first type of checking account?
49 checks or less   [2.5C]

22. Mixtures   Find the cost per pound of a mixture made from 16 lb of chocolate that costs $4.00 per pound and 24 lb of chocolate that costs $2.50 per pound.
$3.10   [2.4A]

23. Uniform Motion   A plane can fly at a rate of 225 mph in calm air. Traveling with the wind, the plane flew 1000 mi in the same amount of time it took to fly 800 mi against the wind. Find the rate of the wind.
25 mph   [8.7B]

24. Physics   The distance d that a spring stretches varies directly as the force f used to stretch the spring. If a force of 20 lb stretches a spring 6 in., how far will a force of 34 lb stretch the spring?
10.2 in.   [8.8A]

25. Carpentry   A carpenter purchased 80 ft of redwood and 140 ft of fir for a total cost of $67. A second purchase, at the same prices, included 140 ft of redwood and 100 ft of fir for a total cost of $81. Find the cost of redwood and of fir.
Redwood: 40¢ per foot
Fir: 25¢ per foot   [5.3B]

26. Investments   The compound interest formula is $A = P(1 + i)^n$, where P is the original value of an investment, i is the interest rate per compounding period, n is the total number of compounding periods, and A is the value of the investment after n periods. Use the compound interest formula to find how many years it will take for an investment of $5000 to double in value. The investment earns 7% annual interest and is compounded semiannually.
10 years   [12.5A]

© iStockphoto.com/David Gomez

FINAL EXAM

1. Simplify:
$12 - 8[3 - (-2)]^2 \div 5 - 3$
-31 [1.3A]

2. Evaluate $\dfrac{a^2 - b^2}{a - b}$ when $a = 3$ and $b = -4$.
-1 [1.4A]

3. Simplify: $5 - 2[3x - 7(2 - x) - 5x]$
$-10x + 33$ [1.4D]

4. Solve: $\dfrac{3}{4}x - 2 = 4$
8 [2.2A]

5. Solve: $8 - |5 - 3x| = 1$
$-\dfrac{2}{3}, 4$ [2.6A]

6. Find the volume of a sphere with a diameter of 8 ft. Round to the nearest tenth.
268.1 ft^3 [3.3A]

7. Graph $2x - 3y = 9$ using the x- and y-intercepts.

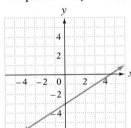

[4.3C]

8. Find the equation of the line containing the points $P_1(3, -2)$ and $P_2(1, 4)$.
$y = -3x + 7$ [4.5B]

9. Find the equation of the line that contains the point $P(-2, 1)$ and is perpendicular to the line $3x - 2y = 6$.
$y = -\dfrac{2}{3}x - \dfrac{1}{3}$ [4.6A]

10. Simplify: $2a[5 - a(2 - 3a) - 2a] + 3a^2$
$6a^3 - 5a^2 + 10a$ [6.3A]

11. Factor: $8 - x^3y^3$
$(2 - xy)(4 + 2xy + x^2y^2)$ [7.4B]

12. Factor: $x - y - x^3 + x^2y$
$(x - y)(1 - x)(1 + x)$ [7.4D]

13. Divide: $(2x^3 - 7x^2 + 4) \div (2x - 3)$
$x^2 - 2x - 3 - \dfrac{5}{2x - 3}$ [6.4B]

14. Divide: $\dfrac{x^2 - 3x}{2x^2 - 3x - 5} \div \dfrac{4x - 12}{4x^2 - 4}$
$\dfrac{x(x - 1)}{2x - 5}$ [8.1C]

15. Subtract: $\dfrac{x - 2}{x + 2} - \dfrac{x + 3}{x - 3}$
$\dfrac{-10x}{(x + 2)(x - 3)}$ [8.2B]

16. Simplify: $\dfrac{\dfrac{3}{x} + \dfrac{1}{x + 4}}{\dfrac{1}{x} + \dfrac{3}{x + 4}}$
$\dfrac{x + 3}{x + 1}$ [8.3A]

Unless otherwise noted, all content on this page is © Cengage Learning.

17. Solve: $\dfrac{5}{x-2} - \dfrac{5}{x^2-4} = \dfrac{1}{x+2}$

$-\dfrac{7}{4}$ [8.4A]

18. Solve $a_n = a_1 + (n-1)d$ for d.

$d = \dfrac{a_n - a_1}{n-1}$ [8.6A]

19. Simplify: $\left(\dfrac{4x^2y^{-1}}{3x^{-1}y}\right)^{-2}\left(\dfrac{2x^{-1}y^2}{9x^{-2}y^2}\right)^3$

$\dfrac{y^4}{162x^3}$ [6.1B]

20. Simplify: $\left(\dfrac{3x^{2/3}y^{1/2}}{6x^2y^{4/3}}\right)^6$

$\dfrac{1}{64x^8y^5}$ [9.1A]

21. Subtract: $x\sqrt{18x^2y^3} - y\sqrt{50x^4y}$

$-2x^2y\sqrt{2y}$ [9.2B]

22. Simplify: $\dfrac{\sqrt{16x^5y^4}}{\sqrt{32xy^7}}$

$\dfrac{x^2\sqrt{2y}}{2y^2}$ [9.3B]

23. Simplify: $\dfrac{3}{2+i}$

$\dfrac{6}{5} - \dfrac{3}{5}i$ [9.5D]

24. Write a quadratic equation that has integer coefficients and has solutions $-\dfrac{1}{2}$ and 2.

$2x^2 - 3x - 2 = 0$ [10.1A]

25. Solve by using the quadratic formula:
$2x^2 - 3x - 1 = 0$

$\dfrac{3+\sqrt{17}}{4}, \dfrac{3-\sqrt{17}}{4}$ [10.2B]

26. Solve: $x^{2/3} - x^{1/3} - 6 = 0$

$-8, 27$ [10.3A]

27. Graph: $f(x) = -x^2 + 4$

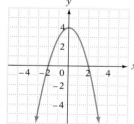

[11.1A]

28. Graph: $f(x) = -\dfrac{1}{2}x - 3$

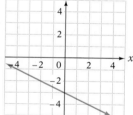

[4.3A]

29. Solve: $\dfrac{2}{x} - \dfrac{2}{2x+3} = 1$

$-2, \dfrac{3}{2}$ [10.3C]

30. Find the inverse of the function $f(x) = \dfrac{2}{3}x - 4$.

$f^{-1}(x) = \dfrac{3}{2}x + 6$ [11.4B]

31. Solve by the addition method:
$3x - 2y = 1$
$5x - 3y = 3$

$(3, 4)$ [5.2A]

32. Simplify: $\sqrt{49x^6}$

$7x^3$ [9.1C]

Unless otherwise noted, all content on this page is © Cengage Learning.

33. Solve: $2 - 3x < 6$ and $2x + 1 > 4$

$\left\{x \mid x > \dfrac{3}{2}\right\}$ [2.5B]

34. Solve: $|2x + 5| < 3$

$\{x \mid -4 < x < -1\}$ [2.6B]

35. Graph the solution set: $3x + 2y > 6$

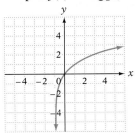

[4.7A]

36. Graph: $f(x) = 3^{-x} - 2$

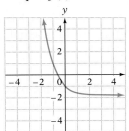

[12.1B]

37. Graph: $f(x) = \log_2 (x + 1)$

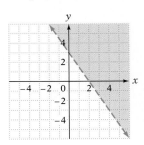

[12.3A]

38. Write $2(\log_2 a - \log_2 b)$ as a single logarithm with a coefficient of 1.

$\log_2 \dfrac{a^2}{b^2}$ [12.2B]

39. Solve for x: $\log_3 x - \log_3 (x - 3) = \log_3 2$

6   [12.4B]

40. Education   An average score of 70–79 in a history class receives a C grade. A student has grades of 64, 58, 82, and 77 on four history tests. Find the range of scores on the fifth test that will give the student a C grade for the course.

$69 \le x \le 100$ [2.5C]

41. Uniform Motion   A jogger and a cyclist set out at 8 A.M. from the same point headed in the same direction. The average speed of the cyclist is two and a half times the average speed of the jogger. In 2 h, the cyclist is 24 mi ahead of the jogger. How far did the cyclist ride in that time?

40 mi   [2.4C]

Unless otherwise noted, all content on this page is © Cengage Learning.

42. Investments   You have a total of $12,000 invested in two simple interest accounts. On one account, a money market fund, the annual simple interest rate is 8.5%. On the other account, a tax-free bond fund, the annual simple interest rate is 6.4%. The total annual interest earned by the two accounts is $936. How much do you have invested in each account?
$8000 at 8.5%; $4000 at 6.4%   [5.1C]

43. Geometry   The length of a rectangle is 1 ft less than three times the width. The area of the rectangle is 140 ft². Find the length and width of the rectangle.
Width: 7 ft; length: 20 ft   [10.4A]

44. The Stock Market   Three hundred shares of a utility stock earn a yearly dividend of $486. How many additional shares of the utility stock would give a total dividend income of $810?
200 additional shares   [8.5A]

45. Travel   An account executive traveled 45 mi by car and then an additional 1050 mi by plane. The rate of the plane was seven times the rate of the car. The total time for the trip was $3\frac{1}{4}$ h. Find the rate of the plane.
420 mph   [8.7B]

46. Physics   An object is dropped from the top of a building. Find the distance the object has fallen when the speed reaches 75 ft/s. Use the equation $v = \sqrt{64d}$, where v is the speed of the object and d is the distance. Round to the nearest whole number.
88 ft   [9.4B]

47. Travel   A small plane made a trip of 660 mi in 5 h. The plane traveled the first 360 mi at a constant rate before increasing its speed by 30 mph. Then it traveled another 300 mi at the increased speed. Find the rate of the plane for the first 360 mi.
120 mph   [8.7B]

48. Light   The intensity (L) of a light source is inversely proportional to the square of the distance (d) from the source. If the intensity is 8 foot-candles at a distance of 20 ft, what is the intensity when the distance is 4 ft?
200 foot-candles   [8.8A]

49. Travel   A motorboat traveling with the current can go 30 mi in 2 h. Against the current, it takes 3 h to go the same distance. Find the rate of the motorboat in calm water and the rate of the current.
Rate of the boat in calm water: 12.5 mph; Rate of the current: 2.5 mph   [5.3A]

50. Investments   An investor deposits $4000 into an account that earns 9% annual interest compounded monthly. Use the compound interest formula $P = A(1 + i)^n$, where A is the original value of the investment, i is the interest rate per compounding period, n is the total number of compounding periods, and P is the value of the investment after n periods, to find the value of the investment after 2 years. Round to the nearest cent.
$4785.65   [12.5A]

Transitioning to Intermediate Algebra

OBJECTIVES

SECTION T.1
A To evaluate a variable expression
B To simplify a variable expression

SECTION T.2
A To solve a first-degree equation in one variable
B To solve an inequality in one variable

SECTION T.3
A To graph points in a rectangular coordinate system
B To graph a linear equation in two variables
C To evaluate a function
D To find the equation of a line

SECTION T.4
A To multiply and divide monomials
B To add and subtract polynomials
C To multiply polynomials
D To divide polynomials
E To factor polynomials of the form $ax^2 + bx + c$

Focus on Success

Reviewing concepts and skills that you have studied is one way to strengthen your ability to use those concepts and skills to solve problems. This transition chapter provides you with an opportunity to review material that you have studied in the Introductory Algebra chapters of this text. As you begin the Intermediate Algebra chapters of the text, also review the study skills that you and other students have used to be successful by reading Chapter A, AIM for Success.

© Carlos E. Santa Maria/Shutterstock.com

Prep Test

Are you ready to succeed in an intermediate algebra course? Take the Prep Test below to find out if you are ready to review some of the skills you will need.

1. Place the correct symbol, $<$ or $>$, between the two numbers.

$-37 < -28$ [1.1A]

2. Write $\frac{9}{100}$ as a decimal.

0.09   [1.2A]

3. Subtract: $\frac{3}{5} - \frac{11}{12}$

$-\frac{19}{60}$ [1.2C]

4. Find the least common multiple (LCM) of 24 and 36.

72   [1.2C]

5. Find the quotient of -6.904 and 1.35.

-5.11 [1.2D]

6. Evaluate $(-2)^3 (-7^2)$.

128   [1.2E]

7. Evaluate $x - y$ when $x = 3005$ and $y = 387$.

2618   [1.4A]

8. Simplify:
$3x^2 - 4x + 1 + 2x^2 - 5x - 7$

$5x^2 - 9x - 6$ [1.4B]

9. Given $3x - 4y = 12$, find the value of x when $y = 0$.

4   [2.2A]

10. Find the greatest common factor (GCF) of $24x^2$ and $36xy$.

$6x$ [7.1A]

SECTION

T.1 | Variable Expressions

OBJECTIVE A | *To evaluate a variable expression*

Whenever an expression contains more than one operation, the operations must be performed in a specified order, as listed below in the Order of Operations Agreement.

1.3A* Order of Operations Agreement

The Order of Operations Agreement

Step 1	Perform operations inside grouping symbols. Grouping symbols include parentheses ( ), brackets [ ], braces { }, the absolute value symbol | |, and fraction bars.
Step 2	Simplify exponential expressions.
Step 3	Do multiplication and division as they occur from left to right.
Step 4	Do addition and subtraction as they occur from left to right.

EXAMPLE 1

Evaluate: $-2(7 - 3)^2 + 4 - 2(5 - 2)$

Solution

$-2(7 - 3)^2 + 4 - 2(5 - 2)$
$= -2(4)^2 + 4 - 2(3)$ • Perform operations inside parentheses.

$= -2(16) + 4 - 2(3)$ • Simplify exponential expressions.

$= -32 + 4 - 2(3)$ • Do multiplication and division from left to right.
$= -32 + 4 - 6$

$= -28 - 6$ • Do addition and subtraction from left to right.
$= -28 + (-6) = -34$

YOU TRY IT 1

Evaluate: $(-4)(6 - 8)^2 - (-12 \div 4)$

Your solution

-13

Solution on p. S35

1.4A Evaluate variable expressions

A **variable** is a letter that represents a quantity that is unknown or that can change, or *vary*. An expression that contains one or more variables is a **variable expression**. $3x - 4y + 7z$ is a variable expression. It contains the variables x, y, and z.

Replacing a variable in a variable expression by a number and then simplifying the resulting numerical expression is called **evaluating the variable expression.** The number substituted for the variable is called the **value of the variable.** The result is called the **value of the variable expression.**

*Review this objective for more detailed coverage of this topic.

EXAMPLE 2

Evaluate $5ab^3 + 2a^2b^2 - 4$ when $a = 3$ and $b = -2$.

Solution

$5ab^3 + 2a^2b^2 - 4$

$5(3)(-2)^3 + 2(3)^2(-2)^2 - 4$ • Replace a by 3
                                        and b by -2.

$= 5(3)(-8) + 2(9)(4) - 4$ • Use the Order of
                                 Operations Agreement
$= -120 + 72 - 4$ to simplify the
$= -48 - 4$ numerical expression.
$= -48 + (-4)$
$= -52$

YOU TRY IT 2

Evaluate $3xy^2 - 3x^2y$ when $x = -2$ and $y = 5$.

Your solution

-210

Solution on p. S35

OBJECTIVE B *To simplify a variable expression*

1.4B
1.4C
Simplify variable
expressions
using the
Properties of
Addition and
Multiplication

The Properties of Real Numbers are used to simplify variable expressions.

EXAMPLE 3

Simplify: $-\dfrac{1}{3}(-3y)$

Solution

$-\dfrac{1}{3}(-3y)$

$= \left[-\dfrac{1}{3}(-3)\right]y$ • Use the Associative Property
                                          of Multiplication to regroup
                                          factors.
$= 1y$ • Use the Inverse Property of
                                          Multiplication.
$= y$ • Use the Multiplication
                                          Property of One.

YOU TRY IT 3

Simplify: $-5(-3a)$

Your solution

$15a$

Solution on p. S35

A variable expression is shown at the right. The expression can be rewritten by writing subtraction as addition of the opposite. Note that the expression has four addends. The **terms** of a variable expression are the addends of the expression. The expression has four terms.

The terms $3x^2$, $-4xy$, and $5z$ are **variable terms.** The term -2 is a **constant term,** or simply a **constant.**

$3x^2 - 4xy + 5z - 2$
$3x^2 + (-4xy) + 5z + (-2)$

$$\overbrace{3x^2 - 4xy + 5z}^{\text{Four terms}} \underbrace{- 2}$$

Variable   Constant
terms   term

Like terms of a variable expression are terms that have the same variable part. The terms $3x$ and $-7x$ are like terms. Constant terms are also like terms. Thus -6 and 9 are like terms.

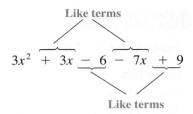

By using the Commutative Property of Multiplication, we can rewrite the Distributive Property as $ba + ca = (b + c)a$. This form of the Distributive Property is used to **combine like terms** of a variable expression by adding their coefficients. For instance,

$$7x + 9x = (7 + 9)x$$
$$= 16x$$

• Use the Distributive Property: $ba + ca = (b + c)a$.

EXAMPLE 4

Simplify: $4x^2 + 5x - 6x^2 - 7x$

Solution

$4x^2 + 5x - 6x^2 - 7x$
$= 4x^2 - 6x^2 + 5x - 7x$
$= (4x^2 - 6x^2) + (5x - 7x)$

• Use the Associative and Commutative Properties of Addition to rearrange and group like terms.

$= -2x^2 + (-2x)$
$= -2x^2 - 2x$

• Use the Distributive Property to combine like terms.

YOU TRY IT 4

Simplify: $2z^2 - 5z - 3z^2 + 6z$

Your solution

$-z^2 + z$

Solution on p. S35

1.4D Simplify variable expressions using the Distributive Property

The Distributive Property also is used to remove parentheses from a variable expression. Here is an example.

$$4(2x + 5z) = 4(2x) + 4(5z)$$

• Use the Distributive Property: $a(b + c) = ab + ac$.

$$= (4 \cdot 2)x + (4 \cdot 5)z$$

• Use the Associative Property of Multiplication to regroup factors.

$$= 8x + 20z$$

• Multiply $4 \cdot 2$ and $4 \cdot 5$.

The Distributive Property can be extended to expressions containing more than two terms. For instance,

$$4(2x + 3y + 5z) = 4(2x) + 4(3y) + 4(5z)$$
$$= 8x + 12y + 20z$$

EXAMPLE 5

Simplify.
a. $-3(2x + 4)$
b. $6(3x - 4y + z)$

Solution

a. $-3(2x + 4)$
  $= -3(2x) + (-3)(4)$ • Use the Distributive
  $= -6x - 12$ Property.

b. $6(3x - 4y + z)$
  $= 6(3x) - 6(4y) + 6(z)$ • Use the Distributive
  $= 18x - 24y + 6z$ Property.

YOU TRY IT 5

Simplify.
a. $-3(5y - 2)$
b. $-2(4x + 2y - 6z)$

Your solution

a. $-15y + 6$
b. $-8x - 4y + 12z$

Solution on p. S36

 Take Note

Recall that the Distributive
Property states that if *a*, *b*,
and *c* are real numbers, then
  $a(b + c) = ab + ac$

To simplify the expression $5 + 3(4x - 2)$, use the Distributive Property to remove the parentheses.

  $5 + 3(4x - 2) = 5 + 3(4x) - 3(2)$ • Use the Distributive Property.

  $= 5 + 12x - 6$

  $= 12x - 1$ • Add the like terms 5 and −6.

EXAMPLE 6

Simplify.
a. $3(2x - 4) - 5(3x + 2)$
b. $3a - 2[7a - 2(2a + 1)]$

Solution

a. $3(2x - 4) - 5(3x + 2)$
  $= 6x - 12 - 15x - 10$ • Use the Distributive
                              Property.
                            • Combine like terms.
  $= -9x - 22$

b. $3a - 2[7a - 2(2a + 1)]$
  $= 3a - 2[7a - 4a - 2]$ • Use the Distributive
                              Property.
                            • Combine like terms.
  $= 3a - 2[3a - 2]$ • Use the Distributive
  $= 3a - 6a + 4$ Property.
                            • Combine like terms.
  $= -3a + 4$

YOU TRY IT 6

Simplify.
a. $7(-3x - 4y) - 3(3x + y)$
b. $2y - 3[5 - 3(3 + 2y)]$

Your solution

a. $-30x - 31y$
b. $20y + 12$

Solution on p. S36

T.1 EXERCISES

OBJECTIVE A   *To evaluate a variable expression*

For Exercises 1 to 9, evaluate the variable expression when $a = 2$, $b = 3$, and $c = -4$.

1. $a - 2c$
10

2. $-3a + 4b$
6

3. $3b - 3c$
21

4. $-3c + 4$
16

5. $16 \div (2c)$
-2

6. $6b \div (-a)$
-9

7. $3b - (a + c)^2$
5

8. $(a - b)^2 + 2c$
-7

9. $(b - 3a)^2 + bc$
-3

For Exercises 10 to 21, evaluate the variable expression when $a = -1$, $b = 3$, $c = -2$, and $d = 4$.

10. $\dfrac{b - a}{d}$
1

11. $\dfrac{d - b}{a}$
-1

12. $\dfrac{2d + b}{-a}$
11

13. $\dfrac{b - d}{c - a}$
1

14. $2(b + c) - 2a$
4

15. $3(b - a) - bc$
18

16. $\dfrac{-4bc}{2a + c}$
-6

17. $\dfrac{abc}{b - d}$
-6

18. $(d - a)^2 - (b - c)^2$
0

19. $(-b + d)^2 + (-a + c)^2$
2

20. $4ab + (2c)^2$
4

21. $3cd - (4a)^2$
-40

For Exercises 22 to 24, evaluate the variable expression when $a = 2.7$, $b = -1.6$, and $c = -0.8$.

22. $c^2 - ab$
4.96

23. $(a + b)^2 - c$
2.01

24. $\dfrac{b^3}{c} - 4a$
-5.68

OBJECTIVE B   *To simplify a variable expression*

For Exercises 25 to 63, simplify.

25. $x + 7x$
$8x$

26. $12y + 9y$
$21y$

27. $8b - 5b$
$3b$

28. $4y - 11y$
$-7y$

29. $-12a + 17a$
$5a$

30. $-15xy + 7xy$
$-8xy$

31. $4x + 5x + 2x$
$11x$

32. $-5x^2 - 10x^2 + x^2$
$-14x^2$

33. $6x - 2y + 9x$
$15x - 2y$

34. $3x - 7y - 6x + 4x$
$x - 7y$

35. $5a + 6a - 2a$
$9a$

36. $2a - 5a + 3a$
0

37. $12y^2 + 10y^2$
$22y^2$

38. $2z^2 - 9z^2$
$-7z^2$

39. $\dfrac{3}{4}x - \dfrac{1}{4}x$
$\dfrac{1}{2}x$

40. $\dfrac{2}{5}y - \dfrac{3}{5}y$
$-\dfrac{1}{5}y$

41. $-4(5x)$
$-20x$

42. $-2(-8y)$
$16y$

43. $(6a)(-4)$
$-24a$

44. $-5(7x^2)$
$-35x^2$

45. $\dfrac{1}{4}(4x)$
x

46. $\dfrac{12x}{5}\left(\dfrac{5}{12}\right)$
x

47. $\dfrac{1}{3}(21x)$
$7x$

48. $-\dfrac{5}{8}(24a^2)$
$-15a^2$

49. $(36y)\left(\dfrac{1}{12}\right)$
$3y$

50. $-(z + 4)$
$-z - 4$

51. $-3(a + 5)$
$-3a - 15$

52. $(4 - 3b)9$
$36 - 27b$

53. $(-2x - 6)8$
$-16x - 48$

54. $3(5x^2 + 2x)$
$15x^2 + 6x$

55. $-5(2y^2 - 1)$
$-10y^2 + 5$

56. $4(x^2 - 3x + 5)$
$4x^2 - 12x + 20$

57. $6(3x^2 - 2xy - y^2)$
$18x^2 - 12xy - 6y^2$

58. $5a - (4a + 6)$
$a - 6$

59. $3 - (10 + 8y)$
$-8y - 7$

60. $12(y - 2) + 3(7 - 4y)$
-3

61. $-5[2x + 3(5 - x)]$
$5x - 75$

62. $-3[2x - (x + 7)]$
$-3x + 21$

63. $-5a - 2[2a - 4(a + 7)]$
$-a + 56$

T.2   Equations and Inequalities

OBJECTIVE A   *To solve a first-degree equation in one variable*

An **equation** expresses the equality of two mathematical expressions. Each of the equations below is a **first-degree equation in one variable.** *First degree* means that the variable has an exponent of 1.

$$x + 11 = 14$$
$$3a + 5 = 8a$$
$$2(6y - 1) = 3$$
$$4 - 3(2n - 1) = 6n - 5$$

A **solution** of an equation is a number that, when substituted for the variable, results in a true equation.

3 is a solution of the equation $x + 4 = 7$ because $3 + 4 = 7$.
9 is not a solution of the equation $x + 4 = 7$ because $9 + 4 \neq 7$.

To **solve an equation** means to find a solution of the equation. In solving an equation, the goal is to rewrite the given equation with the variable alone on one side of the equation and a constant term on the other side of the equation.

$$variable = constant \quad or \quad constant = variable$$

The constant is the solution of the equation.

The following properties of equations are used to rewrite equations in this form.

Properties of Equations

Addition Property of Equations
The same number can be added to each side of an equation without changing its solution. In symbols, the equation $a = b$ has the same solution as the equation $a + c = b + c$.

2.1B
2.1C
Solving equations using the Addition and Multiplication Properties of Equations

Multiplication Property of Equations
Each side of an equation can be multiplied by the same nonzero number without changing the solution of the equation. In symbols, if $c \neq 0$, then the equation $a = b$ has the same solution as the equation $ac = bc$.

 Take Note

Subtraction is defined as addition of the opposite.
$$a - b = a + (-b)$$

The Addition Property of Equations is used to remove a term from one side of the equation by adding the opposite of that term to each side of the equation. Because subtraction is defined in terms of addition, the Addition Property of Equations also makes it possible to subtract the same number from each side of an equation without changing the solution of the equation.

For example, to solve the equation $t + 9 = -4$, subtract the constant term (9) from each side of the equation.

$$t + 9 = -4$$
$$t + 9 - 9 = -4 - 9$$
$$t = -13$$

Now the variable is alone on one side of the equation and a constant term (-13) is on the other side. The solution is the constant. The solution is -13.

To solve $7 = y - 8$, add 8 to each side of the equation.

$$7 = y - 8$$
$$7 + 8 = y - 8 + 8$$
$$15 = y$$

Take Note

Division is defined as multiplication by the reciprocal.

$$a \div b = a \cdot \frac{1}{b}$$

The equation is in the form *constant = variable*. The solution is the constant. The solution is 15.

The Multiplication Property of Equations is used to remove a coefficient by multiplying each side of the equation by the reciprocal of the coefficient. Because division is defined in terms of multiplication, each side of an equation can be divided by the same nonzero number without changing the solution of the equation.

Take Note

When using the Multiplication Property of Equations, multiply each side of the equation by the reciprocal of the coefficient when the coefficient is a fraction. Divide each side of the equation by the coefficient when the coefficient is an integer or a decimal.

For example, to solve the equation $-5q = 120$, divide each side of the equation by the coefficient -5.

$$-5q = 120$$
$$\frac{-5q}{-5} = \frac{120}{-5}$$
$$q = -24$$

Now the variable is alone on one side of the equation and a constant (-24) is on the other side. The solution is the constant. The solution is -24.

2.2B Solve general
2.2C equations

In solving more complicated first-degree equations in one variable, use the following sequence of steps.

Steps for Solving a First-Degree Equation in One Variable

1. Use the Distributive Property to remove parentheses.
2. Combine any like terms on the right side of the equation and any like terms on the left side of the equation.
3. Use the Addition Property to rewrite the equation with only one variable term.
4. Use the Addition Property to rewrite the equation with only one constant term.
5. Use the Multiplication Property to rewrite the equation with the variable alone on one side of the equation and a constant on the other side of the equation.

If one of these steps is not needed to solve a given equation, proceed to the next step.

EXAMPLE 1

Solve.
a. $5x + 9 = 23 - 2x$
b. $8x - 3(4x - 5) = -2x + 6$

Solution

a.
$$5x + 9 = 23 - 2x$$
$$5x + 2x + 9 = 23 - 2x + 2x \quad \bullet \text{ Step 3}$$
$$7x + 9 = 23$$
$$7x + 9 - 9 = 23 - 9 \quad \bullet \text{ Step 4}$$
$$7x = 14$$
$$\frac{7x}{7} = \frac{14}{7} \quad \bullet \text{ Step 5}$$
$$x = 2$$
The solution is 2.

b.
$$8x - 3(4x - 5) = -2x + 6$$
$$8x - 12x + 15 = -2x + 6 \quad \bullet \text{ Step 1}$$
$$-4x + 15 = -2x + 6 \quad \bullet \text{ Step 2}$$
$$-4x + 2x + 15 = -2x + 2x + 6 \quad \bullet \text{ Step 3}$$
$$-2x + 15 = 6$$
$$-2x + 15 - 15 = 6 - 15 \quad \bullet \text{ Step 4}$$
$$-2x = -9$$
$$\frac{-2x}{-2} = \frac{-9}{-2} \quad \bullet \text{ Step 5}$$
$$x = \frac{9}{2}$$
The solution is $\frac{9}{2}$.

YOU TRY IT 1

Solve.
a. $4x + 3 = 7x + 9$
b. $4 - (5x - 8) = 4x + 3$

Your solution

a. -2
b. 1

Solution on p. S36

OBJECTIVE B *To solve an inequality in one variable*

An **inequality** contains the symbol $>$, $<$, $\geq$, or $\leq$. An inequality expresses the relative order of two mathematical expressions. Here are some examples of inequalities in one variable.

$$\left. \begin{array}{c} 4x \geq 12 \\ 2x + 7 \leq 9 \\ x^2 + 1 > 3x \end{array} \right\} \text{Inequalities in one variable}$$

2.5A   Solve an inequality in one variable

A **solution of an inequality in one variable** is a number that, when substituted for the variable, results in a true inequality. For the inequality $x < 4$ shown below, 3, 0, and -5 are solutions of the inequality because replacing the variable by these numbers results in a true inequality.

$$x < 4 \qquad\qquad x < 4 \qquad\qquad x < 4$$
$$3 < 4 \quad \text{True} \qquad 0 < 4 \quad \text{True} \qquad -5 < 4 \quad \text{True}$$

The number 7 is not a solution of the inequality $x < 4$ because $7 < 4$ is a false inequality. Besides the numbers 3, 0, and -5, there is an infinite number of other solutions of the inequality $x < 4$. Any number less than 4 is a solution; for instance, -5.2, $\frac{5}{2}$, π, and 1 are also solutions of the inequality. The set of all the solutions of an inequality is called the **solution set of the inequality.** The solution set of the inequality $x < 4$ is written in set-builder notation as $\{x \mid x < 4\}$. This is read, "the set of all x such that x is less than 4."

The graph of the solution set of $x < 4$ is shown at the right.

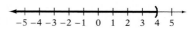

In solving an inequality, the goal is to rewrite the given inequality in the form

$$variable < constant \quad \text{or} \quad variable > constant$$

The Addition Property of Inequalities is used to rewrite an inequality in this form.

Addition Property of Inequalities

The same term can be added to each side of an inequality without changing the solution set of the inequality. Symbolically, this is written

$$\text{If } a < b, \text{ then } a + c < b + c.$$
$$\text{If } a > b, \text{ then } a + c > b + c.$$

This property is also true for an inequality that contains $\leq$ or $\geq$.

The Addition Property of Inequalities is used to remove a term from one side of an inequality by adding the additive inverse of that term to each side of the inequality. Because subtraction is defined in terms of addition, the same number can be subtracted from each side of an inequality without changing the solution set of the inequality.

As shown in HOW TO 1 below, the Addition Property of Inequalities applies to variable terms as well as to constants.

HOW TO 1   Solve $4x - 5 \leq 3x - 2$. Write the solution set in set-builder notation.

$$4x - 5 \leq 3x - 2$$
$$4x - 3x - 5 \leq 3x - 3x - 2 \qquad \bullet \text{ Subtract } 3x \text{ from each side of the inequality.}$$
$$x - 5 \leq -2 \qquad\qquad\qquad \bullet \text{ Simplify.}$$
$$x - 5 + 5 \leq -2 + 5 \qquad\qquad \bullet \text{ Add 5 to each side of the inequality.}$$
$$x \leq 3 \qquad\qquad\qquad\qquad \bullet \text{ Simplify.}$$

The solution set is $\{x \mid x \leq 3\}$.

Unless otherwise noted, all content on this page is © Cengage Learning.

When multiplying or dividing an inequality by a number, the inequality symbol may be reversed, depending on whether the number is positive or negative. Look at the following two examples.

$$3 < 5$$
$$2(3) < 2(5)$$

- Multiply by positive 2. The inequality symbol remains the same.

$$6 < 10$$

- $6 < 10$ is a true statement.

$$3 < 5$$
$$-2(3) > -2(5)$$

- Multiply by negative 2. The inequality symbol is reversed in order to make the inequality a true statement.

$$-6 > -10$$

This is summarized in the Multiplication Property of Inequalities.

The Multiplication Property of Inequalities

Rule 1
If $a > b$ and $c > 0$, then $ac > bc$.
If $a < b$ and $c > 0$, then $ac < bc$.
Rule 2
If $a > b$ and $c < 0$, then $ac < bc$.
If $a < b$ and $c < 0$, then $ac > bc$.

Rule 1 states that when each side of an inequality is multiplied by a *positive* number, the inequality symbol remains the same. Rule 2 states that when each side of an inequality is multiplied by a *negative* number, the inequality symbol must be reversed.

Here are a few more examples of this property.

Rule 1: Multiply by a *positive* number.

$-4 < -2$	$5 > -3$
$-4(2) < -2(2)$	$5(3) > -3(3)$
$-8 < -4$	$15 > -9$

Rule 2: Multiply by a *negative* number.

$7 < 9$	$-2 > -6$
$7(-2) > 9(-2)$	$-2(-3) < -6(-3)$
$-14 > -18$	$6 < 18$

The Multiplication Property of Inequalities is also true for the symbols $\leq$ and $\geq$.

Use the Multiplication Property of Inequalities to remove a coefficient other than 1 from one side of an inequality so that the inequality can be rewritten with the variable alone on one side of the inequality and a constant term on the other side.

Because division is defined in terms of multiplication, when each side of an inequality is divided by a positive number, the inequality symbol remains the same. When each side of an inequality is divided by a negative number, the inequality symbol must be reversed.

Take Note

Solving inequalities in one variable is similar to solving equations in one variable *except* that when you multiply or divide by a negative number, you must reverse the inequality symbol.

HOW TO 2   Solve $-3x < 9$. Write the solution set in set-builder notation.

$$-3x < 9$$

$$\frac{-3x}{-3} > \frac{9}{-3}$$

- Divide each side of the inequality by the coefficient -3 and reverse the inequality symbol.

$$x > -3$$

- Simplify.

$$\{x \mid x > -3\}$$

- Write the answer in set-builder notation.

EXAMPLE 2

Solve $x + 3 > 4x + 6$. Write the solution set in set-builder notation.

Solution

$$x + 3 > 4x + 6$$
$$x - 4x + 3 > 4x - 4x + 6$$
$$-3x + 3 > 6$$
$$-3x + 3 - 3 > 6 - 3$$
$$-3x > 3$$
$$\frac{-3x}{-3} < \frac{3}{-3}$$
$$x < -1$$

- Subtract $4x$ from each side.
- Subtract 3 from each side.
- Divide each side by -3. Reverse the inequality symbol.

$\{x \mid x < -1\}$

Solve $3x - 1 \leq 5x - 7$. Write the solution set in set-builder notation.

Your solution

$\{x \mid x \geq 3\}$

Solution on p. S36

When an inequality contains parentheses, often the first step in solving the inequality is to use the Distributive Property to remove the parentheses.

EXAMPLE 3

Solve $-2(x - 7) > 3 - 4(2x - 3)$. Write the solution set in set-builder notation.

Solution

$$-2(x - 7) > 3 - 4(2x - 3)$$
$$-2x + 14 > 3 - 8x + 12$$

$$-2x + 14 > 15 - 8x$$
$$-2x + 8x + 14 > 15 - 8x + 8x$$
$$6x + 14 > 15$$
$$6x + 14 - 14 > 15 - 14$$
$$6x > 1$$
$$\frac{6x}{6} > \frac{1}{6}$$
$$x > \frac{1}{6}$$

- Use the Distributive Property.
- Combine like terms.
- Add $8x$ to each side.
- Subtract 14 from each side.

- Divide each side by 6.

$\left\{ x \mid x > \dfrac{1}{6} \right\}$

Solve $3 - 2(3x + 1) < 7 - 2x$. Write the solution set in set-builder notation.

Your solution

$\left\{ x \mid x > -\dfrac{3}{2} \right\}$

Solution on p. S36

T.2 EXERCISES

OBJECTIVE A *To solve a first-degree equation in one variable*

For Exercises 1 to 30, solve.

1. $x + 7 = -5$
−12

2. $9 + b = 21$
12

3. $-9 = z - 8$
−1

4. $b - 11 = 11$
22

5. $-48 = 6z$
−8

6. $-9a = -108$
12

7. $-\dfrac{3}{4}x = 15$
−20

8. $\dfrac{5}{2}x = -10$
−4

9. $-\dfrac{x}{4} = -2$
8

10. $\dfrac{2x}{5} = -8$
−20

11. $4 - 2b = 2 - 4b$
−1

12. $4y - 10 = 6 + 2y$
8

13. $5x - 3 = 9x - 7$
1

14. $3m + 5 = 2 - 6m$
$-\dfrac{1}{3}$

15. $6a - 1 = 2 + 2a$
$\dfrac{3}{4}$

16. $5x + 7 = 8x + 5$
$\dfrac{2}{3}$

17. $2 - 6y = 5 - 7y$
3

18. $4b + 15 = 3 - 2b$
−2

19. $2(x + 1) + 5x = 23$
3

20. $9n - 15 = 3(2n - 1)$
4

21. $7a - (3a - 4) = 12$
2

22. $5(3 - 2y) = 3 - 4y$
2

23. $9 - 7x = 4(1 - 3x)$
−1

24. $2(3b + 5) - 1 = 10b + 1$
2

25. $2z - 2 = 5 - (9 - 6z)$
$\dfrac{1}{2}$

26. $4a + 3 = 7 - (5 - 8a)$
$\dfrac{1}{4}$

27. $5(6 - 2x) = 2(5 - 3x)$
5

28. $4(3y + 1) = 2(y - 8)$
−2

29. $2(3b - 5) = 4(6b - 2)$
$-\dfrac{1}{9}$

30. $3(x - 4) = 1 - (2x - 7)$
4

OBJECTIVE B *To solve an inequality in one variable*

For Exercises 31 to 60, solve. Write the answer in set-builder notation.

31. $x - 5 > -2$
$\{x \mid x > 3\}$

32. $5 + n \geq 4$
$\{n \mid n \geq -1\}$

33. $-2 + n \geq 0$
$\{n \mid n \geq 2\}$

34. $x - 3 < 2$
$\{x \mid x < 5\}$

35. $8x \leq -24$
$\{x \mid x \leq -3\}$

36. $-4x < 8$
$\{x \mid x > -2\}$

37. $3n > 0$
$\{n \mid n > 0\}$

38. $-2n \leq -8$
$\{n \mid n \geq 4\}$

39. $2x - 1 > 7$
$\{x \mid x > 4\}$

40. $5x - 2 \leq 8$
$\{x \mid x \leq 2\}$

41. $4 - 3x < 10$
$\{x \mid x > -2\}$

42. $7 - 2x \geq 1$
$\{x \mid x \leq 3\}$

43. $3x - 1 > 2x + 2$
$\{x \mid x > 3\}$

44. $6x + 4 \leq 8 + 5x$
$\{x \mid x \leq 4\}$

45. $8x + 1 \geq 2x + 13$
$\{x \mid x \geq 2\}$

46. $6x + 3 > 4x - 1$
$\{x \mid x > -2\}$

47. $-3 - 4x > -11$
$\{x \mid x < 2\}$

48. $4x - 2 < x - 11$
$\{x \mid x < -3\}$

49. $4x - 2 > 3x + 1$
$\{x \mid x > 3\}$

50. $7x + 5 \leq 9 + 6x$
$\{x \mid x \leq 4\}$

51. $9x + 2 \geq 3x + 14$
$\{x \mid x \geq 2\}$

52. $8x + 1 > 6x - 3$
$\{x \mid x > -2\}$

53. $-5 - 2x > -13$
$\{x \mid x < 4\}$

54. $5x - 3 < x - 11$
$\{x \mid x < -2\}$

55. $4(2x - 1) > 3x - 2(3x - 5)$
$\left\{x \mid x > \dfrac{14}{11}\right\}$

56. $2 - 5(x + 1) \geq 3(x - 1) - 8$
$\{x \mid x \leq 1\}$

57. $3(4x + 3) \leq 7 - 4(x - 2)$
$\left\{x \mid x \leq \dfrac{3}{8}\right\}$

58. $3 + 2(x + 5) \geq x + 5(x + 1) + 1$
$\left\{x \mid x \leq \dfrac{7}{4}\right\}$

59. $3 - 4(x + 2) \leq 6 + 4(2x + 1)$
$\left\{x \mid x \geq -\dfrac{5}{4}\right\}$

60. $12 - 2(3x - 2) \geq 5x - 2(5 - x)$
$\{x \mid x \leq 2\}$

SECTION

T.3 Linear Equations in Two Variables

OBJECTIVE A *To graph points in a rectangular coordinate system*

A **rectangular coordinate system** is formed by two number lines, one horizontal and one vertical. The point of intersection is called the **origin.** The two axes are called the **coordinate axes** or simply the **axes.** Generally, the horizontal axis is labeled the *x*-axis, and the vertical axis is labeled the *y*-axis. In this case, the axes form what is called the *xy*-plane.

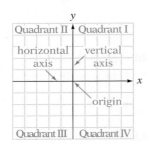

4.1A Points in the plane

Each point in the plane can be identified by a pair of numbers called an **ordered pair.** The first number of the ordered pair measures a horizontal change from the *y*-axis and is called the **abscissa** or *x*-**coordinate.** The second number of the ordered pair measures a vertical change from the *x*-axis and is called the **ordinate** or *y*-**coordinate.** The ordered pair (*x*, *y*) associated with a point is also called the **coordinates** of the point.

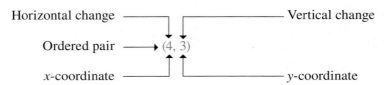

To **graph,** or **plot,** a point means to place a dot at the coordinates of the point. For example, to graph the ordered pair (4, 3), start at the origin. Move 4 units to the right and then 3 units up. Draw a dot. To graph (−3, −4), start at the origin. Move 3 units to the left and then 4 units down. Draw a dot.

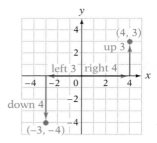

The **graph of an ordered pair** is the dot drawn at the coordinates of the point in the plane. The graphs of the ordered pairs (4, 3) and (−3, −4) are shown above.

The graphs of the points (2, 3) and (3, 2) are different points. The *order* in which the numbers in an *ordered* pair are listed is important.

Unless otherwise noted, all content on this page is © Cengage Learning.

EXAMPLE 1

Graph the ordered pairs $(-4, 2)$, $(3, 4)$, $(0, -1)$, $(2, 0)$, and $(-1, -3)$.

Solution

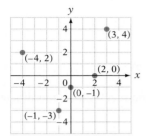

YOU TRY IT 1

Graph the ordered pairs $(-2, 4)$, $(4, 0)$, $(0, 3)$, $(-4, -3)$, and $(5, -1)$.

Your solution

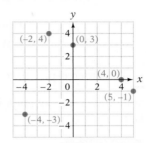

Solution on p. S36

OBJECTIVE B *To graph a linear equation in two variables*

The equations below are examples of equations in two variables.

$$y = 3x - 4$$
$$2x - y = 7$$
$$y = x^2 + 1$$

4.1B Solutions of equations in two variables

A **solution of an equation in two variables** is an ordered pair (x, y) whose coordinates make the equation a true statement.

HOW TO 1 Is the ordered pair $(-4, 9)$ a solution of the equation $y = -2x + 1$?

$y = -2x + 1$

9	$-2(-4) + 1$	• Replace x by -4 and y by 9.
9	$8 + 1$	• Simplify the right side.
$9 = 9$		• Compare the results. If the resulting equation is true, the ordered pair is a solution of the equation. If it is not true, the ordered pair is not a solution of the equation.

Yes, the ordered pair $(-4, 9)$ is a solution of the equation $y = -2x + 1$.

Besides $(-4, 9)$, there are many other ordered pairs that are solutions of the equation $y = -2x + 1$. For example, $(0, 1)$, $(3, -5)$, and $\left(-\frac{3}{2}, 4\right)$ are also solutions. In general, an equation in two variables has an infinite number of solutions. By choosing any value of x and substituting that value into the equation, we can calculate a corresponding value of y.

Unless otherwise noted, all content on this page is © Cengage Learning.

HOW TO 2 Find the ordered-pair solution of $y = \frac{2}{5}x - 4$ that corresponds to $x = 5$.

$$y = \frac{2}{5}x - 4$$

$$y = \frac{2}{5}(5) - 4 \qquad \bullet \text{ We are given that } x = 5. \text{ Replace } x \text{ by 5 in the equation.}$$

$$y = 2 - 4 \qquad \bullet \text{ Simplify the right side.}$$

$$y = -2 \qquad \bullet \text{ When } x = 5, y = -2.$$

The ordered-pair solution of $y = \frac{2}{5}x - 4$ for $x = 5$ is $(5, -2)$.

4.3A Graph an equation of the form $y = mx + b$

Solutions of an equation in two variables can be graphed in the rectangular coordinate system.

HOW TO 3 Graph the ordered-pair solutions of $y = -2x + 1$ for $x = -2, -1,$ 0, 1, and 2.

x	$y = -2x + 1$	y	(x, y)
-2	$-2(-2) + 1$	5	$(-2, 5)$
-1	$-2(-1) + 1$	3	$(-1, 3)$
0	$-2(0) + 1$	1	$(0, 1)$
1	$-2(1) + 1$	-1	$(1, -1)$
2	$-2(2) + 1$	-3	$(2, -3)$

• Use the given values of x to determine ordered-pair solutions of the equation. It is convenient to record these in a table.

The ordered-pair solutions of $y = -2x + 1$ for $x = -2, -1, 0, 1,$ and 2 are $(-2, 5)$, $(-1, 3)$, $(0, 1)$, $(1, -1)$, and $(2, -3)$. These are graphed at the right.

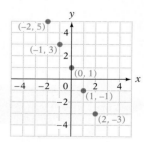

The **graph of an equation in two variables** is a graph of all the ordered-pair solutions of the equation. Consider the equation $y = -2x + 1$ above. The ordered-pair solutions $(-2, 5)$, $(-1, 3)$, $(0, 1)$, $(1, -1)$, and $(2, -3)$ are graphed in the figure above. We can choose values of x that are not integers to produce more ordered pairs to graph, such as $\left(\frac{5}{2}, -4\right)$ and $\left(2\frac{1}{2}, 2\right)$. Choosing still other values of x would result in more and more ordered pairs being graphed. The result would be so many dots that the graph would appear as a straight line, as shown below. This is the graph of $y = -2x + 1$.

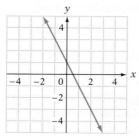

Unless otherwise noted, all content on this page is © Cengage Learning.

The equation $y = -2x + 1$ is an example of a linear equation because its graph is a straight line. It is also called a first-degree equation in two variables because the exponent on each variable is 1.

 Take Note

The equation
$y = x^2 + 2x - 4$ is not
a linear equation in two
variables because there is a
term with a variable squared.

The equation $y = \dfrac{5}{x - 3}$ is
not a linear equation because
a variable occurs in the
denominator of a fraction.

> ### Linear Equation in Two Variables
>
> An equation of the form $y = mx + b$, where m is the coefficient of x and b is a constant, is a linear equation in two variables. The graph of a linear equation in two variables is a straight line.

Examples of linear equations are shown at the right.

$$y = 5x + 3 \qquad (m = 5, b = 3)$$
$$y = x - 4 \qquad (m = 1, b = -4)$$
$$y = -\frac{3}{4}x \qquad \left(m = -\frac{3}{4}, b = 0\right)$$

To graph a linear equation, find ordered-pair solutions of the equation. Do this by choosing any value of x and finding the corresponding value of y. Repeat this procedure, choosing different values for x, until you have found the number of solutions desired. Because the graph of a linear equation in two variables is a straight line, and a straight line is determined by two points, it is necessary to find only two solutions. However, it is recommended that at least three points be used to ensure accuracy.

HOW TO 4   Graph: $y = 2x - 3$

x	$y = 2x - 3$	y
0	$2(0) - 3$	-3
2	$2(2) - 3$	1
-1	$2(-1) - 3$	-5

• Choose any values of x.
  Then find the corresponding
  values of y. The numbers
  0, 2, and -1 were chosen
  arbitrarily for x.

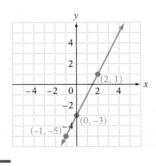

• Graph the ordered-pair
  solutions $(0, -3)$, $(2, 1)$, and
  $(-1, -5)$. Draw a straight
  line through the points.

Remember that a graph is a drawing of the ordered-pair solutions of an equation. Therefore, every point on the graph is a solution of the equation, and every solution of the equation is a point on the graph.

When graphing an equation of the form $y = mx + b$, if m is a fraction, choose values of x that will simplify the evaluation. This is illustrated in Example 2. Note that the values of x chosen are multiples of the denominator, 2.

Unless otherwise noted, all content on this page is © Cengage Learning.

EXAMPLE 2

Graph $y = -\dfrac{3}{2}x - 3$.

Solution

x	y
0	-3
-2	0
-4	3

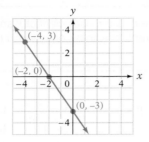

YOU TRY IT 2

Graph $y = \dfrac{3}{5}x - 4$.

Your solution

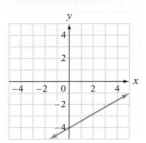

Solution on p. S36

4.3B Graph an equation of the form $Ax + By = C$

An equation of the form $Ax + By = C$ is also a linear equation in two variables. Examples of these equations are shown below.

$$3x + 4y = 12 \qquad (A = 3, B = 4, C = 12)$$
$$x - 5y = -10 \qquad (A = 1, B = -5, C = -10)$$
$$2x - y = 0 \qquad (A = 2, B = -1, C = 0)$$

One method of graphing an equation of the form $Ax + By = C$ involves first solving the equation for y and then following the same procedure used for graphing an equation of the form $y = mx + b$. To solve the equation for y, rewrite the equation so that y is alone on one side of the equation and the term containing x and the constant are on the other side of the equation. The Addition and Multiplication Properties of Equations are used to rewrite an equation of the form $Ax + By = C$ in the form $y = mx + b$.

HOW TO 5   Graph $3x + 2y = 6$.

$$3x + 2y = 6$$
- The equation is in the form $Ax + By = C$.

$$2y = -3x + 6$$
- Solve the equation for y. Subtract $3x$ from each side of the equation.

$$y = -\dfrac{3}{2}x + 3$$
- Divide each side of the equation by 2. Note that each term on the right side is divided by 2.

- Find at least three solutions.

x	y
0	3
2	0
4	-3

Unless otherwise noted, all content on this page is © Cengage Learning.

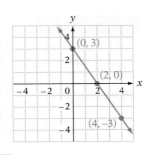

● Graph the ordered pairs (0, 3), (2, 0), and (4, −3). Draw a straight line through the points.

EXAMPLE 3

Graph $2x + 3y = 9$.

Solution

$$2x + 3y = 9$$
$$3y = -2x + 9$$
$$y = -\frac{2}{3}x + 3$$

x	y
−3	5
0	3
3	1

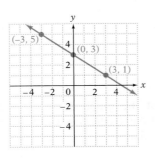

YOU TRY IT 3

Graph $-3x + 2y = 4$.

Your solution

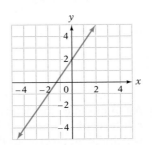

Solution on p. S36

4.4A Slope

The **slope** of a line is a measure of the slant of the line. The symbol for slope is *m*.

For an equation of the form $y = mx + b$, *m* is the slope. Here are a few examples:

The slope of the line $y = -2x + 5$ is -2.

The slope of the line $y = 8x$ is 8.

The slope of the line $y = \frac{3}{4}x - 1$ is $\frac{3}{4}$.

The slope of a line containing two points is the ratio of the change in the *y* values between the two points to the change in the *x* values.

📋 **Take Note**

$$\text{Slope} = m = \frac{\text{change in } y}{\text{change in } x}$$

Slope Formula

The slope of the line containing the two points $P_1(x_1, y_1)$ and $P_2(x_2, y_2)$ is given by

$$m = \frac{y_2 - y_1}{x_2 - x_1}, \ x_1 \neq x_2$$

Unless otherwise noted, all content on this page is © Cengage Learning.

EXAMPLE 4

Find the slope of the line containing the points $(-4, -3)$ and $(-1, 1)$.

Solution

Let $(x_1, y_1) = (-4, -3)$ and $(x_2, y_2) = (-1, 1)$.

$$m = \frac{y_2 - y_1}{x_2 - x_1} = \frac{1 - (-3)}{-1 - (-4)} = \frac{4}{3}$$

The slope is $\frac{4}{3}$.

YOU TRY IT 4

Find the slope of the line containing the points $(-2, 3)$ and $(1, -3)$.

Your solution

-2

Solution on p. S36

4.4B Slope-intercept form of a straight line

An important characteristic of the graph of a linear equation is its *intercepts*. An **x-intercept** is a point at which the graph crosses the x-axis. A **y-intercept** is a point at which the graph crosses the y-axis.

The graph of the equation $y = \frac{1}{2}x - 2$ is shown at the right. The x-intercept of the graph is $(4, 0)$. The y-intercept of the graph is $(0, -2)$.

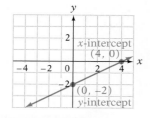

Note that at an x-intercept, the y-coordinate is 0. At a y-intercept, the x-coordinate is 0.

We can find the y-intercept of a linear equation by letting $x = 0$.

HOW TO 6 Find the y-intercept of the graph of the equation $y = 3x + 4$.

$$y = 3x + 4$$
$$y = 3(0) + 4 \qquad \bullet \text{ To find the } y\text{-intercept, let } x = 0.$$
$$y = 0 + 4$$
$$y = 4$$

The y-intercept is $(0, 4)$.

Note that the constant term of $y = mx + b$ is b, and the y-intercept is $(0, b)$.

In general, **for any equation of the form $y = mx + b$, the y-intercept is $(0, b)$.**

Because the slope and the y-intercept can be determined directly from the equation $y = mx + b$, this equation is called the *slope-intercept form of a straight line*.

Unless otherwise noted, all content on this page is © Cengage Learning.

Slope-Intercept Form of a Straight Line

The equation $y = mx + b$ is called the **slope-intercept form of a straight line.** The slope of the line is m, the coefficient of x. The y-intercept is $(0, b)$.

The following equations are written in slope-intercept form.

$$y = -4x + 3 \qquad \text{Slope} = -4, \ y\text{-intercept} = (0, 3)$$

$$y = \frac{2}{5}x - 1 \qquad \text{Slope} = \frac{2}{5}, \ y\text{-intercept} = (0, -1)$$

$$y = -x \qquad \text{Slope} = -1, \ y\text{-intercept} = (0, 0)$$

When an equation is in slope-intercept form, it is possible to quickly draw a graph of the function.

 Take Note

When graphing a line by using its slope and y-intercept, *always* start at the y-intercept.

Take Note

Recall that slope =
$m = \dfrac{\text{change in } y}{\text{change in } x}$.
For HOW TO 7 at the right, $m = -\dfrac{1}{2} = \dfrac{-1}{2} = \dfrac{\text{change in } y}{\text{change in } x}$. Therefore, the change in y is -1 and the change in x is 2.

HOW TO 7   Graph $x + 2y = 4$ by using the slope and y-intercept.

Solve the equation for y.

$$x + 2y = 4$$
$$2y = -x + 4$$
$$y = -\frac{1}{2}x + 2$$

From the equation $y = -\frac{1}{2}x + 2$, the slope is $-\frac{1}{2}$ and the y-intercept is $(0, 2)$.

Rewrite the slope $-\frac{1}{2}$ as $\frac{-1}{2}$.

Beginning at the y-intercept, move down 1 unit (change in y) and then right 2 units (change in x).

The point whose coordinates are $(2, 1)$ is a second point on the graph. Draw a straight line through the points $(0, 2)$ and $(2, 1)$.

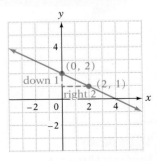

EXAMPLE 5

Graph $y = -\frac{3}{2}x + 4$ by using the slope and the y-intercept.

YOU TRY IT 5

Graph $y = -\frac{2}{3}x + 2$ by using the slope and the y-intercept.

Unless otherwise noted, all content on this page is © Cengage Learning.

Solution

From the equation $y = -\frac{3}{2}x + 4$, the slope is $-\frac{3}{2}$ and the y-intercept is $(0, 4)$.

Rewrite the slope $-\frac{3}{2}$ as $\frac{-3}{2}$.

Place a dot at the y-intercept.

Starting at the y-intercept, move down 3 units (the change in y) and right 2 units (the change in x). Place a dot at that location.

Draw a line through the two points.

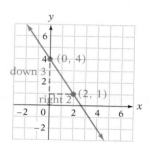

Your solution

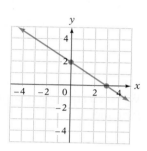

Solution on p. S37

OBJECTIVE C   *To evaluate a function*

4.2A Evaluate a function

A **relation** is a set of ordered pairs. A **function** is a relation in which no two ordered pairs have the same first coordinate.

The relation {(0, 4), (1, 5), (1, 6), (2, 7)} is not a function because the ordered pairs (1, 5), and (1, 6) have the same first coordinate.

The relation {(0, 4), (1, 5), (2, 6), (3, 7)} is a function.

The phrase "y is a function of x," or a similar phrase with different variables, is used to describe those equations in two variables that define functions. To emphasize that the equation represents a function, function notation is used. For example, the square function is written in function notation as follows:

$$f(x) = x^2$$

The process of determining $f(x)$ for a given value of x is called **evaluating the function.** For instance, to evaluate $f(x) = x^2$ when $x = 3$, replace x by 3 and simplify.

$$f(x) = x^2$$
$$f(3) = 3^2 = 9$$

When $x = 3$, the value of the function is 9. An ordered pair of the function is $(3, 9)$.

Take Note

The symbol $f(x)$ is read "the value of f at x" or "f of x."

Unless otherwise noted, all content on this page is © Cengage Learning.

EXAMPLE 6

Evaluate $f(x) = 2x - 4$ when $x = 3$. Use your answer to write an ordered pair of the function.

Solution

$f(x) = 2x - 4$
$f(3) = 2(3) - 4$ • $x = 3$
$f(3) = 6 - 4$
$f(3) = 2$

An ordered pair of the function is $(3, 2)$.

YOU TRY IT 6

Evaluate the function $f(x) = 4 - 2x$ at $x = -3$. Use your answer to write an ordered pair of the function.

Your solution

$f(-3) = 10; (-3, 10)$

Solution on p. S37

OBJECTIVE D   *To find the equation of a line*

4.5A Find the equation of a line given a point and the slope

When the slope of a line and a point on the line are known, the equation of the line can be determined by using the point-slope formula.

> **Point-Slope Formula**
>
> Let m be the slope of a line, and let (x_1, y_1) be the coordinates of a point on the line. The equation of the line can be found using the point-slope formula:
> $$y - y_1 = m(x - x_1)$$

EXAMPLE 7

Find the equation of the line that contains the point $(1, -3)$ and has slope -2.

Solution

$\begin{aligned}
y - y_1 &= m(x - x_1) \\
y - (-3) &= -2(x - 1) \quad \text{• } x_1 = 1, y_1 = -3, m = -2 \\
y + 3 &= -2x + 2 \\
y &= -2x - 1
\end{aligned}$

The equation of the line is $y = -2x - 1$.

YOU TRY IT 7

Find the equation of the line that contains the point $(-2, 2)$ and has slope $-\frac{1}{2}$.

Your solution

$y = -\frac{1}{2}x + 1$

Solution on p. S37

T.3 EXERCISES

OBJECTIVE A *To graph points in a rectangular coordinate system*

For Exercises 1 to 4, graph the ordered pairs.

1. $(2, 3), (4, 0), (-4, 1), (-2, -2)$

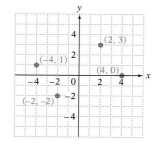

2. $(0, 2), (-4, -1), (2, 0), (1, -3)$

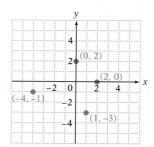

3. $(-2, 5), (3, 4), (0, 0), (-3, -2)$

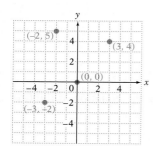

4. $(1, -4), (-2, 0), (-1, -5), (0, 4)$

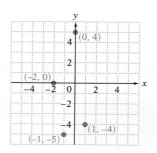

OBJECTIVE B *To graph a linear equation in two variables*

For Exercises 5 to 22, graph.

5. $y = 2x + 1$

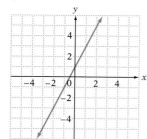

6. $y = 2x - 4$

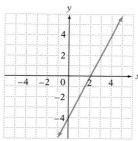

7. $y = -3x + 4$

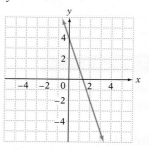

8. $y = -4x + 1$

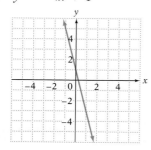

9. $y = 3x$

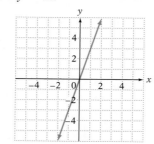

10. $y = 2x$

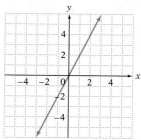

Unless otherwise noted, all content on this page is © Cengage Learning.

11. $y = -\dfrac{4}{3}x$

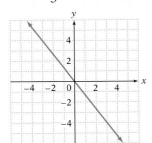

12. $y = -\dfrac{5}{2}x$

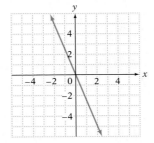

13. $y = \dfrac{3}{2}x - 1$

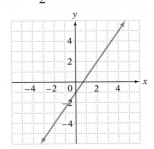

14. $y = \dfrac{2}{3}x + 1$

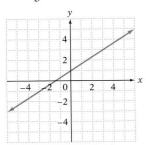

15. $y = -\dfrac{2}{3}x + 1$

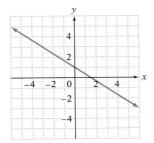

16. $y = -\dfrac{1}{2}x + 3$

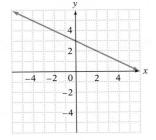

17. $2x + y = -3$

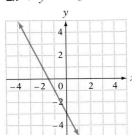

18. $2x - y = 3$

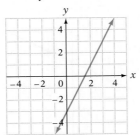

19. $x - 4y = 8$

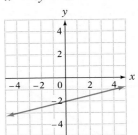

20. $2x + 5y = 10$

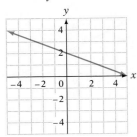

21. $3x - 2y = 8$

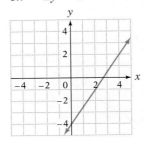

22. $3x - y = -2$

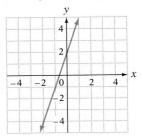

Unless otherwise noted, all content on this page is © Cengage Learning.

For Exercises 23 to 34, find the slope of the line containing the given points.

23. $P_1(2, 1)$, $P_2(3, 4)$

3

24. $P_1(4, 2)$, $P_2(3, 4)$

-2

25. $P_1(-2, 1)$, $P_2(2, 2)$

$\dfrac{1}{4}$

26. $P_1(-1, 3)$, $P_2(2, 4)$

$\dfrac{1}{3}$

27. $P_1(1, 3)$, $P_2(5, -3)$

$-\dfrac{3}{2}$

28. $P_1(2, 4)$, $P_2(4, -1)$

$-\dfrac{5}{2}$

29. $P_1(-1, 2)$, $P_2(-1, 3)$
Undefined

30. $P_1(3, -4)$, $P_2(3, 5)$
Undefined

31. $P_1(5, 1)$, $P_2(-2, 1)$
0

32. $P_1(4, -2)$, $P_2(3, -2)$

0

33. $P_1(3, 0)$, $P_2(2, -1)$

1

34. $P_1(0, -1)$, $P_2(3, -2)$

$-\dfrac{1}{3}$

For Exercises 35 to 37, give the slope and the y-intercept of the graph of the equation.

35. $y = \dfrac{5}{2}x - 4$

$m = \dfrac{5}{2}$, $b = (0, -4)$

36. $y = -3x + 7$

$m = -3$, $b = (0, 7)$

37. $y = x$

$m = 1$, $b = (0, 0)$

For Exercises 38 to 46, graph by using the slope and the y-intercept.

38. $y = \dfrac{2}{3}x - 3$

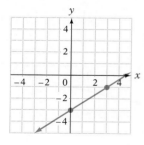

39. $y = \dfrac{1}{2}x + 2$

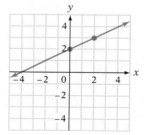

40. $y = \dfrac{3}{4}x$

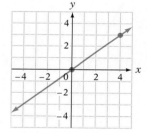

41. $y = -\dfrac{3}{2}x$

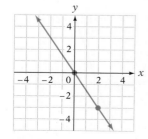

42. $y = \dfrac{2}{3}x - 1$

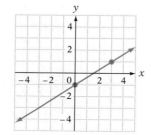

43. $y = -\dfrac{1}{2}x + 2$

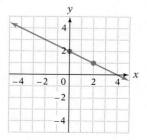

Unless otherwise noted, all content on this page is © Cengage Learning.

44. $3x + 2y = 8$

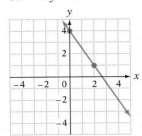

45. $x - 3y = 3$

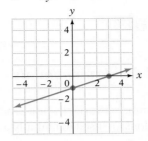

46. $4x + y = 2$

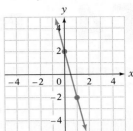

OBJECTIVE C   *To evaluate a function*

For Exercises 47 to 55, evaluate the function for the given value of the variable. Use your answer to write an ordered pair of the function.

47. $g(x) = -3x + 1; x = -4$
13; (−4, 13)

48. $f(x) = 2x + 7; x = -2$
3; (−2, 3)

49. $p(x) = 6 - 8x; x = -1$
14; (−1, 14)

50. $h(x) = 4x - 2; x = 3$
10; (3, 10)

51. $f(t) = t^2 - t - 3; t = 2$
−1; (2, −1)

52. $t(x) = 5 - 7x; x = 0$
5; (0, 5)

53. $h(x) = -3x^2 + x - 1; x = -2$
−15; (−2, −15)

54. $p(n) = n^2 - 4n - 7; n = -3$
14; (−3, 14)

55. $g(t) = 4t^3 - 2t; t = -1$
−2; (−1, −2)

OBJECTIVE D   *To find the equation of a line*

For Exercises 56 to 64, find the equation of the line that contains the given point and has the given slope.

56. $(2, -3); m = 3$
$y = 3x - 9$

57. $(-1, 2); m = -3$
$y = -3x - 1$

58. $(-3, 5); m = 3$
$y = 3x + 14$

59. $(4, -5); m = -2$
$y = -2x + 3$

60. $(3, 1); m = \dfrac{1}{3}$
$y = \dfrac{1}{3}x$

61. $(5, -3); m = -\dfrac{3}{5}$
$y = -\dfrac{3}{5}x$

62. $(4, -2); m = \dfrac{3}{4}$
$y = \dfrac{3}{4}x - 5$

63. $(-3, -2); m = -\dfrac{2}{3}$
$y = -\dfrac{2}{3}x - 4$

64. $(2, 3); m = -\dfrac{1}{2}$
$y = -\dfrac{1}{2}x + 4$

Unless otherwise noted, all content on this page is © Cengage Learning.

T.4 Polynomials

OBJECTIVE A — *To multiply and divide monomials*

6.1A Multiply and
6.1B divide monomials

A **monomial** is a number, a variable, or a product of a number and variables. The following rules and definitions are used to multiply and divide monomials and to write monomials in simplest form.

> ### Rule for Multiplying Exponential Expressions
>
> If m and n are integers, then $x^m \cdot x^n = x^{m+n}$.
>
> ### Rule for Simplifying Powers of Exponential Expressions
>
> If m and n are integers, then $(x^m)^n = x^{mn}$.
>
> ### Rule for Simplifying Powers of Products
>
> If m, n, and p are integers, then $(x^m y^n)^p = x^{mp} y^{np}$.
>
> ### Rule for Dividing Exponential Expressions
>
> If m and n are integers and $x \neq 0$, then $\dfrac{x^m}{x^n} = x^{m-n}$.
>
> ### Rule for Simplifying Powers of Quotients
>
> If m, n, and p are integers and $y \neq 0$, then $\left(\dfrac{x^m}{y^n}\right)^p = \dfrac{x^{mp}}{y^{np}}$.
>
> ### Definition of Zero as an Exponent
>
> If $x \neq 0$, then $x^0 = 1$. The expression 0^0 is undefined.
>
> ### Definition of a Negative Exponent
>
> If n is a positive integer and $x \neq 0$, then $x^{-n} = \dfrac{1}{x^n}$ and $\dfrac{1}{x^{-n}} = x^n$.

HOW TO 1   Simplify: $(3x^4)^2(4x^3)$

$$(3x^4)^2(4x^3) = (3^{1\cdot2}x^{4\cdot2})(4x^3)$$

- Use the Rule for Simplifying Powers of Products to simplify $(3x^4)^2$.

$$= (3^2 x^8)(4x^3)$$
$$= (9x^8)(4x^3)$$
$$= (9 \cdot 4)(x^8 \cdot x^3)$$
$$= 36x^{8+3}$$

- Use the Rule for Multiplying Exponential Expressions.

$$= 36x^{11}$$

An exponential expression is in simplest form when there are no negative exponents in the expression. For example, the expression y^{-7} is not in simplest form; use the Definition of a Negative Exponent to rewrite the expression with a positive exponent: $y^{-7} = \frac{1}{y^7}$. The expression $\frac{1}{c^{-4}}$ is not in simplest form; use the Definition of a Negative Exponent to rewrite the expression with a positive exponent: $\frac{1}{c^{-4}} = c^4$.

HOW TO 2   Simplify: $\dfrac{6x^2}{8x^9}$

$\dfrac{6x^2}{8x^9} = \dfrac{3x^2}{4x^9} = \dfrac{3x^{2-9}}{4}$

- Divide the coefficients by their common factors. Use the Rule for Dividing Exponential Expressions.

$= \dfrac{3x^{-7}}{4} = \dfrac{3}{4} \cdot \dfrac{x^{-7}}{1} = \dfrac{3}{4} \cdot \dfrac{1}{x^7}$

- Rewrite the expression with only positive exponents.

$= \dfrac{3}{4x^7}$

HOW TO 3   Simplify: $\left(\dfrac{a^4}{b^3}\right)^{-2}$

$\left(\dfrac{a^4}{b^3}\right)^{-2} = \dfrac{a^{4(-2)}}{b^{3(-2)}} = \dfrac{a^{-8}}{b^{-6}}$

- Use the Rule for Simplifying Powers of Quotients.

$= \dfrac{b^6}{a^8}$

- Rewrite the expression with positive exponents.

EXAMPLE 1

Simplify.
a. $(-2x)(3x^{-2})^{-3}$

b. $\left(\dfrac{3a^2b^{-1}}{27a^{-3}b^{-4}}\right)^{-2}$

Solution

a. $(-2x)(3x^{-2})^{-3}$

$= (-2x)(3^{-3}x^6)$

$= \dfrac{-2x \cdot x^6}{3^3}$

$= -\dfrac{2x^7}{27}$

- Use the Rule for Simplifying Powers of Products.
- Write the expression with positive exponents.
- Use the Rule for Multiplying Exponential Expressions. Simplify 3^3.

YOU TRY IT 1

Simplify.
a. $(-2ab)(2a^3b^{-2})^{-3}$

b. $\left(\dfrac{2x^2y^{-4}}{4x^{-2}y^{-5}}\right)^{-3}$

Your solution

a. $-\dfrac{b^7}{4a^8}$

b. $\dfrac{8}{x^{12}y^3}$

b. Use the Rule for Simplifying Powers of
Quotients. Then simplify the expression and
write it with positive exponents.

$$\left(\frac{3a^2b^{-1}}{27a^{-3}b^{-4}}\right)^{-2}$$

$$= \left(\frac{a^2b^{-1}}{9a^{-3}b^{-4}}\right)^{-2}$$

$$= \frac{a^{2(-2)}b^{(-1)(-2)}}{9^{1(-2)}a^{(-3)(-2)}b^{(-4)(-2)}}$$

$$= \frac{a^{-4}b^2}{9^{-2}a^6b^8}$$

$$= 9^2a^{-4-6}b^{2-8}$$

$$= 81a^{-10}b^{-6} = \frac{81}{a^{10}b^6}$$

Solution on p. S37

OBJECTIVE B *To add and subtract polynomials*

A **polynomial** is a variable expression in which the terms are monomials. The polynomial
$15t^2 - 2t + 3$ has three terms: $15t^2$, $-2t$, and 3. Note that each of these three terms is
a monomial.

A polynomial of *one* term is a **monomial.** $-7x^2$ is a monomial.
A polynomial of *two* terms is a **binomial.** $4y + 3$ is a binomial.
A polynomial of *three* terms is a **trinomial.** $6b^2 + 5b - 8$ is a trinomial.

6.2B Add polynomials

Polynomials can be added by combining like terms. This is illustrated in Example 2
below.

EXAMPLE 2

Add: $(8x^2 - 4x - 9) + (2x^2 + 9x - 9)$

Solution

$(8x^2 - 4x - 9) + (2x^2 + 9x - 9)$
$= (8x^2 + 2x^2) + (-4x + 9x) + (-9 - 9)$
$= 10x^2 + 5x - 18$

YOU TRY IT 2

Add: $(-4x^3 + 2x^2 - 8) + (4x^3 + 6x^2 - 7x + 5)$

Your solution

$8x^2 - 7x - 3$

Solution on p. S37

6.2B Subtract polynomials

The **additive inverse of the polynomial** $(3x^2 - 7x + 8)$ is $-(3x^2 - 7x + 8)$.

To find the additive inverse of a polynomial, change the sign of each term inside the parentheses.

$$-(3x^2 - 7x + 8) = -3x^2 + 7x - 8$$

To subtract two polynomials, add the additive inverse of the second polynomial to the first.

HOW TO 4   Simplify: $(5a^2 - a + 2) - (-2a^3 + 3a - 3)$

$(5a^2 - a + 2) - (-2a^3 + 3a - 3)$

$= (5a^2 - a + 2) + (2a^3 - 3a + 3)$ • Rewrite subtraction as addition of the additive inverse.

$= 2a^3 + 5a^2 - 4a + 5$ • Combine like terms.

EXAMPLE 3

Subtract: $(7c^2 - 9c - 12) - (9c^2 + 5c - 8)$

Solution

$(7c^2 - 9c - 12) - (9c^2 + 5c - 8)$
$= (7c^2 - 9c - 12) + (-9c^2 - 5c + 8)$
$= -2c^2 - 14c - 4$

YOU TRY IT 3

Subtract:
$(-4w^3 + 8w - 8) - (3w^3 - 4w^2 + 2w - 1)$

Your solution

$-7w^3 + 4w^2 + 6w - 7$

Solution on p. S37

OBJECTIVE C   *To multiply polynomials*

6.3A Multiply a polynomial by a monomial

The Distributive Property is used to multiply a polynomial by a monomial. Each term of the polynomial is multiplied by the monomial.

HOW TO 5   Multiply: $3x^3(4x^4 - 2x + 5)$

$3x^3(4x^4 - 2x + 5)$

$= 3x^3(4x^4) - 3x^3(2x) + 3x^3(5)$ • Use the Distributive Property. Multiply each term of the polynomial by $3x^3$.

$= 12x^7 - 6x^4 + 15x^3$ • Use the Rule for Multiplying Exponential Expressions.

EXAMPLE 4

Multiply: $2xy(3x^2 - xy + 2y^2)$

Solution

$2xy(3x^2 - xy + 2y^2)$
$= 2xy(3x^2) - 2xy(xy) + 2xy(2y^2)$
$= 6x^3y - 2x^2y^2 + 4xy^3$

YOU TRY IT 4

Multiply: $3mn^2(2m^2 - 3mn - 1)$

Your solution

$6m^3n^2 - 9m^2n^3 - 3mn^2$

Solution on p. S37

6.3B Multiply polynomials

A vertical format similar to that used for multiplication of whole numbers is used to multiply two polynomials. The product $(2y - 3)(y^2 + 2y + 5)$ is shown below.

$$\begin{array}{rrrrl} & y^2 & + 2y & + 5 & \\ & & 2y & - 3 & \\ \hline & -3y^2 & - 6y & - 15 & \text{This is } -3(y^2 + 2y + 5). \\ 2y^3 & + 4y^2 & + 10y & & \text{This is } 2y(y^2 + 2y + 5). \text{ Like terms are placed in the same column.} \\ \hline 2y^3 & + y^2 & + 4y & - 15 & \text{Add the terms in each column.} \end{array}$$

EXAMPLE 5

Multiply: $(2a^2 + 4a - 5)(3a + 5)$

Solution

$$\begin{array}{rrrr} & 2a^2 & + 4a & - 5 \\ & & 3a & + 5 \\ \hline & 10a^2 & + 20a & - 25 \\ 6a^3 & + 12a^2 & - 15a & \\ \hline 6a^3 & + 22a^2 & + 5a & - 25 \end{array}$$

- **Align like terms in the same column.**

YOU TRY IT 5

Multiply: $(3c^2 - 4c + 5)(2c - 3)$

Your solution

$6c^3 - 17c^2 + 22c - 15$

Solution on p. S37

6.3B Multiply two binomials

It is frequently necessary to multiply two binomials. The product is computed by using a method called FOIL. The letters of FOIL stand for **F**irst, **O**uter, **I**nner, and **L**ast. The FOIL method is based on the Distributive Property and involves adding the products of the first terms, the outer terms, the inner terms, and the last terms.

The product $(2x + 3)(3x + 4)$ is shown below using FOIL.

First, Last, Inner, Outer

	First terms	Outer terms	Inner terms	Last terms
$(2x + 3) \cdot (3x + 4) =$	$(2x)(3x)$	$+ (2x)(4)$	$+ (3)(3x)$	$+ (3)(4)$

$$= 6x^2 + 8x + 9x + 12$$

$$= 6x^2 + 17x + 12$$

EXAMPLE 6

Multiply: $(4x - 3)(2x + 5)$

Solution

$$\begin{aligned} &(4x - 3)(2x + 5) \\ &\quad = (4x)(2x) + (4x)(5) + (-3)(2x) + (-3)(5) \\ &\quad = 8x^2 + 20x + (-6x) + (-15) \\ &\quad = 8x^2 + 14x - 15 \end{aligned}$$

YOU TRY IT 6

Multiply: $(4y - 7)(3y - 5)$

Your solution

$12y^2 - 41y + 35$

Solution on p. S37

OBJECTIVE D

6.4B Divide polynomials

To divide polynomials

To divide two polynomials, use a method similar to that used for division of whole numbers.

To divide $(x^2 - 5x + 8) \div (x - 3)$:

Step 1

$$\begin{array}{r} x \\ x - 3 \overline{)x^2 - 5x + 8} \\ \underline{x^2 - 3x} \downarrow \\ -2x + 8 \end{array}$$

Think: $x\overline{)x^2} = \dfrac{x^2}{x} = x$

Multiply: $x(x - 3) = x^2 - 3x$

Subtract: $(x^2 - 5x) - (x^2 - 3x) = -2x$

Bring down the $+8$.

Step 2

$$\begin{array}{r} x - 2 \\ x - 3 \overline{)x^2 - 5x + 8} \\ \underline{x^2 - 3x} \\ -2x + 8 \\ \underline{-2x + 6} \\ 2 \end{array}$$

Think: $x\overline{)-2x} = \dfrac{-2x}{x} = -2$

Multiply: $-2(x - 3) = -2x + 6$

Subtract: $(-2x + 8) - (-2x + 6) = 2$

The remainder is 2.

Check: (Quotient $\times$ Divisor) + Remainder = Dividend

$$(x - 2)(x - 3) + 2 = x^2 - 3x - 2x + 6 + 2 = x^2 - 5x + 8$$

$$(x^2 - 5x + 8) \div (x - 3) = x - 2 + \frac{2}{x - 3}$$

EXAMPLE 7

Divide: $(6x + 2x^3 + 26) \div (x + 2)$

Solution

Arrange the terms of the dividend in descending order. There is no x^2 term in $2x^3 + 6x + 26$. Insert $0x^2$ for the missing term so that like terms will be in the same columns.

$$\begin{array}{r} 2x^2 - 4x + 14 \\ x + 2 \overline{)2x^3 + 0x^2 + 6x + 26} \\ \underline{2x^3 + 4x^2} \\ -4x^2 + 6x \\ \underline{-4x^2 - 8x} \\ 14x + 26 \\ \underline{14x + 28} \\ -2 \end{array}$$

Check: $(2x^2 - 4x + 14)(x + 2) - 2$

$= 2x^3 + 6x + 28 - 2 = 2x^3 + 6x + 26$

$(6x + 2x^3 + 26) \div (x + 2) = 2x^2 - 4x + 14 - \dfrac{2}{x + 2}$

YOU TRY IT 7

Divide: $(x^3 - 7 - 2x) \div (x - 2)$

Your solution

$x^2 + 2x + 2 - \dfrac{3}{x - 2}$

Solution on p. S37

OBJECTIVE E *To factor polynomials of the form $ax^2 + bx + c$*

7.1A Factor a monomial from a polynomial

A polynomial is in factored form when it is written as a product of other polynomials. Factoring can be thought of as the reverse of multiplication.

Polynomial   **Factored Form**

$$2x^3 + 6x^2 - 10x = 2x(x^2 + 3x - 5)$$

$$x^2 - 3x - 28 \quad = (x + 4)(x - 7)$$

To factor out a common monomial from the terms of a polynomial, first find the greatest common factor (GCF) of the terms.

The GCF of two or more monomials is the product of the GCF of the coefficients and the common variable factors.

$$10a^3b = \mathbf{2} \cdot 5 \cdot \mathbf{a} \cdot \mathbf{a} \cdot a \cdot \mathbf{b}$$
$$4a^2b^2 = \mathbf{2} \cdot 2 \cdot \mathbf{a} \cdot \mathbf{a} \cdot \mathbf{b} \cdot b$$
$$\text{GCF} = \mathbf{2} \cdot \mathbf{a} \cdot \mathbf{a} \cdot \mathbf{b} = 2a^2b$$

Note that the exponent on each variable in the GCF is the same as the smallest exponent on that variable in either of the monomials.

The GCF of $10a^3b$ and $4a^2b^2$ is $2a^2b$.

HOW TO 6   Factor: $5x^3 - 35x^2 + 10x$

The GCF is $5x$.

• Find the GCF of the terms $5x^3$, $-35x^2$, and $10x$.

$$\frac{5x^3 - 35x^2 + 10x}{5x}$$
$$= x^2 - 7x + 2$$

• Divide each term of the polynomial by the GCF.

$$5x^3 - 35x^2 + 10x$$
$$= 5x(x^2 - 7x + 2)$$

• Write the polynomial as the product of the GCF and the quotient found above.

$$5x(x^2 - 7x + 2)$$
$$= 5x^3 - 35x^2 + 10x$$

• Check the factorization by multiplying.

EXAMPLE 8

Factor: $16x^4y^5 + 8x^4y^2 - 12x^3y$

Solution

The GCF is $4x^3y$.

$$16x^4y^5 + 8x^4y^2 - 12x^3y$$
$$= 4x^3y(4xy^4 + 2xy - 3)$$

YOU TRY IT 8

Factor: $6x^4y^2 - 9x^3y^2 + 12x^2y^4$

Your solution

$3x^2y^2(2x^2 - 3x + 4y^2)$

Solution on p. S37

A **quadratic trinomial** is a trinomial of the form $ax^2 + bx + c$, where a and b are coefficients and c is a constant. Examples of quadratic trinomials are shown below.

$$x^2 + 9x + 14 \qquad a = 1, b = 9, c = 14$$
$$x^2 - 2x - 15 \qquad a = 1, b = -2, c = -15$$
$$3x^2 - x + 4 \qquad a = 3, b = -1, c = 4$$

To **factor a quadratic trinomial** means to express the trinomial as the product of two binomials. For example,

Trinomial   Factored Form
$$2x^2 - x - 1 = (2x + 1)(x - 1)$$
$$y^2 - 3y + 2 = (y - 1)(y - 2)$$

7.2A Factor a trinomial of the form $x^2 + bx + c$

We will begin by factoring trinomials of the form $x^2 + bx + c$, where the coefficient of x^2 is 1.

The method by which factors of a trinomial are found is based on FOIL. Consider the following binomial products, noting the relationship between the constant terms of the binomials and the terms of the trinomial.

Sum of the binomial constants

Product of the binomial constants

$$(x + 4)(x + 5) = x \cdot x + 5x + 4x + 4 \cdot 5 \qquad = x^2 + 9x + 20$$
$$(x - 6)(x + 8) = x \cdot x + 8x - 6x + (-6)(8) \qquad = x^2 + 2x - 48$$
$$(x - 3)(x - 2) = x \cdot x - 2x - 3x + (-3)(-2) \qquad = x^2 - 5x + 6$$

HOW TO 7   Factor: $x^2 - 7x + 10$

Find two integers whose product is 10 and whose sum is -7.

Negative Factors of 10	**Sum**
$-1, -10$	-11
$-2, -5$	**-7**

• These are the correct factors.

$$x^2 - 7x + 10 = (x - 2)(x - 5)$$

• Write the trinomial as a product of factors.

Check:
$$(x - 2)(x - 5) = x^2 - 5x - 2x + 10$$
$$= x^2 - 7x + 10$$

Take Note

Once the correct factors are found, it is not necessary to try the remaining factors.

Take Note

Always check the proposed factorization by multiplying the factors.

EXAMPLE 9

Factor: $x^2 + 6x - 27$

Solution

Two factors of -27 whose sum is 6 are -3 and 9.

$$x^2 + 6x - 27 = (x - 3)(x + 9)$$

YOU TRY IT 9

Factor: $x^2 - 8x + 15$

Your solution

$(x - 3)(x - 5)$

Solution on p. S37

7.3A Factor a trinomial of the form $ax^2 + bx + c$ by using trial factors

To use the trial factor method to factor a trinomial of the form $ax^2 + bx + c$, where $a \neq 1$, use the factors of a and the factors of c to write all of the possible binomial factors of the trinomial. Then use FOIL to determine the correct factorization. To reduce the number of trial factors that must be considered, remember the following guidelines.

Use the signs of the constant term and the coefficient of x in the trinomial to determine the signs of the binomial factors. If the constant term is positive, the signs of the binomial factors will be the same as the sign of the coefficient of x in the trinomial. If the sign of the constant term is negative, the constant terms in the binomials will have different signs.

HOW TO 8   Factor: $2x^2 - 7x + 3$

Because the constant term is positive ($+3$) and the coefficient of x is negative (-7), the binomial constants will be negative.

Positive Factors of 2 ($a = 2$)   **Negative Factors of 3 ($c = 3$)**
1, 2 $-1, -3$

Write trial factors. Use the **O**uter and **I**nner products of FOIL to determine the middle term, $-7x$, of the trinomial.

Trial Factors	**Middle Term**	
$(x - 1)(2x - 3)$	$-3x - 2x = -5x$	
$(x - 3)(2x - 1)$	$-x - 6x = -7x$	• $-7x$ is the middle term.

$2x^2 - 7x + 3 = (x - 3)(2x - 1)$

Check: $(x - 3)(2x - 1) = 2x^2 - x - 6x + 3 = 2x^2 - 7x + 3$

HOW TO 9   Factor: $5x^2 + 22x - 15$

The constant term is negative (-15). The binomial constants will have different signs.

Positive Factors of 5 ($a = 5$)   **Factors of -15 ($c = -15$)**
1, 5 $-1, 15$
                                         $1, -15$
                                         $-3, 5$
                                         $3, -5$

Write trial factors. Use the **O**uter and **I**nner products of FOIL to determine the middle term, $22x$, of the trinomial.

Take Note
It is not necessary to test trial factors that have a common factor. If the trinomial does not have a common factor, then its factors cannot have a common factor.

Trial Factors	**Middle Term**	
$(x - 1)(5x + 15)$	common factor	
$(x + 15)(5x - 1)$	$-x + 75x = 74x$	
$(x + 1)(5x - 15)$	common factor	
$(x - 15)(5x + 1)$	$x - 75x = -74x$	
$(x - 3)(5x + 5)$	common factor	
$(x + 5)(5x - 3)$	$-3x + 25x = 22x$	• $22x$ is the middle term.
$(x + 3)(5x - 5)$	common factor	
$(x - 5)(5x + 3)$	$3x - 25x = -22x$	

$5x^2 + 22x - 15 = (x + 5)(5x - 3)$

Check: $(x + 5)(5x - 3) = 5x^2 - 3x + 25x - 15 = 5x^2 + 22x - 15$

EXAMPLE 10

Factor: $2x^2 + 3x - 5$

Solution

Positive Factors of 2	**Factors of -5**
1, 2	1, -5
	-1, 5

Trial Factors	**Middle Term**
$(x + 1)(2x - 5)$	$-5x + 2x = -3x$
$(x - 5)(2x + 1)$	$x - 10x = -9x$
$(x - 1)(2x + 5)$	$5x - 2x = 3x$
$(x + 5)(2x - 1)$	$-x + 10x = 9x$

$2x^2 + 3x - 5 = (x - 1)(2x + 5)$

Check: $(x - 1)(2x + 5) = 2x^2 + 5x - 2x - 5$
$= 2x^2 + 3x - 5$

YOU TRY IT 10

Factor: $3x^2 - x - 2$

Your solution

$(x - 1)(3x + 2)$

Solution on p. S38

7.4D Factor completely

A polynomial is factored completely when it is written as a product of factors that are nonfactorable over the integers.

The first step in any factoring problem is to determine whether the terms of the polynomial have a common factor. If they do, factor it out first.

EXAMPLE 11

Factor: $5x^2y + 60xy + 100y$

Solution

There is a common factor, $5y$.
Factor out the GCF.

$5x^2y + 60xy + 100y = 5y(x^2 + 12x + 20)$

Factor $x^2 + 12x + 20$. The two factors of 20 whose sum is 12 are 2 and 10.

$5y(x^2 + 12x + 20) = 5y(x + 2)(x + 10)$

$5x^2y + 60xy + 100y = 5y(x + 2)(x + 10)$

Check:
$5y(x + 2)(x + 10) = (5xy + 10y)(x + 10)$
$= 5x^2y + 50xy + 10xy + 100y$
$= 5x^2y + 60xy + 100y$

YOU TRY IT 11

Factor: $4a^3 - 4a^2 - 24a$

Your solution

$4a(a + 2)(a - 3)$

Solution on p. S38

T.4 EXERCISES

OBJECTIVE A *To multiply and divide monomials*

For Exercises 1 to 48, simplify.

1. $z^3 \cdot z \cdot z^4$
z^8

2. $b \cdot b^2 \cdot b^6$
b^9

3. $(x^3)^5$
x^{15}

4. $(b^2)^4$
b^8

5. $(x^2y^3)^6$
$x^{12}y^{18}$

6. $(m^4n^2)^3$
$m^{12}n^6$

7. $\dfrac{a^8}{a^2}$
a^6

8. $\dfrac{c^{12}}{c^5}$
c^7

9. $(-m^3n)(m^6n^2)$
$-m^9n^3$

10. $(-r^4t^3)(r^2t^9)$
$-r^6t^{12}$

11. $(-2a^3bc^2)^3$
$-8a^9b^3c^6$

12. $(-4xy^3z^2)^2$
$16x^2y^6z^4$

13. $\dfrac{m^4n^7}{m^3n^5}$
mn^2

14. $\dfrac{a^5b^6}{a^3b^2}$
a^2b^4

15. $\dfrac{-16a^7}{24a^6}$
$-\dfrac{2a}{3}$

16. $\dfrac{18b^5}{-45b^4}$
$-\dfrac{2b}{5}$

17. $(9mn^4p)(-3mp^2)$
$-27m^2n^4p^3$

18. $(-3v^2wz)(-4vz^4)$
$12v^3wz^5$

19. $(-2n^2)(-3n^4)^3$
$54n^{14}$

20. $(-3m^3n)(-2m^2n^3)^3$
$24m^9n^{10}$

21. $\dfrac{14x^4y^6z^2}{16x^3y^9z}$
$\dfrac{7xz}{8y^3}$

22. $\dfrac{25x^4y^7z^2}{20x^5y^9z^{11}}$
$\dfrac{5}{4xy^2z^9}$

23. $(-2x^3y^2)^3(-xy^2)^4$
$-8x^{13}y^{14}$

24. $(-m^4n^2)^5(-2m^3n^3)^3$
$8m^{29}n^{19}$

25. $4x^{-7}$
$\dfrac{4}{x^7}$

26. $-6y^{-1}$
$-\dfrac{6}{y}$

27. $d^{-4}d^{-6}$
$\dfrac{1}{d^{10}}$

28. $x^{-3}x^{-5}$
$\dfrac{1}{x^8}$

29. $\dfrac{x^{-3}}{x^2}$
$\dfrac{1}{x^5}$

30. $\dfrac{x^4}{x^{-5}}$
x^9

31. $\dfrac{1}{3x^{-2}}$
$\dfrac{x^2}{3}$

32. $\dfrac{2}{5c^{-6}}$
$\dfrac{2c^6}{5}$

33. $(x^2y^{-4})^3$
$\dfrac{x^6}{y^{12}}$

34. $(x^3y^5)^{-4}$
$\dfrac{1}{x^{12}y^{20}}$

35. $(3x^{-1}y^{-2})^2$
$\dfrac{9}{x^2y^4}$

36. $(5xy^{-3})^{-2}$
$\dfrac{y^6}{25x^2}$

37. $(2x^{-1})(x^{-3})$

$\dfrac{2}{x^4}$

38. $(-2x^{-5})(x^7)$

$-2x^2$

39. $\dfrac{3x^{-2}y^2}{6xy^2}$

$\dfrac{1}{2x^3}$

40. $\dfrac{2x^{-2}y}{8xy}$

$\dfrac{1}{4x^3}$

41. $\dfrac{2x^{-1}y^{-4}}{4xy^2}$

$\dfrac{1}{2x^2y^6}$

42. $\dfrac{3a^{-2}b}{ab}$

$\dfrac{3}{a^3}$

43. $(x^{-2}y)^2(xy)^{-2}$

$\dfrac{1}{x^6}$

44. $(x^{-1}y^2)^{-3}(x^2y^{-4})^{-3}$

$\dfrac{y^6}{x^3}$

45. $\left(\dfrac{x^2y^{-1}}{xy}\right)^{-4}$

$\dfrac{y^8}{x^4}$

46. $\left(\dfrac{x^{-2}y^{-4}}{x^{-2}y}\right)^{-2}$

y^{10}

47. $\left(\dfrac{4a^{-2}b}{8a^3b^{-4}}\right)^{2}$

$\dfrac{b^{10}}{4a^{10}}$

48. $\left(\dfrac{6ab^{-2}}{3a^{-2}b}\right)^{-2}$

$\dfrac{b^6}{4a^6}$

OBJECTIVE B   *To add and subtract polynomials*

For Exercises 49 to 64, add or subtract.

49. $(4b^2 - 5b) + (3b^2 + 6b - 4)$
$7b^2 + b - 4$

50. $(2c^2 - 4) + (6c^2 - 2c + 4)$
$8c^2 - 2c$

51. $(2a^2 - 7a + 10) + (a^2 + 4a + 7)$
$3a^2 - 3a + 17$

52. $(-6x^2 + 7x + 3) + (3x^2 + x + 3)$
$-3x^2 + 8x + 6$

53. $(x^2 - 2x + 1) - (x^2 + 5x + 8)$
$-7x - 7$

54. $(3x^2 + 2x - 2) - (5x^2 - 5x + 6)$
$-2x^2 + 7x - 8$

55. $(-2x^3 + x - 1) - (-x^2 + x - 3)$
$-2x^3 + x^2 + 2$

56. $(2x^2 + 5x - 3) - (3x^3 + 2x - 5)$
$-3x^3 + 2x^2 + 3x + 2$

57. $(x^3 - 7x + 4) + (2x^2 + x - 10)$
$x^3 + 2x^2 - 6x - 6$

58. $(3y^3 + y^2 + 1) + (-4y^3 - 6y - 3)$
$-y^3 + y^2 - 6y - 2$

59. $(5x^3 + 7x - 7) + (10x^2 - 8x + 3)$
$5x^3 + 10x^2 - x - 4$

60. $(3y^3 + 4y + 9) + (2y^2 + 4y - 21)$
$3y^3 + 2y^2 + 8y - 12$

61. $(2y^3 + 6y - 2) - (y^3 + y^2 + 4)$
$y^3 - y^2 + 6y - 6$

62. $(-2x^2 - x + 4) - (-x^3 + 3x - 2)$
$x^3 - 2x^2 - 4x + 6$

63. $(4y^3 - y - 1) - (2y^2 - 3y + 3)$
$4y^3 - 2y^2 + 2y - 4$

64. $(3x^2 - 2x - 3) - (2x^3 - 2x^2 + 4)$
$-2x^3 + 5x^2 - 2x - 7$

OBJECTIVE C   *To multiply polynomials*

For Exercises 65 to 101, multiply.

65. $4b(3b^3 - 12b^2 - 6)$
$12b^4 - 48b^3 - 24b$

66. $-2a^2(3a^2 - 2a + 3)$
$-6a^4 + 4a^3 - 6a^2$

67. $3b(3b^4 - 3b^2 + 8)$
$9b^5 - 9b^3 + 24b$

68. $-2x^2(2x^2 - 3x - 7)$
$-4x^4 + 6x^3 + 14x^2$

69. $-2x^2y(x^2 - 3xy + 2y^2)$
$-2x^4y + 6x^3y^2 - 4x^2y^3$

70. $3ab^2(3a^2 - 2ab + 4b^2)$
$9a^3b^2 - 6a^2b^3 + 12ab^4$

71. $(x^2 + 3x + 2)(x + 1)$
$x^3 + 4x^2 + 5x + 2$

72. $(x^2 - 2x + 7)(x - 2)$
$x^3 - 4x^2 + 11x - 14$

73. $(a - 3)(a^2 - 3a + 4)$
$a^3 - 6a^2 + 13a - 12$

74. $(2x - 3)(x^2 - 3x + 5)$
$2x^3 - 9x^2 + 19x - 15$

75. $(-2b^2 - 3b + 4)(b - 5)$
$-2b^3 + 7b^2 + 19b - 20$

76. $(-a^2 + 3a - 2)(2a - 1)$
$-2a^3 + 7a^2 - 7a + 2$

77. $(x^3 - 3x + 2)(x - 4)$
$x^4 - 4x^3 - 3x^2 + 14x - 8$

78. $(y^3 + 4y^2 - 8)(2y - 1)$
$2y^4 + 7y^3 - 4y^2 - 16y + 8$

79. $(y + 2)(y^3 + 2y^2 - 3y + 1)$
$y^4 + 4y^3 + y^2 - 5y + 2$

80. $(2a - 3)(2a^3 - 3a^2 + 2a - 1)$
$4a^4 - 12a^3 + 13a^2 - 8a + 3$

81. $(a - 3)(a + 4)$
$a^2 + a - 12$

82. $(b - 6)(b + 3)$
$b^2 - 3b - 18$

83. $(y - 7)(y - 3)$
$y^2 - 10y + 21$

84. $(a - 8)(a - 9)$
$a^2 - 17a + 72$

85. $(2x + 1)(x + 7)$
$2x^2 + 15x + 7$

86. $(y + 2)(5y + 1)$
$5y^2 + 11y + 2$

87. $(3x - 1)(x + 4)$
$3x^2 + 11x - 4$

88. $(7x - 2)(x + 4)$
$7x^2 + 26x - 8$

89. $(4x - 3)(x - 7)$
$4x^2 - 31x + 21$

90. $(2x - 3)(4x - 7)$
$8x^2 - 26x + 21$

91. $(3y - 8)(y + 2)$
$3y^2 - 2y - 16$

92. $(5y - 9)(y + 5)$
$5y^2 + 16y - 45$

93. $(7a - 16)(3a - 5)$
$21a^2 - 83a + 80$

94. $(5a - 12)(3a - 7)$
$15a^2 - 71a + 84$

95. $(x + y)(2x + y)$
$2x^2 + 3xy + y^2$

96. $(2a + b)(a + 3b)$
$2a^2 + 7ab + 3b^2$

97. $(3x - 4y)(x - 2y)$
$3x^2 - 10xy + 8y^2$

98. $(2a - b)(3a + 2b)$
$6a^2 + ab - 2b^2$

99. $(5a - 3b)(2a + 4b)$
$10a^2 + 14ab - 12b^2$

100. $(2x + 3)(2x - 3)$
$4x^2 - 9$

101. $(4x - 7)(4x + 7)$
$16x^2 - 49$

OBJECTIVE D *To divide polynomials*

For Exercises 102 to 119, divide.

102. $(b^2 - 14b + 49) \div (b - 7)$
$b - 7$

103. $(x^2 - x - 6) \div (x - 3)$
$x + 2$

104. $(2x^2 + 5x + 2) \div (x + 2)$
$2x + 1$

105. $(2y^2 - 13y + 21) \div (y - 3)$
$2y - 7$

106. $(x^2 + 1) \div (x - 1)$
$x + 1 + \dfrac{2}{x - 1}$

107. $(x^2 + 4) \div (x + 2)$
$x - 2 + \dfrac{8}{x + 2}$

108. $(6x^2 - 7x) \div (3x - 2)$
$2x - 1 - \dfrac{2}{3x - 2}$

109. $(6y^2 + 2y) \div (2y + 4)$
$3y - 5 + \dfrac{20}{2y + 4}$

110. $(a^2 + 5a + 10) \div (a + 2)$
$a + 3 + \dfrac{4}{a + 2}$

111. $(b^2 - 8b - 9) \div (b - 3)$
$b - 5 - \dfrac{24}{b - 3}$

112. $(2y^2 - 9y + 8) \div (2y + 3)$
$y - 6 + \dfrac{26}{2y + 3}$

113. $(3x^2 + 5x - 4) \div (x - 4)$
$3x + 17 + \dfrac{64}{x - 4}$

114. $(8x + 3 + 4x^2) \div (2x - 1)$
$2x + 5 + \dfrac{8}{2x - 1}$

115. $(10 + 21y + 10y^2) \div (2y + 3)$
$5y + 3 + \dfrac{1}{2y + 3}$

116. $(x^3 + 3x^2 + 5x + 3) \div (x + 1)$
$x^2 + 2x + 3$

117. $(x^3 - 6x^2 + 7x - 2) \div (x - 1)$
$x^2 - 5x + 2$

118. $(x^4 - x^2 - 6) \div (x^2 + 2)$
$x^2 - 3$

119. $(x^4 + 3x^2 - 10) \div (x^2 - 2)$
$x^2 + 5$

OBJECTIVE E *To factor polynomials of the form $ax^2 + bx + c$*

For Exercises 120 to 182, factor.

120. $8x + 12$
$4(2x + 3)$

121. $12y^2 - 5y$
$y(12y - 5)$

122. $10x^4 - 12x^2$
$2x^2(5x^2 - 6)$

123. $10x^2yz^2 + 15xy^3z$
$5xyz(2xz + 3y^2)$

124. $x^3 - 3x^2 - x$
$x(x^2 - 3x - 1)$

125. $5x^2 - 15x + 35$
$5(x^2 - 3x + 7)$

126. $3x^3 + 6x^2 + 9x$
$3x(x^2 + 2x + 3)$

127. $3y^4 - 9y^3 - 6y^2$
$3y^2(y^2 - 3y - 2)$

128. $2x^3 + 6x^2 - 14x$
$2x(x^2 + 3x - 7)$

129. $x^4y^4 - 3x^3y^3 + 6x^2y^2$
$x^2y^2(x^2y^2 - 3xy + 6)$

130. $4x^5y^5 - 8x^4y^4 + x^3y^3$
$x^3y^3(4x^2y^2 - 8xy + 1)$

131. $16x^2y - 8x^3y^4 - 48x^2y^2$
$8x^2y(2 - xy^3 - 6y)$

132. $x^2 + 5x + 6$
$(x + 2)(x + 3)$

133. $x^2 + x - 2$
$(x + 2)(x - 1)$

134. $x^2 + x - 6$
$(x + 3)(x - 2)$

135. $a^2 + a - 12$
$(a + 4)(a - 3)$

136. $a^2 - 2a - 35$
$(a + 5)(a - 7)$

137. $a^2 - 3a + 2$
$(a - 1)(a - 2)$

138. $a^2 - 5a + 4$
$(a - 1)(a - 4)$

139. $b^2 + 7b - 8$
$(b + 8)(b - 1)$

140. $y^2 + 6y - 55$
$(y + 11)(y - 5)$

141. $z^2 - 4z - 45$
$(z + 5)(z - 9)$

142. $y^2 - 8y + 15$
$(y - 3)(y - 5)$

143. $z^2 - 14z + 45$
$(z - 5)(z - 9)$

144. $p^2 + 12p + 27$
$(p + 3)(p + 9)$

145. $b^2 + 9b + 20$
$(b + 4)(b + 5)$

146. $y^2 - 8y + 32$
Nonfactorable

147. $y^2 - 9y + 81$
Nonfactorable

148. $p^2 + 24p + 63$
$(p + 3)(p + 21)$

149. $x^2 - 15x + 56$
$(x - 7)(x - 8)$

150. $5x^2 + 6x + 1$
$(x + 1)(5x + 1)$

151. $2y^2 + 7y + 3$
$(y + 3)(2y + 1)$

152. $2a^2 - 3a + 1$
$(a - 1)(2a - 1)$

153. $3a^2 - 4a + 1$
$(a - 1)(3a - 1)$

154. $4x^2 - 3x - 1$
$(x - 1)(4x + 1)$

155. $2x^2 - 5x - 3$
$(x - 3)(2x + 1)$

156. $6t^2 - 11t + 4$
$(2t - 1)(3t - 4)$

157. $10t^2 + 11t + 3$
$(2t + 1)(5t + 3)$

158. $8x^2 + 33x + 4$
$(x + 4)(8x + 1)$

159. $10z^2 + 3z - 4$
$(2z - 1)(5z + 4)$

160. $3x^2 + 14x - 5$
$(x + 5)(3x - 1)$

161. $3z^2 + 95z + 10$
Nonfactorable

162. $8z^2 - 36z + 1$
Nonfactorable

163. $2t^2 - t - 10$
$(t + 2)(2t - 5)$

164. $2t^2 + 5t - 12$
$(t + 4)(2t - 3)$

165. $12y^2 + 19y + 5$
$(3y + 1)(4y + 5)$

166. $5y^2 - 22y + 8$
$(y - 4)(5y - 2)$

167. $11a^2 - 54a - 5$
$(a - 5)(11a + 1)$

168. $4z^2 + 11z + 6$
$(z + 2)(4z + 3)$

169. $6b^2 - 13b + 6$
$(2b - 3)(3b - 2)$

170. $6x^2 + 35x - 6$
$(x + 6)(6x - 1)$

171. $3x^2 + 15x + 18$
$3(x + 2)(x + 3)$

172. $3a^2 + 3a - 18$
$3(a + 3)(a - 2)$

173. $ab^2 + 7ab - 8a$
$a(b + 8)(b - 1)$

174. $3y^3 - 15y^2 + 18y$
$3y(y - 2)(y - 3)$

175. $2y^4 - 26y^3 - 96y^2$
$2y^2(y + 3)(y - 16)$

176. $3y^4 + 54y^3 + 135y^2$
$3y^2(y + 3)(y + 15)$

177. $2x^3 - 11x^2 + 5x$
$x(x - 5)(2x - 1)$

178. $2x^3 + 3x^2 - 5x$
$x(x - 1)(2x + 5)$

179. $10t^2 - 5t - 50$
$5(t + 2)(2t - 5)$

180. $16t^2 + 40t - 96$
$8(t + 4)(2t - 3)$

181. $6p^3 + 5p^2 + p$
$p(2p + 1)(3p + 1)$

182. $12x^2y - 36xy + 27y$
$3y(2x - 3)^2$

Appendix A

Keystroke Guide for the TI-84 Plus

Basic Operations

Numerical calculations are performed on the **home screen.** You can always return to the home screen by pressing [2ND] QUIT. Pressing [CLEAR] erases the home screen.

To evaluate the expression $-2(3 + 5) - 8 \div 4$, use the following keystrokes.

Note: There is a difference between the key to enter a negative number, [(-)], and the key for subtraction, [-]. You cannot use these keys interchangeably.

The [2ND] key is used to access the commands in gold writing above a key. For instance, to evaluate $\sqrt{49}$, press [2ND] $\sqrt{}$ 49 [ENTER].

The [ALPHA] key is used to place a letter on the screen. One reason to do this is to store a value of a variable. The following keystrokes give A the value of 5.

5 [STO▸] [ALPHA] A [ENTER]

This value is now available in calculations. For instance, we can find the value of $3a^2$ by using the following keystrokes: 3 [ALPHA] A [x^2]. To display the value of the variable on the screen, press [2ND] RCL [ALPHA] A.

Note: When you use the [ALPHA] key, only capital letters are available on the TI-83 calculator.

Complex Numbers

To perform operations on complex numbers, first press [MODE] and then use the arrow keys to select $a+bi$. Then press [ENTER] [2ND] QUIT.

Addition of complex numbers To add $(3 + 4i) + (2 - 7i)$, use the keystrokes

[(] 3 [+] 4 [2ND] i [)] [+]
[(] 2 [-] 7 [2ND] i [)] [ENTER].

Division of complex numbers To divide $\frac{26 + 2i}{2 + 4i}$, use the keystrokes [(] 26 [+] 2 [2ND] i [)] [÷]
[(] 2 [+] 4 [2ND] i [)] [ENTER].

Note: Operations for subtraction and multiplication are similar.

The descriptions in the margins (for example, Basic Operations and Complex Numbers) are the same as those used in the text and are arranged alphabetically.

Take Note

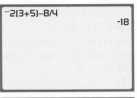

```
-2(3+5)-8/4
                    -18
```

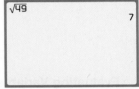

```
√49
                      7
```

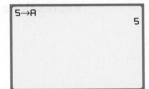

```
5→A
                      5
```

```
3A²
                     75
```

```
Normal  Sci  Eng
Float   0123456789
Radian  Degree
Func   Par  Pol  Seq
Connected     Dot
Sequential    Simul
Real   a+bi  re^θi
Full   Horiz  G-T
```

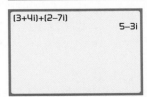

```
(3+4i)+(2-7i)
                   5-3i
```

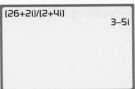

```
(26+2i)/(2+4i)
                   3-5i
```

Unless otherwise noted, all content on this page is © Cengage Learning.

Additional operations on complex numbers can be found by selecting **CPX** under the MATH key.

To find the absolute value of $2 - 5i$, press MATH (scroll to **CPX**) (scroll to **abs**) ENTER 2 − 5 2ND i ENTER .

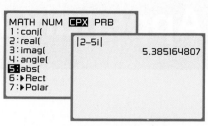

Evaluating Functions

There are various methods of evaluating a function but all methods require that the expression be entered as one of the ten functions Y_1 to Y_0. To evaluate $f(x) = \dfrac{x^2}{x - 1}$ when $x = -3$, enter the expression into, for instance, Y_1, and then press VARS ▸ 11 ((−) 3) ENTER .

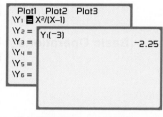

Note: If you try to evaluate a function at a number that is not in the domain of the function, you will get an error message. For instance, 1 is not in the domain of $f(x) = \dfrac{x^2}{x - 1}$. If we try to evaluate the function at 1, the error screen at the right appears.

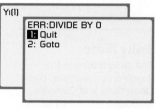

Take Note

Use the down arrow key to scroll past Y_7 to see Y_8, Y_9, and Y_0.

Evaluating Variable Expressions

To evaluate a variable expression, first store the values of each variable. Then enter the variable expression. For instance, to evaluate $s^2 + 2sl$ when $s = 4$ and $l = 5$, use the following keystrokes.

4 STO▸ ALPHA S ENTER 5 STO▸ ALPHA L ENTER ALPHA S x^2 + 2 ALPHA S ALPHA L ENTER

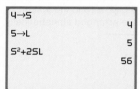

Graph

To graph a function, use the Y= key to enter the expression for the function, select a suitable viewing window, and then press GRAPH. For instance, to graph $f(x) = 0.1x^3 - 2x - 1$ in the standard viewing window, use the following keystrokes.

Y= 0.1 X,T,θ,n ^ 3 ▸ − 2 X,T,θ,n − 1 ZOOM (scroll to 6) ENTER

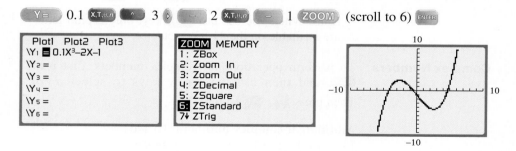

Note: For the keystrokes above, you do not have to scroll to 6. Alternatively, use ZOOM 6. This will select the standard viewing window and automatically start the graph. Use the WINDOW key to create a custom window for a graph.

Graphing Inequalities

To illustrate this feature, we will graph $y \leq 2x - 1$. Enter $2x - 1$ into Y_1. Because $y \leq 2x - 1$, we want to shade below the graph. Move the cursor to the left of Y_1 and press ENTER three times. Press GRAPH .

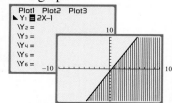

Unless otherwise noted, all content on this page is © Cengage Learning.

Note: To shade above the graph, move the cursor to the left of Y₁ and press ENTER two times. An inequality with the symbol ≤ or ≥ should be graphed with a solid line, and an inequality with the symbol < or > should be graphed with a dashed line. However, a graphing calculator does not distinguish between a solid line and a dashed line.

To graph the solution set of a system of inequalities, solve each inequality for y and graph each inequality. The solution set is the intersection of the two inequalities. The solution set of

$$3x + 2y > 10$$
$$4x - 3y \leq 5$$

is shown at the right.

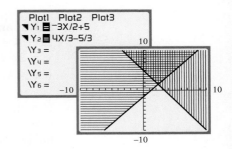

Intersect

The INTERSECT feature is used to solve a system of equations. To illustrate this feature, we will use the system of equations
$$\begin{array}{l} 2x - 3y = 13 \\ 3x + 4y = -6 \end{array}.$$

Note: Some equations can be solved by this method. See the section "Solve an equation" on the next page. Also, this method is used to find a number in the domain of a function for a given number in the range. See the section "Find a domain element" on the next page.

Solve each of the equations in the system of equations for y. In this case, we have $y = \frac{2}{3}x - \frac{13}{3}$ and $y = -\frac{3}{4}x - \frac{3}{2}$.

Use the Y-editor to enter $\frac{2}{3}x - \frac{13}{3}$ into Y₁ and $-\frac{3}{4}x - \frac{3}{2}$ into Y₂. Graph the two functions in the standard viewing window. (If the window does not show the point of intersection of the two graphs, adjust the window until you can see the point of intersection.)

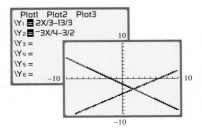

Press 2ND CALC (scroll to 5, **intersect**) ENTER.

Alternatively, you can just press 2ND CALC 5.

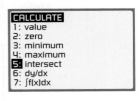

First curve? is shown at the bottom of the screen and identifies one of the two graphs on the screen. Press ENTER.

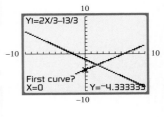

Second curve? is shown at the bottom of the screen and identifies the second of the two graphs on the screen. Press ENTER.

Guess? shown at the bottom of the screen asks you to use the left or right arrow key to move the cursor to the *approximate* location of the point of intersection. (If there are two or more points of intersection, it does not matter which one you choose first.) Press ENTER.

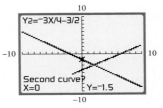

Unless otherwise noted, all content on this page is © Cengage Learning.

The solution of the system of equations is $(2, -3)$.

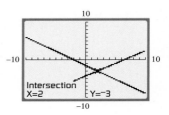

Solve an equation To illustrate the steps involved, we will solve the equation $2x + 4 = -3x - 1$. The idea is to write the equation as the system of equations

$y = 2x + 4$
$y = -3x - 1$ and then use the steps for solving a system of equations.

Use the Y-editor to enter the left and right sides of the equation into Y₁ and Y₂. Graph the two functions and then follow the steps for Intersect.

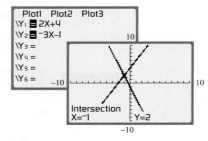

The solution is -1, the x-coordinate of the point of intersection.

Find a domain element For this example, we will find a number in the domain of $f(x) = -\frac{2}{3}x + 2$ that corresponds to 4 in the range of the function. This is like solving the system of equations $y = -\frac{2}{3}x + 2$ and $y = 4$.

Use the Y = editor to enter the expression for the function in Y₁ and the desired output, 4, in Y₂. Graph the two functions and then follow the steps for Intersect.

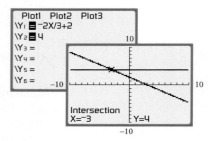

The point of intersection is $(-3, 4)$. The number -3 in the domain of f produces an output of 4 in the range of f.

Math   Pressing (MATH) gives you access to many built-in functions. The following keystrokes will convert 0.125 to a fraction: .125 (MATH) 1 (ENTER).

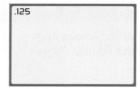   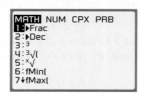   

Additional built-in functions under (MATH) can be found by pressing (MATH) ◊. For instance, to evaluate $-|-25|$, press ((-)) (MATH) ◊ 1 ((-)) 25 (ENTER).

   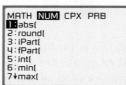   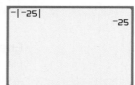

See your owner's manual for assistance with other functions under the (MATH) key.

Unless otherwise noted, all content on this page is © Cengage Learning.

Min and Max   The local minimum and the local maximum values of a function are calculated by accessing the CALC menu. For this demonstration, we will find the minimum value and the maximum value of $f(x) = 0.2x^3 + 0.3x^2 - 3.6x + 2$.

Enter the function into Y₁. Press 2ND CALC (scroll to 3 for **minimum** of the function) ENTER.

Alternatively, you can just press 2ND CALC 3.

Left Bound? shown at the bottom of the screen asks you to use the left or right arrow key to move the cursor to the *left* of the minimum. Press ENTER.

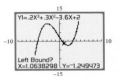

Right Bound? shown at the bottom of the screen asks you to use the left or right arrow key to move the cursor to the *right* of the minimum. Press ENTER.

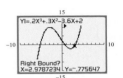

Guess? shown at the bottom of the screen asks you to use the left or right arrow key to move the cursor to the *approximate* location of the minimum. Press ENTER.

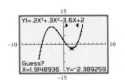

The minimum value of the function is the *y*-coordinate. For this example, the minimum value of the function is -2.4.

The *x*-coordinate for the minimum is 2. However, because of rounding errors in the calculation, it is shown as a number close to 2.

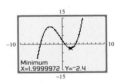

To find the maximum value of the function, follow the same steps as above except select **maximum** under the CALC menu. The screens for this calculation are shown below.

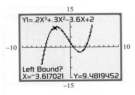 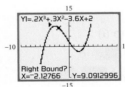 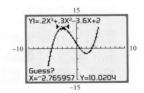 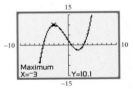

The maximum value of the function is 10.1.

Radical Expressions   To evaluate a square-root expression, press 2ND √.

For instance, to evaluate $0.15\sqrt{p^2 + 4p + 10}$ when $p = 100{,}000$, first store 100,000 in P. Then press 0.15 2ND √ ALPHA P x^2 + 4 ALPHA P + 10 ENTER.

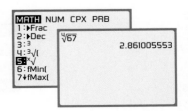

To evaluate a radical expression other than a square root, access $\sqrt[x]{}$ by pressing MATH. For instance, to evaluate $\sqrt[4]{67}$, press 4 (the index of the radical) MATH (scroll to 5) ENTER 67 ENTER.

Unless otherwise noted, all content on this page is © Cengage Learning.

Scientific Notation To enter a number in scientific notation, use 2ND EE. For instance, to find $\frac{3.45 \times 10^{-12}}{1.5 \times 10^{25}}$, press 3.45 2ND EE (-) 12 ÷ 1.5 2ND EE 25 ENTER. The answer is 2.3×10^{-37}.

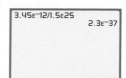

Sequences and Series The terms of a sequence and the sum of a series can be calculated by using the 2ND LIST feature.

Store a sequence A sequence is stored in one of the lists L_1 through L_6. For instance, to store the sequence 1, 3, 5, 7, 9 in L_1, use the following keystrokes.

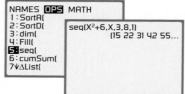

2ND { 1 , 3 , 5 , 7 , 9 2ND } STO► 2ND L1 ENTER

Display the terms of a sequence The terms of a sequence are displayed by using the function seq(expression, variable, begin, end, increment). For instance, to display the 3rd through 8th terms of the sequence given by $a_n = n^2 + 6$, enter the following keystrokes.

2ND LIST ◊ (scroll to 5)

ENTER X,T,θ,n x² + 6

, X,T,θ,n , 3 , 8

, 1 ENTER STO► 2ND L1 ENTER

The keystrokes STO► 2ND L1 ENTER store the terms of the sequence in L_1. This is not necessary but is sometimes helpful if additional work will be done with that sequence.

Find a sequence of partial sums To find a sequence of partial sums, use the cumSum(function. For instance, to find the sequence of partial sums for 2, 4, 6, 8, 10, use the following keystrokes.

2ND LIST ◊ (scroll to 6)

ENTER 2ND { 2 , 4 , 6

, 8 , 10 2ND }) ENTER

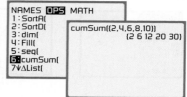

If a sequence is stored as a list in L_1, then the sequence of partial sums can be calculated by pressing 2ND LIST ◊ (scroll to 6 [or press 6]) ENTER 2ND L1) ENTER.

Find the sum of a series The sum of a series is calculated using sum<list, start, end>. For instance, to find $\sum_{n=3}^{6} (n^2 + 2)$, enter the following keystrokes.

2ND LIST ◊ ◊ (scroll to 5)

ENTER 2ND LIST ◊ (scroll to 5 [or press 5])

ENTER X,T,θ,n x² + 2 ,

X,T,θ,n , 3 , 6 , 1) ENTER

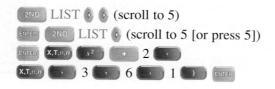

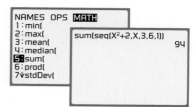

Unless otherwise noted, all content on this page is © Cengage Learning.

Table   There are three steps in creating an input/output table for a function. First use the  editor to input the function. The second step is setting up the table, and the third step is displaying the table.

To set up the table, press **2ND** TBLSET. **TblStart** is the first value of the independent variable in the input/output table. $\triangle$**Tbl** is the difference between successive values. Setting this to 1 means that, for this table, the input values are $-2, -1, 0,$ $1, 2, \ldots$ If $\triangle$**Tbl** $= 0.5$, then the input values are $-2, -1.5,$ $-1, -0.5, 0, 0.5, \ldots$

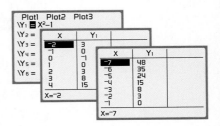

Indpnt is the independent variable. When this is set to **Auto**, values of the independent variable are automatically entered into the table. **Depend** is the dependent variable. When this is set to **Auto**, values of the dependent variable are automatically entered into the table.

To display the table, press **2ND** TABLE. An input/output table for $f(x) = x^2 - 1$ is shown at the right.

Once the table is on the screen, the up and down arrow keys can be used to display more values in the table. For the table at the right, we used the up arrow key to move to $x = -7$.

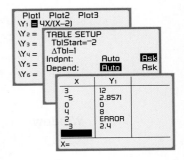

An input/output table for any given input can be created by selecting **Ask** for the independent variable. The table at the right shows an input/output table for

$f(x) = \dfrac{4x}{x-2}$ for selected values of x. Note the word **ERROR** when 2 was entered. This occurred because f is not defined when $x = 2$.

Note: Using the table feature in **Ask** mode is the same as evaluating a function for given values of the independent variable. For instance, from the table at the right, we have $f(4) = 8$.

Test   The TEST feature has many uses, one of which is to graph the solution set of a linear inequality in one variable. To illustrate this feature, we will graph the solution set of $x - 1 < 4$. Press **Y=** **X,T,θ,n** **−** 1 **2ND** TEST (scroll to 5) **ENTER** 4 **GRAPH**.

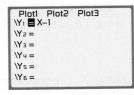  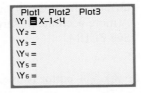 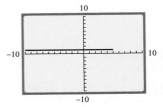

Trace   Once a graph is drawn, pressing **TRACE** will place a cursor on the screen, and the coordinates of the point below the cursor are shown at the bottom of the screen. Use the left and right arrow keys to move the cursor along the graph. For the graph at the right, we have $f(4.8) = 3.4592$, where $f(x) = 0.1x^3 - 2x + 2$ is shown at the top left of the screen.

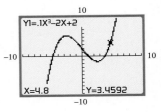

Unless otherwise noted, all content on this page is © Cengage Learning.

In TRACE mode, you can evaluate a function at any value of the independent variable that is within Xmin and Xmax. To do this, first graph the function. Now press (TRACE) (the value of x) (ENTER). For the graph at the left below, we used $x = -3.5$. If a value of x is chosen outside the window, an error message is displayed.

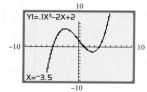 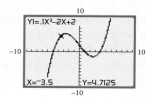 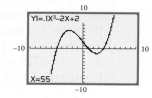 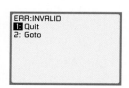

In the example above in which we entered -3.5 for x, the value of the function was calculated as 4.7125. This means that $f(-3.5) = 4.7125$. The keystrokes (2ND) QUIT (VARS) (▷) 11 (MATH) 1 (ENTER) will convert the decimal value to a fraction.

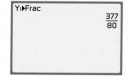

When the TRACE feature is used with two or more graphs, the up and down arrow keys are used to move between the graphs. The graphs below are for the functions $f(x) = 0.1x^3 - 2x + 2$ and $g(x) = 2x - 3$. By using the up and down arrows, we can place the cursor on either graph. The right and left arrows are used to move along the graph.

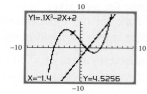 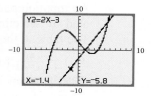

Window
The viewing window for a graph is controlled by pressing (WINDOW). Xmin and Xmax are the minimum value and maximum value, respectively, of the independent variable shown on the graph. Xscl is the distance between tic marks on the x-axis. Ymin and Ymax are the minimum value and maximum value, respectively, of the dependent variable shown on the graph. Yscl is the distance between tic marks on the y-axis. Leave Xres as 1.

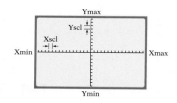

Note: In the standard viewing window, the distance between tic marks on the x-axis is different from the distance between tic marks on the y-axis. This will distort a graph. A more accurate picture of a graph can be created by using a square viewing window. See ZOOM.

The (Y=) editor is used to enter the expression for a function. There are ten possible functions, labeled Y_1 to Y_0, that can be active at any one time. For instance, to enter $f(x) = x^2 + 3x - 2$ as Y_1, use the following keystrokes.

Unless otherwise noted, all content on this page is © Cengage Learning.

Note: If an expression is already entered for Y₁, place the cursor anywhere on that expression and press ⌨CLEAR.

To enter $s = \dfrac{2v - 1}{v^3 - 3}$ into Y₂, place the cursor to the right of the

equals sign for Y₂. Then press ⌨(2 ⌨X,T,θ,n ⌨− 1 ⌨)
⌨÷ ⌨(⌨X,T,θ,n ⌨^ 3 ⌨(⌨− 3 ⌨) ⌨ENTER.

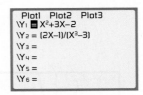

Ploti Plot2 Plot3
\Y₁ ▉ X²+3X−2
\Y₂ ▉ (2X−1)/(X³−3)
\Y₃ =
\Y₄ =
\Y₅ =
\Y₆ =

Note: When we enter an equation, the independent variable, v in the expression above, is entered using ⌨X,T,θ,n. The dependent variable, s in the expression above, is one of Y₁ to Y₀. Also note the use of parentheses to ensure the correct order of operations.

Observe the black rectangle that covers the equals sign for the two examples we have shown. This rectangle means that the function is "active." If we were to press ⌨GRAPH, then the graph of both functions would appear. You can make a function inactive by using the arrow keys to move the cursor over the equals sign of that function and then pressing ⌨ENTER. This will remove the black rectangle. We have done that for Y₂, as shown at the right. Now if ⌨GRAPH is pressed, only Y₁ will be graphed.

Ploti Plot2 Plot3
\Y₁ ▉ X²+3X−2
\Y₂ = (2X−1)/(X³−3)
\Y₃ =
\Y₄ =
\Y₅ =
\Y₆ =

It is also possible to control the appearance of the graph by moving the cursor on the ⌨Y= screen to the left of any Y. With the cursor in this position, pressing ⌨ENTER will change the appearance of the graph. The options are shown at the right.

Ploti Plot2 Plot3
\Y₁ = Default graph line
\Y₂ = Bold graph line
▼Y₃ = Shade above graph
▲Y₄ = Shade below graph
-□Y₅ = Draw path of graph
□Y₆ = Travel path of graph
\Y₇ = Dashed graph line

Zero   The ZERO feature of a graphing calculator is used for various calculations: to find the x-intercepts of a function, to solve some equations, and to find the zero of a function.

***x*-intercepts** To illustrate the procedure for finding x-intercepts, we will use $f(x) = x^2 + x - 2$.

First, use the Y-editor to enter the expression for the function and then graph the function in the standard viewing window. (It may be necessary to adjust this window so that the intercepts are visible.) Once the graph is displayed, use the keystrokes below to find the x-intercepts of the graph of the function.

Press ⌨2ND CALC (scroll to 2 for **zero** of the function) ⌨ENTER.

CALCULATE
1: value
2: zero
3: minimum
4: maximum
5: intersect
6: dy/dx
7: ∫f(x)dx

Alternatively, you can just press ⌨2ND CALC 2.

Left Bound? shown at the bottom of the screen asks you to use the left or right arrow key to move the cursor to the *left* of the desired x-intercept. Press ⌨ENTER.

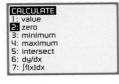

Right Bound? shown at the bottom of the screen asks you to use the left or right arrow key to move the cursor to the *right* of the desired x-intercept. Press ⌨ENTER.

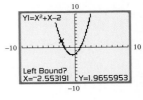

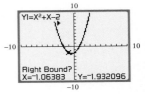

Unless otherwise noted, all content on this page is © Cengage Learning.

Guess? shown at the bottom of the screen asks you to use the left or right arrow key to move the cursor to the *approximate* location of the desired *x*-intercept. Press ⏎.

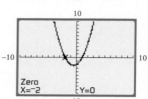

The *x*-coordinate of an *x*-intercept is −2. Therefore, an *x*-intercept is (−2, 0).

To find the other *x*-intercept, follow the same steps as above. The screens for this calculation are shown below.

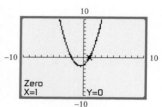

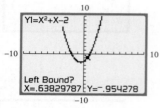 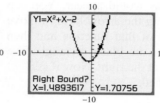 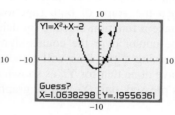 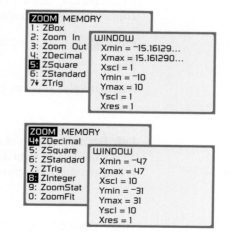

A second *x*-intercept is (1, 0).

Solve an equation To use the ZERO feature to solve an equation, first rewrite the equation with all terms on one side. For instance, one way to solve the equation $x^3 - x + 1 = -2x + 3$ is first to rewrite it as $x^3 + x - 2 = 0$. Enter $x^3 + x - 2$ into Y_1 and then follow the steps for finding *x*-intercepts.

Find the real zeros of a function To find the real zeros of a function, follow the steps for finding *x*-intercepts.

Zoom Pressing **ZOOM** allows you to select some preset viewing windows. This key also gives you access to **ZBox, Zoom In,** and **Zoom Out.** These functions enable you to redraw a selected portion of a graph in a new window. Some windows used frequently in this text are shown below.

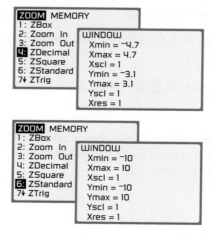

Unless otherwise noted, all content on this page is © Cengage Learning.

Appendix B

Proofs and Tables

Proofs of Logarithmic Properties

In each of the following proofs of logarithmic properties, it is assumed that the Properties of Exponents are true for all real number exponents.

The Logarithm Property of the Product of Two Numbers

For any positive real numbers x, y, and b, $b \neq 1$, $\log_b xy = \log_b x + \log_b y$.

Proof: Let $\log_b x = m$ and $\log_b y = n$.

Write each equation in its equivalent exponential form.
Use substitution and the Properties of Exponents.

$$x = b^m \qquad y = b^n$$
$$xy = b^m b^n$$
$$xy = b^{m+n}$$

Write the equation in its equivalent logarithmic form.
Substitute $\log_b x$ for m and $\log_b y$ for n.

$$\log_b xy = m + n$$
$$\log_b xy = \log_b x + \log_b y$$

The Logarithm Property of the Quotient of Two Numbers

For any positive real numbers x, y, and b, $b \neq 1$, $\log_b \dfrac{x}{y} = \log_b x - \log_b y$.

Proof: Let $\log_b x = m$ and $\log_b y = n$.

Write each equation in its equivalent exponential form.

$$x = b^m \qquad y = b^n$$

Use substitution and the Properties of Exponents.

$$\frac{x}{y} = \frac{b^m}{b^n}$$
$$\frac{x}{y} = b^{m-n}$$

Write the equation in its equivalent logarithmic form.

$$\log_b \frac{x}{y} = m - n$$

Substitute $\log_b x$ for m and $\log_b y$ for n.

$$\log_b \frac{x}{y} = \log_b x - \log_b y$$

The Logarithm Property of the Power of a Number

For any real numbers x, r, and b, $b \neq 1$, $\log_b x^r = r \log_b x$.

Proof: Let $\log_b x = m$.

Write the equation in its equivalent exponential form.
Raise both sides to the r power.

$$x = b^m$$
$$x^r = (b^m)^r$$
$$x^r = b^{mr}$$

Write the equation in its equivalent logarithmic form.
Substitute $\log_b x$ for m.

$$\log_b x^r = mr$$
$$\log_b x^r = r \log_b x$$

Table of Symbols

$+$	add	$<$	is less than
$-$	subtract	$\leq$	is less than or equal to
$\cdot$, $\times$, $(a)(b)$	multiply	$>$	is greater than
$\dfrac{a}{b}$, $\div$	divide	$\geq$	is greater than or equal to
$(\ )$	parentheses, a grouping symbol	(a, b)	an ordered pair whose first component is a and whose second component is b
$[\ ]$	brackets, a grouping symbol	$^\circ$	degree (for angles)
π	pi, a number approximately equal to $\frac{22}{7}$ or 3.14	$\sqrt{a}$	the principal square root of a
$-a$	the opposite, or additive inverse, of a	$\varnothing$, $\{\ \}$	the empty set
$\dfrac{1}{a}$	the reciprocal, or multiplicative inverse, of a	$\|a\|$	the absolute value of a
		$\cup$	union of two sets
$=$	is equal to	$\cap$	intersection of two sets
$\approx$	is approximately equal to	$\in$	is an element of (for sets)
$\neq$	is not equal to	$\notin$	is not an element of (for sets)

Table of Measurement Abbreviations

U.S. Customary System

Length		Capacity		Weight		Area	
in.	inches	oz	fluid ounces	oz	ounces	in²	square inches
ft	feet	c	cups	lb	pounds	ft²	square feet
yd	yards	qt	quarts				
mi	miles	gal	gallons				

Metric System

Length		Capacity		Weight/Mass		Area	
mm	millimeter (0.001 m)	ml	milliliter (0.001 L)	mg	milligram (0.001 g)	cm²	square centimeters
cm	centimeter (0.01 m)	cl	centiliter (0.01 L)	cg	centigram (0.01 g)	m²	square meters
dm	decimeter (0.1 m)	dl	deciliter (0.1 L)	dg	decigram (0.1 g)		
m	meter	L	liter	g	gram		
dam	decameter (10 m)	dal	decaliter (10 L)	dag	decagram (10 g)		
hm	hectometer (100 m)	hl	hectoliter (100 L)	hg	hectogram (100 g)		
km	kilometer (1000 m)	kl	kiloliter (1000 L)	kg	kilogram (1000 g)		

Time

h	hours	min	minutes	s	seconds

Table of Properties

Properties of Real Numbers

The Associative Property of Addition

If a, b, and c are real numbers, then
$(a + b) + c = a + (b + c)$.

The Associative Property of Multiplication

If a, b, and c are real numbers, then
$(a \cdot b) \cdot c = a \cdot (b \cdot c)$.

The Commutative Property of Addition

If a and b are real numbers, then
$a + b = b + a$.

The Commutative Property of Multiplication

If a and b are real numbers, then
$a \cdot b = b \cdot a$.

The Addition Property of Zero

If a is a real number, then
$a + 0 = 0 + a = a$.

The Multiplication Property of One

If a is a real number, then
$a \cdot 1 = 1 \cdot a = a$.

The Multiplication Property of Zero

If a is a real number, then
$a \cdot 0 = 0 \cdot a = 0$.

The Inverse Property of Multiplication

If a is a real number and $a \neq 0$, then
$a \cdot \dfrac{1}{a} = \dfrac{1}{a} \cdot a = 1$.

The Inverse Property of Addition

If a is a real number, then
$a + (-a) = (-a) + a = 0$.

Distributive Property

If a, b, and c are real numbers, then
$a(b + c) = ab + ac$.

Properties of Equations

Addition Property of Equations

If $a = b$, then $a + c = b + c$.

Multiplication Property of Equations

If $a = b$ and $c \neq 0$, then $a \cdot c = b \cdot c$.

Properties of Inequalities

Addition Property of Inequalities

If $a > b$, then $a + c > b + c$.
If $a < b$, then $a + c < b + c$.

Multiplication Property of Inequalities

If $a > b$ and $c > 0$, then $ac > bc$.
If $a < b$ and $c > 0$, then $ac < bc$.
If $a > b$ and $c < 0$, then $ac < bc$.
If $a < b$ and $c < 0$, then $ac > bc$.

Properties of Exponents

If m and n are integers, then $x^m \cdot x^n = x^{m+n}$.
If m and n are integers, then $(x^m)^n = x^{mn}$.

If $x \neq 0$, then $x^0 = 1$.

If m and n are integers and $x \neq 0$, then $\dfrac{x^m}{x^n} = x^{m-n}$.

If m, n, and p are integers, then $(x^m \cdot y^n)^p = x^{mp}y^{np}$.
If n is a positive integer and $x \neq 0$, then

$x^{-n} = \dfrac{1}{x^n}$ and $\dfrac{1}{x^{-n}} = x^n$.

If m, n, and p are integers and $y \neq 0$, then $\left(\dfrac{x^m}{y^n}\right)^p = \dfrac{x^{mp}}{y^{np}}$.

Principle of Zero Products

If $a \cdot b = 0$, then $a = 0$ or $b = 0$.

Properties of Radical Expressions

If a and b are positive real numbers, then $\sqrt{ab} = \sqrt{a}\sqrt{b}$. If a and b are positive real numbers, then $\sqrt{\dfrac{a}{b}} = \dfrac{\sqrt{a}}{\sqrt{b}}$.

Property of Squaring Both Sides of an Equation

If a and b are real numbers and $a = b$, then $a^2 = b^2$.

Properties of Logarithms

If x, y, and b are positive real numbers and $b \neq 1$, then
$\log_b(xy) = \log_b x + \log_b y$.
If x, y, and b are positive real numbers and $b \neq 1$, then

$\log_b \dfrac{x}{y} = \log_b x - \log_b y$.

If x and b are positive real numbers, $b \neq 1$, and r is any real number, then $\log_b x^r = r \log_b x$.
If x and b are positive real numbers and $b \neq 1$, then
$\log_b b^x = x$.

Table of Algebraic and Geometric Formulas

Slope of a Line

$$m = \frac{y_2 - y_1}{x_2 - x_1}, \; x_1 \neq x_2$$

Point-slope Formula for a Line

$$y - y_1 = m(x - x_1)$$

Quadratic Formula

$$x = \frac{-b \pm \sqrt{b^2 - 4ac}}{2a}$$

$$\text{discriminant} = b^2 - 4ac$$

Perimeter and Area of a Triangle, and Sum of the Measures of the Angles

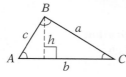

$$P = a + b + c$$

$$A = \frac{1}{2}bh$$

The sum of the measures of the angles in a triangle is 180°.

Pythagorean Theorem

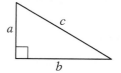

$$a^2 + b^2 = c^2$$

Perimeter and Area of a Rectangle

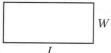

$$P = 2L + 2W$$
$$A = LW$$

Perimeter and Area of a Square

$$P = 4s$$
$$A = s^2$$

Area of a Trapezoid

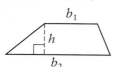

$$A = \frac{1}{2}h(b_1 + b_2)$$

Circumference and Area of a Circle

$$C = 2\pi r$$
$$A = \pi r^2$$

Volume and Surface Area of a Rectangular Solid

$$V = LWH$$
$$SA = 2LW + 2LH + 2WH$$

Volume and Surface Area of a Sphere

$$V = \frac{4}{3}\pi r^3$$

$$SA = 4\pi r^2$$

Volume and Surface Area of a Right Circular Cylinder

$$V = \pi r^2 h$$
$$SA = 2\pi r^2 + 2\pi rh$$

Volume and Surface Area of a Right Circular Cone

$$V = \frac{1}{3}\pi r^2 h$$

$$SA = \pi r^2 + \pi rl$$

Unless otherwise noted, all content on this page is © Cengage Learning.

Solutions to "You Try It"

Solutions to Chapter 1 "You Try It"

SECTION 1.1

You Try It 1 Replace z by each of the elements of the set and determine whether the inequality is true.

$z > -5$

$-10 > -5$ False

$-5 > -5$ False

$6 > -5$ True

The inequality is true for 6.

You Try It 2 **a.** Replace d in $-d$ by each element of the set and determine the value of the expression.

$-d$

$-(-11) = 11$

$-(0) = 0$

$-(8) = -8$

b. Replace d in $|d|$ by each element of the set and determine the value of the expression.

$|d|$

$|-11| = 11$

$|0| = 0$

$|8| = 8$

You Try It 3 $100 + (-43) = 57$

You Try It 4 $-51 + 42 + 17 + (-102)$

$= -9 + 17 + (-102)$

$= 8 + (-102)$

$= -94$

You Try It 5 $-8 + 7 = -1$

You Try It 6 $-9 - (-12) - 17 - 4$

$= -9 + 12 + (-17) + (-4)$

$= 3 + (-17) + (-4)$

$= -14 + (-4)$

$= -18$

You Try It 7 $-11 - (-12) = -11 + 12 = 1$

You Try It 8 $8(-9)10 = -72(10)$

$= -720$

You Try It 9 $(-2)3(-8)7 = -6(-8)7$

$= 48(7)$

$= 336$

You Try It 10 $-9(34) = -306$

You Try It 11 $(-135) \div (-9) = 15$

You Try It 12 $\dfrac{-72}{4} = -18$

You Try It 13 $-\dfrac{36}{-12} = -(-3)$

$= 3$

You Try It 14 $\dfrac{-72}{-8} = 9$

You Try It 15

Strategy

To find the average daily low temperature:

▶ Add the seven temperature readings.

▶ Divide the sum by 7.

Solution

$-6 + (-7) + 0 + (-5) + (-8) + (-1) + (-1) = -28$

$-28 \div 7 = -4$

The average daily low temperature was $-4°C$.

SECTION 1.2

You Try It 1

$$9)\overline{\,4.000\,} \quad \begin{array}{r} 0.444 \\ \hline -3\,6 \\ \hline 40 \\ -36 \\ \hline 40 \\ -36 \\ \hline 4 \end{array}$$

$\dfrac{4}{9} = 0.\overline{4}$

You Try It 2 $125\% = 125\left(\dfrac{1}{100}\right) = \dfrac{125}{100} = \dfrac{5}{4}$

$125\% = 125(0.01) = 1.25$

You Try It 3 $\dfrac{1}{3} = \dfrac{1}{3}(100\%)$

$$= \dfrac{100}{3}\% = 33\dfrac{1}{3}\%$$

You Try It 4 $0.043 = 0.043(100\%) = 4.3\%$

You Try It 5 The LCM of 8, 6, and 4 is 24.

$$-\dfrac{7}{8} - \dfrac{5}{6} + \dfrac{3}{4} = -\dfrac{21}{24} - \dfrac{20}{24} + \dfrac{18}{24}$$

$$= \dfrac{-21}{24} + \dfrac{-20}{24} + \dfrac{18}{24}$$

$$= \dfrac{-21 - 20 + 18}{24}$$

$$= \dfrac{-23}{24} = -\dfrac{23}{24}$$

You Try It 6 $16.127 - 67.91$

$$= 16.127 + (-67.91)$$

$$= -51.783$$

You Try It 7
The quotient is positive.

$$-\dfrac{3}{8} \div \left(-\dfrac{5}{12}\right) = \dfrac{3}{8} \div \dfrac{5}{12} = \dfrac{3}{8} \cdot \dfrac{12}{5}$$

$$= \dfrac{3 \cdot 12}{8 \cdot 5}$$

$$= \dfrac{3 \cdot \overset{1}{2} \cdot \overset{1}{2} \cdot 3}{2 \cdot \underset{1}{2} \cdot \underset{1}{2} \cdot 5} = \dfrac{9}{10}$$

You Try It 8 The product is negative.

$$\begin{array}{r} 5.44 \\ \times\ 3.8 \\ \hline 4352 \\ 1632 \\ \hline 20.672 \end{array}$$

$$-5.44(3.8) = -20.672$$

You Try It 9 $-6^3 = -(6 \cdot 6 \cdot 6) = -216$

You Try It 10 $(-3)^4 = (-3)(-3)(-3)(-3) = 81$

You Try It 11
$(3^3) \cdot (-2)^3 = (3)(3)(3) \cdot (-2)(-2)(-2)$
$$= 27(-8) = -216$$

You Try It 12
$$\left(-\dfrac{2}{5}\right)^2 = \left(-\dfrac{2}{5}\right)\left(-\dfrac{2}{5}\right) = \dfrac{4}{25}$$

You Try It 13 $-3(0.3)^3 = -3(0.3)(0.3)(0.3)$
$$= -0.9(0.3)(0.3)$$
$$= -0.27(0.3) = -0.081$$

You Try It 14
$-5\sqrt{32} = -5\sqrt{16 \cdot 2} = -5\sqrt{16}\sqrt{2}$
$$= -5 \cdot 4\sqrt{2} = -20\sqrt{2}$$

You Try It 15
$\sqrt{216} = \sqrt{36 \cdot 6} = \sqrt{36}\sqrt{6} = 6\sqrt{6}$

You Try It 16

Strategy To find the difference:
- ▸ Read the numbers from the graph that correspond to the numbers of barrels of oil imported each day (13.2, 5.4, 3.2, 2.1, 2.3).
- ▸ Add the numbers.
- ▸ Subtract the sum from the total number of barrels consumed each day, calculated in Example 16 (37.7 million).

Solution $13.2 + 5.4 + 3.2 + 2.1 + 2.3 = 26.2$
$37.7 - 26.2 = 11.5$
The difference is 11.5 million barrels of oil per day.

SECTION 1.3

You Try It 1
$18 - 5[8 - 2(2 - 5)] \div 10$
$= 18 - 5[8 - 2(-3)] \div 10$ • Perform operations inside grouping symbols.
$= 18 - 5[8 + 6] \div 10$
$= 18 - 5[14] \div 10$
$= 18 - 70 \div 10$ • Do multiplication and division from left to right.
$= 18 - 7$
$= 11$ • Do addition and subtraction from left to right.

You Try It 2 $(6.97 - 4.72)^2 \cdot 4.5 \div 0.05$
$= (2.25)^2 \cdot 4.5 \div 0.05$
$= 5.0625 \cdot 4.5 \div 0.05$
$= 22.78125 \div 0.05$
$= 455.625$

SECTION 1.4

You Try It 1 $\dfrac{a^2 + b^2}{a + b}$

$$\dfrac{5^2 + (-3)^2}{5 + (-3)} = \dfrac{25 + 9}{5 + (-3)}$$

$$= \dfrac{34}{2}$$

$$= 17$$

You Try It 2 $x^3 - 2(x + y) + z^2$

$(2)^3 - 2[2 + (-4)] + (-3)^2$
$= 8 - 2(-2) + 9$
$= 8 + 4 + 9$
$= 12 + 9$
$= 21$

You Try It 3 $3a - 2b - 5a + 6b = -2a + 4b$

You Try It 4 $-3y^2 + 7 + 8y^2 - 14 = 5y^2 - 7$

You Try It 5 $-5(4y^2) = -20y^2$

You Try It 6 $-7(-2a) = 14a$

You Try It 7 $-\dfrac{3}{5}\left(-\dfrac{7}{9}a\right) = \dfrac{7}{15}a$

You Try It 8 $(3a - 1)5 = 15a - 5$

You Try It 9 $-8(-2a + 7b) = 16a - 56b$

You Try It 10 $3(12x^2 - x + 8) = 36x^2 - 3x + 24$

You Try It 11 $3y - 2(y - 7x) = 3y - 2y + 14x$
$$= y + 14x$$

You Try It 12
$$-2(x - 2y) - (-x + 3y) = -2x + 4y + x - 3y$$
$$= -x + y$$

You Try It 13
$$3y - 2[x - 4(2 - 3y)] = 3y - 2[x - 8 + 12y]$$
$$= 3y - 2x + 16 - 24y$$
$$= -2x - 21y + 16$$

You Try It 14   the speed of the older model: s
the speed of the new jet plane is twice the speed of the older model: $2s$

You Try It 15   the length of the longer piece: y
the length of the shorter piece: $6 - y$

SECTION 1.5

You Try It 1 $A = \{-9, -7, -5, -3, -1\}$

You Try It 2 $A = \{1, 3, 5, \ldots\}$

You Try It 3 $A \cup B = \{-2, -1, 0, 1, 2, 3, 4\}$

You Try It 4 $C \cap D = \{10, 16\}$

You Try It 5 $A \cap B = \varnothing$

You Try It 6   **a.** $\{x \mid x \le 3\}$ is the set of real numbers less than or equal to 3. This set extends forever in the negative direction. In interval notation, this set is written $(-\infty, 3]$.

   b. $\{x \mid -5 \le x \le -3\}$ is the set of real numbers between -5 and -3, including -5 and -3. In interval notation, this set is written $[-5, -3]$.

You Try It 7   **a.** The interval $(-3, \infty)$ is the set of real numbers greater than -3. In set-builder notation, this set is written $\{x \mid x > -3\}$.

   b. The interval $[0, 4)$ is the set of real numbers between 0 and 4, including 0 and excluding 4. In set-builder notation, this set is written $\{x \mid 0 \le x < 4\}$.

You Try It 8   **a.** The graph is the set of real numbers between -4 and 4, including -4 and 4. Use brackets at -4 and 4.

   b. The graph is the set of real numbers greater than -3. Use a parenthesis at -3.

You Try It 9   The graph is the set of real numbers between 2 and 5, including 2 and 5. Use brackets at 2 and 5.

Solutions to Chapter 2 "You Try It"

SECTION 2.1

You Try It 1

$$10x - x^2 = 3x - 10$$

$10(5) - (5)^2$	$3(5) - 10$
$50 - 25$	$15 - 10$
$25 \ne 5$	

No, 5 is not a solution.

You Try It 2
$$26 = y - 14$$
$$26 + 14 = y - 14 + 14$$
$$40 = y - 0$$
$$40 = y$$

The solution is 40.

You Try It 3
$$-\dfrac{2x}{5} = 6$$
$$\left(-\dfrac{5}{2}\right)\left(-\dfrac{2}{5}x\right) = \left(-\dfrac{5}{2}\right)(6)$$
$$x = -15$$

The solution is -15.

You Try It 4
$$4x - 8x = 16$$
$$-4x = 16$$
$$\dfrac{-4x}{-4} = \dfrac{16}{-4}$$
$$x = -4$$

The solution is -4.

Unless otherwise noted, all content on this page is © Cengage Learning.

You Try It 5

$$P \cdot B = A$$

$$\frac{1}{6}B = 18 \qquad \bullet \ 16\frac{2}{3}\% = \frac{1}{6}$$

$$6 \cdot \frac{1}{6}B = 6 \cdot 18$$

$$B = 108$$

18 is $16\frac{2}{3}\%$ of 108.

You Try It 6

Strategy Use the basic percent equation. $B = 310$, the U.S. population; $A = 162.9$, the number of people who watched the game; P is the unknown percent.

Solution $P \cdot 310 = 162.9$

$$\frac{P \cdot 310}{310} = \frac{162.9}{310}$$

$$P \approx 0.525$$

Approximately 52.5% of the U.S. population watched Super Bowl XLV.

You Try It 7

Strategy
To find how much Clarissa must deposit into the account:
▶ Find the amount of interest earned on the municipal bond by solving $I = Prt$ for I using $P = 1000$, $r = 6.4\% = 0.064$, and $t = 1$.
▶ Solve $I = Prt$ for P using the amount of interest earned on the municipal bond as I. $r = 8\% = 0.08$, and $t = 1$.

Solution
$I = Prt$
$\ \ = 1000(0.064)(1) = 64$
The interest earned on the municipal bond was $64.

$$I = Prt$$
$$64 = P(0.08)(1) \qquad \bullet \ I = 64, r = 0.08, t = 1$$
$$64 = 0.08P$$
$$\frac{64}{0.08} = \frac{0.08P}{0.08}$$
$$800 = P$$

Clarissa must invest $800 in the account.

You Try It 8

Strategy To find the number of ounces of cereal in the bowl, solve $Q = Ar$ for A using $Q = 2$ and $r = 25\% = 0.25$.

Solution
$$Q = Ar$$
$$2 = A(0.25) \qquad \bullet \ Q = 2, r = 0.25$$
$$\frac{2}{0.25} = \frac{A(0.25)}{0.25}$$
$$8 = A$$

The bowl contains 8 oz of cereal.

You Try It 9

Strategy To find the distance, solve the equation $d = rt$ for d. The time is 3 h. Therefore, $t = 3$. The plane is moving against the wind, which means the headwind is slowing the actual speed of the plane. 250 mph $-$ 25 mph $= 225$ mph. Thus $r = 225$.

Solution $d = rt$

$$d = 225(3) \qquad \bullet \ r = 225, t = 3$$
$$\ \ = 675$$

The plane travels 675 mi in 3 h.

SECTION 2.2

You Try It 1

$$5x + 7 = 10$$
$$5x + 7 - 7 = 10 - 7 \qquad \bullet \text{ Subtract 7.}$$
$$5x = 3$$
$$\frac{5x}{5} = \frac{3}{5} \qquad \bullet \text{ Divide by 5.}$$
$$x = \frac{3}{5}$$

The solution is $\frac{3}{5}$.

You Try It 2

$$2 = 11 + 3x$$
$$2 - 11 = 11 - 11 + 3x \qquad \bullet \text{ Subtract 11.}$$
$$-9 = 3x$$
$$\frac{-9}{3} = \frac{3x}{3} \qquad \bullet \text{ Divide by 3.}$$
$$-3 = x$$

The solution is -3.

You Try It 3

$$\frac{5}{8} - \frac{2x}{3} = \frac{5}{4}$$
$$\frac{5}{8} - \frac{5}{8} - \frac{2}{3}x = \frac{5}{4} - \frac{5}{8} \qquad \bullet \text{ Recall that } \frac{2x}{3} = \frac{2}{3}x.$$
$$-\frac{2}{3}x = \frac{5}{8}$$
$$-\frac{3}{2}\left(-\frac{2}{3}x\right) = -\frac{3}{2}\left(\frac{5}{8}\right) \qquad \bullet \text{ Multiply by } -\frac{3}{2}.$$
$$x = -\frac{15}{16}$$

The solution is $-\frac{15}{16}$.

You Try It 4

$$\frac{2}{3}x + 3 = \frac{7}{2}$$

$$6\left(\frac{2}{3}x + 3\right) = 6\left(\frac{7}{2}\right)$$

$$6\left(\frac{2}{3}x\right) + 6(3) = 6\left(\frac{7}{2}\right)$$ • **Distributive Property**

$$4x + 18 = 21$$

$$4x + 18 - 18 = 21 - 18$$ • **Subtract 18.**

$$4x = 3$$

$$\frac{4x}{4} = \frac{3}{4}$$ • **Divide by 4.**

$$x = \frac{3}{4}$$

The solution is $\frac{3}{4}$.

You Try It 5 $x - 5 + 4x = 25$

$$5x - 5 = 25$$

$$5x - 5 + 5 = 25 + 5$$

$$5x = 30$$

$$\frac{5x}{5} = \frac{30}{5}$$

$$x = 6$$

The solution is 6.

You Try It 6

$$5x + 4 = 6 + 10x$$

$$5x - 10x + 4 = 6 + 10x - 10x$$ • **Subtract 10x.**

$$-5x + 4 = 6$$

$$-5x + 4 - 4 = 6 - 4$$ • **Subtract 4.**

$$-5x = 2$$

$$\frac{-5x}{-5} = \frac{2}{-5}$$ • **Divide by -5.**

$$x = -\frac{2}{5}$$

The solution is $-\frac{2}{5}$.

You Try It 7

$$5x - 10 - 3x = 6 - 4x$$

$$2x - 10 = 6 - 4x$$ • **Combine like terms.**

$$2x + 4x - 10 = 6 - 4x + 4x$$ • **Add 4x.**

$$6x - 10 = 6$$

$$6x - 10 + 10 = 6 + 10$$ • **Add 10.**

$$6x = 16$$

$$\frac{6x}{6} = \frac{16}{6}$$ • **Divide by 6.**

$$x = \frac{8}{3}$$

The solution is $\frac{8}{3}$.

You Try It 8

$$5x - 4(3 - 2x) = 2(3x - 2) + 6$$

$$5x - 12 + 8x = 6x - 4 + 6$$ • **Distributive Property**

$$13x - 12 = 6x + 2$$

$$13x - 6x - 12 = 6x - 6x + 2$$ • **Subtract 6x.**

$$7x - 12 = 2$$

$$7x - 12 + 12 = 2 + 12$$ • **Add 12.**

$$7x = 14$$

$$\frac{7x}{7} = \frac{14}{7}$$ • **Divide by 7.**

$$x = 2$$

The solution is 2.

You Try It 9

$$-2[3x - 5(2x - 3)] = 3x - 8$$

$$-2[3x - 10x + 15] = 3x - 8$$ • **Distributive Property**

$$-2[-7x + 15] = 3x - 8$$

$$14x - 30 = 3x - 8$$

$$14x - 3x - 30 = 3x - 3x - 8$$ • **Subtract 3x.**

$$11x - 30 = -8$$

$$11x - 30 + 30 = -8 + 30$$ • **Add 30.**

$$11x = 22$$

$$\frac{11x}{11} = \frac{22}{11}$$ • **Divide by 11.**

$$x = 2$$

The solution is 2.

You Try It 10

Strategy   Given: $F_1 = 45$
$F_2 = 80$
$d = 25$

Unknown: x

Solution

$$F_1 x = F_2(d - x)$$

$$45x = 80(25 - x)$$

$$45x = 2000 - 80x$$

$$45x + 80x = 2000 - 80x + 80x$$

$$125x = 2000$$

$$\frac{125x}{125} = \frac{2000}{125}$$

$$x = 16$$

The fulcrum is 16 ft from the 45-pound force.

SECTION 2.3

You Try It 1

The total of three times the smaller number and six	amounts to	seven less than the product of four and the larger number

Strategy

The smaller number: n

The larger number: $12 - n$

Solution

$$3n + 6 = 4(12 - n) - 7$$
$$3n + 6 = 48 - 4n - 7$$
$$3n + 6 = 41 - 4n$$
$$3n + 4n + 6 = 41 - 4n + 4n$$
$$7n + 6 = 41$$
$$7n + 6 - 6 = 41 - 6$$
$$7n = 35$$
$$\frac{7n}{7} = \frac{35}{7}$$
$$n = 5$$
$$12 - n = 12 - 5 = 7$$

The smaller number is 5.

The larger number is 7.

You Try It 2

Strategy
► First integer: n
  Second integer: $n + 1$
  Third integer: $n + 2$
► The sum of the three integers is -6.

Solution
$$n + (n + 1) + (n + 2) = -6$$
$$3n + 3 = -6$$
$$3n = -9$$
$$n = -3$$
$$n + 1 = -3 + 1 = -2$$
$$n + 2 = -3 + 2 = -1$$

The three consecutive integers are -3, -2, and -1.

You Try It 3

Strategy
To find the number of tickets purchased, write and solve an equation using x to represent the number of tickets purchased.

$3.50 plus $17.50 for each ticket	is	$161

Solution
$$3.50 + 17.50x = 161$$
$$3.50 - 3.50 + 17.50x = 161 - 3.50$$
$$17.50x = 157.50$$
$$\frac{17.50x}{17.50} = \frac{157.50}{17.50}$$
$$x = 9$$

You purchased 9 tickets.

You Try It 4

Strategy
To find the length, write and solve an equation using x to represent the length of the shorter piece and $22 - x$ to represent the length of the longer piece.

The length of the longer piece	is	4 in. more than twice the length of the shorter piece

Solution
$$22 - x = 2x + 4$$
$$22 - x - 2x = 2x - 2x + 4$$
$$22 - 3x = 4$$
$$22 - 22 - 3x = 4 - 22$$
$$-3x = -18$$
$$\frac{-3x}{-3} = \frac{-18}{-3}$$
$$x = 6$$
$$22 - x = 22 - 6 = 16$$

The length of the shorter piece is 6 in.

The length of the longer piece is 16 in.

SECTION 2.4

You Try It 1

Strategy
► Pounds of $.75 fertilizer: x

	Amount	Cost	Value
$.90 fertilizer	20	0.90	0.90(20)
$.75 fertilizer	x	0.75	0.75x
$.85 fertilizer	$20 + x$	0.85	0.85(20 + x)

► The sum of the values before mixing equals the value after mixing.

Solution
$$0.90(20) + 0.75x = 0.85(20 + x)$$
$$18 + 0.75x = 17 + 0.85x$$
$$18 - 0.10x = 17$$
$$-0.10x = -1$$
$$x = 10$$

10 lb of the $.75 fertilizer must be added.

You Try It 2

Strategy
► Liters of 6% solution: x

	Amount	Percent	Quantity
6% solution	x	0.06	0.06x
12% solution	5	0.12	5(0.12)
8% solution	$x + 5$	0.08	0.08(x + 5)

► The sum of the quantities before mixing equals the quantity after mixing.

Unless otherwise noted, all content on this page is © Cengage Learning.

Solution

$$0.06x + 5(0.12) = 0.08(x + 5)$$
$$0.06x + 0.60 = 0.08x + 0.40$$
$$-0.02x + 0.60 = 0.40$$
$$-0.02x = -0.20$$
$$x = 10$$

The pharmacist adds 10 L of the 6% solution to the 12% solution to get an 8% solution.

You Try It 3

Strategy

▸ Rate of the first train: r
Rate of the second train: $2r$

	Rate	Time	Distance
1st train	r	3	$3r$
2nd train	$2r$	3	$3(2r)$

▸ The sum of the distances traveled by the two trains equals 288 mi.

Solution

$$3r + 3(2r) = 288$$
$$3r + 6r = 288$$
$$9r = 288$$
$$r = 32$$
$$2r = 2(32) = 64$$

The first train is traveling at 32 mph.
The second train is traveling at 64 mph.

You Try It 4

Strategy

▸ Time spent flying out: t
Time spent flying back: $5 - t$

	Rate	Time	Distance
Out	150	t	$150t$
Back	100	$5 - t$	$100(5 - t)$

▸ The distance out equals the distance back.

Solution

$$150t = 100(5 - t)$$
$$150t = 500 - 100t$$
$$250t = 500$$
$$t = 2 \qquad \text{(The time out was 2 h.)}$$
The distance out $= 150t = 150(2)$
$$= 300$$
The parcel of land was 300 mi away.

SECTION 2.5

You Try It 1

$$2x - 1 < 6x + 7$$
$$-4x - 1 < 7 \qquad \bullet \text{ Subtract } 6x \text{ from each side.}$$
$$-4x < 8 \qquad \bullet \text{ Add 1 to each side.}$$
$$\frac{-4x}{-4} > \frac{8}{-4} \qquad \bullet \text{ Divide each side by } -4.$$
$$x > -2$$
$$\{x \mid x > -2\}$$

You Try It 2

$$6 - 3(2x + 1) \le 8 - 4x$$
$$6 - 6x - 3 \le 8 - 4x$$
$$3 - 6x \le 8 - 4x$$
$$3 \le 8 + 2x$$
$$-5 \le 2x$$
$$-\frac{5}{2} \le x$$
$$\left[-\frac{5}{2}, \infty \right)$$

You Try It 3

$$-2 \le 5x + 3 \le 13$$
$$-2 - 3 \le 5x + 3 - 3 \le 13 - 3 \qquad \bullet \text{ Subtract 3 from each of the three parts.}$$

$$-5 \le 5x \le 10$$
$$\frac{-5}{5} \le \frac{5x}{5} \le \frac{10}{5} \qquad \bullet \text{ Divide each of the three parts by 5.}$$
$$-1 \le x \le 2$$
$$[-1, 2]$$

You Try It 4

$$\begin{array}{lll}
2 - 3x > 11 & \text{or} & 5 + 2x > 7 \\
-3x > 9 & & 2x > 2 \\
x < -3 & & x > 1 \\
\{x \mid x < -3\} & & \{x \mid x > 1\}
\end{array}$$
$$\{x \mid x < -3\} \cup \{x \mid x > 1\}$$

You Try It 5

Strategy

To find the maximum height, substitute the given values into the inequality
$$\frac{1}{2}bh < A \text{ and solve.}$$

Solution

$$\frac{1}{2}bh < A$$
$$\frac{1}{2}(12)(x + 2) < 50$$
$$6(x + 2) < 50$$
$$6x + 12 < 50$$
$$6x < 38$$
$$x < \frac{19}{3}$$

The largest integer less than $\frac{19}{3}$ is 6.

$$x + 2 = 6 + 2 = 8$$

The maximum height of the triangle is 8 in.

Unless otherwise noted, all content on this page is © Cengage Learning.

You Try It 6

Strategy To find the range of scores, write and solve an inequality using N to represent the score on the fifth test.

Solution

$$80 \le \frac{72 + 94 + 83 + 70 + N}{5} \le 89$$

$$80 \le \frac{319 + N}{5} \le 89$$

$$5 \cdot 80 \le 5\left(\frac{319 + N}{5}\right) \le 5 \cdot 89$$

$$400 \le 319 + N \le 445$$

$$400 - 319 \le 319 + N - 319 \le 445 - 319$$

$$81 \le N \le 126$$

Because 100 is the maximum score, the range of scores that will give Luisa a B grade is $81 \le N \le 100$.

SECTION 2.6

You Try It 1

$|2x - 3| = 5$

$$
\begin{array}{ll}
2x - 3 = 5 & 2x - 3 = -5 \\
2x = 8 & 2x = -2 \qquad \bullet \text{ Add 3.} \\
x = 4 & x = -1 \qquad \bullet \text{ Divide by 2.}
\end{array}
$$

The solutions are 4 and -1.

You Try It 2 $\quad |x - 3| = -2$

There is no solution to this equation because the absolute value of a number must be nonnegative.

You Try It 3 $\quad 5 - |3x + 5| = 3$

$$
\begin{array}{ll}
-|3x + 5| = -2 & \bullet \text{ Subtract 5.} \\
|3x + 5| = 2 & \bullet \text{ Multiply by } -1.
\end{array}
$$

$$
\begin{array}{ll}
3x + 5 = 2 & 3x + 5 = -2 \\
3x = -3 & 3x = -7 \\
x = -1 & x = -\dfrac{7}{3}
\end{array}
$$

The solutions are -1 and $-\dfrac{7}{3}$.

You Try It 4 $\quad |3x + 2| < 8$

$$-8 < 3x + 2 < 8$$

$$-8 - 2 < 3x + 2 - 2 < 8 - 2$$

$$-10 < 3x < 6$$

$$\frac{-10}{3} < \frac{3x}{3} < \frac{6}{3}$$

$$-\frac{10}{3} < x < 2$$

$$\left\{ x \,\middle|\, -\frac{10}{3} < x < 2 \right\}$$

You Try It 5 $\quad |3x - 7| < 0$

The absolute value of a number must be non-negative.

The solution set is the empty set.

$\varnothing$

You Try It 6 $\quad |2x + 7| \ge -1$

The absolute value of a number is nonnegative.

The solution set is the set of real numbers.

You Try It 7

$|5x + 3| > 8$

$$
\begin{array}{lll}
5x + 3 < -8 & \text{or} & 5x + 3 > 8 \\
5x < -11 & & 5x > 5 \\
x < -\dfrac{11}{5} & & x > 1
\end{array}
$$

$$
\left\{ x \,\middle|\, x < -\frac{11}{5} \right\} \qquad \{x \mid x > 1\}
$$

$$
\left\{ x \,\middle|\, x < -\frac{11}{5} \right\} \cup \{x \mid x > 1\}
$$

You Try It 8

Strategy

Let b represent the diameter of the bushing, T the tolerance, and d the lower and upper limits of the diameter. Solve the absolute value inequality $|d - b| \le T$ for d.

Solution

$$|d - b| \le T$$

$$|d - 2.55| \le 0.003$$

$$-0.003 \le d - 2.55 \le 0.003$$

$$-0.003 + 2.55 \le d - 2.55 + 2.55 \le 0.003 + 2.55$$

$$2.547 \le d \le 2.553$$

The lower and upper limits of the diameter of the bushing are 2.547 in. and 2.553 in.

Solutions to Chapter 3 "You Try It"

SECTION 3.1

You Try It 1

$$QR + RS + ST = QT$$

$$24 + RS + 17 = 62 \qquad \bullet \; QR = 24, \; ST = 17, \; QT = 62$$

$$41 + RS = 62 \qquad \bullet \text{ Add 24 and 17.}$$

$$RS = 21 \qquad \bullet \text{ Subtract 41 from each side.}$$

$$RS = 21 \text{ cm}$$

You Try It 2

$$AC = AB + BC$$

$$AC = \frac{1}{4}(BC) + BC \qquad \bullet \; AB \text{ is one-fourth } BC.$$

$$AC = \frac{1}{4}(16) + 16 \qquad \bullet \; BC = 16$$

$$AC = 4 + 16$$

$$AC = 20$$

$$AC = 20 \text{ ft}$$

You Try It 3

Strategy   Supplementary angles are two angles whose sum is 180°. To find the supplement, let x represent the supplement of a 129° angle. Write an equation and solve for x.

Solution $x + 129° = 180°$
$ x = 51°$

The supplement of a 129° angle is a 51° angle.

You Try It 4

Strategy   To find the measure of $\angle a$, write an equation using the fact that the sum of the measure of $\angle a$ and 68° is 118°. Solve for $\angle a$.

Solution $\angle a + 68° = 118°$
$ \angle a = 50°$

The measure of $\angle a$ is 50°.

You Try It 5

Strategy   The angles labeled are adjacent angles of intersecting lines and are therefore supplementary angles. To find x, write an equation and solve for x.

Solution $(x + 16°) + 3x = 180°$
$ 4x + 16° = 180°$
$ 4x = 164°$
$ x = 41°$

You Try It 6

Strategy $3x = y$ because corresponding angles have the same measure. $y + (x + 40°) = 180°$ because adjacent angles of intersecting lines are supplementary angles. Substitute $3x$ for y and solve for x.

Solution $3x + (x + 40°) = 180°$
$ 4x + 40° = 180°$
$ 4x = 140°$
$ x = 35°$

You Try It 7

Strategy   ▶ To find the measure of angle b, use the fact that $\angle b$ and $\angle x$ are supplementary angles.
▶ To find the measure of angle c, use the fact that the sum of the interior angles of a triangle is 180°.
▶ To find the measure of angle y, use the fact that $\angle c$ and $\angle y$ are vertical angles.

Solution $\angle b + \angle x = 180°$
$\angle b + 100° = 180°$
$ \angle b = 80°$

$\angle a + \angle b + \angle c = 180°$
$45° + 80° + \angle c = 180°$
$ 125° + \angle c = 180°$
$ \angle c = 55°$

$\angle y = \angle c = 55°$

You Try It 8

Strategy   To find the measure of the third angle, use the fact that the measure of a right angle is 90° and the fact that the sum of the measures of the interior angles of a triangle is 180°. Write an equation using x to represent the measure of the third angle. Solve the equation for x.

Solution $x + 90° + 34° = 180°$
$ x + 124° = 180°$
$ x = 56°$

The measure of the third angle is 56°.

SECTION 3.2

You Try It 1

Strategy   To find the perimeter, use the formula for the perimeter of a square. Substitute 60 for s and solve for P.

Solution $P = 4s$
$P = 4(60)$
$P = 240$

The perimeter of the infield is 240 ft.

You Try It 2

Strategy   To find the length of molding needed, use the formula for the perimeter of a rectangle. Substitute 12 for L and 8 for W, and solve for P.

Solution $P = 2L + 2W$
$P = 2(12) + 2(8)$
$P = 24 + 16$
$P = 40$

The length of decorative molding needed to edge the tops of the walls is 40 ft.

You Try It 3

Strategy   To find the circumference, use the circumference formula that involves the diameter. Leave the answer in terms of π.

Solution $C = \pi d$
$C = \pi(9)$
$C = 9\pi$

The circumference is 9π in.

You Try It 4

Strategy

To find the number of rolls of wallpaper to be purchased:

▶ Use the formula for the area of a rectangle to find the area of one wall.

▶ Multiply the area of one wall by the number of walls to be covered (2).

▶ Divide the area of wall to be covered by the area one roll of wallpaper will cover (30).

Solution

$A = LW$

$A = 12 \cdot 8 = 96$ • **The area of one wall is 96 ft².**

$2(96) = 192$ • **The area of the two walls is 192 ft².**

$192 \div 30 = 6.4$

Because a portion of a seventh roll is needed, 7 rolls of wallpaper should be purchased.

You Try It 5

Strategy   To find the area, use the formula for the area of a circle. An approximation is asked for; use the π key on a calculator. $r = 11$

Solution $A = \pi r^2$

$A = \pi (11)^2$

$A = 121\pi$

$A \approx 380.13$

The area is approximately 380.13 cm².

SECTION 3.3

You Try It 1

Strategy   To find the volume, use the formula for the volume of a cube. $s = 2.5$

Solution $V = s^3$

$V = (2.5)^3 = 15.625$

The volume of the cube is 15.625 m³.

You Try It 2

Strategy   To find the volume:

▶ Find the radius of the base of the cylinder. $d = 8$

▶ Use the formula for the volume of a cylinder. Leave the answer in terms of π.

Solution $r = \dfrac{1}{2}d = \dfrac{1}{2}(8) = 4$

$V = \pi r^2 h = \pi (4)^2(22) = \pi (16)(22) = 352\pi$

The volume of the cylinder is 352π ft³.

You Try It 3

Strategy   To find the surface area:

▶ Find the radius of the base of the cylinder. $d = 6$

▶ Use the formula for the surface area of a cylinder. An approximation is asked for; use the π key on a calculator.

Solution $r = \dfrac{1}{2}d = \dfrac{1}{2}(6) = 3$

$SA = 2\pi r^2 + 2\pi rh$

$SA = 2\pi (3)^2 + 2\pi (3)(8)$

$ = 2\pi (9) + 2\pi (3)(8)$

$ = 18\pi + 48\pi$

$ = 66\pi$

$ \approx 207.35$

The surface area of the cylinder is approximately 207.35 ft².

You Try It 4

Strategy   To find the surface area, use the formula for the surface area of a cube. $s = 10$

Solution $SA = 6s^2$

$SA = 6(10)^2$

$ = 6(100)$

$ = 600$

The surface area of the cube is 600 cm².

Solutions to Chapter 4 "You Try It"

SECTION 4.1

You Try It 1

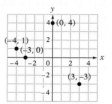

You Try It 2

$A(4, -2)$, $B(-2, 4)$

The abscissa of D is 0.

The ordinate of C is 0.

You Try It 3

$$\begin{array}{c|c} x - 3y = -14 & \\ \hline -2 - 3(4) & -14 \\ -2 - 12 & -14 \\ -14 = -14 & \end{array}$$

Yes, $(-2, 4)$ is a solution of $x - 3y = -14$.

Unless otherwise noted, all content on this page is © Cengage Learning.

You Try It 4 $x + 2y = 4$
$$2y = -x + 4$$
$$y = -\frac{1}{2}x + 2$$

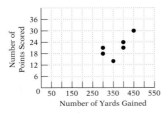

x	y
-4	4
-2	3
0	2
2	1

You Try It 5

Strategy   Graph the ordered pairs on a rectangular coordinate system, where the horizontal axis represents the number of yards gained and the vertical axis represents the number of points scored.

Solution

You Try It 6   Locate 20 in. on the x-axis. Follow the vertical line from 20 to a point plotted in the diagram. Follow a horizontal line from that point to the y-axis. Read the number where the horizontal line intersects the y-axis.

The ordered pair is $(20, 37)$, which indicates that when the space between seats was 20 in., the evacuation time was 37 s.

SECTION 4.2

You Try It 1
The domain is the set of first coordinates. The range is the set of second coordinates.
Domain: $\{-1, 3, 4, 6\}$
Range: $\{5\}$

You Try It 2 $G(x) = \dfrac{3x}{x + 2}$
$$G(-4) = \frac{3(-4)}{-4 + 2} = \frac{-12}{-2} = 6$$

You Try It 3 $f(x) = x^2 - 11$
$$f(3h) = (3h)^2 - 11$$
$$= 9h^2 - 11$$

You Try It 4
Because $3x^2 - 5x + 2$ evaluates to a real number for any value of x, the domain of $f(x) = 3x^2 - 5x + 2$ is all real numbers, or $\{x | -\infty < x < \infty\}$.

Unless otherwise noted, all content on this page is © Cengage Learning.

SECTION 4.3

You Try It 1

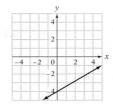

You Try It 2

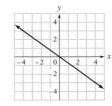

You Try It 3 $-3x + 2y = 4$
$$2y = 3x + 4$$
$$y = \frac{3}{2}x + 2$$

You Try It 4 $y - 3 = 0$
$$y = 3$$

- The graph of $y = 3$ goes through the point with coordinates $(0, 3)$.

You Try It 5

x-intercept:
$$3x - y = 2$$
$$3x - 0 = 2 \quad \bullet \text{ Let } y = 0.$$
$$3x = 2$$
$$x = \frac{2}{3}$$

x-intercept: $\left(\dfrac{2}{3}, 0\right)$

y-intercept:
$$3x - y = 2$$
$$3(0) - y = 2 \quad \bullet \text{ Let } x = 0.$$
$$-y = 2$$
$$y = -2$$

y-intercept: $(0, -2)$

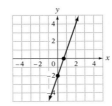

You Try It 6

$$g(x) = 4 + \frac{2}{3}x$$

$$0 = 4 + \frac{2}{3}x \quad \bullet \text{ Let } g(x) = 0.$$

$$-\frac{2}{3}x = 4 \quad \bullet \text{ Solve for } x.$$

$$x = -6$$

The zero is -6.

You Try It 7

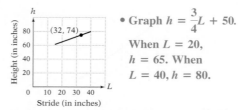

$\bullet$ Graph $h = \frac{3}{4}L + 50$.

When $L = 20$,
$h = 65$. When
$L = 40, h = 80$.

The ordered pair $(32, 74)$ means that a person with a stride of 32 in. is 74 in. tall.

SECTION 4.4

You Try It 1   Use $P_1(4, -3)$ and $P_2(2, 7)$.

$$m = \frac{y_2 - y_1}{x_2 - x_1} = \frac{7 - (-3)}{2 - 4} = \frac{10}{-2} = -5$$

The slope is -5.

You Try It 2   Use $P_1(6, -1)$ and $P_2(6, 7)$.

$$m = \frac{y_2 - y_1}{x_2 - x_1} = \frac{7 - (-1)}{6 - 6} = \frac{8}{0}$$

Division by zero is not defined.
The slope of the line is undefined.

You Try It 3   Choose $P_1(5, 25{,}000)$ and $P_2(2, 55{,}000)$.

$$m = \frac{y_2 - y_1}{x_2 - x_1}$$

$$= \frac{55{,}000 - 25{,}000}{2 - 5} \quad \bullet \begin{array}{l}(x_1, y_1) = (5, 25{,}000), \\ (x_2, y_2) = (2, 55{,}000)\end{array}$$

$$= \frac{30{,}000}{-3}$$

$$= -10{,}000$$

A slope of $-10{,}000$ means the value of the recycling truck is decreasing by \$10,000 per year.

You Try It 4
$$2x + 3y = 6$$
$$3y = -2x + 6$$
$$y = -\frac{2}{3}x + 2$$
$$m = -\frac{2}{3} = \frac{-2}{3}$$

y-intercept $= (0, 2)$

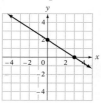

You Try It 5   Locate $P(-3, -2)$.

$$m = 3 = \frac{3}{1}$$

SECTION 4.5

You Try It 1 $m = -\frac{1}{3}$ $(x_1, y_1) = (-3, -2)$

$$y - y_1 = m(x - x_1)$$

$$y - (-2) = -\frac{1}{3}[x - (-3)]$$

$$y + 2 = -\frac{1}{3}(x + 3)$$

$$y + 2 = -\frac{1}{3}x - 1$$

$$y = -\frac{1}{3}x - 3$$

The equation of the line is
$$y = -\frac{1}{3}x - 3.$$

You Try It 2 $m = -3$ $(x_1, y_1) = (4, -3)$

$$y - y_1 = m(x - x_1)$$
$$y - (-3) = -3(x - 4)$$
$$y + 3 = -3x + 12$$
$$y = -3x + 9$$

The equation of the line is $y = -3x + 9$.

You Try It 3   Use $P_1(2, 0)$ and $P_2(5, 3)$.

$$m = \frac{y_2 - y_1}{x_2 - x_1} = \frac{3 - 0}{5 - 2} = \frac{3}{3} = 1$$

$$y - y_1 = m(x - x_1)$$
$$y - 0 = 1(x - 2)$$
$$y = 1(x - 2)$$
$$y = x - 2$$

The equation of the line is $y = x - 2$.

Unless otherwise noted, all content on this page is © Cengage Learning.

You Try It 4   Use $P_1(2, 3)$ and $P_2(-5, 3)$.

$$m = \frac{y_2 - y_1}{x_2 - x_1} = \frac{3 - 3}{-5 - 2} = \frac{0}{-7} = 0$$

The line has zero slope, so the line is a horizontal line.
All points on the line have an ordinate of 3.
The equation of the line is $y = 3$.

You Try It 5
Strategy   ▸ Select the independent and dependent variables. The function will predict the Celsius temperature, so that quantity is the dependent variable, y. The Fahrenheit temperature is the independent variable, x.
▸ From the given data, two ordered pairs are (212, 100) and (32, 0). Use these ordered pairs to determine the linear function.

Solution   Choose $P_1(32, 0)$ and $P_2(212, 100)$.

$$m = \frac{y_2 - y_1}{x_2 - x_1} = \frac{100 - 0}{212 - 32} = \frac{100}{180} = \frac{5}{9}$$

$$y - y_1 = m(x - x_1)$$

$$y - 0 = \frac{5}{9}(x - 32)$$

$$y = \frac{5}{9}(x - 32), \text{ or } C = \frac{5}{9}(F - 32)$$

The linear function is

$$f(F) = \frac{5}{9}(F - 32).$$

SECTION 4.6

You Try It 1

$$m_1 = \frac{1 - (-3)}{7 - (-2)} = \frac{4}{9}$$
• $(x_1, y_1) = (-2, -3)$,
$(x_2, y_2) = (7, 1)$

$$m_2 = \frac{-5 - 1}{6 - 4} = \frac{-6}{2} = -3$$
• $(x_1, y_1) = (4, 1)$,
$(x_2, y_2) = (6, -5)$

$$m_1 \cdot m_2 = \frac{4}{9}(-3) = -\frac{4}{3}$$

No, the lines are not perpendicular.

You Try It 2 $5x + 2y = 2$
$$2y = -5x + 2$$
$$y = -\frac{5}{2}x + 1$$
$$m_1 = -\frac{5}{2}$$
$$5x + 2y = -6$$
$$2y = -5x - 6$$
$$y = -\frac{5}{2}x - 3$$
$$m_2 = -\frac{5}{2}$$
$$m_1 = m_2 = -\frac{5}{2}$$

Yes, the lines are parallel.

Unless otherwise noted, all content on this page is © Cengage Learning.

You Try It 3 $x - 4y = 3$
$$-4y = -x + 3$$
$$y = \frac{1}{4}x - \frac{3}{4}$$
$$m_1 = \frac{1}{4}$$
$$m_1 \cdot m_2 = -1$$
$$\frac{1}{4} \cdot m_2 = -1$$
$$m_2 = -4$$
$$y - y_1 = m(x - x_1)$$
$$y - 2 = -4[x - (-2)]$$ • $(x_1, y_1) = (-2, 2)$
$$y - 2 = -4(x + 2)$$
$$y - 2 = -4x - 8$$
$$y = -4x - 6$$

The equation of the line is $y = -4x - 6$.

SECTION 4.7

You Try It 1 $x + 3y > 6$
$$3y > -x + 6$$
$$y > -\frac{1}{3}x + 2$$

You Try It 2 $y < 2$

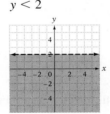

Solutions to Chapter 5 "You Try It"

SECTION 5.1

You Try It 1

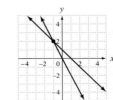

• Find the coordinates of the point of intersection of the graphs of the equations.

The solution is $(-1, 2)$.

You Try It 2

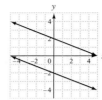

• Graph the two equations.

The lines are parallel and therefore do not intersect. The system of equations is inconsistent. The system of equations has no solution.

You Try It 3

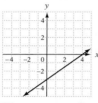

• Graph the two equations.

The two equations represent the same line. The system of equations is dependent. The solutions are the ordered pairs $\left(x, \frac{3}{4}x - 3\right)$.

You Try It 4

(1) $3x - y = 3$
(2) $6x + 3y = -4$

Solve Equation (1) for y.

$3x - y = 3$
$-y = -3x + 3$
$y = 3x - 3$

Substitute into Equation (2).

$6x + 3y = -4$
$6x + 3(3x - 3) = -4$
$6x + 9x - 9 = -4$
$15x - 9 = -4$
$15x = 5$
$x = \dfrac{5}{15} = \dfrac{1}{3}$

Substitute the value of x into Equation (1).

$3x - y = 3$
$3\left(\dfrac{1}{3}\right) - y = 3$
$1 - y = 3$
$-y = 2$
$y = -2$

The solution is $\left(\dfrac{1}{3}, -2\right)$.

You Try It 5

(1) $y = 2x - 3$
(2) $3x - 2y = 6$

$3x - 2y = 6$
$3x - 2(2x - 3) = 6$
$3x - 4x + 6 = 6$
$-x + 6 = 6$
$-x = 0$
$x = 0$

Substitute the value of x into Equation (1).

$y = 2x - 3$
$y = 2(0) - 3$
$y = 0 - 3$
$y = -3$

The solution is $(0, -3)$.

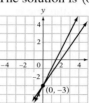

• Graph the two equations.

You Try It 6

(1) $6x - 3y = 6$
(2) $2x - y = 2$

Solve Equation (2) for y.

$2x - y = 2$
$-y = -2x + 2$
$y = 2x - 2$

Substitute into Equation (1).

$6x - 3y = 6$
$6x - 3(2x - 2) = 6$
$6x - 6x + 6 = 6$
$6 = 6$

The system of equations is dependent. The solutions are the ordered pairs $(x, 2x - 2)$.

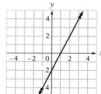

• Graph the two equations.

You Try It 7

Strategy

▶ Amount invested at 4.2%: x
   Amount invested at 6%: y

	Principal	*Rate*	*Interest*
Amount at 4.2%	x	0.042	$0.042x$
Amount at 6%	y	0.06	$0.06y$

▶ The total investment is $13,600:
   $x + y = 13,600$
   The two accounts earn the same interest:
   $0.042x = 0.06y$

Solution

(1) $x + y = 13,600$
(2) $0.042x = 0.06y$

$x = \dfrac{10}{7}y$ • Solve Equation (2) for x.

Unless otherwise noted, all content on this page is © Cengage Learning.

Substitute $\dfrac{10}{7}y$ for x in Equation (1) and solve for y.

$$x + y = 13{,}600$$

$$\frac{10}{7}y + y = 13{,}600$$

$$\frac{17}{7}y = 13{,}600$$

$$y = 5600$$

$$x + 5600 = 13{,}600$$

$$x = 8000$$

$8000 must be invested at 4.2%, and $5600 must be invested at 6%.

SECTION 5.2

You Try It 1

(1) $2x + 5y = 6$
(2) $3x - 2y = 6x + 2$

Write Equation (2) in the form $Ax + By = C$.

$$3x - 2y = 6x + 2$$
$$-3x - 2y = 2$$

Solve the system:
$$2x + 5y = 6$$
$$-3x - 2y = 2$$

Eliminate y.

$$2(2x + 5y) = 2(6)$$
$$5(-3x - 2y) = 5(2)$$

$$\begin{array}{r} 4x + 10y = 12 \\ -15x - 10y = 10 \\ \hline -11x = 22 \end{array}$$ • Add the equations.

$$x = -2$$ • Solve for x.

Replace x in Equation (1).

$$2x + 5y = 6$$
$$2(-2) + 5y = 6$$
$$-4 + 5y = 6$$
$$5y = 10$$
$$y = 2$$

The solution is $(-2, 2)$.

You Try It 2

$$2x + y = 5$$
$$4x + 2y = 6$$

Eliminate y.

$$-2(2x + y) = -2(5)$$
$$4x + 2y = 6$$

$$\begin{array}{r} -4x - 2y = -10 \\ 4x + 2y = 6 \\ \hline 0x + 0y = -4 \end{array}$$ • Add the equations.

$$0 = -4$$

This is a false equation.
The system is inconsistent and therefore has no solution.

Unless otherwise noted, all content on this page is © Cengage Learning.

You Try It 3

(1) $x - y + z = 6$
(2) $2x + 3y - z = 1$
(3) $x + 2y + 2z = 5$

Eliminate z. Add Equations (1) and (2).

$$\begin{array}{r} x - y + z = 6 \\ 2x + 3y - z = 1 \\ \hline 3x + 2y = 7 \end{array}$$ • Equation (4)

Multiply Equation (2) by 2 and add to Equation (3).

$$\begin{array}{r} 4x + 6y - 2z = 2 \\ x + 2y + 2z = 5 \\ \hline 5x + 8y = 7 \end{array}$$ • Equation (5)

Solve the system of two equations.

(4) $3x + 2y = 7$
(5) $5x + 8y = 7$

Multiply Equation (4) by -4 and add to Equation (5). Solve for x.

$$\begin{array}{r} -12x - 8y = -28 \\ 5x + 8y = 7 \\ \hline -7x = -21 \end{array}$$

$$x = 3$$

Replace x by 3 in Equation (4). Solve for y.

$$3x + 2y = 7$$
$$3(3) + 2y = 7$$
$$9 + 2y = 7$$
$$2y = -2$$
$$y = -1$$

Replace x by 3 and y by -1 in Equation (1). Solve for z.

$$x - y + z = 6$$
$$3 - (-1) + z = 6$$
$$4 + z = 6$$
$$z = 2$$

The solution is $(3, -1, 2)$.

SECTION 5.3

You Try It 1
Strategy
▶ Rate of the rowing team in calm water: t
Rate of the current: c

	Rate	Time	Distance
With current	$t + c$	2	$2(t + c)$
Against current	$t - c$	2	$2(t - c)$

▶ The distance traveled with the current is 18 mi.
The distance traveled against the current is 10 mi.

$$2(t + c) = 18$$
$$2(t - c) = 10$$

Solution

$$2(t + c) = 18 \qquad \frac{1}{2} \cdot 2(t + c) = \frac{1}{2} \cdot 18$$

$$2(t - c) = 10 \qquad \frac{1}{2} \cdot 2(t - c) = \frac{1}{2} \cdot 10$$

$$\begin{aligned} t + c &= 9 \\ t - c &= 5 \\ \hline 2t &= 14 \\ t &= 7 \end{aligned}$$

$$\begin{aligned} t + c &= 9 \\ 7 + c &= 9 \qquad \bullet \text{ Substitute 7 for } t. \\ c &= 2 \end{aligned}$$

The rate of the rowing team in calm water is 7 mph.
The rate of the current is 2 mph.

You Try It 2
Strategy

▶ Amount invested at 9%: x
Amount invested at 7%: y
Amount invested at 5%: z

	Principal	Rate	Interest
Amount at 9%	x	0.09	$0.09x$
Amount at 7%	y	0.07	$0.07y$
Amount at 5%	z	0.05	$0.05z$

▶ The amount invested at 9% (x) is twice the amount invested at 7% (y): $x = 2y$
The sum of the interest earned by all three accounts is $1300:
$0.09x + 0.07y + 0.05z = 1300$
The total amount invested is $20,000:
$x + y + z = 20,000$

Solution

$$\begin{aligned} (1) & \qquad x = 2y \\ (2) & \quad 0.09x + 0.07y + 0.05z = 1300 \\ (3) & \qquad x + y + z = 20,000 \end{aligned}$$

Solve the system of equations. Substitute $2y$ for x in Equation (2) and Equation (3).

$$0.09(2y) + 0.07y + 0.05z = 1300$$
$$2y + y + z = 20,000$$

$$\begin{aligned} (4) \quad 0.25y + 0.05z &= 1300 \qquad \bullet \ 0.09(2y) + 0.07y = 0.25y \\ (5) \qquad 3y + z &= 20,000 \qquad \bullet \ 2y + y = 3y \end{aligned}$$

Solve the system of equations in two variables by multiplying Equation (5) by -0.05 and adding to Equation (4).

$$\begin{aligned} 0.25y + 0.05z &= 1300 \\ -0.15y - 0.05z &= -1000 \\ \hline 0.10y &= 300 \\ y &= 3000 \end{aligned}$$

Substituting the value of y into Equation (1), $x = 6000$.
Substituting the values of x and y into Equation (3), $z = 11,000$.

The investor placed $6000 in the 9% account, $3000 in the 7% account, and $11,000 in the 5% account.

SECTION 5.4

You Try It 1 Shade above the solid line graph of $y = 2x - 3$.
Shade above the dashed line graph of $y = -3x$.
The solution set of the system is the intersection of the solution sets of the individual inequalities.

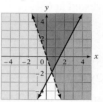

You Try It 2 $3x + 4y > 12$
$$4y > -3x + 12$$
$$y > -\frac{3}{4}x + 3$$

Shade above the dashed line graph of $y = -\frac{3}{4}x + 3$.

Shade below the dashed line graph of $y = \frac{3}{4}x - 1$.

The solution set of the system is the intersection of the solution sets of the individual inequalities.

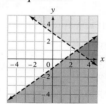

Solutions to Chapter 6 "You Try It"

SECTION 6.1

You Try It 1
$$\begin{aligned} (-3a^5b^7)^4 &= (-3)^4a^{20}b^{28} \\ &= 81a^{20}b^{28} \end{aligned}$$

You Try It 2
$$\begin{aligned} (-3a^2b^4)(-2ab^3)^4 &= (-3a^2b^4)[(-2)^4a^4b^{12}] \\ &= (-3a^2b^4)(16a^4b^{12}) \\ &= -48a^6b^{16} \end{aligned}$$

Unless otherwise noted, all content on this page is © Cengage Learning.

You Try It 3

$$(-4ab^4)^2(2a^4b^2)^4 = [(-4)^2a^2b^8][2^4a^{16}b^8]$$
$$= [16a^2b^8][16a^{16}b^8]$$
$$= 256a^{18}b^{16}$$

You Try It 4

$$\frac{20r^{-2}t^{-5}}{-16r^{-3}s^{-2}} = -\frac{4 \cdot 5r^{-2-(-3)}s^2t^{-5}}{4 \cdot 4}$$
$$= -\frac{5rs^2}{4t^5}$$

You Try It 5

$$\frac{(9u^{-6}v^4)^{-1}}{(6u^{-3}v^{-2})^{-2}} = \frac{9^{-1}u^6v^{-4}}{6^{-2}u^6v^4}$$
$$= 9^{-1} \cdot 6^2u^0v^{-8}$$
$$= \frac{36}{9v^8}$$
$$= \frac{4}{v^8}$$

You Try It 6 $942{,}000{,}000 = 9.42 \times 10^8$

You Try It 7 $2.7 \times 10^{-5} = 0.000027$

You Try It 8

$$\frac{5{,}600{,}000 \times 0.000000081}{900 \times 0.000000028}$$
$$= \frac{5.6 \times 10^6 \times 8.1 \times 10^{-8}}{9 \times 10^2 \times 2.8 \times 10^{-8}}$$
$$= \frac{(5.6)(8.1) \times 10^{6+(-8)-2-(-8)}}{(9)(2.8)}$$
$$= 1.8 \times 10^4 = 18{,}000$$

You Try It 9

Strategy To find the number of arithmetic operations:
- Find the reciprocal of 9.74×10^{-16}, which is the number of operations performed in 1 s.
- Write the number of seconds in 1 min (60) in scientific notation.
- Multiply the number of arithmetic operations per second by the number of seconds in 1 min.

Solution

$$\frac{1}{9.74 \times 10^{-16}} \approx 1.03 \times 10^{15}$$
$$60 = 6.0 \times 10$$
$$(1.03 \times 10^{15})(6.0 \times 10) = 6.18 \times 10^{16}$$

The computer can perform 6.18×10^{16} operations in 1 min.

SECTION 6.2

You Try It 1

$R(x) = -2x^4 - 5x^3 + 2x - 8$
$R(2) = -2(2)^4 - 5(2)^3 + 2(2) - 8$ • Replace x by 2.
$= -2(16) - 5(8) + 4 - 8$ • Simplify.
$= -32 - 40 + 4 - 8$
$= -76$

You Try It 2

x	$y = f(x)$
-4	5
-3	0
-2	-3
-1	-4
0	-3
1	0
2	5

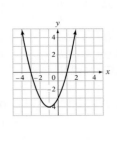

You Try It 3

x	$y = f(x)$
-4	-40
-3	0
-2	18
-1	20
0	12
1	0
2	-10
3	-12
4	0
5	32

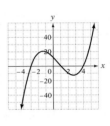

You Try It 4

$$\begin{array}{r} -3x^2 - 4x + 9 \\ -5x^2 - 7x + 1 \\ \hline -8x^2 - 11x + 10 \end{array}$$

You Try It 5 Add the additive inverse of $6x^2 + 3x - 7$ to $-5x^2 + 2x - 3$.

$$\begin{array}{r} -5x^2 + 2x - 3 \\ -6x^2 - 3x + 7 \\ \hline -11x^2 - x + 4 \end{array}$$

You Try It 6
$P(x) - R(x) = (4x^3 - 3x^2 + 2) - (-2x^2 + 2x - 3)$
$= 4x^3 - x^2 - 2x + 5$

SECTION 6.3

You Try It 1
$(2b^2 - 7b - 8)(-5b)$
$= 2b^2(-5b) - 7b(-5b) - 8(-5b)$ • Distributive
$= -10b^3 + 35b^2 + 40b$ Property

You Try It 2 $x^2 - 2x[x - x(4x - 5) + x^2]$
$= x^2 - 2x[x - 4x^2 + 5x + x^2]$
$= x^2 - 2x[6x - 3x^2]$
$= x^2 - 12x^2 + 6x^3$
$= 6x^3 - 11x^2$

Unless otherwise noted, all content on this page is © Cengage Learning.

You Try It 3

$$-\ 2b^2 +\ 5b - 4$$
$$\underline{-\quad 3b + 2}$$
$$\underline{-\ 4b^2 + 10b - 8} \qquad \bullet\ 2(-2b^2 + 5b - 4)$$
$$\underline{6b^3 - 15b^2 + 12b} \qquad \bullet\ -3b(-2b^2 + 5b - 4)$$
$$6b^3 - 19b^2 + 22b - 8$$

You Try It 4

$$(3x - 4)(2x - 3) = 6x^2 - 9x - 8x + 12 \qquad \bullet\ \text{FOIL}$$
$$= 6x^2 - 17x + 12$$

You Try It 5

$$(3ab + 4)(5ab - 3) = 15a^2b^2 - 9ab + 20ab - 12$$
$$= 15a^2b^2 + 11ab - 12$$

You Try It 6

$$(3x - 7)(3x + 7) \qquad \bullet\ \text{The sum and differ-}$$
$$= 9x^2 - 49 \qquad\qquad \text{ence of two terms}$$

You Try It 7

$$(2ab + 7)(2ab - 7) \qquad \bullet\ \text{The sum and differ-}$$
$$= 4a^2b^2 - 49 \qquad\qquad \text{ence of two terms}$$

You Try It 8

$$(3x - 4y)^2 \qquad \bullet\ \text{The square}$$
$$= 9x^2 - 24xy + 16y^2 \qquad \text{of a binomial}$$

You Try It 9

$$(5xy + 4)^2 \qquad \bullet\ \text{The square of}$$
$$= 25x^2y^2 + 40xy + 16 \qquad \text{a binomial}$$

You Try It 10

Strategy To find the area, replace the variables b and h in the equation $A = \frac{1}{2} bh$ by the given values, and solve for A.

Solution

$$A = \frac{1}{2} bh$$
$$A = \frac{1}{2} (2x + 6)(x - 4)$$
$$A = (x + 3)(x - 4)$$
$$A = x^2 - 4x + 3x - 12 \qquad \bullet\ \text{FOIL}$$
$$A = x^2 - x - 12$$

The area is $(x^2 - x - 12)$ ft^2.

You Try It 11

Strategy To find the volume, subtract the volume of the small rectangular solid from the volume of the large rectangular solid.

Large rectangular solid:
Length $= L_1 = 12x$
Width $= W_1 = 7x + 2$
Height $= H_1 = 5x - 4$

Small rectangular solid:
Length $= L_2 = 12x$
Width $= W_2 = x$
Height $= H_2 = 2x$

Solution

$V = $ Volume of large rectangular solid $-$ volume of small rectangular solid
$$V = (L_1 \cdot W_1 \cdot H_1) - (L_2 \cdot W_2 \cdot H_2)$$
$$V = (12x)(7x + 2)(5x - 4) - (12x)(x)(2x)$$
$$= (84x^2 + 24x)(5x - 4) - (12x^2)(2x)$$
$$= (420x^3 - 336x^2 + 120x^2 - 96x) - 24x^3$$
$$= 396x^3 - 216x^2 - 96x$$

The volume is $(396x^3 - 216x^2 - 96x)$ ft^3.

You Try It 12

Strategy To find the area, replace the variable r in the equation $A = \pi r^2$ by the given value, and solve for A.

Solution
$$A = \pi r^2$$
$$A \approx 3.14(2x + 3)^2$$
$$= 3.14(4x^2 + 12x + 9)$$
$$= 12.56x^2 + 37.68x + 28.26$$

The area is $(12.56x^2 + 37.68x + 28.26)$ cm^2.

SECTION 6.4

You Try It 1
$$\frac{4x^3y + 8x^2y^2 - 4xy^3}{2xy} = \frac{4x^3y}{2xy} + \frac{8x^2y^2}{2xy} - \frac{4xy^3}{2xy}$$
$$= 2x^2 + 4xy - 2y^2$$

Check: $2xy(2x^2 + 4xy - 2y^2) = 4x^3y + 8x^2y^2 - 4xy^3$

You Try It 2

$$\begin{array}{r} 5x - 1 \\ 3x + 4 \overline{)15x^2 + 17x - 20} \\ \underline{15x^2 + 20x} \\ -3x - 20 \\ \underline{-3x - 4} \\ -16 \end{array}$$

$$\frac{15x^2 + 17x - 20}{3x + 4} = 5x - 1 - \frac{16}{3x + 4}$$

You Try It 3

$$\begin{array}{r} x^2 + 3x - 1 \\ 3x - 1 \overline{)3x^3 + 8x^2 - 6x + 2} \\ \underline{3x^3 - x^2} \\ 9x^2 - 6x \\ \underline{9x^2 - 3x} \\ -3x + 2 \\ \underline{-3x + 1} \\ 1 \end{array}$$

$$\frac{3x^3 + 8x^2 - 6x + 2}{3x - 1} = x^2 + 3x - 1 + \frac{1}{3x - 1}$$

You Try It 4

$$\begin{array}{r} 3x^2 - 2x + 4 \\ x^2 - 3x + 2)\overline{3x^4 - 11x^3 + 16x^2 - 16x + 8} \\ \underline{3x^4 - 9x^3 + 6x^2} \\ -2x^3 + 10x^2 - 16x \\ \underline{-2x^3 + 6x^2 - 4x} \\ 4x^2 - 12x + 8 \\ \underline{4x^2 - 12x + 8} \\ 0 \end{array}$$

$$\frac{3x^4 - 11x^3 + 16x^2 - 16x + 8}{x^2 - 3x + 2} = 3x^2 - 2x + 4$$

You Try It 5

$$\begin{array}{r|rrr} -2 & 6 & 8 & -5 \\ & & -12 & 8 \\ \hline & 6 & -4 & 3 \end{array}$$

$$(6x^2 + 8x - 5) \div (x + 2)$$
$$= 6x - 4 + \frac{3}{x + 2}$$

You Try It 6

$$\begin{array}{r|rrrr} 2 & 5 & -12 & -8 & 16 \\ & & 10 & -4 & -24 \\ \hline & 5 & -2 & -12 & -8 \end{array}$$

$$(5x^3 - 12x^2 - 8x + 16) \div (x - 2)$$
$$= 5x^2 - 2x - 12 - \frac{8}{x - 2}$$

You Try It 7

$$\begin{array}{r|rrrrr} 3 & 2 & -3 & -8 & 0 & -2 \\ & & 6 & 9 & 3 & 9 \\ \hline & 2 & 3 & 1 & 3 & 7 \end{array}$$

$$(2x^4 - 3x^3 - 8x^2 - 2) \div (x - 3)$$
$$= 2x^3 + 3x^2 + x + 3 + \frac{7}{x - 3}$$

You Try It 8

$$\begin{array}{r|rrr} 2 & 2 & -3 & -5 \\ & & 4 & 2 \\ \hline & 2 & 1 & -3 \end{array}$$

$$P(2) = -3$$

You Try It 9

$$\begin{array}{r|rrrr} -3 & 2 & -5 & 0 & 7 \\ & & -6 & 33 & -99 \\ \hline & 2 & -11 & 33 & -92 \end{array}$$

$$P(-3) = -92$$

Solutions to Chapter 7 "You Try It"

SECTION 7.1

You Try It 1 The GCF is $7a^2$.
$$14a^2 - 21a^4b = 7a^2(2) + 7a^2(-3a^2b)$$
$$= 7a^2(2 - 3a^2b)$$

You Try It 2 The GCF is 9.
$$27b^2 + 18b + 9$$
$$= 9(3b^2) + 9(2b) + 9(1)$$
$$= 9(3b^2 + 2b + 1)$$

You Try It 3
The GCF is $3x^2y^2$.
$$6x^4y^2 - 9x^3y^2 + 12x^2y^4$$
$$= 3x^2y^2(2x^2) + 3x^2y^2(-3x) + 3x^2y^2(4y^2)$$
$$= 3x^2y^2(2x^2 - 3x + 4y^2)$$

You Try It 4
$$2y(5x - 2) - 3(2 - 5x)$$
$$= 2y(5x - 2) + 3(5x - 2) \quad \bullet \; 5x - 2 \text{ is the common}$$
$$= (5x - 2)(2y + 3) \qquad\qquad \text{factor.}$$

You Try It 5
$$a^2 - 3a + 2ab - 6b$$
$$= (a^2 - 3a) + (2ab - 6b)$$
$$= a(a - 3) + 2b(a - 3) \quad \bullet \; a - 3 \text{ is the common factor.}$$
$$= (a - 3)(a + 2b)$$

You Try It 6
$$2mn^2 - n + 8mn - 4$$
$$= (2mn^2 - n) + (8mn - 4)$$
$$= n(2mn - 1) + 4(2mn - 1) \quad \bullet \; 2mn - 1 \text{ is the common}$$
$$= (2mn - 1)(n + 4) \qquad\qquad \text{factor.}$$

You Try It 7
$$3xy - 9y - 12 + 4x$$
$$= (3xy - 9y) - (12 - 4x) \quad \bullet \; -12 + 4x = -(12 - 4x)$$
$$= 3y(x - 3) - 4(3 - x) \quad \bullet \; -(3 - x) = (x - 3)$$
$$= 3y(x - 3) + 4(x - 3) \quad \bullet \; x - 3 \text{ is the common factor.}$$
$$= (x - 3)(3y + 4)$$

SECTION 7.2

You Try It 1
Find the positive factors of 20 whose sum is 9.

Factors	Sum
1, 20	21
2, 10	12
4, 5	**9**

$$x^2 + 9x + 20 = (x + 4)(x + 5)$$

Unless otherwise noted, all content on this page is © Cengage Learning.

You Try It 2

Find the factors of -18 whose sum is 7.

Factors	Sum
+1, −18	−17
−1, +18	17
+2, −9	−7
−2, +9	**7**
+3, −6	−3
−3, +6	3

$x^2 + 7x - 18 = (x + 9)(x - 2)$

You Try It 3

The GCF is $-2x$.

$-2x^3 + 14x^2 - 12x = -2x(x^2 - 7x + 6)$

Factor the trinomial $x^2 - 7x + 6$. Find two negative factors of 6 whose sum is -7.

Factors	Sum
−1, −6	**−7**
−2, −3	−5

$-2x^3 + 14x^2 - 12x = -2x(x - 6)(x - 1)$

You Try It 4

The GCF is 3.

$3x^2 - 9xy - 12y^2 = 3(x^2 - 3xy - 4y^2)$

Factor the trinomial.

Find the factors of -4 whose sum is -3.

Factors	Sum
+1, −4	**−3**
−1, +4	3
+2, −2	0

$3x^2 - 9xy - 12y^2 = 3(x + y)(x - 4y)$

SECTION 7.3

You Try It 1

Factor the trinomial $2x^2 - x - 3$.

Positive factors of 2: 1, 2   Factors of -3: $+1, -3$
                                                $-1, +3$

Trial Factors	Middle Term
$(x + 1)(2x - 3)$	$-3x + 2x = -x$
$(x - 3)(2x + 1)$	$x - 6x = -5x$
$(x - 1)(2x + 3)$	$3x - 2x = x$
$(x + 3)(2x - 1)$	$-x + 6x = 5x$

$2x^2 - x - 3 = (x + 1)(2x - 3)$

You Try It 2

The GCF is $-3y$.

$-45y^3 + 12y^2 + 12y = -3y(15y^2 - 4y - 4)$

Factor the trinomial $15y^2 - 4y - 4$.

Positive factors of 15: 1, 15   Factors of -4:   1, −4
                      3, 5   −1,   4
                                                      2, −2

Trial Factors	Middle Term
$(y + 1)(15y - 4)$	$-4y + 15y = 11y$
$(y - 4)(15y + 1)$	$y - 60y = -59y$
$(y - 1)(15y + 4)$	$4y - 15y = -11y$
$(y + 4)(15y - 1)$	$-y + 60y = 59y$
$(y + 2)(15y - 2)$	$-2y + 30y = 28y$
$(y - 2)(15y + 2)$	$2y - 30y = -28y$
$(3y + 1)(5y - 4)$	$-12y + 5y = -7y$
$(3y - 4)(5y + 1)$	$3y - 20y = -17y$
$(3y - 1)(5y + 4)$	$12y - 5y = 7y$
$(3y + 4)(5y - 1)$	$-3y + 20y = 17y$
$(3y + 2)(5y - 2)$	$-6y + 10y = 4y$
$(3y - 2)(5y + 2)$	$6y - 10y = \mathbf{-4y}$

$-45y^3 + 12y^2 + 12y = -3y(3y - 2)(5y + 2)$

You Try It 3

Factors of -14 $[2(-7)]$	Sum
+1, −14	−13
−1, +14	**13**
+2, −7	−5
−2, +7	5

$$
\begin{aligned}
2a^2 + 13a - 7 &= 2a^2 - a + 14a - 7 \\
&= (2a^2 - a) + (14a - 7) \\
&= a(2a - 1) + 7(2a - 1) \\
&= (2a - 1)(a + 7)
\end{aligned}
$$

Unless otherwise noted, all content on this page is © Cengage Learning.

You Try It 4

The GCF is $5x$.

$15x^3 + 40x^2 - 80x = 5x(3x^2 + 8x - 16)$

Factors of -48 $[3(-16)]$	Sum
$+1, -48$	-47
$-1, +48$	47
$+2, -24$	-22
$-2, +24$	22
$+3, -16$	-13
$-3, +16$	13
$+4, -12$	-8
$-4, +12$	$\mathbf{8}$

$$\begin{aligned}
3x^2 + 8x - 16 &= 3x^2 - 4x + 12x - 16 \\
&= (3x^2 - 4x) + (12x - 16) \\
&= x(3x - 4) + 4(3x - 4) \\
&= (3x - 4)(x + 4)
\end{aligned}$$

$$\begin{aligned}
15x^3 + 40x^2 - 80x &= 5x(3x^2 + 8x - 16) \\
&= 5x(3x - 4)(x + 4)
\end{aligned}$$

SECTION 7.4

You Try It 1

$$\begin{aligned}
x^2 - 36y^4 &= x^2 - (6y^2)^2 \quad &\bullet \text{ Difference of} \\
&= (x + 6y^2)(x - 6y^2) \quad &\text{two squares}
\end{aligned}$$

You Try It 2

$9x^2 + 12x + 4 = (3x + 2)^2$ • **Perfect-square trinomial**

You Try It 3

$$\begin{aligned}
(a + b)^2 - (a - b)^2 \quad &\bullet \text{ Difference of two squares} \\
= [(a + b) + (a - b)][(a + b) - (a - b)] \\
= (a + b + a - b)(a + b - a + b) \\
= (2a)(2b) = 4ab
\end{aligned}$$

You Try It 4

$$\begin{aligned}
a^3b^3 - 27 &= (ab)^3 - 3^3 \quad &\bullet \text{ Difference of two cubes} \\
&= (ab - 3)(a^2b^2 + 3ab + 9)
\end{aligned}$$

You Try It 5

$$\begin{aligned}
8x^3 + y^3z^3 &= (2x)^3 + (yz)^3 \quad &\bullet \text{ Sum of two cubes} \\
&= (2x + yz)(4x^2 - 2xyz + y^2z^2)
\end{aligned}$$

You Try It 6

$$\begin{aligned}
(x - y)^3 &+ (x + y)^3 \quad \bullet \text{ Sum of two cubes} \\
&= [(x - y) + (x + y)] \\
&\quad \times [(x - y)^2 - (x - y)(x + y) + (x + y)^2] \\
&= 2x[x^2 - 2xy + y^2 - (x^2 - y^2) + x^2 + 2xy + y^2] \\
&= 2x(x^2 - 2xy + y^2 - x^2 + y^2 + x^2 + 2xy + y^2) \\
&= 2x(x^2 + 3y^2)
\end{aligned}$$

You Try It 7   Let $u = x^2$.

$$\begin{aligned}
3x^4 + 4x^2 - 4 &= 3u^2 + 4u - 4 \\
&= (u + 2)(3u - 2) \\
&= (x^2 + 2)(3x^2 - 2)
\end{aligned}$$

You Try It 8

$$\begin{aligned}
18x^3 - 6x^2 - 60x &= 6x(3x^2 - x - 10) \quad &\bullet \text{ The GCF is } 6x. \\
&= 6x(3x + 5)(x - 2) \quad &\bullet \text{ Factor the trinomial.}
\end{aligned}$$

You Try It 9

$$\begin{aligned}
4x - 4y - x^3 + x^2y \quad &\bullet \text{ Factor by grouping.} \\
= (4x - 4y) - (x^3 - x^2y) \\
= 4(x - y) - x^2(x - y) \\
= (x - y)(4 - x^2) \quad &\bullet \text{ Difference of two} \\
&\quad\text{squares} \\
= (x - y)(2 + x)(2 - x)
\end{aligned}$$

SECTION 7.5

You Try It 1

$2x(x + 7) = 0$

$$\begin{array}{ll}
2x = 0 \qquad x + 7 = 0 & \bullet \text{ Principle of Zero} \\
x = 0 \qquad\quad x = -7 & \text{Products}
\end{array}$$

The solutions are 0 and -7.

You Try It 2

$$\begin{aligned}
4x^2 - 9 &= 0 \quad &\bullet \text{ Difference of two squares} \\
(2x - 3)(2x + 3) &= 0
\end{aligned}$$

$$\begin{array}{ll}
2x - 3 = 0 \qquad 2x + 3 = 0 & \bullet \text{ Principle of Zero Products} \\
2x = 3 \qquad\quad 2x = -3 & \\
x = \dfrac{3}{2} \qquad\quad x = -\dfrac{3}{2} &
\end{array}$$

The solutions are $\dfrac{3}{2}$ and $-\dfrac{3}{2}$.

You Try It 3

$$\begin{aligned}
(x + 2)(x - 7) &= 52 \\
x^2 - 5x - 14 &= 52 \\
x^2 - 5x - 66 &= 0 \\
(x + 6)(x - 11) &= 0
\end{aligned}$$

$$\begin{array}{ll}
x + 6 = 0 \qquad x - 11 = 0 & \bullet \text{ Principle of Zero Products} \\
x = -6 \qquad\quad x = 11 &
\end{array}$$

The solutions are -6 and 11.

You Try It 4

Strategy   First consecutive positive integer: n
Second consecutive positive integer: $n + 1$
The sum of the squares of the two consecutive positive integers is 61.

Solution

$$\begin{aligned}
n^2 + (n + 1)^2 &= 61 \\
n^2 + n^2 + 2n + 1 &= 61 \\
2n^2 + 2n + 1 &= 61 \\
2n^2 + 2n - 60 &= 0 \\
2(n^2 + n - 30) &= 0 \\
2(n - 5)(n + 6) &= 0
\end{aligned}$$

$$\begin{array}{ll}
n - 5 = 0 \qquad n + 6 = 0 & \bullet \text{ Principle of} \\
n = 5 \qquad\quad n = -6 & \text{Zero Products}
\end{array}$$

Unless otherwise noted, all content on this page is © Cengage Learning.

Because -6 is not a positive integer, it is not a solution.

$n = 5$
$n + 1 = 5 + 1 = 6$

The two integers are 5 and 6.

You Try It 5

Strategy   Width $= x$
Length $= 2x + 4$

The area of the rectangle is 96 in^2.
Use the equation $A = L \cdot W$.

Solution
$A = L \cdot W$
$96 = (2x + 4)x$
$96 = 2x^2 + 4x$
$0 = 2x^2 + 4x - 96$
$0 = 2(x^2 + 2x - 48)$
$0 = 2(x + 8)(x - 6)$

$\begin{aligned} x + 8 &= 0 & x - 6 &= 0 \\ x &= -8 & x &= 6 \end{aligned}$ • **Principle of Zero Products**

Because the width cannot be a negative number, -8 is not a solution.

$x = 6$
$2x + 4 = 2(6) + 4 = 12 + 4 = 16$

The length is 16 in. The width is 6 in.

Solutions to Chapter 8 "You Try It"

SECTION 8.1

You Try It 1

$\dfrac{6x^5y}{12x^2y^3} = \dfrac{\overset{1}{2} \cdot \overset{1}{3} \cdot x^5y}{\underset{1}{2} \cdot 2 \cdot \underset{1}{3} \cdot x^2y^3} = \dfrac{x^3}{2y^2}$

You Try It 2

$\dfrac{x^2 + 4x - 12}{x^2 - 3x + 2} = \dfrac{(x - 2)(x + 6)}{(x - 1)(x - 2)} = \dfrac{x + 6}{x - 1}$

You Try It 3

$\dfrac{x^2 + 2x - 24}{16 - x^2} = \dfrac{\overset{-1}{(x - 4)}(x + 6)}{\underset{1}{(4 - x)}(4 + x)}$ • $\dfrac{x - 4}{4 - x} = \dfrac{x - 4}{-1(x - 4)} = -1$

$= -\dfrac{x + 6}{x + 4}$

You Try It 4

$\dfrac{12x^2 + 3x}{10x - 15} \cdot \dfrac{8x - 12}{9x + 18} = \dfrac{3x(4x + 1)}{5(2x - 3)} \cdot \dfrac{4(2x - 3)}{9(x + 2)}$

$= \dfrac{\overset{1}{3}x(4x + 1) \cdot 4(2x - 3)}{5(2x - 3) \cdot \underset{1}{3} \cdot 3(x + 2)}$

$= \dfrac{4x(4x + 1)}{15(x + 2)}$

You Try It 5

$\dfrac{x^2 + 2x - 15}{9 - x^2} \cdot \dfrac{x^2 - 3x - 18}{x^2 - 7x + 6}$

$= \dfrac{(x - 3)(x + 5)}{(3 - x)(3 + x)} \cdot \dfrac{(x + 3)(x - 6)}{(x - 1)(x - 6)}$ • **Factor.**

$= \dfrac{\overset{-1}{(x - 3)}(x + 5) \cdot \overset{1}{(x + 3)}\overset{1}{(x - 6)}}{\underset{1}{(3 - x)}\underset{1}{(3 + x)} \cdot (x - 1)(x - 6)} = -\dfrac{x + 5}{x - 1}$

You Try It 6

$\dfrac{a^2}{4bc^2 - 2b^2c} \div \dfrac{a}{6bc - 3b^2}$

$= \dfrac{a^2}{4bc^2 - 2b^2c} \cdot \dfrac{6bc - 3b^2}{a}$ • **Multiply by the reciprocal.**

$= \dfrac{a^2 \cdot 3\overset{1}{b}(2c - b)}{2\underset{1}{bc}(2c - b) \cdot a} = \dfrac{3a}{2c}$

You Try It 7

$\dfrac{3x^2 + 26x + 16}{3x^2 - 7x - 6} \div \dfrac{2x^2 + 9x - 5}{x^2 + 2x - 15}$

$= \dfrac{3x^2 + 26x + 16}{3x^2 - 7x - 6} \cdot \dfrac{x^2 + 2x - 15}{2x^2 + 9x - 5}$ • **Multiply by the reciprocal.**

$= \dfrac{\overset{1}{(3x + 2)}(x + 8) \cdot \overset{1}{(x + 5)}\overset{1}{(x - 3)}}{\underset{1}{(3x + 2)}\underset{1}{(x - 3)} \cdot (2x - 1)\underset{1}{(x + 5)}} = \dfrac{x + 8}{2x - 1}$

SECTION 8.2

You Try It 1

The LCM is $(2x - 5)(x + 4)$.

$\dfrac{2x}{2x - 5} = \dfrac{2x}{2x - 5} \cdot \dfrac{x + 4}{x + 4} = \dfrac{2x^2 + 8x}{(2x - 5)(x + 4)}$

$\dfrac{3}{x + 4} = \dfrac{3}{x + 4} \cdot \dfrac{2x - 5}{2x - 5} = \dfrac{6x - 15}{(2x - 5)(x + 4)}$

You Try It 2

$2x - x^2 = x(2 - x) = -x(x - 2)$;
$3x^2 - 5x - 2 = (x - 2)(3x + 1)$
The LCM is $x(x - 2)(3x + 1)$.

$\dfrac{2x - 7}{2x - x^2} = -\dfrac{2x - 7}{x(x - 2)} \cdot \dfrac{3x + 1}{3x + 1} = -\dfrac{6x^2 - 19x - 7}{x(x - 2)(3x + 1)}$

$\dfrac{3x - 2}{3x^2 - 5x - 2} = \dfrac{3x - 2}{(x - 2)(3x + 1)} \cdot \dfrac{x}{x} = \dfrac{3x^2 - 2x}{x(x - 2)(3x + 1)}$

You Try It 3

The LCM is ab.

$\dfrac{2}{b} - \dfrac{1}{a} + \dfrac{4}{ab} = \dfrac{2}{b} \cdot \dfrac{a}{a} - \dfrac{1}{a} \cdot \dfrac{b}{b} + \dfrac{4}{ab}$

$= \dfrac{2a}{ab} - \dfrac{b}{ab} + \dfrac{4}{ab} = \dfrac{2a - b + 4}{ab}$

You Try It 4
The LCM is $a(a-5)(a+5)$.

$$\dfrac{a-3}{a^2-5a}+\dfrac{a-9}{a^2-25}$$

$$=\dfrac{a-3}{a(a-5)}\cdot\dfrac{a+5}{a+5}+\dfrac{a-9}{(a-5)(a+5)}\cdot\dfrac{a}{a}$$

$$=\dfrac{(a-3)(a+5)+a(a-9)}{a(a-5)(a+5)}$$

$$=\dfrac{(a^2+2a-15)+(a^2-9a)}{a(a-5)(a+5)}$$

$$=\dfrac{a^2+2a-15+a^2-9a}{a(a-5)(a+5)}$$

$$=\dfrac{2a^2-7a-15}{a(a-5)(a+5)}=\dfrac{(2a+3)(a-5)}{a(a-5)(a+5)}$$

$$=\dfrac{(2a+3)\cancel{(a-5)}}{a\cancel{(a-5)}(a+5)}=\dfrac{2a+3}{a(a+5)}$$

You Try It 5
The LCM is $(x-4)(x+1)$.

$$\dfrac{2x}{x-4}-\dfrac{x-1}{x+1}+\dfrac{2}{x^2-3x-4}$$

$$=\dfrac{2x}{x-4}\cdot\dfrac{x+1}{x+1}-\dfrac{x-1}{x+1}\cdot\dfrac{x-4}{x-4}+\dfrac{2}{(x-4)(x+1)}$$

$$=\dfrac{2x(x+1)-(x-1)(x-4)+2}{(x-4)(x+1)}$$

$$=\dfrac{(2x^2+2x)-(x^2-5x+4)+2}{(x-4)(x+1)}$$

$$=\dfrac{x^2+7x-2}{(x-4)(x+1)}$$

SECTION 8.3

You Try It 1
The LCM of the denominators in the numerator and denominator is $x-3$.

$$\dfrac{2x+5+\dfrac{14}{x-3}}{4x+16+\dfrac{49}{x-3}}=\dfrac{2x+5+\dfrac{14}{x-3}}{4x+16+\dfrac{49}{x-3}}\cdot\dfrac{x-3}{x-3}$$

$$=\dfrac{(2x+5)(x-3)+\dfrac{14}{x-3}(x-3)}{(4x+16)(x-3)+\dfrac{49}{x-3}(x-3)}$$

$$=\dfrac{2x^2-x-15+14}{4x^2+4x-48+49}=\dfrac{2x^2-x-1}{4x^2+4x+1}$$

$$=\dfrac{(2x+1)(x-1)}{(2x+1)(2x+1)}=\dfrac{\cancel{(2x+1)}(x-1)}{\cancel{(2x+1)}(2x+1)}=\dfrac{x-1}{2x+1}$$

You Try It 2
The LCM of the denominators of 1, 2, and $\dfrac{1}{x}$ is x.

$$2-\dfrac{1}{2-\dfrac{1}{x}}=2-\dfrac{1}{2-\dfrac{1}{x}}\cdot\dfrac{x}{x}$$

$$=2-\dfrac{1\cdot x}{2\cdot x-\dfrac{1}{x}\cdot x}=2-\dfrac{x}{2x-1}$$

The LCM of the denominators is $2x-1$.

$$2-\dfrac{x}{2x-1}=2\cdot\dfrac{2x-1}{2x-1}-\dfrac{x}{2x-1}$$

$$=\dfrac{4x-2}{2x-1}-\dfrac{x}{2x-1}$$

$$=\dfrac{4x-2-x}{2x-1}=\dfrac{3x-2}{2x-1}$$

SECTION 8.4

You Try It 1

$$\dfrac{x}{x+6}=\dfrac{3}{x}$$ • The LCM is $x(x+6)$.

$$\dfrac{x(x+6)}{1}\cdot\dfrac{x}{x+6}=\dfrac{x(x+6)}{1}\cdot\dfrac{3}{x}$$ • Multiply by the LCM.

$$x^2=(x+6)3$$ • Simplify.
$$x^2=3x+18$$
$$x^2-3x-18=0$$ • Standard form
$$(x+3)(x-6)=0$$ • Factor.
$$x+3=0\qquad x-6=0$$ • Principle of Zero Products
$$x=-3\qquad x=6$$

Both -3 and 6 check as solutions.

The solutions are -3 and 6.

You Try It 2

$$\dfrac{5x}{x+2}=3-\dfrac{10}{x+2}$$ • The LCM is $x+2$.

$$\dfrac{(x+2)}{1}\cdot\dfrac{5x}{x+2}=\dfrac{(x+2)}{1}\left(3-\dfrac{10}{x+2}\right)$$ • Clear denominators.

$$\dfrac{x+2}{1}\cdot\dfrac{5x}{x+2}=\dfrac{x+2}{1}\cdot3-\dfrac{x+2}{1}\cdot\dfrac{10}{x+2}$$

$$5x=(x+2)3-10$$ • Solve for x.
$$5x=3x+6-10$$
$$5x=3x-4$$
$$2x=-4$$
$$x=-2$$

-2 does not check as a solution.

The equation has no solution.

SECTION 8.5

You Try It 1

A.
$$\frac{2}{x+3} = \frac{6}{5x+5}$$

$$\frac{(x+3)(5x+5)}{1} \cdot \frac{2}{x+3} = \frac{(x+3)(5x+5)}{1} \cdot \frac{6}{5x+5}$$

$$\frac{\overset{1}{\cancel{(x+3)}}(5x+5)}{1} \cdot \frac{2}{\cancel{x+3}} = \frac{(x+3)\overset{1}{\cancel{(5x+5)}}}{1} \cdot \frac{6}{\cancel{5x+5}}$$

$$(5x+5)2 = (x+3)6 \qquad \text{• Solve for } x.$$
$$10x + 10 = 6x + 18$$
$$4x + 10 = 18$$
$$4x = 8$$
$$x = 2$$

The solution is 2.

B.
$$\frac{5}{2x-3} = \frac{10}{x+3}$$

$$\frac{(2x-3)(x+3)}{1} \cdot \frac{5}{2x-3} = \frac{(2x-3)(x+3)}{1} \cdot \frac{10}{x+3}$$

$$\frac{\overset{1}{\cancel{(2x-3)}}(x+3)}{1} \cdot \frac{5}{\cancel{2x-3}} = \frac{(2x-3)\overset{1}{\cancel{(x+3)}}}{1} \cdot \frac{10}{\cancel{x+3}}$$

$$(x+3)5 = (2x-3)10 \qquad \text{• Solve for } x.$$
$$5x + 15 = 20x - 30$$
$$15 = 15x - 30$$
$$45 = 15x$$
$$3 = x$$

The solution is 3.

You Try It 2

Strategy To find the ounces of medication required, write and solve a proportion using M to represent the ounces of medication.

Solution
$$\frac{3}{120} = \frac{M}{180} \qquad \text{• Write a proportion.}$$

$$(180)(120)\left(\frac{3}{120}\right) = (180)(120)\left(\frac{M}{180}\right) \qquad \text{• Clear denominators.}$$

$$540 = 120M$$
$$4.5 = M$$

4.5 oz of medication are required for a 180-pound adult.

You Try It 3

Strategy To find DE, write and solve a proportion.

Solution
$$\frac{DE}{AB} = \frac{DF}{AC} \qquad \text{• Write a proportion.}$$

$$\frac{DE}{9} = \frac{9}{5} \qquad \text{• } AB = 9, DF = 9, AC = 5$$

$$9\left(\frac{DE}{9}\right) = 9\left(\frac{9}{5}\right) \qquad \text{• Multiply by 9.}$$

$$DE = \frac{81}{5}$$
$$DE = 16.2$$

DE is 16.2 in.

You Try It 4

Strategy To find the area of triangle AOB:
▶ Solve a proportion to find AO (the height of triangle AOB).
▶ Use the formula for the area of a triangle. $\overline{AB}$ is the base and $\overline{AO}$ is the height.

Solution
$$\frac{CD}{AB} = \frac{DO}{AO} \qquad \text{• Write a proportion.}$$

$$\frac{4}{10} = \frac{3}{AO} \qquad \text{• Substitute.}$$

$$10 \cdot AO \cdot \frac{4}{10} = 10 \cdot AO \cdot \frac{3}{AO}$$

$$4(AO) = 30$$
$$AO = 7.5$$

$$A = \frac{1}{2}bh \qquad \text{• Area of a triangle}$$

$$= \frac{1}{2}(10)(7.5) \qquad \text{• Substitute.}$$

$$= 37.5$$

The area of triangle AOB is 37.5 cm^2.

SECTION 8.6

You Try It 1
$$5x - 2y = 10$$
$$5x - 5x - 2y = -5x + 10 \qquad \text{• Subtract } 5x.$$
$$-2y = -5x + 10$$
$$\frac{-2y}{-2} = \frac{-5x + 10}{-2} \qquad \text{• Divide by } -2.$$
$$y = \frac{5}{2}x - 5$$

You Try It 2
$$s = \frac{A+L}{2}$$

$$2 \cdot s = 2\left(\frac{A+L}{2}\right) \qquad \text{• Multiply by 2.}$$
$$2s = A + L$$
$$2s - A = A - A + L \qquad \text{• Subtract } A.$$
$$2s - A = L$$

You Try It 3
$$S = a + (n-1)d$$
$$S = a + nd - d \qquad \text{• Distributive Property}$$
$$S - a = a - a + nd - d \qquad \text{• Subtract } a.$$
$$S - a = nd - d$$
$$S - a + d = nd - d + d \qquad \text{• Add } d.$$
$$S - a + d = nd$$
$$\frac{S - a + d}{d} = \frac{nd}{d} \qquad \text{• Divide by } d.$$
$$\frac{S - a + d}{d} = n$$

You Try It 4

$$S = rS + C$$
$$S - rS = rS - rS + C \quad \bullet \text{ Subtract } rS.$$
$$S - rS = C$$
$$(1 - r)S = C \quad \bullet \text{ Factor.}$$
$$\frac{(1 - r)S}{1 - r} = \frac{C}{1 - r} \quad \bullet \text{ Divide by } 1 - r.$$
$$S = \frac{C}{1 - r}$$

SECTION 8.7

You Try It 1

Strategy   ▸ Time for one printer to complete the job: t

	Rate	Time	Part
1st printer	$\dfrac{1}{t}$	2	$\dfrac{2}{t}$
2nd printer	$\dfrac{1}{t}$	5	$\dfrac{5}{t}$

▸ The sum of the parts of the task completed must equal 1.

Solution
$$\frac{2}{t} + \frac{5}{t} = 1$$
$$t\left(\frac{2}{t} + \frac{5}{t}\right) = t \cdot 1$$
$$2 + 5 = t$$
$$7 = t$$

Working alone, one printer takes 7 h to print the payroll.

You Try It 2

Strategy   ▸ Rate sailing across the lake: r
Rate sailing back: $3r$

	Distance	Rate	Time
Across	6	r	$\dfrac{6}{r}$
Back	6	$3r$	$\dfrac{6}{3r}$

▸ The total time for the trip was 2 h.

Solution
$$\frac{6}{r} + \frac{6}{3r} = 2$$
$$3r\left(\frac{6}{r} + \frac{6}{3r}\right) = 3r(2) \quad \bullet \text{ Multiply by the LCM, } 3r.$$
$$3r \cdot \frac{6}{r} + 3r \cdot \frac{6}{3r} = 6r$$
$$18 + 6 = 6r \quad \bullet \text{ Solve for } r.$$
$$24 = 6r$$
$$4 = r$$

The rate sailing across the lake was 4 km/h.

SECTION 8.8

You Try It 1

Strategy   To find the distance:
▸ Write the basic direct variation equation, replace the variables by the given values, and solve for k.
▸ Write the direct variation equation, replacing k by its value. Substitute 5 for t and solve for s.

Solution
$$s = kt^2 \quad \bullet \text{ Direct variation equation}$$
$$64 = k(2)^2 \quad \bullet \text{ Replace } s \text{ by 64 and } t \text{ by 2.}$$
$$64 = k \cdot 4$$
$$16 = k$$

$$s = 16t^2 \quad \bullet \text{ Direct variation equation}$$
$$= 16(5)^2 = 400 \quad \bullet \text{ Find } s \text{ when } t = 5.$$

The object will fall 400 ft in 5 s.

You Try It 2

Strategy   To find the resistance:

▸ Write the basic inverse variation equation, replace the variables by the given values, and solve for k.
▸ Write the inverse variation equation, replacing k by its value. Substitute 0.02 for d and solve for R.

Solution

$$R = \frac{k}{d^2} \quad \bullet \text{ Inverse variation equation}$$

$$0.5 = \frac{k}{(0.01)^2} \quad \bullet \text{ Replace } R \text{ by 0.5 and } d \text{ by 0.01.}$$

$$0.5 = \frac{k}{0.0001}$$

$$0.00005 = k$$

$$R = \frac{0.00005}{d^2} = \frac{0.00005}{(0.02)^2} = 0.125 \quad \bullet \ k = 0.00005, d = 0.02$$

The resistance is 0.125 ohm.

You Try It 3

Strategy   ▸ Write the basic combined variation equation, replace the variables by the given values, and solve for k.
▸ Write the combined variation equation, replacing k by its value. Substitute 4 for W, 8 for d, and 16 for L. Solve for s.

Solution

$$s = \frac{kWd^2}{L}$$ • Combined variation equation

$$1200 = \frac{k(2)(12)^2}{12}$$ • Replace s by 1200, W by 2, d by 12, and L by 12.

$$1200 = 24k$$

$$50 = k$$

$$s = \frac{50Wd^2}{L}$$ • Replace k by 50 in the combined variation equation.

$$= \frac{50(4)8^2}{16}$$ • Replace W by 4, d by 8, and L by 16.

$$= 800$$

The strength of the beam is 800 lb.

Solutions to Chapter 9 "You Try It"

SECTION 9.1

You Try It 1 $\quad 16^{-3/4} = (2^4)^{-3/4}$ • $16 = 2^4$
$$= 2^{-3}$$
$$= \frac{1}{2^3} = \frac{1}{8}$$

You Try It 2 $\quad (x^{3/4}y^{1/2}z^{-2/3})^{-4/3} = x^{-1}y^{-2/3}z^{8/9}$
$$= \frac{z^{8/9}}{xy^{2/3}}$$

You Try It 3
$$\left(\frac{16a^{-2}b^{4/3}}{9a^4b^{-2/3}}\right)^{-1/2} = \left(\frac{2^4a^{-6}b^2}{3^2}\right)^{-1/2}$$ • Use the Rule for Dividing Exponential Expressions.
$$= \frac{2^{-2}a^3b^{-1}}{3^{-1}}$$ • Use the Rule for Simplifying Powers of Products.
$$= \frac{3a^3}{2^2b} = \frac{3a^3}{4b}$$

You Try It 4 $\quad (2x^3)^{3/4} = \sqrt[4]{(2x^3)^3}$
$$= \sqrt[4]{8x^9}$$

You Try It 5 $\quad -5a^{5/6} = -5(a^5)^{1/6}$
$$= -5\sqrt[6]{a^5}$$

You Try It 6 $\quad \sqrt[3]{3ab} = (3ab)^{1/3}$

You Try It 7 $\quad \sqrt[4]{x^4 + y^4} = (x^4 + y^4)^{1/4}$

You Try It 8 $\quad \sqrt[3]{-8x^{12}y^3} = -2x^4y$

You Try It 9 $\quad -\sqrt[4]{81x^{12}y^8} = -3x^3y^2$

SECTION 9.2

You Try It 1 $\quad \sqrt[5]{x^7} = \sqrt[5]{x^5 \cdot x^2}$ • x^5 is a perfect fifth power.
$$= \sqrt[5]{x^5}\,\sqrt[5]{x^2}$$
$$= x\,\sqrt[5]{x^2}$$

You Try It 2
$$\sqrt[3]{-64x^8y^{18}} = \sqrt[3]{-64x^6y^{18}(x^2)}$$ • $-64x^6y^{18}$ is a perfect third power.
$$= \sqrt[3]{-64x^6y^{18}}\sqrt[3]{x^2}$$
$$= -4x^2y^6\sqrt[3]{x^2}$$

You Try It 3
$$3xy\sqrt[3]{81x^5y} - \sqrt[3]{192x^8y^4}$$
$$= 3xy\sqrt[3]{27x^3 \cdot 3x^2y} - \sqrt[3]{64x^6y^3 \cdot 3x^2y}$$
$$= 3xy\sqrt[3]{27x^3}\sqrt[3]{3x^2y} - \sqrt[3]{64x^6y^3}\sqrt[3]{3x^2y}$$
$$= 3xy \cdot 3x\sqrt[3]{3x^2y} - 4x^2y\sqrt[3]{3x^2y}$$
$$= 9x^2y\sqrt[3]{3x^2y} - 4x^2y\,\sqrt[3]{3x^2y} = 5x^2y\sqrt[3]{3x^2y}$$

SECTION 9.3

You Try It 1
$$5\sqrt{2}(\sqrt{6} + \sqrt{24})$$
$$= 5\sqrt{12} + 5\sqrt{48}$$ • Distributive Property
$$= 5\sqrt{4 \cdot 3} + 5\sqrt{16 \cdot 3}$$ • Simplify each radical.
$$= 5 \cdot 2\sqrt{3} + 5 \cdot 4\sqrt{3}$$
$$= 10\sqrt{3} + 20\sqrt{3} = 30\sqrt{3}$$

You Try It 2
$$(4 - 2\sqrt{7})(1 + 3\sqrt{7})$$
$$= 4 + 12\sqrt{7} - 2\sqrt{7} - 6(\sqrt{7})^2$$ • The FOIL method
$$= 4 + 10\sqrt{7} - 6 \cdot 7$$
$$= 4 + 10\sqrt{7} - 42 = -38 + 10\sqrt{7}$$

You Try It 3
$$(4 - \sqrt{2x})^2$$
$$= (4 - \sqrt{2x})(4 - \sqrt{2x})$$
$$= 16 - 4\sqrt{2x} - 4\sqrt{2x} + (\sqrt{2x})^2$$ • The FOIL method
$$= 16 - 8\sqrt{2x} + 2x$$
$$= 2x - 8\sqrt{2x} + 16$$

You Try It 4
$$\sqrt{\frac{5}{6}} = \frac{\sqrt{5}}{\sqrt{6}}$$ • Quotient Property of Radicals
$$= \frac{\sqrt{5}}{\sqrt{6}} \cdot \frac{\sqrt{6}}{\sqrt{6}} = \frac{\sqrt{30}}{6}$$ • Rationalize the denominator.

You Try It 5
Note that $3x^2 \cdot 9x = 27x^3$, a perfect third power.
$$\frac{3x}{\sqrt[3]{3x^2}} = \frac{3x}{\sqrt[3]{3x^2}} \cdot \frac{\sqrt[3]{9x}}{\sqrt[3]{9x}}$$ • Rationalize the denominator.
$$= \frac{3x\sqrt[3]{9x}}{\sqrt[3]{27x^3}} = \frac{3x\sqrt[3]{9x}}{3x} = \sqrt[3]{9x}$$

You Try It 6

$$\frac{3}{2 - \sqrt{x}} = \frac{3}{2 - \sqrt{x}} \cdot \frac{2 + \sqrt{x}}{2 + \sqrt{x}}$$

• Rationalize the denominator.

$$= \frac{3(2 + \sqrt{x})}{2^2 - (\sqrt{x})^2} = \frac{6 + 3\sqrt{x}}{4 - x}$$

SECTION 9.4

You Try It 1

$$\sqrt{x} - \sqrt{x + 5} = 1$$

• Add $\sqrt{x + 5}$ to each side.

$$\sqrt{x} = 1 + \sqrt{x + 5}$$

$$(\sqrt{x})^2 = (1 + \sqrt{x + 5})^2$$

• Square each side.

$$x = 1 + 2\sqrt{x + 5} + x + 5$$

$$-6 = 2\sqrt{x + 5}$$

$$-3 = \sqrt{x + 5}$$

$$(-3)^2 = (\sqrt{x + 5})^2$$

• Square each side.

$$9 = x + 5$$

$$4 = x$$

4 does not check as a solution. The equation has no solution.

You Try It 2

$$\sqrt[4]{x - 8} = 3$$

$$(\sqrt[4]{x - 8})^4 = 3^4$$

• Raise each side to the fourth power.

$$x - 8 = 81$$

$$x = 89$$

Check:

$$\frac{\sqrt[4]{x - 8} = 3}{\sqrt[4]{89 - 8} \;\big|\; 3}$$

$$\sqrt[4]{81} \;\big|\; 3$$

$$3 = 3$$

The solution is 89.

You Try It 3

Strategy   To find the diagonal, use the Pythagorean Theorem. One leg is the length of the rectangle. The second leg is the width of the rectangle. The hypotenuse is the diagonal of the rectangle.

Solution

$$c^2 = a^2 + b^2$$

• Pythagorean Theorem

$$c^2 = (6)^2 + (3)^2$$

• Replace a by 6 and b by 3.

$$c^2 = 36 + 9$$

• Solve for c.

$$c^2 = 45$$

$$(c^2)^{1/2} = (45)^{1/2}$$

• Raise each side to the $\frac{1}{2}$ power.

$$c = \sqrt{45}$$

• $a^{1/2} = \sqrt{a}$

$$c \approx 6.7$$

The diagonal is approximately 6.7 cm.

You Try It 4

Strategy   To find the height, replace d in the equation with the given value and solve for h.

Solution

$$d = \sqrt{1.5h}$$

$$5.5 = \sqrt{1.5h}$$

• Replace d by 5.5.

$$(5.5)^2 = (\sqrt{1.5h})^2$$

• Square each side.

$$30.25 = 1.5h$$

$$20.17 \approx h$$

The periscope must be approximately 20.17 ft above the water.

You Try It 5

Strategy   To find the distance, replace the variables v and a in the equation by their given values, and solve for s.

Solution

$$v = \sqrt{2as}$$

$$88 = \sqrt{2 \cdot 22s}$$

• Replace v by 88 and a by 22.

$$88 = \sqrt{44s}$$

$$(88)^2 = (\sqrt{44s})^2$$

• Square each side.

$$7744 = 44s$$

$$176 = s$$

The distance required is 176 ft.

SECTION 9.5

You Try It 1

$$\sqrt{-45} = i\sqrt{45} = i\sqrt{9 \cdot 5} = 3i\sqrt{5}$$

You Try It 2

$$-b + \sqrt{b^2 - 4ac}$$

$$-(6) + \sqrt{(6)^2 - 4(1)(25)} = -6 + \sqrt{36 - 100}$$

$$= -6 + \sqrt{-64} = -6 + i\sqrt{64}$$

$$= -6 + 8i$$

You Try It 3

$$(-4 + 2i) - (6 - 8i) = -10 + 10i$$

You Try It 4

$$(16 - \sqrt{-45}) - (3 + \sqrt{-20})$$

$$= (16 - i\sqrt{45}) - (3 + i\sqrt{20})$$

$$= (16 - i\sqrt{9 \cdot 5}) - (3 + i\sqrt{4 \cdot 5})$$

$$= (16 - 3i\sqrt{5}) - (3 + 2i\sqrt{5})$$

$$= 13 - 5i\sqrt{5}$$

You Try It 5

$$(-3i)(-10i) = 30i^2 = 30(-1) = -30$$

You Try It 6

$$-\sqrt{-8} \cdot \sqrt{-5} = -i\sqrt{8} \cdot i\sqrt{5}$$

$$= -i^2\sqrt{40}$$

$$= -(-1)\sqrt{40} = \sqrt{4 \cdot 10} = 2\sqrt{10}$$

You Try It 7

$$-6i(3 + 4i) = -18i - 24i^2$$

• Distributive Property

$$= -18i - 24(-1)$$

$$= 24 - 18i$$

You Try It 8

$$(4 - 3i)(2 - i) = 8 - 4i - 6i + 3i^2$$

• FOIL

$$= 8 - 10i + 3i^2$$

$$= 8 - 10i + 3(-1)$$

$$= 5 - 10i$$

You Try It 9

$$(5 - 3i)^2 = (5 - 3i)(5 - 3i)$$
$$= 25 - 15i - 15i + 9i^2$$
$$= 25 - 30i + 9(-1)$$
$$= 25 - 30i - 9 = 16 - 30i$$

You Try It 10

$$\frac{2 - 3i}{4i} = \frac{2 - 3i}{4i} \cdot \frac{i}{i}$$
$$= \frac{2i - 3i^2}{4i^2}$$
$$= \frac{2i - 3(-1)}{4(-1)}$$
$$= \frac{3 + 2i}{-4} = -\frac{3}{4} - \frac{1}{2}i$$

You Try It 11

$$\frac{2 + 5i}{3 - 2i} = \frac{2 + 5i}{3 - 2i} \cdot \frac{3 + 2i}{3 + 2i} = \frac{6 + 4i + 15i + 10i^2}{3^2 + 2^2}$$
$$= \frac{6 + 19i + 10(-1)}{9 + 4}$$
$$= \frac{-4 + 19i}{13}$$
$$= -\frac{4}{13} + \frac{19}{13}i$$

Solutions to Chapter 10 "You Try It"

SECTION 10.1

You Try It 1

$$2x^2 = 7x - 3$$
$$2x^2 - 7x + 3 = 0$$ • Write in standard form.
$$(2x - 1)(x - 3) = 0$$ • Solve by factoring.

$$2x - 1 = 0 \qquad x - 3 = 0$$
$$2x = 1 \qquad\qquad x = 3$$
$$x = \frac{1}{2}$$

The solutions are $\frac{1}{2}$ and 3.

You Try It 2

$$(x - r_1)(x - r_2) = 0$$
$$(x - 3)\left[x - \left(-\frac{1}{2}\right)\right] = 0$$ • $r_1 = 3,\ r_2 = -\frac{1}{2}$
$$(x - 3)\left(x + \frac{1}{2}\right) = 0$$
$$x^2 - \frac{5}{2}x - \frac{3}{2} = 0$$
$$2\left(x^2 - \frac{5}{2}x - \frac{3}{2}\right) = 2 \cdot 0$$
$$2x^2 - 5x - 3 = 0$$

You Try It 3

$$2(x + 1)^2 - 24 = 0$$
$$2(x + 1)^2 = 24$$
$$(x + 1)^2 = 12$$ • Solve for $(x + 1)^2$.
$$\sqrt{(x + 1)^2} = \sqrt{12}$$ • Take the square root of each side of
$$x + 1 = \pm\sqrt{12}$$ the equation.
$$x + 1 = \pm2\sqrt{3}$$

$$x + 1 = 2\sqrt{3} \qquad\qquad x + 1 = -2\sqrt{3}$$ • Solve for x.
$$x = -1 + 2\sqrt{3} \qquad\qquad x = -1 - 2\sqrt{3}$$

The solutions are $-1 + 2\sqrt{3}$ and $-1 - 2\sqrt{3}$.

SECTION 10.2

You Try It 1

$$4x^2 - 4x - 1 = 0$$
$$4x^2 - 4x = 1$$ • Write in the form $ax^2 + bx = -c$.
$$\frac{1}{4}(4x^2 - 4x) = \frac{1}{4} \cdot 1$$ • Multiply each side by $\frac{1}{a}$.
$$x^2 - x = \frac{1}{4}$$
$$x^2 - x + \frac{1}{4} = \frac{1}{4} + \frac{1}{4}$$ • Complete the square.
$$\left(x - \frac{1}{2}\right)^2 = \frac{2}{4}$$ • Factor.
$$\sqrt{\left(x - \frac{1}{2}\right)^2} = \sqrt{\frac{2}{4}}$$ • Take square roots.
$$x - \frac{1}{2} = \pm\frac{\sqrt{2}}{2}$$

$$x - \frac{1}{2} = \frac{\sqrt{2}}{2} \qquad\qquad x - \frac{1}{2} = -\frac{\sqrt{2}}{2}$$ • Solve for x.
$$x = \frac{1}{2} + \frac{\sqrt{2}}{2} \qquad\qquad x = \frac{1}{2} - \frac{\sqrt{2}}{2}$$

The solutions are $\dfrac{1 + \sqrt{2}}{2}$ and $\dfrac{1 - \sqrt{2}}{2}$.

You Try It 2

$$x^2 + 4x + 8 = 0$$
$$x^2 + 4x = -8$$
$$x^2 + 4x + 4 = -8 + 4$$ • Complete the square.
$$(x + 2)^2 = -4$$ • Factor.
$$\sqrt{(x + 2)^2} = \sqrt{-4}$$ • Take square roots.
$$x + 2 = \pm2i$$

$$x + 2 = 2i \qquad\qquad x + 2 = -2i$$ • Solve for x.
$$x = -2 + 2i \qquad\qquad x = -2 - 2i$$

The solutions are $-2 + 2i$ and $-2 - 2i$.

You Try It 3

$$x^2 = 2x - 10$$
$$x^2 - 2x + 10 = 0 \qquad \text{• Write in standard form.}$$
$$a = 1, b = -2, c = 10$$

$$x = \frac{-b \pm \sqrt{b^2 - 4ac}}{2a}$$

$$= \frac{-(-2) \pm \sqrt{(-2)^2 - 4(1)(10)}}{2 \cdot 1}$$

$$= \frac{2 \pm \sqrt{4 - 40}}{2} = \frac{2 \pm \sqrt{-36}}{2}$$

$$= \frac{2 \pm 6i}{2} = 1 \pm 3i$$

The solutions are $1 + 3i$ and $1 - 3i$.

You Try It 4

$$4x^2 = 4x - 1$$
$$4x^2 - 4x + 1 = 0 \qquad \text{• Write in standard form.}$$
$$a = 4, b = -4, c = 1$$

$$x = \frac{-b \pm \sqrt{b^2 - 4ac}}{2a}$$

$$= \frac{-(-4) \pm \sqrt{(-4)^2 - 4(4)(1)}}{2 \cdot 4}$$

$$= \frac{4 \pm \sqrt{16 - 16}}{8} = \frac{4 \pm \sqrt{0}}{8}$$

$$= \frac{4}{8} = \frac{1}{2}$$

The solution is $\frac{1}{2}$.

You Try It 5

$$3x^2 - x - 1 = 0$$
$$a = 3, b = -1, c = -1$$
$$b^2 - 4ac =$$
$$(-1)^2 - 4(3)(-1) = 1 + 12 = 13$$
$$13 > 0$$

Because the discriminant is greater than zero, the equation has two unequal real number solutions.

SECTION 10.3

You Try It 1

$$x - 5x^{1/2} + 6 = 0$$
$$u^2 - 5u + 6 = 0 \qquad \text{• Let } u = x^{1/2}. \text{ Then } u^2 = x.$$
$$(u - 2)(u - 3) = 0$$

$$u - 2 = 0 \qquad u - 3 = 0$$
$$u = 2 \qquad u = 3$$

Replace u by $x^{1/2}$.

$$x^{1/2} = 2 \qquad\qquad x^{1/2} = 3$$
$$\sqrt{x} = 2 \qquad\qquad \sqrt{x} = 3$$
$$(\sqrt{x})^2 = 2^2 \qquad (\sqrt{x})^2 = 3^2$$
$$x = 4 \qquad\qquad x = 9$$

The solutions are 4 and 9.

You Try It 2

$$\sqrt{2x + 1} + x = 7 \qquad \text{• Solve for the}$$
$$\sqrt{2x + 1} = 7 - x \qquad\quad \text{radical.}$$
$$(\sqrt{2x + 1})^2 = (7 - x)^2 \qquad \text{• Square each side.}$$
$$2x + 1 = 49 - 14x + x^2$$
$$0 = x^2 - 16x + 48 \qquad \text{• Write in standard form.}$$
$$0 = (x - 4)(x - 12) \qquad \text{• Factor.}$$
$$x - 4 = 0 \qquad x - 12 = 0 \qquad \text{• Solve for } x.$$
$$x = 4 \qquad\quad x = 12$$

4 checks as a solution.
12 does not check as a solution.
The solution is 4.

You Try It 3

$$\sqrt{2x - 1} + \sqrt{x} = 2$$

Solve for one of the radical expressions.

$$\sqrt{2x - 1} = 2 - \sqrt{x}$$
$$(\sqrt{2x - 1})^2 = (2 - \sqrt{x})^2 \qquad \text{• Square each side.}$$
$$2x - 1 = 4 - 4\sqrt{x} + x$$
$$x - 5 = -4\sqrt{x} \qquad \text{• Solve for the radical.}$$
$$(x - 5)^2 = (-4\sqrt{x})^2 \qquad \text{• Square each side.}$$
$$x^2 - 10x + 25 = 16x$$
$$x^2 - 26x + 25 = 0 \qquad \text{• Write in standard form.}$$

$$(x - 1)(x - 25) = 0 \qquad \text{• Factor.}$$
$$x - 1 = 0 \qquad x - 25 = 0$$
$$x = 1 \qquad\quad x = 25 \qquad \text{• Solve for } x.$$

1 checks as a solution.
25 does not check as a solution.
The solution is 1.

You Try It 4

$$3y + \frac{25}{3y - 2} = -8$$

$$(3y - 2)\left(3y + \frac{25}{3y - 2}\right) = (3y - 2)(-8) \qquad \text{• Multiply each side by } 3y - 2.$$

$$(3y - 2)(3y) + (3y - 2)\left(\frac{25}{3y - 2}\right) = (3y - 2)(-8)$$

$$9y^2 - 6y + 25 = -24y + 16$$
$$9y^2 + 18y + 9 = 0 \qquad \text{• Write in standard form.}$$

$$9(y^2 + 2y + 1) = 0 \qquad \text{• Factor.}$$
$$9(y + 1)(y + 1) = 0$$
$$y + 1 = 0 \qquad y + 1 = 0 \qquad \text{• Solve for } y.$$
$$y = -1 \qquad\quad y = -1$$

The solution is -1.

SECTION 10.4

You Try It 1
Strategy

► The unknown time for Olivia working alone: t
► The unknown time for William working alone: $t + 3$
► The time working together: 2 h

	Rate	·	Time	=	Part
Olivia	$\dfrac{1}{t}$	·	2	=	$\dfrac{2}{t}$
William	$\dfrac{1}{t+3}$	·	2	=	$\dfrac{2}{t+3}$

► The sum of the parts of the task completed by Olivia and William equals 1.

Solution
$$\frac{2}{t} + \frac{2}{t+3} = 1$$
$$t(t+3)\left(\frac{2}{t} + \frac{2}{t+3}\right) = t(t+3) \cdot 1$$
$$2(t+3) + 2t = t^2 + 3t$$
$$2t + 6 + 2t = t^2 + 3t$$
$$0 = t^2 - t - 6$$
$$0 = (t+2)(t-3)$$

$t + 2 = 0 \qquad t - 3 = 0$
$\quad t = -2 \qquad\quad t = 3$

The time cannot be negative. It takes Olivia 3 h, working alone, to detail the car. The time for William to detail the car is $t + 3$.

$t + 3$
$3 + 3 = 6$

It takes William 6 h, working alone, to detail the car.

You Try It 2
Strategy

► Unknown rate of the wind: r

	Distance	÷	Rate	=	Time
With the wind	1200	÷	$250 + r$	=	$\dfrac{1200}{250+r}$
Against the wind	1200	÷	$250 - r$	=	$\dfrac{1200}{250-r}$

► The time with the wind was 2 h less than the time against the wind.

Solution
$$\frac{1200}{250+r} = \frac{1200}{250-r} - 2$$
$$(250+r)(250-r)\left(\frac{1200}{250+r}\right) = (250+r)(250-r)\left(\frac{1200}{250-r} - 2\right)$$
$$(250-r)1200 = (250+r)1200 - 2(250+r)(250-r)$$
$$300{,}000 - 1200r = 300{,}000 + 1200r - 2(62{,}500 - r^2)$$
$$300{,}000 - 1200r = 300{,}000 + 1200r - 125{,}000 + 2r^2$$
$$0 = 2r^2 + 2400r - 125{,}000$$
$$0 = 2(r^2 + 1200 - 62{,}500)$$
$$0 = 2(r + 1250)(r - 50)$$

$r + 1250 = 0 \qquad\qquad r - 50 = 0$
$\quad r = -1250 \qquad\qquad\quad r = 50$

The rate cannot be negative. The rate of the wind was 50 mph.

You Try It 3
Strategy

► Height of the triangle: h
► Base of the triangle: $2h + 4$

► Use the formula $A = \dfrac{1}{2}bh$.

Solution
$$A = \frac{1}{2}bh$$
$$35 = \frac{1}{2}(2h + 4)h$$
$$35 = h^2 + 2h$$
$$0 = h^2 + 2h - 35$$
$$0 = (h + 7)(h - 5)$$

$h + 7 = 0 \qquad h - 5 = 0$
$\quad h = -7 \qquad\quad h = 5$

The height cannot be negative.
The height of the triangle is 5 in.

SECTION 10.5

You Try It 1
$$2x^2 - x - 10 \le 0$$
$$(2x - 5)(x + 2) \le 0$$

$2x - 5$ $\;---\,|\,----------\,|\,+++$
$x + 2$ $\;\;\;---\,|\,++++++++++\,|\,+++$

$\quad\quad$ −3 −2 −1 0 1 2 3

−5 −4 −3 −2 −1 0 1 2 3 4 5

$$\left\{x \,\middle|\, -2 \le x \le \frac{5}{2}\right\}$$

Unless otherwise noted, all content on this page is © Cengage Learning.

Solutions to Chapter 11 "You Try It"

SECTION 11.1

You Try It 1

x-coordinate of vertex:

$$-\frac{b}{2a} = -\frac{4}{2(4)} = -\frac{1}{2}$$

y-coordinate of vertex:

$$y = 4x^2 + 4x + 1$$

$$= 4\left(-\frac{1}{2}\right)^2 + 4\left(-\frac{1}{2}\right) + 1$$

$$= 1 - 2 + 1$$

$$= 0$$

Vertex: $\left(-\frac{1}{2}, 0\right)$

Axis of symmetry: $x = -\dfrac{1}{2}$

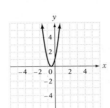

You Try It 2

x-coordinate of vertex:

$$-\frac{b}{2a} = -\frac{-2}{2(-1)} = -1$$

y-coordinate of vertex:

$$f(x) = -x^2 - 2x - 1$$

$$f(-1) = -(-1)^2 - 2(-1) - 1$$

$$= -1 + 2 - 1$$

$$= 0$$

Vertex: $(-1, 0)$

The domain is $\{x | -\infty < x < \infty\}$. The range is $\{y | y \le 0\}$.

You Try It 3

$$y = x^2 + 3x + 4$$

$$0 = x^2 + 3x + 4$$

$$x = \frac{-b \pm \sqrt{b^2 - 4ac}}{2a}$$

$$= \frac{-3 \pm \sqrt{3^2 - 4(1)(4)}}{2 \cdot 1} \qquad \bullet\ a = 1, b = 3, c = 4$$

$$= \frac{-3 \pm \sqrt{-7}}{2}$$

$$= \frac{-3 \pm i\sqrt{7}}{2}$$

The equation has no real number solutions and therefore no x-intercepts.

You Try It 4

$$g(x) = x^2 - x + 6$$

$$0 = x^2 - x + 6$$

$$x = \frac{-b \pm \sqrt{b^2 - 4ac}}{2a}$$

$$= \frac{-(-1) \pm \sqrt{(-1)^2 - 4(1)(6)}}{2(1)} \qquad \bullet\ a = 1, b = -1, c = 6$$

$$= \frac{1 \pm \sqrt{1 - 24}}{2}$$

$$= \frac{1 \pm \sqrt{-23}}{2} = \frac{1 \pm i\sqrt{23}}{2}$$

The zeros of the function are $\dfrac{1}{2} + \dfrac{\sqrt{23}}{2}i$ and $\dfrac{1}{2} - \dfrac{\sqrt{23}}{2}i$.

You Try It 5

$$y = x^2 - x - 6$$

$$a = 1, b = -1, c = -6$$

$$b^2 - 4ac$$

$$(-1)^2 - 4(1)(-6) = 1 + 24 = 25$$

Because the discriminant is greater than zero, the parabola has two x-intercepts.

You Try It 6

$$f(x) = -3x^2 + 4x - 1$$

$$x = -\frac{b}{2a} = -\frac{4}{2(-3)} = \frac{2}{3} \qquad \bullet \text{ The } x\text{-coordinate of the vertex}$$

$$f(x) = -3x^2 + 4x - 1$$

$$f\left(\frac{2}{3}\right) = -3\left(\frac{2}{3}\right)^2 + 4\left(\frac{2}{3}\right) - 1 \qquad \bullet\ x = \frac{2}{3}$$

$$= -\frac{4}{3} + \frac{8}{3} - 1 = \frac{1}{3}$$

Because a is negative, the function has a maximum value.

The maximum value of the function is $\dfrac{1}{3}$.

You Try It 7

Strategy

▶ To find the time it takes the ball to reach its maximum height, find the t-coordinate of the vertex.

▶ To find the maximum height, evaluate the function at the t-coordinate of the vertex.

Solution

$$t = -\frac{b}{2a} = -\frac{64}{2(-16)} = 2 \qquad \bullet \text{ The } t\text{-coordinate of the vertex}$$

The ball reaches its maximum height in 2 s.

$$s(t) = -16t^2 + 64t$$

$$s(2) = -16(2)^2 + 64(2) = -64 + 128 = 64 \qquad \bullet\ t = 2$$

The maximum height is 64 ft.

You Try It 8

Strategy

$$P = 2x + y$$
$$100 = 2x + y \qquad \bullet\ P = 100$$
$$100 - 2x = y \qquad \bullet\ \text{Solve for } y.$$

Express the area of the rectangle in terms of x.

$$A = xy$$
$$A = x(100 - 2x) \qquad \bullet\ y = 100 - 2x$$
$$A = -2x^2 + 100x$$

▶ To find the width, find the x-coordinate of the vertex of $f(x) = -2x^2 + 100x$.

▶ To find the length, replace x in $y = 100 - 2x$ by the x-coordinate of the vertex.

Solution

$$A = -2x^2 + 100x \qquad \bullet\ a = -2, b = 100$$
$$x = -\frac{b}{2a} = -\frac{100}{2(-2)} = 25$$

The width is 25 ft.

$$100 - 2x = 100 - 2(25)$$
$$= 100 - 50 = 50$$

The length is 50 ft.

SECTION 11.2

You Try It 1

The graph of K is the graph of h shifted 3 units to the left and 2 units down.

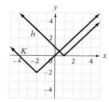

You Try It 2

The graph of $y = f(-x)$ is the mirror image about the y-axis of the graph of f. For instance, the point $(2, 0)$ on $y = f(x)$ reflects to $(-2, 0)$ on $y = f(-x)$.

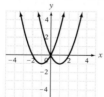

SECTION 11.3

You Try It 1

$$(f + g)(-2) = f(-2) + g(-2)$$
$$= [(-2)^2 + 2(-2)] + [5(-2) - 2]$$
$$= (4 - 4) + (-10 - 2)$$
$$= -12$$
$$(f + g)(-2) = -12$$

You Try It 2

$$(f \cdot g)(3) = f(3) \cdot g(3)$$
$$= (4 - 3^2) \cdot [3(3) - 4]$$
$$= (4 - 9) \cdot (9 - 4)$$
$$= (-5)(5)$$
$$= -25$$
$$(f \cdot g)(3) = -25$$

You Try It 3

$$\left(\frac{f}{g}\right)(4) = \frac{f(4)}{g(4)}$$
$$= \frac{4^2 - 4}{4^2 + 2 \cdot 4 + 1}$$
$$= \frac{16 - 4}{16 + 8 + 1}$$
$$= \frac{12}{25}$$
$$\left(\frac{f}{g}\right)(4) = \frac{12}{25}$$

You Try It 4

$$g(x) = x^2$$
$$g(-1) = (-1)^2 = 1 \qquad \bullet\ \text{Evaluate } g \text{ at } -1.$$
$$f(x) = 1 - 2x \qquad \bullet\ \text{Evaluate } f \text{ at}$$
$$f[g(-1)] = f(1) = 1 - 2(1) = -1 \qquad \quad g(-1) = 1.$$
$$f[g(-1)] = -1$$

You Try It 5

$$M(s) = s^3 + 1$$
$$M[L(s)] = (s + 1)^3 + 1 \qquad \bullet\ \text{Evaluate } M \text{ at } L(s).$$
$$= s^3 + 3s^2 + 3s + 1 + 1$$
$$= s^3 + 3s^2 + 3s + 2$$
$$M[L(s)] = s^3 + 3s^2 + 3s + 2$$

SECTION 11.4

You Try It 1

Because any horizontal line intersects the graph at most once, the graph is the graph of a 1–1 function.

You Try It 2

$$f(x) = \frac{1}{2}x + 4$$

$$y = \frac{1}{2}x + 4$$

$$x = \frac{1}{2}y + 4$$

$$x - 4 = \frac{1}{2}y$$

$$2x - 8 = y$$

$$f^{-1}(x) = 2x - 8$$

The inverse of the function is given by
$$f^{-1}(x) = 2x - 8.$$

You Try It 3

$$f[g(x)] = 2\left(\frac{1}{2}x - 3\right) - 6$$
$$= x - 6 - 6 = x - 12$$

No, $g(x)$ is not the inverse of $f(x)$.

You Try It 4

$$s(x) = \frac{3}{5}x + 1125$$

$$y = \frac{3}{5}x + 1125 \qquad \bullet \text{ Replace } s(x) \text{ by } y.$$

$$x = \frac{3}{5}y + 1125 \qquad \bullet \text{ Interchange } x \text{ and } y.$$

$$x - 1125 = \frac{3}{5}y \qquad \bullet \text{ Solve for } y.$$

$$\frac{5}{3}(x - 1125) = y$$

$$\frac{5}{3}x - 1875 = y$$

$$\frac{5}{3}x - 1875 = s^{-1}(x) \qquad \bullet \text{ Replace } y \text{ by } s^{-1}(x).$$

$s^{-1}(x) = \frac{5}{3}x - 1875$. The equation gives the temperature of the air given the speed of sound.

Solutions to Chapter 12 "You Try It"

SECTION 12.1

You Try It 1

$$f(x) = \left(\frac{2}{3}\right)^x$$

$$f(3) = \left(\frac{2}{3}\right)^3 = \frac{8}{27} \qquad \bullet \; x = 3$$

$$f(-2) = \left(\frac{2}{3}\right)^{-2} = \left(\frac{3}{2}\right)^2 = \frac{9}{4} \qquad \bullet \; x = -2$$

You Try It 2

$$f(x) = 2^{2x+1}$$

$$f(0) = 2^{2(0)+1} = 2^1 = 2 \qquad \bullet \; x = 0$$

$$f(-2) = 2^{2(-2)+1} = 2^{-3} = \frac{1}{2^3} = \frac{1}{8} \qquad \bullet \; x = -2$$

You Try It 3

$$f(x) = e^{2x-1}$$

$$f(2) = e^{2 \cdot 2 - 1} = e^3 \approx 20.0855 \qquad \bullet \; x = 2$$

$$f(-2) = e^{2(-2)-1} = e^{-5} \approx 0.0067 \qquad \bullet \; x = -2$$

You Try It 4

x	$y = 2^{\frac{1}{2}x}$
-4	4
-2	2
0	1
2	$\frac{1}{2}$
4	$\frac{1}{4}$

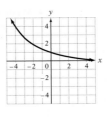

You Try It 5

x	$y = 2^x + 1$
-2	$\frac{5}{4}$
-1	$\frac{3}{2}$
0	2
1	3
2	5

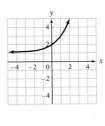

You Try It 6

x	$y = 2^{-x} + 2$
-2	6
-1	4
0	3
1	$\frac{5}{2}$
2	$\frac{9}{4}$

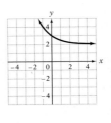

You Try It 7

x	$y = e^{-x/2}$
-4	7.39
-3	4.48
-2	2.72
-1	1.65
0	1.00
1	0.61
2	0.37
3	0.22

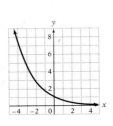

Unless otherwise noted, all content on this page is © Cengage Learning.

SECTION 12.2

You Try It 1

$\log_4 64 = x$ • Write the equivalent
$64 = 4^x$ exponential form.
$4^3 = 4^x$ • $64 = 4^3$
$3 = x$ • The bases are the same.
$\log_4 64 = 3$ The exponents are equal.

You Try It 2

$\log_2 x = -4$
$2^{-4} = x$ • Write the equivalent
$\dfrac{1}{2^4} = x$ exponential form.
$\dfrac{1}{16} = x$

The solution is $\dfrac{1}{16}$.

You Try It 3

$\ln x = 3$ • Write the equivalent
$e^3 = x$ exponential form.
$20.0855 \approx x$ • Use a calculator.

You Try It 4

$\log_8 \sqrt[3]{xy^2} = \log_8 (xy^2)^{1/3} = \dfrac{1}{3} \log_8 (xy^2)$ • Power Property

$= \dfrac{1}{3} (\log_8 x + \log_8 y^2)$ • Product Property

$= \dfrac{1}{3} (\log_8 x + 2 \log_8 y)$ • Power Property

$= \dfrac{1}{3} \log_8 x + \dfrac{2}{3} \log_8 y$ • Distributive Property

You Try It 5

$\dfrac{1}{3} (\log_4 x - 2 \log_4 y + \log_4 z)$

$= \dfrac{1}{3} (\log_4 x - \log_4 y^2 + \log_4 z)$

$= \dfrac{1}{3} \left(\log_4 \dfrac{x}{y^2} + \log_4 z \right) = \dfrac{1}{3} \left(\log_4 \dfrac{xz}{y^2} \right)$

$= \log_4 \left(\dfrac{xz}{y^2} \right)^{1/3} = \log_4 \sqrt[3]{\dfrac{xz}{y^2}}$

You Try It 6 $\log_3 0.834 = \dfrac{\ln 0.834}{\ln 3} \approx -0.1652$

You Try It 7 $\log_7 6.45 = \dfrac{\log 6.45}{\log 7} \approx 0.9579$

SECTION 12.3

You Try It 1

$f(x) = \log_2 (x - 1)$
$y = \log_2 (x - 1)$ • $f(x) = y$
$2^y = x - 1$ • Write the equivalent
$2^y + 1 = x$ exponential equation.

$x = 2^y + 1$	y
$\dfrac{5}{4}$	-2
$\dfrac{3}{2}$	-1
2	0
3	1
5	2

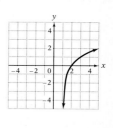

You Try It 2

$f(x) = \log_3 (2x)$
$y = \log_3 (2x)$ • $f(x) = y$
$3^y = 2x$ • Write the equivalent
$\dfrac{3^y}{2} = x$ exponential equation.

$x = \dfrac{3^y}{2}$	y
$\dfrac{1}{18}$	-2
$\dfrac{1}{6}$	-1
$\dfrac{1}{2}$	0
$\dfrac{3}{2}$	1
$\dfrac{9}{2}$	2

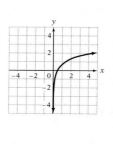

You Try It 3

$f(x) = -\log_3 (x + 1)$
$y = -\log_3 (x + 1)$ • $f(x) = y$
$-y = \log_3 (x + 1)$ • Multiply both sides by -1.
$3^{-y} = x + 1$ • Write the equivalent
$3^{-y} - 1 = x$ exponential equation.

$x = 3^{-y} - 1$	y
8	-2
2	-1
0	0
$-\dfrac{2}{3}$	1
$-\dfrac{8}{9}$	2

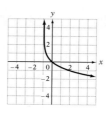

Unless otherwise noted, all content on this page is © Cengage Learning.

SECTION 12.4

You Try It 1

$$(1.06)^n = 1.5$$
$$\log (1.06)^n = \log 1.5 \qquad \text{• Take the log of each side.}$$
$$n \log 1.06 = \log 1.5 \qquad \text{• Power Property}$$
$$n = \frac{\log 1.5}{\log 1.06} \qquad \text{• Divide both sides by log 1.06.}$$
$$n \approx 6.9585$$

The solution is 6.9585.

You Try It 2

$$4^{3x} = 25$$
$$\log 4^{3x} = \log 25 \qquad \text{• Take the log of each side.}$$
$$3x \log 4 = \log 25 \qquad \text{• Power Property}$$
$$3x = \frac{\log 25}{\log 4} \qquad \text{• Divide both sides by log 4.}$$
$$3x \approx 2.3219$$
$$x \approx 0.7740 \qquad \text{• Divide both sides by 3.}$$

The solution is 0.7740.

You Try It 3

$$\log_4 (x^2 - 3x) = 1$$
$$4^1 = x^2 - 3x \qquad \text{• Write in exponential form.}$$
$$4 = x^2 - 3x$$
$$0 = x^2 - 3x - 4$$
$$0 = (x + 1)(x - 4)$$
$$x + 1 = 0 \qquad\qquad x - 4 = 0$$
$$x = -1 \qquad\qquad x = 4$$

The solutions are −1 and 4.

You Try It 4

$$\log_3 x + \log_3 (x + 3) = \log_3 4$$
$$\log_3 [x(x + 3)] = \log_3 4 \qquad \text{• Product Property}$$
$$x(x + 3) = 4 \qquad \text{• 1–1 Property of Logarithms}$$
$$x^2 + 3x = 4$$
$$x^2 + 3x - 4 = 0$$
$$(x + 4)(x - 1) = 0 \qquad \text{• Solve by factoring.}$$
$$x + 4 = 0 \qquad\qquad x - 1 = 0$$
$$x = -4 \qquad\qquad x = 1$$

−4 does not check as a solution. The solution is 1.

You Try It 5

$$\log_3 x + \log_3 (x + 6) = 3$$
$$\log_3 [x(x + 6)] = 3 \qquad \text{• Product Property}$$
$$x(x + 6) = 3^3 \qquad \text{• Write the equivalent exponential equation.}$$
$$x^2 + 6x = 27$$
$$x^2 + 6x - 27 = 0$$
$$(x + 9)(x - 3) = 0 \qquad \text{• Solve by factoring.}$$
$$x + 9 = 0 \qquad\qquad x - 3 = 0$$
$$x = -9 \qquad\qquad x = 3$$

−9 does not check as a solution. The solution is 3.

SECTION 12.5

You Try It 1

Strategy To find the hydrogen ion concentration, replace pH by 2.9 in the equation $\text{pH} = -\log (\text{H}^+)$ and solve for H^+.

Solution
$$\text{pH} = -\log (\text{H}^+)$$
$$2.9 = -\log (\text{H}^+)$$
$$-2.9 = \log (\text{H}^+) \qquad \text{• Multiply by −1.}$$
$$10^{-2.9} = \text{H}^+ \qquad \text{• Write the equivalent exponential equation.}$$
$$0.00126 \approx \text{H}^+$$

The hydrogen ion concentration is approximately 0.00126.

You Try It 2

Strategy Substitute 0.20 for p and 6 for k into the equation $\log p = -kd$ and solve for d.

Solution
$$\log p = -kd$$
$$\log 0.20 = -6d$$
$$\frac{\log 0.20}{-6} = d$$
$$0.116 \approx d$$

The opaque glass must be 0.116 m thick.

Solutions to Chapter T "You Try It"

SECTION T.1

You Try It 1

$$(-4)(6 - 8)^2 - (-12 \div 4) = -4(-2)^2 - (-3)$$
$$= -4(4) - (-3)$$
$$= -16 - (-3)$$
$$= -16 + 3$$
$$= -13$$

You Try It 2

$$3xy^2 - 3x^2y$$
$$3(-2)(5)^2 - 3(-2)^2(5) = 3(-2)(25) - 3(4)(5)$$
$$= -6(25) - 3(4)(5)$$
$$= -150 - 3(4)(5)$$
$$= -150 - 12(5)$$
$$= -150 - 60$$
$$= -150 + (-60)$$
$$= -210$$

You Try It 3

$$-5(-3a) = [-5(-3)]a$$
$$= 15a$$

You Try It 4

$$2z^2 - 5z - 3z^2 + 6z = 2z^2 - 3z^2 - 5z + 6z$$
$$= (2z^2 - 3z^2) + (-5z + 6z)$$
$$= -1z^2 + z$$
$$= -z^2 + z$$

You Try It 5

a. $-3(5y - 2) = -3(5y) - (-3)(2)$
$\qquad\qquad\quad = -15y + 6$

b. $-2(4x + 2y - 6z) = -2(4x) + (-2)(2y) - (-2)(6z)$
$\qquad\qquad\qquad\quad\; = -8x - 4y + 12z$

You Try It 6

a. $7(-3x - 4y) - 3(3x + y) = -21x - 28y - 9x - 3y$
$\qquad\qquad\qquad\qquad\qquad = -30x - 31y$

b. $2y - 3[5 - 3(3 + 2y)] = 2y - 3[5 - 9 - 6y]$
$\qquad\qquad\qquad\qquad\quad = 2y - 3[-4 - 6y]$
$\qquad\qquad\qquad\qquad\quad = 2y + 12 + 18y$
$\qquad\qquad\qquad\qquad\quad = 20y + 12$

SECTION T.2

You Try It 1

a.
$$4x + 3 = 7x + 9$$
$$4x - 7x + 3 = 7x - 7x + 9$$
$$-3x + 3 = 9$$
$$-3x + 3 - 3 = 9 - 3$$
$$-3x = 6$$
$$\frac{-3x}{-3} = \frac{6}{-3}$$
$$x = -2$$
The solution is -2.

b.
$$4 - (5x - 8) = 4x + 3$$
$$4 - 5x + 8 = 4x + 3$$
$$-5x + 12 = 4x + 3$$
$$-5x - 4x + 12 = 4x - 4x + 3$$
$$-9x + 12 = 3$$
$$-9x + 12 - 12 = 3 - 12$$
$$-9x = -9$$
$$\frac{-9x}{-9} = \frac{-9}{-9}$$
$$x = 1$$
The solution is 1.

You Try It 2
$$3x - 1 \le 5x - 7$$
$$3x - 5x - 1 \le 5x - 5x - 7$$
$$-2x - 1 \le -7$$
$$-2x - 1 + 1 \le -7 + 1$$
$$-2x \le -6$$
$$\frac{-2x}{-2} \ge \frac{-6}{-2}$$
$$x \ge 3$$
$$\{x | x \ge 3\}$$

You Try It 3
$$3 - 2(3x + 1) < 7 - 2x$$
$$3 - 6x - 2 < 7 - 2x$$
$$1 - 6x < 7 - 2x$$
$$1 - 6x + 2x < 7 - 2x + 2x$$
$$1 - 4x < 7$$
$$1 - 1 - 4x < 7 - 1$$
$$-4x < 6$$
$$\frac{-4x}{-4} > \frac{6}{-4}$$
$$x > -\frac{3}{2}$$
$$\left\{ x \middle| x > -\frac{3}{2} \right\}$$

SECTION T.3

You Try It 1

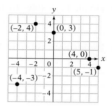

You Try It 2 $\quad y = \dfrac{3}{5}x - 4$

x	y
5	-1
0	-4
-5	-7

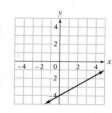

You Try It 3
$$-3x + 2y = 4$$
$$2y = 3x + 4$$
$$y = \frac{3}{2}x + 2$$

x	y
2	5
0	2
-2	-1

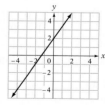

You Try It 4 $\quad$ Let $(x_1, y_1) = (-2, 3)$ and $(x_2, y_2) = (1, -3)$.

$$m = \frac{y_2 - y_1}{x_2 - x_1} = \frac{-3 - 3}{1 - (-2)} = \frac{-6}{3} = -2$$

The slope is -2.

Unless otherwise noted, all content on this page is © Cengage Learning.

You Try It 5 $y = -\dfrac{2}{3}x + 2$

$m = -\dfrac{2}{3}$

$= \dfrac{-2}{3}$

$y\text{-intercept} = (0, 2)$

Place a dot at the y-intercept.
Starting at the y-intercept, move down 2 units (the change in y) and to the right 3 units (the change in x).
Place a dot at that location.
Draw a line through the two points.

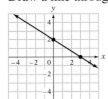

You Try It 6 $f(x) = 4 - 2x$
$f(-3) = 4 - 2(-3)$
$f(-3) = 4 + 6$
$f(-3) = 10$
An ordered pair of the function is $(-3, 10)$.

You Try It 7 $(x_1, y_1) = (-2, 2),\ m = -\dfrac{1}{2}$

$y - y_1 = m(x - x_1)$

$y - 2 = -\dfrac{1}{2}[x - (-2)]$

$y - 2 = -\dfrac{1}{2}(x + 2)$

$y - 2 = -\dfrac{1}{2}x - 1$

$y = -\dfrac{1}{2}x + 1$

SECTION T.4

You Try It 1

a. $(-2ab)(2a^3b^{-2})^{-3} = (-2ab)(2^{-3}a^{-9}b^6)$
$= (-2^{-2})a^{-8}b^7$
$= -\dfrac{b^7}{2^2 a^8}$
$= -\dfrac{b^7}{4a^8}$

b. $\left(\dfrac{2x^2 y^{-4}}{4x^{-2}y^{-5}}\right)^{-3} = \left(\dfrac{x^2 y^{-4}}{2x^{-2}y^{-5}}\right)^{-3}$

$= \dfrac{x^{2(-3)}y^{(-4)(-3)}}{2^{1(-3)}x^{(-2)(-3)}y^{(-5)(-3)}}$

$= \dfrac{x^{-6}y^{12}}{2^{-3}x^6 y^{15}}$

$= 2^3 x^{-6-6}y^{12-15}$

$= 2^3 x^{-12}y^{-3}$

$= \dfrac{8}{x^{12}y^3}$

You Try It 2
$(-4x^3 + 2x^2 - 8) + (4x^3 + 6x^2 - 7x + 5)$
$= (-4x^3 + 4x^3) + (2x^2 + 6x^2) + (-7x) + (-8 + 5)$
$= 8x^2 - 7x - 3$

You Try It 3
$(-4w^3 + 8w - 8) - (3w^3 - 4w^2 + 2w - 1)$
$= (-4w^3 + 8w - 8) + (-3w^3 + 4w^2 - 2w + 1)$
$= (-4w^3 - 3w^3) + 4w^2 + (8w - 2w) + (-8 + 1)$
$= -7w^3 + 4w^2 + 6w - 7$

You Try It 4
$3mn^2(2m^2 - 3mn - 1)$
$= 3mn^2(2m^2) - 3mn^2(3mn) - 3mn^2(1)$
$= 6m^3 n^2 - 9m^2 n^3 - 3mn^2$

You Try It 5
$$
\begin{array}{r}
3c^2 - 4c + 5 \\
2c - 3 \\
\hline
-9c^2 + 12c - 15 \\
6c^3 - 8c^2 + 10c \\
\hline
6c^3 - 17c^2 + 22c - 15
\end{array}
$$

You Try It 6
$(4y - 7)(3y - 5)$
$= (4y)(3y) + (4y)(-5) + (-7)(3y) + (-7)(-5)$
$= 12y^2 - 20y - 21y + 35$
$= 12y^2 - 41y + 35$

You Try It 7
$(x^3 - 7 - 2x) \div (x - 2)$

$$
\require{enclose}
\begin{array}{r}
x^2 + 2x + 2 \\
x - 2 \enclose{longdiv}{x^3 + 0x^2 - 2x - 7} \\
\underline{x^3 - 2x^2} \\
2x^2 - 2x \\
\underline{2x^2 - 4x} \\
2x - 7 \\
\underline{2x - 4} \\
-3
\end{array}
$$

Check: $(x^2 + 2x + 2)(x - 2) - 3$
$= x^3 - 2x - 4 - 3$
$= x^3 - 2x - 7$

$(x^3 - 7 - 2x) \div (x - 2)$

$= x^2 + 2x + 2 - \dfrac{3}{x - 2}$

Unless otherwise noted, all content on this page is © Cengage Learning.

You Try It 8 The GCF is $3x^2y^2$.
$$6x^4y^2 - 9x^3y^2 + 12x^2y^4$$
$$= 3x^2y^2(2x^2 - 3x + 4y^2)$$

You Try It 9 Two factors of 15 whose sum is -8 are -3 and -5.
$$x^2 - 8x + 15 = (x - 3)(x - 5)$$

You Try It 10
$3x^2 - x - 2$

Positive Factors of 3	Factors of -2
1, 3	1, -2
	$-1, 2$

Trial Factors	Middle Term
$(x + 1)(3x - 2)$	$-2x + 3x = x$
$(x - 2)(3x + 1)$	$x - 6x = -5x$
$(x - 1)(3x + 2)$	$2x - 3x = -x$
$(x + 2)(3x - 1)$	$-x + 6x = 5x$

$3x^2 - x - 2 = (x - 1)(3x + 2)$

Check: $(x - 1)(3x + 2) = 3x^2 + 2x - 3x - 2$
$$= 3x^2 - x - 2$$

You Try It 11
$4a^3 - 4a^2 - 24a$

There is a common factor, $4a$. Factor out the GCF.

$4a^3 - 4a^2 - 24a = 4a(a^2 - a - 6)$

Factor $a^2 - a - 6$. The two factors of -6 whose sum is -1 are 2 and -3.

$4a(a^2 - a - 6) = 4a(a + 2)(a - 3)$
$4a^3 - 4a^2 - 24a = 4a(a + 2)(a - 3)$

Check: $4a(a + 2)(a - 3) = (4a^2 + 8a)(a - 3)$
$$= 4a^3 - 12a^2 + 8a^2 - 24a$$
$$= 4a^3 - 4a^2 - 24a$$

Unless otherwise noted, all content on this page is © Cengage Learning.

Answers to Selected Exercises

Answers to Chapter 1 Selected Exercises

PREP TEST

1. 127.16   **2.** 46,514   **3.** 4517   **4.** 11,396   **5.** 508   **6.** 24   **7.** 4   **8.** $3 \cdot 7$ **9.** $\dfrac{2}{5}$ **10.** iv

SECTION 1.1

1a. left   **b.** right   **3.** absolute value   **9.** $8 > -6$ **11.** $-12 < 1$ **13.** $42 > 19$ **15.** $0 > -31$ **17.** $53 > -46$
19. $-23, -18$ **21.** 21, 37   **23.** i   **25.** -4 **27.** 9   **29.** 28   **31.** 14   **33.** -77 **35.** 0   **37.** 74   **39.** -82
41. -81 **43.** $|-83| > |58|$ **45.** $|43| < |-52|$ **47.** $|-68| > |-42|$ **49.** $|-45| < |-61|$ **51.** $19, 0, -28$ **53.** $-45, 0, -17$
55. True   **57.** -11 **59.** -5 **61.** -83 **63.** -46 **65.** 0   **67.** -5 **69.** 9   **71.** 1   **73.** 8   **75.** -7 **77.** -9
79. 9   **81.** -3 **83.** 18   **85.** -10 **87.** -41 **89.** -12 **91.** 0   **93.** -9 **95.** 11   **97.** -18 **99.** 0   **101.** 2
103. -138 **105.** -8 **107.** -12 **109.** Negative   **111.** Positive   **113.** 42   **115.** -28 **117.** 60   **119.** -253
121. -238 **123.** -114 **125.** -2 **127.** 8   **129.** -7 **131.** -12 **133.** -6 **135.** -7 **137.** 11   **139.** -14
141. 15   **143.** -16 **145.** 0   **147.** -29 **149.** Undefined   **151.** -11 **153.** Undefined   **155.** -105 **157.** 252
159. 420   **161.** 0   **163.** Negative   **165.** The difference is 153°F.   **167.** After a rise of 5°C, the temperature is -14°C.
169. The difference in height is 20,370 m.   **171a.** $0, -4, -2$ **b.** Duke's score for the first three days was -6.   **c.** His final score was -10.
173. The average daily high temperature for the week was -4°F.   **175.** False   **177.** The average score of the 10 golfers was -2.
179. The student scored 71 points.   **181.** The largest difference is 25.   **183a.** True   **b.** True   **187.** $-32, 64, -128$

SECTION 1.2

1. 3; 4; terminating   **3.** 0.01   **5.** equivalent; common denominator   **7.** reciprocal   **9.** $(-5)^6$ **11.** 0.125   **13.** $0.\overline{2}$
15. $0.1\overline{6}$ **17.** 0.5625   **19.** $0.58\overline{3}$ **21.** 0.525   **25.** $\dfrac{2}{5}, 0.40$ **27.** $\dfrac{22}{25}, 0.88$ **29.** $\dfrac{8}{5}, 1.6$ **31.** $\dfrac{87}{100}, 0.87$ **33.** $\dfrac{9}{2}, 4.50$

35. $\dfrac{3}{70}$ **37.** $\dfrac{3}{8}$ **39.** $\dfrac{1}{400}$ **41.** $\dfrac{1}{16}$ **43.** $\dfrac{23}{400}$ **45.** 0.091   **47.** 0.167   **49.** 0.009   **51.** 0.0915   **53.** 0.1823

55. 37%   **57.** 2%   **59.** 12.5%   **61.** 136%   **63.** 0.4%   **65.** 83%   **67.** $37\dfrac{1}{2}\%$ **69.** $44\dfrac{4}{9}\%$ **71.** 45%   **73.** 250%

75. Greater than 1%   **77.** $\dfrac{5}{26}$ **79.** $\dfrac{11}{8}$ **81.** $\dfrac{1}{12}$ **83.** $\dfrac{7}{24}$ **85.** 0   **87.** $\dfrac{3}{8}$ **89.** -1.06 **91.** -23.845 **93.** -10.7893

95. 19.61   **97.** -0.88 **99.** 1   **101.** $\dfrac{2}{9}$ **103.** $-\dfrac{2}{9}$ **105.** $-\dfrac{2}{27}$ **107.** $\dfrac{3}{2}$ **109.** $-\dfrac{8}{9}$ **111.** $\dfrac{2}{3}$ **113.** 4.164

115. 4.347   **117.** -2.22 **119.** 0.75   **121.** 2.32   **123.** 3.83   **125a.** Less than 1   **b.** Greater than 1   **127.** 2401

129. -64 **131.** -8 **133.** -125 **135.** $-\dfrac{27}{64}$ **137.** 3.375   **139.** -1 **141.** -8 **143.** -6750 **145.** Negative

147. Positive   **149.** 4   **151.** 7   **153.** $4\sqrt{2}$ **155.** $2\sqrt{2}$ **157.** $18\sqrt{2}$ **159.** $10\sqrt{10}$ **161.** $\sqrt{15}$ **163.** $\sqrt{29}$
165. $-54\sqrt{2}$ **167.** $3\sqrt{5}$ **169.** 0   **171.** $48\sqrt{2}$ **173.** 15.492   **175.** 16.971   **177.** 16   **179.** -11 and -10
181. 2 and 3   **183.** The difference between the temperatures was 55.57°C.   **185.** The difference is 35.438°C.
187a. The difference is 4.3 million barrels of oil per day.   **b.** The predicted increase is 1.1 million barrels of oil per day.

189. The chef should use $1\dfrac{1}{8}$ c of butter.   **191.** The box contains 16 servings.   **193.** Answers will vary. For example, **a.** 0.15, **b.** 1.05,

c. 0.001.   **197.** $a = 2, b = 3, c = 6$

SECTION 1.3

3. 0   **5.** -11 **7.** 20   **9.** -10 **11.** 20   **13.** 29   **15.** 11   **17.** 7   **19.** -11 **21.** 6   **23.** 15   **25.** 4
27. 5   **29.** -1 **31.** 4   **33.** 0.51   **35.** 1.7   **37.** Answers will vary. For example, $\dfrac{17}{24}$ and $\dfrac{33}{48}$.   **39.** Answers will vary.
For example, **a.** $\dfrac{1}{2}$, **b.** 1, **c.** 2.   **41.** Column A: $1,000,000 = 100^3$

CHECK YOUR PROGRESS: CHAPTER 1*

1. $\{1, 2, 3, 4, 5, 6, 7, 8\}$ [1.1A]   **2.** $-7, 0$ [1.1A]   **3.** 13   [1.1B]   **4.** 44; -18 [1.1B]   **5.** $|31| > |-13|$ [1.1B]
6. -24 [1.1C]   **7.** 16   [1.1C]   **8.** -1 [1.1C]   **9.** 52   [1.1C]   **10.** -32 [1.1D]   **11.** 315   [1.1D]   **12.** -10 [1.1D]
13. -16 [1.1D]   **14.** Undefined   [1.1D]   **15.** 0.6875   [1.2A]   **16.** $0.\overline{63}$ [1.2A]   **17.** $\dfrac{9}{20}$; 0.45   [1.2B]   **18.** $\dfrac{29}{200}$ [1.2B]
19. 87.5%   [1.2B]   **20.** 8%   [1.2B]   **21.** 1   [1.2C]   **22.** $-\dfrac{1}{24}$ [1.2C]   **23.** -37.19 [1.2C]   **24.** $-\dfrac{3}{2}$ [1.2D]
25. $-\dfrac{8}{33}$ [1.2D]   **26.** -0.32 [1.2D]   **27.** $6\sqrt{2}$ [1.2F]   **28.** $3\sqrt{3}$ [1.2F]   **29.** -144 [1.2E]   **30.** 9   [1.3A]
31. -31 [1.3A]   **32.** 0   [1.3A]   **33.** The temperature is 5°C.   [1.1E]   **34.** The average daily low temperature for the week is -5°C.   [1.1E]   **35.** The day began at a temperature of -4.6°F.   [1.2G]

SECTION 1.4

1. $2x^2, 5x, \underline{-8}$ **3.** $-a^4, 6$ **5.** $12, -8, -1$ **7.** reciprocal (or multiplicative inverse)   **11.** less than, quotient   **13.** $25 - x$
15. -9 **17.** 41   **19.** -7 **21.** 13   **23.** 41   **25.** 5   **27.** 1   **29.** 57   **31.** 5   **33.** 8   **35.** -3 **37.** -2
39. -4 **41.** Positive   **43.** Negative   **45.** $14x$ **47.** $5a$ **49.** $7 - 3b$ **51.** $5a$ **53.** $5xy$ **55.** 0
57. $-\dfrac{5}{6}x$ **59.** $6.5x$ **61.** $0.45x$ **63.** $7a$ **65.** $-14x^2$ **67.** $-\dfrac{11}{24}x$ **69.** $17x - 3y$ **71.** $-2a - 6b$ **73.** $-3x - 8y$
75. $-4x^2 - 2x$ **77.** iv and v   **79.** $60x$ **81.** $-10a$ **83.** $30y$ **85.** $72x$ **87.** $-28a$ **89.** $108b$ **91.** $-56x^2$ **93.** x^2
95. x **97.** a **99.** b **101.** x **103.** n **105.** $2x$ **107.** $-2x$ **109.** $-15a^2$ **111.** $6y$ **113.** $3y$ **115.** $-2x$
117. $-9y$ **119.** $8x - 6$ **121.** $-2a - 14$ **123.** $-6y + 24$ **125.** $-x - 2$ **127.** $35 - 21b$ **129.** $2 - 5y$
131. $15x^2 + 6x$ **133.** $2y - 18$ **135.** $-15x - 30$ **137.** $-6x^2 - 28$ **139.** $-6y^2 + 21$ **141.** $3x^2 - 3y^2$
143. $-4x + 12y$ **145.** $-6a^2 + 7b^2$ **147.** $4x^2 - 12x + 20$ **149.** $\dfrac{3}{2}x - \dfrac{9}{2}y + 6$ **151.** $-12a^2 - 20a + 28$
153. $12x^2 - 9x + 12$ **155.** $10x^2 - 20xy - 5y^2$ **157.** $-8b^2 + 6b - 9$ **159.** iii   **161.** $a - 7$ **163.** $-11x + 13$
165. $-4y - 4$ **167.** $-2x - 16$ **169.** $14y - 45$ **171.** $a + 7b$ **173.** $6x + 28$ **175.** $5x - 75$ **177.** $4x - 4$
179. $2x - 9$ **181.** $1.24x + 0.36$ **183.** $-0.01x + 40$ **185.** $\dfrac{x}{18}$ **187.** $x + 20$ **189.** $11x - 8$ **191.** $40 - \dfrac{x}{20}$
193. $x^2 + 2x$ **195.** $10(x - 50); 10x - 500$ **197.** $x - (x + 3); -3$ **199.** $(2x - 4) + x; 3x - 4$ **201.** $3x + x; 4x$
203. $(x + 6) + 5; x + 11$ **205.** $x - (x + 10); -10$ **207.** Let M be the number of visitors to the Metropolitan Museum of Art; the number of visitors to the Louvre is $M + 3,800,000$.   **209.** Let d be the noise level, in decibels, of a car horn; the noise level of an ambulance siren is $d + 10$.   **211.** Let T be U2's concert ticket sales; Bruce Springsteen and the E Street Band's concert ticket sales are $T - 28,500,000$.
213. Let N be the number of bones in your body; the number of bones in your foot is $\dfrac{1}{4}N$.
215. Let B be the attendance at major league basketball games; the attendance at major league baseball games is $B + 50,000,000$.
217. Let N be the number of U.S. undergraduate students; the number of U.S. undergraduate students who attend two-year colleges is $0.46N$.
219. Let L be the measure of the largest angle; the measure of the smallest angle is $\dfrac{1}{2}L - 10$.
221. s represents the number of students enrolled in fall-term science classes.   **223.** $\dfrac{1}{4}x$ **227a.** Yes   **b.** No
229. i, ii, iv, and v are equivalent; they are all equal to $10x + 4$.

SECTION 1.5

1. roster; set-builder; interval   **3.** $A = \{16, 17, 18, 19, 20, 21\}$ **5.** $A = \{9, 11, 13, 15, 17\}$ **7.** $A \cup B = \{3, 4, 5, 6\}$
9. $A \cup B = \{-10, -9, -8, 8, 9, 10\}$ **11.** $A \cup B = \{1, 3, 7, 9, 11, 13\}$ **13.** $A \cap B = \{4, 5\}$ **15.** $A \cap B = \varnothing$
17. $A \cap B = \{c, d, e\}$ **19.** $\{x | x > -5, x \in \text{negative integers}\}$ **21.** $\{x | x > 30, x \in \text{integers}\}$ **23.** $\{x | x > 8\}$
25. $(1, 2)$ **27.** $(3, \infty)$ **29.** $[-4, 5)$ **31.** $(-\infty, 2]$ **33.** $[-3, 1]$ **35.** $\{x | -5 < x < -3\}$ **37.** $\{x | x \le -2\}$
39. $\{x | -3 \le x \le -2\}$ **41.** $\{x | x \le 6\}$ **43.**   **45.**

Note: The numbers in brackets following the answers in the Check Your Progress are a reference to the objective that corresponds to that problem. For example, the reference [1.2A] stands for Section 1.2, Objective A. This notation will be used for answers to Prep Tests, Check Your Progress, Chapter Reviews, Chapter Tests, and Cumulative Reviews throughout the text.

Unless otherwise noted, all content on this page is © Cengage Learning.

47.   **49.**   **51.**   **53.**

55.   **57.**   **59.** None   **61.** $m \geq 250$ **63.** True

65. Answers will vary. For example, $A = \{1, 2, 3, 4\}$ and $B = \{1, 2, 3, 4\}$.

CHAPTER 1 REVIEW EXERCISES

1. $-4, 0$ [1.1A]   **2.** 4 [1.1B]   **3.** -5 [1.1B]   **4.** -13 [1.1C]   **5.** 1 [1.1C]   **6.** -42 [1.1D]

7. -20 [1.1D]   **8.** 0.28 [1.2A]   **9.** 0.062 [1.2B]   **10.** 62.5% [1.2B]   **11.** $\frac{7}{12}$ [1.2C]   **12.** -1.068 [1.2C]

13. $-\frac{72}{85}$ [1.2D]   **14.** -4.6224 [1.2D]   **15.** $\frac{16}{81}$ [1.2E]   **16.** 12 [1.2F]   **17.** $-6\sqrt{30}$ [1.2F]

18. 31 [1.3A]   **19.** 29 [1.4A]   **20.** $8a - 4b$ [1.4B]   **21.** $36y$ [1.4C]   **22.** $10x - 35$ [1.4D]

23. $7x + 46$ [1.4D]   **24.** $-90x + 25$ [1.4D]   **25.** $\{1, 3, 5, 7\}$ [1.5A]   **26.** $A \cap B = \{1, 5, 9\}$ [1.5A]

27. [1.5B]   **28.** [1.5B]   **29.** $(-4, \infty)$ [1.5B]

30. The student's score is 98. [1.1E]   **31.** 59.0% of those surveyed opposed abolishing the penny. [1.2G]

32. $2x - \frac{1}{2}x; \frac{3}{2}x$ [1.4E]   **33.** Let A be the number of American League player cards; the number of National League player cards is $5A$. [1.4E]

CHAPTER 1 TEST

1. $-2 > -40$ [1.1A; Example 1]   **2.** 7 [1.1B; Example 2]   **3.** -4 [1.1B; HOW TO 1]   **4.** -14 [1.1C; HOW TO 3]

5. -16 [1.1C; Example 4]   **6.** 4 [1.1C; HOW TO 3]   **7.** 17 [1.1D; Example 11]   **8.** $0.\overline{7}$ [1.2A; HOW TO 1]

9. $\frac{9}{20}, 0.45$ [1.2B; Example 2]   **10.** $\frac{1}{15}$ [1.2C; HOW TO 3]   **11.** -5.3578 [1.2D; HOW TO 8]   **12.** $-\frac{1}{2}$ [1.2D; HOW TO 7]

13. 12 [1.2E; Example 13]   **14.** $-6\sqrt{5}$ [1.2F; You Try It 14]   **15.** 17 [1.3A; HOW TO 3]   **16.** 22 [1.4A; HOW TO 1]

17. $5x$ [1.4B; HOW TO 2]   **18.** $2x$ [1.4C; HOW TO 6]   **19.** $-6x^2 + 21y^2$ [1.4D; Example 10]

20. $-x + 6$ [1.4D; Example 11]   **21.** $-7x + 33$ [1.4D; Example 13]   **22.** $\{-2, -1, 0; 1, 2, 3\}$ [1.5A; Example 1]

23. $\{x | x < -3, x \in \text{real numbers}\}$ [1.5B; HOW TO 4]   **24.** $A \cup B = \{1, 2, 3; 4, 5, 6, 7, 8\}$ [1.5A; Example 3]

25. [1.5B; Example 8b]   **26.** [1.5B; You Try It 9]

27. $10(x - 3); 10x - 30$ [1.4E; HOW TO 16]   **28.** Let s be the speed of the catcher's return throw; the speed of the fastball is $2s$. [1.4E; You Try It 14]   **29a.** The balance of trade increased from the previous year in 1981, 1988, 1989, 1990, 1991, and 1995. **b.** The difference was \$288.6 billion. **c.** The difference was greatest between 1999 and 2000. **d.** The trade balance was approximately 4 times greater in 1990 than in 1980. **e.** The average trade balance per quarter for the year 2000 was $-\$92.425$ billion. [1.2G; Example 16]   **30.** The difference is 215.4°F. [1.2G; You Try It 16]

Answers to Chapter 2 Selected Exercises

PREP TEST

1. 0.09 [1.2B]   **2.** 75% [1.2B]   **3.** 63 [1.4A]   **4.** $0.65R$ [1.4B]   **5.** $\frac{7}{6}x$ [1.4B]   **6.** $9x - 18$ [1.4D]

7. $1.66x + 1.32$ [1.4D]   **8.** $5 - 2n$ [1.4E]   **9.** Let s be the speed of the old card. The speed of the new card is $5s$. [1.4E]   **10.** The length of the shorter piece in terms of x is $5 - x$. [1.4E]

SECTION 2.1

1a. Equation   **b.** Expression   **c.** Expression   **d.** Equation   **e.** Expression   **3.** i, ii, and iv are equations of the form $x + a = b$; you would subtract a from both sides.   **5.** Amount: 30, base: 40   **7.** unknown; 30; 24   **9.** Keith had the greater average speed.

11. Yes   **13.** No   **15.** No   **17.** Yes   **19.** No   **21.** Yes   **23.** Yes   **25.** No   **29.** 2   **31.** 15   **33.** 6   **35.** 3

37. 0   **39.** -7 **41.** -7 **43.** -12 **45.** -5 **47.** 15   **49.** 9   **51.** 14   **53.** -1 **55.** 1   **57.** $-\frac{1}{2}$ **59.** $-\frac{7}{12}$

61. 0.6529   **63.** 9.257   **65.** -3 **67.** 0   **69.** -2 **71.** 180   **73.** 0   **75.** 6   **77.** -10 **79.** 12   **81.** -12 **83.** 0

85. -24 **87.** $\frac{1}{3}$ **89.** 4.745   **91.** 2.06   **93.** 7   **95.** 4   **97.** 3   **99.** Positive   **101.** Negative   **103.** 28   **105.** 0.72

107. 64   **109.** 24%   **111.** 7.2   **113.** 400   **115.** 9   **117.** 25%   **119.** 200%   **121.** 400   **123.** 7.7   **125.** 200

127. 400   **129.** 30   **131.** Less than   **133.** 97.9% of the participants who started the course finished the race.   **135.** 12% of the accidental deaths were not attributed to motor vehicle accidents.   **137.** 82.1% of the vacation cost is charged on a credit card.

139. The annual interest rate is 9%. **141.** Sal will earn $240 from the two accounts after one year. **143.** Makana earned $63 in one year. **145.** The interest rate on the combined investment is between 6% and 9%. **147.** The percent concentration of hydrogen peroxide is 2%. **149.** Apple Dan has the greater concentration of apple juice. **151.** 12.5 g of the cream are not glycerine. **153.** The percent concentration of salt in the remaining solution is 12.5%. **155a.** The distance biked by Emma is equal to the distance biked by Morgan. **b.** The time spent biking by Emma is less than the time spent biking by Morgan. **157.** The dietician's average rate of speed is 30 mph. **159.** Marcella's average rate of speed is 36 mph. **161.** It would take Palmer 2.5 h to walk the same course. **163.** The two joggers will meet in 40 min. **165.** The two cyclists are 8.5 mi apart. **167.** The two trains are 30 mi apart. **169.** -15 **171.** 5 **173.** 6 **175.** No; the prices are lower than the original prices. **179.** One possible answer is $x + 7 = 9$. **181.** $\dfrac{7}{11}$ **183a.** 18.1% of the U.S. population lives in the Northeast; 21.8% of the U.S. population lives in the Midwest; 36.8% of the U.S. population lives in the South; 23.4% of the U.S. population lives in the West. **b.** The region with the largest population is the South. The largest percent of the population lives in the South. **c.** 12.3% of the U.S. population lives in California. **d.** Approximately 520,000 residents live in Wyoming. **e.** Answers will vary.

SECTION 2.2

1a. i **b.** iii **c.** ii **d.** iv **3.** 5; 8 **5.** True **7.** Subtract $2x$ from each side. **9.** 3 **11.** 6 **13.** -1 **15.** -3 **17.** 2 **19.** 2 **21.** 5 **23.** -3 **25.** 6 **27.** 3 **29.** 1 **31.** 6 **33.** -7 **35.** 0 **37.** $\dfrac{3}{4}$ **39.** $\dfrac{4}{9}$ **41.** $\dfrac{1}{3}$ **43.** $-\dfrac{1}{2}$ **45.** $-\dfrac{3}{4}$ **47.** $\dfrac{1}{3}$ **49.** $-\dfrac{1}{6}$ **51.** 0 **53.** 0.15 **55.** $-\dfrac{3}{2}$ **57.** 18 **59.** 8 **61.** -16 **63.** 25 **65.** $\dfrac{3}{4}$ **67.** $\dfrac{3}{8}$ **69.** $\dfrac{16}{9}$ **71.** $\dfrac{1}{18}$ **73.** $\dfrac{15}{2}$ **75.** $-\dfrac{18}{5}$ **77.** 2 **79.** 3 **81.** Negative **83.** Negative **85.** $x = 7$ **87.** -1 **89.** 3 **91.** -2 **93.** -3 **95.** 2 **97.** -2 **99.** -0.2 **101.** 0 **103.** -2 **105.** -2 **107.** -2 **109.** 4 **111.** $\dfrac{3}{4}$ **113.** $\dfrac{3}{2}$ **115.** -14 **117.** 7 **119.** ii **121.** 1 **123.** 4 **125.** -1 **127.** -1 **129.** 24 **131.** 495 **133.** $\dfrac{1}{2}$ **135.** $-\dfrac{1}{3}$ **137.** $\dfrac{10}{3}$ **139.** $-\dfrac{1}{4}$ **141.** 0 **143.** The customer was driven 6 mi. **145a.** The fulcrum is 5 ft from the other person. **b.** The person who is 3 ft from the fulcrum is heavier. **c.** No, the seesaw will not balance. **147.** The fulcrum must be placed 10 ft from the child. **149.** The fulcrum must be placed 4.8 ft from the 90-pound child. **151.** The force on the lip of the can is 1770 lb. **153.** The break-even point is 260 barbecues. **155.** The break-even point is 520 recorders. **157.** The oxygen consumption is 54.8 ml/min. **159.** 4 **161.** No solution **165.** 6 **167.** Hampton's population at the beginning of the 1990s was 30,000 people.

SECTION 2.3

1. True **3.** True **5.** Equals **7.** 1; 2; 2 **9.** $x - 15 = 7$; 22 **11.** $9 - x = 7$; 2 **13.** $5 - 2x = 1$; 2 **15.** $2x + 5 = 15$; 5 **17.** $4x - 6 = 22$; 7 **19.** $3(4x - 7) = 15$; 3 **21.** $3x = 2(20 - x)$; 8, 12 **23.** $2x - (14 - x) = 1$; 5, 9 **25.** 15, 17, 19 **27.** -1, 1, 3 **29.** 4, 6 **31.** iii **33.** The length of the Golden Gate Bridge is 1280 m. **35.** The U. S. gross national product in 1937 was $91 billion. **37.** The lengths of the sides of the triangle are 6 ft, 6 ft, and 11 ft. **39.** The intensity of the sound of a jet engine is 140 decibels. **41.** The area of Greenland is 840,000 mi². **43.** The number of kilowatt-hours used is 515 kWh. **45.** The executive used the phone for 951 min. **47.** The customer pays $.15 per text message over 300 messages. **49.** The perimeter of the larger square is 8 ft. **51.** The cyclist will complete the entire trip in $\dfrac{1}{3}$ additional hour. **53.** The integers are -12, -10, -8, and -6. **55.** Any three consecutive odd integers **57.** even **59.** even **61.** even **63.** even **65.** odd

CHECK YOUR PROGRESS: CHAPTER 2

1. Yes [2.1A] **2.** -11 [2.1B] **3.** 9 [2.1C] **4.** 72 [2.1D] **5.** 4 [2.2A] **6.** Yes [2.1A] **7.** $-\dfrac{1}{2}$ [2.1B] **8.** 1 [2.2C] **9.** 150 [2.1D] **10.** -1 [2.2B] **11.** No [2.1A] **12.** 100 [2.1B] **13.** $\dfrac{4}{5}$ [2.1C] **14.** 28 [2.2A] **15.** 25% [2.1D] **16.** $\dfrac{4}{3}$ [2.1C] **17.** 3 [2.2C] **18.** -7 [2.2B] **19.** 1 [2.2C] **20.** 9 [2.1C] **21.** $\dfrac{15}{x} = -3$; -5 [2.3A] **22.** 3, 5, 7, 9 [2.3A] **23.** The average daily consumption of calories today is 2199 calories. [2.1D] **24.** The trip lasts 6 h. [2.1E] **25.** No, the seesaw is not balanced. [2.2D]

SECTION 2.4

1. $10.50 **3.** $.76 **5.** 100 **7.** True **9.** False **13.** 2 lb of dog food and 3 lb of vitamin supplement should be used to make the 5-pound mixture. **15.** 8 lb of chamomile tea must be used. **17.** The cost per pound of the mixture is $6.98. **19.** The amount of herbs costing $1 per ounce is 20 oz. **21.** The amount of pepper cheese is 1.5 kg; the amount of Pennsylania Jack is 3.5 kg. **23.** The amount of meal costing $.80 per pound is 300 lb. **25.** 37 lb of almonds and 63 lb of walnuts were used. **27.** The cost per pound of the breakfast cereal

Unless otherwise noted, all content on this page is © Cengage Learning.

is $1.40.   **29.** The parks department bought 8 bundles of seedlings and 6 bundles of container-grown plants.   **31.** The cost per ounce of the sunscreen is $3.   **33.** iv   **35.** The resulting mixture contains $33\frac{1}{3}\%$ tomato juice.   **37.** 80 lb of chicken feed that is 50% corn must be used.   **39.** $1\frac{2}{3}$ gal of the lighter green paint must be used.   **41.** The chemist should use 20 ml of the 13% solution and 30 ml of the 18% solution.   **43.** The percent concentration of the resulting alloy is 50%.   **45.** 10 lb of the 40% rye grass is used in the mixture.   **47.** 55 kg of pure silk thread and 20 kg of 85% silk thread must be woven together.   **49.** 12.5 gal of ethanol must be added.   **51.** 150 oz of pure chocolate must be added.   **53.** False   **55.** The first plane is flying at 105 mph and the second plane is flying at 130 mph.   **57.** The second skater will overtake the first skater 40 s after the second skater starts.   **59.** Michael's boat will be alongside the tour boat 2 h after the tour boat leaves.   **61.** The distance from the airport to the corporate offices is 120 mi.   **63.** The sailboat traveled 36 mi in the first 3 h.   **65.** The passenger train is traveling at 50 mph and the freight train is traveling at 30 mph.   **67.** It takes 1 h for the second ship to catch up to the first ship.   **69.** The rate of the faster car is 95 km/h.   **71.** The second car will not overtake the first car.   **73.** The bus overtakes the car 180 mi from the starting point.   **75.** The plane flew 2 h at 115 mph and 3 h at 125 mph.   **77.** The mixture contains 10 lb of walnuts and 20 lb of cashews.   **79.** The chemist used 3 L of pure acid and 7 L of water.   **81.** 85 adults and 35 children attended the performance.   **83.** The campers turned around downstream at 10:15 A.M.   **85.** 3.75 gal of 20% antifreeze must be drained from the radiator and replaced by pure antifreeze.   **87.** The cyclist's average speed for the trip was $13\frac{1}{3}$ mph.   **89.** The round trip was 8 mi.

SECTION 2.5

3. i, iii   **5.** $<$ **7.** $\{x|x < 5\}$ **9.** $\{x|x \le 2\}$

11. $\{x|x < -4\}$ **13.** $\{x|x > 3\}$ **15.** $\{x|x > 4\}$ **17.** $\{x|x \le 2\}$ **19.** $\{x|x > -2\}$

21. $\{x|x \ge 2\}$ **23.** $\{x|x > -2\}$ **25.** $\{x|x \le 3\}$ **27.** $\{x|x < 2\}$ **29.** $\{x|x < -3\}$ **31.** $\{x|x \le 5\}$

33. $\left\{x\middle|x \ge -\frac{1}{2}\right\}$ **35.** Both positive and negative numbers   **37.** Both positive and negative numbers   **39.** $\left(-\infty, -\frac{2}{3}\right)$

41. $\left(-\infty, \frac{8}{3}\right)$ **43.** $(1, \infty)$ **45.** $\left(\frac{14}{11}, \infty\right)$ **47.** $(-\infty, 1]$ **49.** $\left(-\infty, \frac{7}{4}\right]$ **51.** $\left[-\frac{5}{4}, \infty\right)$ **53.** $(-\infty, 2]$ **55.** $[-2, 4]$

57. $(-\infty, 3) \cup (5, \infty)$ **59.** $(-4, 2)$ **61.** $(-\infty, -4) \cup (6, \infty)$ **63.** $(-\infty, -3)$ **65.** $\varnothing$ **67.** $(-2, 1)$ **69.** All real numbers

71. Two intervals of real numbers   **73.** $\{x|x < -2\} \cup \{x|x > 2\}$ **75.** $\{x|2 < x < 6\}$ **77.** $\{x|-3 < x < -2\}$

79. $\{x|x > 5\} \cup \left\{x\middle|x < -\frac{5}{3}\right\}$ **81.** $\varnothing$ **83.** The set of real numbers   **85.** $\left\{x\middle|\frac{17}{7} \le x \le \frac{45}{7}\right\}$ **87.** $\left\{x\middle|-5 < x < \frac{17}{3}\right\}$

89. The set of real numbers   **91.** $\varnothing$ **93.** $t \le 42$ **95.** $t \le 42$ **97.** The maximum width of the rectangle is 11 cm.
99. The advertisement can run for a maximum of 104 days.   **101.** The homeowner can pay a maximum of $19 per gallon of paint.
103. The temperature range is $32° < F < 86°$.   **105.** George's amount of sales must be $44,000 or more.   **107.** The company must produce a minimum of 900 gal.   **109.** The range of scores is $58 \le n \le 100$.   **111. a.** -3 **b.** 7   **113.** True   **115.** True

SECTION 2.6

1. Yes   **3.** Yes   **5.** 7 and -7 **7.** 6 and -6 **9.** No solution   **11.** No solution   **13.** $\{x|x < -3\} \cup \{x|x > 3\}$

15. $|x - 2| < 5$ **17.** 1 and -5 **19.** 8 and 2   **21.** 2   **23.** No solution   **25.** $-\frac{3}{2}$ and 3   **27.** $\frac{3}{2}$ **29.** No solution

31. 7 and -3 **33.** 2 and $-\frac{10}{3}$ **35.** 1 and 3   **37.** $\frac{3}{2}$ **39.** No solution   **41.** $\frac{11}{6}$ and $-\frac{1}{6}$ **43.** $-\frac{1}{3}$ and -1 **45.** No solution

47. 3 and 0   **49.** No solution   **51.** 1 and $\frac{13}{3}$ **53.** No solution   **55.** $\frac{7}{3}$ and $\frac{1}{3}$ **57.** $-\frac{1}{2}$ **59.** $-\frac{1}{2}$ and $-\frac{7}{2}$ **61.** $-\frac{8}{3}$ and $\frac{10}{3}$

63. No solution   **65.** Two positive solutions   **67.** Two negative solutions   **69.** $\{x|x > 1\} \cup \{x|x < -3\}$ **71.** $\{x|4 \le x \le 6\}$

73. $\{x|x \ge 5\} \cup \{x|x \le -1\}$ **75.** $\{x|-3 < x < 2\}$ **77.** $\{x|x > 2\} \cup \left\{x\middle|x < -\frac{14}{5}\right\}$ **79.** $\varnothing$ **81.** The set of real numbers

83. $\left\{x\middle|x \le -\frac{1}{3}\right\} \cup \{x|x \ge 3\}$ **85.** $\left\{x\middle|-2 \le x \le \frac{9}{2}\right\}$ **87.** $\{x|x = 2\}$ **89.** $\{x|x < -2\} \cup \left\{x\middle|x > \frac{22}{9}\right\}$ **91.** $\left\{x\middle|-\frac{3}{2} < x < \frac{9}{2}\right\}$

93. $\{x|x < 0\} \cup \left\{x\middle|x > \frac{4}{5}\right\}$ **95.** $\{x|x > 5\} \cup \{x|x < 0\}$ **97.** All negative solutions   **99.** The desired dosage is 3 ml.
The tolerance is 0.2 ml.   **101.** The lower and upper limits of the diameter of the bushing are 1.742 in. and 1.758 in.
103. The lower and upper limits of the length of the piston rod are $9\frac{19}{32}$ in. and $9\frac{21}{32}$ in.   **105.** The lower and upper limits of the percent of voters who felt the economy was the most important election issue are 38% and 44%.   **107.** The lower and upper limits of the resistor are 28,420 ohms and 29,580 ohms.   **109a.** $\{x|x \ge -3\}$ **b.** $\{a|a \le 4\}$ **111.** 4   **113.** $-\frac{6}{5}$ and 14   **115.** No solution

CHAPTER 2 REVIEW EXERCISES

1. 21   [2.1B]   **2.** 10   [2.2C]   **3.** 7   [2.2A]   **4.** No   [2.1A]   **5.** 20   [2.1C]   **6.** $\left(\dfrac{5}{3}, \infty\right)$ [2.5A]   **7.** 250%   [2.1D]

8. 4   [2.2B]   **9.** -1 [2.2C]   **10.** No solution   [2.6A]   **11.** $\{x|1 < x < 4\}$ [2.6B]   **12.** $\left\{x|-3 < x < \dfrac{4}{3}\right\}$ [2.5B]

13. $(-\infty, \infty)$ [2.5B]   **14.** $\{x|x \geq 2\} \cup \left\{x|x \leq \dfrac{1}{2}\right\}$ [2.6B]   **15.** $\dfrac{8}{5}$ [2.2B]   **16.** 3   [2.2B]   **17.** 9   [2.2C]   **18.** 8   [2.2C]

19. $-\dfrac{8}{5}$ [2.6A]   **20.** $\varnothing$ [2.6B]   **21.** The force is 24 lb.   [2.2D]   **22.** The average speed on the winding road was 32 mph.   [2.4C]

23. The amount of cranberry juice is 7 qt; the amount of apple juice is 3 qt.   [2.4A]   **24.** The three integers are $-1, 0,$ and 1.   [2.3A]

25. $5n - 4 = 16; 4$ [2.3A]   **26.** The height of the Eiffel Tower is 1063 ft.   [2.3B]   **27.** The jet overtakes the propeller-driven plane 600 mi from the starting point.   [2.4C]   **28.** The lower and upper limits of the diameter are 2.747 in. and 2.753 in.   [2.6C]   **29.** The mixture is 14% butterfat.   [2.4B]   **30.** The island is 16 mi from the dock.   [2.4C]

CHAPTER 2 TEST

1. -5 [2.2B; HOW TO 4]   **2.** -5 [2.1B; HOW TO 2]   **3.** -3 [2.2A; Example 1]   **4.** 2   [2.2C; HOW TO 5]

5. No   [2.1A; Example 1]   **6.** 5   [2.2A; Example 2]   **7.** 0.04   [2.1D; HOW TO 6]   **8.** $-\dfrac{1}{3}$ [2.2C; HOW TO 5]

9. 2   [2.2B; Example 7]   **10.** -12 [2.1C; HOW TO 4]   **11.** $(-1, \infty)$ [2.5A; HOW TO 6]   **12.** $\{x|x > -2\}$ [2.5B; HOW TO 10]

13. $\varnothing$ [2.5B; Example 4]   **14.** 3 and $-\dfrac{9}{5}$ [2.6A; HOW TO 1]   **15.** 7 and -2 [2.6A; Example 3]

16. $\left\{x\left|\dfrac{1}{3} \leq x \leq 3\right.\right\}$ [2.6B; Example 4]   **17.** The amount of rye is 10 lb; the amount of wheat is 5 lb.   [2.4A; HOW TO 1]

18. The numbers are 10, 12, and 14.   [2.3A; HOW TO 2]   **19.** 1.25 gal of water must be added.   [2.4B; HOW TO 2]
20. $3x - 15 = 27; 14$ [2.3A; HOW TO 1]   **21.** The rate of the snowmobile was 6 mph.   [2.4C; HOW TO 3]
22. The company makes 110 LCD flat-panel TVs each day.   [2.3B; Example 3]   **23.** The smaller number is 8; the larger number is 10.   [2.3A; Example 1]   **24.** The distance between the airports is 360 mi.   [2.4C; You Try It 4]   **25.** The final temperature is 60°C.   [2.2D; Example 10]   **26.** It costs less to rent from Gambelli Agency if the car is driven less than 72 mi.   [2.5C; Example 5]
27. The lower and upper limits of the diameter of the bushing are 2.648 in. and 2.652 in.   [2.6C; You Try It 8]

CUMULATIVE REVIEW EXERCISES

1. 6   [1.1C]   **2.** -48 [1.1D]   **3.** $-\dfrac{19}{48}$ [1.2C]   **4.** -2 [1.2D]   **5.** 54   [1.2E]   **6.** 24   [1.3A]   **7.** 6   [1.4A]

8. $-17x$ [1.4B]   **9.** $-5a - 2b$ [1.4B]   **10.** $2x$ [1.4C]   **11.** $36y$ [1.4C]   **12.** $2x^2 + 6x - 4$ [1.4D]

13. $6x - 34$ [1.4D]   **14.** $A \cap B = \{-4, 0\}$ [1.5A]   **15.** [number line from -5 to 5 with open interval]   [1.5B]   **16.** Yes   [2.1A]

17. 19.2   [2.1D]   **18.** -25 [2.1C]   **19.** -3 [2.2A]   **20.** 3   [2.2A]   **21.** 13   [2.2C]   **22.** 120   [2.1D]

23. -3 [2.2B]   **24.** $\dfrac{1}{2}$ [2.2B]   **25.** $\{x|x \leq 1\}$ [2.5A]   **26.** $\{x|-4 \leq x \leq 1\}$ [2.5B]   **27.** -1 and 4   [2.6A]

28. $\{x|x > 2\} \cup \left\{x|x < -\dfrac{4}{3}\right\}$ [2.6B]   **29.** $\dfrac{11}{20}$ [1.2B]   **30.** 103%   [1.2B]   **31.** The final temperature is 60°C.   [2.2D]

32. $12 - 5x = -18; 6$ [2.3A]   **33.** The area of the garage is 600 ft^2.   [2.3B]   **34.** 20 lb of oat flour are needed for the mixture.   [2.4A]   **35.** 25 g of pure gold must be added.   [2.4B]   **36.** The length of the track is 120 m.   [2.4C]

Unless otherwise noted, all content on this page is © Cengage Learning.

Answers to Chapter 3 Selected Exercises

PREP TEST

1. 56 [1.3A]   **2.** 43 [2.1B]   **3.** 51 [2.1B]   **4.** 56.52 [1.4A]   **5.** 113.04 [1.4A]   **6.** 120 [1.4A]

SECTION 3.1

1. 12; 5; x; 4   **3.** 160°; 140°; 360°   **5.** a; b **7.** c; d; 180°   **9a.** $\angle a$, $\angle b$, and $\angle c$ **b.** $\angle y$ and $\angle z$ **c.** $\angle x$ **11.** 40°; acute
13. 115°; obtuse   **15.** 90°; right   **17.** 28°   **19.** 18°   **21.** 14 cm   **23.** 28 ft   **25.** 30 m   **27.** 86°   **29.** 30°   **31.** 36°
33. 71°   **35.** 127°   **37.** 116°   **39.** 20°   **41.** 20°   **43.** 20°   **45.** 141°   **47.** 90° − x **49.** 106°   **51.** 11°
53. $\angle a = 38°$, $\angle b = 142°$ **55.** $\angle a = 47°$, $\angle b = 133°$ **57.** 20°   **59.** False   **61.** True
63. The measure of $\angle x$ is 155° and the measure of $\angle y$ is 70°.   **65.** The measure of $\angle a$ is 45° and the measure of $\angle b$ is 135°.
67. The measure of the third angle is 60°.   **69.** The measure of the third angle is 35°.   **71.** True   **73.** Point D, or 3.5, is halfway between two other points.

SECTION 3.2

1. hexagon   **3.** pentagon   **5.** scalene   **7.** equilateral   **9.** obtuse   **11.** acute   **13.** 56 in.   **15.** 14 ft   **17.** 47 mi
19. 8π cm; 25.13 cm   **21.** 11π mi; 34.56 mi   **23.** 17π ft; 53.41 ft   **25.** The perimeter is 17.4 cm.   **27.** The perimeter is 8 cm.
29. The perimeter is 24 m.   **31.** The perimeter is 17.5 in.   **33.** The perimeter is 48.8 cm.   **35.** The circumference is 1.5π in.
37. The circumference is 226.19 cm.   **39.** 60 ft of fencing should be purchased.   **41.** 44 ft of carpet must be nailed down.
43. The length of the playground is 120 ft.   **45.** The length of the third side is 10 in.   **47.** Each side of the frame is 12 in.
49. A diameter of the circle is 2.55 cm.   **51.** The length of molding needed is 13.19 ft.   **53.** The bicycle travels 50.27 ft.
55. The circumference of Earth is 39,935.93 km.   **57.** A square whose side is 1 ft has the greater perimeter.   **59.** 60 ft^2 **61.** 20.25 in^2
63. 546 ft^2 **65.** 16π cm^2; 50.27 cm^2 **67.** 30.25π mi^2; 95.03 mi^2 **69.** 72.25π ft^2; 226.98 ft^2 **71.** The area is 156.25 cm^2.
73. The area is 570 in^2.   **75.** The area is 192 in^2.   **77.** The area is 13.5 ft^2.   **79.** The area is 330 cm^2.   **81.** The area is 25π in^2.
83. The area of the reserve in the Congo is approximately 10,500 mi^2.   **85.** The area watered by the irrigation system is 2500π ft^2.
87. The area of the patio is 72.25 m^2.   **89.** 7500 yd^2 of artificial turf must be purchased.   **91.** The width is 10 in.
93. The length of the base is 20 m.   **95.** You should buy 2 qt of stain.   **97.** The cost to wallpaper the two walls is $148.
99. You should select the 10 × 20 unit.   **101.** The increase in area is 339.29 cm^2.   **103.** You will spend $59.96 on paint.
105. 80 ft^2 of material must be purchased to make the drapes.   **107.** The area of the resulting rectangle is 4 times larger.
109. A rectangle that is 5 units by 5 units will have the greatest possible area.

CHECK YOUR PROGRESS: CHAPTER 3

1. 20 ft [3.1A]   **2.** 168° [3.1A]   **3.** $\angle b = 138°$, $\angle c = 42°$, $\angle d = 138°$ [3.1B]   **4.** The third angle measures 67°. [3.1C]
5. The area of the square quilt is 1600 in^2. [3.2A]   **6.** 20° [3.2B]   **7.** The circumference is 37.70 cm. [3.2A]
8. $\angle a = 135°$, $\angle b = 45°$ [3.1B]   **9.** The length of the base is 5 m. [3.2B]   **10.** The area is 98 m^2. [3.2B]
11. The length of a side is 9.5 in. [3.2A]   **12.** $\angle x = 24°$, $\angle y = 156°$ [3.1C]   **13.** The length of the rectangle is 16 m. [3.2B]
14. The perimeter is 36 in. [3.2A]   **15.** The area is 1.96π m^2, or approximately 6.16 m^2. [3.2B]   **16.** The area is 72 cm^2. [3.2B]
17. The length of decorative molding needed is 37 ft. [3.2A]

SECTION 3.3

1a. cone   **b.** cube   **c.** sphere   **d.** cylinder   **3.** $s^2 + 2sl$; l; s **5.** 840 in^3 **7.** 15 ft^3 **9.** 4.5π cm^3; 14.14 cm^3
11. The volume is 34 m^3.   **13.** The volume is 15.625 in^3.   **15.** The volume is 36π ft^3.   **17.** The volume is 8143.01 cm^3.
19. The volume is 392.70 cm^3.   **21.** The volume is 216 m^3.   **23.** The height of the aquarium is 8.5 in.
25. The height of the cylinder is 15.01 cm.   **27.** The volume of the portion of the silo that is not being used for storage is 1507.96 ft^3.
29. The volume of the lock is 6,600,000 ft^3.   **31.** Yes   **33.** No   **35.** The volume of the guacamole is 172,800 ft^3.   **37.** 94 m^2
39. 56 m^2 **41.** 96π in^2; 301.59 in^2 **43.** The surface area is 184 ft^2.   **45.** The surface area is 69.36 m^2.
47. The surface area is 225π cm^2.   **49.** The surface area is 402.12 in^2.   **51.** The surface area is 6π ft^2.   **53.** The surface area is 297 in^2.
55. The width is 3 cm.   **57.** 3217 ft^2 of fabric was used to construct the balloon.   **59.** 456 in^2 of glass is needed to make the fish tank.
61a. Always true   **b.** Never true   **c.** Sometimes true   **63.** The cone should be filled with water three times.

CHAPTER 3 REVIEW EXERCISES

1. $\angle x = 22°$, $\angle y = 158°$ [3.1C]   **2.** 26 ft [3.2A]   **3.** 168 in^3 [3.3A]   **4.** 68° [3.1B]   **5.** 63.585 cm^2 [3.2B]   **6.** 125.66 m^2 [3.3B]
7. 44 cm [3.1A]   **8.** 19° [3.1A]   **9.** 32 in^2 [3.2B]   **10.** 96 cm^3 [3.3A]   **11.** 42 in. [3.2A]   **12.** $\angle a = 138°$, $\angle b = 42°$ [3.1B]
13. 220 ft^2 [3.3B]   **14.** 3 [3.1A]   **15.** The volume is 42.875 in^3. [3.3A]   **16.** The supplement is 148°. [3.1A]
17. The volume is 39 ft^3. [3.3A]   **18.** The third angle measures 95°. [3.1C]   **19.** The length of the base is 8 cm. [3.2B]
20. The volume is 288π mm^3. [3.3A]   **21.** The length of each side of the frame is 21.5 cm. [3.2A]
22. 4 cans of paint must be purchased in order to paint the cylinder. [3.3B]   **23.** 208 yd of fencing are needed to surround the park. [3.2A]
24. The area of the patio is 90.25 m^2. [3.2B]   **25.** The area of the walkway is 276 m^2. [3.2B]

CHAPTER 3 TEST

1. 168 ft³ [3.3A; You Try It 1]   **2.** 143° [3.1B; HOW TO 4]   **3.** The area is 111 m². [3.2B; HOW TO 5]

4. The area is 42 ft². [3.2B; HOW TO 8]   **5.** The volume is $\dfrac{784\pi}{3}$ cm³. [3.3A; You Try It 2]   **6.** The surface area is 75 m². [3.3B; Example 3]

7. 4618.14 cm³   [3.3A; You Try It 2]   **8.** 159 in²   [3.2B; HOW TO 9]   **9.** 20°   [3.1B; HOW TO 4]   **10.** 75 m² [3.3B; Example 3]
11. 34°   [3.1B; Example 5]   **12.** Octagon [3.2A; HOW TO 1]   **13.** 500π cm² [3.3B; You Try It 3]   **14.** 61° [3.1C; Example 7]
15. The perimeter is 20 m. [3.2A; HOW TO 3]   **16.** The perimeter is 26 cm. [3.2A; HOW TO 2]   **17.** The supplement is 139°.
[3.1A; Example 3]   **18.** The third angle measures 102°. [3.1C; Example 8]   **19.** 90° and 58°   [3.1C; You Try It 8]
20. The bicycle travels approximately 73.3 ft in 10 revolutions.   [3.2A; You Try It 3]   **21a.** The total area of window glass that must
be cleaned is 1600 ft².   [3.2B; You Try It 4]   **b.** The window in the exhibit fills 1,440,000 in³.   [3.3A; Example 1]
22. The volume of the silo is approximately 1144.53 ft³.   [3.3A; You Try It 2]   **23.** The area is 11 m².   [3.2B; HOW TO 8]
24. The cross-sectional area is 103.82 ft².   [3.2B; Example 5]   **25a.** The area of the floor of a cell was 45 ft². [3.2B; HOW TO 5]
b. The volume of a cell was 315 ft³. [3.3A; Example 1]

CUMULATIVE REVIEW EXERCISES

1. −3, 0, and 1   [1.1A]   **2.** 0.089   [1.2B]   **3.** 35%   [1.2B]   **4.** $-\dfrac{2}{3}$ [1.2D]   **5.** −24.51   [1.2D]   **6.** $-5\sqrt{5}$ [1.2F]

7. −28   [1.3A]   **8.** −8   [1.4A]   **9.** $-3m + 3n$ [1.4B]   **10.** $21y$ [1.4C]   **11.** $7x + 9$ [1.4D]

12. $\{-2, -1\}$ [1.5A]   **13.** $C \cup D = \{-10, 0, 10, 20, 30\}$ [1.5A]   **14.** 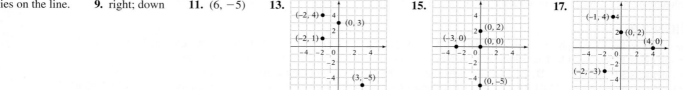   [1.5B]

15. 5   [2.2B]   **16.** $\dfrac{1}{2}$ [2.2C]   **17.** $(-\infty, -4)$ [2.5A]   **18.** $\{x | x \geq 2\}$ [2.5A]   **19.** $\{x | x < -3\} \cup \{x | x > 4\}$ [2.5B]

20. $\{x | 2 \leq x \leq 6\}$ [2.5B]   **21.** $1, -\dfrac{1}{3}$ [2.6A]   **22.** $\{x | 6 \leq x \leq 10\}$ [2.6B]   **23.** $\angle x = 131°$ [3.1B]

24. $4x - 10 = 2; x = 3$ [2.3B]   **25.** The third angle measures 122°.   [3.1C]   **26.** Michael will earn $312.50 from the
two accounts in one year.   [2.1D]   **27.** The third side measures 4.5 m.   [3.2A]   **28.** The height of the box is 3 ft. [3.3A]
29. The depth is 40 ft. [2.2D]   **30.** You sent or received 22 text messages. [2.3B]

Answers to Chapter 4 Selected Exercises

PREP TEST

1. $-4x + 12$ [1.4D]   **2.** 10   [1.2F]   **3.** −2   [1.3A]   **4.** 11   [1.4A]   **5.** 2.5   [1.4A]   **6.** 5   [1.4A]   **7.** 1   [1.4A]
8. 4   [2.2A]

SECTION 4.1

1. Quadrant II   **3.** y-axis   **5.** Answers will vary. For example, $(-3, 2)$ and $(5, 2)$ **7.** Answers will vary. Any point with x-coordinate 1
lies on the line.   **9.** right; down   **11.** $(6, -5)$ **13.**   **15.**   **17.**

19. $A(2, 3)$, $B(4, 0)$, $C(-4, 1)$, $D(-2, -2)$ **21.** $A(-2, 5)$, $B(3, 4)$, $C(0, 0)$, $D(-3, -2)$ **23a.** 2, −4   **b.** 1, −3
25a. I   **b.** II   **c.** IV   **d.** III   **27.** Yes   **29.** No   **31.** No   **33.** Negative
35.   **37.**   **39.**

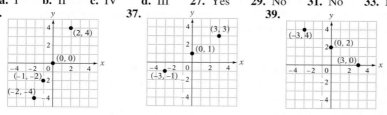

Unless otherwise noted, all content on this page is © Cengage Learning.

41.

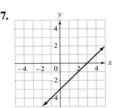

43.

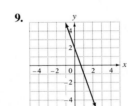

45. The record time for the 800-meter race was 200 s.　**47.** The point will be graphed with an *x*-coordinate of 1200 and a *y*-coordinate equal to the record time for the 1200-meter race. The graph will have an additional point.　**49a.** The highway fuel usage for the car was 15 mpg.
b. The city fuel usage for the car was 13 mpg.　**51.** 4 units　**53.** 2 units　**55.** 5 units　**57.**

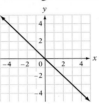

SECTION 4.2

1. 5　**3.** Domain: $\{-3, -2, -1, 1, 2, 3\}$; Range: $\{1, 4, 9\}$　**5.** Yes　**7.** Function; Domain: $\{0, 2, 3, 4, 5\}$; Range: $\{0, 4, 6, 8, 10\}$
9. Function; Domain: $\{-4, -2, 0, 3\}$; Range: $\{-5, -1, 5\}$　**11.** Function; Domain: $\{-2, -1, 0, 1, 2\}$; Range: $\{-3, 3\}$
13. Not a function; Domain: $\{1, 4, 9\}$; Range: $\{-2, -1, 1, 2, 3\}$　**15a.** Yes　**b.** \$1.05　**c.** \$.65　**d.** \$.45　**17.** True
19. $f(3) = 11$　**21.** $f(0) = -4$　**23.** $G(0) = 4$　**25.** $G(-2) = 10$　**27.** $q(3) = 5$　**29.** $q(-2) = 0$　**31.** $F(4) = 24$
33. $F(-3) = -4$　**35.** $H(1) = 1$　**37.** $H(t) = \dfrac{3t}{t+2}$　**39.** $s(-1) = 6$　**41.** $s(a) = a^3 - 3a + 4$　**43.** $4h$

45. The power produced will be 50.625 watts.　**47.** The minimum annual cost savings is \$114.29.　**49a.** \$4.75 per game
b. \$4.00 per game　**51.** $\{x|x \neq 1\}$　**53.** $\{x|-\infty < x < \infty\}$　**55.** $\{x|-\infty < x < \infty\}$　**57.** $\{x|-\infty < x < \infty\}$
59. $\{x|-\infty < x < \infty\}$　**61.** $\{x|x \neq -2\}$　**63a.** $\{(-2, -8), (-1, -1), (0, 0), (1, 1), (2, 8)\}$　**b.** Yes, the set of ordered pairs defines a
function because each member of the domain is assigned to exactly one member of the range.　**67a.** The speed of the paratrooper 11.5 s after
beginning the jump is 36.3 ft/s.　**b.** One second after jumping, the paratrooper is falling at a speed of approximately 30 ft/s.
69a. The runner's heart rate is 110 beats/min.　**b.** The runner's heart rate is 75 beats/min.

SECTION 4.3

3. 0; 0　**5.** The pressure is 49.5 atm.　**7.**

9.

11.

13.

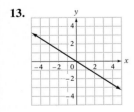

15.

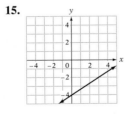

17.

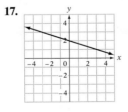

19.

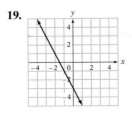

21.

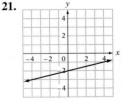

23.

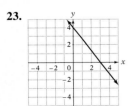

25.

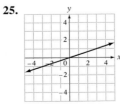

27.

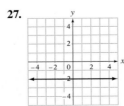

29.

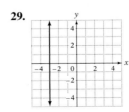

31.

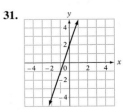

33.

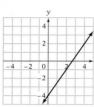

35. No. If $B = 0$, it is not possible to solve $Ax + By = C$ for y.

37. x-intercept: $(-4, 0)$; y-intercept: $(0, 2)$

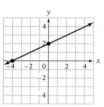

39. x-intercept: $\left(\dfrac{9}{2}, 0\right)$; y-intercept: $(0, -3)$

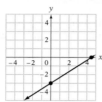

41. x-intercept: $\left(\dfrac{3}{2}, 0\right)$; y-intercept: $(0, 3)$

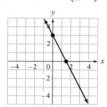

43. x-intercept: $\left(\dfrac{4}{3}, 0\right)$; y-intercept: $(0, 2)$

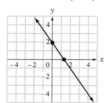

45. No. The graph of the equation $x = a$, $a \neq 0$, is a vertical line and has no y-intercept. **47.** The zero is -2. **49.** The zero is 12.

51. The zero is 0. **53.** The zero is $\dfrac{8}{3}$. **55.** The heart of the hummingbird will beat 8400 times in 7 min of flight.

57. Marlys receives \$165 for tutoring 15 h.

59. The cost of manufacturing 50 pairs of skis is \$9000.

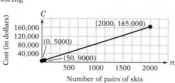

61.

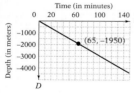

After 65 min, *Alvin* is 1950 m below sea level.

67.

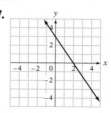

Unless otherwise noted, all content on this page is © Cengage Learning.

CHECK YOUR PROGRESS: CHAPTER 4

1. [4.1A]   **2.** 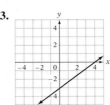 [4.3A]   **3.** 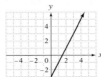 [4.3C]

4. $(-1, 5), (0, 2), (1, -1), (2, -4)$ [4.1B]   **5.** The graph is a vertical line passing through $(-5, 0)$.   [4.3B]

6. Domain: $\{-3, -2, -1\}$; Range: $\{-2, -1, 0\}$; yes   [4.2A]   **7.** 0   [4.2A]   **8.** Yes   [4.1B]   **9.** Yes   [4.2A]

10. Function; Domain: $\{-5, 0, 1, 2\}$; Range: $\{1, 2, 3\}$ [4.2A]   **11.** $\{x \mid x \neq 0\}$ [4.2A]   **12.** $s(-3) = -40$ [4.2A]

13. $f(2 - a) = 1 - 2a$ [4.2A]   **14.** 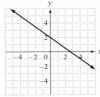 [4.3A]   **15.** 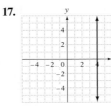 [4.3A]

16. 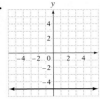 [4.3B]   **17.** [4.3B]   **18.** x-intercept: $(5, 0)$; y-intercept: $(0, -4)$ [4.3C]

19. The zero is -2.   [4.3C]   **20.** A 3-mile taxi ride costs \$10.60.   [4.3D]

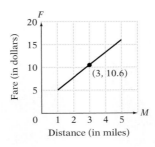

SECTION 4.4

1. increases   **3.** 3; $(0, 4)$ **5.** -1; $(0, 1)$ **7.** -1 **9.** $\dfrac{1}{3}$ **11.** $-\dfrac{2}{3}$ **13.** $-\dfrac{3}{4}$ **15.** Undefined   **17.** $\dfrac{7}{5}$ **19.** 0

21. $-\dfrac{1}{2}$ **23.** Undefined   **25.** a and c **27.** $m = 40$. The average speed of the motorist is 40 mph.   **29.** $m = -5$. The temperature of the oven is decreasing 5°/min.   **31.** $m = -0.05$. For each mile the car is driven, approximately 0.05 gal of fuel is used.   **33.** $m \approx 352.4$. The average speed of the runner was 352.4 m/min.   **35a.** No   **b.** Yes   **37.** $m = -\dfrac{19}{30}$. The water in the lock decreases by $0.6\overline{3}$ million gallons each minute.

39. 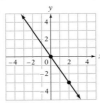   **41.** 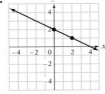   **43.** 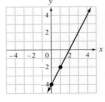   **45.** 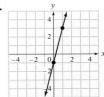   **47.**

49.   **51.** 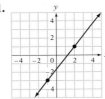   **53.** 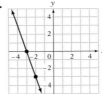   **55.** 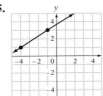   **57a.** Below   **b.** Negative

59. Increases by 2

61. Increases by 2

65. $k = -4$

Unless otherwise noted, all content on this page is © Cengage Learning.

SECTION 4.5

1. One   **3.** A point on the line.   **7.** $(0, 0)$ **9.** $y = 2x + 5$ **11.** $y = \dfrac{1}{2}x + 2$ **13.** $y = \dfrac{5}{4}x + \dfrac{21}{4}$ **15.** $y = -\dfrac{5}{3}x + 5$

17. $y = -3x + 9$ **19.** $y = -3x + 4$ **21.** $y = \dfrac{2}{3}x - \dfrac{7}{3}$ **23.** $y = \dfrac{1}{2}x$ **25.** $y = 3x - 9$ **27.** $y = -\dfrac{2}{3}x + 7$

29. $y = -x - 3$ **31.** $y = \dfrac{7}{5}x - \dfrac{27}{5}$ **33.** $y = -\dfrac{2}{5}x + \dfrac{3}{5}$ **35.** $x = 3$ **37.** $y = -\dfrac{5}{4}x - \dfrac{15}{2}$ **39.** $y = -3$

41. $y = -2x + 3$ **43.** $x = -5$ **45.** Check that the coordinates of each given point are a solution of your equation.

47. $y = x + 2$ **49.** $y = -2x - 3$ **51.** $y = \dfrac{2}{3}x + \dfrac{5}{3}$ **53.** $y = \dfrac{1}{3}x + \dfrac{10}{3}$ **55.** $y = \dfrac{3}{2}x - \dfrac{1}{2}$ **57.** $y = -\dfrac{3}{2}x + 3$

59. $y = -1$ **61.** $y = x - 1$ **63.** $y = -x + 1$ **65.** $y = -\dfrac{8}{3}x + \dfrac{25}{3}$ **67.** $y = \dfrac{1}{2}x - 1$ **69.** $y = -4$ **71.** $y = \dfrac{3}{4}x$

73. $y = -\dfrac{4}{3}x + \dfrac{5}{3}$ **75.** $x = -2$ **77.** $y = x - 1$ **79.** $y = \dfrac{4}{3}x + \dfrac{7}{3}$ **81.** $y = -x + 3$ **83a.** $f(x) = 1200x, 0 \le x \le 26\dfrac{2}{3}$

b. The height of the plane 11 min after takeoff is 13,200 ft.   **85a.** $f(x) = 0.625x - 1170.5$ **b.** In 2020, 92% of trees at 2600 ft will be hardwoods.   **87a.** $f(x) = -0.032x + 16, 0 \le x \le 500$ **b.** After driving for 150 mi, 11.2 gal are left in the tank.
89a. $f(x) = -20x + 230,000$ **b.** 60,000 motorcycles will be sold at $8500 each.   **91a.** $f(x) = 63x$ **b.** There are 315 Calories in a 5-ounce serving of lean hamburger.   **93.** Substitute 15,000 for $f(x)$ and solve for x.   **95.** $f(x) = x + 3$ **97.** 0   **99a.** -10 **b.** 6
101. Answers will vary. Possible answers include $(0, 4), (3, 2)$, and $(9, -2)$.   **103.** $y = -2x + 1$ **105.** The steepness is 3° up.

SECTION 4.6

3. slope   **5.** -5 **7.** $-\dfrac{1}{3}$ **9.** $\dfrac{2}{3}$ **11.** Yes   **13.** No   **15.** Yes   **17.** Yes   **19.** Yes   **21.** No   **23.** Yes

25. $y = 2x - 8$ **27.** $y = \dfrac{3}{2}x + 2$ **29.** $y = \dfrac{2}{3}x - \dfrac{8}{3}$ **31.** $y = \dfrac{1}{3}x - \dfrac{1}{3}$ **33.** $y = -\dfrac{5}{3}x - \dfrac{14}{3}$ **35.** $y = -2x + 5$

37. $y = -2x + 15$

SECTION 4.7

3. Yes   **5.** No   **7.** 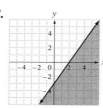   **9.** 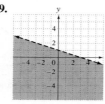   **11.**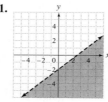

13. 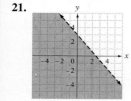   **15.**   **17.**   **19.**

21. 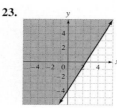   **23.**   **25.** Quadrant I   **29.**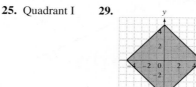

Unless otherwise noted, all content on this page is © Cengage Learning.

CHAPTER 4 REVIEW EXERCISES

1a. 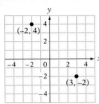   **2.** 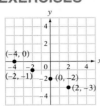   **3.** $(4, 2)$ [4.1B]   **4.** $P(-2) = -2; P(a) = 3a + 4$ [4.2A]

[4.1B]

b. -2

c. -4 [4.1A]

5. The zero is 6.   [4.3C]   **6.** $\{x \mid x \neq -4\}$ [4.2A]   **7.** $y = -\dfrac{5}{2}x + 16$ [4.5A]   **8.** Domain $= \{-1, 0, 1, 5\}$; Range $= \{0, 2, 4\}$ [4.2A]

9. x-intercept: $(-3, 0)$ [4.3A]   **10.** x-intercept: $(-2, 0)$ [4.3C]

y-intercept: $(0, -2)$ y-intercept: $(0, -3)$

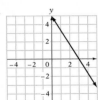   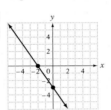

11. [4.3A]   **12.** [4.3B]   **13.** -1 [4.4A]   **14.** $y = \dfrac{5}{2}x + \dfrac{23}{2}$ [4.5A]

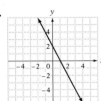   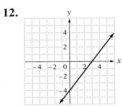

15. [4.3B]   **16.** [4.3B]   **17.** [4.4B]   **18.** [4.4B]

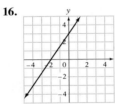   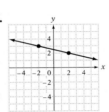   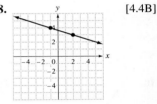

19. $y = -4x - 5$ [4.6A]   **20.** $y = \dfrac{5}{2}x + 8$ [4.6A]   **21.** [4.3B]   **22.** [4.3B]

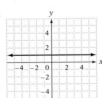   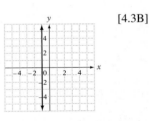

23. $y = -\dfrac{2}{3}x + 1$ [4.5A]   **24.** $y = \dfrac{1}{4}x + 4$ [4.5B]   **25.** [4.7A]   **26.** [4.7A]

    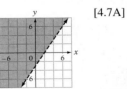

27. $y = -\dfrac{7}{6}x + \dfrac{5}{3}$ [4.5B]   **28.** $y = 2x$ [4.6A]   **29.** $y = -3x + 7$ [4.6A]   **30.** $y = \dfrac{3}{2}x + 2$ [4.6A]

31. [4.1C]   **32a.** $f(x) = -x + 295; 0 \leq x \leq 295$ **b.** When the rate is \$120, 175 rooms will be occupied.   [4.5C]

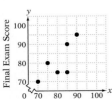

Unless otherwise noted, all content on this page is © Cengage Learning.

33. After 4 h, the car has traveled 220 mi. [4.3D]

34. The slope is 20. The manufacturing cost is $20 per calculator. [4.4A]

35a. $f(x) = 80x + 25,000$ **b.** The house will cost $185,000 to build. [4.5C]

CHAPTER 4 TEST

1. $(3, -3)$ [4.1B; HOW TO 2] **2.**

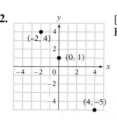

[4.1B; HOW TO 3] **3.**

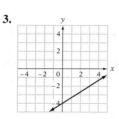

[4.3A; HOW TO 1]

4.

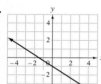

[4.3B; HOW TO 3] **5.** The zero is 3. [4.3C; HOW TO 7] **6.** $f(2) = 6$ [4.2A; HOW TO 2]

7. $-\dfrac{1}{6}$ [4.4A; Example 1] **8.** $P(2) = 9$ [4.2A; HOW TO 2] **9.**

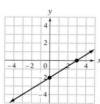

[4.3C; Example 5]

10.

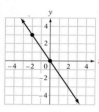

[4.4B; Example 5] **11.** $y = \dfrac{2}{5}x + 4$ [4.5A; HOW TO 2] **12.** $\{x \mid x \neq 0\}$ [4.2A; Example 4]

13. $y = -\dfrac{7}{5}x + \dfrac{1}{5}$ [4.5B; HOW TO 4] **14.** $(2, 0), (0, -3)$ [4.3C; HOW TO 5] **15.** Domain = $\{-4, -2, 0, 3\}$;

Range = $\{0, 2, 5\}$ [4.2A; Example 1] **16.** $y = -\dfrac{3}{2}x + \dfrac{7}{2}$ [4.6A; HOW TO 2] **17.** $y = 2x + 1$ [4.6A; HOW TO 6]

18.

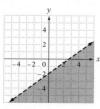

[4.7A; Example 1] **19.** After 1 s, the ball is traveling 96 ft/s. [4.3D; HOW TO 8]

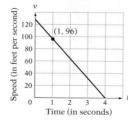

20. The slope is $-10,000$. The value of the house decreases by $10,000 each year. [4.4A; You Try It 3]

21a. $f(x) = -\dfrac{3}{10}x + 175$ **b.** When the tuition is $300, 85 students will enroll. [4.5C; Example 5]

Unless otherwise noted, all content on this page is © Cengage Learning.

CUMULATIVE REVIEW EXERCISES

1. The Commutative Property of Multiplication　[1.4C]　**2.** $\dfrac{9}{2}$　[2.2A]　**3.** $9\sqrt{5}$　[1.2F]　**4.** $-\dfrac{1}{14}$　[2.2C]

5. $\{x \mid x < -1\} \cup \left\{x \mid x > \dfrac{1}{2}\right\}$　[2.5B]　**6.** $-\dfrac{3}{2}$ and $\dfrac{5}{2}$　[2.6A]　**7.** $\left\{x \mid 0 < x < \dfrac{10}{3}\right\}$　[2.6B]　**8.** 8　[1.3A]

9. -4.5　[1.4A]　**10.** 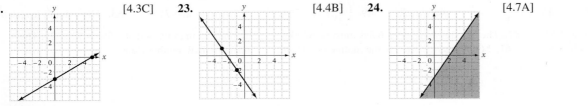　[1.5B]　**11.** $\dfrac{19}{18}$　[2.2C]　**12.** $\dfrac{1}{15}$　[1.2B]

13. $\varnothing$　[2.5B]　**14.** 14　[4.2A]　**15.** $(-8, 13)$　[4.1B]　**16.** $-\dfrac{7}{4}$　[4.4A]　**17.** $y = \dfrac{3}{2}x + \dfrac{13}{2}$　[4.5A]

18. $y = -\dfrac{5}{4}x + 3$　[4.5B]　**19.** $y = -\dfrac{3}{2}x + 7$　[4.6A]　**20.** $y = -\dfrac{2}{3}x + \dfrac{8}{3}$　[4.6A]　**21.** The zero is 3.　[4.3C]

22. 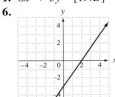　[4.3C]　**23.** 　[4.4B]　**24.** 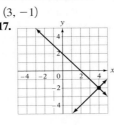　[4.7A]

25. The first plane is traveling at 200 mph and the second plane is traveling at 400 mph.　[2.4C]
26. The mixture consists of 40 lb of $9 coffee and 20 lb of $6 coffee.　[2.4A]
27a. $y = -5000x + 30,000$　**b.** The value of the truck decreases by $5000 per year.　[4.5C]

Answers to Chapter 5 Selected Exercises

PREP TEST

1. $6x + 5y$　[1.4D]　**2.** 7　[1.4A]　**3.** 0　[2.2A]　**4.** -3　[2.2C]　**5.** 1000　[2.2C]
6. 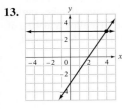　[4.3B]　**7.** 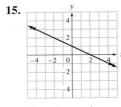　[4.7A]

SECTION 5.1

1. Yes　**3.** No　**5.** Dependent　**7.** $(4, -1)$　**9.**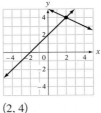

$(3, -1)$　$(2, 4)$

13. 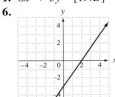　**15.** 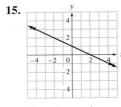　**17.**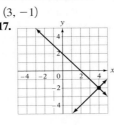

$(4, 3)$　$\left(x, -\dfrac{1}{2}x + 1\right)$　$(4, -2)$

19.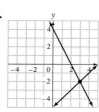

$(3, -2)$

21.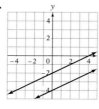

No solution

23.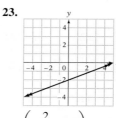

$\left(x, \dfrac{2}{5}x - 2\right)$

25. $(1, 2)$ **27.** $\left(-\dfrac{5}{2}, -4\right)$ **29.** $(-1, 4)$ **31.** $(-2, 5)$ **33.** $(1, 5)$ **35.** $(2, 0)$ **37.** $\left(\dfrac{2}{3}, -1\right)$ **39.** No solution

41. $(-2, 5)$ **43.** $(0, 0)$ **45.** $\left(x, \dfrac{1}{2}x - 4\right)$ **47.** $(1, 5)$ **49.** $\left(\dfrac{2}{3}, 3\right)$ **51.** $\dfrac{2}{3}$ **53.** 5.5% and 7.2%   **55.** The amount invested

in the insured bond fund is $5500.   **57.** The amount invested at 6.5% must be $4000.   **59.** The amount invested at 3.5% is $23,625. The amount invested at 4.5% is $18,375.   **61.** $6000 was invested in the mutual bond fund.   **63.** Any real number except 3   **65.** Answers will vary.

SECTION 5.2

1. Answers may vary. Possible answers are 6 and -5.   **3.** $(6, 1)$ **5.** $(1, 1)$ **7.** $(2, 1)$ **9.** $(-2, 1)$ **11.** $(x, 3x - 4)$ **13.** $\left(-\dfrac{1}{2}, 2\right)$ **15.** No solution   **17.** $(-1, -2)$ **19.** $(-5, 4)$ **21.** $(2, 5)$ **23.** $\left(\dfrac{1}{2}, \dfrac{3}{4}\right)$ **25.** $(0, 0)$ **27.** $(-1, 3)$

29. $\left(\dfrac{2}{3}, -\dfrac{2}{3}\right)$ **31.** $(-2, 3)$ **33.** $(2, -1)$ **35.** $(10, -5)$ **37.** $\left(-\dfrac{1}{2}, \dfrac{2}{3}\right)$ **39.** $\left(\dfrac{5}{3}, \dfrac{1}{3}\right)$ **41.** No solution   **43.** $(1, -1)$

45. $(2, 1, 3)$ **47.** $(1, -1, 2)$ **49.** $(1, 2, 4)$ **51.** $(2, -1, -2)$ **53.** $(-2, -1, 3)$ **55.** No solution   **57.** $(1, 4, 1)$

59. $(1, 3, 2)$ **61.** $(1, -1, 3)$ **63.** $(0, 2, 0)$ **65.** $(1, 5, 2)$ **67.** $(-2, 1, 1)$ **69a.** iii   **b.** ii   **c.** i   **71.** $(1, -1)$

73. $(1, -1)$ **75.** $A = 2, B = 3, C = -3$

CHECK YOUR PROGRESS: CHAPTER 5

1.

$(1, -1)$ [5.1A]   **2.** $(x, -2x + 6)$ [5.1A]

3. $(3, -2)$ [5.1B]   **4.** No solution   [5.1B]   **5.** $\left(\dfrac{44}{13}, -\dfrac{3}{13}\right)$ [5.1B]   **6.** $(-2, 4)$ [5.2A]   **7.** $\left(x, -\dfrac{2}{5}x + 2\right)$ [5.2A]

8. $(2, -1, 3)$ [5.2B]   **9.** $(0, 4, -2)$ [5.2B]   **10.** $(1, -1, -3)$ [5.2B]   **11.** $12,000 is invested at 6%, and $8000 is invested at 4.5%.   [5.1C]

SECTION 5.3

1. The speed of the plane is 450 mph.   **3.** $50x + 100y$ **5.** Less than   **7.** The rate of the motorboat in calm water is 15 mph. The rate of the current is 3 mph.   **9.** The rate of the plane in calm air is 502.5 mph. The rate of the wind is 47.5 mph.   **11.** The rate of the team in calm water is 8 km/h. The rate of the current is 2 km/h.   **13.** The rate of the plane in calm air is 180 mph. The rate of the wind is 20 mph.   **15.** The rate of the plane in calm air is 110 mph. The rate of the wind is 10 mph.   **17.** Greater than   **19.** The cost of the redwood is $3.30/ft. The cost of the pine is $1.10/ft.   **21.** The cost of the wool carpet is $52/yd.   **23.** The company plans to manufacture 25 mountain bikes during the week.   **25.** The owner drove 322 mi in the city and 72 mi on the highway.   **27.** The chemist should use 480 g of the first alloy and 40 g of the second alloy.   **29.** The Model VI computer costs $4000.   **31.** The investor placed $10,400 in the 8% account, $5200 in the 6% account, and $9400 in the 4% account.   **33.** The measures of the two angles are 35° and 145°.   **35.** $d_1 = 6$ in., $d_2 = 3$ in., $d_3 = 9$ in.

SECTION 5.4

1. ii   **3.** 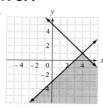   **5.** 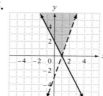   **7.** 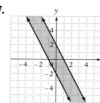   **9.**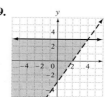

Unless otherwise noted, all content on this page is © Cengage Learning.

11. 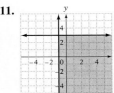 **13.** 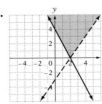 **15.** 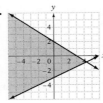 **17.**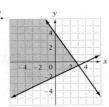

19. The region between the parallel lines $x + y = a$ and $x + y = b$ **21.** **23.** ii, iii

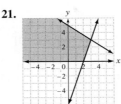

CHAPTER 5 REVIEW EXERCISES

1. $\left(6, -\dfrac{1}{2}\right)$ [5.1B] **2.** $(-4, 7)$ [5.2A] **3.** [5.1A] **4.** [5.1A]

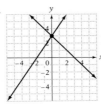

$(0, 3)$

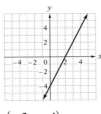

$(x, 2x - 4)$

5. $\left(x, -\dfrac{1}{4}x + \dfrac{3}{2}\right)$ [5.1B] **6.** $\left(x, \dfrac{1}{3}x - 2\right)$ [5.2A] **7.** $(3, -1, -2)$ [5.2B] **8.** $(5, -2, 3)$ [5.2B] **9.** Yes [5.1A]

10. $\left(\dfrac{1}{2}, -\dfrac{5}{2}\right)$ [5.1B] **11.** $(3, -1)$ [5.1B] **12.** $(3, 2)$ [5.2A] **13.** $(-1, -3, 4)$ [5.2B] **14.** $(2, 3, -5)$ [5.2B]

15. [5.4A] **16.** [5.4A]

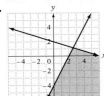

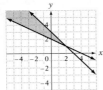

17. The rate of the cabin cruiser in calm water is 16 mph. The rate of the current is 4 mph. [5.3A] **18.** The rate of the plane in calm air is 175 mph. The rate of the wind is 25 mph. [5.3A] **19.** On Friday evening, 100 children attended. [5.3B] **20.** The amount invested at 3% is $5000. The amount invested at 7% is $15,000. [5.1C]

CHAPTER 5 TEST

1. $(3, 4)$ [5.1A; Example 1] **2.** No solution [5.1A; Example 2]

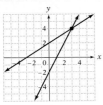

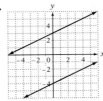

3. [5.4A; HOW TO 1] **4.** [5.4A; HOW TO 1] **5.** $\left(\dfrac{3}{4}, \dfrac{7}{8}\right)$ [5.1B; Example 4]

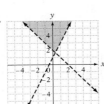

Unless otherwise noted, all content on this page is © Cengage Learning.

6. $(-3, -4)$ [5.1B; HOW TO 3] **7.** $(2, -1)$ [5.1B; You Try It 5] **8.** $(-2, 1)$ [5.2A; HOW TO 1] **9.** No solution
[5.2A; You Try It 2] **10.** $(1, 1)$ [5.2A; Example 1] **11.** No solution [5.2B; HOW TO 5] **12.** $(2, -1, -2)$ [5.2B; HOW TO 6]
13. $\left(-\dfrac{1}{3}, -\dfrac{10}{3}\right)$ [5.1B; HOW TO 3] **14.** Yes [5.1A; HOW TO 1] **15.** $\left(\dfrac{1}{5}, -\dfrac{6}{5}, \dfrac{3}{5}\right)$ [5.2B; HOW TO 4] **16.** The rate of the plane
in calm air is 150 mph. The rate of the wind is 25 mph. [5.3A; Example 1] **17.** The cost of the cotton is $9/yd. The cost of the wool
is $14/yd. [5.3B; HOW TO 2] **18.** The amount invested at 2.7% is $9000. The amount invested at 5.1% is $6000. [5.1C; HOW TO 5]

CUMULATIVE REVIEW EXERCISES

1. $-\dfrac{11}{28}$ [2.2C] **2.** $y = 5x - 11$ [4.5B] **3.** $3x - 24$ [1.4D] **4.** -4 [1.4A] **5.** $(-\infty, 6)$ [2.5B]
6. $\{x | -4 < x < 8\}$ [2.6B] **7.** $\{x | x < -1\} \cup \{x | x > 4\}$ [2.6B] **8.** $f(-3) = -98$ [4.2A]
9. The domain is $\{x | -\infty < x < \infty\}$. [4.2A] **10.** 1 [4.2A] **11.** $3h$ [4.2A] **12.** 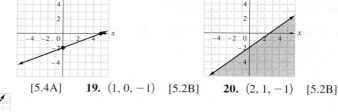 [1.5B]

13. $y = -\dfrac{2}{3}x + \dfrac{5}{3}$ [4.5A] **14.** $y = -\dfrac{3}{2}x + \dfrac{1}{2}$ [4.6A] **15.** [4.4B] **16.** [4.7A]

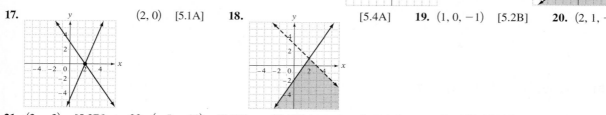

17. $(2, 0)$ [5.1A] **18.** [5.4A] **19.** $(1, 0, -1)$ [5.2B] **20.** $(2, 1, -1)$ [5.2B]

21. $(2, -3)$ [5.2B] **22.** $(-5, -11)$ [5.1B] **23.** The amount of water that must be added is 60 ml. [2.4B]
24. The rate of the wind is 12.5 mph. [5.3A] **25.** The cost of 1 lb of steak is $6. [5.3B] **26.** The lower and upper limits of the resistor
are 10,200 ohms and 13,800 ohms. [2.6C] **27.** The slope of the line is 40. The commission rate of the executive is $40 for every
$1000 worth of sales. [4.4A]

Answers to Chapter 6 Selected Exercises

PREP TEST
1. 1 [1.1C] **2.** -18 [1.1D] **3.** $\dfrac{2}{3}$ [1.1D] **4.** 48 [1.4A] **5.** 0 [1.1D] **6.** No [1.4B]
7. $5x^2 - 9x - 6$ [1.4B] **8.** 0 [1.4B] **9.** $-6x + 24$ [1.4D] **10.** $-7xy + 10y$ [1.4D]

SECTION 6.1
1. i, ii, iv **3.** No; the expression inside the parentheses is a sum, not a product. **5.** $-1, 1$ **7.** a^4b^4 **9.** $-18x^3y^4$ **11.** x^8y^{16}
13. $81x^8y^{12}$ **15.** $729a^{10}b^6$ **17.** $64a^{24}b^{18}$ **19.** x^5y^{11} **21.** $-45a^7b^5$ **23.** $-192x^{14}y$ **25.** $42a^6b^2c^5$ **27.** $54a^{13}b^{17}$
29. $-432a^7b^{11}$ **31.** $-6x^4y^4z^5$ **33.** $-6x^5y^5z^4$ **35.** $-12a^2b^9c^2$ **37.** 33 **39.** $\dfrac{1}{16}$ **41.** 128 **43.** $\dfrac{3}{a^5}$ **45.** $\dfrac{a^7}{3}$
47. $\dfrac{a^3b^2}{4}$ **49.** $\dfrac{x}{y^4}$ **51.** $\dfrac{1}{2}$ **53.** $-\dfrac{1}{9}$ **55.** $\dfrac{1}{x^6y^{10}}$ **57.** $\dfrac{15}{a^6b}$ **59.** $\dfrac{4z^2}{9y^9}$ **61.** $\dfrac{x^{13}}{4y^{18}}$ **63.** $\dfrac{3}{4x^3}$ **65.** $-\dfrac{1}{2x^2}$ **67.** $\dfrac{1}{y^8}$
69. $\dfrac{a^8}{b^9}$ **71.** $-\dfrac{1}{243a^5b^{10}}$ **73.** $\dfrac{3a^4}{4b^3}$ **75.** $\dfrac{16x^2}{y^6}$ **77.** $\dfrac{ab}{72}$ **79.** $\dfrac{8b^{15}}{3a^{18}}$ **81.** 0 **83.** 4.67×10^{-6} **85.** 1.7×10^{-10}
87. 2×10^{11} **89.** 0.000000123 **91.** 8,200,000,000,000,000 **93.** 0.039 **95.** 150,000 **97.** 20,800,000 **99.** 0.000000015
101. 0.0000000000178 **103.** 11,000,000 **105.** 0.000008 **107.** Greater than **109.** It would take a spaceship 2.6×10^{11} years to cross
the galaxy. **111.** The mass of a proton is 1.83664508×10^3 times larger than the mass of an electron. **113.** The radio signal travels
1.12×10^7 mi/min. **115.** The mass of the sun is 3.3898305×10^5 times larger than the mass of Earth. **117.** The lodgepole pine trees
released

$1.\overline{6} \times 10^2$ seeds for each surviving seedling. **119a.** 81 **b.** 19,683 **c.** 1 **d.** 81 **121.** x^{7n} **123.** $\dfrac{y^{4m}}{x^{2n}}$ **125a.** $\dfrac{8}{5}$ **b.** $\dfrac{5}{4}$

Unless otherwise noted, all content on this page is © Cengage Learning.

SECTION 6.2

1a. Binomial **b.** Binomial **c.** Monomial **d.** Trinomial **e.** None **f.** Monomial **3.** All real numbers
5. Polynomial: **a.** -1 **b.** 8 **c.** 2 **7.** Not a polynomial **9.** Polynomial: **a.** -1 **b.** 0 **c.** 6 **11.** $P(3) = 13$
13. $R(2) = 10$ **15.** $f(-1) = -11$ **17.** **19.** **21.**

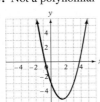

23. $f(c) - g(c) > 0; f(c) - g(c) < 0$ **25.** $6x^2 - 6x + 5$ **27.** $-x^2 + 1$ **29.** $5y^2 - 15y + 2$ **31.** $7a^2 - a + 2$
33. $3x^2 - 3xy - 2y^2$ **35.** $8x^2 - 2xy + 5y^2$ **37.** -2 **39.** 2 **43.** The graph of k is the graph of f moved 2 units down.

CHECK YOUR PROGRESS: CHAPTER 6

1. $-72a^{10}b^4$ [6.1A] **2.** $12x^5$ [6.1B] **3.** $48x^{14}$ [6.1A] **4.** $\dfrac{16a^{18}}{9b^{20}}$ [6.1B] **5.** $\dfrac{x^3}{y^4}$ [6.1B] **6.** $\dfrac{x^2}{2}$ [6.1B]

7. $\dfrac{b^4}{2a^3}$ [6.1B] **8.** $\dfrac{1}{72x^3y^5}$ [6.1B] **9.** 6.83×10^{-7} [6.1C] **10.** 0.002607 [6.1C] **11.** 2,140,000 [6.1C]

12. -19 [6.2A] **13.** $5x^2 - 5x - 2$ [6.2B] **14.** $-11x^2 - 1$ [6.2B] **15.** [6.2A]

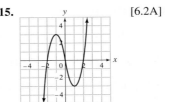

SECTION 6.3

1. Multiply $2(2x + 1) = 4x + 2$. By the Order of Operations Agreement, do multiplication before addition. **5.** $2x^2 - 6x$
7. $6x^4 - 3x^3$ **9.** $6x^2y - 9xy^2$ **11.** $-12x^2y^2 + 15xy^3$ **13.** $-4b^2 + 10b$ **15.** $-6a^4 + 4a^3 - 6a^2$ **17.** $-3y^4 - 4y^3 + 2y^2$
19. $-20x^2 + 15x^3 - 15x^4 - 20x^5$ **21.** $-2x^4y + 6x^3y^2 - 4x^2y^3$ **23.** $-3x^3 - 12x^2 + 28x$ **25.** $5y^2 - 11y$ **27.** $6y^2 - 31y$
29. $9b^5 - 9b^3 + 24b$ **31.** $x^2 + 5x - 14$ **33.** $8y^2 + 2y - 21$ **35.** $4a^2 - 7ac - 15c^2$ **37.** $25x^2 - 70x + 49$
39. $8x^2 + 8xy - 30y^2$ **41.** $x^2y^2 + xy - 12$ **43.** $2x^4 - 15x^2 + 25$ **45.** $10x^4 - 15x^2y + 5y^2$ **47.** $x^3 + 2x^2 - 11x + 20$
49. $10a^3 - 27a^2b + 26ab^2 - 12b^3$ **51.** $4x^4 - 13x^2 + 16x - 15$ **53.** $4x^5 - 16x^4 + 15x^3 - 4x^2 + 28x - 45$
55. $x^4 - 3x^3 - 6x^2 + 29x - 21$ **57.** $2a^3 + 7a^2 - 43a + 42$ **59.** $6x^3 + 5x^2 - 61x - 60$ **61.** $2y^5 - 10y^4 - y^3 - y^2 + 3$
63. mn **65.** $9x^2 - 4$ **67.** $36 - x^2$ **69.** $4a^2 - 9b^2$ **71.** $9a^2b^2 - 16$ **73.** $x^4 - 1$ **75.** $x^2 - 10x + 25$
77. $9a^2 + 30ab + 25b^2$ **79.** $x^4 - 6x^2 + 9$ **81.** $4x^4 - 12x^2y^2 + 9y^4$ **83.** $9m^2n^2 - 30mn + 25$ **85.** $-x^2 + 2xy$
87. $-4xy$ **89.** False **91.** ft^2 **93.** $(3x^2 + 10x - 8) \text{ ft}^2$ **95.** $(x^2 + 3x) \text{ m}^2$ **97.** $(x^3 + 9x^2 + 27x + 27) \text{ cm}^3$
99. $(2x^3) \text{ in}^3$ **101.** $(78.5x^2 + 125.6x + 50.24) \text{ in}^2$ **103a.** $k = 5$ **b.** $k = 1$ **105.** $2x^2 + 11x - 21$
107. Answers will vary. For example, $(x + 3)(2x^2 - 1)$.

SECTION 6.4

1. 4 **5.** $x - 2$ **7.** $-x + 2$ **9.** $xy + 2$ **11.** $x^2 + 3x - 5$ **13.** $3b^3 + 4b^2 + 2b$ **15.** $a^4 - 6a^2 + 1$
17. $6x^3 + 21x^2 - 15x$ **19.** $x + 8$ **21.** $x^2 + 3x + 6 + \dfrac{20}{x - 3}$ **23.** $3x + 5 + \dfrac{3}{2x + 1}$ **25.** $5x + 7 + \dfrac{2}{2x - 1}$

27. $4x^2 + 6x + 9 + \dfrac{18}{2x - 3}$ **29.** $3x^2 + 1 + \dfrac{1}{2x^2 - 5}$ **31.** $x^2 - 3x - 10$ **33.** $x^2 - 2x + 1 - \dfrac{1}{x - 3}$ **35.** $2x^3 - 3x^2 + x - 4$

37. $2x + \dfrac{x + 2}{x^2 + 2x - 1}$ **39.** $x^2 + 4x + 6 + \dfrac{10x + 8}{x^2 - 2x - 1}$ **41.** $x^2 + 4 + \dfrac{3}{2x + 1}$ **43.** $6x^3 + 27x^2 + 18x - 30$ **45.** 3

47. $2x - 8$ **49.** $3x - 8$ **51.** $3x + 3 - \dfrac{1}{x - 1}$ **53.** $2x^2 - 3x + 9$ **55.** $4x^2 + 8x + 15 + \dfrac{12}{x - 2}$ **57.** $2x^2 - 3x + 7 - \dfrac{8}{x + 4}$

59. $3x^3 + 2x^2 + 12x + 19 + \dfrac{33}{x - 2}$ **61.** $3x^3 - x + 4 - \dfrac{2}{x + 1}$ **63.** $3x + 1 + \dfrac{8}{x - 2}$ **65.** $x - 3$ **67.** $P(3) = 8$
69. $R(4) = 43$ **71.** $P(-2) = -39$ **73.** $Z(-3) = -60$ **75.** $Q(2) = 31$ **77.** $F(-3) = 178$ **79.** $P(5) = 122$
81. $R(-3) = 302$ **83.** $Q(2) = 0$ **85.** 3 **87.** -1 **89.** 1, 5; 2, 4; 3, 3 **91.** Yes

Unless otherwise noted, all content on this page is © Cengage Learning.

CHAPTER 6 REVIEW EXERCISES

1. $21y^2 + 4y - 1$ [6.2B]   **2.** $5x + 4 + \dfrac{6}{3x - 2}$ [6.4B]   **3.** $144x^2y^{10}z^{14}$ [6.1B]   **4.** $25y^2 - 70y + 49$ [6.3C]

5. $\dfrac{b^6}{a^4}$ [6.1B]   **6.** 1 [6.4D]   **7.** $4x^2 - 8xy + 5y^2$ [6.2B]   **8.** $4b^4 + 12b^2 - 1$ [6.4A]   **9.** $-\dfrac{1}{2a}$ [6.1B]

10. $-6a^3b^6$ [6.1A]   **11.** $\dfrac{2x^3}{3}$ [6.1B]   **12.** $4x^2 + 3x - 8 + \dfrac{50}{x + 6}$ [6.4B/6.4C]   **13.** -7 [6.2A]

14. $13y^3 - 12y^2 - 5y - 1$ [6.2B]   **15.** $b^2 + 5b + 2 + \dfrac{7}{b - 7}$ [6.4B/6.4C]   **16.** $12x^5y^3 + 8x^3y^2 - 28x^2y^4$ [6.3A]

17. $2ax - 4ay - bx + 2by$ [6.3B]   **18.** $8b^2 - 2b - 15$ [6.3B]   **19.** $33x^2 + 24x$ [6.3A]   **20.** $x^4y^8z^4$ [6.1A]

21. $16x^2 - 24xy + 9y^2$ [6.3C]   **22.** $x^3 + 4x^2 + 16x + 64 + \dfrac{252}{x - 4}$ [6.4B/6.4C]   **23.** $2x^2 - 5x - 2$ [6.2B]

24. $70xy^2z^6$ [6.1B]   **25.** $-\dfrac{x^3}{4y^2z^3}$ [6.1B]   **26.** 9.48×10^8 [6.1C]   **27.** 2×10^{-6} [6.1C]   **28.** 68 [6.4D]

29. $4x^4 - 2x^2 + 5$ [6.4A]   **30.** $2x - 3 - \dfrac{4}{6x + 1}$ [6.4B]   **31.** $a^{3n+3} - 5a^{2n+4} + 2a^{2n+3}$ [6.3A]

32. $x^4 + 3x^3 - 23x^2 - 29x + 6$ [6.3B]   **33.** $-8x^3 - 14x^2 + 18x$ [6.3A]   **34.** $6y^3 + 17y^2 - 2y - 21$ [6.3B]
35. $16u^{12}v^{16}$ [6.1A]   **36.** $2x^3 + 9x^2 - 3x - 12$ [6.2B]   **37.** $2x^2 + 3x - 8$ [6.2B]   **38.** $a^2 - 49$ [6.3C]

39. $100a^{15}b^{13}$ [6.1A]   **40.** 14,600,000 [6.1C]   **41.** $-108x^{18}$ [6.1A]   **42.** $2y - 9$ [6.4B]   **43.** $-\dfrac{1}{16}$ [6.1B]

44. $10a^2 + 31a - 63$ [6.3B]   **45.** $-x + 2 + \dfrac{1}{x + 3}$ [6.4B/6.4C]   **46.** 1.27×10^{-7} [6.1C]   **47.** $-4y + 8$ [6.4A]

48. $\dfrac{c^{10}}{2b^{17}}$ [6.1B]   **49.** $6x^3 - 29x^2 + 14x + 24$ [6.3B]   **50.** $\dfrac{x^4y^6}{9}$ [6.1B]   **51.** $-54a^{13}b^5c^7$ [6.1A]

52. $25a^2 - 4b^2$ [6.3C]   **53.** 0.00254 [6.1C]   **54.** $8a^3b^3 - 4a^2b^4 + 6ab^5$ [6.3A]   **55.** [6.2A]

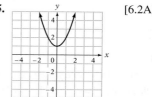

56a. 3 **b.** 8 **c.** 5 [6.2A]   **57.** The mass of the moon is 8.103×10^{19} tons. [6.1D]   **58.** The area is
$(9x^2 - 12x + 4)$ in^2. [6.3D]   **59.** The Great Galaxy of Andromeda is 1.291224×10^{19} mi from Earth. [6.1D]
60. The area is $(10x^2 - 29x - 21)$ cm^2. [6.3D]

CHAPTER 6 TEST

1. $4x^3 - 6x^2$ [6.3A; HOW TO 1]   **2.** -8 [6.4D; HOW TO 6]   **3.** $-\dfrac{4}{x^6}$ [6.1B; Example 4]   **4.** $-6x^3y^6$ [6.1A; HOW TO 1]

5. $x - 1 + \dfrac{2}{x + 1}$ [6.4B; Example 3]   **6.** $x^3 - 7x^2 + 17x - 15$ [6.3B; HOW TO 5]   **7.** $-8a^6b^3$ [6.1A; Example 1]

8. $\dfrac{9y^{10}}{x^{10}}$ [6.1B; Example 5]   **9.** $a^2 + 3ab - 10b^2$ [6.3B; Example 4]   **10.** -3 [6.2A; HOW TO 3]

11. $x + 7$ [6.4C; HOW TO 4]   **12.** $6y^4 - 9y^3 + 18y^2$ [6.3A; HOW TO 1]   **13.** $-4x^4 + 8x^3 - 3x^2 - 14x + 21$ [6.3B; HOW TO 5]

14. $16y^2 - 25$ [6.3C; HOW TO 8]   **15.** $6x^3 + 3x^2 - 2x$ [6.4A; HOW TO 1]   **16.** $8ab^4$ [6.1B; Example 4]   **17.** $\dfrac{2b^7}{a^{10}}$ [6.1B;

Example 5]   **18.** $-5a^3 + 3a^2 - 4a + 3$ [6.2B; HOW TO 8]   **19.** $4x^2 - 20x + 25$ [6.3C; HOW TO 9]

20. $x^2 - 5x + 10 - \dfrac{23}{x + 3}$ [6.4C; Example 6]   **21.** $10x^2 - 43xy + 28y^2$ [6.3B; Example 4]   **22.** $3x^3 + 6x^2 - 8x + 3$
[6.2B; HOW TO 5]   **23.** 3.02×10^{-9} [6.1C; Example 6]   **24.** There are 6.048×10^6 s in 10 weeks. [6.1D; Example 9]
25. The area of the circle is $(\pi x^2 - 10\pi x + 25\pi)$ m^2. [6.3D; Example 12]

CUMULATIVE REVIEW EXERCISES

1. 4 [1.3A]   **2.** $-\dfrac{5}{4}$ [1.4A]   **3.** The Inverse Property of Addition [1.4B]   **4.** $-18x + 8$ [1.4D]   **5.** $-\dfrac{1}{6}$ [2.2A]

6. $-\dfrac{11}{4}$ [2.2B]   **7.** $x^2 + 3x + 9 + \dfrac{24}{x - 3}$ [6.4B/6.4C]   **8.** -1 and $\dfrac{7}{3}$ [2.6A]   **9.** $P(-2) = 18$ [4.2A]

10. $\{x \mid x \neq -2\}$ [4.2A]   **11.** $\dfrac{4}{3}$ [4.3C]   **12.** The slope is $-\dfrac{1}{6}$. [4.4A]   **13.** $y = -\dfrac{3}{2}x + \dfrac{1}{2}$ [4.5A]

14. $y = \dfrac{2}{3}x + \dfrac{16}{3}$ [4.6A]   **15.** $(-1, -2)$ [5.1B]   **16.** $\left(-\dfrac{9}{7}, \dfrac{2}{7}, \dfrac{11}{7}\right)$ [5.2B]

Unless otherwise noted, all content on this page is © Cengage Learning.

17. 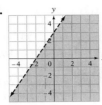 [4.3C]   **18.** 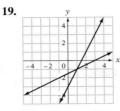 [4.7A]   **19.** 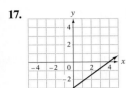 $(1, -1)$ [5.1A]

20. 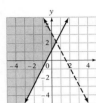 [5.4A]   **21.** $\dfrac{b^5}{a^8}$ [6.1B]   **22.** $\dfrac{y^2}{25x^6}$ [6.1B]   **23.** $\dfrac{21}{8}$ [6.1B]   **24.** $4x^3 - 7x + 3$ [6.3B]

25. The integers are 9 and 15. [2.3A]   **26.** \$2000 is invested in the 4% account. [5.1C]   **27.** The slower cyclist travels at 5 mph and the faster cyclist at 7.5 mph. [2.4C]   **28.** 40 oz of pure gold must be mixed with the alloy. [2.4A]   **29.** $m = 50$. The average speed is 50 mph. [4.4A]   **30.** The vehicle will reach the moon in 12 h. [6.1D]

Answers To Chapter 7 Selected Exercises

PREP TEST
1. $2 \cdot 3 \cdot 5$ [1.2D]   **2.** $-12y + 15$ [1.4D]   **3.** $-a + b$ [1.4D]   **4.** $-3a + 3b$ [1.4D]   **5.** 0 [2.1C]
6. $-\dfrac{1}{2}$ [2.2A]   **7.** $x^2 - 2x - 24$ [6.3B]   **8.** $6x^2 - 11x - 10$ [6.3B]   **9.** x^3 [6.1B]   **10.** $3x^3y$ [6.1B]

SECTION 7.1
1. 4 **3a.** x **b.** $2x - 1$ **5.** $(2x^3 - x^2) + (6x - 3)$ **7.** $5(a + 1)$ **9.** $8(2 - a^2)$ **11.** $4(2x + 3)$ **13.** $x(7x - 3)$
15. $a^2(3 + 5a^3)$ **17.** $2x(x^3 - 2)$ **19.** $2x^2(5x^2 - 6)$ **21.** $4a^5(2a^3 - 1)$ **23.** $xy(xy - 1)$ **25.** $3xy(xy^3 - 2)$
27. $3x(x^2 + 2x + 3)$ **29.** $2x^2(x^2 - 2x + 3)$ **31.** $2x(x^2 + 3x - 7)$ **33.** $y^3(2y^2 - 3y + 7)$ **35.** $xy(x^2 - 3xy + 7y^2)$
37. $5y(y^2 + 2y - 5)$ **39.** $3b^2(a^2 - 3a + 5)$ **41.** x^c **43.** $(b + 4)(x + 3)$ **45.** $(y - x)(a - b)$ **47.** $(x - 2)(x - y)$
49. $(2m - 3n)(8c - 1)$ **51.** $(x + 2)(x + 2y)$ **53.** $(p - 2)(p - 3r)$ **55.** $(a + 6)(b - 4)$ **57.** $(2z - 1)(z + y)$
59. $(2x - 5)(x - 3y)$ **61.** $(y - 2)(3y - a)$ **63.** $(3x - y)(y + 1)$ **65.** $(3s + t)(t - 2)$ **67.** -1 **69.** $b - 3a$

SECTION 7.2
1. -8 **3.** -2 and 6 **5.** Different   **7.** $(x + 1)(x + 2)$ **9.** $(x + 1)(x - 2)$ **11.** $(a + 4)(a - 3)$ **13.** $(a - 1)(a - 2)$
15. $(a + 2)(a - 1)$ **17.** $(b - 3)(b - 3)$ **19.** $(b + 8)(b - 1)$ **21.** $(y + 11)(y - 5)$ **23.** $(y - 2)(y - 3)$
25. $(z - 5)(z - 9)$ **27.** $(z + 8)(z - 20)$ **29.** $(p + 3)(p + 9)$ **31.** $(x + 10)(x + 10)$ **33.** $(b + 4)(b - 5)$
35. $(y + 3)(y - 17)$ **37.** $(p + 3)(p - 7)$ **39.** Nonfactorable over the integers   **41.** $(x - 5)(x - 15)$ **43.** $(p + 3)(p + 21)$
45. $(x + 2)(x + 19)$ **47.** Nonfactorable over the integers   **49.** $(a + 4)(a - 11)$ **51.** $(a - 3)(a - 18)$ **53.** $(z + 21)(z - 7)$
55. $(c + 12)(c - 15)$ **57.** $(p + 9)(p + 15)$ **59.** $(c + 2)(c + 9)$ **61.** $(x + 15)(x - 5)$ **63.** $(x + 25)(x - 4)$
65. $(b - 4)(b - 18)$ **67.** $(a + 45)(a - 3)$ **69.** $(b - 7)(b - 18)$ **71.** $(z + 12)(z + 12)$ **73.** $(x - 4)(x - 25)$
75. $(x + 16)(x - 7)$ **77.** Positive   **79.** $3(x + 2)(x + 3)$ **81.** $-(x + 6)(x - 2)$ **83.** $a(b + 8)(b - 1)$ **85.** $x(y + 3)(y + 5)$
87. $-2a(a + 1)(a + 2)$ **89.** $4y(y + 6)(y - 3)$ **91.** $2x(x^2 - x + 2)$ **93.** $6(z + 5)(z - 3)$ **95.** $3a(a + 3)(a - 6)$
97. $(x + 7y)(x - 3y)$ **99.** $(a - 5b)(a - 10b)$ **101.** $(s + 8t)(s - 6t)$ **103.** Nonfactorable over the integers
105. $z^2(z + 10)(z - 8)$ **107.** $b^2(b + 2)(b - 5)$ **109.** $3y^2(y + 3)(y + 15)$ **111.** $-x^2(x - 12)(x + 1)$ **113.** $3y(x + 3)(x - 5)$
115. $-3x(x - 3)(x - 9)$ **117.** $(x - 3y)(x - 5y)$ **119.** $(a - 6b)(a - 7b)$ **121.** $(y + z)(y + 7z)$ **123.** $3y(x + 21)(x - 1)$
125. $3x(x + 4)(x - 3)$ **127.** $2(t - 5s)(t - 7s)$ **129.** $3(a + 3b)(a - 11b)$ **131.** $5x(x + 2y)(x + 4y)$ **133a.** Yes   **b.** No
135. $-2x$ **137.** $y(x + 6)(x - 9)$ **139.** $3p(p + 8)(p - 4)$ **141.** $19, 11, 9, -9, -11, -19$ **143.** $15, 9, -9, -15$ **145.** $6, 10, 12$
147. $6, 10, 12$ **149.** $4, 6$

SECTION 7.3
1. $2x + 5$ **3.** $4x - 3$ **5.** $4, -5$ **7.** $-2x - 6x$ **9.** $(x + 1)(2x + 1)$ **11.** $(y + 3)(2y + 1)$ **13.** $(a - 1)(2a - 1)$
15. $(b - 5)(2b - 1)$ **17.** $(x + 1)(2x - 1)$ **19.** $(x - 3)(2x + 1)$ **21.** $(t + 2)(2t - 5)$ **23.** $(p - 5)(3p - 1)$
25. $(3y - 1)(4y - 1)$ **27.** Nonfactorable over the integers   **29.** $(2t - 1)(3t - 4)$ **31.** $(x + 4)(8x + 1)$
33. Nonfactorable over the integers   **35.** $(3y + 1)(4y + 5)$ **37.** $(z - 14)(2z + 1)$ **39.** $(p + 8)(3p - 2)$ **41.** $2(x + 1)(2x + 1)$
43. $5(y - 1)(3y - 7)$ **45.** $x(x - 5)(2x - 1)$ **47.** $b(a - 4)(3a - 4)$ **49.** Nonfactorable over the integers

51. $-3x(x + 4)(x - 3)$ **53.** $4(4y - 1)(5y - 1)$ **55.** $z(2z + 3)(4z + 1)$ **57.** $y(2x - 5)(3x + 2)$ **59.** $5(t + 2)(2t - 5)$
61. $p(p - 5)(3p - 1)$ **63.** $2(z + 4)(13z - 3)$ **65.** $2y(y - 4)(5y - 2)$ **67.** $yz(z + 2)(4z - 3)$ **69.** $3a(2a + 3)(7a - 3)$
71. $y(3x - 5y)(3x - 5y)$ **73.** $xy(3x - 4y)(3x - 4y)$ **75.** Odd   **77.** $(2x - 3)(3x - 4)$ **79.** $(b + 7)(5b - 2)$
81. $(2a - 3)(3a + 8)$ **83.** $(z + 2)(4z + 3)$ **85.** $(2p + 5)(11p - 2)$ **87.** $(y + 1)(8y + 9)$ **89.** $(3t + 1)(6t - 5)$
91. $(b + 12)(6b - 1)$ **93.** $(3x + 2)(3x + 2)$ **95.** $(2b - 3)(3b - 2)$ **97.** $(3b + 5)(11b - 7)$ **99.** $(3y - 4)(6y - 5)$
101. $(3a + 7)(5a - 3)$ **103.** $(2y - 5)(4y - 3)$ **105.** $(2z + 3)(4z - 5)$ **107.** Nonfactorable over the integers
109. $(2z - 5)(5z - 2)$ **111.** $(6z + 5)(6z + 7)$ **113.** $(x + y)(3x - 2y)$ **115.** $(a + 2b)(3a - b)$ **117.** $(y - 2z)(4y - 3z)$
119. $-(z - 7)(z + 4)$ **121.** $-(x - 1)(x + 8)$ **123.** $3(x + 5)(3x - 4)$ **125.** $4(2x - 3)(3x - 2)$ **127.** $a^2(5a + 2)(7a - 1)$
129. $5(b - 7)(3b - 2)$ **131.** $(x - 7y)(3x - 5y)$ **133.** $3(8y - 1)(9y + 1)$ **135.** One positive, one negative
137. One positive, one negative   **139.** $(x - 2)(x + 3)$ **141.** $y(y + 1)$ **143.** $(3a + 2)(a + 3)$ **145.** $2y(y - 3)(4y - 1)$
147. $ab(a + 4)(a - 6)$ **149.** $7, -7, 5, -5$ **151.** $7, -7, 5, -5$ **153.** $11, -11, 7, -7$

CHECK YOUR PROGRESS: CHAPTER 7

1. $5(4b + 1)$ [7.1A]   **2.** $(b + 7)(2x - y)$ [7.1B]   **3.** $(x + 10)(x + 10)$ [7.2A]   **4.** $y(x + 4)(x - 6)$ [7.2B]
5. $-(x - 7)(x + 5)$ [7.2B]   **6.** Nonfactorable over the integers   [7.2A]   **7.** $(7x + 2y)(3x - 7)$ [7.1B]   **8.** $3a(2b + 3)$ [7.1A]
9. $(y - 4)(5y - 2)$ [7.3A]   **10.** $(3x + 1)(4x + 9)$ [7.3B]   **11.** $x(9 - 5x)$ [7.1A]   **12.** $(2x + 1)(x + y)$ [7.1B]
13. $(2a + b)(4a - 3b)$ [7.3B]   **14.** $(b + 4)(b + 5)$ [7.2A]   **15.** $2a(a + 9)(a + 3)$ [7.2B]   **16.** $(a - 5)(11a + 1)$ [7.3A]
17. $4(9y + 1)(10y - 1)$ [7.3B]   **18.** $y(14y^2 + 5y + 11)$ [7.1A]   **19.** $(x - 2)(x - 5)$ [7.2A]
20. Nonfactorable over the integers   [7.2B]   **21.** $(b + 8)(b + 5)$ [7.2A]   **22.** $(2x - 5)(x - 3y)$ [7.1B]
23. $xy(x - y^2 + x^2)$ [7.1A]   **24.** $(b + 4)(3b + 4)$ [7.3A]   **25.** $(x + 3)(x - 14)$ [7.2A]

SECTION 7.4

1a. No   **b.** Yes   **c.** No   **d.** Yes   **3a.** Yes   **b.** No   **c.** Yes   **d.** No   **5a.** $2x^3$ **b.** $3y^5$ **c.** $4a^2b^6$ **d.** $5c^4d$
7a. Yes   **b.** No   **c.** Yes   **d.** Yes   **9.** $(x + 4)(x - 4)$ **11.** $(2x + 1)(2x - 1)$ **13.** $(4x + 11)(4x - 11)$
15. $(1 + 3a)(1 - 3a)$ **17.** $(xy + 10)(xy - 10)$ **19.** Nonfactorable over the integers   **21.** $(5 + ab)(5 - ab)$
23. $(x - 6)^2$ **25.** $(b - 1)^2$ **27.** $(4x - 5)^2$ **29.** Nonfactorable over the integers   **31.** Nonfactorable over the integers
33. $(x + 3y)^2$ **35.** $(5a - 4b)^2$ **37.** $(x - 7)(x - 1)$ **39.** $(x - y - a - b)(x - y + a + b)$
41. $(x - 3)(x^2 + 3x + 9)$ **43.** $(2x - 1)(4x^2 + 2x + 1)$ **45.** $(x - y)(x^2 + xy + y^2)$ **47.** $(m + n)(m^2 - mn + n^2)$
49. $(4x + 1)(16x^2 - 4x + 1)$ **51.** $(3x - 2y)(9x^2 + 6xy + 4y^2)$ **53.** $(xy + 4)(x^2y^2 - 4xy + 16)$ **55.** Nonfactorable over the
integers   **57.** Nonfactorable over the integers   **59.** $(5 - c)(25 + 5c + c^2)$ **61.** No, polynomials cannot have square roots as variable
terms.   **63.** $(xy - 5)(xy - 3)$ **65.** $(xy - 12)(xy - 5)$ **67.** $(x^2 - 3)(x^2 - 6)$ **69.** $(b^2 + 5)(b^2 - 18)$
71. $(x^2y^2 - 2)(x^2y^2 - 6)$ **73.** $(\sqrt{x} + 1)(\sqrt{x} + 2)$ **75.** $(3xy - 5)(xy - 3)$ **77.** $(2ab - 3)(3ab - 7)$ **79.** $(x^2 + 1)(2x^2 - 15)$
81. $(x^3 + 2)(x^3 - 3)$ **83.** $(2xy^2 - 3)^2$ **85.** $3(2x - 3)^2$ **87.** $a(3a - 1)(9a^2 + 3a + 1)$ **89.** $5(2x + 1)(2x - 1)$
91. $y^3(y + 11)(y - 5)$ **93.** $(4x^2 + 9)(2x + 3)(2x - 3)$ **95.** $2a(2 - a)(4 + 2a + a^2)$ **97.** $b^3(ab - 1)(a^2b^2 + ab + 1)$
99. $2x^2(2x - 5)^2$ **101.** $(x^2 + y^2)(x + y)(x - y)$ **103.** $(x^2 + y^2)(x^4 - x^2y^2 + y^4)$ **105.** Nonfactorable over the integers

SECTION 7.5

1a. Yes   **b.** No   **c.** Yes   **3a.** Yes   **b.** Yes   **c.** No   **d.** Yes   **e.** No   **f.** Yes   **5.** $-3, -2$ **7.** $7, 3$ **9.** $0, 5$
11. $0, 9$ **13.** $0, -\dfrac{3}{2}$ **15.** $0, \dfrac{2}{3}$ **17.** $-2, 5$ **19.** $9, -9$ **21.** $\dfrac{7}{2}, -\dfrac{7}{2}$ **23.** $\dfrac{1}{3}, -\dfrac{1}{3}$ **25.** $-4, -2$ **27.** $2, -7$
29. $-\dfrac{1}{2}, 5$ **31.** $-\dfrac{1}{3}, -\dfrac{1}{2}$ **33.** $0, 3$ **35.** $0, 7$ **37.** $-1, -4$ **39.** $2, 3$ **41.** $\dfrac{1}{2}, -4$ **43.** $\dfrac{1}{3}, 4$ **45.** $3, 9$
47. $9, -2$ **49.** $-1, -2$ **51.** $5, -9$ **53.** $4, -7$ **55.** $-2, -3$ **57.** $-8, 9$ **59.** $1, 4$ **61.** $-5, 2$ **63.** Less than
65. The number is 6.   **67.** The numbers are 2 and 4.   **69.** ii   **71.** The numbers are 4 and 5.   **73.** The numbers are 15 and 16.
75. The base of the triangle is 18 ft. The height is 6 ft.   **77.** The length of the rectangle is 18 ft. The width is 8 ft.
79. The length of a side of the original square is 4 m.   **81.** The radius of the original circle is 3.81 in.   **83.** The dimensions are 4 in. by 7 in.
85. The width of the lane is 16 ft.   **87.** The object will hit the ground in 4 s.   **89.** There are 15 consecutive natural numbers beginning with
1 that will give a sum of 120.   **91.** There are 10 teams in the league.   **93.** The golf ball will return to the ground in 3 s.   **95.** $\dfrac{3}{2}, -4$
97. $-1, -9$ **99.** $0, 7$ **101.** $18, 1$ **103.** 2 or -128 **107.** The length of the piece of cardboard is 20 in. The width is 10 in.

CHAPTER 7 REVIEW EXERCISES

1. $(b - 3)(b - 10)$ [7.2A]   **2.** $(x - 3)(4x + 5)$ [7.1B]   **3.** Nonfactorable over the integers   [7.3A]   **4.** $(7x^2y^2 + 3)(3x^2y^2 + 2)$ [7.4C]
5. $7y^3(2y^6 - 7y^3 + 1)$ [7.1A]   **6.** $(y - 4)(y + 9)$ [7.2A]   **7.** $(2x - 7)(3x - 4)$ [7.3A]   **8.** $3ab(4a + b)$ [7.1A]
9. $(a^3 + 10)(a^3 - 10)$ [7.4A]   **10.** $n^2(n + 1)(n - 3)$ [7.2B]   **11.** $(6y - 1)(2y + 3)$ [7.3A]   **12.** $2b(3b - 4)(2b - 7)$ [7.4D]
13. $(3y^2 + 5z)(3y^2 - 5z)$ [7.4A]   **14.** $(c + 6)(c + 2)$ [7.2A]   **15.** $(6a - 5)(3a + 2)$ [7.3B]   **16.** $\dfrac{1}{4}, -7$ [7.5A]

17. $4x(x - 6)(x + 1)$ [7.2B]   **18.** $(4a - 3b)(16a^2 + 12ab + 9b^2)$ [7.4B]   **19.** $(2a + 5)(a - 12)$ [7.3B]
20. $-3, 7$ [7.5A]   **21.** $(3a - 5b)(7x + 2y)$ [7.1B]   **22.** $(6x^4 - 1)(6x^4 - 5)$ [7.4C]   **23.** $(2x + 5)(5x + 2y)$ [7.1B]
24. $5(x + 2)(x - 3)$ [7.2B]   **25.** $3(x + 6)^2$ [7.4D]   **26.** $(3x - 2)(x - 5)$ [7.3B]   **27.** The length is 100 yd.

The width is 60 yd.   [7.5B]   **28.** The distance is 20 ft.   [7.5B]   **29.** The width of the frame is 1.5 in. or $1\frac{1}{2}$ in.   [7.5B]
30. The two integers are 4 and 5.   [7.5B]

CHAPTER 7 TEST
1. $(b + 6)(a - 3)$ [7.1B; Example 6]   **2.** $2y^2(y + 1)(y - 8)$ [7.2B; Example 3]   **3.** $4(x + 4)(2x - 3)$ [7.3B; Example 4]
4. $(2x + 1)(3x + 8)$ [7.3A; HOW TO 2]   **5.** $(a - 3)(a - 16)$ [7.2A; Example 1]   **6.** $2x(3x^2 - 4x + 5)$ [7.1A; HOW TO 2]

7. $(x + 5)(x - 3)$ [7.2A; Example 2]   **8.** $\frac{1}{2}, -\frac{1}{2}$ [7.5A; Example 2]   **9.** $5(x^2 - 9x - 3)$ [7.1A; HOW TO 2]

10. $(p + 6)^2$ [7.4A; HOW TO 2]   **11.** $3, 5$ [7.5A; Example 3]   **12.** $3(x + 2y)^2$ [7.4D; Example 8]
13. $(b + 4)(b - 4)$ [7.4A; You Try It 1]   **14.** $3y^2(2x + 1)(x + 1)$ [7.3B; Example 4]   **15.** $(3x - 2)(9x^2 + 6x + 4)$ [7.4B; HOW TO 4]
16. $(2a^2 - 5)(3a^2 + 1)$ [7.4C; You Try It 7]   **17.** $(p + 1)(x - 1)$ [7.1B; Example 4]   **18.** $3(a + 5)(a - 5)$ [7.4D; HOW TO 9]
19. Nonfactorable over the integers   [7.3B; HOW TO 6]   **20.** $(x + 3)(x - 12)$ [7.2A; HOW TO 2]   **21.** $(2a - 3b)^2$ [7.4A; Example 2]

22. $(2x + 7y)(2x - 7y)$ [7.4A; HOW TO 1]   **23.** $\frac{3}{2}, -7$ [7.5A; HOW TO 1]   **24.** The two numbers are 3 and 7.   [7.5B; Example 4]

25. The length is 15 cm. The width is 6 cm.   [7.5B; You Try It 5]

CUMULATIVE REVIEW EXERCISES
1. 7 [1.1C]   **2.** 4 [1.3A]   **3.** -7 [1.4A]   **4.** $15x^2$ [1.4C]   **5.** 12 [1.4D]   **6.** $\frac{2}{3}$ [2.1C]   **7.** $\frac{7}{4}$ [2.2B]

8. 3 [2.2C]   **9.** 45 [2.1D]   **10.** 1 [4.2A]   **11.**

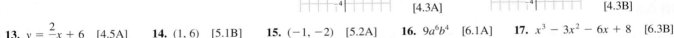

[4.3A]   **12.** [4.3B]

13. $y = \frac{2}{3}x + 6$ [4.5A]   **14.** $(1, 6)$ [5.1B]   **15.** $(-1, -2)$ [5.2A]   **16.** $9a^6b^4$ [6.1A]   **17.** $x^3 - 3x^2 - 6x + 8$ [6.3B]
18. $4x + 8 + \frac{21}{2x - 3}$ [6.4B]   **19.** $\frac{y^6}{x^8}$ [6.1B]   **20.** $(a - b)(3 - x)$ [7.1B]   **21.** $5xy^2(3 - 4y^2)$ [7.1A]
22. $(x - 7y)(x + 2y)$ [7.2A]   **23.** $\frac{5}{2}, 4$ [7.5A]   **24.** $3a(2a + 5)(3a + 2)$ [7.4D]   **25.** $(6a - 7b)(6a + 7b)$ [7.4A]
26. $\frac{2}{3}, -7$ [7.5A]   **27.** The shorter piece is 4 ft long. The longer piece is 6 ft long.   [2.3B]   **28.** The measure of $\angle a$ is 72°.
The measure of $\angle b$ is 108°.   [3.1B]   **29.** The distance to the resort is 168 mi.   [2.4C]   **30.** The integers are 10, 12, and 14.   [2.3A]
31. The length of the base of the triangle is 12 in.   [7.5B]

Answers to Chapter 8 Selected Exercises

PREP TEST
1. 50 [1.2C]   **2.** $-\frac{1}{6}$ [1.2D]   **3.** $-\frac{3}{2}$ [1.2D]   **4.** $\frac{1}{24}$ [1.2C]   **5.** $\frac{5}{24}$ [1.2C]   **6.** $\frac{1}{3}$ [1.4A]   **7.** -2 [2.2C]

8. $\frac{10}{7}$ [2.2C]   **9.** The rates of the planes are 110 mph and 130 mph.   [2.4C]

SECTION 8.1
5. $\frac{3}{4x}$ **7.** $\frac{1}{x + 3}$ **9.** -1 **11.** $\frac{2}{3y}$ **13.** $-\frac{3}{4x}$ **15.** $\frac{a}{b}$ **17.** $-\frac{2}{x}$ **19.** $\frac{y - 2}{y - 3}$ **21.** $\frac{x + 5}{x + 4}$ **23.** $\frac{x + 4}{x - 3}$ **25.** $-\frac{x + 2}{x + 5}$
27. $\frac{2(x + 2)}{x + 3}$ **29.** $\frac{2x - 1}{2x + 3}$ **31.** $-\frac{x + 7}{x + 6}$ **33.** $\frac{2}{3xy}$ **35.** $\frac{8xy^2ab}{3}$ **37.** $\frac{2}{9}$ **39.** $\frac{y^2}{x}$ **41.** $\frac{y(x + 4)}{x(x + 1)}$ **43.** $\frac{x^3(x - 7)}{y^2(x - 4)}$
45. $-\frac{y}{x}$ **47.** $\frac{x + 3}{x + 1}$ **49.** $\frac{x - 5}{x + 3}$ **51.** $-\frac{x + 3}{x + 5}$ **53.** $-\frac{x + 3}{x - 12}$ **55.** $\frac{x + 2}{x + 4}$ **57.** 1 **59.** 1 **61.** $\frac{7a^3y^2}{40bx}$ **63.** $\frac{4}{3}$

Unless otherwise noted, all content on this page is © Cengage Learning.

65. $\dfrac{3a}{2}$ **67.** $\dfrac{x^2(x + 4)}{y^2(x + 2)}$ **69.** $\dfrac{x(x - 2)}{y(x - 6)}$ **71.** $-\dfrac{3by}{ax}$ **73.** $\dfrac{(x + 6)(x - 3)}{(x + 7)(x - 6)}$ **75.** 1 **77.** $-\dfrac{x + 8}{x - 4}$ **79.** $\dfrac{2n + 1}{2n - 3}$ **81.** Yes

83. No **85.** 5, −1 **87.** $\dfrac{4}{25}$ **89.** $\dfrac{x + 4}{x - 3}$ and $\dfrac{2x - 1}{3x + 1}$ or $\dfrac{x + 4}{3x + 1}$ and $\dfrac{2x - 1}{x - 3}$

SECTION 8.2

1a. $14(x + 3)$ **b.** $(x + 4)(x - 6)$ **c.** $(x - 2)(x + 2)(x + 5)$ **3a.** Six **b.** Four **5.** $\dfrac{9y^3}{12x^2y^4}, \dfrac{17x}{12x^2y^4}$

7. $\dfrac{2x^2 - 4x}{6x^2(x - 2)}, \dfrac{3x - 6}{6x^2(x - 2)}$ **9.** $\dfrac{3x - 1}{2x(x - 5)}, -\dfrac{6x^3 - 30x^2}{2x(x - 5)}$ **11.** $\dfrac{6x^2 + 9x}{(2x + 3)(2x - 3)}, \dfrac{10x^2 - 15x}{(2x + 3)(2x - 3)}$

13. $\dfrac{2x}{(x + 3)(x - 3)}, \dfrac{x^2 + 4x + 3}{(x + 3)(x - 3)}$ **15.** $\dfrac{6}{6(x + 2y)(x - 2y)}, \dfrac{5x + 10y}{6(x + 2y)(x - 2y)}$ **17.** $\dfrac{3x^2 - 3x}{(x + 1)(x - 1)^2}, \dfrac{5x^2 + 5x}{(x + 1)(x - 1)^2}$

19. $-\dfrac{x - 3}{(x - 2)(x^2 + 2x + 4)}, \dfrac{2x - 4}{(x - 2)(x^2 + 2x + 4)}$ **21.** $\dfrac{2x^2 + 6x}{(x - 1)(x + 3)^2}, -\dfrac{x^2 - x}{(x - 1)(x + 3)^2}$

23. $-\dfrac{12x^2 - 8x}{(2x - 3)(2x - 5)(3x - 2)}, \dfrac{6x^2 - 9x}{(2x - 3)(2x - 5)(3x - 2)}$ **25.** $\dfrac{5}{(3x - 4)(2x - 3)}, -\dfrac{4x^2 - 6x}{(3x - 4)(2x - 3)}, \dfrac{3x^2 - x - 4}{(3x - 4)(2x - 3)}$

27. $\dfrac{2x^2 + 10x}{(x + 5)(x - 3)}, -\dfrac{2x - 6}{(x + 5)(x - 3)}, -\dfrac{x - 1}{(x + 5)(x - 3)}$ **29.** True **31.** $\dfrac{1}{2x^2}$ **33.** $\dfrac{1}{x + 2}$ **35.** $\dfrac{12ab - 9b + 8a}{30a^2b^2}$

37. $\dfrac{5 - 16b + 12a}{40ab}$ **39.** $\dfrac{7}{12x}$ **41.** $\dfrac{2xy - 8x + 3y}{10x^2y^2}$ **43.** $-\dfrac{a(2a - 13)}{(a - 2)(a + 1)}$ **45.** $\dfrac{5x^2 - 6x + 10}{(2x - 5)(5x - 2)}$ **47.** $\dfrac{a}{b(a - b)}$

49. $\dfrac{a^2 + 18a - 9}{a(a - 3)}$ **51.** $\dfrac{17x^2 + 20x - 25}{x(6x - 5)}$ **53.** $\dfrac{6}{(x - 3)^2(x + 3)}$ **55.** $-\dfrac{2(x - 1)}{(x + 2)^2}$ **57.** $-\dfrac{5x^2 - 17x + 8}{(x + 4)(x - 2)}$

59. $\dfrac{x^2 - 52x + 160}{4(x + 3)(x - 3)}$ **61.** $\dfrac{3x - 1}{4x + 1}$ **63.** $\dfrac{2(5x - 3)}{(x + 4)(x - 3)(x + 3)}$ **65.** $\dfrac{x - 2}{x + 3}$ **67.** 1 **69.** $\dfrac{x + 1}{2x - 1}$ **71.** $\dfrac{1}{2x - 1}$

73. $\dfrac{1}{x^2 + 4}$ **75.** $-\dfrac{2x^2 + 5x - 2}{(x + 2)(x + 1)}$ **77.** $\dfrac{b - a}{b + 2a}$

SECTION 8.3

3. $\dfrac{5}{23}$ **5.** $\dfrac{2}{5}$ **7.** $\dfrac{x}{x - 1}$ **9.** $-\dfrac{a}{a + 2}$ **11.** $\dfrac{a}{2a - 1}$ **13.** $\dfrac{(x - 1)(x + 1)}{x^2 + 1}$ **15.** $-\dfrac{a - 1}{a + 1}$ **17.** $\dfrac{2}{5}$ **19.** $-\dfrac{1}{2}$ **21.** $\dfrac{x + 1}{x - 4}$

23. $\dfrac{x - 2}{x - 3}$ **25.** $\dfrac{x - 3}{x + 4}$ **27.** $\dfrac{x + 2}{x + 5}$ **29.** $\dfrac{x + 2}{x - 1}$ **31.** $-\dfrac{x + 4}{2x + 3}$ **33.** $\dfrac{x + y}{x - y}$ **35.** $\dfrac{x^2 - x - 1}{x^2 + x + 1}$ **37.** $-\dfrac{2(a + 1)}{7a - 4}$

39. $-\dfrac{2x}{x^2 + 1}$ **41.** $\dfrac{a(a^2 + a + 1)}{a^2 + 1}$ **43.** $\dfrac{a - 1}{a}$ **45.** $\dfrac{2y}{x}$ **47.** $-\dfrac{1}{x(x + h)}$ **49a.** $P(x) = \dfrac{Cx(x + 1)^{60}}{(x + 1)^{60} - 1}$

b. The monthly payment is $405.53.

SECTION 8.4

1. Multiplication Property of Equations **3.** We can clear denominators in an *equation*, as in part (a), but not in an *expression*, as in part (b).

5. −1, 2 **7.** 0, 9 **9.** 1 **11.** 3 **13.** 2 **15.** 2 **17.** $\dfrac{2}{3}$ **19.** 4 **21.** −3 **23.** $\dfrac{3}{4}$ **25.** 7 **27.** −7

29. −1 **31.** −1 **33.** No solution **35.** 2, −6 **37.** $-\dfrac{2}{3}, 5$ **39.** −1, 6 **41.** 0 **43.** $0, -\dfrac{2}{3}$

45. The intensity of the illumination is 4 lm. **47.** A 320-candela light source is needed. **49.** The light source must be placed 2 m from the desk surface.

CHECK YOUR PROGRESS: CHAPTER 8

1. $\dfrac{x + 2}{x - 4}$ [8.1A] **2.** $\dfrac{x - 8}{3x - 8}$ [8.1A] **3.** $\dfrac{(x - 6)(x - 5)}{(x + 9)(x - 8)}$ [8.1B] **4.** $\dfrac{2x + 5}{3x + 9}$ [8.1B] **5.** $\dfrac{b^4x^2y}{a^2}$ [8.1C]

6. $\dfrac{(x + 7)(x - 3)}{(4x + 1)(x + 5)}$ [8.1C] **7.** $x(x + 4)(x + 5)$ [8.2A] **8.** $(x - 2)(x + 2)(x + 4)$ [8.2A] **9.** $\dfrac{5x^2 + 3x - 29}{(x - 1)(x - 4)}$ [8.2B]

10. $\dfrac{x^2 + 3x - 77}{(x - 3)(x - 9)}$ [8.2B] **11.** $-\dfrac{2x^2 + 3x + 3}{(3x + 4)(x + 1)}$ [8.2B] **12.** $-\dfrac{x^2 + 10x + 26}{(x + 9)(3x - 7)}$ [8.2B] **13.** $\dfrac{x}{x - 3}$ [8.3A]

14. $\dfrac{19x + 6}{8x - 15}$ [8.3A] **15.** −4 [8.4A] **16.** No solution [8.4A]

SECTION 8.5

1. A ratio is the quotient of two quantities that have the same unit. A rate is the quotient of two quantities that have different units.
3a. Rate, $\dfrac{25\ \text{ft}}{2\ \text{s}}$ **b.** Ratio, $\dfrac{4}{3}$ **c.** Rate, $\dfrac{10\ \text{mi}}{1\ \text{h}}$ **d.** Ratio, $\dfrac{1}{6}$ **5a.** YZ **b.** $\angle R$ **7.** 9   **9.** 12   **11.** 7   **13.** 6   **15.** 1

17. -6 **19.** 4   **21.** $-\dfrac{2}{3}$ **23.** 20,000 people voted in favor of the amendment.   **25.** 45 million Americans do not have health insurance.
27. 140 air vents are required for the office building.   **29.** The length of the sea scorpion's claw was 1.23 ft.   **31.** There are approximately 800 fish in the lake.   **33.** 180 panels are needed to provide 600 watts of power.   **35.** Yes, the shipment will be accepted.   **37.** The height of the person is 67.5 in.   **39.** There are 750 mi between the two cities.   **41.** The engine burns 127,500 lb of fuel in 45 s.   **43.** 22.5 gal of yellow paint are needed.   **45.** 10 additional acres must be planted.   **47.** AC is 6.7 cm.   **49.** The height is 2.9 m.
51. The perimeter is 22.5 ft.   **53.** The area is 48 m^2.   **55.** BC measures 6.25 cm.   **57.** DA measures 6 in.   **59.** OP measures 13 cm.

61. True   **63.** The width of the river is 35 m.   **65.** The number is 5 or $\dfrac{1}{5}$.   **67.** The player made 210 foul shots.

69a. Eratosthenes calculated the circumference of Earth to be 24,960 mi.   **b.** The difference is 86 mi.

SECTION 8.6

1. True   **3.** R **5.** $t = \dfrac{d}{r}$ **7.** $T = \dfrac{PV}{nR}$ **9.** $l = \dfrac{P - 2w}{2}$ **11.** $b_1 = \dfrac{2A - hb_2}{h}$ **13.** $h = \dfrac{3V}{A}$ **15.** $S = C - Rt$

17. $P = \dfrac{A}{1 + rt}$ **19.** $w = \dfrac{A}{S + 1}$ **21.** $y = -3x + 10$ **23.** $y = 4x - 3$ **25.** $y = -\dfrac{3}{2}x + 3$ **27.** $y = \dfrac{2}{5}x - 2$

29. $y = -\dfrac{2}{7}x + 2$ **31.** $y = -\dfrac{1}{3}x + 2$ **33.** $x = -3y + 6$ **35.** $x = \dfrac{1}{3}y + 1$ **37.** $x = -\dfrac{5}{2}y + 5$ **39.** $x = 2y - 1$

41a. Yes   **b.** Yes   **43.** $R_2 = \dfrac{RR_1}{R_1 - R}$

SECTION 8.7

3. $\dfrac{1}{x}$ **5.** Jen has the greater rate of work.   **7a.** Chris's rate of work is $\dfrac{1}{x}$ of the job per hour.   **b.** Chris can lay $\dfrac{3}{x}$ of the floor in 3 h.
9. The faster printer printed $\dfrac{h}{5}$ of the brochures.   **11a.** The speed of the boat traveling with the current is 12 mph.   **b.** The speed of the boat traveling against the current is 4 mph.   **13.** Row 1: $\dfrac{1}{10}, t, \dfrac{t}{10}$; Row 2: $\dfrac{1}{12}, t, \dfrac{t}{12}$ **15.** It will take 2 h to fill the fountain with both sprinklers operating.   **17.** With both skiploaders working together, it would take 3 h to transfer the earth.   **19.** It would take both computers, working together, 30 h to solve the problem.   **21.** With both air conditioners working, it would take 24 min to cool the room 5°F.   **23.** It would take the second welder, working alone, 15 h to complete the task.   **25.** It would take the second pipeline, working alone, 90 min to fill the tank.
27. It would take 3 h to harvest the field using only the older reaper.   **29.** It will take the second technician 3 h to complete the wiring.
31. It would take the small unit, working alone, $14\dfrac{2}{3}$ h to heat the pool.   **33.** It will take the apprentice, working alone, 3 h to complete the repairs.   **35.** t is less than k.   **37a.** Row 1: 1440, $380 - r$, $\dfrac{1440}{380 - r}$; Row 2: 1600, $380 + r$, $\dfrac{1600}{380 + r}$ **b.** $\dfrac{1440}{380 - r}, \dfrac{1600}{380 + r}$
39. The camper hiked at 4 mph.   **41.** The rate of the jet is 360 mph.   **43.** The rate of the boat for the first 15 mi was 7.5 mph.
45. The technician traveled at 20 mph through the congested traffic.   **47.** The family can travel 21.6 mi down the river and still return the boat in 3 h.
49. The rate of the river's current is 5 mph.   **51.** The rate of the freight train is 30 mph, and the rate of the express train is 50 mph.
53. The rate of the current is 2 mph.   **55.** The rate of the jet stream is 50 mph.   **57.** The rate of the current is 5 mph.   **59.** It would take
$1\dfrac{1}{19}$ h to fill the tank with all three pipes operating.   **61.** The amount of time spent traveling by canoe was 2 h.   **63.** The bus usually travels at 60 mph.

SECTION 8.8

1. $y = kx$ **3.** $z = kxy$ **5.** The profit is $80,000.   **7.** The pressure is 6.75 lb/in^2.   **9.** In 10 s, the object will fall 1600 ft.
11. It would take five computers 100 s to solve the problem.   **13.** The load on the sail is 360 lb.
15. Approximately 4.8 billion barrels leaked during the spill.   **17.** The current is 7.5 amps.
19. The intensity is 48 foot-candles when the distance is 5 ft.   **21.** inversely   **23.** y is doubled.   **25.** y is divided by 4.

CHAPTER 8 REVIEW EXERCISES

1. $\dfrac{b^3 y}{10ax}$ [8.1C]   **2.** $\dfrac{7x + 22}{60x}$ [8.2B]   **3.** $\dfrac{2xy}{5}$ [8.1B]   **4.** $\dfrac{2xy}{3(x + y)}$ [8.1C]   **5.** $\dfrac{x - 2}{3x - 10}$ [8.3A]   **6.** $-\dfrac{x + 6}{x + 3}$ [8.1A]

7. $\dfrac{2x^4}{3y^7}$ [8.1A]   **8.** 62 [8.5A]   **9.** $\dfrac{(3y - 2)^2}{(y - 1)(y - 2)}$ [8.1C]   **10.** $x = \dfrac{5}{3a - 1}$ [8.6A]   **11.** 8 [8.4A]   **12.** $\dfrac{x^2 + 3y}{xy}$ [8.2B]

13. $y = -\dfrac{5}{4}x + 5$ [8.6A]   **14.** $\dfrac{by^3}{6ax^2}$ [8.1B]   **15.** $\dfrac{x}{x-7}$ [8.3A]   **16.** $\dfrac{3x^2 - x}{(2x+3)(6x-1)(3x-1)}, \dfrac{24x^3 - 4x^2}{(2x+3)(6x-1)(3x-1)}$ [8.2A]

17. $a = \dfrac{T - 2bc}{2b + 2c}$ [8.6A]   **18.** 2 [8.4A]   **19.** $\dfrac{2x+1}{3x-2}$ [8.3A]   **20.** $\dfrac{x^2 + 5}{(x-5)(x-2)}$ [8.2B]   **21.** $c = \dfrac{100m}{i}$ [8.6A]

22. No solution [8.4A]   **23.** $\dfrac{1}{x^2}$ [8.1C]   **24.** $\dfrac{2y-3}{5y-7}$ [8.2B]   **25.** $\dfrac{1}{x+3}$ [8.2B]   **26.** $(5x-3)(2x-1)(4x-1)$ [8.2A]

27. $y = -\dfrac{4}{9}x + 2$ [8.6A]   **28.** $\dfrac{2x+1}{x+2}$ [8.1B]   **29.** 5 [8.4A]   **30.** $\dfrac{3x-1}{x-5}$ [8.2B]   **31.** 10 [8.5A]   **32.** 12 [8.5A]

33. The length of QO is 15 cm. [8.5B]   **34.** The area is $\dfrac{256}{3}$ in^2. [8.5B]   **35.** It would take 6 h to fill the pool. [8.7A]

36. The rate of the car is 45 mph. [8.7B]   **37.** The rate of the wind is 20 mph. [8.7B]   **38.** The pitcher's ERA is 1.35. [8.5A]
39. The current is 2 amps. [8.8A]   **40.** 48 mi would be represented by 12 in. [8.5A]   **41.** The stopping distance for a car traveling at 65 mph is 287.3 ft. [8.8A]   **42.** It would take the apprentice 104 min to complete the job working alone. [8.7A]

CHAPTER 8 TEST

1. $\dfrac{x^2 - 4x + 5}{(x-2)(x+3)}$ [8.2B; Example 5]   **2.** -1 [8.5A; You Try It 1B]   **3.** $\dfrac{(x-5)(2x-1)}{(x+3)(2x+5)}$ [8.1B; Example 5]

4. $\dfrac{2x^3}{3y^3}$ [8.1A; Example 1]   **5.** $t = \dfrac{d - s}{r}$ [8.6A; Example 3]   **6.** 2 [8.4A; Example 2]   **7.** $-\dfrac{x+5}{x+1}$ [8.1A; Example 3]

8. $3(2x-1)(x+1)$ [8.2A; HOW TO 2]   **9.** $\dfrac{5}{(2x-1)(3x+1)}$ [8.2B; HOW TO 4]   **10.** $\dfrac{x+5}{x+4}$ [8.1C; Example 7]

11. $\dfrac{x-3}{x-2}$ [8.3A; HOW TO 1]   **12.** $\dfrac{3x+6}{x(x-2)(x+2)}, \dfrac{x^2}{x(x-2)(x+2)}$ [8.2A; Example 2]   **13.** $\dfrac{2}{x+5}$ [8.2B; Example 5]

14. $y = \dfrac{3}{8}x - 2$ [8.6A; Example 1]   **15.** No solution [8.4A; HOW TO 2]   **16.** $\dfrac{x+1}{x^3(x-2)}$ [8.1B; Example 4]   **17.** CE is 12.8 ft. [8.5B; Example 3]   **18.** 6 lb of salt are needed. [8.5A; Example 2]   **19.** It would take 4 h to fill the pool. [8.7A; HOW TO 1]
20. The rate of the wind is 20 mph. [8.7B; HOW TO 2]   **21.** 54 sprinklers are needed for a 3600-square-foot lawn. [8.5A; You Try It 2]
22. The intensity is 128 decibels. [8.8A; HOW TO 3]

CUMULATIVE REVIEW EXERCISES

1. $\dfrac{31}{30}$ [1.3A]   **2.** 21 [1.4A]   **3.** $5x - 2y$ [1.4B]   **4.** $-8x + 26$ [1.4D]   **5.** $-\dfrac{9}{2}$ [2.2A]   **6.** -12 [2.2C]

7. 10 [2.1D]   **8.** $\{x|x \le 8\}$ [2.5A]   **9.** $y = \dfrac{3}{2}x + 2$ [4.6A]   **10.** $(2, 1, -1)$ [5.2B]   **11.** a^3b^7 [6.1A]

12. $a^2 + ab - 12b^2$ [6.3B]   **13.** $x^2 + 2x + 4$ [6.4B]   **14.** $(4x+1)(3x-1)$ [7.3A/7.3B]   **15.** $\dfrac{3}{7}$ [4.2A]

16. $a(2a-3)(a+5)$ [7.3A/7.3B]   **17.** $4(b+5)(b-5)$ [7.4D]   **18.** $-3, \dfrac{5}{2}$ [7.5A]   **19.** $\dfrac{2x^3}{3y^5}$ [8.1A]   **20.** $-\dfrac{x-2}{x+5}$ [8.1A]

21. 1 [8.1C]   **22.** $\dfrac{3}{(2x-1)(x+1)}$ [8.2B]   **23.** $\dfrac{x+3}{x+5}$ [8.3A]   **24.** 4 [8.4A]   **25.** 3 [8.5A]   **26.** $t = \dfrac{f-v}{a}$ [8.6A]
27. $5x - 13 = -8; x = 1$ [2.3A]   **28.** The 120-gram alloy is 70% silver. [2.4B]   **29.** The cost of a $5000 policy is $80. [8.5A]
30. The resistance is 0.4 ohm. [8.6A]   **31.** It would take both pipes 6 min to fill the tank. [8.7A]   **32.** The rate of the current is 2 mph. [8.7B]

Answers to Chapter 9 Selected Exercises

PREP TEST

1. 16 [1.1D]   **2.** 32 [1.2E]   **3.** 9 [1.2D]   **4.** $\dfrac{1}{12}$ [1.2C]   **5.** $-5x - 1$ [1.4D]   **6.** $\dfrac{xy^5}{4}$ [6.1B]

7. $9x^2 - 12x + 4$ [6.3C]   **8.** $-12x^2 + 14x + 10$ [6.3B]   **9.** $36x^2 - 1$ [6.3C]   **10.** -1 and 15 [7.5A]

SECTION 9.1

1. 125   **3.** -32 **5.** 9   **7.** 3   **9.** i, iii   **11.** 2   **13.** 27   **15.** $\dfrac{1}{9}$ **17.** 4   **19.** Not a real number   **21.** $\dfrac{343}{125}$

23. x **25.** $y^{1/2}$ **27.** $x^{1/12}$ **29.** $a^{7/12}$ **31.** $\dfrac{1}{a}$ **33.** $\dfrac{1}{y}$ **35.** $y^{3/2}$ **37.** $\dfrac{1}{x}$ **39.** $\dfrac{1}{x^4}$ **41.** a **43.** $x^{3/10}$ **45.** a^3

47. $\dfrac{1}{x^{1/2}}$ **49.** $y^{1/9}$ **51.** $\dfrac{x}{y^2}$ **53.** $\dfrac{x^{3/2}}{y^{1/4}}$ **55.** $\dfrac{x^2}{y^8}$ **57.** $\dfrac{1}{x^{11/12}}$ **59.** $\dfrac{1}{y^{5/2}}$ **61.** $\dfrac{1}{b^{7/8}}$ **63.** a^5b^{13} **65.** $\dfrac{x^2}{4y^{3/2}}$ **67.** $\dfrac{1}{3xy^{11}}$

69. $\dfrac{b^4}{2a}$ **71.** $\dfrac{y^{17/2}}{x^3}$ **73.** $\dfrac{16b^2}{a^{1/3}}$ **75.** $y^2 - y$ **77.** $a - a^2$ **79.** False **81.** $\sqrt[4]{3}$ **83.** $\sqrt{a^3}$ **85.** $\sqrt{32t^5}$ **87.** $-2\sqrt[3]{x^2}$

89. $\sqrt[3]{a^4b^2}$ **91.** $\sqrt[5]{a^6b^{12}}$ **93.** $\sqrt[4]{(4x-3)^3}$ **95.** $\dfrac{1}{\sqrt[3]{x^2}}$ **97.** $14^{1/2}$ **99.** $x^{1/3}$ **101.** $x^{4/3}$ **103.** $b^{3/5}$ **105.** $(2x^2)^{1/3}$

107. $-(3x^5)^{1/2}$ **109.** $3xy^{2/3}$ **111.** $(a^2-2)^{1/2}$ **113.** Positive **115.** Not a real number **117.** x^8 **119.** $-x^4$ **121.** xy^3

123. $-x^5y$ **125.** $4a^2b^6$ **127.** Not a real number **129.** $3x^3$ **131.** $-4x^3y^4$ **133.** $-x^2y^3$ **135.** x^4y^2 **137.** $3xy^5$

139. $2ab^2$

SECTION 9.2

1. Neither **3.** Perfect square **5.** Perfect cube **7.** Perfect cube **9.** No **11.** No **13.** $x^2yz^2\sqrt{yz}$ **15.** $2ab^4\sqrt{2a}$

17. $3xyz^2\sqrt{5yz}$ **19.** $2xyz\sqrt[4]{3yz^2}$ **21.** $a^5b^2\sqrt[3]{ab^2}$ **23.** $-5y\sqrt[3]{x^2y}$ **25.** $abc^2\sqrt[3]{ab^2}$ **27.** $2x^2y\sqrt[4]{xy}$ **29.** True **31.** $-6\sqrt{x}$

33. $-2\sqrt{2x}$ **35.** $3\sqrt{2b} + 5\sqrt{3b}$ **37.** $-2xy\sqrt{2y}$ **39.** $6ab^2\sqrt{3ab} + 3ab\sqrt{3ab}$ **41.** $-\sqrt[3]{2}$ **43.** $8b\sqrt[3]{2b^2}$ **45.** $3a\sqrt[4]{2a}$

47. $17\sqrt{2} - 15\sqrt{5}$ **49.** $5b\sqrt{b}$ **51.** $-8xy\sqrt{2x} + 2xy\sqrt{xy}$ **53.** $2y\sqrt[3]{2x}$ **55.** $-4ab\sqrt[4]{2b}$ **57.** $(-\infty, 3]$

59. $(-\infty, \infty)$

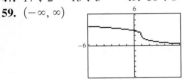

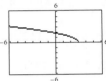

SECTION 9.3

1. If $\sqrt[n]{a}$ and $\sqrt[n]{b}$ are real numbers, then $\sqrt[n]{a} \cdot \sqrt[n]{b} = \sqrt[n]{ab}$. **3.** $3 - \sqrt{5}$ **5.** $4 + 3\sqrt{11}$ **7.** i, iii **9.** 16 **11.** $2\sqrt[3]{4}$

13. $xy^3\sqrt{x}$ **15.** $8xy\sqrt{x}$ **17.** $2x^2y\sqrt[3]{2}$ **19.** $2ab\sqrt[4]{3a^2b}$ **21.** 6 **23.** $x - \sqrt{2x}$ **25.** $4x - 8\sqrt{x}$ **27.** $-4 - \sqrt{5}$

29. $29 - 7\sqrt{7}$ **31.** 12 **33.** -3 **35.** $-10x + 14\sqrt{2x} + 3$ **37.** $x + 4\sqrt{x} + 4$ **39.** $3x - 10\sqrt{3x} + 25$

41. $2x - 8\sqrt{2x+1} + 17$ **43.** True **47.** $\sqrt[3]{4x}$ **49.** $\sqrt{3} + x$ **51.** $y\sqrt{5y}$ **53.** $b\sqrt{13b}$ **55.** $\dfrac{\sqrt{2}}{2}$ **57.** $\dfrac{2\sqrt{3y}}{3y}$

59. $\dfrac{3\sqrt{3a}}{a}$ **61.** $\dfrac{\sqrt{2y}}{2}$ **63.** $\dfrac{5\sqrt[3]{3}}{3}$ **65.** $\dfrac{5\sqrt[3]{9y^2}}{3y}$ **67.** $\dfrac{2\sqrt[4]{9x^3}}{3}$ **69.** $3x\sqrt[5]{9x^4}$ **71.** $\dfrac{b\sqrt{2a}}{2a^2}$ **73.** $\dfrac{\sqrt{15x}}{5x}$ **75.** $2 + 2\sqrt{2}$

77. $-\dfrac{12 + 4\sqrt{2}}{7}$ **79.** $-\dfrac{10 + 5\sqrt{7}}{3}$ **81.** $-\dfrac{7\sqrt{x} + 21}{x-9}$ **83.** $-\sqrt{6} + 3 - 2\sqrt{2} + 2\sqrt{3}$ **85.** $-\dfrac{17 + 5\sqrt{5}}{4}$

87. $\dfrac{8a - 10\sqrt{ab} + 3b}{16a - 9b}$ **89.** $\dfrac{3 - 7\sqrt{y} + 2y}{1 - 4y}$ **91.** $2\sqrt{2}$ **93.** $5 - 2\sqrt{6}$ **95.** $\dfrac{1}{\sqrt{9+h} + 3}$ **97.** $\dfrac{9 - 3\sqrt[3]{2} + \sqrt[3]{4}}{29}$

CHECK YOUR PROGRESS: CHAPTER 9

1. 16 [9.1A] **2.** $\dfrac{1}{8}$ [9.1A] **3.** $\dfrac{2}{3}$ [9.1A] **4.** $\dfrac{729}{512}$ [9.1A] **5.** $x^{1/4}$ [9.1A] **6.** $4x^4$ [9.1A] **7.** $z^{1/12}$ [9.1A]

8. $\dfrac{a^2c^4}{b^9}$ [9.1A] **9.** $3x^6$ [9.1C] **10.** $-2ab^3$ [9.1C] **11.** $6ab^5\sqrt{2a}$ [9.2A] **12.** $2x^2yz^3\sqrt[3]{2xy^3z^2}$ [9.2A]

13. $-4ab\sqrt{2b}$ [9.2B] **14.** $3\sqrt{2}$ [9.2B] **15.** $2x^2\sqrt{3x}$ [9.2B] **16.** $2xy\sqrt[3]{3y^2}$ [9.3A]

17. $6x^2\sqrt{2} - 6x$ [9.3A] **18.** $23 + 19\sqrt{5}$ [9.3A] **19.** $4x - 12\sqrt{x} + 9$ [9.3A]

20. $\dfrac{3\sqrt{2}}{2}$ [9.3B] **21.** $\dfrac{\sqrt[3]{6}}{2}$ [9.3B] **22.** $\dfrac{14\sqrt{3} - 21}{3}$ [9.3B] **23.** $\dfrac{2x + 4\sqrt{x}}{x-4}$ [9.3B] **24.** $\dfrac{17 + 13\sqrt{2}}{7}$ [9.3B]

SECTION 9.4

1. Sometimes true **3.** No **5.** -2 **7.** 9 **9.** No solution **11.** 35 **13.** 27 **15.** -12 **17.** 3 and 11 **19.** 9 **21.** 2

23. 2 **25.** 1 **29.** On the moon, an object will fall 24.75 ft in 3 s. **31.** The height of the water is 5.0 ft after the valve has been opened for 10 s. It takes 27.7 s to empty the tank. **33.** The length of the pendulum is 7.30 ft. **35.** The HDTV screen is approximately 7.15 in. wider. **37.** The longest pole that can be placed in the box is 5.4 ft. **39.** No

SECTION 9.5

3. Real part: 3; imaginary part: 7   **5.** Real part: 7; imaginary part: 0   **7.** $5i$ **9.** $7i\sqrt{2}$ **11.** $3+i$ **13.** $\dfrac{3}{2}-\dfrac{5\sqrt{2}}{2}i$

15. $-4+2i$ **17.** $4+8i$ **19.** $8+2i\sqrt{2}$ **21.** $-2+6i\sqrt{3}$ **23.** $-5+i\sqrt{23}$ **25.** $-4+2i\sqrt{14}$ **27.** $8-i$

29. $-8+4i$ **31.** $6-6i$ **33.** $11-4i$ **35.** $-6-2i$ **37.** The real parts of the complex numbers are additive inverses.   **39.** 63

41. -4 **43.** 29   **45.** $-4+12i$ **47.** $-3-4i$ **49.** $17-i$ **51.** $8+27i$ **53.** $-21+20i$ **55.** 1   **57.** True

59. $-3i$ **61.** $\dfrac{3}{4}+\dfrac{1}{2}i$ **63.** $\dfrac{10}{13}-\dfrac{2}{13}i$ **65.** $\dfrac{4}{5}+\dfrac{2}{5}i$ **67.** $-i$ **69.** $\dfrac{12}{17}+\dfrac{3}{17}i$ **71.** $\dfrac{3}{10}-\dfrac{11}{10}i$ **73.** $\dfrac{6}{5}+\dfrac{7}{5}i$ **75.** True

77. No   **79.** Yes   **81.** 1   **83.** i

CHAPTER 9 REVIEW EXERCISES

1. $20x^2y^2$ [9.1A]   **2.** 7 [9.4A]   **3.** $39-2i$ [9.5C]   **4.** $7x^{2/3}y$ [9.1B]   **5.** $6\sqrt{3}-13$ [9.3A]   **6.** -2 [9.4A]

7. $\dfrac{1}{x^5}$ [9.1A]   **8.** $\dfrac{8\sqrt{3y}}{3y}$ [9.3B]   **9.** $-2a^2b^4$ [9.1C]   **10.** $2a^2b\sqrt{2b}$ [9.2B]   **11.** $4+\sqrt{2}$ [9.3B]

12. $\dfrac{2}{3}-\dfrac{5}{3}i$ [9.5D]   **13.** $3ab^3\sqrt{2a}$ [9.2A]   **14.** $2+12i$ [9.5B]   **15.** $5x^3y^3\sqrt[3]{2x^2y}$ [9.2B]   **16.** $4xy^2\sqrt[3]{x^2}$ [9.3A]

17. $7+3i$ [9.5C]   **18.** $\dfrac{a^{2/3}}{2b^5}$ [9.1A]   **19.** $-2ab^2\sqrt[5]{2a^3b^2}$ [9.2A]   **20.** $-2+7i$ [9.5D]   **21.** $-6\sqrt{2}$ [9.5C]

22. 30 [9.4A]   **23.** $3a^2b^3$ [9.1C]   **24.** $\dfrac{3\sqrt[3]{9x^2}}{x}$ [9.3B]   **25.** $-12+10i$ [9.5B]   **26.** $2x+4\sqrt{2x-1}+3$ [9.3A]

27. $6x^2\sqrt{3y}$ [9.2B]   **28.** $\dfrac{1}{3}$ [9.1A]   **29.** $\dfrac{1}{a^{10}}$ [9.1A]   **30.** $-7x^3y^8$ [9.1C]   **31.** $4\sqrt[3]{a^2}$ [9.1B]   **32.** $\dfrac{x}{3}$ [9.1A]

33. $xy^2z^2\sqrt[4]{x^2z^2}$ [9.2A]   **34.** $5\sqrt{6}$ [9.2B]   **35.** $4x^2\sqrt{3xy}-4x^2\sqrt{5y}$ [9.2B]   **36.** 40 [9.3A]   **37.** $3x+3\sqrt{3x}$ [9.3A]

38. $5x\sqrt{x}$ [9.3B]   **39.** $\dfrac{-9-16\sqrt{7}}{29}$ [9.3B]   **40.** $6i$ [9.5A]   **41.** $8+6i$ [9.5A]   **42.** $-2+4i\sqrt{2}$ [9.5A]

43. $9-i$ [9.5B]   **44.** -11 [9.3A]   **45.** $-4-9i$ [9.5B]   **46.** $15-8i$ [9.5C]   **47.** $6i$ [9.5D]

48. $\dfrac{14}{5}+\dfrac{7}{5}i$ [9.5D]   **49.** 8 [9.4A]   **50.** -24 [9.4A]   **51.** The width is 5 in. [9.4B]   **52.** The amount of power is 120 watts. [9.4B]   **53.** The distance required is 242 ft. [9.4B]   **54.** The distance is 6.63 ft. [9.4B]

CHAPTER 9 TEST

1. $\dfrac{1}{2}x^{3/4}$ [9.1B; HOW TO 8]   **2.** $-2x^2y\sqrt[3]{2x}$ [9.2B; Example 3]   **3.** $3\sqrt[5]{y^2}$ [9.1B; Example 5]

4. $18+16i$ [9.5C; HOW TO 7]   **5.** $4x+12\sqrt{x}+9$ [9.3A; Example 3]   **6.** $r^{1/6}$ [9.1A; Example 3]

7. 4 [9.4A; HOW TO 3]   **8.** $2a^2b^2\sqrt[4]{2b^2}$ [9.3A; HOW TO 1]   **9.** $-4x\sqrt{3}$ [9.3A; HOW TO 2]

10. $-3+2i$ [9.5B; HOW TO 3]   **11.** $4x^2y^3\sqrt{2y}$ [9.2A; HOW TO 2]   **12.** $14+10\sqrt{3}$ [9.3A; HOW TO 3]

13. -10 [9.5C; HOW TO 5]   **14.** 2 [9.3B; Example 6]   **15.** $8a\sqrt{2a}$ [9.2B; Example 3]

16. $2a-\sqrt{ab}-15b$ [9.3A; HOW TO 3]   **17.** $\dfrac{64x^3}{y^6}$ [9.1A; HOW TO 6]   **18.** $2\sqrt[3]{25x}$ [9.3B; You Try It 5]

19. $-\dfrac{4}{5}+\dfrac{7}{5}i$ [9.5D; HOW TO 9]   **20.** -3 [9.4A; HOW TO 2]   **21.** $\dfrac{b^3}{8a^6}$ [9.1A; HOW TO 6]

22. $3abc^2\sqrt[3]{ac}$ [9.2A; Example 2]   **23.** $\dfrac{4x^2}{y}$ [9.3B; HOW TO 5]   **24.** $\sqrt{5x}$ [9.3B; HOW TO 6]

25. The distance is 576 ft. [9.4B; Example 4]

CUMULATIVE REVIEW EXERCISES

1. 92 [1.3A]   **2.** 56 [1.4A]   **3.** $-10x+1$ [1.4D]   **4.** $\dfrac{3}{2}$ [2.2A]   **5.** $\dfrac{2}{3}$ [2.2C]   **6.** $\{x \mid x>1\}$ [2.5A]

7. $\dfrac{1}{3}, \dfrac{7}{3}$ [2.6A]   **8.** $\{x \mid -6 \le x \le 3\}$ [2.6B]   **9.** The area is 187.5 cm². [3.2B]   **10.** The volume is 14 ft³. [3.3A]

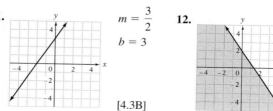

11. $m=\dfrac{3}{2}$, $b=3$ [4.3B]   **12.** [4.7A]   **13.** $y=\dfrac{1}{3}x+\dfrac{7}{3}$ [4.5B]

Unless otherwise noted, all content on this page is © Cengage Learning.

14. $(3, 2)$ [5.2A]   **15.** $2x^2y^2$ [6.1B]   **16.** $(9x + y)(9x - y)$ [7.4A]   **17.** $x(x^2 + 3)(x + 1)(x - 1)$ [7.4D]

18. $C = R - nP$ [8.6A]   **19.** $\dfrac{y^5}{x^4}$ [9.1A]   **20.** $-x\sqrt{10x}$ [9.2B]   **21.** $13 - 7\sqrt{3}$ [9.3A]   **22.** $\sqrt{6} + \sqrt{2}$ [9.3B]

23. $-\dfrac{1}{5} + \dfrac{3}{5}i$ [9.5D]   **24.** -20 [9.4A]   **25.** The length of side DE is 27 m. [8.5C]   **26.** \$2500 is invested at 3.5%. [5.1C]

27. The rate of the plane is 250 mph. [8.7B]   **28.** The time is 1.25 s. [6.1D]   **29.** The height of the periscope is 32.7 ft. [9.4B]
30. $m = 0.08$. The annual interest income is 8% of the investment. [4.4A]

Answers to Chapter 10 Selected Exercises

PREP TEST

1. $3\sqrt{2}$ [1.2F]   **2.** $3i$ [9.5A]   **3.** $\dfrac{2x - 1}{x - 1}$ [8.2B]   **4.** 8 [1.4A]   **5.** Yes [7.4A]   **6.** $(2x - 1)^2$ [7.4A]

7. $(3x + 2)(3x - 2)$ [7.4A]   **8.** [1.5B]   **9.** -3 and 5 [7.5A]   **10.** 4 [8.5A]

SECTION 10.1
1. i, iii   **5.** -4 **7.** 0 and 4   **9.** 0 and 4   **11.** -5 and 5   **13.** -2 and 3   **15.** 3   **17.** 0 and 2   **19.** -2 and 5

21. 2 and 5   **23.** $-\dfrac{3}{2}$ and 6   **25.** $\dfrac{1}{4}$ and 2   **27.** -4 and $\dfrac{1}{3}$ **29.** $-\dfrac{2}{3}$ and $\dfrac{9}{2}$ **31.** -4 and $\dfrac{1}{4}$ **33.** -2 and 9

35. -2 and $-\dfrac{3}{4}$ **37.** -5 and 2   **39.** -4 and $-\dfrac{3}{2}$ **41.** $x^2 - 7x + 10 = 0$ **43.** $x^2 + 6x + 8 = 0$ **45.** $x^2 - 5x - 6 = 0$

47. $x^2 - 9 = 0$ **49.** $x^2 - 8x + 16 = 0$ **51.** $x^2 - 5x = 0$ **53.** $2x^2 - 7x + 3 = 0$ **55.** $4x^2 - 5x - 6 = 0$

57. $3x^2 + 11x + 10 = 0$ **59.** $6x^2 - 5x + 1 = 0$ **61.** $10x^2 - 7x - 6 = 0$ **63.** $8x^2 + 6x + 1 = 0$ **65.** 0   **67.** -7 and 7

69. $-2i$ and $2i$ **71.** -2 and 2   **73.** $-\dfrac{9}{2}$ and $\dfrac{9}{2}$ **75.** $-7i$ and $7i$ **77.** $-4\sqrt{3}$ and $4\sqrt{3}$ **79.** $-3i\sqrt{2}$ and $3i\sqrt{2}$

81. -5 and 7   **83.** -7 and 3   **85.** 0 and 1   **87.** $-5 - \sqrt{6}$ and $-5 + \sqrt{6}$ **89.** $3 - 3i\sqrt{5}$ and $3 + 3i\sqrt{5}$

91. $-\dfrac{2 - 9\sqrt{2}}{3}$ and $-\dfrac{2 + 9\sqrt{2}}{3}$ **93.** Two complex solutions   **95.** Two equal real solutions   **97.** $x^2 - 2 = 0$ **99.** $x^2 + 1 = 0$

101. $x^2 - 6x + 7 = 0$ **103.** $x^2 - 10x + 26 = 0$ **105.** $-\sqrt[4]{7}$ and $\sqrt[4]{7}$ **107.** $-\sqrt[6]{2}$ and $\sqrt[6]{2}$ **109.** No   **111.** Yes   **113.** Yes

SECTION 10.2
1a. No   **b.** Yes   **c.** Yes   **d.** No   **3.** Yes   **5.** -1 and 5   **7.** 3   **9.** $-2 - \sqrt{11}$ and $-2 + \sqrt{11}$

11. $3 - \sqrt{2}$ and $3 + \sqrt{2}$ **13.** $\dfrac{3 - \sqrt{5}}{2}$ and $\dfrac{3 + \sqrt{5}}{2}$ **15.** $3 - \sqrt{13}$ and $3 + \sqrt{13}$ **17.** $\dfrac{1 - \sqrt{17}}{2}$ and $\dfrac{1 + \sqrt{17}}{2}$

19. $1 - i$ and $1 + i$ **21.** $2 - 3i$ and $2 + 3i$ **23.** $-3 - 2i$ and $-3 + 2i$ **25.** $\dfrac{2 - \sqrt{14}}{2}$ and $\dfrac{2 + \sqrt{14}}{2}$

27. $\dfrac{1}{2} - i$ and $\dfrac{1}{2} + i$ **29.** $\dfrac{1}{3} - \dfrac{1}{3}i$ and $\dfrac{1}{3} + \dfrac{1}{3}i$ **31.** $1 - \sqrt{5}$ and $1 + \sqrt{5}$ **33.** $\dfrac{1}{2}$ and 5   **35.** $2 - \sqrt{5}$ and $2 + \sqrt{5}$

37. -3.236 and 1.236 **39.** 0.293 and 1.707 **41.** $c \le 4$ **45.** -2 and 5   **47.** $4 - \sqrt{7}$ and $4 + \sqrt{7}$

49. $3 - 2\sqrt{7}$ and $3 + 2\sqrt{7}$ **51.** $7 - 3\sqrt{5}$ and $7 + 3\sqrt{5}$ **53.** $\dfrac{1 - \sqrt{3}}{2}$ and $\dfrac{1 + \sqrt{3}}{2}$ **55.** $\dfrac{5 - 2\sqrt{2}}{2}$ and $\dfrac{5 + 2\sqrt{2}}{2}$

57. $-1 - i$ and $-1 + i$ **59.** $1 - 2i$ and $1 + 2i$ **61.** $2 - 3i$ and $2 + 3i$ **63.** $\dfrac{1}{2} - 2i\sqrt{2}$ and $\dfrac{1}{2} + 2i\sqrt{2}$

65. $\dfrac{1 - 6\sqrt{2}}{3}$ and $\dfrac{1 + 6\sqrt{2}}{3}$ **67.** $\dfrac{1 - \sqrt{11}}{2}$ and $\dfrac{1 + \sqrt{11}}{2}$ **69.** -3 and 1   **71.** -8 and 1

73. $-3 - i\sqrt{5}$ and $-3 + i\sqrt{5}$ **75.** $1 - i\sqrt{5}$ and $1 + i\sqrt{5}$ **77.** $\dfrac{1}{4} - \dfrac{\sqrt{19}}{4}i$ and $\dfrac{1}{4} + \dfrac{\sqrt{19}}{4}i$ **79.** 0.394 and 7.606

81. -4.236 and 0.236 **83.** -1.351 and 1.851 **85.** Two complex number solutions   **87.** Two equal real number solutions
89. Two unequal real number solutions   **91.** $\sqrt{4ac}$ **93.** $\{p \mid p < 9\}$ **95.** $(1, \infty)$ **97.** $-3i$ and $5i$
99. No. The ball will have gone only 197.2 ft when it hits the ground.

SECTION 10.3

1. i, ii, iii, iv, v   **3.** Yes   **5.** No   **7.** $-3, -2, 2,$ and 3 **9.** $-2, -\sqrt{2}, \sqrt{2},$ and 2 **11.** 1 and 4   **13.** 16

15. $-1, 1, -2i,$ and $2i$ **17.** $-2, 2, -4i,$ and $4i$ **19.** 16   **21.** 1 and 512   **23.** $-1, -\dfrac{2}{3}, \dfrac{2}{3},$ and 1

25. Exercises 30, 31, 32, 36, 37, 38, 40, 41, 44   **27.** 3   **29.** 9   **31.** -1 and 2   **33.** 0 and 2   **35.** $-\dfrac{1}{2}$ and 2   **37.** -2 **39.** 1

41. 1   **43.** -3 **45.** $y + 2$ **47.** -1 and 10   **49.** -3 and 1   **51.** -1 and 0   **53.** -4 **55.** $-\dfrac{2}{3}$ and 6   **57.** $\dfrac{4}{3}$ and 3

59. 4   **61a.** $\{x \mid -\sqrt{29.7366} \le x \le \sqrt{29.7366}\}$ **b.**    The $\pm$ symbol occurs in the equation so that the graph pictures the entire shape of the football.   **c.** The radius is 2.7592 in.

CHECK YOUR PROGRESS: CHAPTER 10

1. $-\dfrac{2}{3}$ and 4   [10.1A]   **2.** -2 and 6   [10.1A]   **3.** $5x^2 + 12x - 9 = 0$ [10.1A]   **4.** $-3 - 2\sqrt{5}$ and $-3 + 2\sqrt{5}$ [10.1B]

5. $4 - 2i$ and $4 + 2i$ [10.1B]   **6.** $-1 - 5\sqrt{2}$ and $-1 + 5\sqrt{2}$ [10.2A]   **7.** $-\dfrac{3}{2} - i\sqrt{3}$ and $-\dfrac{3}{2} + i\sqrt{3}$ [10.2A]

8. $\dfrac{1 - 4\sqrt{2}}{2}$ and $\dfrac{1 + 4\sqrt{2}}{2}$ [10.2B]   **9.** $-4 - 3i$ and $-4 + 3i$ [10.2B]   **10.** $-\sqrt{2}, \sqrt{2}, -i\sqrt{10}, i\sqrt{10}$ [10.3A]   **11.** 24   [10.3B]

12. $-\dfrac{5}{7}$ and 4   [10.3C]

SECTION 10.4

1. $\dfrac{1}{t}$ **3.** Down: $r + 2$; up: $r - 2$ **5.** The maximum speed is 33 mph.   **7.** The rocket takes 12.5 s to return to Earth.

9. The length of a side of the square base is 177 m.   **11.** It would take the larger pipe 6 min to fill the tank. It would take the smaller pipe 12 min to fill the tank.   **13.** It would take the faster computer 2 h to run the program.   **15.** It would take the apprentice carpenter 6.8 h to install the floor.   **17.** The rate of the wind was approximately 108 mph.   **19.** The rate of the jet stream is 93 mph.   **21.** At $1600 per month, 96 units will be rented. At $9600 per month, 16 units will be rented.   **23.** The piece of cardboard needed must be 90 cm by 90 cm.   **25.** The dimensions of the work area will be 10 ft by 30 ft.   **27.** The bottom of the scoop is 2.3 in. from the bottom of the cone.

SECTION 10.5

3a. True   **b.** False   **c.** False   **5a.** True   **b.** False   **c.** True   **7.** $\{x \mid x < -2\} \cup \{x \mid x > 4\}$

9. $\{x \mid x \le 1\} \cup \{x \mid x \ge 2\}$ **11.** $\{x \mid -3 < x < 4\}$

13. $\{x \mid x < -2\} \cup \{x \mid 1 < x < 3\}$ **15.** $\{x \mid -4 \le x \le 1\} \cup \{x \mid x \ge 2\}$

17. $\{x \mid x < -2\} \cup \{x \mid x > 4\}$ **19.** $\{x \mid -1 < x \le 3\}$

21. $\{x \mid x \le -2\} \cup \{x \mid 1 \le x < 3\}$ **23.** $\{x \mid x > 4\} \cup \{x \mid x < -4\}$ **25.** $\{x \mid -3 \le x \le 12\}$

27. $\left\{x \mid \dfrac{1}{2} < x < \dfrac{3}{2}\right\}$ **29.** $\{x \mid x < 1\} \cup \left\{x \mid x > \dfrac{5}{2}\right\}$ **31.** $\{x \mid x < -1\} \cup \{x \mid 1 < x \le 2\}$ **33.** $\left\{x \mid \dfrac{1}{2} < x \le 1\right\}$

35. $\{x \mid x > 5\} \cup \{x \mid -4 < x < -1\}$ **37.**   **39.**

41.   **43.** The arrow will be more than 200 m high between 4 s and 10 s.

CHAPTER 10 REVIEW EXERCISES

1. 0 and $\dfrac{3}{2}$ [10.1A]   **2.** -2 and $\dfrac{1}{2}$ [10.1A]   **3.** $-4\sqrt{3}$ and $4\sqrt{3}$ [10.1B]   **4.** $-\dfrac{1}{2} - 2i$ and $-\dfrac{1}{2} + 2i$ [10.1B]

5. -3 and -1 [10.2A]   **6.** $\dfrac{7 - 2\sqrt{7}}{7}$ and $\dfrac{7 + 2\sqrt{7}}{7}$ [10.2A]   **7.** $\dfrac{3}{4}$ and $\dfrac{4}{3}$ [10.2B]   **8.** $\dfrac{1}{2} - \dfrac{\sqrt{31}}{2}i$ and $\dfrac{1}{2} + \dfrac{\sqrt{31}}{2}i$ [10.2B]

9. $x^2 + 3x = 0$ [10.1A]   **10.** $12x^2 - x - 6 = 0$ [10.1A]   **11.** $1 - i\sqrt{7}$ and $1 + i\sqrt{7}$ [10.2A]   **12.** $-2i$ and $2i$ [10.2A]

13. $\dfrac{11 - \sqrt{73}}{6}$ and $\dfrac{11 + \sqrt{73}}{6}$ [10.2B]   **14.** Two complex number solutions   [10.2B]   **15.** $\left\{x \mid -3 < x < \dfrac{5}{2}\right\}$ [10.5A]

Unless otherwise noted, all content on this page is © Cengage Learning.

16. $\{x|x \le -4\} \cup \left\{x\left|-\dfrac{3}{2} \le x \le 2\right.\right\}$ [10.5A]   **17.** -64 and 27 [10.3A]   **18.** $\dfrac{5}{4}$ [10.3B]   **19.** -1 and 3 [10.3C]

20. -1 [10.3C]   **21.** ⟨number line −5 to 5⟩ $\left\{x\left|x < \dfrac{3}{2}\right.\right\} \cup \{x|x \ge 2\}$ [10.5A]   **22.** ⟨number line −5 to 5⟩

$\{x|x \le -3\} \cup \left\{x\left|\dfrac{1}{2} \le x < 4\right.\right\}$ [10.5A]   **23.** 4 [10.3B]   **24.** 5 [10.3B]   **25.** $\dfrac{-3 - \sqrt{249}}{10}$ and $\dfrac{-3 + \sqrt{249}}{10}$ [10.3C]

26. $\dfrac{-11 - \sqrt{129}}{2}$ and $\dfrac{-11 + \sqrt{129}}{2}$ [10.3C]   **27.** $3x^2 + 8x - 3 = 0$ [10.1A]   **28.** -5 and $\dfrac{1}{2}$ [10.1A]

29. $-1 - 3\sqrt{2}$ and $-1 + 3\sqrt{2}$ [10.1B]   **30.** $-3 - i$ and $-3 + i$ [10.2B]   **31.** 2 and 3 [10.3C]

32. $-5, 5, -\sqrt{3}$, and $\sqrt{3}$ [10.3A]   **33.** $\dfrac{25}{18}$ [10.3B]   **34.** -8 and $\dfrac{1}{8}$ [10.3A]   **35.** 2 [10.3B]

36. $5 - 3\sqrt{2}$ and $5 + 3\sqrt{2}$ [10.2A]   **37.** $-\dfrac{4}{9}$ and 5 [10.3C]   **38.** $\dfrac{1 - \sqrt{5}}{6}$ and $\dfrac{1 + \sqrt{5}}{6}$ [10.2B]   **39.** $\dfrac{5}{4}$ [10.3B]

40. $\dfrac{-13 - \sqrt{217}}{2}$ and $\dfrac{-13 + \sqrt{217}}{2}$ [10.3C]   **41.** Two unequal real number solutions   [10.2B]   **42.** $\{x|-2 \le x \le 5\}$ [10.5A]

43. The sculling crew's rate of rowing in calm water is 6 mph.   [10.4A]   **44.** The width of the rectangle is 5 cm. The length of the rectangle is 12 cm.   [10.4A]   **45.** Working alone, the new computer can print the payroll in 12 min.   [10.4A]
46. The rate of the first car is 40 mph. The rate of the second car is 50 mph.   [10.4A]

CHAPTER 10 TEST

1. -4 and $\dfrac{2}{3}$ [10.1A; HOW TO 1]   **2.** $-\dfrac{2}{3}$ and $\dfrac{3}{2}$ [10.1A; HOW TO 1]   **3.** $x^2 - 9 = 0$ [10.1A; HOW TO 3]

4. $2x^2 + 7x - 4 = 0$ [10.1A; Example 2]   **5.** $2 - 2\sqrt{2}$ and $2 + 2\sqrt{2}$ [10.1B; Example 3]

6. $3 - \sqrt{11}$ and $3 + \sqrt{11}$ [10.2A; HOW TO 3]   **7.** $\dfrac{3 - \sqrt{15}}{3}$ and $\dfrac{3 + \sqrt{15}}{3}$ [10.2A; HOW TO 4]

8. $\dfrac{1 - \sqrt{3}}{2}$ and $\dfrac{1 + \sqrt{3}}{2}$ [10.2B; HOW TO 6]   **9.** $-2 - 2i\sqrt{2}$ and $-2 + 2i\sqrt{2}$ [10.2B; HOW TO 7]

10. $\dfrac{1}{4}$ [10.3A; HOW TO 1]   **11.** $-\sqrt{3}, -1, \sqrt{3}$, and 1 [10.3A; Example 1]   **12.** 4 [10.3B; HOW TO 2]

13. No solution   [10.3B; Example 3]   **14.** -9 and 2 [10.3C; HOW TO 3]   **15.** ⟨number line −5 to 5⟩

$\{x|x < -4\} \cup \{x|2 < x < 4\}$ [10.5A; HOW TO 2]   **16.** ⟨number line −5 to 5⟩ $\left\{x\left|-4 < x \le \dfrac{3}{2}\right.\right\}$ [10.5A; HOW TO 3]

17. Two equal real number solutions   [10.2B; HOW TO 6]   **18.** The ball hits the basket 1.88 s after it is released.   [10.4A; HOW TO 1]
19. It would take Cora 6 h to stain the bookcase.   [10.4A; Example 1]   **20.** The rate of the canoeist in calm water is 4 mph.   [10.4A; Example 2]

CUMULATIVE REVIEW EXERCISES

1. 14 [1.4A]   **2.** 54π m^3 [3.3A]   **3.** $-\dfrac{3}{2}$ [4.4A]   **4.** $y = x + 1$ [4.6A]   **5.** $-3xy(x^2 - 2xy + 3y^2)$ [7.1A]

6. $(2x - 5)(3x + 4)$ [7.3A/7.3B]   **7.** $(x + y)(x - 2)$ [7.4D]   **8.** $x^2 - 3x - 4 - \dfrac{6}{3x - 4}$ [6.4B]   **9.** $\dfrac{x}{2}$ [8.1B]

10. 16 cm   [8.5C]   **11.** $b = \dfrac{2S - an}{n}$ [8.6A]   **12.** $-8 - 14i$ [9.5C]   **13.** $1 - a$ [9.1A]   **14.** $\dfrac{x\sqrt[3]{4y^2}}{2y}$ [9.3B]

15. $-\dfrac{3}{2}$ and -1 [10.3C]   **16.** $-\dfrac{1}{2}$ [8.4A]   **17.** $-2, -\sqrt{2}, \sqrt{2}$, and 2 [10.3A]   **18.** 0 and 1 [10.3B]

19. $\left\{x\left|-2 < x < \dfrac{10}{3}\right.\right\}$ [2.6B]   **20.** $\left(\dfrac{5}{2}, 0\right)$ and $(0, -3)$ [4.3C]   **21.** ⟨graph⟩   [5.4A]

22. $(1, -1, 2)$ [5.2B]   **23.** $-\dfrac{7}{3}$ [4.2A]   **24.** $\{x|x \ne -3, 5\}$ [4.2A]

Unless otherwise noted, all content on this page is © Cengage Learning.
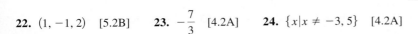

25. $\{x|x < -3\} \cup \{x|0 < x < 2\}$ 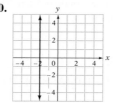 [10.5A]

26. $\{x|-3 < x \le 1\} \cup \{x|x \ge 5\}$ [10.5A]   **27.** The lower limit is $9\frac{23}{64}$ in. The upper limit is $9\frac{25}{64}$ in.   [2.6C]

28. The area is $(x^2 + 6x - 16)$ ft^2.   [6.3D]   **29.** Two complex number solutions   [10.2B]

30. $m = -\dfrac{25,000}{3}$. The building depreciates $\dfrac{\$25,000}{3}$, or about \$8333, each year.   [4.4A]

Answers to Chapter 11 Selected Exercises

PREP TEST

1. 1 [1.4A]   **2.** -7 [1.4A]   **3.** 30 [4.2A]   **4.** $h^2 + 4h - 1$ [4.2A]   **5.** $-\dfrac{2}{3}$ and 3 [10.1A]

6. $2 - \sqrt{3}$ and $2 + \sqrt{3}$ [10.2B]   **7.** $y = \dfrac{1}{2}x - 2$ [8.6A]   **8.** Domain: $\{-2, 3, 4, 6\}$; Range: $\{4, 5, 6\}$ The relation is a function.   [4.2A]

9. $\{x|x \ne 8\}$ [4.2A]   **10.**   [4.3B]

SECTION 11.1

1. i, iii   **5.** Zero, one, or two   **7.** $(-1, 0), (3, 0)$ **11.**

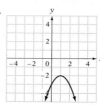

Vertex: $(1, -5)$

Axis of symmetry: $x = 1$

13.

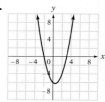

Vertex: $(1, -2)$

Axis of symmetry: $x = 1$

15.

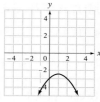

Vertex: $\left(\dfrac{1}{2}, -\dfrac{25}{4}\right)$

Axis of symmetry: $x = \dfrac{1}{2}$

17.

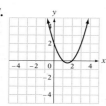

Vertex: $\left(\dfrac{3}{2}, -\dfrac{1}{4}\right)$

Axis of symmetry: $x = \dfrac{3}{2}$

19.

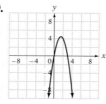

Vertex: $\left(\dfrac{3}{2}, \dfrac{9}{2}\right)$

Axis of symmetry: $x = \dfrac{3}{2}$

21.

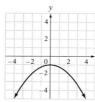

Vertex: $(0, -1)$

Axis of symmetry: $x = 0$

23.

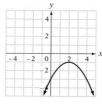

Vertex: $(2, -1)$

Axis of symmetry: $x = 2$

25.

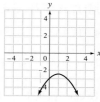

Vertex: $\left(1, -\dfrac{5}{2}\right)$

Axis of symmetry: $x = 1$

27. Domain: $\{x|-\infty < x < \infty\}$
Range: $\{y|y \ge 2\}$ **29.** Domain: $\{x|-\infty < x < \infty\}$
Range: $\{y|y \ge -5\}$ **31.** Domain: $\{x|-\infty < x < \infty\}$
Range: $\{y|y \le 0\}$ **33.** Domain: $\{x|-\infty < x < \infty\}$
Range: $\{y|y \ge -7\}$

35. $(3, 0)$ and $(-3, 0)$ **37.** $(0, 0)$ and $(-2, 0)$ **39.** $(4, 0)$ and $(-2, 0)$ **41.** $\left(-\dfrac{1}{2}, 0\right)$ and $(3, 0)$

43. $\left(-2 + \sqrt{7}, 0\right)$ and $\left(-2 - \sqrt{7}, 0\right)$ **45.** No x-intercepts   **47.** -1 and 5   **49.** $-\dfrac{4}{3}$ and 2   **51.** $\dfrac{1}{2}$ **53.** 0 and $\dfrac{4}{3}$

55. -2 and 2   **57.** $-3\sqrt{3}$ and $3\sqrt{3}$ **59.** $1 - 3\sqrt{2}$ and $1 + 3\sqrt{2}$ **61.** $-2 - i$ and $-2 + i$ **63.** $-2 - 3i$ and $-2 + 3i$

Unless otherwise noted, all content on this page is © Cengage Learning.

65. Two   **67.** One   **69.** No x-intercepts   **71.** Two   **73.** No x-intercepts   **75.** Two   **77a.** $a > 0$ **b.** $a = 0$ **c.** $a < 0$

79. Minimum value: -2 **81.** Maximum value: -3 **83.** Maximum value: $\dfrac{9}{8}$ **85.** Minimum value: $-\dfrac{11}{4}$ **87.** Maximum value: $\dfrac{9}{4}$

89. Minimum value: $-\dfrac{1}{12}$ **91.** A price of $250 will give the maximum revenue.   **93.** The numbers are 10 and 10.

95. The vertex of the parabola is approximately (42, 8500), so the maximum height of the plane is about 8500 m.

97. The stream reaches its maximum height after 0.9375 s. The maximum height is 14.0625 ft.   **99.** The manager should rent 56 units. The maximum revenue is $313,600.   **101.** The length is 100 ft and the width is 50 ft.   **103.** The dimensions are 350 ft by 525 ft.

105. -2 and 3   **107.** $h = r = \dfrac{50}{\pi + 4}$, or about 7.00 ft

SECTION 11.2

1. left   **3.** up   **5.** $(-3, 5)$ **7.** 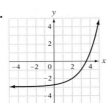   **9.** 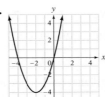   **11.**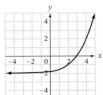

13. 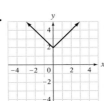   **15.** 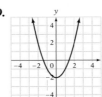   **17.** 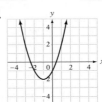   **19.**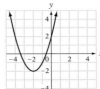

21. 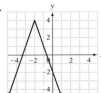   **23.** $(0, 9)$ **25.** 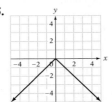   **27.** 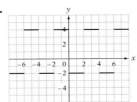   **29.**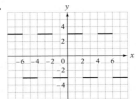

31. 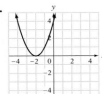   **33.** 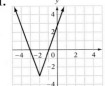   **35.** 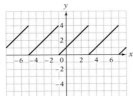   **37.**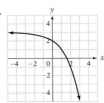

39. 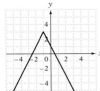   **41.**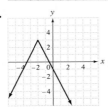

CHECK YOUR PROGRESS: CHAPTER 11

1. 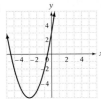   **2.** 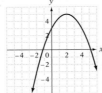   **3.** $(-3, 0), \left(\dfrac{1}{2}, 0\right)$ [11.1B]

Vertex: $(-3, -6)$
Axis of symmetry: $x = -3$ [11.1A]

Vertex: $(2, 5)$
Axis of symmetry: $x = 2$ [11.1A]

Unless otherwise noted, all content on this page is © Cengage Learning.

4. $(1 - \sqrt{2}, 0)$ and $(1 + \sqrt{2}, 0)$ [11.1B] **5.** -6 and 5 [11.1B] **6.** $2 - 6i$ and $2 + 6i$ [11.1B] **7.** -36 [11.1C]
8. 4 [11.1C] **9.** [11.2A] **10.** [11.2A] **11.** [11.2B]

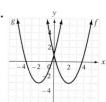

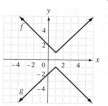

12. [11.2B] **13.** The maximum height the rock will attain is 140 ft. [11.1D]

SECTION 11.3

1a. Yes **b.** Yes **c.** Yes **d.** No **3a.** No **b.** Yes **5.** 5 **7.** 1 **9.** 0 **11.** $-\dfrac{29}{4}$ **13.** 2 **15.** 39 **17.** -8
19. Undefined **21.** -2 **23.** 7 **25.** -13 **27.** -29 **29.** $8x - 13$ **31.** 4 **33.** 3 **35.** $x + 4$ **37.** 1 **39.** 5
41. $x^2 + 1$ **43.** 5 **45.** 11 **47.** $3x^2 + 3x + 5$ **49.** -3 **51.** -27 **53.** $x^3 - 6x^2 + 12x - 8$
55a. $S[M(x)] = 80 + \dfrac{16{,}000}{x}$ **b.** \$83.20 **c.** When 5000 digital cameras are manufactured, the camera store sells each camera for \$83.20.
57a. $I[n(m)] = 50{,}000m$ **b.** \$150,000 **c.** The garage's income from doing conversions for 3 months is \$150,000.
59a. $d[r(p)] = 0.90p - 1350$ **b.** $r[d(p)] = 0.90p - 1500$ **c.** $r[d(p)]$ (The final price is less.) **61.** 0 **63.** -2 **65.** 7
67. $h^2 + 6h$ **69.** $2 + h$ **71.** $2a + h$ **73.** -1 **75.** -6 **77.** $6x - 13$

SECTION 11.4

3a. Yes **b.** No **7.** ii **9.** Yes **11.** No **13.** Yes **15.** No **17.** No **19.** No **21.** $\{(0, 1), (3, 2), (8, 3), (15, 4)\}$
23. No inverse **25.** $\{(-2, 0), (5, -1), (3, 3), (6, -4)\}$ **27.** No inverse **29.** $f^{-1}(x) = \dfrac{1}{4}x + 2$ **31.** $f^{-1}(x) = \dfrac{1}{2}x - 2$
33. $f^{-1}(x) = 2x + 2$ **35.** $f^{-1}(x) = -\dfrac{1}{2}x + 1$ **37.** $f^{-1}(x) = \dfrac{3}{2}x - 6$ **39.** $f^{-1}(x) = -3x + 3$ **41.** $f^{-1}(x) = \dfrac{1}{2}x + \dfrac{5}{2}$
43. $f^{-1}(x) = \dfrac{1}{5}x + \dfrac{2}{5}$ **45.** $f^{-1}(x) = \dfrac{1}{6}x + \dfrac{1}{2}$ **47.** $\dfrac{5}{3}$ **49.** 3 **51.** Yes; Yes **53.** Yes **55.** No **57.** Yes **59.** Yes
61. $f^{-1}(x) = 16x$; The inverse function converts pounds to ounces. **63.** $f^{-1}(x) = x - 30$; The inverse function converts a dress size in France
to a dress size in the United States. **65.** $f^{-1}(x) = \dfrac{1}{90}x - \dfrac{13}{18}$; The inverse function gives the intensity of training percent for a given target heart
rate for the athlete. **67.** $f^{-1}(x) = 0.008307x$ **73.** **75.** **77.**

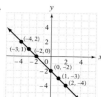

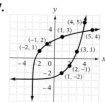

CHAPTER 11 REVIEW EXERCISES

1. [11.2A] **2.** Yes [11.4A] **3.** [11.2A] **4.** [11.2B]

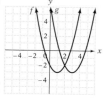

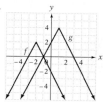

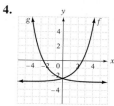

Unless otherwise noted, all content on this page is © Cengage Learning.

5. [11.2B] 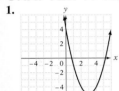 **6.** 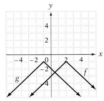 **7.** Two x-intercepts [11.1B]   **8.** No x-intercepts [11.1B]

Vertex: $(1, 2)$
Axis of symmetry: $x = 1$ [11.1A]

9. $(0, 0)$ and $(-3, 0)$ [11.1B]   **10.** $(3 - \sqrt{2}, 0)$ and $(3 + \sqrt{2}, 0)$ [11.1B]   **11.** $-\dfrac{3}{2}$ and 5 [11.1B]   **12.** $1 - 3i$ and $1 + 3i$ [11.1B]

13. 3 [11.1C]   **14.** $-\dfrac{17}{4}$ [11.1C]   **15.** 5 [11.3B]   **16.** 10 [11.3B]   **17.** $12x^2 + 12x - 1$ [11.3B]

18. $6x^2 + 3x - 16$ [11.3B]   **19.** 7 [11.3A]   **20.** -9 [11.3A]   **21.** 70 [11.3A]   **22.** $\dfrac{12}{7}$ [11.3A]

23. $f^{-1}(x) = -\dfrac{1}{6}x + \dfrac{2}{3}$ [11.4B]   **24.** $f^{-1}(x) = \dfrac{3}{2}x + 18$ [11.4B]   **25.** Yes [11.4B]   **26.** No [11.4B]

27. $p^{-1}(x) = 2.5x - 37.5$; The inverse function gives a diver's depth below the surface of the water for a given pressure on the diver. [11.4B]
28. Fifty gloves should be made each month. The maximum monthly profit is $5000. [11.1D]   **29.** The dimensions are 7 ft by 7 ft. [11.1D]

CHAPTER 11 TEST

1. 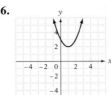 **2.** **3.** 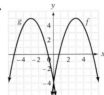 **4.**

Vertex: $(3, -5)$
Axis of symmetry: $x = 3$
[11.1A; HOW TO 3]   [11.2A; HOW TO 2]   [11.2A; Example 1]   [11.2B; You Try It 2]

5. No x-intercepts [11.1B; HOW TO 7]   **6.** $\left(-\dfrac{2}{3}, 0\right)$ and $(3, 0)$ [11.1B; HOW TO 4]   **7.** 9 [11.1C; HOW TO 8]

8. Domain: $\{x|-\infty < x < \infty\}$; Range: $\{y \mid y \geq -7\}$ [11.1A; You Try It 2]   **9.** -2 [11.3A; Example 1]

10. 234 [11.3A; Example 2]   **11.** $-\dfrac{13}{2}$ [11.3A; Example 3]   **12.** -5 [11.3A; Example 1]   **13.** -17 [11.3B; Example 4]

14. 31 [11.3B; Example 4]   **15.** $3x^2 - 1$ [11.3B; HOW TO 4]   **16.** $f[g(x)] = 2x^2 - 4x - 5$ [11.3B; HOW TO 4]

17. No [11.4B; HOW TO 1]   **18.** $\{(6, 2), (5, 3), (4, 4), (3, 5)\}$ [11.4B; HOW TO 1]   **19.** $f^{-1}(x) = \dfrac{1}{4}x + \dfrac{1}{2}$ [11.4B; HOW TO 2]

20. $f^{-1}(x) = 4x + 16$ [11.4B; You Try It 2]   **21.** Yes [11.4B; Example 3]   **22.** Yes [11.4B; Example 3]
23. No [11.4A; Example 1]   **24.** $C^{-1}(x) = 0.8x - 4$; The inverse function gives the number of miles to the location for a given cost.
[11.4B; Example 4]   **25.** Twenty-five speakers should be made. The minimum daily production cost is $50. [11.1D; Example 7]
26. 14 and 14 [11.1D; Example 8]   **27.** Dimensions of 50 cm by 50 cm will give a maximum area of 2500 cm^2. [11.1D; HOW TO 9]

CUMULATIVE REVIEW EXERCISES

1. $-\dfrac{23}{4}$ [1.4A]   **2.** [1.5B]   **3.** 3 [2.2C]   **4.** $\{x|x < -2\} \cup \{x|x > 3\}$ [2.5B]

5. The set of all real numbers [2.6B]   **6.** $-\dfrac{a^{10}}{12b^4}$ [6.1B]   **7.** $2x^3 - 4x^2 - 17x + 4$ [6.3B]   **8.** $\left(\dfrac{1}{2}, 3\right)$ [5.2A]

9. $xy(x + 3y)(x - 2y)$ [7.2B]   **10.** -3 and 8 [7.5A]   **11.** $\{x|x < -3\} \cup \{x|x > 5\}$ [10.5A]

12. $\dfrac{5}{2x - 1}$ [8.2B]   **13.** -2 [8.4A]   **14.** $-3 - 2i$ [9.5D]   **15.** 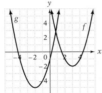 Vertex: $(0, 0)$ [11.1A]
Axis of symmetry: $x = 0$

Unless otherwise noted, all content on this page is © Cengage Learning.

16. 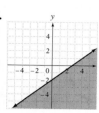 [4.7A]   **17.** $y = -2x - 2$ [4.5B]   **18.** $y = -\dfrac{3}{2}x - \dfrac{7}{2}$ [4.6A]

19. $\dfrac{1}{2} - \dfrac{\sqrt{3}}{6}i$ and $\dfrac{1}{2} + \dfrac{\sqrt{3}}{6}i$ [10.2B]   **20.** 3   [10.3B]   **21.** -3 [11.1C]   **22.** $\dfrac{4}{3}$ [4.3C]   **23.** Yes   [4.2A]

24. 2   [9.4A]   **25.** 10   [11.3B]   **26.** $f^{-1}(x) = -\dfrac{1}{3}x + 3$ [11.4B]   **27.** The cost per pound of the mixture is $3.96.   [2.4A]

28. 25 lb of the 80% copper alloy must be used.   [2.4B]   **29.** An additional 4.5 oz of insecticide are required.   [8.5B]

30. It would take the larger pipe 4 min to fill the tank.   [10.4A]   **31.** A force of 40 lb will stretch the spring 24 in.   [8.8A]

32. The frequency is 80 vibrations/min.   [8.8A]

Answers to Chapter 12 Selected Exercises

PREP TEST

1. $\dfrac{1}{9}$ [6.1B]   **2.** 16   [6.1B]   **3.** -3 [6.1B]   **4.** 0; 108   [4.2A/6.2A]   **5.** -6 [2.2B]   **6.** -2 and 8   [10.1A]

7. 6326.60   [1.4A]   **8.** 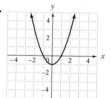   [11.1A]

SECTION 12.1

3. iii, v, vi   **5a.** 9   **b.** 1   **c.** $\dfrac{1}{9}$ **7a.** 16   **b.** 4   **c.** $\dfrac{1}{4}$ **9a.** 1   **b.** $\dfrac{1}{8}$ **c.** 16   **11a.** 7.3891   **b.** 0.3679

c. 1.2840   **13a.** 54.5982   **b.** 1   **c.** 0.1353   **15a.** 16   **b.** 16   **c.** 1.4768   **17a.** 0.1353   **b.** 0.1353   **c.** 0.0111

19. $f(a) > f(b)$ **21.** 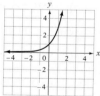   **23.** 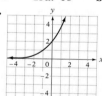   **25.** 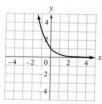   **27.**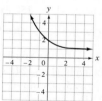

29. 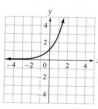   **31.** 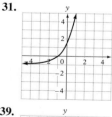   **33.** The speed of the object after 4 s is 31.41 ft/s.   **35.** i and iii, ii and iv

37. 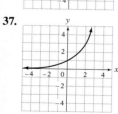   **39.** 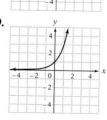   **41.** e

SECTION 12.2

3. $\log_5 25 = 2$ **5.** $\log_4\left(\dfrac{1}{16}\right) = -2$ **7.** $3^4 = 81$ **9.** $e^q = p$ **11.** False   **13.** True   **15.** True   **17.** 4   **19.** 7   **21.** 2

23. 3   **25.** 0   **27.** 4   **29.** 9   **31.** 64   **33.** $\dfrac{1}{7}$ **35.** 1   **37.** 316.23   **39.** 0.02   **41.** 7.39   **43.** 0.61   **45.** 1

49. False   **51.** 0   **53.** 1   **55.** x **57.** v **59.** $x^2 + 1$ **61.** $x^2 - x - 1$ **63.** $\log_8 x + \log_8 z$ **65.** $5 \log_3 x$

67. $\log_b r - \log_b s$ **69.** $2 \log_3 x + 6 \log_3 y$ **71.** $3 \log_7 u - 4 \log_7 v$ **73.** $2 \log_2 r + 2 \log_2 s$ **75.** $2 \ln x + \ln y + \ln z$

77. $\log_5 x + 2 \log_5 y - 4 \log_5 z$ **79.** $2 \log_8 x - \log_8 y - 2 \log_8 z$ **81.** $\dfrac{3}{2} \log_4 x + \dfrac{1}{2} \log_4 y$ **83.** $\dfrac{3}{2} \log_7 x - \dfrac{1}{2} \log_7 y$

85. $\log_3 t - \dfrac{1}{2} \log_3 x$ **87.** $\log_3 (x^3 y^2)$ **89.** $\ln \dfrac{x^4}{y^2}$ **91.** $\log_7 x^3$ **93.** $\ln(x^3 y^4)$ **95.** $\log_4 (x^2 y^2)$ **97.** $\log_3 \dfrac{x^2 z^2}{y}$ **99.** $\ln \dfrac{x}{y^2 z}$

101. $\log_6 \sqrt{\dfrac{x}{y}}$ **103.** $\log_4 \dfrac{s^2 r^2}{t^4}$ **105.** $\ln \dfrac{x}{y^2 z^2}$ **107.** $\log_4 \sqrt{\dfrac{x^3 z}{y^2}}$ **109.** $\log_2 \dfrac{\sqrt{xz}}{\sqrt[3]{y^2}}$ **111.** 0.8617   **113.** 2.1133   **115.** -0.6309

117. 0.2727   **119.** 1.6826   **121.** 1.9266   **123.** 0.6752   **125.** 2.6125   **127.** $\log_5 x = \dfrac{\log x}{\log 5}$ **129.** 19,683   **131.** 3

133. Because $x = 4$, $x - 5 = -1$. The logarithm of a negative number is undefined.   **135a.** 2.3219281   **b.** Less   **c.** Less
d. 0; Because this system contains only one species, there is no diversity in the system.

SECTION 12.3

3. $x = 2^{y/3}$ **5.** $x = e^{(y - 2)/3}$

7. 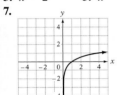   **9.** 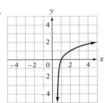   **11.** 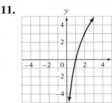   **13.**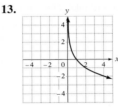

15. 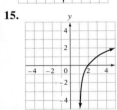   **17.** 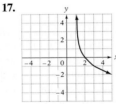   **19.** $(1, 0)$ **21.** 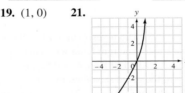   **23.**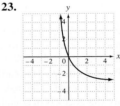

25a. 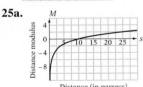   **b.** The ordered pair (25.1, 2) means that a star that is 25.1 parsecs from Earth has a distance modulus of 2.
c. The distance is 6.3 parsecs.

CHECK YOUR PROGRESS: CHAPTER 12

1. 81 [12.1A]   **2.** $\dfrac{1}{8}$ [12.1A]   **3.** $\dfrac{1}{4}$ [12.1A]   **4.** [12.1B]   **5.** [12.3A]

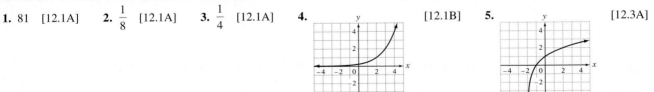

6. 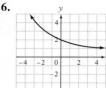 [12.1B]   **7.** 4 [12.2A]   **8.** −3 [12.2A]   **9.** −1 [12.2A]   **10.** 33 [12.2B]   **11.** 625 [12.2A]

12. $\dfrac{1}{27}$ [12.2A]   **13.** 0 [12.2A]   **14.** 0.0001 [12.2A]   **15.** $2\log_7 x + 5\log_7 y$ [12.2B]   **16.** $\log_8 x - 3\log_8 y$ [12.2B]

17. $2\log_3 x - \dfrac{1}{2}\log_3 y - \dfrac{1}{2}\log_3 z$ [12.2B]   **18.** $\log_3 \dfrac{x^3}{y^4}$ [12.2B]   **19.** $\ln \dfrac{xz^5}{y^2}$ [12.2B]   **20.** $\log\sqrt{xy}$ [12.2B]   **21.** 2.2619 [12.2C]

22. −1.4307 [12.2C]   **23.** 0.8271 [12.2C]

SECTION 12.4

3. $x < 0$ **5.** $-\dfrac{1}{3}$ **7.** −3   **9.** 1   **11.** 6   **13.** $\dfrac{4}{7}$ **15.** $\dfrac{9}{8}$ **17.** 1.1133   **19.** −1.7626   **21.** 0.7740   **23.** −2.8074

25. 3.5850   **27.** 1.1309   **29.** −2.5850   **31.** 1.2224   **33.** 4.1166   **35.** 4.4629   **37.** $\dfrac{1}{2}$ **39.** No solution   **41.** 3

43. 2   **45.** $\dfrac{11}{2}$ **47.** 17.5327   **49.** 1   **51.** $\dfrac{5}{3}$ **53.** No solution   **55.** $x - 2 < x$, and therefore $\log(x - 2) < \log x$.
This means that $\log(x - 2) - \log x < 0$ and cannot equal the positive number 3.   **57.** −6.1285   **59.** 3.0932
61a. 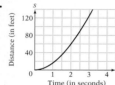   **b.** It will take 2.64 s for the object to fall 100 ft.

SECTION 12.5

3. The value of the investment after 2 years is $1172.   **5.** The investment will be worth $15,000 in approximately 18 years.
7a. After 3 h, the technetium level is 21.2 mg.   **b.** The technetium level will reach 20 mg after 3.5 h.   **9.** The half-life is 2.5 years.
11. The intensity of the earthquake was $794{,}328{,}235 I_0$.   **13.** The Honshu earthquake was approximately 3.2 times stronger than the Quetta earthquake.   **15.** The magnitude of the earthquake for the given seismogram is 5.3.   **17.** The magnitude of the earthquake for the given
seismogram is 5.6.   **19.** The pH of the baking soda solution is 8.4.   **21.** The range of hydrogen ion concentrations is 2.5×10^{-7} to 5.0×10^{-6}.
23. A blue whale emits 188 decibels of sound.   **25.** The sound intensity of a cat's purr is 3.16×10^{-14} watts/cm^2.   **27.** 79.4% of the light will
pass through the glass.   **29.** The thickness of the copper is 0.4 cm.   **31a.** Decay   **b.** Growth   **c.** Decay   **d.** Growth
33a. The value of the investment after 3 years is $5986.09.   **b.** The investor needs an annual interest rate of 11.2%.   **35a.** When $v = 160$,
$f(375) \approx 0.43$ ft.   **b.** When $v = 166.4, f(375) \approx 24.4$ ft.   **c.** The greatest distance the fence could be placed is 385 ft from home plate.

CHAPTER 12 REVIEW EXERCISES

1. 1 [12.1A]   **2.** $5^2 = 25$ [12.2A]   **3.** 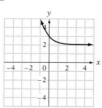 [12.1B]   **4.** 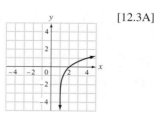 [12.3A]

5. $\dfrac{2}{5}\log_3 x + \dfrac{4}{5}\log_3 y$ [12.2B]   **6.** $\log_3 \dfrac{x^2}{y^5}$ [12.2B]   **7.** −12 [12.4A]   **8.** $\dfrac{1}{4}$ [12.4B]   **9.** 1.7251 [12.2C]

10. 32 [12.2A]   **11.** 79 [12.4B]   **12.** 1000 [12.2A]   **13.** $\log_7 \sqrt[3]{xy^4}$ [12.2B]   **14.** $\dfrac{5}{2}\log_8 x - 3\log_8 y$ [12.2B]

15. $\log_2 32 = 5$ [12.2A]   **16.** 0.4278 [12.2C]   **17.** −0.535 [12.4A]   **18.** $\dfrac{3}{2}$ [12.1A]   **19.** $\dfrac{11}{7}$ [12.4B]

Unless otherwise noted, all content on this page is © Cengage Learning.

20. 3 [12.4B]   **21.** 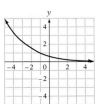 [12.1B]   **22.**   [12.3A]   **23.** 2 [12.2A]

24. $\log_2 \sqrt[3]{\dfrac{x}{y}}$ [12.2B]   **25.** 1 [12.4A]   **26.** 1.5938 [12.4A]   **27.** $\dfrac{1}{5}$ [12.2A]   **28.** $\log_3 81 = 4$ [12.2A]

29. 5 [12.4B]   **30.** $\dfrac{2}{3}\log_5 x + \dfrac{1}{3}\log_5 y$ [12.2B]   **31.** -0.5207 [12.4A]   **32.** $\dfrac{1}{7}$ [12.1A]   **33.** 4 [12.2A]

34. 2 [12.4B]   **35.** 2.3219 [12.2C]   **36.** 3 [12.4A]   **37.** 625 [12.2A]   **38.** $\log_b \dfrac{x^3}{y^7}$ [12.2B]

39. 5 [12.1A]   **40.** 3.2091 [12.4A]   **41.** The value of the investment after 2 years is \$4692. [12.5A]
42. The Richter scale magnitude of the earthquake is 9.2. [12.5A]   **43.** The half-life is 27 days. [12.5A]
44. The sound emitted from the busy street corner is 107 decibels. [12.5A]

CHAPTER 12 TEST

1. $f(0) = 1$ [12.1A; Example 1]   **2.** $f(-2) = \dfrac{1}{3}$ [12.1A; Example 2]   **3.** 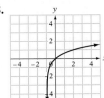 [12.1B; Example 5]

4.   [12.1B; You Try It 5]   **5.** 2 [12.2A; You Try It 1]   **6.** $\dfrac{1}{9}$ [12.2A; HOW TO 4]

7.   [12.3A; You Try It 2]   **8.**   [12.3A; Example 3]

9. $\dfrac{1}{2}\log_6 x + \dfrac{3}{2}\log_6 y$ [12.2B; You Try It 4]   **10.** $\log_3 \sqrt{\dfrac{x}{y}}$ [12.2B; Example 5]   **11.** $\ln x - \dfrac{1}{2}\ln z$ [12.2B; HOW TO 11]

12. $\ln \dfrac{x^3}{y\sqrt{z}}$ [12.2B; HOW TO 12]   **13.** -2 [12.4A; HOW TO 1]   **14.** -3 [12.4A; HOW TO 2]

15. 2.5789 [12.4A; HOW TO 3]   **16.** 6 [12.4B; You Try It 4]   **17.** 3 [12.4B; HOW TO 7]
18. 1.3652 [12.2C; HOW TO 14]   **19.** 2.6801 [12.2C; HOW TO 14]   **20.** 3.1826 [12.4A; HOW TO 4]
21. 6.0163 [12.4A; Example 2]   **22.** 5 [12.4B; Example 5]   **23.** The shard is 3099 years old. [12.5A; HOW TO 3]
24. The intensity of the sound is 3.16×10^{-9} watts/cm². [12.5A; Example 1]   **25.** The half-life is 33 h. [12.5A; HOW TO 2]

CUMULATIVE REVIEW EXERCISES

1. $\dfrac{8}{7}$ [2.2C]   **2.** $y = 2x - 6$ [4.6A]   **3.** $(4x^2 + 3)(x^2 + 1)$ [7.3A/7.3B]   **4.** $\dfrac{x-3}{x+3}$ [8.3A]   **5.** $\dfrac{x\sqrt{y} + y\sqrt{x}}{x - y}$ [9.3B]

6. $2 - \sqrt{10}$ and $2 + \sqrt{10}$ [10.2A]   **7.** $3x^2 + 8x - 3 = 0$ [10.1A]   **8.** 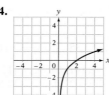   [5.4A]   **9.** $(0, -1, 2)$ [5.2B]

10. $-\dfrac{2x^2 - 17x + 13}{(x - 2)(2x - 3)}$ [8.2B]   **11.** $\{x | -5 \le x \le 1\}$ [10.5A]   **12.** $\{x | 1 \le x \le 4\}$ [2.6B]

13.   [12.1B]   **14.** 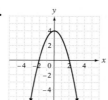   [12.3A]   **15.** 4   [12.1A]

16. 125   [12.2A]   **17.** $\log_b \dfrac{x^3}{y^5}$ [12.2B]   **18.** 1.7712   [12.2C]   **19.** 2   [12.4A]   **20.** $\dfrac{1}{2}$ [12.4B]

21. The customer can write at most 49 checks.   [2.5C]   **22.** The cost per pound of the mixture is \$3.10.   [2.4A]
23. The rate of the wind is 25 mph.   [8.7B]   **24.** The spring will stretch 10.2 in.   [8.8A]   **25.** The cost of redwood is \$.40 per foot. The cost of fir is \$.25 per foot.   [5.3B]   **26.** In approximately 10 years, the investment will be worth \$10,000.   [12.5A]

FINAL EXAM

1. -31 [1.3A]   **2.** -1 [1.4A]   **3.** $-10x + 33$ [1.4D]   **4.** 8   [2.2A]   **5.** $4, -\dfrac{2}{3}$ [2.6A]   **6.** The volume is

268.1 ft^3.   [3.3A]   **7.** 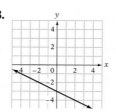   [4.3C]   **8.** $y = -3x + 7$ [4.5B]   **9.** $y = -\dfrac{2}{3}x - \dfrac{1}{3}$ [4.6A]

10. $6a^3 - 5a^2 + 10a$ [6.3A]   **11.** $(2 - xy)(4 + 2xy + x^2y^2)$ [7.4B]   **12.** $(x - y)(1 + x)(1 - x)$ [7.4D]

13. $x^2 - 2x - 3 - \dfrac{5}{2x - 3}$ [6.4B]   **14.** $\dfrac{x(x - 1)}{2x - 5}$ [8.1C]   **15.** $\dfrac{-10x}{(x + 2)(x - 3)}$ [8.2B]   **16.** $\dfrac{x + 3}{x + 1}$ [8.3A]

17. $-\dfrac{7}{4}$ [8.4A]   **18.** $d = \dfrac{a_n - a_1}{n - 1}$ [8.6A]   **19.** $\dfrac{y^4}{162x^3}$ [6.1B]   **20.** $\dfrac{1}{64x^8y^5}$ [9.1A]   **21.** $-2x^2y\sqrt{2y}$ [9.2B]

22. $\dfrac{x^2\sqrt{2y}}{2y^2}$ [9.3B]   **23.** $\dfrac{6}{5} - \dfrac{3}{5}i$ [9.5D]   **24.** $2x^2 - 3x - 2 = 0$ [10.1A]   **25.** $\dfrac{3 + \sqrt{17}}{4}, \dfrac{3 - \sqrt{17}}{4}$ [10.2B]

26. $-8, 27$ [10.3A]   **27.** [11.1A]   **28.** [4.3A]

29. $-2, \dfrac{3}{2}$ [10.3C]   **30.** $f^{-1}(x) = \dfrac{3}{2}x + 6$ [11.4B]   **31.** $(3, 4)$ [5.2A]   **32.** $7x^3$ [9.1C]

33. $\left\{x | x > \dfrac{3}{2}\right\}$ [2.5B]   **34.** $\{x | -4 < x < -1\}$ [2.6B]

Unless otherwise noted, all content on this page is © Cengage Learning.

35. 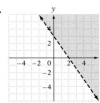 [4.7A] **36.** 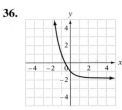 [12.1B] **37.** 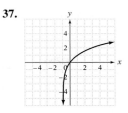 [12.3A]

38. $\log_2 \dfrac{a^2}{b^2}$ [12.2B] **39.** 6 [12.4B] **40.** The range of scores is $69 \le x \le 100$. [2.5C]

41. The cyclist rode 40 mi. [2.4C] **42.** There is $8000 invested at 8.5% and $4000 invested at 6.4%. [5.1C]
43. The length is 20 ft. The width is 7 ft. [10.4A] **44.** An additional 200 shares are needed. [8.5A] **45.** The rate of
the plane was 420 mph. [8.7B] **46.** The object has fallen 88 ft when the speed reaches 75 ft/s. [9.4B] **47.** The rate
of the plane for the first 360 mi was 120 mph. [8.7B] **48.** The intensity is 200 foot-candles. [8.8A] **49.** The rate of
the boat in calm water is 12.5 mph. The rate of the current is 2.5 mph. [5.3A] **50.** The value of the investment after
2 years is $4785.65. [12.5A]

Answers to Chapter T Selected Exercises

PREP TEST
1. $-37 < -28$ [1.1A] **2.** 0.09 [1.2A] **3.** $-\dfrac{19}{60}$ [1.2C] **4.** 72 [1.2C] **5.** -5.11 [1.2D] **6.** 128 [1.2E]

7. 2618 [1.4A] **8.** $5x^2 - 9x - 6$ [1.4B] **9.** 4 [2.2A] **10.** $6x$ [7.1A]

SECTION T.1
1. 10 **3.** 21 **5.** -2 **7.** 5 **9.** -3 **11.** -1 **13.** 1 **15.** 18 **17.** -6 **19.** 2 **21.** -40 **23.** 2.01

25. $8x$ **27.** $3b$ **29.** $5a$ **31.** $11x$ **33.** $15x - 2y$ **35.** $9a$ **37.** $22y^2$ **39.** $\dfrac{1}{2}x$ **41.** $-20x$ **43.** $-24a$

45. x **47.** $7x$ **49.** $3y$ **51.** $-3a - 15$ **53.** $-16x - 48$ **55.** $-10y^2 + 5$ **57.** $18x^2 - 12xy - 6y^2$
59. $-8y - 7$ **61.** $5x - 75$ **63.** $-a + 56$

SECTION T.2
1. -12 **3.** -1 **5.** -8 **7.** -20 **9.** 8 **11.** -1 **13.** 1 **15.** $\dfrac{3}{4}$ **17.** 3 **19.** 3 **21.** 2 **23.** -1

25. $\dfrac{1}{2}$ **27.** 5 **29.** $-\dfrac{1}{9}$ **31.** $\{x|x > 3\}$ **33.** $\{n|n \ge 2\}$ **35.** $\{x|x \le -3\}$ **37.** $\{n|n > 0\}$ **39.** $\{x|x > 4\}$

41. $\{x|x > -2\}$ **43.** $\{x|x > 3\}$ **45.** $\{x|x \ge 2\}$ **47.** $\{x|x < 2\}$ **49.** $\{x|x > 3\}$ **51.** $\{x|x \ge 2\}$ **53.** $\{x|x < 4\}$

55. $\left\{x \middle| x > \dfrac{14}{11}\right\}$ **57.** $\left\{x \middle| x \le \dfrac{3}{8}\right\}$ **59.** $\left\{x \middle| x \ge -\dfrac{5}{4}\right\}$

SECTION T.3
1.   **3.** 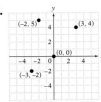 **5.** 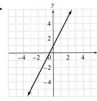 **7.**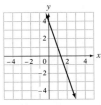

9. 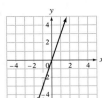 **11.** 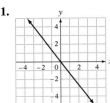 **13.** 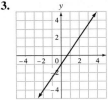 **15.**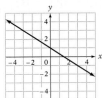

Unless otherwise noted, all content on this page is © Cengage Learning.

17. 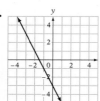 **19.** 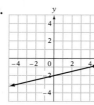 **21.** 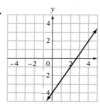 **23.** 3 **25.** $\dfrac{1}{4}$ **27.** $-\dfrac{3}{2}$

29. Undefined **31.** 0 **33.** 1 **35.** $m = \dfrac{5}{2}, b = (0, -4)$ **37.** $m = 1, b = (0, 0)$ **39.**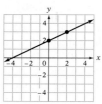

41. **43.** 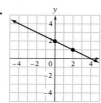 **45.** 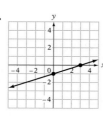 **47.** $13; (-4, 13)$ **49.** $14; (-1, 14)$

51. $-1; (2, -1)$ **53.** $-15; (-2, -15)$ **55.** $-2; (-1, -2)$ **57.** $y = -3x - 1$ **59.** $y = -2x + 3$ **61.** $y = -\dfrac{3}{5}x$

63. $y = -\dfrac{2}{3}x - 4$

SECTION T.4

1. z^8 **3.** x^{15} **5.** $x^{12}y^{18}$ **7.** a^6 **9.** $-m^9n^3$ **11.** $-8a^9b^3c^6$ **13.** mn^2 **15.** $-\dfrac{2a}{3}$ **17.** $-27m^2n^4p^3$

19. $54n^{14}$ **21.** $\dfrac{7xz}{8y^3}$ **23.** $-8x^{13}y^{14}$ **25.** $\dfrac{4}{x^7}$ **27.** $\dfrac{1}{d^{10}}$ **29.** $\dfrac{1}{x^5}$ **31.** $\dfrac{x^2}{3}$ **33.** $\dfrac{x^6}{y^{12}}$ **35.** $\dfrac{9}{x^2y^4}$ **37.** $\dfrac{2}{x^4}$

39. $\dfrac{1}{2x^3}$ **41.** $\dfrac{1}{2x^2y^6}$ **43.** $\dfrac{1}{x^6}$ **45.** $\dfrac{y^8}{x^4}$ **47.** $\dfrac{b^{10}}{4a^{10}}$ **49.** $7b^2 + b - 4$ **51.** $3a^2 - 3a + 17$ **53.** $-7x - 7$

55. $-2x^3 + x^2 + 2$ **57.** $x^3 + 2x^2 - 6x - 6$ **59.** $5x^3 + 10x^2 - x - 4$ **61.** $y^3 - y^2 + 6y - 6$

63. $4y^3 - 2y^2 + 2y - 4$ **65.** $12b^4 - 48b^3 - 24b$ **67.** $9b^5 - 9b^3 + 24b$ **69.** $-2x^4y + 6x^3y^2 - 4x^2y^3$

71. $x^3 + 4x^2 + 5x + 2$ **73.** $a^3 - 6a^2 + 13a - 12$ **75.** $-2b^3 + 7b^2 + 19b - 20$ **77.** $x^4 - 4x^3 - 3x^2 + 14x - 8$

79. $y^4 + 4y^3 + y^2 - 5y + 2$ **81.** $a^2 + a - 12$ **83.** $y^2 - 10y + 21$ **85.** $2x^2 + 15x + 7$ **87.** $3x^2 + 11x - 4$

89. $4x^2 - 31x + 21$ **91.** $3y^2 - 2y - 16$ **93.** $21a^2 - 83a + 80$ **95.** $2x^2 + 3xy + y^2$ **97.** $3x^2 - 10xy + 8y^2$

99. $10a^2 + 14ab - 12b^2$ **101.** $16x^2 - 49$ **103.** $x + 2$ **105.** $2y - 7$ **107.** $x - 2 + \dfrac{8}{x + 2}$ **109.** $3y - 5 + \dfrac{20}{2y + 4}$

111. $b - 5 - \dfrac{24}{b - 3}$ **113.** $3x + 17 + \dfrac{64}{x - 4}$ **115.** $5y + 3 + \dfrac{1}{2y + 3}$ **117.** $x^2 - 5x + 2$ **119.** $x^2 + 5$ **121.** $y(12y - 5)$

123. $5xyz(2xz + 3y^2)$ **125.** $5(x^2 - 3x + 7)$ **127.** $3y^2(y^2 - 3y - 2)$ **129.** $x^2y^2(x^2y^2 - 3xy + 6)$ **131.** $8x^2y(2 - xy^3 - 6y)$

133. $(x + 2)(x - 1)$ **135.** $(a + 4)(a - 3)$ **137.** $(a - 1)(a - 2)$ **139.** $(b + 8)(b - 1)$ **141.** $(z + 5)(z - 9)$

143. $(z - 5)(z - 9)$ **145.** $(b + 4)(b + 5)$ **147.** Nonfactorable over the integers **149.** $(x - 7)(x - 8)$ **151.** $(y + 3)(2y + 1)$

153. $(a - 1)(3a - 1)$ **155.** $(x - 3)(2x + 1)$ **157.** $(2t + 1)(5t + 3)$ **159.** $(2z - 1)(5z + 4)$ **161.** Nonfactorable over the integers

163. $(t + 2)(2t - 5)$ **165.** $(3y + 1)(4y + 5)$ **167.** $(a - 5)(11a + 1)$ **169.** $(2b - 3)(3b - 2)$ **171.** $3(x + 2)(x + 3)$

173. $a(b + 8)(b - 1)$ **175.** $2y^2(y + 3)(y - 16)$ **177.** $x(x - 5)(2x - 1)$ **179.** $5(t + 2)(2t - 5)$ **181.** $p(2p + 1)(3p + 1)$

Unless otherwise noted, all content on this page is © Cengage Learning.

Glossary

abscissa The first number of an ordered pair; it measures a horizontal distance and is also called the first coordinate of an ordered pair. [4.1]

absolute value equation An equation that contains the absolute-value symbol. [2.6]

absolute value inequality An inequality that contains the absolute-value symbol. [2.6]

absolute value of a number The distance of the number from zero on the number line. [1.1, 2.6]

acute angle An angle whose measure is between 0° and 90°. [3.1]

acute triangle A triangle that has three acute angles. [3.2]

addend In addition, a number being added. [1.1]

addition The process of finding the total of two numbers. [1.1]

addition method An algebraic method of finding an exact solution of a system of linear equations. [5.2]

additive inverse of a polynomial The polynomial with the sign of every term changed. [6.2]

additive inverses Numbers that are the same distance from zero on the number line but lie on different sides of zero; also called opposites. [1.1, 1.4]

adjacent angles Two angles that share a common side. [3.1]

alternate exterior angles Two nonadjacent angles that are on opposite sides of the transversal and lie outside the parallel lines. [3.1]

alternate interior angles Two nonadjacent angles that are on opposite sides of the transversal and lie between the parallel lines. [3.1]

analytic geometry Geometry in which a coordinate system is used to study relationships between variables. [4.1]

angle Figure formed when two rays start from the same point. [3.1]

area A measure of the amount of surface in a region. [3.2]

axes The two number lines that form a rectangular coordinate system; also called coordinate axes. [4.1]

axis of symmetry of a parabola A line of symmetry that passes through the vertex of the parabola and is parallel to the y-axis for an equation of the form $y = ax^2 + bx + c$ or parallel to the x-axis for an equation of the form $x = ay^2 + by + c$. [11.1]

balance of trade The difference between the value of a nation's imports and the value of its exports over a stated period of time. [1.1]

base In an exponential expression, the number that is taken as a factor as many times as indicated by the exponent. [1.2]

base of a parallelogram Any side of a parallelogram may be designated the base. [3.2]

base of a triangle Any side of a triangle may be designated the base. [3.2]

basic percent equation Percent times base equals amount. [2.1]

binomial A polynomial of two terms. [6.2]

center of a circle The central point that is equidistant from all the points that make up a circle. [3.2]

circle Plane figure in which all points are the same distance from its center. [3.2]

circumference The perimeter of a circle. [3.2]

clearing denominators Removing denominators from an equation that contains fractions by multiplying each side of the equation by the LCM of the denominators. [2.2]

coefficient The number part of a variable term. [1.4]

combined variation A variation in which two or more types of variation occur at the same time. [8.8]

combining like terms Using the Distributive Property to add the coefficients of like variable terms; adding like terms of a variable expression. [1.4]

common logarithms Logarithms to the base 10. [12.2]

complementary angles Two angles whose measures have the sum 90°. [3.1]

completing the square Adding to a binomial the constant term that makes it a perfect-square trinomial. [10.2]

complex fraction A fraction whose numerator or denominator contains one or more fractions. [8.3]

complex number A number of the form $a + bi$, where a and b are real numbers and $i = \sqrt{-1}$. [9.5]

composite number A natural number that is not a prime number. [1.1]

composition of two functions The operation on two functions f and g denoted $f \circ g$. The value of the composition of f and g is given by $(f \circ g)(x) = f[g(x)]$. [11.3]

compound inequality Two inequalities joined with a connective word such as *and* or *or*. [2.5]

compound interest Interest that is computed not only on the original principal but also on the interest already earned. [12.5]

conjugates Binomial expressions that differ only in the sign of a term. The expressions $a + b$ and $a - b$ are conjugates. [9.5]

consecutive even integers Even integers that follow one another in order. [2.3]

consecutive integers Integers that follow one another in order. [2.3]

consecutive odd integers Odd integers that follow one another in order. [2.3]

constant function A function given by $f(x) = b$, where b is a constant. Its graph is a horizontal line passing through $(0, b)$. [4.3]

constant of proportionality k in a variation equation; also called the constant of variation. [8.8]

constant of variation k in a variation equation; also called the constant of proportionality. [8.8]

constant term A term that includes no variable part; also called a constant. [1.4, 6.2]

coordinate axes The two number lines that form a rectangular coordinate system; also called axes. [4.1]

coordinates of a point The numbers in the ordered pair that is associated with the point. [4.1]

corresponding angles Two angles that are on the same side of the transversal and are both acute angles or are both obtuse angles. [3.1]

cube A rectangular solid in which all six faces are squares. [3.3]

cube root of a perfect cube One of the three equal factors of the perfect cube. [7.4]

cubic function A third-degree polynomial function. [6.2]

decimal notation Notation in which a number consists of a whole-number part, a decimal point, and a decimal part. [1.2]

degree A unit used to measure angles. [3.1]

degree of a monomial The sum of the exponents of the variables. [6.1]

degree of a polynomial The greatest of the degrees of any of the polynomial's terms. [6.2]

dependent system of equations A system of equations whose graphs coincide. [5.1]

dependent variable In a function, the variable whose value depends on the value of another variable known as the independent variable. [4.2]

descending order The terms of a polynomial in one variable are arranged in descending order when the exponents of the variable decrease from left to right. [6.2]

diameter of a circle A line segment with endpoints on the circle and going through the center. [3.2]

diameter of a sphere A line segment with endpoints on the sphere and going through the center. [3.3]

difference of two perfect squares A polynomial in the form $a^2 - b^2$. [7.4]

difference of two perfect cubes A polynomial in the form $a^3 - b^3$. [7.4]

direct variation A special function that can be expressed as the equation $y = kx$, where k is a constant called the constant of variation or the constant of proportionality. [8.8]

discriminant For an equation of the form $ax^2 + bx + c = 0$, the quantity $b^2 - 4ac$ is called the discriminant. [10.2]

distance between two points For two points on a number line, the absolute value of the difference between the coordinates of the two points. [2.6]

domain The set of the first coordinates of all the ordered pairs of a relation. [4.2]

double root When a quadratic equation has two solutions that are the same number, the solution is called a double root of the equation. [10.1]

elements of a set The objects in the set. [1.1, 1.5]

empty set The set that contains no elements; also called the null set. [1.5]

endpoint In geometry, the point at which a ray starts. [3.1]

equation A statement of the equality of two mathematical expressions. [2.1]

equilateral triangle A triangle in which all three sides are of equal length. [3.2]

equivalent equations Equations that have the same solution. [2.1]

evaluating a function Replacing x in $f(x)$ with some value and then simplifying the numerical expression that results. [4.2]

evaluating a variable expression Replacing each variable by its value and then simplifying the resulting numerical expression. [1.4]

even integer An integer that is divisible by 2. [2.3]

exponent In an exponential expression, the raised number that indicates how many times the factor, or base, occurs in the multiplication. [1.2]

exponential equation An equation in which the variable occurs in the exponent. [12.4]

exponential form The expression 2^6 is in exponential form. Compare *factored form*. [1.2]

exponential function The exponential function with base b is defined by $f(x) = b^x$ where b is a positive real number not equal to 1. [12.1]

export A good or service produced in one's own country and sold for consumption in another country. [1.1]

exterior angle An angle adjacent to an interior angle of a triangle. [3.1]

extraneous solution When each side of an equation is raised to an even power, the resulting equation may have a solution that is not a solution of the original equation. Such a solution is called an extraneous solution. [9.4]

factor In multiplication, a number being multiplied. [1.1]

factored form The multiplication $2 \cdot 2 \cdot 2 \cdot 2 \cdot 2 \cdot 2$ is in factored form. Compare *exponential form*. [1.2]

factoring a polynomial Writing the polynomial as a product of other polynomials. [7.1]

factoring a quadratic trinomial Expressing the trinomial as the product of two binomials. [7.2]

favorable balance of trade Occurs when the value of a nation's exports is greater than that of its imports. [1.1]

FOIL A method of finding the product of two binomials. The letters stand for First, Outer, Inner, and Last. [6.3]

formula A literal equation that states a rule about measurement. [8.6]

function A relation in which no two ordered pairs that have the same first coordinate have different second coordinates. [4.2]

function notation A function designated by $f(x)$, which is the value of the function at x. [4.2]

graph of a function A graph of the ordered pairs that belong to the function. [4.3]

graph of an integer A heavy dot directly above the number on the number line. [1.1]

graph of an ordered pair The dot drawn at the coordinates of the point in the plane. [4.1]

graphing a point in the plane Placing a dot at the location given by the ordered pair; also called plotting a point in the plane. [4.1]

greater than A number that lies to the right of another number on the number line is said to be greater than that number. [1.1]

greatest common factor (GCF) The greatest common factor of two or more integers is the greatest integer that is a factor of all the integers. [7.1]

greatest common factor (GCF) of two or more monomials The greatest common factor of two or more monomials is the product of the GCF of the coefficients and the common variable factors. [7.1]

grouping symbols Parentheses ( ), brackets [ ], braces { }, the absolute value symbol, and the fraction bar. [1.3]

half-plane The solution set of a linear inequality in two variables. [4.7]

height of a parallelogram Found by drawing a line segment perpendicular to the base from the opposite side. [3.2]

height of a triangle Found by drawing a line perpendicular to the base from the vertex opposite the base. [3.2]

horizontal-line test A graph of a function represents the graph of a one-to-one function if any horizontal line intersects the graph at no more than one point. [11.4]

hypotenuse In a right triangle, the side opposite the 90° angle. [9.4]

imaginary number A number of the form ai, where a is a real number and $i = \sqrt{-1}$. [9.5]

imaginary part of a complex number For the complex number $a + bi$, b is the imaginary part. [9.5]

import A good or service bought from another country and consumed in one's own country. [1.1]

inconsistent system of equations A system of equations that has no solution. [5.1]

independent system of equations A system of equations whose graphs intersect at only one point. [5.1]

independent variable In a function, the variable that varies independently and whose value determines the value of the dependent variable. [4.2]

index In the expression $\sqrt[n]{a}$, n is the index of the radical. [9.1]

inequality An expression that contains the symbol >, <, ≥ (is greater than or equal to), or ≤ (is less than or equal to). [1.1]

infinity symbol A symbol used to indicate an interval that extends forever in the positive direction. [1.5]

integers The numbers . . . , $-3, -2, -1, 0, 1, 2, 3, \ldots$. [1.1]

interior angle of a triangle Angle within the region enclosed by a triangle. [3.1]

intersecting lines Lines that cross at a point in the plane. [3.1]

intersection of two sets The set that contains all elements that are common to both of the sets. [1.5]

interval notation A method of designating a set by enclosing the endpoints of the interval in brackets and/or parentheses. A bracket indicates that the endpoint is included in the set. A parenthesis indicates that the endpoint is not included in the set. [1.5]

inverse of a function The set of ordered pairs formed by reversing the coordinates of each ordered pair of the function. [11.4]

inverse variation A function that can be expressed as the equation $y = \dfrac{k}{x}$, where k is a constant. [8.8]

irrational number The decimal representation of an irrational number never terminates or repeats and can only be approximated. [1.2, 9.2]

isosceles triangle A triangle that has two sides of equal length; the angles opposite the equal sides are of equal measure. [3.2]

joint variation A variation in which a variable varies directly as the product of two or more variables. A joint variation can be expressed as the equation $z = kxy$, where k is a constant. [8.8]

leading coefficient In a polynomial, the coefficient of the variable with the largest exponent. [6.2]

least common denominator The smallest number that is a multiple of each denominator in question. [1.2, 8.2]

least common multiple (LCM) The LCM of two or more numbers is the smallest number that is a multiple of each of those numbers. [1.2]

least common multiple (LCM) of two or more polynomials The simplest polynomial of least degree that contains the factors of each polynomial. [8.2]

leg In a right triangle, one of the two sides that are not opposite the 90° angle. [9.4]

less than A number that lies to the left of another number on the number line is said to be less than that number. [1.1]

like terms Terms of a variable expression that have the same variable part. Having no variable part, constant terms are like terms. [1.4]

line Having no width, it extends indefinitely in two directions in a plane. [3.1]

linear equation in three variables An equation of the form $Ax + By + Cz = D$, where A, B, and C are coefficients of the variables and D is a constant. [5.2]

linear equation in two variables An equation of the form $y = mx + b$, where m is the coefficient of x and b is a constant; also called a linear function. [4.3]

linear function A function that can be expressed in the form $y = mx + b$. Its graph is a straight line. [4.3, 6.2]

linear inequality in two variables An inequality of the form $y > mx + b$ or $Ax + By > C$. The symbol $>$ could be replaced by $\geq$, $<$, or $\leq$. [4.7]

line segment Part of a line; it has two endpoints. [3.1]

literal equation An equation that contains more than one variable. [8.6]

logarithm For b greater than zero and not equal to 1, the statement $y = \log_b x$ (the logarithm of x to the base b) is equivalent to $x = b^y$. [12.2]

lower limit In a tolerance, the lowest acceptable value. [2.6]

magic square A square grid in which the numbers in every row, column, and diagonal sum to the same number. [1.3]

maximum value of a function The greatest value that the function can take on. [11.1]

minimum value of a function The least value that the function can take on. [11.1]

monomial A number, a variable, or a product of a number and variables; a polynomial of one term. [6.1, 6.2]

multiplication The process of finding the product of two numbers. [1.1]

multiplicative inverse The multiplicative inverse of a nonzero real number a is $\dfrac{1}{a}$; also called the reciprocal. [1.4]

natural exponential function The function defined by $f(x) = e^x$, where $e \approx 2.71828$. [12.1]

natural logarithm When e (the base of the natural exponential function) is used as the base of a logarithm, the logarithm is referred to as the natural logarithm and is abbreviated $\ln x$. [12.2]

natural numbers The numbers 1, 2, 3 . . . ; also called the positive integers. [1.1]

negative infinity symbol A symbol used to indicate an interval that extends forever in the negative direction. [1.5]

negative integers The numbers . . . , -3, -2, -1. [1.1]

negative reciprocal The negative reciprocal of a nonzero real number a is $-\dfrac{1}{a}$. [4.6, 11.1]

negative slope The slope of a line that slants downward to the right. [4.4]

nonfactorable over the integers A polynomial is nonfactorable over the integers if it does not factor using only integers. [7.2]

nth root of a A number b such that $b^n = a$. The nth root of a can be written $a^{1/n}$ or $\sqrt[n]{a}$. [9.1]

null set The set that contains no elements; also called the empty set. [1.5]

numerical coefficient The number part of a variable term. When the numerical coefficient is 1 or -1, the 1 is usually not written. [1.4]

obtuse angle An angle whose measure is between 90° and 180°. [3.1]

obtuse triangle A triangle that has one obtuse angle. [3.2]

odd integer An integer that is not divisible by 2. [2.3]

one-to-one function In a one-to-one function, given any y, there is only one x that can be paired with the given y. [11.4]

opposites Numbers that are the same distance from zero on the number line but lie on different sides of zero; also called additive inverses. [1.1]

Order of Operations Agreement A set of rules that tells us in what order to perform the operations that occur in a numerical expression. [1.3]

ordered pair A pair of numbers expressed in the form (a, b) and used to locate a point in the plane determined by a rectangular coordinate system. [4.1]

ordered triple Three numbers expressed in the form (x, y, z) and used to locate a point in the xyz-coordinate system. [5.2]

ordinate The second number of an ordered pair; it measures a vertical distance and is also called the second coordinate of an ordered pair. [4.1]

origin The point of intersection of the two number lines that form a rectangular coordinate system. [4.1]

parabola The graph of a quadratic function is called a parabola. [11.1]

parallel lines Lines that never meet; the distance between them is always the same. In a rectangular coordinate system, parallel lines have the same slope and thus do not intersect. [3.1, 4.6]

parallelogram Four-sided plane figure with opposite sides parallel. [3.2]

percent Parts of 100. [1.2]

percent mixture equation The equation $Q = Ar$, where Q represents the quantity of a substance in a solution, A is the amount of the solution, and r is the percent concentration of the substance. [2.1]

perfect cube The product of the same three factors. [7.4]

perfect square The product of a term and itself. [1.2, 7.4]

perfect-square trinomial The square of a binomial. [7.4]

perimeter The distance around a plane geometric figure. [3.2]

perpendicular lines Intersecting lines that form right angles. The slopes of perpendicular lines are negative reciprocals of each other. [3.1, 4.6]

plane A flat surface that extends indefinitely. [3.1]

plane figure A figure that lies entirely in a plane. [3.1]

plotting a point in the plane Placing a dot at the location given by the ordered pair; also called graphing a point in the plane. [4.1]

point A basic unit of geometry, symbolized by a dot. [3.1]

point-slope formula The equation $y - y_1 = m(x - x_1)$, where m is the slope of a line and (x_1, y_1) is a point on the line. [4.5]

polygon A closed figure determined by three or more line segments that lie in a plane. [3.2]

polynomial A variable expression in which the terms are monomials. [6.2]

positive integers The numbers 1, 2, 3 . . . ; also called the natural numbers. [1.1]

positive slope The slope of a line that slants upward to the right. [4.4]

prime number A natural number greater than 1 that is evenly divisible only by itself and 1. [1.1]

prime polynomial A polynomial that is nonfactorable over the integers. [7.2]

principal square root The positive square root of a number. [1.2, 9.1]

product In multiplication, the result of multiplying two numbers. [1.1]

product of the sum and difference of two terms A polynomial that can be expressed in the form $(a + b)(a - b)$. [6.3, 7.4]

proportion An equation that states the equality of two ratios or rates. [8.5]

protractor A device used to measure angles. [3.1]

Pythagorean Theorem The square of the hypotenuse of a right triangle is equal to the sum of the squares of the two legs. [9.4]

quadrant One of the four regions into which a rectangular coordinate system divides the plane. [4.1]

quadratic equation An equation of the form $ax^2 + bx + c = 0$, where a and b are coefficients, c is a constant, and $a \neq 0$; also called a second-degree equation. [7.5, 10.1]

quadratic equation in standard form A quadratic equation written in descending order and set equal to zero. [7.5]

quadratic formula A general formula, derived by applying the method of completing the square to the standard form of a quadratic equation, used to solve quadratic equations. [10.2]

quadratic function A function that can be expressed by the equation $f(x) = ax^2 + bx + c$, where a is not equal to zero. [6.2, 11.1]

quadratic inequality An inequality that can be written in the form $ax^2 + bx + c < 0$ or $ax^2 + bx + x > 0$, where a is not equal to zero. The symbols $\leq$ and $\geq$ can also be used. [10.5]

quadratic trinomial A trinomial of the form $ax^2 + bx + c$, where a and b are nonzero coefficients and c is a nonzero constant. [7.4]

quadrilateral A four-sided closed figure. [3.2]

quotient In division, the result of dividing the divisor into the dividend. [1.1]

radical equation An equation that contains a variable expression in a radicand. [9.4]

radical sign The symbol $\sqrt{}$, which is used to indicate the positive, or principal, square root of a number. [1.2, 9.1]

radicand In a radical expression, the expression under the radical sign. [1.2, 9.1]

radius of a circle Line segment from the center of the circle to a point on the circle. [3.2]

radius of a sphere A line segment going from the center to a point on the sphere. [3.3]

range The set of the second coordinates of all the ordered pairs of a relation. [4.2]

rate The quotient of two quantities that have different units. [8.5]

rate of work That part of a task that is completed in one unit of time. [8.7]

ratio The quotient of two quantities that have the same unit. [8.5]

rational expression A fraction in which the numerator or denominator is a polynomial. [8.1]

rational number A number of the form $\frac{a}{b}$, where a and b are integers and b is not equal to zero. [1.2]

rationalizing the denominator The procedure used to remove a radical from the denominator of a fraction. [9.2]

ray Line that starts at a point and extends indefinitely in one direction. [3.1]

real numbers The rational numbers and the irrational numbers taken together. [1.2, 9.1]

real part of a complex number For the complex number $a + bi$, a is the real part. [9.5]

reciprocal The reciprocal of a nonzero real number a is $\frac{1}{a}$; also called the multiplicative inverse. [1.2, 1.4]

reciprocal of a rational expression The rational expression with the numerator and denominator interchanged. [8.1]

rectangle A parallelogram that has four right angles. [3.2]

rectangular coordinate system A coordinate system formed by two number lines, one horizontal and one vertical, that intersect at the zero point of each line. [4.1]

rectangular solid A solid in which all six faces are rectangles. [3.3]

regular polygon A polygon in which each side has the same length and each angle has the same measure. [3.2]

relation A set of ordered pairs. [4.2]

repeating decimal A decimal number in which the decimal part contains a block of digits that repeats infinitely. [1.2]

right angle An angle whose measure is 90°. [3.1]

right triangle A triangle that contains a 90° angle. [3.1]

roster method A method of designating a set by enclosing a list of its elements in braces. [1.5, 1.1]

scalene triangle A triangle that has no sides of equal length; no two of its angles are of equal measure. [3.2]

scatter diagram A graph of collected data as points in a coordinate system. [4.1]

scientific notation Notation in which a number is expressed as the product of a number between 1 and 10 and a power of 10. [6.1]

second-degree equation An equation of the form $ax^2 + bx + c = 0$, where a and b are coefficients, c is a constant, and $a \neq 0$; also called a quadratic equation. [10.1]

set A collection of objects. [1.1, 1.5]

set-builder notation A method of designating a set that makes use of a variable and a certain property that only elements of that set possess. [1.5]

similar objects Similar objects have the same shape but not necessarily the same size. [8.5]

simple interest equation The equation $I = Prt$, where I is the simple interest; P is the principal, or amount invested; r is the simple interest rate; and t is the time.

simplest form of a rational expression A rational expression is in simplest form when the numerator and denominator have no common factors. [8.1]

slope A measure of the slant, or tilt, of a line. The symbol for slope is m. [4.4]

slope-intercept form of a straight line The equation $y = mx + b$, where m is the slope of the line and $(0, b)$ is the y-intercept. [4.4]

solution of a system of equations in three variables An ordered triple that is a solution of each equation of the system. [5.2]

solution of a system of equations in two variables An ordered pair that is a solution of each equation of the system. [5.1]

solution of an equation A number that, when substituted for the variable, results in a true equation. [2.1]

solution of an equation in three variables An ordered triple (x, y, z) whose coordinates make the equation a true statement. [5.2]

solution of an equation in two variables An ordered pair whose coordinates make the equation a true statement. [4.1]

solution set of a system of inequalities The intersection of the solution sets of the individual inequalities. [5.4]

solution set of an inequality A set of numbers, each element of which, when substituted for the variable, results in a true inequality. [2.5]

solving an equation Finding a solution of the equation. [2.1]

sphere A solid in which all points are the same distance from point O, which is called the center of the sphere. [3.3]

square A rectangle with four equal sides. [3.2]

square of a binomial A polynomial that can be expressed in the form $(a + b)^2$. [6.3]

square root A square root of a positive number x is a number a for which $a^2 = x$. [1.2]

square root of a perfect square One of the two equal factors of the perfect square. [7.4]

standard form of a quadratic equation A quadratic equation is in standard form when the polynomial is in descending order and equal to zero. [7.5, 10.1]

straight angle An angle whose measure is 180°. [3.1]

substitution method An algebraic method of finding an exact solution of a system of linear equations. [5.1]

sum In addition, the total of two or more numbers. [1.1]

supplementary angles Two angles whose measures have the sum 180°. [3.1]

synthetic division A shorter method of dividing a polynomial by a binomial of the form $x - a$. This method uses only the coefficients of the variable terms. [6.4]

system of equations Two or more equations considered together. [5.1]

system of inequalities Two or more inequalities considered together. [5.4]

terminating decimal A decimal that is formed when division of the numerator of its fractional counterpart by the denominator results in a remainder of zero. [1.2]

terms of a variable expression The addends of the expression. [1.4]

tolerance of a component The amount by which it is acceptable for the component to vary from a given measurement. [2.6]

translate To change a sentence written in words to a mathematical equation. [2.3]

transversal A line that intersects two other lines at two different points. [3.1]

triangle A three-sided closed figure. [3.1]

trinomial A polynomial of three terms. [6.2]

undefined slope The slope of a vertical line is undefined. [4.4]

unfavorable balance of trade Occurs when the value of a nation's imports is greater than that of its exports. [1.1]

uniform motion The motion of an object whose speed and direction do not change. [2.1, 8.7]

uniform motion equation The equation $d = rt$, where d is the distance traveled, r is the rate of travel, and t is the time spent traveling. [2.1]

union of two sets The set that contains all elements that belong to either of the sets. [1.5]

upper limit In a tolerance, the greatest acceptable value. [2.6]

value of a function The value of the dependent variable for a given value of the independent variable. [4.2]

value mixture equation The equation $AC = V$, where A is the amount of an ingredient, C is the cost per unit of the ingredient, and V is the value of the ingredient. [2.4]

variable A letter of the alphabet used to stand for a number that is unknown or that can change. [1.1, 1.4]

variable expression An expression that contains one or more variables. [1.4]

variable part In a variable term, the variable or variables and their exponents. [1.4]

variable term A term composed of a numerical coefficient and a variable part. When the numerical coefficient is 1 or -1, the 1 is usually not written. [1.4]

vertex Point at which the rays that form an angle meet. [3.1]

vertex of a parabola The point on the parabola with the smallest y-coordinate or the largest y-coordinate. [11.1]

vertical angles Two angles that are on opposite sides of the intersection of two lines. [3.1]

vertical-line test A graph defines a function if any vertical line intersects the graph at no more than one point. [12.1]

volume A measure of the amount of space inside a closed surface. [3.3]

whole numbers The natural numbers and zero. [1.1]

x-coordinate The abscissa in an xy-coordinate system. [4.1]

x-intercept The point at which a graph crosses the x-axis. [4.3]

y-coordinate The ordinate in an xy-coordinate system. [4.1]

y-intercept The point at which a graph crosses the y-axis. [4.3]

zero of a function A value of x for which $f(x) = 0$. [4.3, 11.1]

zero slope The slope of a horizontal line. [4.4]

Index

TI-30X IIS

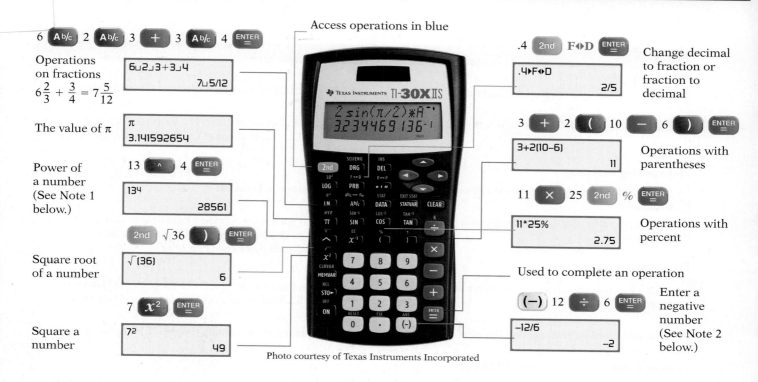

Access operations in blue

Operations on fractions
$6\frac{2}{3} + \frac{3}{4} = 7\frac{5}{12}$

6 A b/c 2 A b/c 3 + 3 A b/c 4 ENTER =

6⌴2⌴3+3⌴4
7⌴5/12

The value of π

π
3.141592654

Power of a number (See Note 1 below.)

13 ^ 4 ENTER =

13⁴
28561

Square root of a number

2nd √ 36) ENTER =

√(36)
6

Square a number

7 x² ENTER =

7²
49

Photo courtesy of Texas Instruments Incorporated

Change decimal to fraction or fraction to decimal

.4 2nd F◆D ENTER =

.4▸F◆D
2/5

Operations with parentheses

3 + 2 (10 — 6) ENTER =

3+2(10–6)
11

Operations with percent

11 × 25 2nd % ENTER =

11*25%
2.75

Used to complete an operation

Enter a negative number (See Note 2 below.)

(—) 12 ÷ 6 ENTER =

–12/6
–2

---

fx-300MS

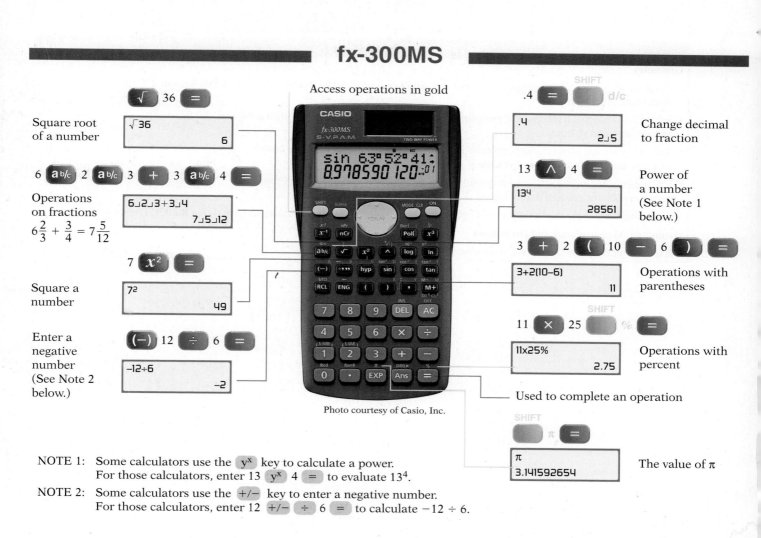

Access operations in gold

Square root of a number

√ 36 =

√36
6

Operations on fractions
$6\frac{2}{3} + \frac{3}{4} = 7\frac{5}{12}$

6 a b/c 2 a b/c 3 + 3 a b/c 4 =

6⌴2⌴3+3⌴4
7⌴5⌴12

Square a number

7 x² =

7²
49

Enter a negative number (See Note 2 below.)

(—) 12 ÷ 6 =

–12÷6
–2

Photo courtesy of Casio, Inc.

Change decimal to fraction

.4 = SHIFT d/c

.4
2⌴5

Power of a number (See Note 1 below.)

13 ^ 4 =

13⁴
28561

Operations with parentheses

3 + 2 (10 — 6) =

3+2(10–6)
11

Operations with percent

11 × 25 SHIFT % =

11x25%
2.75

Used to complete an operation

The value of π

SHIFT π =

π
3.141592654

NOTE 1: Some calculators use the yˣ key to calculate a power. For those calculators, enter 13 yˣ 4 = to evaluate 13^4.

NOTE 2: Some calculators use the +/– key to enter a negative number. For those calculators, enter 12 +/– ÷ 6 = to calculate $-12 \div 6$.

FUNCTIONS

Functions

- Define, evaluate, and compare functions.

  4.1 The Rectangular Coordinate System; 4.2 Introduction to Functions

- Use functions to model relationships between quantities.

  4.5 Finding Equations of Lines; 11.1 Properties of Quadratic Functions; 12.5 Applications of Exponential and Logarithmic Functions

Interpreting Functions

- Understand the concept of a function and use function notation.

  4.2 Introduction to Functions

- Interpret functions that arise in applications in terms of the context.

  4.3 Linear Functions; 4.5 Finding Equations of Lines; 11.1 Properties of Quadratic Functions; 12.5 Applications of Exponential and Logarithmic Functions

Building Functions

- Build a function that models a relationship between two quantities.

  4.5 Finding Equations of Lines; 11.1 Properties of Quadratic Functions; 12.5 Applications of Exponential and Logarithmic Functions

- Build new functions from existing functions.

  11.2 Translating and Reflecting Graphs; 11.3 Algebra of Functions

Linear, Quadratic, and Exponential Models

- Construct and compare linear, quadratic, and exponential models and solve problems.

  Chapter 4 Linear Functions and Inequalities in Two Variables; 11.1 Properties of Quadratic Functions; 12.5 Applications of Exponential and Logarithmic Functions

GEOMETRY

Geometry

- Graph points on the coordinate plane to solve real-world and mathematical problems.

  4.1 The Rectangular Coordinate System

- Solve real-world and mathematical problems involving area, surface area, and volume.

  3.2 Plane Geometric Figures; 3.3 Solids

- Solve real-life and mathematical problems involving angle measure, area, surface area, and volume.

  Chapter 3: Geometry

- Understand congruence and similarity using physical models, transparencies, or geometry software.

  8.5 Ratio and Proportion

- Understand and apply the Pythagorean Theorem.

  9.4 Solving Equations Containing Radical Expressions

Geometric Measurement and Dimension

- Explain volume formulas and use them to solve problems.

  3.3 Solids

ALGEBRA

Operations and Algebraic Thinking

- Write and interpret numerical expressions.

1.1 Introduction to Integers; 1.2 Rational and Irrational Numbers

Expressions and Equations

- Apply and extend previous understandings of arithmetic to algebraic expressions.

1.4 Variable Expressions

- Reason about and solve one-variable equations and inequalities.

2.1 Introduction to Equations; 2.2 General Equations; 2.5 First-Degree Inequalities; 2.6 Absolute Value Equations and Inequalities

- Use properties of operations to generate equivalent expressions.

1.4 Variable Expressions; 6.2 Introduction to Polynomial Functions; 6.3 Multiplication of Polynomials; 6.4 Division of Polynomials; 7.1 Common Factors; 7.2 Factoring Polynomials of the Form $x^2 + bx + c$; 7.3 Factoring Polynomials of the Form $ax^2 + bx + c$; 7.4 Special Factoring

- Solve real-life and mathematical problems using numerical and algebraic expressions and equations.

2.1 Introduction to Equations; 2.2 General Equations; 2.3 Translating Sentences into Equations; 2.4 Applications: Mixture and Uniform Motion Problems; Chapter 3: Geometry

- Work with radicals and integer exponents.

6.1 Exponential Expressions; 9.1 Rational Exponents and Radical Expressions

- Understand the connections between proportional relationships, lines, and linear equations.

4.3 Linear Functions; 4.4 Slope of a Straight Line; 4.5 Finding Equations of Lines; 4.6 Parallel and Perpendicular Lines; 8.5 Ratio and Proportion

- Analyze and solve linear equations and pairs of simultaneous linear equations.

4.3 Linear Functions; Chapter 5 Systems of Linear Equations and Inequalities

Seeing Structure in Expressions

- Interpret the structure of expressions.

1.4 Variable Expressions

- Write expressions in equivalent forms to solve problems.

1.4 Variable Expressions

Arithmetic with Polynomials and Rational Expressions

- Perform arithmetic operations on polynomials.

6.2 Introduction to Polynomial Functions; 6.3 Multiplication of Polynomials; 6.4 Division of Polynomials; 7.1 Common Factors; 7.2 Factoring Polynomials of the Form $x^2 + bx + c$; 7.3 Factoring Polynomials of the Form $ax^2 + bx + c$; 7.4 Special Factoring

- Understand the relationship between zeros and factors of polynomials.

4.3 Linear Functions; 6.4 Division of Polynomials; 7.5 Solving Equations; 11.1 Properties of Quadratic Functions

- Rewrite rational expressions.

8.1 Multiplication and Division of Rational Expressions; 8.2 Addition and Subtraction of Rational Expressions; 8.3 Complex Fractions

Creating Equations

- Create equations that describe numbers or relationships.

1.4 Variable Expressions; 2.1 Introduction to Equations; 2.2 General Equations; 2.3 Translating Sentences into Equations; 2.4 Applications: Mixture and Uniform Motion Problems; 8.4 Solving Equations Containing Fractions; 8.8 Variation; 10.4 Applications of Quadratic Equations

Reasoning with Equations and Inequalities

- Solve equations and inequalities in one variable.

2.1 Introduction to Equations; 2.2 General Equations; 2.5 First-Degree Inequalities; 7.5 Solving Equations; Chapter 10: Quadratic Equations

- Solve systems of equations.

5.1 Solving Systems of Linear Equations by Graphing and by the Substitution Method; 5.2 Solving Systems of Linear Equations by the Addition Method; 5.4 Solving Systems of Linear Inequalities

Common Core Correlation Guide

THE NUMBER SYSTEM	
• Understand and apply properties of operations and the relationship between addition and subtraction.	1.1 Introduction to Integers; 1.2 Rational and Irrational Numbers
• Work with addition and subtraction equations.	2.1 Introduction to Equations
• Represent and solve problems involving multiplication and division.	1.1 Introduction to Integers; 1.2 Rational and Irrational Numbers
• Find a percent of a quantity as a rate per 100 (30% of a quantity means 30/100 times the quantity); solve problems involving finding the whole, given a part and the percent.	2.1 Introduction to Equations
Numbers and Operations in Base Ten	
• Perform operations with multi-digit whole numbers and with decimals to hundredths	1.2 Rational and Irrational Numbers
Numbers and Operations—-Fractions	
• Use equivalent fractions as a strategy to add and subtract fractions.	1.2 Rational and Irrational Numbers
• Apply and extend previous understandings of multiplication and division to multiply and divide fractions.	1.2 Rational and Irrational Numbers
• Apply and extend previous understandings of multiplication and division to divide fractions by fractions.	1.2 Rational and Irrational Numbers
• Compute fluently with multi-digit numbers and find common factors and multiples.	1.2 Rational and Irrational Numbers
• Apply and extend previous understandings of numbers to the system of rational numbers	1.2 Rational and Irrational Numbers
• Apply and extend previous understandings of operations with fractions to add, subtract, multiply, and divide rational numbers.	1.2 Rational and Irrational Numbers
• Know that there are numbers that are not rational, and approximate them by rational numbers.	1.2 Rational and Irrational Numbers
The Real Number System	
• Extend the properties of exponents to rational exponents.	9.1 Rational Exponents and Radical Expressions
• Use properties of rational and irrational numbers.	1.2 Rational and Irrational Numbers; 8.1 Multiplication and Division of Rational Expressions; 8.2 Addition and Subtraction of Rational Expressions; 8.3 Complex Fractions
Quantities	
• Reason quantitatively and use units to solve problems.	1.1 Introduction to Integers; 1.2 Rational and Irrational Numbers; 3.2 Plane Geometric Figures; 3.3 Solids
The Complex Number System	
• Perform arithmetic operations with complex numbers.	9.5 Complex Numbers
• Use complex numbers in polynomial identities and equations.	Chapter 10 Quadratic Equations
Ratios and Proportional Relationships/Measurement and Data	
• Understand ratio concepts and use ratio reasoning to solve problems.	8.5 Ratio and Proportion
• Analyze proportional relationships and use them to solve real-world and mathematical problems.	8.4 Solving Equations Containing Fractions; 8.5 Ratio and Proportion